ENCYCLOPEDIA OF

ETHICAL, LEGAL,
— AND —
POLICY ISSUES IN BIOTECHNOLOGY

VOLUME 1

# WILEY BIOTECHNOLOGY ENCYCLOPEDIAS

Encyclopedia of Bioprocess Technology: Fermentation, Biocatalysis, and Bioseparation
Edited by Michael C. Flickinger and Stephen W. Drew

Encyclopedia of Molecular Biology
Edited by Thomas E. Creighton

Encyclopedia of Cell Technology
Edited by Raymond E. Spier

Encyclopedia of Ethical, Legal, and Policy Issues in Biotechnology
Edited by Thomas H. Murray and Maxwell J. Mehlman

ENCYCLOPEDIA OF ETHICAL, LEGAL, AND POLICY ISSUES IN BIOTECHNOLOGY
EDITORIAL BOARD

ENCYCLOPEDIA OF

ETHICAL, LEGAL,
AND
POLICY ISSUES IN

BIOTECHNOLOGY

VOLUME 1

Thomas H. Murray
The Hastings Center
Garrison, New York

Maxwell J. Mehlman
Case Western Reserve University
Cleveland, Ohio

A Wiley-Interscience Publication

John Wiley & Sons, Inc.

New York / Chichester / Weinheim / Brisbane / Singapore / Toronto

Copyright © 2000 by John Wiley & Sons, Inc. All rights reserved.

Published simultaneously in Canada.

For ordering and customer service, call 1-800-CALL-WILEY.

Library of Congress Cataloging in Publication Data:
Murray, Thomas H. (Thomas Harold), 1926–
    Encyclopedia of ethical, legal, and policy issues in biotechnology / Thomas Murray, Maxwell J. Mehlman.
        p.   cm.
    Includes index.
    ISBN 0-471-17612-5 (cloth : alk. paper)
    1. Biotechnology — Moral and ethical aspects — Encyclopedias. 2. Biotechnology industries — Law and legislation — Encyclopedias. 3. Biotechnology — government policy — United States — Encyclopedias. I. Mehlman, Maxwell J. II. Title.

TP248.16.M87   2000
174′.96606 — dc21   00-021383

Printed in the United States of America.

10 9 8 7 6 5 4 3 2 1

PREFACE

The Wiley Biotechnology Encyclopedias, composed of the *Encyclopedia of Molecular Biology*; the *Encyclopedia of Bioprocess Technology: Fermentation, Biocatalysis, and Bioseparation*; the *Encyclopedia of Cell Technology*; and the *Encyclopedia of Ethical, Legal, and Policy Issues in Biotechnology* cover very broadly four major contemporary themes in biotechnology. The series comes at a fascinating time in that, as we move into the twenty-first century, the discipline of biotechnology is undergoing striking paradigm changes.

Biotechnology is now beginning to be viewed as an informational science. In a simplistic sense there are three types of biological information. First, there is the digital or linear information of our chromosomes and genes with the four-letter alphabet composed of G, C, A, and T (the bases guanine, cytosine, adenine, and thymine). Variation in the order of these letters in the digital strings of our chromosomes or our expressed genes (or mRNAs) generates information of several distinct types: genes, regulatory machinery, and information that enables chromosomes to carry out their tasks as informational organelles (e.g., centromeric and telomeric sequences).

Second, there is the three-dimensional information of proteins, the molecular machines of life. Proteins are strings of amino acids employing a 20-letter alphabet. Proteins pose four technical challenges: (*1*) Proteins are synthesized as linear strings and fold into precise three-dimensional structures as dictated by the order of amino acid residues in the string. Can we formulate the rules for protein folding to predict three-dimensional structure from primary amino acid sequence? The identification and comparative analysis of all human and model organism (bacteria, yeast, nematode, fly, mouse, etc.) genes and proteins will eventually lead to a lexicon of motifs that are the building block components of genes and proteins. These motifs will greatly constrain the shape space that computational algorithms must search to successfully correlate primary amino acid sequence with the correct three-dimensional shapes. The protein-folding problem will probably be solved within the next 10–15 years. (*2*) Can we predict protein function from knowledge of the three-dimensional structure? Once again the lexicon of motifs with their functional as well as structural correlations will play a critical role in solving this problem. (*3*) How do the myriad of chemical modifications of proteins (e.g., phosphorylation, acetylation, etc.) alter their structures and modify their functions? The mass spectrometer will play a key role in identifying secondary modifications. (*4*) How do proteins interact with one another and/or with other macromolecules to form complex molecular machines (e.g., the ribosomal subunits)? If these functional complexes can be isolated, the mass spectrometer, coupled with a knowledge of all protein sequences that can be derived from the complete genomic sequence of the organism, will serve as a powerful tool for identifying all the components of complex molecular machines.

The third type of biological information arises from complex biological systems and networks. Systems information is four dimensional because it varies with time. For example, the human brain has 1,012 neurons making approximately 1,015 connections. From this network arise systems properties such as memory, consciousness, and the ability to learn. The important point is that systems properties cannot be understood from studying the network elements (e.g., neurons) one at a time; rather the collective behavior of the elements needs to be studied. To study most biological systems, three issues need to be stressed. First, most biological systems are too complex to study directly, therefore they must be divided into tractable subsystems whose properties in part reflect those of the system. These subsystems must be sufficiently small to analyze all their elements and connections. Second, high-throughput analytic or global tools are required for studying many systems elements at one time (see later). Finally, the systems information needs to be modeled mathematically before systems properties can be predicted and ultimately understood. This will require recruiting computer scientists and applied mathematicians into biology — just as the attempts to decipher the information of complete genomes and the protein folding and structure/function problems have required the recruitment of computational scientists.

I would be remiss not to point out that there are many other molecules that generate biological information: amino acids, carbohydrates, lipids, and so forth. These too must be studied in the context of their specific structures and specific functions.

The deciphering and manipulation of these various types of biological information represent an enormous technical challenge for biotechnology. Yet major new and powerful tools for doing so are emerging.

One class of tools for deciphering biological information is termed high-throughput analytic or global tools. These tools can be used to study many genes or chromosome features (genomics), many proteins (proteomics), or many cells rapidly: large-scale DNA sequencing, genomewide genetic mapping, cDNA or oligonucleotide arrays, two-dimensional gel electrophoresis and other global protein separation technologies, mass spectrometric analysis of proteins and protein fragments, multiparameter, high-throughput cell and chromosome sorting, and high-throughput phenotypic assays.

A second approach to the deciphering and manipulating of biological information centers around combinatorial strategies. The basic idea is to synthesize an informational string (DNA fragments, RNA fragments, protein fragments, antibody combining sites, etc.) using all combinations of the basic letters of the corresponding alphabet, thus creating many different shapes that can be used to activate, inhibit, or complement the biological functions of designated three-dimensional shapes (e.g., a molecule in a signal transduction pathway). The power of combinational chemistry is just beginning to be appreciated.

A critical approach to deciphering biological information will ultimately be the ability to visualize the functioning of genes, proteins, cells, and other informational elements within living organisms (in vivo informational imaging).

Finally, there are the computational tools required to collect, store, analyze, model, and ultimately distribute the various types of biological information. The creation presents a challenge comparable to that of developing new instrumentation and new chemistries. Once again this means recruiting computer scientists and applied mathematicians to biology. The biggest challenge in this regard is the language barriers that separate different scientific disciplines. Teaching biology as an informational science has been a very effective means for breeching these barriers.

The challenge is, of course, to decipher various types of biological information and then be able to use this information to manipulate genes, proteins, cells, and informational pathways in living organisms to eliminate or prevent disease, produce higher-yield crops, or increase the productivity of animals for meat and other foods.

Biotechnology and its applications raise a host of social, ethical, and legal questions, for example, genetic privacy, germline genetic engineering, cloning of animals, genes that influence behavior, cost of therapeutic drugs generated by biotechnology, animal rights, and the nature and control of intellectual property.

Clearly, the challenge is to educate society so that each citizen can thoughtfully and rationally deal with these issues, for ultimately society dictates the resources and regulations that circumscribe the development and practice of biotechnology. Ultimately, I feel enormous responsibility rests with scientists to inform and educate society about the challenges as well as the opportunities arising from biotechnology. These are critical issues for biotechnology that are developed in detail in the *Encyclopedia of Ethical, Legal, and Policy Issues in Biotechnology*.

The view that biotechnology is an informational science pervades virtually every aspect of this science, including discovery, reduction to practice, and societal concerns. These Encyclopedias of Biotechnology reinforce the emerging informational paradigm change that is powerfully positioning science as we move into the twenty-first century to more effectively decipher and manipulate for humankind's benefit the biological information of relevant living organisms.

Leroy Hood
University of Washington

LIST OF ARTICLES

CONTRIBUTORS

Terrence F. Ackerman, *University of Tennessee, Memphis, Tennessee,* Human Subjects Research, Ethics, Compensation of Subjects for Injury

Sheri Alpert, *University of Notre Dame, Notre Dame, Indiana,* Genetic Information, Ethics, Ethical Issues in Tissue Banking and Human Subject Research in Stored Tissues

Mary R. Anderlik, *University of Houston, Houston, Texas,* Genetic Information, Legal, Genetic Privacy Laws; Genetic Information, Legal, Genetics and the Americans with Disabilities Act

Mary Jane Angelo, *St. John's River Water Management District, Palatka, Florida,* Agricultural Biotechnology, Law, and EPA Regulation

Adrienne Asch, *Wellesley College, Wellesley, Massachusetts,* Reproduction, Ethics, Prenatal Testing, and the Disability Rights Critique

Richard E. Ashcroft, *Imperial College, London, England,* International Aspects: National Profiles, United Kingdom

Patricia Backlar, *Portland State University, Oregon Health Sciences University, Portland, Oregon,* Human Subjects Research, Ethics, Research on Vulnerable Populations

Dianne M. Bartels, *University of Minnesota, Center for Bioethics, Minneapolis, Minnesota,* Reproduction, Ethics, Sex Selection

Françoise Bieri, *Swiss Priority Programme Biotechnology of the Swiss National Science Foundation, Basel, Switzerland,* International Aspects: National Profiles, Switzerland

Barbara Bowles Biesecker, *National Human Genome Research Institute, Bethesda, Maryland,* Reproduction, Ethics, the Ethics of Reproductive Genetic Counseling: Nondirectiveness

Michael R. Bleavins, *Warner-Lambert Company, University of Michigan Medical School, Ann Arbor, Michigan,* Pharmacogenetics

David Blumenthal, *Massachusetts General Hospital, Harvard Medical School, Boston, Massachusetts,* Academic Industry Relationships in Biotechnology, Overview

Mary Anne Bobinski, *University of Houston Law Center, Houston, Texas,* Genetic Information, Legal, ERISA Preemption, and HIPAA Protection

Robert A. Bohrer, *California Western School of Law, San Diego, California,* Federal Regulation of Biotechnology Products for Human Use, FDA, Orphan Drug Act

Andrea L. Bonnicksen, *Northern Illinois University, DeKalb, Illinois,* Gene Therapy, Ethics, and International Perspectives

Jeffrey R. Botkin, *University of Utah School of Medicine, Salt Lake City, Utah,* Reproduction, Law, Wrongful Birth, and Wrongful Life Actions

Dan W. Brock, *Brown University, Providence, Rhode Island,* Cloning, Ethics

Paul Brodwin, *University of Wisconsin, Milwaukee, Wisconsin,* Professional Power and the Cultural Meanings of Biotechnology

Baruch Brody, *Baylor College of Medicine, Houston, Texas,* Human Subjects Research, Law, FDA Rules

William P. Browne, *Central Michigan University, Mount Pleasant, Michigan,* Agricultural Biotechnology, Law, and Social Impacts of Agricultural Biotechnology

Jeffrey Burkhardt, *University of Florida, Gainesville, Florida,* Agricultural Biotechnology, Ethics, Family Farms, and Industrialization

David L. Cameron, *Willamette University College of Law, Salem, Oregon,* Scientific Research, Policy, Tax Treatment of Research and Development

Courtney S. Campbell, *Oregon State University, Corvallis, Oregon,* Patents and Licensing, Ethics, Moral Status of Human Tissues: Sale, Abandonment or Gift; Religious Views on Biotechnology, Protestant

Eric G. Campbell, *Massachusetts General Hospital, Harvard Medical School, Boston, Massachusetts,* Academic Industry Relationships in Biotechnology, Overview

Benjamin Capps, *University of Bristol, Bristol, England,* International Aspects: National Profiles, United Kingdom

Catherine Chabert, *Alain Bensoussan — Avocats, Lyon, France,* International Aspects: National Profiles, France

R. Cole-Turner, *Pittsburgh Theological Seminary, Pittsburgh, Pennsylvania,* Patents and Licensing, Ethics, Organizations with Prominent Positions on Gene Patenting

Robert Mullan Cook-Deegan, *Georgetown University, Annapolis, Maryland,* Medical Biotechnology, United States Policies Influencing its Development

George C. Cunningham, *State of California-Department of Health Services, Berkeley, California,* Genetic Information, Legal, Regulating Genetic Services

Elliot N. Dorff, *University of Judaism, Bel Air, California,* Religious Views on Biotechnology, Jewish

Daniel W. Drell, *U.S. Department of Energy, Washington, District of Columbia,* Informing Federal Policy on Biotechnology: Executive Branch, Department of Energy

Elisa Eiseman, *RAND Science and Technology Policy Institute, Washington, District of Columbia,* Cloning, Policy Issues

Susan S. Ellenberg, *U.S. Food and Drug Administration, Washington, District of Columbia,* Human Subjects Research, Ethics, Stopping Rules for Randomized Clinical Trials

Lisa M. Gehrke, *Nilles & Nilles, S.C., Milwaukee, Wisconsin,* Ownership of Human Biological Material

Oreste Ghisalba, *Swiss Priority Programme Biotechnology of the Swiss National Science Foundation, Basel, Switzerland,* International Aspects: National Profiles, Switzerland

Paul Giannelli, *Case Western Reserve University, School of Law, Cleveland, Ohio,* Genetic Information, Law, Legal Issues in Law Enforcement DNA Databanks

Henry T. Greely, *Stanford University Law School, Stanford, California,* Human Genome Diversity Project

Ronald Green, *Dartmouth College, Hanover, New Hampshire,* Human Subjects Research, Ethics, Research on Human Embryos

Louis M. Guenin, *Harvard Medical School, Boston, Massachusetts,* Patents, Ethics, Human Life Forms

Kathi E. Hanna, *Science and Health Policy Consultant, Prince Frederick, Maryland,* Federal Policy Making for Biotechnology, Executive Branch, National Human Genome Research Institute; Cloning, Overview of Human Cloning

Mats G. Hansson, *Uppsala University, Uppsala, Sweden,* International Aspects: National Profiles, Scandinavia

Sarah C. Hensel, *Mayo Clinic, Rochester, Minnesota,* Animal Medical Biotechnology, Policy, Would Transgenic Animals Solve the Organ Shortage Problem?

Göran Hermerén, *Lund University, Lund, Sweden,* Patents and Licensing, Ethics, International Controversies

Sharona Hoffman, *Case Western Reserve University, School of Law, Cleveland, Ohio,* Genetic Information, Law, Legal Issues in Law Enforcement DNA Databanks

Rachelle D. Hollander, *National Science Foundation, Arlington, Virginia,* Scientific Research, Ethics, Values in Science

Cynthia M. Ho, *Loyola University of Chicago School of Law, Chicago, Illinois,* International Intellectual Property Issues for Biotechnology

Anny Huang, *Wachtell, Lipton, Rosen, and Katz, New York, New York,* Genetic Information, Legal, FDA Regulation of Genetic Testing

Richard Huxtable, *University of Bristol, Bristol, England,* International Aspects: National Profiles, United Kingdom

Gail H. Javitt, *Johns Hopkins School of Hygiene and, Public Health, Baltimore, Maryland,* Gene Therapy, Law and FDA Role in Regulation

Eric T. Juengst, *Case Western Reserve University, Cleveland, Ohio,* Genetic Information, Ethics, Family Issues; Human Enhancement Uses of Biotechnology, Ethics, the Ethics of Enhancement

Marshall B. Kapp, *Wright State University School of Medicine, Dayton, Ohio,* Human Subjects Research, Law, Law of Informed Consent

Othmar Käppeli, *Swiss Priority Programme Biotechnology of the Swiss National Science Foundation, Basel, Switzerland,* International Aspects: National Profiles, Switzerland

Nancy E. Kass, *Johns Hopkins University, Baltimore, Maryland,* Human Subjects Research, Ethics, Informed Consent in Research

Dana Katz, *University of Pennsylvania, Philadelphia, Pennsylvania,* Patents and Licensing, Policy, Patenting of Inventions Developed with Public Funds

Matthias Kettner, *Cultural Studies Center of Northrhein-Westfalia (KWI), Essen, Germany,* International Aspects: National Profiles, Germany

John Kleinig, *City University of New York, New York, New York,* Research on Animals, Ethics, Principles Governing Research on Animals

Bartha Maria Knoppers, *University of Montreal, Montreal, Canada,* Human Subjects Research, Ethics, and International Codes on Genetic Research

Loretta M. Kopelman, *Brody School of Medicine at East Carolina University, Greenville, North Carolina,* Human Subjects Research, Ethics, and Research on Children

William Lacy, *University of California, Davis, California,* Agricultural Biotechnology, Socioeconomic Issues, and the Fourth Criterion

Robert P. Lawry, *Case Western Reserve University, Cleveland, Ohio,* University-Industry Research Relationships, Ethics, Conflict of Interest

Susan K. Lehnhardt, *Morrison & Foerster LLP, New York, New York,* Transferring Innovations from Academic Research Institutions to Industry: Overview

Josh Lerner, *Harvard University and National Bureau of Economic Research, Boston, Massachusetts,* Strategic Alliances and Technology Licensing in Biotechnology

Bonnie S. Leroy, *University of Minnesota, Institute of Human Genetics, Minneapolis, Minnesota,* Reproduction, Ethics, Sex Selection

Robert J. Levine, *Yale University School of Medical, New Haven, Connecticut,* Human Subjects Research, Ethics, Principles Governing Research with Human Subjects; Human Subjects Research, Ethics, Stopping Rules for Randomized Clinical Trials

Susan Lewis, *Case Western Reserve University, University Hospitals of Cleveland, Cleveland, Ohio,* Human Subjects Research, Ethics, Family, and Pedigree Studies

Darryl R.J. Macer, *Institute of Biological Sciences, University of Tsukuba, Tsukuba Science City, Japan,* International Aspects: National Profiles, Japan

David R. Mackenzie, *University of Maryland, College Park, Maryland,* Agricultural Biotechnology, Law, APHIS Regulation

David Magnus, *University of Pennsylvania, Philadelphia, Pennsylvania,* Eugenics, Ethics

Michael J. Malinowski, *Widener University School of Law, Wilmington, Delaware,* FDA Regulation of Biotechnology Products for Human Use

Wendy K. Mariner, *Boston University, Boston, Massachusetts,* Human Subjects Research, Law, Common Law of Human Experimentation

Cynthia R. Marling, *Ohio University, Athens, Ohio,* Human Enhancement Uses of Biotechnology, Ethics, Cognitive Enhancement

Anne L. Matthews, *Case Western Reserve University, Cleveland, Ohio,* Genetic Counseling

Charles R. McCarthy, *Kennedy Institute of Ethics, Georgetown University, Washington, District of Columbia,* Human Subjects Research, Law, HHS Rules

Jean E. McEwen, *National Human Genome Research Institute, Bethesda, Maryland,* Genetic Information, Ethics, and Information Relating to Biological Parenthood

Glenn McGee, *University of Pennsylvania, Philadelphia, Pennsylvania,* Eugenics, Ethics

Joseph D. McInerney, *Foundation for Genetic Education and Counseling, Baltimore, Maryland,* Education and Training, Public Education about Genetic Technology

Gerald P. McKenny, *Rice University, Houston, Texas,* Gene Therapy, Ethics, Religious Perspectives; Human Enhancement Uses of Biotechnology, Ethics, Therapy vs. Enhancement

Maxwell J. Mehlman, *Case Western Reserve University, Cleveland, Ohio,* Human Enhancement Uses of Biotechnology, Law, Genetic Enhancement, and the Regulation of Acquired Genetic Advantages

Richard A. Merrill, *University of Virginia, Charlottesville, Virginia,* Gene Therapy, Law and FDA Role in Regulation

Jon F. Merz, *University of Pennsylvania, Philadelphia, Pennsylvania,* Patents and Licensing, Policy, Patenting of Inventions Developed with Public Funds

Henry I. Miller, *Stanford University, Stanford, California,* Agricultural Biotechnology, Law, and Food Biotechnology Regulation

Barbara Mishkin, *Hogan Hartson L.L.P., Washington, District of Columbia,* Scientific Research, Law, and Penalties for Scientific Misconduct

Ronald Y. Nakasone, *Graduate Theological Union, Berkeley, California,* Religious Views on Biotechnology, Buddhism

Curtis Naser, *Fairfield University, Fairfield, Connecticut,* Genetic Information, Ethics, Ethical Issues in Tissue Banking and Human Subject Research in Stored Tissues

Dorothy Nelkin, *New York University, New York, New York,* Media Coverage of Biotechnology

Robert M. Nelson, *University of Pennsylvania, Philadelphia, Pennsylvania,* Professional Power and the Cultural Meanings of Biotechnology; Gene Therapy, Ethics, Germ Cell Gene Transfer

Marion Nestle, *New York University, New York, New York,* Agricultural Biotechnology, Policy, and Nutrition

Gilbert S. Omenn, *University of Michigan, Ann Arbor, Michigan,* Federal Policy Making for Biotechnology, Executive Branch, Office of Management and Budget

F. Barbara Orlans, *Kennedy Institute of Ethics, Georgetown University, Washington, District of Columbia,* Research on Animals, Law, Legislative, and Welfare Issues in the Use of Animals for Genetic Engineering and Xenotransplantation

Stuart M. Pape, *Patton Boggs, LLP, Washington, District of Columbia,* Animal Biotechnology, Law, FDA Regulation of Genetically Modified Animals for Human Food Use

Erik Parens, *The Hastings Center, Garrison, New York,* Reproduction, Ethics, Prenatal Testing, and the Disability Rights Critique

Kayhan P. Parsi, *Institute for Ethics, American Medical Association, Chicago, Illinois,* Reproduction, Law, Regulation of Reproductive Technologies

Rebecca D. Pentz, *M.D. Anderson Cancer Center, Houston, Texas,* Genetic Information, Legal, Genetic Privacy Laws

Jeffrey L. Platt, *Mayo Clinic, Rochester, Minnesota,* Animal Medical Biotechnology, Policy, Would Transgenic Animals Solve the Organ Shortage Problem?

Cynthia Powell, *University of North Carolina, Chapel Hill, North Carolina,* Reproduction, Ethics, Prenatal Testing, and the Disability Rights Critique

Madison Powers, *Georgetown University, Washington, District of Columbia,* Genetic Information, Ethics, Privacy and Confidentiality: Overview

Kimberly A. Quaid, *Indiana University School of Medicine, Indianapolis, Indiana,* Genetic Information, Ethics, Informed Consent to Testing and Screening

Philip R. Reilly, *Shriver Center for Mental Retardation, Inc., Waltham, Massachusetts,* Eugenics, Ethics, Sterilization Laws

Herbert Reutimann, *Swiss Priority Programme Biotechnology of the Swiss National Science Foundation, Basel, Switzerland,* International Aspects: National Profiles, Switzerland

Laura F. Rothstein, *Louis D. Brandeis School of Law, University of Louisville, Louisville, Kentucky,* Reproduction, Law, Is Infertility a Disability?

Maria Ruano, *Alain Bensoussan—Avocats, Lyon, France,* International Aspects: National Profiles, France

Paul D. Rubin, *Patton Boggs, LLP, Washington, District of Columbia,* Animal Biotechnology, Law, FDA Regulation of Genetically Modified Animals for Human Food Use

Lilly-Marlene Russow, *Purdue University, West Lafayette, Indiana,* Research on Animals, Ethics, and the Moral Status of Animals

Kenneth J. Ryan, *Harvard Medical School, Boston, Massachusetts,* Scientific Research, Ethics, Scientific Misconduct

Ani B. Satz, *Princeton University, Princeton, New Jersey, University of Michigan Law School, Ann Arbor, Michigan,* Disability and Biotechnology

Jennifer Scott, *Case Western Reserve University, University Hospitals of Cleveland, Cleveland, Ohio,* Human Subjects Research, Ethics, Family, and Pedigree Studies

Harold T. Shapiro, *Princeton University, Princeton, New Jersey,* Federal Policy Making for Biotechnology, Executive Branch, National Bioethics Advisory Commission

Michael H. Shapiro, *University of Southern California Law School, Los Angeles, California,* Human Enhancement Uses of Biotechnology, Policy, Technological Enhancement, of Human Equality

Michael M. Shi, *Warner-Lambert Company, University of Michigan Medical School, Ann Arbor, Michigan,* Pharmacogenetics

Anita Silvers, *San Francisco State University, San Francisco, California,* Disability and Biotechnology

Rivers Singleton, Jr., *University of Delaware, Newark, Delaware,* Transgenic Animals: An Overview

Dominique Sprumont, *University of Neuchâtel, Neuchâtel, Switzerland,* Human Subjects Research, Ethics, and International Codes on Genetic Research

Bonnie Steinbock, *SUNY Albany, Albany, New York,* Reproduction, Ethics, Moral Status of the Fetus

Michele Svatos, *Iowa State University, Ames, Iowa,* Patents and Licensing, Ethics, Ownership of Animal, and Plant Genes

Carol A. Tauer, *College of St. Catherine, St. Paul, Minnesota,* Gene Therapy, Ethics, Gene Therapy for Fetuses and Embryos; Human Enhancement Uses of Biotechnology, Ethics, Human Growth Hormone

Paul B. Thompson, *Purdue University, West Lafayette, Indiana,* Agricultural Biotechnology, Ethics, Food Safety, Risk, and Individual Consent

Françoise Touraine-Moulin, *Immunology Laboratory—Hôpital Neurologique Pierre Wertheimer, Lyon, France,* International Aspects: National Profiles, France

Andrew Trew, *John Carroll University, Cleveland, Ohio,* Animal Medical Biotechnology, Legal, Laws and Regulations Governing Animals as Sources of Human Organs

Gary Varner, *Texas A&M University, College Station, Texas,* Cloning, Overview of Animal Cloning

Marissa Vicari, *La Tour de Salvagny, Lyon, France,* International Aspects: National Profiles, France

Robert Wachbroit, *University of Maryland, College Park, Maryland,* Genetic Determinism, Genetic Reductionism, and Genetic Essentialism; Human Enhancement Uses of Biotechnology: Overview

Wendy E. Wagner, *Case Western Reserve University, Cleveland, Ohio,* Federal Policy Making for Biotechnology, Congress, EPA; Genetic Information, Law, Legal Issues in Law Enforcement DNA Databanks

LeRoy Walters, *Kennedy Institute of Ethics, Georgetown University, Washington, District of Columbia,* Federal Policy Making for Biotechnology, Executive Branch, RAC; Gene Therapy: Overview

David T. Wasserman, *University of Maryland, College Park, Maryland,* Behavioral Genetics, Human

Charles Weiner, *Massachusetts Institute of Technology, Cambridge, Massachusetts,* Recombinant DNA, Policy, Asilomar Conference

Dorothy C. Wertz, *The Shriver Center, Waltham, Massachusetts,* Public Perceptions: Surveys of Attitudes Toward Biotechnology

Peter J. Whitehouse, *Case Western Reserve University, Cleveland, Ohio,* Human Enhancement Uses of Biotechnology, Ethics, Cognitive Enhancement

Georgia L. Wiesner, *Case Western Reserve University, University Hospitals of Cleveland, Cleveland, Ohio,* Human Subjects Research, Ethics, Family, and Pedigree Studies

Michael D. Witt, *Technology Enterprise Development Company, Troy, Michigan,* Transferring Innovations from Academic Research Institutions to Industry: Overview

Nelson A. Wivel, *University of Pennsylvania School of Medicine, Philadelphia, Pennsylvania,* Gene Therapy, Ethics, Somatic Cell Gene Therapy

Audra Wolfe, *University of Pennsylvania, Philadelphia, Pennsylvania,* Federal Policy Making for Biotechnology, Executive Branch, ELSI

INTRODUCTION

If future generations mark the third millennium as the age of biotechnology, they may well regard the dawn of the twenty-first century as its birth. The mapping and sequencing of the human genome is almost complete. Products made with biotechnology, ranging from drugs to pest-resistant crops, are becoming commonplace. Human gene therapy is poised to make clinically significant strides.

In the midst of these astounding developments, a growing cadre of scientists and scholars are struggling to understand the profound ethical, legal, and social implications. Never before has humanity been able so directly to manipulate the code of life. The result is both opportunity and danger on an unprecedented scale. How safe is biotechnology? In what ways will genetic information affect our conceptions of who we are and our relationships with each other? Will the promise of eradicating genetic disease lead us to take untoward experimental risks, or to impose these risks on vulnerable subjects? Can democracy exist in a society in which only the wealthy obtain access to new genetic advances? At what point does a genetically modified human being cease to be a member of the human species?

Our objective in producing the *Encyclopedia of Ethical, Legal, and Policy Issues in Biotechnology* is to bring together the best minds to describe these issues, analyze their implications, and present public policy options. The Encyclopedia contains 112 entries arranged in broad alphabetical order. Related entries are cross-referenced. Each entry includes a list of sources. Virtually all of the entries have undergone peer review by two independent reviewers. For the most part, we have asked authors to be objective. Some entries, however, reflect partisan viewpoints due to their subject matter and the background of their authors. We hope we have made sure that, whenever this occurs, it is obvious to the reader.

The potential audience for this Encyclopedia is extremely broad, ranging from individuals with substantial knowledge and experience in these fields to those just embarking on their journey of understanding. We have endeavored to make all of the entries useful to the former while still accessible to the latter. Yet many entries address complex, technically demanding subjects, and we apologize to readers who find specific entries either too elementary or too abstruse. In addition, completing a project of this scope takes time. We recognize that, in a field as dynamic as biotechnology, descriptions of developments in science, ethics, law, and public policy rapidly become out-of-date. We have undertaken to make entries as current as possible.

Our ultimate goal has been to make this a comprehensive reference work. We began by forming an advisory board of renowned experts, headed by David Blumenthal. With their assistance, we created an exhaustive list of topics and identified potential contributors. We aimed for learned, highly respected authors, individuals who are actively involved in their fields and consequently in great demand. It was not always possible to engage their participation. As a result, there are gaps in coverage, which we mention here to dispel the notion that we simply failed to identify important topics for inclusion. For example, while individual entries describe a number of key government agencies and offices that affect biotechnology policy, we were unable to secure entries for some other government agencies, government offices, industry groups, and interest groups. We include profiles of a number of countries with extensive involvement in biotechnology, but were unable to obtain profiles for some of the countries on our list. The Encyclopedia contains discussions of the views of a number of religions toward biotechnology, but we were unable to obtain discussions for Islam or Roman Catholicism. Moreover, we originally intended to include a separate entry on each topic from an ethical, legal, and public policy perspective. This was not always possible. Nevertheless, we feel that, for the most part, entries that we were able to obtain from one or two of the perspectives provide adequate coverage of the major issues from the other viewpoints.

We wish to thank the authors, the members of the advisory board, our editors at Wiley, our staffs, and our families for their dedication and support.

Thomas H. Murray
Maxwell J. Mehlman

A

ACADEMIC INDUSTRY RELATIONSHIPS IN BIOTECHNOLOGY, OVERVIEW

DAVID BLUMENTHAL
ERIC G. CAMPBELL
Massachusetts General Hospital
Harvard Medical School
Boston, Massachusetts

OUTLINE

Introduction
Recent History of AIRs in Biotechnology
Definitions of AIRs in Biotechnology
Prevalence and Magnitude of AIRs
Benefits of AIRRs
Risks of AIRRs
AIRRs in Genetics (A Special Case)
AIRRs in Biotechnology, 1985–1995
Implications for Policy and Management
Bibliography

INTRODUCTION

Relationships between academic institutions and industries have become central to the biotechnology enterprise, and indeed, to all the life sciences in the United States. Academic–industry relationships (AIRs) in biotechnology are thought to serve a variety of purposes for participating academic institutions (universities, their medical schools, and their associated clinical teaching and research facilities and their faculty), for industries, and for the larger society. It is widely believed that AIRs facilitate the transfer of new knowledge from the academic to the industrial sector, and thereby the application of that knowledge to the practical needs of human beings in this country and around the globe. Funds from industry and from the commercial sales of intellectual properties owned by academic institutions support research and training in those institutions, and thus may spur the development of new knowledge and young investigators needed by both universities and industries. Industry, in turn, benefits not only from the profits realized from transferred academic intellectual property but also from increased productivity in its internal research resulting from enhanced opportunities to recruit talented scientists from academia and from exposure of its investigators to ongoing academic research.

The apparent growth of AIRs has also generated ongoing concerns about their risks. The greatest potential risks affect the academic enterprise. There are fears that AIRs will retard scientific progress through a number of effects: by involving the nation's most talented academic investigators in commercially relevant work, thus distracting them from the pursuit of fundamental questions whose answers will set the stage for the next biological revolution; by promoting secrecy in academic science, which will undermine scientific exchange that is essential to optimal progress in the life sciences; by involving young investigators in commercial projects of lesser scientific interest and import, and thereby compromising the quality of their training. Industrial partners of AIRs, of course, also face potential downsides, though these are primarily business risks of a type that is routine in any for-profit enterprise. Academic institutions may turn out to unproductive partners either because cultural differences between universities and industries cannot be successfully bridged, or because academic work produces little commercializable intellectual property, or because garrulous academics prove unable to protect industrial secrets until they can be commercialized.

Given the stakes involved, it is not surprising that AIRs in biotechnology and the life sciences generally continue to attract considerable attention in the popular press and to be the subject of both praise and deprecation (1–4). This article reviews the status of AIRs in biotechnology at the close of the twentieth century. We cover the following relevant topics:

1. The history of AIRs in biotechnology.
2. The definition of AIRs.
3. Their current prevalence.
4. Evidence of their benefits and risks.
5. Evidence concerning their evolution over time.
6. The policy implications of these findings.

RECENT HISTORY OF AIRs IN BIOTECHNOLOGY

Relationships between academic life scientists and industrial organizations have existed through much of the twentieth century (5). Prior to the 1970s these interactions consisted predominantly of consulting by academicians retained by pharmaceutical companies to help with the solution of particular research problems. Academically based clinicians also participated in clinical trials to test newly developed industrial products. Furthermore universities had been commercializing their intellectual property on a modest scale since the 1920s, when the University of Wisconsin created the Wisconsin Alumni Research Foundation to hold the patent on the technology for irradiating dairy products to instill them with vitamin D (6).

In the late 1970s and early 1980s, however, events in science, law and public policy combined to expand the potential value of AIRs in biotechnology. First and most important was the biotechnology revolution itself. This revolution represented the flowering of research investments by the federal government over 30 years following the end of the Second World War. The signature breakthrough heralding the new era in biotechnology was, of course, the development of recombinant DNA technology by Cohen and Boyer in 1973, but parallel breakthroughs were occurring in monoclonal antibody technologies, large-scale fermentation, genetic sequencing, and genetic synthesis. The arrival of these new techniques suggested that academic research might have much more commercial relevance, both short term and long term, than had been supposed in the past, and also suggested that the next generation of dramatic pharmacological breakthroughs might be based in the biological rather than, as previously, the chemical sciences. During much of the early history of the pharmaceutical industry, the primary source of new agents was the chemical isolation and synthesis of naturally occurring compounds that had been found to have biological activity. Thus major pharmaceutical companies had developed deep expertise in chemistry, but were ill-prepared to take advantage of the biological revolution occurring in university laboratories. It became a pressing business priority therefore for the pharmaceutical industry to develop relationships with universities. AIRs would allow pharmaceutical companies to capture new intellectual property arising in universities, and would facilitate retraining of current industry investigators and/or recruitment of new talent who could then work within industry to exploit the new biotechnologies.

The dawn of the biotechnology revolution coincided with policy developments that facilitated AIRs in biotechnology. The 1970s were a time of deep economic anxiety for the United States (7,8). Rising oil prices and decades of complacency on the part of major U.S. industries had combined to produce rapid inflation and stagnant productivity at a time when Japanese industries were thriving. A crisis of confidence in the U.S. economy ensued, and the U.S. government began examining strategies for restoring the vitality of the country's economy. Policy makers concluded that as one such strategy, the United States should take better advantage of its large investment in university research (9). They also concluded that lack of incentives for universities and their scientists to exploit the commercial potential of their work was a major impediment to the success of this strategy. Congress in 1980 enacted the Bayh-Dole Act, which enabled universities to claim ownership to intellectual property resulting from federally sponsored research. The law also required that inventors within universities receive a share of the gains from commercialization of these properties. At least in theory, Bayh-Dole gave universities and their scientists a financial motive to cooperate with industrial partners.

Still another development, this one in patent law, gave an additional boost to the development of AIRs. In 1980 the Supreme Court ruled in the *Diamond v. Chakrabarty* case that it was legal to patent new life forms created as a result of biotechnological manipulation. Since many of the most promising products of the biotechnology revolution resulted from the creation of novel cells and organisms that yielded valuable biological agents, the Chakrabarty case reassured industries and universities that the intellectual property likely to result from AIRs (and other biotechnology endeavors) could be protected under existing patent law.

Following these changes in science, policy, and law, a number of highly publicized AIRs ensued. One of the first, the subject of critical congressional hearings, was an arrangement between the Massachusetts General Hospital and a German chemical and drug company, Hoechst A.G. Hoechst funded not only research at the MGH but also the creation of a new department of genetics and the construction of a research building. Other large relationships developed in the early 1980s as well: between Harvard Medical School and the Dupont Co., Monsanto and the Washington University, Bristol-Myers and Yale University, and others. Equally interesting was the emergence of small, biotech start-up companies founded by and with university faculty members. These included a number of enterprises that have survived to this day: Amgen, Biogen, Genentech, Immunex, and Chiron, to name a few. It is estimated that university faculty participated in the founding of 500 such companies over the 1980s and 1990s (8).

The participation of university faculty in founding new companies was hardly unprecedented. During the 1960s a number of engineers had left universities such as Massachusetts Institute of Technology, Stanford University, and California Institute of Technology to found the computer companies that gave rise to Silicon Valley and Route 128. However, for the most part, these professors cut their formal ties to the university, becoming full-time business persons. In contrast, many university-based biotechnology entrepreneurs wanted to maintain their faculty positions even while they held major roles in start-up companies. Some faculty became major equity holders in for-profit enterprises, and then conducted university-based research funded by those companies. This reflected not only the business aspirations of faculty but also the wishes of the venture capital companies that funded their new start-ups. In many cases venture capitalists felt that the success of the new businesses was dependent on providing faculty with strong incentives to stay involved and support the work of the start-up. Whatever the cause, the participation of faculty in founding new companies created a new and more intimate form of AIR.

There was also ample precedent prior to the biotechnology revolution for the funding of research in universities by industry. Research relationships between universities and companies were common in chemistry and engineering disciplines prior to the biotechnology revolution, and even after the arrival of the new biotechnologies, AIRs in chemistry and engineering were more prevalent than in the biological sciences. In 1985, 43 percent of faculty principal investigators in chemistry and engineering had research relationships with industries, compared with 23 percent involved in biotechnology (10).

Nevertheless, AIRs provoked greater controversy and more soul-searching in the life sciences than had AIRs in other fields. There were several reasons for this. First and most important were the potential implications of AIRs in biotechnology for health care services. The products of AIRs were likely to yield pharmaceuticals and devices that would be used in the treatment of patients. Observers worried that conflicts of interest on the part of university researchers might cause research bias that would ultimately hurt patients. More immediately, some research sponsored by biotechnology companies required the use of human subjects. Concerns arose about whether university-based researchers would let financial interests compromise their management of research subjects in ways that could adversely affect their health in the short term.

Second, university research in biomedicine benefits from billions of dollars annually in federal support. Universities and researchers were very anxious that AIRs, which support only about 14 percent of research in U.S. academic health centers compared to 67 percent from the federal government (11), not in any way compromise the federal government's or voter's trust in academic biomedical research (12). The federal government, through the National Institutes of Health (NIH), was further concerned that the participation of faculty in AIRs not reduce the effectiveness of federal investments in research or training (13).

Third, the fact that university researchers were founding companies at a great rate, and staying in the university, seemed to set AIRs in biotechnology apart from AIRs in other fields. AIRs in biotechnology were creating a cadre of faculty entrepreneurs who continued to have teaching and administrative responsibilities in universities, and continued to participate in federal peer review and consulting roles, as though nothing had changed. The conflicts of interest and conflicts of commitment among such entrepreneurs were likely to be particularly intense, but there was no way to ensure, under then existing university policies, that colleagues or outside clients would be aware of the existence of such conflicts.

Fourth, the above concerns focused attention on the fact that there was virtually no data in field of biotechnology or any other academic discipline on the extent and consequences of AIRs in universities. The balance of this article explores these issues.

DEFINITIONS OF AIRs IN BIOTECHNOLOGY

Before trying to describe the extent and consequences of AIRs in biotechnology, it is useful to define more precisely what AIRs consist of from our standpoint: What exactly we are describing. For the purposes of this discussion, AIRs consist of arrangements between for-profit corporations and academic institutions (or their faculty, staff, and trainees) in which something of value is exchanged. Commonly universities provide a service (e.g., research or training) or intellectual property (in the form of a patent, license, or advice) in return for financial considerations of various types (research support, honoraria, consulting fees, royalties, or equity) (14).

AIRs in the health sciences can assume a variety of forms. The following types of AIRs are among the most common but by no means exhaust the alternatives:

1. Academic-industry research relationships (AIRRs): the support by industry (through grant or contract) of university-based research.
2. Consulting relationships: the compensated provision of advice or information, usually by individual faculty members, to commercial organizations.
3. The sale or licensing of patents by university to industries.
4. The participation by academic institutions or their faculty in the founding and/or ownership of new companies commercializing university based research: AIRs of this type often occur when cash-poor start-up companies use small amounts of equity to compensate faculty for consulting or other services. However, academic institutions or their faculty may also participate in the founding of new commercial entities, sometimes taking much larger amounts of equity in return for contributions of intellectual property (14–16).
5. Academic-industry training relationships (AITRs): industries provide support for the research or educational expenses of graduate students or postdoctoral fellows, or contract with academic institutions to provide various educational experiences (e.g., seminars or fellowships) to industrial employees.

These and other forms of AIRs may occur singly or in combination. The mixed forms of AIRs (e.g., those involving AIRRs and consulting or equity holding) often raise the most troubling concerns about conflict of interest because multiple relationships often involve more money (both real and potential) than single forms of AIRs.

Most current information on the dimensions and consequences of the AIR phenomenon in biotechnology concerns AIRRs, and much of that information pertains to AIRRs in the life sciences generally rather than specifically in the subfield of biotechnology. For that reason the ensuing discussion focuses particularly on AIRRs, and often references the field of life sciences generally. Nevertheless, where data specific to the biotechnology area and data concerning other types of relationships are available, we convey these as well.

PREVALENCE AND MAGNITUDE OF AIRs

The most recent nationally representative data on the prevalence and magnitude of AIRs stems from surveys of industries and faculty members in 1994 and 1995 (15,17). A 1994 survey of senior executives in a representative sample of life-sciences companies revealed that over 90 percent participated in some form of AIR. The most prevalent form was retention of university faculty as consultants (88 percent). Fifty-nine percent participated in AIRRs and 38 percent in AITRs. Seven percent of companies reported that faculty members were significant equity holders in their companies (17).

A contemporaneous survey of 2052 faculty members at the 50 most research intensive U.S. universities revealed that 28 percent of respondents reported receiving some research support from industrial sources (15). The prevalence of support was greater for clinical (36 percent) than nonclinical (21 percent) departments. Among a subgroup of faculty whose research involved what the authors defined as "biotechnologies" (recombinant DNA technology, monoclonal antibodies, gene synthesis, gene sequencing, tissue culture, enzymology, and large-scale fermentation), 21 percent of principal investigators on research grants reported receiving research support from industry. This subsample was specifically chosen to be comparable to a 1986 survey of faculty using the same biotechnology techniques (see below) (10,15). Thus, as of the middle of the 1990s, between 20 and 30 percent of life-sciences and biotechnology faculty in research-intensive U.S. universities participated in AIRRs. There is no reason to suppose that this number has declined since that time, so current levels of faculty participation in AIRRs are likely to be at least that high.

Characteristics of AIRRs in the life sciences suggest that relationships tend on average to be small in size and short in duration. Industry respondents indicated that 71 percent of AIRRs in 1994 and 1995 were funded at less than $100,000 a year. Only 6 percent of responding firms provided annual funding of $500,000 or more. For 84 percent of respondents whose firms had relationships with academe, the typical relationship lasted two years or less. The generally short duration of AIRRs and the small funding levels suggest that at that time, the research they supported tended to be targeted—that is, applied rather than fundamental (17). AIRRs also constituted a relatively small proportion of the total research funding to universities in the mid-1990s: about 12 percent of the total. (The 14 percent figure cited above refers to academic health centers, which include teaching hospitals and which are more heavily weighted toward clinical departments, where the prevalence of AIRRs is higher).

Recent data about the prevalence of other types of AIRs are scarce. The 1985 survey of biotechnology faculty cited above indicated that 7 percent held equity in a biotechnology company related to their own work, while 47 percent consulted to industry. In a separate survey of nearly 700 graduate students and fellows in life-sciences departments at six leading universities, 19 percent reported receiving some research support from industry (18). Krimsky et al. showed in 1988 that as many as 31 percent of scientists in certain life-sciences departments had some form of link to outside firms (19).

BENEFITS OF AIRRs

The best documentation of the benefits that result from the relationships between academic institutions and industry derive from studies of AIRRs. For this reason we primarily focus AIRRs; however, we will comment on the benefits that derive from other forms of AIRs when such data are available.

The most obvious benefit of AIRRs is that these relationships provide funds to support the research conducted in academic institutions. In a 1994 survey of senior research executives at 306 life-sciences companies in the United States, respondents reported that their companies supported more than 1500 academe based research projects at a cost of over $340 million (17). Based on these reports it was estimated that the life-sciences industry as a whole supported more than 6000 life-sciences, projects and expended $1.5 billion for academic research in the life sciences.

Receipt of industry funds is not associated with detectable adverse effects on academic productivity.

Indeed, if anything, AIRRs are associated with enhanced productivity on the part of involved university investigators. Faculty involved AIRRs exhibit higher levels of research productivity than faculty without such relationships. In a 1994-95 survey of over 2000 life-sciences faculty in the 50 most research intensive universities, faculty with funding from industry published significantly more articles in peer-reviewed journals in the previous three years than faculty without AIRRs (15). Faculty benefit from increased publications, since articles in peer-reviewed journals represent one of the main criteria by which faculty are awarded the trappings of academic success including promotions, tenure, prizes, future research grants, positions in professional organizations, and ultimately a place in the history of the scientific endeavor (21). At an institutional level more publications by faculty translate into greater prestige and, perhaps, an increased ability to attract top students, faculty, and future research funding for universities.

In addition to publications, AIRRs are associated with an increased likelihood of commercial activities on the part of faculty and their institutions. Blumenthal and colleagues (1996) found that compared to faculty without AIRRs, those with industry funding were significantly more likely to report that they had applied for a patent (24 vs. 42 percent), had a patent granted (12.6 vs. 25 percent), had a patent licensed (8.7 vs. 18.5 percent), a product under review (5.5 vs. 26.7 percent) a product on the market (10.8 vs. 26.1 percent), or a start-up company (6.0 vs. 14.3 percent) (15). A number of benefits may accrue to faculty as a result of the commercial opportunities that are associated with AIRRs, including financial returns, the opportunity to see the results of their research developed into useful products and services, and perhaps enhanced career opportunities in the industrial sector. Universities benefit from faculty commercialization, since their polices often provide the institution with the option to participate in commercial ventures such as supporting the costs of filing a patent in exchange for a portion of the licensing revenues or by providing venture capital funding for a start-up in exchange for a share of the future profits of that firm.

Several other benefits accrue to universities, faculty, and students as a result of AIRRs. For example, 66 percent of faculty reported that research grants and contracts from industry involved less "red tape" than those from federal sources, 60 percent felt that AIRRs enhanced career opportunities for students, 49 percent felt AIRRs increased the prestige of their department, 37 percent felt AIRRs contributed to their promotion and tenure, and 34 percent reported that these relationships led to salary increases (15).

Like universities and their faculty, companies also benefit from AIRRs. In our 1994 survey of life-sciences companies 60 percent of firms with AIRRs have realized patents, products and sales as direct result of those relationships (17). In addition to direct benefits such as patents, products, and sales, companies receive access to ideas, knowledge, and a pool of talented potential researchers. For example, 56 percent of companies with AIRRs reported that they depend very much on these relationships to "keep staff current" with important research, 53 percent depend on them to provide ideas for new products and services, and 37 percent to aid in recruiting able researchers. Only 29 percent reported that they rely somewhat or very much on AIRRs to invent the products that the company will license (17).

Perhaps one of the most important benefits of AIRRs to industry may be that these relationships provide sponsors with access to the most recent research results of faculty—often months or years ahead of competitors. It is common for most AIRRs to allow a sponsor 30 to 90 days to review the results of the research they sponsored prior to submission for publication. An executive of a company said that in his field the published literature is "miles behind the front line of what is happening in universities" (16). For companies, especially those rapidly developing fields such as human genetics, using AIRRs as a means of access to new knowledge that is not yet public may constitute a considerable competitive advantage over companies without AIRRs.

Society ultimately benefits from AIRRs in terms of the increased flow of research results from universities into the industrial setting—a process often called *technology transfer*. A study by Cohen, Florida, and Goe (1994) suggested that collaborative research and development and other forms of intimate interactions between university researchers and industry personnel were more effective in transferring information into the industrial sector than communication through traditional academic channels such as publications and presentations (22). Cohen and his colleagues studied the extent, characteristics, and consequences of university–industry research centers (UIRCs). Based on a national survey of UIRC directors (response rate 48 percent), they found that the overwhelming majority of UIRCs were created with government support (70 percent) and continued to derive about 86 percent of their research funds from governmental sources. On average UIRCs derived 46 percent of their funds from the federal government compared to only 31 percent from industrial sources. These data suggested the important role of federal funding in creating and maintaining UIRCs as a potential mechanism for technology transfer and indicated that federal and industrial research coexist in close proximity to university settings. Cohen and his colleagues also found that UIRCs reported generating 211 patents in 1990. UIRC patent productivity per dollar invested was about one-third of that observed in industrial research and development (R&D) laboratories. Patent productivity tended to be higher in small UIRCs, those predominately funded by industry and UIRCs in the fields of biotechnology and advanced materials. The biotechnology field was also most productive among all scientific fields of new products from UIRCs. Twenty-two percent of UIRCs reported spin-off companies resulting from their work (22).

Broader evidence of the localized benefits of AIRRs have resulted from studies of what economists refer to as "spill over." In a seminal study Jaffe (1989) found a positive association between university-based and industrial innovation among companies in the same state, as measured by the number of patents issued to the firms between 1972 and 1986. In the drug industry, a 1 percent increase in university-based, biomedical research was associated with a 0.28 percent increase in the number of patents issued to drug firms (23). Additional work by Jaffe, Trajtenberg, and Henderson (1993) provided further confirmation of spillovers in a study of whether patents were more likely to cite earlier patents that originated in the same geographic location compared to earlier patents originating in other geographic settings (24). They found that later patents were significantly more likely to cite earlier patents that originated nearby than they were to cite a control group of patents from the same field and technological area that resulted from work done in a different state, standard metropolitan statistical area, or county.

In addition there are real, visible effects of AIRRs on local economies. For example, academic researchers have played a seminal role establishment of high tech industries. Etzkowitz (1988), Etzkowitz and Peters (1991), and Dorfman (1993) have documented the role of investigators at the Massachusetts Institute of Technology, Stanford University, and the University of California in founding and staffing local biotechnology and electronics companies such as Raytheon, Data General, Digital Equipment Corporation, Genetics Institute, and Biogen in the Route 128 area and Genentech in the Silicon Valley (25–27). Creation of these firms/industries no doubt included the creation of high-paying technical and professional jobs, an increased source of tax revenues for local economies, and an increased inflow of venture capital and other research-related services into the local economies.

RISKS OF AIRRs

As with all forms of individual and institutional behavior AIRRs have risks that must be addressed. According to Derek Bok, former president of Harvard University, AIRs may "...divert the faculty. Graduate students may be drawn into projects in ways that sacrifice their education for commercial gain. Research performed with an eye towards profit may lure investigators into conflicts of interest or cause them to practice forms of secrecy that hamper scientific progress. Ultimately, corporate ties may undermine the university's reputation for objectivity" (16). This quote encapsulates many of the worst fears about the potential negative influences of AIRs on the academic enterprise.

A frequently cited risk of AIRs is the potential for increased secrecy in academic science. Secrecy in scientific research can take a number of forms including delaying publications for an extended period of time, faculty refusing to share research results and materials when asked by other academic scientists, and the keeping of trade secrets. Blumenthal and colleagues (1997) found in their 1994–95 survey of academic life scientists that 27.2 percent of researchers with AIRRs delayed publication of their research longer than 6 months compared to 16.5 percent of faculty without industry funding ($p < 0.001$) (15). Faculty with AIRRs were significantly more likely than those without AIRRs to report having denied other university faculty access to their research results or biomaterials such as cell lines, tissues, reagents, and so on (11 vs. 8 percent, $p < 0.01$). Participation in trade secrecy, defined as information kept secret to protect its proprietary value was significantly more prevalent among researchers with AIRRs (14.5 percent) compared faculty without AIRRs (4.7 percent, $p < 0.001$).

While secrecy may stem from individual scientists' motivations and career aspirations, it is likely that the policies of universities and corporate research sponsors encourage secrecy as well. According to an NIH study in the early 1990s, 20 percent of AIRRs permitted companies to delay publications for longer than six months, so that companies can review findings and secure rights to commercializable products (28). More than 80 percent of life science companies supporting research in universities in 1994 reported that their agreements sometimes require academic researchers to keep research results secret prior to filing a patent (17). Cohen and colleagues found that 41 percent of UIRCs had restrictions on their ability to communicate their research results to the general public, 29 percent on their communication with faculty at other universities, 21 percent on sharing information with faculty in their own institution ,and 13 percent on sharing information with scientists within their own research center (22).

Another risk mentioned by Bok was that AIRRs may have a negative impact on scientists in training. A 1985 survey of 693 advanced trainees in the life sciences at 6 universities found that 34 percent of respondents whose projects were supported by industry felt constrained in discussing their research results with other scientists (18). Further this study found that graduate students and post-doctoral fellows whose projects were supported by industry reported significantly fewer publications on average (2.62) than those with no industry support (3.67).

A third risk of AIRRs is that these relationships may lure faculty away from basic research, which has long been the mainstay of academics, towards research that has commercial applications. Unpublished research by Blumenthal and colleagues conducted in 1994 found that more than half (54 percent) of all life scientists felt AIRRs created pressures on faculty to "spend too much time on commercial activities" (29). Blumenthal et al. (1996) found that faculty members with industrial support were significantly more likely than those without AIRRs to report that their choice of research topics had been influenced somewhat or greatly by the likelihood that the results would have commercial application (35 vs. 14 percent, $p < 0.001$) (15).

A fourth risk associated with AIRRs is that too much funding from industry may be associated with lower research productivity on the part of involved faculty. Faculty who received more than 66 percent of their funding from industry published significantly fewer articles over a three-year period, published in less influential journals, and were less likely to report commercial outcomes from their research than faculty with less support from industry (15). This finding may reflect the faculty that faculty with more than two-thirds of their funding from industry are less able than others to attract peer-reviewed

support from governmental and nonindustrial funding sources.

As with all business relationships there is the risk that AIRRs may not produce the outcome(s) industrial sponsors had predicted or hoped for. There are some data to suggest that the behavior of faculty may cause some AIRRs to be less useful to companies than when they were conceived. In 1995, 33 percent of life science firms with AIRRs reported that academic scientists had changed the direction of research conducted under an AIRR to the extent that the results were less useful to the corporate sponsor than had been originally expected (17).

Bok articulated the greatest potential risk of AIRRs when he wrote that these relationships may "undermine the university's reputation for objectivity." The public's generous support for research is founded on the belief that the results of research represent faculties' best effort to detect the truth, untainted by commercial interests. Recent research regarding the effects of AIRs on the outcomes of studies that examined the efficacy and safety of calcium-channel antagonists in the treatment of cardiovascular disorders suggests there may be cause for some concern (30). Between March 1995 and September 1996 more than 70 studies were published that were either supportive, neutral, or critical with respect to the safety and efficacy of using calcium-channel antagonists in the clinical setting (30). Stellfox and colleagues surveyed the authors of these 70 papers about their relationships with companies producing the antagonists or with companies producing competing products. He found that 96 percent of authors whose research was supportive of the use of calcium-channel antagonists had financial relationships with companies that produced antagonists, compared with only 60 percent of those whose research was neutral, or 37 percent of those whose findings were critical. Further he found that despite the fact that 44 out of the 70 authors had AIRs only 2 of the studies disclosed the authors' relationships with industry. As Stellfox and colleagues wrote, "We wonder how the public would interpret the debate over calcium channel antagonists if it knew that most of the authors participating in the debate had undisclosed financial ties with pharmaceutical manufacturers. ...Full disclosure of relationships between physicians and pharmaceutical manufacturers is necessary to affirm the integrity of the medical profession and maintain public confidence" (30).

Despite concerns, several of the risks associated with AIRRs have not been substantiated. First, there is no evidence that AIRRs have resulted in a diversion of faculty effort from academic and administrative commitments — so-called conflicts of commitment. Data from Blumenthal and colleagues (1996) show that faculty with AIRRs spent as much time per week teaching undergraduates, graduate students, and postdoctoral fellows as those without AIRRs (15). Also AIRRs were associated with increased rather than decreased service activities on the part of faculty to their institution and their discipline. Faculty with AIRRs were significantly more likely than those without AIRRs to have been chairs of a departments, universitywide administrators, members of review panels for federal agencies, editors or editorial board members of journals, the heads or associate heads of research institutes, chairs of a universitywide committees, or officers of professional associations (15).

AIRRs IN GENETICS (A SPECIAL CASE)

AIRRs in genetics differ significantly from the other life-sciences fields in terms of their prevalence, magnitude, benefits and risks. First, AIRRs are significantly more prevalent in the field of genetics than the other life-sciences fields. Based on a survey of 210 life-sciences companies in the United States, Blumenthal and colleagues (1997) found that after controlling for firm size, companies conducting genetics research were significantly more likely than nongenetics firms to support research in universities (69 vs. 45 percent, $p < 0.005$) (31). Also genetics firms were significantly more likely to support research training than other life science firms (46 vs. 33 percent, $p < 0.005$).

Second, AIRRS in genetics are longer in duration and involve more money that AIRRs in other life-sciences fields (31). Among genetics firms, 19 percent of AIRRs lasted three years or more compared to 14 percent among nongenetics companies. Agreements of one year or less were significantly less common in among genetics firms than nongenetics firms (15 vs. 34 percent, $p < 0.05$). Among large companies the median amount of research support provided to universities by genetics companies was $102,000 compared to $70,000 for large, nongenetics companies.

Third, there is some evidence suggesting that AIRRs in genetics have greater benefits in some respects than AIRRs in other fields (31). Among faculty with AIRRs, genetics researchers reported publishing more articles in peer reviewed in the preceding three years (18 vs. 14.5), participating in more service related activities within their institution or discipline and publishing in more influential journals than nongeneticists. Also genetics researchers with industry support were significantly more likely than other researchers with AIRRs to have applied for a patent, received a patent, licensed a patent, or started a new company.

Fourth, AIRRs in genetics are more prone to the risks of data-withholding than those in the other life sciences (31). Genetics firms with AIRRs were more likely than nongenetics firms with AIRRs to report that their agreements with universities required researchers to keep results secret beyond the time to file a patent. Also genetics researchers with AIRRs were significantly more likely to report that trade secrets resulted from their university research, to have delayed publication of their results in order to file for a patent, and to have denied direct requests from other scientists for access to their research results and materials than nongenetics researchers with AIRRs.

AIRRs IN BIOTECHNOLOGY, 1985–1995

Since the mid-1980s the rate of faculty participation in AIRRs has remained about the same (10,15). Based on 1985 and an 1995 survey of academic biotechnology researchers (faculty using recombinant DNA, monoclonal antibodies, gene synthesis, gene sequencing, cell tissue and culture, enzymology, and large-scale fermentation), 23 percent of biotechnology faculty in 1985 reported that they were principal investigators on research grants or projects funded by industry compared to 21 percent in 1995. For these faculty, industry supplied 7.4 percent of their total research budgets in 1985, as compared with 5.8 percent in 1995.

The experiences of faculty members in 1985 and 1995 were similar in other ways as well. From 1985 to 1995 the percentage of biotechnology researchers with AIRRs who reported that trade secrets had resulted from their research increased slightly from 12 percent to 17.2 percent. However, among those without AIRRs the percentage who had engaged in trade secrecy doubled from 3 percent in 1985 to 6.6 percent in 1995. Similar results were found regarding biotechnology researchers' choice of investigational topics. In 1985 and 1995, 30 percent of those with AIRRs reported that their choice of research topics had been influenced to some extent or to a great extent by the likelihood that the results would have commercial application. However, among those without AIRRs the percentage who reported that their choice of research topics had been influenced to some extent or to a great extent by the likelihood that the results would have commercial application doubled from 7 percent in 1985 to 14 percent in 1995.

IMPLICATIONS FOR POLICY AND MANAGEMENT

Persistent uncertainties about the scope and consequences of AIRs in biotechnology somewhat complicate the tasks of regulating and managing these relationships. Nevertheless, it is possible to draw some reasonable conclusions concerning these issues from existing data on the prevalence, magnitude, benefits, risks, and historical development of AIRs.

First, AIRs in biotechnology and the other life sciences have documented benefits that constitute a persuasive argument for continuing and even promoting AIRs of certain types and in certain situations. These benefits are best demonstrated for AIRRs, patent and licensing arrangements, and academe–industry training relationships, and include increased funding for academic research, possible increases in rates of patenting and publishing on the part of academic investigators, income from patents and licenses, and the apparent enhancement of the educational experiences of trainees.

Second, many types of AIRs also pose real risks for the academic institutions that participate in them. These risks include reductions in the openness of communication among investigators, channeling of research in more applied directions, and threats to the public credibility of the life sciences. With the exception of certain limited situations, however, these risks seem not to present a clear and immediate danger to the conduct of science in universities or to their educational missions and do not justify at this time limiting the freedom of academic institutions to establish AIRs. This conclusion could change as further information emerges concerning the long-term benefits and risks of AIRs both for universities and for the scientific enterprise. For the time being, however, policy and management should emphasize disclosure on the part of life scientists, vigilance on the part of academic and public administrators, and further research into the positive and negative effects of these relationships. In addition it would seem prudent for academic institutions to avoid excessive dependence on industrial relationships for research support. This will minimize the chances that the effects of AIRs on the norms and behaviors of universities will have significant, lasting impact on the character of life science research.

Third, certain forms or combinations of AIRs pose qualitatively greater concerns than others. Intense conflicts of interest arise in equity-holding AIRs (especially those where potential gains to investigators are large) and in major, sustained consulting arrangements in which faculty members derive appreciable amounts of their annual income from one particular company or a small number of companies. Public and private managers are justified in subjecting these relationships to a higher level of scrutiny, in limiting the size and number of such relationships among their faculty, and, in certain cases, in forbidding them altogether.

Another situation requiring different treatment is AIRs in which patients are directly involved, such as research involving living human subjects. When academic clinical investigators have financial relationships with companies (usually in the form of substantial equity positions or major consulting income) that may benefit from their clinical research, the resulting conflicts of interest create the appearance or reality that the interests of human subjects may in some way be compromised. This could occur, for example, if financially involved clinical investigators failed to fully inform prospective research subjects of the benefits and risks of the clinical protocols, inappropriately pressed subjects who wished to withdraw to remain in a research protocol, or engaged in biased patient recruitment to increase the chances of a successful outcome. The chances of such occurrences are less for large-scale, multicenter clinical trials than for more exploratory types of clinical investigation. Nevertheless, even large-scale clinical trials may not be exempt from such concerns when they occur in a discipline with a small number of leading investigators who may all have relationships with sponsoring companies.

One final situation that raises special issues is when academic administrators develop personal financial relationships with outside firms with which their faculty are also involved. Such relationships have no documented benefits but jeopardize the real or apparent ability of universities to manage objectively their faculty members' AIRs.

A fourth conclusion is that it is neither practical nor desirable for the federal government to dictate detailed rules for management of AIRs in biotechnology or the other life sciences across the United States. Past experience with oversight of research involving human subjects suggests the feasibility of permitting individual institutions to take responsibility for overseeing AIRs. However, experience with institutional review boards also suggests the need for continuing federal supervision and review of universities as they attempt self-regulation of sensitive ethical and policy issues related to research. Such supervision and review will undoubtedly be required for federal sponsors of research in the area of AIRs.

With these conclusions in mind, the following specific recommendations for universities and for federal research sponsors seem appropriate. For academic institutions:

1. All universities conducting biotechnology and health science research should require regular disclosure by faculty (including those not receiving federal funds) and senior administrators of financial relationships with companies that have life-sciences or health care interests. These disclosures should be reviewed carefully and confidentially by academic managers. Disclosure constitutes the minimal acceptable response of academic institutions to the demonstrated risks posed by AIRs in biotechnology and other fields. It is also impossible for academic institutions to learn from experience with AIRs if they do not know they exist.

2. Academic institutions should develop explicit policies for deciding which AIRs in biotechnology are desirable and undesirable. In this regard it would be prudent for universities to prohibit research involving living human subjects on the part of investigators with major financial interests in companies that may benefit from the results of that research.

3. Academic institutions should avoid excessive dependence on industrially sponsored research, given the proven risks of such relationships. The definition of *excessive* will undoubtedly vary from institution to institution, but a reasonable rule would be to keep industrial research support below one-third of total funds for biotechnology and other life-sciences research. Given dramatic recent increases in NIH funding, meeting this target will be appreciably easier than it was in the 1990s.

For federal sponsors of research:

1. The federal government should not fund clinical research when the principal investigator has a personal financial relationship with a company that may be affected by the outcome of that research. Exceptions may be made in cases where the relationships are minimal according to standards defined by the federal government.

2. The federal government should not fund research at institutions that do not have formal policies and procedures governing academe–industry relationships, or where such policies and procedures cannot be fairly and effectively enforced. The latter circumstance would arguably exist when there is no effective enforcement process, or where academic administrators themselves share financial interests in companies in which faculty members are also involved.

Academic industry relationships in biotechnology and the life sciences generally are part of the modern life-sciences economy. They cannot and should not be prevented. But their benefits should not be exaggerated, nor their risks minimized. Academic institutions are priceless resources whose integrity and independence are critical to the long-term health of the American people and the American economy.

BIBLIOGRAPHY

1. E. Lau, *Washington Times*, D2 (1999).
2. R.L. Hotz, *L.A. Times*, A1 (1999).
3. R. Saltus, *Boston Globe*, A6 (1998).
4. C.G. Stolberg, *N.Y. Times*, A17 (1998).
5. J.P. Swan, *Academic Scientists and the Pharmaceutical Industry*, The Johns Hopkins University Press, Baltimore, MD, 1988.
6. D. Blumenthal, S. Epstein, and J. Maxwell, *N. Engl. J. Med.* **314**, 1621–1626 (1986).
7. W.E. Deming, *Out of the Crisis*, MIT Press, Cambridge, MA, 1982.
8. M. Kenney, *Biotechnology: The University-Industry Complex*, Yale University Press, New Haven, CT, 1988.
9. D.J. Prager and G.S. Omenn, *Science* **207**, 379–84 (1980).
10. D. Blumenthal et al., *Science* **232**, 1361–1366 (1986).
11. The Commonwealth Fund Task Force on Academic Health Centers, *From Bench to Bedside: Preserving the Mission of Academic Health Centers*, The Commonwealth Fund, New York, 1999.
12. Hearings Before the Human Resources and Intergovernmental Relations Subcommittee, 101st Congr., 1st Sess., Is Science for Sale? Conflicts of Interest v. the Public Interest, 1989.
13. General Accounting Office, *Controlling Inappropriate Access to Federally Funded Research Results*, U.S. General Accounting Office, Washington, DC, 1992.
14. D. Blumenthal, *J. Am. Med. Assoc.* **268**, 3334–3339 (1992).
15. D. Blumenthal et al., *N. Engl. J. Med.* **335**, 1734–1739 (1996).
16. N. Bowie, *University-Business Partnerships: An Assessment*, Lanham: Rowan & Littlefield Publishers, London, 1994.
17. D. Blumenthal et al., *N. Engl. J. Med.* **334**, 368–373 (1996).
18. M. Gluck, D. Blumenthal, and M.A. Stoto, *Res. Policy* **16**, 327–366 (1987).
19. S. Krimsky, J.G. Ennis, and R. Weissman, *Sci. Technol. Hum. Values* **16**, 275–287 (1991).
20. D. Blumenthal et al., *Science* **231**, 242–246 (1986).
21. M.F. Fox, in J.C. Smart, ed., *Higher Education: Handbook of Theory and Research*, Agathon Press, New York, 1985, pp. 255–282.
22. W.R. Cohen, R. Florida, and W.R. Goe, *University-industry Research Centers in the United States*, Carnegie Mellon University, Pittsburgh, PA, 1994.
23. A.B. Jaffe, *Am. Econ. Rev.* **79**, 957–990 (1989).
24. A.B. Jaffe, R. Trajtenberg, and R. Henderson, *Q. J. Econ.* **108**, 557–598 (1993).
25. N.S. Dorfman, *Res. Policy* **12**, 299–316 (1983).
26. H. Etzkowitz, in E. Mendelsohn, M.R. Smith, and P. Weingart, eds., *Science, Technology and the Military*, Kluwer Academic Press, Boston, MA, 1988, pp. 515–540.
27. H. Etzkowitz and L. Peters, *Minerva* (Summer) 133–66 (1991).
28. B. Healy, *N. Engl. J. Med.* **329**, 725–727 (1993).
29. D. Blumenthal, Unpublished results from a 1995 survey of life science faculty, 1999.
30. H.T. Stelfox et al., *N. Engl. J. Med.* **339**, 101–106 (1998).
31. D. Blumenthal, N. Causino, and E.G. Campbell, *Nat. Genet.* **16**, 104–108 (1997).

See other entries TRANSFERRING INNOVATIONS FROM ACADEMIC RESEARCH INSTITUTIONS TO INDUSTRY: OVERVIEW; UNIVERSITY-INDUSTRY RESEARCH RELATIONSHIPS, ETHICS, CONFLICT OF INTEREST.

AGRICULTURAL BIOTECHNOLOGY, ETHICS, FAMILY FARMS, AND INDUSTRIALIZATION

JEFFREY BURKHARDT
University of Florida
Gainesville, Florida

OUTLINE

INTRODUCTION

After nearly a century of neglect, agriculture is receiving increased attention from intellectuals concerned with social ethics. Along with the ethical dimensions of agriculture's impact on the environment and nonhuman animals, one fundamental ethical concern is the plight of the family farm, the traditional farming unit around the globe. The family farm is considered by many ethicists to have special moral, cultural, or political-economic significance, and various social and economic forces—especially new agricultural technologies—appear to threaten to drive the family farm to extinction. The increased reliance on high technology approaches to farming, and the increasing dependence of agriculture on other sectors of the economy, especially manufacturing and petrochemical refining, is referred to as the industrialization of agriculture. While the industrialization of agriculture has been a century-long trend, the emergence of agricultural biotechnology in the 1970s heightened concerns that the family farm was becoming even more threatened. This is because agricultural biotechnology was thought to benefit primarily (or exclusively) large, already industrialized farms. If, as some people argue, society has some ethical obligation to protect or save family farms, then industrialization overall, and the new agricultural biotechnologies in particular, are cause for serious ethical concern. Independent of religious or environmental objections to biotechnology, there are three "family farm critiques" of agricultural biotechnology. These are based on the potential damage biotechnology might inflict on (1) an important political-economic entity, (2) a cherished symbol if not the embodiment of basic moral values, and (3) the solution to long-term natural resource problems.

INDUSTRIALIZATION OF AGRICULTURE

Industrialization of Agriculture: A Brief Overview

The "industrialization of agriculture" is a catchword for a broad set of changes that have occurred in agriculture in the United States and other developed countries over the last one hundred years. Its main features are as follows:

- The transition from animal powered farming techniques to machine power, with its attendant need for electrical or petrochemical fuel.
- The transition from the use of inputs (seed, fertilizers, and pest control measures) produced on the farm and reused yearly to the purchase of inputs from nonfarm sources, such as seed companies.
- The transition from small- to medium-sized farms, worked by a farm family and a few hired hands, to large-scale farms where all workers are hired and the farm manager may not even be the farm owner.
- The transition from localized and seasonal farm markets to regional, national and international markets, and "seasonal" produce available year-round.
- The transition from numerous independent farm producers supplying markets and processing firms to a relatively small number of large farm producers supplying commodities under contract to processing firms.
- The overall integration of agriculture into the larger industrial society with farming conceived of as "just another industry" akin to manufacturing or retail.

One could argue that the "industrialization" of agriculture began when humans first began using fabricating tools for farming, such as stone hoes, metal plows, or leather harnesses for draft animals. Most observers of agriculture, however, identify the beginnings of industrialization with the emergence of agricultural chemicals, particularly fertilizers, in the late 1800s. A more significant change in agriculture occurred with the development and widespread adoption of mechanical technologies, especially the gasoline-powered tractor, in the 1920s in the United States. Refinements and major breakthroughs in agricultural machinery (combines, harvesters, and postharvest storage technologies) and agricultural chemicals (pesticides, herbicides, and animal pharmaceuticals) continued throughout the middle twentieth century. After World War II the pace of these developments accelerated. In recent decades additional improvements in machines and chemicals have been accompanied by the growing use of computers in agriculture. Now we are witnessing what some see as the culmination of technological industrialization, the introduction of various biotechnologically produced products for agriculture.

The industrialization of agriculture is not simply the transition from a primarily manual-labor-based enterprise to a more technologically based production system. The biological, economic, and sociological dimensions of agricultural have changed in important ways as well. For example, the basic biological unit of crop production—the seed—has undergone significant changes. Soybeans and corn are no longer produced from the best seed saved from the farmer's previous year's harvest. Seeds are now biologically altered by scientists, patented (or given other patent-like protections) and sold to farmers annually. Hybridization, while augmenting desired traits in plants (e.g., salt tolerance and drought resistance), has made farmers dependent on external agents—seed stores, seed companies, and researchers in agricultural colleges—for the acquisition of their basic production input. Similarly chemicals and gas-powered machinery have to be purchased, often at considerable cost. Computer usage requires expensive operating programs and training. Products of biotechnology such as bioengineered seed or genetically altered animals also have to be purchased. Each of these changes in farm technology carries the attending consequence that agriculture increasingly must rely on nonfarm sources for the biological and mechanical inputs necessary for farming in its current form.

These changes have brought socioeconomic consequences. Three in particular stand out: (1) The size of farms has steadily increased since the turn of the twentieth century, (2) the number of farms in the United States has rapidly declined since around 1930, and (3) there has been a sharp decline in the number of people involved in agriculture—both owners and hired labor. The trend in farm size is directly attributable to the nature of the technological changes that have occurred in farming. It is a generally recognized fact that technologies are not what economists refer to as "scale-neutral." Certain technologies tend to favor production systems or enterprises of a particular size or scale. In agriculture, most of the technologies introduced over the last 40 years tend to favor large farm operations. Cost is a factor. Purchasing chemicals or seed in large quantities reduces the per unit or marginal cost of those expenses. Large operations are typically in a better financial position to buy mass qualitites and thereby realize a savings. Cost is not the only factor, however. While a large 400 horsepower grain harvester is expensive, it is also better suited to large fields. Most of the important new technologies introduced into farming since World War II have tended to be (and it is frequently claimed were intended to be) more useful to large-scale farming operations. As large-scale farms have grown larger, the total number of farms has declined along with the number of people involved in farming. There are several reasons for these related trends. The departure of people from agriculture and the decline in the number of farms from 1930 to 1945 has been attributed to the economic hardships farmers faced during the Great Depression. Many lost their farms because of an inability to purchase critical inputs and/or receive credit from failed banks. Around 1940 many young farmers and male children of older farmers went to serve in the war. Of those who returned home, many preferred to forsake the hard life of farming for alternative employment in newly sub-urbanizing America. Others pursued advanced education with the help of the GI Bill. The result was a rapid decline in the total number of farms and farm-based employment.

Even 50 years after World War II farm numbers and farm employment continue to decline. Urbanization, loss of farmland, and the less-than-glorious nature of farm work undoubtedly continue to contribute to this decline. It has been argued that there has been a systematic attempt by the federal government and nonfarm agribusiness industries to consolidate production into fewer and larger farms. The development of non-scale-neutral technologies have contributed to that. Many small farms are now what amounts to "hobby" farms that do not contribute in a significant way to the total output of agricultural as a whole. Middle-sized farms are forced either to get big or be reduced to hobby-type farms.

Current Structure of Agriculture

The term "structure of agriculture" refers to the overall nature of the agricultural industry in terms of its economic, sociological, and demographic features. The industrialization of agriculture has resulted in a restructuring of the nature of farming in the United States and other developed nations. The composition of farming, for example, the number and size of farms, who farms, and what the relationship is between farms and nonfarm enterprises, has changed. Currently the United States has what is called a "bipolar" agricultural structure. Of the nearly two million farms in the United States, fewer than 25 percent produce more than 80 percent of foodstuffs Americans consume. About 15 percent of farms account for three quarters of total agricultural sales (in dollars). The average income for roughly 75 percent of U.S. farms is around $17,000. Nearly all U.S. farms supplement their income with non-farm income, and for most farms the nonfarm income exceeds farm income. However, for most small farms, the nonfarm income actually covers net farm losses. In terms of size, nearly a million and a half of U.S. farms are less than 80 acres, whereas some farms in California, Texas, and Florida are as large as 200,000 acres. Ownership patterns have changed as well. While most small- and mid-size farms are single-family proprietorships or family-owned corporations, many of the largest and greatest revenue-producing farms are owned by corporations whose primary enterprise is not agricultural. For example, many farms are owned by petrochemical companies, restaurant chains, and a consortium of urban-based investors. A legitimate question that can be raised

is why agricultural industrialization or the increasingly bipolar structure of agriculture should be of any concern, given that (1) less than 3 percent of the U.S. population is engaged in farming and (2) Americans spend the smallest portion of their income (15 to 18 percent) on food compared with the rest of the world. One could argue that the structural changes occurring are simply the logical result of the technological transformation of agriculture. As long as consumers continue to spend a relatively small portion of their income on agricultural products and remain happy with the outcome of agriculture's industrialization, there is little cause for concern, economic, ethical, or otherwise. Some people believe, however, that there is cause for concern.

Ethics and Industrialization

Consumers are generally satisfied with the relatively low price of food and fiber products in the United States. This establishes, for some, an ethical justification for the industrialization of agriculture and its attendant structural effects. Utilitarian ethical theory holds that actions or policies are ethically sound if they result in "the greatest good of the greatest number," which many utilitarian ethicists interpret to mean "most satisfied preferences." The longstanding goals of agricultural policy and practices — including the development of agricultural technologies — have been (1) enough food, (2) a safe food supply, and (3) inexpensive food (the cost of the food should be at a rate that will return a reasonable profit to farmers). These three goals can be referred to as the "QQP" outcome: sufficient quantity, adequate quality, affordable (and profitable) price. It is generally held that the U.S. food system has met these goals and any utilitarian evaluation of the system would conclude that the current nature of agriculture is not only acceptable but also "ethically best." However, this line of thinking precludes customer awareness of agricultural processes. Consumers may never see the process or even think about the process that their food goes through before it reaches their table. Some would say as long as people get their food it does not matter to them what is going on in the agricultural process. Therefore agriculture and the technologies employed in food and fiber production can remain "invisible" tools for the satisfaction of consumer preferences and still be justified on utilitarian grounds.

According to utilitarian reasoning, it is only the increasing visibility of some of the results of industrialization (agricultural or otherwise) which accounts for questions about the ethics of contemporary practices. The noticeable environmental effects of agriculture processes, for instance, water pollution, soil erosion, or even the smell of diary or swine production, have begun to erode, though not completely undermine, the "satisfied preferences" basis for industrialized agriculture's legitimacy. In the 1960s heavy air pollution and fouled rivers led to concerns about the legitimacy of the smokestack industry and forced changes in practices and policies in that industry. If the impacts of industrialization on small farms were to become as apparent to the public as they are to residents of rural communities, one might expect a change in attitudes and policies toward the industrialization of agriculture. For the present, however, it appears that industrialization will be endorsed despite the attempts of contemporary Agrarian philosophers and small-farm activists to alert the public to the negative impacts of industrialization. After all, in the utilitarian calculus, the livelihoods of less than 2 million people weigh lightly against the satisfaction of 260 million others. Even so, there are strong ethical arguments about why we should be concerned about industrialization and the restructuring of agriculture.

ETHICS AND THE FAMILY FARM

Why Focus on the Family Farm?

In the 1960s and early 1970s, people concerned with social ethics were alerted to the significance of agricultural practices and policies through the publication of Rachel Carson's *Silent Spring* (1) and Ruth Harrison's *Animal Machines, the New Factory Farming Industry* (2). During this time the focus of agricultural ethics was on the environmental consequences of agriculture and the treatment of nonhuman animals. The scope of agricultural ethics began to broaden in the late 1970s, and its focus began to shift. Films and television documentaries in the United States highlighted the conditions of farm labor. The public's attention was caught by the successful unionization of California farm labor. Food safety became a matter of larger public concern as questions were raised about the carcinogenicity of artificial food additives. U.S. foreign aid, trade, and development policies came under increased scrutiny after the sale of U.S. grain surpluses to the Soviet Union and after the famine in Bangladesh. By the early 1980s ethical analysis and critique of agriculture was well underway. It was Wendell Berry's *The Unsettling of America* (3), however, that ignited the most focused philosophical and ethical works in agricultural ethics. Berry articulated and critiqued the whole philosophy of modern agriculture. His work prompted social scientists, ethicists, and philosophers to study the trends, meanings, and normative implications of the modern agricultural production with a focus on its impact on traditional family farms.

Today agricultural or food system ethics encompasses a broad range of ethical concerns—chemical use, farm labor and management practices, impacts on animals, environmental pollution and resource depletion, the health of the land, food safety, and issues relating to international aid, trade, and development. Yet, for many agricultural ethicists, the key to understanding the ethical depth and complexity of agriculture—and how many of the practices of contemporary farming are ethically objectionable—is the family farm. Berry argued that the traditional family farm represents a way of life that precludes its contributing to environmental or cultural problems associated with industrialized agriculture. Most agricultural ethicists see the family farm as Berry characterized it: a standard against which most ethical issues or problems in agriculture can be "tested." There are three ways in which the family farm serves as this standard: (*1*) from a political-economic perspective, the question of whether family farms can continue to

exist represents a test of the fairness or justness of democratic, market-based societies, (*2*) from a cultural or moral-value perspective, the question of whether values associated with traditional family farms continue to be viable serves as a test of the moral or spiritual health of modern society, and (*3*) from the perspective of our responsibility to future generations, the question of whether environmentally sustainable practices of family farms can be employed in modern farming tests the legitimacy of current production practices. Acceptance of the family farm as an ethical standard in any of these three domains gives agricultural ethicists reasons to take issue with biotechnology: Agricultural biotechnology may threaten or undermine the family farm. As such, these three perspectives together form what can be referred to as "the family farm critique" of agricultural biotechnology. Before proceeding to that matter, a brief examination of the arguments supporting the ethical standing of the family farm is in order. It should be noted that the following sections, the terms *agrarian populism* and *agrarian traditionalism*, when referring to contemporary philosophies, are derived from Thompson et al. (4).

Agrarian Populism and Farmers' Rights

Agrarianism is a philosophy that holds that farming is an important social good. Farming is a profession or occupation that ought to be respected, and in the legal-political realm, protected. The United States has had a long history of political and social concern for the plight of farmers. In his *Notes on the State of Virginia*, Thomas Jefferson argued that small-scale agricultural freeholders are good citizens and essential for the political-cultural life of the new nation being formed. They embody the "spirit of independence," but more important, a class of land-based, geographically dispersed, independent laborers serve as a political check against powerful urban interests and other threats to democracy. According to Jefferson, farmers and farm communities are to be encouraged and protected. Jefferson's ideas continued to be voiced in the cultural practices and government policies of the United States for more than a hundred years. It was not until the late 1860s that small-scale farmers realized that the Jeffersonian agrarian vision was being undermined. It was during this time, that the combined effect of U.S. Department of Agriculture programs, federal monetary policies, and the growing political strength of banking and manufacturing interests began to place small farmers in political and economic jeopardy.

At the turn of the century, Agrarian *Populism* (after the People's Party, circa 1870–1920) resurrected political-cultural arguments concerning the importance of small-scale farms. According to Populists, farmers' fundamental rights (property rights and the right to self-determination) were being deliberately violated or threatened by large business enterprises and government programs. Populists argued that small-scale farmers were entitled to "equal protection under the law," at the very least. Indeed, big business and big government at the time were acting to undermine basic principles of democracy and free-market capitalism. Populists demanded reforms and found some concessions. Nevertheless, the Agrarian

Populist movement failed to protect small farms over the longer term.

If the Populists had succeeded in all their demands regarding the protection of small farms, industrialization and structural change would not be an issue. Indeed, it is partly the result of the *failure* of Agrarian Populism as a political reform movement that Agrarian Populist arguments remains philosophically significant position to this day. As things stand, the philosophical tenets of Agrarian Populism are the basis for an ethical analysis and critique of contemporary agricultural practices and policies, including biotechnology and biotechnology policy. According to one contemporary spokesperson for Agrarian Populism, Jim Hightower (5), the family farm remains an important ethical, political, and economic entity: The family farm is one of the last, if not the last, holdouts in the attempt to secure fundamental values of democratic societies — freedom, self-determination, and equality of opportunity. While it may be no more important than, for example, hardware stores or plumbers, family farms are at the very least entitled to not be discriminated against in markets or in public policy (including research and development policies). The claim (like that of the turn-of-the-century Agrarians) is that the government and big business have conspired to drive family farms out of business and out of existence. Government policies encourage large farms, and large-scale agribusiness firms are always waiting in the wings to scavenge the remains of small farms unable to stay in production. It is unethical that individual farmers' rights and opportunities are systematically being violated, whether it is deliberate or simply the result of the socioeconomic system. It is also a harbinger of the death of democracy and free-market capitalism. According to Hightower and fellow Populist Marty Strange (6), family farms are potentially economically and politically viable despite what defenders of current agricultural practices claim. This is true, however, only if the government takes steps to protect family farms. Family farms may have no intrinsic or special ethical value to some, but they are valuable to the greater society. As a matter of justice, governments should guarantee that markets and policies are fair.

Agrarian Traditionalism and the Moral Value of Farms

Somewhat in contrast to Populism, Agrarian *Traditionalism*, exemplified in Wendell Berry's work, holds that traditional family farms have intrinsic or special ethical significance. Traditionalists agree with Populists that family farms should be preserved and protected but for different reasons. According to Agrarian Traditionalism, the family farm is at once the embodiment as well as the symbol of a set of values and virtues that have inherent worth. Among these values or virtues are self-reliance, community, and communion with nature. Traditionalists argue that the traditional family farm, by the very nature of the activities that occur thereon, promotes those virtues. Family farms engaged in the difficult labor of harnessing nature's power in order to survive, foster strength of will, courage, and self-determination among family members. The nature of farm work and the need for members of the larger farm community to help each other

during times of adversity fosters a sense of community values — sharing benefits and burdens, joys and sufferings. Those values and virtues are self-reinforcing and above all healthy for body and soul. Traditionalists argue that the modernization of farming — agriculture as a business — degrades and threatens farming as a "way of life." There are differences within Traditionalism concerning the precise meaning of the family farm as a moral ideal in this regard. To some, family farms, even in the present age, embody and promote those values or virtues. If family farms are intrinsically good in so doing, an implicit moral judgment might follow that everyone should engage in family farming. Since this is not possible in modern society, some traditionalists hold that while family farms do embody these virtues, the more important point is that they serve as symbols or paradigms for how fundamental ethical virtues such as community and respect for nature should be regarded. The family farm, in other words, is a metaphor for the good life, ethically conceived, rather than a profession or occupation to which all people ought to devote themselves.

Whether Agrarian Traditionalism is understood as advocating for the intrinsic value of actual family farms or suggesting that the family farm serves as a metaphor for the ethical life, a critique of modern agriculture or modern society necessarily follows from its tenets. Traditionalism holds that modern agricultural practices are essentially inimical to self-reliance, moral character, family values, and communities. Modern society is business-oriented and materialistic, and decidedly out of sync with basic human needs, especially spirituality. Moreover modern agriculture and society are insensitive to the rhythms of nature and the organic or holistic features of the traditional family farm. If there are fundamental ecological, psychological, and philosophical truths, modern agriculture and society are alienated from these. Perhaps it is not possible for everyone to experience these by actually farming. They are nevertheless fundamental ethical goods that would, if followed, lead to significant changes in the way modern people live their lives.

Family Farms and the Future

Agrarian Populists and Agrarian Traditionalists share the belief that since family farms have political-economic or intrinsic ethical value, public policies should at least preserve and protect family farms, if not actively promote or encourage them. Agrarians' arguments are increasingly being joined by proponents of "sustainable agriculture," who find the structure of traditional family farms and their farming techniques to be more consistent with long-term environmental stewardship which is essential for a sustainable planet. According to proponents of sustainability, human beings have fundamental ethical obligations to the future. Among these obligations is the duty to not exploit natural resources to the point that future generations will be unable to sustain themselves though food and fiber production. Given that many if not most modern industrial agricultural practices are resource-depleting there is an ethical obligation to change those practices. Taking this one step further, it has been argued that present people have an obligation to leave for posterity sound democratic institutions and a heritage of deep cultural and moral values. In each case, then, sustainability advocates strike notes similar to the Agrarians: Save the family farm, for political-economic, cultural-moral, and environmental reasons. This also is a matter of intergenerational justice.

In each of the positions described above, the ethical argument implies a critique of many if not most contemporary agricultural practices and policies. Agriculture has become in most developed nations a fairly large-scale, high technology, inputs-dependent industry. Family farm advocates have challenged large-scale corporate farmers, government policy makers, and agribusiness inputs manufacturers (chemical, mechanical, biotechnological) on ethical grounds. They argue that not only should family farms be protected or promoted, but many of the practices and practices associated with modern agriculture must be rejected. Family farm proponents have begun to target corporate and government/university actors involved in research and development for foisting ever-increasing industrialization on the farm sector. In essence, Agrarians and advocates of sustainable agriculture have come to find high technology — machines, chemicals, computers, and now biotechnology — ethically indictable in the apparent continuing demise of the traditional family farm. There is a shared critique and a shared vision of what is wrong with modern, industrial agriculture.

BIOTECHNOLOGY, INDUSTRIAL AGRICULTURE, AND THE FAMILY FARM

Emergence of Agricultural Biotechnology

The first applications of biotechnology were primarily in the medical/pharmaceutical field, such as in producing bioengineered insulin. The earliest agriculturally related products to emerge from genetic engineering were enzymes for fermentation (mainly for cheese) and agents for the biological control of pests, for example, *bacillus thuringiensis* (Bt). These products of biotechnology were not exactly the breakthroughs that the biotechnology community and agriculturalists have envisioned in the early 1970s (7). Rather, they were simply organisms whose commercial use value was enhanced through biotechnology, since genetic engineering made their large-scale production more efficient and less costly.

In 1970 the commercial prospects for agricultural biotechnology were made more attractive to the scientific community and to industry with the passage of the Plant Variety Protection Act (PVPA). PVPA provided for patentlike protections for novel plant species, whether the new or improved species was produced through conventional plant breeding or through bioengineering. PVPA's legal protections were extended as a result of two U.S. Supreme Court decisions (1980 and 1985), which allowed complete patent protection of novel organisms and plant species. The U.S. Patent and Trademark Office's decision (1988) to allow researchers at Harvard University to patent a mouse that had been generically altered, and this brought to completion the legal protections necessary for researchers involved in biotechnology to forge ahead with the development and commercial release of biotechnology products and processes, including those related to agriculture.

Agricultural researchers interviewed in a 1983 U.S. National Science Foundation-sponsored study believed that genetically engineered varieties of tomatoes and wheat, for example, would be in farmers' fields within five to ten years (7). In fact those bioengineered species are only now becoming commercially widespread. The overall process of bringing commercial agricultural biotechnologies to the marketplace or to the farm has been very slow. While pharmaceuticals and "biologics" (e.g., growth hormones) for animal agriculture are becoming increasingly available, the pace of the introduction of these has been far less rapid than early proponents of agricultural biotechnology (researchers and corporate marketing agents) had predicted and promoted. At present, the actual number of products from biotechnology currently being used in agriculture is not large. Nevertheless, given patents and patentlike protections, and the prospects for large profits from bioengineered agricultural products, that number is likely to increase at a rapid rate. Estimates are that market for U.S. agricultural biotechnology will grow from approximately $400 million in 1998 to over $2 billion by 2008 (8,9).

Despite the risks generally associated with biotechnology, there has been little public controversy over agricultural biotechnology. Although the most sophisticated application of genetic engineering to date (1998), cloning, was performed on an agricultural animal, controversies about cloning focused the possibilities of cloning human beings and not on agricultural applications. Perhaps the most contentious development in agricultural biotechnology per se has been the introduction of bovine somatotropin (bST), the nonsteroidal hormone capable of increasing milk production in dairy cattle. Yet even the main objections to bST, mostly at the public policy level and initiated by consumer groups, have had to do with its safety relative to human consumption of milk from bST-treated cows. Its role in agricultural industrialization and restructuring, while noted by some activist groups, has received little public or governmental attention. This is understandable given the relatively little attention the larger public devotes to *anything* related to agriculture — unless it directly affects human health or the environment.

Reach of Agricultural Biotechnology

The industrialization of agriculture has had the effect of blurring lines between research activities, industries, and governmental activities that formerly were separable and relatively isolated from each other. Agricultural animal science research is now closely tied to human medicinal and pharmaceutical research. Petrochemical and pharmaceutical firms own seed companies and animal-breeding facilities. Food safety oversight and regulation now includes involvement from the Department of Agriculture, the Food and Drug Administration, and the Environmental Protection Agency. Each of these connections suggest that agricultural production has become more and more integrated into the larger industrial society.

One major goal of industrialization in any sector of an economy is *control*. Industrialization requires and enhances control over a production-distribution system, whether it is the factory, the transportation of products or marketing. This goal has permeated industrial agriculture. Farmers and the firms supplying agricultural technologies have always looked for ways of increasing control over the production of food and fiber. Part of the appeal of agricultural *bio*technology is the promise of even greater control over farming. Farmers and agribusiness concerns differ, however, in what kinds of control they wish to exercise over farming.

Agricultural biotechnologies can be categorized in terms of what sorts of control they allow the farmer over the production process. The most basic units in agriculture are, for crops, the soil, water, and solar resources and the plants or seeds to be grown. For animal agriculture, they are the animals (swine, cattle, chickens, etc.) and the foodstuffs necessary for producing animal products. The longstanding goal of traditional plant and animal breeding has been the introduction of traits into plants and animals that would allow them to be more productive relative to the conditions under which they are grown, which includes both natural circumstances (soil, water, etc.) as well as inputs such as feed. Biotechnology, at least in theory, makes those goals more attainable because genetic engineering is quicker and more precise in the transference of the genes controlling those traits. Plant varieties that are pest resistant, drought resistant, better able to absorb nutrients in the soil, and the like, are a desirable outcomes of crop biotechnology. Similarly animals bioengineered to withstand hostile climatic conditions or resist diseases while continuing to produce milk, eggs, or lean muscle tissue are important goals for animal biotechnology. Plant varieties or animal species so engineered are in less need of constant management of inputs and external conditions. Therefore biotechnologically improved plants and animals should give agricultural producers more control over their production processes. The most desirable products of biotechnology for industrialized farms thus are bioengineered plant and animal species.

In the absence of a plant variety that is high-yielding and resistant to pests or climatic stresses, a second level of agricultural biotechnologies is desirable. These are genetically engineered organisms or biotechnologically produced substances that assist the plant in resisting pests or diseases or taking up nutrients from the soil. *Bacillus thuringiensis* (Bt, mentioned above) is one such product. Bt is a bacterium, engineered so as to be deadly to various species of caterpillar that are harmful to vegetable (tomato, bean) plants but not harmful to the plant nor other species (including humans). Organisms that prevent frost damage to fragile young plants (e.g., potatoes, strawberries) have also been bioengineered. For crop farming, these "external" control agents (sprayed on crops) are obviously less desirable than a crop with pest resistance or frost tolerance "built in," but they are nevertheless useful in helping control the environment in which production occurs.

For animal agriculture, the second tier in biotechnology is similar: organisms or chemical substances that might be injected or added to feed in order to help prevent disease, augment nutrition, and increase control of milk or egg or meat production. *Bovine somatotrophin* (bST, also called bovine growth hormone (BGH)) is one of these. When bST is administered to dairy cattle, it increases milk output without a corresponding need for increased animal feed intake.

A third tier of agricultural biotechnology is concerned with postharvest control over the products of agriculture, whether plant or animal. Though some of these directly benefit farmers, most are designed to reduce spoilage in vegetables and grains and keep animal products fresh and safe during transportation and marketing. Some are designed to help control or speed up processing, for example, bioengineered enzymes that are more efficient than traditionally used biochemicals in the fermentation of cheeses or beer. While some of these third-tier products have become available for commercial use in transportation, processing, and marketing, most postharvest biotechnology products are still in the developmental stage.

In fact most of the desired agricultural biotechnology products and processes are still in the developmental stage, and many will never reach commercial applicability. Nevertheless, the energies and monies invested in agricultural biotechnology suggest that the long-touted "biotechnological revolution" in agriculture is quite possible. Given the longstanding goal of increased productivity and its corollary, increased control in the production–processing–distribution system, increased agricultural biotechnology is desirable at least for its corporate producers. This suggests an additional issue of control associated with agricultural biotechnology. That is the potential control of agriculture, including both large and small farms, associated with the corporate entities who are by and large the major proponents and suppliers of agricultural biotechnology.

Biotechnology, Corporations, and the Family Farm

Technological change and economics of scale have placed small- and medium-sized farms (both of which most likely to be family farms) in a precarious position. According to the theory of the "technology treadmill," farms must be able to adopt the latest, efficiency-enhancing technologies as these technologies emerge from the research and development process. Large industrialized farms are most likely to be able to keep up with the acquisition of "new and improved" technologies. Middle-sized and small farms must either figure out ways to move toward adoption, or fall off the treadmill. Small farms are most likely to fall.

The issue with agricultural biotechnology is the same. Some of biotechnology's real and potential products and processes for agriculture may be scale neutral, equally capable of benefiting small and large farms. However, it has been claimed that most, including, for example, bST, are not scale neutral, and appear to be *designed* for large farming operations. Research and development of agricultural biotechnology in universities (especially in colleges of agriculture) has been criticized for being financially and politically influenced by large-scale farm operators. What large farms desire is what is produced.

While the intention may not be to deliberately harm small or medium family farms, the nature of the products and processes developed does so anyway by exacerbating the technology treadmill.

More significant is the fact that final product development, production, and marketing of biotechnology products and processes is in the hands of large, in many cases, multinational corporations. In fact much of the university-based research is being funded by those corporations, which include pharmaceutical firms, seed companies, petrochemical giants, and global food-and-fiber products distributors. The primary concern of these agribusiness firms is to increase profits and market shares for their products, and one way of doing so is to bring "new and improved" agricultural biotechnology to the market as quickly, and as often, as possible. In this regard corporate producers of agricultural biotechnology can control the speed and direction of the technology treadmill. In so doing, these corporations effectively control the further industrialization of agriculture, with attendant implications for the restructuring of agriculture and the fate of the family farm.

There are two likely outcomes of the way in which agricultural biotechnology is currently being developed and commercialized: Only large-scale farming operations will be able to afford them, and only large-scale enterprises will be able to efficiently and effectively employ the products. The structural effects of these outcomes will be reinforced by other current trends. For example, in the broiler chicken industry, producers must purchase inputs (chicks, feed) from a given firm and sell their products back to that firm under a contractual tie between corporate actors and on-farm producers. The small grower who is unable either to afford the inputs or able to guarantee a particular quantity at a particular weight at a fixed price is effectively excluded from the market. With no viable market for its products, the smaller farm will fall out of agricultural production (other things being equal). In the case of some of the new agricultural biotechnology products, a similar contractual arrangement is taking place. Growers who want to plant a particularly high-yielding soybean hybrid bioengineered for tolerance to a particular herbicide must agree in writing to purchase both seed and the herbicide from the same corporation. This may be contrary to growers' best interests, given the fact that the herbicide's patent protection has run out and other companies are producing identical herbicides at a lower price. Moreover the seed-herbicide "package" is expensive, effectively limiting its market to large producers.

The nature of agricultural biotechnology's products then, combined with the fact that these products and processes will be the domain of large corporate actors in the food system, does not portend well for family farms. Indeed, it may raise concerns for all of agriculture as even large-scale farms find themselves increasingly tied to nonfarm agribusiness corporations. Biotechnology per se may not be the ultimate cause of further industrialization or increasing bipolarization in the structure of agriculture. Agricultural biotechnology may not be the cause of the ultimate demise of the family farm in the United States.

Nevertheless, as a further and soon to be more pervasive tool in the toolbox of agricultural researchers, on-farm producers, and corporate producers and suppliers, it is and will continue to be a contributing factor to the decrease in the competitiveness and viability of small family farms. Given the fact that only 20 percent or so of the large farm producers account for over 80 percent of farm output and sales in the United States, there is little incentive for the producers and marketers of agricultural biotechnology to focus their research and development efforts on small farms' needs or interests. Instead, the trend is likely to continue that the big will get bigger and the small will be placed in jeopardy if not driven to extinction. This trend will be aided by the researchers and corporate actors who have a vested interest in the success of the "biotechnological revolution" in agriculture, that is, if the trend is not halted by public policy or changes in consumer tastes and preferences.

CONCLUSION

Agricultural ethics is the systematic application of disciplinary tools from philosophy and ethics to the problems and issues of agriculture. It focuses on specific problems as well as on the ethical aspects of agriculture or the agricultural/food system as a whole. Agricultural ethics, as in other areas of applied ethics (business ethics, medical ethics, etc.), is not simply about identifying issues and concerns, however. That would be only half its philosophical task, that is, the *description* of problems, conflicts, values, and orientations in agriculture — *descriptive ethics*. The other more important task is prescriptive (proscriptive), the work of *normative ethics*. Although agricultural ethicists may be in no authoritative position to tell farmers, agribusinesses, consumers, or public policy makers what to do concerning such things as using pesticides, managing livestock, or adopting new technologies, it is nevertheless part of the responsibility of ethicists to articulate the normative implications of actual or potential decisions. If, for example, there is general agreement that people have an obligation to future generations to leave a habitable environment (and there seems to be some such consensus), then when an ethicist shows how a particular agricultural or natural resource-related practice potentially endangers the environment, the ethicist "proves" that the practice is unethical and therefore must be stopped. The difficulties are (1) finding the consensus we individually and collectively might have regarding ethical obligations and (2) accurately describing and analyzing the facts and ethical implications of specific actions or general practices.

In the discussion of the utilitarian justification for industrial agriculture, for example, a conclusion was drawn that "quantity, qaulity, and price" represent widely held and ethically acceptable goals for agriculture. This is a legitimate inference, because with only a few exceptions related to the environment or food safety, public actions and public policy have not challenged agriculture so long as QQP has been achieved. The task for the agricultural ethicist in this case is to carefully analyze actions or developments in agriculture to see if there are any that may be inconsistent with QQP in the present (e.g., the use of some pesticides threaten food quality even if unbeknownst to consumers) or in the future (e.g., the control over agricultural production by a few corporations raises the prospect of monopoly pricing). The normative implications of finding such inconsistencies should be clear: From a utilitarian perspective, inconsistencies between practices or trends and QQP (or whatever goals the public has for the food system) entail that those practices or trends are ethically wrong and should be corrected or stopped.

Rights-based analysis involves similar reasoning. Suppose that we can identify a widely held rule or principle concerning peoples' rights as human beings or as participants in a democratic, market-based society, This rights-defining rule or principle may be implicit in current policies or public thinking, or may be only very general and vague, for example, "People should have equal opportunities" or "People should be treated fairly." However implicit or vague, if we can also show that something is happening in agriculture that may undermine or infringe on those rights or entitlements, we have arrived at a basis for judging that this particular action, practice, or trend is ethically unacceptable.

In sum, it is not the practice of ethics or agricultural ethics to preach or dictate. Rather, it is to show that *if* there are dimensions of agriculture — including how agriculturalists are themselves treated by nonagricultural actors including governments — that are at odds with basic ethical principles such as utility maximization, fairness, or respect for the future, *then* these dimensions should be seen for what they are — unethical. It is up to the relevant actors to decide on the basis of this moral knowledge whether or not they will do the right thing.

In any event, there is one thing normative ethical analysis, in agriculture or elsewhere, a priori rejects: Retreat to the claim "that's the way it is." Reviewing past occurrences can give us perspective as to how present circumstances have developed. However, we must be careful to see "developments" or "trends" for what they are: the collective results of individual decisions. Sometimes those collective results are unintended. A farmer purchasing a tractor or a sack of hybrid corn seed in the 1930s could not have predicted that 50 years hence the development of the tractor or hybrid seed would be early occurrences in the process of transforming farming into a high technology, international, corporate-controlled agribusiness system. Nevertheless, in purchasing that tractor and seed, the farmer made a decision that affected the development of agriculture into the system before as today. While normative ethicists cannot ask decision makers to acquire predictive powers in all their actions, ethicists can ask, indeed demand, that each of us be more circumspect in our decisions and actions about the potential grand-scale outcomes of small decisions. Trends begin and end when individuals begin or end them.

From a normative ethical perspective, we are left with a set of questions about the family farm, the industrialization and restructuring of agriculture, and the role of biotechnology and the purveyors of biotechnology.

- Has the industrialization of agriculture over the last century been a good thing? For whom? According to what ethical criteria (rights, utility, etc.)?
- Is the continuing industrialization a good thing? Again, for whom and on what ethical basis?
- Is it ethically justifiable that small family farms have been the major losers in the process of the industrialization of agriculture? On what basis?
- Indeed, does society have any obligations to small family farms?
- Are the new agricultural biotechnologies (individually or as a whole) good for agriculture? Are they good for society? On what grounds?
- Are there ethical problems associated with the research, development, or marketing goals or strategies of the biotechnology entrepreneurs?
- What should society's ethical judgment be regarding the new agricultural biotechnologies and their governmental and corporate "sponsors"? What should the public's response be?

These are the big questions to consider as the industrialization of agriculture nears completion in the United States and most other developed nations. They merit ethical reflection.

BIBLIOGRAPHY

1. R. Carson, *Silent Spring*, Houghton Mifflin, Boston, MA, 1962.
2. R. Harrison, *Animal Machines: The New Factory Farming Industry*, V. Smart, London, 1964.
3. W. Berry, *The Unsettling of America*, Sierra Club Books, San Francisco, CA, 1977.
4. P.B. Thompson, R.J. Matthews, and E.O. Van Ravenswaay, *Ethics, Public Policy and Agriculture*, Macmillan, New York, 1994.
5. J. Hightower, *Eat Your Heart Out: Food Profiteering in America*, Crown Publishers, New York, 1975.
6. M. Strange, *Family Farming: A New Economic Vision*, University of Nebraska Press, Lincoln, 1988.
7. L. Busch, W.B. Lacy, J. Burkhardt, and L.R. Lacy, *Plants, Power, and Profit: Social, Economic and Ethical Consequences of the New Biotechnologies*, Basil Blackwell, London, 1991.
8. Biotechnology Industrial Organization (BIO), 1998. Industry website: *http://www.bio.org*
9. BioTech Navigator, 1998. Industry online newsletter: *http://www.biotechnav.com*

See other entries AGRICULTURAL BIOTECHNOLOGY; ANIMAL BIOTECHNOLOGY, LAW, FDA REGULATION OF GENETICALLY MODIFIED ANIMALS FOR HUMAN FOOD USE.

AGRICULTURAL BIOTECHNOLOGY, ETHICS, FOOD SAFETY, RISK, AND INDIVIDUAL CONSENT

PAUL B. THOMPSON
Purdue University
West Lafayette, Indiana

OUTLINE

INTRODUCTION

Agricultural biotechnology has spawned heated controversy. While human and biomedical applications raise clear ethical issues, the ethics of food biotechnology are less obvious, and concern may appear entirely misguided. Changes to food stimulate reactions of intense emotion, resentment, and resistance among some consumers. These reactions are themselves complex, combining feelings based on religious beliefs about genetic technology in general, environmental impact, sympathy for animals, or solidarity with small farm, organic and sustainable agriculture movements with genuine concern about the hazards and uncertainties associated with the consumption of genetically modified foods and food products. For some individuals, fear and resentment about human gene technologies may be manifest in attitudes toward food biotechnology (1,2). Whatever the ultimate basis, these concerns about agricultural biotechnology often surface as concerns about consuming food (3).

Scientifically and analytically trained observers of the debate apply a narrower set of criteria to the evaluation of food safety. Even when concern about environmental impact is seen as valid, for example, experts do not translate this into a legitimate basis for concern about the consumption of food. As such, experts often dismiss the broader public's reaction to food biotechnology as muddled, uninformed, and even irrational (4,5). Yet it should not be surprising that public resistance to genetic technology might center on food. Most people will rarely or never face a personal choice about the more exotic and ethically troubling applications of genetic technology, and few may be willing to undertake what is necessary to influence policy in a political forum. Everyone, however, makes food choices everyday. It is natural that many of those who experience the greatest anxiety and moral opposition to genetic technology would express their concern most vehemently with respect to food.

The inconsistency between expert and lay approaches to the risk of genetically modified foods is the overarching philosophical issue for food safety. Experts understand food safety almost exclusively as a problem framed by the conditional probability of injury or disease as a result of consuming a food product, either once or over a lifetime. For the expert, issues are relevant to food safety only to the extent that they affect these probability estimates. To be sure, the expert approach to risk entails ethical questions of its own (discussed below). Nevertheless, it is reasonable to conclude that the likelihood of injury or disease from eating genetically modified foods is low, and can be reduced to minimal levels with a modicum of cautionary practices. To most experts, this conclusion establishes a sound basis for repudiating many of the lay public's concerns about consuming genetically modified food. However, a different philosophical interpretation of risk would countenance a much broader array of factors, and would make each individual the sovereign judge of whether the interests threatened by genetically engineered food are vital. Such an approach would emphasize an individual's right to choose whether or not to consume genetically engineered food. Safety, in the sense of probable harm, would clearly be relevant in this alternative view, but the appropriate ethical response to risk might be to secure conditions of individual consent, rather than to minimize probability of injury or disease.

There has been comparatively little discussion of these issues by professional bioethicists. Some of the key philosophical questions can be highlighted, and a sketch of each of the two contrasting approaches to risk can be made. Perhaps the crucial policy questions revolve around labeling of genetically modified foods, but legal and economic issues contend with ethical concerns in fixing labeling policies. Even the definition of a "genetically engineered food" is open to debate.

BIOTECHNOLOGY AND FOOD

Recombinant DNA techniques have many applications in the food system, and many of the key terms are ambiguous. The phrase "whole foods" indicates foods that have not been combined or adulterated; other foods are "processed foods." Apples, beef, whole-wheat flour, and whole milk are whole foods. Sausages, breads, and cheeses are processed. However, some foods, such as fruit juices, vegetable oils, or skimmed milk, fit uncomfortably in this simple dichotomy. The term "constituent" will be used to identify the various parts (fat, fiber, vitamins, etc.) of whole foods, while "ingredient" will be used to indicate a food or additive used in processed food. "Contaminants" such as pesticide residue, insect parts, or fecal matter are unintentionally incorporated in both whole and processed foods. "Adulterates" are impurities that have been intentionally introduced, generally to enhance the bulk or appearance of a commodity. Regulatory agencies approve certain substances (e.g., dyes) as additives. Nevertheless, philosophical controversies arise because one person's additive is another's adulteration.

The term "genetically modified organism" (GMO) is used commonly to include whole foods from plants or animals whose germ plasm has been modified using recombinant DNA techniques. The expression "genetically modified food" would normally be given a broader interpretation to include processed foods with GMO ingredients. Such processed foods would involve consumption of a GMO as an ingredient, though the chemical properties of the GMO may be substantially affected by processing. The use of genetic engineering in the food chain, however, is often indirect, as in the case of recombinant rennet, where a bacteria is modified to produce an enzyme used in making cheese. The use of terminology in such cases can be controversial, but here we will refer to all cases with the phrase "GMO." However, all foods are "genetically modified" in a broad sense. Virtually all plants and animals consumed for human food are the product of crossbreeding and genetic selection. Such practices of genetic modification have been used since antiquity, and for our purposes they do not count as creating a GMO.

Key ethical and philosophical judgments depend on the interpretation of terminology. One use of genetic engineering is to develop DNA probes designed to increase the accuracy and efficiency of meat and produce inspection. The availability of such probes will significantly change the procedures for monitoring food pathogens and enforcing food safety regulations. These changes in procedure may, in turn, have an impact on food safety. DNA probes are, indeed, intended to enhance food safety, though whether this result will occur can be disputed. However, DNA probes do not in themselves alter the composition of foods on which they are used. At the opposite end of the spectrum are genetically modified plants and animals. Maize and soybean, for example, have been modified through genetic engineering to make the crops resistant to plant viruses or the use of herbicide, and to synthesize the naturally occurring substance bacillus thuringensus (Bt), which is toxic to leipedoptera. The grain and oil derived from these genetically engineered crops are consumed directly by humans, and indirectly when the crops are fed to animals who in turn produce milk and meat products.

A number of other food technologies stand in between diagnostics, which do not alter the composition of food, and genetically engineered crops or food animals, which clearly have been altered. For example, recombinant rennet, the enzyme essential to cheese making, was one of the first commercial products from genetic engineering outside of medicine. In nature, rennet is produced within the gut of calves. Traditional rennet for cheese making has been harvested from the entrails of slaughtered calves. Genetically engineered bacteria modified to synthesize the enzyme produce recombinant rennet in a process similar to fermentation. Other bacteria have been modified to synthesize bovine somatotropin (BST), which can be administered to lactating dairy cows as a stimulant to milk production. Such modifications can reasonably be interpreted as changing the constituents of food, though they do not involve genetic manipulation of the organism that is thought of as the food source itself.

DNA itself is present in virtually all foods and food ingredients, excepting only minerals (e.g., salt) and water. The high temperatures used in many forms of food preparation destroy DNA. Fresh fruits or vegetables and raw milk or meat contain DNA in the forms in which they are normally consumed. New food biotechnologies thus do not introduce DNA into the human food chain. DNA is nontoxic and is thoroughly metabolized in normal human food consumption. Some applications of food biotechnology result in direct human consumption of altered genetic material. Some do not. Alterations in genetic material are normally made to affect an organism's production of proteins, or to affect the regulation of the organism's cellular processes. Such changes affect the organism at a phenotypic level. Phenotypic alterations could affect the suitability of an organism for use as food.

Whether some, none, or all of these applications trigger ethical issues depends on pragmatic circumstances. In regulatory contexts, authority may be constrained by legislation or bureaucracy. Regulatory decision makers must decide whether a DNA probe or an intermediate product (e.g., recombinant rennet or BST) is defined as a food, an additive, an ingredient, or a contaminant in order to know which regulatory criteria to apply. Often such a judgment determines which agency or division has jurisdiction. If reducing public exposure to injury or disease is the overriding objective, applications that affect food safety inspection procedures are more significant (for good or ill) than genetically modified crops such as herbicide tolerant soybeans or Bt maize. An entirely different set of criteria may be appropriate when the circumstances that lead to individual consumption of a genetically modified substance are at issue, rather than general public health. For example, a person who believes that religious dietary rules prohibit the consumption of genetic materials derived from specific animals will be far more interested in the source of genes that ultimately find their way into the food chain than in the material impact of biotechnology on the probability of disease or injury.

FOOD: SAFE, PURE, AND WHOLESOME

Dietary rules have occupied a minor place in religious and philosophical discourse from time immemorial. Semitic dietary rules are well known. Christian rules have been associated with specific rituals and seasons such as Lent. In ancient Greece the Pythagorean cult of which Plato was a member had a rule against eating beans. In addition to explicit rules, all human cultures adopt implicit beliefs about what is and is not considered edible, and in what combinations or at what seasons edibles may be eaten. Though the basis and meaning for food regimes is a subject of debate among anthropologists, there is no doubt that such regimes fulfill a minimal social function. Every human society must have some means for avoiding poisons. Such knowledge directs ordinary food choices to plants and animals that are not acutely toxic, and encourages practices of harvest, storage, and preparation that minimize risk. Furthermore, since the type and availability of edibles will vary according to season, location, and climatic conditions of drought or pestilence, this knowledge must be reproduced from day to day, year to year, and generation to generation. Any successful human society will have developed a food regime that satisfies these conditions (6,7).

Food regimes thus represent an implicit knowledge system of enormous complexity, and one that is highly sensitive to technological transformation. Claude Lévi-Strauss attributed deep significance to the emergence of cooking, suggesting that a culture's entire system of signs is rooted in this fundamental food technology. Lévi-Strauss's structuralism represents a view that might be interpreted to entail hidden meanings wrapped up in food beliefs and the potential for serious ethical considerations when that system of belief is challenged. Yet even without structuralism's backing, it is easy to see why beliefs about what is and is not food would become deeply interwoven with a culture's ideas about purity, hierarchy, and the sacred. It is also easy to see why such beliefs would become a minor battleground when distinct cultures come in contact with one another and why minorities would nurture food regimes as components of cultural identity. Given this background, the emergence of the modern system for state regulation of foods is one of industrialization's more remarkable achievements.

Pythagorean rules on beans aside, philosophers have largely neglected food regimes, and the history of food safety during industrialization has yet to be written in a definitive fashion. Any concise overview is thus necessarily speculative to some degree. Over a few centuries, industrialized societies evolved a conception of food safety in which impurities were thought to be the primary cause of food-borne illness, a view that evolved into the germ theory of disease in the nineteenth century. Avoiding germ contamination was consistent with a traditional emphasis on pure and wholesome foods. In both traditional and early industrial food regimes, pure and wholesome foods are "good for you" in the broadest sense, meaning that they promoted physical health, a positive mental outlook, and were socially acceptable. In traditional societies, this was assured by following implicit or culturally based rules about what could and could not be eaten. Under industrialization, "pure and wholesome" foods were those not contaminated or adulterated by germs, residues or foreign substances (7).

In the emerging industrial food regime, science was deployed first as a means of identifying impurities, and then of measuring the risk associated with them. Regulatory policy was to protect the food supply from adulteration by contaminants, and to weigh the benefit and risk from additives and residues, removing offending substances that posed significant risk. Many members of the lay public probably retain this mental picture of food safety today. It shares two important features with traditional food regimes. First, the determination that a substance is or is not food is the primary basis for a judgment of purity. Second, risk is associated with a compromise in purity. Until comparatively recently, most people would probably have also presumed a strong link between purity and wholesomeness, or between nonadulteration of a food and its nutritional or social acceptability.

Toxicologists, nutritionists, and other experts on food have so modified this picture during the last quarter of the twentieth century as to constitute its gradual abandonment. Scientists have put complex interaction among the chemicals that make up whole foods in place of the idea that purity is equivalent to safety (8). Dietary induced cancer is now thought to be caused by interaction between two groups of food constituents, mutagens and antimutagens, along with the genetic disposition of the individual consumer. Furthermore mutagens and antimutagens are thought to be found in virtually all foods. The more general link between nutrition and health has also shifted from eating pure foods derived from groups, to the proper balance of fats, protein, carbohydrate, fiber, and other food constituents. Most whole foods contain a combination of all these constituents. This reductionist view of food safety has made some experts skeptical of regulatory approaches that require identification and elimination of alleged carcinogens. It plays an important role in the reasoning of scientists who evaluate the safety of genetically modified foods.

At the same time physicians are diagnosing more and more individuals as suffering from allergic reaction to specific foods or to chemical sensitivities. True allergies cause toxic and sometimes fatal reactions, usually within a few hours of consuming the allergen. Food sensitivities such as lactose intolerance cause less critical reactions such as gastrointestinal distress. Individuals with specific allergies or sensitivities have reasons to avoid foods of uncertain origin, and they have raised a number of questions about the effects of genetic modification. However, some experts question the rising tide of allergy diagnoses, claiming that most people with sensitivities can consume moderate amounts of the given food, especially in combination with other foods. There is thus controversy as to whether the increase in allergy diagnosis is biologically, psychologically, or culturally based, and an individual's view on this question often colors their assessment of genetically modified food.

In summary, the starting point for any discussion of food safety and biotechnology is an environment where science has outrun cultural attitudes on purity and wholesomeness, attitudes that have evolved as the primary social basis for food safety over centuries, if not millennia. The science itself is dynamic, and debates are frequent. With this background in mind, it is remarkable that there is relatively little debate among food safety experts about the low probability of injury or disease from consuming GMOs. Yet it is not surprising that the broader public debate over GMOs should be affected by cultural and technical disputes that attend food safety generally.

RISK, SAFETY, AND DELIBERATIVE RATIONAL CHOICE

Experts approach food safety as a problem of risk assessment framed by the parameters of deliberative rational choice. Some experts also presume a public health philosophy that reflects a utilitarian approach to public policy. Deliberative rational choice presumes that decision makers see each course of action open to them as a means for bringing about consequences. These consequences can then evaluated in such a way that each course of action can be understood as having an expected value. A given course of action commonly has two or more possible outcomes. In such cases the expected value of a given course of action is derived by considering both the expected value of each possible outcome, and the likelihood or probability that it will occur. In common parlance, the *risk* of a given course of action is a function of the probability that unwanted consequences will occur, and the harm or loss associated with those consequences (9).

Deliberative rational choice can be applied to many different kinds of choice situation. At the policy level, the choice might concern whether to allow *any* genetically modified foods on the market or whether to allow a specific application of the technology (e.g., recombinant rennet) on the market. In either case, the *risk* of the policy would be found by assessing both the likelihood of unwanted consequences and the harm or loss expected to be associated with those consequences. Individuals might also apply deliberative rational choice in making individual food decisions. Here each food purchase or consumption decision might be evaluated as having an expected value (10–12). The risk of purchasing or consuming a genetically modified food would, for the individual consumer, be a function of the probability that unwanted consequences would occur, and of the harm or loss associated with those consequences, should they occur.

Regulators, scientists, and others who have contributed to the literature on food safety appear to be applying a framework of deliberative rational choice, though few say so explicitly. They assume that consumers either apply or intend to apply such a framework in their food choices. Experts contributing to this debate rarely question the assumption that consumers see food choice in terms of instrumental rationality. Policy choices that end in general dietary health and no increase in the rate of dietary diseases or disorders are assumed to meet all relevant ethical criteria. Such policies are seen as consistent with public health and consumer wishes, and there is rarely any acknowledgment that these two criteria could diverge.

Approaches to Food Safety

Given this general approach to risk, one might develop any of several approaches to food safety, but three paradigms dominate debate over food safety. *Risk thresholds* are routinely applied to determine criteria for food safety in most industrial countries. In the United States the Food and Drug Administration applies threshold criteria to genetically modified food. *Risk-benefit* averaging adapts broadly utilitarian criteria to administrative decision making. It differs dramatically from a threshold approach in conceptual terms, but practical differences consist in the fact that benefits are taken into consideration along with risk. A third approach would allow *market forces* to determine acceptable levels of risk.

Threshold Approaches. A food might be deemed safe if the risk of any policy decision to allow it on markets falls below a given threshold. Note that the level of risk might be driven below a threshold by many different characteristics.

- An event that occurs relatively frequently but with trivial harm or loss might be considered "low risk."
- An event that occurs very infrequently may be considered "low risk," even if harm or loss in those infrequent occurrences is comparatively significant. In particular, an event likely to affect only a small percentage of the population may be assessed as "low risk," even when it is very likely to affect that small minority.

- Even the potential for catastrophic effects may not bring a risk above the threshold, if the probability of catastrophe is sufficiently small.

In practice, however, food safety policy makers have adopted very conservative approaches to setting thresholds. The U.S. debate over thresholds has centered on the Delaney Clause, which stipulates that no carcinogenic substances may be used as additives. Regulators have interpreted the Delaney Clause as requiring "zero risk," or a threshold of zero for acceptable risks.

The zero-risk threshold has created enormous problems for regulators. Given the general parameters of deliberative rational choice, regulators must consider any scenario that leads to an unwanted outcome in making a risk assessment. The only way that such a scenario can be found acceptable under a zero-risk threshold is to prove that there is a zero probability of its occurrence. But statistical methods do not support such a proof. As such, regulators have treated zero risk as "no measurable risk." Even this approach has become problematic as the general philosophy of food safety has become more reductionistic. It is increasingly difficult to design experiments that could establish meaningful probabilities, while eliminating the confounding effects of other mutagens and antimutagens that are natural constituents of food. Regulators and experts share a general consensus that the zero-risk threshold is impossible to support with modern scientific methods (see (13) for a related discussion on workplace hazards).

Risk–Benefit. Frustration with thresholds has led to a surge of interest in risk–benefit criteria. In this approach a decision maker must weigh both risks and benefits in assessing policy. Economists and nutritionists would be consulted to assess the benefits of a genetically modified food. Foods that return economic benefits to consumers, farmers and the food industry, or that improve nutritional quality might be found acceptably safe, even when risks are nonzero, or even above *de minimus* levels. Such a standard is currently used to assess risks associated with chemicals regulated under the U.S. Federal Insecticide, Fungicide and Rodenticide Act (FIFRA). A risk–benefit approach to genetically engineered foods could be given one of several interpretations:

- Safety could be deemed acceptable whenever benefits outweigh risks.
- Policies could be required to produce the optimal ratio of benefit over risk.
- Policy could reflect a "mixed mode" of thresholds and consideration of benefits, so options entailing significant risk would not be considered, no matter what level of benefit might be associated with them.

Additional technical parameters would have to be specified before a formal risk–benefit can begin, as well.

Risk–benefit is attractive in part because it seems more consistent with the utilitarian philosophy that some associate with deliberative rational choice. Though utilitarians differ over how to assess the expected value (utility) of a choice, they would advocate an approach of predicting consequences for all affected parties, assigning utility or value to these consequences, then selecting the course of action that maximizes utility. The emphasis on including consequences for all affected parties provides a rationale for considering benefit as well as risk. The struggle over which way to interpret risk–benefit requirements applied to GMOs would be seen as part of a larger problem in interpreting the maximization requirement: Is it maximal total utility, is it average utility, or is there some hybrid notion?

Market Solutions. Both approaches described above assume that administrative decision makers will assess risk and apply a policy decision rule to determine when and whether consumers will be exposed to risks associated with GMOs. An alternative approach would stress providing individual consumers with information about risks, and then letting market forces determine the level of acceptable risk. In practice, such an approach has many pitfalls. It would require difficult and contestable judgments about how to provide information on the probability of disease or injury that might be associated with any consumer's food decision. Furthermore, since consumers are in the habit of assuming that foods on grocery shelves meet standards of food safety, it is likely that they would be slow to apply a more critical approach to risk in their individual food choices. There is also doubt that the average consumer has the ability to make appropriate risk judgments.

For all these reasons, market solutions are often treated more as a foil in technical debates over food safety than as a serious alternative. However, experts tend to presume that those who advocate labels for GMOs are advocating a market solution (14–17). As such, they evaluate the question of labels in light of standards derived from deliberative rational choice. Labeling is seen as a policy option that should be evaluated in light of whether individuals use information on labels to satisfy their personal values with respect to risk, safety, price, taste, and the other characteristics affecting the expected value of a food decision. The mere occurrence of a label might lead some consumers to assume that a food is risky, even when the substantive information would support a different comparison. Labels for GMOs thus have a clear potential for to suboptimal policy performance from a deliberative rational choice perspective. This theme will be revisited again in the discussion below.

Assessing Risk

Deliberative rational choice demands an assessment of the probability and value of the consequences that might be expected to ensue from general public consumption of GMOs. Such assessments require epistemological and methodological judgments that have ethical implications. Although only a few authors have addressed the questions of food safety for GMOs, there is an extensive philosophical literature on risk and probability that is relevant to this general problem. First, risk assessors must settle questions about the *interpretation of probability*. Second, they must have a general philosophy for distinguishing

risk and uncertainty, and must have norms for responding to each. Third, they must make assumptions about minimizing *type I and type II errors*. Finally, they must decide between *formal and informal* risk assessment.

Interpretation of Probability. Three general approaches to probability can be found in the literature, and considerable innovation in the philosophy of probability has taken place in the last decade. Classical probabilities are derived from formal properties of the systems under study. An ordinary die has six sides, and a "fair die" can be defined as one for which the likelihood for each face turning up on a given role is equal, or 1/6. From this one can derive probabilities for combinations arising from several dice thrown at once. Relative frequency is a probability stated as the frequency with which a given outcome occurs in a given population of trials. Subjective probability refers to the confidence or expectation that a given person has that a predicted event will occur, or that a given proposition obtains.

Classical probability provides a formal specification of probability that permits substantial development of statistical theory. One may then treat both relative frequency measurements and subjective probabilities as situations where the analyst simply lacks an adequate specification of the formal system. Bayes's theorem provides a way to combine relative frequencies with subjective judgments, as well as to update results from several trials. Uncertainty then reflects a measure of how likely it is that any given statistical measurement is wrong, given available evidence. Although classical probability is generally thought to be inapplicable in most problems of empirical risk assessment, statistical theory and methods generally render philosophical questions about which theory of probability is the true one moot in practical situations.

Nevertheless, the potential for subjective interpretations of probability opens the door for dispute about the legitimacy of any given risk assessment. Whose subjective judgments are to count? Why should expert judgments supplant lay judgments? These are easy questions to answer in a purely instrumental context. Experts generally make more reliable judgments in virtue of their expertise. Yet the questions may be asked as a challenge to the authority of experts in making regulatory decisions that affect the lay public. Here the above-mentioned assumption that consumers are themselves deploying a deliberative rational choice model becomes crucial, for if that is so, relying on expert judgment may result in better choices than if individuals make their own judgments about risk. Even a resort to relative frequency does not settle this issue, for there are value judgments embedded in how to construct and evaluate the populations on which relative frequency trials are based. Such concerns might be more productively pursued within the framework of a general rejection of some assumptions common in deliberative rational choice (18).

Risk and Uncertainty. Although statistical theory provides one approach to uncertainty, it demands evidence or assumptions about the relationship between observations and the total population of instances for which observations might be made. In practical terms, this means that one must have at least speculative knowledge of the mechanism or correlation thought to lie at the root of a risk. This approach to uncertainty does nothing to address the possibility that there has been some scenario or possibility that no one has thought of. Yet entirely unknown risks do materialize. When scientists developed the feeding strategies that are now thought to have led to the variant of encephalopathy associated with mad cow disease, prions were not known as potential risk agents. It is clear that many who raise questions about uncertainty associated with human consumption of GMOs through the food chain are referring to this kind of uncertainty, rather than the sort addressed through standard statistical approaches.

Though it would be difficult to say how such uncertainty should be measured, it is common to apply relative quantitative judgments to the uncertainty that surrounds GMOs as compared to traditional foods. Common practice would also lead to the judgment that practices associated with great uncertainty are thereby "risky." Given the vagueness and unmeasurability of this uncertainty, it must be approached through subjective probability, if it is to be incorporated into deliberative rational choice at all. Here, again, the problem of whose judgments to use (discussed immediately above), and whether Bayesian techniques may be applied, reasserts itself. In general, philosophers have concluded that experts are prone to dismiss uncertainties too quickly, and to conclude that highly novel practices are inherently risky. For their part, risk experts have tended to demand at least a plausible scenario for how an unwanted event might transpire before they will seriously consider reviewing it in a risk assessment (18).

Type I and Type II Errors. In standard scientific research, conservative practice demands that a result be rejected unless uncertainty (in the technical sense) can be reduced to the point that the research is 95 percent certain that the result is true. Accepting a false result is a type I statistical error. A type II error occurs when one fails to accept a result that is, in fact, true. In risk assessment, early results often indicate risk but are not adequate to corroborate that result with 95 percent confidence. Should scientists minimize the chance of a type I error, and withhold judgment until further studies are done? Or should they minimize the chance of a type II error, and announce that a risk is present, even though future studies may well show that it is not?

Within the regulatory system for genetically engineered foods, decision makers would certainly apply a principle of caution (minimizing type II error) prior to the approval or release of a given product. The situation can be far more difficult if evidence for risk appears after a substance is already on the market. Here the potential for needless panic and economic loss often leads regulators to regard relatively unconfirmed results pointing toward risk as premature. Though no clear cases of type I/type II dilemmas have yet occurred in the food safety regulation of GMOs, uncertainty is itself treated as evidence for type II error by some critics. It is not clear how regulatory agencies would or should handle such cases (19).

Formal or Informal Risk Assessment. Risks for pesticides, food additives and drugs are subjected to laboratory and clinical research trials that establish measures of risk. Many have argued that such formal procedures introduce unnecessary costs in the development of a GMO. Instead, they argue that adequate risk assessment can be done simply by reviewing the nature of the planned alteration subjectively. Reviewers would note that interventions involving known allergens or that had a known potential to create new proteins in food should undergo formal risk assessment, but other GMOs would pass directly into the food chain. Reviews would be done by individual researchers or by Institutional Review Boards (IRBs) at universities, laboratories, and within the food industry. This is, in effect, the procedure currently used for assessing the food safety risk of consuming organisms developed through conventional plant and animal breeding. Critics of agricultural biotechnology have vociferously opposed the informal approach.

These two views split two key value assumptions of deliberative rational choice. On the one hand, deliberative rational choice is supposed to be deliberative, which would imply careful review and objective assessment of all risks for all affected parties. On the other hand, rational choice is supposed to produce the best outcome, and if otherwise acceptable products are never developed because process of policy approval is too costly, that is a result hardly in keeping with its spirit, (5,20–22).

RISK, PURITY, AND CONSENT

Traditional approaches to food choice combine the safety of food with culturally based judgments about purity. Anthropologists have studied these approaches, but no philosophical literature articulates the ethical principles on which they rest. One possibility is that people are attempting to emulate deliberative rational choice through their food traditions. An alternative possibility is that individuals and groups have been thought to have the right to apply whatever standards of purity they deem appropriate in making food choices. This alternative view finds philosophical support in two complementary positions. First, purity norms may function to promote rational ends, but through a nondeliberative mechanism. If so, it may be rational to rely on purity rules, even when deliberative calculations indicate otherwise. Second, many bioethicists have long argued that risks may only be imposed on subjects with their consent. Together, purity rules and consent criteria establish a procedural burden of proof for the safety of genetically engineered foods and food products.

Risk and Purity

Anthropologist Mary Douglas has approached risk from the standpoint of cultural norms that establish the most basic categories of acceptable behavior. In any society certain patterns of conduct are established as accepted and unexceptional. Cultural norms and expectations determine the boundaries for accepted conduct. Since behavior that falls within these bounds is expected, it does not occasion special consideration or deliberation. Douglas notes that food regimes and purity rules constitute an important part of the implicit norms that provide background rules for acceptable conduct in any society. Conduct that challenges these boundaries is defined as risky. Such conduct will either be repressed, or it will require justification according to burdens of proof that are also culturally determined (23,24).

From the standpoint of rational choice, cultural norms function as pre-deliberative filters that limit the circumstances in which deliberation will be applied. Although the range of deliberative choices in industrial societies is quite broad, it is impossible for individuals or organizations to apply the calculation and ranking entailed by deliberative rational choice to every potential choice situation. People simply do not have enough time and mental energy to weigh the consequences of every possible action. As such, rational behavior presupposes the existence of cognitive filters that sort life's options into categories. Some actions require no deliberative attention, others do. Douglas's purity rules function as pre-rational filters that sort life's options into the unexceptional and the risky. Risky actions require further consideration; they fall under a bias that demands proof of their acceptability. Actions that are consistent with purity norms do not trigger these additional burdens of proof, which is to say that they are not risky.

Although this approach is tantalizingly close to the rational choice paradigm, it is important to see that it utilizes an altogether different conception of risk. Crucially, there is no contradiction in saying that a given action has a nonzero likelihood of causing harm, but is not risky in virtue of the fact that it would not call for deliberative consideration. To say that an action entails risk in Douglas's sense is simply to say that it is out of the ordinary. Many daily activities—walking down stairs, making a pot of coffee—are undertaken without deliberative, conscious calculation. Bad things can happen as a result of walking down stairs or making coffee, but we do not apply a calculation of the probability of bad consequences in ordinary daily pursuit of these activities.

A culturally based set of food purity rules would have to be functionally rational in the sense that they would have to limit the number of cases of accidental poisoning as well as short-term disorders and dietary deficiencies. A society with too many such incidents would experience any number of weaknesses that would threaten its survival. Indeed, if a food regime appears to functioning adequately, it might be irrational for the cultural elite to expend time and energy on a deliberative review of it. However, anyone who violates these rules or who challenges them in any way would be engaging in conduct that *does* call for deliberative review, and quite possibly sanction. Such conduct would be classified as risk (18).

The time-honored response to risk is to repress it, to ban risky actions altogether. Responses to risk in complex societies are more varied. One response is to initiate the conscious, deliberative review of choice that leads one eventually to an assessment of the probability that a given course of action will result in harm or loss. However, other responses may also be reasonable, including policy norms that permit risky actions when affected parties have been given the opportunity to give or withhold their consent.

New foods would thus not constitute a risk to an existing food regime so long as practitioners of traditional purity rules could continue their usual practices, experimenting with novelty only under conditions of consent.

Purity rules have two important philosophical ramifications. First, they blur the distinction that experts draw between safety understood as probability of harm or loss and broader, culturally based views about what is or is not acceptable food. When an observant Jew or Muslim violates dietary laws, their conduct challenges tradition in a manner that might be called "risky," even though it may have little objective probability of causing illness or injury. The social, cultural, and individual objectives being served through dietary rules may be much broader than the expert's conception of safety, and the attempt to supplant them with a reductive approach to risk may itself be perceived as a challenge to the cultural integrity of a food regime (25). The policy debate that occurs in response to this situation leads to political and ethical issues about how expected value assessments of food safety should be deployed, and whether they are consistent with consent criteria, discussed below.

Second, while functionally rational, purity rules suggest an approach to risk that is conceptually incompatible with deliberative rational choice. It is meaningful to claim that a given food or diet poses "no risk" on this view. Advocates of deliberative rational choice translate this as "zero risk." They go on to assert that the zero risk goal is irrational, implying that anyone who continues to address dietary choice through a framework of purity rules is irrational, hysterical, or at least profoundly misinformed. In itself, this may not represent a deep philosophical problem. We have two ways to use our general concept of risk, one in a general classificatory sense, the other to specifically call attention to the probabilistic dimension of unintended consequences. Pragmatic and contextual circumstances should determine which sense is in play at any given moment. However, this pragmatic or contextualist approach to risk is itself challenged by many who defend a more essentialist analysis of risk. Here genetically engineered foods become a case for a larger philosophical dispute (26).

Risk and Consent

Literature on risk and consent emphasizes situations in which individuals will be exposed to hazards as a condition of employment, medical treatment, or involvement in scientific research. In the almost unanimous opinion of scholars who have written on the subject, such risks may not be imposed on conscious and competent individuals without their consent. Individuals may claim a right to the information needed to evaluate the likely consequences of such risks, and to withhold consent, or *exit* from the risky situation. When such risks are imposed without consent, the party imposing risk may be held responsible for damages, and it is morally culpable for imposing risk, even when damages do not actually occur.

Classic environmental risks from air or water pollution present an altogether different situation. The entire population in a region is typically exposed to such risks, and it is in no position to claim a right of exit. There is debate as to the significance of these cases for consent criteria. On one view, such risks differ in their nature, and must be evaluated according to criteria of general public good, rather than consent. On another view, pollution risks differ only in virtue of the fact that it is difficult to identify the source of pollution, and hence difficult to discern who should be culpable for imposing risk without consent (27–29).

Food safety risks introduced through the food chain bear some similarity to environmental risks. Once a genetically engineered variety of corn or soybean is combined with other bulk crops, it is very difficult to trace the source and content of any given commodity lot. Processed foods with variable ingredients have long been sold, and it would be impossible to argue that consumers have been able to claim a right to reject food constituents on an ingredient by ingredient basis. However, it is also true that religious minorities have successfully maintained an ability to exercise dietary rules. Individuals with very idiosyncratic beliefs about diet and health have been able make food choices based on those beliefs, mainly by eating a diet consisting in whole foods. GMOs challenge their opportunity to do this, for there may be no way to determine whether any given whole food commodity may contain grains, meats, or milk products derived from GMOs (30).

Food biotechnology's challenge to consent is procedural. One need not show that GMOs pose measurable risk to individual consumers in order to show that they challenge an individual's right of consent. As noted above, this right is typically exercised against a broad range of challenges to an existing dietary regime, not simply against the probability that the individual will suffer from injury or disease. As such it is useful to review some of the reasons why individuals might prefer not to eat genetically engineered foods.

1. *Religious objections.* Genetic engineering raises religious issues for many individuals. At least three types of religious concern may be relevant to food.

   - *Genetic engineering is wrong.* Clearly, individuals who believe that all forms of genetic engineering are wrong may have a legitimate reason to avoid genetically engineered food.
   - *Dietary rules.* The question of whether a particular food biotechnology is consistent with a given sect or congregation's interpretation of dietary rules (e.g., *kashrut*) must be left to religious authorities.
   - *Sanctity of life.* Some critics of biotechnology have extended a religiously based concern about commercial exploitation of genetic technology to animals and plants.

2. *Mistrust of science.* Many do not trust scientists or scientific pronouncements in the wake of well-publicized mistakes and deceptions. This mistrust takes at least three forms.

   - *Safety concerns.* Though no evidence suggests that GMOs increase the probability of disease or injury, many are unwilling to rely on existing studies or the word of scientists.
   - *Reflexive risk inferences.* Some people infer that if scientists and regulators are unwilling to provide information through food labeling (discussed below), the technology must be dangerous.
   - *Increasing power of scientific elites.* Some may resist genetic technology because the see it as one instance of a general loss of individual autonomy in complex society.

3. *Broader consequences.* Consumers may feel that food choices provide them the best opportunity to voice concerns about environmental, social, or animal impacts of food biotechnology.

4. *The yuk factor.* Many find genetic engineering aesthetically repulsive. Since individual aesthetics are an intrinsic dimension of food choice, it is reasonable for individuals to cite their aversion as a basis for withholding consent.

Any of these reasons might provide an individual with a reason to reject food biotechnology as a personal choice, and to regard it as "impure" or "unwholesome" (31).

FOOD LABELS

The larger philosophical issue is that experts and a segment of the lay public may be applying different philosophical frameworks to food safety. Experts define food safety quite narrowly, with respect to the conditional probability of disease or injury associated with the consumption of genetically modified foods. Technical problems of risk assessment aside, they see little ethical basis for concern about GMOs. Some in the lay public are applying a form of purity rules or have one or more of the concerns noted above. They claim a right to exit from the emerging food regime that includes GMOs. This tension is manifesting itself as a dispute over the need for labeling of genetically modified foods.

The labeling dispute replicates issues in the larger dispute. On the one hand, food labels can be seen as instruments that enable rational choice. In this view, information placed on the label of a food has ethical significance to the extent that it helps consumers realize the optimizing objectives of deliberative rational choice. Information must be true, but it must also be usable. It must help consumers reach the goals they seek to implement. To the extent that these goals are limited to health concerns, the only basis for distinguishing one product from another is when there is some reason to associate a measurable probability of disease or harm. The paradigm cases in the United States have been tobacco and saccharin, both subjected to mandatory labels in the wake of scientific studies. Lacking studies that identify risk from GMOs, there is no basis for requiring labels.

On the other hand, those who claim a right to know whether any given food is the product of genetic engineering or other forms of food biotechnology are demanding a right of exit. They see food labels as mechanisms for securing consent, without regard to whether or how they will use the information on labels in making food choices. Even those who plan to eat GMOs may believe that the right of consent should be protected. Their mistrust of food biotechnology may, ironically, be based on suspicions that arise in the wake of resistance to labeling that arises on the part of the expert community.

The philosophical perspectives of rational choice and of consent thus present radically different burdens of proof for evaluating a policy on the labeling of GMOs, just as they do for evaluating the broad questions of risk and safety themselves. The likely consequences of labeling may be largely irrelevant to an advocate of consent, and the rational choice view that labels must be constructed so as to enable better choices will be seen as paternalistic. On the other side of the controversy, experts fear that poorly informed individuals will misinterpret labels, that they will make unwise food choices, and that labels will stigmatize GMOs without basis (32).

BIBLIOGRAPHY

1. P. Sparks, R. Shepherd, and L.J. Frewer, *Agric. Hum. Values* **11**(1), 19–28 (1994).

2. L.J. Frewer, R. Shepherd, and P. Sparks, *Br. Food J.* **9**, 26–33 (1994).

3. T.J. Hoban, in D. Maurer and J. Sobal, eds., *Eating Agendas: Food and Nutrition as Social Problems*, de Gruyter, New York, 1995, pp. 189–209.

4. I.C. Munro and R.L. Hall, in J.F. MacDonald, ed., *Agricultural Biotechnology, Food Safety, and Nutritional Quality for the Consumer*, National Agricultural Biotechnology Council, Ithaca, NY, 1991, pp. 64–73.

5. S. Huttner, in R.D. Weaver, ed., *U.S. Agricultural Research: Strategic Opportunities and Options*, Agricultural Research Institute, Bethesda, MD, 1993, pp. 155–168.

6. N. Elias, *The Civilizing Process*, Basil Blackwell, Oxford, UK, 1976.

7. W.C. Whit, *Food and Society: A Sociological Approach*, General Hall, Dix Hills, NY, 1995.

8. B. Ames, *Science* 1256–1263 (1983).

9. P.B. Thompson, in T.B. Mepham, G.A. Tucker, and J. Wiseman, eds., *Issues in Agricultural Bioethics*, Nottingham University Press, Nottingham, England, 1995, pp. 31–45.

10. S. Stich, *Philos. Public Affairs* **7**, 187–205 (1978); reprinted in M. Ruse, ed., *Philosophy of Biology*, Macmillan, New York, 1989, pp. 229–243.

11. D. Johnson, *Agric. Hum. Values* **3**(1&2), 171–179 (1986).

12. R. Wachbroit, in M.A. Levin and H.S. Strauss, eds., *Risk Assessment in Genetic Engineering*, McGraw-Hill, New York, 1991, pp. 368–377.

13. J.D. Graham, L.C. Green, and M.J. Roberts, *In Search of Safety: Chemicals and Cancer Risk*, Harvard University Press, Cambridge, MA, 1988.

14. J. Halloran, *Agric. Hum. Values* **3**(1&2), 5–9 (1986).

15. M.P. Doyle and E.H. Marth, in B. Baumgardt and M. Martin, eds., *Agricultural Biotechnology: Issues and Choices*, Purdue Research Foundation, Lafayette, IN, 1991, pp. 55–67.

16. H.R. Cross, in J.F. MacDonald, ed., *Animal Biotechnology: Opportunities and Challenges*, National Agricultural Biotechnology Council, Ithaca, NY, 1992.

17. M.J. Dobbins et al., *J. Nutr. Educ.* **26**(2), 69–72 (1994).

18. P.B. Thompson, *Food Biotechnology in Ethical Perspective*, Chapman & Hall, London, 1997.

19. C.F. Cranor, *Regulating Toxic Substances: A Philosophy of Science and the Law*, Oxford University Press, New York, 1993.

20. A.L. Caplan, *Agric. Hum. Values* **3**(1&2), 180–190 (1986).

21. G.L. Johnson and P.B. Thompson, in B. Baumgardt and M. Martin, eds., *Agricultural Biotechnology: Issues and Choices*, Purdue Research Foundation, Lafayette, IN, 1991, pp. 121–137.

22. H.I. Miller, *Trends Biotechnol.* **13**, 123–125 (1995).

23. M. Douglas, *Implicit Meanings*, Routledge, London, 1975.

24. M. Douglas and A. Wildavsky, *Risk and Culture*, University of California Press, Berkeley, 1982.

25. K. Taylor, in J.F. MacDonald, ed., *Agricultural Biotechnology, Food Safety, and Nutritional Quality for the Consumer*, National Agricultural Biotechnology Council, Ithaca, NY, 1991, pp. 96–102.

26. R. von Schomberg, in R. von Schomberg, ed., *Contested Technology: Ethics, Risk and Public Debate*, International Centre for Human and Public Affairs, Tilburg, 1995, pp. 13–28.

27. D. MacLean, in *Values at Risk*, Rowman & Allanheld, Totawa, NJ, 1986, pp. 17–30.

28. H. Shue, *Agric. Hum. Values* **3**(1&2), 191–200 (1986).

29. K. Shrader-Frechette, *Risk and Rationality*, University of California Press, Berkeley, 1991.

30. J. Burkhardt, in H.D. Hafs and R.G. Zimbelman, eds., *Low-Fat Meats: Design Strategies and Human Implications*, Academic Press, San Diego, CA, 1994, pp. 87–111.

31. P.B. Thompson, *Hastings Cent. Rep.* **27**(4), 34–38 (1997).

32. P.B. Thompson, *Choices: Mag. Food, Farm, Resour. Issues*, First Quarter, 11–13 (1996).

See other entries AGRICULTURAL BIOTECHNOLOGY; ANIMAL BIOTECHNOLOGY, LAW, FDA REGULATION OF GENETICALLY MODIFIED ANIMALS FOR HUMAN FOOD USE.

AGRICULTURAL BIOTECHNOLOGY, LAW, AND EPA REGULATION

MARY JANE ANGELO
St. John's River Water Management District
Palatka, Florida

OUTLINE

INTRODUCTION

As we cross the threshold into the new millennium, agriculture is on the brink of a new technological era. Biotechnology is dramatically changing the landscape of modern agriculture. In the mid-1990s there were more than 1200 biotechnology companies in the United States (1). This number is expected to increase dramatically in the new millennium. In the year 2000, farm-level sales of biotechnology products are expected to be in the tens of billions of dollars (2). One of the most promising areas of agricultural biotechnology is the area of pest resistance. Biotechnology has made it now possible to produce naturally occurring proteins that act as pesticides in quantity in microbial organisms or plants. The ability of a microbe or crop plant to produce pest resistance may obviate the need for harsher and less selective synthetic pesticides (3). In addition to pest resistance, a variety of crops have been engineered to produce increased levels of desired nutrients or to impart them with other desirable characteristics such as cold tolerance (3). Regulatory agencies, such as the U.S. Environmental Protection Agency (EPA) have been struggling to keep up with these rapidly developing technologies and to establish regulatory programs that promote the beneficial uses of the new products, while attempting to protect the public health and the environment from the associated potential risks. This article focuses on EPA's regulation of agricultural biotechnology under the Federal Insecticide, Fungicide, and Rodenticide Act (FIFRA) (4) and the Federal Food, Drug, and Cosmetic Act (FFDCA) (5).

AGRICULTURAL BIOTECHNOLOGY

Since 1962, when Rachel Carson's seminal work, *Silent Spring* (6), first awakened the country to the risks of chemical pesticides, the public has been skeptical of the government's ability to protect them and their environment from the hazards of pesticide use. While in recent years public attention has increasingly focused on the risks of pesticides in food, particularly the risks to small children, many pesticides may also pose significant risks to farm workers, consumers, and the natural environment. Within three decades following the publication of *Silent Spring*, environmentalists and consumer groups repeatedly called upon the government, in particular EPA, to push for the reduction of pesticide use. Despite these efforts, in 1991 alone, approximately 4.1 billion dollars worth of pesticides, roughly 320 million kilograms of pesticides, were used in the United States (7). EPA has registered over 19,000 pesticides, containing 913 different active ingredients (8).

In the 1990s dramatic changes began to take place, especially with regard to an ever-increasing number of pesticide products derived from biotechnology processes. The promise of these pesticides stems from the potentially lower risk to humans and the environment. This is due to their greater specificity to the target pest than chemical pesticides, their tendency to have lower toxicity than chemical pesticides, and the tendency of many biotechnology pesticides to have limited persistence in the environment. The universe of biotechnology pesticides is large and diverse, and it includes microorganisms such as bacteria, fungi, algae, protozoa, and viruses, which act as pesticides by producing toxins, acting as parasites, or acting through competition and macroorganisms such as parasitic wasps or plants that produce substances that exert a pesticidal effect.

What is Biotechnology? Biotechnology, in its broadest sense, is the use of living organisms, be they plants, animals, or microorganisms, to make or modify products. While there does not appear to be one standard definition of "biotechnology," most definitions are broad enough to cover a wide array of processes including genetic engineering and more traditional processes such as plant breeding and fermentation. The U.S. government has defined "biotechnology" as the "use of various biological processes, both traditional and newly devised, to make products and perform services from living organisms or their components" (9). For centuries, biotechnology has been used to manufacture products such as bread, beer, wine, yogurt, and cheese (10). For thousands of years traditional plant breeding has enabled the production of crop plants with desired traits such as high seed yields and increased resistance of pests and environmental stresses (11). For example, early farmers are believed to have created wheat over 5000 years ago by combining traits from three different species (12). By repeatedly selecting plants that exhibit the desired traits and crossbreeding them with closely related plants over several generations, traditional plant breeders were able to create plants with a desired combination of traits (11). However, traditional plant breeding is limited by two major constraints: (*1*) Removing undesirable traits from the original cross can take generations and often takes years; (*2*) only closely related plant species can be directly bred together, severally limiting the gene pool available. Genetic engineering enables plants to be developed that cannot be produced through traditional plant breeding (12).

In recent years, through the use of recombinant DNA (rDNA), researchers have been able to "genetically engineer" organisms by moving genes from one organism to another. Recombinant DNA technology allows the isolation and characterization of specific pieces of DNA from one organism and transfer of the DNA sequences into another organism. The term "genetic engineering" generally refers to the use of recombinant DNA (rDNA), cell fusion, or other novel bioprocessing techniques (13). Recombinant DNA technology has dramatically increased the speed of inserting a desired trait into a plant (11). Moreover rDNA techniques eliminate the problem of undesirable traits being introduced into the plant along with the desired genes (14). Additionally rDNA techniques can be used to move desired genes from virtually any types of living organism, be it plant, animal, or microorganism, into the plant (14).

How is Biotechnology Used in Agriculture? For the past 10 years, EPA has exercised regulatory oversight over genetically engineered microbial organisms that act as pesticides. Because microbial pesticides are living organisms and have the potential to reproduce and spread on their own in the environment, they pose the potential for unique risks. Thus EPA's regulatory scheme for microbial pesticides is somewhat different from that for conventional chemical pesticides. Nevertheless, microbial pesticides are similar to conventional pesticides in that they are "applied to" crops, and thus they are in many ways regulated like traditional pesticides. In the past several years, however, EPA has been faced with a completely new class of genetically engineered pesticidal products that poses a new set of regulatory challenges. In the 1990s significant technological advances were made in altering plants to produce pesticidal substances. That rDNA technology has advanced to the point where researchers are able to more easily move genes from microorganisms, animals, or other plants into agricultural crop plants. For example, through these new rDNA technologies, plants can be made to produce toxins normally produced only by microorganisms such as the *Bacillus thuringiensis* (Bt) insecticidal delta-endotoxin. Bt acts by forming a protein crystal, referred to as the delta endotoxin, that becomes toxic upon ingestion by the insect. EPA considers the pesticidal substances produced by plants and the genetic material necessary to produce them to be "plant-pesticides." Although EPA does not yet have in place a final comprehensive regulatory scheme to address plant-pesticides, the biotechnology industry has advanced to the point where it is commercializing products, and thus, for the past several years, EPA has been regulating these products under a proposed regulatory scheme. EPA has issued several Experimental Use Permit (EUP) applications and registration applications for the Bt delta-endotoxin produced in various plants. (See section on FIFRA below.) EPA also has granted applications for tolerance exemptions for residues of pesticidal substances produced in plants as a result of genetic engineering.

Some of the agricultural crops that have been developed through this type of genetic modification in recent years include corn, cotton, and potato plants that have been genetically modified to contain a bacteria gene that leads to the production of the Bt insecticidal toxin, squash that has been genetically modified to contain a virus gene that leads to the production of a viral coat protein to make the plant resistant to infection by viruses, and cotton and soybean plants have been genetically modified to contain bacteria genes that cause the plants to tolerate herbicides that are applied to the plant. Recombinant DNA techniques also are being used to produce a new class of animal hormones, the somatotropins, such as the bovine somatotropin (BST) hormone, which was approved by the Food and Drug Administration (FDA) in 1993 for use in lactating dairy cows to produce more milk.

EPA received the first EUP application for the Bt toxin produced by a genetically engineered plant, cotton, in November 1991. In the years following the Bt in cotton EUP, EPA has granted EUPs for Bt in potatoes and corn and a number of registrations for Bt in cotton, potatoes, and corn. EPA has also granted exemptions from the requirement of a tolerance for a number of Bt plant-pesticides. In addition EPA has registered and granted a

tolerance exemption for the potato leafroll virus resistance gene and has granted a number of tolerance exemptions for a variety of viral coat proteins in raw agricultural commodities. Currently EPA is reviewing a number of EUP registration, and tolerance exemption applications for other plant-pesticide products. It is anticipated that number of applications for EUPs, registrations, and tolerances for plant-pesticides will continue to grow at a rapid pace.

What is the Significance of Agricultural Biotechnology?

Risks. Many of the risk considerations for biotechnology pesticides are similar to, if not the same as, those for traditional chemical pesticides. In general, EPA has expressed a view that traditional chemical pesticides pose greater environmental risks than biochemical, microbial, or plant-pesticides (15). As with any pesticide risk assessment, the underlying considerations for analyzing risks posed by biotechnology pesticides are the potential for humans and other nontarget organisms to be exposed to the pesticide, and the hazard (usually toxicity) of the pesticide to nontarget organism, humans, and the environment. For biotechnology pesticides, as with other pesticides, hazard will be determined by the chemical and toxicological properties of the pesticidal substance. Exposure, on the other hand, will be determined somewhat differently for biotechology pesticides than for traditional chemical pesticides. For traditional pesticides the primary factors in determining exposure is the amount of chemical that is introduced into the environment and the likelihood that humans or other nontarget organisms will come into contact with the chemical. Because microbial and plant-pesticides are produced by living organisms, however, exposure issues are more complex for these substances and are dependent, in large part, on the biological characteristics of the organism itself. For example, exposure to a plant-pesticide could be determined by factors such as whether the production of the plant-pesticide is limited to particular plant parts (e.g., leaves, stems, fruit, or roots) and what organisms consume or are associated with those plant parts.

Moreover one of the most significant exposure considerations for microbial pesticides and plant-pesticides not seen for chemical pesticides is the potential for spread of the living organism or the organism's genetic material. For example, plants can reproduce sexually and/or asexually, and as a result the genetic material that was introduced into the plant and that enables the plant to produce plant-pesticides could spread through agricultural or natural ecosystems. Thus, if a plant that produces a plant-pesticide has the capacity to spread in the environment, or to spread its genetic material to other plants, there would be a greater potential for increased exposure to nontarget organisms than there would be for a plant-pesticide produced in a plant that can only grow in a limited geographic area or does not have the ability to cross-fertilize with other plants in the environment. This is a particular concern for plant-pesticides produced in plants that have wild relatives in the United States. If these wild relatives acquire the ability to produce the plant-pesticide, through cross-fertilization, many additional nontarget organisms could potentially be exposed to the pesticide.

The potential for a genetically modified organism (GMO) or its genetic material to spread from one plant to another raises additional risk issues beyond those of exposure to humans and nontarget organisms. One potential risk of biotechnology products parallels the risk of the introduction of any nonnative species into a new environment (15). Small genetic manipulations can result in significant changes in an organism's ability to survive and flourish in a particular ecosystem (12). There are dozens of examples of the disastrous, but unpredicted, effects of the introduction of nonnative species into the environment displacing native species (13). Genetically modified organisms introduced into the environment could have similar impacts (13). The risks that appear to be most significant are that a genetically engineered plant might become a weed or pest itself or that it might outcross with related species to create new weeds or pests (13). Once released into the environment, the spread of a GMO may be difficult, if not impossible, to control (13). One of the most cited concerns about plant-pesticides is the concern over the potential for the development of "superweeds" through the outcrossing of plants producing plant-pesticides to wild relatives. If the ability to produce a plant-pesticide that, for example, makes a plant resistant to insect or viral pests is spread to a wild relative and passed on to subsequent generations of that relative, there is the potential that the wild relative, by virtue of its newly acquired ability to resist insects or viruses, could become a hardy weed. Development of such a weed has the potential to disrupt agricultural or natural ecosystems. For a transgenic plant to transfer its genes to related existing weed species, however, wild relatives of the transgenic plant must grow in the geographic areas where the transgenic plant is introduced (13). Most crops grown in the United States are of foreign origin. Thus the risk of hybridization between transgenic crops and wild relatives is unlikely in the United States. Many domestic crops including soybeans, corn, and wheat have been bred to the point where they have lost their ability to compete with wild species in the environment. Thus these crops are unlikely to become weeds when genetically altered (12).

Another issue that has received considerable attention is the potential for plant-pesticides in foods to pose a risk of allergenicity to humans. The primary concern appears to be that if a gene that leads to the production of a plant-pesticide is moved from one plant, for example, a peanut, into another plant, for example, corn, people who know they are allergic to peanuts will not know to avoid the corn plant. Thus, if the plant-pesticide derived from the peanut plant contains an allergen from the peanut plant, allergic consumers could be put at risk (14,15). Other areas of potential adverse effects on the environment center on specific plant-pesticides or categories of plant-pesticides. For example, some environmental organizations have expressed their concern that engineering plants to produce viral coat proteins has the potential to result in the develop of new unintended viruses.

In addition to the risk concerns described above, public interest organizations have articulated other concerns that are more philosophical, ethical, or religious in nature. For example, the movement of genes from animals to plants may be of concern to subpopulations of people with special dietary preferences such as vegetarians or persons who observe kosher (Jewish) or halal (Muslim) laws (17). Other philosophical issues that have been raised include a concern that the prospect of "human-made" organisms, even if they pose no risk to humans or the environment, may threaten the concepts of "wildness" and "wilderness" (18, p. 33). Some argue that while biotechnology pesticidal products may be environmentally preferable to traditional chemical pesticides, the focus on developing these products may be diverting attention from the more important goal of developing a system of sustainable agriculture (19, p. 67).

Probably the most significant concern with agricultural biotechnology stems from the fact that the risks of biotechnology are uncertain. Although the risk of a genetically modified organism released into the environment creating a new superweed or disrupting the balance of natural ecosystems may be small, the consequences could be disastrous and potentially irreversible (15). The precise nature and magnitude of the risk is difficult to predict because of the almost infinite variety of potential genetically modified organisms, the reproductive ability of GMO's, the complexity of the natural balance of ecosystems and the dearth of long-term data (15).

Benefits. To many, agricultural biotechnology products hold the promise of a less risky substitute for traditional chemical agricultural products. The use of rDNA technologies has enabled organisms, particularly plant varieties, to be developed that either could not have been developed through traditional plant breeding or could only be developed through traditional techniques with a great amount of time and difficulty. Chemical pesticides often are of relatively high toxicity. Many, but not all, traditional chemical pesticides are toxic to a broad range of organisms, including humans. In addition the manner in which traditional pesticides are applied—often sprayed over large areas—could result in significant exposure to nontarget organisms. Biotechnology pesticides, on the other hand, are generally of low toxicity, target-specific, and produced in relatively small quantities in the organism. Because plant-pesticides are generally produced in small amounts in the plant, nontarget organisms are not as likely to be exposed to these pesticides as they are to pesticides that are sprayed over large areas. Moreover, even if nontarget organisms are exposed to plant-pesticides, because these pesticides are often of low toxicity and are generally target specific, nontarget organisms are not as likely to be adversely affected by these pesticides as they are with pesticides that are more highly toxic or toxic to a broad spectrum of organisms. For example, the Bt. toxin is specific to specific groups of insects (e.g., Lepidoptera) and is not toxic to humans or other mammals.

One example of where a plant-pesticide is believed to have the potential for significant environmental benefits, is viral coat protein-mediated resistance. By genetically modifying plants to produce certain viral coat proteins, researchers have been able to produce plants that are resistant to infection by particular viruses. For viruses spread by vectors such as insects, the most common agricultural practice for preventing viral attack is the use of chemical pesticides to control the insect vector that spreads the virus. It is believed that the use of viral coat protein-mediated resistance would reduce the need for these chemical pesticides. In addition to the environmental benefits of viral coat protein-mediated resistance, there is a high potential for significant economic benefits. Another potential environmental benefit is the reduction of runoff of agricultural chemicals such as pesticides and fertilizers, which can contaminate surface and ground water (11). For example, rDNA technique may be used to create plants with improved photosynthetic and nitrogen fixation capabilities, thereby reducing the need to apply fertilizers (11). Moreover, some of the most environmentally friendly herbicides with relatively low toxicity, low soil mobility, and rapid biodegradation are also the herbicides that are the most nonselective, and thus the most likely to kill crops plants along with the weed (11). Farmers are often forced to apply more selective and more toxic herbicides as a result (11). The use of genetically modified herbicide tolerant crop plants may benefit the environment by causing a reduction in the use of highly toxic herbicides. Environmental organizations have expressed concerns that plants that have been genetically modified to be tolerant to herbicides could actually result in an increase in herbicide use because herbicides would be able to be directly applied to crop plants without killing them (18). Industry groups, however, assert that these plants will enable farmers to reduce the number of herbicide applications by allowing farmers to target the timing of herbicide application to after the plant has emerged, when herbicides are most needed (20,21). Herbicide tolerant plants are not considered to produce plant-pesticides because the substances they produce are not intended to prevent, destroy, or repel pests. Thus these products would not be covered by the plant-pesticide policy or rules. EPA is planning to develop a separate policy for these plants. Other potential benefits of plant-pesticides may not yet be apparent. Nevertheless, many scientists believe that the technological advances in this area hold out great promise for the future.

Public Perception. The intensity of the public response to the 1992 FDA policy on foods derived from new plant varieties, as well as the public concerns surrounding FDA's approval of the BST milk, illustrates the important function that public perception will play in defining the role of agricultural biotechnology in the marketplace. As others have pointed out, while many new technologies will soon be commercially viable, they all will not automatically be put to use—consumers will be the ultimate judge of emerging technologies (22). Key to the success or failure of new biotechnology products will be the ability of the government agencies responsible for regulating these technologies, such as EPA, to effectively communicate to the public the risks and benefits of these products and the public's resulting acceptance or nonacceptance. Many people are skeptical of any new technology. This skepticism is even more pronounced with biotechnology, which could be difficult for the layperson to understand

because it is surrounded by many uncertainties. A recent survey conducted to gather information on consumer attitudes about the use of biotechnology in agriculture and food production concluded that one of the most important factors influencing public perception of biotechnology is the perceived credibility of public policies and regulations. This survey found that while most consumers supported the use of biotechnology in agriculture and food production (22), they also favored an active role for government agencies in establishing biotechnology regulations that ensure environmental protection and food safety (22). Thus EPA must be mindful that the public will be looking to it, not only to evaluate the risks and benefits of biotechnology pesticides in order to develop a regulatory program that will protect humans and the environment but also to effectively communicate with the public on these issues. Possibly the most serious public concern over agricultural biotechnology is the use of the technology in the production of food crops (23).

HOW IS AGRICULTURAL BIOTECHNOLOGY REGULATED?

The Coordinated Framework

The U.S. government's first systematic attempt to address the regulation of biotechnology in a comprehensive fashion was with the publication of the 1984 document entitled *Proposal for a Coordinated Framework for Regulation of Biotechnology* (24). The purpose of this document was "to provide a concise index to U.S. laws related to biotechnology, to clarify the policies of the major regulatory agencies that will be involved in reviewing research and products of biotechnology, to describe a scientific advisory mechanisms for assessment of biotechnology issues, and to explain how the activities of the Federal agencies in biotechnology will be coordinated." In 1986 the Office of Science and Technology Policy (OSTP) published in the Federal Register a *Coordinated Framework for Regulation of Biotechnology; Announcement of Policy and Notice for Public Comment* (the Coordinated Framework) (25). This document made clear that the Executive Branch believed it could adequately regulate biotechnology under its existing authorities and did not intend to seek new legislation to address emerging technologies. The Coordinated Framework described in detail the roles of the five federal agencies with significant involvement in the regulation of biotechnology: FDA, the United States Department of Agriculture (USDA), EPA, the National Institutes of Health (NIH), and the Occupational Safety and Health Administration (OSHA). The Coordinated Framework was created to harmonize the regulation of biotechnology between several federal agencies and to address gaps and overlaps between and among agencies (15). The Coordinated Framework contains four major conclusions: (1) Existing federal statutes are sufficient to regulate biotechnology, (2) federal agencies should regulate "products" rather than the "process," (3) the safety of biotechnology products should be addressed on a case-by-case basis, and (4) the efforts of all agencies involved in regulating biotechnology should be coordinated (13). The Coordinated Framework gave

EPA the primary responsibility over the environmental regulation of biotechnology.

Under the Coordinated Framework, the regulatory approach taken by U.S. regulatory agencies, including EPA, has been to rely on existing statutes and to focus on the "product" rather than the "process" used to create the product (15). The thinking underlying this approach is that rDNA technology in itself does not create risk, and thus does not necessitate an entirely new regulatory system (15). Instead, certain types of products of biotechnology may pose risks that can be addressed in the same fashion as risks posed by traditional "chemical" products.

EPA's Statutory Authority

EPA's primary authority for regulating agricultural biotechnology products can be found in two statutes: the Federal Insecticide, Fungicide, and Rodenticide Act (FIFRA), 7 U.S.C. §136–136y, and the Federal Food, Drug, and Cosmetic Act (FFDCA), 21 U.S.C. §301 *et seq*. Since EPA was created in 1970, it has had responsibility for the regulation of pesticides under both laws. Reorganization Plan No. 3 of 1970, 84 Stat. 2086. Under FIFRA, EPA is responsible for regulating the distribution, sale, use, and testing of pesticides to prevent unreasonable adverse effects to humans and the environment. In evaluating a pesticide, EPA balances the potential human and environmental risks against the potential benefits to society of using that pesticide. Under FFDCA, EPA has the authority to set tolerances for pesticide chemical residues in or on food. In establishing tolerances, EPA evaluates the impacts of human dietary exposure to the pesticide residues. EPA also regulates biologicals and biotechnology products that are not pesticides, food, or drugs under the Toxic Substances Control Act (TSCA), 15 U.S.C. §§2601–2692. TSCA grants EPA the authority to screen new chemical substances and impose controls to prevent unreasonable risks and, through rule making, to acquire information and impose restrictions to prevent unreasonable risks on existing chemical substances. Although some agricultural biotechnology products may fall within the purview of TSCA, the majority of agricultural biotechnology products regulated by EPA are considered pesticides under EPA's broad definition of the term, and thus are regulated under FIFRA and FFDCA.

FIFRA. Section 2(u) of FIFRA defines the term "pesticide" as "(*1*) any substance or mixture of substances intended for preventing, destroying, repelling, or mitigating any pest, (*2*) any substance or mixture of substances intended for use as a plant regulator, defoliator, or desiccant . . . ." This definition is very broad and can include living organisms and substances produced by living organisms as well as traditional chemical pesticides. The definition of "pesticide" in FIFRA does not depend on the process by which a particular pesticide is produced. EPA has interpreted this definition to include biological pesticides and genetically engineered pesticides.

Section 3 of FIFRA provides that no person may distribute or sell in the United States any pesticide that is not registered under the Act. FIFRA Section 3(c)(5) requires that before a pesticide can be registered, it

must be shown that when used in accordance with widespread and commonly recognized practice, it will not generally cause "unreasonable adverse effects on the environment." The term "unreasonable adverse effects on the environment" is defined in FIFRA Section 2(bb) as any unreasonable risk to humans or the environment, taking into account the economic, social, and environmental costs and benefits of the use of any pesticide or human dietary risk resulting from pesticide residues in food inconsistent with the safety standard under Section 408 of FFDCA. Thus FIFRA involves a balancing of the risks presented by the use of the pesticide against the benefits associated with the use of that pesticide. The procedures governing the regulation of pesticides are set forth in 40 CFR Parts 152 through 172. One of the most important requirements is that the registrant or applicant submit data in support of registration. 40 CFR Part 158 sets forth data requirements for conventional pesticides and microbial pesticides (specifically at 40 CFR 158.740), and provides for the submission of comprehensive health and environmental effects data. EPA has not yet established specific data requirements for plant-pesticides. In addition to submitting required data, an applicant for registration must submit all proposed labeling with the registration application. FIFRA Section 2(p) defines the term "label" as the written, printed, or graphic matter on, or attached to, the pesticide. The term "labeling" under FIFRA includes the label as well as all other written, printed, or graphic matter that accompanies the pesticide or to which reference is made on the label. Registered pesticide products must bear a label or labeling that contains certain information, including precautionary statements, warnings, directions for use of the product, and an ingredient statement. FIFRA requires users of pesticides to follow all label directions. A product whose label or labeling does not contain the information required by EPA or that sets forth false or misleading information is misbranded pursuant to FIFRA Sections 2(q) and 12(a)(1)(E). For conventional pesticides, many risk reduction measures are achieved through labeling restrictions. As discussed below, however, many of these types of restrictions may not be appropriate for plant-pesticides.

FIFRA also provides EPA with a number of other regulatory tools beyond the registration authority. For example, large-scale field testing of pesticides is necessary to evaluate the efficacy of a potential product and to obtain data needed to support registration under FIFRA Section 3. This large-scale testing is regulated under Section 5 of FIFRA. Under this section, EPA is authorized to issue experimental use permits (EUPs) for limited use of an unregistered product for an unregistered use. Before an EUP is issued, EPA must determine that the field test will not cause an "unreasonable adverse effect" on the environment. For most new pesticides, EPA grants conditional registration while it continues to evaluate whether the pesticide product qualifies for full registration. Under Section 3(c)(7) of FIFRA, conditional registration is granted when EPA lacks sufficient data to make a final determination on a full registration. Finally, under FIFRA Section 25(b), EPA may exempt from some or all requirements of FIFRA, by regulation, any pesticide determined to be (1) adequately regulated by another federal agency, or (2) of a character that is unnecessary to be subject to the Act in order to carry out the purposes of the Act.

FFDCA. EPA regulates pesticide residues in or on food under the authority of Section 408 of FFDCA. Under FFDCA Section 408, any pesticide chemical residue in or on food is deemed to be unsafe unless a tolerance, or an exemption from the requirement of a tolerance, for such pesticide is established and the pesticide is within the tolerance limits. The term "pesticide chemical" is defined in Section 201(q) of FFDCA as "any substance that is a pesticide within the meaning of [FIFRA] . . . ."

Thus pesticide chemicals subject to Section 408 of FFDCA are defined by reference to the definition of pesticide under FIFRA. Section 408(b) of FFDCA authorizes EPA to promulgate regulations to establish tolerances for pesticide chemical residues in or on food if EPA determines that the tolerance is "safe." This section goes on to provide that "safe" means that EPA has determined that there is a reasonable certainty that no harm will result from aggregate exposure to the pesticide chemical residue, including all anticipated dietary exposures and all other exposures for which there is reliable information. Thus unlike FIFRA, FFDCA only addresses human dietary risks. FFDCA Section 408(c) authorizes EPA to promulgate regulations exempting any pesticide chemical residue from the necessity of a tolerance when the exemption is safe.

In 1996 both FIFRA and FFDCA were amended by the Food Quality Protection Act (FQPA). FQPA amended FIFRA such that a registration cannot be issued for a pesticide to be used in or on food unless the residues of the pesticide in the food qualify for a tolerance or an exemption from tolerance. FQPA also modified FIFRA Section 2bb by incorporating the FFDCA safety standard into the test for determining whether a pesticide poses an unreasonable adverse effect. FQPA also amended Section 408 of FFDCA to require EPA to give special consideration to exposure of infants and children to pesticide residues.

EPA REGULATION OF AGRICULTURAL BIOTECHNOLOGY

Microbial Pesticides

EPA has regulated naturally occurring microbial pesticides, such as Bt, for many years. Microbial pesticides are regulated in much the same way as traditional pesticides at the large-scale testing and registration stages. For the past 10 years, however, EPA has been concerned about the potential for adverse effects associated with small-scale environmental testing of certain microbial pesticides, both naturally occurring and genetically engineered. Small-scale testing of most traditional pesticides generally is considered to pose very limited risks, and thus is not usually regulated by EPA. Because microbial pesticides are living organisms that have the potential to reproduce and spread in the environment, however, even small-scale testing has the potential to present unreasonable adverse effects on the environment.

Section 5 of FIFRA authorizes EPA to issue EUPs for the testing of new pesticides or new uses of existing pesticides. Under EPA's existing regulations at 40 CFR Part 172, EUPs are generally issued for large-scale testing of pesticides. A large-scale test under Part 172 includes any terrestrial application on a cumulative acreage of more than 10 acres of land or any aquatic application on more that 1 acre of surface water. EPA has generally presumed that tests conducted on 10 acres or less of land or 1 acre or less of water (small-scale tests) would not require EUPs. The Agency has determined, however, that small-scale tests conducted with certain naturally occurring and genetically engineered microbial pesticides may pose sufficiently different risk considerations from tests conducted with convention chemical pesticides so that a closer evaluation at the small-scale testing stage is warranted.

In October 1984, EPA published a policy statement entitled *Microbial Pesticides: Interim Policy on Small Scale Field Testing* (26). In June 1986, EPA reiterated the provisions of the Interim Policy Statement as part of the Office of Science and Technology Policy's Coordinated Framework for Regulation of Biotechnology. These policy statements described EPA's concern about the potential for adverse effects associated with small-scale environmental testing of certain microbial pesticides. To address this situation, these statements required that EPA be notified prior to initiation of small-scale testing of all nonindigenous and genetically engineered microbial pesticides. The purpose of the notification was to allow EPA to screen these small-scale tests and determine whether the tests should be carried out under an EUP that allows EPA oversight. In addition the 1986 Policy stated EPA's plan for future rule making in order to codify the interpretation set out in the policy.

After almost 10 years of deliberation and a series of EPA and federal government-wide policy statements that were made available to EPA's Scientific Advisory Panel (SAP) and the Biotechnology Science Advisory Committee (BSAC), on January 14, 1993, the EPA issued a proposed rule that was a somewhat revised version of the 1986 policy (27). The rule would codify the early screening procedure in the Coordinated Framework by requiring notification before the initiation of small-scale field testing of certain microbial pesticides in order to determine whether an EUP is necessary. Under the proposed rule, testing conducted in facilities designed and operated to adequately contain the microbial pesticide would not be subject to the notification requirements.

EPA received comments in response to the proposed rule making from trade associations, business firms, public interest groups, scientific researchers, and state and federal agencies. Perhaps the most controversial issue that arose during the lengthy development of this rule was the issue of what constitutes the appropriate scope of regulation. The proposal identified three options for defining the scope of genetically modified microbial pesticides subject to notification requirements. EPA's preferred option provided the most clear-cut scope of regulation — namely microbial pesticides whose pesticidal properties have been imparted or enhanced by the introduction of genetic material that has been deliberately modified. EPA developed this option based on comments from the public in response to earlier Federal Register announcements, the SAP subpanel, the BSAC, and other agencies including USDA. The Agency preferred this option because it believed that it covers the appropriate microbial pesticides and has a high degree of regulatory utility. The majority of comments supported this option. The commenters generally agreed that EPA's preferred option was more clear-cut and that the decision of whether notification is necessary should not be left solely to the judgment of the researcher. In 1994 EPA issued the final microbial rule (28). The final rule included EPA's preferred option for the scope of regulation from the proposed rule. The final rule also included a mechanism to exempt, by rule making, additional microbial pesticides or categories of microbial pesticides from the requirement for notification as data and experience permit.

One other issue that was somewhat controversial was that of whether EPA should require notification for "nonindigenous" microbial pesticides. Under EPA's 1984 Policy Statement and the 1986 Coordinated Framework, EPA had been requiring notifications to be submitted for all small-scale testing of nonindigenous organisms. In all of the scope options presented in the proposal, EPA proposed to no longer require notifications for any nonindigenous microbial pesticides that have not been genetically modified. EPA based this decision on its belief that continued imposition of the notification requirement on these microbial pesticides would constitute duplicative oversight because the USDA's Animal and Plant Health Inspection Service (APHIS) already regulates small-scale testing of these organisms. Some commenters supported EPA's decision to exclude nonindigenous microbial pesticides from notification, while others believed that EPA should regulate any nonindigenous microbial pesticide that is not regulated by another federal agency. EPA responded to these comments by stating that it continues to believe that the vast majority, if not all, nonindigenous microbial pesticides are reviewed by APHIS. However, to address the concerns of some commenters that there might be a regulatory gap, EPA revised the language in the final rule to state that only those nonindigenous microbial pesticides that have not been acted upon by APHIS (i.e., either by issuing or denying a permit or determining that a permit is unnecessary; or when a permit is not pending with APHIS) are exempt from the notification requirement.

The final rule also contains several provisions that were not very controversial and were not changed significantly from what was proposed. In the final rule, testing conducted in facilities designed and operated to adequately contain the microbial pesticide would not be subject to the notification requirements. Records describing containment, however, would be required to be developed and maintained. The final rule also includes provisions that will enable EPA to address situations where small-scale testing results in unanticipated and untoward effects. Section 172.57 requires persons using microbial pesticides in small-scale tests to submit any information they obtain concerning the potential for unreasonable adverse effects from the microbial pesticide,

and Section 172.59 enables EPA to take immediate actions to prevent use of a microbial pesticide if such use would create an imminent threat of substantial harm to health or the environment. Finally, the rule amends 40 CFR §172.3 to clarify its rationale for presuming that an EUP is not required prior to small-scale testing with most pesticides. As explained in the preamble to the final rule, Section 172.3 is modified to clarify that the determination of whether an EUP is required would be based on risk considerations, rather than on a definitional presumption about whether a substance is a pesticide. This clarification has general applicability to all pesticides and is not limited to microbial pesticides.

Plant-Pesticides

EPA's first attempt to describe its plans to regulate plant-pesticides was in early 1994. On January 21, 1994, EPA held a joint meeting of a subgroup of the Agency's SAP and BSAC to address certain scientific issues related to the regulation of pesticidal substances produced in plants. For the meeting, EPA made available to the public a draft proposal of a comprehensive policy and four draft proposed rules (together referred to as the "draft proposal") that were developed under FIFRA and FFDCA. On November 23, 1994, EPA published in the Federal Register somewhat modified versions of these draft documents (together referred to as "the proposal") (29–33). The proposal is intended to clarify the status of plant-pesticides under FIFRA and FFDCA and outline the scope of what types of plant-pesticides EPA believes warrant regulation based on risk–benefit considerations. Under the proposal many plant-pesticides would not be subject to regulation because they pose a low potential for risk to humans and/or the environment. Others would be subject to regulation but would be regulated somewhat differently than conventional pesticides because of the unique nature of plant-pesticides. The proposal outlines how EPA intends to assess plant-pesticides at different stages of environmental testing and at the sale and distribution stage. In developing this policy, EPA worked closely with two other federal agencies that also have regulatory jurisdiction over agricultural biotechnology products, to integrate the three agencies' regulatory programs and minimize duplicative regulation. Those agencies are APHIS, which regulates certain genetically modified plants, including plants that are modified to produce pesticidal substances, and FDA, which regulates nonpesticidal substances in food plants as food additives under FFDCA.

As described above, FIFRA defines the term "pesticide" very broadly, and under this definition both the "plant" and the pesticidal substances produced in the plant are considered to be "pesticides." However, in 1982 EPA promulgated a regulation under FIFRA Section 25(b) that exempted all biological control agents from the requirements of FIFRA, except for certain microorganisms (34). This exemption was promulgated because EPA found that macroorganisms used as biological control agents were adequately regulated by other federal agencies such as APHIS. Plants, as biological control agents, were implicitly exempted from regulation under FIFRA through this exemption. EPA does not believe is it necessary to revoke this exemption for the plant itself. Instead EPA intends to focus on the pesticidal substance produced by the plant. This is consistent with EPA's past actions. For example, EPA does not regulate chrysanthemums, but it regulates the pesticidal substance pyrethrum that is produced by the chrysanthemum when it is extracted from the plant and applied onto other plants as an insecticide. However, prior to 1994, EPA had not clearly stated its policies for regulating pesticidal substances that are produced in living plants and not extracted from the plants (i.e., substances produced in plants naturally, or through genetic engineering or other technologies, that actually exert their pesticidal effect while still in the plant). It is these substances that EPA considers to be plant-pesticides and that are the subjects of the proposal.

One point that should be emphasized is that in the proposal, EPA has defined the pesticidal active ingredient as including not only the substance that is produced in the plant for the purpose of inducing the pesticidal effect but also the genetic material necessary for the production of that substance. To understand why EPA is including this genetic material in the definition of active ingredient, it is necessary to understand how a plant-pesticide is created. There are three primary steps involved in creating a plant-pesticide: (1) isolating the gene to be transferred from the source organism to the plant, (2) adding regulatory DNA sequences to the gene so that it will be properly expressed in the gene (these regulatory DNA sequences typically are derived from plant viruses), and (3) moving the gene to the plant. The last step can be accomplished by using physical methods such as microinjection and biolistic delivery (firing very small metal particles coated with DNA into plant tissues) or by using biological vectors such as the soil bacterium, *agrobacterium*. There are several reasons why EPA included the genetic material as part of the active ingredient. First, it is the genetic material that is actually added to the plant and that leads to the production of the substance that ultimately results in the pesticidal effect. Moreover EPA is not only concerned with the environmental risks associated with the pesticidal substance itself but also with potential environmental impacts associated with the spread of genetic material. Finally, from a practical standpoint, it may be easier to detect, for monitoring or enforcement purposes, the genetic material in a plant than the pesticidal substance itself.

Under EPA's definition of plant-pesticide, all substances produced by plants and intended for a pesticidal purpose are within EPA's jurisdiction, whether the plant is genetically engineered or not. However, just because a substance is considered to be a plant-pesticide, it does not necessarily mean that EPA will regulate it under FIFRA. The Agency believes there are many plant-pesticides that do not warrant any regulation under FIFRA because they pose a low probability of risk and will not cause unreasonable adverse effects on the environment. One category of plant-pesticides that EPA believes does not warrant regulation are those that will not cause new exposures to nontarget organisms. EPA is proposing to exempt from FIFRA regulation those plant pesticides that are not new to the plant (i.e., derived from closely related plants). Thus

the Bt delta-endotoxin would not be exempt when it is produced in corn, for example, because the delta-endotoxin is derived from a bacterium rather than from a plant that is closely related to corn. A pesticidal substance that is naturally produced by a certain variety of corn and is introduced into another variety of corn, however, would be exempt. Another category that EPA is proposing to exempt are those plant-pesticides that would not be expected to adversely affect nontarget organisms because they are less likely to be directly toxic because of their mechanism of action. This category consists of plant-pesticides that act primarily by affecting the plant so that pests are inhibited from attaching to the plant, penetrating the plant's surface, or invading the plant's tissue. Thus a substance that acts by causing a structural barrier to pest penetration in the plant would be exempt. EPA also believes that coat proteins from viruses pose low risks and do not warrant regulation under FIFRA.

In addition to the low potential for risk associated with these categories of plant-pesticides, EPA believes that these plant-pesticides may have significant benefits associated with them because they could be used as alternatives to more toxic and persistent conventional pesticides.

Although EPA scientists and the members of the SAP and BSAC that have evaluated these exemptions believe that the plant-pesticides proposed for exemption pose low risks, many environmentalists are concerned that the exemptions are too broad. These concerns seem to stem, in large part, from the uncertainty surrounding many of the issues and the historical lack of experience with plant-pesticides. Some have suggested that EPA should require ongoing monitoring of exempt plant-pesticides. In response to this concern, EPA is considering proposing a regulation that would require reporting of adverse effects information for exempt plant-pesticides. This regulation would be similar to FIFRA Section 6(a)(2), which requires reporting of unreasonable adverse effects information for all registered pesticides. If EPA does impose such a requirement, the next issue to consider is how EPA will react if it finds that a particular plant-pesticide, or category of plant-pesticides, is riskier than EPA believed when it exempted it. Currently, under FIFRA Section 25(b), to exempt a pesticides, EPA must go through notice and comment rule making. It follows that to repeal an exemption, EPA also may be required to go through rule making. Rule making can be a lengthy process, particularly when coupled with the FIFRA requirement of submittal of all proposed and final regulations to the SAP and USDA for comment. A statutory amendment that would authorize EPA to repeal exemptions with a more abbreviated process would enable EPA to more quickly gain regulatory control over plant-pesticides found to pose unreasonable adverse effects.

Under the proposal, once it is determined that a substance is a plant-pesticide subject to FIFRA regulation, the regulatory process is similar to, with some modification, the regulatory process for all pesticides. Prior to sale or distribution, if a crop is to be used as food or feed at any test acreage, an EUP would be required. For crops that will not be used as food or feed, and if subject to the authority of the Plant Pest Act, an EUP would be required when environmental testing will be on greater than 10 acres of land or greater than one surface acre of water. Currently, for all pesticides, the 10-acres requirement is triggered when the cumulative acreage of environmental tests reaches ten acres. In the proposal, EPA indicates that it is considering changing this requirement for plant-pesticides so that an EUP is required when a single environmental test exceeds 10 acres. EPA is also considering a number of other options for EUP triggers. One option is to utilize APHIS's determination that a plant is no longer a regulated article as the point at which regulatory responsibility is handed off from APHIS to EPA. If a plant-pesticide is not subject to the authority of the Plant Pest Act, an EUP would be required at first introduction into the environment regardless of acreage. If a producer has been granted an exemption by APHIS from permitting requirements under the Plant Pest Act, an EUP would be required at the time the exemption is granted.

Before sale or distribution of a plant-pesticide, a producer must obtain a registration under FIFRA Section 3 if the plant-pesticide is not otherwise exempt. Where there is food or feed use at sale or distribution, the potential registrant must further fulfill the necessary FFDCA obligations. FIFRA Section 3 requires that all registered pesticides be labeled. Labeling includes both the written, printed, or graphic material on, or attached to, the pesticide or any of its containers or wrapper and all other written, printed or graphic material accompanying the pesticide at any time. An improperly labeled pesticide is considered to be misbranded and in violation of FIFRA. As noted earlier, EPA generally relies on labeling requirements to impose risk reduction measures on the use of traditional pesticide products. For example, EPA regulations at 40 CFR 156.10 contain extensive labeling requirements dealing with, among other things, warnings and precautionary statements and directions for use. Other labeling restrictions are imposed, case by case, through the registration process. Restrictive labeling may include anything from requirements that personal protective equipment such as gloves and respirators be used to reduce the risk to pesticide users, to the requirement that a buffer zone be provided around fields to prevent risks to bystanders from spray-drift, to geographic restrictions on the use of certain pesticides to reduce the risk to endangered species or other beneficial organisms that occur in a limited geographical area. These labeling restrictions are translated into use restrictions via FIFRA Section 12(a)(2)(G), which provides that it is unlawful for any person to use any registered pesticide in a manner inconsistent with its labeling. EPA has stated that it recognizes that many types of restrictive labeling that it relies on to regulate traditional chemical pesticides may not be appropriate for plant-pesticides. For example, geographical limitations on the use of the plant-pesticide may not be meaningful if the plant that produces the pesticide can reproduce and spread in the environment beyond those geographical limits. Similarly other use restrictions (e.g., prohibiting use within 50 feet of a stream, river, or lake) may not be effective if seeds

from plants that produce plant-pesticides are saved and planted during subsequent growing seasons. Such seeds would not be labeled, and the farmers using these seeds might not even be aware that the seeds were from plants that had been engineered to produce a plant-pesticide. Although EPA recognizes that the more typical labeling restrictions may not be meaningful for plant-pesticides, it is not yet clear how EPA will adapt its regulatory practice to these new forms of pesticides. The success of EPA's plant-pesticide program will depend, in large part, on EPA's ability to diverge from its historical reliance on labeling restrictions to achieve risk reduction. Because traditional restrictive labeling is not likely to result in true risk reduction for plant-pesticides, EPA will need to consider whether registrations should not be granted for plant-pesticides that pose significant risks in the absence of meaningful risk reduction. Despite the problems with traditional risk reduction labeling, EPA recognizes that other forms of labeling may be useful for plant-pesticides. Specifically, EPA is considering requiring labeling on bags of seeds containing plant-pesticides that inform farmers or other users of the type of pesticide that the plants will produce and against which pest it is active. This information could help prevent unnecessary application of additional pesticides to the plants that already produce plant-pesticides.

If a plant-pesticide is being used in food or feed, EPA has two options in its regulation under FFDCA: It can set a tolerance for the plant-pesticide, or it can exempt the plant-pesticide from the requirements of a tolerance. FIFRA and FFDCA are independent statutes: A plant-pesticide that is exempt from regulation under the proposed scope for FIFRA is not necessarily exempt from regulation under FFDCA. Moreover, the two Acts have different, but overlapping, purposes: Under FIFRA, EPA considers *all* environmental and human health risks, whereas, under FFDCA, EPA focuses on the risks posed by human dietary consumption. In the proposal, under FFDCA Section 408(c), EPA would exempt certain categories of plant-pesticides from the requirement of a tolerance. The plant-pesticides that EPA believes warrant review, and thus would not be exempted, are those that are most likely to result in new or different dietary exposures. The proposal would exempt the following:

1. Plant-pesticides produced in food and derived from closely related food or nonfood plants.
2. Plant-pesticides produced in food and derived from food plants that are not closely related to the recipient food plant and would not result in significantly different dietary exposure when produced in the recipient food plant. "Results in significantly different dietary exposure" can be interpreted in a number of ways:
   a. The pesticidal substance is produced in inedible portions of the source food plant, but in the recipient plant, the pesticidal substance is present in the plant's edible portions.
   b. The pesticidal substance is produced in the immature but not in the mature edible portions of the source food plant, but in the recipient plant,

the pesticidal substance is present in the mature edible portions.

c. The pesticidal substance is from a source food plant normally cooked or processed and is produced in a recipient plant that is not normally cooked or processed prior to consumption.

d. The pesticidal substance is derived from a source food plant that is not a major crop for human dietary consumption (i.e, not wheat, corn, soybeans, potatoes, oranges, tomatoes, grapes, apples, peanuts, rice, and beans or any other crop that EPA has determined is a major crop for human dietary consumption) and is introduced into a recipient plant that is a major crop for human dietary consumption.

EPA also is proposing to exempt from the requirement of a tolerance the coat proteins from plant viruses and nucleic acids. EPA believes that tolerances are not necessary for coat proteins from viruses because virus-infected plants have always been a part of the human diet without any known adverse human health effects. It is necessary for EPA to address nucleic acids under FFDCA because they are considered part of the pesticidal active ingredient. EPA plans to exempt these substances from the requirement of a tolerance, however, because nucleic acids are present in the cells of every living organism and thus are ubiquitous in the food supply. Because of their ubiquity in the food supply and because they lack any toxicity when consumed in food, EPA does not believe tolerances for nucleic acids are necessary to protect the public health.

During the five years since it first published its plant-pesticide proposal, EPA has issued a number of registrations and granted several tolerance exemptions for a variety of plant-pesticides. EPA has not, however, completed or published a set of final regulations governing plant-pesticides. The reason for the delay most likely stems from the controversies surrounding the plant-pesticide debate. One of the most significant controversies involves the strongly opposing views on whether genetically engineered food should be required to be labeled. Many, particularly in the European Community, believe that all genetically modified foods should be labeled so that consumers are fully informed. Thus EPA's position on whether to require labeling may have serious implications also with regard to international trade. In addition serious concerns have arisen regarding the risk that plants producing pesticidal substances such as the Bt toxin on a continual basis may hasten the development of pest resistance to these beneficial pesticides. This is a very difficult issue that EPA has not yet come to grips with. These and other issues will have to be resolved before EPA's plant-pesticide program will be fully in place. Perhaps only time and experience will tell how to address these difficult and uncertain issues.

ACKNOWLEDGMENT

Portions of this article are reprinted from the previously published article by the author, *Genetically Engineered Plant Pesticides: Recent Developments in the EPA's Regulation of Biotechnology*, 7 UFJL & Pub. Pol'y 257 (1996), with the permission of the University of Florida Journal of Law and Public Policy.

BIBLIOGRAPHY

1. R.A. Bohrer, *U. Pitt. Law Rev.* **55**, 607–610 (1994).
2. K. Bosselmann, *Colo. J. Int. Environ. Law Policy* **7**, 111–148 (1996).
3. D.L. Burk, *U. Pitt. Law Rev.* **55**, 611–633 (1994).
4. 7 U.S.C. §136–136y (1988); Supp. (1988).
5. 21 U.S.C. §301 et seq. (1988).
6. R. Carson, *Silent Spring*, 1962.
7. Pimmentel et al., *Bioscience* **41**, 402 (1991).
8. S. Matten, W. Schneider, B. Slutsky, and E. Milewski, in L. Kim, ed., *Advanced Engineered Pesticides*, 1993.
9. 57 Fed. Reg. 6753 (1992).
10. S.H. Mantell, J.A. Matthews, and R.A. McKee, *Principles of Plant Biotechnology: An Introduction to Genetic Engineering in Plants*, 5, Blackwell Scientific, Oxford, UK, 1985.
11. P.J. Goss, *Calif. Law Rev.* **84**, 1395–1435 (1996).
12. D.J. Earp, *Environ. Law* **24**, 1633 (1994).
13. J.J. Kim, *Fordham Int. Law J.* **16**, 1160 (1993).
14. J.E. Beach, *Food Drug Law J.* **53**, 181–191 (1998).
15. C.M. Steen, *Ariz. Law Rev.* **38**, 763–791 (1996).
16. U.S. General Accounting Office, Food Safety and Quality, *Innovative Strategies May be Needed to Regulate New Food Technologies*, U.S. Government Printing Office, Washington, DC, 1993.
17. Environmental Defense Fund, A Mutable Feast: Assuring food safety in the era of genetic engineering, *New York*, October 1, 1991.
18. M. Mellon, *Biotechnology and the Environment: A Primer on the Environmental Implications of Genetic Engineering*, National Wildlife Federation, Washington, DC, 1988.
19. M. Mellon, in *Agricultural Biotechnology at the Crossroads: Biological and Social Institutional Concerns*, National Agricultural Biotechnology Council Report No. 3, Washington, DC, 1991.
20. W. Fehr, R. Goldberg, and J. Fearn, in *Agricultural Biotechnology at the Crossroads: Biological and Social Institutional Concerns*, National Agricultural Biotechnology Council Report No. 3, Washington, DC, 1991.
21. W. Fehr, in *Agricultural Biotechnology at the Crossroads: Biological and Social Institutional Concerns*, National Agricultural Biotechnology Council Report No. 3, Washington, DC, 1991.
22. T. Hoban and P. Kendall, in *Agricultural Biotechnology: A Public Conversation about Risk* (National Agricultural Biotechnology Council Report No. 5, Washington, DC, 1993 (citing Office of Technology Assessment, A New Technological Era for American Agriculture, Summary Report 1992).
23. R.A. Bohrer, *U. Pitt. Law Rev.* **55**, 653–679 (1994).
24. 49 Fed. Reg. 50856 (1984).
25. 51 Fed. Reg. 23302 (1986).
26. 49 Fed. Reg. 40,659 (1984).
27. 58 Fed. Reg. 5878 (1993).
28. 59 Fed. Reg. 45,600 (1994).
29. 59 Fed. Reg. 60,496 (1994).
30. 59 Fed. Reg. 60,519 (1994).
31. 59 Fed. Reg. 60,535 (1994).

32. 59 Fed. Reg. 60,542 (1994).
33. 59 Fed. Reg. 60,545 (1994).
34. 40 CFR §152.20.

See other entries AGRICULTURAL BIOTECHNOLOGY; ANIMAL BIOTECHNOLOGY, LAW, FDA REGULATION OF GENETICALLY MODIFIED ANIMALS FOR HUMAN FOOD USE; FEDERAL POLICY MAKING FOR BIOTECHNOLOGY, CONGRESS, EPA.

AGRICULTURAL BIOTECHNOLOGY, LAW, AND FOOD BIOTECHNOLOGY REGULATION

HENRY I. MILLER
Stanford University
Stanford, California

OUTLINE

Introduction
Food Biotech's Venerable Past and Promising Future
FDA's Oversight of New Crop Varieties
    The 1992 Policy
    FDA's Volte-face on Biotech Food Policy
OECD's Biotech Food Policy
Other International and Supranational Approaches
    The European Union
    Japan
Bibliography

INTRODUCTION

Governmental oversight of the foods made with the techniques of the new biotechnology offers examples of the spectrum of possible approaches and their ripple effects. FDA's official risk-based policy has served the public interest well, but the agency appears to be retreating to a more politically correct, defensive posture that is closer to the European approach. The European experience provides striking illustrations of what can happen when regulatory policy is built on a foundation of invalid scientific assumptions, gratuitous controversy, and political and ideological goals. The outcome is expensive, expansive, and irrational regulation — which leads, in turn, to narrower application of the technology, and fewer benefits. The arguments that have been marshaled during the policy "debate" over biotechnology regulation are revealing. Those who would encourage unnecessary regulation sometimes argue that, in the face of uncertainty, it is only prudent to "err on the side of safety," to avoid taking any chances, and act instead on the basis of the worst-case scenario — the "precautionary principle." A related argument is that even if only a handful of adverse events (e.g., toxicity caused by the consumption of genetically engineered plants) are prevented by government oversight regimens, not to act would be unconscionable and would amount to putting a price on human life. But the principles of "erring on the

side of safety" and the pricelessness of life do not withstand rigorous scrutiny. What appears to be the "safe" choice may, upon analysis, actually pose greater risk. One must consider the risks of various alternative courses of action; forgoing new technology can put lesser theoretical risks ahead of known, palpable, existing ones.

Despite their many advantages, gene-spliced organisms — sometimes called genetically modified (GM) — are controversial in some parts of the world. In Europe, for instance, there has been widespread public opposition to importing gene-spliced corn and soybeans. Foods produced through gene splicing must be labeled as such, and most major supermarket chains and food producers have said they will not sell them. Threats by antitechnology activists of boycotts and hostile publicity have induced several companies doing business in the United States to reject gene-spliced ingredients used to make their products. For instance, the Japanese breweries Kirin and Sapporo, whose beer is popular in the United States, have announced that they will phase out their use of gene-spliced corn. Two of the largest U.S. producers of baby food, Gerber and Heinz, have promised not to use gene-spliced materials — even if what they use instead is nutritionally inferior or less safe. As an example, they will reject materials from corn plants modified so that they do not need to be sprayed with toxic chemical insecticides.

Scientists around the world agree that introducing genes from other organisms does not make plants less safe either to the environment or for humans to eat. Dozens of new plant varieties produced through hybridization and other traditional methods of genetic modification enter the marketplace each year, without special scientific review or labeling for consumers. Moreover many of these foods on the market are from "wide crosses," hybridizations in which genes are moved from one species or one genus to another to create a variety of plant that does not exist in nature. While such changes may sound dramatic, the results are as mundane as a tomato that is more resistant to disease, or that has a thicker skin that won't be damaged during mechanical picking. Plants that have undergone those slight but important alterations have been an integral part of European and American diets for decades. However, these scientific and commercial realities have often been lost in the nether world of regulatory politics.

FOOD BIOTECH'S VENERABLE PAST AND PROMISING FUTURE

The almost unimaginably wide spectrum of foods consumed throughout the world owes its existence to both the ingenuity of history's cooks and the practitioners of biotechnology. Microorganisms were creating and improving humans' food and drink long before anyone knew that microorganisms existed. In time — but still without knowing what was happening biologically — early practitioners of biotechnology learned to exploit the fermentative action of microorganisms to produce such things as cheese, bread, and alcoholic beverages. Still later, food producers began to isolate favored microbial cultures with highly descriptive names such as *Penicillium roqueforti* (used to make

Roquefort cheese) and *Lactobacillus san francisco* (for San Francisco-style sourdough bread). It is no coincidence that some of the past century's most sophisticated microbiology applied to beverage production has been performed in the laboratories of companies like Guinness, Carlsberg, Kirin, and Bass. Vastly popular regional foods produced by fermentation include milk products (yoghurt, sour cream, buttermilk, kefir), preserved vegetables (cabbage, olives), tempeh, sufu, tofu, soy sauce, and natto.

The modification of crop plants has been performed ever since ancient agriculturists selected and cross-bred plants with desirable traits, often creating domesticated relatives of wild species. The rediscovery in 1900 of Mendel's concepts of inheritance ushered in the scientific application of genetic principles to crop improvement. Since then, each scientific advance has increased our ability to improve predictably the *genotype* (and more important for the farmer, food manufacturer, and consumer, the *phenotype*). Currently a combination of several techniques is routinely used to improve plants. For example, an existing plant might have been modified by many generations of classical breeding and selection, and more recently by techniques developed during the past half-century, including somaclonal variation and wide crosses with embryo rescue. These plants are being further improved by the newer molecular techniques such as recombinant DNA (rDNA), or "gene splicing," and can then be reintroduced into a classical breeding program from which its descendants eventually will be released into commerce.

In this century plant breeders have increasingly used hybridization to transfer genes from certain noncultivated plant species to a variety of a different (but closely related) species. These "interspecific" transfers of traits from wild species to domesticated relatives in the same genus stimulated attempts at even wider crosses, including those between members of different genera. These "wide crosses," that transcend natural barriers to mating, have been facilitated by "embryo rescue" or culture techniques in which a sexual cross yielding a viable embryo but abnormal endosperm is "rescued" by culturing the embryo. This is done by providing the hybrid embryo with the life support normally supplied in the early stages of development by maternal tissue and the endosperm. A number of plants resulting from wide crosses have been used in further breeding, extensively field tested, and marketed in the United States and elsewhere. These plants include commonly available varieties of tomatoes, potatoes, corn, oats, sugarbeets, rice, and bread and durum wheat (1).

The use of molecular techniques for the genetic manipulation of plants enables scientists to direct the movement of specific and useful segments of genetic material readily between unrelated organisms. These techniques offer several advantages and complement existing breeding efforts by increasing the diversity of genes and germ plasm available for incorporation into crops. The numerous molecular techniques for genetic manipulation of plants can be divided into two main types — vectored and nonvectored. Vectored modifications rely on the use of biologically active agents, such as plasmids and viruses, to facilitate the entry of the foreign gene into the plant cell. Nonvectored modifications rely on the foreign genes being physically inserted into the plant cell by such methods as electroporation, microinjection, or particle guns. In both kinds of approaches, new DNA enters the plant's genome and is stably maintained and expressed. A landmark report from the U.S. National Research Council concluded that:

> Recombinant DNA methodology makes it possible to introduce pieces of DNA, consisting of either single or multiple genes, that can be defined in function and even in nucleotide sequence. With classical techniques of gene transfer, a variable number of genes can be transferred, the number depending on the mechanism of transfer; but predicting the precise number or the traits that have been transferred is difficult, and we cannot always predict the phenotypic expression that will result. With organisms modified by molecular methods, we are in a better, if not perfect, position to predict the phenotypic expression (2, p. 13).

> Crops modified by molecular and cellular methods should pose risks no different from those modified by classical genetic methods for similar traits. As the molecular methods are more specific, users of these methods will be more certain about the traits they introduce into the plants (2, p. 3).

Far from eliciting concern, techniques that yield a better-characterized and more predictable plant variety should be welcomed as a means for improving food. The new biotechnology lowers even further the already minimal risk associated with introducing new plant varieties into the field and the food supply. The use of the latest biotechnology techniques makes the final product even safer, as it is now possible to introduce pieces of DNA that contain one or a few well-characterized genes. In contrast, the older genetic techniques transferred a variable number of genes haphazardly. Users of the new techniques can be more certain about the traits they introduce into the plants and about the presence of unwanted, deleterious genetic changes. Thousands of products from plant varieties crafted with the older techniques have entered the marketplace in the last three or four decades, and only three products (two squash varieties and one potato type) had unsafe levels of toxins; in addition one celery variety caused allergic skin reactions in some farm and supermarket workers. But today's more precise gene-splicing techniques mitigate against any repetition. A group of chefs who announced a boycott of biotechnology-produced foods in 1990, lacked perspective on the new products' pedigree — that is, on the continuum between conventional and new biotechnology. They were against the use of plants engineered with the newest, most precise, and sophisticated techniques, while they lacked any scruples about using the mutant peaches we call "nectarines," the genetic hybrid (of tangerine and grapefruit) known as "tangelos," or the genetically improved oats, rice, and other plants that have resulted from wide crosses (3).

Many of the improvements introduced by gene-splicing enable plants to grow with less agricultural chemicals such as pesticides and herbicides, and in regions that have salty or other low-quality water. They offer increased yields with lower inputs, and diminish the runoff of chemicals into waterways, so they are favorable to the environment. It is difficult, however, to pass on inflated regulatory costs for these kinds of improvements whose value is obscure to consumers.

FDA'S OVERSIGHT OF NEW CROP VARIETIES

FDA, which is responsible for the safety and wholesomeness of the nation's food supply (excepting only most meats, which are regulated by the U.S. Department of Agriculture, USDA), monitors the continuing progress of new biotechnology-derived products. The agency's policy on foods derived from new plant varieties — whether crafted with conventional or new biotechnology or other techniques — was published in 1992. It set out a carefully considered, scientific, and "transparent" — that is, clear and predictable — regulatory approach.

The 1992 Policy

FDA's approach to safety assessment of new varieties of crops developed by both traditional and newer methods of genetic modification (4) (Fig. 1) is based on the agency's long-standing oversight of "old" biotech-derived varieties commonly introduced into the U.S. marketplace (5). Foods derived from new plant varieties are not routinely

subjected to extensive scientific tests for safety, although there are exceptions. The usual practices employed by plant breeders — such as chemical and visual analyses and taste testing — are generally recognized as adequate for ensuring food safety. Additional tests, however, are performed when required by the product's history of use or scientific judgment. For example, potatoes are tested for the glycoalkaloid solanine, because this natural toxicant has been present at toxic levels in some new potato varieties.

FDA's 1992 regulatory approach identifies scientific and regulatory issues related to characteristics of foods that raise safety questions and that elicit a higher level of FDA review. Such characteristics, which are further discussed below, include the presence in the new variety of a substance that is completely new to the food supply, the presence of an allergen in an unusual or unexpected milieu, or increased levels of toxins that are normally found in foods. Consistent with scientific consensus about recombinant DNA techniques, the use of any particular technique(s) of genetic manipulation does not in itself determine the need for or the level of governmental review.

The "Guidance to Industry" section of the 1992 policy statement instructs developers to consider initially the characteristics of the host plant that is being modified, the donor organism that is contributing genetic information,

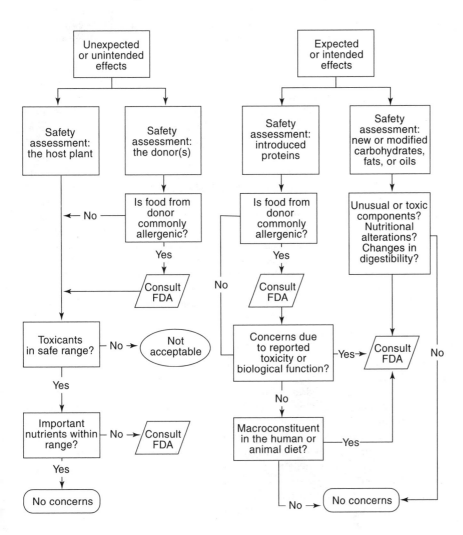

Figure 1. Food and Drug Administration (FDA) safety assessment of new varieties of crops.

and the genetic material and substances being introduced or modified. The guidance section also provides criteria that enable developers to determine whether a substance intentionally introduced or altered by genetic modification will require premarket approval as a food additive. In general, premarket review is not required for introduced or modified proteins of known function if they are derived from food sources or are substantially the same as existing food substances, are not known to be toxic or to raise food safety concerns, and will not be a major constituent of the diet. New carbohydrates with unusual structural or functional groups or oils that contain new, unusual fatty acids may require premarket approval as food additives.

The question whether biotech-derived foods should be labeled as such has received a disproportionate amount of attention. The primary reason is not ambiguity in either the science or law; rather, it is that the issue has been a focus of activists' attention. Their attention to this issue is yet another manifestation of anti-biotechnology activists' attempts to use overregulation to retard the progress of every stage of biotechnology R&D. They have lobbied against laboratory research, then against scale-up and against testing and commercial uses of products. Finally, thwarted in their desire to get the FDA to require clinical trials or case by case evaluation of biotech-derived foods, activists have retreated to demanding labeling that would reveal when new biotechnology techniques were used in a food's manufacture. The ostensible rationale for such a requirement is that information is power and that consumers can never know too much about the products they buy. Especially for foods, the more information the better, goes the mantra. But that's not necessarily true. A message can mislead and confuse consumers if it is irrelevant, unintelligible, or crafted to tell only part of the truth. Moreover a requirement for labeling carries added production expenses and raises costs to both producers and consumers that can constitute a barrier to the development of and access to new products.

To serve the consumer best, regulation should focus on genuine risks and should require only disclosure of information about a food's origin or use that is relevant to safety and that supports informed choice. Mandatory labeling of all biotech foods would achieve none of this.

Labeling has become a key issue for gene-splicing applied to food production. Labels that warn of gene-spliced ingredients in food are fundamentally different from the labels currently on food—that list calories, fat content, and so on—because these represent a modest, one-time expense for food producers. By contrast, gene-spliced fruits, vegetables, and grains would have to be continually segregated through all phases of production—planting, harvesting, processing, and distribution—adding costs and compromising economies of scale. The need to segregate gene-spliced foods, especially the thousands of processed foods that contain small amounts of derivatives of corn or soybeans, would raise production costs and pose a particular disadvantage to products in this competitive market with low profit margins.

FDA's long-standing commonsense approach to food labeling has been that label information must be both accurate and "material." FDA does not require a "product of biotechnology" or "genetically engineered" label for foods from plants or animals that have been improved with rDNA techniques. In the 1992 food policy statement, the FDA said that labeling is required "if a food derived from a new plant variety differs from its traditional counterpart such that the common or usual name no longer applies to the new food, or if a safety or usage issue exists to which consumers must be alerted." The statement of policy also emphasized that, as for other foods derived from new plant varieties, no premarket review or approval is required unless the characteristics of the new biotechnology products raise explicit safety issues. It noted that these safety issues could be raised by food from new plant varieties however they were created. The safety issues include the introduction of a substance that is new to the food supply (and, hence, lacks a history of safe use), increased levels of a natural toxicant, changes in the levels of a major dietary nutrient, and transfer of an allergen to a milieu where a consumer would not expect to find it (e.g., a peanut protein transferred to a potato).

FDA clarified that if a new food raises any of these safety issues, it could be subject to FDA regulations for premarket testing, product labeling, or removal from the marketplace. FDA cited the example of new allergens in a food as a possible material fact whose omission could make a label misleading. The agency reiterated that the genetic method used in the development of a new plant variety is *not* considered to be material information because there is no evidence that new biotech foods are different from other foods in ways related to safety. Therefore FDA said that product labeling will not be required to include the method of development of a new plant variety. Biotech foods would not be required to be labeled as such.

The 1992 FDA policy statement has already been tested and validated. A 1996 report in the *New England Journal of Medicine* reported that allergenicity common to Brazil nut proteins was transferred into soybeans by genetic engineering and was readily identified by routine procedures (6). The plant breeder, Pioneer Hi-Bred International, was required to and did consult with the FDA during product development. During the course of consultation and subsequent analysis, the allergenicity was identified. Confronted with the dual prospects of potential product liability and the costs of labeling all products derived from the new plant variety, the company abandoned all plans for using the new soybeans in consumer products. Not a single consumer was exposed to or injured by the newly allergenic soybeans. In what might be considered a "positive control," the system worked.

The approach taken by FDA in its 1992 policy statement is consistent with scientific consensus that the risks associated with new biotechnology-derived products are fundamentally the same as for other products (2). Dozens of new plant varieties modified with traditional genetic techniques (e.g., hybridization and mutagenesis) enter the marketplace every year without premarketing regulatory review or special labeling (7). As discussed above, many of these products are from "wide crosses" in which genes have been moved across natural breeding barriers (without rDNA techniques). None of these plants exists in nature. None requires or gets a premarket review by a government agency. (Safety tests by plant breeders are primarily taste and appearance and, in the case of plants with high levels of known intrinsic toxicants — e.g., tomato and potato — levels of certain alkaloids.) Nonetheless, they have become an integral, familiar, and safe part of our diet: wheat, corn, rice, oats, black currants, pumpkins, tomatoes, and potatoes.

There are other reasons why special regulations and labeling requirements are often not in the best interest of consumers. As food producers know well, requiring a label can add significantly to the production costs of certain foods, particularly those that are produced from pooled fresh fruits and vegetables. To maintain the accuracy of labels, recombinant DNA-modified fruits and vegetables would have to be segregated through all phases of production — planting, harvesting, processing, and distribution — which adds costs and eliminates economies of scale. Added production costs, in turn, raise consumer prices and disadvantage products in the highly competitive, low profit-margin marketplace of processed foods.

Superfluous labeling requirements for new biotech products would constitute, in effect, an unwarranted and punitive tax on the use of a new, superior technology. The requirement would exact excess costs and reduce profits to plant breeders, farmers, food processors, grocers, and others in the distribution pathway. The power of regulatory disincentives is such that this burden could virtually eliminate new biotechnology tools from food research, development and production. For example, as required in the European Union, the United Kingdom introduced mandatory labeling of gene-spliced foods in 1998, which Britain's agriculture minister called "a triumph for consumer rights to better information," and which the country's additives and novel-foods chief regulator characterized as "a question of choice, of consumer choice." But as a direct result of the labeling law, there's hardly any choice there now at all: The new law sparked a stampede by manufacturers, retailers, and restaurant chains to rid their products of any genetically modified ingredients so that they would not have to alter their labels and risk losing sales.

It is unclear how far these scientifically dubious food label requirements will extend. Will special labels be required for foods such as pizza or burritos containing cheese made with new biotech-produced chymosin (rennin), for chickens raised on feed from new biotech-manipulated corn, and for cattle vaccinated with a new biotech vaccine? Will labels be required if highly sensitive analytical techniques detect one part per million of recombinant DNA in a food?

An analysis of the economic impacts of a labeling requirement for new-biotech foods by the California Department of Consumer Affairs (CDCA) predicted that the additional costs would be "substantial," and that "while the American food processing industry is large, it is doubtful that it would be either willing or able to absorb most of the additional costs associated with labeling biotech foods" (7). The analysis concluded that "there is cause for concern that consumers will be unwilling to pay even the increased price for biotech foods necessary to cover biotechnology research and development, much less the additional price increases necessary to cover the costs associated with labeling biotech food."

The CDCA assessment implies another outcome of unwarranted but compulsory labeling. Overregulation will reduce competition and, therefore, increase prices, and overpriced biotech products would be limited to upscale, higher-income markets. Wealthier consumers would be able to pay more for the "boutique" products, while the less affluent would simply do without them.

It is noteworthy that FDA's current approach to labeling was upheld indirectly by a federal appeals court, which found in a pivotal 1996 decision regarding another product of biotechnology that food labeling cannot be compelled just because some consumers wish to have the information. That case, *International Dairy Foods Association v. Amestoy* (92 F. 3rd 67; 2nd Cir. 1996), involved a Vermont state law requiring labeling of dairy products from cows treated with a gene-spliced protein that increases the productivity of dairy cows. In overturning the law, the appeals court found that such regulation merely to satisfy the public's alleged "right to know" is a constitutional violation of commercial free speech. "Were consumer interest alone sufficient, there is no end to the information that states could require manufacturers to disclose about their production methods," the court wrote.

Although FDA's 1992 food biotech policy is scientifically defensible and favors the public interest, as discussed below, the agency has shown a willingness to ignore scientific consensus, bow to political pressure, and accommodate activists' whims.

FDA's Volte-face on Biotech Food Policy

In 1993, only a year after publishing the progressive and scientific food biotech policy described above, FDA informally announced plans to require registration of all new biotechnology foods. FDA never published a proposal, but various agency officials announced repeatedly that one was being prepared. The new policy would have directly contradicted the widely praised 1992 policy statement that specified that new biotechnology foods would be treated in the same manner as other, similar foods. The ostensible rationale for this *volte-face* in policy was the gratuitous "controversy" over biotech foods in Europe and the United States, fueled by activists who are ideologically opposed to the new biotechnology (8).

The actions of antitechnology activists have shown that their agenda is neither a good-faith attempt to air issues of technological risk, nor an attempt to offer innovators and consumers a greater spectrum of choices. Rather, they wish to control what research is performed, what techniques are permitted, and what products are brought to market. Academic freedom, industrial innovation, free markets, and consumer choice are among their victims. The controversies about the new biotechnology are only a microcosm of that struggle. Activists' minds will not be changed by scientifically reasonable arguments, by assertions of the primacy of empirical evidence and the scientific method, or by invoking the benefits to the public of new products and choices. The activists' modest success at discouraging prospective end-users of new biotechnology products from accepting them is worrisome, for if end-users such as food producers and consumers do not want the products, plant biologists, breeders, and farmers will stop developing and growing them, and the technology will no longer be widely used. The major exceptions will be in which the inflated price of the final product can offset the cost of making it. Inevitably, with lowered demand for and acceptance of gene-spliced products, there will be less interest in and resources for basic research on plants for food, fiber, and medicines.

Overregulation, in the form of additional risk assessment (e.g., case-by-case review) or risk management (e.g., labeling) is interpreted by the public as a warning of potential risk. This is evident from the observations of Barbara Keating-Edh, representing the consumer group *Consumer Alert*, before the National Biotechnology Policy board, a group established by Congress and located within the NIH, September 20, 1991:

> For obvious reasons, the consumer views the technologies that are *most* regulated to be the *least* safe ones. Heavy involvement by government, no matter how well intended, inevitably sends the wrong signals. Rather than ensuring confidence, it raises suspicion and doubt (9, emphasis in original).

Since 1993 FDA has generally been receptive to activists' demands. The agency has seemed more concerned about placating activists than ensuring access by consumers to the fullest array of improved foods and providing opportunities to American companies for product innovation.

At the direction of Vice President Gore's domestic policy staff, in 1993 FDA announced a policy that would require selected foods to be registered with the agency before being sold to consumers (10). Extra government scrutiny certainly makes sense when there is uncertainty about health risks or a reason to suspect a problem such as the presence of toxins or allergenic components. In this proposal, however, the FDA decided to require registration only of those foods made with the most precise state-of-the-art biotechnology techniques. While in the first instance there was to be only a requirement for *registration* of new products at FDA's Center for Food Safety and Nutrition, officials could, of course, request additional information and testing for individual products. Such a policy would confer no public-health advantage but would discourage research that could produce foods of better quality and greater variety. The inevitable result of such regulation would have been more responsibility and bureaucracy for FDA but more limited choices for consumers at the supermarket or greengrocer.

Scientific and professional groups — including the Chicago-based Institute of Food Technologists and the University of California's Systemwide Biotechnology Program — objected strenuously to the unscientific basis of the proposal. Only after the Republican-controlled 104th Congress made pointed inquiries about the plan did FDA officially withdraw the proposal — deleting it in early 1995 from the agency's regular report to OMB of regulations in preparation. The proposal was gone but not forgotten.

In July 1996 FDA began, in a peculiar and surreptitious way, to circulate the news of new requirements. A seven-page document, "Foods Derived from New Plant Varieties:

Consultation Procedures," went out to state officials (11). In it, FDA adopts a pretense that the new policy is applicable to all new plant varieties and not just those produced with the new biotechnology. The agency's intent is transparent, however: Oversight of the consultation process rests with the *Biotechnology* Evaluation Team, and the degree of detail requested from the plant's developer would only be available for those crafted with new biotechnology.

Again signaling apparent willingness to retreat from its scientifically sound 1992 policy, in late 1999 the agency held a series of public meetings around the country to inform the public "about current FDA policy for assuring the safety of bioengineered foods [and to ask] whether this policy should be modified," according to the agency's press release. This was a thinly veiled invitation for anti-biotech activists to stuff the ballot box and demand more stringent regulation — and that is exactly what happened at the meetings, held in Chicago, Washington, DC, and Oakland.

In Chicago, three hundred prospective speakers showed up, the vast majority of them from radical environmental groups, to denounce gene-spliced food as, variously, unproven, dangerous, worthless, unnatural, and anti-religious. Interestingly many members of these anti-biotech organizations registered merely as "consumer," presumably so their groups could in effect have multiple representatives. There were also two panels (whose members were selected by FDA), with long-time antagonists of biotechnology heavily represented. Outside the meeting, many of the activists mugged for the cameras, staging mini-morality-plays in which, for example, children costumed as monarch butterflies fled in mock terror from a figure dressed as a huge ear of gene-spliced corn.

These most recent attempts by FDA to regulate biotech foods in a discriminatory way reverse the agency's 15-year-old guiding principle for the oversight of biotechnology: Regulation should focus on real risks and should not turn on the use of one technique or another. These tenets have provided effective oversight for thousands of new biotechnology products, including drugs, vaccines, diagnostic tests, and foods. Ironically, as discussed above, as recently as May 1992 FDA formally reiterated this policy for foods, affirming that new biotechnology foods would be treated no differently from those produced with other techniques and that oversight would be risk-based.

In several ways the FDA's new policy will discourage the application of biotechnology to foods. The data requirements are substantial; FDA lists nine categories of obligatory information:

1. Name of the bioengineered food and the crop from which it was derived.

2. Description of the various intended uses of the bioengineered food, including animal feed uses.

3. Information concerning the sources and functions of introduced genetic material.

4. Information on the purpose or intended technical effects of the modification, and its expected effects on the composition or characteristics of the food or feed.

5. Information about the identity and function of the newly introduced genetic material and new gene-expression products, including an estimate of the concentration of any expression product in the bioengineered crop and the food derived from it.

6. Comparison of the composition and characteristics of the bioengineered food with the food derived from the parental variety or other commonly consumed varieties.

7. Information about the identity and levels of toxicants that occur naturally in the food.

8. Discussion of the available information concerning the potential for altered allergenicity (ability to elicit an allergic reaction) in the bioengineered food.

9. Any other information relevant to the safety and nutritional assessment of the bioengineered food.

The detail is far greater than would be required for food products made with less-precise, less-sophisticated techniques; if applied to traditionally crafted plants these new, draconian requirements would spell the end of new varieties of apples, pears, strawberries, or wheat, for example. Imagine trying to determine the function of poorly characterized genes situated on whole chromosomes and newly introduced into a new cultivar of wheat by the wide cross-hybridization of wheat and a wild grass to which it is distantly related.

FDA's new policy will entail significant costs for the government and industry, and by extension, the public, in those few instances where food producers actually decide to apply biotechnology to foods and attempt to negotiate the new regulatory hoops. According to FDA's description of the new regulatory scheme the Biotechnology Evaluation Team will always consist of no fewer than *six* FDA staff, drawn from different parts of the agency. There will be endless and conflicting demands for information about each product, causing delay and uncertainty among manufacturers. Another disadvantage of the new regulatory regime is that every biotechnology product will be placed squarely in the sights of anti-biotechnology activists: according to FDA, the results of consultations with industry will be available on the Internet.

The policy is intentionally murky about whether developers of new biotech foods are *required* to consult with the agency, although in private conversations Dr. James Maryanski, the biotechnology coordinator in FDA's Center for Food Safety and Nutrition, has protested that these "requirements" are "only suggestions." However, the reality is unequivocal: A "suggestion" from the nation's most ubiquitous and draconian regulator is akin to an armed mugger "suggesting" you turn over your wallet. In practice, this is mandatory premarket regulation—applied uniquely to biotech-derived foods. And it is extra-legal regulation, in the sense that it is not the product of the rule making required by law.

The bottom line is that the policy will, in effect, impose a tax on the use of biotechnology for food production. This discriminatory treatment will discourage research on more varied, appetizing, and nutritious foods—research that has given us low-saturated-fat oils, seedless grapes,

tangelos, and the like. American farmers and food processors will be less competitive, consumers will be deprived of new choices and the price of biotech-derived foods will be inflated.

OECD'S BIOTECH FOOD POLICY

The Paris-based Organization for Economic Cooperation and Development (OECD) has endorsed a policy for biotech foods similar to FDA's 1992 approach. In its 1993 publication, *Safety Evaluation of Foods Derived by Modern Biotechnology*, OECD invoked the concept of "substantial equivalence" (borrowed from the U.S. FDA's medical device regulations), the crux of which is that new foods that are "substantially equivalent" to other varieties should be regulated "in the same manner as their analogous conventional counterparts" (12). In other words, no additional regulatory requirements, such as notification, review, or labeling, should be required. *Safety Evaluation of Foods Derived by Modern Biotechnology* spells out clearly the rationale, theory, and practice for applying the principle of "substantial equivalence":

Historically, foods prepared and used in traditional ways have been considered to be safe on the basis of long-term experience, even though they may have contained natural toxicants or anti-nutritional substances. In principle, food has been presumed to be safe unless a significant hazard was identified.

Modern biotechnology broadens the scope of the genetic changes that can be made in food organisms, and broadens the scope of possible sources of foods. This does not inherently lead to foods that are less safe than those developed by conventional techniques. Therefore, evaluation of foods and food components obtained from organisms developed by the application of the newer techniques does not necessitate a fundamental change in established principles, nor does it require a different standard of safety.

[T]he precision inherent in the use of certain molecular techniques for developing organisms for use as food should enable direct and focused assessment of safety where such assessment is desired. Knowledge obtained using these methods might also be used to approach safety assessment of new foods or food components from organisms developed by traditional methods.

For foods and food components from organisms developed by the application of modern biotechnology, the most practical approach to the determination of safety is to consider whether they are *substantially equivalent* to analogous conventional food product(s), if such exist. ...The concept of substantial equivalence embodies the idea that existing organisms used as food, or as a source of food, can be used as the basis for comparison when assessing the safety of human consumption of a food or food component that has been modified or is new.

If one considers a modified traditional food about which there is extensive knowledge on the range of possible toxicants, critical nutrients or other relevant characteristics, the new product can be compared with the old in simple ways. These ways can include, *inter alia*, appropriate traditionally performed analytical measurements (for example, alkaloid levels in potatoes, cucurbatin in vegetable squash cultivars, and psoralens in celery) or crop-specific markers, for comparative purposes. The situation becomes more complex as the origins/composition/exposure experience decreases, or if the new products lack similarity to old established products or, in fact, have no conventional counterpart.

A demonstration of substantial equivalence takes into consideration a number of factors, such as:

- knowledge of the composition and characteristics of the traditional or parental product or organism;
- knowledge of the characteristics of the new component(s) or trait(s) derived, as appropriate, from information concerning: the component(s) or traits(s) as expressed in the precursor(s) or parental organism(s); transformation techniques (as related to understanding the characteristics of the product) including the vector(s) and any marker genes used; possible secondary effects of the modification; and the characterization of the component(s) or trait(s) as expressed in the new organism; and
- knowledge of the new product/organism with the new component(s) or trait(s), including the characteristics and composition [i.e. the amount of the component(s) or the range(s) of expression(s) of the new trait(s)] as compared with the conventional counterpart(s) (i.e. the existing food or food component).

The OECD document elaborated "principles for the application of substantial equivalence to the assessment of organisms developed by the application of biotechnology":

- If the new or modified food or food component is determined to be substantially equivalent to an existing food, then further safety or nutritional concerns are expected to be insignificant.
- Such foods, once substantial equivalence has been established, are treated in the same manner as their analogous conventional counterparts.
- Where new foods or classes of new foods or food components are less well-known, the concept of substantial equivalence is more difficult to apply; such new foods or food components are evaluated taking into account the experience gained in the evaluation of similar materials (for example, whole foods or food components such as proteins, fats or carbohydrates).
- Where a product is determined not to be substantially equivalent, the identified differences should be the focus of further evaluations.
- Where there is no basis for comparison of a new food or food component, that is, where no counterpart or similar materials have been previously consumed as food, then the new food or food component should be evaluated on the basis of its own composition and properties.

The consideration of safety may include the need to evaluate possible effects occurring through cooking or other processing. For example, trypsin inhibitors from certain leguminous plants, such as the cowpea trypsin inhibitor, have a long history of safe consumption when properly cooked. However, if the cowpea trypsin inhibitor is expressed in other plants, the safety question relates to whether the normal use of these plants as food involves cooking sufficient for its inactivation.

Another consideration [related to whether a new food is substantially equivalent to another] is the influence of the newly introduced modification(s) on the nutritional value of the food or food components(s). For the majority of modifications being carried out, such changes are unlikely.

Nonetheless, when modifications are directed at metabolic pathways of key macro- or micro-nutrients, the possibility of an impact on nutritional value is increased. Such impacts are of potential significance in cases where the modified food or food component may become a major dietary source of the nutrient affected.

It is obvious from the foregoing that in order to apply substantial equivalence generally, as well as to specific cases, judgments by regulators are necessary. And therein lies what has become in practice an anomaly: Contrary to the concept as conceived at the OECD, national regulators and others have often defined virtually *any* change wrought by molecular techniques as yielding a food or food component that falls outside the realm of substantial equivalence and that, therefore, requires more extensive review and evaluation. Although FDA does not apply the term "substantial equivalence" to its oversight of food, the concept is implicit. As described above, FDA's 1992 official policy defines certain safety-related characteristics of new foods that, if present, define "non-substantial-equivalence" and require greater scrutiny by the agency. These include the presence of a substance that is completely new to the food supply, an allergen presented in an unusual or unexpected way (e.g., a peanut protein transferred to a potato), changes in the levels of major dietary nutrients, and increased levels of toxins normally found in foods. (The absence of such characteristics, in effect, defines foods that are substantially equivalent to antecedent products.) Foods lacking characteristics that raise these safety issues need not be subject to premarket FDA review.

OTHER INTERNATIONAL AND SUPRANATIONAL APPROACHES

The European Union

The European Union (EU) announced controversial rules for the labeling and sale of new biotechnology-derived foods in December 1996, after months of acrimonious debate (13). The now-mandatory labels will add significantly to the costs of processed foods made from fresh fruits and vegetables. The precise costs will vary according to the product. But a company using a gene-spliced, higher-solids, less-watery tomato (more favorable for processing), for example, must bear the additional costs of segregating the product at all levels of planting, harvesting, shipping, processing and distribution. Labels must appear on minestrone soup, indicating the presence of gene-spliced tomato, potato or other products (at least any amount above an arbitrary one percent threshold). The added production costs are a particular disadvantage to products in this competitive, low profit-margin market segment, and at best, will likely relegate many gene-spliced products to the status of expensive "boutique" foods, out of the reach of less affluent consumers (15). As discussed above, labeling requirements have virtually eliminated biotech foods from the shelves of European retailers.

The EU compromise was reached after five years of negotiations by a joint committee of the European

Parliament and the EU Council of Ministers (which represents the 15 states). The Council and the European Commission had preferred to require labeling only when the new food or ingredient was "significantly different" from its predecessors; but the European Parliament had its way, and labeling is required for "live" genetically modified products — those that could, in theory, grow if put in soil, such as tomatoes or potatoes. The compromise has not mollified the radicals. A new "technical amendment" to the regulation adopted by the European Commission on April 2, 1997, would require the labeling of seed products that will give rise to transgenic plants. The European Novel Food Regulation, with or without the April 1997 amendment, is irrelevant to public health. A label that says "genetically modified" provides no useful or material information to consumers — but at significant economic and societal cost.

Quite apart from gene-splicing considerations, other parts of the regulation also fail to take scientific principles and precedents into account. For example, new varieties of wheat improved by the introduction of genes from hardy grasses (a common plant-improvement strategy) might be deemed "different in comparison with a conventional food or food ingredient." Under which circumstances, says the regulation, the varieties are "no longer equivalent" to preexisting foods or ingredients, and would require special — and costly — labeling. (Consider, also, that using sophisticated analytical techniques, the chemical composition of English potatoes can easily be distinguished from those grown in Italy; under the regulation, the two varieties — even if the same species and cultivar — would arguably be nonequivalent and need to be distinguished by labeling.)

As often happens with political compromises by poorly informed, paternalistic politicians, the citizenry are compromised by an outcome that makes neither scientific nor economic sense. The unnecessary and arbitrary novel food regulation constitutes, in effect, a punitive "tax" on regulated products or activities, which, in turn, creates a potent disincentive to product development and use. Finally, the EU's regulation and its tax are incompatible with the U.S. policies and may well precipitate trade conflicts or even a trade war — corollaries of the law of unintended consequences.

Japan

The Japanese government has made no pretense of adopting policies that are consistent with the scientific consensus that the new biotechnology is an extension, or refinement, of older genetic techniques, or with the spirit of the OECD's "substantial equivalence." Rather, the Japanese Ministry of Health and Welfare (MHW) has imposed a strict regulatory regime specific to foods and food additives manufactured with rDNA techniques (16). This regime captures virtually all rDNA-derived products for case by case review and subjects them to extraordinarily stringent standards for manufacture, documentation, record keeping, characterization of the source organism and the actual food products, and so on — a far higher standard than any other class of food, except perhaps for the preparation of *Fugu*, a fish much favored in Japan that contains a potent and potentially fatal neurotoxin. Because food products' profit margins are low, discriminatory and unnecessary regulation is a strong disincentive to using a new technology. Predictably, regulatory disincentives have prevented Japan from exploiting its experience and traditional strengths in agriculture and the production of fermented foods and beverages.

In 1986 MHW promulgated guidelines concerning the manufacture of new-biotechnology products, loosely based on the OECD 1986 report, "Recombinant DNA Safety Considerations" (18), but failed to incorporate the spirit of "substantial equivalence." MHW also has regulatory responsibility for food additives, and in 1991 issued a policy statement, "Basic Principles on Safety Assurance for Foods and Food Additives Produced by Biotechnology" (17). A year later the MHW issued two guidelines for the new biotechnology used in food production: a Manufacturing Guideline (GMP, or Good Manufacturing Practice) and a Safety Assessment Guideline. In December 1999 MHW announced that beginning in April 2001 mandatory tests on the potential health risks of genetically modified (GM) foods would replace voluntary testing, and also that products approved and considered "safe" would be identified by labels as having been produced by the new biotechnology (18). However, as of March 2000, there remained uncertainty about how this would be accomplished and whether the labels would, indeed, indicate "safety." The confusion lies in the existence of conflicting (August 1999) draft regulations of the Ministry of Agriculture, Forestry, and Fisheries, which require 30 food ingredients containing gene-spliced ingredients to be labeled as such, and that products with a mixture of gene-spliced and non-gene-spliced ingredients be labeled as "undifferentiated" (18).

In theory, the Japanese adhere to the concept of "substantial equivalence" as articulated by the OECD (*vide supra*), but in practice their primary regulatory trigger is process based — that is, the use of recombinant DNA techniques — and there are unprecedented and irrational admonitions that "recombinants themselves are not to be consumed" (19).

BIBLIOGRAPHY

1. R.M. Goodman et al., *Science* **236**, 48–54 (1987).
2. U.S. National Research Council, *Field Testing Genetically Modified Organisms: Framework for Decisions*, National Academy Press, Washington, DC, 1989.
3. H.I. Miller, *San Francisco Chronicle*, May 31, A17 (1994).
4. Fed. Regist. **57**, 22984–23005 (May 29, 1992).
5. H.I. Miller, *J. Am. Med. Assoc.* **269**, 910–914 (1993).
6. J.A. Nordler et al., *N. Engl. J. Med.* **334**, 688–699 (1996).
7. V.J. Taylor, Memorandum to the Labeling Subcommittee, California Interagency Biotechnology Task Force, April 15, 1994.
8. H.I. Miller, *Policy Controversy in Biotechnology: An Insider's View*, R.G. Landes, Austin, TX, 1997, pp. 43–51.
9. Statement of Barbara Keating-Edh, President of Consumer Alert, before the National Biotechnology Policy Board, September 20, 1991.
10. Anon., *Food Chemical News*, January 31, p. 59 (1994).

11. R. Arbaugh, FDA Office of Regulatory Affairs, Memorandum to State Feed Control Officials, July 1, 1996.

12. Organization for Economic Cooperation and Development, *Safety Evaluation of Foods Derived by Modern Biotechnology*, OECD, Paris, 1993.

13. D. Butler, *Nature* **384**, 502–503 (1996).

14. H.I. Miller, *Financial Times (London)*, April 22 (1997).

15. H.I. Miller, *Financial Times (London)*, November 29 (1996).

16. Japanese Ministry of Health and Welfare, *Guidelines for Foods and Food Additives Produced by Recombinant DNA Techniques*, Tokyo, Japan, 1992.

17. Organization for Economic Cooperation and Development, *Recombinant DNA Safety Considerations*, OECD, Paris, 1986.

18. Y. Uozum, *Nature* **402**, 846 (1999).

19. Y. Suzuki, in *Proc. Japan–USA Workshop on Scientific Political and Social Aspects of Recombinant DNA*, Japanese Science and Technology Agency, Tokyo.

See other entries AGRICULTURAL BIOTECHNOLOGY; ANIMAL BIOTECHNOLOGY, LAW, FDA REGULATION OF GENETICALLY MODIFIED ANIMALS FOR HUMAN FOOD USE.

AGRICULTURAL BIOTECHNOLOGY, LAW, AND SOCIAL IMPACTS OF AGRICULTURAL BIOTECHNOLOGY

WILLIAM P. BROWNE
Central Michigan University
Mount Pleasant, Michigan

OUTLINE

INTRODUCTION

Biotechnology in agriculture has gained as much attention for its social and political controversies as it has for its science and its promise for food and fiber production. Those controversies as they pertain to different types of scientific products and processes are the main subject of this article. One polar view holds that agricultural biotechnology is the primary key to providing for future world food needs. The opposite view is that biotechnology products subject agriculture, people, and food availability all to potentially grave risks — too much so as some see them. These opposing positions as well as myriad centrist views have led to legal, regulatory, and other political activity and conflicts that are described here. In the process of engendering conflict, they also have led to a complex and growing maze of statutes, regulations, and international agreements that govern agricultural biotechnology to the satisfaction of almost no one. Those institutional rules and dissatisfactions are also covered here.

LEGAL FRAMEWORK

Two types of genetically engineered products have received the greatest attention in agriculture. The first is recombinant products such as hormones that are added to living organisms through application or injection. The second is transgenic manipulation, in which genes of one organism are engineered into another, often of a very different species. Since the two types of products are quite distinct in technology and in approach, they have evoked unique interests within the governing process in recent years. Considerable irony exists in these unique interests because, at least in the United States, the essence of legislation actually proceeded the emergence of agricultural biotechnology products. Precedents were set long before the new technology arrived.

U.S. Domestic Legislation: Protecting Property

The generic legal framework for biotechnology plant products is the Plant Patent Act (PPA) of 1930 (1). It was essentially a matter of protecting profits for those building an industry on new products. So was subsequent legislation. The intention was to provide protection of intellectual property developed primarily by agribusiness companies (2). The Act allowed patent-style protection for new plant innovations, which sets those important legal precedents. PPA, however, had limited utility since it only covered asexually reproducing plants, which included grafting but not new seeds. That obviously failed to cover much of the U.S. plant industry, as became startlingly clear in the 1940s with the introduction of corn hybridization.

As a corrective, the Plant Variety Protection Act (PVPA) of 1970 extended important features of the 1930 legislation. Specifically, it gave intellectual property protection for 20 years to sexually reproducing plants that hold over time their uniform characteristics. Far more products were able to get certificates of protection. PVPA, however, had significant exemptions, for research efforts to develop new products and for crop growing purposes, or farmer-to-farmer sales. These exemptions meant that plant industries still had relatively limited product control as well as incentives to litigate. In 1994 these two problems with PVPA were marginally narrowed by Congress, but not anywhere to the satisfaction of industry.

Much of the logic for plant protection was touched on but not actually extended to livestock and other animals in the Animal Patent Act of 1986 (3). It, like a subsequent act in 1988, failed. Nonetheless, arguments made in the congressional hearings combined with federal genetics privacy legislation which protects DNA, provided the eventual basis for certifying animal innovations as well as plant species (4). But federal administrators rather than Congress were to become the actual arbitrators.

Evolution: Litigation and Regulation in the United States

The congressional emphasis on property protection rather than the direct topic of biotechnology products is not surprising. Historically U.S. federal legislation develops only in the vaguest sense and with minimal explicitness. Then bureaucracy takes over. Many issues simply get avoided for political reasons, although administrators are often forced to eventually rule as part of their legal duties. Either that or the courts rule. Those reasons lie in part with the complex and shared governing powers within the policy-making process. Powers are both divided and separated as well as shared from one institution to the next. This brings very different and more compromised results than the far more explicit and integrated policy processes that characterize parliamentary forms of national government, which dominant in most other democratic regimes. With more executive control in keeping agencies accountable, governing in parliamentary nations is far less fragmented.

Within the United States, Congress disposes of policy initiatives half-heartedly. The presence of both active federal courts and of technical administrative agencies explains why. With their involvement Congress understands full well that most of its policy decisions will be refined further through judicial litigation and through agency regulation when enabling authority exists. Thus Congress has rarely tried to resolve all relevant issues in statutory law. That, of course, is why congressional initiatives on protecting plant and animal properties were necessarily to be evolving rather than dealing with all possible new products and resulting legal concerns (5).

It was clear from legislative hearings on all of the major acts that property protection was not the only issue up for contest. Most of those who testified as critics in the hearings addressed side issues such as animal well-being, environmental impact, human health as linked to new products, and such economic and social consequences of biotechnology as resulting financial concentration in farming (6). Some of those concerns led to the aforementioned plant variety exemptions for research and for growers.

But for the most part these things were left to be at least partially resolved in other, mostly nonpatent federal institutions. So they evolved in those places to set legal case law. Federal courts were involved at first, and again most explicitly in interpreting property protection since that had been the emphasis of legislation.

While litigation has been prolific, only three cases merit specific comment for their precedent setting nature for biotechnology. A California state case, *Moore v. Regents of the University of California*, ruled that biotechnology industries can utilize genetic raw materials freely and to their own purposes, regardless of donor circumstances.

Even more landmark, however, was the earlier U.S. Supreme Court ruling in *Diamond v. Chakabarty*. The Chakabarty decision affirmed that traditional patent law is indeed still patent law, that there are no differences in law between the animate and inanimate innovation. The legal distinction is rather between what nature produces and what humans engineer. This case not only supported the biotechnology industry, it also dealt with living bacterium and thus extended plant protections to animals. The federal patent office moved under this authority in 1988 to certify the famous Harvard mouse. *Ex parte Hibberb* then moved to clarify which products could receive utility patents. The courts ruled that when Congress did not explicitly exclude plants from such patents, it left intact all forms of procedural plant protection. Armed with that interpretation, and confident that it is also extended to animals, industry firms moved successfully to prosecute numerous patent violations as well as to patent more and more products (7). There, however, still remained a downside for industry: Patent protection for plants and animals is hard to obtain because of the nature of both these products and the utility patent process. To gain protection, inventors must prove novel and useful application, describe the invention well enough to let others with necessary skills recreate it, and demonstrate true innovation as opposed to a simply logical extension of past ideas. Protecting trade secrets is allowed within that legal context.

The regulatory arena has been no less active than the courts in promulgating legal precedents that shape the law for biotechnology, indeed far more so. Using and interpreting patenting and other regulatory processes shaped by Congress through enabling legislation, administrators have become prominent and consequential public policymakers. Both the U.S. Patent and Trademark Office (PTO) within the Department of Commerce and the U.S. Department of Agriculture (USDA) share specific patent responsibilities which lead to frequent interpretive differences. USDA in the 1990s had very explicit opposition, for example, to patenting an entire species, which its officials argue cannot represent a new discovery. PTO disagrees, leaving resolution to continue into the future.

Points of administrative confusion and conflict are far more than just over patent interpretations since myriad other nonpatent agencies have agricultural biotechnology jurisdictions or at least interests. This opens a whole Pandora's Box of American public policymaking for product review. Most notable of the agencies are the federal Food and Drug Administration (FDA) and the Environmental Protection Agency (EPA). USDA's traditional agency in this area of food safety also is important but inclined to be less critical of biotechnology. The Animal, Plant, Health Inspection Service (APHIS) nonetheless adds to the regulatory muddle through PPA. Under the long existing Public Health Service Act and other lesser statutes, FDA has clear authority to regulate and ensure the safety of foods derived from new plant varieties and, therefore, new techniques of production. This entails what can be an extensive product safety review. EPA has contrasting and less clear authority to mandate and review environmental impact statements. It has been at this juncture where transgenic manipulation has been separated in impact from recombinant technologies. EPA has been concerned that new products with variable gene structures may contaminate existing and especially wild or natural settings. This could threaten existing strains of plants and animals ranging from commercial canola to migratory salmon. To respond to this problem, EPA has moved to regulate and test all new agricultural varieties which promise plant protection under the Federal Insecticide, Fungicide, and Rodenticide Act (FIFRA).

EPA's response infuriated most of the active U.S. scientific community, including eliciting a direct and angry critique from 11 scientific societies (8). This was, they argued, neither EPA's job nor was it a scientifically sound conclusion (9). The case nonetheless shows a great deal about the effect American politics has on a fragmented and barely integrated regulatory process that opens its official doors to nearly any and all interest groups and citizen complaints. If complainants fail to be content with decisions in one place, they simply look for other agencies with which to complain. When APHIS says no to a complaint, for example, the complainant goes on to EPA—and looks and hopes for a different response on different grounds. In effect, the United States regulates agricultural biotechnology to ease public fears more than to bring forward sound and lasting scientific decisions. Dissenting social forces move, for example, from PTO to USDA patent officials to FDA to EPA to APHIS. Animal Patent Law hearings, as noted, were places where many nongermane issues were raised.

A good example of cultivating the public can be seen in another task of USDA's Office of Agricultural Biotechnology (10). Since 1987 this office has taken a neutral stance on product innovations and on biotechnology itself. Yet it further opens up decision-making units within government to intense lobbying and media coverage from any dissenting social faction. Federalism, or the separation of American governments into three levels of government, does even more of the same. When no federal agencies respond, agricultural biotechnology critics become politically active within the American states and even localities. Chicago provides a superb illustration with its decision to insist on consumer labeling for any food product that may be adulterated by agricultural biotechnology, either through recombinant or transgenic processes. Quite clearly, those who wish to promote this biotechnology industry in the United States face a costly, cumbersome, and uncertain structure of governance. Yet biotechnology critics and opponents love that structure for the many obstacles it provides.

Agricultural Biotechnology in Other Countries

As the above makes abundantly clear, the United States can hardly have a systematic and supportive public policy to promote agricultural biotechnology. Things within governing structures are just too cumbersome for that to be the case. Yet, on a comparative basis, U.S. governments have been generally friendlier to agricultural biotechnology than have those of almost all foreign governments (11). There are some exceptions. Germany,

Australia, and Japan actively support the industry, because in large part both agriculture and high levels of industrialization matter to their political economies. That support, however, has not quelled loud and extensive public criticism in those countries. China with its major worries over food supplies and Brazil with its aspirations as a leading world food producer tend also to be supportive yet not particularly activist in their biotechnology political agendas.

Much of Europe as well as countries of the Southern Hemisphere are far more skeptical. France as a recent convert to conservation policy links agricultural biotechnology to those concerns, particularly in protecting indigenous germ plasm. Industry is also not advantaged in the United Kingdom, where both government and public skeptics about technology's effects on a prominent farming sector prevail. With such splits as between France and Germany, and with family farming revered throughout most of Europe, no common laws on anything of substance have passed. The European Parliament has, despite the splits, passed a nonbinding resolution against animal patenting, seen the introduction of several resolutions attacking plant patents, failed to come up with a proposal for a European initiative supporting and regulating agricultural biotechnology, and condemned the international European Patent Office for actually approving an animal patent as it did in 1992 (12). Europe, as a consequence, lags several years behind the United States in actual public policy supports for the industry. Moreover its federated European Union (EU) governing structure is proving to be even as cumbersome and fragmented as the United States on biotechnology matters in agriculture, only just not *within* the individual countries.

Despite serious food and trade needs as well as rapidly growing human populations, countries of the Southern Hemisphere also act skeptically. Culture plays a major role in structuring negative responses as does the importance of farming in each nation's politics. The greatest reluctance, however, comes from fears of further domination by the United States and other highly developed industrial nations. Exploitation has long been a dramatic political issue in Asia, Africa, and Latin America. Its importance has only escalated as industry firms have moved into underdeveloped countries in major initiatives to research as well as to market biotechnology products. Pioneer Hi-Bred International, with its offices in over 30 countries, has become for many an archetype of the fearsome multinational corporation. Thus, while governments around the world could quite easily promote agricultural biotechnology because of their more integrated parliamentary structures, these countries have demonstrated the ease with which they can also react reluctantly.

International Cooperation

Agriculture has long been a policy area marked by substantial efforts at international cooperation. In part, this is because agriculture and now agricultural biotechnology are subject to agreements between so many participant nations within several international agencies. From world food needs to nutrition problems of the poor, the United

Nations has played an activist role in fostering various agreements. That has worked in part because there has been much to agree on in promoting solutions to world farm, food, and fiber problems. Humanitarianism is a common goal, one easily favored. Agricultural biotechnology, however, brings no such consensus, as those nation-by-nation differences discussed above would obviously suggest. Disagreements abound and limit any efforts to think that global cooperation on this subject will ever take on either a voluntary or a legal status. There was passed in 1992, though, a UN Convention on Biological Diversity. While the United States did not ratify the convention, it still participates actively in its frequent and ongoing conferences. The biggest controversy surrounding the convention has been over the issue of a legally binding protocol governing the release of genetically modified organisms in developing countries. While an earlier UN report by 15 experts saw no need for such a protocol, it still passed but has not yet been adopted. Opposition of major agricultural biotechnology producer countries favored a voluntary protocol and numerous future meetings were planned and held to facilitate discussions (13). No substance has yet emerged. That both irritates friends and foes of the binding protocol and reveals again the lack of authority to be found in international cooperation.

The UN, however, is not the sole international agency to be involved with or split by controversy. Global standards on health and environmental protection are developed by the World Trade Organization (WTO). On those grounds WTO voted to delay action on rBGH, which meant that the EU's ban on the product remained in place. The World Health Organization (WHO) advocated strategies for assessing foods produced by biotechnology. And the Organization for Economic Cooperation and Development (OECD) has directed its mostly European member countries to carefully scrutinize agricultural biotechnology products. As a member of all three organizations, the United States stood as a biotechnology proponent and faced formidable opposition from many developing nations and several European countries. Once again, none of these three organizations have actually changed global production or trade conditions. They have, however, increased negative attention to biotechnology as applied to food products.

Much of that negative attention goes back to the UN as well as its relatively independent suborganizations. Two strong suborganizations have along with central UN deliberations had the strongest impact. The UN Food and Agriculture Organization has both recognized the promise of agricultural biotechnology and issued conditions under which its application should continue. FAO wants that biotechnology to be used only for highly planned special circumstances and specified problems, to be adapted to local infrastructure needs, to wrestle with the complexity and equality of property rights issues, and to ensure food and environmental safety along with fostering biodiversity. FAO clearly holds that what it calls "novel foods" should be subject to use with extreme caution. Such concerns impose further checks on the actions of agricultural biotechnology producer countries, further limiting and restricting their market inroads.

The same is true of the UN Environmental Programme (UNEP). Along with FAO, UNEP produces extensive research and information. But unlike FAO, UNEP takes a more activist and indeed interventionist position as well, especially through its Environmental Law Programme (ELP). The ELP promotes the development of international legal instruments, develops international environmental law aimed at sustainable development, provides technical and legal assistance to countries with developing and transitional economies, and exercises leadership in implementing environmental law both internationally and nationally. It has become clear from recent actions of the 1990s that ELP/UNEP has concerned itself extensively and provocatively with food biotechnology products and practices. Its partner in all such instances has been OECD. That partnership, like OECD more generally, has focused more negative than positive attention on such agricultural issues. Thus the balanced result of voluntary member organizations has proved more negative than positive for the whole of worldwide agricultural biotechnology.

SOCIAL PROMISE OF AGRICULTURAL BIOTECHNOLOGY

There is no mystery why agricultural biotechnology has been subject to so much social and political concern. Two related reasons explain it. World politics has long been influenced by Malthusian fears. Economist Thomas Malthus determined in the early 1830s that food production was increasing at a far slower rate than was the rate for the human population. Yet industrialization of agriculture occurred in the twentieth century, and the food production rate of growth outpaced the population rate of growth by 3 to 2 between 1950 and century's end. Still, however, fears of food shortages remain real, for a very good reason. The world's population of 6 billion people is expected to peak at nearly 11 billion by 2050 (14).

These contrasting observations have affected politics and social values at two levels. First, Malthusian fears have led at least indirectly to massive government intervention in agriculture over time. Nations have supported agricultural expansion, policies to bring about a more educated and productive farm population, and research that has hastened farm industrialization and modernization (15). All that has been especially true of the United States. But the U.S. model of agricultural education, research, and outreach or extension assistance to farmers has been exported to other nations worldwide. Thus proactive and interventionist government in agriculture has prevailed. No country trusts agriculture only to the private market. The contrasting belief that the world will not experience long-term food shortages has produced a rather different political scenario in more recent years. While food availability fears still exist, governments have become more skeptical of their own agricultural expenditures and involvement. A view exists that Malthusians have always been wrong and will continue to be wrong in the future. As a consequence fear no longer drives, though it does still in tandem influence, agricultural policy initiatives. The private sector, or market, also gets much of the credit so far for having avoided food shortages. Some trust in business has then emerged from some parts of governments. This enhances the influence of agribusinesses. However, since much of the social base of fears of a food-short world are gone, more stern critics of agriculture also have emerged. These new critics are much less inclined than previous generations of critics to support commercial agriculture and research at any social and economic cost. Indeed, the new critics want instead to closely scrutinize any costs of a further developing agriculture, especially one led by market considerations which may not take into account human safety, cultural values, or essentials of nutrition.

Those contrasting views of advocates and skeptics both get played out daily in agricultural policy making and in the media. Nowhere is that more true than for transgenic agricultural biotechnology. Without the fear factor of Malthus, agricultural biotechnology has been the subject of intense calls for regulation, as can be surmised from the previous section. Government therefore intervenes in what is often less than a supportive fashion for industry. At the same time, few interests want agricultural biotechnology to truly fail—but rather just to be safe (16). Exceptions do exist, mostly from those widely labeled to be modern Luddites. The general view of both industry and most critics of agricultural biotechnology, though, is much the same: Support for or opposition to the social promises and fearful circumstances that emerging technologies seem possibly to produce. Agricultural technology then is the subject of both severe complaints as well as valued for what it may offer in the near- and long-time future. This makes for conditions that fail to bring about extensive and comprehensive public policy. No politicians wish to decide.

That social promise, as well as the diverse fears that come with it, need to be specifically examined in order to make sense of this rather dizzying politics, for the promises are immense and of far more than marginal social value. So too are the fears. Promised contributions of agricultural biotechnology include the obvious one of finalizing the closing of the gap between food availability and food needs. What began with commercial industrialization of agriculture and moved from a green revolution of hybridization into molecular biology now can be turned toward engineering plants and animals whose genetic characteristics make for more food. By identifying those genes, by marking them, and by transferring them to host organisms, better products can result. "Better," however, does not just mean more world food to those doing agricultural research. "Better" also means foods that are, for instance, more nutritious, build in pesticide control genes, and foster a resulting ecological improvement from reduced agricultural chemical use. That value-laden term of the advocates of "better" also implies social gains as new biotechnology products use nutrients more efficiently and so lead to higher yields that also promise to limit destruction of old growth wildernesses for farming and ranching (5). Even with the increased financial costs of the technologies, they might bring economies of production scale as well and therefore at least some lower food costs. "Better" also means food products that are less prone to spoilage, have longer shelf lives, and are less subject to bacterial contamination. The social concern over wasted foodstuffs may thus be addressed. Perhaps "better" also means foods

that have more appealing taste and can find usable markets where only limited ones existed before (17). That has compounding importance when these more tasty and now marketable products can substitute for foodstuffs grown in short supply on environmentally fragile lands or by using high levels of degrading chemical inputs.

The "better" arguments are countered by fears of worsening food conditions. To opponents, "worse" also means many things. The range of possible difficulties go from the creation of pesticide resistant bacteria and other pests to the loss of genetically proven and strong seed and animal strains, to the contervailing view that food will be less nuitrious over time. A major disagreement over "worse" owes to what will be the nature of a specific food product. What are the religious implications of a Muslim or Jew eating plants expressing pig genes? What is the likelihood of a fish allergic consumer being ravaged by eating a tomato expressing trout genes? Both the public and the media pay widespread attention to such fears for quite obvious reasons.

With these conditions in mind, U.S. government regulators have approved the following sorts of products, which in total still number less than two dozen by FDA. A few more are approved by other regulatory agencies depending on their jurisdiction. There are far fewer approvals worldwide. Recombinant bovine somatotrophin (rBST) came first. It falls into the category of more product, or milk, with more efficient nutrient use by cows. Adoption in the United States has become relatively high but plagued with significant milk producer expense. European countries still await approval, which is hardly surprising since they have—for trade reasons—long banned imported meat raised with growth hormones. Beyond rBST, there are varieties of marketed biotechnology substances. Calgene brings a transgenically altered and more lasting and transportable tomato. Large firms, mostly Monsanto, have introduced products to deal with agricultural chemicals and use, both for plant survival and to reduce applications. There are on the market soybeans and canola that can withstand herbicide application, canola that produces improved oils, and corn that resists some insects. Despite these innovative products, however, there is little to conclude from them about the future successes of agricultural biotechnology. These are simply too few of these innovations yet on the market to comment on and assume success of the technology in attaining its social promises. Yet from the above illustration two fears about "worse" have come to life: the example of a tomato expressing trout genes and the death of monarch butterflies eating too near pesticide resistant crops.

All that anticipated value then is still but the visionaries' promise of agricultural biotechnology: more food, less future starvation, a more secure environment, and more choices of affordable products. Small wonder that these new food technologies have been widely championed, especially with some fears coming real. The scientific and industrial advocacy, however, is enhanced further by the accompanying advent of new mechanisms for business and profit. There exists substantial room for firms to make money by developing new products with all those desirable characteristics (18). Some are large firms such as Monsanto and Pioneer that have set up extraordinarily large scientific enterprises to foster the new technologies. Monsanto, for example, has invested more than four billion dollars in agricultural biotechnology. Others are far smaller firms set up by entrepreneurs who organize around a single product, often with state government backing to enhance area economic development. Both types have in general found appeal among finance capitalists and stock market investors, which further drives advocacy of the technology. Industry and scientific jobs and technically sound advancement are, of course, powerful social motivators for investors and for governments. Combined with all the visionary promises, profits and their additional potential add real social luster to agricultural biotechnology and lend it greater social legitimacy. The technology is therefore a dramatic social and political force. Yet that economic promise is still to be proved. Monsanto supports its agricultural biotechnology science by selling more herbicides, many corporate stock prices have experienced declines, some firms have failed, and some states have divested their investments as far too costly to state budgets.

BUT THERE ARE THOSE LOOMING CRITICS

These broad-ranging promises, however, are only predictions premised on the good guesses of a very able but always limited scientific community. So too are social fears. No one can be exact as to what will actually occur, either as a benefit or as a consequence of agricultural biotechnology. That uncertainty produces, first of all, considerable perceived risk and, second, opens up political and economic doors to every advocate who holds a contrary position. To the great frustration of public policy makers, no one can give either outright assurances or absolute reasons for resistance to biotechnology. Thus, for scientists and industry, there are a great many opposing and competing views from formidable critics that especially plague them. Some common and recurring critical assessments that filter down to and move the national publics at large need highlighting.

These can be typed and analyzed in five general but not mutually exclusive categories: the philosophy-and-ethics-of-risk problem, the agricultural-sustainability problem, the farm problem, the Jeremy Rifkin problem, and the opposing-coalition problem. Each will be explained in full, not because of the importance of any one of them. Rather the explanations are sequenced according to the formal logic and indeed science of their advocates' opposition. The most scientifically logical problems come earliest, Then, in descending order, the problems become even more purely political and driven by the competing social values of and myths held out by the opponents. To an important extent, however, a degree of both logic and politics can be found within each type of critical disagreement because there are so few hard, or perfect, pieces of evidence.

The Philosophy-and-Ethics-of-Risk Problem

At the center of the debate for those critics are the growing relationships between scientists, the universities

that employ them, and the industries that offer grants and contracts to university budgets and researchers (19). The role that government plays in encouraging applied research, cutting university budgets, and supporting economic growth and development also enters in according to these critics. The basic skepticism results from beliefs that as these forces became more interdependent, scientific inquiry lost its former neutral ethics of research. Ethical values, for reasons of scientific self-preservation, had to give way to science's move to the center of social and economic service (20). As a consequence scientists lost much of their traditional culture of critical and objective inquiry. This, of course, is a harsh view.

The harshest charge of those who see this problem goes beyond that perversion-of-science charge. It emphasizes that in pursuing social and economic relationships, scientists fail to adequately consider even the possibilities of the negative consequences of their actions. What they learn might hurt. In drawing an analogy that simplifies the problem, one critic raised the specter of Dr. Frankenstein. Frankenstein was not scientifically wrong for what he did in creating the monster. He was wrong in not anticipating what might result (21). Thus he failed to take any action to prevent disasters that may occur. Scientists at the center of society are thus myopic, inclined not to analyze everything that should be subject to inquiry. Finding likely consequences may be a threat. At least some of that myopia is tied to budget constraints but even more is linked to doing whatever best serves the new scientific culture of interdependent rather than independent relationships with those who fund research.

Beyond this generic concern there are few agreements as to what are the specific risks of biotechnology and what should be done about them. An extreme position is that science should ensure that any genetic creations that escape to the environment should quickly die and never become ecological contaminants (22). Doing so certainly makes avoiding risk the highest priority of agricultural biotechnology. Many such proponents favor no-risk tolerance laws. These critics look to China as the object lesson for this concern over preserving natural or indigenous germ plasm, species, and plant varieties. Due to massive manipulation of plants and animals, China has not seen the semblance of a normal environment for over 1000 years. Major loss of varieties has occurred.

Others, however, have less comprehensive and restrictive concerns. Some will even accept low levels of risk. Biotechnology, according to others, should not be injurious to animals as a basic ethical standard (23). An ironic expression of risk is the likelihood of producing too much food and destroying in the process social and institutional stability in the world order (24). On an entirely different level, communication specialists argue for getting citizens involved in assessing risks (25). With increased public involvement of citizens, risk-minimizing benefits of two kinds accrue. First, the likelihood of public opinion having too much faith in that culturally changing science is minimized. Increased social scrutiny results. A competing concern comes from the came critics. They note that despite perhaps misplaced faith in science, the public also has simultaneously developed a highly critical view of government and other social institutions. Should those views be moved further by food system or environmental failure laid to agricultural biotechnology, public opinion would become irrationally cynical of these new technologies and lead to hostile and even backward controls.

Thus, with a philosophical framework intact of minimizing risk, ethics and developing standards still mean several different things to several different critics. That does not imply, however, that these critics have not made their mark. The EPA takes a strongly risk-averse approach, even in confronting organized scientists. USDA's Office of Agricultural Biotechnology publicly and politically holds a neutral position to ensure confidence and avoid charges of product advocacy over that of the public interest. Finally, USDA's APHIS has also taken an innovative stand by establishing standards for releasing information and encouraging a dialogue with various segments of society (10). Avoiding risks of various kinds has certainly penetrated the public policy process and the attendant debate over regulatory standards. That, of course, complicates U.S. national politics, especially in bringing hypothetical fears to a cautious public that might well revolt. European politics is hardly immune to the same problems.

Agricultural-Sustainability Problem

While discussions of agricultural biotechnology have been a mix of effusiveness, critical thought, and social attention, other innovative approaches to the Malthusian fear of world food shortage have been in evidence. A prominent one is sustainable agriculture, the search for methods of production and food products that will not lead either gradually or even in a crisis to production disasters. Lester Brown's Worldwatch argues first and foremost for long-term sustainability over biotechnical advancement as do Rodale industries, both two organizations that reach countless Americans and a great many residents of other countries worldwide. The sustainability debate in agriculture is often linked to better and more productive agricultural biotechnology (26). This linkage is but more of the idea that biotechnology offers food and production needs almost limitless social value. The logic, given earlier, is obvious but unproved.

Not all observers agree with the positive linkage, though, including many in USDA's original office of sustainable agriculture. The reasons for the disagreement are complex and not as intuitively obvious as ones that in a positive way link biotechnology and sustainability (27). Part of the reason for obfuscation of problems lies in the origins of sustainable agriculture. Although not all of its advocates agree, sustainability has its origins in organic farming. As such, much of both its rhetoric and science are anti-chemical input with a championing of healthy ecological conditions. As sustainability developed its own goals, though, it moved its position from no chemicals to reduced chemicals. Sustainability advocates, accordingly, mostly want food that is healthier to consume and healthier for long-term soil and water resources. Thus not all these advocates are extremists in their beliefs. They also want food that is less expensive for producers to grow as the amount of expensive agrochemical inputs are decreased. The latter point has gained these

advocates a growing number of supporters among farmers and within several federal agencies in the United States. Research support is established and considerable. Sustainability has become popular enough that most USDA agencies try to promote it at least at the margins. The Extension Service promotes it more aggressively through its farmer assistance outreach in some regions of the country. The U.S. South is especially sustainability conscious. Sustainable advocates in general are also more impressed with precision farming innovations rather than biotechnology, which holds up another roadblock for many.

Opposition from some sustainability segments nonetheless charges agricultural biotechnology interests as bringing forth several production negatives. They tend to argue the following: that biotechnology is owned and controlled by large industries that have records of environmental abuse rather than of contribution, that biotechnology in agriculture can have no life without this corporate control, that diversity in agriculture will further give way worldwide to specialized and large-scale crop and animal production, that all of this takes decision making away from local farmers who now often creatively address their own sustainability and environmental needs, and finally, that the sum total effect is to distance people and producers from what is grown and how it is grown.

Stewardship over natural resources, as a consequence, is likely to disappear as a human value. These are another set of very harsh and especially cynical views. The cynicism and the targets of attack make them less potent than are the arguments of advocates against risk. Despite what seems the marginal status of the most extreme sustainability critics, these views have enough attention in public, farm, and policy-making circles worldwide to be a considerable factor in evaluating policy options. The 1999 move of the Henry Wallace Institute for Alternative Agriculture, with its sustainability emphasis, to the worldly acclaimed Winrock International has enhanced that research identification. The anti-business concerns are nonetheless particularly persuasive in bringing opposition in developing nations to the policy wants of heavily industrialized, biotechnology-producing agricultural nations.

The Farm Problem

It seems unlikely at first thought that farmers, producers, and growers would present agricultural biotechnology with an obstacle. The farm problem, however, exists as a major one for industry and science. Production agriculture, even through the present, is treated both in the United States and in much of the world as a unique economic sector deserving special government assistance (28). Even as farming loses its economic uniqueness, a raft of institutional structures can still be found that protect producers against market failure, natural disaster, and even economic loss. The reason owes to another irony of production agriculture: Just after Malthus was spreading his fears, the United States entered a period through today and into the foreseeable future that has brought food supplies that are too large, farm prices that are too low, and accompanying failures and farm losses among producers. Technological innovations kept that spiral in place (29).

In a nation that was settling its frontiers and attracting massive development, failures of this magnitude among so many national agricultural institutions were found to be politically unacceptable.

First, considerable social and political investment had been made in increasing the number of producers, getting them to frontiers, keeping them productive, and even giving them land. Nations as diverse as Japan and Mexico passed legal structures either to keep farm prices artificially high and away from a worldwide market or simply to give expropriated lands to peasants. The belief was and still remains the simple one that producers as providers of food and stewards of both natural resources and national security deserve special treatment. As Jeffersonian Democracy spread in the United States, a myth emerged that family farmers were important and should be preserved for reasons of protecting basic social values (30). This played out as the widespread social belief that family farming structures of production deserved protecting against large industrializing farm and ranch holdings (31). Other nations were not immune to this notion of farming as a basic social value, and protectionism became a common worldwide practice.

Farm failures were unacceptable from a second perspective as well. Not only was national development important and existing on a powerful base of social values and myths, farmers were also acquiring vast political importance in agrarian and in evolving nations. Farm power was institutionalized in the United States in its own structures within Congress and in administrative agencies. Farmers felt entitled to assistance, and government officials treated them as if they indeed were. Homestead laws were passed, railroad transportation was regulated, a huge agricultural establishment of research and education was put in place, marketing supports were advanced, and, as farm failures continued, direct government payments to farmers were made. Farmers and organized farm interests zealously protected all this. Congress proved the best vehicle for protection because congressional members from rural districts and states were anxious to serve local constituents, especially where they could effectively destabilize regional politics (32). Similar ventures worked worldwide to play off and protect myths of agrarians as special and often unassailable interests.

Thus, when agribusiness industry took on economic and political significance, farmers won concessions frequently and with a political flair. In the United States, patent laws allowed farmers to sell certified products to others as these were left over from individual production efforts. Until patent law cases were more effective, farmer actions were often abusive. Yet only in 1994 did the courts limit farmers against industry. In *Asgrow v. Winterboer*, the U.S. Supreme Court ruled that it was quite excessive for a single family farm to stockpile and sell enough soybean seed to plant 400,000 acres. As long as farmer interests work to compromise with those other social and economic interests in their nation, however, farmers still hold potent influence. Environmental and farm trades among otherwise competing interests are a good example, one directed harshly and in tandem against biotechnology industries.

The emergence of opposition to agricultural biotechnology shows that influence with clarity, at least relative to countering industry and science. Dairy farmers delayed the opposition of rBST for years. They argued, through to the present, that the cost of the technology was too high for smaller-scale farmers to bear. Economic benefits could go primarily to larger-sized producers who enjoyed economies of scale in distributing costs throughout large herds. This smaller-scale producer fear argued that rBST would further increase escalating farm size and disrupt the widely valued structure of family farm production. Several other biotechnology products have been opposed for similar reasons of costs and farm structure. These controversial products range from strawberries that can withstand lower temperatures and not freeze as well as transgenic innovations that at considerable expense added preservation qualities and had longer shelf lives. Opposition also was strong against adding transgenic innovations in farm animals. In all these cases farmers wanted protection from changes and competitive circumstances in their own operations. They wanted to preserve their own current animal breeding plans, keep personal costs low, resist greater agribusiness dependency, and continue to produce with well-understood and familiar practices. As a result much of the farm problem over biotechnology has been directed against agribusiness corporations and production structures that are far from the traditions of family farming. This same farm problem perspective has also been directed against other than biotechnology producers, for instance, against those proposing irradiation of food for longer shelf life. This rather populist rhetoric finds considerable public support as well, and it has gained a number of limited-technology champions within Congress and numerous state legislatures.

Agricultural biotechnology thus comes constantly against prevailing social sentiments that are protective of family farming. The tendency is for liberal, progressive, and populist elements in agriculture to generate much of this daily opposition. A long and enduring tradition of ideological farm protest has always split farm sectors in many nations, including the United States (33). The farm problem, however, is not restricted only to populist opposition. Many mainstream general and commodity farm interest groups have cooperated in challenging some biotechnology products and attacking agribusiness projects. What to support and what to oppose is in many of these cases a matter of primarily what farmers like and what they do not. Farmers, for obvious reasons, like and support new pesticide resistant crops. Yet they dislike many other products because these disrupt existing and widely accepted production and even marketing practices. That opposition results from all types of producers, regardless of political ideology and farm size. It gets played out in all of the institutions of government that have long provided farm services. This indicates that the farm problem will continue to plague at least some biotechnology innovations, as well as others, into the more than immediate future.

The Jeremy Rifkin Problem

It may seem unfair to personalize a critical problem with the name of a single human being. Yet Jeremy Rifkin and his unrelenting and vitriolic opposition to agricultural biotechnology deserve special note for personalizing and spreading this dissenting style of politics. Rifkin is a guru, the one who founded the word of warning. His works have spread internationally and he would probably welcome the guru label, since he has personally and continuously contested genetic research since the mid-1970s. Labeling the problem after Rifkin also calls attention to the widespread growth of public interest groups that lobby against the unintended externalities of agricultural biotechnology and for alternatives to genetic experimentation. For example, Greenpeace has been a Rifkin disciple organization in Europe, bringing this same type of opposition and same rhetoric to nearly the whole of that continent's extreme Green politics.

In this and other issue areas, individuals have a strong tendency to be better identifiable than their own organizations. That is, Rifkin matters more than Greenpeace, or even more than his own group. As entrepreneurs, such people have founded numerous organized political interests, sought grants and donations to get and keep them going, made extensive contacts in policy-making and media circles, and lobbied as near free agents with their own personal agendas (34). Their organizations are often quite small, sometimes with no members. In opposing agricultural biotechnology, scientists and industry have often been confounded by the importance of these personalities, ranging from the aforementioned Lester Brown to the Land Institute's Wes Jackson and including the agrarian poet from Kentucky, Wendell Berry. As well-recognized entities, these individuals keep the attention of much of society because they personally make claims that followers enthusiastically disseminate. They matter less because of what they say than how and to whom they routinely say things.

Rifkin epitomizes this approach and came to command ongoing media coverage after the publication of his highly critical and controversial book, *Algeny* (35). Politically, even as a distinct outsider from government, Rifkin remains a major contact or at least reference for those who wish to investigate what might be the seamy side of agricultural biotechnology. Having only very small financial resources compared to the largest agribusiness firms, the Rifkins of international politics raise enough money to mount obstacles through law suits, frightening scenarios released to the press, and ongoing exposure to electronic media hosts who include them in news reports. With political and social institutions around the world both skeptical of new technologies and lacking integregation, Rifkin and other public interest entrepreneurs have proved to be influential in both starting and formulating debates. By articulating the semblance of public policy plans and by always having a story that reporters enjoy and pick up, public interest groups like that Rifkin follower Greenpeace often appear far better at basic politics than do advocates of emerging agricultural biotechnology. That explains why in the United States these people have won permanent friends in such agencies as FDA, EPA, and USDA's Food Safety Inspection Service. The same is true in the UN. In Europe they have won even more policy-making friends and have been very influential

in delaying product introductions. This is true even in countries that are strong in biotechnology research. Yet even some developing nations are homes to the critic entrepreneurs, especially by using international grants. Others from developed countries also influence domestic policies in such developing countries. Robert Rodale of Rodale Press gained frequent attention for being the worldwide exporter of organic farming. And his successors have recently used Rodale's publications to link organic farming to biotechnology skepticism.

The Coalition Problem

Closely tied to the Jeremy Rifkin problem is that of coalition politics. Political coalitions are formed when different organized interests and social segments cooperate on public policy questions (36). The basic objective of a coalition is bringing together various interests so that they can share resources of money, political and media contacts, and even skill in order to win their agreed-upon common goals. Proponents of causes related to agricultural biotechnology have formed coalitions in order to gain government cooperation. Critics of biotechnology have organized coalitions in order to delay or prevent product introduction and regulatory approval. The coalition problem results because critics are more advantaged by such cooperation than is science and industry. That, for instance, makes Rodale Press opposition to biotechnology especially formidable when it is linked to the idea of organic production.

The explanation as to the opponents' advantage is quite simple. First, critics aim to stop something. Second, stopping or halting a product or technology is but a matter of creating doubts. Third, critics need not share all of the same values as reasons for their common opposition. They only need to agree to cooperatively bring impediments into the policy process. Thus minimal cooperation is necessary, and no plans for the details of future policy need be formulated. Their politics is really only about saying "no" in a unified voice. The more who say "no" together, the stronger is the coalition.

Proponents of biotechnology have contrary disadvantages. Given the uniqueness of products and technology, scientists and agribusiness firms must decide whether or not any cooperation is worthwhile. Cooperation on securing a patent or on approving a single pesticide is a task for the stockholder firm, not the entire industry. Moreover, too much cooperation between advocates of agricultural biotechnology may release trade secrets and give away firm advantages in the marketplace. That explains why most universities, research facilities, and corporations lobby on their own. And it explains why they also support more generically active common trade associations, the Biotechnology Industry Organization or the National Agricultural Biotechnology Council, when obvious policy questions such as binding international protocols unite the diverse proponents. This quite clearly is neither as simple or as easily agreed upon a cooperation as that of the opponents collectively and loudly saying "no" (37). Far more strategic options need be considered by proponents in order to pursue collective ends or even merely share information. As a consequence opposing coalitions tend to grow to the largest possible numbers while coalitions of proponents tend to be kept far smaller, and therefore more internally conciliatory and politically weakened.

The net result is that coalitions tend then to be especially favored by Rifkin-style public interest groups. When Rifkin's Foundation on Economic Trends worked to stop the introduction of rBST, coalition politics was the obvious and favored route. All of the other three types of critical opponents were brought together to force greater study and long and costly industry delays. Sustainable agriculture specialists raised their questions under Rifkin's direction. Populist, smaller-scale dairy producers engaged with Rifkin in active social protest in favor of family farming. Academics who studied philosophy of science and policy ethics willingly shared their skeptical information (38). On the other side, as more industries that were once hoping to produce BST disassociated themselves from the innovation, Monsanto was left nearly alone to lobby for regulatory approval and against legislative roadblocks. In a more recent and publicly inspiring effort to capture attention, Rifkin helped organize the Pure Food Campaign. Nationally and internationally prominent chefs with important local followings gained Rifkin's views extensive publicity, even on cable TV's *The Food Channel*. The Beyond Beef Coalition was similarly conceived and attended.

The conclusion as to the impact of the coalition problem on science and industry is easy to draw: When it comes time to challenge agricultural biotechnology in public policy making, the interests are easily merged of those who represent each of the opposition problems. The critics do not exist in political isolation nor have secrets to necessarily withhold from one another. Thus critics of agricultural biotechnology may each be relatively small and resourceless and anti-science, but they still exist in cooperation with one another, with a resulting considerable political influence over corporate and research innovations.

SOCIAL IMPACT ON AGRICULTURAL BIOTECHNOLOGY

The social impact of agricultural biotechnology is full of both promise and fear, but for the foreseeable future, that impact is problematic. Sound sciences and capable industries make projections that must be far from reliable forecasts. Intelligent critics raise questions that are logical and rationally derived. As a consequence these emerging technologies may yield great benefits but they might also offer several social and political problems. No one knows. All extrapolations are only guesses as to probable impact.

The more immediate question for the early decades of the twenty-first century is on what social and political impact affects agricultural biotechnology. Answers exist for this query. First, any world order for governing agricultural technology's future will be fragmented and full of inconsistencies nation to nation. Second, what people believe even mythically in each society will exert strong political pressures that influence product and technology innovations. Rational science will not preordain adoption outcomes because near perfect information is never possible to have. National

politics, as affected by international events, will prove determinate. Third, existing evidence indicates that a skeptical public, responsive public officials, and numerous critical positions on this technology will put distinctive and either unfortunate or fortunate limits on its ability to offer the social values it otherwise might. There exists a general suspicion both of new technologies and of corporate enterprises, especially in developing countries (39). That explains why public interest groups, as well as many academics, so often predict food scares that seldom, but certainly could, bring about the anticipated disasters (40). The rhetoric seems believable. The unfortunate plight of the monarch butterfly as it feeds close to genetically altered crops is an example that adds to the believability of critical commentaries. So too is the Muslim who confronts a tomato having a pig gene. People and politicians worldwide like change to be slow; and they can from that value position be easily influenced and mobilized by social and political critics to impose impediments from an anti-science perspective. Agricultural biotechnologies for these reasons will not soon escape either its controversies or its difficult politics.

BIBLIOGRAPHY

1. N.J. Seay, *Aipla Q. J.* **16**, 418–422 (1955).
2. P.J. Goss, *California Law R.* **84**, 1395–1836 (1996).
3. W.H. Lesser, *Animal Patents: The Legal, Economic, and Social Issues*, Stockton Press, New York, 1989.
4. M.M.J. Lin, *Am. J. Law Med.* **22**, 109–134 (1996).
5. J. Doyle, *Altered Harvests: Agriculture, Genetics, and the Fate of the World's Food Supply*, Viking Penguin, New York, 1985.
6. P.B. Thompson, *Agricul. Hum. Values* **14**, 11–27 (1997).
7. N.D. Hamilton, *Tulsa Law J.* **28**, 587–596 (1993).
8. S. Schuff, *Agri-Pulse* **13**, 1–2 (1997).
9. J.A. Nettleton, *Food Tech.* **50**, 28 (1996).
10. A.L. Young and M.R. Asner, *J. Nutr.* **126**, 1010S–1012S (1996).
11. L. Busch et al., *Making Nature, Shaping Cultures: Plant Biodiversity in Global Context*, University of Nebraska Press, Lincoln, NE, 1995.
12. E.I. Tsevdos, R.A. Chadwick, and G. Matthews, *Nat. Law J.* **19**, C1, C27–C31 (1997).
13. B. Hileman, *Chem. Eng. News* **73**, 8–17 (1995).
14. *The Economist* **341**, 21–23 (1996).
15. D.F. Hadwiger and W.P. Browne, in D.F. Hadwiger and W.P. Browne, eds., *Public Policy and Agricultural Technology: Adversity Despite Achievement*, St. Martin's Press, New York, 1987, pp. 1–13.
16. D. Avery, *Biodiversity: Saving Species with Biotechnology*, Hudson Institute, Indianapolis, IN, 1993.
17. D.D. Jones and A.L. Young, in G.R. Takeoka et al., eds., *Biotechnology for Improved Foods and Flavors*, American Chemical Society, Washington, DC, 1996, pp. 1–13.
18. L. Grant, *Forbes* **135**, 116–118 (1997).
19. W.B. Lacy, L.R. Lacy, and L. Busch, *Agricul. Hum. Values* **5**, 3–14 (1988).
20. K.P. Ruscio, *Sci. Tech. Hum. Values* **19**, 205–222 (1994).
21. B.E. Rollin, in S.M. Gendel et al., eds., *Agricultural Bioethics: Implications of Agricultural Biotechnology*, Iowa State University Press, Ames, IA, 1985, pp. 292–308.

22. M. Sagoff, *Agricul. Hum. Values* **3**, 26–35 (1988).

23. A. Gore, Jr., *Gen. Eng. News* **7**, 4 (1987).

24. G. Comstock, *Agricul. Hum. Values* **3**, 36–52 (1988).

25. S.H. Priest, *J. Comm.* **45**, 39–54 (1995).

26. H.A. Schneiderman and W.D. Carpenter, *Env. Sci. Tech.* **24**, 472 (1990).

27. J. Kloppenburg, Jr. and B. Burrows, *The Ecologist* **26**, 61–67 (1996).

28. D. Paarlberg, *Farm and Food Policy: Issues of the 1980s*, University of Nebraska Press, Lincoln, NE, 1980, pp. 5–13.

29. W.W. Cochrane, *The Development of American Agriculture: A Historical Analysis*, University of Minnesota Press, Minneapolis, MN, 1979, pp. 378–395.

30. W.P. Browne et al., *Sacred Cows and Hot Potatoes: Agrarian Myths in Agricultural Policy*, Westview Press, Boulder, CO, 1992, pp. 5–16.

31. J. Burkhardt, *Agricul. Hum. Values* **3**, 53–60 (1988).

32. W.P. Browne, *Cultivating Congress: Constituents, Issues, and Interests in Agricultural Policymaking*, University Press of Kansas, Lawrence, KA, 1995, pp. 22–39.

33. P.H. Mooney and T.J. Majka, *Farmers' and Farm Movements: Social Protest in American Agriculture*, Twayne Publishers, New York, 1995.

34. W.P. Browne, *Private Interests, Public Policy, and American Agriculture*, University Press of Kansas, Lawrence, KA, 1998.

35. J. Rifkin, *Algeny*, Viking Press, New York, 1983.

36. K. Hula, in A.J. Cigler and B.A. Loomis, eds., *Interest Group Politics*, 4th ed., Congressional Quarterly Press, Washington, DC, 1995, pp. 239–258.

37. W.H. Riker, *The Theory of Political Coalitions*, Yale University Press, New Haven, CT, 1962.

38. W.P. Browne and L.G. Hamm, *Policy Studies J.* **17**, 181–192 1988.

39. A. Wildavsky, *But Is It True? A Citizen's Guide to Environmental Health and Safety Issues*, Harvard University Press, Cambridge, MA, 1995.

40. S.C. Phillips, *CQ Researcher* **4**, 675–691 (1994).

See other entries AGRICULTURAL BIOTECHNOLOGY; ANIMAL BIOTECHNOLOGY, LAW, FDA REGULATION OF GENETICALLY MODIFIED ANIMALS FOR HUMAN FOOD USE.

AGRICULTURAL BIOTECHNOLOGY, LAW, APHIS REGULATION

DAVID R. MACKENZIE
University of Maryland
College Park, Maryland

OUTLINE

INTRODUCTION

Agriculture in the United States is extremely productive today because of extraordinary natural resources and technological efficiency. The present capacity of the U.S. agricultural production system (from field to fork) is a result of scientific contributions made by both the private and public sectors. It rests on a foundation of basic and applied research outcomes delivered to the intended users through multiple technology transfer mechanisms.

Science in general, and U.S. agricultural research in particular, has been for the most part exempted from regulatory oversight, with some notable exceptions. Research scientists have for some time been expected to comply with federal regulations regarding the handling of radiological materials, and they are required to obtain institutional permission to conduct research with human subjects. Federal law also sets strict standards for the care of research animals. In recent times research scientists have been required by federal law to obtain certification to handle registered pesticides, although many types of hazardous chemicals (including experimental pesticides) have long been exempted from federal regulation in small-scale tests.

One notable exception to this regulatory pattern for research activities is the strict federal requirement on the interstate shipment of plant pest, pathogens and noxious weeds. To move a "regulated article," from one state to another, or to import a "regulated article" requires a permit under the Federal Plant Pest Act. The Plant Pest Act is administered by the United States Department of Agriculture (USDA) through the Animal and Plant Health Inspection Service (APHIS), a division of the USDA's Food Safety and Inspection Service (FSIS) (1).

Thus, for the most part, during the first three-quarters of the twentieth century agricultural researchers in the public and private sectors were mostly free to manipulate the genetics of animals and plants, and even microorganisms, with little government oversight or regulatory attention. The freedom to investigate all types of organisms for applications to crop and livestock improvement came to be a common expectation. New technologies emerged and flourished, as did U.S. agricultural production.

In the United States this virtual absence of scientific regulation seemed to work, and when something untoward happened, solutions were modest. This point is perhaps best typified by the federal government's response to the 1969 release of a potato clone named Lenape. This cultivar was soon discovered to have a significant concentration of poisonous glycoalkaloids in the tubers, under most commercial growing conditions. Lenape was the product of a conventional breeding program that had used some wild species as parents to obtain superior potato chip processing quality. It has been presumed that the wild parentage brought to the progeny high glycoalkaloid concentrations that later required its withdrawal from commercial production. The federal government's response to public concerns was to establish, through the U.S. Food and Drug Administration (FDA), requirements for establishing the safety of newly released cultivars. The agency's Generally Regarded As Safe (GRAS) guidelines asked plant breeders to give self-assurances that what was about to be released was at least as safe as the cultivar(s) to be replaced. This "self-policing" approach to the problem was generally well received within the scientific community. And it was fairly typical of the federal government's hands-off approach to research-related concerns for environmental, public health, and food safety issues. The GRAS guidelines are still in place, but there is no GRAS police force.

AGRICULTURAL BIOTECHNOLOGY

In the early 1970s, the relationship between society and the scientific community began to change as a result of research breakthroughs in the ability to manipulate the cells, tissues, and the genetic code of plants and animals (2). These collective technologies were subsequently called *biotechnology*. Understanding the technology's components is necessary for gaining an understanding of the shift that occurred in the regulation of biotechnology in general, and agricultural biotechnology research in particular, by the federal government.

Tissue and Cell Culture

The culture of living tissues, and subsequently of individual living cells, has long been available to research scientists. Plant tissue culture dates back six decades. But in the 1970s scientists discovered how to chemically dissolve cell walls and grow single protoplasts (naked plant cells). This represented a technological advancement of significance to plant researchers. Similarly, in the animal science research community, the ability to culture animal tissue and individual cells opened up new areas of investigation. At the time, none of these discoveries were considered to represent a risk worthy of federal regulation.

Regeneration

Subsequent research discoveries allowed the regeneration of individual cells and tissues into once again whole organisms, through a phenomenon called *totipotency* (which is

actually very poorly understood). This remarkable biological characteristic means that certain types of individual living cells have all of the genetic information necessary to become a complex organism, and can do so when given the right culture conditions.

Regeneration of plants and animals also was not considered worthy of federal regulatory attention. This was true until recently when the cloning of some higher animals (notably with the cloning of the sheep named Dolly) raised public concerns regarding the ethics of a technology that might lead to the cloning of humans. Still, no federal regulations for sheep cloning are being seriously considered.

Transformation

The third emerging technology is genetic transformation which was first described in bacteria. This discovery had obvious applications to higher organisms, once it was understood that complex organisms could be reduced to a single cell (or small group of cells), transformed, and then regenerated. Taken together, tissue or cell culture, genetic transformation, and regeneration have become the tools of genetic engineering (a.k.a. biotechnology).

PLANT GENETIC ENGINEERING

The genetic engineering of plants first occurred in 1982 at the University of Wisconsin. Researchers there genetically engineered a protein from sunflower into a bean plant. This "genetic transformation" technology later became the standard for genetically engineering plants. Involved in genetic transformation processes are the following steps.

- Identification of a specific *genetic sequence*
- Isolation of the sequence by the use of *restriction enzymes*
- Matching the result to a *promoter sequence*
- Application of a gene *vector*
- Insertion into a host plant's *genome*
- *Regeneration*.

Both the vector and the promoter DNA sequences are essential to the transformation process. A *marker gene* (e.g., a gene for antibiotic resistance) is also desirable in the *genetic construct* to more easily identify the successfully transformed cells or tissues.

The terminology associated with genetic engineering requires some explanation. Restriction enzymes were first named because when they are applied to cultures of a bacterophage, the plague's (i.e., infected bacteria's) growth is "restricted." Subsequent research showed that the restricted infection was due to the presence of enzymes, but it was not known why they restricted plague growth. In the end it turned out to be the result of ribonucleases (RNA enzymes). These enzymes were cutting the DNA at specific sites in the DNA sequence, thus restricting infectivity. The proteins came to be termed "restriction enzyme" (because the bacteria's growth was restricted). Technically they are more accurately called endonucleases, since they cut the DNA sequences internally (endo).

Gene vectors come in many forms, but the most commonly used (and highly efficient) vector in plants was found in the plant pathogen *Agrobacterium tumafasciens*. This pathogen has a wide host range and induces a cancer like growth near the soil line, which accounts for the common name for the disease, crown gall (3).

Significant research funding in the 1970s from the National Institutes of Health (NIH) for investigation of crown gall was likely a consequence of their interest in the cancer-like growth promoting characteristic of the pathogen. Thus, as a benefit of cancer research, plant pathologists were able to discover that a plasmid (an extra-chromosomal ring of DNA) harbored a sequence for tumor induction (commonly referred to as the TI plasmid) (4). The plasmid had the capacity to insert itself into host DNA, so it could serve as a vector for attached DNA. The attachment of cloned DNA to the TI plasmid came to be called a *construct* and targeted sequences could be used to transform plants that were susceptible to *Agrobacterium tumafasciens*.

It was soon discovered that in order to get the DNA sequences to express themselves effectively, a promoter sequence had to precede the gene of interest. The commonly used promoter sequence during the early years of plant research was also from a plant pathogen, the cauliflower mosaic virus.

All of this is important to subsequent regulation of biotechnology by the Animal and Plant Health Inspection Service (APHIS) of the U.S. Department of Agriculture (USDA) under the Plant Pest Act, as we will see later.

HISTORY OF BIOTECHNOLOGY REGULATION

In 1974 a group of about 300 concerned scientists met at the Asilomar Conference Center in Monterey, California, to discuss their concerns for the safety of research with recombinant (a.k.a. rDNA) organisms (5–7). At that time virtually all genetic engineering was being done with bacteria. There were no biosafety standards for the handling of these biological materials in research laboratories. Moreover much of the biotechnology research at the time was medically related, often involving human pathogens. Fear of an epidemiological disaster that might result from an unintended release of a genetically engineered human pathogen prompted the Asilomar conference. From that conference emerged a set of voluntary guidelines that set levels of containment for the handling of recombinant organisms, primarily based on human disease hazards (8,9). That is, the disease risk of the research organisms being handled helped determine the levels of containment, which increased with the level of concern for human pathogenicity. The guidelines described handling and containment protocols for various types of research microorganisms.

NIH Guidelines

The Recombinant DNA Guidelines were to be administered by NIH, and they were to be voluntary (10–12). In response, the NIH created the Office of Recombinant DNA Activities (ORDA). ORDA relied on a panel of experts to review application and make recommendations to the director of NIH on proposed research protocols.

The NIH guidelines were more than self-policing. The awarding of federal funds for research projects required compliance with the guidelines. It was commonly asserted at the time that voluntary compliance by the private sector and nonfederally sponsored public sector research was virtually 100 percent. Some individuals concluded that the threat of civil penalties for negligence induced many public and private laboratories to adopt the NIH rDNA guidelines, even though technically they were not required to do so. Although never tested in court, it is presumed that with NIH approval the researcher was doing what a reasonable and prudent scientist would have done. Negligence would then be hard to prove. Thus the endorsement by the NIH's Recombinant DNA Advisory Committee (RAC) became a standard for research protocols for new and novel recombinant organisms, but only for contained laboratory research.

By the early 1980s it became obvious that biotechnology would have many applications in agricultural science (13). What was not known at the time was how to provide biosafety assurances for field-testing (i.e., tests to be conducted outside of a contained laboratory or greenhouse) with organisms of recombinant parentage (14–17). The NIH guidelines were strictly for controlled laboratory facilities, and inasmuch as they were primarily based on human pathogenic traits, the 1982 announcement of transgenic sunbeans drew attention to biosafety issues, rightly or wrongly. Many felt that something had to be done by the USDA to ensure the safety of organisms being handled in research activities outside of containment (at least for the projects if funded). Thus began USDA's two-track approach to biosafety assurance.

USDA's Two Track Approach

The USDA is a complex structure with responsibilities that required it both to promote and to regulate scientific activities. At the time (early 1980s) the Assistant Secretary for Science and Education, Dr. Orville Bentley, foresaw a need to extend the NIH recombinant DNA guidelines to cover field experimentation with plants and animals. This would be done in ways to provide public assurances of the safety of research funded through his office. At the same time, NIH was expressing little interest in assuming responsibility for safety assurances for the environmental release of recombinant DNA organisms. And thus was born the idea for an Agricultural Biotechnology Research Advisory Committee (ABRAC), patterned after the NIH RAC. (To be factually accurate, the Committee on Biotechnology of the National Association of State Universities and Land-Grant Colleges first proposed to the USDA that it should use the NIH rDNA guidelines model for the oversight of field tests with recombinant organisms.) Additionally ABRAC was given staff support by the Office of Agricultural Biotechnology (OAB), which was obviously patterned after the NIH ORDA. But ABRAC/OAB was to address the biosafety (i.e., the environmental and public health) questions arising from research tests conducted outside of containment facilities.

Simultaneously, the regulatory arm of the USDA (which houses APHIS) began exploring its authority under the Federal Plant Pest Act to regulate genetically engineered plants as plant pests. This created a dual track situation and set up an interesting dynamic between the research side and the regulatory side of the same federal department.

Meanwhile, other federal regulatory agencies were giving thought to asserting their authorities to regulate biotechnology in ways that would impact agricultural research. The Food and Drug Administration (FDA) initiated notable regulatory activities, under the U.S. Food, Drug, and Cosmetic Act. Also the U.S. Environmental Protection Agency (EPA) became active under the Federal Insecticide, Fungicide, and Rodenticide Act and the Toxic Substances Control Act.

Coordinated Framework

By 1983 it was becoming obvious that a plan was needed for assuring the safety of biotechnology (18). Some interest groups argued that regulation was needed at the federal level, to coordinate better regulatory actions, and perhaps to preclude state-by-state, or county-by-county, or even city-by-city regulation of biotechnology research and commercial development. At the time one of the greatest fears of the technology's champions was a patchwork quilt of federal, state, and local government regulations that would make product commercialization not feasible.

In response to this need the White House Office of Science and Technology Policy (OSTP) called together federal regulatory and research agencies to map out a coordinated plan for the regulation of biotechnology. In June 1984 the OSTP published in the Federal Register a proposed Federal Coordinated Framework for Regulation of Biotechnology (herein after referred to as the Federal Coordinated Framework) (19,20). Intense public debate ensued, and the debate still has not subsided.

There were two fundamental principles of the Federal Coordinated Framework:

- No new laws were needed to regulate biotechnology, because existing laws were adequate.
- All regulations would be based on the product, not the process, of biotechnology (21).

Both points need some explanation.

Leading proponents of biotechnology, particularly the pharmaceutical industry, argued strongly at the time that sufficient regulatory authority already existed to ensure the safety of biotechnology (research and commercialization) and that no new laws were needed. It was argued, particularly strongly, that the pharmaceutical, drug, and medical device industries were already subject to extensive clinical trials and product registration. Adding another layer of regulatory oversight would unfairly and unnecessarily slow the development of the commercial products from this new and exciting technology.

Opponents of the Federal Coordinated Framework came from many sectors, including the scientific community that was, as noted above, unaccustomed to regulatory oversight for research activities heretofore not regulated. Environmental interest groups saw the Coordinated Framework as inadequate, and they registered their concerns during the public comment period. Interestingly the U.S. Department of Interior, which historically has responsibilities for aquatic organisms through the Fish and Wildlife Service, was not invited by OSTP to participate in the first rounds of the Federal Coordinated Framework's development. This later turned out to be a major oversight, as aquatic species, especially genetically engineered fish, soon became one of the major environmental safety concerns. No provision for regulatory oversight of fish or shellfish, or even informal research guidelines, were proposed through the Federal Coordinated Framework. Additionally the USDA's Food Safety and Inspection Service (FSIS), and FDA have long sought for themselves regulatory authority over fish and shellfish. Thus, with no existing law and no place at the table, a biosafety void was created.

Following a substantial period of public criticism, the Federal Coordinated Framework was formally announced in June of 1986 as a plan for biosafety assurance (22). Lead agencies were identified with specific responsibilities. Regulations were to be built on existing legal authorities, and with a focus on the *product*, not the process of biotechnology.

The second point of consideration became an issue of contention when defining the scope of regulatory authority, especially for obtaining White House approval to implement the regulations (23). Regulation of the products (not the processes) required that any regulatory wording could not single out biotechnology as a process. The regulations had to identify the product that was being regulated, and only if it specifically represented a biosafety hazard requiring federal regulation. This seemingly logical approach to regulation fit well with the needs of the pharmaceutical industry that had long been required to verify product safety, quality, and efficacy. Under the new rules this assurance was irrespective of whether the product was recombinant or conventional. Policy makers argued that it does not matter whether insulin was to be provided as a recombinant product or not. The process was irrelevant. It was, they said, the product that was important for making biosafety assurances.

As one might guess, significant problems occurred in several areas of scientific research from this policy since heretofore many research activities, for the most part, were not regulated. As noted above, this was particularly true for agricultural science. Many scientists shuddered to think that a federal regulatory apparatus would be imposed on their activities, on an experiment-by-experiment basis.

In the worse case scenarios, some critics lamented, a new, perhaps unforeseen hazardous consequence might occur in field tests with rDNA organisms. Would the transferred genes be stable? Would the traits be expressed in ways that heretofore were not seen? Would pleiotropic effects be expressed? Would recombinant organisms have a superior ecological advantage or greater fitness over natural types? Would unknown or unanticipated characteristics cause environmental, public health, or safety problems?

An additional biosafety issue that needed to be addressed was the capacity of biological organisms to reproduce and disseminate, once released into the environment. Would the Genie, once out of the bottle, become an unretrievable problem? Could experiments be designed to contain or mitigate an organism released into the environment? And who would do the biological monitoring?

To implement the Federal Coordinated Framework, USDA decided to move ahead with its two-track approach (24). This led to some interesting consequences.

APHIS began, in quick succession, its proposed rule making, followed 30 days later with final rules to regulate the "products" of plant genetic engineering, using its Federal Plant Pest Act authority (25). The final rule clearly states "genetic engineering" as the regulatory trigger, thus seemingly violating the product-not-the-process terms of the Federal Coordinated Framework. The intended regulatory targets were the plant pathogen-derived vectors and promoters of genetically engineered plants. Through this regulatory strategy the plants themselves would become the regulated articles as "plant pests," under the Federal Plant Pest Act. The White House policy reviewers allowed this curious strategy. This way APHIS regulatory authority was put in place for plant biotechnology.

Simultaneously USDA's Science and Education Office drafted guidelines for field testing recombinant organisms (26). The guidelines met with mixed results.

First, the ABRAC guidelines attempted to prescribe physical and biological "confinement" practices in anticipation of field experiments yet to be proposed. That task proved to be too daunting. The deployment of recombinant DNA confinement methods on so many forms of organisms was evolving faster than the advisory committee could come up with procedures that could be accepted as reasonably safe. Notably, drafts of the ABRAC guidelines met with severe criticism, leading to more revisions and considerable frustration by ABRAC members.

Second, federally funded scientists were required to get both a permit from APHIS and an ABRAC review of their research protocol. The science community saw this as double jeopardy (27).

Also applicants expressed frustration with the delays in obtaining their reviews. Some ABRAC members began to question the validity of their own decision-making process, and the legitimacy of using peer review for biosafety risk assessments (28).

Meanwhile, USDA's Science and Education office created the National Biological Impact Assessment Program (NBIAP) with responsibility for facilitating safe agricultural and environmental biotechnology research (29). NBIAP was to monitor progress, foster biosafety communication, and focus research activities on priority biosafety issues. As an independent office, the NBIAP could work across agency lines and with institutions external to USDA to identify emerging biosafety issues and expedite solutions. During a very short period of time, and with limited funding, the USDA's Cooperative State Research Service, working with several Land-Grant University partners, developed a protocol:

- A biotechnology information bulletin board (a forerunner World Wide Web)
- A compilation of all APHIS generated Environmental Assessments for field testing permits as a CD-ROM
- An expert system for assisting scientists with completing an application for an APHIS permit
- An annual international conference, co-sponsored with U.S. EPA and Environment Canada, on biosafety research results
- A $1.7 million competitive grants program in biotechnology risk assessment research

To this last point Section 1664 of the 1990 Farm Bill established a competitive grants program in biotechnology risk assessment to begin answering some of the risk assessment questions brought forward by the critics of agricultural biotechnology.

NBIAP continues today, renamed as the Information Systems for Biotechnology (ISB), with modest USDA funding to Virginia Polytechnic Institute and State University (Virginia Tech).

THE APHIS REGULATORY PROCESS

As noted earlier, under the requirements of the Federal Coordinated Framework, APHIS could not seek new legal authority. APHIS had to use the provisions of the Plant Pest Act to write regulations to assure the safety of recombinant plants being tested outside of containment. This required some significant reinterpretations of the Plant Pest Act, including promulgating provisions to regulate the movement of articles intrastate as well as interstate, and the use of on-site inspections to verify the conditions stated in the application for a permit. Some individuals said that the statutory authorities that APHIS had to claim for these activities exceeded those assigned in the Plant Pest Act, including the regulation of genetically engineered plants as a "plant pest."

The National Environmental Policy Act (NEPA) requires a federal agency, when making a decision that may have a significant impact on the environment, to conduct an Environmental Assessment (EA) of the various options considered, and to post the reasons for the final determinations. Although some said it was unnecessary, APHIS decided to comply with the NEPA by requiring each permit application, and the agency's own determination of plant pest status, to undergo its own EA, on a permit-by-permit basis. This decision was no doubt driven by the experience of other federal agencies (e.g., the Department of Health and Human Services, DHHS) challenged in federal court by the Foundation on Economic trends on procedural missteps.

Implementing the provisions in the NEPA caused APHIS to hire or reassign considerable staffing to prepare the lengthy documents assessing the environmental consequences of a proposed experiment with a genetically engineered plant. This in turn required that the applications for an APHIS permit had to provide information sufficient for the agency to make its determination and complete an EA. From this emerged the NBIAP idea of designing an expert system that would build on the experiences of past permit applications and the associated EAs. The expert system was to ensure the design of safe experiments with genetically engineered organisms and to facilitate the drafting of similar permit applications. Secondarily, the expert system was to facilitate, to some extent, compliance with ABRAC guidelines. NBIAP software was developed and distributed in 1988, and updated versions were periodically made available free of charge to scientists upon request. Sample paragraphs were offered by the expert system for adoption, revisions, or technical correction, based on previously successful applications. Applicants made responses to a series of organism-relevant questions relating to the permit decision-making process. Key biosafety questions were developed to identify high-risk situations. Responses to some of the critical expert questions resulted in advice to the applicant that under no circumstances would APHIS be likely to issue them a permit. The NBIAP expert system software was well received by some, but it was not extensively used by the scientific community, for reasons that were never well understood.

Following the preparation of EA by APHIS, one of two determinations was made, each requiring more documentation by APHIS. If there was a Finding of No Significant Impact (FONSI), the document was filed with the EA, and the availability of the EA and FONSI documents was then announced in the Federal Register. When more than 300 Environmental Assessment/FONSIs had accumulated, NBIAP assembled the documentation into a searchable database and issued it on a CD-ROM, as a service to the research community and biotechnology interest groups.

If the EA made a finding of a significant impact, the agency would have been required, under NEPA, to prepare an Environmental Impact Statement (EIS). An EIS is often an enormously large document and always requires considerable technical detail, large amounts of information, and sophisticated analysis. Some critics note that to their surprise, APHIS never made a finding of significant impact for a proposed field test with a recombinant plant, and thus was never required to write an EIS. In response to this criticism, APHIS countered that if a permit application was submitted that would have led to a conclusion of a significant environmental impact, the applicant was so notified and was provided an opportunity either to redesign the experiment or to withdraw the application. Inasmuch as the application process was not a public record, little information exists beyond FONSIs.

In the 10 years following the initiation of the APHIS permit process, more than 3000 field trials were conducted in the United States, and more than 30 products were commercialized. This high level of success is related to a number of legal, policy, and political issues that needed to be resolved to make the APHIS permit process work.

LEGAL, POLICY, AND POLITICAL ISSUES

APHIS had a number of issues that needed to be resolved for the successful implementation of the agency's field-testing permit system. In the end APHIS was successful

in resolving most of these issues, and this required them to expend a lot of time and attention on their resolution.

Conflicting Legal Authorities

In 1990 Congress proposed to hold hearings on the Omnibus Biotechnology Bill (OBB) that would have standardized the regulation of the process of biotechnology research and the commercialization of the resulting products through one federal agency (30,31). Although not specifically named in the bill, it was presumed at the time that the U.S. Environmental Protection Agency would be assigned the authority to administer the OBB.

The OBB had both opponents and proponents who were equally outspoken. A congressional hearing placed those issues on the table and the divisions were clearly evident. Proponents liked the idea of "one-stop shopping" through a centralized regulatory process that would end "shopping around" and would promote the consistent application of biotechnology regulation with an environmental perspective.

Opponents of the proposed bill (primarily the biotechnology industry) were very content with their regulatory experience, particularly with the APHIS office that issued the field-testing permits, because they understood the procedure. Under the careful guidance of attorney Terry Medley, the unit gave careful and courteous attention to their "customers" (in the sense of Ed Deming's Total Quality Management). Although criticized as being too helpful, the unit became known as an office that returned its phone calls and answered correspondence in a timely manner. As a consequence permit applicants became supporters and strong defenders of the existing APHIS regulatory process, and the services provided. (An interesting historical comparison could be made with the EPA regulatory office. EPA was viewed at the time, as an adversary of the technology and less than helpful to their applicants.)

Thus, when Congress presented the OBB for public comment, strong industry support for the existing regulatory procedures defended the status quo, and the Federal Coordinated Framework continued as originally devised.

Policy Conflicts

Regulatory policies that were in conflict with other agencies were addressed by APHIS through ongoing conversations, staff exchange, and high-level consultations. This open dialogue with other agencies avoided conflicts (and the consequent interference) with APHIS's regulatory decisions. When questions arose regarding lead-agency responsibilities, APHIS was quick to move to a satisfactory resolution. For example, when EPA had overlapping authority with APHIS on plant pathogenic microorganisms with pesticidal properties, they conferenced to work out the differences. As a result of these patterns, it became relatively easy for APHIS staff to understand their authorities vis-à-vis other regulatory agencies, and to act decisively.

Political Issues

APHIS was particularly effective in providing information through congressional hearings during the uncertain early years of regulating agricultural biotechnology. Even during highly charged hearings, APHIS presented carefully thought out arguments for why it was necessary to apply the Plant Pest Act for the regulation of plant biotechnology products.

It is important to note that the political milieu of the federal government during this period was a Republican White House and a Democratically controlled House of Representatives. Traditionally it is said that Republicans favor industry and business, while Democrats favor environmental stewardship, equity, and other issues that may be at odds with commerce. This placed APHIS in an awkward situation, seemingly in need of representing, through the USDA, a pro-industry approach to biosafety assurance. It provided opportunity for special interest groups to prod members of Congress to hold hearings, some of which were designed to embarrass the administration. Thus a spotted congressional record exists on APHIS's accomplishment, probably reflecting the political situation of the time more than the merits of the accomplishments of APHIS.

Special Interest Groups

The Environmental Defense Fund, the National Wildlife Federation, and the Foundation on Economic Trends provided a continuing challenge to the APHIS regulatory process (32,33). This effort met with mixed results (34).

Retrospectively, these special interest groups probably provided an important public service as watchdogs to the biotechnology regulatory process. They no doubt deserve credit for serving as a public conscience for agricultural biotechnology. And they served well as an information funnel for APHIS. Many of their arguments were founded on solid science and careful investigation. However, more often their arguments showed an absence of any scientific knowledge, one way or the other. This left many observers to question the validity of regulatory decisions being made by APHIS. But this was an important contribution as well, as it helped identify areas needing more research to uncover the scientific facts required for better regulatory decision making. And it was a constant reminder that not all things that matter can be quantified.

Section 1664 of the 1990 Farm Bill established a Biotechnology Risk Assessment Research program, as a 1 percent set aside of USDA's outlays for biotechnology research (yielding about $1.7 million annually). It was understood at the time that this provision in the Farm Bill was placed into law at the insistence of environmental interest groups. The funding allowed establishment of a targeted competitive grants program to begin answering some of the risk assessment questions brought forward by both critics and the regulators of agricultural biotechnology. This competitive grants program, although opposed by some influential leaders in the scientific community (who saw it as an unnecessary admission of biotechnology's risks), was able to resolve some of the questions raised by outspoken critics of agricultural biotechnology. The program continues today, but with less than enthusiastic support from either the USDA or the agricultural research community.

Trade Groups

The Biotechnology Industry Organization (BIO) and the National Association of State Universities and Land-Grant Colleges' Committee on Biotechnology were the primary trade groups interacting with APHIS during the early years of agricultural biotechnology regulation (35,36). The trade groups tended to operate as a counterbalance to the special interest groups by seeking to voice the concerns of the public and private sectors over unneeded and unnecessary regulatory burdens. APHIS was responsive to this perspective by periodically reviewing its permit application procedures, by disseminating information on permits issued and eventually, by the mid-1990s, converting to a notification procedure for six crops (and later, additional ones) that eliminated the need to apply for a permit under fairly broad circumstances. The APHIS notification procedure greatly facilitated field experimentation in prescribed areas, which was justified by the Agency's accumulated experience and knowledge derived by that time from issuing hundreds of permits for field tests, which were all carried out safely.

An additional influential trade group was the National Agricultural Biotechnology Council (NABC) that performed a different service for APHIS (37). NABC was a coalition of universities that annually sponsored a forum for dialogue primarily focused on the biosafety issues of agricultural biotechnology. The proceedings of their annual meetings clearly depict a pattern of an evolving consensus on how to approach the biosafety questions of agricultural biotechnology research.

Science Community

The scientific community became divided on the issues of biosafety, mostly along scientific discipline lines. Plant pathologists gave arguments that the Federal Plant Pest Act was, scientifically, not a legitimate legal authority for regulating the products of agricultural biotechnology (38). They used contemporary scientific information to question the supposition that a disarmed TI vector or a promoter sequence from a virus could in any way lead to a plant becoming a plant pest. To many plant pathologists this supposition was an absurdity (39,40).

To the molecular biologists, the APHIS regulatory approach seemed overly heavy-handed, unnecessary and probably an impediment to the agricultural applications of biotechnology (41,42). Molecular biologists appeared at the time to be giving little attention to the environmental consequences of field releases with genetically modified organisms, even though problems might occur in very low frequency.

Ecologists had a very different perspective. They foresaw that severe consequences could result from "releases" of recombinant DNA, based on other "environmental disasters," such as the gypsy moth, Dutch elm disease, and the kudzu vine (43,44). Several serious and scholarly treatments of the issues of biosafety were published during this period. The failure to resolve the differences among the plant pathologists, molecular biologists, and ecologists stemmed from the absence of a factual basis for decision making.

It must be noted that APHIS never asserted that its regulatory authority was based on risk. It was merely a determination of whether or not the organism was a plant pest, and therefore a regulated article. At the time many individuals in the scientific community seemed to have been arguing for a risk-based determination of the safety of recombinant organisms being tested in the environment. This would have been a major challenge for APHIS, since much of the necessary information was missing (and still is) for a thorough risk assessment of proposed field tests with recombinant DNA organisms. Moreover the fundamental risk analysis paradigm, as presented in the 1983 National Research Council's "Red Book," proposed a conceptual separation between risk assessment and risk management (45). Risk assessment is a science-based, stepwise process, for which the first step is hazard identification. Part of the biosafety controversy was over arguments that biotechnology in and of itself represents a hazard, and therefore the whole process needed to be regulated. Others argued that biotechnology was not in and of itself a hazard, and thus not a risk. Out of this paradigm difference came the product versus the process debate. The Federal Coordinated Framework sided with the second view (i.e., regulating the product). But the argument continues today as to whether biotechnology itself represents a hazard, and thus should be subject to risk assessment (46–48). APHIS redirected this argument by asking whether or not the organism was a plant pest, and thus a regulated article.

Another complication regarding the application of the 1984 Red Book approach to risk analysis was the identification of the appropriate authority for risk management. According to the Red Book, a firewall needs to be created between risk assessment and risk management. The first activity is a scientific process used to derive a science-based recommendation, while the second process is policy implementation that takes into account scientific fact and public consideration. The scientific community in general has long been resistant to the notion that anything other than scientific fact should be used to decide the safety of an organism (49). However, in a democracy, sometimes the best scientific evidence is not sufficient to gain social or market acceptance.

The APHIS strategy to focus on a determination of pest status completely avoided the risk assessment/risk management question, but much of this finesse was lost on the scientific community. The science-oriented critics of APHIS were focused on an all together different set of questions.

Industry and Academe

The transactional cost of complying with the federal regulatory permit application process segmented industry from academe. Many of the larger companies that were investing heavily in agriculture biotechnology (e.g., Monsanto, Calgene, CIBA-Geigy) could afford staff assistants and legal support to help in the preparation of lengthy permit applications for field tests. Universities, on the other hand, argued that they could not afford to hire such staff. Thus the regulatory burden fell upon the shoulders of the individual faculty member wishing to conduct the field research. During the early years of APHIS permit issuance, the ratio of private to public sector was approximately 9 to 1. This caused considerable alarm, as it appeared that the regulatory apparatus was interfering with the normal flow of research activity from the laboratory to the field, at least at the public institutions and federal research laboratories.

It was this point of concern that caused the NBIAP to conduct a national survey in 1990 on the impact of regulating agricultural and environmental biotechnology research in both the public and private sectors (50). The survey method was face-to-face interviews with open-ended questions. The questions were designed to determine the degree of interference with the research process being caused by regulatory requirements, with a particular focus on APHIS.

The results of the NBIAP survey were surprising, from two perspectives. First it was discovered that there was not a big backlog of research results awaiting field testing for lack of permits. Interviews with university scientists mostly noted their apprehensions, but little specific biological material could be cited as ready but not yet tested. Moreover a significant portion of the private sector's permits had been issued by APHIS to private sector and university scientist working in partnerships to jointly conduct field tests with recombinant plants. But, inasmuch as the permit was issued to the private company, the participation of the public institution was lost in the calculated ratio of private to public sector permits.

Second, when asked about the transactional costs for university scientists, an APHIS officer described a permit request they received from a university scientist that had been handwritten on yellow paper, with an attached hand-drawn diagram of the genetic construct and a crude plot map of the proposed test site. To APHIS's credit, they began to work with the scientist to develop an adequate permit application that had the required information. APHIS eventually issued a field test permit to the institution. This test became the university's first sanctioned field test of a recombinant plant. In defense of APHIS, and contrary to alleged burdens of the permit application process, it is to their credit that they did not summarily reject the application but instead worked with the scientist to assemble a proper application for a permit.

The results of the NPIAP survey clearly demonstrated high levels of satisfaction with APHIS's permit application process, services, and the professionalism of the agency's staff. As noted earlier, this was very likely the consequence of their customer-focused attention, in the sense of Ed Deming's Total Quality Management, a then-contemporary model for the federal government. Ironically this also led to criticism that APHIs was too friendly with its permit applicants. One then needs to ask, what is the proper balance between government agency responsiveness and regulatory adversity?

Trading Partners

U.S. agriculture has maintained for decades a positive trade balance that is derived from its reliability as a source of many types of commodities. Among these commodities are the basic grain and oil crops that are traded internationally as a very profitable business.

Early in the process of identifying the benefits of agricultural biotechnology, it became apparent that the acceptance of recombinant commodities by U.S. trading partners would become a key issue, if U.S. agricultural biotechnology-derived commodities were to be accepted in global commerce. Recombinant corn, soybean, and cotton are just some of the commodities that are now a point of contention in international trade markets. This pattern will undoubtedly continue to grow unless some significant change comes about. What is not known is the market receptivity and consumer acceptance of those recombinantly derived commodities. One could conclude that the eventual consumer receptivity of the products of agricultural biotechnology will depend more on the regulation of the process (as Europe has done), rather than regulation of the products (as the United States has done).

Early on, APHIS saw the need to begin discussions with our economic trading partners. This was done through the Paris-based Organization of Economic Cooperation and Development (OECD). Through this forum APHIS began the process of finding common ground and agreement on biotechnology regulations impacting agricultural products. The OECD was in a good position to provide a forum for the discussion of how to harmonize biosafety assurances. Working through the U.S. Department of State, and by coordinating USDA activities, APHIS gave early leadership to the development of documents that set out concepts and expectations for regulatory requirements. This turned out to be difficult in Europe, which is noted for its resistance to the application of biotechnology to food products. After 15 years of discussion and consensus building, there are no formal agreements on the acceptance of the products of U.S. agriculture biotechnology. Some farm organizations are now calling for farmers not to plant seed of biotechnology-derived cultivars. They say the prospects for the loss of international markets are too great of a risk.

State and Local Governments

APHIS actively sought the cooperation and compliance of state and local governments that were expressing an interest in creating competing regulatory mechanisms under their own jurisdiction. Concerns that a "patchwork quilt" of regulatory requirements could emerge no doubt motivated APHIS to give extra attention to these governmental units. This was done through listening sessions, workshops, and shared documentation. APHIS contained earlier proposals to create subfederal-level regulatory requirements for biotechnology. Had this conflict not been resolved, the pace of U.S. biotechnology commercialization would undoubtedly have been slowed.

THE ROAD MAP

APHIS co-sponsored with several other federal regulatory agencies a forum for discussions on the public and private sector's development of a road map to bring the products of biotechnology forward to commerce. This discussion proved enlightening, as it soon became clear that the conflicting requirements of the various agencies were going to slow the process of product commercialization. Out of these discussions came interagency agreements for better coordination of the product commercialization process.

A particularly difficult issue for APHIS was the process of no longer regulating the products of agricultural biotechnology as they entered commercialization. This was approached by APHIS in several ways, including the eventual certification that a trait was not a plant pest (51). This certification allowed an owner to take a product to commerce, certified by APHIS as an unregulated article.

APHIS also made the determination that it was the genetic trait, not the specific cultivar, that would be certified as not a plant pest. This became important to plant-breeding programs. If a transformation trait could be subsequently moved by conventional plant-breeding methods, once certified by APHIS as not a plant pest, it would not require another round of permit applications and certification. That is, if a recombinant potato plant carrying the Bt endotoxin were to be subsequently crossed by conventional plant-breeding methods to another potato cultivator, the progeny would not be an APHIS-regulated article. This opened significant opportunity to exploit plant genetic engineering, relatively free from federal regulatory oversight, using conventional plant-breeding methods.

A RETROSPECTIVE VIEW

During the first 15 years of APHIS regulating the products of biotechnology, the following significant outcomes have occurred:

- Plant genetic engineering in both the public and private sector is now mostly conducted under a notification process that represents minimal burden to the scientific community.
- More than 3000 field tests have been conducted without a significant incident.
- Thirty commercial products of agricultural biotechnology research are now a market reality. These include genetically engineered corn, soybeans, potatoes, tomatoes, squash, cotton, tobacco, and papaya, which carry in various combinations resistance to plant viruses, tolerance to stress, herbicide tolerance, improved product quality, insect resistance, and male sterility (for making hybrid seed).

Meanwhile ABRAC was decommissioned and the Office of Agricultural Biotechnology no longer exists (52,53). The National Biological Impact Assessment Program's functions have been distributed within the USDA, although a modest special research grant to Virginia Tech continues to provide biotechnology information services to the public and private sector scientific communities. The NASULGC Committee on Biotechnology was disbanded in 1997, after a 15-year history of institutional services and information exchange (54).

Two distinct areas remain in need of resolution. First, the Federal Coordinated Framework's approach to regulating the products (not the process) of biotechnology has left a regulatory incongruency that will not be easily resolved with our European trading partners. The European Community (EC) has shown a preference for regulating the process of biotechnology apart from other regulatory requirement for the products of commerce. This distinction rose to white-hot intensity in the early 1990s when our European trading partners wanted to establish a "fourth criterion" for registering the products of biotechnology. Heretofore the three standard criteria for registering a product of commerce were safety, efficacy, and quality. Europeans at the time were proposing the fourth criterion that would answer the question, "Do we need it?" In the United States there is no such regulatory authority, since the marketplace is expected to determine whether or not the product has market value or meets a social need. The EC, on the other hand, through its fourth criterion, was proposing that a social-need standard should be imposed on the products of biotechnology. This proposal still seems to be floating in the gap between the U.S. approach to regulating biotechnology and that of our European trading partners.

Second, U.S. policy has established that the products of biotechnology will not be required to be labeled unless the product has a distinctly identified allergen, such as a peanut protein. Our European trading partners, on the other hand, are approaching a consensus that all the products of biotechnology must be labeled. This represents a major challenge for U.S. agriculture production inasmuch as many of our grains and oil seeds are commonly blended from multiple sources prior to shipping to Europe. Maintaining product identity seems, at least to the export/shipping industry, not to be feasible. Labeling everything as "may contain recombinant DNA" is simply uninformative.

LESSONS LEARNED

In the years that have passed since the development of the Federal Coordinated Framework a few biotechnology regulation-lessons have been learned. We now know that:

- Assigning regulatory functions based on existing authorities (as was done by the Federal Coordinated Framework) probably hastened the commercialization of the products of biotechnology. This is the conclusion drawn in a study for the OECD that compared the progress of agricultural biotechnology in those countries with a specifically implemented biotechnology regulatory authority. In these countries (e.g., Germany, the Netherlands, and Belgium) commercialization of the products of agricultural biotechnology seems to be going slower (55).

- Divergent views on the safety of biotechnology had to be expressed through public dialogue, and NABC met this need very well.

- Communication of perspectives was essential for establishing regulatory positions and aligning public policies that allowed safe field testing with recombinant organisms.

- The experience with ABRAC/OAB indicates that the success of the NIH–RAC was not transferable for biosafety assessments of field testing protocols for recombinant organisms.

- It would seem that the designation of lead regulatory agencies is necessary to avoid multiple reviews, which can become inefficient and unworkable.

- Research projects focused on biosafety questions can fill information gaps and thus help to resolve otherwise contentious public health and environmental protection issues.

What remains to be determined is the level of consumer acceptance of the products of biotechnology (56). Is there a linkage between the type of regulatory approval that is used to provide biosafety assurance and the level of consumer product acceptance? This question needs to be resolved.

BIBLIOGRAPHY

1. U.S. Department of Agriculture, Animal and Plant Health Inspection Service: *www.aphis.usda.gov/biotech/OECD/ ustegs.htm#usdalaw*

2. S.N. Cohen et al., *Proc. Nat. Acad. Sci.* **70**(11), 3240–3244 (1973).

3. E.F. Smith and C.O. Townsend, *Science* **25**, 671–673 (1907).

4. M.D. Chilton et al., *Cell* **11**, 263–271 (1977).

5. P. Berg et al., *Science* **185**, 303 (1974).

6. M.N. Oxman, in A. Hellman, M.N. Oxman, and R. Pollack, eds., *Proc. Biohazards in Biological Research*, (Asilomar I), Cold Spring Harbor Laboratory, Pacific Grove, CA, 1973.

7. P. Berg et al., *Science* **188**, 991 (1975).

8. National Institutes of Health, *Fed. Reg.*, No. 48, p. 16459, April 15 (1983).

9. National Institutes of Health, 39 Fed. Reg., 39306 (November 6, 1974).

10. National Institutes of Health, 41 Fed. Reg., 27902–27943 (July 7, 1976).

11. National Institutes of Health, 41 Fed. Reg., 38426–38483 (September 9, 1976).

12. D.S. Fredrickson, *Recomb. DNA Techn. Bull.* **2**(2), 87–90 (1979).

13. *Genetic Engineering of Plants: Agricultural Research Opportunities and Policy Concerns*, National Academy Press, Washington, DC, 1984.

14. S.A. Tolin, *Recomb. DNA Techn. Bull.* **4**(4), 156–159 (1981).

15. E. Milewski and S.A. Tolin, *Recomb. DNA Techn. Bull.* **7**(3), 114–124 (1984).

16. E. Milewski, *Recomb. DNA Techn. Bull.* **8**(3), 102–108 (1985).

17. E. Milewski, in J.R. Fowle, III, ed., *Application of Biotechnology: Environmental and Policy Issues*, Westview Press Boulder, CO, 1987, pp. 55–90.

18. U.S. Congress, House of Representatives, Committee on Science and Technology, Subcommittee on Investigations and Oversight, June 22, Hearing No. 98/36, *The Environmental Implications of Genetic Engineering*, U.S. Government Printing Office, Washington, DC, 1983.

19. Office of Science and Technology Policy and Executive Office of the President, 49 Fed. Reg., 50856–50907 (December 31, 1984).

20. Office of Science and Technology Policy and Executive Office of the President, 50 Fed. Reg., 47174–47195 (November 14, 1985).

21. H.I. Miller, in D. Brauer, ed., *Biotechnology*, VCH, Weinheim, 1995.

22. Office of Science and Technology Policy and Executive Office of the President, 51 Fed. Reg., 23302–23350 (June 26, 1986).

23. 57 Fed. Reg., 6753–6762 (1992).

24. U.S. Department of Agriculture, 50 Fed. Reg., 29367–29368 (July 19, 1985).

25. U.S. Department of Agriculture Animal and Plant Health Inspection Service, 52 Fed. Reg., 22892–22915 (1987).

26. *Guidelines recommended to USDA by the ABRAC*, USDA Doc. No. 91-04, March (1992).

27. U.S. General Accounting Office, Report to the House Committee on Science and Technology, GAO/RCED-86-59, GAO, Washington, DC, March, 1986.

28. *Minutes of the Agricultural Biotechnology Research Advisory Committee*, USDA Doc. No. 91-01, pp. 25–30 (1991).

29. J.F. Fulkerson, *BioSci.* **37**(3), 187–189 (1987).

30. U.S. Congress, House of Representatives, Committee on Agriculture and Subcommittee on Department Operations Research and Foreign Agriculture, *Review of Current and Proposed Agricultural Biotechnology Regulatory Authority and the Omnibus Biotechnology Act of 1990*, Hearing 101/75, U.S. Government Printing Office, Washington, DC, October 2, 1990.

31. G. Simon, *Science* **252**, 629–630 (1991).

32. J. Doyle, *Altered Harvest: Agriculture, Genetics, and the Fate of the World's Food Supply*, Viking Penguin, New York, 1985.

33. *Crossfire*, with Tom Braden and Bob Novak, John Fulkerson, USDA, versus Jeremy Rifkin, FET. CNN-TV, April 20, Washington, DC, 1987.

34. I. Rabino, *Sci. Technol. Hum. Values* **16**, 70–87 (1991).

35. Committee on Biotechnology, Division of Agriculture and NASULGC, National Association of State Universities and Land-Grant Colleges, Progress Rep. 5, November (1986).

36. Committee on Biotechnology, Division of Agriculture and NASULGC, National Association of State Universities and Land-Grant Colleges, Progress Rep. 6, November (1987).

37. J. Fessenden Mac Donald, ed., *Agricultural Biotechnology: A Conversation about Risk*, Rep. 5, National Agricultural Biotechnology Council, Ithaca, NY, 1993.

38. *Field Testing Genetically Modified Organisms: Framework for Decisions*, National Academy Press, Washington, DC, 1989.

39. *Introduction of Recombinant DNA-Engineered Organisms into the Environment: Key Issues*, National Academy Press, Washington, DC, 1987.

40. S.A. Tolin and A.K. Vidaver, *Annu. Rev. Phytopathol.* **27**, 551–581 (1989).

41. D. Kennedy, *MBL Sci.* (Winter), 5–9 (1988, 1989).

42. C. Arntzen, *Science* **257**, 1327 (1992).

43. J.M. Tiedje et al., *Ecol.* **70**(2), 297–315 (1989).

44. U.S. Congress, Office of Technology Assessment, *New Developments in Biotechnology — Field-Testing Engineered Organisms: Genetic and Ecological Issues*, OTA-BA-350, U.S. Government Printing Office, Washington, DC, 1998.

45. National Research Council, *Risk Assessment in the Federal Government: Managing the Process*, National Academy Press, Washington, DC, 1983.

46. H.I. Miller et al., *Science* **250**, 490–491 (1990).

47. H.I. Miller, *Science* **252**, 1599–1600 (1991).

48. H.I. Miller, *Policy Controversy in Biotechnology: An Insider's View*, R.G. Landes Co., Austin, TX, 1997.

49. V.T. Covello and J.R. Fiksel, Rep. NSF/PRA 8502286, National Science Foundation, Washington, DC, 1985.

50. M. Ratner, *Bio/Technology* **8**, 196–198 (1990).

51. U.S. Department of Agriculture Animal and Plant Health Inspection Service, 58 Fed. Reg., 17044–17059 (1993).

52. Office of Agricultural Biotechnology and USDA, Mission Accomplished: Final Report of the Office of Agricultural Biotechnology, Washington, DC, 1996.

53. J. Kaiser, *Science* **270**, 1911 (1995).

54. Committee on Biotechnology, Division of Agriculture and NASULGC, National Association of State Universities and Land-Grant Colleges. Progress Rep. 15 (November 1996). Available at: *www.nbiap.vt.edu*

55. Lisa Zannoni, Personal Communication, 1998.

56. S. Kilman and H. Cooper, *Wall Street J.* May 11, A1, A10 (1999).

See other entries AGRICULTURAL BIOTECHNOLOGY; ANIMAL BIOTECHNOLOGY, LAW, FDA REGULATION OF GENETICALLY MODIFIED ANIMALS FOR HUMAN FOOD USE.

AGRICULTURAL BIOTECHNOLOGY, POLICY, AND NUTRITION

MARION NESTLE
New York University
New York, New York

OUTLINE

INTRODUCTION

Despite the widely recognized potential of agricultural biotechnology to improve the quantity and nutritional quality of the world's food supply, the first commercial products elicited extraordinary levels of controversy over issues of human and environmental safety as well as social values. This article examines the reasons why so potentially useful an application of molecular techniques to improving the nutritional status of populations has proved so controversial in the United States, Europe, and elsewhere. Focusing on the situation in the United States, it reviews key policy issues related to nutrition — economics, marketing, risk, and regulation — that have affected acceptance of the first genetically engineered food products. It reviews evidence indicating that early public acceptance of genetically modified foods was product-specific; people were willing to accept products believed beneficial, safe, and consistent with personal values. Because the failure to label genetically modified foods undermined public trust in industry as well as in government, the chapter addresses implications of the present controversy for future product development, industry actions, and public policies.

FOOD BIOTECHNOLOGY: PROMISE OR REALITY

Food biotechnology — the use of recombinant deoxyribonucleic acid (rDNA) and cell fusion techniques ("genetic engineering") to confer selected characteristics upon food plants, animals, and microorganisms — is well understood as a means to increase agricultural productivity, especially in the developing world. The great *theoretical* promise of biotechnology is that it will help solve world food problems by creating a more abundant and more nutritious food supply (1). Despite this widely recognized and undisputed promise, food biotechnology has elicited extraordinary levels of controversy. In the United States and especially in Europe, the first commercial food products of genetic engineering have been greeted with suspicion by the public, vilified by the press, confonted with boycotts and legislative prohibitions, and threatened with trade barriers. Such reactions reflect widespread concerns about the safety of the products, as well as about their economic impact, environmental effects, ethical implications, and social value (2). The reactions also reflect public fears of the unknown dangers of genetic engineering, along with deep distrust of the biotechnology industry and its governmental regulators. Biotechnology industry leaders and their supporters, however, have tended to dismiss such concerns as antiscientific and irrational, to consider "biotechnophobia" as the single most serious threat to research and commercialization, and to identify as their most important challenge the need to convince the public that the products are safe as well as beneficial (3).

The divergent viewpoints derive directly from the conflict between the two fundamental goals of the food biotechnology industry: to benefit humanity by developing agricultural products that will improve the nutritional status of populations, and to benefit the industry itself through successful marketing of products. Although the new products might well be expected to meet both goals, such is not always the case. The lack of a viable market constitutes a major barrier to research on food problems of the developing world, and the industry's need for rapid returns on investment drives virtually all decisions related to research and development. Indeed, financial imperatives have caused industry leaders to view legitimate public questions about the use, safety, or social consequences of particular food products as threats to the entire biotechnology enterprise and to make defensive marketing decisions that have only undermined public trust.

Theoretical Potential

There seems little doubt that biotechnology holds great promise for addressing world food and nutrition problems, most notably the overall shortfall in food production now expected early in the twenty-first century. No theoretical barriers impede the use of the techniques of molecular and cellular genetics to improve the quantity and quality of the food supply, increase food and environmental safety, and reduce food costs. Table 1 lists the wide range of potentially beneficial applications of food biotechnology now under investigation or theoretically possible. Such applications

Table 1. Theoretical and Current Applications of Food Biotechnology

- Improve the nutrient content, flavor, texture, or freshness of fruit and vegetables
- Increase levels of vitamins, protein, and other nutrients in plant food crops
- Modify seed storage proteins to increase concentrations of limiting amino acids
- Reduce saturated fatty acids in plant seed oils
- Increase plant production of specialty chemicals such as sugars, waxes, phytooxidants, or pharmaceutically active chemicals
- Enable fruits and vegetables to remain fresh during processing, transport, and storage
- Decrease levels of caffeine or other undesirable substances in plant food crops
- Increase resistance of crops to damage by insect, fungal, or microbial pests
- Increase resistance of crops to "stress" by frost, heat, salt, or heavy metals
- Develop herbicide-resistant plants to improve weed control
- Enable crop plants to be grown under conditions of low input of fertilizers, pesticides, or water
- Enable major crop plants to fix atmospheric nitrogen
- Develop plant foods that contain "vaccine" antigens
- Increase the efficiency of growth and reproduction of food-producing animals
- Create disease-resistant animals
- Develop animal veterinary vaccines and diagnostic tests
- Enable cows to produce milk containing human milk proteins
- Alter mosquitoes so they prefer animal blood to human, or convey vaccines
- Create microorganisms, enzymes, and other biological products useful in food processing
- Develop microorganisms capable of converting environmental waste products — plastics, oil, pesticides, or PCBs — into usable animal feeds

could well increase world food production, especially given the conditions of climate, soil, and environmental degradation characteristic of many developing countries. The potential benefits constitute the principal basis for industry arguments that biotechnology is not only the most important scientific tool to affect food production in the history of the world, but the *only* solution to the expanding global needs for food (4). These promises, however, have not yet been fulfilled, nor are they likely to be achieved in the immediate future, not least because many of the listed applications pose technical problems of formidable complexity. The slow progress of biotechnology in addressing world food problems should not imply that such problems cannot be solved. Given sufficient time, commitment, and funding support, technical barriers can be overcome. Whether doing so will help feed the world, however, is a matter of considerable debate.

Economic Realities

Investment Demands. Rather than technical problems, the most important barriers to addressing world food problems derive from the industry's need to recover research costs and maximize returns on investment. The potential returns are enormous; worldwide sales of genetically-modified crops could reach $300 billion by 2010 (5). To date, however, stock market returns have not reflected such projections. Although the food biotechnology industry increased in sales, revenues, and numbers of companies and employees in the 1990s, net losses also increased. The Monsanto company has been a notable exception; its stock prices rose rapidly in the mid-1990s and its agricultural products continue to be highly profitable (4). The generally poor performance of other food biotechnology stocks has been attributed to uneven management, corporate shortsightedness, and product failures. More recently international resistance to genetically modified foods has affected sales and confidence in the industry. Low levels of government investment also have impeded industry growth, as most federal biotechnology funding has supported drug rather than agricultural research. Only recently has the U.S. government begun to support agricultural biotechnology research to any significant extent.

The immediate need for returns on investment requires the industry to focus development efforts on products that are technically feasible and economically productive rather than on those that might be more useful to the public or to developing countries (6). Thus product development efforts concentrate on traits that most benefit agricultural producers and processors: control of weeds, plant diseases, ripening, insects, or herbicide resistance (7). For example, the Monsanto company's research budget (which is more than twice that of all of the public sector tropical research institutes combined) is applied almost entirely to temperate-zone agricultural problems (8). The company's principal agricultural products are soybeans and other plants genetically modified to resist the company's flagship herbicide, "Roundup," and "Bt corn" containing a insect-inhibiting toxin naturally produced by the soil bacterium *Bacillus thuringiensis.* Monsanto began selling "Roundup Ready" soybeans in 1996; by 1998 they had been planted

on one-third of U.S. soybean farmland and covered 25 million acres (10.1 million hectares). By 1999 more than 35 percent of U.S. corn and 45 percent of soybean acres were grown from genetically modified seeds, and total worldwide acres devoted to such crops were expected to triple within the next five years (9). Monsanto's research "pipeline" mainly emphasizes "Roundup Ready" crops designed for animal feed, although a high-carotenoid canola oil designed to prevent vitamin A deficiency is a rare nutritionally focused exception (4). Projects to improve the nutritional content of basic food sources are expensive to produce and unlikely to be affordable by the populations they would most benefit. Although Third World agricultural problems and their biotechnological solutions are well defined, and many sources of private and public funding are available to support such projects, the sources are fragmented and poorly coordinated and often favor the priorities of the donors rather than recipients (10). Despite recent advances in cassava biotechnology, for example, nearly all international budgets for such research have been reduced (11). Ultimately biotechnology may well improve the world's food supply but to date it has not done so to any appreciable extent.

Marketing Barriers. To ensure adequate returns on investment, the biotechnology industry must create and sell new products. Because the United States vastly overproduces food, new products must compete in a market that is already highly competitive. In 1997, for example, the U.S. food supply provided an average of 3,800 kcal (18.2 MJ) per day for every man, woman, and child in the country, an increase of 500 kcal (2.0 MJ) per day since 1970 (12); most adults require one-half to two-thirds that amount and children much less. Because the amount of energy that any one person can consume is finite, such overproduction implies that a choice of any one food product will preclude the choice of another, making the food-marketing system extremely competitive. Food marketers compete for consumer purchases through two principal means: advertising and new product development. Retail sales of food and beverages generate about $800 billion annually in the United States, and food marketers spend $11 billion on direct consumer advertising, and about twice that amount on retail promotion. They introduce about 15,000 food products into the marketplace every year (13). Nevertheless, the food-processing sector rarely grows by more than 1 percent a year, a rate considered stagnant by comparison to that of other industries. In so competitive an environment, biotechnology is viewed as a critically important process for developing new products that will increase economic returns.

SAFETY ISSUES

From their inception, gene cloning experiments elicited safety concerns, mainly focused on the potential hazards of releasing new organisms with unknown properties into the environment. At a conference in 1975, scientists suggested stringent guidelines for research studies employing

Table 2. Safety Issues Raised by Food Biotechnology

- Adverse changes in the composition, absorption, or metabolism of key nutrients
- Unanticipated health effects resulting from genetic changes
- Increases in levels of naturally occurring toxins or allergens
- Activation of dormant toxins or allergens
- Introduction of known or new toxins, allergens, or antinutrients
- Induction of resistance to useful antibiotics through use of antibiotic marker genes
- Adverse environmental effects on wildlife and ecosystems
- Adverse changes in the nutrient content of animal feed
- Increased levels of toxins in plant byproducts fed to animals

rDNA techniques. The following year, the National Institutes of Health required researchers to follow similar guidelines. In subsequent years, as understanding of the techniques improved, concerns about safety diminished and the guidelines were modified accordingly. From the standpoint of the biotechnology industry and its supporters, genetically engineered foods are no different from foods produced by conventional genetic crosses; If they induce any risks at all, these are small and greatly outweighed by benefits. Nevertheless, the common genetic techniques for modifying foods, especially those involving bacteria that cause plant diseases (e.g., crown gall), antibiotics as part of the selection process, and genes from one living species inserted into another, continue to elicit debate. Table 2 summarizes the principal safety issues raised by the use of food biotechnology. Although most such concerns remain theoretical, some that could affect human nutritional status and health have a limited basis in observation or experiment, as discussed below.

Unintended Consequences: Tryptophan Supplements

Critics of food biotechnology insist that without prior experience, the techniques raise safety concerns that are difficult to define, predict, or quantify. As an example, they point to the demonstrable hazards of genetically engineered nutritional supplements of the amino acid tryptophan. Tryptophan is a normal constituent of all body proteins that is sometimes sold as self-medication for insomnia and other conditions. In 1989 health officials linked tryptophan supplements from a single manufacturer to *eosinophilia-myalgia syndrome*, an unusual syndrome of muscle pain, weakness, and increased blood levels of certain white blood cells (eosinophils). Eventually, more than 1500 cases of illness and nearly 40 deaths were attributed to the supplements. Because tryptophan is an essential component of body proteins, investigators believed that the amino acid itself could not have caused harm, but that toxic contaminants must have developed during the manufacturing process. This process involved genetically modifying a strain of bacteria to produce unusually high levels of tryptophan, and then concentrating, collecting, and purifying the amino acid. To date, the toxin remains incompletely characterized. Although the genetic techniques do not appear to be directly at fault, their use in modifying a strain of bacteria created a situation — albeit inadvertently — that favored the formation of

toxic products (14). This example suggests that concerns about the unknown hazards of biotechnology cannot be dismissed out of hand.

Allergenicity

Because genes encode proteins, and proteins are allergenic, the introduction of allergenic proteins into previously nonallergenic foods could be another unintended consequence of plant biotechnology (15). In support of this idea, a biotechnology company transferred an allergenic protein from Brazil nuts to soybeans, and researchers confirmed that people who are allergic to Brazil nuts react similarly to soybeans containing the transgenic Brazil-nut protein (16). The company had developed the Brazil-nut soybeans as a means to increase the content of methionine, a sulfur-containing amino acid, in poultry feeds. Feathers contain high levels of sulfur-containing amino acids, and poultry feeds must be supplemented with methionine — at additional cost — to promote optimal growth. Because the Brazil-nut protein is especially rich in methionine, its gene was a logical choice as donor. Nuts, however, are often allergenic; the researchers happened to have collected serum samples from people known to be allergic to Brazil nuts. Thus they had in place all components necessary to test for allergies to Brazil-nut proteins, a situation that is rarely the case for other food allergens.

True allergies to food proteins can be documented in less than 2 percent of the adult population, but many more people might be expected to develop food sensitivities as proteins are increasingly added to commercially-prepared foods. Soy proteins, for example, already are very widely used in processed foods, and genetically modified soy ingredients already are widely prevalent in the food supply (17). Most biotechnology companies use microorganisms rather than food plants as gene donors, however, and their proteins do not appear to share sequence similarities with known food allergens. Few have as yet entered the food supply, but their allergenic potential is uncertain, unpredictable, and untestable (18).

As discussed below, allergenicity raises complex regulatory issues. Under a policy developed by the U.S. Food and Drug Administration (FDA) in 1992, company scientists were required to — and did — consult agency staff about the need for premarket testing. Because testing demonstrated transmission of the allergenic protein, the company would have had to label its soybeans as genetically modified. Because the company could not guarantee that people would not eat soybeans intended for animal feed, it wisely withdrew the transgenic soybeans from the market. Supporters of the FDA policy interpreted these events as a clear demonstration of the policy's effectiveness. Others, however, argued that the policy failed to protect the public against less well-studied transgenic allergens to which they might be sensitive and therefore favored industry. Critics were especially concerned about the lack of a requirement for labeling, as avoidance is often the only effective way to prevent allergic reactions. In 1993 the FDA requested public comment on whether and how to label food allergens in transgenic foods and later proposed rules to help resolve safety issues related to allergenicity, but by mid-2000 had taken no action on the matter.

Antibiotic Resistance

Genes for antibiotic resistance are used as part of selection processes. They can be expressed in transgenic plants when under the control of genes taken from higher organisms (but not bacteria), raising the possibility that they could be transferred to bacteria in the human intestine. Most experts consider this possibility remote but not entirely impossible, and antibiotic resistance is a principal concern of critics of genetically modified foods, especially in Europe. To avoid this possibility, the FDA suggests that companies evaluate the risks of transferring resistance to the antibiotics they are using, avoid using antibiotics effective against human diseases, and especially avoid using antibiotics uniquely effective against certain conditions (e.g., vancomycin and staphylococcal infections) (19).

REGULATORY ISSUES

Current debates about the regulation of food biotechnology center on the conflict between issues of safety on the one hand and a broad range of ecological, ethical, and societal issues on the other (20). For the industry and its supporters, safety is the only issue of relevance; because science supports the safety of most genetically engineered products, unnecessarily restrictive regulations appear to create barriers to further research and economic growth. Critics, however, view regulations as needed to protect the public not only against known safety risks but also against those that cannot yet be anticipated. They view safety as only one component of a far broader range of concerns about the impact of biotechnology on individuals, society, and the environment — issues that might also demand regulatory intervention. For government officials, biotechnology regulation must find the proper balance between oversight and encouragement of industry efforts to develop and market new food products (21). Current U.S. regulatory policies affect three aspects of food biotechnology directly — food safety, environmental protection, and intellectual property rights — and affect international trade indirectly. Thus far, these policies have achieved a balance that neither satisfies industry nor consumer groups.

Food Safety

In 1986 the U.S. White House developed a "Coordinated Framework" for regulating biotechnology based on the premise that the techniques led to products no different from those developed through conventional genetic methods. Thus existing laws and agencies should be sufficient for regulatory purposes. At the time regulation of food biotechnology involved no less than 3 offices reporting directly to the president; 4 major federal agencies; 8 centers, services, offices, or programs within agencies; and 5 federal committees — all operating under the authority of 10 distinct Acts of Congress (1). As might be expected, critics identified obvious flaws in this regulatory framework, among them lack of coordination, duplication

of effort, overlapping responsibility, and gaps in oversight. The principal laws affecting food safety preceded the use of genetic engineering, however, and did not necessarily apply to the new methods.

This uncertain regulatory status caused the food biotechnology industry to demand more precise guidance from FDA. In response, FDA developed a formal policy for the regulation of genetically-modified plant foods (22). The policy presumed that foods produced through recombinant techniques raised no new safety or nutritional issues and therefore could be regulated by FDA's existing policies for foods considered Generally Recognized as Safe (GRAS). Instead, safety evaluation would focus on changes in the "objective characteristics" of foods — changes in nutrient composition or new substances, toxins, or allergens. FDA would invoke requirements for premarket safety evaluation, premarket approval, or labeling only when those characteristics were sufficiently altered. The biotechnology industry welcomed this policy as a strong incentive for investors, but consumer groups judged it inadequate not least because the foods would not be labeled. As early as 1992 it became evident that consumer choice in the marketplace would influence acceptance of genetically modified foods (23), and a federal study recommended a review of the entire regulatory framework in order to establish a more equitable balance between promotion of industry and protection of the public (21). By late 1994 the FDA had approved the marketing of tomatoes genetically altered to reach optimal ripening after harvest, milk from cows treated with recombinant growth hormone, virus-resistant squash, insect-resistant potatoes, and herbicide-resistant cotton (used to make seed oil for animal feed), and soybeans, none of which addressed nutritional characteristics directly.

Environmental Impact

The "Coordinated Framework" affirmed that the U.S. Department of Agriculture (USDA) and the Environmental Protection Agency (EPA) were the primary agencies for regulating agricultural biotechnology. The EPA was to regulate recombinant plants developed to control insects and other pests, and it did so by requiring biotechnology companies to obtain permits prior to the manufacture or release of their agricultural products. The EPA policies were designed to address concerns that widespread agricultural use of new kinds of living species might present direct risks to human health — risks generally agreed to be minimal. Instead, environmentalists were concerned that transgenic crop plantings might pose *ecological* risks — displace existing plants and animals, create new plant pathogens, disrupt ecosystems, or reduce crop diversity. They predicted that widespread use of genetically modified crops such as those containing the gene for Bt toxin might undermine ongoing efforts to promote sustainable agricultural practices by selecting for Bt-resistance (24). They also argued that increased planting of herbicide-resistant crops would increase reliance on toxic chemicals to manage pests (25). In 1994 the EPA proposed to extend pesticide laws to transgenic "plant-pesticides" such as Bt and to require their registration, meaning that manufacturers would have to conduct tests of their nutritional and ecological impact as well as label them (26). Despite the EPA's assurance that the regulations would resolve uncertainties and attract investors, the rules appeared to favor large, established companies but discourage small, innovative companies.

Environmentalists were concerned that the proposals did not place enough emphasis on crops designed to resist chemical herbicides. Their ecological concerns have been encouraged by subsequent observations and research. Since 1996 researchers have reported preliminary signs of Bt-resistance in cotton plants and among moths and tobacco budworms, transmission of herbicide resistance from oilseed rape (canola) to related weeds, and higher mortality rates among bees fed proteins isolated from genetically modified rapeseed. Most famously, monarch butterfly larvae consuming pollen from Bt corn were reported to grow more slowly and die more quickly than larvae not exposed to such pollen (27). Although these observations are preliminary and require confirmation, they suggest a rational basis for environmentalists' fears that weeds and insects can develop resistance to currently available control methods, transgenic toxins can kill "friendly" insects such as bees and monarch butterflies, and genetically modified pollen can cross-fertilize conventional and organic crop plants. By late 1999 the EPA had not yet issued final rules on its plant-pesticide proposals.

Intellectual Property Rights

The U.S. intellectual property laws grant rights to patent owners to exclude everyone else from making, using, or selling the protected product for at least 17 years. Patents were first granted for plant varieties developed through asexual propagation in 1930. In 1970 Congress extended these rights to new varieties of plants developed through traditional genetic methods of sexual propagation. In 1980 the Supreme Court granted patent rights to microorganisms developed through recombinant techniques, and the Patent Office issued the first patent for such an organism. Patent rights were further extended to transgenic plants in 1985 and to animals in 1988 (28). The patenting of transgenic microorganisms and plants provided a major incentive for the growth of the food biotechnology industry. Within just a few years, however, industry and government officials in the United States, Canada, and Europe began challenging patent awards. By 1995, however, the U.S. Patent Office had issued 112 patents for genetically engineered plants. Among these were exclusive patent rights to one company for all forms of bioengineered cotton and to another for all uses of "antisense" genes such as those used to create tomatoes with a long shelf life (discussed below). The breadth of such patents seemed excessive and various groups soon filed lawsuits. Patent issues are especially pressing for Monsanto, as its U.S. patent for the Roundup herbicide expires in 2000.

The patenting of animals has generated even greater debate, particularly from animal-rights organizations and other groups who believe that the genetic engineering of farm animals might adversely affect family farmers, be cruel to animals, and endanger other living species.

Table 3. Principal Arguments for and Against the Patenting of Transgenic Animals

Arguments in favor

- Patent laws regulate inventiveness, not commercial uses
- Patenting is an incentive to research and development
- Patenting enables the biotechnology industry to compete in international markets
- Patenting is preferable to trade secrets
- Patenting rewards innovation and entrepreneurship

Arguments opposed

- Metaphysical and theological considerations make patenting untenable
- Patenting involves inappropriate treatment of animals
- Patenting reflects inappropriate human control over animal life
- Patenting disturbs the sanctity and dignity of life
- Most other countries do not permit patenting of animals
- Patenting could cause adverse economic effects on developing countries
- Patenting promotes environmentally unsound policies
- Animal patents will increase costs to consumers and producers
- Animal patents will result in further concentration in agricultural production
- Patent holders will derive unfair benefits from royalties on succeeding generations of patented animals

The successful cloning of a sheep ("Dolly") in 1997 (29) only heightened concerns about the ethics of "tampering" with animal life. The principal arguments for and against the patenting of genetically engineered animals are summarized in Table 3. Perhaps in response to concerns about such issues, the Patent Office ceased issuing patents for transgenic animals in 1988. In 1993 it resumed processing of the nearly 200 animal patent applications that had accumulated during that "self-imposed moratorium," but fewer companies were attempting to patent farm animals by that time, largely because persistent technical problems and costs had encouraged them to shift to more profitable areas of research.

PUBLIC PERCEPTIONS

Because food is overproduced, the food industry is so competitive, and overseas sales of American products are so important to the economic viability of agricultural producers, biotechnology companies have long viewed consumer acceptance of genetically modified products as critical to the industry. Thus various agencies in the United States and many other countries have conducted surveys of public perceptions of food biotechnology. In the United States these surveys cover a 15-year period. Despite differences in methods, year, and subjects, they have produced remarkably consistent information over time, and they reveal an internally consistent logic of considerable predictive value. They also explain why the responses to genetically modified foods in the United States have differed so substantially from those in Europe. Although U.S. survey respondents have only a limited understanding of science and technology, they hold high expectations that food biotechnology will produce benefits for them and for society as a whole. Surveys find respondents to be concerned about the potential and unknown dangers of genetically modified foods, but believe that the benefits outweigh risks. For example, 75 percent answer yes to the question, Do you feel that biotechnology will provide benefits for you or your family within the next five years? The surveys indicate clearly that respondents prefer some transgenic food products to others, most favoring products that appear beneficial to health or society, save money or time, are safe, or improve the environment. For example, 77 percent of U.S. respondents say they would be likely to buy genetically modified foods that protected against insect damage or required fewer pesticide applications (30).

Safety considerations, although often the focus of biotechnology debates, do not emerge in these surveys as the most important public concern. Instead, respondents appear most troubled by *ethical* issues related to food biotechnology. They are more willing to accept genetically modified foods that involve plants rather than animals, that do not harm animals, and that do not involve the transfer of animal genes into plants (31). These views derive from value systems that encompass issues that extend beyond food safety to include fundamental social, cultural, and religious beliefs. The surveys also reveal substantial public distrust of government credibility in safety matters, its ability to regulate food biotechnology appropriately, and the ability of the biotechnology industry to make decisions in the public interest. Perhaps for these reasons, surveys invariably find a large majority of respondents to want genetically modified foods to be labeled. Demands for labeling of genetically modified foods are especially prominent in Europe where nearly everyone (96 percent in Great Britain) favors such action (32).

If industry leaders view public opinion as irrational and as evidence for the need to educate consumers about the safety and benefits of biotechnology, they are missing the most strikingly useful conclusion to be drawn from these surveys. Prior to 1996, consumer attitudes toward food biotechnology in the United States and in Europe were largely product specific. People were willing to accept genetically modified products perceived as valuable to public health and welfare (33). From the surveys it should have been evident to industry that most consumers would accept genetically modified products if they were demonstrably beneficial to the public as well as to the industry's economic interests. The failure of genetically modified foods to address needs perceived by the public as important, and the industry's refusal to label the products, explain much of the subsequent resistance to the products, especially in Europe.

PREDICTIVE IMPLICATIONS: RECOMBINANT PRODUCTS

For the first recombinant products approved for sale in the United States, survey results suggested an analytical framework — based on the value of a product, its safety, and its ethical value — for predicting the degree of difficulty a product might experience with

Table 4. Analytical Framework for Predicting Public Acceptance of a Genetically Modified Food Product

1. Is the food safe for people and for the environment?

2. Is the food valuable? Will it:

- Increase nutrient content?
- Increase food availability?
- Decrease food cost?
- Improve food taste?
- Grow better under difficult conditions?
- Reduce use of herbicides and pesticides?

3. Is the food "ethical"? Does it avoid:

- Harm to animals?
- Insertion of animal genes into plants?
- Economic harm to small farms or businesses?
- Economic harm to populations in developing countries?

public acceptance. Table 4 outlines the questions that comprise this framework. The more positive the answers, the more likely consumers were to accept the product. To the extent that the answers were negative or equivocal, consumer resistance was likely to increase. This framework predicted the degree of acceptance of products released through 1995. Beginning with the release of recombinant soybeans and corn in 1996, however, policies regarding labeling, food safety, and international trade also influenced public acceptance.

Pharmaceuticals: Insulin

By the early 1990s the FDA had approved at least 15 recombinant drugs for use in human subjects. Recombinant insulin, for example, received approval in 1982 and was of unquestionable utility (6). It solved problems of scarcity and quality; it could be produced in unlimited amounts, and its amino acid structure was identical to that of human insulin and therefore superior to insulin obtained from the pancreas of pigs or cows. It was safe and raised no ethical considerations. Recombinant insulin readily met all three criteria for consumer acceptance, and it is neither surprising nor inconsistent that it and other recombinant drugs were accepted without protest in the United States as well as in Europe.

Enzymes: Chymosin

Recombinant enzymes used in food manufacture also were readily accepted. Chymosin, an enzyme used to coagulate milk to make cheese, was traditionally extracted from the stomachs of calves and sold as part of a mixture called rennet. It was difficult to extract, varied in quality, and was scarce and expensive. Through genetic techniques, the gene for chymosin could be transferred to bacteria that produced the enzyme in large quantities. The recombinant enzyme was approved for food use in 1990 (6). This action elicited no noticeable complaints from biotechnology critics, perhaps because the manufacturer did not publicize the enzyme's recombinant origins but also because obtaining the nontransgenic enzyme required the slaughter of baby calves. Transgenic chymosin also met the three criteria for consumer acceptance: It was more useful, ethical, and just as safe as the enzyme it replaced.

Hormones: Bovine Somatotropin (rbST)

The history of recombinant bovine somatotropin (rbST), the first product to be approved by the FDA under its 1992 food biotechnology policy, best illustrates how issues of safety, societal benefit, and ethics contribute to consumer resistance. The product, a growth hormone that increases milk production in cows by at least 10 to 20 percent, elicited considerable debate in the United States in the mid-1990s and, more recently, in Canada and Europe. Its name reflects the controversy: proponents generally use the scientific name, rbST, whereas critics call it Bovine Growth Hormone (rBGH). For purposes of consistency, this article uses rbST. The Monsanto company developed rbST in the mid-1980s and promoted it as a method to increase the efficiency of dairy farming. Although such efficiency would seem to be of great benefit to consumers, critics soon raised questions about the product's effects on human health, animal welfare, and the economic viability of small dairy farms (6). They were especially concerned that rbST-treated milk would not be labeled as such. When the FDA approved Monsanto's rbST as a new animal drug in 1993, it ruled that labeling would be misleading as treated and untreated milk were indistinguishable by available methods. The level of protest against rbST was extraordinary; Supermarket chains announced that they would not carry milk from rbST-treated cows, and several states enacted legislation banning the hormone. Some dairy companies, concerned about consumer reactions, began to label their products as "BGH-free," but industry groups challenged the legality of this practice. The FDA permitted that designation to be used if accompanied by a disclaimer: "No significant difference has been shown between milk derived from rbST-treated and non-rbST-treated cows" (34). Protest against rbST could easily have been anticipated, as rbST raised safety, value, and ethics issues addressed by the questions in Table 4.

Safety Issues. Bovine somatotropin stimulates milk production, and the natural hormone is always present in cow's milk in low concentrations. Milk from rbST-treated cows contains both the natural hormone and rbST; these are almost identical. The hormone itself is unlikely to be harmful to humans, even though its concentration is higher in milk from treated cows. Its protein structure differs from that of the human hormone and is not biologically active in people. Like all proteins, the cow hormone is largely broken down in the human intestinal tract. In 1990 Monsanto-sponsored scientists reported that rbST milk was safe for human consumption and that the FDA studies had answered all safety questions. That same year FDA scientists reviewed more than 130 studies of the effects of rbST on cows, rats, and humans and concluded that the hormone did not adversely affect human health. The publication of this last report in a prestigious scientific journal was judged "unprecedented," as it appeared that the FDA was favoring a drug it had not yet approved. However, other expert groups also concluded that milk

from rbST-treated cows was essentially the same—and as safe—as milk from untreated cows (35).

Despite this evidence critics continued to raise safety concerns about two factors that might be present in rbST milk: antibiotics and insulinlike growth factor-I (IGF-I). The concern about antibiotics derives from observations that cows treated with rbST develop udder infections (mastitis) more frequently than untreated cows (36). Because the infections are treated with antibiotics that linger in milk and meat, it is theoretically possible that consumed antibiotics could contribute to human antibiotic resistance. Although federal regulations require testing for antibiotic residues in milk, the FDA tests for only a small fraction of animal drugs in common use—just 4 out of 82 in one study—leading to charges that the agency lacks a comprehensive strategy for monitoring such drugs.

The IGF-I issue derives from concerns that the increased concentration of this factor in milk from rbST-treated cows might stimulate premature growth of infants or cancers in adults. Although IGF-I appears to be denatured in infant formulas and seems unlikely to be absorbed in significant amounts by the human digestive tract, the factor is readily absorbed from milk, is biologically active in rats, and is associated in epidemiological studies with increased risk of prostate cancer in men (37) and breast cancer in premenopausal (but not postmenopausal) women (38). Although the clinical significance of these observations is unknown, they have encouraged dairy and consumer groups to demand further examination of the data and to file suit against the FDA; they also have been used as a basis for refusal to license rbST in Canada and Europe.

Value Issues. For many years milk production in the United States exceeded demand, resulting in large surpluses of dairy products. The use of rbST was expected to further increase milk production. Biotechnology companies contended that use of the hormone would reduce farm costs because equivalent amounts of milk could be produced by fewer cows. Although it might seem logical that creation of more surplus milk would lead to lower prices, dairy prices are tightly linked to federal support programs and unlikely to change very much. Given this situation, rbST offered no evident cost benefits to consumers.

Ethical Issues. Because rbST-treatment of cows increases milk production, concerns have been raised about effects of the drug on the health and reproductive ability of animals. The more milk cows produce, the more likely they are to develop mastitis. In addition rbST is delivered through injection and can cause localized reactions at the point of entry. Despite industry assertions that appropriate herd-management practices can minimize such problems, they were reported regularly. In addition, increasing the supply of milk might be expected to accelerate long-standing trends toward the elimination of small dairy farms, and most commentors believed that at least some dairy farmers would be forced out of business. For these reasons, answers to some of the ethical questions are negative or equivocal.

Consumer Reactions. Taken together, public scepticism about rbST related to nearly all of the areas of concern listed in Table 4, suggesting that this product was an unfortunate first choice for commercialization. U.S. farmers already overproduced milk, and rbST offered no evident benefit to consumers in availability, price, or quality. That the product affected milk also was unfortunate, as this food often is promoted as conveying an image of purity. The primary beneficiaries of rbST therefore appeared to be its manufacturers and the large dairy farmers who were best able to exploit its use. Despite considerable resistance from farmers, the company has said that rbST broke even in 1996 and has been profitable ever since, but its annual report does not permit independent verification of this assertion (4). Because the use of rbST to produce commercial milk is not labeled, public acceptance of the hormone in the U.S. remains uncertain. One indicator of public opinion is the spectacular growth in sales of organic ("BGH-free") milk since 1996 (39). In Europe, use of the hormone is prohibited at least until the end of 1999. Although several international committees have reaffirmed the safety of rbST, European Community members have not yet reached consensus on its approval (40). In Canada, applications to market rbST have been pending for more than 15 years, largely because of conflicting opinions about the safety of the products for both cows and people. These events also could have been predicted from the questions in Table 4.

Foods: The "Flavr Savr" Tomato

Americans expect tomatoes to be available on a year-round basis. In 1997 farmers produced nearly 16 pounds (7.3 kg) per capita of fresh tomatoes and another 73 pounds (33.1 kg) for processing (12), but supermarket tomatoes, bred for disease resistance, appearance, and durability, have long been the bane of consumers longing for "backyard" taste and freshness. Beginning in the mid-1980s, Calgene, a California-based biotechnology company, invested $25 million and 8 years of effort to develop a tomato with a reversed (and therefore blocked) gene for ripening that would allow it to be picked and marketed at a more mature stage of taste (6). Calgene expected this "Flavr Savr" tomato to capture at least 15 percent of the market for fresh tomatoes as soon as it became available, and the company planned to sell—and label—it as genetically engineered to taste better. As the first company to develop a genetically modified food, Calgene worked closely with FDA to determine the tomato's regulatory status. FDA insisted that review committees focus exclusively on the tomato's *safety* and judged concerns about ethical issues or labeling as irrelevant. The agency decided in 1994 that all safety and nutritional questions about the new tomato had been resolved and approved its marketing. Although some groups threatened boycotts and "dumpings," most analysts believed that consumers would accept the tomato if its improved taste seemed worth the premium price, initially expected to be twice that of conventional tomatoes. From the answers to the questions in Table 4, some consumer resistance should have be expected. Although

the Flavr Savr was as safe and nutritious as market tomatoes, it raised issues related to impact on small growers, and its benefit to the public was restricted to taste. Its higher costs, however, identified the Flavr Savr as a luxury product targeted to an upscale market. To Calgene, the tomato was well worth the huge investment of time, money, and effort as it paved the way for subsequent approval of the company's seed oils and other genetically modified crops. Eventually the Flavr Savr proved impossible to grow and ship in adequate quantities, and its acceptance in the markeplace could not be tested (41).

Food Crops: Soybeans and Corn

Experts predict that within five years virtually *all* of U.S. agricultural exports—worth $50 billion annually—will be transgenic or combined with genetically modified bulk commodities. Any resistance to acceptance of transgenic soybeans and corn would pose a serious economic threat (42). In the United States these products encountered little public opposition, perhaps because they are mainly fed to animals and their environmental hazards seem geographically remote from most people. Alternatively, the lack of protest reflected ignorance of the extent to which genetically modified ingredients pervaded the food supply (17). Also, until quite recently, most press reports about genetically modified crops appeared exclusively in business pages. Europeans, however, could not help but be informed; at the peak of coverage early in 1999, the seven largest British daily newspapers ran nearly 2000 column inches (5,000 cm) of copy about genetically modified food, nearly all of it unfavorable (43).

The controversy in Europe began in 1996 with the first marketing of unlabeled recombinant soybeans and corn. Since then the European Union and various member countries have issued outright bans, prohibitive regulations, or labeling requirements. Food producers and retailers have refused to use genetically modified ingredients in their products and have withdrawn products containing them from sale (9). The intense resistance in Europe, so much more extreme than in the United States, is nevertheless related to a similar set of consumer issues. Many Europeans have long-standing traditions of animal welfare, vegetarianism, and other value systems that might affect attitudes toward food biotechnology, as well as memories of Nazi eugenic experiments during World War II. In addition concerns about the spread of antibiotic resistance, fears generated by the 1996 food safety crisis over "mad cow" disease, and, more recently, alarms raised by outbreaks caused by foodborne pathogens have reduced public trust in government as well as in industry. European consumers, also were reacting to the perceived arrogance of American officials and companies who seemed to be forcing U.S. exports "down their throats" (44). In this context the aggressive marketing of unlabeled genetically modified soybeans and corn by American biotechnology companies only intensified public resistance. An advertising campaign by the Monsanto company emphasizing environmental and nutritional benefits of biotechnology through slogans such as "While we'd never claim to have solved world hunger at a stroke, biotechnology provides one means to feed the world more effectively," and "Food labelling. It has Monsanto's full backing," produced the opposite of the effect intended and only increased public suspicion of genetically modified foods (45). Also inciting protest was the company's investment in "terminator" technology, a method for ensuring that genetically modified crops could not produce viable seeds. This technology would protect the company's proprietary rights to the seeds but adversely affect Third World farmers who grow 15 to 20 percent of the world's food from saved seed (46). Calls for an international ban on terminator research also brought negative public attention to food biotechnology.

By the late 1990s European surveys revealed nearly unanimous public support for labeling of food products containing genetically modified ingredients. Labeling requires strict segregation of genetically modified from conventional corn or soybeans and the use of genetic tests to distinguish them. Because of cross-pollination, some mixing of genetically modified with conventional field crops is inevitable, and authorities have yet to agree on the lowest level of "contamination" permissible for products to be labeled "GM-free." Labeling also has trade implications. Any differences between the food regulations of one country and another must be "harmonized" by the international Commission (Codex Alimentarius) that sets food standards. For several years, the Commission has considered standards for mandatory labeling of genetically modified foods but has yet to reach consensus (47).

POLICY IMPLICATIONS

Although consumers in the United States have been slow to oppose genetically modified foods, organized opposition appears to be growing. When the USDA proposed that standards for foods designated as "organic" could include those that had been genetically modified, 275,000 people wrote letters of protest and the agency was forced to withdraw the suggestion. Although news articles critical of biotechnology had been published for years in environmentalist and other specialist magazines, mainstream publications have begun to feature the issue and to raise public awareness. Coalitions of consumer groups, religious groups, chefs, and scientists have filed lawsuits and organized petition campaigns to force the FDA to require labeling and safety testing for transgenic products. One consumer organization collected nearly 500,000 signatures on a petition calling for mandatory labeling. By the end of 1999, 68 percent of American adults polled by Gallup wanted genetically modified foods to be labeled even if it meant paying higher prices (48). Feeling the pressure from organic competitors, the Gerber's and Heinz baby food companies have announced that they will not allow genetically modified ingredients in their products, as has the Archer-Daniels-Midland Company. In response to European rejection of genetically modified corn, the American Corn Growers Association advised its members to consider planting only conventional seeds in spring 2000. Labeling of genetically modified products seems likely to occur in the United States as well as in Europe.

In promoting public acceptance of the first genetically modified products, industry leaders focused on safety as the sole basis for discussion and characterized other concerns as unscientific or irrational. The industry's dismissal of concerns other than safety and its opposition to labeling missed an important point: Initial views of food biotechnology were product specific and, as such, were consistent and predictable. If the marketplace was to be allowed to determine the success of genetically modified products — as it does with all others — the products would have to be labeled. If the products were valuable to consumers, the label should have encouraged purchases as well as trust in the industry (18). The failure to label could well have been the single factor most responsible for the hostile European reception to genetically modified soybeans and corn.

The controversy over food biotechnology derives directly from the conflict between the industry's need to be profitable and the desire of consumers for products that are economically and socially valuable, as well as safe. To frame the debate in terms of rational science versus an irrational public is to do a disservice to both. Biotechnology is not inherently dangerous, and it should be capable of doing much good. The public is not inherently irrational and should be capable of judging whether genetically modified products are worth buying. This analysis suggests that the public will continue to be unconvinced that genetically modified foods are necessary or safe as long as the principal beneficiaries of the technology are the companies themselves. The current debates about genetically modified foods offer the industry an opportunity to address consumers' concerns about the credibility, safety, and ethical implications of the products. To improve credibility, the industry must bring its rhetoric more in line with reality. If industry leaders continue to state that food biotechnology is necessary to solve world food problems, they should be investing substantial resources into research on those problems. Companies, for example, could institute tithing programs that apply 10 percent of income to Third World research and development projects that might never prove profitable. Such programs might help convince the public that the industry recognizes its own conflict of interest and distinguishes its societal from its investment goals. Of course, the most effective way the industry can achieve credibility is to be credible. If biotechnology companies want to convince the public that their products are beneficial, they should develop beneficial products — those that promote public goals and truly promote sustainable agriculture, prevent environmental degradation, and improve nutritional quality.

ACKNOWLEDGMENTS

This article is an extensive revision and update of "Food biotechnology: politics and policy implications," in K.F. Kiple and C.K. Ornelas-Kiple, eds., *The Cambridge World History of Food and Nutrition*, Cambridge University Press, Cambridge, UK, 2000: used with permission.

The work was supported by research challenge grants from New York University and its School of Education, and by expert research assistance from Monica Bhagwan.

BIBLIOGRAPHY

1. Office of Technology Assessment, *A New Technological Era for American Agriculture*, OTA-F-475, U.S. Congress, Washington, DC, 1992.

2. T.J. Hoban, in D. Maurer and J. Sobal, eds., *Eating Agendas: Food and Nutrition as Social Problems*, de Gruyter, New York, 1995, pp. 189–212.

3. G.E. Gaull and R.A. Goldberg, eds., *New Technologies and the Future of Food and Nutrition*, Wiley, New York, 1991.

4. Monsanto, *1998 Annual Report* Online: *http://www.monsanto.com/*

5. J. Enriquez and R. Goldberg, *Gene Research, the Mapping of Life and the Global Economy Case N9-599-016*, Harvard Business School, Boston, MA, 1998.

6. U.S. Congress, Office of Technology Assessment, *Biotechnology in a Global Economy*, OTA-BA-494, U.S. Government Printing Office, Washington, DC, 1991.

7. Z.S. Olempska-Beer, P.M. Kuznesof, M. DiNovi, and M.J. Smith, *Food Technol.* **47**, 64–72 (1993).

8. J. Sachs, *The Economist*, August 14, pp. 17–20 (1999).

9. *The Economist*, June 19, pp. 19–21 (1999).

10. E. Messer, *Adv. Technol. Assess. Syst.* **9**, 371–378 (1992).

11. A.M. Thro et al., *Nat. Biotechnol.* **16**, 428–430 (1998).

12. J.J. Putnam and J.E. Allshouse, *Food Consumption, Prices and Expenditures, 1970–97*, USDA Stat. Bull. No. 965, U.S. Department of Agriculture, Washington, DC, 1999.

13. A.E. Gallo, *The Food Marketing System in 1996*, USDA Agric. Inform. Bull. No. 743, U.S. Department of Agriculture, Washington, DC, 1998.

14. R.M. Philin et al., *Am. J. Epidemiol.* **138**, 154–159 (1993).

15. D.D. Metcalfe et al., eds., *Crit. Rev. Food Sci. Nutr.* **36**(Suppl.), s1–s186 (1996).

16. J.A. Nordlee et al., *N. Engl. J. Med.* **334**, 688–689 (1996).

17. *Consumer Rep.*, September, pp. 41–46 (1999).

18. M. Nestle, *Nutr. Today* **33**(1), 6–12 (1998).

19. Food and Drug Administration (FDA), *Guidance for Industry: Use of Antibiotic Resistance Marker Genes in Transgenic Plants: Draft Guidance*, FDA, Washington, DC, 1998. Online: *http://vm.cfsan.fda.gov/*

20. P.B. Thompson, *Food Biotechnology in Ethical Perspective*, Blackie Academic & Professional, London, 1997.

21. U.S. General Accounting Office, *Food Safety and Quality: Innovative Strategies may be Needed to Regulate New Food Technologies*, GAO/RCED-93-142. GAO, Washington, DC, 1993.

22. D.A. Kessler et al., *Science* **256**, 1747–1749, 1832 (1992).

23. M. Nestle, *Bio/Technology* **10**, 1056 (1992).

24. J. Rissler and M. Mellon, *Perils Amidst the Promise*, Union of Concerned Scientists, Washington, DC, 1993.

25. R. Goldburg, J. Rissler, H. Shand, and C. Hassebrook, *Biotechnology's Bitter Harvest*, Biotechnology Working Group, Washington, DC, 1990.

26. Environmental Protection Agency, *Fed. Regist.* **59**, 60496–60518 (1994).

27. J.E. Losey, L.S. Rayor, and M.E. Carter, *Nature (London)* **399**, 214 (1999).

28. Office of Technology Assessment, *New Developments in Biotechnology: Patenting Life — Special Report*, OTA-BA-370, U.S. Congress, Washington, DC, 1989.

29. I. Wilmut et al., *Nature (London)* **385**, 810–813 (1997).

30. International Food Information Council, *U.S. Consumer Attitudes Toward Food Biotechnology*, Wirthlin Group Quorum Survey, Washington, DC, 1999.

31. T.J. Hoban, *J. Food Distrib. Res.* **27**, 1–10 (1996).

32. *The Economist*, June 13–19, pp. 113–114 (1998).

33. G. Gaskell, M.W.S. Bauer, J. Durant, and N.C. Allum, *Science* **285**, 384–387 (1999).

34. Food and Drug Administration, *Fed. Regist.* **59**, 6279–6280 (1994).

35. Office of Medical Applications of Research, *Bovine Somatotropin: National Institutes of Health Technology Assessment Conference Statement*, U.S. Government Printing Office, Washington, DC, 1990.

36. U.S. General Accounting Office, *Recombinant Bovine Growth Hormone: FDA Approval Should Be Withheld Until the Mastitis Issue Is Resolved*, GAO/PEMD-92-26, GAO, Washington, DC, 1992.

37. J.M. Chan et al., *Science* **279**, 563–566 (1998).

38. S.E. Hankinson et al., *Lancet* **351**, 1393–1396 (1998).

39. S. Gilbert, *N.Y. Times*, January 19, p. F7 (1999).

40. Food and Agriculture Organization of the United Nations, *Press Release*, FAO/UN, Rome, 1999. Online: *http://www.fao.org/*

41. M. Graves, *Los Angeles Times*, August 18, p. C1 (1997).

42. P.J. Longman, *U.S. News & World Rep.*, July 26, pp. 38–41 (1999).

43. E. Dorey, *Nat. Biotechnol.* **17**, 631 (1999).

44. Y.M. Ibrahim, *N.Y. Times*, November 7, pp. D1, D24 (1996).

45. S. Kilman and H. Cooper, *Wall Street J.*, May 11, pp. A1, A10 (1999).

46. R.A. Steinbrecher and P.R. Mooney, *The Ecologist* **28**(5), 276–279 (1998).

47. E. Masood, *Nature* **398**, 641 (1999).

48. S. Kilman, *Wall Street J.*, October 7, pp. A1, A15 (1999).

See other entries AGRICULTURAL BIOTECHNOLOGY; ANIMAL BIOTECHNOLOGY, LAW, FDA REGULATION OF GENETICALLY MODIFIED ANIMALS FOR HUMAN FOOD USE.

AGRICULTURAL BIOTECHNOLOGY, SOCIOECONOMIC ISSUES, AND THE FOURTH CRITERION

WILLIAM LACY
University of California
Davis, California

OUTLINE

Introduction

The Fourth Criterion

Farmers, Rural Communities, and the Food System

Agribusiness and Industry

Consumers

INTRODUCTION

The world economy is currently undergoing major structural changes. A central factor in these changes has been the development and diffusion of fundamentally new kinds of technologies, in particular, computers and the new biotechnologies. Social and economic changes that result from these profoundly enhanced capacities in science and technology are visible in every sphere of human life from health, transportation and communication to agriculture and the food system. However, each change is associated not only with new benefits but also with new risks, latent complications, and long-term consequences that are often poorly understood. Some have argued that the new biotechnologies may be the most radical experiment humankind has ever carried out on the natural world, in many ways representing our fondest hopes and aspirations as well as our darkest fears and misgivings. The technology according to some actually touches the core of our self-definition (1).

The new tools are arguably the ultimate expression of human control both helping us to shape and define nature itself as well as our very sense of self and society. The changes brought by biotechnology could deeply affect our individual and collective consciousness, the future of our civilization, and the biosphere itself. Until recently far more public attention has been focused on the other great technology revolution of this century, computers and telecommunications. However, after nearly 40 years of parallel development, the information in life sciences is slowly beginning to fuse into a single technological and economic force. Computers are increasingly being used to decipher, manage, and organize the vast genetic information as a new resource of the emerging biotechnology economy. This new field called "bioinformatics" is being used to download the genetic information of millions of years of evolution and thereby creating a powerful new genre of biological databanks. This new genetic information database may be used by researchers to remake the natural world. As Jeremy Rifkin has noted "These changes represent a turning point for civilization. We are in the throes of one of the great transformations in world history" (1, p. 4).

In the area of agricultural biotechnology, the president of a large U.S. public university noted, "Our society has moved into the era of high technology ... as we move into the new millennium, we will see more technological changes than we have experienced over the entire history of our nation. It promises to be one of the most exciting and challenging times in the history in mankind." He further observed, "Biotechnology, genetic manipulation and engineering research will have tremendous impact on the crops and animals we grow for food, affecting agriculture in ways never before dreamed possible" (2, p. 3). According to Rifkin (1), global agriculture could find

itself in the midst of a great transition in world history, with an increasing volume of food and fiber being grown in enclosed tissue culture vats. The shift to indoor agriculture could bring significantly low prices, more abundant supplies of food, and massive displacement of millions of farmers in both the developing and developed world.

Despite the enormous optimism in the scientific community, national and state governments, and the private commercial sector, the applications of biotechnology have been fraught with concern and controversy within the both the scientific community and the broader public. Much of the initial public concern has centered on human and animal health and environmental safety issues. Issues of the environmental impact of the creation, mass production and wholesale release of thousands of genetically altered life forms into the environment and the potential irreversible change and wholesale reseeding of the earth's biosphere has been raised. Many groups of scientists and environmentalists and the local citizen groups have raised safety issues regarding the unexpected but possible consequences of introducing new life forms such as the production of a toxic secondary metabolite or protein toxin or the undesired self-perpetuation and spread of the organism. In agriculture, some have suggested that living natural inputs may be even more dangerous to society than the artificial products they replace. In addition there is concern that genetically engineered crops bearing resistance to nature herbicides may become weeds. Some fear that these herbicide-resistant crops may even cross with weedy relatives and spread resistance into sectors of the weed flora. In the area of human safety there has also been concern raised about the safety and human health issues surrounding genetically modified foods and pharmaceuticals. In the case of human health, these include potential allergenic or toxic effects resulting from genetic changes that are not completely understood.

THE FOURTH CRITERION

The debate, however, cannot be reduced to a simple risk controversy that focuses on health and environmental safety issues. While the three standard criteria often utilized to evaluate and approve new products and processes have been (1) human safety, (2) animal and environmental safety, and (3) efficacy, increasingly over the last couple of decades, a fourth criterion or fourth hurdle for product approval and regulation has been proposed. This refers to the social and economic effects of the product or a technology. With many citizens in both the United States and Europe the issue of food and agriculture modified from modern molecular genetics and biotechnology elicit deeper concerns about the relationship to the natural environment where there are strong dimensions of social and political risk. Increasingly it is recognized that the issues of agricultural biotechnology are not purely technical but also concern the balance between the different worldviews and values that enter the scene of each national regulation and product approval process. Assessing the risk associated with agricultural biotechnology therefore becomes a complex problem that is being heatedly debated. Many efforts have been directed against the use of the fourth criterion on the basis that it inhibits trade and is in violation of a number of global trade agreements. Despite the explicit prohibition of the use of the fourth criterion to inhibit trade, the social criteria are being implicitly or explicitly included in a number of policy debates and decision-making processes. As a result of these developments and experiences with previous technologies, an increasingly accepted position among technology assessment professionals is that (1) all technologies have multiple effects, (2) many of these effects are potentially harmful and require conscious decisions, and (3) these critical decisions entail social, economic and moral as well as scientific analysis (3,4).

In the early 1990s this broader fourth criterion was employed in the European Common Market's ban on growth hormones in food products. At that time the Advocate General of the Court of Justice in the European Communities released an opinion on the legality of the hormone stating that "it was appropriate and justifiable to prohibit the administration of the five substances for fattening purposes, even in the absence of scientific evidence showing that they were harmful. A total prohibition was the only solution which could bring an end to the distortions of competition and barriers to intra-Community trade in meat, eliminate risks to public health, even if they were purely hypothetical ones, and avoid a further reduction in consumption" (5, p. 1).

Similarly, in the Austrian biotechnology regulations, the social and economic criteria are clearly present. Their regulations state that products containing or consisting of genetically engineered organisms must not cause any social unsustainability, no unbalanced burden on society or any social group that is unacceptable for economic, social, or moral reasons (6). Arguments regarding health or environmental risk are frequently countered with fears of socioeconomic hazards. In practice, the two perspectives are exceedingly difficult to separate. The absence of risk to environment, life, and limb appears to be the necessary, but inadequate, condition for acceptance among increasingly larger percentages of the Austrian population. As a consequence Austria passed paragraph 63 of the Genetic Engineering Act of 1994 which required that the genetically engineered products be assessed for the possible risk of social unsustainability before they could be licensed. These provisions seem to contradict the European Union (EU) Commission's ruling because they include aspects of assessment according to socioeconomic criteria that had been clearly rejected as the fourth hurdle to licensing. According to the EU Commission, socioeconomic criteria cannot be assessed by scientifically clear criteria and therefore would likely provoke legal issues and lead to a drain of capital.

Despite the EU's policies and prior to the passage of the Act, the Austrian parliament voted unanimously to set up a parliamentary inquiry commission to investigate technology assessment issues based on the example of genetic engineering. In its report in 1992, the commission demanded that social sustainability be taken into consideration in addition to the ethical requirements and environmental impact. One member of the commission indicated that the concept of social sustainability should entail safeguarding the balance of interests

and the maintenance of consensual value orientations. Moreover, the inquiry commission recommended information for the public, participation by the public, mandatory disclosure (annual report on all genetic engineering activities), and measures to promote the public discussion of genetic engineering. Paragraph 63 of the Austrian Genetic Engineering Act specifically stated that genetically engineered products must not lead to social unsustainability and would not be approved for use "if it may be assumed on a technical basis that such products would lead to an unbalanced burden on society or on social groups, and if this burden no longer appears acceptable to the population for economic, social or moral reasons" (6, p. 303).

To implement the socioeconomic criteria for product approval, several models were explored. One focused on creating an interdisciplinary expert panel to design future scenarios and test them scientifically for their compatibility with the Constitution and societal values. The decision on introducing technology was to reside in diverse politically authorized groups. Criticisms of this scheme included the fact that experts would not be able to obtain consensus because of the pluralistic values of these groups. A second model involved the idea that anything that is accepted by society is socially sustainable. However, this meant turning the agendum into research on societal acceptance and focusing on the manipulation of public opinion and the creation of acceptance. The third approach focused on the participatory process where social sustainability would remain a transparent and preliminary working term only put into practice if a situation demanded it. The requirement that certain agricultural biotechnology products must not cause social unsustainability made them a mandate for public discussion and negotiation. The state's function was to provide a framework for the public dialogue to occur and to guarantee that framework from a legal point of view. However, it is unlikely that anyone was considering participatory procedures when social sustainability was conceptualized. It was something viewed instead as an expertise type of administrative procedure. Nevertheless, in time meaningful participatory procedures were developed for technology assessments, and they are now in place in Austria.

The compromise in Europe has been to introduce the precautionary principle to address uncertainty without specific reference to the fourth criterion or socioeconomic concerns. In the recent European Union's directive on genetic engineering and international laws, the precautionary principle was upheld. While it comes in many interpretations, the principle is that whenever there is a serious threat of irreversible damage, any lack of scientific evidence must lead to postponement or avoidance of the biotechnology. Built into this conclusion is time for deliberation. The use of precautionary measures allows for selection among possible risks and consideration of the severity of those risks. The decision regarding risks depend on who is affected in which ways and who derives which benefits, and this ultimately takes on socioeconomic dimensions.

The fourth criterion has also entered the discussion and negotiations for an international biosafety protocol. In 1998 an ad hoc working group on biosafety met and heard concerns from many members of developing countries about the impact of transgenic crops on their farming communities, and they voiced their support for adding socioeconomic issues to the protocol (7). The ability to implement certain new biotechnology products varies greatly across countries and regions (e.g., the enforcement of obligatory resistance management strategies for Bt crops will be difficult in a country like India, and there are legitimate concerns that where such strategies fail, the commercial usefulness of Bt transgenic crops will be limited to just a few years). Those advocating for socioeconomic factors in the biosafety protocol argue that it is important for individual countries to be free to examine, case by case, the social and economic impacts of an imported biotechnology. The decision to subjugate national interest to that of free trade is particularly problematic in the case of a developing country's agriculture where impact of biotechnology on farming systems will differ greatly from that in industrialized countries. Moreover the regulatory interests of countries that will primarily export genetically modified crops and animals will differ from those that do not have a domestic biotechnology industry. The immediate economic benefits of a burgeoning agribiotechnology market unencumbered by regulatory controls pose potential barriers to free trade, and these benefits will accrue to corporations based in countries that are in a position to export the genetically modified products. Therefore individual countries must be allowed to consider the socioeconomic impact of importing genetically modified plants and animals on a case by case basis in order to ameliorate the effects of disparities (7).

As noted earlier, some countries of the European Union, in particular, Austria, Denmark, Sweden, and Finland, have expressed the need to interpret the European directive on the deliberate release of living modified organisms in order to permit them to consider the socioeconomic impacts. The Norwegian Gene Technology Act developed by a non-EU country goes even further. The Act states that in "Deciding whether or not to grant the application, significant emphasis shall also be placed on whether the deliberate release represents a benefit to the community and a contribution to sustainable development" (7, p. 698). The Norwegian interpretation takes the discussion one step further: If there is any question that a negative impact may arise, this doubt must comply with environmental and the social factors that take priority and override the particular application of the biotechnology.

In another recent example of the potential utilization of broader socioeconomic criteria, Switzerland was faced with a Gene Protection Initiative that demanded the government to outlaw the release of genetically altered organisms into the environment as well as the patenting of transgenic animals and plants, of their components, and of the relevant processes. This initiative, defeated in a national vote, demanded that experiments of all genetically modified organisms require proof of the lack of alternatives and a statement of ethical responsibility. Schatz (8) noted that despite Switzerland being the corporate headquarters of a number of multinational biologically based corporations, the public has begun to

view companies around the world as heartless giants and choose the high tech products of those giants as targets of their frustration. The Swiss basically have begun to doubt that their elected representatives can reign in the international conglomerates.

In contrast, the U.S. Executive Branch concluded a review of literature on the social consequences of rBST with this revealing sentence: "At no time in the past has the U.S. Federal Government prevented a technology from being adopted on the basis of socioeconomic consequences" (9). Despite this statement the recent report from the U.S. National Research Council (4) entitled *Understanding Risk: Informing Decisions in a Democratic Society*, has moved the U.S. debate on technology impact from a narrow scientific discussion of risk assessment to the broader issue of risk characterization. The committee was asked to review "the appropriateness of including in risk characterization such considerations as economic factors, equity issues, risk mitigation and tradeoffs, and technical control feasibility as well as environmental-equity issues and other issues of social context" (4, p. x). They were also charged to examine ways for improving public participation and building trust. As a consequence the study addressed such issues as the social, behavioral, economic, and ethnical aspects of risk that were viewed as relevant to the content or process of risk characterization (4).

This brief discussion illustrates that the social and economic concerns about biotechnology have become part of the policy and regulatory process. To date, these concerns have only selectively become a fourth criterion or fourth hurdle. The extent to which these concerns are made explicit may vary, but they are inherent in the debates in both the developed and developing nations. Therefore it is important for scientists, regulators, policy makers, and citizens to understand and evaluate not only elements of human safety, animal safety, environmental risks, and efficacy but also the range of socioeconomic impacts and concerns.

The potential social and economic impacts of agricultural biotechnology on the food and fiber system and society are just emerging. Consequently the proposed implications of biotechnology for the system represent only possible scenarios. The socioeconomic effects may include impacts on: (1) farmers, rural communities, and the food system; (2) the organization and structure of agribusiness and industry; (3) consumers; (4) science and technology transfer; and (5) the global economy and developing nations. The social impacts and consequences of any technology are likely to be dispersed in both time and space and occur through a wide variety of mechanisms. The social impacts of technology are controversial, in part, because the mechanisms that link technological innovations to their eventual consequences are generally opaque to both developers of the innovation, scholars of technology, policy makers, and citizens.

FARMERS, RURAL COMMUNITIES, AND THE FOOD SYSTEM

The impact of agricultural change on rural communities is largely proportional to the level of local dependence on agriculture. Today nationwide, fewer than 40 congressional districts have more than 20 percent of their population living on farms. The overwhelming majority of farms that once existed in the United States no longer exist, and production is highly concentrated among the remaining farms characterized by productivity-enhancing technology. In 1978 there were 2.3 million farms in the United States while there were less than 2.0 million by 1999. Only 6 percent of U.S. farms, involving primarily the super-large farms, receive the majority of farm receipts (10). Over the last century, agricultural technologies have emerged that use ever greater levels of capital to enable fewer people to produce the nation's food. As a result income and opportunities have shifted from farms to the companies that produce and sell goods to farmers. As farmers focused on producing undifferentiated raw commodities, food system profit and opportunities have shifted to the companies that sell the farm inputs, and process, package, and market food. Consequently from 1910 to 1990 the share of the agricultural economy received by farmers dropped from 21 to 5 percent (11). Agricultural biotechnology will likely continue this trend with the profits accruing to the industries developing the biotechnology products. Finally, and importantly, a trend that appears in all sectors of American agriculture is a widening spread between what farmers receive for their production and what consumers pay at the supermarket.

The industrialization of agriculture in the United States has almost one-third of the total value of production of U.S. farms generated under contractual arrangements and mostly under market contracts. Large agricultural integrators tend to avoid capital investment in the means of production and pass the risk and costs on to their contract growers or to society at large. Under these conditions farmers contract to sell their products to a specific processor or contractor, but the farmer owns the product and the risks until the product is sold and makes all the managerial and production decisions. Production contracts are also increasing with the contractor owning the livestock or crop and paying the producer a flat fee plus additional payments for performance-based incentives. Under these conditions the producer or farmer becomes very similar to industrial laborers. The poultry industry is perhaps the most industrialized subsector of agriculture with 89 percent of poultry farms using contracts and about 86 percent of the total value of poultry production grown under contract. Competition in the hog, cattle, and lamb industries has been declining even before the recent rise in livestock contracting with the proportion of the market controlled by the four largest steer and heifer slaughter firms increasing from 36 percent in 1980 to 72 percent in 1990 and 82 percent in 1994 (10). The vast majority of small farms, however, are now buffered from the effects of technological change, since the farm is no longer the primary source of income for their owners. Consequently biotechnology will probably have less impact on the total number of farms in the United States and developed countries than previous mechanical and chemical technologies adopted by farmers during the last 50 to 75 years. Moreover it is likely biotechnology will not greatly exaggerate the decline in the number of farms,

although it will certainly maintain present trends, which indicate that farming will continue to be one of the fastest declining occupations.

To better understand the potential impact of technologies such as agricultural biotechnologies, it may be important to briefly review the concept of the technological treadmill. Numerous scholars have argued that new production technology allows farmers to reduce the costs of production with early adopters of the technology reaping substantial profits. They produce more than their neighbors can with a comparable investment of time, labor, and capital. As more and more people adopt the new technology, however, total production rises and prices begin to fall. Those operating with the old technologies find themselves operating at a loss, and they often go out of business. On the other hand, those who adopt the new technology, find that higher profits disappear and they are producing more food to retain the same income level. However, the treadmill is more than a technology transfer. It also accounts for how societies consisting of many independent, owner-operated farms become societies that consist of a small number of land and capital-owning investors with masses of workers relegated to wage labor. The social transition described by the technological treadmill process is thus a change in social structure. The fundamental change that occurs is a shift from owner-operators, each with control over their work activity and relatively equal opportunity to succeed, to a society of owners and managers of capital who control the work life of laborers and who determine future directions of society through their investment policies and practices (12).

Genetic engineering is likely not the most important technology implicated in this transition, particularly in industrialized countries where this transition occurred long before the development of biotechnology. This technology may be far more important in affecting the social structure of agricultural economies in the developing world. Small farmers in those nations constitute a significant if not majority portion of the population, and they will become displaced or marginalized as urban populations begin to rely increasingly on industrialized agriculture from Europe, North America, Japan, and Australia. Even the successful few and large producers will have to share a larger portion of their farm profits with the companies that produce the biotechnology and become more dependent on those companies in a manner not unlike those of wage laborers dependent on their employers.

Other changes and impacts on family farms and rural communities include the shift in returns on production from labor to capital. Capitalists use technology to gain a larger share of the value of their product at the expense of labor. A new technology lowers costs and eventually dominates the industry. Those who work in the industry then are forced to accept wages offered by the owners of the technology. Thompson and others argue that this shift violates a farmer's right, for it reduces the farmer's autonomy and control in disposing of his primary assets, land and labor. Kloppenburg and his colleagues (13) have proposed a further impact that concerns the loss of a moral economy associated with traditional agriculture. They argue that the institution of alienable property rights in land introduced commercial practices into food production that have inexorably undercut the moral economy. They propose that rural and urban people can invigorate a moral economy for contemporary agriculture that will reverse the commodifying influence of technology and its attendant impact on the human condition. The underlying assumption here is that if capitalism systematically consigns labor to a situation of wage servitude, it cannot be considered morally legitimate.

The extent of biotechnology's influence on the trend toward fewer and larger farms depends, in part, on how adoption effects the cost structure of farms. If biotechnology development significantly alters costs, returns, competitive positions, and the special location of production, and if certain trade and farm policies are implemented, the potential impact of biotechnology could be relatively important. It has been argued that these new technologies, like those of previous generations, will be adopted by well-financed, innovative farmers who are presumed capable to run the competitively large farms. However, others have argued that biotechnology innovations will provide widespread benefits to the full range of farmers because new technologies will be used in traditional ways. Regardless of which farmers are likely to benefit, however, biotechnology will probably increase the value added off farm at the expense of value added on farm.

Other significant changes in the farming community may result if the information and products of this technology bypass the Cooperative Extension system and the agricultural cooperatives. Previous products and information of biological research have been disseminated through the Cooperative Extension system. However, the development of new seed and chemical packages through biotechnology has emerged from private research. Public sector scientists may have limited knowledge with which to support extension education programs, with a consequence that extension, and potentially agricultural cooperatives, may gradually be reduced to playing a secondary role in farm change. Moreover many agriculturally based rural communities will continue the ongoing process of shrinkage and consolidation, as producers, and local supply and marketing firms continue to decline in numbers. Biotechnology may also accelerate the trend noted above regarding the integration of contract farming, already common in the United Status, where commodities such as poultry and most processed vegetables are produced on contract. These arrangements will further reduce the autonomy of farmers and will certainly reduce their contact with and need for extension education, agricultural cooperatives, and local farm suppliers. The new biotechnologies may also restructure the relationship between farmers and researchers. Until very recently farmers were seen as a primary clientele of public sector research. However, the entry of molecular biology into agricultural research has increasingly been accompanied by the insertion of the agribusiness sector between farmers and researchers. As a result it is quite possible that the interests of agribusiness sector will dominate the agenda setting in the public research arena (2).

Michael Gertler (14), participating in a National Agricultural Biotechnology Council annual meeting, focused on several reasons why he thought agricultural biotechnologies may become, or should be, social issues in rural agricultural communities. He noted that with the advent of the agricultural biotechnology products, some of which are proposed as environmentally friendly such as herbicide-resistant crops, the farmer is likely to incur increased costs and risks without assurance of gains. The use of expensive genetically engineered seeds do not guarantee a commensurate increase in yields. Furthermore supply companies and firms licensing particular generically engineered organisms are adept at charging what markets bear with the economic benefits arising from these technologies likely to be accrued by those holding the patents.

A second concern Gertler observed is the industrialization and accelerated structural change in the farm sector noted earlier. Biotechnology is being introduced in the context of increasing the industrialization of farming. This technology is an important development in responding to environmental, agronomic, and veterinary problems encountered when industrializing livestock and crop production. Gertler argues that it may permit further industrial development without addressing fundamental contradictions and efficiencies of this system. Although biotechnology may appear to be scale neutral, the level of investment required, the increased risk, and need for higher levels of management mean that larger and more capitalized farmers will likely benefit disproportionately. Moreover biotechnologies may create deeper divisions between farmers subscribing to different models or systems of production, between farmers and nonfarm rural populations, and between farmers and the nonrural public. For example, organic farmers may feel even more marginalized as the traditional industrialized agricultural food system embraces these new technologies and public section research is directed more extensively to production systems that overlook their needs. The organic growers, however, may experience increased demand from consumers distrustful of these new genetically engineered food products. Another potential impact is on the self-esteem of farmers as they are transformed from practitioners into the objects of agricultural practice. The proliferation of new genetically engineered products and processes may inhibit the ability of farmers to make educated choices with respect to crops and inputs appropriate to their regions and cropping systems. Gertler concludes by observing that farmers may eventually become more like consumers, less able to distinguish quality because of product proliferation, lack of information, and misinformation.

Perhaps the broadest analysis of potential social and economic impacts of modern technology and the new agricultural biotechnologies on rural communities focuses on the sweeping challenges to democratic rights that it poses. Langdon Winner argues that technical changes have the social effect that are equivalent to legal or constitutional changes. Here citizens would not tolerate such sweeping changes coming through government without due process, but scientists and business leaders are able to bring about wrenching social change through the process of the introduction of new technologies that is totally isolated from public influence or participation. For Winner and others this amounts to a total usurpation of the most fundamental democratic rights and has been the basis for proposing the need for a fourth criterion to regulate technology (12).

Others such as Wendell Berry (15) have argued that industrialization undermines the moral meaning of work which is considered both the formation and expression of personal identity. The hard work that is necessary for traditional farming has the effect of providing the farmer a well-developed sense of self, an identity that attaches naturally to a set of interests arising from work. In contrast, Berry argues that the factory pattern of life encourages people to identify with leisure activities and to acquire interests that are not related to their identity or self-expression. This ecological perspective of work is embedded in his vision of community. Farmers depend not only on each other but on tradespeople, merchants and other members of the rural town. These constitute particular nonuniversal dependencies that establish strong moral bonds to specific community members. In such a community the farmer is linked to others in the community by their work activities that form their personalities and identities. As a consequence, Berry argues, community becomes meaningful as an ethical concept (15). The extent to which biotechnology continues a process of industrializing farming, its impact, in Berry's terms, goes far beyond the farm itself.

AGRIBUSINESS AND INDUSTRY

Many business and government leaders view biotechnology as a force to not only restructure farming and rural communities but also to catalyze a major change in the structure of worldwide agribusiness. They note that the application of molecular biology permits the various segments of the world's largest industrial sector, agribusiness, to form logical linkages to other economic sectors as was never before practical. This $1.3 trillion agribusiness sector, not counting feed and fiber, consists of four basic elements: input suppliers, growers, processors, and consumers. The system has experienced mechanical and chemical eras that contributed to increased productivity and efficiency and will likely continue to make significant contributions in the future. However, according to a number of business leaders, the new biological and biotechnological era will further increase both efficiency and productivity along with the ability to change the quality of food and feed. It will lead to consolidation and new forms of vertical and horizontal integration of the food industry (2,16).

The formation of new biotechnology companies increased dramatically, starting with the founding of Genentech in 1976, with more than 250 small venture capital biotechnology firms founded in the United States over the next decade. Proliferation of these risk-tasking companies helped raise billions of dollars from private investors and gave the United States a comparative lead in the early stages of biotechnology commercialization.

By the late 1980s the number of these firms had grown to over 600. Despite consolidation which began in the industry with mergers, bankruptcies, and major multinational corporation investments, the number of companies have continued to grow. By 1998 the Genetic Engineering News, *Guides to Biotechnology Companies*, listed over 3500 companies worldwide and approximately 1500 in the United States with a substantial number involved in agriculture (17). However, multinational corporations now clearly dominate biodevelopment. These corporations presently control nearly a third of the fledgling bioindustry, a figure predicted to rise to 50 percent in the near future.

During the 1980s and continuing throughout the 1990s, these multinational corporations began diversifying into every field or specialty that uses living organisms as a means of production. The new biotechnologies appear to further reduce the distinctions among the traditional industrial sectors, rendering corporate boundaries virtually unlimited. These large multinational corporations specializing in chemicals, food, and pharmaceuticals have taken the leadership in agricultural biotechnology research and development (e.g., American Cyanamid, Agr Evo, Dow Chemical, Dupont, Eli Lilly, Merck, and Monsanto). At the forefront is Novartis, formed in 1996 by the merger of Sandoz and Ciba-Geigy. Novartis is now the world's largest agrochemical company, the second largest seed firm, the third largest pharmaceutical firm and the fourth largest veterinarian medicine company (16). At the same time companies like Monsanto moved rapidly to expand and consolidate their market share of several key crops with their genetically engineered seeds. For example, in 1997, 15 percent of the U.S. soybean crop was grown from genetically engineered seeds. By 1998 this had grown to 44 percent of the soybean crop and 36 percent of the nations corn crop with Monsanto's Round-up Ready (herbicide resistant) seeds controlling a majority of the market. In 1998 U.S. farmers planted more than 50 million acres of genetically modified crops about six times the acreage planted with such crops just two years earlier (18). Worldwide it was estimated that in 1999 GM crops were grown on over 100 million acres (19). Indeed, two-thirds of the genetically engineered crops available in 1999 are designed specifically to increase the sale of herbicides and pesticides produced by the companies selling the genetically engineered seeds (20). Lappe and Bailey conclude their analyses of the U.S. and international agricultural biotechnology developments by noting that never before in the history of the world has such a rapid and large-scale revolution occurred in a nation's food supply. According to bioindustry analysts, by the year 2025 some 70 percent of the industrial economy and 40 percent of the entire global economy will have, at its base, some form of biotechnology.

Michael Pollan, a writer for *The New York Times*, recently noted that with the advent of biotechnology, agriculture is entering the information age with a small number of multinational corporations positioned to become another Microsoft, supplying the proprietary operating systems to run the new generation of plants and animals (21). Most analysts predict biotechnology will continue and accelerate the trend toward increasing concentration of power in a small number of large multinational corporations. Consequently development and commercial control of agricultural biotechnology will be in the hands of corporations that transcend geography boundaries and hold limited national allegiance. Within this context people question how we can ensure that democratic participation will occur in the decision-making processes surrounding the development and commercialization of biotechnology. This is difficult within the national boundaries and generally prohibited internationally given current government structures.

CONSUMERS

For consumers, the new biotechnologies could mean dramatic improvements in the productivity and efficiency of food production and processing, and the expansion and extension of food and nonfood uses of raw agricultural commodities. Consumers could benefit in the form of reduced prices, increased food safety, and more nutritional foods. Products in the pipeline, for example, could produce plants that lack allergenic proteins or have a healthier oil composition and may also provide benefits for developing countries such as the pro-vitamin A and iron-enriched rice (22). The new technologies also have a potential to change the very nature of food itself and to expand a range of possible food products. It is now possible to consider the production of new fabricated foods in which basic foods are broken down into their component parts (e.g., starch, fat, and sugar) and recombined into wholly new types of food. However, to date, the new food products and processes have been met with mixed reactions. On one level are the concerns about food safety from unexpected allergenic or toxic foods resulting from the insertion of a foreign gene into crops, food, or animals. At issue is whether a foreign gene can activate the expression of a latent toxic gene. On a more subtle level, the new biotechnologies may make it far more difficult in the future for the consumer to determine the composition of the food and to maintain a balanced diet.

Another impact of biotechnology has been the stimulation of new moral and ethical debates among consumers and the general public regarding the limits of science. Public concern about a range of scientific developments, including biotechnology, are resulting in a decline in public confidence in science and increasing public perception of the likelihood of environmental risks from genetically altered bacteria, plants, and animals. Krimsky and Wrubel argue that on the basis of their analysis the level of pre-market public scrutiny of some of the first products of agricultural biotechnology has been unprecedented. Citizens are demanding earlier entry points and broader participation in technological decisions (23). The development of biotechnology is stimulating a wider range of public concerns about science that extend beyond human health, environmental risks, food safety, and animal health issues and includes such concerns as negative socioeconomic consequences and the morality of tampering with nature and life itself. Many environmental and consumer groups view transgenic food as a symbol of the assault on traditional sources of food. At issue here is the dignity of the food supply even more than its safety.

In 1999 reacting to the escalating public concern, several European supermarket chains banned GM (genetically modified) products from their house brands. Moreover in Britain, Unilever and Nestle announced that they would phase out genetically modified ingredients in their products. Meanwhile the European Union decided that products in which more than 1 percent of one of the ingredients was transgenic should be labeled and that the introduction of new GM crops would be suspended for several years. Even in the U.S. the Federal Drug Administration chose to hold public hearings around the country on whether it should adjust its role in regulating GM crops. These hearings were often confronted by consumer protests against "foods created by altering genes" (frequently characterized as "Frankenfoods"). Finally, leading food manufacturers in the U.S., Gerber and Heinz announced that they would permit no GM foods in their products. Many analysts have concluded that the next few years will be crucial for the future of GM crops and that in the end consumers, rather than the farmers that the industry had long considered its primary customers, will decide the fate of GM foods (22).

While, a small number of groups oppose any form of genetically modified food, the general focus of public policy consumer concerns has turned towards the question of labeling. Krimsky and Wrubel (24) note that labeling transgenic food would enable consumers to express social values in their food preferences, which is consistent with the trend towards "green consumerism." In 1999 labeling genetically modified food was high on the agenda of the Codex Alimentarius Commission with members of the European Union who were conscious of public pressure over GM foods; they argued that any food containing detectable GM ingredients should be labeled (24). From recent analyses of public and consumer concerns about biotechnology in Europe, people seem prepared to accept some risks as long as there is a perception of usefulness and no moral concern. But crucially, moral doubts act as a veto irrespective of people's views on use and risk. In one study the finding that risk is less significant than moral acceptability in shaping public perceptions of biotechnology held true for each European country and across all six specific applications of biotechnology (genetic testing, medicines, crop plants, food production, research animals, and xenotransplants). In the same survey, 74 percent of the respondents consider that genetically modified foods should be labeled and 60 percent believe that there should be public consultation about new developments in biotechnology (25).

In a column in *Nature*, the editor stated that the genetically modified foods debate needs a recipe for restoring trust. While there is no simple institutional formula for achieving this, some of the principles include (1) acceptance for the need to ensure the regulation of genetically modified foods based on the soundest possible science, (2) acknowledgment of the current limits to scientific certainty, (3) the need to find ways of facilitating public access to credible scientific information and of communicating it in a responsible form, (4) the need for honest brokers, and (5) taking into account in food regulations broad public concerns. The editorial concluded if labeling all foods produced by genetically modified techniques turns out to be a necessary step in regaining trust on both sides, it should be a small price to pay (24).

SCIENCE AND TECHNOLOGY TRANSFER

Perhaps the most dramatic, immediate impact of the new biotechnologies is on science itself. During the last 20 years, the convergence of a number of new scientific techniques and biotechnologies, legal policy, and commercial developments has had a major impact on the way in which knowledge is generated and commercialized and on the evolution of agriculture and our food system. While some argue that biotechnology is a continuation of the application of biological techniques to improve plants, animals, and microorganisms, many biologists contend that biotechnology has revolutionized the field. The knowledge and tools generated by molecular biology and biotechnology have stimulated a great deal of enthusiasm and redirected large sums of money in an effort to pursue knowledge in this area. At the federal level, financial support for biotechnology has grown steadily since the mid-1980s and reached several billion dollars annually in the 1990s. While 80 percent of the federal nonmilitary research budget has been devoted to the National Institutes of Health (NIH) program, support for agricultural biotechnology has been relatively meager, constituting less than 3 percent of federal expenditures for biotechnology.

The techniques and tools of biotechnology are facilitating basic research efforts to understand the intricate, complex, functioning of living organisms at the molecular and cellular level. This reductionist approach, often called logical positivism, continues and extends the basic methods and approaches of modern science. Modern biology attempts to reduce nature to small, definable pieces, subject to human manipulation, and separated from broader questions of value. From this perspective, scientists control, measure, reduce, and divide nature in order to generate knowledge. Biotechnology, particularly in agriculture, may truncate both the time and space required to develop new plant, animal, and food products. However, one concern is that this approach, while providing important but only partial knowledge, is rapidly becoming the dominant epistemology, often to the exclusion of important alternative ways of knowing. As a consequence lack of adequate support has occurred for critical complementary research to molecular biology and genomics agendas, and this includes whole-plant and animal-level research (e.g., traditional plant breeding), systems-level research programs (e.g., agroecology, farming systems), social assessments, and indigenous knowledge (2).

Another development stimulated by agricultural biotechnology with implications for the generation of knowledge is the increased concentration of research funds, scientific talent, and intellectual property at a small number of public and private institutions. In the public sector, every U.S. state could afford and has had conventional soils, breeding, and pathology programs. Every state

cannot afford and will not be able to have a comprehensive agricultural biotechnology program. By the late 1980s and early 1990s, for example, eight states accounted for over half of the State Experiment Station expenditures and nearly half of all science years for agricultural biotechnology research (26). It is unclear how the absence of diverse and heterogeneous institutions and groups of scientists will affect the generation and dissemination of knowledge.

The new agricultural biotechnologies are also contributing to a changing collaborative relationship between the universities and industries. While partnerships between universities and industries have existed for several decades, the new types of university and industry relationships in biotechnology are generally more varied, wider in scope, more aggressive and experimental, and more publicly visible than the relationships of the past. The legal/contractual bases for these relationships depend on the goals and institutional characteristics of the partners, and consequently involve diverse approaches including: large grants and contracts between companies and universities in exchange for patent rights and exclusive licenses to discoveries; programs and centers organized with industrial funds at major universities, that give participating private firms privileged access to university resources and a role in shaping research agendas; professors, particularly in the biomedical sciences serving in extensive consulting capacities on scientific advisory boards or in managerial positions in the firms; faculty receiving research funds from private corporations in which they hold significant equity; and public universities establishing for profit corporations to develop and market innovations arising from research (27).

A notable example of these new types of collaborative arrangements between universities and industry is the five-year $25 million "strategic" research alliance announced in late 1998 between the University of California, Berkeley's College of Natural Resources and a unit at the Swiss biotechnology giant, Novartis. While large multimillion dollar industry grants to universities are not unheard of, this agreement applies not to a single researcher or team focusing on a specific topic but rather to the entire department of plant and microbiology. Under the agreement the Novartis until will provide funds and access to proprietary technology to Berkeley faculty members and graduate students, and in return it will receive first rights to negotiate licenses up to one-third of the inventions that result. Novartis is also considering the development of a facility on or near the Berkeley campus for 20 to 30 of its own scientists who would be available to work with university researchers and to share equipment and space (28).

The university and the private sector have very different goals for research and ways of pursuing those goals. When collaborating, the consequences of these two distinct and complementary research communities can be both positive and negative. In the United States, for example, university–industry collaboration may bring useful products to market more rapidly and promote U.S. technological leadership in a changing world economy. Second, in light of funding stagnation within U.S. Department of Agriculture (USDA) and in many states, such collaborations are a means of raising new funds for university research and support for graduate education. Third, these joint efforts may expand the scientific network, increasing communication between some university and industry scientists and provide some university scientists access to cutting-edge research tools, proprietary materials, and vast databases owned by the particular company (29).

However, a number of concerns have been voiced regarding the impact of these new relationships. First, long-term research, previously a major emphasis of the public sector, may decline. The private sector has short-term proprietary goals, and as a consequence funding for research is also generally short term, spanning one or two years. In contrast, nearly all the federal NIH extramurally funded programs and USDA Hatch-based funded projects are for three years or longer. Moreover dependence on private sector funds will generally change not only the time frame but also the stability of funding. It seems unlikely that these university–industry relationships will provide stable long-term funding, nor will they significantly address the capital needs of the universities. For example, in a study of executives of 210 agricultural, chemical, and pharmaceutical corporations, 59 percent reported supporting university research totaling $340 million for more than 1500 projects. However, most said their support lasts two years or less and involves research contracts for less than $100,000 (30).

Universities are also concerned about ensuring that research projects are generally originated by faculty members and not adopted as a result of outside pressure, either implicit or explicit. If a sufficiently large and influential number of academic scientists and engineers become involved with industry, a whole range of research agendas, traditionally the purview of the university community, might be de-emphasized. Furthermore the scientific community could become desensitized to the environmental or social impacts of proprietary research. Some research that lacks commercial application could be neglected entirely.

As noted earlier, with increased focus on knowledge and technology as intellectual property, particularly in the biological arena, there has been an enormous increase in patents, licensing, and material transfer agreements. Many analysts suggest that these new practices and processes may impede or limit the pace and direction of scientific efforts, restrict scientific communication, or undermine an academic scientist's ability to carry out research. The potential restriction of communication is particularly true of university scientists with private sector grants, who often must delay public discussion of work, or its results, pending review by the sponsoring company. Even some scientists with public funding feel inhibited about discussing their work, for fear that some private company with the money, equipment, and time will utilize their ideas and perform the experimental work before they can (29). Companies sponsoring university biomedical research often ask scientists to go beyond the standard secrecy requirements needed to obtain a patent for products related to their research. While NIH calls for a delay of only one or two months while an application

is filed, 58 percent of the companies in a recent survey ask researchers to keep data secret for more than six months (31). The net effect of these various developments appears to be a reduction of the free flow of information. Many have argued that open communication and the freedom of thought is the best path towards realizing the social benefit from science (12).

A final impact involves potential conflicts of interest and/or scientific misconduct. In interviews, public and private sector scientists alike stress the potentially detrimental effects of restrictive agreements between the universities and corporations. These effects include favoritism, unwarranted financial advantages through privileged use of information or technology derived from the publicly funded research, and shelving of research of interest to the public but not to the corporation (2). In a recent article entitled "University–industry research must get closer scrutiny," Mildred Cho observed that "one major reason for concern is that if faculty members are profiting financially from their research either through royalties from, or as investors in, companies that market products based on their discoveries, the outcome or direction of their work may be affected. They might, for example, be tempted (consciously or unconsciously) to design studies that are more likely than not to have an outcome favorable to the product" (31, p. B4). A recent study (32), for example, designed to measure how drug company money might influence scientists, points directly toward a need for disclosure of industry relations and funding sources. This study found that 96 percent of the researchers who wrote favorable articles about a controversial class of drugs for treating hypertension and angina also had financial ties to the marketers of those drugs. In contrast, among those who published articles critical of the drugs, only 37 percent had financial ties. Conflicts were disclosed in only 2 of the 70 papers. As one researcher at George Washington Medical Center noted, researchers like to think that they are not influenced by their financial ties, "but the pressures may be too subtle for them to realize" (33, p. A41). These divided loyalties and conflicts of interest betray the public trust. According to Krimsky, the most significant social consequence of change within scientific institutions is "the disappearance of a critical mass of elite, independent and commercially unaffected scientists to whom we turn for vision and guidance when we are confronted by technological choices" (3, p. 79). Thompson (12) has further noted that as little as the public might care about the institutional effects of biotechnology within science, they may well be among the most far-reaching.

DEVELOPING COUNTRIES AND THE GLOBAL ECONOMY

The new technologies offer the hope of increasing crop yields where population growth is outstripping the food supply. Microbiology in conjunction with plant propagation and breeding is already creating more drought and resistant varieties of cassava, oil palms, and groundnuts. Yet despite biotechnologies' great promise for feeding the world's rapidly growing population, particularly in developing countries, science and policy makers admit it will not be easy to ensure that this technology has the desired positive effects. Much that has been said about the social consequences for family farming and rural communities in industrialized countries applies more dramatically to the resource poor farmers in developing countries. Several analysts have predicted that biotechnology will have an unfavorable impact on the rural poor in Africa, Asia, and Latin America while benefiting relatively better-off farmers in those regions. As farms become larger and fewer, more people both in absolute numbers and as a percentage of the population in developing countries are being affected. Those who are affected are much worse off to begin with and are more vulnerable to displacement.

There is legitimate concern arising that the developed nations will use the new technology to undercut traditional Third World exports, such as vanilla, sugar, cocoa butter, and other important cash crops. Genetic engineering processes are being used to transform the production of certain agricultural commodities into industrial processes. In principle, any commodity that is consumed in an undifferentiated or highly processed form, could be produced using new biotechnological processes, and product substitutions could be easily introduced. Similarly, although with greater difficulty, tissue culture techniques could be used to produce edible plant parts in vitro. Several companies are now capable of final production of a natural vanilla product in the laboratory. A genetic modification of oilseed plants to convert cheap oils (e.g., palm or soybean oil) into high-quality cocoa butter is well advanced. Biotechnology is also being used to produce substitutes for sugar as an industrial sweetener. Even moderate success in realizing these product substitutions would have profound effects around the world, most immediate and important would be the restructuring of global markets (34–36).

Another concern is that biotechnology will increase the disparities between the developed and developing nations. With the shift in applied research and associated product development from the public to the private sector, the benefits from the new biotechnologies may become less widely available. Furthermore the products developed are unlikely to be ones that are important to the poor developing countries, particularly in the tropics. Only a small amount of the estimated $2.5 billion dollars of research spending on agricultural biotechnology around the globe is carried out in the developing world. According to Robert Herdt, from the Rockefeller Foundation, between $50 and $75 million per year is spent on agricultural biotechnology in the developing world, with about half of that conducted by the Consultative Group for International Agricultural Research (CGIAR) centers. However, the financial support for that system has weakened in significant part because of declining U.S. support. While the International Service for Applied Agricultural Agrobiotechnology data on field trials of genetically engineered crops reveal over 3700 trials of genetically engineered crops through the end of 1995, none of the field trials to date has been directed specifically at increasing output. Instead, most of the work has been done on transforming crops to be herbicide resistant (40 percent) and insect resistant (22 percent). Research

on some of the traits most needed in the developing world such as the ability to tolerate low soil fertility, the ability to tolerate soil salinity or alkalinity, and techniques for producing biological pesticides has gone unstudied. This could further widen the gap between the agricultural production methods in the North and the less developed practices in the South (37).

A further concern revolves around the controversy over the property rights in genetic resources. Biotechnology is playing a key role in conserving genetic diversity worldwide at the same time as it is accelerating the privatization of these genetic resources. Biotechnology is providing new incentives to patent commercially valuable genetic resources as well as providing the means to both enhance those resources and protect them from patent infringement. It has been argued that native genetic resources (i.e., germ plasm or seeds) are owned by indigenous farmers, by their governments, or collectively, by the whole society and considered the common heritage of humankind. Intellectual property rights may deprive farmers in developing countries of something they currently have. Farmers are losing the right to plant seed freely from land races or other publicly available varieties (12). This is probably unlikely since most legal codes protect any existing uses of the raw materials from which new seed varieties or plants are derived. For example, the convention on Biological Diversity, which came into force in December 1993, provides an internationally legally binding instrument that explicitly recognizes national sovereign rights over the genetic resources existing within a country's territory (38). However, although indigenous farmers may have a legal right to use plants in traditional ways, they lack the resources and knowledge to protect those rights. Moreover many countries, especially developing countries rich in genetic diversity, have called for measures to protect their interests and to insure that they share in the benefits derived from the use of their resources by others (39). Issues of ownership, access to genetic resources, and intellectual property protection of genetically engineered products are currently receiving considerable attention in various international forums. These complex and highly politically charged debates will likely continue into the future.

Finally, many developing countries have no basic and limited applied research capacity, marginal capabilities to adapt biotechnological advances to local conditions, and few resources to attract transnational corporations to conduct their own research. In conclusion, agricultural biotechnology may shift the geographic location of agricultural production from one Third World country to another or from the Third World to the First World. The literature on social consequences of agricultural biotechnology for developing countries includes very little in the way of detailed ex ante studies. However, for many Third World countries that are dependent on one or two agricultural commodities for their continued viability, this production and market restructuring, and increased productivity gaps, could result in a collapse in existing markets. Significant numbers of farmers and farm workers could find themselves with no products to sell. This could increase the already high Third World debt and exacerbate the deficit imbalance of payments in Third World countries. If this were to occur, political instability, already a problem in the developing world, would doubtless increase.

STRATEGIES FOR INCORPORATING THE FOURTH CRITERION

This article has discussed a number of both positive and negative social and economic impacts biotechnology may have on (1) farmers, rural communities, and the food system; (2) the structure and organization of agribusiness and industry; (3) consumers; (4) science and technology transfer; and (5) developing countries and the global economy. Genetic engineering and biotechnology are areas in which nontechnical decision-making inputs, on initially technical issues, are found to be increasingly important. It was noted that reduction of risks to health and environment are usually emphasized as the task of any regulation but that goals based on social and ethical principles also appear within the scope of genetic engineering laws and regulations. Indeed, it is argued that the health and environmental risks, as well as the socioeconomic hazards, are difficult to separate. Moreover the issues of risk to environment, life and limb appear to be a necessary but inadequate conditions for public acceptance. Although the socioeconomic criteria, often referred to as the fourth hurdle or fourth criterion, have been clearly rejected in licensing, it is equally clear that socioeconomic criteria are being taken into account in numerous contexts (e.g., the Austrian Genetic Engineering Act; the Norwegian Gene Technology Act; the UNEP Technical Guidelines for Safety and Biotechnology; release applications for genetically engineered organisms in Denmark, Sweden, and Finland; the Swiss Gene Protection Initiative).

As a consequence a number of alternatives have emerged for incorporating the socioeconomic issues into a broader public discussion and eventually into decision making. One process or procedural approach described by Thompson (12) is discourse ethics where certain morally relevant constraints on discourse must be met for it to be reasonably successful. First, discourse must be open to all competent speakers whose interests will be affected. Second, people must be free to construe the issue and their own interests in whatever terms they deem appropriate. Third, participants must be free of rigid inflexibility that precludes them from adopting a hypothetical stance toward their own and others' interests and values. Fourth, the process must be free of external coercion and, fifth, statements must be focused exclusively on establishing the best reasons for accepting a prescription or conclusion (12). Thompson admits that since these conditions are rarely met, actual public debates over biotechnology are unlikely to reach an ethically defensible consensus. However, the debates will be greatly enhanced if informed by efforts to approximate these ideal discourse conditions—if not in a public forum, then at least under some controlled circumstances in which these issues can be seriously pursued.

Seifert and Torgensen (6) also examined various mechanisms whereby socioeconomic criteria can be included in the decision-making process for new technologies such as genetic engineering. Their discussion of the Austrian efforts focused on incorporating social sustainability in consideration of regulations for genetic engineering. They suggested three forms to enhance broader participation in the debates. The first is discursive meditation whose main emphasis is on clarifying controversial issues. The best example of this process in the United States is the annual public forum conducted by the National Agricultural Biotechnology Council that brings together diverse participants and stakeholders in the arena of agricultural biotechnology to discuss and clarify concerns surrounding agricultural biotechnology. Consensus conferences are a second form that serves the task of a development of a political will. Institutional innovations to provide for independent mediation among competing positions on issues like biotechnology is a third form. For example, the Dutch consumer protection organization Consumer and Biotechnology is an example of such an institutional innovation. The function of this institution is to mediate among nongovernmental organizations, the biotechnology industry, the authorities, agricultural associations, and consumer organizations.

All these processes and procedures, however, have two fundamental problems. First, the results are generally not binding and require the approval of the decision makers for their implementation. To partially overcome this limitation in Denmark, for example, the institution of consensus conferences was established in association with the parliament so that the results of the procedures and process could be considered in parliamentary decision making as quickly as possible. The second problem concerns the vulnerability of key parties to manipulate the results. In all the processes there is a risk of being co-opted or overwhelmed by powerful interests and ultimately serving only as justification for the implementation of these new technologies. Therefore the goals and motives of the organizers, the credibility of spokespersons who both raise and discuss the range of issues surrounding biotechnology, the funding sources for these procedures and processes, the process by which content, scope, and audience for these processes are made, and how this information and the outcomes of the conferences will be pursued must be kept as transparent as possible.

Barling and his colleagues (40) in examining social aspects of food and agricultural biotechnology in Europe have observed that public concern about these new technologies are primarily focused on issues of trust, choice and care for a sustainable society of natural balance. They have recommended improving consumer choice and promoting greater public involvement in decision making. Complementing labeling, they propose transparency right through the food chain. This would entail the use of a comprehensive system of segregation and certification of genetically modified crops and their products from nongenetically modified crops at each stage of the food chain, and for this to be reflected in the final labeling information. This degree of transparency, they argue, would allow consumers to make more fully informed choice of foodstuffs in line with their more deeply felt values on such issues and would provide for a more democratic and participatory basis for transparency. Finally, in attempting to predict the effects of the use of genetically modified organisms in agriculture and food production, they propose integrating the precautionary principle more actively into risk management. As noted earlier, the precautionary principle is applied in circumstances of scientific uncertainty reflecting the need to take action in the face of potentially serious harm in the absence of scientific proof. As a consequence the precautionary principle is not simply a matter of science but is socially and politically informed. They argue that the incorporation of wider social concerns, as articulated by different social actors, should be included in risk analysis to produce a more socially embedded and accepted process of risk analysis of the applications of modern biotechnology. This would build greater trust in and acceptance of the regulatory process and provide a more socially responsible, plural, and accountable form of decision making. Implementation, of course, remains quite complex and highly problematic.

While efforts to enhance the dialogue and to ensure wider participation in the debates is laudable, Youngberg (41) at a National Agricultural Biotechnology Council annual meeting challenged the participants to move beyond preoccupation with the dialogue process and to begin to explore new and innovative ways to involve the broader society such as the sustainable agriculture community in the biotechnology decision-making process itself. He noted that there is a critical difference between broad participation in the dialogue about biotechnology, and the actual involvement in planning and decision-making phases of agricultural biotechnology research development and the introduction of these technologies into the marketplace. He continued by suggesting that the time has come for the biotechnology industry to begin exploring the principles and processes of participatory decision making and to initiate a serious assessment of ways to implement concrete decision-making opportunities involving all elements of the agricultural biotechnology constituency, including farmers, public interest group representative, and other citizens. He concluded by noting that the dialogue offers only sporadic, short-term opportunities for interaction while ongoing relatively intimate interactions characteristic of participatory decision making would create authentic opportunities to directly influence the biotechnology agenda. This action would likely create greater trust, result in more comprehensive and enlightened planning that includes meaningful consideration of the socioeconomic issues, potentially save money, time, and resources, and make possible endorsement not mere acceptance of the new agricultural biotechnologies.

CONCLUSION

Although introduced as the fourth criterion, it may be more appropriate in evaluating public research agendas as well as the regulation and approval of new agricultural biotechnology products and processes to consider the broader socioeconomic effects as the first criterion.

As most scientists and policy analysts acknowledge, biotechnologies are the tools and means to achieve particular socioeconomic goals. These biotechnologies are options, albeit compelling options, among many and involve choices. As such they should be framed and evaluated in terms of local, regional, national, and international social goals and values. In the final analysis the key question is not how we learn to accept and live with the new biotechnologies but rather under what conditions, price, costs, gains, and benefits on a personal, community, national, and global scale. In a democracy the public has an obligation and a right to be informed, to participate, and to shape the developments of technology in terms of the broader socioeconomic values of their respective society. The effective public representation of the public interest also ensures that society avoids the potential abuses of power by those with vested interests (16). In the case of agricultural biotechnology as we have seen, the public is increasingly exercising their obligations and rights.

Thompson concludes his thoughtful analysis of the ethics of food biotechnology by looking at the relationship among science, trust, and democracy. He notes that to the extent "that democracy is understood as a form of government distinctive for its receptivity to participation and resting upon consent of the governed, the events that turn ordinary people into enemies of science can be seen to compromise government, rather than science" (12, p. 237). How we balance scientific criteria embedded in traditional risk analysis and the social criteria embedded in the fourth criterion may well determine the extent to which the future of biotechnology is able to live up to its promise as the first important science of the twenty-first century.

BIBLIOGRAPHY

1. J. Rifkin, *The Biotech Century: Harnessing the Gene and Remaking the World*, Tarcher/Putnam, New York, 1998.
2. L. Busch, W.B. Lacy, J. Burkhardt, and L.R. Lacy, *Plants, Power and Profit: Socioeconomic and Ethical Consequences of the New Biotechnologies*, Basil Blackwell, Oxford, UK, 1991.
3. S. Krimsky, *Biotechnics and Society: The Rise of Industrial Genetics*, Praeger, New York, 1991.
4. P.C. Stern and H.B. Fineberg, eds., *Understanding Risks: Informing Decisions in a Democratic Society*, National Academy Press, Washington, DC, 1996.
5. G. Weber, Extension Biotechnology Coordinator's Memo, USDA-ES, Washington, DC, March 24, 1990.
6. F. Seifert and H. Torgersen, *Public Understand. Sci.* **6**, 301–327 (1997).
7. T. Crompton and T. Wakeford, *Nature Biotechnol.* **16**, 697–698 (1998).
8. G. Schatz, *Science* **281**, 810–811 (1998).
9. U.S. Executive Office of the President, *Use of Bovine Somatotropin BST in the United States: Its Potential Effects*, Washington, DC, 1994, pp. 35–36.
10. U.S. Department of Agriculture, *A Time to Act: A Report of the USDA National Commission on Small Farms*, USDA, MP-1545, Washington, DC, 1998.
11. S. Smith, *Choices* **7**(1), 1–10 (1992).
12. P. Thompson, *Food Biotechnology in Ethical Perspective*, Blackie Academic and Professional, London, 1997.

13. J.J. Kloppenburg, J. Henrickson, and S. Stevenson, *Agric. Hum. Values* **13**(3), 33–42 (1996).

14. M.E. Gertler, in R.W.F. Hardy, J.B. Segelken, and M. Voronmaa, eds., *Resource Management in Challenged Environments*, National Agricultural Biotechnology Council, Ithaca, NY, 1998, pp. 137–145.

15. W. Berry, in W. Vitek and W. Jackson, eds., *Rooted in the Land: Essays on Community and Place*, Yale University Press, New Haven, CT, 1996, pp. 76–84.

16. R. Hindmarsh, G. Lawrence, and J. Norton, in R.G. Hindmarsh, G. Lawrence and J. Norton, eds., *Altered Genes: Reconstructing Nature, the Debate*, Allen & Unwin, St. Leonards, Australia, 1998, pp. 3–23.

17. Genetic Engineering News, *Guides to Biotechnology Companies*, GEN Publishing, Larchmont, NY, 1998.

18. D. Strickland, *Genetic Eng. News* **19**(21), 1, 12, 62, 72 (1999).

19. D. Ferber, *Science* **286**(5448), 1662–1666 (1999).

20. M. Lappe and B. Bailey, *Against the Grain: Biotechnology and the Corporate Takeover of Your Food*, Common Courage Press, Monroe, ME, 1998.

21. M. Pollan, *N.Y. Times Mag.* October 25, 1998, pp. 44–51, 62, 63, 82, 92, 93.

22. M. Enerink, *Science* **286**(5448), 1666–1668 (1999).

23. S. Krimsky and R. Wrubel, *Agricultural Biotechnology in the Environment: Science, Policy, and Social Issues*. University of Illinois Press, Urbana, IL, 1996.

24. *Nature* **398**(6729), 639 (1999).

25. Biotechnology and the European Public Concerted Action Group, *Nature* **387**, 845–847 (1997).

26. W.B. Lacy and L. Busch, in J.F. MacDonald, ed., *Agricultural Biotechnology at the Crossroads: Biological, Social and Institutional Concerns*, National Agricultural Biotechnology Council, Ithaca, NY, 1991, pp. 153–168.

27. W.B. Lacy, L.R. Lacy, and L. Busch, in M.C. Hallberg, ed., *Bovine Somatotropin and Emerging Issues*, Westview Press, Boulder, CO, 1992, pp. 3–32.

28. G. Blumenstyk, *Chron. Higher Educ.*, December 11, 1998, p. A56.

29. W.B. Lacy, in S. Wolfe, ed., *Knowledge Generation and Transfer: Implications for the 21st Century*, forthcoming.

30. D. Blumenthal, E.G. Campbell, N. Causino, and K.S. Locus, *N. Eng. J. Med.* **335**, 1734–1739 (1996).

31. M.K. Cho, *Chron. Higher Educ.*, August 1, 1997, pp. B4–B5.

32. H.T. Stelfox, G. Chua, K. O'Rourke, and A.S. Detsky, *N. Eng. J. Med.* **338**(2), 101–106 (1998).

33. G. Blumenstyle 1998b. *Chron. Higher Educ.*, May 22, 1998, pp. A40–A41.

34. N.P. Barbosa, *Int. J. Sociol. Agric. Food* **4**, 161–168 (1996).

35. R.M.A.A. Galhardi, *Futures* **27**(6), 641–656 (1995).

36. H.J. Shand, in J.F. MacDonald, ed., *Agricultural Biotechnology and the Public Good*, National Agricultural Biotechnology Council, Ithaca, NY, 1994, pp. 73–86.

37. R. Herdt, in R.W.F. Hardy, J.B. Segelken, and M. Doionmaa, eds., *Resource Management and Challenged Environment*, National Agricultural Biotechnology Council, Ithaca, NY, 1998, pp. 33–40.

38. G. Hawtin, in R.W.F. Hardy, J.B. Segelken, and M. Doionmaa, eds., *Resource Management in Challenged Environments*, National Agricultural Biotechnology Council, Ithaca, NY, 1998, pp. 113–121.

39. W.B. Lacy, *Crop Sci.* **35**(2), 335–345 (1995).

40. D. Barling et al., *Environ. Toxicol. Pharmacol.* **7**, 89–93 (1999).

41. G.I. Youngberg, in J.F. MacDonald, ed., *Agricultural Biotechnology and the Public Good*, National Agricultural Biotechnology Council, Ithaca, NY, 1994, pp. 147–150.

See other entries AGRICULTURAL BIOTECHNOLOGY; ANIMAL BIOTECHNOLOGY, LAW, FDA REGULATION OF GENETICALLY MODIFIED ANIMALS FOR HUMAN FOOD USE.

ANIMAL BIOTECHNOLOGY, LAW, FDA REGULATION OF GENETICALLY MODIFIED ANIMALS FOR HUMAN FOOD USE

STUART M. PAPE
PAUL D. RUBIN
Patton Boggs, LLP
Washington, District of Columbia

OUTLINE

Background
Benefits of Animal Biotechnology for Human Food Use
    Biotechnology Overview
    Examples of Animal Biotechnology
Regulatory Categories of Animal Biotechnology
    Biopharm Animals
    Somatic Cell Therapy
    Transgenic Breed Modification
U.S. Regulatory Agencies: Overview
FDA Regulation of Biotechnology
    Background: FDA Regulation of Biotechnology in General
    FDA Regulation of Genetically Modified Animals Intended for Food Use
Regulation of Biotechnology by Agencies Other Than the FDA
    USDA
    Environmental Issues and EPA Involvement
Conclusion
Bibliography

BACKGROUND

Since 1981, when scientists at Ohio University were credited with producing the first genetically modified animals by transferring genes from other animals into mice, there have been significant scientific advancements in the field of animal biotechnology. Despite these advancements, genetically modified animals intended for human food use have not yet been commercially distributed in the United States. The use of animal biotechnology for human food use, however, is expected to become prevalent within the next decade.

To date the primary focus of biotechnology policy has been directed toward agricultural biotechnology products—which have been marketed in the United States for a number of years. Agricultural biotechnology products have been highly controversial and have been subject to significant criticism from a variety of interest groups, scientists, and members of the public (particularly in Europe). Many observers expect the controversy and criticism directed toward genetic modification of animals to be even more intense.

Although not actually involving genetic manipulation of animals, the public's reaction toward the use of recombinant bovine somatotropin (rBST), also known as recombinant bovine growth hormone (rBGH), in cows is instructive (1). In 1993, after a comprehensive review of safety and efficacy, the Food and Drug Administration (FDA) approved a new animal drug application (NADA) for a product, called Posilac, that contained rBST. The genetically engineered hormone, which was found by FDA to be identical to natural pituitary-derived bovine growth hormone, was approved for injection into cows to increase milk production. (Although the cows received the rBST injections, they were not genetically modified (2).)

Subsequently, based on the alleged impact on cow health and milk, numerous challenges were made to FDA's finding that food products from cows treated with rBST are safe for human consumption (3–6). Challengers argued that the possible adverse health effects of Posilac were not addressed, in part because long-term toxicology studies to ascertain human health safety were not required by FDA or conducted by the NADA applicant, Monsanto.

In response to these challenges, the FDA conducted a comprehensive audit of the human food safety sections of the NADA supporting the drug approval (2). The audit reviewed all of the studies relied upon to determine the human food safety of rBST. FDA concluded that an examination had not been performed on antibody data during the course of the original review of the Monsanto application (2). FDA subsequently reconsidered all of the studies and concluded that there were no new scientific concerns regarding the safety of milk derived from cows treated with the drug (2). The determination that long-term studies were not necessary for assessing the safety of rBST was based on studies that demonstrated that rBST is biologically inactive in humans even if injected, and that rBST and pituitary-derived bovine growth hormone are biologically indistinguishable.

The public's reaction toward the use of rBST in cows provides evidence of the extensive controversy that may result when changes are made to animals that affect the human food supply. Not surprisingly, the controversy is expected to be even more intense if genetic modifications are involved.

Compared with agricultural biotechnology, however, the visceral reaction opposed to the use of biotechnology techniques may be tempered in the United States by the fundamental difference between the current regulation by FDA of agricultural biotechnology and genetically modified animals intended for human food use. Whereas most genetically modified agricultural products are currently only subject to a voluntary notification system by FDA, genetically modified animals intended for human food use

are subject to extensive agency regulation. FDA is currently reevaluating its policy toward agricultural biotechnology and is considering adopting a mandatory review policy. The mandatory and arduous nature of animal biotechnology regulation may satisfy the "safety" concerns raised by critics of biotechnology as applied to agricultural products, and could conceivably mitigate opposition to genetically modified animals. Other issues, however, including religious and/or ethical-based concerns, could subject animal biotechnology to the same public backlash that has recently plagued agricultural biotechnology.

BENEFITS OF ANIMAL BIOTECHNOLOGY FOR HUMAN FOOD USE

Biotechnology Overview

The Genetic Code. Living organisms contain cells that contain DNA (deoxyribonucleic acid) in their chromosomes. DNA contains the genetic code, or the genome, for an organism. The genetic code is derived from a four-letter alphabet, A,C,G,T (Adenine, Cytosine, Guanine, and Thymine), and based on the sequence and number of genes, the hereditary traits and characteristics of the living organism is determined.

The chemical and physical composition of DNA does not vary from organism to organism. It is only the sequence and number of letters in the genome that create differences between different living organisms—the physical and chemical composition of the actual DNA, and letters themselves, are constant from organism to organism. In practice, this means that DNA from any organism is capable of functioning even if it is transferred into a different organism.

Definition of Biotechnology. Biotechnology has been defined as the manipulation of the DNA molecules of living organisms. By means of selective breeding of animals and crops, humans have practiced basic biotechnology for centuries. The transfer of DNA occurs naturally through sexual reproduction and has been utilized in plant and animal breeding programs for centuries in order to alter the traits in living organisms.

Modern biotechnology techniques, however, are technologically superior to traditional breeding in that (1) precise genetic manipulation can alter specific genes while leaving others unchanged and (2) genes not native to an organism may be added from a distinguishable organism (i.e., a distinguishable species).

Intragenetic combinations involve the transfer of genes that are native to a species (e.g., adding an additional growth hormone gene that is ordinarily found in the species). Intergenetic combinations, which involve the addition of a nonnative gene from one species to another (e.g., adding a nonnative disease resistance gene from one species to another), allow the transfer of desirable traits found in nature from one organism to another. In other words, rather than relying solely upon conventional breeding, which can be structured to exploit traits that exist naturally within a breed, intergenetic combinations permit the breeder to add genes from outside the breed. Furthermore unlike conventional breeding, specific genes can be added or modified within a breed without potentially altering other genes as well. In other words, biotechnology permits the transfer of DNA from one species to another; DNA may be exchanged between plants, animals, bacteria, or viruses in order to alter the genetic information contained in the genome.

Examples of Animal Biotechnology

Fish. Fish comprise a significant portion of the diet and are a major source of protein for people throughout the world. At present, the majority of animal biotechnology research is being conducted on fish. Seafood has been the focal point of animal biotechnology research due to the simpler biological make-up of fish compared with farm animals. Although no transgenic fish have yet been approved for food use in the United States, investigations of transgenic fish are being conducted throughout the world at the present time.

Research on seafood biotechnology is currently focused on a variety of genetic changes, including (1) improving the growth rate of fish, (2) increasing fish size, (3) improving the food conversion capabilities of fish, (4) improving the nutritional profile of fish, (5) altering the color, flavor, or texture of fish, (6) improving disease resistance of fish, (7) improving temperature resistance of fish, and (8) using fish as "biopharm animals" to create drugs or other chemicals for human use.

Meat and Poultry. In comparison with fish biotechnology, genetic modification of farm animals is still in its infancy. Genetic modification of farm animals is generally far more complex than for plants or fish due to the genetic complexity of the organisms, and difficulties with embryo transfer. Biotechnological developments ultimately may be capable of increasing the muscle mass of cattle (i.e., cattle with less fat), increase growth, improve digestive capabilities, increase disease resistance, and improve the nutritional profile of meat and poultry products (including eggs).

Meat and poultry are currently produced by numerous breeds and varieties based on years of domestication and selection. Genetic modifications, however, are capable of adding traits and characteristics that are currently impossible to develop under conventional breeding techniques. For example, it may be possible to genetically modify poultry to improve digestive function—such that the poultry would be capable of digesting lower-quality animal feeds (which are less expensive and readily available even in developing nations). Animals may also be genetically altered to produce a human or veterinary drug or biologic, food additive, or other product that can be harvested from the milk, blood, or other tissues of the animal (i.e., a "biopharm animal").

Animals may also be genetically altered to improve disease resistance to specified pathogens. Such genetic alterations could complement animal vaccination programs, and decrease the human health risk associated with ingestion of meat and poultry. In addition genetic modification could reduce or eliminate the need to use antibiotics, and thus may help to address the potential problem of antibiotic resistance.

Finally, poultry could also be genetically modified to improve the nutritional profile of whole eggs for human ingestion. For example, in the near future it may be possible to create eggs with high levels of protein and lower levels of cholesterol.

REGULATORY CATEGORIES OF ANIMAL BIOTECHNOLOGY

For regulatory purposes FDA has characterized uses of animal biotechnology into three major categories, each subject to special regulatory issues: (1) biopharm animals, (2) somatic cell therapy, and (3) transgenic gene modification.

Biopharm Animals

Biopharm animals are animals that have been genetically modified to produce a human or veterinary drug or biologic, food additive, or other product that can be harvested from the milk, blood, or other tissues of the animal (7). The genetic modification is designed to harness the metabolic capabilities of the animal to produce a product in lieu of using chemical synthesis or other traditional production methods (7). Although biopharm animals that are salvaged may also end up in the human food supply, the vast majority of biopharm animals are not in themselves intended for human food use. Rather, only the products harvested from the milk, blood, or other tissues of the animal are ordinarily intended for human use.

In general, products derived via biopharm animals will be regulated by the regulatory agency with experience in regulating that type of derived product, regardless of the breed of the biopharm animal or the method used to genetically modify the animal (8).

For example, if a genetically modified animal produces a human biologic (which is expected to account for the majority of products derived from biopharm animals), FDA's Center for Biologics Evaluation and Research (CBER) would conduct the safety and efficacy review (although if the biopharm animal were also used for human food use, FDA's Center for Veterinary Medicine (CVM) would be expected to consult with CBER regarding food salvage and safety issues).

A guidance document issued by CBER in 1995, entitled *Points to Consider in the Manufacture and Testing of Therapeutic Products for Human Use Derived from Transgenic Animals*, provides an overview of the FDA regulatory considerations associated with biopharm animals. The guidance document outlines FDA's concerns with regard to the use of transgenic animals to produce FDA-regulated drugs and biological products for human use.

Among the issues CBER will evaluate are (1) the generation and characterization of the transgene construct, (2) creation and characterization of the transgenic founder animal, including genetic stability and expression, (3) establishment of a reliable and continuous source of transgenic animals, (4) generation and selection of the production herds, (5) maintenance of the transgenic animals, including monitoring, feeding, and use of by-products, and (6) purification and characterization of the transgenic product.

Somatic Cell Therapy

Somatic cell therapy involves treating somatic cells, cells of the body that compose the tissues and organs, with new DNA to change the function of the recipient somatic cell. Somatic cell therapy may be accomplished via individual animal injections that modify specified cells in the body in order to express a protein, hormone, or enzyme. Individual steers, for example, could be modified to produce more muscle mass without having to modify the breeding herd (thereby avoiding calving difficulties that might be caused by additional muscle mass in a brood cow) (9). Somatic cell therapy does not ordinarily change the heritable traits of the animal.

Somatic cell therapy is expected to be ordinarily regulated by the FDA and subject to NADAs, unless the purpose is to prepare animals to be used in the production of regulated biopharm animal products (8). In addition, if the genetic modification more appropriately falls within the regulatory category of a food additive or color additive, FDA is expected to review the product under its food-related regulatory requirements rather than its drug-related requirements.

Transgenic Breed Modification

Transgenic breed modification involves germ line modifications made to affect growth characteristics or quality of food products derived from the target animal. The modifications are made to eggs or sperm and are heritable. FDA has indicated that animals derived from traditional breeding and selection, including artificial insemination and in vitro fertilization (IVF), would be excluded from the definition of transgenic breed modification animals (8).

As with somatic cell therapy, transgenic breed modifications are expected to be ordinarily regulated by the FDA and subject to NADAs. As with somatic cell therapy, if the modifications more appropriately fall within the food additive or color additive requirements, FDA is expected to review the applicable products under those regulatory requirements rather than its drug regulatory requirements.

U.S. REGULATORY AGENCIES: OVERVIEW

The FDA is the primary U.S. regulatory agency responsible for regulating genetically modified animals intended for human food use. Although other regulatory agencies, such as the U.S. Department of Agriculture (USDA) and Environmental Protection Agency (EPA) may share jurisdiction over animal biotechnology in certain circumstances, FDA will be the focal point in the United States for establishing regulatory policy over animal biotechnology.

FDA is composed of five separate centers, and the center responsible for regulating a specific product derived via animal biotechnology will vary depending on the product type rather than the process used to create the genetic modification. It is anticipated that CVM will be responsible for regulating the majority of products derived via animal biotechnology. CVM is responsible for regulating "animal drugs," which are defined as including most products intended to improve (1) animal growth, (2) animal feed efficiency and digestive capabilities, (3) animal carcass characteristics, and (4) animal disease resistance (i.e., an antibiotic effect).

Other centers within the FDA, however, may have primary responsibility for regulating certain types of animal biotechnology products. It is anticipated that FDA's Center for Food Safety and Applied Nutrition (CFSAN), for example, would have primary responsibility for regulating animal biotechnology intended to improve the nutritional profile of food used for human use. Genetically modified fish, such as that intended to increase the level of omega-3 fatty acids present in the fish (in order to improve the nutritional profile of the fish when ingested by humans), would likely be regulated as a "food additive" by CFSAN. It is anticipated, however, that CVM may also play a significant role in reviewing the product due to CVM's expertise in evaluating animal health issues. CFSAN would also be expected to have primary responsibility over animal biotechnology used to improve the color profile of animals intended for human use. It is anticipated that salmon, genetically modified to contain increased pink muscle tissue, would primarily be regulated by CFSAN as a "color additive."

FDA's Center for Drug Evaluation and Research (CDER), which is primarily responsible for reviewing the safety and efficacy of drugs intended for human use, would have primary jurisdiction over drugs produced for human use from biopharm animals (i.e., animals genetically modified to produce drugs for human use that are harvested from animal milk, blood, or other tissue). Similarly CBER, which is primarily responsible for reviewing the safety and efficacy of biologics (e.g., vaccines and many products derived via biotechnology) intended for human use, would have primary jurisdiction over biologics produced for human use from biopharm animals.

USDA's Animal and Plant Health Inspection Service (APHIS) is responsible for regulating animal "biologics." It is therefore anticipated that a genetic modification of an animal intended to produce a vaccine-type response to a disease in the animal (an immune-response) would be primarily regulated by APHIS. If applicable, APHIS would also conduct a food safety review for animal vaccines.

For products of animal biotechnology regulated by CDER and CBER, however, if the genetically modified animals are also intended for human food use, CVM would consult with the appropriate FDA center regarding the safety of the genetically modified animal for human food use.

Finally, numerous federal and state regulatory agencies, including the U.S. Fish and Wildlife Service and EPA, may have partial regulatory responsibility to review the environmental effects of animal biotechnology (e.g., introduction of genetically modified fish into the environment). For animal biotechnology applications that are managed by CVM, it would be expected that the assessment of potential environmental effects of such products would be coordinated by CVM under the National Environmental Policy Act (NEPA). Table 1 identifies the regulatory agencies expected to assert primary jurisdiction over various types of animal biotechnology products.

Table 1. U.S. Regulatory Agencies: Expected Primary Jurisdiction Over Biotechnology Products

Type of Product	Primary Agency Jurisdiction
Biopharm animal	
Produces a human drug	FDA, CDER
Produces a human biologic (e.g., vaccine)	FDA, CBER
Produces a food additive for use in human food	FDA, CFSAN
Produces a color additive for use in human food	FDA, CFSAN
Produces an animal drug	FDA, CVM
Produces an animal biologic (e.g., vaccine)	USDA, APHIS
Somatic cell therapy or transgenic gene modification	
Increases animal muscle mass	FDA, CVM (animal drug)
Increases animal growth	FDA, CVM (animal drug)
Reduces the amount of fat present in the animal	FDA, CVM (animal drug)
Improves digestive capabilities of the animal	FDA, CVM (animal drug)
Improves animal disease resistance via a vaccine antibody/antigen response	USDA, APHIS (animal biologic)
Improves the nutritional profile of an animal for improved nutrition upon ingestion by humans[a]	FDA, CFSAN (food additive)
Modifies the color of the animal for improved appearance for human food use[b]	FDA, CFSAN (color additive)

[a] One example would be the genetic modification of fish to increase omega-3 fatty acid content.

[b] One example would be the genetic modification of fish to increase the amount of pink muscle, which would be more aesthetically pleasing when intended for human food use. This would be a complex issue, however, as the "color additive" review would not assess animal health. Accordingly, it would not be surprising if CVM consulted with CFSAN regarding animal health issues. In addition, FDA could attempt to regulate such fish via the NADA process.

FDA REGULATION OF BIOTECHNOLOGY

Background: FDA Regulation of Biotechnology in General

Congress has not enacted any new statutory provisions specifically governing products derived via biotechnology processes. In 1986 the FDA issued a General Statement of Policy on Biotechnology (10) that indicates that because FDA regulates products rather than processes, products of biotechnology may be regulated under existing statutory authority. The Policy established the following general principles that should be followed in determining the safety of food produced by biotechnology (11):

1. The cloned DNA, as well as the vector DNA, should be properly identified.
2. The details of construction of the production organism should be available.
3. There should be information documenting that the inserted DNA is well characterized (i.e., the exact nucleotide sequence of the insert and any flanking nucleotides should be characterized) and is free from sequences that code for harmful products.
4. Food produced should be purified, characterized, and standardized.

The Policy also included a variety of considerations that should be evaluated for determining the safety of food produced by microbial isolation that has been genetically manipulated (11):

1. The microbial isolate used for production should be identified taxonomically, and if the strain of the isolate has been genetically manipulated, it should be determined whether each strain contributing genetic information to the production strain is identified.
2. The cultural purity and the genetic stability of the isolate should be maintained.
3. Fermentation should be performed with a pure culture and monitored for purity.
4. It should be determined whether the microbial isolate used for production also produces antibiotics or toxins.
5. It should be determined whether the isolates are pathogenic.
6. It should be determined whether viable cells of the production strain are present in the final product.

In addition a 1992 federal government oversight document confirms that federal government regulation should focus on the characteristics and risks of the biotechnology product—not the process by which it is created (12). Where oversight is warranted, the extent and type of oversight measure(s) must be commensurate with the type of risk being addressed, must maximize the net benefits of oversight by choosing the oversight measure that achieves the greatest risk reduction benefit at the least cost, and must consider the effect that additional oversight could have on existing safety incentives (12).

FDA Regulation of Genetically Modified Animals Intended for Food Use

Overview. The FDA has determined that existing FDA regulatory requirements are capable of ensuring the safety and efficacy of genetically modified animals intended for food use. FDA officials have noted that regulatory determinations will focus on the resulting product of the biotechnology method, rather than the process (9), and that, accordingly, most genetically modified animals intended for human food use will be regulated as new animal drugs (9).

In 1994 FDA published a status report summarizing proposed regulatory approaches and issues for FDA regulation of animal biotechnology (8). The status report identifies several considerations underlying FDA's oversight of animal biotechnology products. First, FDA will seek consistent regulation of similar products. Second, the use of existing statutory authority and administrative processes is expected to save resources, offer a measure of consistency, and minimize disruption by taking advantage of many existing regulations, FDA's scientific surveillance staff, and existing FDA guidelines. Third, a mechanism should be established to inform the public which animal biotechnology products are regulated by FDA. There is no centralized program currently regulating investigations and field trials of transgenic animals, though there is a program for transgenic plants. Fourth, clear lines are needed to define those types of animal biotechnology where governmental oversight is required and those where it is not. Fifth, scientific flexibility will be required by FDA, particularly with regard to safety assessments.

FDA Regulation of Animal Drugs. The vast majority of genetically modified animals intended for human food use are expected to be regulated by FDA as animal drugs. As noted, FDA focuses on the effect of a product in determining regulatory jurisdiction, rather than the process used to produce the product. For example, growth hormones may be delivered to animals via injection, somatic cell therapy, or transgenic breed modification. Regardless of which method is used, the animals receive additional growth hormone. Accordingly, each of these methods would be regulated by FDA under its animal drug regulatory requirements (9).

Drugs are defined under the FFDCA as including (1) articles intended for use in the diagnosis, cure, mitigation, treatment or prevention of disease in man or other animals (13), and (2) articles (other than food) intended to affect the structure or function of the body of man or other animals (e.g., production drugs and hormones, anesthetics, contraception drugs). Animal drugs do not include vaccines designed to prevent animal disease (which are regulated as veterinary biologics by APHIS), or food or color additives (e.g., genetic modifications that change the color of fish, or improve the nutrition of animal meat for human food consumption). Finally, genetic modifications developed via traditional breeding techniques are not regulated by FDA as animal drugs.

It is interesting that animal clones are currently not expected to be regulated as animal drugs. If a clone is identical to a traditional animal (i.e., it is not a clone of a transgenic animal), FDA is not likely to assert jurisdiction since there would be no distinction between the cloned animal and the traditional animal. FDA's Center for Veterinary Medicine regulates products (not processes), and cloning is a process.

If, however, cloning is used to "produce" transgenic animals that produce an animal drug, CVM would regulate the production of the animal drug and require NADA approval. It is unclear, however, if FDA would regulate the safety of food derived from cloned animals that are not genetically modified. Theoretically the cloned animals would be indistinguishable from their noncloned parents — and therefore should not present a food safety concern (14).

For all animal drugs, FDA is responsible for evaluating (1) the safety and efficacy of the drug on the target animal, (2) labeling and promotional claims for the animal drug product, (3) environmental safety, (4) manufacturing and quality controls (ensuring that the product may be consistently manufactured to comply with established specifications under good manufacturing practices), and (5) the safety profile of the animal drug when provided to animals that are ultimately ingested by humans (e.g., toxicity and potential for adverse health effects).

Ordinarily, with regard to human food safety concerns based upon animal drug residues, FDA is expected to focus primarily upon the effect of potential chronic low level exposure to the drug residues. Drug residues are ordinarily not expected to produce acute toxicity in humans. For traditional animal drugs, FDA ordinarily relies upon toxicological studies to determine the "no observed effect level" (NOEL) — which is defined as the highest dose at which the drug produces no adverse effect. Based upon the NOEL, FDA utilizes a safety factor to establish an "acceptable daily intake" (ADI) — which is defined as the highest amount of drug residue that should be safely allowed in the edible tissues of the target animal.

As part of the NADA review process, FDA also stringently reviews all drug residues remaining in human food. For instance, an FDA regulation provides the following with regard to analytical methods used to identify and evaluate drug residues (15):

> Applications shall include a description of practicable methods for determining the quantity, if any, of the new animal drug in or on food, and any substance formed in or on food because of its use, and the proposed tolerance or withdrawal period or other use restrictions to ensure that the proposed use of this drug will be safe. When data or other adequate information establish that it is not reasonable to expect the new animal drug to become a component of food at concentrations considered unsafe, a regulatory method is not required.

FDA has also provided the following examples of the types of studies that may be used to evaluate the existence and safety profile of drug residues potentially found in food-producing animals (15):

1. Complete experimental protocols should be employed for determining the drug residue levels in the edible products (including residues present in muscle, liver, kidney, fat, and possibly skin, milk, and eggs), and the length of time required for residues to be eliminated from such products following the drug's use (studies should be conducted under appropriate conditions of dosage, time, and route of administration to show the levels of the drug and any metabolites in test animals during and upon cessation of drug treatment).

2. If an animal drug is provided via animal feed or water, appropriate consumption records of the medicated feed or water and appropriate performance data for the treated animal should be evaluated.

3. If an animal drug is to be used in more than one species, drug residue studies or appropriate metabolic studies should be conducted for each species that is food-producing.

4. If residues of the animal drug are suspected or known to be present in litter from treated animals, it may be necessary to obtain data with respect to such residues becoming components of other agricultural commodities because of use of litter from treated animals.

For genetically modified animals, FDA will evaluate how different the transgenic animal is from the traditional animal. FDA will ordinarily conduct a case-by-case assessment based on molecular biology research, toxicological studies, and perceived stability of the gene pool. As noted, however, the statutory human food safety requirements for genetically modified animals are the same as those for other animal drugs. FDA is therefore expected to require the food products produced from genetically modified animals to be as safe as those from nontransgenic animals.

However, the standard battery of toxicology studies used to establish the safety of "traditional" animal drugs are not expected to be appropriate for assessing the safety of a transgene in a genetically modified animal. Unlike traditional drugs, the genetically modified genes will not be eliminated from the animal, and therefore the concept of a "withdrawal period" would not apply. Accordingly the safety profile of the genetically modified gene and any expression products must be evaluated — and if safety issues arise due to expression products, FDA may be required to establish an appropriate tolerance for a level of acceptable use.

In evaluating the safety profile of genetically modified animals, FDA may also take into consideration the fact that mammals are often important indicators of their own safety, since adverse consequences of introduced genetic material will generally be reflected in the growth, development, and reproductive abilities of the mammals. Accordingly, if genetically modified mammals are healthy, FDA should be expected to take this fact into consideration when conducting its scientific analysis.

Fish, however, are known to produce toxins that are harmful to humans — but for which the fish have developed a natural resistance. Accordingly, unlike mammals, healthy fish may impose safety issues when intended for human food use — and the FDA is therefore unlikely to rely on the health of genetically modified fish as demonstrative of safety for human health.

REGULATION OF BIOTECHNOLOGY BY AGENCIES OTHER THAN THE FDA

USDA

Animal and Plant Health Inspection Service. APHIS regulates animal biologics, plants, plant pests, and nonhuman animal pests. The Animal Virus, Serum, and Toxin Act of 1913, which provides for the regulation of veterinary biological products (16), defines a veterinary biological product as including:

all viruses, serums, toxins (excluding substances that are selectively toxic to microorganisms, *e.g.*, antibiotics), or analogous products at any stage of production, shipment, distribution, or sale, which are intended for use in the treatment of animals and which act primarily through the direct stimulation, supplementation, enhancement, or modulation of the immune system or immune response. The term "biological products" includes but is not limited to vaccines, bacterins, allergens, antibodies, antitoxins, toxoids, immunostimulants, certain cytokines, antigenic or immunizing components of live organisms, and diagnostic components, that are of natural or synthetic origin or that are derived from synthesizing or altering various substances or components of substances such as microorganisms, genes or genetic sequences, carbohydrates, proteins, antigens, allergens, or antibodies (17).

USDA issues licenses for biological products and establishments if the applicant meets standards designed to ensure the safety, purity, potency, and efficacy of the product (18). Animal biologics derived from biotechnology are expected to be regulated in the same manner as products that are prepared via conventional techniques (19). APHIS has also published a guidance document describing its method for conducting a risk analysis for veterinary biologics (20).

Food Safety Inspection Service. The Food Safety Inspection Service (FSIS) regulates products prepared from domestic livestock and poultry pursuant to the Federal Meat Inspection Act (FMIA) (21) and the Poultry Products Inspection Act (PPIA) (22). The FSIS is required to inspect meat and poultry products intended for human food to ensure they are wholesome, not adulterated, and properly marked, labeled, and packaged. Although both FDA and USDA share adulteration and misbranding jurisdiction over meat and poultry products, FSIS has primary jurisdiction over general compliance issues (22).

FSIS also regulates the slaughter for food use of livestock and poultry involved in biotechnology experiments under its regulations for livestock and poultry involved in research (23). FSIS regulates the slaughter of genetically modified animals (including experimental animals) somewhat differently than conventional animals. Specifically, FSIS has noted: "For nontransgenic livestock or poultry derived from transgenic experiments, the data should be submitted to FSIS, and would have to show that the animals to be slaughtered for food use do not have the experimental transgene, and consequently are equivalent to the parental line and, thus, are not adulterated as a result of the experiment" (24). If an animal with a transgenic modification is to be slaughtered, review and approval in accordance with the regulations would be required (25).

In general, for genetically modified meat and poultry, FSIS is expected to consult with FDA regarding safety and labeling issues. The FDA is expected to have primary responsibility for evaluating the safety and efficacy of the genetically modified animals intended for human food use.

FSIS, however, would still be responsible for conducting the inspections of genetically modified meat and poultry.

Environmental Issues and EPA Involvement

EPA regulates all pesticides under the Federal Insecticide, Fungicide, and Rodenticide Act (FIFRA) and controls the use of genetically engineered microorganisms under the Toxic Substances Control Act (TSCA). EPA is not principally involved in regulating genetically modified animals intended for human food use, but rather it is expected to consult with FDA and USDA in evaluating environmental issues.

Significant environmental concerns may arise when genetically modified species are released into the environment. A genetically modified gene may spread more widely than anticipated, and environmental and ecological changes may occur as a result of competition between the transgenic variety of the species and the natural variety. Accordingly environmental issues and biocontainment strategies are expected to be evaluated in addition to the more traditional safety and efficacy review.

CONCLUSION

The majority of genetically modified animals intended for human food use are expected to be regulated by CVM as animal drugs. CVM does not currently intend to issue a standard set of guidelines on how the food safety determination for transgenic animals should be conducted. Accordingly CVM advises companies seeking approval for genetically modified animals to consult with CVM in order to develop appropriate protocols for evaluating human food safety issues.

The legal and regulatory climate appears hospitable toward the development of genetically modified animals for human food use. The federal food and drug laws provide the FDA with a degree of regulatory flexibility, and within the next decade, as FDA conducts its safety evaluations for these products, modifications to the existing regulatory regime may be implemented without substantial difficulties. The genetic modification of animals for human food use is still in its infancy, however, and the public and political reaction to such products is largely unknown at the present time.

BIBLIOGRAPHY

1. Executive Branch Study, *Use of Bovine Somatotropin in the United States: Its Potential Effects*, U.S. Government Printing Office, Washington, DC, January, 1994.
2. Food and Drug Administration, *1999 Report on the Food and Drug Administration's Review of the Safety of Recombinant Bovine Somatotropin*, FDA, Washington, DC, 1999.
3. *Ben and Jerry's Homemade, Inc. v. Lumpkin*, 1996 U.S. Dist. LEXIS 12469 (N.D. Ill. Aug. 27, 1996).
4. *International Dairy Foods Ass'n v. Amestoy*, 92 F.3d 67 (2nd Cir. 1996).
5. *Stauber v. Shalala*, 895 F. Supp. 1178 (W.D. Wis. 1995).
6. *Barnes v. Shalala*, 865 F. Supp. 550 (W.D. Wis. 1994).
7. G.A. Mitchell et al., Regulatory Issues in Agricultural Biotechnology, Conference on Urban/Rural Environmental, Food and Agricultural Issues, November 14, 1997, Citing L.M. Houdebine, in *Transgenic Animals Generation and Use*.
8. J.C. Matheson, *Possible Approaches and Issues for FDA Regulation of Animal Biotechnology: A Status Report*, Presentation at 1994 Workshop at Davis, CA, April 6, 1994.
9. J.C. Matheson, *FDA Veterinarian* (May–June) (1999).
10. 51 Fed. Reg. 23309 (1986).
11. 51 Fed. Reg. 23313 (1986).
12. 57 Fed. Reg. 6753 (1992).
13. FFDCA, §201(g)(1)(B).
14. M.A. Miller and J.C. Matheson, III, *FDA Veterinarian* (March–April) (1998).
15. 21 CFR 514.1(b)(7) (1999).
16. Pub. L. No. 62-430, 31 Stat. 828 (1913) (codified at 21 USC §151).
17. 62 Fed. Reg. 31326, 31328 (1997) (9 CFR §101.2).
18. 9 CFR Part. 102 (1999).
19. 51 Fed. Reg. 23336, 23340 (1986).
20. Animal and Plant Health Inspection Service, Guidance Document: *Risk Analysis for Veterinary Biologics*, APHIS, Washington, DC, 1994.
21. Federal Meat Inspection Act, 21 U.S.C. §§601–695.
22. Poultry Products Inspection Act, 21 U.S.C. §§451–470.
23. 9 CFR §309.17 (1999); 9 CFR §381.75 (1999).
24. 56 Fed. Reg. at 67054, 67055 (1991).
25. 9 CFR §§309.17 and 381.75 (1999).

See other entries AGRICULTURAL BIOTECHNOLOGY; FDA REGULATION OF BIOTECHNOLOGY PRODUCTS FOR HUMAN USE.

ANIMAL MEDICAL BIOTECHNOLOGY, LEGAL, LAWS AND REGULATIONS GOVERNING ANIMALS AS SOURCES OF HUMAN ORGANS

ANDREW TREW
John Carroll University
Cleveland, Ohio

OUTLINE

INTRODUCTION

How far do we need to regulate xenotransplantation, the use of animal organs, tissue, and cells for transplantation into human subjects? Xenotransplantation offers life-saving potential and brings hope of an end to waiting lists for transplants, and current ethical dilemmas surrounding who should receive transplants, at a time when demand for organs outstrips supply. However, xenotransplantation also involves fears about crossing the barriers between the species which are the result of gradual evolutionary changes over millions of years. This creates a real risk that human patients may be harmed by infections, transferring with the animal organs, cells, or tissue. Retroviral infection of transplant patients might take years before emerging as new diseases and could meanwhile spread to wider populations, creating public health implications. Regulatory provisions, nationally and internationally, address many of these fears.

In the United States, regulatory policy is based mainly on 1996 Federal Drug Administration (FDA) Guidelines. These provide for patient protection, through informed consent and patient autonomy provisions. There are stringent biocontainment and long-term post-operative monitoring provisions that involve major restrictions on civil liberties. The Guidelines also supplement existing Animal Welfare Laws, and provide detailed risk management strategies regarding the use of donor animals.

Patent laws are also central to regulatory frameworks for xenotransplantation. Genetically engineered (transgenic) animals are likely to be used as donors. For example, transgenic pigs have been designed to include human genetic material in order to reduce rejection problems associated with the use of animal organs (e.g., transgenic pig hearts). Transgenic techniques (in effect, transgenic animals) have been patented, by corporations who own patents to the associated immunosuppressive drugs. The advantages of using transgenics include a possible reduction in acute rejection. The risk of viral infection

crossing from animal to human remains. Moreover, the use of drugs to suppress the rejection of organs, also suppresses the immune system, and may open the way for viruses, contained in the donor animal's DNA, to transfer more easily into human patients. There are also questions about the long-term viability of xenografts within human beings, including the ability of animal organs to perform effectively within another species.

There are also concerns about rearranging the DNA and genome of animals, and using patent laws to claim ownership over animals and thus control their availability or use in xenotransplantation. There are deep-seated legal concerns in Europe that conceptually the patent regime does not, or ought not to, extend to life forms. A movement toward the treatment of life as a commodity is central to this concern. Corporate "creators" of transgenic animals are likely to impose controls on the centers chosen for xenotransplantation clinical trials, which will promote corporate and not public interests. Transplant centers wishing to advance research in xenotransplantation, and provide greater equality of access to the new techniques, may be tempted to proceed using "natural" animals and develop new immunosuppressive drugs. This is permitted within the terms of the FDA Draft Guidelines. However, the risk of zoonosis (infectious agents crossing from animal to human) might be equally high whether transgenics or natural animals were used as donors in xenotransplantation procedures.

The lack of federal legislation, or an advisory agency, may possibly lead to a two-tier development of clinical trials, one using transgenics and the other using natural animals. Given the unquantifiable nature of any viral infection of the patient, and/or the public at large, there is growing national concern about the appropriate level of regulatory intervention. Xenotransplantation may continue to exist in a largely unregulated setting similar to existing provisions concerning assisted reproduction and human-to-human transplants. Policing of the FDA Guidelines, through the Institutional Review Board (IRB), at a local level, is clearly inadequate. Long-term monitoring of patients to detect viral infections may raise insurmountable restrictions on civil liberties. On the other hand, if there were to be viral infection of the public by xenotransplant recipients, it may be too late to regulate.

In the international context, the World Health Organization (WHO), the Council of Europe, the European Union, and the Organization for Economic Cooperation and Development (OECD), have all addressed the risks and benefits of xenotransplantation, emphasizing the need for international cooperation on appropriate public health policies and regulation. International concerns center on issues of justice, equality of access, and the availability of xenotransplantation in Third World countries. International conventions may be needed to avoid the possibility of forum shopping, as has occurred in the context of human organs for transplants. Forum shopping involves patients searching out a market where organs from wild, poorly screened animals are procured for use by transplant centers, in lucrative safe havens provided by countries where no xenotransplant regulations exist. These centers would offer tempting solutions to desperate patients. Any resulting viral infections would not respect political boundaries.

Internationally there may also be an inadequate scientific base on which to decide whether xenotransplantation ought to go ahead and what level of regulatory supervision would be appropriate to minimize risks.

Recent reports that infections may be able to cross over from animals into human cells add fuel to the view of some experts that a moratorium is appropriate. A moratorium on human cloning was proposed by the National Bioethics Advisory Commission, and accepted by President Clinton. However, it would appear xenotransplantation is not currently on the Commission's agenda. There is therefore increasing need for public debate and informed dialogue between scientists and lawyers. Discussion might center on regulatory polices that balance the benefits of research into new medical treatments, against the risks of emergent retroviral infection in patients, that could develop into public health concerns. Ought we to wait, or to go ahead now to regulate or permit a largely free market in xenotransplantation? Do the perceived risks outweigh the benefits of this frontier surgery? Or should we be looking to alternative sources for transplant material, such as embryonic stem cells or cloned material? Perhaps we should change the laws on organ and tissue donation, or develop more artificial spare parts for human patients? There is clearly a need for further public debate.

REGULATORY CONTEXT

Encouraging Increased Organ Donation

There is a severe shortage of organs from human donors for transplantation. The 1998 United Network for Organ Sharing (UNOS) statistics indicate a huge waiting list of patients, many of whom will die before a suitable organ is available. For UNOS statistics see: www.unos.org. As of December 1999 there were 69,550 patients on the national waiting lists. During 1999 only 21,941 transplants were carried out, spread over 272 medical institutions. The United States government has proposed measures to allocate organs to those most in need, which UNOS has opposed. Other countries experience similar shortages (e.g., statistics from the OECD in Ref. 1). Current legislation, the National Organ Transplant Act 1984 (2), relies on donations that are voluntary. Donor status depends on statements in driver's licences, in living wills, or donor cards, and in practice, the consent of relatives or proxies of the deceased. In the case of renewable organs or tissue, donations depend on individuals' altruistic conduct. Shortages may partly be the result of cultural or religious prohibitions. There are also deep-seated taboos. These are associated with respect for the dead or revulsion at removing organs and interfering with the integrity of a recently loved human being. Attempts to reduce organ shortages center on (1) developing artificial organs such as the electric heart, which may form a "bridge" until a human organ is available, (2) creating new criteria for legal brain death and/or creating a legal presumption of donation (as in the laws of Belgium and France), and (3) creation of a market for the sale of organs and tissue (3). There are two more recent proposals. First, organs might be developed from cloned human stem cells. This might enable genetically compatible transplantation, subject to removing any inherent genetic defects present in the donor's body. Second, organs may be taken from nonhuman animals to transplant into human subjects. This is known as xenotransplantation, as opposed to allotransplantation where the donor is human.

Regulatory Policy

What role might the law play in the development of xenotransplantation? Transplantation from animals to humans represents a turning point in medicine. However, it has not received the same publicity as animal and human cloning. The prospect of replicating humans has raised ethical and legal debate worldwide, leading to a ban in many countries (4). Replicating parts of humans, by creating transgenic animals with human genes, for example, donor pigs, has caused less outcry. Public awareness of the issues is lower, and experts have called for more open public debate (5) and education. Discussion focuses on appropriate ethical and regulatory policies to deal with public health risks potentially associated with xenotransplantation. Ought xenotransplants to be banned in the same way as many jurisdictions have outlawed human cloning research? The response in the United States has been a "no" in the form of FDA Draft Guidelines (6). These envisage controlled use of the donor animal, which is either "natural" or transgenic. However, internationally, WHO (7) and other international bodies are concerned about emerging infectious diseases that "cross over" from animal to human species. The central regulatory issue for public health is how far to control the possible spread of such retroviral infections, both to patients, and by patients, who have received transplanted organs or tissue from animal sources. The question is similar to the HIV/AIDS experience. How far can the potential spreading of a retroviral infection justify restrictions on patients' civil liberties?

There are existing warning signs that infections do indeed transfer from animals to humans. Recent cases off CJD (Creutzfeldt Jakob Disease) have occurred in Britain (8), which appear to originate in a form of crossover infection from cattle with BSE (Bovine Spongeoform Encephalopathy). Recent evidence suggests a strong connection between HIV and a form of simian immune deficiency (SIV) (9). Outbreaks of Hanta virus and Dengue fever are also warning precedents about the reality of such risks. The problem is that if xenotransplants are banned by law, research that may be vital will not proceed. Regulatory concerns therefore involve effective ways to strike a balance between the risks and benefits of xenotransplantation and conflicting interests.

In general, the United States is reluctant to legislate in the area of biotechnology and genetics, preferring to allow the marketplace, self-regulation, and existing patent regulation to govern. Extensive use of patents has led to the primacy of the interests of biotechnology corporations. For example, the U.S. government prefers a free market approach; it has refused to adopt the 1999 Protocol on Biosafety, which regulates trade in genetically engineered plants and animals, despite its adoption by over 120 countries (10).

Xenotransplantation is heavily dependant on the manufacture of genetically engineered animals (the animal of choice being the pig) and on associated immunosuppressive drugs, which control rejection of animal organs by humans. Both the animals and drugs are subject to patent protection. A central issue of justice is how far ought corporations to own the newly engineered species of transgenic animals, which in turn provides ownership rights to DNA and the building blocks of life (see the discussion below). A small number of patent owners in effect control access to xenotransplantation under patent licences granted to medical centers who wish to pioneer these new treatments.

Regulatory policy also needs to address how far existing concepts of informed consent and patient autonomy are adequate to protect the patients who consent to xenotransplant protocols. Given current organ shortages, patients may be tempted to opt for xenotransplants as a life-saving possibility, without full appreciation of the risks and outcomes. It might be suggested that a national body be established to guide Institutional Review Boards (IRBs) in design of protocols.

DEFINING XENOTRANSPLANTATION

Xenotransplantation is an emerging field. The scope of definition may change. Currently xenotransplantation covers a range of animal types as sources in the "xeno" part of the definition, and a range of procedures within the "transplantation" part of the definition. (See FDA Guidelines 1996, Para. 1.2.) The key issue is the use of 'live' tissue organs or cells. Other animal material used in treatments is regulated separately (see the discussion below).

"Xeno"

The "xeno," or nonhuman source of transplant material, covers three animal types. First, there are animals that arise in nature. Pigs, baboons, and even sheep have been used in pioneering transplants. The animals may be taken from the wild, bred in captivity, or reared under laboratory conditions. Second, there are transgenic animals. Such animals are genetically engineered to include genes from another species. These include transgenic pigs that have been genetically modified to include a small number of human genes to overcome specific aspects of the rejection problem when animal organs are transplanted into human patients (11). Third, there are cloned animals (12). Animals may be cloned from laboratory raised natural animals, for example, baboons or pigs. Such clones might be created following the successful use of a natural animal in transplant procedures, in order to ensure consistency of donor quality. Clones may also be created from transgenic animal donors identified as optimum specimens. There are other possibilities in the future. Animal stem cells may be used to "grow" an organ rather than the whole animal. Such techniques (13), if applied to human genetic material, may largely dispense with the need for xenotransplants in the future. Organs may also be cloned from human stem cells in the future. The organs could provide a compatible match for human patients, if genetically engineered to remove inherent defects that initially caused the patient's need for transplantation.

"Transplantation"

Transplantation refers to any procedure that involves the use of live cells, tissue, and organs, from a nonhuman animal source ("xeno" as above), whether transgenic or nontransgenic, transplanted or implanted into a human, or used for ex vivo perfusion. Use of a bioartificial liver support system might be included, where live sterile pig liver cells (hepatocytes) form part of liver dialysis treatment. Equally, animal neural cells are used in treatments for Parkinson's and Huntington's diseases, and baboon bone marrow is used as a treatment for AIDS.

Associated Definitions

The use of natural and transgenic animals as sources for treatments may include nonliving animal products. Such products have been used for some time without unforseen risks arising through transfer from animal to human. These nonliving materials are regulated separately. Porcine heart valves are classified as medical devices. Porcine insulin, on the other hand, is classified as a drug. Bovine serum albumin is classified as a biologic.

FEDERAL REGULATION IN THE UNITED STATES

Animal Regulatory Framework

There is no federal statue specifically regulating the use of animals as sources of organ donation. The National Organ Transplant Act (1984) only applies to human donors. However, animals used as donors will be subject to the Animal Welfare Act 1966, as amended in 1985 (14). Regulatory authority under the Act vests in the Secretary of the U.S. Department of Agriculture (USDA). The law regulates handling, sale, transportation, and humane treatment of a wide range of animals intended for research or experimentation, including a list of animals determined by the Secretary of Agriculture. The list covers warm-blooded animals, such as nonhuman primates, dogs, and rabbits. Legal provisions extend to health certification of animals prior to sale or transport, treatment of animals, humane care, and documentation requirements, together with appropriate use of anesthetics and other drugs.

The Food Security Act 1985, Subtitle F-Animal Welfare (15), provides for institutional supervision of protocols, standards of housing, and humane care of animals to be undertaken by the IACUC (Institutional Animal Care and Use Committee). The Act also expands the provision of humane and ethical care for, and use of, laboratory animals. The National Institutes of Health (NIH) Office for Protection from Research Risks (OPRR) provides an IACUC Guidebook, detailing procedures for the review of protocols which involve animal use (16). The Guidebook covers issues such as minimization of pain, euthanasia and research methodology, including the numbers of animals required, animal health and husbandry, facilities record keeping, and the health and safety of workers.

The IACUC, in the context of xenotransplantation, will need to address the difference, from a legal perspective, between the use of transgenic animals and natural animals. Transgenics, being regulated by patent law (17), are subject to contractual and licensing restrictions as to their use and disposal, especially restrictions on the use of gametes or on breeding or cloning. Biosafety provisions in laboratories are crucial to avoid legal actions based on licencing infringements or negligence. Management should design procedures governing the secure physical and biological containment of transgenic animals and material, especially in relation to escape of animals or pathogens into the environment. Welfare issues specific to transgenic animals include possible suffering caused by exposure to infectious animals and physical abnormalities caused by mistakes in the genetic manipulation that leads to the creation of the transgenics. Similar, more familiar restrictions are likely to apply to naturally arising animals from controlled herds. Physical conditions of animals generally are regulated, including standards governing heating, ventilation, space, and sanitation. The need for physical exercise and for psychological well being are recognized. Procedures involving pain and distress to animals are defined and are to be minimized. Detailed regulations on animal welfare (18) ensure compliance with the Animal Welfare Act and its various amendments. The regulations incorporate provisions on licencing and registration, research facilities, record keeping, containment policies, holding periods, minimizing the numbers of animals used in research protocols, destruction of animals, and inspection of premises and records. The regulations set standards that cover the health of warm-blooded animals, breeding, and the operation of animal facilities. A valuable *'Guide to the Care and Use of Laboratory Animals'* is published by the National Research Council (19). This details the housing, management and medical care of animals, including euthanasia of animals, together with institutional policies and laboratory hazard containment, health and safety provisions. The Guide also provides details of national accreditation standards for animal facilities.

FDA Regulatory Supervision

In addition to the Draft Guidelines (discussed below), the FDA is also involved through IND (Investigational New Drug) approvals of clinical trials. These regulations require sponsors of transgenic experiments to obtain written FDA approval. One example of such approval is the use of neural cells from pigs for treatment of degenerative neurological diseases such as Parkinson's and Huntington's diseases. This is an alternative to the use of human fetal cells in the treatment of these conditions. Other trials have involved the use of livers from pigs as a bridge to human organ transplantation and the use of living sterile pig liver cells (hepatocytes) in a dialysis machine. These trials are all subject to FDA regulations in 21 CFR part 312 (20). IND applications are handled by the FDA Center for Biologics Evaluation and Research (CBER). Its first IND applications relating to xenotransplantation date from 1995. CBER had expressed concern about the possibility of infectious retroviruses from porcine blood derived cells infecting human patients. This was confirmed in the literature in 1998 (20). CBER publishes *Vision*, a valuable guide to its role and current activities, on the Internet (21).

FDA Draft Guidelines on Xenotransplantation

The 1996 Draft Public Health Service Guidelines as amended in May 2000, provide a regulatory framework for xenotransplantation (22), addressing the central risks to patients, medical teams, and the public at large. A final version of the Guidelines had not been produced by February 2000. However, detailed discussion with experts did take place in January 1998 (23) about the advances in research relating to viral risks as well as regulatory, ethical, and public health concerns. In general, the Guidelines attempt to contain such risks, including possible viral infection (zoonosis), and long-term retroviral risks to patients and their contacts.

Issues raised by the Guidelines can be conveniently divided into four main headings: (*1*) the medical team, (*2*) institutional requirements, (*3*) animal donors, and (*4*) patient concerns.

The Medical Team. Transplanting animal organs, tissues or cells into human patients requires interdisciplinary expertise. The Guidelines envisage a team comprising infectious disease and zoonosis specialists (both physicians and veterinarians), an immunologist, the director of a microbiology/virology laboratory, and an infection control specialist. Since there are currently some 270 transplant centers in the United States, it is clear that such teams may not be available at smaller sites. For all health care workers, including veterinarians and laboratory staff, there is a need for a continuing health education program to alert workers to the risks associated with both the handling of animal tissues and organs, and to provide support for xenotransplantation patients (para. 4.3.3). Educational materials should be developed to include instruction on safety procedures, appropriate protective clothing, and detailed risk management strategies. The seriousness of potential long-term risks of infection to workers is recognized in detailed provisions about worker surveillance (collection of baseline sera), postexposure protocols for monitoring or treatment of infected personnel, and maintenance of long term records regarding nosocomial (hospital/laboratory) based exposures to health risks.

Institutional Requirements. Centers engaging in solid organ transplants are required to be members of the OPTN (Organ Procurement and Transplantation Network) and to comply with legislation regarding the housing and treatment of laboratory animals, and also with the Public Health Service Act (24). Protocols will need to be reviewed by the center's IRB, IACUC, and Biosafety Committee. The Guidelines require members of the various boards reviewing protocols to have high levels of expertise in assessing and evaluating potential risks of infection. The risks could affect animals and laboratory workers; patients, their families, and contacts; and the health care team and the public at large. However, multidisciplinary expertise is unlikely to be available except in larger institutional settings. There is clearly a need for institutions to share expertise in the design and approval of protocols, and avoid duplication of research. Self-education is also vital to ensure awareness of latest research and developments in xenotransplantation trials.

In addition there is a case for establishing a national xenotransplantation advisory agency to integrate matters concerning the institutional needs of transplant centers, oversight of regulatory issues, and an ongoing review of risks to patients, health care workers, and the public. This would also be valuable in coordinating the collection and comparison of health care records.

The Guidelines provide that records of the progress and outcomes of xenotransplantation should be kept indefinitely. An Institutional Xenotransplantation Record would contain full details of the procedures, animal sources, and all those concerned with the patient's care. The risks of hospital acquired (nosocomial) infection are to be minimized by following infection control policies. Where appropriate, isolation precautions should be followed. Hence a permanent Nosocomial Health Exposure Log is to be kept. This should track risks to employees of potential transmission of xenogeneic infections. A third permanent document, covering individual health care records, should follow the patient throughout the clinical stage and record the results of postoperative surveillance.

Although a national registry is envisaged, as well as a serum and tissue archive, there is currently no provision for an integrated federal advisory resource. Those who advocate a federal agency might consider a recent precedent in Great Britain where a regulatory agency, the United Kingdom Xenotransplantation Interim Regulatory Authority (UKXIRA) was established in 1998. A regulatory or advisory agency could service institutional needs in this context and include a central registry for statistics, avoidance of duplication in early trials, and sharing of both research data and details of xenogeneic infections. Equally, there is a need for a national training resource to provide the IRB, IACUC, and indeed hospital legal departments and management with updated information about scientific, legal, and ethical concerns. Without some form of regulatory or advisory agency, equality of access, to both the xenografts and to research findings, may be restricted by the vested interests of the patent owners, who control access to transgenic animals and associated immunosuppressive drugs. This may lead institutions who are not chosen as initial centers by corporate patent owners, to pursue the use of "natural" animals, such as pigs or baboons as organ or tissue donors, and attempt to overcome the problems associated with early xenotransplant experiments.

Animal Donors. Central to the Guidelines on Xenotransplantation are detailed provisions regulating the use of animals as sources of organs, tissue or cells. The animals should not be procured from the wild (para. 3.1.4) from abattoirs (para. 3.1.7), or be imported from overseas or be immediate offspring of such imports (para. 3.1.5). However, imported animals of a species not available in the United States, may provide donor material if their history is properly documented. For example, provision is made to ensure imported animals are free from transmissible diseases (e.g., BSE-Mad Cow Disease). The animals bred and reared in captivity should have documented lineage to ensure disease-free animal donors. Failure to follow the FDA Guidelines as a standard of care could result

in negligence liability. The use of transgenic animals (e.g., pigs) may reduce some risks of rejection but still carry the potential for retroviral infection for transplant recipients (para. 5.5.). However, their use is not specifically addressed in the text of the Guidelines. Whatever type of animal is used as donor, the Guidelines provide detailed safety provisions governing animal housing, feeding, screening, and surveillance (para. 3.2).

Generally, any facility that houses animals used in xenotransplantation procedures must comply with the National Accreditation Standards (detailed in Ref. 19). The animals and humans entering the facility would be monitored and screened for diseases and infections. Quarantine for animals (para. 3.5) is to be provided for at least three weeks before animals are used in xenotransplantation procedures. Animals should be screened for infectious agents and viral agents, which research shows can infect human cells in vivo or in vitro. Routine serum samples should be obtained from the animals, and animal deaths investigated thoroughly. As in patient records, documentation about the health of the animal donors and the herds (if any) from which they are drawn from, be maintained indefinitely. Equally important is control of animal feed (para. 3.2.1.3). Recycled and rendered animal materials may be a risk factor associated with prion associated diseases. These include BSE in cattle and CJD (Mad Cow Disease) in humans. Other methods of risk containment cover screening for infectious agents (para. 3.3) and include preclinical studies to identify the nature of species specific endogenous retroviruses and their potential to transfer to human patients as infections. Such studies might extend to an assessment of the potential of samples from xenografts to infect human cells.

What is missing from the FDA Draft Guidelines is any attempt to prohibit the use of certain animals as donors. There is clearly an animal welfare issue here. More especially there is a threat of species-crossing infection, posed by the use of nonhuman primates such as baboons. The Guidelines fail to provide a detailed assessment of the risks of using natural animals in general. Clearly, the extensive treatment of animal issues in the Guidelines illustrates the degree of concern about the possible spread to patients of animal viral infections that might be contained in the xenograft. This raises a central issue of regulatory policy. If the FDA, having the benefit of national and international expertise in drawing up the Guidelines, provides such detailed risk management procedures, might it be that the risks are substantial enough for there to be a strong case for enforceable federal legislation, or even for a moratorium? There are clearly international regulatory problems, since patients may readily spread their infections travelling abroad. Concerns about animals ultimately reflect a fundamental concern for the impact of xenotransplantation on patients.

Patient Concerns. Xenotransplantation is an invasive and experimental procedure. The patient is entitled to standard legal protections governing informed consent to risks. These are ensured by patient autonomy under the Patient Self-determination Act 1990, IRB supervision of protocols, and through leading court decisions such as

Canterbury v. Spence (25). Informed consent also protects medical teams against actions for the tort of battery. Risks communicated to the patient include "inherent and potential hazards of the proposed treatment, the alternatives ... and the results likely if the patient remains untreated" (25, p. 776). Both the risks and outcomes of the surgery should be explained to the patient in lay language and on a signed consent form (26). When reviewing protocols IRB members have a responsibility to include easily understood language to describe the surgery.

The Guidelines envisage that the issues specific to xenotransplantation would be fully explained to patients prior to their signing consent documents. Explanations would cover risks of both known and unknown infections being passed from the animal donor to the patient. Patients would be made aware that any infection may not occur until some time after the surgery and that it may be possible to infect both partners and third parties. Because of the possible outcomes, patients should be prepared to be monitored, possibly for life, to undertake regular tests, and to provide up to date information about changes of address. Further restrictions include not donating blood, gametes, tissue, or body parts for use by human beings. Moreover patients face the risks of passing infection to those involved in any future sexual relationships. The Guidelines advise patients to treat the risk of infecting third parties in the same way as transmission of HIV, suggesting that the use of barriers during intercourse may minimize risks to partners. All these provisions reflect attempts to monitor future unquantifiable risks of infection to patients, their families, or the general public. These restrictions clearly involve severe incursions into traditional civil liberties and raise questions of how far they could be enforced as a contractual obligation rather than a simple consent form. How could they actually be monitored? Although the patient will be reliant on drug or other therapies for a long period of time after the xenograft, compliance with "guidelines" of conduct will depend on individual responsibility. Otherwise, the early patients could well be condemned to a life of quarantine isolation.

How patients can give meaningful "informed" consent to such a range of risks and restrictions on personal freedom is uncertain. Indeed, absent federal legislation, how could a patient be restrained by an institution and his or her liberties be curtailed? Suppose that the patient is making a good recovery but must submit to the long-term surveillance required by the Guidelines. Over what time period should this extend? What if any retroviral infection takes years to emerge (as in the case of HIV)? The Guidelines suggest warning the patient that xenogeneic infectious agents may be transmitted by unprotected sexual activity. How could that be monitored? If we treat this as a case where risks and benefits need to be balanced, how could patients realistically be given advice about outcomes or future potential for viral infections when so little is known? In reality, "consent" is meaningless here. Moreover the institution that foresees possible public health risks might well be held accountable for initiating a chain of events involving risks to third parties. A xenotransplant patient post operatively is in effect a walking source of novel and unknown infection. Knowing

the patient is the source of foreseeable risk of harm, how might this impact on negligence liability? On a product liability basis, we might argue that the "manufacture" and sale of a transgenic animal, with foreseeable potential to harbor retroviruses, could lead to third party liability. This may indeed justify federal intervention. At the minimum greater public discussion is needed (27).

PUBLIC HEALTH RISKS

Identifying public health risks has been central to the Guidelines from the outset. In the absence of a national regulatory agency for xenotransplantation, the heavy burden to monitor patients and health care workers indefinitely, and to maintain tissue and data archives, falls on individual transplant centers at the point where an infection threatens public health. Public health risks could derive from (*1*) patients, (*2*) animal donors, or (*3*) health care workers.

Risks from Patients

The consent section of xenotransplantation protocols require patients to be apprised of the risks of transmitting infections to family or close contacts "with whom the recipient participates in activities that could result in exchange of bodily fluids" (para. 2.5.3). Educating the patient about risks of infection is left to the transplant center. It is unclear how this might be done outside the framework of xenotransplant protocols and consent forms. Counseling may be appropriate, and clearly centers will need to consider their obligation to ensure informed consent is not subject to any pressure. Education of a patient's close contacts about the uncertainty regarding risks, and of the need to tell their doctor about any unexplained illness, is seen as the individual responsibility of the patient (para. 4.2). Suitable indemnity clauses may need to be inserted into the consent section of protocols. Conduct that might spread infections, such as needle sharing, unprotected intercourse, donation of blood should be prohibited. In the event of a clinical episode, which might lead to infection, it is envisaged that the state health department, the Center for Disease Control) (CDC) in Atlanta and the FDA will be notified (para. 4.1.1.6).

Every transplant center is obliged to maintain permanent archives of biological specimens relating to the patient, taken before the xenotransplant, immediately after the operation, and then at regular intervals for the rest of the patient's life. The intervals could be regulated by individual protocols (para. 4.1.1.2). Similar data of potential public health value would include the provision of permanent archiving of major organs and tissue samples from autopsies of patients who have died after xenotransplantation. Secure storage facilities would have to be provided within the transplant centers. All of these obligations raise the need for detailed in-house policies in collecting and maintaining long-term archival data on patients. So for there is no provision for early warning, or sharing of information, that might alert to potential public health concerns. The Guidelines do recognize the need for a national registry of data to track common features among xenograft recipients and to assess long-term safety (para. 5.1). However, it is not clear how or when such an agency could be created.

Risks from Animals

When it comes to animals as a public health concern, the Guidelines provide for detailed systematic record keeping to link biological samples from donor animals to transplant recipients. This is "essential for public health investigation and containment of emergent xenogeneic infections." (para. 3.7). Once again, maintaining the integrity and security of long term cryogenic storage of samples is the responsibility of individual transplant centers. Transplant centers are expected to provide education, and suitable educational materials, for health care workers who face risks involved with animal donors or human recipients of xenografts.

Risks from Health Care Workers

Occupational health care programs should provide details of risks, ensure that standard precautions are followed, including protective clothing and disposal of waste, and monitor for possible infection among workers. Workers forming part of a xenotransplant team that handles donor animal material or provides care for patients would be subject to health checks including the archiving of baseline sera, obtained from workers, before exposure to xenografts or recipients. In the event of exposure to infection, or unexplained illness after exposure, there is provision for reporting and recording incidents (para. 4.3.3.).

Risk Assessment and Wider Concerns

The provisions relating to public health thus center on specific risk management provisions. However, the Guidelines contain no provision to ensure maintenance of safety provisions, no means to provide effective exchange of know how, no ways to standardize data collection, analysis, and exchange among individual centers. No effective provision exists to ensure uniformity of standards regarding protocol approval and self-education of the members of IRB or IACUC. Without objective consideration of risks, there is a danger that institutional review of protocols will initially be driven by commercial interests. In those centers where non-patented animals are used, there is no accessible national resource to provide expert advice to IRBs and research groups. Appropriate federal laws or regulatory agency could provide this.

Risk Reduction, "Natural" and Transgenic Animals

Overall, assessment or reduction of risks is difficult absent clinical trials with real patients. Recent research has indicated that the potential for infection of human cells by animal donor material is considerable, and this sounds a warning note. The use of transgenic organs genetically modified to include some five to ten human genes would help to reduce problems of rejection by human patients. However, retroviral transfer would remain a serious risk to patients. Already the pig genome is known to contain retroviruses that are endogenous, essentially integrated into the genetic makeup of the pig. If retroviral transfer

does take place, the threat to public health from infected patients could be considerable. Against any evidence of foreseeable harm to both the patient and public, to proceed with clinical trials would seem at best negligent. An alternative attempt may be to transplant porcine material into monkeys in order to obtain the clues as to the long-term viability of placing a pig organ in a human.

The Guidelines do not ban the use of nonhuman primates as donors, although there are restrictions as to their procurement. For example, wild caught animals ought not to be used, nor captive free-range animals, in order to minimize some infectious risks. In the past a variety of animals had been used in the early xenotransplantation experiments. Since 1963 chimpanzees, baboons, sheep, and pigs have been used. In the so-called Baby Fae case, in 1984, surgeons at Loma Linda Hospital in California transplanted the liver of a seven-month-old baboon into a newborn human infant. The baby survived for 20 days and then died of kidney failure. More recently the University of Pittsburgh Medical Center transplanted baboon livers into two patients who were dying from Hepatitis B. One, aged 35, lived for 71 days in 1992. The other patient aged 62 died after 26 days. Baboon bone marrow was transplanted into an AIDS patient in 1995. These early cases do not indicate how far the risk of acquiring infections might be significantly reduced by breeding and rearing nonhuman primates in captivity and isolation. Moreover there are widespread ethical concerns about breeding baboons, monkeys, or other animals to farm their organs and tissue for human use.

Experts seem to agree that the pig offers a better choice of donor, whether transgenically engineered or not. However, although pig heart valves and insulin have been used in human patients, inserting an animal organ into a human patient raises novel issues. Although breeding pigs in isolation may reduce some infectious risks, the fear is that genetically inherited retroviruses in the makeup of the pig genome may transfer to the human patient. Even if transgenic pigs were used, it seems that "human proteins expressed on the surface of transgenic pig cells can act as receptors for viruses." Opinion is divided; see Refs. 5 and 21 (28). Producing a virus-free transgenic would require identifying viruses as yet unknown. Indeed, the impact on humans of viruses that are of no danger to pigs raises the familiar argument that a virus like that in Michael Chrichton's *Andromeda Strain* may mutate into a wider population. It is also conceivable that transplant recipients may infect other animals. Clearly, concerns about public health are the basis for extreme caution.

Individual patients may in desperation agree to become guinea pigs in early surgery in having accepted the likelihood of an uncertain future or death. But suppose that the surgery were reasonably successful using transgenic pigs and new immunosuppressive treatments. How would it be possible in a democracy to control the sexual behavior and personal freedom of xenotransplant recipients, as envisaged in the Guidelines, so that there were minimum risk to public health? It is surely unrealistic not to be deeply concerned about a scenario where the public became infected with new animal-to-human viral infections. There are some precedents already. In Britain and other parts of Europe, outbreaks of CJD (Mad Cow Disease) in humans was transmitted from cattle infected with BSE. In Australia, the deaths of veterinarians examining an infected horse were attributed to cross-species infection. Worldwide, in the case of the HIV virus, the likelihood is that it transferred to humans from the sooty mangabey and from chimpanzees (9), originating in strains of SIV (Simian Immunodeficiency Virus). These examples offer cautionary tales for those who advocate going ahead with xenotransplantation. For some it may be a question of the "unnatural" way in which species barriers are crossed. Is nature showing us a red light here? Jean Rostand, the French biologist, wrote soon after the discovery of the DNA sequence in the 1950s that in Nature there may be some defined "no trespassing" signs (29), which we cross at unknown peril.

A XENOTRANSPLANTATION MORATORIUM?

Could xenotransplantation take us to the point where we should pause before crossing into new territory in biotechnology? The evidence of risks is strong. If there is a public health implication, the public has an interest in being informed about the risks and benefits of xenotransplantation. The transfer of organs from animals raises a possible but unquantifiable risk of xenotropic organisms. These may cause infections of an unknown kind in humans that were not harmful to donor animals. Endogenous retroviruses may transfer with the animal organ, with risks of new types of infection. Were the viruses to mutate in humans, it is conceivable that animal donor species could be infected through contact with humans. Research into the interplay of human and porcine viruses in the influenza epidemic of 1918 lends support to this idea. Moreover the patient will be more open to infection generally after immunosuppressive drug treatments, and consequently react in a different way to infections derived from animal donors than would a healthy person. A moratorium on xenotransplantation might be based, either on a version of the so-called Asilomar agreement, which involved scientists curtailing their own use of science through self-regulation, or on a federal ban on funding, as is proposed in the case of human cloning research. The issue whether a moratorium is required has so far not been addressed by the President's National Bioethics Advisory Commission and is unlikely to be the subject of federal legislation in the near future. Unless the issues of public and patient risks are raised to a higher profile within the media, public debate on the future of xenotransplantation regulatory policy is likely to remain largely uninformed.

DRAFTING CONSENT FORMS

It may be too late for a moratorium. There are now proposals being presented to IRBs in medical centers for clinical trials with backing from major pharmaceutical companies. If transplants are to proceed, it is vital that attorneys representing medical centers be aware of the public health risks and monitoring issues in the Guidelines and in the literature. Drafting consent forms within protocols ought

not to be left to the medical/research team. Patients are "consenting" both to life-saving surgery and to becoming an additional risk of infection for others after surgery, which is unusual. Thus attorneys need to consider novel contractual and negligence liability clauses. They should be aware of the competing interests represented by the commercial sponsors anxious to promote patented commercial advantage, the patients who desperately seek potentially life-saving surgery but lack expert independent advice, and the ultimate "consumers" of infections, the general public. There may be a need to create a forum of legal and ethical experts, for group discussion and education of patients' families or partners about potential risks, and to explore issues of liability. Public indemnity insurance may also require revision. Positive undertakings would also be required. Perhaps patients would need to submit to a form of quarantine, the length of which seems difficult to stipulate. Expert scientific opinion ought to be invoked. However, the basis of medical care is consensual. Unless the risk to public health is certified by some state or federal agency, how could patients legally be detained against their will without the hospital facing false imprisonment litigation? Without a clear federal enforcement mechanism, the FDA Guidelines are really a paper tiger.

FDA REVISED GUIDELINES

After considering concerns from scientists and ethicists, the FDA issued Revised Draft Guidelines at the end of May 2000 (29a). These provide the FDA with closer regulatory oversight of clinical trials involving xenotransplantation. No clinical trials are allowed unless investigators have first submitted applications for prior FDA authorization. The FDA will propose further regulations and guidance on protocols for industry. There are further limitations on the use of nonhuman primates, coupled with FDA acceptance that there is insufficient information about risks from the use of animals as donors in xenotransplantation. Risk minimization procedures are strengthened regarding the screening and surveillance of animals, and those coming into contact with them, including patients.

Taking into account the needs to counsel patients, their families and close contacts about risks, the Guidelines place new obligations on sponsors for the design and monitoring of clinical trials and for counseling. This might seem inadequate, given the vested interests of sponsors. However, the FDA expects a coordinating national role will be played by the Secretary's Advisory Committee on Xenotransplantation, soon to be established. In part, the Committee will oversee protocol designs, and evaluate wider scientific, medical, and public health issues related to xenotransplantation. To meet wider public health monitoring concerns, the FDA will now require maintaining of both records and specimens from animal donors and patients for a period of fifty years.

The revised Guidelines go some way toward meeting the concerns raised since the 1996 Guidelines were published. However, there remain real concerns about the ways in which long term monitoring of patients might be achieved, and the preventative measures necessary to address any widespread infection of the public at large. Moreover, there is a growing need to view the problems of risk management and surveillance as an international and not merely an U.S. policy concern.

INTERNATIONAL REGULATORY ACTION

The regulation of xenotransplantation is of growing international concern. Equality of access to pioneering surgery tends to grow slowly over a period of time. The pace of medical advances may be deterred by overregulation in a given country. There are cultural differences with regard to health care coverage and approaches to the use of animals in surgery. Further, there is a general concern that if patent owners control the medical market for the use of transgenic animals, some countries may be kept out of the early exchange of research data and clinical expertise. Health, unlike the environment, has not been the subject of major international regulation. Nonetheless, both international organizations and individual countries are engaged in promoting regulatory initiatives, in some cases involving a moratorium on clinical trials pending further expert and public debate. These regulatory initiatives indicate a shared international perception about the risks from animal to humans of viral infections that could threaten public health as transplant recipients travel from country to country.

World Health Organization

WHO is concerned about the emergence of new communicable diseases. Although lacking regulatory authority, WHO reports provide a catalyst for international research and cooperation, leading in turn to international treaties and conventions. In 1998 WHO published a Consultation on Xenotransplantation (7,30), providing a framework for international policy. International agreement is urged "to ensure that xenotransplantation is developed in conformity with accepted ethical and legal standards, based on the need to respect human dignity and individual rights together with community interests" (para. 8.1.14.). The WHO report is neutral on whether to advocate xenotransplantation. Recommendations (para. 8) provide a check list of issues which countries thinking of undertaking xenotransplantation need to address. These cover maximizing individual and public health by managing risks of zoonoses; procurement of healthy animals; risk assessment, counseling, and monitoring of transplant recipients and their contacts; developing archives of human recipient and animal donor tissue; and ensuring animal welfare. Informed public debate should include sensitivity to national, cultural, and religious norms. Equality of access should be encouraged through information exchange, nationally and internationally. A multidisciplinary xenotransplantation review body should be established on regulatory policy, patient welfare, archives, and global exchange of information.

Council of Europe

The Council of Europe, which represents some 40 states (31), has adopted a Convention on Human Rights and Biomedicine (32). This provides the basis (under Articles 19 and 20) for the current development of a common regulatory policy on organ transplants (32). A Second Protocol on Xenotransplantation may be added to the Convention following the proposal in 1999 to establish a moratorium on xenotransplantation. In 1997 the Committee of Ministers passed a Recommendation on Xenotransplantation, aimed at regulations to minimize "the risk of transmission of known or unknown diseases and infections to either the human or animal population" (33). Regulations would cover issues such as research and clinical trials, sources and welfare of donor animals, and long-term review of animal donors and transplant recipients. Subsequently, in 1998, the Parliamentary Assembly of the Council of Europe received two Reports on xenotransplantation prepared by member representatives (34). Both endorsed a "legally binding moratorium on clinical xenotransplantation," based on risks of a major pandemic affecting public health. The moratorium is linked to proposals to stimulate research into risks of infections to humans and animals, and full examination of appropriate ethical and legal policies.

Most recently, in 1999, the Parliamentary Assembly adopted a Recommendation on xenotransplantation (35), formally urging the Committee of Ministers to work for rapid introduction of a legally binding moratorium, to consider development of a Second Protocol to the Convention on Human Rights and Medicine, and to plan an international strategy with WHO to balance ethical, legal, medical, and public health aspects of xenotransplantation. In addition the ministers are urged to lobby for a worldwide legally binding agreement for a moratorium. Despite the optimistic tone of all theses documents, they are unlikely to be supported by the United States, given the reluctance to legislate on biotechnology issues. However, declarations by some 40 countries might add weight to the argument that it is time for the President's National Bioethics Advisory Commission to address xenotransplantation regulatory policy in detail.

The European Union

Within the 15-member states of the European Union (EU), there has been a long and heated discussion about appropriate legal policies to balance the availability of patent protection to biotechnology companies against the philosophy, prevalent in Europe, that patenting life forms raises serious ethical and moral objections. Safeguarding the dignity and integrity of the person is seen as a central legal issue. In 1998 the EU adopted a Legal Protection of Biotechnological Inventions directive for implementation by the member states no later than July 30, 2000 (36). In general, the new law relates to existing uncertainties about the scope of Article 53(b) of the European Patent Convention, as applied to genetically modified animals. Article 53 (b) excludes from patentability "animal varieties or essentially biological processes for the production of . . . animals." Specifically, this complex directive includes legislation of relevance to xenotransplantation. The law identifies the difference between inventions and discoveries. Attempts to patent a mere DNA sequence or partial sequence will not be permitted, as being discoveries about natural phenomena. On the other hand, even inventions may be unpatentable if they fall within Article 6 "where their commercial exploitation would be contrary to public order or morality." The list of inventions includes "uses of embryos for industrial or commercial purposes," which might prevent patents on embryonic stem cell usage to create organs or tissue as an alternative to xenografts." Furthermore, "processes for modifying the genetic identity of animals" are unpatentable, if "likely to cause them suffering without any substantial medical benefit to man or animal," as are "animals resulting from such processes." Would the transgenic pigs, engineered to contain some human genes, fall within this Article? The "substantial medical benefit" of xenotransplantation, as we have seen, is subject to risks of substantial harm to patients from the use of genetically altered animal donors.

Methods of National Regulation

Some countries, such as India, are adopting a similar model of regulation to that proposed by the Council of Europe. India has declared a strict ban on organ transplantation including xenotransplantation clinical trials. A doctor who attempted to transplant a pig's liver into a patient with severe heart problems was prosecuted in 1997 (37). At the same time India's Council of Medical Research is drawing up guidelines. Another model for regulation is the establishment of a national regulatory authority. This has been adopted in countries such as Britain, where there is a xenotransplantation regulatory authority, UKXIRA. The Canadian government has initiated proposals for a National Advisory Board on Xenotransplantation based on recommendations of a National Forum on Xenotransplantation exploring clinical ethical and regulatory issues, which was convened by Health Canada in November 1997. Canada's proposals are likely to involve a standard-based regulatory system linked to the Food and Drugs regulations, and the National Animal Care Committee. In France, a similar expert advisory committee oversees safety issues, and legislation is under review. In Sweden, a Committee on Xenotransplantation was established during 1998; public opinion was tested through a questionnaire and public hearings held on xenotransplantation. In Australia, the government has taken initial steps to establish a regulatory body. The Australian Health Ethics Committee has been developing guidelines to assist local Institutional Ethics Committees on scientific and ethical issues. The context for the xenotransplant debates in Australia was a proposal to commence diabetes treatments using pancreatic islet cells from pigs. Another model is to include xenotransplantation in constitutional reforms covering organ transplants. Switzerland is considering a constitutional amendment to cover organ, tissue, and cell transplants from both human and animal donors. The Swiss Science Council is also researching the ethics and risks of xenotransplantation.

Establishing National Regulatory Bodies

The main regulatory shortcoming of the current FDA Guidelines in the United States remains the lack of enforceability. The role of the law might be seen as standard-setting here. Enforcement of the Occupational

Health and Safety laws provides a model that might be valuable in xenotransplantation policy. The regulatory framework in Britain offers an alternative models to the current FDA Guidelines.

British research pioneered the techniques for creating transgenic pigs as an alternative to "natural" donor animals such as baboons. UKXIRA offers a national focus for monitoring research and proposed clinical trials, a framework for regulatory issues, and a means of assessing the scientific evidence about risks and current techniques (38). Established in 1998, UKXIRA promotes links with other scientific and regulatory bodies across the globe. Ideally such an agency would encourage exchange and minimize duplication of data between government departments, and provide a cross-departmental focus. Regulation is to be provided through legally binding Codes of Practice and Health Service Circulars. As for animal donors, existing legislation, the Animal (Scientific Procedures) Act 1986, provides for an Inspectorate to oversee licensed use of animals in scientific or medical procedures and experiments. A Code of Practice governing the welfare of donor animals used in xenotransplantation is being developed by the Inspectorate, and UKXIRA is initiating parallel provisions on biosecurity. Clinical trials are governed by a legally binding Health Service Circular (39) which contains detailed guidance provisions, in the form of Directions to the National Health Service. Health authorities cannot undertake xenotransplantation treatments without prior written approval by the Secretary of State for Health under Section 17, National Health Service Act 1977. The guidance sections address information about the medical team, sources of animals, their housing and welfare, infectious risk controls, patient consent and monitoring. The role of UKXIRA is to advise the Secretary of State for Health on "the acceptability of specific applications to proceed with xenotransplantation in humans." Thus the British model clearly ties together the general regulatory framework governing health provision and animal welfare, with binding specific regulations administered by a national expert regulatory agency. In this way coordination of information and coherent national standards for xenotransplantation clinical trials are ensured.

REGULATORY LIMITS TO THE USES OF ANIMALS AS DONORS

What is the future for animals as donors? The patenting of animals and other life forms continues to be an ongoing legal controversy central to xenotransplantation.

Comparative Legal Issues

Since 1980, when the Supreme Court decided the *Diamond v. Chakrabarty* case, the way has been open to engineer life forms and gain patent protection in the United States. The original patent law was amended in 1995 after the Trade in Related Aspects of Intellectual Property (TRIP) agreement at the Uraguay Round of GATT extended protection to 20 years from the date of application (40). The Chakrabarty case turned on the genetic manipulation of bacteria that might be used to break up oil slicks, which was thus a rearrangement of life forms. However, the Court viewed this as a narrow case. It did not seem to foresee the huge impact that would result from extending patent protection, beyond the purposes envisaged by Thomas Jefferson's understanding of "inventions," and the concepts adopted since the original Act of 1793 which exclude from patentability "discoveries" about Nature. After all, the discovery of the structure of the DNA "double helix" was not patented by Crick and Watson in the 1950s. Extension of U.S. patent laws to protect the discovery, rather than invention, of what seem to be the rules of Nature governing the sequences, arrangement, and rearrangement of DNA has caused animated controversy in the European Patent Office. As a result there was reluctance to register the U.S. Oncomouse, as the first patent on a genetically engineered mouse used in human cancer research (41). Initially the argument prevailed, both in Europe and in the United States, that the techniques for sequencing and manipulating genetic material are inventions.

In 1998, however, legislation in the form of a directive (36), was passed by the European Union (EU), covering the legal protection of biotechnological inventions. This attempts to deliniate boundaries between discoveries and inventions. For example, a "mere DNA sequence without indication of a function" is not patentable because it is treated as a discovery. Although inventions would normally be patentable, as discussed earlier, the directive excludes certain classes of inventions if "their commercial exploitation offends against *ordre morale* and morality." The inventions are detailed in Article 6, and include "processes for modifying the genetic identity of animals." The United States continues to pursue policies permitting patents on a wider basis. The driving force may be considerations of trade. There are billions of dollars invested in biotechnology companies, based on potential revenues from genetic patents, including those on animal life and human stem cells.

Biodiversity and Animal Donors

Could the Chakrabarty case be challenged? The Supreme Court decision seems to contain a fundamental conceptual error. If genetically "created" bacteria can be patented, then any life forms, including human beings, might be altered and patented, producing possible adverse effects on biodiversity. Control over patenting of transgenic plants, such as wheat and tomatoes, or transgenic animals, such as the Oncomouse or the pig engineered to contain human genes, proved a logical extension of Chakrabarty. The ability to selectively breed and "own" hybrids of plants or animals is familiar to us. Explorers over centuries have rearranged the geographical location and enabled exposure to other species of countless varieties of flora and fauna. All this could be seen as interfering with "natural" biodiversity, and in 1992 the Biodiversity Treaty attempted both to encourage and regulate biodiversity. Nonetheless, the United States refused to sign the subsequent Biosafety Protocol on the trade in genetically altered plants and animals, endorsed by more than 120 countries, in February 1999 (10). The Protocol is graphic indication of the spread of genetically manufactured life forms. Biotechnology enables us to bypass evolution and genetically engineer plants and animals and potentially

human beings. This in turn enables us to impose a new evolutionary shift on plants, animals, and eventually humans, and it sets in motion a chain of events over which we have little control. Controling the nature of Nature may thus lead to harmful ecological changes. Many would argue that genetically engineered animals, plants, or seeds can effectively improve the quality of agricultural output, promote resitance to disease, and reduce the risk of famine. Yet trade in and use of genetically modified organisms may result in major adaptation to the food chain, mingling wild and genetically engineered species. Within a period of years, biodiversity and evolutionary changes may produce instability as a result of our short-term perception of agricultural problems and their solutions.

The recent refusal of the United States to ratify the Biosafety Protocol shows the pressures that biotechnology industries can exert in this area of patents. Countries in which initial development of the new genetics occurred, notably Britain, have supported a more global perspective through a regulatory interventionist policy. By rejecting the Protocol, the United States will encounter barriers to trade and restricted access to markets overseas in which to license patented plants and animals.

Conflict of Laws Issues

Research and overseas investment by American biotechnology companies may be particularly affected by conflict between the U.S. Patent regime and the 1998 EU directive on biotechnology patents adopted by the 15 EU countries. On the other hand, xenotransplantation advances within the EU may be hindered if reciprocal recognition of U.S. patents on transgenics is denied within its countries. American companies would presumably be unable to protect overseas licencing restrictions within the EU jurisdiction. EU countries wishing to pursue xenotransplantation, using U.S. originated transgenic pigs, may not have access to these animals and may develop unpatented transgenics or possibly use "natural" animals for transplantation, coupled with developing new types of immunosuppressive drugs.

In the United States, patents continue to be registered that cover human genetic material, whether or not incorporated into animals. Ought we not to be concerned about human or animal life being owned and commodified through intellectual property rights? The ongoing Human Genome Project, which aims to map all human genes, was certainly not intended to lead to the patenting of gene sequences and human or animal genetic material. Rather, it was a public project, a multinational research exercise, to map the human genome, backed by many governments, including the United States. In the event, American-based corporate interests have successfully patented individual human gene sequences and a variety of transgenic animals and the techniques for their "manufacture," after overtaking the slower pace of discovery in state-funded research projects. Why is it not contrary to law to own and control the use of manufactured animal life? If the animal were a human, it would be unconstitutional to own a complete person, on grounds of slavery (42). Yet it is not unlawful through intellectual property rights to own or as it were to "enslave" the genetic building blocks that make up both animals and humans.

BIBLIOGRAPHY

1. E.g. Statistics from the OECD in E. Ronchi, *OCDE/GD(96)* **167**, 5–9 (1996).

2. 42U.S.C.273 (1984); *Xenotransplantation: Science, Ethics & Public Policy*, Institute of Medicine, Washington, DC, 1996.

3. A. Trew, *Toledo Law Rev.* **29**, 271, 306–315.

   R.M. Arnold and S. Youngner, *Transplant. Proc.* **27**, 2913 (1995).

4. E.g. Council of Europe, *Convention on Human Rights and Biomedicine*, ETS No. 164, Council of Europe, 1997.

   Protocol (Banning Human Cloning), ETS No. 168 (1998), adopted by 23 European states; *Human Fertilization and Embryology Act*, UK, 1990.

5. R.A. Weiss, *Br. Med. J.* **317**, 931–934 (1998).

   D. Butler, *Nature (London)* **391**, 320–325 (1998).

6. FDA Draft Guidelines, *Fed. Regist.* **61**, 49920–49932 (1996).

7. World Health Organization (WHO), *Xenotransplantation: Guidance on Infectious Disease Prevention and Management*, WHO/ZOO/98.1, WHO, Geneva, 1998.

8. Department of Health Advisory Group of the Ethics of Transplantation, *Animal Tissues into Humans*, H.M. Stationery Office, London, 1997.

9. F. Gao, *Nature (London)* **397**, 436–441 (1999).

10. *N.Y. Times*, February 25, Sect. A, p. 1 (1999).

11. D. Dickson, *Nature (London)* **377**, 185 (1995).

    H. Vanderpool, *Lancet* **351**, 1347–1350 (1998).

12. R. Wilmut, *Nature (London)* **385**, 810–813 (1997).

13. U.S. Pat. 5,843,280 (1998) approved for primate embryonic stem cells. See advice by President's National Bioethics Advisory Commission November 20, 1998: *www.bioethics.gov*

14. PL89-544. The amended text is in 7USC 2131–2156.

15. PL99-198.

16. *NIH Publ.* **92-3415.**

17. U.S. Constitution Article 1, Section 8; Title 35 USC para. 101.

18. CFR Title 9, Chapter 1, Subchapter A-"Animal Welfare."

19. National Research Council, *Guide to the Care and Use of Laboratory Animals*, National Academy Press, Washington, DC, 1996.

20. 21CFR part 312.

21. The Website for CBER is at *www.fda.gov*

    C. Wilson, *J. Virol.* **72**(4), 3082–3087 (1998).

22. *Fed. Regist.* **61**(185), 49919–49932 (1996); Health & Human Services Fact Sheet 26th May 2000: *www.hhs.gov/news/ress/2000/pres/20000526.html*. Test of revised Guidelines: *www.fda.gov/cber/xap.xap.htm*

23. U.S. Department of Health and Human Services, *Developing US Public Health Care Policy in Xenotransplantation*, US DHHS, Bethesda, MD, 1998. For text, use "Search" at Web site: *www.fda.gov*

24. 42 USC 262,264.

25. *Canterbury v. Spence*, 464 F2d. 772 DCCircuit (1972).

26. 45 CFR 46; 21 CFR Part 50.

27. F. Bach, *Nat. Med.* **4**, 141–144 (1998).

28. W. Heinene, *Lancet* **352**, 695–699 (1988).

United Kingdom Xenotransplantation Interim Regulatory Authority, *Workshops on Porcine Endogenous Retroviruses*, UK Xenotransplant, Interim Regul. Auth., 1999.

29. J. Rostand, *Can Man be Modified?* (J. Griffin, transl.), Basic Books, New York, 1959.

    Institute of Medicine Report, Washington, DC, 1996, Ch. 2 (note 6 above).

29a. FDA Revised Guidelines on Xenotransplantation Full Text (May 2000): *www.fda.gov/cber/gdlns/xeno0500.txt*

30. *Economic Aspects of Biotechnology Related to Human Health*, OECD Report, 1998, DSTI/STP/BIO(98)8/FINAL.

31. Council of Europe Website is at *www.stars.coe.fr*

32. *Explanatory Report to the Convention on Human Rights and Biomedicine*, Council of Europe, 1997, DIR/JUR(97)1.

33. Council of Europe Recommendation R(97)15.

34. Council of Europe, 1998, Doc.8166; Doc.8264.

35. Council of Europe, *Xenotransplantation*, Recommendation 1399, Council of Europe, 1999, (1).

36. European Union, 1998, Directive 98/44/EC.

37. G. Mudur, *Br. Med. J.* **318**, 79 (1999).

38. UKXIRA, *First Annual Report, May 1997, August 1998*, Department of Health, UK, 1998.

39. Department of Health, HCS1998/126; Web site: *www.open.gov.uk/doh/coinh.htm*

40. *Diamond v. Chakrabarty* et al., 65 L ed 2d 144 (1980); TRIPS amendment to USC Title 35, para. 154 (1995); Utility patents defined: U.S. Constitution Article 1, Section 8; USC Title 35, para. 101–103.

41. U.S. Pat. 4,736,866 (1988).

    A. Kimbrell, *The Human Body Shop*, Harper Collins, New York, 1995, pp. 188–202.

    U.S. Office of Technology Assessment Report, *Patenting Life*, U.S. Office of Technology, Washington, DC, 1990.

42. U.S. Constitution, Amendment XIII: Neither slavery nor involuntary servitude ... shall exist within the United States.

See other entries ANIMAL, MEDICAL BIOTECHNOLOGY, POLICY, WOULD TRANSGENIC ANIMALS SOLVE THE ORGAN SHORTAGE PROBLEM?; FDA REGULATION OF BIOTECHNOLOGY PRODUCTS FOR HUMAN USE; see also RESEARCH ON ANIMALS entries; TRANSGENIC ANIMALS: AN OVERVIEW.

ANIMAL MEDICAL BIOTECHNOLOGY, POLICY, WOULD TRANSGENIC ANIMALS SOLVE THE ORGAN SHORTAGE PROBLEM?

JEFFREY L. PLATT
SARAH C. HENSEL
Mayo Clinic
Rochester, Minnesota

OUTLINE

Introduction

Rationale for Xenotransplantation

Selection of Animal Sources for Xenotransplantation

The Hurdles to Xenotransplantation

INTRODUCTION

The transplantation of organs or tissues between individuals of different species, such as the transplanting of a baboon heart into a human, is called xenotransplantation. Developing the technologies needed to safely carry out xenotransplants has been a long-standing goal, as it would allow the widespread application of organ transplantation and, potentially, other benefits, as discussed below. Successful application of xenotransplantation, however, has been impeded by several hurdles. Recent progress in identifying the molecular basis of the hurdles to xenotransplantation, however, has allowed the genetic engineering of source animals to address these hurdles, those being (1) the immune response of the recipient against the organ graft, (2) the functional limitations of the foreign tissue, and (3) the possibility of a xenograft introducing an infectious disease into the recipient. This article summarizes the rationale and hurdles to xenotransplantation, the potential application of genetic engineering to address these hurdles, and the ethical issues related to xenotransplantation. The reader is referred to other reviews for more detailed consideration of these topics (1).

RATIONALE FOR XENOTRANSPLANTATION

The field of transplantation began during the early years of the twentieth century, when the development of the vascular anastomosis, a surgical technique allowing the suturing of the cut ends of blood vessels, provided the means for transplanting organs from one individual to another (2,3). The transplanting of human organs, also known as *allotransplantation*, was not being undertaken at that time, however, because it was not clear how human organs could be obtained in an ethical manner in the absence of brain death laws that are now in place today, and allow for the harvesting of human organs for transplantation. For instance, because the organs of the recently deceased contained living cells, the organs could be argued to be living, and thus surgical removal was considered unethical. Accordingly, the first efforts to transplant organs into humans were xenotransplants, the organs originating from sheep and pigs, thereby avoiding that particular ethical conflict (4).

Today there are at least three reasons for interest in transplanting animal organs into humans. The first reason is that animal organs would be able to supplement the very limited supply of human organs available for transplantation. Indeed, the shortage of human organs for transplantation is widely seen as the most urgent problem in the clinical treatment of patients with organ failure. So great is the shortage of donor organs, that in the United States, as few as 5 percent of the organs needed ever become available (5), with the problem being even worse in countries where use of human organs is discouraged by law or custom. Furthermore, the shortage of human organs for transplantation sometimes forces the allocation of organs based on social rather than purely medical criteria. While improvements in immunosuppressive therapy and the introduction of immunological tolerance might lessen the shortage of organs by increasing the duration of organ graft survival, this advantage will likely be stifled by the ever-increasing demand for organs and tissues.

A second reason for interest in xenotransplantation stems from the possibility that an animal organ might offer some advantages over the use of a human organ for transplantation. For example, use of an animal organ would allow the transplant procedure to be planned well in advance. The advance planning of a transplant, in turn, allows for the pre-treatment of the patient in ways that more effectively suppress immune responses. Advance planning also allows screening of the donor to minimize infectious risks, and it reduces injury to the graft. The use of animals as a source of organ transplants may also allow the application of transplantation in parts of the world such as Asia, where the use of human tissue is discouraged for cultural reasons. Finally, the use of animal tissues may avert infection of the transplant with human viruses. Such was the motivation behind recent attempts to transplant baboon livers into patients with hepatitis B infection, and baboon bone marrow into an individual with AIDS.

A third reason for interest in xenotransplantation is the possibility that the use of animals would allow for the genetic or biochemical manipulation of the donor, so as to lower the risk of rejection of the organ, or even to express new genes or biochemical processes that benefit the patient. Genetic engineering of large animals such as pigs, to alleviate rejection, has already been undertaken. As a further step, one can envision genetic engineering being carried out to improve the function of an organ transplant, or even to impart new and novel functions on the transplant, such as the secretion of a needed protein, by a patient, to treat a disease.

SELECTION OF ANIMAL SOURCES FOR XENOTRANSPLANTATION

The most important characteristics of the animals used as sources of xenografts are summarized in Table 1. While it might seem intuitive that animals genetically similar to humans, such as chimpanzees and baboons, would be preferable as sources of xenografts, there are a number of factors that argue against this approach and suggest that nonprimates, especially pigs, are preferable. First, and most important, the animals most genetically similar

Table 1. Preferred Characteristics of Animals Used as Sources of Xenografts

Characteristic	Goal	Example
Available in large numbers	Unlimited source	Pig
Size suitable for human	Organs large enough for full sized adults	Pig
Genetically close to human	Minimize immune and biochemical incompatibility	Nonhuman primate
Minimal risk of zoonosis	Ideal donor is from species in regular contact with humans and with well characterized flora	Pig
Ease of breeding	Short gestation time and large litters	Pig
Ease of genetic engineering	Ability to introduce transgenes	Pig

to humans, nonhuman primates, such as baboons, are not available in large numbers, and their relatively small size may prevent the use of some organs in humans. In contrast, the availability of pigs is unlimited, and the size of pig organs is appropriate for transplantation into humans. Second, nonhuman primates frequently harbor viruses, such as the herpes B virus, that are difficult to detect and can be fatal if transferred to humans. Pigs harbor relatively few infectious agents that are harmful to humans, and these agents can generally be eliminated from herds. Third, owing to short gestation time and large litters, pigs can be easily bred, and genes can be introduced into lines of pigs using transgenic techniques. Nonhuman primates are difficult to breed and cannot be engineered genetically with such ease.

THE HURDLES TO XENOTRANSPLANTATION

While various factors favor the use of lower animals, such as pigs, as sources of organs for transplantation into humans, there are still formidable hurdles to carrying out such transplants. The most daunting hurdle to the clinical application of xenotransplantation is the immune response of the recipient against the graft, leading to graft rejection. Recent years have brought about new information regarding the immunology of xenograft rejection, as well as highly specific techniques, such as the genetic engineering of transplant donors, to deal with rejection. In fact, recent successes in dealing with the immune mechanisms of rejection have given rise to two other issues, the physiological limitations of a xenogeneic organ transplant and the possibility that a xenogeneic organ transplant might serve as a vector for introducing novel pathogens into human populations, a process known as *zoonosis*.

IMMUNE RESPONSES TO XENOTRANSPLANTATION

The immune responses to an organ xenograft consist of both "natural" (or innate) immunity, and elicited immune responses. Natural immunity, primarily consisting of natural antibodies and complement but also, in some cases, of natural killer cells, exists without prior exposure to foreign cells or substances. It normally provides an immediate, highly potent defense against extracellular microorganisms. Elicited immune responses, including T cell mediated responses and production of antibodies, following immunization with a foreign cell or substance, provide a defense against intracellular organisms, or organisms less virulent than those targeted by natural immunity.

Natural and elicited immune responses have important roles in the rejection of different types of transplants. While natural immunity is not generally involved in the rejection of organs transplanted between individuals of the same species, that is, allografts (except when the transplants are carried out across blood groups), it is the first and most severe type of immunity directed against organ xenografts. Elicited immune responses contribute to the rejection of both allografts and xenografts.

Hyperacute Rejection

An organ transplanted from a pig into a human would first be subject to hyperacute rejection (6), the most severe and violent immune response known. Hyperacute rejection begins almost immediately upon perfusion of the graft with the blood of the recipient. The graft, initially appearing pink and normal, becomes mottled, then deep red. The flow of blood to the graft declines sharply, then ceases, and the graft is destroyed over a period of minutes to hours. Microscopic analysis of the graft reveals that blood has leaked through small blood vessels, and clots, consisting predominantly of platelets, are formed. Hyperacute rejection of porcine xenografts by primates is mediated by xenoreactive natural antibodies and the complement system of the recipient. Research within the past decade has shed light on the molecular mechanisms by which natural antibodies and complement cause hyperacute rejection, with this knowledge leading to the development of new and insightful therapeutic approaches to this problem.

Xenoreactive Natural Antibodies. Natural antibodies exist in the circulation without a known history of immunization with a foreign substance or organism. For example, some natural antibodies recognize the blood group A or B substances and, thus, define the human blood groups. Some other natural antibodies recognize the cells of foreign species and, as such, are referred to as xenoreactive natural antibodies.

Hyperacute rejection of porcine organ xenografts by primates is initiated by the binding of xenoreactive natural antibodies to the endothelial cells that line blood vessels in the newly transplanted xenograft (7,8). Xenoreactive natural antibodies are present in the circulation of all normal individuals without a known history of exposure to animal cells (9,10). Xenoreactive natural antibodies in humans are mainly directed against a saccharide, Galα1-3Gal (11,12). The importance of Galα1-3Gal as the primary barrier to xenotransplantation was recently demonstrated by the observation that removal of anti-Galα1-3Gal antibodies from baboons prevents the hyperacute rejection of the pig organs transplanted into the treated baboons (13).

Although the identification of the relevant antigen for pig-to-primate xenotransplantation allows specific depletion of the offending antibodies, more enduring and less intrusive forms of therapy would be preferable. One approach to overcoming the antibody–antigen reaction is to develop lines of pigs with low levels of expression of Galα1-3Gal (14,15). The most obvious approach to developing such lines of xenograft donors might be to "knock out" the enzyme that synthesizes the critical sugar, α1,3-galactosyl transferase. Unfortunately, embryonic stem cells are not available for pigs, so this approach cannot be applied with current technology. The first tested method for modifying expression of α1,3-galactosyl transferase involved expression of a glycosyl transferase, which would terminate sugar chains with a sugar other than Galα1-3Gal (16,17). Transgenic mice and pigs expressing the H-transferase have increased expression of H antigen, as expected, and decreased expression of Galα1-3Gal. Whether the expression of a glycosyl transferase, such as H transferase, will prove sufficient to eliminate hyperacute rejection is yet uncertain, but studies using isolated porcine cells have demonstrated that with H-transferase expression, complement mediated lysis significantly decreases.

Complement Activation. The complement system consists of more than 20 proteins that can assemble into complexes that can facilitate the engulfing of foreign organisms by inflammatory cells and cause the death of foreign organisms and foreign cells. Through these functions, the complement system provides the most potent line of defense against severe infections. In addition to helping in defense against infection, the complement system is involved in the development of various immunological diseases, among which is the rejection of xenografts.

Hyperacute rejection is caused by the activation of the recipient's complement system on donor blood vessels (18,19). Triggered by the binding of xenoreactive antibodies to graft endothelium (6,18), complement activation causes loss of the integrity of the endothelial lining of blood vessels and induces abnormal functions in blood vessels.

Xenografts are especially susceptible to complement-mediated injury. The basis for this susceptibility is an intrinsic incompatibility of the recipient's complement system with complement regulatory proteins expressed in the graft. Complement activation is amplified in a xenograft because the mechanisms that prevent complement activation from causing inadvertent injury to normal cells fail to protect the xenograft from the foreign complement of the recipient (20).

Under normal conditions the complement system is inhibited by proteins in the plasma and on the surface of cells (21). These proteins, called *complement regulatory proteins*, protect normal cells from inadvertent injury during the activation of complement (some of the reactions of the complement cascade occur in the complement deposition of cells residing in the vicinity of infectious organisms). Complement regulatory proteins function in a species-restricted fashion, which is to say, they inhibit complement of the same species far more effectively than they inhibit complement of foreign species. Accordingly, the complement regulatory proteins expressed in a xenograft are ineffective at controlling the complement cascade of the recipient, and thus the graft is subject to severe complement-mediated injury (20,22).

Therapeutic Considerations. One approach to preventing hyperacute rejection of xenografts is to administer complement inhibitors to the recipient, such as cobra venom factor, a complement inhibitory protein found in the venom of cobra snakes, or soluble complement receptor type 1, a recombinant protein that inhibits complement at the level of C3 (23), both of which have been found highly effective. One impediment to using complement inhibitors such as these for the prevention and treatment of xenograft rejection is that the inhibitors also prevent the utilization of the complement system to protect the recipient against infectious organisms.

To address this problem, and to overcome the incompatibility of complement regulatory proteins discussed above, several biotechnology groups have developed lines of pigs transgenic for human complement regulatory proteins and, thus, are able to control human complement reactions occurring in transplanted organs (20,24,25). These efforts have focused on expression of human decay accelerating factor, which regulates complement at the level of C3, and CD59, which regulates complement at the level of C8 and C9 (25,26) or decay accelerating factor alone (24). Recent studies have demonstrated that the expression of even low levels of human decay accelerating factor and CD59 in porcine organs prevents hyperacute rejection by primates (25,26). These results, and the dramatic prolongation of xenograft survival brought about by expressing higher levels of human decay accelerating factor in the pig (27), underscore the importance of complement regulation as a determinant of xenograft outcome.

While hyperacute rejection was once considered to be the most daunting hurdle to the clinical application of xenotransplantation, the development of methods for the specific depletion of xenoreactive antibodies and the inhibition of complement have shown that hyperacute rejection can be reliably prevented. This progress has also disclosed another type of rejection — acute vascular rejection — that may occur when hyperacute rejection is prevented.

Acute Vascular Rejection

If hyperacute rejection of a xenograft is averted, a xenograft is subject to the development of acute vascular rejection (28,29). Acute vascular rejection (sometimes called *delayed xenograft rejection*) may begin within 24 hours of connection to the recipient's circulation, and it leads to graft destruction over a period of days to weeks (28,30). It is characterized by injury to endothelial cells and widespread formation of intravascular clots. Although the cause of acute vascular rejection is not completely understood, there is growing evidence that it is triggered, at least in part, by the binding of xenoreactive antibodies to the graft. The importance of xenoreactive antibodies in triggering acute vascular rejection is suggested by the findings that (1) anti-donor antibodies are present in the circulation of allograft recipients with acute vascular rejection, (2) depletion of anti-donor antibodies delays or prevents the occurrence of acute vascular rejection (31), and (3) administration of drugs that inhibit the synthesis of anti-donor antibodies delays or prevents the onset of acute vascular rejection (27). Recent studies suggest that the antibodies that trigger acute vascular rejection might include antibodies directed against Galα1-3Gal. Other factors that may contribute to the development of acute vascular rejection include complement, endothelial cell activation, and natural killer cells.

Therapeutic Considerations. The physical removal of anti-donor antibodies from the xenograft recipient, or inhibition of antibody synthesis, as described above, may effectively prevent acute vascular rejection of experimental xenografts. However, both types of treatments expose the graft recipient to potential complications. Furthermore, while hyperacute rejection can be prevented by temporary treatments, acute vascular rejection poses a more significant hurdle because therapies need to be provided on an ongoing basis. Accordingly, it would be desirable to address this problem through genetic engineering. The various approaches which might be used to deal with acute vascular rejection include removal of xenoreactive antibodies combined with administration of immunosuppressive drugs to limit synthesis of new antibody, induction of immunologic tolerance, and genetic engineering to decrease expression of Galα1-3Gal. In addition to lowering antigen expression, it is likely that expression of human complement regulatory proteins will be helpful in preventing complement mediated injury from contributing to graft injury. Also under consideration is the expression of molecules that inhibit either endothelial injury or the abnormal functions exhibited by injured endothelium.

Accommodation. Fortunately, the presence of anti-donor antibodies in the circulation of a graft recipient does not inevitably trigger acute vascular rejection. Some years ago it was found that if anti-donor antibodies are temporarily depleted from a recipient, an organ transplant may be established so that rejection does not ensue when the anti-donor antibodies return to the circulation (20). This phenomenon was referred to as "accommodation" (20). Accommodation, if it can be established, may be especially important in xenotransplantation, as it would obviate the need for certain ongoing interventions. One potential approach to accommodation may be the use of genetic engineering to reduce the susceptibility of an organ transplant to acute vascular rejection and endothelial cell activation associated with it (32).

Elicited Immune Responses. Organ transplants are subject to elicited immune responses leading to rejection. In contrast to the natural immune response, which exists without prior exposure to foreign cells or antigens, elicited immune responses are brought about by exposure to these entities. For example, the immune responses engendered by administration of vaccines are "elicited" responses.

The elicited immune responses that cause the rejection of transplants between individuals of the same species (allotransplants) can be effectively controlled by conventional immunosuppressive therapy, using drugs such as cyclosporine. There is concern, however, that the immune responses elicited by xenotransplantation will be more severe than the immune responses elicited in allotransplantation and that, accordingly, the responses to xenografts may not be agreeable with conventional immunosuppressive therapy. One reason the immune response to a xenograft may be especially intense is that xenografts have a great variety of antigenic proteins, and that introduction of these proteins may lead to recruitment of a diverse set of "xenoreactive" T lymphocytes. Another reason is that the binding of xenoreactive antibodies to the graft, and activation of the complement system, may lead to amplification of elicited immune responses (33). For example, activation of complement in a graft may cause activation of antigen presenting cells, in turn, stimulating the T cell responses that lead to cellular rejection.

Xenografts may be especially susceptible to immune responses mediated by natural killer cells. Natural killer cells are lymphocytes that recognize and kill tumor and virus-infected cells. The recognition of abnormal cells occurs through the function of two types of cellular receptors. One type of receptor recognizes abnormal carbohydrates and, upon doing so, up regulates natural killer cells' activities. The other type of receptor recognizes major histocompatibility antigens and down regulates natural killer activity (major histocompatibility antigens may be expressed at a decreased level or in an abnormal way by tumor cells or cells infected with viruses). The reason natural killer cells may be active against xenografts is that their carbohydrates recognize Galα1-3Gal (34) and their major histocompatibility receptors may fail to recognize the major histocompatibility antigens of the donor species (35). Thus, human natural killer cells may be especially active against xenogeneic cells because of stimulation and a failure to down regulate natural killer cell functions.

Still another type of elicited immune response to a xenograft might be the production of antibodies against foreign substances in the graft. As already mentioned, natural antibodies are specific for Galα1-3Gal and are produced in increased amounts after xenotransplantation. In addition to these antibodies, there also occurs the production of antibodies against "new" antigens (36). The identity of these antigens and the importance of the antibodies directed against them are still unclear. However, experience with experimental xenografts would suggest that the production of these antibodies can be suppressed by measures that inhibit T lymphocytes, an observation that is explained by the likelihood that antibody secretion by B lymphocytes depends on the function of T lymphocytes.

Another significant hurdle may be the elicited humoral response to xenotransplantation. Such a response in allotransplants, occurring over a period of months, is typically directed against major histocompatibility antigens and has been thought to cause acute vascular rejection. There is every reason to believe that this will be true of xenografts as well. Thus far, studies in xenografts over periods of a few weeks to months suggest that the major immune responses in recipients given immunosuppressive therapy are directed against Galα1-3Gal (36,37), appearing to be an enhancement of the natural immune response. However, humoral responses against other antigens might be elicited by xenotransplantation. If elicited humoral immune responses occur, the responses may be addressed by immunosuppressive therapies or by the development of immunological tolerance. On the other hand, development of genetic approaches to dealing with elicited humoral immune responses must await the identification of the antigens recognized.

Therapeutic Considerations. One major question in xenotransplantation is whether or not the immunosuppressive drugs currently available which so effectively hold the rejection of allografts in check, or those in development, would be able to control the elicited immune responses to xenografts. Yet another question is whether the xenograft can be genetically modified to delimit the elicited immune response. Efforts to control the natural immune barriers to xenotransplantation will likely contribute to limiting the elicited immune response. How a xenogeneic donor could be further modified to limit elicited immune responses is still uncertain but an important area of investigation.

PHYSIOLOGICAL HURDLES TO XENOTRANSPLANTATION

Progress in addressing some of the immune hurdles to xenotransplantation has brought into focus the question of the extent to which a xenotransplant would function optimally in a foreign host. It is known that pig kidneys and pig lungs can replace the most important functions of the primate kidney and primate lung, respectively (38–41). However, subtle defects in physiology across species may exist. Furthermore, organs such as the liver that secrete a variety of proteins and depend on complex enzymatic cascades may prove incompatible with a primate host. Accordingly one important application of genetic engineering in xenotransplantation may be the amplification or modulation of xenograft function to allow for more complete establishment of physiologic function or to overcome critical defects. For example, recent studies by Akhter et al. (42) and Kypson et al. (43) showed that the function of cardiac allografts can be improved by expression of genes encoding beta-adrenergic receptors. Expression of these genes could be adapted to the xenotransplant in order to improve cardiac function. The key question, then, is which of the many potential defects actually need to be repaired.

Another potential hurdle is the possibility that the xenograft may disturb normal metabolic and physiologic functions in the recipient. For example, Lawson has shown that porcine thrombomodulin fails to interact adequately with human thrombin and Protein C to generate activated Protein C (44), a molecule that is critical for the control of coagulation. This defect could lead to a propensity for formation of blood clots in the graft. Of even greater concern is the possibility that a xenogeneic organ, such as the liver, might release substances that would promote abnormal clotting of the blood of the recipient. While a great many physiologic defects can be detected at a molecular level, the critical question will be which of these defects are of importance for the well-being of the recipient, and must therefore be repaired by pharmaceutical or genetic means, and which are innocent defects.

In addition to using genetic engineering to overcome physiological incompatibilities, there is the possibility that genetic engineering might be used to impart novel functions on the xenograft. The animal organ or tissue would then be used as a vehicle for introducing a foreign gene. The genetic engineering of pigs, through transgenic techniques, and the use of these transgenic tissues and organs, as grafts, has certain advantages over conventional gene therapy. First, the genetic material introduced into the genome of the pig can be expressed at high levels in all cells of a given type. Second, since the genetic material is expressed in stem cells, it is passed on to subsequent generations, and therefore the genetic manipulation need not be carried out repeatedly.

ZOONOSIS

The emerging success of experimental xenotransplants and the impending therapeutic trials also bring to the fore the question of zoonosis, that is, infectious disease introduced from the graft into the recipient. The transfer of infectious agents from the graft to the recipient is a well-known complication of allotransplantation. However, the extent of this particular risk of transplantation can generally be estimated so that a decision can be made based on the eventual risk versus the potential benefits conferred by the transplant. The greatest concern of zoonosis in xenotransplantation is not so much the risk to the transplant recipient as it is the risk that an infectious agent will be transferred from the recipient to the population at large. Fortunately, all microbial agents known to infect the pig can be detected by screening and can be potentially eliminated from a population of xenotransplant donors. There is concern, however, that the pig may harbor viruses, such as retroviruses, that might become activated and transferred to the cells of the recipient. For example, Patience et al. recently reported that a C type retrovirus, endogenous to the pig, could be activated in pig cells, leading to release of particles that can infect human cell lines (45). Whether or not this virus or other endogenous viruses can actually infect across species, and whether or not such infection would lead to disease, is still not certain, but recent studies suggest that the risk of infection may be low (46).

If cross-species infection does prove to be a significant hurdle, genetic therapies might also be used to address this problem with further possibility of introducing genes that might target the infectious agent. The simplest genetic therapy would involve "breeding out" the organism, but this approach might fail if the organism were widespread in the pig populations or integrated at multiple loci. Some genetic therapies have been developed to potentially control HIV (47). While these therapies have not generally succeeded, because it has been difficult or impossible to gain expression of the transferred genes in stem cells and at levels sufficient to deal with high viral loads, the application of such therapies might be much easier in xenotransplantation because the therapeutic genes could be carried by all of the animal cells.

Another concern is that infection of the xenograft might fail to engender an immune response in the recipient that would otherwise lead to control of that infection, because antigenic peptides of the infecting organism would be presented in association with xenogeneic MHC molecules. If this problem proves limiting in xenotransplantation, the problem could potentially be overcome by eliminating those microorganisms through mechanisms described above, or by the introduction of "generic" MHC molecules, which the host might recognize.

A SCENARIO FOR THE CLINICAL APPLICATION OF XENOTRANSPLANTATION

The past few years have brought significant progress in defining the immunological, physiological, and infectious disease hurdles to xenotransplantation. This information has been exploited in the development of incisive new therapies for overcoming these hurdles. Perhaps the most exciting of these therapeutic strategies is the development of genetically engineered animals for use as a source of organs for transplantation. To this point, research on genetic manipulation has focused on the most severe of the immune hurdles, the action of complement on graft blood vessels. This effort will no doubt continue while new work will seek to introduce genes for dealing with the effects of cellular immunity or physiological defects or infectious agents. With these, it can be envisioned that xenotransplantation will enter the clinical arena in a stepwise fashion. First, there will occur free tissue xenografts and extracorporeal use of xenogeneic organs. Limited clinical trials of this sort are in progress (48–50) and there is encouraging early evidence that neural cells may endure in a human recipient (50). Next, xenogeneic organs will probably be used as "bridge" or temporary transplants. Bridge transplants will not solve the problem of organ shortage, but the transplants will allow the gathering of vital information regarding the remaining immune and biological hurdles. Third, there will be the use of porcine organs as permanent replacements, but this use will probably be restricted to patients who cannot receive a human organ allograft. Only with further refinements may there eventually be a fourth and final step, in which xenotransplantation is used as an alternative to allotransplantation.

ETHICAL ISSUES IN XENOTRANSPLANTATION

The ethical issues as related to xenotransplantation can be considered to be of three categories: (1) animal welfare, (2) clinical experimentation, and (3) societal implications. With the exception of the third category, the issues are, for the most part, shared with many fields of experimental medicine and are not unique to xenotransplantation. Some

of the ethical aspects of xenotransplantation have been reviewed (51–59).

Animal Welfare

One aspect of the ethics of xenotransplantation relates to the use of animals as a source of organs and tissues for transplantation into humans. The ethical question is whether or not it is proper or just to "use" animals for the benefits of human beings. Some believe that animals should not be used for such purposes, while others believe such use of animals should be permissible. The question, then, is whether society should regulate the use of animals according to the wishes of one group or to the other. One approach to this question has been to reason that if it is permissible to "use" animals for food, it should be permissible to "use" animals as a source of organs or tissues for transplantation. Thus, to the extent that society is willing to countenance the use of animals for food, it would seem difficult to justify development of laws prohibiting the "use" of animals for other purposes that are beneficial to humans provided that the generally accepted facets of animals welfare are addressed.

If it is not unethical to use animals for organ transplantation, ethical issues still remain pertaining to animal welfare or treatment. Similar to the broader subject of animal welfare, these issues include how the source animals are raised, how they will be handled, the potential occurrence of suffering, and the subjection of animals to genetic engineering. For example, to keep animals as free as possible from infectious agents, it will likely be necessary to raise them in isolation. Is this unethical, and, if so, how might this issue be addressed? As another example, it is possible that genetic modification may impair an animal's health and cause suffering.

Of the issues discussed above, one of particular interest is genetic engineering. Genetic engineering involves inserting genes into the germ line (accomplished by various means, e.g., a direct injection of genetic material into a fertilized egg), thereby passing that gene on to subsequent generations of animals. There seems to be a fear, among some, of the changes that may occur as a result of this type of deliberate introduction of genetic material, for instance, in the case of genetic engineering of plants and livestock. Some question whether or not the species will change as a result of genetic manipulation or whether the animal will become more "human." Perhaps some of this fear stems from the lack of knowledge that genetic material can be naturally exchanged between species as a result of infection by retroviruses.

Clinical Experimentation

A second category of ethical issues relates to the use of human subjects in clinical xenotransplantation trials. The recipient of a xenotransplant, as the subject of an experimental procedure, must be adequately informed about that procedure, including the anticipated risks and the limitations of knowledge regarding the anticipated benefits. However, these aspects of the use of human subjects do not differ in substantive ways from any use of human subjects in medical research. For example, similar weighing of uncertain risks and benefits might also occur with the use of an experimental device or an experimental device or an experimental human-to-human transplant. Some of the matters of ethics to be addressed include (1) how to obtain consent for an experimental procedure that will surely require long-term follow-up or monitoring, (2) how the right of the patient to withdraw will impact on the individual, (3) how to avoid encouraging patients to accept greater-than-usual risks for the benefit of future patients, (4) how the experiment might affect quality of life — as an increase in risk of organ rejection means a need for greater immunosuppressive therapies to an unknown end or risk, and (5) how to forecast what the possible psychological effects of such an experiment might be.

Ethics of Society

While the ethics of animal use and human experimentation are not unique to xenotransplantation, the social implications may be so. The most important ethical question for society would seem to relate to the possibility that the recipient of xenotransplant might become infected by microbial agents contained in the transplant and that agent might then be spread more broadly in the population. The possibility that a medical or surgical procedure could have implications for the broader community is a relatively novel aspect of medical ethics, but one that does exist outside of xenotransplantation. For example, the use of antibiotics by an individual patient may alter the types of organisms present in a hospital and, thus, impose a risk on other individuals. Similarly, the use of inhaled substances could impact on those in contact with a given patient. While it is now thought by many experts that the risk of an epidemic being caused by a xenotransplant is very low, the potential impact of an epidemic keeps this issue at the fore.

There are other ethical questions, the impact or scope of which might extend beyond the recipient of a xenotransplant. For example, it is thought that family and/or sexual contacts of the recipient may have to provide consent, since they may be at risk. As another example, there is the question of whether or how the recipient of a xenotransplant might withdraw consent after a xenotransplant is in place; such withdrawal would lead to cessation of monitoring for potential infectious agents, which would, in turn, place society at risk. Yet another issue involves the possible costs of xenotransplantation to society, for, although transplantation has proven to be less costly than other chronic care for organ failure, it is uncertain whether this advantage will apply to xenotransplantation. A further ethical question includes how to achieve justice in distribution of human versus animal organs, assuming human organs are preferred. Furthermore, this preference may result from another issue, that of the possibility of a low level of public acceptance, at least at the outset, as many may perceive xenotransplantation as "unnatural" or tampering with nature or God. Important questions then become: Does the use of animal organs change how we define ourselves as human? What is or isn't "natural"? Is it not natural that we evolve and obtain the knowledge that enables us to move forward in ways such as this? Finally, to what extent

should the advantages and risks of xenotransplantation be subject to public discussion?

In addition, there are some issues that may be more political than ethical. One such issue is whether or not regulation occurring in one country, though not in another, would be meaningful. Another issue involves the role of the private sector, which will likely play a major role in the research and creation of transgenic breeds of porcine. Relative questions become: What is the role of the private sector in support of research or commercialization? Will patents be given for transgenic breeds and, if so, to whom? Furthermore, are the ethics of business too vastly different from those of science and medicine, and while their motivation is surely different, does this necessarily pose any ethical dilemmas or risk to the public? It would seem logical that though their motivation is, in fact, different, they would have a vested interest in ensuring a safe, quality "product," if not simply to ensure their own survival.

CONCLUSIONS

While ethical issues are salient, many do not differ greatly from the ethical issues raised in any experiments involving animal or human subjects. What makes xenotransplantation somewhat unique is that transplantation has a high profile, and it raises issues for society, although it is far from the only human endeavor to do so. Throughout history (e.g., with the advent of the railroad train) there has been a quandary about how advances in science will affect society and the state. Clearly, some of these issues relating to xenotransplantation will most likely remain unsettled, raising one further question: To what extent can one proceed, in activities that impact on society, without the existence of consensus?

ACKNOWLEDGMENT

I thank the National Institutes of Health for supporting my work in this area.

BIBLIOGRAPHY

1. J.L. Platt, *Nature* **392**(suppl.), 11 (1998).
2. A. Carrel, *Lyon Med.* **98**, 858 (1902).
3. C.C. Guthrie, *Blood-Vessel Surgery and Its Applications*, Longmans, Green & Co., New York, 1912.
4. E. Ullman, *Ann. Surg.* **60**, 195 (1914).
5. R.W. Evans, C.E. Orians, and N.L. Ascher, *J. Am. Med. Assoc.* **267**, 239 (1992).
6. J.L. Platt, *Hyperacute Xenograft Rejection*, CRC Press, Austin, TX, 1995.
7. R.J. Perper and J.S. Najarian, *Transplantation* **4**, 377 (1966).
8. G.R. Giles et al., *Transpl. Proc.* **2**, 522 (1970).
9. S.V. Boyden, *Adv. Immunol.* **5**, 1 (1964).
10. W. Parker et al., *J. Clin. Immunol.* **17**, 311 (1997).
11. A.H. Good et al., *Transpl. Proc.* **24**, 559 (1992).
12. M.S. Sandrin et al., *Proc. Natl. Acad. Sci. USA* **90**, 11391 (1993).
13. S.S. Lin et al., *Transpl. Immunol.* **5**, 212 (1997).

14. R.L. Geller, P. Rubinstein, and J.L. Platt, *Transplantation* **58**, 272 (1994).

15. C.G. Alvarado et al., *Transplantation* **59**, 1589 (1995).

16. M.S. Sandrin et al., *Nat. Med.* **1**, 1261 (1995).

17. A. Sharma et al., *Proc. Natl. Acad. Sci. USA* **93**, 7190 (1996).

18. J.L. Platt et al., *Transplantation* **52**, 214 (1991).

19. A.P. Dalmasso et al., *Am. J. Pathol.* **140**, 1157 (1992).

20. J.L. Platt et al., *Immunol. Today* **11**, 450 (1990).

21. D. Hourcade, V.M. Holers, and J.P. Atkinson, *Adv. Immunol.* **45**, 381 (1989).

22. A.P. Dalmasso et al., *Transplantation* **52**, 530 (1991).

23. H.F. Weisman et al., *Science* **249**, 146 (1990).

24. D. White and J. Wallwork, *Lancet* **342**, 879 (1993).

25. K.R. McCurry et al., *Nat. Med.* **1**, 423 (1995).

26. G.W. Byrne et al., *Transplantation* **63**, 149 (1997).

27. E. Cozzi et al., in D.K.C. Cooper et al., eds., *Xenotransplantation: The Transplantation of Organs and Tissues between Species*, Springer, Berlin, Germany, 1997, p. 665.

28. J.R. Leventhal et al., *Transplantation* **56**, 1 (1993).

29. W. Parker et al., *Immunol. Today* **17**, 373 (1996).

30. J.C. Magee et al., *J. Clin. Invest.* **96**, 2404 (1995).

31. S.S. Lin et al., *J. Clin. Invest.* **101**, 1745 (1998).

32. F.H. Bach et al., *Nat. Med.* **3**, 196 (1997).

33. D.T. Fearon and R.M. Locksley, *Science* **272**, 50 (1996).

34. L. Inverardi et al., *Transpl. Proc.* **28**, 552 (1996).

35. L.L. Lanier and J.H. Phillips, *Immunol. Today* **17**, 86 (1996).

36. K.R. McCurry et al., *Hum. Immunol.* **58**, 91 (1997).

37. A.H. Cotterell et al., *Transplantation* **60**, 861 (1995).

38. G.P.J. Alexandre, in M.A. Hardy, ed., *Xenograft 25*, Elsevier Science Publishers, New York, 1989, p. 259.

39. T. Sablinski et al., *Xenotransplantation* **2**, 264 (1995).

40. C.W. Daggett et al., *Surg. Forum* **47**, 413 (1996).

41. C.W. Daggett et al., *J. Thor. Cardiovas. Surg.* **115**, 19 (1998).

42. S.A. Akhter et al., *Proc. Natl. Acad. Sci. USA* **94**, 12100 (1997).

43. A.P. Kypson et al., *J. Thor. Cardiovas. Surg.* **115**, 623 (1998).

44. J.H. Lawson and J.L. Platt, *Transplantation* **62**, 303 (1996).

45. C. Patience, Y. Takeuchi, and R.A. Weiss, *Nat. Med.* **3**, 282 (1997).

46. K. Paradis et al., *Science* **285**, 1236 (1999).

47. B.A. Sullenger et al., *Cell* **63**, 601 (1990).

48. C.G. Groth et al., *Lancet* **344**, 1402 (1994).

49. R.S. Chari et al., *N. Engl. J. Med.* **331**, 234 (1994).

50. T. Deacon et al., *Nat. Med.* **3**, 350 (1997).

51. The Nuffield Council on Bioethics, London, 1996.

52. R. Downie, *J. Med. Ethics* **23**, 205 (1997).

53. M. First, *Kidney Int.* **58**, S46 (1997).

54. A. Daar and D. Phil, *World J. Surg.* **21**, 975 (1997).

55. J. Moatti, *An. N.Y. Acad. Sci.* **862**, 188 (1998).

56. B. Miranda and R. Matesanz, *An. N.Y. Acad. Sci.* **862**, 129 (1998).

57. J. Hughes, *J. Med. Ethics* **24**, 18 (1998).

58. A. Daar, *Bull. World Health Organ* **77**, 54 (1999).

59. G. Beauchamp, *Can. J. Surg.* **42**, 5 (1999).

See other entries ANIMAL, MEDICAL BIOTECHNOLOGY, LEGAL, LAWS AND REGULATIONS GOVERNING ANIMALS AS SOURCES OF HUMAN ORGANS; FDA REGULATION OF BIOTECHNOLOGY PRODUCTS FOR HUMAN USE; see also RESEARCH ON ANIMALS entries; TRANSGENIC ANIMALS: AN OVERVIEW.

B

BEHAVIORAL GENETICS, HUMAN

DAVID T. WASSERMAN
University of Maryland
College Park, Maryland

OUTLINE

INTRODUCTION

A variety of contemporary research programs in biology, psychology, and medicine investigate the magnitude and character of genetic influence on human behavior, affect, and cognition. Heritability studies attempt to tease out genetic from environmental effects, largely by comparisons involving twins and adoptees; association and linkage studies seek markers, and ultimately genes, associated with behaviors or traits of interest; neurobiological research explores causal pathways by which genetic variations may affect behavior. While specific research programs differ greatly in method and ambition, they share the assumption that it makes scientific sense to look for genetic contributions to important mental and behavioral differences among people.

Statistical techniques for teasing apart genetic and environmental influence date back to Francis Galton, who was—not coincidentally, critics insist—the founder of eugenics. In the half-century after Galton proposed the systematic study of heritable differences among people and the improvement of the genetic stock of humanity, heritability research evolved in close association with eugenic policies. The research, while not devoid of scientific achievement, proceeded in ignorance of the complex patterns and actual mechanisms of inheritance, and produced such retrospective embarrassments as Charles Davenport's work on the genetics of seafaring, which argued, in effect, that the love of the sea was a simple Mendelian trait. The policies, while supported in part by liberals as well as conservatives, included the restrictive immigration laws and sterilization campaigns in the United States and led, indirectly, to the programs of mass sterilization and "euthanasia" of the unfit in Nazi Germany (1,2).

Research on the genetics of human behavior and psychology underwent a period of understandable quiescence after the Second World War. In the past two decades, though, it has made a dramatic comeback. Two developments have arguably contributed to its resurgence: first, the advent of sophisticated techniques for isolating and manipulating genetic material; second, increasing public dissatisfaction with the optimistic environmentalism that supposedly dominated the social policy of the postwar era.

Much of the debate between critics and proponents of genetic research centers on the interpretation of these two trends. Proponents see the resurgence of genetic research on human behavior as driven by its scientific progress and a growing appreciation of the complexity of human behavior; they see the public as rejecting the environmental determinism of the postwar era much as it had rejected the genetic determinism of the pre-war era. They argue that judging contemporary behavioral human genetics in terms of the genetic research and eugenic policies of the first half of this century is "somewhat akin to attempting to explain the behavior of a butterfly by studying the caterpillar (or for that matter, understanding the fruitfly by studying the maggot)" (3).

Critics of behavioral genetics doubt the claimed metamorphosis; they are less inclined to see it as emerging from a cocoon than crawling out of an overturned rock. They see the research as exploiting the general retreat from an environmentalism that policy makers never seriously pursued; they regard the advances in molecular genetics, however useful in other areas of research, as serving a largely cosmetic role in behavioral genetics, lending a veneer of hard science to an enterprise that remains essentially confused and speculative.

This clash of interpretations helps frame the central issue this article will address: Do advances in molecular biology make it significantly more likely that behavioral genetics will avoid the scientific pitfalls and social abuses of earlier human genetic research? Or do behavioral geneticists misunderstand the scientific and social lessons of the past, making them likely to repeat it?

This article will not attempt to adjudicate the conflict between behavioral genetics and its critics so much as to clarify the sources of disagreement between them. Their disagreement, it will suggest, does not concern the specific methods, assumptions or findings of the research so much as its explanatory and practical value. Researchers believe that much can be learned about the causes and control of significant human traits and behavior by isolating genetic features associated with individual differences in those traits and behaviors, while critics believe that even if such features can be isolated, they will yield limited insight into causal and developmental processes, while creating a substantial risk of scientific oversimplification and social abuse.

SCOPE OF THE ARGUMENT

Some objections to human behavioral genetics are based on a denial that human conduct is subject to, or can be described by, the kind of law-like generalizations found in other sciences. These objections are directed against the generalizations offered by sociology no less than those offered by behavioral genetics. Other objections to behavioral genetics are to biological, as opposed to social or psychological explanations of behavior; still others are to individual, as opposed to situational explanations. Some of these more general objections, e.g., to the attempt to reduce human behavior and cognition to biology, are addressed in other articles in this Encyclopedia. This article will discuss these general objections only to the extent that they figure in the debate over genetics and behavior.

One such general objection is that biological accounts of mental traits and social behavior inevitably oversimplify or distort the phenomena they seek to explain. Thus, for example, a vast array of social behavior and interaction is lumped together as "criminality," a classification that assumes a common trait or disposition underlying the myriad of criminal activities and styles of transgression. Similarly, the richly varied forms of sexual attraction, play, and intimacy are dichotomized as hetero- and homosexuality. Researchers would respond that simplification and abstraction are essential to scientific progress. Even if the initial categories employed by the researchers are vague or coarse, they will be refined by additional research, and they may have significant heuristic value in generating hypotheses and developing theories. Simplification and abstraction are critical for sociological as biological generalizations about behavior.

Another general objection is that explanations based on individual differences divert attention from broader structural or institutional causes. For example, it is claimed that the attempt to attribute learning failures to individual factors—whether biological, like Attention Deficit Disorder, or social, like family dysfunction—diverts attention from the appalling state of public schools and public education. The attribution to biological differences may appear to be especially diversionary, since those differences may appear, unlike family dysfunction, to have no relation to structural or institutional causes. Researchers may insist in good faith that biological or genetic variations are only one type of causal factor among many, and that the effect of those variations is mediated by broader social factors. But policy makers will inevitably single out biological or genetic factors, since it is far easier to identify "bad apples" than to change a rotten system.

Other objections are specific to genetic as opposed to other biological causes—to genetic variations as opposed, say, to birth trauma. Those who regard genetic explanations as especially reductive or prone to abuse often assume that genetic causes have less mutable effects than other biological causes, and that the effects of genetic causes are essential to, or constitutive of, an individual in a way that the effects of trauma or insult are not. Both these assumptions, however, are mistaken. The effects of many genetic conditions are, or may be, largely or completely preventable, such as PKU, a single-gene disorder, causes severe retardation, but that effect can be avoided or mitigated by a modified diet. And if genetic effects are not immutable, neither are they essential or constitutive. Even if possessing all or most of one's genome is a necessary condition for personal identity, acquiring all or most of the traits associated with that genome in standard environments is not. We hardly think that a phenolallenine-free diet alters the identity of a child with PKU, even if the absence of the PKU mutation would (and that is debatable). But these beliefs may be as ingrained and recalcitrant as they are mistaken, and they may have a profound effect on the social reception of behavioral genetic research.

Behavioral genetics confronts all of the above objections—to individual, biological, and genetic explanations—because its defining feature is its focus on the effects of individual differences in genetic constitution. Humans beings share 99 percent of their genome with chimpanzees and more with each other; behavioral genetics is concerned with the relatively small portion that differs from person to person. Many critics believe this emphasis is misplaced; they question whether attempts to explain differences in human behavior and personality in terms of genetic differences are scientifically promising or socially beneficial. Some critics argue from an evolutionary perspective that while all human beings, or human males, may have genetic propensities for, say, violent or antisocial behavior, individual differences in the strength of those propensities are unlikely to have a genetic source. Even if genetic differences among individuals made them more or less prone to violent or antisocial behavior, they are likely to contribute far less to the understanding of human violence than genetic commonalities. Other critics fear that the search for individual genetic differences, whether or not scientifically justified, will stigmatize those individuals thought to be genetically predisposed, and the social groups to which they belong.

While behavioral genetics looks at a wide range of traits and behaviors, this article will focus on predispositions to criminal, violent, aggressive, impulsive and antisocial behavior. There are two reasons for this focus: first, research in these areas has engendered more public attention and controversy over the past decade than research on other traits, even intelligence; second, and more importantly, genetic research on antisocial and criminal behavior has taken much more of a molecular turn than genetic research on intelligence or other traits, bringing to behavioral genetics the powerful hopes and fears raised by the Human Genome Project.

CLASSICAL AND CONTEMPORARY BEHAVIORAL GENETICS

Heritability Research

Until techniques were developed in the 1970s for isolating and manipulating genetic and other molecular material, human behavioral genetics was largely confined to heritability studies. Heritability research takes off from the scarcely debatable observation that many behaviors and psychological traits "travel in families." But parents usually confound the assessment of genetic influence by providing their children with rearing environments as well as genes; the transmission of behavior from one generation to the next can be attributed to either one.

Behavioral genetics exploits two processes which tend to tease apart these genetic and environmental contributions: twinning and adoption. Although the logic of these comparisons has been understood and debated for over a century, the implementation of careful studies only became possible with the systematic government record-keeping and refined statistical techniques of the twentieth century (4).

Twinning produces offspring that share either half their genes (dizygotic/DZ), the same proportion as in normal siblings, or all their genes (monozygotic/MZ). If the rearing environments of DZ twins can be assumed to be as much alike as those of MZ twins, and if other, less controversial assumptions are satisfied, then any greater similarity (concordance) in the behavior of the MZ twins can be attributed to their greater genetic commonality, and differences in the concordance of MZ and DZ twins can be used to estimate the heritability of the behavior or trait in the general population.

The second process that tends to separate out genetic and environmental contributions is adoption. A true experiment would randomly assign children at birth (or better yet, at conception) to other parents, and compare their traits and behaviors with those of their biological and adopted parents. Social practice very roughly approximates such an experiment by somewhat fortuitously if nonrandomly assigning children at some point after birth to adoptive parents. The greater the similarity of the children to their biological parents in the trait or behavior studied, the stronger the evidence of a genetic contribution. In addition to providing potentially corroborating evidence for twin studies, adoption studies help identify specific environmental factors and provide insight into the role of genetic factors in shaping the rearing environment, such as by evoking parental responses (5). Researchers can combine twin and adoption studies by examining the similarities of MZ twins reared apart — a prized commodity in human behavioral genetics.

Studies conducted over the past 40 years have reported significant heritabilities for a variety of psychiatric and behavioral conditions, including schizophrenia, intelligence, neuroticism, antisocial behavior, and property crimes. Interestingly, they have failed to find significant heritabilities for other behavior, such as violent crime, which the public assumes to have a heavy genetic loading, but such negative findings rarely receive the same press as positive ones (6,7). Among the strongest, and most surprising, results from twin and adoption studies have been negative — the consistent finding that "shared environment," the congeries of parental, domestic, and other local factors affecting all siblings equally, makes virtually no contribution to most behavioral and psychological states that have been investigated. The environment that appears to matter is the idiosyncratic environment of each sibling (8).

There has been much discussion about the validity of the assumptions on which these findings rest: For example, are the rearing environments of DZ twins as much alike as those of MZ twins, or are identical twins environmentally as well as genetically more alike, down to their identical wardrobes? Does the time adoptees spend with their biological parents, or do the adoption agencies' nonrandom placement practices, help explain the behavioral similarities of the adoptees and their biological parents? The researchers themselves are keenly aware of weaknesses in the assumptions and methods of previous studies, although they tend to be more optimistic than the critics about the prospects for dispensing with controversial assumptions or controlling for their violation.

There is a broader, more important debate about what to make of findings that a given trait or behavior has high heritability. Heritability studies do not identify the genetic or chromosomal variations responsible for traits or behavior. Rather, they calculate the population variance on some measure of a trait or behavior (e.g., IQ score, number of arrests), then use the assumptions about genetic similarity referred to earlier to assign percentages of that variance to heredity and environment. Thus a typical finding of an MZ/DZ twin study might be that "60 percent of the variance in IQ score is attributable to heredity." There are several critical points to make about this type of finding, about which researchers and their critics do not disagree.

First, heritability studies assess genetic contributions to differences in traits or behaviors. Features shared by virtually everyone in the population have little or no heritability; for example, the trait of "having two arms" will have very low heritability in a country where almost everyone does, and where most of those who do not are missing limbs because of dismemberment or infection rather than genetic disease or mutation (9). It is important to distinguish a trait's heritability from its genetic basis, a distinction critical to the evolutionary critique of behavioral genetics discussed below.

Second, in human heritability studies, the "environment" covers everything besides the genetic endowment to which variance in a trait or behavior can be attributed, from prenatal stress to adolescent peer pressure. Human studies do not plot the trait or behavior against an array of distinct environments, defined by the presence, absence, or level of quantified variables manipulated by the experimenter — the standard methodology in plant and animal behavioral genetics. The differences among the environments where human genetic variation is expressed are largely unmeasured and often unknown (10,11).

Third, heritability studies attempt to account for population variance in the trait or behavior; they reveal nothing directly about the comparative importance of genetic and biological factors in any individual. A particular level of heritability could reflect the effect of different genetic and environmental factors in different segments of the population, and could arise from the large effects of a small number of genes or the small effects of a large number. The claim, for example, that heredity accounts for 60 percent of the variance in IQ does not mean that 60 percent of an individual's IQ score, or deviation from the mean IQ, is genetically determined (if such claims are even intelligible) (9).

Fourth, heritability studies do not investigate or distinguish among possible causal pathways from gene to trait or behavior. Thus, to take an example from Christopher Jencks, people with red hair might have below-average IQ solely because they are neglected and mistreated as children throughout the society. Common sense would treat this as an environmental explanation of their lower IQ. Nevertheless, because there might be little or no variation in the treatment or the performance of red-haired children across the environments actually studied, low IQ could be counted as genetic in a heritability study. The extent to which genes lower IQ by evoking discrimination is not reflected in the proportion of the variance assigned to heredity. The example becomes less fanciful if we substitute dark skin for red hair (11).

The indirect manner in which behaviors and mental traits are typically measured gives such alternative explanations further plausibility. If the researcher adopts arrests or convictions as a proxy for criminal behavior or disposition, or score on a written test as a measure of task proficiency, the causal pathway from the genes to the measured variable may not run through the trait or behavior at all: What may be transmitted genetically is not a propensity to commit crimes, but a lack of talent for concealment or evasion; not proficiency in a specific cognitive task, but a general facility with written tests.

Again, behavioral geneticists are well aware that the analysis of variance is quite different from the analysis of causation; that heritability studies face inherent limits in what they can reveal about the ways genes affect traits and behavior; and that observed heritability of a trait or behavior may vary with the measure chosen. Thus the American Society for Human Genetics nicely summarized the limitations of heritability research in a 1997 statement on the state of the research in behavioral genetics:

> The concept of heritability refers to the ratio of the genetic variance to the overall phenotypic variance. It is based on a specific situation involving a particular phenotype in a population with some array of genetic and environmental factors at a given time. It can differ from population to population and from time to time. It can change with age during development. It is important to keep in mind that heritability is a descriptive statistic of a trait in a particular population, not of a trait in an individual (5, p. 1266).

But researchers and critics disagree about the significance of these limitations. For researchers, heritability studies make a valuable contribution despite these limitations, by identifying traits and behaviors that are good candidates for genetic influence in light of the strong evidence that something is being genetically transmitted. Critics, in contrast, argue that the very attempt to partition causes of variance into environmental and genetic factors is misleading, if not meaningless. Given the indefinite variety of ways in which causally inert genetic traits such as skin color interact with powerful social forces such as racism to produce destructive social behavior, there is no reason to think that interesting or informative genetic explanations lie behind findings of high heritability even if they are valid on their own terms (5,10).

This difference in outlook is illustrated in the response to reports of high heritabilities for traits such as religious affiliation, where any genetic effect is obviously mediated by a vast array of cellular, somatic, and social variables (12). While researchers see such findings as intriguing (though clearly in need of a more complete causal account), critics regard them as a reductio ad absurdum: if religious affiliation is heritable, then heritability can have only the most tenuous relation to genetic causation.

Twin and adoption studies were not the only tools in the behavioral geneticist's armamentarium before the introduction of molecular genetics. During the 1970s some sophisticated mathematical techniques were developed for teasing out genetic and environmental contributions to behavior from a wider range of family data. These "model-fitting" techniques employ data from a variety of family relationships and a variety of populations. They have proved useful in overcoming one of the principal limitations of heritability estimates based on standard twin and adoption studies — their restriction to a particular population (5). These techniques have also been useful in incorporating the associations found between behavioral and psychological traits and specific genetic locations into models of gene–environment interaction, integrating the old behavioral genetics and the new.

Out with Heritability, in with Molecular Genetics

The clearest difference in outlook between behavioral geneticists and their critics is not in their evaluation of past or present research but in their assessment of the prospects for future research. Researchers and critics agree that heritability research will play a diminishing role in the behavioral genetics of the next century. But they disagree about the scientific potential of the techniques that are superseding it. These techniques are designed to link behavioral and mental traits to specific genes and to move beyond heritability research in several respects: to identify what is transmitted genetically, trace causal pathways from genes to traits or behavior, and thereby move partway from the population to the individual. But they also maintain the focus of heritability research on the genetic basis for differences among individuals.

Researchers expect that molecular and neurogenetic research will resolve the ambiguities about causation by tracing the complex pathways through which specific genes affect traits or behavior. This work, they claim, will place behavioral genetics on a solid biological foundation, and it will be less susceptible to oversimplification than prior research, since it will replace global estimates of heritability with narrow and testable causal hypotheses. They point to the progress made by medical genetics in identifying genes associated with a variety of disorders, many of which had not previously been regarded as "genetic." At the same time they recognize that even with the new molecular techniques, genetic contributions will be harder to identify, in part because genes may make a more modest and complex contribution to behavioral and psychological disorders than somatic disorders.

While researchers expect genetic effects on psychology and behavior to be modest and complex, they also expect them to be sufficiently strong and interpretable to yield insight into the causes of important human traits and behavior. For example, Gregory Carey and Irving Gottesman speculate:

> The current generation of molecular genetic research is likely to uncover polymorphisms associated with behavior, and some of these loci will probably be correlated with antisocial behavior. No one is banking on a major "crime gene." Instead, most suspect that there will be a number of loci of small effect that partly influence temperament, motivation, and cognition. ...The statistical predictability of these loci may be quite small. They may, however, prove quite important for unraveling the complicated psychology and neurobiology behind behavior (13, p. 89).

Moreover researchers expect that the identification of specific genes, and of the cellular, somatic and social variables that mediate between those genes and behavior, will increase the possibilities for humane and effective intervention, and put to rest the public misconception that genetic causes have immutable effects.

Critics, on the other hand, think that molecular genetics will help to sustain the false hopes that they believe have always informed behavioral genetics. They fear that neurogenetic research focused on individual differences is unlikely to yield much understanding of the causes of antisocial and violent behavior. They agree that the research is likely to find markers and genes that are loosely associated with that behavior. But they expect that most genetic contributions to differences in mental traits and behavior will be slight and oblique — all interaction and no main effects — difficult to interpret and of very limited theoretical interest. On the rare occasions when genetic contributions are significant, they are likely to be the work of mutations that cause major dysfunctions in small numbers of people. On this point, Evan Balaban contends that "the biochemical equivalent of hitting a subject on the head with a club may explain a pattern of pathology in a small number of individuals but will not be very enlightening for either the scientific study of behavioral biology or for general societal problems involving crime and violence" (14, p. 87).

Critics fear that the discovery of molecular markers for behavior will be less to advance scientific understanding than to increase social control. The danger of abuse is much greater for research employing sophisticated genomic technologies and enjoying the cachet of molecular genetics. Because the markers and genes will be easy to detect, and will have the appearance of hard scientific data, they are likely to be employed in programs of screening and preemptive intervention.

From XYY to MAO

These conflicting expectations about the scientific and social value of molecular genetic research are reflected in the divergent lessons that critics and researchers draw from the history of the first microbiological marker linked with human behavior, the XYY karyotype. In 1965 researchers found an apparently high incidence of that karyotype among prison inmates in Britain (15). They assumed that this incidence was higher than in the general population, an assumption since confirmed, and they speculated that men with an extra Y chromosome tended to be hyperaggressive, a hypothesis that subsequent research failed to support. It is now widely believed that if an extra Y chromosome leads to prison, it is by an indirect route. XYY individuals do not appear to be more aggressive or violent than average, but they may be taller, less intelligent, and more impulsive (16,17). Their increased risk of arrest may reflect a greater likelihood of getting caught, not heightened aggressiveness or a greater disregard of social norms.

Both researchers and critics regard the XYY story as a cautionary tale. But where critics see an illustration of the risks inherent in any inquiry into biological markers for social behavior, researchers see a modest triumph of scientific self-correction. Critics observe that the early XYY investigators, in their rush to find a direct link between genes and behavior, assumed that an extra "male" chromosome would make a specific contribution to violence or aggression, instead of having the generally impairing effects typically associated with an extra chromosome. Researchers, on the other hand, note that it was behavioral geneticists who ruled out any association between the XYY karyotype and violence or aggression (while confirming the high incidence of XYY in prisons and other institutions).

In the 25 years since the XYY controversy, the techniques for identifying biological markers for behavior have changed far more than the issues concerning their interpretation. With the development of recombinant DNA technology in the late 1970s, researchers were able to identify and manipulate individual genes and genetic material. That technology became directly relevant to behavioral genetics with the discovery of genetic "markers" for a variety of diseases and traits — highly variable ("polymorphic") but functionally inert DNA segments found to be associated with various phenotypic traits, presumably because they were located in close proximity to genes that actually contributed to those traits.

In the late 1980s and early 1990s, markers, and in some cases genes, were identified for a number of diseases known or suspected by their inheritance patterns to have a significant genetic component. Behavioral geneticists were quick to adopt the same methods, hoping to replicate the dramatic successes of medical genetics. They were soon reporting markers for a number of psychiatric and behavioral conditions, including bipolar disorder, schizophrenia, and alcoholism. Early findings in the first two areas had to be retracted, however, and findings in the third remain mired in controversy (5,18). A decade after the first application of molecular genetic techniques to psychiatric and behavioral disorders, there has yet to be a single consistently replicated, generally accepted association between a specific gene or marker and a common behavioral or psychiatric disorder. Although the yield of medical genetic research has also been more modest than its enthusiasts had predicted, the contrast between the two areas of research is still striking.

In 1993 researchers did find a marker, then a gene, associated with violence and aggression, in the male members of a Dutch family (19). This was an unexpected and somewhat awkward finding; it looked like the kind of "major 'crime gene'" that researchers had not been expecting to discover. While the family studied was atypical in several relevant respects, the study had enormous impact, in part because the affected gene was known to produce a protein, MAO, involved in regulating the metabolism of serotonin, a neurotransmitter thought to play a critical role in mediating between genes and behavior: Serotonin has been implicated in psychiatric and behavioral conditions ranging from depression to impulsive violence.

A comparison of the MAO and XYY studies suggests both significant advances in scientific technique and persisting issues of interpretation. The connection between genotype and phenotype was closer in several respects for MAO than for XYY. First, a statistically significant association was found between MAO and aggressive and violent behavior in one family; in contrast, no association was established between XYY and any form of criminal behavior until a decade after the karyotype was identified (17). Second, there appeared to be better prospects for finding a specific causal pathway to antisocial behavior from a mutation in the MAO gene, which helps regulate the inhibitory mechanisms of the central nervous system, than from an extra copy of the entire chromosome linked with male gender. Over 50 studies have found an association between aggressive, violent, antisocial, or suicidal behavior and the serotonin metabolite, CSF 5-HIAA, that MAO helps to produce (20). Third, the measure of behavior was more direct in the MAO study than in the original XYY study—observation by the researchers or reports from close relatives, as opposed to inferences from official records.

And yet critics have argued that the link between MAO and violence and aggression, even in the one family studied, is much more tenuous than the researchers suggest, and that their work reveals some of the same inferential leaps that characterized early XYY research. In seeking a genetic cause for the high incidence of violent and aggressive behavior among male family members, the researchers may have paid insufficient attention to more global effects of the MAO mutation:

> Since a primary characteristic of the affected subjects was lowered IQ, it is unclear why the subjects' aggression received more emphasis than their cognitive deficits, and why there was no mention of the possibility that these cognitive deficits may have contributed to behavioral pathologies. . . .Perhaps these acts of violence are secondary to some more widespread defect in affect or cognition (20, p. 18).

Although some of the men in the family studied engaged in clearly violent and antisocial acts, that conduct, like the convictions on which the XYY researchers relied, may well have reflected a more general deficit. The mutant-MAO males, like the XYY males, may not have been more aggressive but less intelligent, lacking constructive outlets for their aggression or clever ways of concealing it.

The MAO researchers acknowledged that their findings did not support "a simple causal relationship between the metabolic abnormality and the behavioral disturbance" they observed, and that "borderline mental retardation" was also associated with the MAO mutation (21). Critics, however, complained that they did not fully investigate alternative explanations for the observed effects of MAO on

behavior, nor adequately examined the range of behavior that might have been affected. Some critics (10) also question the claimed link between low serotonin and aggressive and antisocial behavior, which gives the MAO finding its threshold plausibility. These critics claim that the misconduct of serotonin-deficient individuals may reflect little more than the broadly debilitating impact of low serotonin levels on mental function, a breadth suggested by the sheer range of conditions in which serotonin deficits are implicated. They suggest that the fixation of serotonin researchers on specific psychiatric or behavioral pathologies has obscured such more general effects.

Researchers might respond by citing other evidence, or other studies, which tend to rule out more global explanations of the association between serotonin and violence, or which support a more direct causal pathway. Or they might regard this challenge merely as posing a legitimate question for further research. But they might also insist that these concerns, however reasonable, are hardly unique to behavioral genetics. If medical genetics has enjoyed greater success, it has also been beset by many of the same interpretive difficulties.

Thus medical geneticists also confront the problem of genetic pleiotropy—the multiple effects of a single gene and the multiplicity of possible causal pathways from gene to trait. The same kind of ambiguity arises in studies that search for the genes associated with various physiological effects, and in studies that probe the physiological effects of "candidate" genes (22). The association between a gene mutation and a disease revealed in a linkage or targeting study may arise indirectly, from more global effects of the gene on the organism or from the attempt of other genes or bodily systems to compensate for the loss of the gene's standard function. Because researchers will rarely be able to track the full range of functions and interactions a gene can have, they will rarely be warranted in claiming a direct causal relationship between that gene and a disease or other condition on the basis of a linkage or targeting study. Such claims will only be justified when researchers have acquired enough knowledge about developmental and physiological processes to narrow the range of plausible causal pathways. But that knowledge cannot come from molecular genetics alone.

If similar inferential difficulties confront medical and behavioral genetics, however, why should critics be so much more skeptical about the findings of the latter than of the former? One reason is that they may doubt the possibility of any causal generalizations concerning human behavior (a skepticism briefly discussed at the outset of this article). This blanket skepticism would give explanations based on cognitive limitations greater threshold plausibility than explanations based on behavioral predispositions, such as heightened aggression. A second reason is that critics may believe that genes contribute less, and less directly, to behavior than to physiology, and that the ways in which human behavior is defined and classified compound the difficulties of finding causal relationships. Because of the vast array of social and environmental forces shaping variations in human behavior and psychology, any genetic effect on behavior or psychology will not only be modest but very difficult to track. And because many important behavioral and psychological categories are socially constructed or imposed, they may not be amenable to genetic or other biological explanation.

Thus researchers looking for genetic contributions to criminal behavior must confront the fact that such behavior is defined by legislators and "ascertained" by police, prosecutors, judges, and juries. The problem is not that genes cannot affect voluntary behavior—scientific critics readily concede that they can—but that social types may not correspond with biological types, so there may be no one type of behavior to be explained. We should not expect much in common psychologically or neurobiologically, let alone genetically, between a child-abuser, a pickpocket, a mob boss, and a political terrorist; between the bank robber prosecuted by one regime and the bank founder prosecuted by another. It is unlikely that any genetic feature distinguishes the members of such an eclectic rogues' gallery from the general population, and if there were, it would be unlikely to have much explanatory value.

Researchers acknowledge that the genetic contribution to behavior may well be more subtle and elusive than the genetic contribution to physiology, but they deny that there is less scientific value in finding smaller or less direct effects—Carey and Gottesman, for example, concede that "some of these loci may have little to do with an internal biology of antisocial behavior" as opposed to a general (in)sensitivity to environmental influences (13). Researchers also regard the heterogeneity and social construction of human behavior as part of the challenge of their work: either to discover unity in heterogeneity, through such underlying traits as impulsivity, antisocial personality, or novelty-seeking (the general tendency toward disobedience or risk) or to develop refined typologies of criminal behavior and look for different kinds of genetic influence on different types of crime (serial killing, leadership of an urban drug ring, embezzlement or tax evasion to finance a second home or third car) (23).

A final reason for the divergent expectations of researchers and critics is that the latter doubt the potential of genetic differences to explain behavioral differences. Behavioral geneticists using a wide variety of methods are united in the belief that small difference in genetic constitution can cause large differences in behavior in a fairly direct way. This belief is challenged by critics convinced that genes contribute little or nothing to behavior except its neuromuscular, sensory, and cognitive prerequisites. But it is also challenged by evolutionary biologists and psychologists who believe that genetic commonalities explain more than genetic differences, that differences in human behavior are generally explained by complex contingency plans encoded in genetic material almost all of us share. From this evolutionary perspective the genetic variations in important traits are likely to be preserved over time only if they are adaptive, that is, if they enhance reproductive fitness, thereby increasing their representation in subsequent generations. It is not clear, however, that genetic differences in consequential behavioral traits such as aggressiveness would be adaptive. Thus Allan Gibbard contrasts genetic differences affecting skin color from genetic differences affecting behavior:

> Survival and vitality are reproduction-enhancing, other things equal, and when there is much sun, dark pigment promotes survival and vitality, whereas with little sun, survival and vitality are promoted by having low levels of skin pigment. One could imagine that the same kind of pattern could hold for certain kinds of behavior.
>
> Might such effects be substantial, though? There are grounds for thinking not. The pattern requires genetic selection of different characteristics in different environments—as opposed to contingency plans for the same individual's having one set of characteristics if in one kind of environment and another set in another. This requires special conditions: the difference in which characteristic is more advantageous must last, on average, over many generations. Climate can make such stable differences, and account for such things as differences in skin color. With violence, relevant differences in life circumstances are likely to be much more volatile. When violence "pays" reproductively and when it doesn't will depend chiefly on characteristics of one's society and one's position in it. Whether social differences will be stable enough to produce different selection pressures is doubtful (24).

Gibbard, somewhat like Balaban, suspects that the genetic differences most likely to increase violence are maladaptive mutations that affect violence only indirectly (10).

Although these differences in expectation are in a broad sense empirical, they are unlikely to be resolved by the findings of one study, or even a series of studies. Behavioral geneticists and their critics are likely to interpret the findings of specific studies quite differently, and to have sharply divergent views about the comparative plausibility of alternative hypotheses.

There is also likely to be disagreement about whether the kind of genetic differences most likely to be identified could be said to predispose a person to a particular type of behavior. Can a genetic constitution that makes a person less intelligent be said to predispose him to crime or violence, if he is more likely to engage in crime or violence only because of more limited opportunities for education and employment? Can a person be said to be predisposed to violence if he acts violently only in extreme enviroments, such as a concentration camp, a space station, or a burnt-out urban neighborhood? Does it depend on how common such environments are? These are not narrowly scientific questions; they concern our understanding of causation and ascription of causal responsibility.

SOCIAL MEANING AND IMPACT

It is ironic that one of the strongest challenges to behavioral genetics comes from evolutionary biology and psychology. Those fields are only recently descended from sociobiology, a discipline that has been attacked by many critics of behavioral genetics for the same offense: that it "naturalizes" social injustice by attempting to ground it in biological reality. Although an evolutionary perspective may appear to rationalize patriarchy, infidelity, and other objectionable social practices, in suggesting a genetically based predisposition to engage in them, it regards all people, or at least all males, as created equally predisposed. In contrast, behavioral genetics is dedicated to the proposition that some people are more predisposed genetically than others to destructive behavior. It is this controversial proposition that sets behavioral genetics apart from other disciplines committed to finding biological explanations for social phenomena. There has been more debate over the social implications of this proposition than over its scientific plausibility.

The debate has focused on the historical legacy of genetic-difference research and the social context in which it now takes place. Critics of behavioral genetics not only see more continuity between the old genetic-difference research and the new, they also see more continuity in their social settings; they do not believe that the "bad old days" are over. In light of the role that research on human genetic differences has played in justifying racism and inequality, they fear that even research focused solely on individual differences, and apolitical on its face, will be taken to justify racial stereotypes and used to justify coercive policies.

Behavioral geneticists see their research as having made a more complete break with the past socially as well as scientifically. They argue that contemporary society has a far greater capacity to use the results of their research humanely or at least to protect it from abuses. They see in their work the promise of more effective and less coercive solutions to recalcitrant social problems, and deny that it is a vehicle for a repressive political agenda or the latest incarnation of scientific racism.

We can conveniently divide the conflicts over the social context and impact of the research into issues concerning the past and the present.

The Eugenic Legacy

Behavioral geneticists tend to identify the cardinal sins of eugenics as coercion and group discrimination. Programs of forced sterilization were objectionable primarily because they were *forced*, employing physical compulsion, threats, or deception to induce people to limit their fertility. They were further objectionable because they relied on crude racial, ethnic, and class generalizations to decide whose fertility to limit. Most contemporary researchers emphatically reject both features of the old eugenics. They intend their findings to be used only by individuals for their own benefit or the benefit of their families, they see the added information they provide as enhancing rather than limiting reproductive choice, and they eschew generalizations for (nonfamily) groups. That is, they do not investigate, and do not expect to find, racial or ethic differences in significant behavioral and mental traits. Most researchers also oppose restrictions on reproductive freedom and support restrictions on third-party access to, or use of, behavioral genetic information.

Critics respond to these avowals in two ways. First, they think that the specter of coercion and discrimination is still present, albeit in subtler form. They fear that eugenic values have become internalized in the decisions made by individual reproductive agents; that our society

is embracing a "backdoor" eugenics more insidious, if less oppressive, than the old front-door variety (25). The social and economic pressures to prevent the birth of "defective" fetuses are very strong, and they are felt across the whole population, not merely by the marginal groups once targeted for "improvement" or elimination. And while behavioral geneticists, with a few notorious exceptions, are not interested in investigating genetic differences among racial, ethnic, or other social groups, they are widely perceived to be, and their research methods can be easily adapted to that purpose.

Critics also point out that coercion and discrimination were not the only objectionable features of the old eugenics. Equally central was the belief that social problems such as poverty and crime persist in part because of the failure to design social institutions to take account of constitutional differences among individuals. However emphatically current genetic-difference researchers reject the racism or the political conservatism of their predecessors, they share that underlying conviction. Critics argue that that conviction also lies behind the public enthusiasm for the resurgence of behavioral genetics (26).

The Current Climate

Critics of behavioral genetics attribute its renewed popularity to the perceived failure of the movements of social reform that began with the New Deal and ended with the Great Society. Having made desultory efforts to improve conditions for the worst-off members of society and witnessed few dramatic successes, our society has been all too eager to return to policies that blame "defective" individuals and groups for the social problems that persist. Behavioral genetics provides a scientific pretext for this retrenchment.

For contemporary behavioral geneticists, the apparent failure of many of the social reforms of the post–World War II reflects scientific ignorance more than excessive optimism. The failure of public schools to educate or prisons to reform does not mean that students or immates were stupid or incorrigible, but that policy makers did not know how to teach or rehabilitate them. They failed to recognize individual differences in cognitive and behavioral disposition and to tailor their interventions accordingly. In offering one-size-fits-all programs, they slighted the needs of those with the greatest difficulties in learning or conforming. Behavioral genetics suggests ways of customizing interventions that actually work. Any stigma associated with the identification of individuals requiring special intervention is far milder than the stigma of failure or recalcitrance that those individuals would otherwise bear.

Deficient Serotonin and Heightened Aggression

Long before the advent of molecular genetics, it was widely suspected that the most prolific and recalcitrant offenders differed genetically from the rest of us. That suspicion was memorably captured in Maxwell Anderson's play *Bad Seed*, written in the 1950s, a time when the crime rate was relatively low and public confidence in environmental interventions relatively high. Almost fifty years later those suspicions are far stronger. Although the unprecedented rise in violent crime after 1960 is often attributed to broad social conditions, from the decline of traditional families to the increase in economic and social inequality, there has been a growing conviction among researchers and policy makers that a large share of the increase is due to a small number of individuals, whose identification and incapacitation will go a long way toward reducing violent crime (27). While the primary means of identifying such individuals has been by their recorded deeds — arrests and convictions — some researchers expect to find chemical and genetic "indicators" that would permit early identification and preemptive intervention. This prospect places the hopes and fears about the social impact of the new behavioral genetics into sharp focus.

Researchers see the promise of more humane and effective treatment: "prenatal and postnatal care, rehabilitation for pregnant drug abusers, educational enrichment programs, parenting education, conflict resolution tactics, media restrictions for violent programming, and gun control" (28). Critics see the threat of further racial polarization and social control. They argue that it would be reckless to develop biological techniques for predicting and "treating" juvenile delinquency or adult criminality, when those techniques would be placed in the hands of educational and criminal justice institutions that have shown so little capacity to serve their clients humanely. They imagine genetic testing and medical treatment being made a condition of probation or parole, or "offered" to disruptive or unreliable employees as an alternative to termination. Other pharmacological treatments, from anabuse to lithium, have been introduced in this way, setting a strong and ominously seductive precedent.

Individual Differences and Racial Generalizations

Researchers and critics disagree about the *racial implications* of individual-differences research. Researchers insist that their work is free of the racial taint of previous research, since it does not even look for group differences and studies mainly white population subgroups with large families or good archives, from Finns to Mormons. Critics, however, argue that the distinction between individual and group differences will prove to be untenable because claims of individual differences in genetic predisposition will almost inevitably lend support to claims of group differences. There are several ways in which this might happen.

First, if some researchers find genes or markers associated with criminal behavior in individuals, other researchers may try to compare the incidence of those genes or markers in racial and ethnic groups. (The data for such comparisons will be found in studies on individual predispositions if those studies involve multiracial populations and code subjects by race.) There are good reasons to doubt that researchers would find significant group differences, or that any differences they found would correspond to differences in present offense rates. Still the discovery of individual differences in genetic predisposition could have an adverse impact on minority groups even if, as seems likely, those groups are *not* found to have a higher incidence of the relevant genes

or markers. There will be considerable pressure to use those genes or markers for detecting criminal tendencies in young children and assessing dangerousness in convicted offenders. Universal screening is very unlikely, but the selective screening of those who look "vulnerable" by dint of misbehavior is well within the realm of social possibility. Those screened are likely to be drawn disproportionately from the predominantly black and Hispanic inner cities, since these are the areas in which violent crime is concentrated (29).

Even without population research or screening programs, evidence of differences in genetic predisposition within groups will be widely taken as evidence of differences between groups, however unwarranted that inference may be. The discovery of genes or markers associated with criminal behavior may be publicly perceived as implicating the black community, at a time when an alarmingly high proportion of African-American males are involved in the criminal justice system (30).

Whether or not the critics' apprehensions are more realistic than the researchers' hopes, these different expectations about the social impact of research on genetics and crime reflect broader differences in outlook, about the fairness and reliability of our institutions of social control, and about the pervasiveness and recalcitrance of racial bias in contemporary society. Defenders and critics of genetic research on crime may differ less in their political values than in their levels of trust and optimism.

A Marker for Homosexuality?

The same year the Dutch study on MAO and aggression was published, researchers claimed to find a marker associated with homosexual orientation (31). Although there has been an unusual degree of controversy over the methods used to ascertain the phenotype of sexual orientation, as well as over the statistical design that yielded the linkage, what was most striking about the study was the enthusiasm with which it was received by some members of the gay community: It was the first time that a large segment of a minority community had warmly embraced a claim of genetic predisposition (32).

For the researchers who located the marker, the discovery was important socially as well as scientifically. It showed that behavioral genetics could be liberating rather than oppressive, legitimizing socially deviant conditions by revealing that they had a biological basis (33). This optimistic view was shared by some members of the gay community, who saw the research as helping to place sexual orientation in the constitutionally protected category of "immutable characteristics," thereby giving homosexuals a legal defense against many forms of discrimination to which they were currently subject.

For critics of behavioral genetics, however, as well as for other representatives of the gay community, the research was simplistic empirically and naive politically. It distorted the complexity and plasticity of sexual attraction and activity, transforming shifting "preferences" into deep, immutable "orientations," dichotomizing a multidimensional array of preferences and behaviors into hetero and homosexuality, and falsely associating effeminacy with homosexuality (34). Moreover critics regard genetic vindication as a Faustian bargain. To embrace the claim that sexual orientation is genetically influenced is to treat it as something that needs to be excused. At the same time genetic predisposition would furnish at most an incomplete excuse: No researcher would claim that sexual orientation, let alone sexual conduct, is fully determined genetically. And even if sexual orientation were seen as genetically determined, it would be unlikely to furnish an excuse the homophobic public would accept—someone born gay might, if anything, be seen as more deeply flawed than someone made gay by a domineering mother—the most popular "scientific" explanation a generation ago. As Diane Paul argues:

> [T]here is little reason to think that social policy would change even if it were generally agreed that homosexuals were in no way blameworthy. Those who consider genetic (or other biological) explanations necessarily liberating are, in my view, naive. Research results acquire social meaning only in the context of other assumptions, for example, how repugnant and threatening others find the behavior. . . .The history of eugenics provides a warning to those with a socially progressive agenda: to stress that our fate is in our genes is a very risky strategy (34, p. 99).

Finally, the mitigating effects of a genetic excuse might well be undermined or offset by the use of prenatal testing for suspected markers or genes to prevent the birth of gay children. While it is unlikely that such testing will ever become accurate enough to substantially reduce the incidence of homosexuality, many parents would seek testing as a way to avoid the frustration and embarrassment of having a gay child (35). The mere availability of testing would contribute to the perception of homosexuality as a "preventable" disorder, like Down syndrome or Tay Sachs disease. Life would be especially difficult for the "false negatives"—those children who tested straight but became gay—who would risk condemnation for embracing a deviant orientation against their genetic grain. Matters might be even worse if genetically based treatments for homosexuality became available: Those who declined treatment would be seen as recalcitrant as well as deviant, ratifying the sexual orientation that they had been acquitted of choosing in the first place. What genetic diagnosis would give by way of excuse, genetic therapy would take away.

Researchers might respond that a deeper understanding of the biological roots of sexual orientation would increase tolerance by making differences in orientation appear not only involuntary but natural. Instead of being seen as a disease or disorder, homosexuality might be seen as a normal variation, such as red hair. They could cite historical precedent: The very first scientific attempts to explain homosexuality in biological terms were made by researchers who regarded all combinations of male and female attributes as natural and sought to decriminalize same-sex sexual behavior. Critics, however, would see this precedent as double-edged. The work that sought to normalize homosexuality was soon built upon by researchers seeking to pathologize it; the hypothesis that

homosexuality had a biological or genetic basis became the cornerstone of eugenic policies against homosexuals (31).

The researchers might respond that the fact that biological research had been misappropriated in this way hardly meant that it would inevitably be turned to eugenic or repressive purposes. They might agree with Diane Paul that the way genetic explanations are received "depends on how repugnant and threatening others find the behavior," but insist that public attitudes toward homosexuality in Western societies have become increasingly accepting, and that genetic explanations will only accelerate the trend (34). As in the case of criminal behavior, divergent expectations about the social impact of genetic research into sexual orientation appear to arise from broader differences in outlook about the social maturity and political fairness of the societies in which the research takes place.

CONCLUSION

It would be naive to expect a single finding, or series of findings, to resolve doubts about the scientific value of human behavioral genetics, or to expect a single application of the research, or series of applications, to resolve doubts about its social utility. At the same time it would be dogmatic to insist that no conceivable findings or applications could vindicate the researchers' confidence and optimism, or their critics' doubts and pessimism. It is, however, quite likely that an article on human behavioral genetics written for the next edition of this Encyclopedia will reflect some of the same uncertainty and ambivalence as this one.

BIBLIOGRAPHY

1. D.J. Kevles, *In the Name of Eugenics: Genetics and the Uses of Human Heredity*, Knopf, New York, 1985.

2. D. Paul, *Controlling Human Heredity: 1865 to the Present*, Humanities Press, Atlantic Highlands, NJ, 1995.

3. D. Goldman, *Politics Life Sci.* **15**(1), 97–98 (1996).

4. R. Plomin et al., *Behavioral Genetics*, 3rd ed., W.F. Freeman, New York, 1997.

5. S.L. Sherman et al., *Am. J. Hum. Genet.* **60**, 1265–1275 (1997).

6. R. Plomin, M.J. Owen, and P. McGuffin, *Science* **264**, 1733–1739 (1994).

7. G. Carey, in A.J. Reiss, Jr. and J.A. Roth, eds., *Understanding and Preventing Violence: Biobehavioral Influences on Violence*, vol. 2, National Academy Press, Washington, DC, 1994.

8. D.C. Rowe and K.C. Jacobson, in R.A. Carson and M.A. Rothstein, eds., *Behavioral Genetics: The Clash of Biology and Culture*, Johns Hopkins University Press, Baltimore, MD, 1999, pp. 12–34.

9. E. Sober, in D. Wasserman and R. Wachbroit, eds., *Genetics and Criminal Behavior: Methods, Meanings, and Morals*, Cambridge University Press, New York, forthcoming.

10. E. Balaban, in Singh et al., eds., *Thinking about Evolution: Historical, Philosophical, and Political Perspectives*, Cambridge University Press, Cambridge, UK, forthcoming.

11. C. Jencks, *New York Rev. Books*, February 12, pp. 33–41, 34 (1987).

12. L.J. Eaves, N.G. Martin, and A.C. Heath, *Beh. Genet.* **20**, 1–22 (1990).

13. G. Carey and I. Gottesman, *Politics Life Sci.* **15**(1), 88–90 (1996).

14. E. Balaban, *Politics Life Sci.* **15**(1), 86–88 (1996).

15. P. Jacobs et al., *Nature* **208**, 1351–1352 (1965).

16. H.A. Witkin et al., *Science* **196**, 547–555 (1976).

17. M. Rutter, H. Giller, and A. Hagell, *Antisocial Behavior by Young People*, Cambridge University Press, New York, 1998.

18. I.D. Waldman, in R.A. Carson and M.A. Rothstein, eds., *Behavioral Genetics: The Clash of Biology and Culture*, Johns Hopkins University Press, Baltimore, MD, 1999, pp. 35–59.

19. H.G. Brunner et al., *Science* **262**, 578–580 (1993).

20. E. Balaban, J. Alper, and Y.L. Kasamon, *J. Neurogenet* **11**, 1–43 (1996).

21. H.G. Brunner, in G.R. Bock and J.A. Goode, eds., *Genetics of Criminal and Antisocial Behavior*, Wiley, New York, 1996, pp. 156, 159.

22. S. Culp, *Philos. Sci. (Proceedings)* **64**, S268–S278 (1997).

23. I. Gottesman and H.H. Goldsmith, in C.A. Nelson, ed., *Threats to Optimal Development: Integrating Biological, Psychological, and Social Risk Factors*, Lawrence Erlbaum Associates, Hillsdale, NJ, 1994.

24. A. Gibbard, in D. Wasserman and R. Wachbroit, eds., *Genetics and Criminal Behavior: Methods, Meanings, and Morals*, Cambridge University Press, New York, forthcoming.

25. T. Duster, *Backdoor to Eugenics*, Routledge, New York, 1990.

26. R. Lewontin, S. Rose, and L. Kamin, *Not in Our Genes: Biology, Ideology, and Human Nature*, Pantheon, New York, 1984.

27. J.Q. Wilson and R. Herrnstein, *Crime and Human Nature*, Simon & Schuster, New York, 1985.

28. D. Fishbein, *Politics Life Sci.* **15**(1), 91–94, 93 (1996).

29. D. Wasserman, in Robert Post, ed., *Censorship and Silencing: Practices of Cultural Regulation*, Getty Research Institute, Santa Monica, CA, 1998, pp. 169–193.

30. J. Miller, *Search and Destroy: African-American Males in the Criminal Justice System*, Cambridge University Press, New York, 1996.

31. D. Hamer et al., *Science* **261**, 321–327 (1993).

32. D. Hamer, *The Search for the Gay Gene and the Biology of Behavior*, Simon & Schuster, New York, 1994.

33. J. DeCecco and M. Shively, eds., *Sex, Cells, and Same-Sex Desire: The Biology of Sexual Preference*, Haworth, New York, 1995.

34. D. Paul, in *Politics Life Sci.* **15**(1), 99–100 (1996).

35. E. Stein, *Bioethics* **12**(1), 1–48 (1998).

See other entries GENETIC DETERMINISM, GENETIC REDUCTIONISM, AND GENETIC ESSENTIALISM; see also HUMAN ENHANCEMENT USES OF BIOTECHNOLOGY entries; PUBLIC PERCEPTIONS: SURVEYS OF ATTITUDES TOWARD BIOTECHNOLOGY.

C

CLONING, ETHICS

Dan W. Brock
Brown University
Providence, Rhode Island

OUTLINE

INTRODUCTION

One of the most dramatic recent advances in biotechnology was the successful cloning of a sheep from a single cell of an adult sheep. The world of science and the public at large were both shocked and fascinated by this accomplishment of Ian Wilmut and his colleagues (1). Scientists were in part surprised because many had believed that after the very early stage of embryo development at which differentiation of cell function begins to take place it would not be possible to achieve the cloning of an adult mammal by nuclear transfer. In this process the nucleus from the cell of an adult mammal is inserted into an ennucleated ovum, and the resulting embryo develops following the complete genetic code of the mammal from which the inserted nucleus was obtained. But some scientists and much of the public were troubled or apparently even horrified at the prospect that if adult mammals such as sheep could be cloned, then cloning of adult humans by the same process would likely be possible as well. Of course, the process is far from perfected even with sheep—it took 276 failures by Wilmut and his colleagues to produce Dolly, their one success, and whether the process can be successfully replicated in other mammals, much less in humans, is not now known. But those who were horrified at the prospect of human cloning were not assuaged by the fact that the science with humans is not yet there, for it looked to them now perilously close.

The response of most scientific and political leaders to the prospect of human cloning, indeed of Dr. Wilmut as well in testimony before Congress in March 1997, was immediate and strong condemnation. In the United States President Clinton immediately banned federal financing of human cloning research and asked privately funded scientists to halt such work until the newly formed National Bioethics Advisory Commission could review the "troubling" ethical and legal implications. The Director-General of the World Health Organization characterized human cloning as "ethically unacceptable as it would violate some of the basic principles which govern medically assisted reproduction. These include respect for the dignity of the human being and the protection of the security of human genetic material" (2). Around the world similar immediate condemnation was heard as human cloning was called a violation of human rights and human dignity. Even before Wilmut's announcement, human cloning had been made illegal in nearly all countries in Europe and had been condemned by the Council of Europe (3).

A few more cautious voices were heard both suggesting some possible benefits from the use of human cloning in limited circumstances and questioning its too quick prohibition, but they were a clear minority. In the popular media, nightmare scenarios of laboratory mistakes resulting in monsters, the cloning of armies of Hitlers, the exploitative use of cloning for totalitarian ends as in Huxley's *Brave New World*, and the murderous replicas of the film Blade Runner all fed the public controversy and uneasiness. A striking feature of these early responses was that their strength and intensity seemed far to outrun the arguments and reasons offered in support of them—they seemed often to be "gut level" emotional reactions rather than considered reflections on the issues. Such reactions should not be simply dismissed, both because they may point to important considerations otherwise missed, and not easily articulated, and because they often have a major impact on public policy. But the formation of public policy should not ignore the ethical reasons and arguments that bear on the practice of human cloning—these must be articulated in order to understand and inform people's more immediate emotional responses. This article will evaluate critically the main moral considerations and arguments for and against human cloning. Though many people's religious beliefs inform their views on human cloning, and it is often difficult to separate religious from secular positions, this article is restricted to arguments and reasons that can be given a clear secular formulation and does not take up explicitly religious positions and arguments pro or con. This article is also concerned principally with cloning by nuclear transfer, which permits cloning of an adult, not cloning by embryo splitting, although some of the issues apply to both (4).

I begin by noting that on each side of the issue there are two distinct kinds of moral arguments brought forward. On the one hand, some opponents claim that human cloning would violate fundamental moral or human rights, while some proponents argue that its prohibition would violate such rights. On the other hand, both opponents and proponents also cite the likely harms and benefits, both to individuals and to society, of the practice. While moral and even human rights need not be understood as absolute, that is, as morally requiring people to respect them no matter how great the costs or bad consequences of doing so, actions that would violate them cannot be justified by a mere balance of benefits over harms. For example, the rights of human subjects in research must be respected even if the result is that some potentially beneficial research is more difficult or cannot be done, and the right of free expression prohibits the silencing of unpopular or even abhorrent views; in Ronald Dworkin's striking formulation, rights trump utility (5). This article will take up the moral rights implicated in human cloning, as well as its more likely significant benefits and harms, because none of the rights as applied to human cloning is sufficiently uncontroversial and strong to settle decisively the morality of the practice one way or the other. But because of their strong moral force, the assessment of the moral rights putatively at stake is especially important. A further complexity here is that it is sometimes controversial whether a particular consideration is merely a matter of benefits and harms, or is instead a matter of moral or human rights. We begin with the arguments in support of permitting human cloning, although with no implication that it is the stronger or weaker position.

MORAL ARGUMENTS IN SUPPORT OF HUMAN CLONING

Is There a Moral Right to Use Human Cloning?

What moral right might protect at least some access to the use of human cloning? Some commentators have argued that a commitment to individual liberty, as defended by J.S. Mill, requires that individuals be left free to use human cloning if they so choose and if their doing so does not cause significant harms to others, but liberty is too broad in scope to be an uncontroversial moral right (6,7). Human cloning is a means of reproduction (in the most literal sense), so the most plausible moral right at stake in its use is a right to reproductive freedom or procreative liberty (8,9). Reproductive freedom includes not only the familiar right to choose not to reproduce, for example, by means of contraception or abortion, but also the right to reproduce. The right to reproductive freedom is properly understood to include as well the use of various assisted reproductive technologies, such as in vitro fertilization (IVF), and oocyte donation. The reproductive right relevant to human cloning is a negative right, that is, a right to use assisted reproductive technologies without interference by the government or others when made available by a willing provider. The choice of an assisted means of reproduction, such as surrogacy, can be defended as included within reproductive freedom even when it is not the only means for individuals to reproduce, just as the choice among different means of preventing conception is protected by reproductive freedom. However, the case for permitting the use of a particular means of reproduction is strongest when that means is necessary for particular individuals to be able to procreate at all. Sometimes human cloning could be the only means for individuals to procreate while retaining a biological tie to the child created, but in other cases different means of procreating would also be possible.

It could be argued that human cloning is not covered by the right to reproductive freedom because, whereas current assisted reproductive technologies and practices covered by that right are remedies for inabilities to reproduce sexually, human cloning is an entirely new means of reproduction; indeed, its critics see it as more a means of manufacturing humans than of reproduction. Human cloning is a different means of reproduction than sexual reproduction, but it is a means that can serve individuals' interest in reproducing. It cannot be excluded from the moral right to reproductive freedom merely because it is a new means of reproducing, but rather only if it has other objectionable moral features, such as eroding human dignity or uniqueness; we will evaluate these other ethical objections to it below.

When individuals have alternative means of procreating, human cloning typically would be chosen because it replicates a particular individual's genome. The reproductive interest in question then is not simply reproduction itself, but a more specific interest in choosing what kind of children to have. The right to reproductive freedom is usually understood to cover at least some choice about the kind of children one will have; for example, genetic testing of an embryo or fetus for genetic disease or abnormality, together with abortion of an affected embryo or fetus, is now used to avoid having a child with that disease or abnormality. Genetic testing of prospective parents before conception to determine the risk of transmitting a genetic disease is also intended to avoid having children with particular diseases. Prospective parents' moral interest in self-determination, which is one of the grounds of a moral right to reproductive freedom, includes the choice about whether to have a child with a condition that is likely to place severe burdens on them, and to cause severe burdens to the child itself.

The more a reproductive choice is not simply the determination of oneself and one's own life but the determination of the nature of another, as in the case of human cloning, the more moral weight the interests of that other person, that is the cloned child, should have in decisions that determine its nature (10). But even then parents are typically taken properly to have substantial, but not unlimited, discretion in shaping the persons their children will become, for example, through education and other child-raising decisions. Even if not part of reproductive freedom, the right to raise one's children as one sees fit, within limits mostly determined by the interests of the children, is also a right to determine within limits what kinds of persons one's children will become. This right includes not just preventing certain diseases or harms to children, but selecting and shaping desirable features and traits in one's children. The use of human cloning is one way to exercise that right.

Its worth pointing out that current public and legal policy permits prospective parents to conceive, or to carry a conception to term, when there is a significant risk, or even certainty, that the child will suffer from a serious genetic disease. Even when others think the risk or presence of genetic disease makes it morally wrong to conceive, or to carry a fetus to term, the parents' right to reproductive freedom permits them to do so. Most possible harms to a cloned child to be considered below are less serious than the genetic harms with which parents can now permit their offspring to be conceived or born.

To conclude our discussion of a moral right to use human cloning, there is good reason to accept that a right to reproductive freedom presumptively includes both a right to select the means of reproduction, as well as a right to determine what kind of children to have, by use of human cloning. However, the particular reproductive interest of determining what kind of children to have is less weighty than other reproductive interests and choices whose impact falls more directly and exclusively on the parents rather than the child. Accepting a moral right to reproductive freedom that includes the use of human cloning does not settle the moral issue about human cloning, however, since there may be other moral rights in conflict with this right, or serious enough harms from human cloning to override the right to use it; this right can be thought of as establishing a serious moral presumption supporting access to human cloning.

There is a different moral right that might be thought to be at stake in the dispute about human cloning—the right to freedom of scientific inquiry and research in the acquisition of knowledge. If there is such a right, it would presumably be violated by a legal prohibition of research on human cloning, although the government could still permissibly decide not to spend public funds to support such research. Leaving aside for the moment human subject ethical concerns, research on human cloning might provide valuable scientific or medical knowledge beyond simply knowledge about how to carry out human cloning. Whether or not there is a moral right to freedom of scientific inquiry, for example, as part of a right to free expression, prohibiting and stopping scientific research and inquiry is a serious matter and precedent that should only be undertaken when necessary to prevent grave violations of human rights or to protect fundamental human interests. But even for opponents of human cloning the fundamental moral issue is not acquiring the knowledge that would make it possible, but using that knowledge to do human cloning. Since it is possible to prohibit human cloning itself, without prohibiting all research on it, it is not necessary to limit the freedom of scientific inquiry in order to prevent human cloning from taking place. But this means as well that a right to freedom of scientific inquiry could only protect research on human cloning, not the use of human cloning. For this reason the fundamental moral right which provides presumptive moral support for permitting the use of human cloning is the right to reproductive freedom, not the right to freedom of scientific inquiry. In what follows, the discussion will principally concern the moral issues in the use of human cloning, not those restricted to research on it.

What Individual or Social Benefits Might Human Cloning Produce?

Largely Individual Benefits. The literature on human cloning by nuclear transfer, as well as the literature on embryo splitting where it is relevant to the nuclear transfer case, contain a few examples of circumstances in which individuals might have good reasons to want to use

human cloning. However, human cloning does not appear to be the unique answer to any great or pressing human need or social problem, and its benefits would likely be at most limited. What are the principal benefits of human cloning that might give persons good reasons to want to use it?

Human Cloning Would be a New Means to Relieve the Infertility Some Persons Now Experience. Human cloning would allow women who have no ova or men who have no sperm to produce an offspring that is biologically related to them (8,11–13). Embryos might also be cloned, either by nuclear transfer or embryo splitting, in order to increase the number of embryos for implantation and improve the chances of successful conception (14). While the moral right to reproductive freedom creates a presumption that individuals should be free to choose the means of reproduction that best serves their interests and desires, the benefits from human cloning to relieve infertility are greater the more persons there are who cannot overcome their infertility by any other means acceptable to them; data are not now available about the numbers of persons who could only relieve their infertility by human cloning.

It is not enough to point to the large number of children throughout the world possibly available for adoption as a solution to infertility, unless we are prepared to discount as illegitimate the strong desire many persons, fertile and infertile, have for the experience of pregnancy and for having and raising a child biologically related to them. While not important to all infertile (or fertile) individuals, it is important to many and is respected and met through other forms of assisted reproduction that maintain a biological connection when that is possible; there seems no good reason to refuse to respect and respond to it when human cloning would be the best or only means of overcoming individuals' infertility.

Human Cloning Would Enable Couples in Which One Party Risks Transmitting a Serious Hereditary Disease, a Serious Risk of Disease, or an Otherwise Harmful Condition to an Offspring, to Reproduce Without Doing So (8). Of course, by using donor sperm or egg donation, such hereditary risks can generally be avoided now without the use of human cloning. These procedures may be unacceptable to some couples, however, or at least considered less desirable than human cloning because they introduce a third party's genes into their reproduction, instead of giving their offspring only the genes of one of them. Thus in some cases human cloning would be a means of preventing genetically transmitted harms to offspring. Here too there are no data on the likely number of persons who would wish to use human cloning for this purpose instead of either using other available means of avoiding the risk of genetic transmission of the harmful condition or accepting the risk of transmitting the harmful condition.

Human Cloning a Later Twin Would Enable a Person to Obtain Needed Organs or Tissues for Transplantation (12,15,16). Human cloning would solve the problem of finding a transplant donor who is an acceptable organ or tissue match and would eliminate, or drastically reduce, the risk of transplant rejection by the host. The availability of human cloning for this purpose would amount to a form of insurance policy to enable treatment of certain kinds of medical needs. Of course, often the medical need would be too urgent to permit waiting for the cloning, gestation and development of the later twin necessary before tissues or organs for transplant could be obtained. In other cases the need for an organ that the later twin would him or herself need to maintain life, such as a heart or a liver, would preclude cloning and then taking the organ from the later twin.

Such a practice has been criticized on the ground that it treats the later twin not as a person valued and loved for his or her own sake, as an end in itself in Kantian terms, but simply as a means for benefiting another. This criticism assumes, however, that only this one motive would determine the relation of the person to his or her later twin. The well-know case some years ago in California of the Ayala's, who conceived in the hopes of obtaining a source for a bone marrow transplant for their teenage daughter suffering from leukemia illustrates the mistake in this assumption. They argued that whether or not the child they conceived turned out to be a possible donor for their daughter, they would value and love the child for itself, and treat it as they would treat any other member of their family. That one reason it was wanted was as a means to saving their daughter's life did not preclude its also being loved and valued for its own sake; in Kantian terms, it was treated as a possible means to saving their daughter, but not *solely as a means*, which is what the Kantian view proscribes.

Indeed, when people have children, whether by sexual means or with the aid of assisted reproductive technologies, their motives and reasons for doing so are typically many and complex, and include reasons less laudable than obtaining life-saving medical treatment, such as having a companion or someone who needs them, enabling them to live on their own, and qualifying for public or government benefit programs. While these other motives for having children sometimes may not bode well for the child's upbringing and future, public policy does not assess prospective parents motives and reasons for procreating as a condition of their doing so.

One commentator has proposed human cloning for obtaining even life-saving organs (15). After cell differentiation some of the brain cells of the embryo or fetus would be removed so that it could then be grown as a brain dead body for spare parts for its earlier twin. This body clone would be like an anencephalic newborn or presentient fetus, neither of whom arguably can be harmed because of their lack of capacity for consciousness. Most people would likely find this practice appalling and immoral, in part because here the cloned later twin's capacity for conscious life is destroyed *solely as a means* for the benefit of another. Yet, if one pushes what is already science fiction quite a bit further in the direction of science fantasy, and imagines the ability to clone and grow in an artificial environment only the particular life-saving organ a person needed for transplantation, then it is far from clear that it would be morally impermissible to do so.

Human Cloning Would Enable Individuals to Clone Someone Who Had Special Meaning to Them, Such as a Child Who Had Died (12). There is no denying that if human cloning were available, some individuals would want to use it in order to clone someone who had special meaning to them, such as a child who had died, but that desire usually would be based on a deep confusion. Cloning such a child would not replace the child the parents had loved and lost, but rather would create a new different child, though with the same genes. The child they loved and lost was a unique individual who had been shaped by his or her environment and choices, not just his or her genes, and, more important, who had experienced a particular relationship with them. Even if the later cloned child could have not only the same genes but also be subjected to the same environment, which is in fact impossible, it would remain a different child than the one they had loved and lost because it would share a different history with them (17). Cloning the lost child might help the parents accept and move on from their loss, but another already existing sibling or another new child that was not a clone might do this equally well; indeed, it might do so better since the appearance of the cloned later twin would be a constant reminder of the child they had lost. Nevertheless, if human cloning enabled some individuals to clone a person who had special meaning to them and doing so had deep meaning and satisfaction for them, that would be a benefit to them even if their reasons for wanting to do so, and the satisfaction they in turn received, were based on a confusion.

Largely Social Benefits

Human Cloning Would Enable the Duplication of Individuals of Great Talent, Genius, Character, or Other Exemplary Qualities. The first four reasons for human cloning considered above all looked to benefits to specific individuals, usually parents, from being able to reproduce by means of human cloning. This reason looks to benefits to the broader society from being able to replicate extraordinary individuals—a Mozart, Einstein, Gandhi, or Schweitzer (18,19). Much of its appeal, like much thinking both in support of and in opposition to human cloning, rests on a confused and mistaken assumption of genetic determinism, that is, that one's genes fully determine what one will become, do, and accomplish. What made Mozart, Einstein, Gandhi, and Schweitzer the extraordinary individuals they were was the confluence of their particular genetic endowments with the environments in which they were raised and lived and the particular historical moments they in different ways seized. Cloning them would produce individuals with the same genetic inheritances (nuclear transfer does not even produce 100 percent genetic identity, although for the sake of exploring the moral issues we follow here the common assumption that it does), but neither by cloning, nor by any other means, would it be possible to replicate their environments or the historical contexts in which they lived and their greatness flourished. We do not know, either in general or with any particular individual, the degree or specific respects in which their greatness depended on their "nature" or their "nurture," but we do know in all cases that it depended on an interaction of them both. Thus human cloning could never replicate the extraordinary accomplishments for which we admire individuals like Mozart, Einstein, Gandhi, and Schweitzer.

If we make a rough distinction between the extraordinary capabilities of a Mozart or an Einstein and how they used those capabilities in the particular environments and historical settings in which they lived, it would also be a mistake to assume that human cloning could at least replicate their extraordinary capabilities, if not the accomplishments they achieved with them. Their capabilities too were the product of their inherited genes and their environments, not of their genes alone, so it would be a mistake to think that cloning them would produce individuals with the same capabilities, even if they would exercise those capabilities at different times and in different ways. In the case of Gandhi and Schweitzer, whose extraordinary greatness lies more in their moral character and commitments, we understand even less well the extent to which their moral character and greatness was produced by their genes.

None of this is to deny that Mozart's and Einstein's extraordinary musical and intellectual capabilities, nor even Gandhi's and Schweitzer's extraordinary moral greatness, were produced in part by their unique genetic inheritances. Cloning them might well produce individuals with exceptional capacities, but we simply do not know how close their clones would be in capacities or accomplishments to the great individuals from whom they were cloned. Even so, the hope for exceptional, even if less and different, accomplishment from cloning such extraordinary individuals might be a reasonable ground for doing so.

The examples above are of individuals whose greatness is widely appreciated and largely uncontroversial, but if we move away from such cases we encounter the problem of whose standards of greatness would be used to select individuals to be cloned for the benefit of society or humankind in general. This problem inevitably connects with the important issue of who would control access to and use of the technology of human cloning, since those who controlled its use would be in a position to impose their standards of exceptional individuals to be cloned. This issue is especially worrisome if particular groups or segments of society, or if government, controlled the technology for we would then risk its use for the benefit of those groups, segments of society, or governments under the cover of benefiting society or humankind.

Human Cloning and Research on Human Cloning Might Make Possible Important Advances in Scientific Knowledge, for Example About Human Development (20,21). While important potential advances in scientific or medical knowledge from human cloning or human cloning research have frequently been cited in some media responses to Dolly's cloning, there are at least three reasons why these possible benefits are highly uncertain. First, there is always considerable uncertainty about the nature and importance of the new scientific or medical knowledge that a dramatic new technology like human cloning will lead to; the road to that new knowledge is never mapped in advance and takes many unexpected turns. Second, we do not know what new knowledge from human cloning or human cloning research could also be gained by other methods and research that do not have the problematic moral features of human cloning to which its opponents

object. Third, what human cloning research would be compatible with ethical and legal requirements for the use of human subjects in research is complex, controversial, and largely unexplored. For example, in what contexts and from whom would it be necessary, and how would it be possible, to secure the informed consent of parties involved in human cloning? No human cloning should ever take place without the consent of the person cloned and the woman receiving a cloned embryo, if they are different. But we could never obtain the consent of the cloned later twin to being cloned, so research on human cloning that produces a cloned individual might be barred by ethical and legal regulations for the use of human subjects in research (22). Moreover, creating human clones solely for the purpose of research would be to use them solely for the benefit of others without their consent, and so unethical. Of course, once human cloning was established to be safe and effective, then new scientific knowledge might be obtained from its use for legitimate, nonresearch reasons. How human subjects regulations would apply to research on human cloning, needs further exploration to help clarify how significant and likely the potential gains are in scientific and medical knowledge from human cloning research and human cloning.

Although there is considerable uncertainty concerning most of the possible individual and social benefits of human cloning discussed above, and although no doubt it may have other benefits or uses that we cannot yet envisage, it does appear reasonable to conclude that human cloning at this time does not promise great benefits or uniquely meet great human or social needs. Nevertheless, a case can be made that scientific freedom supports permitting research on human cloning to go forward and that freedom to use human cloning is protected by the important moral right to reproductive freedom. We must therefore assess what moral rights might be violated, or harms produced, by research on or use of human cloning.

MORAL ARGUMENTS AGAINST HUMAN CLONING

Would the Use of Human Cloning Violate Important Moral Rights?

Many of the immediate condemnations of any possible human cloning following Wilmut's cloning of an adult sheep claimed that it would violate moral or human rights, but it was usually not specified precisely, or often even at all, what the rights were that would be violated. We will consider here two possible candidates for such a right: a right to have a unique identity and a right to ignorance about one's future or to an open future. The former right is cited by many commentators, but even if any such a right exists, it seems not to be violated by human cloning. The latter right has only been explicitly defended by two commentators, and in the context of human cloning, only by Hans Jonas; it supports a more promising, even if ultimately unsuccessful, argument that human cloning would violate an important moral or human right.

Is there a moral or human right to a unique identity, and if so would it be violated by human cloning? For human cloning to violate a right to a unique identity, the relevant sense of identity would have to be genetic identity, that is a right to a unique unrepeated genome. This would be violated by human cloning, but is there any such right? It might be thought there could not be such a right because it would be violated in all cases of identical twins, yet no one claims in such cases that the moral or human rights of each of the twins have been violated. However, this consideration is not conclusive (14,23). It is commonly held that only deliberate human actions can violate others' rights, but that outcomes that would constitute a rights violation if done by human action are not a rights violation if a result of natural causes; if Arthur deliberately strikes Barry on the head so hard as to cause his death, he violates Barry's right not to be killed, but if lightening strikes Cheryl causing her death, then we would not say that her right not to be killed has been violated. The case of twins does not show there could not be a right to a unique genetic identity.

What is the sense of identity that each person might have a right to have uniquely? What constitutes the special uniqueness of each individual (24,25)? Even with the same genes, two individuals, for example, homozygous twins, are numerically distinct and not identical, so what is intended must be the various properties and characteristics that make each individual qualitatively unique and different than others. Does having the same genome as another person undermine that unique qualitative identity? Only on the crudest genetic determinism, a genetic determinism is possible according to which an individual's genes completely and decisively determine everything else about the individual, all his or her other nongenetic features and properties, together with the entire history or biography that will constitute his or her life. But there is no reason whatever to believe that kind of genetic determinism. Even with the same genes, as we know from the case of genetically identical twins, while there may be many important similarities in the twins' psychological and personal characteristics, differences in these develop over time together with differences in their life histories, personal relationships, and life choices. This is true of identical twins raised together, and the differences are still greater in the case of identical twins raised apart; sharing an identical genome does not prevent twins from each developing a distinct and unique personal identity of their own.

We need not pursue what the basis or argument in support of a moral or human right to a unique identity might be—such a right is not found among typical accounts and enumerations of moral or human rights—because even if we grant that there is such a right, sharing a genome with another individual as a result of human cloning would not violate it. The idea of the uniqueness, or unique identity, of each person historically predates the development of modern genetics and the knowledge that except in the case of homozygous twins, each individual has a unique genome. A unique genome thus could not be the ground of this long-standing belief in the unique human identity of each person.

Would human cloning violate instead what Hans Jonas called a right to ignorance, or what Joel Feinberg called a right to an open future (26,27)? Jonas argued that human cloning in which there is a substantial time gap between the beginning of the lives of the earlier and later twin is fundamentally different from the simultaneous beginning of the lives of homozygous twins that occur in nature. Although contemporaneous twins begin their lives with the same genetic inheritance, they also begin their lives or biographies at the same time, and so in ignorance of what the other who shares the same genome will by his or her choices make of his or her life. To whatever extent one's genome determines one's future, each begins ignorant of what that determination will be and so remains as free to choose a future, to construct a particular future from among open alternatives, as are individuals who do not have a twin. Ignorance of the effect of one's genome on one's future is necessary for the spontaneous, free, and authentic construction of a life and self.

A later twin created by human cloning, Jonas argues, knows, or at least believes he or she knows, too much about him or herself. For there is already in the world another person, one's earlier twin, who from the same genetic starting point has made the life choices that are still in the later twin's future. It will seem that one's life has already been lived and played out by another, that one's fate is already determined, so the later twin will lose the spontaneity of authentically creating and becoming his or her own self. One will lose the sense of human possibility in freely creating one's own future. It is tyrannical, Jonas claims, for the earlier twin to try to determine another's fate in this way. And even if it is a mistake to believe the crude genetic determinism according to which one's genes determine one's fate, what is important for one's experience of freedom and ability to create a life for oneself is whether one thinks one's future is open and undetermined, and so still to be determined by one's own choices.

One might try to interpret Jonas' objection so as not to assume either genetic determinism or a belief in it. A later twin might grant that he is not determined to follow in his earlier twin's footsteps, but claim nevertheless that the earlier twin's life would always haunt him, standing as an undue influence in shaping his life in ways to which others' lives are not vulnerable. But the force of the objection still seems to rest on a false assumption that having the same genome as his earlier twin unduly restricts his freedom to choose a different life than the earlier twin chose. A family environment also importantly shapes children's development, but there is no force to the claim of a younger sibling that the existence of an older sibling raised in that same family is an undue influence on his freedom to make a life for himself in that environment. Indeed, the younger twin or sibling might gain the benefit of being able to learn from the older twin's or sibling's mistakes.

In a different context, and without applying it to human cloning, Joel Feinberg has argued for a child's right to an open future. This requires that others raising a child not close off the future possibilities that the child would otherwise have so as to eliminate a reasonable range of opportunities for the child to choose autonomously and construct his or her own life. One way this right to an open future would be violated is to deny even a basic education to the child, and another way might be to create the child as a later twin so that he will believe his future has already been set for him by the choices made and the life lived by his earlier twin.

A central difficulty in evaluating the implications for human cloning of a right either to ignorance or to an open future is whether the right is violated merely because the later twin may be likely to *believe* that its future is already determined, even if that belief is clearly false and supported only by the crudest genetic determinism. If the twin's future in reality remains open and his to freely choose, then someone's acting in a way that unintentionally leads him to believe that his future is closed and determined seems not to have violated his right to ignorance or to an open future. Consider an analogous case of causing a false belief that one's right has been violated. Suppose that you drive down the twin's street in your new car that is just like his, knowing that when he sees you he is likely to believe that you have stolen his car and therefore to abandon his driving plans for the day. You have not violated his property right to his car even though he may feel the same loss of opportunity to drive that day as if you had in fact stolen his car. In each case he is mistaken that his open future or car has been taken from him, and so no right of his to them has been violated. If we know that the twin will believe that his open future has been taken from him as a result of being cloned, even though in reality it has not, then we know that cloning will cause him psychological distress but not that it will violate his right. Thus Jonas's right to ignorance, and our employment of Feinberg's analogous right of a child to an open future, turn out not to be violated by human cloning, though they do point to psychological harms that a later twin may be likely to experience and that we will take up below.

The upshot of our consideration of a moral or human right either to a unique identity or to ignorance and an open future is that neither would be violated by human cloning. Perhaps there are other possible rights that would make good the charge that human cloning is a violation of moral or human rights, but it is not clear what they might be. We turn now to consideration of the harms that human cloning might produce.

What Individual or Social Harms Might Human Cloning Produce?

There are many possible individual or social harms that have been posited by one or another commentator, and we will consider only the more plausible and significant of them.

Largely Individual Harms

Human Cloning Would Produce Psychological Distress and Harm in the Later Twin. This is perhaps the most serious individual harm that opponents of human cloning foresee, and we have just seen that even if human cloning is no violation of rights, it may nevertheless cause psychological distress or harm. No doubt knowing the path in life taken by one's earlier twin might in many cases have several bad psychological effects (13,24,28–32). The later twin may feel, even if mistakenly, that his or her fate has already been substantially laid out, and so have difficulty freely and spontaneously taking responsibility for and making

his or her own fate and life. The later twin's experience or sense of autonomy and freedom may be substantially diminished, even if in actual fact they are diminished much less than it seems to him or her. Together with this might be a diminished sense of one's own uniqueness and individuality, even if once again these are in fact diminished little or not at all by having an earlier twin with the same genome. If the later twin is the clone of a particularly exemplary individual, perhaps with some special capabilities and accomplishments, he or she may experience excessive pressure to reach the very high standards of ability and accomplishment of the earlier twin (31). All of these psychological effects may take a heavy toll on the later twin and be serious burdens under which he or she would live. One commentator has also cited special psychological harms to the first, or first few, human clones from the great publicity that would attend their creation (13). While public interest in the first clones would no doubt be enormous, medical confidentiality should protect their identity. Even if their identity became public knowledge, this would be a temporary effect only on the first few clones and the experience of Louise Brown, the first child conceived by IVF, suggests this publicity could be managed to limit its harmful effects.

While psychological harms of these kinds from human cloning are certainly possible, some would argue even likely, they remain at this point only speculative, since we have no experience with human cloning and the creation of earlier and later twins. With naturally occurring identical twins, while they sometimes struggle to achieve their own identity, a struggle shared by many people without a twin, there is typically a very strong emotional bond between the twins. Such twins are, if anything, generally psychologically stronger and better adjusted than nontwins (8). It is even possible that being a later twin would confer a psychological benefit on the twin; for example, having been deliberately cloned to have his or her specific genes might make the later twin feel especially wanted for the kind of person he or she is. Nevertheless, if experience with human cloning confirmed that serious and unavoidable psychological harms typically occurred to the later twin, that would be a serious moral reason to avoid the practice.

In the discussion above of potential psychological harms to a later twin, it was assumed that one later twin is cloned from an already existing adult individual. Cloning by means of embryo splitting, as carried out and reported by Hall and colleagues at George Washington University in 1993, has limits on the number of genetically identical twins that can be cloned (33). Cloning by nuclear transfer, however, has no technological limit to the number of genetically identical individuals who might be cloned. Intuitively many of the psychological burdens and harms noted above seem more likely and serious for a clone who is only one of many identical later twins cloned from one original source, whereby the clone might run into an identical twin around every street corner. This prospect could be a good reason to place sharp limits on the number of twins that can be cloned from any one source.

There is one argument that has been used by several commentators to undermine the apparent significance of potential psychological harms to a later twin (8,24,25). The point derives from a general problem, called the nonidentity problem posed by the philosopher Derek Parfit and not originally directed to human cloning (34). Here is the argument. Even if all the psychological burdens and pressures from human cloning discussed above can not be avoided for any later twin, they are not harms to the twin, and so not reasons not to clone the twin. That is because the only way for the twin to avoid the harms is never to be cloned and so never to exist at all. But no one claims that these burdens and stresses, hard though they might be, are so bad as to make the twin's life, all things considered, not worth living — that is, to be worse than no life at all. So the later twin is not harmed by being given a life with these burdens and stresses, since the alternative of never existing at all is arguably worse — he or she loses a worthwhile life — but certainly not better for the twin. And if the later twin is not harmed by having been created with these unavoidable burdens and stresses, then how could he or she be wronged by having been created with them? And if the later twin is not wronged, then why is any wrong being done by human cloning? This argument has considerable potential import, for if it is sound it will undermine the apparent moral importance of any bad consequence of human cloning to the later twin that is not so serious as to make the twin's life all things considered not worth living.

Parfit originally posed the nonidentity problem, but he does not accept the above argument as sound. Instead, he believes that if one could have a *different* child without these psychological burdens (e.g., by using a different method of reproduction that does not result in a later twin), there is as strong a moral reason to do so as there would be not to cause similar burdens to an already existing child; this position has been defended in the general case of genetically transmitted handicaps or disabilities (35). The theoretical philosophical problem is to formulate the moral principle that implies this conclusion and that also has acceptable implications in other cases involving bringing people into existence, such as issues about population policy. The issues are too detailed and complex to pursue here and the nonidentity problem remains controversial and not fully resolved, but suffice it to say, what is necessary is a principle that permits comparison of the later twin with these psychological burdens and a different person who could have been created instead, for example, by a different means of reproduction, without such burdens. Choosing to create the later twin with serious psychological burdens instead of a different person who would be free of them, without a weighty overriding reason for choosing the former, would be morally irresponsible or wrong, even if doing so does not harm or wrong the later twin who could only exist with the burdens. At the least, the argument for disregarding the psychological burdens to the later twin because he or she could not exist without them is controversial, and it does not justify ignoring unavoidable psychological burdens to later twins among reasons against human cloning. Such psychological harms, as we will continue to call them, do remain speculative, but they should not be disregarded because of the nonidentity problem.

Human Cloning Procedures Would Carry Unacceptable Risks to the Clone. One version of this objection to human cloning concerns the research necessary to perfect the procedure, the other version concerns the later risks from its use. Wilmut's group had 276 failures before their success with Dolly, indicating that the procedure is far from perfected even with sheep. Further research on the procedure with animals is clearly necessary before it would be ethical to use the procedure on humans. But even assuming that cloning's safety and effectiveness was established with animals, research would need to be done to establish its safety and effectiveness for humans. Could this research be ethically done (36)? There would be little or no physical risk to the donor of the cell nucleus to be transferred, and his or her informed consent could and must always be obtained. There might be greater risks for the woman to whom a cloned embryo is transferred, but these should be comparable to those associated with IVF procedures and the woman's informed consent too could and must be obtained.

What of the risks to the cloned embryo itself? Judging by the experience of Wilmut's group in their work on cloning a sheep, the principal risk to the embryos cloned was their failure successfully to implant, grow, and develop. Comparable risks to cloned human embryos would apparently be their death or destruction long before most people or the law consider them to be persons with moral or legal protections of their life. Moreover artificial reproductive technologies now in use, such as IVF, have a known risk that some embryos will be destroyed or will not successfully implant and will die. It is premature to make a confident assessment of what the risks to human subjects would be of establishing the safety and effectiveness of human cloning procedures, but there are no unavoidable risks apparent at this time that would make the necessary research clearly ethically impermissible.

Could human cloning procedures meet ethical standards of safety and efficacy? Risks to an ovum donor (if any), a nucleus donor, and a woman who receives the embryo for implantation would likely be ethically acceptable with the informed consent of the involved parties. But what of the risks to the human clone if the procedure in some way goes wrong, or unanticipated harms come to the clone; for example, Harold Varmus, director of the National Institutes of Health, has raised the concern that a cell many years old from which a person is cloned could have accumulated genetic mutations during its years in another adult that could give the resulting clone a predisposition to cancer or other diseases of aging (37). Moreover it is impossible to obtain the informed consent of the clone to his or her own creation, but of course, no one else is able to give informed consent for their creation either.

It is too soon to say with any confidence whether unavoidable risks to the clone would make human cloning unethical. At a minimum, further research on cloning animals, as well as research to better define the potential risks to humans, is needed. For the reasons given above, we should not set aside risks to the clone on the ground that the clone would not be harmed by them, since its only alternative is not to exist at all; that is a bad argument. But we should not insist on a standard that requires risks to be lower than those we accept in sexual reproduction, or in other forms of assisted reproduction. It is not possible now to know when, if ever, human cloning will satisfy an appropriate standard limiting risks to the clone.

Largely Social Harms

Human Cloning Would Lessen the Worth of Individuals and Diminish Respect For Human Life. Unelaborated claims to this effect in the media were common after the announcement of the cloning of Dolly. Ruth Macklin has explored and criticized the claim that human cloning would diminish the value we place on, and our respect for, human life because it would lead to persons being viewed as replaceable (24). Just as with a supposed right to a unique identity, only on a confused and indefensible notion of human identity is a person's identity determined solely by his or her genes. Instead, an individual's identity is determined by the interaction of his or her genes over time with his or her environment, including the choices the individual makes and the important relations he or she forms with other persons. This means in turn that no individual could be fully replaced by a later clone possessing the same genes. Ordinary people recognize this clearly. For example, parents of a 12-year-old child dying of a fatal disease would consider it insensitive and ludicrous if someone told them they should not grieve for their coming loss because it is possible to replace their dying child by cloning him; it is *their child who is dying* whom they love and value, and that child and his importance to them could never be replaced by a cloned later twin. Even if they would also come to love and value a later twin as much as their child who is dying, that would be to love and value that *different child* who could never replace the child they lost. Ordinary people are typically quite clear about the importance of the relations they have to distinct, historically situated individuals with whom over time they have shared experiences and their lives, and whose loss to them would therefore be irreplaceable.

A different version of this worry is that human cloning would result in persons' worth or value seeming diminished because we would now see humans as able to be manufactured or "hand-made." This demystification of the creation of human life might reduce our appreciation and awe of it and of its natural creation. It would be a mistake, however, to conclude that a human being created by human cloning is of less value or is less worthy of respect than one created by sexual reproduction. At least outside of some religious contexts, it is the nature of a being, not how she is created, that is the source of her value and makes her worthy of respect. Moreover, for many people gaining a scientific understanding of the extraordinary complexity of human reproduction and development increases, instead of decreases, their awe of the process and its product.

A more subtle route by which the value we place on each individual human life might be diminished could come from the use of human cloning with the aim of creating a child with a particular genome, either the genome of another individual especially meaningful to those doing the cloning or an individual with exceptional talents, abilities, and accomplishments. The child might then be valued only for his or her genome, or at least

for his or her genome's expected phenotypic expression, and no longer be recognized as having the intrinsic equal moral value of all persons, simply as persons. For the moral value and respect due all persons to come to be seen as resting only on the instrumental value of individuals, or of individuals' particular qualities, to others would be to fundamentally change the moral status accorded to persons. Everyone would lose their moral standing as full and equal members of the moral community, replaced by the different instrumental value each of us has to others.

Such a change in the equal moral value and worth accorded to persons should be avoided at all costs, but it is far from clear that such a change would be the unavoidable result of permitting human cloning. Parents, for example, are quite capable of distinguishing their children's intrinsic value, just as individual persons, from their instrumental value based on their particular qualities or properties. The equal moral value and respect due all persons just as persons is not incompatible with the different instrumental value of people's particular qualities or properties; Einstein and an untalented physics graduate student have vastly different value as scientists, but share and are entitled to equal moral value and respect as persons. It would be a confusion and a mistake to conflate the two kinds of value and respect. Making a large number of clones from one original person might be more likely to foster this confusion and mistake in the public, and if so that would be a further reason to limit the number of clones that could be made from one individual.

Human Cloning Would Divert Resources from Other More Important Social and Medical Needs (13,28). As we saw in considering the reasons for, and potential benefits from, human cloning, in only a limited number of uses would it uniquely meet important human needs. There is little doubt that in the United States, and certainly elsewhere, there are more pressing unmet human needs, both medical or health needs and other social or individual needs. This is a reason for not using public funds to support human cloning, at least not if the funds will in fact be redirected to more important ends and needs. It is not a reason, however, either to prohibit other private individuals or institutions from using their own resources for research on human cloning or for human cloning itself, or to prohibit human cloning or research on human cloning.

The other important point about resource use is that it is not now clear how expensive human cloning would ultimately be, for example in comparison with other means of relieving infertility. The procedure itself is not extremely complex scientifically or technologically and might prove not to require a significant commitment of resources.

Human Cloning Might be Used by Commercial Interests for Financial Gain. Both opponents and proponents of human cloning agree that cloned embryos should not be able to be bought and sold. In a science fiction frame of mind, one can imagine commercial interests offering genetically certified and guaranteed embryos for sale, perhaps offering a catalog of different embryos cloned from individuals with a variety of talents, capacities, and other desirable properties. This would be a fundamental violation of the equal moral respect and dignity owed to all persons, treating them instead as objects to be differentially valued, bought, and sold in the marketplace. Even if embryos are not yet persons at the time they would be purchased or sold, they would be being valued, bought, and sold for the persons they will become. The moral consensus against any commercial market in embryos, cloned or otherwise, should be enforced by law whatever public policy ultimately is on human cloning. It has been argued that the law may already forbid markets in embryos on grounds that they would violate the thirteenth amendment prohibiting slavery and involuntary servitude (38).

Human Cloning Might be Used by Governments or Other Groups for Immoral and Exploitative Purposes. In *Brave New World*, Aldous Huxley imagined cloning individuals who have been engineered with limited abilities and conditioned to do, and to be happy doing, the menial work that society needed done (39). Selection and control in the creation of people was exercised not in the interests of the persons created, but in the interests of the society and at the expense of the persons created. Any use of human cloning for such purposes would exploit the clones solely as means for the benefit of others, and would violate the equal moral respect and dignity they are owed as full moral persons. If human cloning is permitted to go forward, it should be with regulations that would clearly prohibit such immoral exploitation.

Fiction contains even more disturbing and bizarre uses of human cloning, such as the Nazi, Joseph Mengele's creation of many clones of Hitler in Ira Levin's *The Boys from Brazil*, Woody Allen's science fiction cinematic spoof *Sleeper* in which a dictator's only remaining part, his nose, must be destroyed to keep it from being cloned, and the contemporary science fiction film *Blade Runner*. Nightmare scenarios like Huxley's or Levin's may be quite improbable, but their impact should not be underestimated on public concern with technologies like human cloning. Regulation of human cloning must assure the public that even such farfetched abuses will not take place.

Human Cloning Used on a Very Widespread Basis Would Have a Disastrous Effect on the Human Gene Pool by Reducing Genetic Diversity and Our Capacity to Adapt to New Conditions (11). This is not a realistic concern, since there is little if any reason to believe that human cloning would be used on a wide enough scale to have the feared effect on the gene pool. The vast majority of humans seem quite satisfied with traditional sexual means of reproduction; if anything, from the standpoint of worldwide population, we could do with a bit less enthusiasm for it. Programs of eugenicists like Herman Mueller earlier in the century to impregnate thousands of women with the sperm of exceptional men, as well as the more recent establishment of sperm banks of Nobel laureates, have met with little or no public interest or success (40). People prefer sexual means of reproduction and they prefer to keep their own biological ties to their offspring.

CONCLUSION

Human cloning has until now received little serious and careful ethical attention because it was typically dismissed as science fiction, and it stirs deep, but difficult to articulate, uneasiness and even revulsion in many people. Any ethical assessment of human cloning at this point must be tentative and provisional. Fortunately the science and technology of human cloning are not yet in hand, so a public and professional debate is possible without the need for a hasty, precipitate policy response.

The ethical pros and cons of human cloning seem at this time to be sufficiently closely balanced and uncertain that there is not an ethically decisive case either for or against permitting it or doing it. Access to human cloning can plausibly be brought within a moral right to reproductive freedom, but it is not a central component of a moral right to reproductive freedom. The circumstances in which its use would have significant benefits appear at this time to be few and infrequent, and it is not the solution to any major or pressing individual or social needs. On the other hand, contrary to the pronouncements of many of its opponents, human cloning seems not to be a violation of any moral or human rights. While it does risk some significant individual or social harms, most are based on common confusions about genetic determinism, human identity, and the effects of human cloning. Because most moral reasons against doing human cloning remain speculative, they seem insufficient to warrant a complete legal prohibition of either research on or later use of human cloning. Legitimate moral concerns about the use and effects of human cloning, however, underline the need for careful public oversight of research on its development, together with a continued and wider public debate and review before cloning is used on human beings.

BIBLIOGRAPHY

1. I. Wilmut et al., *Nature* **385**, 810–813 (1997).
2. WHO (World Health Organization) Press Office, Geneva, Switzerland, March 11, 1997.
3. Council of Europe, *Recommendation 1046 on the Use of Human Embryos and Fetuses for Diagnostic, Therapeutic, Scientific, Industrial and Commercial Purposes*, 1988.
4. J. Cohen and G. Tomkin, *Kennedy Inst. Ethics J.* **4**, 193–204 (1994).
5. R. Dworkin, *Taking Rights Seriously*, Duckworth, London, 1978.
6. J.S. Mill, *On Liberty*, Bobbs-Merrill Publishing, Indianapolis, IN, 1859.
7. R. Rhodes, *Cambridge Q. Healthcare Ethics* **4**, 285–290 (1995).
8. J.A. Robertson, *Children of Choice: Freedom and the New Reproductive Technologies*, Princeton University Press, Princeton, NJ, 1994.
9. D.W. Brock, in D. Thomasma and J. Monagle, eds., *Health Care Ethics: Critical Issues for Health Professionals*, Aspen Publishers, Gaithersburg, MD, 1994.
10. G.J. Annas, *Kennedy Inst. Ethics J.* **4**(3), 235–49 (1994).
11. L. Eisenberg, *J. Med. Philos.* **1**, 318–331 (1976).
12. J.A. Robertson, *Hastings Center Rep.* **24**, 6–14 (1994).
13. M. LaBar, *Thought* **57**, 318–333 (1984).
14. NABER (National Advisory Board on Ethics in Reproduction), *Kennedy Inst. Ethics J.* **4**, 251–282 (1994).
15. C. Kahn, *Free Inquiry* **9**, 14–18 (1989).

16. J. Harris, *Wonderwoman and Superman: The Ethics of Biotechnology*, Oxford University Press, Oxford, UK, 1992.

17. L. Thomas, *N. Engl. J. Med.* **291**, 1296–1297 (1974).

18. J. Lederberg, *Am. Naturalist* **100**, 519–531 (1966).

19. R. McKinnell, *Cloning: A Biologist Reports*, University of Minnesota Press, Minneapolis, MN, 1979.

20. W.A.W. Walters, W.A.W. Walters, and P. Singer, eds., *Test-Tube Babies*, Oxford University Press, Oxford, UK, 1982.

21. G.P. Smith, *University of New South Wales Law J.* **6**, 119–132 (1983).

22. P. Ramsey, *Fabricated Man: The Ethics of Genetic Control*, Yale University Press, New Haven, CT, 1970.

23. L. Kass, *Toward a More Natural Science*, Free Press, New York, 1985.

24. R. Macklin, *Kennedy Inst. Ethics J.* **4**, 209–226 (1994).

25. R.F. Chadwick, *Philos.* **57**, 201–209 (1982).

26. H. Jonas, *Philosophical Essays: From Ancient Creed to Technological Man*, Prentice-Hall, Englewood Cliffs, NJ, 1974.

27. J. Feinberg, in *Parental Authority, and State Power, Whose Child? Children's Rights*, W. Aiken and H. LaFollette, eds., Rowman and Littlefield, Totowa, NJ, 1980.

28. D. Callahan, *Los Angeles Times*, November 12, p. B7 (1993).

29. R. McCormick, *Notes on Moral Theology: 1965 through 1980*, University Press of America, Washington, DC, 1981.

30. A. Studdard, *Man Med. J. Values Ethics Health Care* **3**, 109–114 (1978).

31. J.D. Rainer, "Commentary." *Man Med. J. Values Ethics Health Care* **3**, 115–117 (1978).

32. A.D. Verhey, *Kennedy Inst. Ethics J.* **4**, 227–234 (1994).

33. J.L. Hall et al., *1993 Abstracts of the Scientific Oral and Poster Sessions*, Program Suppl. Abstract 0-001, S1, American Fertility Society/Canadian Fertility and Andrology Society, 1993.

34. D. Parfit, *Reasons and Persons*, Oxford University Press, Oxford, UK, 1984.

35. D.W. Brock, *Bioethics* **9**, 269–275 (1995).

36. R. Pollack, *N.Y. Times*, November 17, 1993.

37. R. Weiss, *Int. Herald Trib*, March 7, 1997.

38. P.O. Turner, *University of Detroit J. Urban Law* **58**, 459–487 (1981).

39. A. Huxley, *Brave New World*, Chalto & Winders, London, 1932.

40. M. Adams, ed., *The Well-Born Science*, Oxford University Press, Oxford, UK, 1990.

See other CLONING entries; FEDERAL POLICY MAKING FOR BIOTECHNOLOGY, EXECUTIVE BRANCH, NATIONAL BIOETHICS ADVISORY COMMISSION.

CLONING, OVERVIEW OF ANIMAL CLONING

GARY VARNER
Texas A&M University
College Station, Texas

OUTLINE

INTRODUCTION

Animal cloning is an area where the ethics and the science are both just coming into focus at the start of the twenty-first century. Although the cloning of embryos had been used for years in production agriculture, prior to the late 1990s, many or most scientists doubted that it would soon be possible to clone an adult animal of a complex species. Dolly's appearance in 1997 sparked renewed interest in cloning, especially the implications of applying the technique to human beings. However, many questions about the cloning of animals are just beginning to be addressed. On the scientific front, it is still unclear how to clone adults reliably and efficiently, and it is still unclear what the effects of the process are on the DNA of the resulting clones. On the philosophical front, although much has been published on the ethics of human cloning since Dolly's appearance, almost nothing has been published on the implications of adult cloning for animal welfare and animal rights (on the science of cloning, see K. Hanna, this volume).

ANIMAL CLONING: SCIENTIFIC QUESTIONS

In February 1997 a team led by Scottish scientist Ian Wilmut announced that it had successfully cloned a six-year-old ewe. Dolly, as they named the sheep clone, became a media celebrity overnight. Never before had a clone been produced from an adult animal. Cattle and sheep embryos had been cloned for years, but most scientists had doubted that it would be possible to clone adults in the foreseeable future.

The first sheep clone was produced in 1979 and the first bovine clone the next year, but these were produced by separating the cells of blastomeres. Blastomeres are very early embryos in which no differentiation of cells has yet occurred. Although still an expensive technique, by the 1990s blastomere separation had come to play an important role in production agriculture. For instance, an embryonic bull or cow of a promising breed line can be cloned, and the clones can then be frozen and shipped to farmers or stock breeders for implantation in surrogate mothers.

The technique used to produce Dolly was radically different. Although almost every cell of a mature animal contains its entire genetic code, in all somatic cells (i.e., body cells, as opposed to "germ" cells—i.e., sperm and eggs) the vast majority of these genes have been somehow "turned off." The nucleus of a bone cell, for instance, contains all of the genetic "instructions" necessary to build the entire animal, but only those needed to produce a bone cell are "active." To clone an adult animal from a somatic cell it is necessary to somehow "reactivate" all this unused DNA, but since no one yet understood how genes are turned on and off, most scientists were skeptical that a clone of an adult animal would be produced within the foreseeable future. Indeed, when Dolly's existence was announced, some scientists publicly expressed skepticism that she was really a clone.

The Wilmut team used a process called nuclear transplantation, in which the nucleus of an egg is replaced by the nucleus of another cell. This technique had first been used in the 1980s, but until 1997 had only been known to work when the transferred nucleus came from an embryo cell. Attempts to clone adult frogs from skin cells, for instance, resulted in some tadpole births but no further development (1). The Wilmut team tried transferring nuclei from the mammary glands of an adult ewe and then manipulating them so that the cells first quit growing and then, after a quiescent phase, began dividing again. They reasoned that something about this manipulation would "facilitate reprogramming of gene expression" (2, p. 811). Apparently they were right. In the summer of 1998, DNA analysis confirmed that Dolly was indeed a clone of the six-year-old ewe the Wilmut team had claimed was the nucleus donor (3,4), and over the next 18 months a number of other teams announced successes using various adult cells as nucleus donors and using variations on the Wilmut team's process for manipulating the resulting blastocysts through a quiescent phase and back into growth. A Japanese team of scientists announced that it had successfully cloned large numbers of mice using cumulus cells (which surround eggs inside the ovaries), and had even made clones of adult clones (5). Then labs in France and Texas announced that they had cloned cattle from skin cells (6–8).

Although cloning of adult animals using the somatic cell nuclear transfer technique — which, following (9), we can simply call "somacloning" — is an acknowledged reality at the beginning of the twenty-first century, the process is still unreliable and questions remain about the health of somaclones. Media coverage of the early work described above emphasized the high ratio of nuclear transfers to live births and a range of concerns about the health of somaclones.

It took 277 nuclear transfers to produce Dolly, and the Wakayama team did almost 1400 transfers on the way to producing their first 31 live births. In some teams' work there was a high rate of lost pregnancies and/or perinatal deaths (6,7,9). And since mitochondria outside the nucleus of the egg into which the donor nucleus is inserted carry some of the egg donor's DNA, somaclones might face problems of compatibility between the mitochondrial DNA of the egg donor and the DNA of the nucleus donor.

One of the most discussed potential problems was premature aging due to telomere shortening. Telomeres are sequences of DNA at the ends of all chromosomes. Although they do not code for any genes, they function to protect the gene-coding portion of the chromosome from being lost during cell division, which strips several base pairs of DNA from each end of the chromosome every time the cell divides. Telomere shortening is widely suspected of being tied to senescence, so a clone produced from a 21-year-old steer might be expected to age suddenly and prematurely. In May 1999 Dolly's telomeres were found to be significantly shorter than age-matched ewes produced from embryonic donor nuclei using the nuclear transfer method (10), but Dolly had by then delivered two healthy lambs herself and the Wakayama team's second generation mouse clones had themselves reared healthy offspring. Of particular interest in this respect was Texas A&M University's cloning of a 21-year-old steer named Chance, whose clone (Second Chance) was born after Chance had died of old age. However, that bull is still quite young at the time of this writing, and at present, then, the jury is still out on the effects of telomere shortening in somaclones. However, even if it turns out that current somaclones suffer from telomere shortening, it may be possible in the future to treat the problem using telomerase, the enzyme which "resets" telomere length in sperm and eggs, allowing naturally produced embryos to begin life with full-length telomeres.

ANIMAL CLONING: ETHICAL AND PHILOSOPHICAL QUESTIONS

The scientific questions discussed in the preceding section are directly relevant to some of the ethical concerns commonly raised in early media coverage of somacloning. A spate of philosophical work on cloning followed Dolly's appearance, but almost none of the articles focused on animal cloning as anything other than a harbinger of, or a slippery slope toward, cloning humans. Accordingly this section addresses the ethical issues most commonly raised in media treatments of somacloning and considers the significance of these concerns from the perspective of animal rights and animal welfare philosophies as those were developed in the last quarter of the twentieth century by philosophers concerned primarily with more traditional uses of animals.

The most commonly raised concern about animal cloning, that it would lead to human cloning, will not be considered here. (For a summary of ethical concerns about human cloning, see D. Brock, this volume.) There is only a limited overlap between the concerns commonly expressed about human cloning and those expressed about animal cloning. Since even the most cognitively sophisticated nonhuman animals presumably lack the ability to understand that they are clones, worries commonly expressed about human clones resenting the fact that they are clones, or suffering from knowing that their earlier-born twins have made certain decisions or suffered certain inherited diseases are irrelevant to nonhuman animals. And because animal breeding has been used for millennia to "improve breeding stock," concerns about cloning of "gifted" individuals and about cloning reducing the worth of individuals seem at least less intense than the analogous concerns about human clones. There are legitimate economic and environmental (and thus social) concerns about increasing genetic homogeneity in farm animals, and people who are worried about human cloning are concerned that the practice of animal cloning puts us on a slippery slope to that, but considered in and of itself, the cloning of animals has been objected to primarily on two grounds. The first is that it is unnatural or amounts to "playing God." The second is that it violates animals' rights and/or may have dramatic adverse impacts on their welfare. This section discusses these objections in turn, along with a related philosophical question about the moral significance of unactualized potentials.

The "It's Unnatural" and "Playing God" Objections

As Rolston (11) notes, there are two "absolute" and antithetically opposed senses of the term "natural." On one extreme, anything that happens in accordance with natural law (the laws of physics, biology, gravity, etc.) is natural, but in this sense nothing that human beings could do would be unnatural. If cloning works, it is consistent with the laws of biological nature, whether or not anyone presently understands how "reprogramming of gene expression" works. On the other extreme is an "artifactual" sense of the unnatural, according to which a process is natural to the extent that it is not guided by human intention or deliberation. In this sense, somacloning is extremely unnatural, but so are highways, computers, and good sanitation. And, Rolston notes, even the attempt to act naturally is in this sense unnatural, insofar as it is intentional or deliberate. So it is difficult to see how either of these "absolute" senses of "unnatural" helps us articulate a clear ethical concern about somacloning.

Varner (12) identifies two other senses of the "natural." First, the term is sometimes used to pick out what is distinctive or unique about a species. Aristotle employed this sense of "natural" in a famous argument in his *Nicomachean Ethics*. Aristotle used the Greek term *ergon* (translated variously as "work" or "function") to refer to whatever capacity the members of a taxon share with no "lower" organisms (13). He concluded that nutrition and growth are *ergon* for plants, sense perception and motion are *ergon* for animals, and reason is *ergon* for humans (14). However, it is unclear that this in any way helps us sort out ethical concerns about cloning, because cloning research and applications are paradigm cases of reasoned inquiry, and thus they are, in Aristotle's sense, perfectly natural things for humans to do. Also, in light of growing scientific evidence that animals engage in reasoning of various kinds, a more plausible candidate for the Aristotelean *ergon* of humans would be some kind of moral reasoning, so Aristotle's *ergon* argument at most raises the question of whether or not cloning is morally acceptable, without providing any part of the answer.

Varner notes that in yet another sense of the term, a process or activity is "natural" for a given species to the extent that it was characteristic of the species in its evolutionary past and/or it played an important role in natural selection's producing the various traits that are now characteristic of the species. Hunting is often claimed to be natural for humans in this sense,

insofar as distinctively human traits like language and the very sophisticated group planning and coordination, which language makes possible, appear to have evolved because they were adaptive for our distant ancestors. Is cloning unnatural in this sense of the term? Many adult plants and animals reproduce by having clones "bud off" from them, and even in mammals, clonal reproduction occurs whenever identical twins are produced. So it is at most *somacloning* that is unnatural for humans and for the other mammals involved in research to date. However, it is unclear why this should make somacloning a bad thing, for plenty of things are unnatural in this sense without being bad, for instance, vegetarian diets and living without engaging in hunting.

Thus claims about what is natural and unnatural do not seem likely to be helpful in articulating the substance of ethical concerns about cloning, and similar things can be said about the objection that by engaging in cloning we are "playing God." Ruth Chadwick observes that claims about "playing God" are sometimes used to express fear of unforeseen consequences. The Greek concept of *hubris* involves people overestimating their power and suffering severe, unforeseen consequences, sometimes via the wrath of the gods. But, Chadwick argues, if the fear is of unforeseen consequences, it would be more clear to say so and "discard the concept of 'playing God'" (15, p. 203).

More generally, basing morality on God's will is philosophically problematic and basing public policy on religion is politically problematic in a pluralist society. Philosophically, the problem can be posed in terms of what is sometimes called "the Euthyphro Question" after a similar question posed about piety in a Platonic dialogue (16): *Is what is right right because God wills it, or does God will what He wills because it is right.*

In the former case, anything would be right if God did will it, including putatively unjust practices like slavery, sexism, wars of aggression, and so on. Also, if doing the right thing means nothing more than acting in accordance with God's will, then God's goodness is trivialized. For on such a view, "John is a good person" seems to mean roughly that "John does what God wills." But by parity of reasoning, on such a view, "God is good" would mean nothing more than "God does what God wills."

The second answer to the Euthyphro question, that "God wills what He wills because it is right," preserves the significance of claims about God's goodness. On such a view, it is at least possible, in principle, to investigate moral claims independently of God's will (since morality is logically prior to God's will on this view). In practice, doing so would be necessary for two reasons: first, to resolve ambiguities in the evidence about what God wills, and second, to determine what God would will about issues (like somacloning, presumably) that are not addressed in the available evidence regarding God's will. For on this view, God is like a wise and benevolent elder who can be depended upon for reliable advice about what is right and what is wrong, but when evidence about what God wills is lacking or ambiguous, the only way to determine what God wills would be to determine, independently, what would be the right thing for Him to will.

So just like claims that cloning is unnatural, claims that cloning is "playing God" seem unlikely to help elucidate and evaluate ethical concerns about cloning. If the concerns cannot be expressed clearly and persuasively without relying on religious assumptions based only in faith, then most people would conclude that the concerns cannot justify sweeping restrictions on cloning in a pluralist society (17).

Animal Welfare and Animal Rights

Since the publication of Peter Singer's book *Animal Liberation* in 1975, philosophers have shown a keen interest in the areas of "animal rights" and "animal welfare." In her contribution to this volume, "Research on Animals, Ethics, and the Moral Status of Animals," Lilly-Marlene Russow questions the usefulness of this distinction. Some of the reasons she gives mirror points raised below, but the received interpretation of the distinction still provides a much better template for analyzing ethical concerns about animal cloning than does either talk of its unnaturalness or its being "playing God." The philosophical characterizations of animal welfare and animal rights sketched briefly below are fairly standard. More detail is available in Russow's contribution to this encyclopedia, and in (18–21) as well as the many works referenced there.

The Distinction as Popularly Conceived

Any careful consideration of animal welfare and animal rights concerns must begin with a clear characterization of what counts as one or the other type of view. Much confusion exists about this, because popular accounts of the distinction typically cut the pie very differently than philosophers writing on the topic.

In particular, in media treatments of animal issues, "animal rightists" are typically distinguished from "animal welfarists" in terms of the political ends and means the two endorse, rather than in terms of their underlying philosophical commitments. Accordingly animal rightists are typically portrayed as seeking the abolition of various practices, whereas animal welfarists are said to seek reform of problematic practices rather than elimination of animal use. Animal rightists are portrayed as willing to employ illegal and even violent tactics, whereas animal welfarists are portrayed as "working within the system." And, perhaps because the term seems to have been coined by agriculturalists, medical researchers, hunters, and so forth, whose colleagues felt threatened by self-professed animal rights activists, animal welfarists typically are portrayed as reasonable, well-informed people, in contrast to animal rightists who are often characterized as irrational and poorly informed people acting in the grip of their emotions.

Self-professed animal welfarists are seen to defend some uses of animals, while emphasizing that they take seriously a moral imperative to take animals' interests into account—that they are not neo-Cartesians who deny all significance to animals' lives and suffering. So it is not surprising that they wish to self-consciously distance themselves—in both their colleagues' and in the public's eyes—from self-professed animal rights activists, whose actions have sometimes richly earned

them the above stereotype (21). However, this stark political characterization of the animal rights/animal welfare distinction in the media hides an underlying overlap in philosophical commitments that is crucial to assessing ethical objections to animal cloning.

Animal Welfare, Philosophically Conceived

Among philosophers writing on related issues, the animal welfare stance generally is taken to assume a specific ethical principle, utilitarianism, and a specific understanding of how animals' happiness or well-being, and the opposites of these, are to be assessed.

Utilitarianism is the view that right actions and institutions maximize aggregate happiness. The "classical" utilitarians of the nineteenth century—Jeremy Bentham (22) and John Stuart Mill (23)—championed various social causes on utilitarian grounds, arguing that various penal code, educational, and welfare reforms would increase the happiness of underprivileged peoples more than the marginal cost of such programs to more privileged persons, thus maximizing human happiness in the aggregate.

Bentham and Mill also both endorsed a hedonistic conception of happiness. Both claimed that the only thing that is intrinsically good is pleasure and the only thing that is intrinsically bad is pain. To say that something is intrinsically good is to say that its existence is good in and of itself, independently of its relationship to other things, and independently of its instrumental value in the pursuit of other ends. And since at least some nonhuman animals are capable of feeling pain, both Mill and Bentham concluded that these animals' happiness must be taken into consideration in the utilitarian calculus.

A widely cited passage in Bentham's *An Introduction to the Principles of Morals and Legislation* conveys the logic of this extension of the moral community and, taken in context, underlines an important point about animal welfare views as philosophically construed. Observing that animals have traditionally been mere "things" in the eyes of the law, Bentham adds:

> The day *may* come, when the rest the animal creation may acquire those rights which never could have been withholden from them but by the hand of tyranny. The French have already discovered that the blackness of the skin is no reason why a human being should be abandoned without redress to the caprice of a tormentor. It may come one day to be recognized, that the number of the legs, the villosity of the skin, or the termination of the *os sacrum*, are reasons equally insufficient for abandoning a sensitive being to the same fate. What else is it that should trace the insuperable line? Is it the faculty of reason, or, perhaps, the faculty of discourse? But a full-grown horse or dog is beyond comparison a more rational, as well as a more conversable animal, than an infant of a day, or a week, or even a month, old. But suppose the case were otherwise, what would it avail? The question is not, Can they *reason*? nor, Can they *talk*? but, Can they *suffer*? (22, p. 412).

In this passage Bentham clearly holds that sentience, rather than some more sophisticated cognitive capacity or species membership, is the criterion of moral standing, and a criterion which many animals clearly meet. (Although "sentience" means, generally, consciousness or awareness of something, in discussions of animal welfare and animal rights the term is usually taken to mean consciousness of pain [and/or pleasure] specifically.) However, immediately preceding this widely quoted passage, Bentham states that

> If the being eaten were all, there is very good reason why we should be suffered to eat such of them as we like to eat; we are the better for it, and they are never the worse. They have none of those long-protracted anticipations of future misery which we have. The death they suffer in our hands commonly is, and always may be, a speedier and by that means a less painful one, than that which would await them in the inevitable course of nature (22, pp. 411–412).

That is, in hedonistic utilitarian terms, a sufficiently humane form of slaughter is perfectly permissible. (Which probably explains why only the portion of the passage quoted first above appears in animal rights tracts.)

Self-styled animal welfarists usually share Bentham's and Mill's basic philosophical commitments regarding animals. That is, they think in utilitarian terms (at least when it comes to animals — see the next subsection regarding their attribution of rights to humans), and they think of individual happiness in hedonistic terms (at least when it comes to animals). So it is common to hear animal welfarists defend humane agricultural systems in roughly Bentham's own terms, medical research on animals on the ground that the harms to animals are greatly outweighed by the suffering humans (and other animals) are spared when cures are found for debilitating diseases, and hunting in terms of reducing animal suffering on the whole. And in doing their utilitarian calculus, self-professed animal welfarists commonly speak as if animal happiness consists entirely in feeling pleasure while avoiding pain so that, for instance, if all pain and suffering is removed from the protocol of an experiment with important clinical implications, all ethical concerns about the experiment have been addressed.

Philosophically speaking, adopting the animal welfare perspective means using hedonistic utilitarian thinking to assess various human uses of animals, and so conceived, both defenders of animal research, agriculture, and hunting, and some prominent critics of these practices, are animal welfarists. Significantly Singer (24), whom the popular, political characterization of the distinction places squarely in the animal rights camp, is himself a utilitarian and, with some important qualifications (see below), a hedonistic utilitarian. How, then, does Singer reach such different conclusions regarding the very practices which self-professed animal welfarists defend? The answer is that while sharing certain deep philosophical commitments about animals, the two differ in their understanding of what those commitments imply in practice. Singer (19,24) argues that among other things, defenders of these practices exaggerate the magnitude and the certainty of benefits to humans and other animals, and underestimate the costs to animals involved in agriculture, research, and hunting.

Animal Rights, Philosophically Conceived

Philosophically construed, a true animal *rights* view must be nonutilitarian. This subsection briefly sketches a paradigm example: Tom Regan's view in *The Case for Animal Rights* (18).

Regan argues that whether or not the defenders of animal agriculture, research, and hunting are right that the aggregate benefits outweigh the aggregate costs, these practices are still wrong from a perspective that extends respect for individuals from humans to animals. In day-to-day talk about ethics, appeals to moral "rights" are often used to claim such respect for individual humans. To say that someone "has a moral right" to something usually means that we would not be justified in infringing that right on purely utilitarian grounds. For instance, an aggressive fundamentalist preacher's right to free speech may not be overridden just because the aggregate suffering of listeners and passers-by (in terms of anger, frustration, and hurt feelings) outweighs the joy the preacher and his followers get from him speaking. Self-styled animal welfarists commonly invoke such rights claims regarding humans, while denying that rights are relevant where animals are concerned. Thus one might claim that while it would never be permissible to experiment on humans without their consent, the utilitarian calculus suffices to justify this in the case of nonhuman animals.

In *The Case for Animal Rights* Regan examined the consequences of extending such respect for individuals from humans to animals. Specifically, Regan's view is that respect for individuals involves treating them as more than mere "utility receptacles," that if an individual "has moral rights," then we cannot justify harming him or her (or it) on the sole basis that doing so maximizes aggregate happiness. Respect for individuals requires nonutilitarian reasons for involuntarily imposing harm, or significant risk of harm.

Which animals deserve this kind of respect? Regan argues that given the range of humans who are usually thought to deserve it, consistency requires us to extend it to all animals who are "subjects of a life," by which he means, roughly, that they have memories, a sense of their own future, and desires or preferences about their future (18, p. 243). He argues that available empirical evidence shows convincingly that at least all normal adult mammals have these capacities, and that birds probably do too (18).

One of the obvious questions for anyone who, like Regan, construes rights claims as "trump cards" against utilitarian arguments is what to do where rights conflict. Regan defends two principles for use in such cases (18, pp. 301–312). For situations involving comparable harms to various individuals, Regan defends what he calls "the miniride" principle. This requires that where rights violations are inevitable, we minimize the violations of rights (hence the name of the principle). Although Regan admits that where it applies, this principle would have the same implications as utilitarianism, he denies that it is utilitarian in spirit or justification. Regan's other principle applies to cases involving dramatically unequal degrees of harm. This principle requires that we avoid harming "the worse-off individual," which means whatever individual under one option would be harmed significantly more than any individual who would be harmed under any other possible option.

Obviously the concept of harm is central to Regan's view—it plays a prominent role both in his general characterization of "having moral rights" and in his two principles regarding conflicting rights. We saw in the preceding section that the classical utilitarians were hedonists, assessing harm in terms of felt pain and lost opportunities for pleasure. Regan endorses a very different, nonhedonistic view of harm, one that many interpreters of Mill have thought was implicitly at work in his thinking, and one that Peter Singer endorses with regard to at least humans and other mammals. On this alternative view, an individual is harmed to the extent that he or she fails to achieve an integrated satisfaction of his or her preferences. A lot is built into the notion of "an integrated satisfaction" of preferences, including, usually, a qualification about weeding out irrational or imprudent preferences. Without going into further detail, however, we can see how such a nonhedonistic conception of happiness would be attractive, at least when it comes to humans. For the death of a normal adult human in the prime of his or her life seems tragic in a way that is difficult to capture from the hedonistic perspective, which would assess the harm purely in terms of any pain felt while dying plus lost opportunities for pleasure in the future. But when a normal adult human dies prematurely, a whole network of preferences for the future—all of one's plans, projects, hopes, dreams, wishes, and so on—go unsatisfied whether one dies painlessly or not, and it is something about this impact on one's preferences that makes the human's death tragic.

Both Regan and Singer hold that this nonhedonistic conception of harm is applicable to many nonhuman animals, including at least all normal mammals. This is one of the important qualifications on Singer's view, alluded to above. Although Singer is a hedonistic utilitarian when it comes to animals who do not have a robust sense of their future, he is a *preference* utilitarian with regard to those who do, including but not limited to human beings. Singer's own view thus differs in two important respects from self-professed animal welfarists. First, he endorses a nonhedonistic conception of happiness for some animals (mammals) in addition to humans. But second, he is a thorough-going utilitarian, unlike those animal welfarists who would attribute rights to at least human beings Singer (24) denies that even humans hold "trump cards" against utilitarian arguments. So it is only the hedonistic utilitarian portion of Singer's thinking about animals which strictly corresponds to the standard philosophical conception of animal welfare.

Assessing Somacloning: Health Risks to Somaclones

Media reports on the early results of somacloning emphasized the low success rate and possible health risks to the resulting clones. Assessing the low success rates raises a complex philosophical question that will be discussed in the next subsection. This subsection describes the differing ways animal welfare and animal rights views would treat the potential health risks to somaclones after birth. First, however, it is necessary to recall the deep scientific uncertainties which still remain on the topic at the time of this writing.

Early work with genetic engineering or gene splicing produced some dramatically deformed animals. The so-called Beltsville pigs, who had a human growth hormone gene inserted into their DNA, suffered from numerous debilitating health problems (25). Because it is still not understood how DNA is re-programmed during the somacloning process, it is natural to worry that similar problems might result. For just as gene splicing alters a complex and well-evolved genetic code, so too might somacloning, if the DNA re-programming process was somehow incomplete or haphazard. Also, because mitochondrial DNA may play a role in gene activation and the enucleated egg cell into which the donor nucleus is transferred contains many mitochondria in its cytoplasm, somacloning by nuclear transfer might cause abnormalities if the two sources of DNA are somehow incompatible.

The number of somaclones produced to date is so small that it is hard to evaluate these risks. By January 2000 only a few dozen somaclones, most of them mice, are reported in the literature. A few health problems in somaclones are reported (6), but Dolly is reported to be healthy and to have delivered two healthy lambs (10) and second generation somacloned mice (somaclones of somaclones) are reported to have reared healthy offspring (5).

Probably the most widely discussed health risk to somaclones involves premature aging due to telomere shortening. At something over one year of age, Dolly's telomeres were found to be significantly shorter than age-matched ewes who had also been produced by nuclear transfer, but using embryonic rather than adult cells. However, "considering the large size distribution of sheep TRFs [terminal restriction fragments or telomeres], it remains to be seen whether a critical length will be reached during the animal's lifetime" (10, p. 317). So it may be that telomere shortening will only affect clones of clones, or perhaps only clones made from already elderly animals. In these respects the Wikayama team's multigeneration mice and Second Chance (the clone produced from an elderly steer at Texas A&M) are of particular interest, but it is just too soon to tell how important telomere shortening will be from published reports available at the time of this writing (5,7). Moreover, even if telomere shortening turns out to be a significant problem, it may be treatable, since in adult mammals and birds, sperm and eggs are produced with fully "repaired" telomeres, and the reagent involved, telomorase, has already been identified.

All in all, then, the magnitude and nature of health risks to somaclones are unclear. There is reason to be cautiously optimistic that somaclones will be of average health, but it is still quite possible that significant health problems will be found. In light of that uncertainty, what can be said about animal welfare and animal rights perspectives on somacloning research?

From the animal welfare perspective, any suffering that results from health problems counts as a negative in the moral ledger. In utilitarian terms, these costs must be weighed against the likely benefits in assessing the research in question. The complex problem of assessing probabilities cannot be addressed here; in particular, whether or not it is even possible to rate experimental protocols in terms of their contributing to the development of valuable clinical applications is a very controversial issue. However, as a general line of research, somacloning seems likely to have a large range of valuable applications which a utilitarian would have to balance against whatever suffering somaclones endure, and if these benefits are very significant, then they could easily outweigh the still unsubstantiated risks somaclones face, at least from a utilitarian perspective.

How many and what kind of applications somacloning will have depends largely on what technical hurdles scientists are able to clear and on some related policy issues. It is all but certain that adult humans can be cloned by the nuclear transfer technique, paving the way for at least certain new reproductive technologies. With somacloning succeeding in mice, sheep, and cattle, whether and when these benefits of somacloning are realized is a matter of political rather than scientific uncertainty.

Other benefits of somacloning will only be realized if some significant remaining technical problems are solved. For instance, medical researchers envision cloning tissue from adults to produce rejection-free grafts of homogenous tissues like skin, bone, and liver. Such tissues can already be grown in culture, but since no one yet understands how DNA reprogramming occurs when embryos are manipulated after nuclear transfer, no one can yet say how difficult it will be to control the process so that instead of producing a whole new individual, an embryonic somaclone can be made to produce just one specific type of tissue cells. More ambitious treatments in the same vein would involve producing whole organs in vitro for rejection-free transplant into the DNA donor, but this vision of "organs in vats" still seems far-fetched.

Still other uses of somacloning are proposed besides such clinical applications in human medicine. The Missyplicity Project (26) is funded by a wealthy couple seeking to clone their pet dog, Missy, but the project has collateral goals with far-reaching implications (9). The project aims to establish commercial dog-cloning services which could be used to reproduce the best "service" dogs — guide, drug sniffing, and search and rescue dogs — whose largely genetically determined abilities can only be discovered through expensive training processes. The project is also intended to make the technology available free of charge to programs benefiting endangered canid species. In the latter case, somacloning could be used to preserve individual genotypes that would otherwise be lost to old age or when unsuccessfully released into the wild, and to introduce genotypes from the wild into captive breeding populations without removing the donor individuals from the wild.

With somacloning promising this diverse array of significant benefits, the health risks to somaclones would have to be very significant in order to outweigh the benefits in a utilitarian calculus. And when we recall that the standard animal welfare perspective assumes a hedonistic conception of happiness (at least

where nonhuman animals are concerned), it seems even less likely that somacloning could be condemned from that perspective. Suppose, for instance, that telomere shortening necessarily reduces a somaclone's life span by a significant amount, but without causing it any kind of suffering apart from that involved in the normal aging process. Because the animal welfare perspective evaluates premature death purely in terms of lost opportunities for pleasure, it is hard to see this as tragic, especially since the somaclone itself would never have existed at all without this shortened life span. When the question is whether or not to bring into the world an individual with deformities or a disease, a utilitarian well might oppose doing so because there is the possibility of instead bringing into existence a similar individual without the deformity or disease. However, when the question is whether or not to go forward with research on somacloning, if some health problems for somaclones are inevitable, then there is no way to achieve the result in question without the attendant costs, and in light of the above consideration of potential uses of the technology, the result in question appears very important from a utilitarian perspective. So unless health problems facing somaclones cause fairly significant amounts of suffering, it is hard to see how this research could be opposed from an animal welfare perspective.

From an animal rights perspective, however, the situation is very different, for two reasons. First, recall that Regan's paradigm animal rights view assesses harm to individuals in terms of lost opportunities to form and satisfy desires. This means that for Regan, even a painless death can be a significant harm to an individual. Probably it is true, as Regan himself says, that mice, sheep, cattle, and even dogs have a significantly less robust sense of their own futures than do humans, but still the same medical problems are likely to translate into a different kind and degree of harm done to an animal when viewed from a perspective like Regan's rather than from the purely hedonistic perspective of animal welfare.

The second reason animal rights philosophies are likely to condemn somacloning research is that rights views are essentially antiutilitarian. Remember that Regan's work explores the implications of extending the kind of respect commonly attributed to human individuals — rights as "trump cards" against utilitarian arguments for harming them — to animals, and this involves denying that harm to the individual can be justified by appealing to the aggregate benefits of doing so. The array of benefits likely to emerge from cloning research thus appears *irrelevant* from an animal rights perspective. If the individuals involved are harmed and they have not consented (or cannot consent), then the research violates their rights, whether the individuals in question are humans or animals, and whatever the magnitude of the harms in question.

One obvious objection to Regan's blanket opposition to animal experimentation is suggested by what was said above while summarizing his view. Regan endorses two nonutilitarian principles for deciding whose rights to violate when rights violations are inevitable, and one of these, the worse-off principle, arguably implies that some animal experimentation would be not only permissible but morally mandatory. For instance, suppose (not implausibly) that it would be possible to save more human burn victims in the future if somacloned skin grafts are developed. On Regan's view, the harm that death is to an individual is a function of how much capacity for desire formation and satisfaction that individual has, and Regan admits that this makes death to a human being (at least in the prime of his or her life) a significantly greater harm than death to any nonhuman animal, whose sense of its own future is much less robust. But then wouldn't the worse-off principle imply that the research ought to be done, since the future burn victims who would die without it would be harmed more than any nonhuman animal involved in the cloning research?

Regan responds to this sort of argument in *The Case for Animal Rights* (18, pp. 363–394) by claiming, in effect, that the worse-off principle only applies where background conditions of justice or fairness have been maintained, and that because the research involuntarily imposes risks that the experimental subjects would not otherwise face, that is not the case. That is, if either a human or an animal were going to die no matter what anyone did, then, Regan claims, the appropriate way to decide who should die would be using the worse-off principle — indeed, to decide on any other basis would be to fail to respect the two individuals equally. However, in the case of biomedical research, the risks of disease and death faced by the experimental animals have been imposed upon them without their consent. So, Regan argues, it seriously misrepresents the situation to paint it as a choice between saving one or the other life one of those lives is in jeopardy only because researchers have unfairly chosen to put it there, without the animal's consent.

Regan's response may encounter a problem when applied to somacloning research specifically (and some kinds of genetic engineering), however. Most animals involved in biomedical research only face special health risks if and when researchers choose to expose them to disease, injury, or drugs. Yet if somacloned animals necessarily face special risks, such as shortened life spans, then it is simply impossible for the individual somaclones to exist without facing that risk. And, if it is impossible for the individual in question to have existed without facing the risk in question, does it make sense to say that the individual is harmed by being brought into existence facing that risk? This raises complex philosophical issues in personal identity (some of which are touched on in Brock's contribution to this volume).

Assessing Somacloning: The Question of Potentiality

In addition to potential health risks to somaclones, media reports on early work in somacloning stressed that the success rate was abysmally low. For instance, Wilmut et al. (2) did 277 nuclear transfers to get the one pregnancy that resulted in Dolly. However, in assessing animal rights and animal welfare views on somacloning, it is important to note how many embryos failed to result in live births and for what reasons and the related question of these entities' moral status.

Most of the eggs into which donor nuclei were transferred by early researchers were disposed of without

any attempt to implant them in surrogate mothers. For instance, Wilmut et al. only attempted 29 implantations, and from these attempts only one pregnancy resulted. However, at the same time they were trying to produce Dolly, the Wilmut team was using the nuclear transfer technique with embryo cells as the source of donor nuclei, and in this portion of their work, they apparently had several miscarriages and one neonatal death. They report attempting 135 implantations resulting in only 21 pregnancies, based on ultrasonic scans performed about 8 weeks after attempted implantation (around the end of the first trimester in a sheep's pregnancy). Subsequent scans revealed fewer pregnancies remaining, "suggesting either misdiagnosis or foetal loss," and in the end, only 8 live births resulted. Of these, one died within minutes of birth, although "post-mortem analysis failed to find any abnormality or infection" (2, p. 811). So apart from whether the source of the DNA is a somatic cell or a germ cell, the nuclear transfer technique may pose significant risks to the resulting embryos, fetuses, and neonates.

Although some researchers soon began to achieve higher success rates, what is the moral status of the many, many blastocysts disposed of prior to implantation, the spontaneously aborted embryos, and the neonates lost in research like the Wilmut team's? That depends crucially on a philosophical point about the moral significance of merely potential (as yet unactualized) traits or capacities: *If an individual has the potential to develop a trait or capacity that would make it deserving of some kind of respect, or moral rights, does that individual deserve the same kind of respect or rights now?* This we may call the question of potentiality.

Many people say yes to this question, at least when it comes to human beings. This is a secular, philosophical basis for a pro-life position on abortion. Just as many say no, however, and this seems to be a necessary plank in any pro-choice platform. For if we answer yes to the question of potentiality, how could we deny that every human embryo deserves the same respect or rights as any adult, and how could we fail to reach the conclusion that abortion is murder? Indeed, since even a newly fertilized egg has the potential to become a fully functioning adult, understanding the implications of saying yes to the question of potentiality helps us understand why many pro-life advocates could equate several forms of birth control with abortion (and, in turn, murder). For IUDs and "morning after" treatments (like RU 486 and "emergency contraception" with high doses of birth control pills) both function by preventing implantation in the uterus rather than by preventing conception, and even standard birth control pills have this as a side effect and second line of defense.

People often wonder how an animal rights advocate could consistently be pro-choice on the abortion issue, but if one answers no to the question of potentiality, then the moral status of the developing human being changes dramatically from conception through birth. From the animal welfare perspective, sentience (the ability to feel pain) is the criterion for moral standing, and on Regan's rights view, it is being a "subject of a life," by which he means having memories, desires, and a sense of one's own future. (As was noted earlier, Singer thinks that sentience is sufficient for moral standing but also recognizes that having a robust sense of one's future gives one's life some added moral significance.) If sentience is the criterion for moral standing, then developing embryos probably do not qualify at all, since the first brain waves are detectable only at around six weeks and consciousness of pain may not develop until much later. And if having a sense of one's future is the criterion for having moral rights, humans may not qualify until sometime after birth. It is for this very reason that Regan restricts his claim about rights possession to "mentally normal mammals of a year or more" (18, p. 78). He does not deny that neonates have a sense of their future, but he admits this is more controversial than the claim that one-year-olds do.

Similar things can of course be said about developing animals. Although newborn animals are usually more precocious than are newborn humans, presumably the numerous animal embryos discarded before implantation during somacloning research, as well as at least early-term fetuses, are incapable of either feeling pain or thinking about their futures. When in an individual's development these capacities arise is a complex issue involving both scientific and philosophical claims that cannot adequately be addressed here (for more on the issue, see Refs. 12, 27, and 29). However, it is clear how, if we answer no to the question of potentiality, neither an animal welfare nor an animal rights position need find any risks faced by embryos and early fetuses in somacloning research morally problematic, leaving the emphasis on the kind of health risks discussed in the preceding subsection.

Apart from relying on one's antecedent intuitions about the moral status of embryos and newly fertilized eggs, how might one argue for a negative answer to the question of potentiality? A classic argument in favor of a negative answer was offered by S.I. Benn and Joel Feinberg in a pair of classic essays on the general concept of moral rights (29,30). They compare moral rights to legal rights, noting that a potential president does not have the right to command the armed forces until he or she actually becomes president. If moral rights and the more general concept of moral standing (which a utilitarian or animal welfarist would use) are treated analogously, then the potential to develop sentience or become a "subject of a life" does not give the individual the same moral status as actually developing these capacities.

A second classic argument offered for a negative answer is based on the phenomena of twinning and chimeras. At conception, a unique genotype is created, but there is more to personal identity than one's genotype. In nature, genetically identical twins result from fission of early embryos, yet in humans and other conscious animals, we consider identical twins distinct individuals in virtue of their differing psychological states. (Although this may be false when it comes to clonal reproduction in plants and even in animals which are not conscious; see Ref. 12.) The reverse phenomenon, when two early embryos with distinct genetic codes fuse, is also known to happen, albeit very infrequently, in nature. The resulting individual is called a chimera. Although every cell in a chimera's body has a genetic code identical to that of one of the two embryos from which it was produced, in any region of the chimera's body, 50 percent of the cells have the code of one embryo and 50 percent the other embryo's. Humans have been born cross-sexual chimeras, and in the lab, chimeras have been produced from closely related species (e.g., by fusing the embryos of a goat and a sheep). In an essay defending experimentation on early human embryos, Helga Kuhse and Peter Singer (31) claim that the above phenomena show that very early embryos are not properly regarded as specific individuals at all: Identical twins cannot both be identical with the single early embryo from which they sprang, nor can a single chimera be identical with the two early embryos from which it sprang.

The foregoing argument appears flawed, however. David Oderberg counters it with the following thought experiment about a "split brain operation." In the philosophical literature on personal identity, this refers to a hypothetical process by which one brain is split into two, which carry (to begin with) identical memories, personalities, and so on. Oderberg writes:

> Suppose it is certain that in five minutes Jones will undergo a split-brain operation. If Jones, being a person, has moral rights, then he is no less a person, and no less a possessor of moral rights, because of this certainty. ...Similarly, the possibility, even the *certainty*, of division of a human embryo does not *of itself* show that the embryo is fair experimental game (32, pp. 277–278).

So even if the correct inference to draw regarding twins and chimeras is that the early embryos involved in producing them are not identical with the resulting adults, that would not, by itself, prove that early embryos have no moral standing or moral rights. At most, it would prove that those early embryos are not the same individuals as either the twins or the chimera.

CONCLUSION

Media coverage of early work on somacloning emphasized health risks to the resulting clones both before and after birth. From an animal welfare stance, the benefits promised by somacloning seem likely to outweigh any presently foreseeable postpartum harms to the clones. From an animal rights perspective, the magnitude of these promised benefits is irrelevant—any significant harm inflicted upon the clones without their consent violates their rights, although a philosophical problem about identity may short-circuit that objection. From both perspectives, the question of potentiality determines the (in)significance of risks to all embryos and at least early-term fetuses.

BIBLIOGRAPHY

1. J.B. Gurdon, R.A. Laskey, and O.R. Reeves, *J. Embryol. Exp. Morphol.* **34**, 93–112 (1975).
2. I. Wilmut et al., *Nature* **385**, 810–813 (1997).
3. D. Ashworth et al., *Nature* **394**, 329 (1998).
4. E. Singer, Y. Dubrova, and A. Jeffreys, *Nature* **394**, 229–230 (1998).

5. T. Wakayama et al., *Nature* **394**, 369–374 (1998).

6. J.-P. Renard et al., *Lancet* **353**, 1489–1491 (1999).

7. J. Hill et al., *Biol. Reprod.* **62**, 1135–1140 (2000).

8. N. Wade, *N.Y. Times*, January 5, A14 (2000).

9. G. Varner, *Animal Welfare* **8**, 407–420 (1999).

10. P.G. Shiels et al., *Nature* **399**, 316–317 (1999).

11. H. Rolston, III, *Environ. Ethics* **1**(1), 7–30 (1979).

12. G. Varner, *In Nature's Interests? Interests, Animal Rights and Environmental Ethics*, Oxford University Press, Oxford, UK, 1998.

13. R. Kraut, *Can. J. Philos.* **9**(3), 467–478 (1979).

14. Aristotle, *Nicomachean Ethics*, T. Irwin, trans., Hackett Publishing Company, Indianapolis, IN, 1985.

15. R. Chadwick, *Philosophy* **57**, 201–209 (1982).

16. Plato, in E. Hamilton and H. Cairns, eds., *Plato: The Collected Dialogues*, L. Cooper, trans., Princeton University Press, Princeton, NJ, 1961, pp. 169–185.

17. L. Silver, *Cambridge Q. Healthcare Ethics* **7**, 168–172 (1998).

18. T. Regan, *The Case for Animal Rights*, University of California Press, Berkeley, CA, 1983.

19. P. Singer, *Animal Liberation*, 2nd ed., Avon Books, New York, 1990.

20. G. Varner, *Hastings Cent. Rep.* **24**(1), 24–28 (1994).

21. J. Jasper and D. Nelkin, *The Animal Rights Crusade: The Growth of a Moral Protest*, Free Press, New York, 1992.

22. J. Bentham, An Introduction to the Principles of Morals and Legislation, Basil Blackwell, Oxford, UK, 1948. Originally published in 1789.

23. J.S. Mill, *Utilitarianism*, Bobbs-Merril, Indianapolis, 1957. Originally published in 1861.

24. P. Singer, *Practical Ethics*, 2nd ed., Cambridge University Press, Cambridge, UK, 1993.

25. D. Bolt et al., in *Veterinary Perspectives on Genetically Engineered Animals*, 58-61, American Veterinary Medical Association, Schaumburg, IL, 1990.

26. L. Hawthorne, Missyplicity Project, Available at: *www.mis syplicity.com*

27. D. DeGrazia, *Taking Animals Seriously*, Cambridge University Press, Cambridge, UK, 1996.

28. M. Tooley, *Abortion and Infanticide*, Clarendon Press, Oxford, UK, 1983.

29. S.I. Benn, in J. Feinberg, ed., *The Problem of Abortion*, Wadsworth, Belmont, CA, 1984, pp. 145–150.

30. J. Feinberg, in J. Feinberg, ed., *The Problem of Abortion*, Wadsworth, Belmont, CA, 1984, pp. 145–150.

31. H. Kuhse and P. Singer, in P. Singer et al., eds., *Embryo Experimentation*, Cambridge University Press, Cambridge, UK, 1990, pp. 65–75.

32. D.S. Oderberg, *Philos. Public Affairs* **26**, 259–298 (1997).

See other CLONING entries.

CLONING, OVERVIEW OF HUMAN CLONING

KATHI E. HANNA
Science and Health Policy Consultant
Prince Frederick, Maryland

OUTLINE

INTRODUCTION

The term "cloning" is used by scientists to describe many different processes that involve making duplicates of biological material. In most cases isolated genes or cells are duplicated for scientific study, and no new animal results. The experiment that led to the cloning of Dolly the sheep in 1997 was different: It used a cloning technique called *somatic cell nuclear transfer* and resulted in an animal that was a genetic twin—although delayed in time—of an adult sheep. This technique of transferring a nucleus from a somatic cell into an egg that produced Dolly was an extension of experiments that had been ongoing for over 40 years. It built on previous work in which embryos were cloned by a technique called *blastomere separation*, or twinning, and by a method of extracting the nucleus of an embryo and transferring it into another enucleated egg. Although the birth of Dolly was lauded as a success, in fact the procedure is not perfected. It is not yet clear whether Dolly will remain healthy or whether she could have subtle problems that might lead to serious diseases. Therefore the prospect of proceeding to application of this technique in humans is troubling for scientific and safety reasons as well as for additional ethical reasons having to do with family and the order of generations.

DEFINITION OF CLONING

The word "clone" is used in many different contexts in biological research but in its most simple and strict sense it refers to a precise genetic copy of a molecule, cell, plant, nonhuman animal, or human being. The feasibility of cloning varies according to the complexity of the genetic material being used and its status in the plant or animal kingdom.

Genetically identical copies of whole organisms are commonplace in the plant-breeding world (referred to as *varieties* rather than clones) because it is relatively easy to regenerate a complete plant from a small cutting. However, the developmental process in animals does not usually permit cloning as easily as in plants, except in the simplest of invertebrate species. Although a single adult vertebrate cannot naturally generate another whole organism, cloning of vertebrates does occur in nature primarily with the formation of identical twins. Twins occur by chance in humans and other mammals when a single embryo splits into halves at an early stage of development.

At the molecular and cellular level, scientists have been cloning human and animal cells and genes for several decades. Such cloning provides greater quantities of identical cells or genes for study, as each cell or molecule is identical to the others. At the most basic level, biologists routinely make clones of deoxyribonucleic acid (DNA), the molecular basis of genes. DNA fragments containing genes are copied and amplified in a host cell, usually a bacterium. The availability of large quantities of identical DNA makes possible many scientific experiments. This process, often called *molecular cloning*, is the mainstay of recombinant DNA technology and has led to the production of important therapies, such as insulin to treat diabetes, tissue plasminogen activator (tPA) to dissolve clots after a heart attack, and erythropoietin (EPO) to treat anemia associated with dialysis for kidney disease.

In *cellular cloning* copies are made of cells derived from the soma, or body, by growing these cells in culture in a laboratory. The genetic makeup of the resulting cloned cells, called a *cell line*, is identical to that of the original cell. Since molecular and cellular cloning of this sort does not involve germ cells (eggs or sperm), the cloned cells are not capable of developing into a baby.

The third level of cloning aims to reproduce genetically identical animals. Cloning of animals can typically be divided into three distinct techniques, embryo cloning by *blastomere separation* (twinning), embryo cloning by *nuclear transplantation*, and cloning via *somatic cell nuclear transfer*. It is this last type of cloning that has raised so many legal, social, and ethical concerns, particularly after the February 1997 announcement that Ian Wilmut and colleagues had successfully cloned a sheep using somatic cell nuclear transfer (1).

CLONING BY BLASTOMERE SEPARATION (TWINNING)

Blastomere separation is an increasingly common form of cloning used in animal biotechnology. In this technique, the developing embryo is split very soon after fertilization when it is composed of two to eight cells (see Fig. 1) (2). To split the embryo, technicians dissolve the protective covering, or zone pellucida, of an early cleaving embryo and place the embryos in a medium in which they separate into individual cells. The cells, called *blastomeres*, are then placed in another solution that forms an artificial zone and are moved to a culturing medium, where they will begin cleaving (dividing). Each blastomere is able to produce an entirely new, individual organism. This is because blastomeres are considered to be totipotent; that is, they possess the total potential to make an entire new organism. This totipotency allows scientists to split animal embryos into several cells to produce multiple organisms that are genetically identical. This capability has tremendous relevance to breeding cattle and other livestock although it is an expensive technique and it is not yet perfected, having been associated with higher rates of congenital malformations.

Researchers first used blastomere separation to twin sheep embryos in 1979 and cattle embryos in 1980 (3). It is the sole cloning technique to have been attempted experimentally with human embryos. In 1993 investigators at the George Washington University (GWU) Medical Center separated the cells of 17 human embryos and generated 48 embryos (an average of three embryos for each original), demonstrating that one human embryo could be split to create two or more genetically identical embryos (4). In the GWU study the investigators used 22 eggs that were fertilized by more than one spermatozoan during in vitro fertilization (IVF) and were thereby considered ineligible for implantation; that is, they would have been

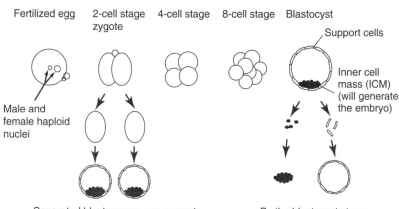

Fertilized egg   2-cell stage   4-cell stage   8-cell stage   Blastocyst
                  zygote

Support cells

Male and female haploid nuclei

Inner cell mass (ICM) (will generate the embryo)

Separated blastomeres can generate blastocysts and ultimatedly develop into 'cloned' animals

By the blastocyst stage separated ICM and support cells cannot regenerate blastocysts

Figure 1. Preimplantation embryo development in mammals (*Source*: Ref. 21).

discarded (5). The researchers allowed the polyspermic eggs to divide into two, four, or eight cells. Cells from 17 of the embryos were placed in another solution that formed an artificial zone and were moved to a culturing medium. An average of three embryos developed from each original embryo, leading to a total of 48 cleaving cells. The GWU experiment showed that twinning is technically feasible, but it created an ethical firestorm and reopened debate about the appropriateness of research conducted on the preimplantation (ex utero) human embryo (6).

Proponents of blastomere separation justify its use as an aid in IVF. If too few embryos are available for transfer to the uterus, IVF success rates are low. Embryo twinning could increase the number of embryos and do so by manipulating (i.e., splitting) the embryos, rather than administering powerful and risky fertility drugs to women in order to stimulate the release of more eggs for fertilization. In addition, blastomere separation could facilitate preimplantation diagnosis, a technique in which a cell is removed from each embryo of a couple at risk for passing a deleterious gene to their offspring. The DNA from the extracted cell is amplified and examined for the presence of the disease-related gene. Only embryos lacking the deleterious gene are transferred. Preimplantation diagnosis remains problematic because in 30 percent of attempts DNA amplification fails due to insufficient quantities of DNA. With blastomere separation more than one cell can be used, which could improve the success rate of amplification and improve the accuracy of diagnoses (7).

Blastomere separation is also proposed as an aid to scientific inquiry. In research studies using human embryos this technique could provide a control embryo for each experimental embryo in order to hold constant as many variables as possible and state with more confidence that differences in the experimental and control groups were due to manipulation rather than to genetic differences in the two groups (8).

Ethical Considerations

One of the most cited objections to the use of twinning in humans is the questionable effect it might have on children. In one application of this technique multiple, identical embryos could be produced and all transferred at the same time to the prospective mother's uterus. If all implanted, the woman would give birth to identical twins, triplets, quadruplets, or more. Aside from the possible hardship this might cause the family, it is unlikely that the children would suffer psychologically as a result. However, if not all embryos were transferred at the same time, unique concerns arise. If the first transfer failed, a second frozen embryo could be thawed and transferred. If the transfer was successful, few would express concern, because only one live born child was produced as a result of multiple, although spaced, transfers.

An issue that has followed IVF since its first use concerns the disposition of frozen embryos. Whether embryos are identical, via blastomere separation or not couples can transfer them at a later date. If the embryos are identical, however, the result would be spaced twins. No one knows what it would be like to be the older or younger of spaced twins but because of the influence of social rearing it is highly unlikely that they would be identical in any way other than genotype. Rather, they would share many of the same environmental influences experienced by children reared in the same household. However, the temptation might exist for parents to unfairly compare the children. Thus concerns about the spacing of identical twins raises questions about the potential adverse psychosocial effects that might occur for the second twin because of unrealistic and unfounded expectations on the part of the parents.

Perhaps of more concern is the possibility that parents would store embryos in the event of the death of a child: The newly transferred identical would provide them with a genetic replica of the deceased child. Also a concern is the possibility that the stored embryos might later be used to generate cell lines for therapeutic transplantation for the living child who has become ill. Even more far-fetched, but possible, would be transgenerational transfer. That is, frozen embryos could be bequeathed to future generations, allowing a woman to give birth to her identical twin or a great grandchild to give birth to the identical twin of her great grandfather.

EMBRYO CLONING BY NUCLEAR TRANSPLANTATION

A more complex cloning technique is the nuclear transplantation of embryo cells to enucleated eggs. This technique was first used in animals in the early 1980s and the first birth in higher animals (lambs) occurred in 1986 when scientists transplanted the nuclei of cells obtained from four- to eight-cell embryos to enucleated eggs (9). The first experiments of this type were successful only when the donor cell was derived from an early embryo. Later, in 1996 and 1997, lambs were produced in this manner, this time using the nuclei of cells from a late stage embryo (1). In 1997 the technique was successfully used in Rhesus monkeys (10). In theory, large numbers of genetically identical animals could be produced through such nuclear transplantation cloning. In practice, the nuclei from embryos which have developed beyond a certain number of cells seem to lose their totipotency, limiting the number of animals that can be produced in a given period of time from a single, originating embryo.

The safety of this technique is questionable. It has not been attempted in humans. Although apparently healthy offspring have been produced in mammals, there is a high rate of fetal and infant loss. In the 1996 experiment with sheep, of the eight fetuses two were spontaneously aborted and two died at birth. Another died at 10 days (11). Other investigators have reported a high rate of congenital abnormalities (20 to 30 percent and larger than normal offspring) (12). The effects of the transfer of genetic material from one organism to another might include a compromised immune system.

Finally, more has to be learned about the role of mitochondrial DNA in development. A cellular component called the *mitochondrion* is the energy-producing component of the cell. Although most of the genes associated with the mitochondrion reside in the nucleus, the mitochondrion itself houses some genes. Thus, in nuclear transfer, mitochondrial genes are not transferred to the enucleated egg along with the nuclear genes. Much has to be learned about the compatibility of nuclear DNA and mitochondrial DNA when such a transfer occurs (13). Preliminary studies of the effects of genetic imprinting suggest that mitochondrial DNA can play an important role in the early activation and function of some genes; 37 genes responsible for energy metabolism in cells are known to be coded by mitochondrial DNA (14). Thus mitochondrial DNA likely plays an important role in some disease expression, a role that must be considered when designing somatic cell nuclear transfer.

Ethical Considerations

Many of the issues raised by blastomere separation also apply to embryo cloning by nuclear transplantation, with some subtle differences. First, embryo cloning by nuclear transplantation would permit the creation of more embryos because it can take place at a later stage of embryonic development (8- to 16-cell stage). Thus many of the concerns posed by twinning are magnified by the potential to create more embryos through nuclear transplantation. Second, the embryos would not be as genetically identical in nuclear transplantation because the nuclei would be fused with the cytoplasm (and therefore the mitochondrial DNA) of the egg donor(s). This might diffuse some of the concern about loss of individuality. However, the safety issues appear to be the most pressing of concerns and perhaps the greatest ethical obstacle to proceeding in this technique.

SOMATIC CELL NUCLEAR TRANSFER CLONING

In the two previous examples of cloning, the genetic material being used in cloning was genetically unique,

that is it was the product of a fusion between an egg and a sperm. The manipulation used to create clones occurred after the fusion. Thus a new individual was being replicated several times. What made Dolly different is that she was not genetically unique, that is, she was genetically identical to an existing six-year-old ewe. She was not created by the union of egg and sperm (see Fig. 2). To appreciate how her creation was possible, it is helpful to understand the science that led up to her birth. Much of it centers on an evolving understanding of how cells develop and differentiate, and how genes are expressed.

Cell Differentiation

Nearly every cell contains a spheroid organelle called the *nucleus*, which houses nearly all the genes of the organism. Genes are composed of DNA, which serve as a set of instructions to the cell to produce particular proteins. Although all somatic cells of an individual contain the same genes in the nucleus, the particular genes that are activated vary by the type of cell. For example, a differentiated somatic cell, such as a neuron, must keep a set of neural-specific genes active and silence those genes specific to the development and functioning of other types of cells such as muscle or liver cells. In contrast, gametes (eggs and sperm) do not differentiate but retain activity necessary to create new life after fusion with egg or sperm.

Investigations that began over 40 years ago sought to determine whether a differentiated somatic cell still contained all genes, even those it did not express. Early experiments in frogs and toads (15,16) provided evidence that the expression potential of the genes in differentiated cells is essentially unchanged from that of the early embryo. Nuclei from donor-differentiated cells were injected into recipient eggs in which the nucleus had been inactivated (Fig. 3). Another carefully controlled

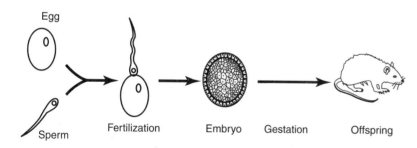

Egg
Sperm
Fertilization
Embryo
Gestation
Offspring

Figure 2. Sexual reproduction (*Source*: Ref. 21).

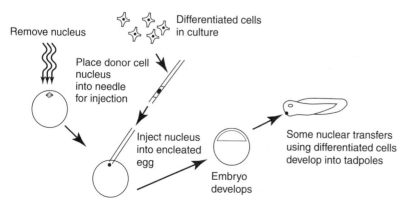

Remove nucleus

Differentiated cells in culture

Place donor cell nucleus into needle for injection

Inject nucleus into encleated egg

Embryo develops

Some nuclear transfers using differentiated cells develop into tadpoles

Figure 3. Nuclear transfer carried out in frogs (*Source*: Ref. 21).

series of experiments used nuclei from adult frog skin cells for transfer to an enucleated egg (17). Four percent of the nuclei transferred eventually gave rise to fully developed tadpoles. These experiments provided evidence that the genes contained in the nuclei of differentiated cells could be reactivated by the cytoplasm of the egg and thus direct normal development, but only up to a certain stage. No viable adult frog ever developed from these tadpoles and there was a decrease in the number of tadpoles born as the age of the transferred nucleus increased. This left open the possibility that complete reactivation of the adult nucleus was prevented by some irreversible change in the genetic material, and that there was a progressive decline in nuclear potential with age. Later analysis, however, suggested that the major reason for developmental failure of the transplanted embryos appeared to be chromosomal abnormalities that occurred during the process of nuclear transplantation itself.

Experiments in mammals also suggested that it is possible to reprogram adult somatic cells. Experiments in mice followed those in frogs and toads. Early development occurs at a slower rate in mammals than amphibians; thus it was believed that reprogramming of the donor nucleus would occur more efficiently in mice than in frogs. It was shown in mice in the 1980s that nuclei could be successfully exchanged between fertilized eggs, with 90 percent reaching the blastocyst stage of embryonic development and beyond, but only if the nuclei were recovered from an embryo at the two-cell stage (18). Many experiments since then have shown that blastomeres up to the early blastocyst stage are still totipotent when combined with other embryonic cells (19). This means that the failure of nuclear reprogramming has to be the result of something other than irreversible changes to the genetic material of the cells.

In 1986, experiments in sheep had results that differed from those in mice. Enucleated eggs from sheep could be fused with blastomeres taken from embryos at the eight-cell stage to provide donor nuclei and viable offspring were produced (20). Apparently the use of early-stage eggs prolongs the period of reprogramming before the donor nucleus has to undergo the first division. The advent in the last few years of electrofusion for both fusion of cells and activation of the egg has been another major advance, because activation and fusion occur simultaneously.

Enucleated eggs used for fusion only proceed to division after activation by some artificial signal, such as the electrical current used in the electrofusion technique. Because these experiments used fusion of two cells and not simple injection of an isolated nucleus, all of the cellular components are transferred. Thus the mitochondria, which contain some genes of their own, are transferred along with the nucleus. Because an enucleated egg also contains mitochondria, the result of a fusion experiment is a cell with a mixture of mitochondria from both the donor and the recipient. Since the mitochondrial genes represent an extremely small proportion of the total number of mammalian genes, mixing of mitochondria per se is not expected to have any major effects on the cell (2,21). However, if one of the donors suffers from a mitochondrial disease, then mixture of the mitochondria may significantly alleviate the disease.

Over the past 10 years or so, there have been numerous reports of successful nuclear transfer experiments in mammals, nearly all of them using cells taken directly from early embryos. The oldest embryonic nucleus that can successfully support development differs among species (22–26).

Reprogramming Gene Expression

More recently, studies suggested that it might be possible to reprogram the gene expression of somatic cells so that they perform a different task. Thus, it should be possible to activate or inactivate almost any gene in a cell, given the right cellular environment containing the appropriate regulatory molecules. To reprogram the gene expression of a somatic cell it is not essential to fuse it with an egg; in some cases re-programming can occur through fusion of two adult cells. Cell fusion experiments have demonstrated that extensive reprogramming of differentiated nuclei can occur (27). The knowledge that regulatory molecules can reprogram an adult nucleus led to the speculation that cloning via somatic cells nuclear transfer was a real possibility.

Synchronization of the Cell Division Cycle

All of the work just described struggled with understanding the relationships between the normal cell division cycle, the age of the embryo, and the ability of the nuclei to be reprogrammed (28,29). Work by Wilmut showed that the phase of cell cycle division at which transfer is attempted is critically important. Thus the need to transfer nuclei in a specific phase of division (called the G1 phase) before replication is initiated is important to avoid the chromosome damage that can occur and prevent development of the embryo into a viable offspring (1).

Cloning of Dolly

In the research that preceded the birth of Dolly, Wilmut and colleagues established cell lines from sheep blastocysts and used these cells as nuclear donors (11). In an attempt to avoid the problems of nuclear transfer of non-G1 nuclei into activated eggs, they starved the donor cell line by removing all nutrients from the medium prior to nuclear transfer. Under these starvation conditions, the cells exit the cell cycle and enter another phase (Gap phase 0), which is similar to the G1 phase in which chromosomes have not replicated. For Wilmut and colleagues, approximately 14 percent of fusions resulted in development of blastocysts, and 4 of the 34 (12 percent) embryos transferred developed into live lambs. Two died shortly after birth. The success rate in sheep and cow experiments was almost identical, and it suggests that division of cells in culture for many days does not inhibit the ability of their nuclei to be reprogrammed by the egg environment. These findings led Wilmut and colleagues to ask if the same would be true of nuclei from fully differentiated somatic cells.

In the most simple of terms, the technique used to produce Dolly the sheep, somatic cell nuclear transplantation cloning, involves removing the nucleus of an egg and replacing it with the diploid nucleus of a somatic cell.

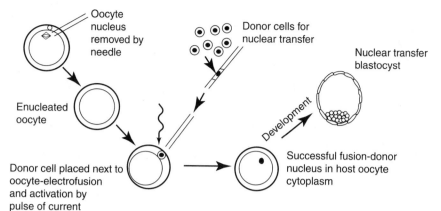

Figure 4. Nuclear transfer via electrofusion in mammals (*Source*: Ref. 21).

In such nuclear transplantation cloning there is a single genetic "parent," unlike sexual reproduction where a new organism is formed when the genetic material of the egg and sperm fuse (see Fig. 4). In addition, this technique differs from blastomere separation and embryo cloning via nuclear transplantation because it does not involve an existing embryo.

In the ground-breaking research reported in February 1997 (1), Wilmut and colleagues took late embryo, fetal cell cultures, and cell cultures derived from the mammary gland of an adult sheep and applied the same approach of synchronizing the cell in the G0 stage prior to nuclear transfer. They reported successful production of live offspring from all three cell types, although only 29 of 277 (11 percent) of successful fusions between adult mammary gland nuclei and enucleated oocytes developed to the blastocyst stage, and only 1 of 29 (3 percent) blastocysts transferred developed into a live lamb. This experiment was, in fact, the first time any fully developed animal had been born following transfer of a somatic cell nucleus.

It should be noted, however, that no attempt was made to document that the donor cells were fully differentiated cells, the genes of which expressed specialized mammary gland proteins. In the earlier experiments with frogs, the fact that the donor cells were fully differentiated was documented in such a manner. In the case of Dolly, it is possible that the cells could have been derived from a less-differentiated cell in the population, such as a mammary stem cell.

Scientific Uncertainties

Several important concerns remain about the science and safety of nuclear transfer cloning using adult cells as the source of nuclei. The first concerns the Dolly experiment itself. Only one animal was produced thus it is not clear that this technique is reproducible even in sheep.

Second, there might be true species differences in the ability to achieve successful nuclear transfer (21). It has been shown that nuclear transfer in mice is much less successful than in larger domestic animals. Part of this difference may reflect the intensity of research in this area in the last 10 years; agricultural interests have meant that more nuclear transfer work has been performed in domestic animals than in mice. But part of the species differences may be real and

not simply reflect the greater recent effort in livestock. For example, in order for a differentiated nucleus to redirect development in the environment of the egg, its constellation of regulatory proteins must be replaced by those of the egg in time for the embryo to use the donor nucleus to direct normal development of the embryo. The species difference may be the result of the different times of embryonic gene activation. In mammals, unlike many other species, the early embryo rapidly activates its genes and cannot survive on the components stored in the egg. The time at which embryonic gene activation occurs varies between species (30,31). The later onset of embryonic gene activation and transcription in sheep provides an additional round or two of cell divisions during which nuclear reprogramming can occur, unlike the rapid genome activation in the mouse. Cross-species comparisons are needed to assess the importance of this difference in the time of genome activation for the success of nuclear transfer experiments (21). In humans, for example, the time period before gene activation is very short, which might not permit the proper reprogramming of genes after nuclear transfer to allow for subsequent normal development.

Third, the phenomenon of genetic imprinting may affect the ability of nuclei from later stages to reprogram development. Genetic imprinting refers to the relative effect on embryonic development of genes inherited from the father (paternal genes) versus those from the mother (maternal genes) (32). Some heritable imprint is established on the chromosomes during the development of the egg and the sperm such that certain genes are expressed only when inherited from the father or mother. Nuclei transferred from diploid cells, whether embryonic or adult, should contain maternal *and* paternal copies of the genome, and thus not have an imbalance between the maternally and paternally derived genes. However, an adult nucleus, if it is to be successfully reprogrammed, must retain intact the chromosomal imprints that under normal conditions would determine whether maternal or paternal gene copies will be active. There is some speculation that some instability of the imprint, particularly in cells in culture, could limit the efficiency of nuclear transfer from somatic cells (21). In addition, it is known that disturbances in imprinting lead to growth abnormalities in mice and are associated with cancer and rare genetic conditions in children.

A fourth concern is whether cellular aging will affect the ability of somatic cell nuclei to program normal development. As somatic cells divide they progressively age and there is normally a defined number of cell divisions that they can undergo before senescence. Part of this aging process involves the progressive shortening of the ends of the chromosomes, the telomeres, and other genetic changes. Germ cells (eggs and sperm) evade telomere shortening by expressing an enzyme, telomerase, that can keep telomeres full length. It seems likely that returning an adult mammalian nucleus to the egg environment will expose it to sufficient telomerase activity to reset telomere length, since oocytes have been found to be potent sources of telomerase activity (33). In 1999 a team of researchers, including those that cloned Dolly, measured the telomeres in her cells as an indication of her actual age and found that her telomeres were 20 percent shorter than would be expected (34). However, it is not yet known whether the shorter telomeres actually make a difference in the physiological age of the cloned sheep. In addition, the 20 percent difference may be within the normal variation for sheep.

The health effects for the resulting liveborn, having been created with an "aged' nucleus, are unknown. Therefore, a fifth concern is raised by the possibility that the mutations that accumulate in somatic cells might affect nuclear transfer efficiency and lead to cancer and other diseases in the offspring. As cells divide and organisms age, mutations in the DNA will inevitably occur and will accumulate with time. If these mistakes occur in the sperm or the egg, the mutation will be inherited in the offspring. Normally mutations that occur in somatic cells affect only that cell and its descendants, which are ultimately dispensable. Nevertheless, such mutations are not necessarily harmless. Transfer of a nucleus from a somatic cell carrying such a mutation to an egg would transform a sporadic somatic mutation into a germline mutation (i.e., transmitted to all of the cells of the body).

POTENTIAL THERAPEUTIC APPLICATIONS OF NUCLEAR TRANSFER CLONING

The demonstration that in mammals as in frogs, the nucleus of a somatic cell can be reprogrammed by the egg environment provides further impetus to studies on how to reactivate embryonic programs of development in adult cells. These studies have exciting prospects for regeneration and repair of diseased or damaged human tissues and organs, and they may provide clues as to how to reprogram adult differentiated cells directly without the need for oocyte fusion. In addition, the use of nuclear transfer has potential application in the field of assisted reproduction.

Potential Applications in Organ and Tissue Transplantation

Some human diseases can be treated effectively by organ or tissue transplantation, including some leukemias, liver failure, and heart and kidney disease. In some instances, the organ required is nonvital, that is, it can be taken from the donor without great risk (e.g., bone marrow, blood, kidney). In other cases, the organ is obviously vital and required for the survival of the individual, such as the heart. All transplantation is imperfect, with the exception of that which occurs between identical twins, because transplantation of organs between individuals requires genetic compatibility.

The application of nuclear transfer cloning to humans could provide a potential source of organs or tissues of a predetermined genetic background. The notion of using human cloning to produce individuals for use solely as organ donors is not only morally repugnant but also illegal as it is unconstitutional and may violate the prohibition against slavery. A morally more acceptable and potentially feasible approach is to direct differentiation along a specific path to produce specific tissues (e.g., muscle or nerve) for therapeutic transplantation rather than to produce an entire individual. Given current uncertainties about the feasibility of this, however, much research would be needed in animal systems before it would be scientifically sound, and therefore potentially morally acceptable, to go forward with this approach.

Potential Applications in Cell-Based Therapies

Another therapeutic possibility raised by cloning is transplantation of cells or tissues not from an individual donor, but from an early embryo or embryonic stem cells. Blastomeres and stem cells are primitive, undifferentiated (or totipotent) cells. This potential application would not require the generation and birth of a cloned individual. Embryonic stem cells provide an interesting model for such studies, since they represent the precursors of all cell lineages in the body. Mouse embryonic stem cells can be stimulated to differentiate in vitro into precursors of the blood, neuronal and muscle cell lineages, among others (35), and they thus provide a potential source of stem cells for regeneration of all tissues of the body.

It might be possible to take a cell from an early blastomere and treat it in such a manner as to direct its differentiation along a specific path. By this procedure it might be possible to generate in the laboratory sufficient numbers of specialized cells, for example, bone marrow stem cells, liver cells, or pancreatic beta-cells (which produce insulin) for transplantation. If even a single tissue type could be generated from early embryonic cells by these methods and used clinically, it would constitute a major advance in transplantation medicine by providing cells that are genetically identical to the recipient.

One could imagine the prospect of nuclear transfer from a somatic cell to generate an early embryo and from it an embryonic stem cell line for each individual human, which would be ideally tissue-matched for later transplant purposes. This might be a rather expensive and far-fetched scenario. An alternative scenario would involve the generation of a few, widely used and well-characterized human embryonic stem cell lines, genetically altered to prevent graft rejection in all possible recipients.

The preceding scenarios depend on using cells of early human embryos, generated either by IVF or nuclear transfer into an egg. Because of ethical and moral concerns raised by the use of embryos for research purposes, it would be far more desirable to explore the direct use of human cells of adult origin to produce specialized cells or tissues for transplantation into patients (2). It may not be necessary to reprogram terminally differentiated cells but rather to stimulate proliferation and differentiation of the quiescent stem cells, which are known to exist in many adult tissues, including even the nervous system (36). Experiments in this area are likely to focus more on the conditions required for direct stimulation of the stem cells in specific tissues, than on actual use of nuclear transfer to activate novel developmental programs. These approaches to cellular repair using adult stem cells will be greatly aided by an understanding of how stem cells are established during embryogenesis.

Another strategy for cell-based therapies would be to identify methods by which somatic cells could be "de-differentiated" and then re-differentiated along a particular path. This would eliminate the need to use cells obtained from embryos. Such an approach would permit the growth of specialized cells compatible with a specific individual person for transplantation. Although at the current time this strategy is highly speculative, ongoing research in animal systems may identify new approaches or new molecular targets that might make this approach feasible.

It will be of great importance to understand through experiments in animals how the environment of the egg reprograms a somatic cell nucleus. What cellular mechanisms can be elucidated? What components are involved in these processes? Can we direct cells along particular developmental pathways in the laboratory and use these cells for therapy? The capacity to grow human cells of different lineages in culture would also dramatically improve prospects for effective somatic gene therapy.

Assisted Reproduction

Another area of medicine where the knowledge gained from animal work has potential application is in the area of assisted reproduction. Assisted reproduction technologies are already widely used and encompass a variety of parental and biological situations, that is, donor and recipient relationships. In most cases an infertile couple seeks remedy through either artificial insemination or IVF using sperm from either the male or an anonymous donor, an egg from the woman or a donor, and in some cases surrogacy. In those instances where both individuals of a couple are infertile or the prospective father has nonfunctional sperm, one might envision using cloning of one member of the couple's nuclei to produce a child.

Although this constitutes an extension of current clinical practice, aside from the serious, moral, and ethical issues surrounding this approach, there are significant technical and medical causes for caution, some of which were described in the research questions enumerated above. In most situations of assisted reproduction, other than the intentional union of the gametes by IVF techniques, the fertilized egg and initial cells of the early embryo are not otherwise manipulated. In some rare cases, such as preimplantation genetic diagnosis, the embryo is manipulated by the removal of one of the identical cells

of the blastomere to test its genetic status. In contrast, if nuclear transfer were to be used as a reproductive option, it would entail substantially more invasive manipulation. Thus far, the animal cloning of Dolly is a singular success, one seemingly normal animal produced from 277 nuclear transfers. Until the experiment is replicated the efficiency, and even the validity, of the procedure cannot be fully determined. It is likely that the mere act of manipulating a nucleus and transferring it into an egg could decrease the percentage of eggs that go on to develop and implant normally, as well as increase the rate of birth defects.

CLONING AND GENETIC DETERMINISM

The announcement of Dolly sparked widespread speculation about a human child being created using somatic cell nuclear transfer. Much of the perceived fear that greeted this announcement centered on the misperception that a child or many children could be produced who would be identical to an already existing person. This fear reflects an erroneous belief that a person's genes bear a simple relationship to the physical and psychological traits that compose that individual. This belief, that genes alone determine all aspects of an individual, is called *genetic determinism* (37). Although genes play an essential role in the formation of physical and behavioral characteristics, each individual is in fact the result of a complex interaction between his or her genes and the environment within which they develop. As social and biological beings we are creatures of our biological, physical, social, political, historical, and psychological environments. Indeed, the great lesson of modern molecular genetics is the profound complexity of both gene–gene interactions and gene–environment interactions in the determination of whether a specific trait or characteristic is expressed.

While the concept of complete genetic determinism is wrong and overly simplistic, genes do play a major role in determining biological characteristics including a predisposition to certain diseases. Moreover, the existence of families in which many members are affected by these diseases suggest that there is a single gene that is passed down with each generation that causes the disease. When such a disease gene is identified, scientists often say they have "cloned the gene for" breast cancer, for instance, implying a direct cause and effect of gene and disease. Indeed, the recent efforts of the Human Genome Project (HGP) have led to the isolation of a large number of genes that are mutated in specific diseases, such as Duchenne muscular dystrophy, and certain types of breast and colon cancer.

However, recent scientific findings have revealed that a "one-gene, one-disease" approach is far too simplistic. Even in the relatively small list of genes currently associated with a specific disease, knowing the complete DNA sequence of the gene does not allow a scientist to predict if a given person will get the disease. For example, in breast cancer there can be many different changes in the DNA, and for some specific mutations there is a calculated risk of developing the disease, while for other changes the risk is unknown. Even when a specific genetic change is identified that "causes" the disease in some people, others may be found who have the same change but do not get the disease. This is because other factors, either genetic or environmental, are altered that mask or compensate for "the" disease gene. Thus even with the most sophisticated understanding of genes, one cannot determine with certainty what will happen to a given person with a single change in a single gene.

Once again, the reason rigid genetic determinism is false is that genes interact with each other and with the environment in extremely complex ways. For example, the likelihood of developing colon cancer, a disease with a strong hereditary component and for which researchers have identified a single "causative" gene, is also strongly influenced by diet. When one considers a human trait that is determined by multiple genes, the situation becomes even more complex. The number of interactions between genes and environment increases dramatically. In fact the ability to predict what a person will be like knowing only their genes becomes virtually impossible because it is not possible to know how the environment and chance factors will influence the outcome.

Thus the idea that one could make through somatic cell nuclear transfer a team of super athletes or a superior military force is simply false. Knowing the complete genetic makeup of an individual would not tell you what kind of person that individual would become. Even identical twins that grow up together and thus share the same genes and a similar home environment have different likes and dislikes, and can have very different talents. The increasingly sophisticated studies coming out of human genetics research are showing that the better we understand gene function, the less likely it is we will ever be able to produce at will a person with any given complex trait.

CONCLUSIONS

Dolly's birth demonstrates the feasibility of somatic cell nuclear transplant cloning. There are many applications that nuclear transfer cloning might have for biotechnology, livestock production, and new medical approaches. Work with embryonic stem cells and genetic manipulation of early embryos in animal species (including nuclear transfer) is already providing unparalleled insights into fundamental biological processes and promises to provide great practical benefit in terms of improved livestock, improved means of producing pharmaceutical proteins, and prospects for regeneration and repair of human tissues.

However, the possibility of using human cloning for the purposes of creating a new individual entails significant scientific uncertainty and medical risk at this time (2). Potential risks include those known to be associated with the manipulation of nuclei and eggs and those yet unknown, such as the effects of aging, somatic mutation, and improper imprinting. These effects could result in high rates of failed attempts at pregnancy as well as the increased likelihood of developmentally and genetically abnormal embryos.

In addition, cloning in this manner would change the nature of procreation so that an individual would not need a partner of the opposite gender to have offspring (although a woman would be needed to carry the pregnancy). What makes this technique distinctive is that for the first time women and men would not need to share genes to create a child. There are various arguments against this form of procreation based on evolutionary biology and the need for a genetic "lottery."

More troubling are the issues of how cloning might feed the egos of the very wealthy, the very powerful, or the less than honorable. It would also necessitate a new way of thinking about relationships because the child would have only one biological parent. This elimination of one biological parent would pose a unique challenge to our existing concepts of parenthood and the order of generations. Finally, it is not clear whether being created in this manner would create adverse psychological effects, and there is reason to believe that there may be significant adverse physical impacts.

The three types of cloning described raise similar ethical concerns about the emotional well-being of the resulting child, effects on human relationships, the safety of the procedures, and the use of artificial means to produce a child. The motivation for the use of each technique, however, may differ dramatically. For example, blastomere separation for immediate transfer as a treatment for infertility raises fewer concerns than the use of somatic cell nuclear transfer to create a child in one's own image.

ACKNOWLEDGMENT

Some of this article is based on "The Science and Application of Cloning," a chapter in *Cloning Human Beings* (National Bioethics Advisory Commission, 1997) for which the author was a writer and consultant. In turn, much of the Commission's scientific analysis relied on material contained in two commissioned papers provided by Janet Rossant, Samuel Lunenfeld Research Institute, Mount Sinai Hospital, Toronto, and by Stuart Orkin, Children's Hospital and Dana Farber Cancer Institute, Boston.

BIBLIOGRAPHY

1. I. Wilmut et al., *Nature (London)* **385**, 810–813 (1997).

2. National Bioethics Advisory Commission, *Cloning Human Beings: Report and Recommendations of the National Bioethics Advisory Commission*, NBAC, Rockville, MD, 1997.

3. H.W. Jones, R.G. Edwards, and G.E. Seidel, *Fertil. Steril.* **61**, 423–426 (1994).

4. J.L. Hall et al., Paper presented at the 1993 Annual Meeting of the American Fertility Society (1993).

5. R.R. Angell, A.A. Templeton, and I.E. Messinis, *Cytogenet. Cell Genet.* **42**, 1–7 (1986).

6. National Institutes of Health (NIH), *Report of the Human Embryo Research Panel*, NIH, Bethesda, MD, 1994.

7. R. Kolberg, *Science* **262**, 652–653 (1993).

8. R. Voelker, *J. Am. Med. Assoc.* **271**, 331–332 (1994).

9. D. Solter, *Nature (London)* **380**, 24–25 (1996).

10. L. Meng and D.P. Wolf, *Abstr. 51st Annu. Meet. Am. Soc. Reprod. Med.* Program Suppl. S236 (1996).

11. K.H.S. Campbell et al., *Nature (London)* **380**, 64–66 (1996).

12. M. Sims and N.L. First, *Proc. Natl. Acad. Sci. U.S.A.* **91**, 6143–6147 (1994).

13. D.S. Rubenstein et al., *Cambridge Q. Healthcare Ethics* **4**(3), 316–339 (1995).

14. J.H. Tanne, *Br. Med. J.* **319**, 593 (1999).

15. J.B. Gurdon, *J. Embryol. Exp. Morphol.* **10**, 622–640 (1962).

16. R. Briggs and T.J. King, *Proc. Natl. Acad. Sci. U.S.A.* **38**, 455–463 (1952).

17. J.B. Gurdon, R.A. Laskey, and O.R. Reeves, *J. Embryol. Exp. Morphol.* **34**, 93–112 (1975).

18. J. McGrath and D. Solter, *Science* **226**, 1317–1319 (1984).

19. J. Rossant and R.A. Pederson, *Experimental Approaches to Mammalian Embryonic Development*, Cambridge University Press, Cambridge, UK, 1986.

20. S.M. Willadsen, *Nature (London)* **320**, 63–65 (1986).

21. J. Rossant, *The Science of Animal Cloning*, commissioned paper prepared for the National Bioethics Advisory Commission, Rockville, MD, 1997.

22. H.T. Cheong, Y. Takahashi, and H. Kanagawa, *Biol. Reprod.* **48**, 958–963 (1993).

23. P. Collas and F.L. Barnes, *Mol. Reprod. Dev.* **38**, 264–267 (1994).

24. L.C. Smith and I. Wilmut, *Biol. Reprod.* **40**, 1027–1035 (1989).

25. R.S. Prather, M.M. Sims, and N.L. First, *Biol. Reprod.* **41**, 414–418 (1989).

26. X. Yang et al., *Biol. Reprod.* **47**, 636–643 (1992).

27. H.M. Blau et al., *Science* **230**, 758–766 (1985).

28. K.H.S. Campbell et al., *Biol. Reprod.* **50**, 385–1393 (1994).

29. S.L. Stice, C.L. Keefer, and L. Mathews, *Mol. Reprod. Dev.* **38**, 61–68 (1994).

30. P. Braude, V. Bolton, and S. Moore, *Nature (London)* **332**, 459–461 (1988).

31. R.M. Schultz, *BioEssays* **15**, 531–538 (1993).

32. D. Solter, *Annu. Rev. Genet.* **22**, 127–146 (1988).

33. L.L. Mantell and C.W. Greider, *EMBO J.* **13**, 3211–3217 (1994).

34. P.G. Shiels et al., *Nature (London)* **399**, 316–317 (1999).

35. M.J. Weiss and S.H. Orkin, *J. Clin. Invest.* **97**, 591–595 (1995).

36. F.H. Gage, J. Ray, and L.J. Fisher, *Annu. Rev. Neurosci.* **18**, 159–192 (1995).

37. P. Ramsey, *Fabricated Man: The Ethics of Genetic Control*, Yale University Press, New Haven, CT, 1970.

See other CLONING entries; FEDERAL POLICY MAKING FOR BIOTECHNOLOGY, EXECUTIVE BRANCH, NATIONAL BIOETHICS ADVISORY COMMISSION.

CLONING, POLICY ISSUES

ELISA EISEMAN
RAND Science and Technology Policy Institute
Washington, District of Columbia

OUTLINE

INTRODUCTION

The birth of Dolly, the first mammal to be cloned from an adult, has brought into sharp focus the future possibility of cloning human beings. Because of the inherent moral, ethical, and legal implications associated with cloning of human beings, policy responses around the world have been intense. In the United States there has been legislative action taken at the state and federal levels. The U.S. Food and Drug Administration (FDA) has also asserted its authority to regulate the cloning of human beings. Similarly several other nations have enacted laws prohibiting the cloning of human beings, and international organizations have issued policy statements calling for a worldwide ban on the cloning of human beings. While the cloning of Dolly has revolutionized science by proving that it is possible to clone an adult mammal and cloning technology may one day transform medicine by providing improved treatments for diseases, there appears to be broad international agreement that cloning of human beings for reproductive purposes should be prohibited.

BACKGROUND

Cloning, which literally means to make a copy, is the asexual reproduction of a precise genetic copy of a molecule, cell, tissue, plant, or animal. However, scientists use the word "cloning" in many different ways. Molecular cloning refers to the copying of DNA fragments. For example, the human gene for insulin has been cloned into bacteria to produce insulin for the treatment of diabetes. In addition, human cells are routinely cloned to study cancer or genetic diseases. These types of cloning are integral tools in biotechnology, and they have been used to produce breakthrough medicines, diagnostics and vaccines to treat heart attacks, cancer, diabetes, hepatitis, cystic fibrosis, and many other diseases.

The cloning of animals was originally conceived of as a way to understand the genetic processes that regulate development and differentiation. The first attempts to clone animals occurred in 1952 with amphibians and in the 1980s with mammals (1). In these experiments, animals were successfully cloned only when cells from embryos were used, and none were cloned from cells of adult animals. It was not until the birth of Dolly on July 5, 1996, that scientists were able to show that it was possible to use genetic material from a single adult mammalian cell to develop a new individual (2). The ability to clone adult animals has moved the prospects for cloning into areas well beyond basic developmental biology, and has paved the way for major advances in biotechnology, reproductive medicine, and cell-based therapies.

In the last four years, since the cloning of Dolly, there have been several scientific advances. The ability to clone mammals other than sheep from adult cells has been reported for cows and mice (3,4). In addition techniques have been developed to produce cloned animals carrying specific genes, providing an efficient means for producing genetically engineered animals that can make proteins in their milk that could then be used for pharmaceutical or clinical purposes (5,6). The last cow of a rare breed has been cloned in an effort to save the breed from extinction (7), and scientists are preparing to clone other rare and endangered animals (8). Nuclear transfer technology has also been used for human applications, including in preimplantation diagnosis during in vitro fertilization (IVF) (9), to try to treat infertility (10), and in an attempt to produce human embryonic stem cells (11). Many of these advances hold the promise of improved treatments for diseases for which there are currently no good alternatives.

While cloning techniques may one day provide improved treatments for diseases, revolutionize the production of biopharmaceuticals, and save endangered species, mammalian cloning does have its risks. In addition to high rates of spontaneous abortion late in pregnancy and death soon after birth, mammalian cloning has been linked to a developmental defect of the immune system (12) and may be associated with premature aging (13). Thus the question of safety remains, and this casts doubt on the future uses of mammalian cloning, especially for human reproductive purposes. Beyond the safety concerns, the prospect of cloning human beings raises several other ethical, social, and legal concerns. These concerns have been addressed at the state, national, and international level.

POLICY AND LEGISLATION IN THE UNITED STATES

Immediately after the announcement of Dolly's birth, President Clinton asked the National Bioethics Advisory Commission (NBAC) for their recommendations on the use of cloning technology. Soon after, legislation was introduced in the 105th Congress and approximately a dozen states, aimed at prohibiting the cloning of human beings (14). President Clinton also transmitted legislation to Congress that would make it illegal for anyone to create a human being through cloning. More recently two bills have been introduced in the 106th Congress. While there was no law in the United States that directly prohibited creating a child through somatic cell nuclear transfer, there were already a variety of state and federal laws and some existing policies that did apply. Summarized in the discussion below are the NBAC's recommendations as well as enacted and pending legislation at the federal and state levels that both directly and indirectly prohibit the cloning of human beings (Tables 1–4).

National Bioethics Advisory Commission

On February 24, 1997, two days after the news about the birth of Dolly, President Clinton asked NBAC to deliver a report to him within 90 days on the legal and ethical issues involved in the cloning of human beings and "possible federal actions to prevent its abuse." On June 9, 1997, NBAC delivered its recommendations to the President. In their report *Cloning Human Beings*, NBAC agreed that the creation of a child by somatic cell nuclear transfer is scientifically and ethically objectionable at this time. The reasons cited were the following: (*1*) the efficiency of nuclear transfer is so low and the chance of abnormal offspring so high that experimentation of this sort in humans was premature, and (*2*) the cloning of an already existing human being may have a negative impact on issues of personal and social well-being such as family relationships, identity and individuality, religious beliefs, and expectations of sameness (14).

NBAC also recommended a continuation of both the moratorium on the use of federal funding in support of any attempt to create a child by somatic cell nuclear transfer, as well as the voluntary moratorium for the private and nonfederally funded sectors (14). NBAC further recommended that federal legislation be enacted to prohibit anyone from attempting, whether in a research or clinical setting, to create a child through somatic cell nuclear transfer cloning (14). Such legislation should include a sunset clause to ensure that Congress reviews the issue after a specified time period (three to five years) in order to decide whether the prohibition continues to be needed (14). In addition any regulatory or legislative actions undertaken to prohibit the creation of a child by somatic cell nuclear transfer should be carefully written so as not to interfere with other important areas of scientific research, such as the cloning of human DNA sequences and cell lines, neither of which raises the scientific and ethical issues that would arise from an attempt to create a child through somatic cell nuclear transfer (14).

Administration Policy

On March 4, 1997, President Clinton released a statement to the heads of executive departments and agencies prohibiting the use of federal funds for cloning of human beings. Even though restrictions already exist on the use of federal funds for the creation of human embryos for research purposes (see Federal Legislation below), these restrictions do not explicitly cover the creation of human embryos for implantation and do not cover all federal agencies. Therefore President Clinton issued his statement "to make it absolutely clear that no federal funds shall be allocated for cloning of human beings." In addition to the ban on the use of federal funds, President Clinton also asked for a voluntary moratorium by researchers funded by private sources.

Acting on NBAC's key recommendation, President Clinton announced the "Cloning Prohibition Act" of 1997 on June 9, 1997. Consistent with NBAC's recommendations, the President's legislative proposals prohibited the use of somatic cell nuclear transfer to create a human being for five years and directed the NBAC to report to the President in four and a half years on whether to continue the ban. The proposal was carefully worded to ensure that it would not interfere with beneficial biomedical and agricultural activities. This legislation therefore would not prohibit the use of somatic cell nuclear transfer techniques to clone DNA in cells and it would not ban the cloning of animals. To date, this legislation has not been signed into law; however, the ban on federal funding the President declared in March remains in effect. In addition the President called upon the private sector to refrain voluntarily from using this technology to attempt to clone a human being.

The Office of Management and Budget released a Statement of Administrative Policy on February 9, 1998, in response to Senator Lott's Human Cloning Prohibition Act (S.1601, see below). The Statement detailed the Administration's position that it did not support the passage of S.1601 in its current from because it was "too far-reaching" and it would "prohibit important biomedical research aimed at preventing and treating serious and life-threatening diseases." Instead, the Administration offered several amendments to S.1601, including: (1) a five-year sunset on the prohibition on human somatic cell nuclear transfer technology to ensure that there is a continuing examination of the risks and benefits, (2) permitting somatic cell nuclear transfer using human cells for the purpose of developing stem cell technology, (3) striking the bill's criminal penalties and instead making any property derived from or used to commit violations of the Act subject to forfeiture to the United States, and (4) striking the provision establishing a new Commission to Promote a National Dialogue on Bioethics, since it would be duplicative of NBAC's mission. The President's proposal would "prohibit any attempt to create a human being using somatic cell nuclear transfer, provide for further review of the ethical and scientific issues associated with the use of somatic cell nuclear transfer, and protect important biomedical research."

Federal Legislation

In fiscal years 1996 and 1997, Congress passed prohibitions on the use of funds appropriated to the Departments of Labor, Health and Human Services, and Education, and Related Agencies for any research that involves exposing embryos to risk of destruction for nontherapeutic research (P.L. 104-91 and P.L. 104-208). The net effect of these appropriation decisions has been to eliminate virtually all

federal funding for research to perfect methods for cloning human beings, including research aimed at initiating pregnancy, since it would probably involve the destruction of many embryos that failed to develop normally (14). This type of research could, however, proceed uninhibited in the private sector.

More recently, language that directly prohibits the use of federal funds for cloning of humans beings has been included in the appropriations legislation for the Departments of Labor, Health and Human Services, and Education, and Related Agencies in fiscal years 1998, 1999, and 2000 (Table 1). These appropriations continue the human embryo research ban in the public sector by prohibiting the use of federal funds for the creation of a human embryo for research purposes or for research in which a human embryo is destroyed, discarded, or knowingly subjected to risk of injury or death greater than that allowed for research on fetuses in utero. By expanding the definition of a human embryo to "include any organism, not protected as a human subject under 45 CFR 46 as of the date of the enactment of this Act, that is derived by fertilization, parthenogenesis, cloning, or any other means from one or more human gametes or human diploid cells" (P.L. 105-78 and P.L. 105-277), these appropriations also effectively prohibit the use of federal funds for the cloning of human beings.

There are two other federal laws and policies that do not directly prohibit cloning, but may have some applicability. The Fertility Clinic Success Rate and Certification Act of 1992 (42 U.S.C.A. Sec, 263a-1 et seq.) requires that clinics using assisted reproduction techniques, such as IVF be monitored. The Act covers all laboratories and treatments that involve manipulation of human eggs and embryos, and requires that pregnancy success rates be reported to the Department of Health and Human Services (DHHS) for publication in a consumer guide. DHHS is also directed to develop a model program to be implemented by the states for the inspection and certification of laboratories that use human embryos. Since any effort to use cloning to create a child would involve manipulation of human eggs and embryos, these requirements would probably also apply to efforts to clone human beings.

The Federal Policy for the Protection of Human Subjects (also called the "Common Rule") describes the requirements for conducting research on human subjects, such as ensuring that human subjects are not exposed to unreasonably risky experiments and are enrolled in research only after giving informed consent (45 CFR Part 46, Subpart A). The Common Rule, promulgated by 17 federal agencies that conduct, support or otherwise regulate human subjects research, governs research that is conducted with federal funds or is performed at institutions that have executed an assurance with the federal government. (These assurances typically promise that any researcher affiliated with the institution will abide by the federal regulations even if that particular researcher is not using federal funds.) Other human subjects regulations codified at Title 45 Part 46 of the Code of Federal Regulations include additional protections pertaining to research involving fetuses, pregnant women, and human IVF. Enforcement of these protections is

Table 1. Enacted Federal Legislation Prohibiting Cloning of Human Beings

Public Law	Title	Synopsis	Status
P.L. 106-113	Health and Human Services FY2000 Appropriations bill (part of the Omnibus Appropriations Bill FY2000)	Continues the ban on the use of federal research funds for human embryo research. This means that federal funds may not be used for the creation of a human embryo for research purposes or for research in which a human embryo is destroyed, discarded or knowingly subjected to risk of injury or death greater than that allowed for research on fetuses in utero. The definition of a human embryo includes "any organism, not protected as a human subject under 45 CFR 46 as of the date of the enactment of this Act, that is derived by fertilization, parthenogenesis, cloning, or any other means from one or more human gametes or human diploid cells."	Sponsor: Rep Istook Introduced as H.R. 3242 by Rep. Young: 11/17/99. Incorporated into H.R. 3194. Became Public Law 106-113: 11/29/99.
P.L. 105-277	Departments of Labor, Health and Human Services, and Education, and Related Agencies Appropriations Act, 1999 (part of the Omnibus Appropriations Bill FY99)	Continues the ban on the use of federal research funds for human embryo research. This means that federal funds may not be used for the creation of a human embryo for research purposes or for research in which a human embryo is destroyed, discarded or knowingly subjected to risk of injury or death greater than that allowed for research on fetuses in utero. The definition of a human embryo includes "any organism, not protected as a human subject under 45 CFR 46 as of the date of the enactment of this Act, that is derived by fertilization, parthenogenesis, cloning, or any other means from one or more human gametes or human diploid cells."	Sponsor: Rep Wolf Introduced as H.R. 4274 by Rep. Porter: 07/20/98. Incorporated into H.R. 4328. Became Public Law 105-277: 10/21/98.
P.L. 105-78	Departments of Labor, Health and Human Services, and Education, and Related Agencies Appropriations Act, 1998	Continues the ban on the use of federal research funds for human embryo research. This means that federal funds may not be used for the creation of a human embryo for research purposes or for research in which a human embryo is destroyed, discarded or knowingly subjected to risk of injury or death greater than that allowed for research on fetuses in utero. Congress also expanded the definition of a human embryo to "include any organism, not protected as a human subject under 45 CFR 46 as of the date of the enactment of this Act, that is derived by fertilization, parthenogenesis, cloning, or any other means from one or more human gametes or human diploid cells."	Sponsor: Rep. Porter Introduced as H.R. 2264: 07/25/97. Became Public Law: 105-78: 11/13/97.

primarily the responsibility of Institutional Review Boards (IRBs), which review experiments before people can be enrolled. Any effort to use federal funds to clone a human being would raise serious questions about the physical harms that might result, making it difficult for an IRB to approve such research.

Just 11 days after the announcement of Dolly, Representative Ehlers introduced two bills, H.R. 922 and H.R. 923, in the House. H.R. 922 would have prohibited the "expenditure of Federal funds to conduct or support any research on the cloning of humans." H.R. 923 would have made it unlawful for any person to use a human somatic cell for the process of producing a human clone, and set forth a civil money penalty. Several other bills were introduced almost a year later, following Richard Seed's announcement that he intended to clone human beings. Altogether, nine bills prohibiting the cloning of human beings were introduced in the 105th Congress, six in the Senate and three in the House.

While no action was taken on any of these bills before the 105th Congress adjourned, one bill, H.R. 3133, was reintroduced in the 106th Congress by Representative Stearns as H.R. 2326 — the Human Cloning Research Prohibition Act. H.R. 2326 prohibits the use of federal funds to "conduct or support any project of research that includes the use of somatic cell nuclear transfer technology to produce an oocyte that is undergoing cell division toward development of a fetus" (Table 2). To date, one other bill prohibiting federal funding of human cloning has been introduced during the 106th Congress. The Human Cloning Prevention Act of 1999, H.R.571, sponsored by Representative Paul, prohibits "federal payments to any business, institution, or organization that engages in human cloning or human cloning techniques" (Table 2).

While both bills expressly prohibit the use of federal funds for research on the cloning of a human being, neither of these bills set forth penalties, such as fines or prison time, for these actions. It is also interesting to note that H.R. 2326 specifically prohibits the use of human somatic cell nuclear transfer technology, while H.R. 571 does not specify any particular technique for cloning. H.R. 2326 contains additional language that strives not to restrict areas of biomedical and agricultural research that use somatic cell nuclear transfer to clone molecules, DNA, cells, tissues, and nonhuman animals. H.R. 2326 also contains language that requires the Director of the National Science Foundation (NSF) to enter into an agreement with the National Research Council (NRC) to

Table 2. Pending Federal Cloning Legislation in the 106th Congress

Bill	Title	Synopsis	Status
H.R.2326.IH	Human Cloning Research Prohibition Act	Prohibits the expenditure of federal funds to conduct or support research on the cloning of humans, and to express the sense of the Congress that other countries should establish substantially equivalent restrictions.	Sponsor: Rep. Stearns Introduced in the House: 06/23/99
H.R.571.IH	Human Cloning Prevention Act of 1999	Prohibits federal payments to any business, institution, or organization that engages in human cloning or human cloning techniques.	Sponsor: Rep. Paul Introduced in the House: 02/04/99

review the implementation of any legislation prohibiting the cloning of human beings. Finally, H.R. 2326 states that foreign countries should establish similar restrictions set forth in the bill to prohibit the cloning of human beings. None of this additional language is included in H.R. 571.

State Legislation

Even though most states do not have legislation directly regulating assisted reproduction techniques, a number of state laws regarding the management of embryos could restrict even privately funded research aimed at cloning human beings (15). Ten states have laws regulating research and/or experimentation on conceptuses, embryos, fetuses, or unborn children that use broad enough language to include early stage conceptuses: Florida, Louisiana, Maine, Massachusetts, Michigan, Minnesota, North Dakota, New Hampshire, Pennsylvania, and Rhode Island (15).

Five states have enacted legislation that directly prohibits cloning of a human being, California, Louisiana, Missouri, Michigan, and Rhode Island (Table 3). California was the first state to enact legislation prohibiting the cloning of human beings. Within three weeks of the announcement of the cloning of Dolly, California introduced a bill into the Senate (SB1344). The bill was signed into law by the governor on October 4, 1997.

Michigan has enacted four separate bills all prohibiting cloning of human beings. Three bills were introduced in the House (HB5475, HB4846, and HB4962), and one bill was introduced in the Senate (SB864). All three of the House bills needed to be enacted into law for each act to take effect. All four bills passed the Senate by a vote of 37 to zero. In addition all four bills were presented to the governor for signature on May 20, 1998, and signed into law by the governor on June 3, 1998.

Rhode Island and Missouri both introduced bills in January 1998. In Rhode Island, the bill became law without the governor's signature on July 7, 1998 (HB7123). In Missouri, the bill was signed into law by the governor on July 10, 1998 (SB722). Louisiana is the latest state to enacted cloning legislation. Legislation banning the cloning of human beings was introduced into Louisiana's Senate on March 29, 1999, and signed into law by the governor on July 2, 1999 (SB825).

In addition, there are four states that have pending legislation to prohibit the cloning of human beings (Table 4). Massachusetts has three bills pending, New York has five bills pending, and New Jersey and Ohio each have one bill pending. Six states, Arkansas, Connecticut,

Illinois, Oregon, South Carolina, and Virginia, introduced proposed legislation; however, bills in these states are now inactive because of the adjournment of their legislatures.

REGULATION IN THE UNITED STATES

Food and Drug Administration

In response to provocative statements by scientist Richard Seed, who announced January 7, 1998, that he plans to clone human beings, FDA announced that it has the authority to regulate human cloning. FDA asserts that human cloning by somatic cell nuclear transfer requires "more than minimal manipulation" of a cell, and therefore requires approval by the FDA under Section 351 of the *Public Health Service Act* (16). In addition cellular products resulting from "more than minimal manipulation" of cells would require approval for safety and efficacy under provisions in the *Public Health Service Act* that regulate products derived from human materials (16). Acting Commissioner Michael Friedman has affirmed that the FDA will take legal action against anyone who attempts to clone a human being without obtaining prior approval from the FDA (17).

In October 1998 Stuart L. Nightingale, the Associate Commissioner for Health Affairs at FDA, distributed a letter detailing FDA's position on the use of cloning technology to create a human being. The purpose of this letter was to confirm to IRBs that FDA has jurisdiction over clinical research involving cloning of human beings, and to inform IRBs of the FDA regulatory process that is required before any investigator can proceed with such a clinical investigation. The letter states:

> Clinical research using cloning technology to create a human being is subject to FDA regulation under the *Public Health Service Act* and the *Federal Food, Drug, and Cosmetic Act*. Under these statutes and FDA's implementing regulations, before such research may begin, the sponsor of the research is required to submit to FDA an Investigational New Drug Application (IND) describing the proposed research plan; to obtain authorization from a properly constituted and functioning IRB; and to obtain a commitment from the investigators to obtain informed consent from all human subjects of the research. Such research may proceed only when an IND is in effect. Since FDA believes that there are major unresolved safety questions pertaining to the use of cloning technology to create a human being, until those questions are appropriately addressed in the IND, FDA would not permit any such investigation to proceed (17a).

Table 3. Enacted State Cloning Legislation

State	Bill	Synopsis	Status
California	SB1344	Prohibits a person from cloning a human being and from purchasing or selling an ovum, zygote, embryo, or fetus for cloning purposes. Penalties for a corporation not to exceed $1,000,000, penalties for an individual $250,000. Those in violation may lose their license. Provisions will be repealed on 1/1/03 unless a later enacted statue deletes or extends that date.	Sponsor: Johnston Introduced: 3/11/97. Passed Assembly: 9/2/97. Passed Senate: 9/10/97. Signed by the governor: 10/4/97. Filed with Secretary of State: 10/6/97.
Louisiana	SB825	Prohibits human cloning.	Sponsor: Hines Introduced: 3/29/99. Passed Senate: 5/11/99. Passed House; to Senate for concurrence: 6/14/99. Signed by governor: 7/2/99.
Michigan	HB4846	Amends the Public Health Code relating to the practice of a health profession by a licensee, a registrant, or an applicant for licensure or regulation. Prohibits a licensee or registrant or other individual from cloning or attempting to clone a human being. A licensee or registrant or other individual who violates this subsection is subject to a civil penalty of $10,000,000.	Sponsor: Profit Introduced: 1/14/98. Passed House: 1/29/98. Passed Senate by a 37-0 vote: 4/28/98. Bill received House concurrence and with SB864, HB5475, HB4962 was presented to governor for signature: 5/20/98. Signed by the governor: 6/3/98.
Michigan	HB4962	Amends the Penal Code to prohibit an individual from cloning or attempting to clone a human being. Provides felony penalties of not more than 10 years imprisonment or a fine of not more than $5,000.00 or both. Definitions of "clone," "cloning," "human somatic cell nuclear transfer," and "somatic cell" are the same as in HB 4846.	Sponsor: McManus Introduced: 1/14/98. Passed House: 1/29/98. Passed Senate by a 37-0 vote: 4/28/98. Bill received House concurrence and with SB864, HB5475, HB4846 was presented to governor for signature: 5/20/98. Signed by the governor: 6/3/98.
Michigan	HB5475	Prohibits the expenditure of state funds to clone a human being or to conduct or to support research on the cloning of human beings. Definitions are the same as HB 4846 and HB 4962. All three bills must be enacted into law for each act to take effect.	Sponsor: Mans Introduced: 1/14/98. Passed the House: 1/29/98. Passed Senate by a 37-0 vote: 4/28/98. Bill received House concurrence and with SB 864, HB4962, HB4846 was presented to governor for signature: 5/20/98. Signed by the governor: 6/3/98.
Michigan	SB864	Prohibits human cloning for a period of 5 years and provides for civil and criminal penalties.	Sponsor: Bennett Introduced: 2/5/98. Passed Senate by a 37-0 vote: 4/28/98. SB864, HB4846, HB4962 & HB5475 were concurred and presented to governor for signature: 5/22/98. Signed by the governor: 6/3/98.
Missouri	SB722	The Senate bill incorporates HB 1316. Section 17 of the bill states that no state funds shall be used for research with respect to the cloning of a human person. For purposes of this section, the term "cloning" means the replication of a human person by taking a cell with genetic material and cultivating such cell through the egg, embryo, fetal, and newborn stages of development into a new human person.	Sponsor: Sims Introduced: 1/14/98. Passed Senate: 3/4/98. Passed House; to the Senate for concurrence: 5/6/98. Senate concurred in House amendments: 5/8/98. Signed by the governor: 7/10/98.
Rhode Island	HB7123	Prohibits cloning of a human being and purchasing or selling of an ovum, zygote, embryo, or fetus for the purpose of cloning a human being. Provides for civil penalties in the amount not to exceed $1,000,000 for a corporation, etc.; not to exceed $250,000 for an individual for violations of the act.	Sponsor: Cambio Introduced: 1/9/98. Passed House: 4/29/98. Passed on Senate Floor: 6/26/98. House concurred with amendment: 6/29/98. Became law without governor's signature: 7/7/98.

Source: From Pharmaceutical Research and Manufacturers of America (PHRMA), Cloning Legislation and Regulation, http://www.phrma.org/genomics/cloning/legislation.html

Table 4. Pending State Legislation

State	Bill	Synopsis	Status
Massachusetts	HB2455	Prohibits science of cloning.	Sponsor: Fallon Introduced 1/6/99.
Massachusetts	HB2462	Regulates the science of cloning.	Sponsor: Harkins Introduced 1/6/99.
Massachusetts	SB1394	Prohibits human cloning.	Sponsor: Magnani Introduced: 1/6/99. Referred to Senate Committee on Science and Technology: 1/6/99.
New Jersey	AB329	States that a person who knowingly engages or assists, directly or indirectly, in the cloning of a human being is guilty of a crime of the first degree. Defines "cloning of a human being" to mean the replication of a human individual by cultivating a cell with genetic material through the egg, embryo, fetal and newborn stages into a new human individual. Amends the Genetic Privacy Act of 1996 to provide that an individual's genetic information is the property of the individual, and deletes exceptions from current New Jersey law where procedures for obtaining informed written consent already are governed by national standards.	Sponsors: Doria and Gill Introduced: 1/13/98. Referred to Assembly Health Committee: 1/13/98. Carryover to 1999 legislative session.
New York	SB2123	Provides that appropriations and reappropriations to the New York State Advisory Commission on Cloning and Genetic Engineering shall be subject to the provisions which apply to all other legislative commissions; creates the temporary state commission on cloning and genetic engineering to examine, evaluate, and make recommendations to the legislature and governor on the scientific, technical, moral, and ethical issues raised by cloning.	Sponsor: Goodman Introduced: 2/3/99. Referred to Senate Committee on Finance: 2/3/99. Withdrawn from Senate Committee on Finance: 3/18/99. Referred to Senate Committee on Corporations, Authorities, and Commissions: 3/18/99.
New York	SB1954	Enacts Cloning Prohibition and Research Protection Act, prohibits cloning of human beings and provides a $250 civil fine for violation of prohibition.	Sponsor: Goodman Introduced: 2/1/99. Referred to Senate Committee on Health: 2/1/99.
New York	SB1179	Prohibits any person from cloning a human and from purchasing or selling an ovum, zygote, embryo, or fetus for the purpose of cloning a human and establishes civil penalties of up to a specified amount.	Sponsor: Marchi Introduced: 1/15/99. Referred to Senate Committee on Health: 1/15/99.
New York	AB6874	Prohibits human cloning and the use of public funds, resources, property, employees, or those of political subdivisions or public corporations in furtherance thereof; makes violation a felony and grounds for license revocation.	Sponsor: Labriola Introduced: 3/10/99. Referred to Assembly Committee on Health: 3/10/99.
New York	AB3026	Prohibits any person from cloning a human being and from purchasing or selling an ovum, zygote, embryo, or fetus for the purpose of cloning a human being; establishes civil penalties; requires the Commissioner of Health to submit a report to the governor and the legislature on the implications of human cloning.	Sponsor: Connelly Introduced: 1/28/99. Referred to Assembly Committee on Health: 1/28/99.
Ohio	SB102	Prohibits cloning a human being. Makes it unlawful to purchase or sell an ovum, zygote, embryo, or fetus for the purpose of cloning. Creates a "Cloning Enforcement Fund" in the state treasury which would consist of moneys from civil penalties. Civil penalty would not exceed $5,000.	Sponsor: Ray Introduced: 3/11/99. Referred to Senate Committee on Reference: 3/11/99. Senate Committee on Reference recommended referral: 3/16/99. Sent to Senate for second reading; read a second time: 3/16/99. Sent to Senate Committee on Judiciary: 3/16/99. Hearing in Senate Committee on Judiciary: 4/21/99.

Source: From Pharmaceutical Research and Manufacturers of America (PHRMA), Cloning Legislation and Regulation, http://www.phrma.org/genomics/cloning/legislation.html

However, the FDA has not specified which provision of current law grants it such authority (17). There are three possible bases for FDA's assertion of jurisdiction over cloning of human beings: (1) classification as a "drug" under Section 201(g) of the Federal Food, Drug, and Cosmetic Act (FDCA), (2) classification as a "medical device" under Section 201(h) of the FDCA, and (3) classification as a "biological product" under Section 351(a) of the Public Health Service Act (PHSA) (17). If human cloning is covered by any of these statutory provisions, the FDA would have authority to require premarket approval and/or licensing based on reasonable clinical assurance of safety and efficacy (17). However, FDA's authority to regulate cloning of human beings has been questioned, and the matter may require a statutory amendment to expand FDA's authority (17).

CLONING LEGISLATION IN OTHER NATIONS

Due to the transnational characteristics of science, there exists a need for international cooperation regarding the conduct of scientific and medical research. Some of this need may be met through legislation adopted on a country-by-country basis, but some international agreement is probably also needed. NBAC recognized the importance of international cooperation in the effort to prohibit the cloning of human beings, and concluded that "[t]he United States Government should cooperate with other nations and international organizations to enforce any common aspects of their respective policies on the cloning of human beings" (14).

The possibility of cloning human beings has prompted responses from several nations. Several countries already had existing legislation that prohibited the cloning of human beings, including Australia, Austria, Denmark, France, Germany, Norway, Slovakia, Spain, Sweden, Switzerland, and the United Kingdom. Three countries, Israel, Malaysia, and Peru, passed cloning legislation in response to the news of Dolly. In addition Argentina, Belgium, Canada, China, Japan, and South Korea have proposed legislation but have not yet passed laws to prohibit the cloning of human beings. Countries that already have laws or have announced plans to pass laws prohibiting the cloning of human beings are discussed below and in Table 5.

Countries With Existing Cloning Legislation Before Dolly Was Cloned

Even though the announcement of the first cloning of an adult mammal seemed to take everyone by surprise, several countries already had existing legislation that prohibited the cloning of human beings. South Australia and Spain have had laws prohibiting the cloning of human beings since 1988. In Victoria, Australia, and the United Kingdom legislation was drafted and implemented based on reports from their national ethics commissions. In addition the ethics commissions in Australia, France, and the United Kingdom provided their respective governments with further recommendations after the cloning of Dolly was announced. In contrast, Austria,

Norway, Slovakia, and Sweden have laws that only implicitly prohibit cloning of human beings.

Australia. Three Australian states, Victoria, South Australia, and Western Australia, already have existing legislation preventing reproductive cloning. In addition, in October 1997, the New South Wales government issued a discussion paper entitled "Review of the Human Tissue Act 1983." In this paper the Minister for Health of New South Wales announced that the government had introduced a law to ban human cloning and trans-species fertilization involving human gametes or embryos. This ban was developed in response to community concern.

The Commonwealth Minister for Health and Aged Care requested the Australian Health Ethics Committee (18) of the National Health and Medical Research Council (NHMRC) to advise him on the need for possible legislation regarding cloning of human beings. In their report of December 18, 1998, entitled "Scientific, Ethical and Regulatory Considerations Relevant to Cloning of Human Beings," the AHEC advised that:

- A basic distinction should be drawn between the cloning of a *whole* human individual and the copying (also referred to as "cloning") of the component *parts* of a human (such as DNA and cells);
- The cloning of individual human beings is prohibited by State legislation in Victoria, South Australia and Western Australia and is prohibited by NHMRC guidelines; and
- Legislation should be introduced in the remaining States and Territories to regulate human embryo research and to prohibit research on human embryos except as it is permitted in the NHMRC's *Ethical guidelines on assisted reproductive technology* (18).

NHMRC's *Ethical guidelines on assisted reproductive technology* already prohibits experimentation with the intent to produce two or more genetically identical individuals, including development of human embryonic stem cell-lines with the aim of producing clones of individuals. Although infringement of these guidelines is not a legal offense, sanctions usually involve loss of access to NHMRC research funds. These guidelines are regarded as national standards of acceptable practice.

In May 1998, the Australian Academy of Science initiated a project on human cloning to contribute to the public debate in this area. In February 1999, the Academy released its position statement "On Human Cloning" (19). In its statement the Academy distinguishes between "reproductive cloning" to produce a human fetus and "therapeutic cloning" to produce human stem cells, tissues and organs, and bases its recommendations on this distinction (19). The Academy's first recommendation states "that reproductive cloning to produce human fetuses is unethical and unsafe and should be prohibited" (19). The statement goes on to say that "human cells, whether derived from cloning techniques, from embryonic stem cell lines, or from primordial germ cells should not be precluded from use in approved research activities in cellular and developmental biology" (19). Based on its recommendations, the Academy concludes that the

Table 5. Legislation in Other Countries Prohibiting Cloning of Human Beings

Country	Law	Date	Synopsis
Argentina		(proposed)	Intend to deter efforts to clone human beings using somatic cell nuclear transfer.
Victoria, Australia	Infertility Treatment Act	1995	Bans cloning of human beings.
South Australia	Reproductive Technology Act	1988	Bans cloning of human beings.
Western Australia	Human Reproductive Technology Act	1991	Bans cloning of human beings.
Austria	Federal Law on Medically Assisted Procreation	1992	Implicitly prohibits cloning of human beings.
Belgium		(proposed)	Legislation covering medical ethics including cloning is currently being considered by Parliament.
Canada	An Act to amend the Criminal Code (genetic manipulation) (Bill C-247)	(proposed)	Would criminalize human cloning and germ-line genetic alteration without prohibiting beneficial scientific research in genetics.
China	Human Reproductive Technology Bill	(proposed)	Intend to deter efforts to clone human beings using somatic cell nuclear transfer.
Denmark	Scientific Ethical Committee System and the Handling of Biomedical Research Projects (Act No. 503)	1992	Research on cloning (production of genetically identical individuals) is forbidden, as is nuclear substitution.
Denmark	Medically Assisted Procreation in Connection with Medical Treatment, Diagnosis and Research (Act No. 460)	1997	Confirms the Danish Parliament's position of January 25, 1995, that treatment can not be initiated in areas where a research ban already exists under the 1992 Act.
France	Federal Bioethics Legislation (Laws 94-653 and 94-654)	1994	Implicitly prohibits human cloning. Bioethics Committee recommended that the ban should be made more explicit when the bioethics legislation is revised in 1999.
Germany	Federal Embryo Protection Act	1990	The creation of an embryo genetically identical to another embryo, fetus, or any living or dead person is an offense.
Israel	Anti-genetic Intervention Law	1998	Places a five year moratorium on any attempt to clone human beings or create a human being through germ-line gene therapy. Does not prohibit research and development of cloning technologies.
Japan		(proposed)	A Committee of the Council for Science and Technology has proposed a ban on human cloning.
Malaysia		1997	Bans the cloning of human beings.
Norway	Medical Use of Biotechnology (Law No. 56)	1994	Implicitly prohibits embryo cloning.
Peru	General Health Law	1997	Prohibits human cloning.
Slovakia	1994 Health Care Law	1994	Implicitly prohibits embryo cloning.
South Korea		(proposed)	Legislators want to ban all human cloning experiments except those that relate to disease research. A proposal before the National Assembly creates a committee of representatives from government, religious groups, research, and industry.
Spain	Spanish Civil Code-Assisted Reproduction Procedures (Law No. 35/1988)	1988	Explicitly prohibits embryo and oocyte cloning with criminal sanctions.
Sweden	Measures for the Purposes of Research or Treatment in Connection with Fertilized Human Oocytes (Law No. 115)	1991	Implicitly prohibits embryo and oocyte cloning with criminal sanctions.

(continued)

Table 5. *Continued*

Country	Law	Date	Synopsis
Switzerland	Law on Reproductive Medicine in Humans	1990	prohibits interventions on the genetic material of gametes, live embryos, and fetuses; prohibits measures aimed at influencing the sex or inherited characteristics of the unborn child; prohibits use of live embryos, fetuses, and parts thereof for research purposes; and prohibits cloning, creation of chimeras, interspecies hybridization, and extracorporeal procreation.
Switzerland	Amendment of Federal Constitution	1992	Legally binding, implicitly prohibits embryo cloning.
Switzerland	Federal Bill on Medically Assisted Procreation	1996	Proposes criminal sanctions for the artificial creation of genetically identical beings.
United Kingdom	Human Fertilisation and Embryology Act	1990	The nuclear substitution of an embryo, or any cell while it forms part of an embryo is expressly prohibited.

Sources: From Refs. 22 and 24.

1996 NHMRC *Ethical guidelines on assisted reproductive technology* and relevant State legislation should be revised to allow research on therapeutic cloning thereby allowing "Australia to participate fully and capture benefits from recent progress in cloning research" (19).

Denmark. Denmark passed *Act No. 503 on a Scientific Ethical Committee System on the Handling of Biomedical Research Projects* in 1992 (20). The 1992 Act forbids research on cloning and nuclear substitution. Cloning is defined as the production of genetically identical individuals. In 1997, *Act No. 460 on Medically Assisted Procreation in Connection with Medical Treatment, and Research* confirms the Danish Parliament's position of January 25, 1995, that treatment cannot be initiated in areas where a research ban already exists under the 1992 Act (20).

France. The Federal Bioethics Legislation, passed in 1994, basically bans embryo research, only allowing research on human embryos if it does not harm the integrity of the embryo. In May 1997, France's national bioethics committee recommended that the ban on human embryo research be loosened to allow for the use of excess embryos from IVF for the development of embryonic stem cells for fundamental and therapeutic research (21). The bioethics committee qualified its recommendation with multiple safeguards including requiring informed consent from the parents of the embryo, as well as bans on the creation of embryos for research, germline modifications, and cloning of human beings. The bioethics committee recommended that the legislature adopt their recommendations during the scheduled revision of France's bioethics legislation in 1999 (21).

Germany. Germany already has some of the world's most restrictive laws on genetic engineering, applying even to food plants such as tomatoes and soybeans. The *Federal Embryo Protection Act 1990* makes the creation

of an embryo genetically identical to another embryo, fetus, or any living or dead person an offense punishable by up to five years imprisonment or by a fine (20). The Act also prohibits alteration of the genetic information of the human germline, and the creation of chimeras and hybrids. In March 1997, the German Parliament passed a resolution calling for a comprehensive international ban on human cloning.

Spain. Spain's law on *Assisted Reproduction Procedures* (Law No. 35/1988), passed in 1988, explicitly prohibits embryo and oocyte cloning with criminal sanctions (20). It also prohibits the fertilization of a human ovum for any other purpose than human procreation. This legislation is sufficiently broad enough to prohibit both embryo twinning and somatic cell nuclear transfer because it concentrates on the result rather than the technique used (22).

Switzerland. Switzerland's *Law on Reproductive Medicine in Humans* of October 18, 1990, prohibits interventions on the genetic material of gametes, live embryos, and fetuses (23). It likewise prohibits measures aimed at influencing the sex or inherited characteristics of the unborn child. Live embryos, fetuses, and parts thereof may not be used for research purposes. Furthermore the following are prohibited: cloning, the creation of chimeras, interspecies hybridization, and extracorporeal procreation. Switzerland's Federal Constitution is itself a legally binding document that implicitly prohibits embryo cloning (20). In 1996 Switzerland proposed the *Federal Bill on Medically Assisted Procreation* that would explicitly prohibit the artificial creation of genetically identical beings by imposing criminal sanctions (22).

United Kingdom. The 1984 Report of the Committee of Inquiry into Human Fertilisation and Embryology (Warnock Report) was commissioned by the government following the 1978 birth of the first baby conceived through IVF. The Warnock report was the basis for The *Human*

Fertilisation and Embryology Act of 1990. The *Human Fertilisation and Embryology Act* makes provisions to regulate and monitor treatment centers and to ensure that research using human embryos is carried out in a responsible way. This is done by means of a licensing system administered through the Human Fertilisation and Embryology Authority (HFEA). Three areas of activity are covered by the Act: (*1*) any fertility treatment that involves the use of donated eggs or sperm, or embryos created outside the body (IVF) — these are referred to as licensed treatments, (*2*) storage of eggs, sperm, and embryos, and (*3*) research on human embryos (20).

The *Human Fertilisation and Embryology Act* expressly prohibits one type of cloning technique, the nuclear substitution of any cell while it forms part of an embryo (20,24). However, it does not expressly prohibit embryo splitting or nuclear transplantation. Since both of these techniques involve the creation of embryos outside the body, a license is required from the HFEA. In 1997 the HFEA announced a policy not to issue licenses for any procedures involving embryo splitting or nuclear transfer to any IVF practice either in the private or public sector.

In response to the cloning of Dolly, the Human Genetics Advisory Commission (HGAC) and the HFEA decided to hold a consultation exercise on cloning. In their report issued December 1998, entitled "Cloning Issues in Reproduction, Science and Medicine," it was concluded that the *Human Fertilisation and Embryology Act* has been effective in dealing with new developments relating to human cloning, and should be extended to ban all human reproductive cloning regardless of the technique used (20). In addition it was recommended that somatic cell nuclear transfer to create embryonic stem cells should be allowed. However, under the *Human Fertilisation and Embryology Act*, laboratory research is allowed on human embryos less than 14 days old only if it is used for research into the treatment of infertility and congenital diseases, but research cannot be aimed at developing replacement tissue. Therefore, the scientists at HGAC and HFEA advised that the Secretary of State for Health should consider specifying in regulations two further purposes for which the HFEA might issue licenses for research, so that potential benefits can clearly be explored: (*1*) the development of methods of therapy for mitochondrial disease, and (*2*) the development of therapeutic treatments for diseased or damaged tissues or organs (20).

Austria, Norway, Slovakia, and Sweden. In contrast to the countries described above that have laws explicitly prohibiting cloning of human beings, the laws in Austria, Norway, Slovakia, and Sweden are implicit. Austria's Federal Law of 1992 regulating Medically Assisted Procreation implicitly prohibits cloning of human beings by stating that assisted reproductive techniques must use "viable cells" to achieve pregnancy (22). Sweden's Law No. 115, passed on March 14, 1991, implicitly prohibits embryo and oocyte cloning with criminal sanctions (20). Section 2 states that "the purpose of experimentation shall not be to develop methods aimed at causing heritable genetic effects." Norway's *Law No. 56 on the Medical Use of Biotechnology 1994* and Slovakia's 1994 *Health Care Law* also implicitly prohibit embryo cloning (20).

Countries That Passed Cloning Legislation in Response to the News of Dolly

Three countries, Israel, Malaysia, and Peru, passed cloning legislation in response to the news of Dolly. Legislation was passed in both Malaysia and Peru because cloning of human beings was viewed as unnatural. However in Malaysia, cloning of animals for scientific purposes is allowed. In Israel, there is a five-year moratorium on cloning of human beings; however, the law allows cloning for medical purposes if the Health Minister deems that it does not violate human dignity.

Israel. The Israeli Knesset unanimously passed an anti-genetic intervention law on December 29, 1998. The law places a five-year moratorium on any attempt to clone human beings. Germ-line gene therapy is also forbidden. The law does allow genetic intervention for medical purposes, such as cloning a healthy organ for donation. However, specific clinical research proposals would only be allowed to proceed if safety and efficacy could be established, and if the Health Minister deemed them not to violate human dignity. The Health Minister will be responsible for deciding how to supervise such intervention. Violation of the ban is punishable by two years in prison (25).

Interestingly, the law does not specifically say that human genetic intervention is opposed to human dignity. During the ban, the law states that the Supreme Helsinki Committee will act as an advisory committee "to follow developments in medicine, science, and biotechnology in the sphere of genetic experimentation on human beings, and to report annually, advise, and make recommendations to the Health Minister as to how to proceed, to continue as is, or to reformulate the law" (26).

Malaysia. The Malaysian Cabinet has banned the cloning of human beings because it is "against nature" (27). The cloning of human beings was seen as unethical and an interference with God's creation. However, cloning of animals is allowed for scientific purposes.

Peru. Peru was the first Latin American nation to ban human cloning in a new General Health Law passed by Congress on June 12, 1997 (28). The Congress' health committee found that cloning of human beings goes against people's individuality. The law's aim is to avoid "creating unnatural procreation."

Countries with Proposed Legislation to Prohibit Cloning

Argentina, Belgium, Canada, China, Japan, and South Korea have proposed legislation but have not yet passed laws to prohibit the cloning of human beings. Until Canada passes legislation, cloning of human beings is subject to a voluntary moratorium introduced by the Minister of Health in July 1995. Even though China and Japan have not yet passed legislation, the Chinese Minister of Health and the Japanese Education Ministry have stated that they will not provide funding for research on cloning human beings. South Korea's Ministry of Health and Social Welfare has proposed an expansion of existing rules that ban the implantation of genetically engineered human embryos to include a prohibition on human cloning. In Argentina and Belgium, legislation to regulate the cloning of human beings is also being considered.

Canada. In its final report the Royal Commission on New Reproductive Technologies concluded that "certain activities conflict so sharply with the values espoused by Canadians and by this Commission, and are so potentially harmful to the interests of individuals and of society, that they must be prohibited by the federal government under threat of criminal sanction." These activities include human zygote or embryo research related to ectogenesis, cloning, animal or human hybrids, and the transfer of zygotes to another species.

Based on the recommendations of the Royal Commission, Canada proposed a comprehensive national policy on the management of human reproductive and genetic technologies in June 1996. The *Human Reproductive and Genetic Technologies Act* would have prohibited 13 unacceptable uses of new reproductive and genetic technologies, including cloning of human embryos, germ-line genetic alteration, and other practices that commercialize reproduction and are contrary to the principles of human dignity, respect for life, and protection of the vulnerable. However, the legislation died on the order paper in April 1997, leaving all research and experiments in Canada subject to a voluntary moratorium introduced by the Minister of Health in July 1995.

On October 9, 1997, Bill C-247, an Act to amend the Criminal Code by adding a section on genetic manipulation was introduced into the House of Commons as a Private Member's Bill. (Private Members' Public Bills, sponsored by a private Member who is not a Minister of the Crown, are public policy initiatives that affect the entire general public or a portion thereof.) Bill C-247 criminalizes human cloning and germ-line genetic alteration without prohibiting beneficial scientific research in genetics. The bill states that

> No person shall knowingly (a) manipulate an ovum, zygote or embryo for the purpose of producing a zygote or embryo that contains the same genetic information as a living or deceased human being or a zygote, embryo or foetus, or implant in a woman a zygote or embryo so produced; or (b) alter the genetic structure of an ovum, human sperm, zygote or embryo if the altered structure is capable of transmission to a subsequent generation.

Violation of the above prohibitions would be a criminal offense punishable by a fine of up to $500,000, imprisonment for up to ten years, or both. This bill was still being considered by the House of Commons as late as February 1999 (29).

In addition, a group composed of three of Canada's major funding bodies, the Medical Research Council, the Natural Sciences and Engineering Research Council, and the Social Sciences and Humanities Research Council, issued a policy statement entitled the "Tri-Council Policy Statement: Ethical Conduct for Research Involving Humans" in August 1998 (30). This policy statement describes standards and procedures for governing research involving human subjects. Included in the policy statement is a section on "Research Involving Human Gametes, Embryos or Foetuses," Article 9.5 of which states:

> It is not ethically acceptable to undertake research that involves ectogenesis, cloning human beings by any means including somatic cell nuclear transfer, formation of animal/human hybrids, or the transfer of embryos between humans and other species.

While these policies only apply to research funded by these three Councils, application of these policies to privately funded research is being considered.

China. In May 1997 the Chinese Academy of Sciences, China's leading institute of scientific research, banned the cloning of human beings, and called for a committee to set standards for cloning animals (31). In response to strong objections to human cloning by both scientists and the Chinese government, legislation similar to that currently being implemented in Hong Kong will probably soon be passed (32). The new *Human Reproductive Technology Bill* will prohibit the cloning of any human embryo, and specifically outlaw cloning by nuclear transfer (32). In Hong Kong, a statutory monitoring committee has been set up together with an ethics committee to exercise tight control of reproductive technology, and a similar body comprising scientists, ethicists, and government agencies has been strongly advocated in mainland China (32). Meanwhile the Chinese Minister of Health has emphasized the "Policy of the Four Nos" toward research on human cloning, No support, No approval, No license, No acceptance (32).

Japan. In March 1997, the Japanese Ministry of Education, Science, Sports, and Culture announced that it would not provide funding for scientific research on cloning human beings. However, this is an administrative guideline that only applies to state run institutions and carries no penalties for violators (33). In August 1998, the Japanese Science Council, an advisory panel of the Ministry of Education, introduced strict controls on cloning research carried out at universities and national research institutes (34). Regulations restrict the application of techniques, such as somatic cell transfer to nonhuman cells, and all cloning projects have to undergo scrutiny by a committee of experts in ethics, medicine, and law. In addition the Council for Science and Technology, the country's principal science policy body, has proposed a legal ban on human cloning (34). According to media reports in Japan, the government is preparing to submit a bill to parliament based on the Council's recommendations, representing the first legal prohibition of life science research in Japan (33).

South Korea. In response to the December 1998 announcement by Korean scientists of the cloning of a human embryo, politicians are working to expand the 1997 rules adopted by the Ministry of Health and Social Welfare that cover genetic research that bans the implantation of genetically engineered human embryos, but not human cloning (35). Therefore South Korea's Parliament is now considering legislation to ban cloning of human cells except for disease research (35). One proposal before the National Assembly gives the task of reviewing such experiments to a committee of representatives from government, religious groups, research, and industry (35).

Argentina and Belgium. Argentina has proposed legislation that is intended to deter efforts to clone human beings using somatic cell nuclear transfer. In Belgium, legislation covering medical ethics including cloning is currently being considered by parliament (20).

POLICY STATEMENTS AND ETHICAL GUIDELINES OF INTERNATIONAL ORGANIZATIONS

The possibility of cloning human beings has prompted responses from several international organizations. The Council of Europe, the World Health Organization (WHO), UNESCO's International Bioethics Committee (IBC), the Human Genome Organization (HUGO), the European Commission's bioethics advisory panel, and the Denver Summit of Eight have all called for a worldwide ban on the cloning of human beings (Table 6). (In addition to the original G7 leaders from the world's leading industrialized countries, Britain, Canada, France, Germany, Italy, Japan, and the United States, the 1997 Denver Summit of the Eight included Russia.) The policy statements of these international organizations are detailed below.

Council of Europe

On April 4, 1997, 21 countries associated with the Council of Europe signed an international convention, the *Convention for the Protection of Human Rights and Dignity with Regard to the Application of Biology and Medicine: Convention on Human Rights and Biomedicine*, which calls for a ban on human cloning (36). (The countries that signed were Denmark, Estonia, Finland, France, Greece, Iceland, Italy, Latvia, Lithuania, Luxembourg, the Netherlands, Norway, Portugal, Romania, San-Marino, Slovakia, Slovenia, Spain, Sweden, Turkey, and Macedonia; http://www.coe.fr/oviedo/index.htm.) In addition, Article 13 of the *Convention on Human Rights and Biomedicine* prohibits interventions seeking to introduce any modification in the genome of any descendants and therefore, implicitly, forbids cloning of human beings including by use of somatic (nonreproductive) cells (36). The Convention is open for signature to the Council's 40 member countries as well as Australia, Canada, Japan, the United States, and the Holy See, which contributed to the drafting process. This text represents the first binding legal instrument ever drafted on an international scale with a view to safeguarding human dignity and fundamental rights against any improper applications of medicine and biology.

On January 12, 1998, representatives from 19 members of the Council of Europe signed an *Additional Protocol to the Convention on Human Rights and Biomedicine on the Prohibition of Cloning Human Beings* that committed their countries to prohibiting by law "any intervention seeking to create human beings genetically identical to another human being, whether living or dead" (37). The Protocol is limited to a ban on the cloning of human beings by embryo splitting or nuclear transfer. It does not prohibit the cloning of cells and it does not deal with the use of embryonic stem cells. Two European countries did not sign the Protocol. Germany claimed that the measure was weaker than a current German law that forbids all research on human embryos (38).

Table 6. Policy Statements and Ethical Guidelines of International Organizations

Organization	Policy/Guideline	Date	Synopsis
Council of Europe	Additional Protocol to the Convention on Human Rights and Biomedicine on the Prohibition of Cloning Human Beings	January 1998	Prohibits any intervention seeking to create human beings genetically identical to another human being, whether living or dead.
World Health Organization (WHO)	Resolution on Human Cloning (WHA50.37)	1997	Affirmed that the use of cloning for the replication of human individuals is ethically unacceptable and contrary to human integrity and morality.
UNESCO's International Bioethics Committee (IBC)	Universal Declaration on the Human Genome and Human Rights (29 C/Resolution 17)	November 1997	Prohibits practices which are contrary to human dignity, such as reproductive cloning of human beings.
Human Genome Organization (HUGO)	Statement on cloning	March 1999	States that there should be no attempt to produce a genetic "copy" of an existing human being by somatic cell nuclear transfer.
European Commission	Meeting at the Hague	June 1997	Called human cloning ethically unacceptable and should be prohibited by law.
Denver Summit of the Eight	Communique: The Denver Summit of the Eight	June 1997	The heads of state for the United States, Japan, Germany, England, France, Italy, and Canada, endorsed a worldwide ban on human cloning.

The United Kingdom did not sign because of its strong tradition of defending the freedoms of scientific research (38). Initially, the Netherlands also refused to sign the Protocol. However, following a debate in the Lower House of the Dutch Parliament, the Dutch government decided to sign the Protocol with the caveat that the term "human being" be defined as humans who are already born (39). The other countries that signed were Denmark, Estonia, Finland, France, Greece, Iceland, Italy, Latvia, Luxembourg, Moldova, Norway, Portugal, Romania, San Marino, Slovenia, Spain, Sweden, Macedonia, and Turkey.

World Health Organization

On March 11, 1997, Dr. Hiroshi Nakajima, the Director-General of WHO, issued a statement condemning human cloning:

> WHO considers the use of cloning for the replication of human individuals to be ethically unacceptable as it would violate some of the basic principles which govern medically assisted procreation. These include respect for the dignity of the human being and protection of the security of human genetic material (40).

However, other uses of cloning technology, such as animal cloning and the routine cloning of human DNA, genes, and cells, should not be banned (40). These uses of cloning technology hold the promise of advancing biomedical research on the diagnosis and treatment of diseases such as cancer, heart disease and diabetes.

In his statement the Director-General also referred to the guiding principles set forth in 1992 by the scientific group convened by the Special Programme of Research, Development and Research Training in Human Reproduction. The role of this group was to review the technical aspects of medically assisted procreation and related ethical issues. The group upheld "the right of everyone to enjoy the benefits of scientific progress and it applications" and the need "to respect the freedom indispensable for scientific research and creative activity" (40). They also stressed that "there is a universal consensus on the need to prohibit extreme forms of experimentation, such as human cloning, interspecies fertilization, the creation of chimeras and, at present, the alteration of germ-cell genome" (40).

In May 1997, at the meeting in Geneva, the Fiftieth World Health Assembly adopted a resolution affirming that "the use of cloning for the replication of human individuals is ethically unacceptable and contrary to human integrity and morality" (41). The Director-General was asked to clarify the potential applications of cloning procedures in human health and their ethical, scientific and social implications. This resolution was affirmed and upheld in 1998 at the Fifty-first World Health assembly (42).

In October 1998, a small working group of independent and government experts met at WHO headquarters to consider a report containing a first draft of guiding principles and recommendations to WHO and its Member States entitled *Cloning in Human Health* (43). The draft guiding principles were inspired by the basic principles of medical ethics, including beneficence, nonmaleficence, confidentiality, autonomy, equity and access to care for all, and were based on fundamental values such as dignity, human rights, and freedom (43). The draft guiding principles included subjects such as the need for public education on genetic research, the interaction of genes and the environment, the right to retain control over one's genetic material and the information derived from it, and gene therapy (43).

United Nations Economic, Scientific, and Cultural Organization

The *Universal Declaration on the Human Genome and Human Rights* was formulated in December 1996 by the United Nations Economic, Scientific and Cultural Organization (44) International Bioethics Committee (IBC). The Declaration received widespread support and was unanimously adopted on November 11, 1997, by UNESCO's 186 member States. On November 19, 1998, the 86 member countries of the United Nations Commission on Human Rights approved the Declaration, and on December 9, 1998, it was adopted by the United Nations General Assembly.

Article 11 of the Declaration addresses the issue of cloning of human beings. Article 11 states:

> Practices which are contrary to human dignity, such as reproductive cloning of human beings, shall not be permitted. States and competent international organisations are invited to co-operate in identifying such practices and in determining, nationally or internationally, appropriate measures to be taken to ensure that the principles set out in this Declaration are respected (44).

Human Genome Organization

In March 1996, about a year before Dolly was cloned, the International Ethics Committee of HUGO issued the *Statement on the Principled Conduct of Genetic Research* (45). The statement is concerned with research under the Human Genome Project (HGP) and Human Genome Diversity Project (HGDP). In its background principles, the statement refers to the "acceptance and upholding of human dignity and freedom." Cloning of human beings would violate these principles. In addition the cloning of a human being would violate a principle referred to in the statement's preamble that is concerned with the "reduction of human beings to their DNA sequences and attribution of social and other human problems to genetic causes."

In March 1999 the HUGO Ethics Committee issued its *Statement on Cloning* that makes specific recommendations on both animal and human cloning (46). The recommendations on human cloning are subdivided according to the purposes for which the cloning is carried out, reproductive cloning, basic research, and therapeutic cloning (46). The HUGO Ethics Committee makes the following recommendations:

- *Animal cloning.* Animal cloning should be subject to the same principles for animal welfare as other experimentation on animals, and possible consequences on biodiversity should be considered.

- *Reproductive cloning.* There should be no attempt to produce a genetic "copy" of an existing human being by somatic cell nuclear transfer. However, the use of somatic cell nuclear transfer may be supported if it is used to avoid a disease, such as an error in mitochondrial DNA.
- *Basic research.* In both humans and animals, cloning techniques should be supported to investigate a wide variety of scientific questions, including the study of gene expression and the study of aging.
- *Therapeutic cloning.* Research to produce cells and tissues for therapeutic transplants should be supported.

HUGO also states that the creation of human embryos should be considered for certain types of research that may be of widespread benefit to humanity, such as the development of embryonic stem cells (46).

European Commission

In June 1997 at a meeting at the Hague, the European Commission's bioethics advisory panel called human cloning ethically unacceptable and said it should be prohibited by law (47). The bioethics panel also specifically rejected the idea of embryo splitting in order to increase the success rate of IVF. However, the panel did recognize that cloning research might have important therapeutic implication such as in the study of aging and cancer, or the development of stem cells that could be used to repair or regenerate human organs. However, the European Commission must leave legislation against such experiments up to its individual member nations.

Denver Summit of the Eight

The Denver Summit of the Eight concluded their 23rd annual summit calling for specific actions on a host of economic, global, and political issues. The 18-page final communiqué, issued June 22, 1997, included a specific article related to cloning. Article 47 of the communiqué states that the G8 "agree on the need for appropriate domestic measures and close international cooperation to prohibit the use of somatic cell nuclear transfer to create a child" (48).

CONCLUSIONS

The cloning of Dolly has paved the way for major advances in biotechnology, reproductive medicine, and cell-based therapies. Before long, the preservation of genetically important strains and mutants of laboratory and farm animals, the preservation and propagation of rare and endangered species, and the unlimited multiplication of elite animals from selected matings will be routine. Combining cloning technology with transgenic techniques will provide an efficient way to produce animals that can make proteins in their milk that could then be used for pharmaceutical or clinical purposes. By genetically engineering cloned animals to express human proteins (e.g., histocompatibility antigens) on the surface of cells and organs, the risk of immune rejection

in xenotransplantation may be significantly reduced. In addition, cloning technology may lead to the development of customized (e.g., autologous) human embryonic stem cells for use as cell and tissue-based therapies that would not be rejected by the patient's immune system. Many of these advances hold the promise of improved treatments for diseases for which there are currently no good alternatives.

While cloning techniques may one day provide improved treatments for diseases, revolutionize the production of biopharmaceuticals, and save endangered species, mammalian cloning does have its risks. In addition to high rates of spontaneous abortion late in pregnancy and death soon after birth, mammalian cloning has been linked to a developmental defect of the immune system and may be associated with premature aging. Thus, the question of safety remains and casts doubt on the future uses of mammalian cloning.

Beyond the safety concerns, the prospect of cloning human beings raises several other ethical concerns. These concerns have prompted calls for worldwide bans. Consequently language that directly prohibits the use of federal funds for cloning of human beings was included in appropriations legislation that prohibits the use of federal funds for human embryo research. In addition, five states have enacted legislation prohibiting cloning of human beings. FDA has also asserted its authority to regulate the cloning of human beings. Similarly, several other nations and international organizations have also enacted laws or issued policy statements prohibiting the cloning of human beings.

There appears to be broad international agreement that cloning of human beings for reproductive purposes should be prohibited. However, there is less agreement as to whether or not the use of cloning technology to develop novel therapeutic applications should be allowed. Some of the legislation and policies have specifically recognized the potential benefits of the use of cloning technology for therapeutic purposes. However, other policies are very broad and essentially prohibit any use of somatic cell nuclear transfer using human cells.

It is clear that the potential benefits that may be realized through the use of cloning technology are many. However, the potential for cloning a child is an issue that we will be grappling with for a long time to come. Therefore, responsible public policy will need to be crafted in such a way as to prevent the use of cloning technology for purposes for which it is found to be ethically unacceptable, while allowing for beneficial uses that hold so much promise for curing human diseases.

BIBLIOGRAPHY

1. N.L. First and R.S. Prather, *Differentiation (Berlin)* **48**, 1–8 (1991).
2. I. Wilmut et al., *Nature (London)* **385**, 810–813 (1997).
3. Y. Kato et al., *Science* **282**, 2095–2098 (1998).
4. T. Wakayama et al., *Nature (London)* **394**, 369–374 (1998).
5. A.E. Schnieke et al., *Science* **278**, 2130–2133 (1997).
6. A. Fitzgerald, *The Detroit News/Associated Press*, January 21, 1998.

7. R. Weiss, *The Washington Post*, August 20, 1998, p. A2.

8. M. Farley, *Los Angeles Times*, September 11, 1998.

9. P. Cohen, *New Sci.* **158**, 6 (1998).

10. R. Weiss, *The Washington Post*, October 9, 1998, p. A1.

11. N. Wade, *N.Y. Times*, November 12, 1998, p. A1.

12. J.P. Renard et al., *Lancet* **353**, 1489–1491 (1999).

13. P.G. Shiels et al., *Nature (London)* **399**, 316–317 (1999).

14. National Bioethics Advisory Commission (NBAC), *Cloning Human Beings: Report and Recommendations of the National Bioethics Advisory Commissions*, NBAC, Rockville, MD, 1997.

15. L. Andrews, *Cloning Human Beings: Report and Recommendations of the National Bioethics Advisory Commission*, vol. II, NBAC, Rockville, MD, 1997.

16. The Bureau of National Affairs (BNA), *Bur. Natl. Affairs* **3** (1998).

17. E.C. Price, *Harv. J. Law Technol.* **11**, 619–641 (1998).

17a. S.L. Nightingale, Letter from the Food and Drug Administration, Rockville, MD, October 26, 1998.

18. Australian Health Ethics Committee (AHEC) of the National Health and Medical Research Council, *Scientific, Ethical and Regulatory Considerations Relevant to Cloning Human Beings*, AHEC, Canberra, Australia, 1998.

19. Australian Academy of Sciences (AAS), *On Human Cloning: A Position Statement*, AAS, Canberra, Australia, 1999.

20. Human Genetics Advisory Commission (HGAC) and Human Fertilization and Embryology Authority (HFEA), *Cloning Issues in Reproduction, Science and Medicine*, HGAC, United Kingdom, 1998.

21. D. Butler, *Nature (London)* **387**, 218 (1997).

22. B. Knoppers, *Cloning Human Beings: Report and Recommendations of the National Bioethics Advisory Commission*, vol. II, NBAC, Rockville, MD, 1997.

23. International Digest of Health Legislation, *Int. Dig. Health Legislation* **44**, 256–257 (1993).

24. Human Genetics Advisory Commission (HGAC) and Human Fertilisation and Embryology Authority (HFEA), *Cloning Issues in Reproduction, Science and Medicine*, HGAC, United Kingdom, 1998.

25. S. Bashi, *The Associated Press*, December 30, 1998.

26. R.H.B. Fishman, *Lancet* **353**, 218 (1999).

27. Reuter, March 19, 1997.

28. Reuter, June 14, 1997.

29. Debates of the House of Commons of Canada, *Edited Hansard*, No. 179, February 11, 1999. Available at: *http://www.parl.gc.ca/36/1/parlbus/chambus/house/debates/179_1999-02-11/toc179-e.htm*

30. Tri-Council Policy Statement, *Ethical Conduct for Research Involving Humans*, MRC, NSERC, and SSHRC, 1998.

31. Associated Press, May 12, 1997.

32. G.K. Becker, *Eubios J. Asian Int. Bioethics* **7**, 175–178 (1997).

33. J. Lamar, *Br. Med. J.* **319**, 1390 (1999).

34. A. Saegusa, *Nat. Med.* **4**, 993 (1998).

35. M. Baker, *Science* **283**, 16–17 (1999).

36. Council of Europe, *Convention for the Protection of Human Rights and Dignity of the Human Being with Regard to the Application of Biology and Medicine: Convention on Human Rights and Biomedicine*, Council of Europe, Oviedo, Spain, 1997.

37. Council of Europe, *Additional Protocol to the Convention for the Protection of Human Rights and Dignity of the Human Being with Regard to the Application of Biology and Medicine, on the Prohibition of Cloning of Human Beings*, Council of Europe, Paris, 1998.

38. J. Schuman, Associated Press, January 12, 1998.

39. B.I. Gordijn, *Professional Option: The Cloning of Human Beings. The Dutch Debate in an International Context*, Occasional Paper No. 5, Centre for Professional Ethics, University of Central Lancashire, Preston, Lancashire, UK, 1999.

40. World Health Organization (WHO), *WHO Director-General Condemns Human Cloning*, Press Release WHO/20; WHO, Geneva, 1997. Available at: *http://www.who.ch/press/1997who20.html*

41. World Health Organization (WHO), *World Health Assembly States its Position on Cloning in Human Reproduction*, Press Release WHA/9; WHO, Geneva, 1997. Available at: *http://www.who.ch/press/1997/wha9.html*

42. World Health Organization (WHO), *Ethical, Scientific and Social Implications of Cloning in Human Health*, WHA51.10, Agenda Item 20; Fifty-First World Health Assembly, WHO, Geneva, 1998.

43. World Health Organization (WHO), *Cloning in Human Health: Report by the Secretariat*, A52/12, Provisional agenda Item 13; Fifty-Second World Health Assembly, WHO, Geneva, 1999.

44. UNESCO, *Universal Declaration on the Human Genome and Human Rights*, UNESCO, Geneva, 1997.

45. Human Genome Organization (HUGO), *HUGO Statement on the Principled Conduct of Genetics Research: HUGO Ethical, Legal, and Social Issues Committee Report to HUGO Council*, HUGO, 1996, Available at: http://www.gene.ucl.ac.uk/hugo/conduct.htm

46. R. Chadwick, *Ebios J. Asian Int. Bioethics* **9**, 70 (1999).

47. R. Herman, *The Washington Post*, June 10, 1997, p. Z19.

48. Denver Summit of the Eight, *Communique: The Denver Summit of the Eight*, 1997.

See other CLONING entries.

D

DISABILITY AND BIOTECHNOLOGY

Ani B. Satz
Princeton University
Princeton, New Jersey

University of Michigan Law School
Ann Arbor, Michigan

Anita Silvers
San Francisco State University
San Francisco, California

OUTLINE

INTRODUCTION

From the medical point of view, people are disabled when they are less functionally proficient than is commonplace for humans, and when their dysfunction is associated with a biological anomaly. Medicine traditionally has aimed at least to reduce, and preferably to cure, such dysfunction, and thus eventually to eliminate disability. There are four main ways in which biotechnology may be expected to help achieve this goal. First, biotechnology may prevent the inception of people who are biologically anomalous. For instance, technology derived from our increasingly accurate understanding of human biology identifies "at-risk" individuals who may be dissuaded from reproducing when apprised of their liability of having a disabled child. Second, biotechnology may prevent or protect people from being biologically anomalous and thereby becoming disabled. Technologies that immunize against disabling diseases, or that delay the disabling effects of aging, use this strategy. Third, biotechnology may repair anomalies by curing them, as do technologies which restore diseased persons to wellness. Fourth, biotechnology may mitigate a biological anomaly's impact by creating compensatory products, such as insulin for people with diabetes, or prosthetics, such as cochlear implants for people with nerve deafness.

Pervasive as the medical view of disability may be, people with disabilities often interpret their situation differently. They understand their limitations in terms of social rather than personal deficits. This social model of disability transforms the notion of "handicapping condition" from a biological state which disadvantages unfortunate individuals to a state of society which disadvantages an oppressed minority. The social model attributes the dysfunction experienced by people with disabilities primarily to hostile social arrangements. On this social view of disability, people who do not function in species-typical ways often are obstructed by socially constructed barriers. These range from discriminatory practice such as disability-based denial of employment to thoughtlessly inaccessible design such as the installation of steps rather than ramps. Sometimes the absence of adequate support services and health care benefits also is construed as a barrier to the effective functioning of people with certain kinds of impairments (1).

From this viewpoint, reforming social arrangements to achieve equitable opportunity and accessibility is the best route to reducing dysfunction in biologically anomalous people. From this perspective, the preeminent strategy of the medical model—namely altering biologically anomalous people to make them species-typical or normal—unfairly disparages personal traits central to the identity of people with disabilities. Further, by placing a premium on medically altering them so as to bring their modes or levels of functioning into better conformity with species-typical functioning, this medicalized approach to disability can be coercive and can expose the disabled to risky or ineffective medical interventions. To illustrate, Deaf Culture advocates believe that implanting cochlear devices in prelingually deaf children hazards their future by supplanting the proven effectiveness of communication in Sign language with a device whose success is unpredictable. They charge that such risky intervention is impelled by a mistaken idea, namely the unfounded assumption that living as a hearing person necessarily is better than living as deaf.

At least four fears converge to make disabled people suspicious about the promise of biotechnology. In the medical domain, the prospect that biotechnology can normalize disabled people may invite inadequately tested or coercive interventions and expose them to otherwise unacceptable levels of medical risk. In the social domain, this prospect may induce policy makers to promote medical strategies for altering individuals with disabilities over social strategies for accommodating them. In the political domain, biotechnology may diminish the proportion of the population who are disabled and thereby may attenuate the political influence of disabled people and endanger the allocation of benefits and special services they need. The fourth fear, a philosophical one, sees biotechnology as an irresistible instrument for promoting homogenization and thereby reducing the diversity in capabilities that is now a feature of humankind.

Running through all four fears is anxiety about practices which assume that there is a biological mandate for functioning in the normal ways, that is, in the modes and at the levels of performance most common to the species. Indeed, this supposed biological mandate often is invoked to argue that restoring anomalous individuals to species-typical functioning is preeminently desirable, and thereby to procure public resources for doing so (2). In the experience of many people with disabilities, however, medical practice traditionally has been dominated by policies that distend the importance of species-typical functioning and consequently damage the disabled by devaluing their differences and discounting their alternative approaches of functioning. To illustrate, individuals in whom thalidomide caused congenital anomalies of the upper limbs report that, throughout their childhood, medical professionals forced them to wear dysfunctional prosthetic hands and disparaged the superior function they could achieve by using their feet to pick up and manipulate objects (3).

The problem is by no means new. Medicine has compiled a mixed record in treating the disabled. The progress medicine has made in reducing mortality rates from disabling disease and accident swells the numbers of people with disabilities. The historical record also reveals that the medical model of disability prompted the institutionalization of many biologically anomalous people merely because they did not seem normal (4). Medicine also imposed unnecessary suffering through worthless treatments aimed at making biologically anomalous people who functioned capably appear more normal. For example, walking, rather than wheeling, is the most common way humans gain mobility. Consequently it has not been unusual for individuals with walking limitations to be discouraged from atypical, but effective, compensatory modes of functioning and to be subjected to dangerous, ineffective surgery merely to try to make them mobile in the species-typical mode.

Biotechnology thus is threatening because it appears to make the pursuit of biological homogeneity an eminently practicable enterprise. In this regard disability advocates have urged that the obligation to protect the collective interests of the disabled, a minority group whose members definitively do not meet the standard of normality, takes precedence over the obligation to develop biotechnology that relieves some individuals of burdens arising from biological anomalies.

Because identifying as disabled quintessentially means experiencing one's self as an exception among normal people, people with disabilities may not agree that being normal is unquestionably valuable. Nevertheless, in the past, people with disabilities often acceded to medicine's traditional aim of normalizing them. From a disability perspective, molecular genetic medicine may make this goal more problematic because it has the capacity to effectuate a much more thorough program of normalizing than traditional drug and surgical interventions could accomplish. From a disability perspective, genetic technology's potential for thoroughly normalizing the population by reducing the natural variety in the functional capacities of humans increases the urgency of weighing its possible dangers.

DISABILITY AND GENETIC TECHNOLOGY

From its inception, the science of human genetics has aimed at enhancing people's lives. Reducing the incidence or impact of inherited disability, and thereby raising the aggregated level of human achievement, seemed an obvious policy for furthering this aim. While the benefits of such a strategy are apparent, many wrongs have been committed in its pursuit. Camouflaged by the claim that they were merely liberating unfortunate individuals from living out a destiny blighted by biologically predetermined disadvantage, eugenics programs targeted individuals with disabilities.

The excesses of eugenics programs committed to reducing disability make some people with anomalous inherited traits suspicious of any technology that facilitates genetic intervention. They are reluctant to permit some current, and many proposed, applications of genetic technology. Nevertheless, there are other people with disabilities who welcome new genetic technologies and invest in them. They believe that with recent developments in molecular genetic medicine, the science of human inheritance now is positioned to fulfill its promise in regard to alleviating the burdens of disability.

The Question About Disability

At the very least, history illustrates how often medical interventions intended to counter disability by elevating corporeal or cognitive capacity have harmed people who did not meet the desired standard. In this regard contemporary bioethical conversations about distinguishing beneficial from detrimental applications of genetic technology are wise to consider such deleterious interventions into the processes of human inheritance as the Nazi program for euthanizing "defective" Germans (5) or the U.S. practice of sterilizing "defective" Americans, notoriously endorsed by the Supreme Court in *Buck v. Bell* (6). Although many bioethicists see little resemblance between these historical incidents and contemporary applications of human genetics, disability activists believe that this history shifts the burden of proof to advocates of expanded usage of genetic technology.

When we move from the level of phenotype to the level of genotype, this debate translates into questions about the impact of molecular genetic medicine on people with disabilities. Specifically, do practices such as the termination of pregnancies because pre-natal testing is positive for genetic deficits in the fetus, or the alteration of fetal chromosomes by inserting genes needed to preclude genetic deficits, fall squarely within the benefits of therapeutic medicine? Or do they instead display the morally problematic aspects of the destructive eugenics programs of the past?

Relevant Genetic Technologies

Scientific literature refers to four categories of genetic technology with implications for disability: drug treatments and pharmacogenetics, cloning, genetic testing, and somatic-cell gene therapy and germ-cell genetic engineering. Advances in recombinant DNA technology have resulted in the capability for mass production of some gene products that may be used to treat disabling conditions. (Recombinant DNA technology allows the insertion of DNA into a bacterial or other host for duplication.) Insulin, for example, which now can be produced cheaply and with better control of the quality of the product, has clinical applications in the treatment of some forms of diabetes. Drugs are being developed to target specific genes for disabling conditions in order to suppress their expression, although such treatments are not yet readily available. Advances in our understanding of human genetics may also contribute to pharmacogenetics, or the understanding of how genes influence drug treatment for disabling and other conditions. Although cloning is commonly thought of as the production of one living organism from another, the technology also may be used to clone cells that will not develop into an organism. Thus, blood, tissue, and organ replacement in the treatment of disabling conditions may be possible through cloning.

Drug treatment enabled by genetic technology and pharmacogenetics, and the use of cloning to replace body parts both appear simply to be new ways to facilitate traditional therapies. Proteins and other substances, as well as organs or tissues, created in vitro can be used for the same treatment purposes as products obtained from human donors and other animals. The last two categories-genetic testing, and gene therapy and genetic engineering-may differ in kind from nongenetic therapies because they have the potential to eliminate disability altogether. These applications may evoke concerns propelled by recollections of the past eugenics programs that victimized people with disabilities.

REDUCING INHERITED DISABILITY: GENETIC TESTING

Genetic testing may be used to detect single gene, multigene (polygenic), or environmentally influenced (multifactorial) genetic conditions that are associated with disability. (Genetic screening refers to genetic testing programs involving either targeted populations or the general population, or to testing programs used to determine the need for further diagnostic testing.) Testing may be direct or through linkage analysis. The latter involves testing family members and identifying certain normal variations in genetic sequences called polymorphisms that serve as markers, indicating the potential presence of a genetic anomaly. The predictive value of information about disabling conditions generated by genetic testing is limited by variances in gene expression, false negatives and positives, and in the case of linkage analysis specifically, by genetic recombination causing the genetic marker or markers to separate from the disease gene. These technological limitations compound concerns about detecting and preventing disability because they indicate that information generated by genetic testing may be inaccurate or of limited predictive value. Despite these known uncertainties and imprecisions, our aversion to disability is so great that people who receive a positive result for a disabling genetic condition may be stigmatized.

Genetic testing is difficult to define because it is conceivable that many diseases have a genetic component. Further, analysis of some nongenetic material, such as metabolites, may furnish strong indication of a genetically anomalous condition. Tests that examine genes for mutations (DNA diagnostic tests), analyze gene products such as RNA, amino acids, proteins, and their associated enzymes, or identify the structure of chromosomes (cytogenetic diagnostic tests) are commonly viewed as genetic tests. Tests for metabolites such as blood sugar and cholesterol level may be viewed as genetic tests when they are highly indicative of mutations in single genes (7). In the United States, the genetic/nongenetic distinction is important for purposes of reimbursement and privacy protection. Currently reimbursement for commonly recognized genetic tests is limited under government and private insurance schemes. Privacy laws in some states afford special protection for information generated from genetic tests, possibly decreasing access to these tests as employers become more reluctant to provide coverage for them in their insurance benefit plans and to expose themselves to greater litigation risk.

There are several occasions on which genetic testing may arise. Each of these has implications for disability. Genetic testing may be used to diagnose (diagnostic testing) or predict a condition associated with disability in embryos (embryonic screening), fetuses (prenatal testing), or living persons (pre-symptomatic testing), or to predict carriers of such a condition (carrier testing). Disabling genetic conditions detected by these tests include Tay-Sachs, Duchenne's muscular dystrophy, cystic fibrosis, fragile-X syndrome, hemophilia A and Down syndrome. (Down syndrome is a congenital rather than a hereditary condition, though it is detected through chromosomal analysis. Risk may be identified prior to such analysis by maternal serum screening or chemical analysis of fetal gene product or biochemical analysis.) Genetic tests may also be used to identify genetic conditions such as deafness or anchondroplasia (the most common kind of dwarfism), which may or may not be viewed as disabilities.

Prenatal testing detects genetic conditions in fetuses. Studies indicate that prenatal testing that reveals certain genetic condition leads to abortion in most cases, although some expecting parents use the information to prepare to care for and support a disabled child. (In a compilation of international surveys, 73–100 percent of individuals chose to abort their fetuses when Down syndrome was detected, 100 percent made this choice for metabolic disorders such as Tay-Sachs, 95–100 percent for thalassemia, 38–63 percent for sex chromosome abnormalities, and 39–54 percent for sickle cell anemia.) (8). Abortion is considered therapeutic for some genetic anomalies such as Tay-Sachs disease, which causes children to suffer pain and severe deterioration of functioning and not to survive past the age of five. On the other hand, parents with conditions such as deafness or achondroplasia sometimes explicitly wish to select for fetuses with the conditions they possess (9). They may believe, for instance, that they can most effectively parent children who are like themselves. Similar selection is possible through embryo screening, which involves testing embryos for conditions prior to implantation during assisted reproduction.

Diagnostic genetic testing confirms a suspected diagnosis or else eliminates the possibility of a genetic condition associated with disability. This contributes to more accurate identification of illnesses and avoids unnecessary or inappropriate drug treatment. By revealing the underlying genetic causes of some disabling conditions, developing diagnostic genetic tests facilitates research that may mitigate or cure such conditions. Diagnostic testing also identifies individuals in whom the effects of such conditions may be cured or mitigated. Newborn screening programs for PKU and congenital hypothyroidism are conducted throughout the United States for this reason. Similar motivations underlie diagnostic testing of adults. Once diagnosed, individuals with hemochromatosis may be stabilized by phlebotomy, individuals with cystic fibrosis may receive antibiotic and pulmonary therapies, and individuals suffering from the copper build-up caused by Wilson's disease may be placed in remission through treatment with chelating agents or zinc.

Presymptomatic genetic testing discovers individuals' predisposition for genetic conditions associated with disability. A negative result may provide comfort and reassurance, while a positive result could offer time for psychological, financial, and familial preparations for the onset of the condition and, when available, the opportunity to take prophylactic measures to delay or prevent the onset of dysfunction. Here again, testing seems to place disabling conditions within the context of medicine by focusing on prevention, preparation, and cure. Thus, where prophylactic or curative measures for a condition are available, testing for members of at-risk families is strongly recommended.

Carrier testing identifies individuals who do not have genetic impairments but who are carriers of certain genes for such impairments. This form of testing usually is conducted in order to allow for more informed family planning. Carrier screening programs—for example, within the Northern European-American population for cystic fibrosis, the Ashkenazy Jewish-American population for Tay-Sachs, the Mediterranean-American population for thalessemia, and the African-American population for sickle trait—have sometimes been implemented when there is an elevated risk of genetic anomaly. As presymptomatic individuals can be identified by carrier testing for certain late-onset conditions, such as Huntington's disease, the two forms of testing could raise similar ethical issues.

Are there Unique Benefits and Harms?

As a technology, genetic testing is sometimes thought to bestow unique benefits, or else to threaten unique harms, for people with disabilities. This is, in part, because genetic testing may occur within a medical setting, predict disability, generates shared information about disability, and have eugenic implications. Many of the genetic conditions associated with disability are the product of molecular anomalies we think of as diseases, but some are not. Acondroplasia and deafness are deemed disabilities, at least for purposes of protection under disability discrimination law, but they are neither diseases nor illnesses. Nevertheless, testing for all genetic anomalies occurs within a medical setting, either at a primary health center, such as a clinic or hospital, or at a clinical genetics laboratory, subsequent to physician referral. Considering genetic anomalies associated with disability to be diseases introduces possibilities of both benefits and burdens for individuals in whom the anomaly is detected, for their families, and for other people with disabilities. These benefits and burdens are not unique, however; they are imposed by genetic and nongenetic diagnostic technologies alike.

The general benefits associated with detecting present or future disabling conditions that are construed specifically as diseases include prevention, prophylaxis, and treatment. Detection within a medical setting may confer indirect benefits of clinical quality controls, genetic counseling, and physician fiduciary obligations. However, when genetic counseling is conducted in a climate of disability prevention, its neutrality may be so compromised that its benefits become questionable. (Consider the mission of genetic testing centers such as the Murdoch Institute in Victoria, Australia, which advertises, "[o]ur aim is to help every child to be born healthy and with normal abilities.") (10).

Three concerns arise in regard to understanding genetic disability as a medical condition. First, testing may promote the medicalizing of genetic characteristics that are not illnesses or impairments but are considered to be weaknesses or are otherwise thought of as undesirable. For instance, red hair might come to be considered a sign of being diseased because it is associated with an elevated risk of skin cancer. If so, having red hair might disable people from being hired for occupations that require them to work out of doors. Likewise, being irresponsible or aggressive or aloof might be counted as biological impairments requiring medical intervention, rather than as personal problems requiring character building, if genetic testing were to correlate them with genetic anomalies.

Second, individuals may feel obligated to prevent, cure, or treat inherited medical conditions because testing for them is available. They might be coerced into testing even if the conditions are valued, as achondroplasia and deafness are, within certain communities or families, and even if it is only in hostile social environments that individuals with these conditions are dysfunctional. They might, for instance, be thought irresponsible or be refused insurance coverage unless they submit to testing.

Third, opportunities to test may result in social pressure to eliminate mildly disabling genetic conditions that do not occasion significant dysfunction. Further compounding this concern is that testing for some conditions, such as Down syndrome and fragile-X syndrome, leaves the severity of the predicted dysfunction unclear. For example, Williams syndrome occasionally severely limits people who have the condition; however, in many cases people with Williams syndrome are better described as different rather than as dysfunctional.

The moral complexities of genetic testing are illustrated by the conflicting considerations weighed by people with achondroplasia. There is a fatal form of achondroplasia that occurs only when both parents are achondroplastic dwarfs; of the 10 to 15 percent of achondroplastic babies that are born to achondroplastic parents, one-fourth have the fatal, homozygous condition (inheriting the dominant achondroplasia gene from both parents), which is 2 to 4 percent of all achondroplasia births. Should genetic testing for achondroplastic fetuses be allowed? Should it be recommended? Because three-quarters of achondroplastic children are born to two average-size parents who are unprepared for them, fetal testing could result in dramatically fewer dwarf children being born. Nevertheless, such testing benefits dwarf couples who can avoid bringing to term high-risk pregnancies where the child does not survive. Some disability activists argue that the benefit genetic testing bestows on individual parents in a very small percentage of all dwarf births is not worth risking the collective future of the dwarf community. Although some people with short stature support this argument against permitting genetic testing, many reject it (11).

Reductionism and the Genetic Identification of Disability

Some commentators argue that testing for the genetic anomalies that cause physical and mental impairments ignores the fact that hostile social environments may contribute to the performance limitations associated with disability. This concern may be understood as a complaint either about reducing disability to biology or, more specifically, reducing disability to genetics (12). The criticism is leveled against procedures that suggest the source of the individual's limitation lies in herself rather than in the unfavorable way society treats people like her.

Claims of this kind fail to withstand further scrutiny. While genetic testing may identify an impairment as arising from a biological or genetic anomaly, it is a non-sequitur to suggest that doing so ignores how the social environment limits people with disabilities. Whether an individual uses a wheelchair because of accident or illness, genetic or nongenetic, is irrelevant to the concern that society limits the opportunities of such people by denying them access to education, employment, transportation, and both public and private places of commerce and accommodation.

Further, the charge that genetic testing promotes genetic reductionism by picturing people with disabilities as the victims of their genes is problematic. Most genetic tests, especially tests for polygenic or multifactorial conditions, do not predict genetic disease with certainty. They only identify predispositions for disease. Some tests for rare, highly penetrative, autosomal dominant conditions, such as Huntington's disease, are 100 percent predictive, though expression and age of onset, both very important to the definition of disease and the social impact of disability, vary. In this regard individuals with the Huntington's disease gene may have 35 to 50 years of existence free of disability. Our new ability to assess the probability of a currently disease-free individual's future disability raises questions about how

to describe such individuals, as well as how to protect them against discrimination. While information that a person will develop Huntington's disease or has a high risk of familial Alzheimer's disease often suffices to stigmatize the individual, the U.S. Supreme Court appears to have narrowed protection against disability discrimination to the class of people who are presently rather than prospectively disabled (13).

Genetic reductionism generally is understood as the broad concept that genes determine who we are or what we will become (14). Even if a genetic test predicted expression and the onset of dysfunction with 100 percent accuracy and precision, it is wrong to imagine that a person's genetic condition necessarily undermines her self-conception, or that it is so inextricably entwined with her self-identity that it determines who she is and will become. While some individuals may identify strongly with their genetic makeup, the majority do not. Thus, even the ability to predict expression with precision does not entail genetic reductionism. It is true that some impairments may be so severe, or may be so socially stigmatizing, that they leave no room for conceptions of self that ignore the dysfunctional state. In such cases, a predictive test that identifies the genetic cause of the condition also reveals the genetic determinants of the disabled person's identity. However, it is the disabling condition itself, not the disclosure of its source or cause, that determines identity.

A remaining concern is that it is harmful to place disabling genetic conditions in a medical context because to do so invites coercive efforts for prevention or cure (15). Concerns of this nature have arisen with regard to deaf parents who prefer having deaf to hearing offspring. In the rich traditions and culture of the Deaf community, deafness, a characteristic deemed an impairment in the medical community, is viewed as a capability by deaf people (16). Deaf people who undergo genetic testing are much less likely to seek to prevent or treat their own deafness or deafness in their offspring than hearing people. Similarly, prospective parents with sickle cell disease are less likely than those who have not lived with the disease to abort fetuses affected with the same condition (17).

Thus genetic testing generates information that can be valuable to individuals, their offspring, and others in their care for many purposes other than prevention or repair. Genetic information may contribute to financial and psychological preparedness and may offer the comfort of taking control of one's current or future health state. It also often suggests the best means for reducing or avoiding the pain and suffering associated with disease.

Because of the high percentage rate of terminations of pregnancies following detection of genetic anomalies, prenatal testing constitutes the most troubling application of genetic testing. Even in this situation, however, the problem does not lie in the fact that the test identifies a genetic condition. Just as a genetic test may disclose that a fetus is likely to develop myotonic dystrophy, a genetic condition that eventually will limit use of the limbs, a nongenetic ultrasound may reveal a fetus with malformed arms and legs that also will limit use of the limbs. Termination of both pregnancies then arises from the same belief, namely that the child will not have normal use of his appendages.

Predicting Disability

Predicting disabling genetic conditions raises the question of the moral relevance of predictive information about disability. Although one could loosely predict other causes for conditions that occasion impairment, for example, that those who do not wear seat belts or bike helmets may have a greater percentage of brain or spinal chord injury, these are at best correlative generalizations. Nongenetic diagnostics such as ultrasound may predict some physical deformities, genetic and nongenetic, but genetic testing generally is the most effective way of predicting disabling genetic conditions associated with individual genomes or familial gene pools. To illustrate, although achondroplasia can be detected by ultrasound, this technique cannot determine whether a fetus with two achondroplastic parents has inherited the gene from both parents, the fatal "double-dominant" condition occurring in a quarter of such pregnancies, or only inherited one parent's achondroplasic gene, a nonfatal condition.

Although predicting impairments may allow for their treatment or prevention, or preparation for the onset of the disabling condition, it has at least two morally troubling aspects. One is the effect of predictive information upon the autonomy of presymptomatic individuals. The second is the use of predictive information to prevent the birth of people with disabilities.

For presymptomatic individuals, predicting impairments may support autonomous behavior or hinder it. Some people find that predictive information preserves their autonomy by allowing them time to plan and prepare for the onset of a disorder to which they are susceptible. Others may find their autonomy is compromised, especially in a future or dispositional sense, because they must alter their life plans and restrict their choices, so as to prepare for life with a disability. The disparity between these alternative responses to the same news suggests that the impact of predictive testing upon autonomy is determined by the character of the recipient of the news, not by the character of the news itself.

Of course, nongenetic tests may also predict disability. Predictions about the effects of environmental carcinogens, exposure to contagious disease, and suffering from domestic violence may be made with respect to individuals or families. These predictions, like genetic predictions, invoke the same concerns over current and future autonomy of individuals of all ages. It is consequently difficult to establish that genetic testing presents a special threat to the autonomy of people with disabilities.

The most pronounced example of the impact of predictive information upon autonomy is demonstrated by testing children to learn whether they will become disabled. Testing the child and revealing the results to the child precludes the opportunity for that child to decide later, as an adult, whether she desires the predictive information. In this sense, testing children may violate both the child's current autonomy, if the child does not want to be tested, as well as the child's future or dispositional autonomy (18). Dispositional autonomy is reduced if the child is preempted from deciding, once she becomes a competent adult, whether she should have this information. Arguably, the best practice balances the detriments of paternalistic intervention against the benefits of prophylactics or available treatment. In instances where the child would reach the age of majority before the onset of disability, or where there are no measures to prevent or ameliorate impairment brought on by the condition, testing the child could be discouraged. The child could be tested if prophylactics or treatments are available.

Carrier and prenatal testing raise a different set of issues with regard to predicting disability. Although carrier testing might reveal genetic predispositions and raise concerns about autonomy as enumerated above, it is usually employed to detect individuals who carry deleterious genes but who do not exhibit the conditions therewith associated. Prenatal genetic testing reveals information about fetuses, which, regardless of debate about other aspects of their moral status, cannot be autonomous.

Disclosing the Results of Tests for Disability

Some of the concerns about carrier and prenatal testing grow out of fears that positive results will expose the subjects of such tests to various forms of social bias against disability. Genetic testing for disabling conditions may reveal shared information among biological relations. Although this is not a unique feature of genetic testing, as nongenetic information about contagious disease or exposure to carcinogens may be shared, it highlights some reasons for hesitating to share information about the results of tests for disabling conditions. Families may be discouraged from supporting their members who seek testing for fear that information about them could become public or that they could receive information about themselves that they do not wish to know.

The benefits and burdens associated with shared genetic information usually are discussed with respect to overlapping autonomy and privacy considerations. Autonomy in this context pertains to the right of an individual to know or not to know information about herself. This form of autonomy is also described as personal privacy. Informing an individual of her own predicted future disability may violate her privacy (19). It may drain her confidence in the feasibility of her life plans and in doing so may enervate her dispositional autonomy. Current and future autonomy may be constrained by the loss of her social identity as a fully-functioning individual (20). Or, if her identity has been formed by the possibility that she may have inherited a familial condition, learning that she has no great likelihood of doing so may demolish the basis on which she has planned to live her life. On the other hand, avoiding predictions of disability for fear that others will gain access to this news may conflict with one's need, as an autonomous individual, to learn enough about one's future to facilitate self-determination.

Certain features of contemporary techniques for predicting genetic disability bring the values of privacy and autonomy into conflict with the value associated with preventing disability. If an individual believes that being tested herself might reveal the genetic condition of a family member who neither wants to know it nor wants anyone else to know it, she may be obligated to refuse to be tested, so as to respect the privacy and autonomy of another person. On the other hand, if the social value of preventing disability, or of acknowledging it, outweighs the individualistic values of privacy and autonomy, the person may be morally obligated to be tested in order to aid a relative who wishes to have linkage analysis for personal or family planning purposes, and may also be morally compelled to disclose information about other at-risk family members for family planning purposes or for the benefit of future caretakers.

These conflicting values raise complex questions about the significance of identifying or preventing familial disability that is occasioned by genetic anomalies, although the questions are not uniquely provoked by disabling genetic conditions. Detecting tuberculosis in one family member, for example, may benefit or harm other family members. It may result in treatment or prevention of a disabling condition, but it may also result in the individual being quarantined or otherwise isolated, or being denied employment or insurance because of her disability. In fact, one of the Supreme Court's earliest disability discrimination cases involved a person who was fired from her job because of a diagnosis of tuberculosis (21). Similarly, detecting cancer in a member of a family exposed to environmental carcinogens may indicate the likelihood of cancer in other family members. In this instance, the testing that confirms the cause of one family member's illness can expose other members to insurance and employment discrimination. So, while diagnosing a genetically occasioned disabling condition in one member of the family may have deleterious personal or social effects on other family members, once again there is no reason to believe that these problems arise from the nature of the test rather than the typical social responses to disability.

Genetic diagnostics may also serve as a vehicle for paternalistic medical intervention in the detection of disabling conditions shared within families. Physicians sometimes provide unsolicited information to individuals about their health status or that of other family members, believing that is in the best interests of the patient or of society. Physicians may suggest testing to one patient for the benefit of her family members who are also patients, where such a suggestion indicates that some or all of them are at risk for a disabling condition. Similarly, communities may urge testing on their members, as Cypriots do for thalessemia and certain Jewish communities do for Tay-Sachs disease. When recommendations for such testing are made in order to dissuade certain kinds of at-risk individuals from reproducing, or to dissuade individuals from carrying certain kinds of at-risk fetuses to term, the specter of negative eugenics is raised.

Genetic Testing as Negative Eugenics

Genetic testing that is aimed at preventing the existence of certain sorts of people carries the suggestion

of negative eugenics. Negative eugenics programs aim to elevate the level of collective human performance by eliminating underachieving performers. Historically, these programs typically targeted people suspected of having inheritable inferior characteristics. The milder ways of practicing negative eugenics denied desirable employment, immigration, and other opportunities to people in categories associated with certain types of physical or mental limitations (22). The more menacing practices prevented these people from reproducing by prohibiting them from marrying, sterilizing them, or forcing them to terminate pregnancies (23). The most malignant practices euthanized people, both children and adults, whose performance limitations or behavioral infelicities were perceived as burdensome to themselves or injurious to society (24).

Genetic testing has consequences for disability in ways that broadly coincide with the familiar objectives of negative eugenics programs. There is no surprise here, for while genetic testing is not the sole way of facilitating eugenics, the advent of this genetic technology could permit these programs to be executed in a greatly refined way. A history of endorsing programs of this sort is the reason eugenics became identified as a genocidal practice conducted by dominant or strong classes and aimed at eliminating people, such as the disabled, who belonged to inferior or weak classes (25).

In the absence of knowledge about the causes of various impairments, virtually all members of certain disability categories were believed to suffer from limitations that would be inherited by their progeny. For example, because blindness was observed to run in some families, people who could not see were sterilized regardless of whether their blindness resulted from retinitis pigmentosa (a genetic condition) or opthalmia (an infection, sometimes acquired during birth). Genetic testing allows for somewhat more accuracy than was possible previously in identifying who will become disabled or will pass along a disability to future generations.

Nevertheless, genetic testing does not seem to have unique negative consequences for disability. Each of the concerns it evokes has an analogue in problems that previously have been found in the practice of medicine. Further, prenatal testing discriminates on the basis of disability only if its predominant use is to eliminate fetuses that are at risk for disability, and fetuses are accorded the moral status of persons (26). Still, putting genetic technology to this use exposes the entire field to the familiar fearful reactions provoked by negative eugenics. As Jonathan Glover remarks, "What is controversial is to eliminate or prevent ... disability by eliminating or preventing the existence of the person who has the disability. This controversial policy is the basis of screening programs" (27). Glover subsequently suggests that, while negative eugenics is destructive, there are positive eugenic applications of genetic technology that are not so. According to this view, uses of genetic technology that transform people with disabilities by alleviating or eliminating their functional differences are incontrovertibly positive and beneficial.

COMPENSATING FOR INHERITED DISABILITY: GENETIC ENGINEERING

Testing is not the only application of genetic technology that can reduce the proportion of the population that is disabled. Gene transfer technology has the potential to transform at least some persons whose disabilities are occasioned by genetic impairments into individuals who are temporarily or permanently free of disabling biological limitations, thereby promoting their existence. This new capability also promises to reduce the proportion of the population that is disabled, but not to do so through preventing the birth of or euthanizing individuals with genetic diseases or genetic anomalies.

Gene Transfer as Positive Eugenics

Programs that manipulate biological inheritance to promote, rather than prevent, the existence of certain sorts of people sometimes are characterized as positive eugenics. Selective breeding programs implement positive eugenics, as do some kinds of interventions that enhance congenital health or raise the levels of human performance. Unless we subscribe to the view that every person's existence substantively prevents the existence of others who otherwise might have replaced her, so that to engender a stronger or smarter child necessarily eliminates the weaker or duller person who otherwise would have been born in her place, positive eugenics does not collapse into negative eugenics. Neither do programs aimed at facilitating the flourishing of one kind of person necessarily disadvantage or damage other kinds of people. Thus, although positive eugenics programs may aim at genetically transforming people who have current or potential disabilities, such practice is not necessarily a form of disability discrimination (28).

Applying genetic technology to make people with inherited disabling conditions healthier can be a form of positive eugenics. Single gene anomalies appear to be obvious candidates for therapies that apply gene transfer technology. To illustrate, there is a hereditary form of retinoblastoma which causes multiple tumors in both eyes. Our current therapeutic interventions may damage the retinas or require removal of the eyes and thus may result in blindness. Moreover, retinoblastoma patients have an increased susceptibility to develop sarcomas in later life, either as sequallae of therapeutic radiation or chemical interventions, or as another manifestation of their genetic anomaly. In this form of genetic impairment, the Rb gene is missing from the chromosome (29). In principle, repairing the chromosome by adding the Rb gene would heighten the probability of long-term good results.

Achondroplasia, cystic fibrosis, Duchenne's muscular dystrophy, hemophilia A and B, Huntington's disease, and sickle cell disease are among the many other disabling conditions that have been attributed to single gene anomalies. As of this writing, the success of most gene therapy trials remains uncertain. However, there are many cases of genetically produced disability where addressing the underlying genetic anomaly conceivably could be more therapeutic than other approaches to curing or mitigating the disadvantages associated with the condition.

What is the ethical status of therapies aimed at normalizing such anomalies by "fixing" chromosomes or by otherwise changing them in order to mitigate their impact? Glover thinks there is a commonsense answer to this question: "When disorders are caused by the absence of a gene or the presence of a 'wrong' gene, it is attractive to think of inserting or deleting genes in embryos as required. ...The day may come when this sort of gene therapy can be performed without harmful side effects. Choosing between a normal baby and one with a disability will become a genuine possibility" (27, pp. 128–129). In contrast to testing programs, Glover thinks, "One kind of intervention against disability is uncontroversially right. This is any treatment that does not prevent the existence of the person with the disability but aims to alleviate or cure the disability" (27, p. 129).

In a similar vein, LeRoy Walters and Julie Gage Palmer comment: "To people with disabilities that are diagnosable at the prenatal or preimplantation stages of development, the message of selective abortion and selective discard may seem ... threatening. The message may be read as, 'If we ... had known you were coming, we would have terminated your development and attempted to find or create a nondisabled replacement.' ...[G]ene-therapy best accords with the health professions' healing role and with the concern to protect rather than penalize individuals who have disabilities" (30, p. 82).

Thus, at first glance, gene transfer therapy may seem to be the antithesis of a technique for eliminating currently low functioning members of the human collective. Far from doing so, the techniques of molecular genetic medicine promise to increase these individuals' presence in the population by countering genetic anomalies that heretofore have prevented the individuals in whom they are expressed from reproducing. Severe Combined Immune Deficiency (SCID) has prevented individuals with this condition from reproducing because it has killed them before puberty. To give another illustration, males with cystic fibrosis usually are infertile. By mitigating such effects, gene transfer therapies would increase the progeny of individuals with these inheritable conditions.

In addition to enabling them to reproduce, applications of gene transfer technology may also enlarge such individuals' opportunities for social participation by enhancing their capabilities for productive performance, for instance, by reducing or eliminating the physical deterioration characteristic of hemophilia, sickle cell disease, muscular dystrophy, and Huntington's disease. Currently the effectiveness of available therapies for such genetic conditions as cystic fibrosis, hemophilia, and sickle cell disease is limited. Rather than cure genetically anomalous individuals by removing the cause of their impairment, current nongenetic therapies merely slow or otherwise reduce the impact of the impairment.

To give another illustration, osteogenesis imperfecta (OI, or brittle bone disease) currently is treated with the traditional medical techniques for healing broken bones. In one form of OI, collagen molecules are not anomalous but fewer than usual are made, while in another a dysfunctional collagen molecule is produced and incorporated into bone tissue. Current gene transfer research includes attempts to remove bone cells, alter them in vitro, and return them to the body; to apply the mechanism found in the first form of OI to reduce the production of dysfunctional molecules found in the second form; and to introduce functional collagen genes into cells (31).

As these illustrations indicate, existing therapeutic techniques for individuals with disabling genetic conditions often do not free them from their role as patients. In contrast, gene transfer technology has the potential to transform genetically impaired individuals into fully functional ones and thus to preclude the divisive conceptualization of strong and weak classes that drives negative eugenics (32). It therefore appears plausible to assess applications of gene transfer technology as potentially complying with the goals of positive eugenics, namely to promote the existence of certain people by making them more capable. In this scenario, the target group whose capabilities are to be enhanced through genetic transformation consists of those at risk for genetically induced disabilities, people who otherwise are in danger of living lives burdensome to themselves as well as others.

Despite the widely held view among medical professionals that gene transfer therapy holds comparatively little threat and some promise for people with disabilities, the fact that disability activists increasingly have mounted protests against at least some instances of it suggests that this technology is not as unproblematic as the authors cited above suppose. The reasons for the activists' alarm are subtle but worth considering because of the influence their views sometimes command, especially outside the United States. For instance, early in the era of recombinant DNA technology the Council of Europe called for "explicit recognition ... of the right to a genetic inheritance which has not been interfered with" (33). Are there precautions and prohibitions that could be put in place to ensure that curative gene transfers do not abridge any rights people with disabilities have to their genetic heritage?

Prohibiting Germ-Line Alteration of People with Disabilities

Initially the obvious response would seem to be to ban any therapeutic protocol with the potential of affecting the patient's germ-line. Tom Murray proposes that "The most important question in the debate over the ethics of gene therapy is whether gene therapy is ethically distinctive from other forms of medical therapy. ...The ethically distinctive element of gene therapy is only characteristic of germ-line manipulation" (34, p. 484). Some commentators agree with Murray. For instance, Glover and Walters and Palmer agree in holding it is always right to repair a disabling genetic impairment. However, Glover believes that we should refrain from any intervention that might alter the germ line, for fear of sliding into eugenics (27, pp. 134–135), while Walters and Palmer believe that we should encourage interventions that alter the germ-line in order to achieve eugenics. For Walters and Palmer, germ-line repair is obligatory if it advances "the effort to cure and prevent serious disease or premature death

... the noblest of all human undertakings" (30, p. 85), while for Glover it raises controversial questions about the characteristics to be encouraged or discouraged, and incurs the risk of making mistakes about matters fundamental to other people's futures (27, p. 134).

Although, as of this writing, the National Institutes of Health endorses only therapeutic protocols that alter somatic cells (35) and the pharmaceutical industry has directed its efforts at somatic-cell alteration, there is no reason why germ-line engineering might not occur — if not intentionally, then by accident during an administration of somatic-cell therapy. Accidents like this can happen. For example, an early effort to repair defective eye color in fruit flies by inserting the gene for a missing enzyme inadvertently created a heritable repair (36). It also has been shown that when a retrovirus vector with a transgene is administered to fetal lambs, the lambs' offspring can inherit it (37). Somatic-cell therapies for conditions like retinoblastoma and adenosine deaminase deficiency might best be administered at the fetal stage because these conditions seriously compromise neonates, but doing so increases the probability of altering germ cells. Furthermore the success of somatic therapies may have outcomes that make subsequent germ-line intervention attractive to cost-conscious policy makers.

As LeRoy Walters observes, successful somatic-cell therapies eventually may permit many more individuals with dysfunctional biological conditions to live to an age at which they can reproduce (38). However, even in the absence of effective somatic-cell gene therapy for these specific conditions, medical advances in areas such as the mechanical assistance of breathing and the control of respiratory infections already have reduced early (pre-reproductive) mortality in individuals with such conditions as cystic fibrosis and spinal muscular atrophy, and presumably will continue to do so. Thus, whether prompted by the advent of successful somatic-cell therapies or by improvements in nongenetic therapies, such an eventuality will necessitate either greater health care resource expenditures due to the increasingly large number of individuals inheriting these conditions, or else the introduction of germ-line alteration that can transmit therapeutic benefits to succeeding generations. As Walters and Palmer argue, germ-line, but not somatic-cell, alterations could reverse this effect and reduce the incidence of inherited disadvantageous anomalies in the human gene pool (30, p. 81).

Parenthetically it sometimes is argued that the most benign approach consistent with such considerations of efficiency would be to eliminate genetically defective embryos, and to provide sperm or egg donation for parents for whom the probability of defective embryos is too high (39). Parents may quite reasonably prefer transmitting their own advantageous characteristics to their offspring, however, rather than risk the child's inheriting a less gratifying aggregate of traits from one or more donors, as long as they can avoid transmitting unfavorable genetic dispositions to their offspring. While germ-line alterations pose uncertain risk (as well as uncertain benefit) to future children, it is becoming clear that egg donation and egg reception procedures also carry risks. Among other problems, we have not had the opportunity for longitudinal study of the effects on women of the various interventions that implement these procedures. Thus we cannot assume that germ-line alteration automatically puts people at more peril, or puts more people at peril, than reproductive technologies that are used for preventing the transmission of genetic defects by individuals who want to have their own (in some sense) biological children.

Contrary to Murray's supposition, then, if gene transfer technology is ethically distinctive from other forms of medical intervention, the ethical issues it raises apply to both somatic-cell and germ-line applications. One is not inherently more problematic than the other. For several reasons, including the pressure the success of somatic-cell fixes might bring to bear to pursue the permanence of germ-line repair, it is difficult to have confidence that the basic arguments for alteration of somatic-cells to treat genetic anomalies will ultimately be less persuasive in regard to the latter application. One is not inherently more or less morally problematic than the other. So, if there are purposes for which altering genes is morally suspect, the assessment is likely to be equally apt with respect to somatic-cell and germ-line intervention.

Pursuing Species-Typicality through Genetic Transformation

Consequently we may ask whether genetic alteration programs really differ from genetic testing programs in their eventual impact on disability. Both contribute to reducing the percentage of people who have disabilities. The prospect of doing so, by whatever method, provokes disability activists who insist that genetically altering individuals so as to surmount inherited biological flaws is tantamount to denying the moral worth of people with disabilities (40). Some also contend that altering their inherited traits changes who they are by denying them their identity as the progeny of ancestors who carried genes for these traits (41).

Further, they charge, programs that aim at adjusting the molecular genetic conditions occasioning these traits cannot help but promote an oppressively exacting standard of biological function. Such a standard, they say, is impossible for anyone with a corporeal or cognitive impairment to meet, and thus it threatens not only people whose genetic heritage incorporates a heightened potential for being corporeally or cognitively impaired (42), but whoever is currently impaired, whether or not the origin of the impairment lies in their genetic material. It is interesting to note here that concerns about the potential of gene transfer technology to elevate standards of human performance are not voiced exclusively by people with disabilities. In 1982 the U.S. President's Commission for the Study of Ethical Problems in Medicine and Biomedical and Behavioral Research assessed "interventions aimed at enhancing 'normal' people." These were considered to be problematic because "the difficulty of drawing a line suggests the danger of drifting toward attempts to 'perfect' human beings once the door of 'enhancement' is opened" (42, pp. 1–3).

The prospect is of a future in which individuals who come by advantageous biological traits as the result of natural processes will be rare. This is a future where it will be common to acquire genetically engineered biological advantages as a result of social forces that determine the development and allocation of the technology. Heretofore, in our democratically organized competitive society in which vigor, industry, and talent supposedly constitute the primary determinants of success, individuals' natural biological and moral endowments have been regarded as relatively impervious to the influence of social position and thereby as offering an antidote to artificially induced social privilege. Gene transfer technology purportedly makes humans' most fundamental biological characteristics so malleable that biological superiority could be realigned so as to be a product of, rather than a constraint upon, the privilege social rank or power bestows. From a perspective influenced by our growing power to uncouple individuals' destiny from their biological endowment, laboring under a biological disadvantage could become identified with inferior social rank. In other words, as the President's Commission for the Study of Ethical Problems in Medicine and Biomedical and Behavioral Research warned in 1982, genetic engineering could transform the "natural lottery" into a new kind of "social lottery," (42) one in which being left with an unrepaired genetic disability (and therapy lacking species-typicality) is the sign of a social loser.

Despite the confidence in the positive effects of gene transfer therapy expressed by such commentators as Glover, and Walters and Palmer, alleviating disability by altering genes continues to be controversial and to provoke concerned response from representatives of the disability community. Such interventions are seen as being motivated by the narrow and coercive standard of species-typicality. They thus are condemned as instruments of a social policy that promotes biological homogeneity to the detriment of people with disabilities. For instance, representatives of disabled people's organizations in 27 countries signed a statement "demanding an end to the bio-medical elimination of diversity, to gene selection based on market forces and to the setting of norms and standards by non-disabled people." The question then comes down to whether molecular genetic medicine must be governed by the goal of promoting species-typicality (43).

Norman Daniels's influential discussions of health care policy exemplify the view of medicine that is objectionable from a disability perspective. Daniels is persuaded that achieving common or normal functioning for all citizens is a moral imperative for medicine. Further, providing the resources to transform people with anomalous biological conditions to species-typicality is a civic obligation grounded in the principles of democratic morality (44). Daniels's interpretation of these principles entails that normality is a preeminent social value and thus that making everyone species-typical is a desirable social policy. In urging that democratic theory requires health care policy designed to ensure that all citizens exhibit baseline physical and mental normality, Daniels intends to promote interventions that improve particular people's lives by adjusting their health. The proposed trigger for such intervention is an individual's deviation from biological normality.

How clear is this account of health, and how fair is it as a standard for authorizing medically induced genetic alterations? To be in a healthy state is assumed by Daniels and many others to equate with being typical of one's group or species and, thereby, with being normal. Further, the functional organization typical of the human species is imagined to be the one best suited to meet our biological goals, so that to fall away from this standard is to be definitively disadvantaged in meeting these goals (42). These ideas lead to designing social arrangements to accommodate normal, not extraordinary, people. These or similar valorizing assumptions about the biological superiority of species-typicality are brought to bear to establish the moral importance of medically transforming biologically anomalous individuals into species-typical ones.

The assumptions promoting species-typicality as the goal of medicine are neither perspicacious nor unbiased. First, as to their clarity, discussions that take species-typicality to be a standard for medical intervention appear to conflate criteria that reflect three different levels of outcomes: standardizing biological states, engendering familiar modes of performing functions, and bringing about typical levels of functional outcome. Thus, it is not clear whether the goal of medical intervention is to standardize human genetic configurations, to ensure that people all can execute functions in the familiar ways, or to enable genetically diverse individuals to exhibit common levels of functional outcomes.

Second, because normal performance is defined with reference to familiar modes and levels of performance, what is thought of as normal often is artificially skewed by patterns of social domination that favor some kinds of performances over others and consequently ensure that those will continue to be the most common and seemingly normal ones. For instance, the domination of individuals for whom text is the most efficient conveyer of information has led to social arrangements that presume the ability to read text is normal. This bias causes people who perform the activities of reading and writing texts badly to be disadvantaged, even when they excel at aural, haptic, or pictorial communication. When this bias is medicalized, energy and expenditures are applied to developing therapies that can alter such individuals so that they can perform as normal readers and writers, whereas it may be more efficient for them to communicate in alternative ways. Similarly, in treating children exposed prenatally to thalidomide, medical professionals biased toward manipulating objects with upper rather than lower extremities insisted, to the disadvantage of their patients, that mechanical hands necessarily are better than fleshly feet for manipulating objects.

To suppose that anomalous performance must be functionally inferior to species-typical levels and modes of performance is to make two mistakes about human biological functioning. It is, first, to assume that human biological organization is functionally rigid, when it is instead immensely adaptive. Second, it is to assume that human social organization also is functionally rigid, when it can be flexible in expanding the opportunities of different kinds of people.

One way of grasping how the drive to normalize can lead to misperceptions about dysfunction is to note that acknowledging or ignoring differences in social context affects whether a genetic anomaly is counted as a genetic disease. To illustrate, it is well known that mild mental retardation does not disable women in low technology environments or simple societies in which a woman's role is to clean, cook, and bear children. Writing about equitable health care resource allocation in the very influential *The Global Burden of Disease*, Christopher Murray insists that, nevertheless, it would be inequitable to allocate resources to rich societies to prevent mental retardation but not to poor ones. Consequently, he thinks, we must assess the burden of mental retardation uniformly from nation to nation, regardless of significant national differences in how mental retardation affects people's lives. Murray acknowledges that the same impairment may be differentially disabling, or not at all disabling, depending on the environment. In fact, "in many cases," Murray says, "allocating resources to avert disability could exacerbate inequalities." Regardless of the realities of functioning in different social environments, he thinks, egalitarianism demands uniformity in assessing the burden of an impairment lest differences in context suggest that exalted functioning is reserved for the most privileged (45).

Contrary to Murray, however, there is no inequity in refraining from intervening where intervention has no benefit. What is inequitable is to treat people as dysfunctional when they are not, and on that biased basis to impose medical treatment that alters them. To the extent that medical perspectives assume species-typicality to be the goal, people with disabilities will be concerned about biased medical use of gene transfer technology.

So it is far from clear that we should build a dominating standard of normalization into the policy that governs the development of medical interventions. What is evident, however, are the dangers attendant upon policies that fail to disentangle natural functional disadvantage from artificial social disadvantage. Public policy should not privilege, and so fix and fortify, common modes of functioning by favoring medical practice that privileges the modes of functioning typical of our species over anomalous but comparably efficient modes. The prevalence of this kind of mistake in discussions about the deployment of medical technology is an understandable cause of wariness among people with genetic disabilities.

They have some reason to fear being subjected to genetic homogenization so as to relieve society from the responsibility of adjusting to their differences, even where their differences do not make them especially ill or dependent. Historically, accepting the dominant class's fashion of functioning as the biologically preferable mode has been a source of negative eugenics. To illustrate, Herbert Spencer, the social-Darwinist sociologist whose writing promoted eugenics, wrote about a group he labeled as "weak" because its members' "defective" biology prevented their functioning in the fashion of the dominant group. Rescuing them from their "natural" subservience would result in "a puny, enfeebled and sickly race," he warned in his repeated explanations of how the biology of the female human made her inferior to human males. Translated into social policy, this perspective on women's differences resulted in an array of discriminatory exclusions attributed to "women's disabilities" (46). Imagine what mischief could have been done had it been possible, a century ago, to treat such disabilities by altering women so that their biological makeup became more like that of men.

Compensation, Enhancement, and Functional Diversity

This illustration suggests that distinguishing benign from baleful genetic alterations requires avoiding interventions made primarily for the purpose of homogenization. In this regard molecular genetic medicine must avoid confusing difference with dysfunction. In other words, interventions that alter genes should be governed not by the aim of imposing species-typicality but instead by the goal of enhancing functionality by whatever strategy is most effective.

The outline of this comprehensively flexible approach to disability was presented by the World Health Organization in the recently released beta revision of the International Categorization of Impairments, Disabilities, and Handicaps. For policy purposes, this document attempts to integrate the medical and social interpretations of disability. Here dysfunction is understood to emerge from a mismatch between the individual's mode and level of biological performance and the demands of the environment. No single strategy is made central to the medical interventions recommended for disability (47). The strategy of repair, which restores dysfunctional individuals to species-typical mode and level of functioning, is acknowledged to be one, but not the only, medical approach to disability. Compensatory strategies are equally recommended. A compensatory strategy differs from a reparative strategy in that effective functioning may be achieved without seeking or accomplishing restoration of species-typicality. In compensatory efforts, anomalies remain, but these need not compromise functional success.

Thus, for example, individuals who experienced successful postpolio rehabilitation often developed an exquisite sense of balance to compensate for one of the disease's typical sequellae, namely the marked disparity of strength between right and left sides, and upper and lower parts of the body. Similarly non-oral communicators often surpass most others in their ability to express themselves in body-language (48). Applications of biotechnology can be similarly compensatory. Gene transfer that boosts the low density lipoprotein receptor above normal range compensates for the effects of familial hypercholesterolemia, and gene transfer that induces capillary formation compensates for the effects of arterial blockage.

Whether all of an individual's modes of functioning are species-typical is not decisive for the individual's capabilities, for biological anomalies are not inherently dysfunctional. While restoring patients to genetic normality once seemed to be the obvious medical goal, responding to anomalous genetic configurations by enhancing the functionality of alternative modes of performance (to compensate for a deficit in another aspect of the individual's functioning) should be recognized as an equally promising strategy. Already of importance at the level of rehabilitative medicine, compensatory interventions appear to be similarly apt at the level of molecular genetic medicine.

There appears to be no medical basis for thinking that gene transfer applications designed to restore individuals to species-typical biological condition have a natural priority over compensatory interventions, that is, over gene transfer applications that improve the level of one biological performance to compensate for dysfunctional deficit in another (49). Further, although disability advocates sometimes fear that engineering their genes would threaten disabled people's identities (41), compensatory biotechnology no more does so than wheelchairs, sign language, or talking computers. All of these enhance the functionality of one kind of performance (arm movement, gesturing, listening) to compensate for limitations in another kind of performance (leg movement, speaking and hearing, seeing). It thus appears that compensatory genetic alterations are compatible with both medical and disability perspectives. However, a question remains about the ethics of using genetic technology to enhance human functioning.

It sometimes is argued that enhancing any biological function above its species-typical level, regardless of whether such an intervention is compensatory, brings about differences that are dangerous. In its influential analysis of the moral and social dimensions of "biologists' newly gained ability to manipulate...the material that is responsible for the different forms of life" the President's Commission for the Study of Ethical Problems in Medicine and Biomedical and Behavioral Research opposed "interventions aimed at enhancing" because these permit people to escape the limits imposed on them by the "natural" lottery (42).

Of course, no one appears prepared to abandon all attempts to apply medical technology to improve on nature. Immunization programs are a familiar example of a medical strategy that enhances our physical performance beyond the level that is native to our species. Immunization compensates for the fact that certain organisms or events can initiate damaging biological processes in the body by enhancing other biological processes so that they become capable of defending against such damage. Although now considered to be unexceptional and to provide the protection deserved by fragile populations like the elderly and the very young, vaccination originally was denounced as unnatural, immoral, and irreligious because it boosted individuals' resistance to disease above the level that then was typical of the species.

Not only have enhancements such as immunization become standard interventions for traditional medicine, but it has become unexceptional for gene therapeutic protocols to attempt to raise performance of one function above the common level to compensate for the adverse impact of other performances. For instance, research into genetic analogues of vaccination abounds (50). To give another illustration, one approved gene transfer protocol enhances the low density lipoprotein receptor to above normal to achieve an acceptably low level of cholesterol in the context of the genetic condition hypercholesterolemia.

This protocol is neither less natural nor more threatening than the SCID protocol that restores a species-typical genetic sequence in patients whose biological inheritance omitted it (51).

Ironically, refraining from interventions that enhance performance at the molecular level may further disadvantage those who have not been favored by the natural lottery. Promoting the primacy of natural limitations over biotechnically engineered enhancements is not necessarily fair to the disabled. To illustrate, at one time lower leg amputees were deemed noncompetitive because their prostheses made them run too slowly. Now, however, new materials and designs have created specially springy sports prostheses that permit their wearers, when very skilled and talented, to run faster than can be done with fleshly feet. So now, using these corrective devices is banned in competitive running to prevent unfairly disadvantaging nondisabled runners in the competition. Of course, some shapes and strengths of fleshly feet are better for running than others; yet athletes who have the best kinds of feet are not banned for disadvantaging the common-footed runner. Arguably, it is unfair to exclude runners when prosthetics render them uncompetitive, and also to exclude them when better prosthetics make them very competitive (52).

Similar issues conceivably can arise in molecular genetic medicine. To illustrate, let us imagine a gene transfer intervention that has been developed to decrease the deformation of erythrocytes in sickle cell anemia and thalessemia, but that also optimizes the hemoglobin in nondeformed cells. As Tom Murray has noted in respect to a similar kind of intervention, the procedure also might be effective when there is no diminution of species-typical functioning. The procedure could serve as a genetic version of blood doping, which temporarily increases the oxygen-carrying capacity of the blood. According to Murray, both blood-doping through infusion and its genetic analogues are ethically suspect in competitive sports (34). Does this consideration generalize so as to cast suspicion on interventions that enable individuals to perform better (on the molecular level) than is typical of the species in one respect, in order to compensate for their deficient performance in another respect?

Let us pursue this question in the context of the gene transfer technique described above. For people with certain genetic conditions, the intervention results in fewer erythrocytes deforming and in the hemoglobin being optimized in some cells that do not. Although more cells are species-typical than in the untreated patient, deformed cells remain, so the patient has not been restored to species-typicality. The erythrocytes have not been restored to species-typical performance because some of them perform below the species-typical level, while others perform above the common level.

Such a procedure is compensatory rather than reparative because it does not restore patients to a species-typical state. Whatever level of functionality is attained will result from the enhanced performance of some cells compensating for the limited performance of others. In other words, the performance of these cells will not be species-typical. Some patients will remain in deficit with respect to the oxygen present in their bodies, although on balance they will be better off, but for others the aggregated oxygen-carrying capacity of their blood will be elevated above that of the average person. Surely, however, the fact that some patients benefit from the intervention by attaining superior rather than species-typical functioning in no way makes it ethically suspect. Indeed, the same elevation of performance that might permit an athlete who is erthyrocytically improved to triumph in a game might give an otherwise biologically typical individual the stamina to save another person's life, or an otherwise biologically compromised individual the energy to be productive in the service of others. In sum, whether prosthetic or genetic technology is in question, intervening medically to enhance human capabilities can be the morally appropriate thing to do.

CONCLUSION

Although bioethicists like Glover and Walters and Palmer assume that genetic interventions that alleviate or cure a disability are uncontroversially good, altering individuals for this purpose can be threatening from a disability perspective. The reason, however, is not that altering genes inherently endangers the disabled but that bioethicists and policy makers seem to focus on reparative applications that promote the standard of species-typical functioning. In the past, this standard has inspired practices oppressive to people with disabilities who, by definition, cannot meet it.

Reparative medical strategies aim to restore people to normal modes and levels of functioning. However, there seem to be no clinical reasons for making such an approach the preeminent strategy in molecular genetic medicine. In contrast to reparative interventions, compensatory medical strategies promote exceptional modes and levels of functioning to secure capabilities that otherwise would be in deficit because of an individual's biological anomalies. Thus the adoption of compensatory strategies in gene transfer interventions counters the homogenization that focusing on reparative genetic alteration tends to bring about. Indeed, as long as reparative approaches do not dominate compensatory ones, molecular genetic medicine is as capable of promoting as it is of preventing the existence of people with genetic anomalies.

From the medical perspective on disability, it is tempting to imagine, as Glover does, that the paradigm for genetic engineering is a reparative process in which people are made species-typical by inserting good genes or deleting bad ones. However, medicine itself is beginning to acknowledge the importance of compensatory medical strategies. The recent revision of the World Health Organization's International Categorization of Impairment, Disability, and Handicap exhibits the compatibility of compensatory medical strategies with disability perspectives. Further, compensatory approaches already constitute a familiar strategy in the development of gene transfer research.

Some bioethicists and medical policy makers discourage compensatory genetic interventions because these may promote exceptional rather than species-typical functioning and may enhance performance above the common level. Their objections distract us from the question most important from a disability perspective, namely whether applications of molecular genetic medicine must be governed by a standard of normality that is inherently dangerous to people with disabilities. As we have seen, however, compensatory genetic alteration can improve disabled people's functioning without normalizing them. Thus gene transfer technology need not promote nor impose species-typicality, and consequently it is not inherently threatening to people with disabilities.

BIBLIOGRAPHY

1. A. Silvers, D. Wasserman, and M. Mahowald, *Disability, Difference, Discrimination: Perspectives on Justice in Bioethics and Public Policy*, Rowman & Littlefield, Lanham, MD, 1998.

2. N. Daniels, in D. Van DeVeer and T. Regan, eds., *Health Care Ethics: An Introduction*, Temple University Press, Philadelphia, PA, 1987.

3. E-mails from Bill Baughn, Theresia Degener, and Gregor Wolbring, January 11, 2000, January 12, 2000, and January 18, 2000//www.egroups.com/subscribe/Bioethics

4. J. Trent, *Inventing the Feeble Mind: A History of Mental Retardation in the United States*, University of California Press, Berkeley, CA, 1994.

5. M. Burleigh, *Death and Deliverance: "Euthanasia" in Germany c. 1900–1945*, Cambridge University Press, Cambridge, England, 1994.

   T. Duster, *Backdoor to Eugenics*, Routledge, New York, 1990, pp. 29–31.

   H.G. Gallagher, *By Trust Betrayed: Patients, Physicians, and the License to Kill in the Third Reich*, Holt, New York, 1990.

6. *Buck v. Bell*, 274 U.S. 200 (1927).

7. N.A. Holtzman and M.S. Watson, eds., *Promoting Safe and Effective Genetic Testing in the United States: Final Report of the Task Force on Genetic Testing*, Johns Hopkins University Press, Baltimore, MD, 1998, p. 6.

8. D.C. Wertz and J.C. Fletcher, *Clinical Obstetrics Gynecol.* **36**, 554 (1993).

9. D.S. Davis, *Hastings Center Rep.* **27**(2) (1997).

   A. Silvers, Hastings Center Rep. **27**(5) (1997).

10. Murdoch Institute, Victoria, Australia, Annual Report, 1997.

11. R. Ricker, Web site of Little People of America, Inc. 1995.

12. H.S. Reinders, *Bioethics Forum*, 3–10 (1996).

13. *Sutton v. United Airlines*, Inc., 119 S.Ct. 2139 (1999).

14. D. Nelkin and M.S. Lindee, *The DNA Mystique*, Freeman, New York, 1995.

15. C. Newell, in G. O'Sullivan, E. Sharman, and S. Short, eds., *Goodbye Normal Gene*, Pluto Press, Lane Cove, 1999, pp. 59–74.

16. R.A. Crouch, *Hastings Center Rep.* **4**, 14–21 (1997).

17. D. Hoar, *Biomedical Ethics and Fetal Therapy*, Wilfred Laurier Press, Calgary, Canada, 1988, p. 56.

18. A.B. Satz, Paper presented at the Australian Bioethics Association Conference on Autonomy, Community, and Justice at the University of Queensland, Australia, September 27, 1995.

19. R. Chadwick, in R. Chadwick, M. Levitt, and D. Shickle, eds., *The Right to Know and the Right not to Know*, Aveberry Brookfied, VT, 1997, p. 19–20.

20. Wiggins et al., *N. Engl. J. Med.* **327**, 1401–1405 (1992).

21. *School Board of Nassau County, v. Arline*, 480 U.S. 273 (1987).

22. M. Miller and Marvin, *Terminating the "Socially Inadequate": The American Eugenicists and the German Race Hygienists, California to Cold Spring Harbor, Long Island to Germany*, Malamud-Rose, Commack, NY, 1996.

23. T. Duster, *Backdoor to Eugenics*, Routledge, New York, 1990.

24. M. Pernick, *The Black Stork: Eugenics and the Death of "Defective" Babies in American Medicine and Motion Pictures since 1915*, Oxford University Press, New York, 1996.

25. D. Wertz, *Gene Lett.* **3**(2) (1999).

26. A.B. Satz, *Monash Bioethics Rev.* **18**(4), 11–22 (1999).

27. J. Glover, in J. Fishkin and P. Laslett, eds., *Justice between Age Groups and Generations*, Yale University Press, New Haven, CT, 1992, pp. 127–143.

28. A. Silvers, in R. Ankeny and L. Parker, eds., *Medical Genetics: Conceptual Foundations and Classic Questions*, Kluwer, Amsterdam, 2001.

29. D.R.J. Macer, *Shaping Genes: Ethics, Law and Science of Using New Genetic Technology in Medicine and Agriculture*, Eubios Ethics Institute, 1990, pp. 10–20.

30. L. Walters and J.G. Palmer, *The Ethics of Human Gene Therapy*, Oxford University Press, New York, 1997.

31. Osteogenesis Imperfecta Foundation Home Page, Profiles in OI Research, *Dr. Rowe Focusing on Genetic Treatments for OI*, copy on file with author, bonelink@oif.org.

32. A. Silvers, in A. Donchin and L. Purdy, eds., *Embodying Bioethics: Feminist Advances*, Rowman & Littlefield, Lanham, MD, 1998, pp. 177–202.

33. Council of Europe Parliamentary Assembly, 1982, Recommendation 934; J.F. Keenam, *America*, October 20, 1990, 262–263.

34. T. Murray, *FASEB J.* **5**(1), 55–60 (1991). Reprinted in J. Arras and B. Steinbock, eds., *Ethical Issues in Modern Medicine*, Mayfield Publishing, Mountain View, CA, 1995, pp. 479–487.

35. Department of Health and Human Services, *Recombinant DNA Techn. Bull.* **9**, 221–242 (1986).

36. Office of Technology Assessment, 1984, 17–18.

37. *Science*, December 9, 1994, 1631; *Economist*, April 25, 1992, pp. 95–97.

38. L. Walters, *Nature* **320**, 225–227 (1986).

39. Arras and Steinbock: 428.

40. D. Kaplan, in K. Rothenberg and E. Thomson, eds., *Women and Prenatal Testing: Facing the Challenges of Genetic Technology*, Ohio State University Press, Columbus, OH, 1994, pp. 49–61.

41. Campaign against Human Genetic Engineering (CAHGE): 1998; e-mail distributed to disability-related listservs.

42. President's Commission for the Study of Ethical Problems in Medicine and Biomedical and Behavioral Research, *Splicing Life: The Social and Ethical Issues of Genetic Engineering with Human Being*, U.S. Government Printing Office, Washington, DC, 1982.

43. Disabled People's International, "The Right to Live and Be Different" Solihull, U.K., February 13, 2000.

44. N. Daniels, *Just Health Care*, Cambridge University Press, Cambridge, England, 1985.

45. C. Murray and A. Lopez, *The Global Burden of Disease*, Harvard School of Health, Cambridge, MA, 1996.

46. R. Miles, *A Women's History of the World*, London, Michael Joseph, London, 1988, p. 180.

47. World Health Organization: 1999, Versions of ICIDH-2, www.who.int/msa/mnh/ems/icidh/icidhtrg/sld033.htm

48. M. Corker, *Deaf or Disabled or Deafness Disabled?*, Open University Press, Buckingham, U.K., 1998, p. 48.

49. A. Silvers, in E. Parens, ed., *Enhancing Human Capacities: Conceptual Complexities and Ethical Implications*, Georgetown University Press, Washington, DC, 1998, pp. 95–123.

50. L. Morgenthaler, *Barron's*, September 20, 1993, pp. 10–11; H. Carey and G. McWilliams, *Business Week*, March 30, 1992, pp. 78–82.

51. Recombinant DNA Advisory Committee (RAC): 1997, "Discussion Regarding the Use of Normal Subjects in Human Gene Transfer Clinical Trials" Minutes of March 6–7, 1997, Bethesda, MD, National Institutes of Health.

52. A. Silvers, in L.P. Francis and A. Silvers, eds., *Americans with Disabilities: Exploring Implications of the Law for Individuals and Institutions*, Routledge, New York, 2000.

See other EUGENICS entries; REPRODUCTION, ETHICS, PRENATAL TESTING, AND THE DISABILITY RIGHTS CRITIQUE; REPRODUCTION, LAW, IS INFERTILITY A DISABILITY?

E

EDUCATION AND TRAINING, PUBLIC EDUCATION ABOUT GENETIC TECHNOLOGY

Joseph D. McInerney
Foundation for Genetic Education and Counseling
Baltimore, Maryland

OUTLINE

INTRODUCTION

Of the many branches of modern science that compete for the public's attention and that claim to have significance for individuals and society, why should we expect that biotechnology, and particularly that aspect of biotechnology concerned with genetics and molecular biology, will be of interest to the average person? Consider the following. At the conclusion of his 1977 book *The First Three Minutes* (1), in which he discusses the nature of the universe immediately after the Big Bang, physicist and Nobel laureate Steven Weinberg makes the following assertion: "The more the universe seems comprehensible, the more it also seems pointless." There is, according to Weinberg, no purpose, no inherent plan, no evidence of design in the universe, even if there is great order amid the complexity. Weinberg's is not an isolated opinion among cosmologists (2), yet his profound challenge to humanity's deeply rooted belief in a guiding, supernatural hand went virtually unremarked in public discourse.

Two decades later, in the wake of voluminous, often sensational media coverage of the cloning of Dolly (3), the now-famous Scottish sheep, Chicago physicist Richard Seed declared his intention to clone a human, even if he had to leave the United States to avoid the applicable legal strictures on such research (4). Seed even offered to clone ABC's Ted Koppel, who interviewed Seed on *Nightline* (5). Despite the lack of any credible evidence that Seed had the knowledge, skills, or resources to pursue his fantasy, the reaction was rapid and strenuous, with scientists and social critics outlining the penalties for hubris and cautioning against "playing God." Even before Seed's pronouncement, the President's National Bioethics Advisory Committee met to prepare guidelines for research on cloning, and specifically placed cloning of humans out of bounds (6), even though the scientific and technical prospects for such research remain remote.

Why was there no public outcry about the challenges that physics presents to long-standing and cherished beliefs, no call to stop research in cosmology, while even the remote prospect of human cloning caused such an uproar? Perhaps the distinction lies in the methods and materials of cosmology as compared with those of biology. The remoteness and immensity of the universe are incomprehensible to most of us, especially when explanations of their origin and nature are packaged, as they often must be, in the complex language of mathematical models. True, each of us is made of "star stuff"—the elements that had their origin in the Big Bang and stellar evolution—but the stars matter little to us personally; we feel no kinship with them. Biology, however, is a different matter. Biology deals with living things, including us. This is stuff we know about, and there is a special quality to it, a quality that we have been taught to respect, even revere, notwithstanding that we cannot define its essence (7).

Public reaction to the steady and stunning progress in biology, and especially genetic technology, reveals a mixture of optimism and anxiety, the former driven perhaps by the promise of benefits to personal and public health, the latter by public ignorance of the technology and the underlying science and by the unsettling sense that progress in this arena consistently challenges many of society's most cherished values and beliefs. Many articles and books on the implications of genetic technology call for public education to help individuals and society understand and accommodate the application of new knowledge in genetics and molecular biology. Indeed, public education has been one of the central objectives of the Human Genome Project's (HGP) Ethical, Legal, and Social Implications (ELSI) program since the project's inception, and the ELSI five-year plan developed in 1998 (8) makes clear the need to expand and improve education for the public (see Table 1). Like most calls for education about genetic technology, however, the new ELSI plan provides little guidance about the content of such education, the instructional approaches, the desired outcomes of educational programs for the public, or even the reasons for educating the public about such a complex area of scientific and technological inquiry. This article provides an overview of those topics by addressing four significant challenges to public education about genetic technology and by recommending how educational programs can meet those challenges.

CHALLENGE 1: SELECTING THE CONTENT

As in other branches of science, the amount of information generated by research in genetics overwhelms even the discipline's practitioners. How then do we make the content and implications of genetics comprehensible to the nonspecialist? The answer to that question requires a clear understanding what we want the public to know about genetics in the first place, and why. Most often, that

Table 1. Ethical, Legal, and Social Implications (ELSI) Research: Excerpts from Goals and Related Research Questions and Education Activities for the Next Five Years of the U.S. Human Genome Project, October 1998

a. Examine the issues surrounding the completion of the human DNA sequence and the study of human genetic variation.

Examples of Research Questions and Education Activities:

- Will the discovery of DNA polymorphisms influence current concepts of race and ethnicity? (*e.g., How will individuals and groups respond to potential challenges to or affirmations of their racial and/or ethnic self-identification, based on new genetic information?*)

- What are the most effective strategies for educating health professionals, policy makers, the media, students, and the public regarding the interpretation and use of information about genetic variation?

b. Examine issues raised by the integration of genetic technologies and information into health care and public health activities.

Examples of Research Questions and Education Activities:

- What are the clinical and societal implications of identifying common polymorphisms that predict disease susceptibility or resistance? (*e.g., Will genetic testing promote risky behavior in persons found to be genetically resistant to particular pathogens, such as HIV, or environmental hazards, such as cigarette smoke?*)

- What are the best strategies for educating health care providers, patients and the general public about the use of genetic information and technologies? (*e.g., What are the most effective mechanisms for educating providers, patients, and the public about the uncertainties inherent in genetic risk information?*)

c. Examine issues raised by the integration of knowledge about genomics and gene–environment interactions into nonclinical settings.

Examples of Research Questions and Education Activities:

- What are appropriate and inappropriate uses of genetic testing in the employment setting? (*e.g., Are there conditions under which it might be ethical and/or legal to use genetic testing to identify those employees who may have a susceptibility to workplace hazards? What implications does the Americans with Disabilities Act have for such testing?*)

- What are the potential uses and abuses of genetic information in educational settings? (*e.g., Is placement of students on the basis of genetic data any more or less beneficial or harmful than tracking on the basis of traditional categories or classifications?*)

d. Explore ways in which new genetic knowledge may interact with a variety of philosophical, theological, and ethical perspectives.

Examples of Research Questions and Education Activities:

- Will continuing research in molecular biology and functional genomics affect how individuals and society view the relationship of humans to one another and to the rest of the living world? (*e.g., As new genetic technologies and information provide additional support for the central role of evolution in shaping the human species, how will society accommodate the challenges that this may pose to traditional religious and cultural views of humanity?*)

Table 1. *Continued*

- What are the implications of behavioral genetics for traditional notions of personal, social and legal responsibility? (*e.g., What role will the discovery of putative genetic predispositions to violent behavior play in criminal prosecutions?*)

e. Explore how socioeconomic factors and concepts of race and ethnicity influence the use and interpretation of genetic information, the utilization of genetic services, and the development of policy.

Examples of Research Questions and Education Activities:

- How is the impact of genetic testing in clinical and nonclinical settings affected by concepts of race and ethnicity and other social or economic factors? (*e.g., Will particular communities and groups be more vulnerable to employment discrimination based on genotype?*)

From Ref. 8.

understanding is stated in the negative: The intent is *not* to turn all members of the public into specialists in genetics. Beyond that, one finds little agreement about the purpose or content of public education, save the assertion that the public should be able to make informed decisions about the personal and societal uses of new genetic knowledge. Some scholars (9), however, question the validity even of that argument as a general rationale for improved public science literacy or a guide for determining content.

Work by the Biological Sciences Curriculum Study (BSCS) provides a framework for defining the possible dimensions of public literacy with respect to genetic technology, and consequently, some guidance on the selection of content (10). According to this framework, *nominal* biological literacy consists simply of recognizing "certain words as belonging to the realm of biology." For genetic technology, nominal literacy would constitute the public's recognizing that terms such as gene, DNA, chromosome, and the polymerase chain reaction (PCR) are related to that area of investigation. In *functional* biological literacy, a person "can define certain biological terms or concepts but has limited understanding of or personal experience with them." The third level, *structural* biological literacy, implies an understanding of major concepts and the ways in which they are related. With respect to genetic technology, for example, a structurally literate person would understand that genes comprise DNA — a virtually universal information molecule in living systems — and, further, that DNA often is organized in chromosomes — cellular structures that help to maintain genetic continuity between generations of cells and generations of organisms. In addition a structurally literate person would understand that the ability to analyze and manipulate DNA with techniques such as PCR can have significant implications for individuals and society.

What does structural literacy require of the public in terms of content knowledge? Most important, the translation for nonscientists of content knowledge in genetic technology requires decisions about the level of detail, and the most difficult decisions concern what to leave out rather than what to include. Should nonspecialists, for example, know the fine structure of DNA? Or is it sufficient that they understand that DNA is an information molecule whose content can be analyzed to provide insights into the construction and expression of human traits? Should education focus on the details of transcription and translation of genetic information, or on the variable expression of that information? Must nonspecialists know the structure of nitrogenous bases, or should public education highlight the extraordinary variation in the sequence of those bases? It is, after all, that variation, in concert with environmental variables, that confounds our ability to make definitive statements about the role of genes in complex human characters and about the eventual role of genetic medicine in improving human health (11,12).

Unfortunately, formal science education worldwide has been plagued by the tendency to subordinate the central concepts of science to a concentration on disarticulated facts. In genetics and molecular biology especially, the rate at which new knowledge is generated is so staggering that precollege and college curricula alike often are overwhelmed by the accretion of isolated details and extensive vocabulary that do little to help students form a conceptual picture of genetics or of biology (10,13,14).

Ironically, the concept most often ignored in public education about genetics is the concept that is at the heart of the discipline: variation. The centrality of variation in understanding genetics is reaffirmed by the HGP's focus on variation in its newest five-year plan (8), yet there is almost no mention of that concept in most educational programs about genetics, largely because of a concentration on relatively rare single-gene Mendelian traits, to the near exclusion of more common, multifactorial traits. In humans most single-gene traits are disorders, and a strict focus on them conveys inappropriately that the study of genetics is related only to disease. In addition an exclusive focus on single-gene traits conveys the misconception that human traits always are straightforwardly qualitative and that there always is a clear, direct relationship between genotype and phenotype. The standard textbook treatment of Mendelian genetics, for example, teaches only that one either has cystic fibrosis or does not, this disorder being a frequently used example of autosomal recessive inheritance. The same approach holds for those disorders generally chosen to illustrate autosomal dominant (Huntington disease) and X-linked (Duchenne muscular dystrophy) inheritance.

Single-gene disorders are not as invariant as most textbook treatments indicate, and the expression and severity of traits such as cystic fibrosis are often highly variable (15). Public education should emphasize that variability rather than leave the impression that all occurrences of a given disorder have the same degree of severity or the same natural history. This shift would help reinforce the message of variation and biochemical individuality (16) that is at the heart of genetics.

Although it is true that there is considerable variation in expression for any of the single-gene disorders cited above, it is nonetheless true that the presence of the mutant allele (or alleles) results in the generally recognized phenotype. Furthermore, in many instances,

geneticists have a very clear sense of the biological relationship between gene and phenotype, and they often know the biochemical details of the relationship. It is clear, for example, that the accumulation of lipids that characterizes Tay-Sachs disease results from a deficiency of the enzyme hexosaminidase A, and that the range of symptoms associated with cystic fibrosis results from impaired transport of chloride ions across the membranes of certain epithelial cells.

It is more difficult to determine the biological relationship between gene and phenotype for multifactorial traits, including many common diseases. Although it is clear, for example, that genetic factors contribute to the risk for early onset heart disease, the exact relationship is yet unclear, as is the relationship between certain genetic markers and the risk of schizophrenia or bipolar disorder. In such common, complex diseases, expression is influenced by the products of multiple genes interacting—throughout one's developmental history—with a host of environmental variables whose influences are difficult to discern.

Given current trends in research, it is important for educational programs to demonstrate the distinction between the relatively rare single-gene disorders and the more common human traits, including many common diseases, that are polygenic and multifactorial (17):

> The term "complex disease" has been coined for conditions that arise from multifaceted interactions of environmental and heritable factors.... The heritability of these disorders deviates in important ways from that of classical (mono) genetic diseases: no simple Mendelian mode of transmission is apparent, and the severity of the disorder shows quantitative, unimodal variation rather than a dichotomous distribution. Complex traits are regarded as polygenic and multifactorial, with the phenotype representing the net effect of all contributing genes and environmental factors.

Educational efforts in genetics also should demonstrate that single-gene disorders are generally severe and often take their toll in infancy, childhood, or adolescence, while common, multifactorial disorders are generally less severe and tend to express themselves later in life. The later onset of common, multifactorial diseases provides opportunities for modification of environmental variables that otherwise magnify the risks inherent in genetic predisposition, and it is this prospect for prevention that is most likely to raise the prominence of genetical thinking among primary-care providers and consumers alike.

Education that addresses complex traits will help to acquaint the public with an important aspect of human genetics and will provide opportunities to help the public understand that genes and environment are both important in the expression of many human characters. That perspective will help counter the simplistic "gene for" approach to causation that often pervades media treatments of genetics (18). Any group of individuals—children, adolescents, or adults—is a living laboratory of human variation manifested as observable multifactorial traits, and educational programs should exploit that variation. Furthermore the quantitative traits manifest in any group of individuals often are those of most interest to the public: height, weight, intelligence, and athletic or artistic ability, for example. Discussions of the expansive range of normal variation for these traits, the complex nature of causation, and societal perceptions of normality itself (19,20), can help prepare the public for the ethical and policy debates that must follow as continued research uncovers genes that are putatively associated with complex and controversial traits such as intelligence, aggression, or sexual orientation (21,22).

Finally, a focus on variation and on quantitative characters in addition to qualitative Mendelian traits may improve the public's ability to understand evolution (23), the central organizing theme of biology. Ernst Mayr (24) has written that perhaps the most important aspect of the Darwinian revolution is the replacement of "typological thinking" with "population thinking," that is, recognition that the members of any given species do not constitute a single, fixed type but instead are highly variable with respect to virtually all traits. Indeed, disease itself is a by-product of the genetic variation required for survival of the species. In certain individuals in certain environments, some variation is expressed as disease. Furthermore familiarity with evolutionary perspectives, and especially those related to human evolution, can help students understand the distribution of disease in human populations. The genetic variations associated with common diseases, which are present in all populations, have an older evolutionary origin than do the variations associated with more rare, single-gene disorders that aggregate in certain groups. The former variations arose before *Homo sapiens* migrated out of Africa and spread across the globe.

Just as genetic variation is the sine qua non of differential selection, population thinking is central to one's understanding of evolution. Because they focus on typology rather than on variation, current practices in genetics education may impede rather than enhance that understanding, and to the extent that genetics is a piece of biology education, the current, incomplete picture of genetics for the public—including pre-college students—may be a disservice. That is especially problematic given that about 95 percent of all students take high school biology and that for most, the high school course will be their last formal exposure to the life sciences (25).

With the foregoing discussion in mind, Table 2 provides a list of basic concepts that might serve to organize public education about genetic technology, in the hope of achieving structural literacy where this field is concerned.

CHALLENGE 2: EXPLAINING THE NATURE OF SCIENCE

Among the more difficult challenges the public faces as it struggles to understand genetics is the portrayal of the discipline in the media. Genetic themes—especially those related to manipulation of DNA—have become increasingly popular, pervading television (*The X-Files*) and movies (*Jurassic Park, GATTACA*) especially, and blurring considerably the lines between fact and fiction, the possible and the fantastic. This trend exposes what is perhaps the most disturbing deficiency in general science literacy: the public's failure to distinguish science as a way

Table 2. Proposed Set of Basic Concepts in Genetic Technology for Public Education

A. Concepts related to biological variation

1. Genetics is the study of biological variation. Genetic medicine is the study of human genetic variation that is associated with mortality and morbidity.

2. Individual genetic variation (biochemical individuality) results from the variable sequence of the four bases that are central components of the DNA molecule. Mutations introduce additional variation, although mutations rarely have biological significance. Some mutations can be deleterious, while other mutations can provide selective advantages that are central to evolution by natural selection. There would be no differential selection, and therefore no evolution, without mutation and variation. This principle helps to explain phenomena such as the emergence of bacterial strains that are resistant to antibiotics.

3. Human biological variation results from the interaction of individual genetic variation with environmental variables (the experiences that one accumulates during one's developmental history, from conception to old age).

4. There is no fixed type—no archetypical individual—in a species, including *Homo sapiens*. A species comprises a population of genetically unique individuals who vary in morphology, physiology, and behavior. Disease is a by-product of the genetic variation required for the survival of our species. Some variations are manifested as disease in some people in some environments.

5. The genotype for a given trait is the gene(s) associated with that trait. The phenotype is the expression of the genotype, which is influenced by the environment.

6. Some human traits result primarily from the action of one gene. Most human traits, however, are multifactorial, resulting form the action of more than one gene in concert with the influence of environmental variables.

7. The phrase "the gene for" can be misleading because it can imply that only genetic influences are responsible for a given trait (discounting the influence of the environment) or that only one gene in particular is associated with a given trait when there may be genetic heterogeneity.

8. Genetic medicine is uniquely positioned to provide insights into prevention because it acknowledges the individuality of each patient and the biological and environmental influences that produce that individuality. Genetic medicine does not focus on the disease, but rather on the individual. It asks, "Why does this person have this disease at this point in his or her life?" (12).

B. Concepts related to cell biology

1. Classic cell theory holds that all life is made of cells and that all cells come from preexisting cells.

2. Cells pass through a series of structural and functional stages know as the cell cycle. The cell cycle is under genetic control. Disruption of that control can lead to disorders such as cancer. In that sense, all cancer is genetic, but not all cancer is hereditary.

3. Cell division is the process that produces new cells.

4. Mitosis, one part of cell division, helps ensure genetic continuity from one generation of somatic cells to the next. Human somatic cells contain 46 chromosomes (the diploid number): 22 pairs of autosomes and one pair of sex chromosomes (X and Y).

Table 2. *Continued*

5. Human germ cells, sperm and ova, contain 23 chromosomes (the haploid number). A special type of cell division — meiosis — occurs in the precursors to germ cells. Meiosis has two major biological effects: it reduces the number of chromosomes from 46 to 23 and it increases genetic variation through the exchange of genetic material (crossing over).

6. In humans (eukaryotes), cells contain a distinct structure (the nucleus) that includes the chromosomes, the carriers of most of the genetic material (DNA).

7. Human cells contain mitochondria. Because mitochondria likely were free-living prokaryotes early in the evolution of life, they carry their own DNA. This DNA in humans has been sequenced completely; mutations in mtDNA can cause health problems, often associated with neuromuscular diseases (because of the role of the mitochondrion in energetics).

C. Concepts related to classical (Mendelian) genetics

1. Our understanding of the behavior of chromosomes during first meiosis allows us to make predictions about genotype and phenotype from one generation to the next.

2. Some traits are inherited through an autosomal dominant pattern of inheritance, others through an autosomal recessive pattern. Still others, those traits associated with genes on the X chromosome, follow somewhat different patterns of transmission because the male has only one X chromosome.

3. Traits, not genes, are dominant or recessive, but we refer to genes as dominant or recessive because it is convenient.

4. Aberrations in the behavior of chromosomes during meiosis can result in structural or numerical alterations that have serious consequences for ontogeny and subsequent growth and development. Some of these aberrations are associated more frequently with advanced maternal age. We can detect many chromosomal aberrations prenatally.

5. Our understanding of the movement of genes through populations allows us to make predictions about the frequency of genes in given population groups and, therefore, about the frequency of disease phenotypes.

6. During the last two decades, research has uncovered genetic mechanisms that extend our understanding of inheritance and that provide biological explanations for heretofore unexplained observations. These mechanisms include imprinting, uniparental disomy, mitochondrial inheritance, and the expansion of trinucleotide repeats.

D. Concepts related to molecular genetics

1. DNA and RNA are information molecules; they store biological information in digital form in a well-defined code.

2. DNA is the primary information molecule for virtually all life on earth; this is evidence for the relatedness of all life by descent with modification.

3. The structure of DNA lends itself to replication. DNA replicates with great fidelity, which is critical to the maintenance of genetic continuity.

4. Sometimes mistakes arise during DNA replication. Evolution has produced mechanisms that repair such mistakes. In fact those mechanisms are conserved evolutionarily all the way back to *Escherichia coli*. When these mechanisms fail, mutations arise. Some cancers result from the failure of DNA repair mechanisms.

Table 2. *Continued*

5. In most biological systems, the flow of information is: DNA–RNA–protein. The process by which this occurs is known as the central dogma of molecular biology: replication, transcription, translation (into protein).

6. DNA is a fragile molecule; it is easily damaged by a host of environmental insults. It is especially important to avoid such insults during pregnancy.

7. The damage that occurs to our DNA during the course of our lives can contribute to the onset of cancer.

8. A gene is a segment of DNA (although the segment may not be contiguous). Some genes code for the production of structural proteins (collagen) or enzymes (lactase). Other genes are regulatory, helping to control such processes as prenatal development.

9. A gene occupies a particular place on a chromosome — a locus. A gene can have two or more alternate forms — alleles.

E. Concepts related to new genetic technologies

1. Advances in technology allow us to analyze and manipulate the genetic material in ways that were not possible even a few years ago. These technologies include PCR, RFLPs, direct and indirect (linkage) analysis of DNA, and quantitative trait loci.

2. These technologies allow us to identify, isolate, and test for genes associated with disease.

3. Like all technologies, genetic technologies are fallible, can have unintended consequences, and often serve the interests of entities apart from the patient.

4. The growth of information technology in concert with the expansion of genetic technology is a great boon to genetic medicine and to basic research, but it also raises concerns about the use of genetic information.

of explaining the natural world from other explanatory systems, particularly those that appeal to supernatural events. This deficiency leaves the public susceptible to fanatics and pseudoscientific charlatans, ranging from radical animal-rights activists and creationists to purveyors of questionable cures for illness and disability. Even Pope John Paul II, in his 1998 encyclical, "Faith and Reason" (26), warns against the abandonment of reason and rational thought: "It is an illusion to think that faith, tied to weak reasoning, might be more penetrating; on the contrary, faith then runs the grave risk of withering into myth or superstition."

Because of the near-homogeneous — and generally incorrect — portrayal of science in science textbooks, the public may view the science as a set of invariant steps in "the scientific method." To help the public make sense of the rapidly moving science of genetic technology and of the ways in which new knowledge supplants old, educational programs must reflect more realistically what science tries to do. Above all, education should emphasize the habits of mind that distinguish scientific inquiry from other ways of knowing and should provide the public with the skills required to make sound judgments about the validity of information they encounter about genetic technology. The following material, excerpted from two programs produced by BSCS (27,28), proposes some major concepts that should pervade public education about genetic technology.

Concept 1: The laboratory seldom provides definitive or final answers to scientific questions.

Although laboratory investigations in molecular biology may provide concrete data, it is a mistake to assume that those data always provide final and immutable answers to complex questions about life on earth. Uncertainty — resulting from indeterminacy and from the emergent properties of organisms at higher levels of biological organization (29,30) — makes simple extrapolations from molecular data to complex characteristics difficult at best. Growing interest in functional genomics (8), which seeks to integrate our understanding of sequence data into the biology of whole organisms, reflects recognition among biologists that it is difficult to derive much helpful information about complex systems from an analysis of DNA sequences alone. One hesitates to enter into a protracted debate about the relative merits of reductionism in the context of educational programs for the public, but it seems only judicious that education about genetic technology include a strong caveat about assuming that we understand life on earth simply because we have access to its constituent molecules.

Concept 2: Scientific explanations are based on empirical observations or experiments.

Scientific inquiry assumes that the universe is explainable without appeals to supernatural phenomena. Evidence in science includes empirical data and existing explanations about related phenomena that are supported by independent data and are publicly observable. The fact that others can confirm or refute what one investigator claims to observe provides a crucial test of the scientific validity of that observation. Rumor, speculation, mystical experiences, and other unsubstantiated information are not accepted as credible scientific data. Public understanding of the distinction between credible and questionable information becomes ever more important as the amount of unvalidated information available via the Internet grows.

Concept 3: Scientific explanations are tentative.

Explanations can and do change. There are no scientific truths in an absolute sense, and scientists often suspend final judgment on the answers to scientific questions. Advances in genetic technology, in fact, have caused geneticists to revise their definition of a gene (31) and to extend their view of inheritance to encompass genetic mechanisms such as extranuclear inheritance, trinucleotide repeat expansions, and genetic imprinting. Additional research likely will result in additional revisions.

One key point to consider is the public's perception of the word "theory." Many people mistakenly think that this word means an ephemeral guess or a hunch; they see a theory as an unsubstantiated idea and, as in the case of the evolution/creation controversy, try to dismiss as "only a theory" what scientists recognize as a powerful explanatory framework. As scientists use the term, a theory refers to a large-scale explanation, or series of explanations, that describe the causes of many natural processes (32). An accepted scientific theory is very well substantiated by evidence, has been built logically upon valid assumptions, and has been tested extensively. A scientific theory is neither established nor refuted on the basis of personal opinion that fails to follow the discipline of scientific methods. Rather, a theory is an explanation of far-reaching significance so well tested and supported by such an abundance of credible evidence that it becomes a broadly accepted and fundamental scientific concept. There are a number of powerful theories in biology: for example, cell theory, chromosome theory, germ theory, and the theory of evolution. Each is supported by overwhelming amounts of evidence.

Concept 4: Scientific explanations are probabilistic.

A statistical view of nature, not an absolute view, is fundamental to science. The probabilistic view of explanations is evident implicitly or explicitly when stating scientific predictions of phenomena or explaining the likelihood of events in actual situations. Geneticists long have faced the problem of conveying a statistical view of nature to the public, and the impending identification of large numbers of human genetic variations will complicate the problem still further as geneticists try to explain to the public the expression of those variations in different environments.

Concept 5: Scientific explanations assume cause–effect relationships.

Cause–effect relationships are fundamental to making sense of phenomena, and much of science is directed toward determining causal relations and developing explanations for interactions and linkages among objects, organisms, and events. Distinctions among causality, correlation, coincidence, and contingency separate science from pseudoscience.

It is not easy to establish causality. People who do not employ scientific reasoning, however, sometimes mistakenly assume that one event is the cause of a second event solely because it precedes the second event. Science requires a much stronger association to establish a casual process. A classic example from folklore states that eating strawberries while pregnant can cause birthmarks on one's newborn child. The timing of the putative cause (eating strawberries) precedes the observed result (birthmark), but there is no credible and relevant evidence to support the first event as a *cause* of the second one.

Concept 6: Pseudoscientific explanations do not meet the requirements of science.

Pseudoscience can be defined as the promotion of unsubstantiated, allegedly scientific opinions. Some of these opinions may be very appealing to the public (33) and thus gain popularity, but they lack supporting, credible evidence. Pseudoscientific ideas have not been tested reliably. Often they are built on inaccurate premises, or they do not follow logically from what is observable. Pseudoscience often involves claims for which it is almost impossible to provide scientific evidence. Just as a poorly formulated hypothesis cannot easily be tested, pseudoscientific claims usually are so vague, ill-formed, and undetailed that they make no specific predictions and cannot be tested through credible experiments.

Ideas based in pseudoscience generally are inconsistent with other, well-tested concepts. Indeed, pseudoscientific claims may not even be internally consistent, but without rigorous criteria for evidence and reasoning, proponents of pseudoscience are not likely to recognize the inconsistencies. We may find ourselves drawn toward the ideas put forth by pseudoscientists because they have an emotional appeal, but when we do so, it is often because we are being intellectually lazy. Science is not easy, but it does provide sound results. New scientific findings, including those in genetics, sometimes challenge comforting, long-standing assumptions about the natural world, but scientists must go where the data take them even if the destination is a bit unsettling.

Concept 7: Science cannot answer all questions.

Some questions simply are beyond the realm of science. Questions involving the meaning of life, ethics, and theology are examples of questions that science cannot answer. Genetic technology, for example, might help us determine the processes by which *Home sapiens* arose from its earliest mammalian ancestors and the chronology of descent with modification, but it is powerless to determine why we are here, in the sense of ultimate causes (34).

It is important, however, to distinguish for the public those questions that science cannot answer from those that science likely can answer, but has not yet. The origin of life is a case in point. Science cannot explain why there is life on earth, and it has yet to provide a complete explanation for life's origin. Failure to provide a completely naturalistic explanation for the latter question does not, however, mean that scientists should throw up their hands in resignation and ascribe the origin of life to supernatural causation. Indeed, such explanations are inadmissable in science, and scientists assume that the problem of life's origins ultimately will yield to the same methods of inquiry and habits of mind that have successfully stripped the mystery from so many of nature's mechanisms.

Concept 8: Science is not authoritarian.

Ecologist Garrett Hardin (35) reminds us that "science is ineluctably married to doubt." Although the public may view disagreements among scientists as evidence that the science in question is somehow flawed, disagreements and multiple competing hypotheses are essential to the health of the scientific enterprise.

The collective nature of science encourages objectivity in the field. Scientists must report their findings, and communication must use standardized descriptions so that results are meaningful to any informed person to whom they are communicated. Science is not monolithic; it attempts to be authoritative, but it is not authoritarian. Certainly recognition of an "authority" in a particular area of study encourages other scientists to pay attention to what is reported by that individual or research group, but the requirements for supporting evidence and all of the other criteria of valid science still apply. A famous name in itself does not establish disciplined methods nor produce the credible evidence required to substantiate an idea presented to the scientific community. The rules

Table 3. Snapshot of the Goals and Methods of Science

1. What does science try to do?

Science tries to provide causal explanations for natural phenomena.

2. How do we know that causal explanations are on the right track?

The causal explanations demonstrate predictive power. Other observations (evidence) rule out competing causal explanations.

3. What counts as evidence in science?

Empirical data are included as evidence in science. Existing explanations about related phenomena are included if they are supported by independent data.

4. What makes science objective?

Science is conducted by a rigorous set of methods. Authority and fame by themselves are not sufficient to establish the scientific validity of an explanation.

5. How and why does scientific knowledge change?

Scientific explanations always are open to change. New evidence may show that an existing explanation is inadequate and that it needs to change.

6. What is the difference between pseudoscience and science?

Pseudoscience fails to meet the intellectually rigorous requirements of science. Internal consistency sets scientific knowledge apart from pseudoscience.

From Ref. 27.

of science are the same for everyone, and anyone who proposes a new explanation for a natural phenomenon ultimately must answer the same two questions from any other scientist: "Where are your data?" and "How do you know they are sound?"

The establishment of valid scientific explanations depends on review and critique by the scientific community at large, but the critique begins with the individual scientist. Scientists are trained to be critics of their own work. Indeed, British biologist Peter Medawar noted that "most of a scientist's wounds are self-inflicted" (36). In addition, at any given time, a minority view on some important scientific concept usually exists. With time, if sufficient evidence is brought to light, even an unpopular view can be validated.

Table 3 provides an overview of the goals and methods of science that can serve as guidance in the development of educational programs in genetic technology.

CHALLENGE 3: THE PRINCIPLES OF TECHNOLOGY

Genetics may be the science of the future, but lay persons almost never will encounter that science directly. They are much more likely to encounter genetically based technologies, ranging from chorionic villus biopsy and PCR to FISH and pre-implantation diagnosis. This distinction is especially true of the Human Genome Project, which is driven by a wide range of technologies, from those employed in mapping and sequencing to those central to the creation and maintenance of international genomic databases (37). The details of the underlying science are likely to remain inaccessible to the average person, partially because those details are extremely complex, but also because they are unimportant for the nonspecialist.

The technology-dependent nature of genetics and of genetic medicine highlights a more recent challenge to public education: the need to include technology as a focus of serious study (13,14,28). If cherished tenets of medical ethics such as informed consent and nondirective counseling are to be more than topics of discussion at genetics meetings, public education must acquaint potential consumers of genetic medicine with some basic principles of technology. The AAAS, in its publication *Benchmarks for Science Literacy* (13), provides helpful guidance about those concepts of technology that should be central to public education. The brief discussion below paraphrases several of those concepts and demonstrates their relationship to education about genetic technology.

Concept 1: Technologies extend our senses.

Many technologies associated with science help us see, hear, or measure objects or phenomena that we would be unable to experience otherwise. We must, however, understand the limitations of our technologies and the role of inference in the interpretations of information we derive from them. For example, we can infer the presence of mutations in the genes for muscular dystrophy or cystic fibrosis on the basis of restriction fragments displayed on Southern blots. We can neither see nor touch the mutations themselves, however, and the accuracy and predictive value of our technologies are partially a function of the soundness of our inferences.

Concept 2: Technologies often have unintended consequences.

Almost all technologies are developed for specific purposes, yet many have side effects that are unintended, and worse, undesired. DNA analysis, for example, provides insights into long-standing biological questions such as gene regulation and evolution and allows us to detect genes associated with disease. But DNA analysis also has raised questions of privacy and discrimination to levels heretofore thought unimaginable (38,39).

Furthermore the impact of unintended consequences multiples rapidly with the introduction of public-health initiatives such as voluntary or mandatory genetic screening and testing. The sickle cell screening programs of the early 1970s, which resulted in discrimination against heterozygotes, demonstrated well the unintended consequences of technologies applied in the absence of thorough planning, education, and counseling. The more recent screening programs for Tay-Sachs disease benefited from those errors and reduced unintended consequences.

Concept 3: All technologies are fallible.

With respect to medical genetics, the public must know that diagnostic, laboratory, and treatment techniques can fail for reasons that range from those inherent in the technologies themselves to those associated with human

error. When technologies fail in the context of personalized genetic medicine, the results can be tragic for individuals and families. When they fail in the context of broadly based public-health initiatives such as genetic screening, the negative consequences are likely to be much more pervasive. The public must be aware that the extent to which society embraces technologies such as genetic screening and testing is the extent to which society also embraces the risk of technological failure.

Concept 4: All technologies serve the interests of particular individuals, groups, or agencies.

Sometimes those interests compete. For example, genetic testing for breast or colon cancer can identify individuals at risk for those diseases, but insurance companies might use the same information to restrict coverage for those found to be at risk (40). Similarly population screening for carriers of mutations for cystic fibrosis can identify at-risk couples. Screening programs, however, can enrich companies that provide the tests and analyses, and those companies might push for wholesale screening before the laboratory tests meet appropriate standards.

CHALLENGE 4: THE PERSONAL AND SOCIAL IMPACT OF SCIENCE AND TECHNOLOGY

Progress in genetic technology demonstrates that we need public education not only because we can do new things (e.g., detect mutations associated with breast cancer or compare base sequences between humans and chimpanzees), but because the new things we can do raise profound, sometimes troubling questions for individuals, families, and society (41). Although the HGP has not necessarily raised issues of ethics and policy that are new to genetics, it has accelerated the rate at which once-hypothetical issues are likely to become reality. Furthermore the HGP is likely to result in more widespread public awareness of genetics in general and in increased use of molecular medicine and its clinical applications (11,42).

Such complex issues challenge scientists and educators to provide serious and rigorous educational treatments of ethics and public policy. These treatments should address concrete applications of genetic technology, such as genetic screening and testing, and as genes come to be associated with specific behaviors, conceptual issues such as notions of normality, societal views of what it is to be human, perceptions of free will, and even biological and cultural perceptions of race (8,43–45).

Many educational programs address ethical and policy issues related to genetic technology, and experience indicates that the most effective instruction includes the features described below.

Feature 1: A clear recognition that controversy is inherent in such instruction, and clear recommendations for dealing with it.

Progress in genetic technology invites controversy almost as a matter of course. Whether the issue is genetic screening for predisposition to breast cancer or the development of chorionic villus sampling, which allows first-trimester detection of certain birth defects, genetic technology challenges traditional values and traditional views of the world. The incredible rate of progress in science and technology ensures that there always will be great disparity between what is possible and what people find acceptable.

The public generally encounters two major categories of scientific controversy: debates within the scientific community and debates about the use of science and technology that extend into society as a whole. The public must understand that debates between scientists are essential, or there will be no scientific progress. Science is a dynamic, self-correcting enterprise that continually tests new information and ideas in open, sometimes confrontational, debates. Debates within the scientific community demonstrate that the concepts under scrutiny are intellectually viable.

The current debate between evolutionary biologists who espouse gradualism and those who espouse punctuated equilibria is a good example, as is the dispute over whether *A. africanus* is ancestral to *A. afarensis,* and vice versa. Neither debate questions the validity of the theory of evolution. The former is a debate over the pace of evolutionary change, the latter, a disagreement over the sequence in which our hominid ancestors diverged. Each debate has a healthy effect on evolutionary biology because the scientists involved must work harder and be more creative and insightful to establish sound arguments.

Although debates within the scientific community may or may not attract public attention, other issues derived from biological progress are certain to do so, and the public encounters a number of those issues with respect to genetic technology. Such issues call into question many long-standing values and moral traditions, but the attendant controversy does not necessarily mean that those values and moral traditions will be found wanting. It does mean, however, that new knowledge and new techniques raise what once were intellectual abstractions to the level of hard, often painful, reality for individuals, families, and policy makers.

Feature 2: A clear conceptual framework for ethical analysis.

Sound instruction about bioethics must specify the underlying criteria for making and evaluating arguments to reduce the likelihood of unproductive discussions. The criteria need not be esoteric, and science educators, in conjunction with specialists in ethics, have developed a number of frameworks for ethical analysis that work effectively with nonspecialists, including the general public and precollege students. For example, the use of competing interests or contrasting goals, rights, and duties can focus analysis and highlight sources of disagreement.

Feature 3: A clear structure for discussions.

Ethical analysis and argument are forms of public discourse, but there is a misperception among nonspecialists that ethical discourse consists of a rather free-form sharing of ideas among the participants. Although that view is incorrect, the prospect of such unstructured discussions often is unsettling, especially to educators who are justifiably uncomfortable with deliberations that may have no structure or resolution. Productive ethical analysis requires structure and an objective. A well-structured model for discussion and analysis can help participants make arguments that lead to important insights and conclusions, while learning that there can be competing, well-made arguments about complex aspects of an area such as genetic technology. Even more important, a well-structured model for discussion can help the public realize that even seemingly intractable issues are amenable to analysis, and that civil, respectful discourse is essential if one wants to understand conflicting views.

Feature 4: A clear understanding that ethical analysis is a form of rational inquiry.

Instruction in bioethics and related policy questions often asserts that there are no correct answers to bioethical dilemmas. That statement may be correct, but it does not go far enough. Well-designed instruction in bioethics conveys clearly that there are well-reasoned and badly reasoned arguments, just as there are in science. Sound ethical analysis, for example, does not permit conclusions that are unsupported by the facts of the case, any more than science would allow such conclusions.

Feature 5: A clear demonstration of the connection between ethics and public policy.

Ethics is vital to public policy because it provides the concepts and terminology for the carefully organized debate that can result in well-reasoned conclusions about what society should or should not do. This inquiry is valuable in and of itself. Once society identifies a well-reasoned conclusion, however, it is reasonable to ask whether it should be enacted into formal public policy. Sometimes the best response is *not* to enact new policies in response to a controversy but rather to allow individuals, institutions, and society to act in the manner they choose.

One example (46) of an educational framework that helps make the connection between ethics and policy uses the conditions of urgency, means, and effectiveness to assess whether any conclusion of a well-reasoned ethical argument should become formal public policy. This analysis (see Table 4) assumes that it is not reasonable to enact new policy if conclusions from ethical arguments do not satisfy those conditions. Instead, ethical inquiry should continue, and public policy should remain at the de facto level.

CONCLUSION

As knowledge of genetics expands, perhaps the greatest challenge to public education will be the integration of new data into a cohesive picture of biology for the average person, much as the greatest challenge of the HGP will be the integration of the complete physical map of three billion bases into a cohesive understanding of the biology of *Homo sapiens.* The scientific community must play a central role in this integration, working with educators to determine how best to make this difficult, sometimes disconcerting discipline understandable to the average

Table 4. Framework for Connecting Ethics to Public Policy

Condition 1.

The situation is urgent: There is immediate risk of serious, far-reaching, and irreversible harm if legislation is not enacted or an existing law is not changed.

Immediate means that there are reasonable scientific grounds to conclude that impairment of interests will occur in the near future, for example, access by unauthorized individuals or institutions to human genome databases will deny or violate individuals' rights of privacy.

Serious means that the risk involves potentially grave injury to interests, for example, some private insurance carriers may deny coverage to an individual who has a genetic predisposition to a particular disease.

Far-reaching means that the impact of severe impairment of interests may be widespread, for example, a public policy of mandatory genetic testing of all pregnant women or children at birth would threaten the interests of millions of individuals.

Irreversible means that the serious damage to interests likely will be permanent, for example, labeling infants as having a genetic predisposition to learning disabilities could well have a permanent impact on their future education.

Conditions 2 and 3.

There are effective means to address the urgency of the situation: Scientifically valid or technologically practical means are available to prevent, reduce, or avoid risk of serious, far-reaching, and irreversible harm. The public policy must work and be enforceable — resources must be available to implement the public policy. An example would be that there are enough scientifically and technologically qualified individuals to carry out populationwide genetic screening. Public policy works when it enjoys broad-based political acceptance, that is, when few, if any, individuals or groups disagree strongly with the public policy. An example of a policy that is controversial in this respect is the requirement that all job applicants submit their individual genetic profile to prospective employers. Public policy is enforceable when few, if any, individuals or groups will disobey the public policy. An example of policy that is controversial in this respect is a requirement that all citizens submit their individual genetic profile to the federal government for inclusion in a national database.

From Ref. 46.

person and to situate genetics in the larger context of life on earth and, especially, of human biology. Because genetics and its associated technologies will continue to raise difficult questions of ethics and public policy, scientists and educators also must develop mechanisms for the rational consideration of those questions and must help develop in the public at large the skills and knowledge essential to informed, dispassionate analysis.

The translation of science for the public has never been easy. We must overcome the inherent difficulty of the information itself and, especially in genetics, the anxiety that arises when new information forces us to confront and perhaps revise long-standing assumptions about what it is to be human. The good news is that the public appears endlessly interested in genetics, and especially in its applications to personal and public health.

BIBLIOGRAPHY

1. S. Weinberg, *The First Three Minutes*, Basic Books, New York, 1977.
2. P. Davies, *The Last Three Minutes: Conjectures About the Ultimate Fate of the Universe*, Basic Books, New York, 1997.
3. I. Wilmut et al., *Nature (London)* **385**, 810–813 (1997).
4. J.B. Elshtain, *The New Republic*, February 9, p. 9 (1998).
5. K. Kestenbaum, *Science* **279**, 315 (1998).
6. National Bioethics Advisory Commission, *Cloning Human Beings: Report and Recommendations*, NBAC, Washington, DC, 1997.
7. E. Mayr, *This Is Biology: The Science of the Living World*, Belknap/Harvard University Press, Cambridge, MA, 1997.
8. F.S. Collins et al., *Science* **282**, 682–689 (1998).
9. J. Gregory and S. Miller, *Science in Public: Communication, Culture, and Credibility*, Plenum Trade, New York, 1998.
10. Biological Sciences Curriculum Study (BSCS), *Developing Biological Literacy*, BSCS, Colorado Springs, CO, 1993.
11. L.B. Andrews, J.E. Fullarton, N.A. Holtzman, and A.G. Motulsky, eds., *Assessing Genetic Risks: Implications for Health and Social Policy*, National Academy Press, Washington, DC, 1994.
12. B. Childs, *Genetic Medicine: A Logic of Disease*, Johns Hopkins University Press, Baltimore, MD, 1999.
13. American Association for the Advancement of Science (AAAS), *Benchmarks for Science Literacy*, Oxford University Press, New York, 1993.
14. National Research Council (NRC), *National Science Education Standards*, National Academy Press, Washington, DC, 1995.
15. L.B. Jorde, J.C. Carey, and R.L. White, *Medical Genetics*, 2nd ed., Mosby-Year Book, St. Louis, MO, 1998.
16. C.R. Scriver and B. Childs, *Garrod's Inborn Factors in Disease*, Oxford University Press, Oxford, UK, 1989.
17. K. Lindpaintner, *N. Engl. J. Med.* **10**(332), 679–680 (1995).
18. D. Nelkin and M.S. Lindee, *The DNA Mystique: The Gene as a Cultural Icon*, Freeman, New York, 1995.
19. M. Rutter and R. Plomin, *Br. J. Psychiatry* **171**, 209–219 (1997).
20. P. Kitcher, *The Lives to Come: The Genetic Revolution and Human Possibilities*, Touchstone Books, New York, 1996.
21. R. Plomin, J.C. Defries, G.E. McClearn, and M. Rutter, *Behavioral Genetics*, 3rd ed., Freeman, New York, 1997.
22. J.D. McInerney, *Q. Rev. Biol.* **71**(1), 81–96 1996.
23. M.U. Smith, H. Siegel, and J.D. McInerney, *Sci. Edu.* **4**, 23–46 (1995).
24. E. Mayr, *The Growth of Biological Thought*, Belknap/Harvard University Press, Cambridge, MA, 1982.
25. Council of Chief State School Officers, *State Indicators of Science and Mathematics Education*, Council of Chief State School Officers, Washington, DC, 1995.
26. Pope John Paul II, *Fides et Ratio*, The Vatican, Rome, 1998.
27. Biological Sciences Curriculum Study (BSCS), *The Puzzle of Inheritance: Genetics and the Methods of Science*, BSCS, Colorado Springs, CO, 1997.
28. Biological Sciences Curriculum Study (BSCS) and Social Sciences Education Consortium (SSEC), *Teaching about the History and Nature of Science and Technology: A Curriculum Framework*, BSCS and SSEC, Colorado Springs, CO, 1992.

29. E. Mayr, *Toward a New Philosophy of Biology: Observations of an Evolutionist*, Belknap/Harvard University Press, Cambridge, MA, 1988.

30. M. Blois, *N. Engl. J. Med.* **318**, 847–851 (1988).

31. P. Portin, *Q. Rev. Biol.* **68**(2), 173–223 (1993).

32. A. Rosenberg, *The Structure of Biological Science*, Cambridge University Press, Cambridge, UK, 1985.

33. C. Sagan, *The Demon-Haunted World: Science as a Candle in the Dark*, Random House, New York, 1995.

34. National Academy of Sciences, *Teaching about Evolution and the Nature of Science*, National Academy Press, Washington, DC, 1998.

35. G. Hardin, *Am. Zool.* **25**, 469–476 (1985).

36. D. Pyke, *Perspect Biol. Med.* **39**(4), 555–568 (1996).

37. N.G. Cooper, ed., *The Human Genome Project: Deciphering the Blueprint of Heredity*, University Science Books, Mill Valley, CA, 1994.

38. M. Rothstein, ed., *Genetic Secrets: Protecting Privacy and Confidentiality in the Genetic Era*, Yale University Press, New Haven, CT, 1997.

39. T.H. Murray, M.A. Rothstein, and R.F. Murray, Jr., eds., *The Human Genome Project and the Future of Health Care*, Indiana University Press, Bloomington, IN, 1996.

40. J. Beckwith and J.S. Alper, *J. Law, Med., Ethics* **26**, 205–210 (1998).

41. B. Gert et al., *Morality and the New Genetics*, Jones & Bartlett, Sudbury, MA, 1996.

42. F.S. Collin, *N. Engl. J. Med.* **341**, 28–37 (1999).

43. J.D. McInerney, *Am. Biol. Teacher* **60**(3), 168–173 (1998).

44. M.R. Gorman, *Perspect. Biol. Med.* **38**, 61–81 (1994).

45. K.J. Kevles and L. Hood, eds., *The Code of Codes: Scientific and Social Issues in the Human Genome Project*, Harvard University Press, Cambridge, MA, 1992.

46. Biological Sciences Curriculum Study (BSCS) and the American Medical Association (AMA), *Mapping and Sequencing the Human Genome: Science, Ethics, and Public Policy*, BSCS, Colorado Springs, CO, 1992.

See other entries FEDERAL POLICY MAKING FOR BIOTECHNOLOGY, EXECUTIVE BRANCH, ELSI; FEDERAL POLICY MAKING FOR BIOTECHNOLOGY, EXECUTIVE BRANCH, NATIONAL BIOETHICS ADVISORY COMMISSION; PROFESSIONAL POWER AND THE CULTURAL MEANINGS OF BIOTECHNOLOGY; PUBLIC PERCEPTIONS: SURVEYS OF ATTITUDES TOWARD BIOTECHNOLOGY.

EUGENICS, ETHICS

GLENN MCGEE
DAVID MAGNUS
University of Pennsylvania
Philadelphia, Pennsylvania

OUTLINE

Introduction

History

Coercive Versus Voluntary

Positive and Negative

Population and Individual

Modern Dilemmas and Eugenics

Conclusion

Bibliography

INTRODUCTION

Francis Galton, an Englishman and cousin of Charles Darwin, who synthesized the word from the Greek *eugenes* (wellborn), coined the term "eugenics" in 1883 (1). Galton described a "science of improving human heredity over time," through the systematic, social and even governmental application of human knowledge about the hereditary roots of desirable and undesirable traits. The history of the twentieth century eugenics movement has been widely chronicled, and it is associated with what were eventually thought of as nefarious political aims. However, policies and arguments that could be described as eugenic both predate and antedate that movement. With every significant advance in reproductive genetic technology, fears of eugenics are revived in social and political institutions around the world. Eugenics is both rooted in the history of biology and tied to contemporary debates, and is thus always complicated by its past and future.

HISTORY

That eugenics is a difficult concept to define is in part a result of its disparate origins and uses. Long before Galton described systematic eugenics, cultures had devised strategies for the regulation of reproductive relationships. Society has always exercised a measure of control over reproduction: in sexual recombination, it takes two to reproduce, and those two choose each other under the influence of family, economic, political and other community values. We learn what counts as attractive, successful, and desirable within the ethos of a community that has models of the successful family and of health. At many times and in various places that ethos has been fairly emphatic. Long before Galton, different cultures were telling families not to have children or with whom reproductive behavior is authorized.

Early versions of what would later be called eugenics were deployed in three ways, each of which was designed to prevent reproduction: exposure or infanticide, abortion, and sterilization. These three techniques are distinguished from milder, more positive means of regulating reproduction in only two ways. First, agreement to terminate or forestall pregnancy is achieved by a political or social instrument, such as a law sanctioning sterilizations or a medical protocol for therapeutic abortion. Second, the techniques involve surgical intervention into the bodies of citizens, rather than acts of verbal coercion.

History reports numerous cases in which societies discussed and employed both clinical and nonclinical techniques. The Spartans left their unwanted offspring to the elements. Plato wrote in the *Republic*: "those of our young men who distinguish themselves ... [should receive] ... more liberal permission to associate with the women, in order that ... the greatest number of children may be the issue of such parents." Contemporary scholars of Greek history argue that while his utopian scheme was not to be realized, systematic control of procreation in Greece was an objective on several occasions.

In the first half of the twentieth century, the eugenics movements in the United States, Great Britain, Germany, and other countries included leaders from across the political spectrum. In 1865 Galton published a series of magazine articles and eventually a book (*Hereditary Genius*) that developed the key theses of early eugenics. First, intellectual and moral qualities could be inherited. Second, through appropriate policies, humans could take advantage of burgeoning (quantitative) biological knowledge to produce an improved version of humanity.

Galton's eugenic ideas were taken up by many of the followers of his biometrical work, including most notably Karl Pearson. In the late nineteenth and early twentieth century, eugenics was largely a movement of scientists and intellectuals, and in fact the history of eugenics and the development of genetics are deeply intertwined. In Great Britain, the group that initially took Galton's mantle (under Pearson's leadership) was largely socialist. Pearson, George Bernard Shaw, and Havelock Ellis used eugenics as the basis of an attack on class distinctions. They argued that these social distinctions created artificial barriers to the "natural" breeding between fitter individuals which otherwise take place. Thus, in the name of good breeding, we should tear down class distinctions.

In the United States, the leader in the eugenics movement was Charles B. Davenport. As in Great Britain, the United States eugenics movement was initially tied to developments in the biological study of heredity. But, whereas in England there was a sharp divide between biometricians (Galton, Pearson) and Mendelians, in the United States, no such deep division occurred. Consequently Davenport was able to draw on both traditions as he helped establish the Cold Spring Harbor Laboratories (one of the most prestigious institutions in the world for biological research to this day) and the Eugenics Record Office (ERO) in Cold Spring Harbor. The ERO carried out studies of "important" character traits and their hereditary pattern on a grand scale accumulating an enormous amount of data (though much of suspect scientific value). Politically Davenport was a conservative and his political approach was to have a far greater impact on eugenic policies in the United States and Great Britain than that of the socialists.

In those early days of eugenics, the possible benefit of science to the improvement of human conditions though better breeding served to unite conservatives like Davenport, radicals like Pearson, and progressives like David Starr Jordan (first president of Stanford University and a leading biologist and politician). Jordan argued that eugenics provided the best possible ground for pacifism. In times of war, countries send out their fittest individuals to die, leaving the unfit behind to reproduce. There were anecdotes that the average height in France had fallen by 6 inches after the Napoleonic wars.

In the early twentieth century there was a parallel discussion of euthenics—the scientific study of how environmental influences could be manipulated to improve humanity, and euthenics was pursued by many of the same people who were prominent in the eugenics movement. However, the focus on the role of heredity eventually led to the popularization of the eugenics movement. This took place not through the positive enthusiasm for eugenics' potential to improve conditions. Rather, a growing fear that society was becoming overrun with the unfit offspring of those with less than desirable hereditary backgrounds motivated the rise of eugenics into a full blown political force. Works such as Henry Goddard's classic on the Kallikak family gave rise to concern about the "menace of the moron." Goddard estimated the number of "feeble-minded" in the United States ranged from several hundred thousand to a million individuals. Given the perceived rapid rate of reproduction among the feeble-minded relative to the disturbingly low reproductive rates among elites, there was a great fear that the country was would be overrun with degenerates. "Fitter family" contests and displays demonstrating the dangers of the breeding of the unfit were regular displays at state fairs (2–4). Estimates of the cost to society of allowing the "least fit" to breed were estimated and became a hot political and social issue. The Jukes and the Kallikaks became household names and policies were soon developed to "solve" the problem of "race suicide" which the popular press increasingly decried.

Increasingly conservative solutions soon became promulgated. Laws allowing the forced sterilization of those deemed less fit became commonplace in most states and resulted in the forced sterilization of over 60,000 individuals. In the famous 1927 Supreme Court decision which upheld these laws (*Buck v. Bell*) Justice Oliver Wendell Holmes argued that "three generations of imbeciles is enough."

At the same time, concern grow that undesirable, inferior immigrants (with large birth rates) were rapidly contributing to the race suicide. The U.S. Immigration Restriction Act of 1924 favored immigration from Northern Europe and greatly restricted the entry of persons from other areas referred to as "biologically inferior." Intelligence testing and the rating of families became a national mania and an important aspect of immigration restriction.

Argentina, Austria, Brazil, Canada, China, Finland, France, Italy, Japan, Mexico, Norway, and Sweden each had eugenic initiatives. In Germany, the fledgling eugenics movement had existed since 1904 with the creation of the Archives of Race-Theory and Social Biology by Dr. Alfred Ploetz, and the creation in 1905 of the German Society of Racial Hygiene. The German movement could only look in envy at the growing movement in the United States and Great Britain (which ironically had now become a way of curbing the growing ranks of the lower classes). That was soon to change as the Nazis to rose to power and began to imitate and eventually "surpass" their American and British counter parts. The Kaiser Wilhelm Institute of Anthropology, Human Heredity, and Eugenics was created in 1927. German sterilization laws were enacted in 1933, requiring compulsory sterilization "for the prevention of progeny with hereditary defects" in cases of "congenital mental defects, schizophrenia, manic-depressive psychosis, hereditary epilepsy ... and severe alcoholism (5).

The counselor of the Reich Interior Ministry called sterilization "an exceptionally important public health initiative ... we go beyond neighborly love; we extend it to future generations" (1, pp. 117–118). Under the Nazi law, physicians reported all "unfit" persons to Hereditary Health Courts, established to determine the sorts of persons who ought not to procreate. Decisions could be appealed to a "supreme" eugenics court, whose decision was final — and could be carried out by force. Within three years, German authorities had sterilized some 325,000 people, more than 10 times that in the previous 30 years in America (6). Marriage or sexual contact between Jews and other Germans was banned. According to Muller-Hill, a few hundred black children and 30,000 German gypsies were sterilized. Eventually the eugenics movement culminated in the Holocaust.

The science behind the eugenics movement was often of a suspect nature. The movement authorized some dramatic, if ill-informed inferences about the hereditary roots of a variety of behaviors and traits. Davenport, for example, argued that family pedigrees established the existence of the trait "thalosophia" or love of the sea. He traced multiple generations of sea faring in families and concluded that this was a simple Mendelian trait — and a sex-linked one (women did not seem to go to sea).

After the German experience, eugenic thought was at its nadir. Eugenics was associated with terrible images and widely discredited in the media, scholarly literature, and policy. However, whether eugenics continued to be practiced or still remains a prominent feature of modern medical genetics and counselling depends on how one understands the term.

Much has changed in the years since the endorsement and rejection of eugenics in overt political and scholarly institutions. Yet eugenics is constantly referenced as a danger of contemporary reproductive and genetic technology. With every advance in the ability of genetic testing to detect disease in adults, fetuses, or germ cells, the likelihood that such technologies might be used in discriminatory ways or as part of a thoughtless or diabolical public campaign is debated. Could eugenics, either in its optimistic or conservative historical incarnations be repeated? Scholars agree that the flowering of early twentieth century eugenics involved a fairly specific set of circumstances (e.g., the rise of the Nazi regime) and a primitive understanding of genetics and inheritance. The present context is very much different, and it bodes much more complex opportunities for misuse of technologies. In addition scholars such as Daniel Kevles have argued that an authoritarian politics, albeit perhaps different than the one seen in Germany, would be a necessary precursor to any state effort at sterilization.

Overt control and planning of reproduction has certainly seen a manifest increase in this century. The development of a birth control pill expanded reproductive control, but it carried new risks and ways of choosing whether to have children. Amniocentesis, ultrasonography, and chorionic villus sampling (CVS) made it possible to look into the womb to check on a fetus's condition. The possibility of doing so without risk to the fetus (amniocentesis and CVS each carried the risk of inducing a spontaneous miscarriage) through sampling of fetal cells circulating in maternal blood has begun to loom large. With the 1973 *Roe v. Wade* decision legalizing abortion, diagnosis of a fetal anomaly entailed the new option of therapeutic abortion. These events enlarged the region of reproductive control for families, physicians, and the community. Parents and health care providers were able to participate in social decisions about the traits that are acceptable in a child before that child is born. As the sensitivity and specificity of reproductive genetic testing has improved, the fetus and donor gamete have been more open to genetic testing. Society exerts influences on parents as they make new decisions about how and when and with what outcome they will reproduce, but such influences are the same sort it exerts on those who are deciding with whom to mate, whom to marry, and when to have children.

In most states, hereditary information is also available to institutions, such as employers and the life insurance industry. Companies whose workers are exposed to chemicals (e.g., Kodak and DuPont) routinely screen for hereditary sensitivities to a particular chemical in the work environment (7–10). Insurance companies and governments have begun to debate the use of detailed genetic information about applicants prior to granting health, life, or annuity policies, and laws banning such use have been passed in many U.S. states.

While sterilization by medical institutions or at the hand of state agencies, authorized to act on behalf of patients, is not presently illegal in many nations or U.S. states, existing laws do emphasize the importance of competent decision making and attenuate the role of the state in actual reproductive decisions that might be aimed at the prevention of disease. The repeal of sterilization laws and the censure of institutions that carried out eugenic policies did not render eugenics illegal, but the trend toward patients' rights and autonomy in reproductive health care have made attempts at comprehensive eugenic policies by governments markedly more difficult and visible (11).

Human genetic testing, gene therapy, choice of gametes, gestational carrying, surrogacy, DNA banking, and cloning all portend a range of possible benefits and hazards, only some of which can usefully be understood in terms of the legacy of eugenics. To understand what might be drawn from this century's experience with eugenics, we here make some distinctions among those ethical issues that have been grouped as "eugenics" and discuss their respective implications.

COERCIVE VERSUS VOLUNTARY

Past policies of eugenics became notorious when their application involved the use of brute force, and particularly against those least able to resist. Most notable were state-directed policies that forced sterilization or institutionalization of those whose reproductive capacity was seen as threatening to the public health. It is for this reason that contemporary genetic counselling has avowed an ethic of noncoercion and even nondirectiveness. As patient autonomy and personal freedom have grown in social importance in the twentieth century, so too has an emphasis on allowing reproductive decision making to rest with the parties involved. Important U.S. regulatory and judicial tests of the rights of patients to make reproductive decisions were the legal battle over abortion rights and the willingness of states to regulate assisted reproductive technology. In both areas the United States has endorsed an enormous amount of personal and familial freedom against intrusion by the state into sexual and procreative activity, while still holding that it is the state's role to protect already born children against abuses by parents and others.

A variety of scholars have argued that these personal liberties are "negative" in nature, meaning that citizens are guaranteed only freedom from procreative interference (12). A positive liberty would entail the obligation of the community to provide procreative aid, akin to the courts' avowed responsibility to provide due process or many states' guarantee of a primary education. A negative liberty interest is obviously much narrower. It entails an emphasis by the state and other institutions on the rights of the sexually active, the procreating parent and the future parent, rather than an emphasis on the embodiment of future generations or their genetic endowment.

Interference by the contemporary state in reproductive decision making does take place. U.S. states and the federal government regulate marriage, prenatal testing, licensure for obstetrical services, and reproductive services for minors. In a few cases courts have required that pregnant women take action to protect a fetus that they intended to bear. However, the likelihood of comprehensive state sterilization or genetic discrimination is less today. The primary questions of reproductive freedom (apart from those discussed below under Population and Individual today concern *coercion*, either by government and other institutions or, more controversially, by social conditions more generally.

In the first case, agencies of government or medicine (or other institutions) may offer incentives or structure the delivery of information about reproductive decisions. Obviously it will be very difficult for those making reproductive decisions to do so carefully if physicians, nurses, employers, insurers, clergy, or the state skew or introduce bias in describing reproductive options. Similarly upstream decisions about which reproductive decisions will be covered by insurance companies or other paying organizations will have a marked effect on the ability of many to make choices. When research monies are allocated to one kind of disease or one technology rather than another, options are similarly constrained for those at the bedside. All these cases suggest some interference with an ideal state in which choices are maximally exercised by consenting and mature adults in a state of informed and reasonable decision making. However, it certainly remains to be seen that such a state ever existed, and in any case the development of upstream research and payment schemes cannot be made in such a way as to allow all choices to be made by all persons toward all desired ends.

At issue is the meaning of voluntariness. Two kinds of challenges to the voluntariness of reproductive decisions are made more likely as technology develops in this area. First, reproductive decisions can be made more difficult and less "free" when in a context of insufficient or coercive information. Second, and perhaps more controversially, economic and social pressures may create situations where reproductive decision making is constrained as if the situation were legislated. As philosopher John Dewey claimed, it makes little sense to say that people have a free choice if only one option is practically available. This is particularly important in the context of contemporary genetic testing. Lack of social support, economic security, or insurance could be important factors in determining whether a woman will abort a fetus at risk for a genetic disorder. Taken at the individual level, this is a threat to the negative right against interference with procreative decision making. More globally, entire groups of people, many of whom would share other ethnic or economic or class distinctions, might find themselves left out of genetic testing options or encouraged to utilize tests or procedures that would lessen the costs to the state. The collective impact of such patterns of economic allocation of services, or pressures toward allocation, might well resemble the impact of early experiments with eugenics. More dangerous, such decisions would not be traceable to a particular policy maker or agency but would be suffused in the economic climate of a market in genetic and reproductive technology. Put another way, if a market in genetic services became the primary mode of distribution, and that market failed to provide opportunities for all to make equal choices, the lack of equality might manifest itself in the appearance of a genetic underclass.

POSITIVE AND NEGATIVE

Another important distinction in the eugenics movement was between efforts primarily aimed at producing more people with desired traits and those aimed at eliminating undesirable traits. Many of the early eugenicists wanted to promote increased production of "geniuses" and people of great talent, through encouraging more scientific selection of mates, and more breeding by the chosen few. This was "positive eugenics." In its most extreme form, German SS officers were encouraged to reproduce with Aryan women, and offspring of the unions were placed with families chosen by the lead scientist of the program. Today positive practices include the selection of sperm and egg donors from highly selective reproductive recruiting pools, and a general emphasis on the importance of genetic relationship in the family (13–15).

In contrast, "negative eugenics" was concerned with eliminating the least fit individuals through reducing or eliminating their reproduction. Sterilization laws that were aimed at eliminating defectives from the population came to be the popular image of eugenics in the United States. In the infamous *Buck v. Bell* Supreme Court case (1927), Justice (and American philosopher) Oliver Wendell Homes articulated for many the felt need "to prevent our being swamped with incompetence." In upholding the eugenic sterilization laws, Holmes wrote, "It is better for all the world if, instead of waiting to execute degenerate offspring for crime, or to let them starve for their imbecility, society can prevent those who are manifestly unfit from continuing their kind.... Three generations of imbeciles is enough." (2, p. 83). As has been mentioned, on the basis of these laws, over 60,000 Americans would be sterilized.

The analysis of contemporary policy in genetics and reproduction reveals little statutory regulation in most nations. Certainly there are few nations who express positive eugenic aims, though many have rules against incest and intrafamilial marriage. However, the increasing use of donated gametes that have been highly selected suggests a growing interest in improvement of particular offspring among those able to afford such improvement. Importantly, the interest in such technologies grows out of a felt need to be responsible for the embodiment of children. Parents who are unable to be linked by genetics to their children seek to replace that seamless bond with responsible decision making about what gifts they can instead give a child. In this way, what begin as a few choices about the health of potential gamete donors can expand to include a wide variety of traits that parents would like to be able to give their child in the absence of their own DNA.

In the background of new positive decision making is an assumption that the transmission of DNA from parent to child through sex and recombination is not only common but also medically normal. In offering treatments for infertility that aim at restoring as much as possible the sexually created DNA bond, medicine and society has embraced the assumption that being "related" is largely a function of sharing genetic information. This context emphasizes both a particular kind of relatedness and, more broadly, the importance of DNA and its stewardship. A \$1 billion infertility expenditure in the United States alone is proof of the importance of this faith in the importance of DNA to the economy, but the real impact is the reification of a particular kind of positive reproductive control in which certain kinds of family life are both socially privileged and made medically normal.

POPULATION AND INDIVIDUAL

The most aggressive policies of eugenics were motivated by a concern for the public health, described in terms of the genetic makeup of the population as a whole. Improvement of the population is the explicit aim of these policies. In contrast, most recent practices that are seen as "eugenic" are motivated largely by a desire to improve people's decision-making options or to improve the health of individuals. Genetic testing and screening is intended primarily to be aids for prospective parents facing difficult decisions. However, it is important to keep in mind that there are collective consequences of individual decisions. The *effect* of individual decisions at the population level may be just as significant as the effect of decisions made with the explicit goal of producing population level changes. Worries about the collective consequences of individual decision making — of leaving choices to "the market" and the good intentions of parents

is what motivates concerns about what sociologist Troy Duster referred to as "backdoor" eugenics, the regulation of reproductive patterns by class, access to medical services, and income (16).

MODERN DILEMMAS AND EUGENICS

In several developing nations amniocentesis is used to determine the sex of the fetus, with the goal of terminating unwanted females. This has resulted in skewed sex ratios in India and China, just one example of what can happen if genetic testing and reproductive technologies are utilized in unregulated or poorly structured ways. Yet how do we distinguish between the moral dilemma of an India that aborts unwanted female fetuses and parents who (perhaps aided by a genetic counselor) choose to abort a fetus destined to die an early painful death? Some patients with Huntington's disease feel that the several healthy decades of life that they have is what really matters. Other genetic traits many cause lesser health problems and risks. Will testing eventually stigmatize all those who are "unhealthy" or "abnormal" in any way? Will parents choose to test for other socially important traits, such as being thin, or tall? Will they test for homosexuality along with a propensity to develop heart disease (17,18)?

Critics of backdoor or market eugenics argue that the same prejudices and values that were problematic in the earlier state-directed eugenics are equally present in the new eugenics. They deny that it is possible to delimit the use of genetic selection and manipulation to any value-free concept of disease. Hence "citizens will end up being engineered in accordance with a dominant set of values after all, and the new eugenics will collapse into the eugenics of old (16,19,20).

In response to this, a number of authors have attempted a defense of market eugenics. Some claim that what is objectionable in the old eugenics is absent from the new eugenics and a significant moral distinction can be made based upon concepts such as disease and disability (21). Other "defenders" of backdoor eugenics argue that while there is no nonnormative basis for distinguishing medical from nonmedical (enhancements) genetic interventions, they deny that any sort of unity in values is likely to result from the marketplace (13,19). Kitcher argues that as long as decisions are left to well-intentioned, well-educated parents trying to do what is best for their children, eugenics is inevitable and unproblematic.

A third set of defenders begins by rejecting the "genetic exceptionalism" they see in criticisms of backdoor eugenics. We currently allow tremendous inequalities in access to environmental and educational circumstances that are far more likely to have a direct, measurable impact on the lives of children and their future expectations and opportunities. Exclusive focus on the potential for genetic inequalities is misleading and unjustified (22,23).

Another problem in genetics is the general lack of education about its meaning and use. Most in the world today do not understand genetic science, let alone the complex fact that genetic probabilities are always understood in terms of particular populations in particular environments. As a result many will not understand how to interpret their risks. For example, testing positive for a BRCA mutation for breast cancer will have different implications for a patient with a family history of breast cancer than for one without such a family history. Yet economic incentives may lead to a push for population level screening before we have a good understanding of the relevant risks for most women.

CONCLUSION

Eugenics is a complex concept with a diverse set of meanings and uses and a rich (if nefarious) history. The ethical issues associated with eugenics depend crucially on the different meanings of the term, and on the particular details of the uses and the context of the reproductive practices under consideration.

BIBLIOGRAPHY

1. D. Kevles, *In the Name of Eugenics*, University of California Press, Berkeley, 1985.
2. D. Paul, *Controlling Human heredity: 1865 to the Present*, Humanities Press, NJ, 1995.
3. G. Allen, *Genome* **31**, 885–889.
4. M. Haller, *Eugenics: Hereditarian Attitudes in American Thought*, Rutgers University Press, New Brunswick, NJ, 1985.
5. B. Muller-Hill, *Murderous Science*, Oxford University Press, Oxford, UK, 1988.
6. R. Proctor, *Racial Hygienie: Medicine under the Nazis*, Harvard University Press, Cambridge, MA, 1988.
7. D. Nelkin, *Dangerous Diagnostics: The Social Power of Biological Information*, Basic Books, New York, 1989.
8. M.A. Rothstein, *Houston Law Rev.* **29**(1), 23–84 (1992).
9. E.V. Lapham, C. Kozma, and J.O. Weiss "Genetic discrimination: perspectives of consumers." *Science* **274**(5287), 621–624 (1996).
10. S.M. Wolf, *J. Law, Med. Ethics* **23**(4), 345–353 (1995).
11. P. Reilly, *The Surgical Solution*, Mcmillan, New York, 1993.
12. D. Heyd, *Genethics*, Cambridge University Press, New York, 1993.
13. P. Kitcher, *The Lives to Come*, Free Press, New York, 1997.
14. J. Glover, *What Kind of People Should Their Be?*, Oxford University Press, New York, 1996.
15. T. Murray, *The Worth of a Child*, University of California Press, Berkeley, CA, 1996.
16. T. Duster, *Backdoor to Eugenics*, Routledge, London, 1990.
17. G. McGee, *The Perfect Baby: A Pragmatic Approach to Genetics*, Rowman & Littlefield, New York, 1997.
18. J. Robertson, *Children of Choice*, Princeton University Press, Princeton, 1994.
19. N. Ager, *Public Affairs Q.* **12**, 137–155 (1998).
20. B. Muller-Hill, in A. Clarke, ed., *Genetic Counselling: Practice and Principles*, Routledge, London, 1994, pp. 133–141.
21. J. Harris, *Wonderwoman and Superman: The Ethics of Human Biotechnology*, Oxford University Press, Oxford, 1992.
22. A. Caplan, G. McGee, and D. Magnus, *B. Med. J.* **319**, 1284–1285 (1999).
23. J. Glover, *What Sort of People Should There Be?*, Penguin, Harmondsworth, UK, 1984.

See other entries Disability and biotechnology; Eugenics, ethics, sterilization laws.

EUGENICS, ETHICS, STERILIZATION LAWS

Philip R. Reilly
Shriver Center for Mental Retardation, Inc.
Waltham, Massachusetts

OUTLINE

Introduction
Early Eugenic Sterilization Laws
    Origins and Rise of the Eugenics Movement
    Impact of Vasectomy
    First Sterilization Laws
Resurgence of Sterilization Laws
    Eugenic Thinking in the 1920s
    Buck v. Bell
    Sterilization Data
    Critics
Europe and Canada
Sterilization in the United States After World War II
    Sterilization in the Courts
Eugenic Laws Today
Conclusion
Bibliography

INTRODUCTION

During the last quarter of the nineteenth century and the first half of the twentieth century, the idea that the instruments of social policy could and should be used to protect the gene pool of a nation's population attracted broad interest in the United States and in several European nations. Beginning in 1907 a number of states and nations began to enact compulsory laws that authorized officials to order the sterilization of institutionalized (allegedly, retarded) persons. Prior to World War I, a number of state supreme courts held such laws to be unconstitutional. In 1927 the United States Supreme Court upheld a revised state involuntary sterilization law as a valid exercise of the police power. The years from 1927 to 1939 mark the zenith of such eugenic programs. In the United States at least 60,000 institutionalized persons were sterilized pursuant to state law. In Germany during the period 1934 to 1945 many hundreds of thousands of person were sterilized pursuant to a law with similar intent and broader reach. Although eugenically rationalized involuntary sterilization programs remained active, particularly in several southern states after World War II, the scope of most state programs rapidly diminished. The major reasons for the decline included the growing sophistication of genetics which revealed the intellectual bankruptcy of most eugenic ideas, revulsion from the revelations concerning events in Nazi Germany, and the growth of the Civil Rights Movement. Eugenic thinking remains widespread in the world today. A prime example is a maternal and infant health law that was enacted in China in 1995, a statute with sections that recall the sterilization laws of the 1930s.

EARLY EUGENIC STERILIZATION LAWS

Origins and Rise of the Eugenics Movement

Although speculations about the perfectibility of humankind date at least as far back as the flowering of philosophy among the ancient Greek city states, the first serious social policy proposal to improve the gene pool of our species that was tied to scientific claims arose in England in the second half of the nineteenth century. Francis Galton, a Victorian polymath who was probably influenced by the revolutionary impact of *The Origin of Species* published in 1859 by his cousin, Charles Darwin, began investigating the inheritance of talent among eminent English families about 1864 (1). He coined the term eugenics (from the Greek, fusing words for good and birth) in 1883 in *Inquiries into Human Faculty and its Development* (2). Public interest in the notion that success and failure in life might be closely tied to the germ plasm that one inherited grew rapidly, especially in England and the United States, and the new word enjoyed a certain vogue. In 1904, when he made a major financial gift to the Eugenics Record Office at the University of London, Galton drafted an official definition of "natural eugenics" as "the study of the agencies under social control that may improve or impair the racial qualities of future generations either physically or mentally" (1).

In the United States eugenic policies, which flowered in the early twentieth century, germinated in the climate of progressive reform that took root in the last quarter of the nineteenth century. From 1850 to 1880, the states had built many prisons, hospitals, insane asylums, and colonies for the mentally retarded. Initial enthusiasm faded as funding problems arose and conditions declined. Richard Dugdale, a well-to-do Englishman who made New York City home, was an ardent social reformer and one of many who sought to improve such facilities. It was while inspecting prisons in upstate New York that he discovered a large family many of whose members seemed to be inhabiting one state facility or another (3). His book, *The Jukes* (4), based on an exhaustive study of the family, detailed the cost to taxpayers of their incarceration and support. He also championed two ideas that would become core beliefs among many Americans within a decade: that feeblemindedness, epilepsy, drunkenness, criminality, and insanity had strong hereditary influences, and that affected individuals tended to produce larger than average numbers of offspring.

Interest in eugenics was widespread in the United States just as biologists rediscovered the laws of inheritance that Gregor Mendel had postulated in 1865, but which had been reported in an obscure local scientific journal and been little noticed. Charles Davenport, a talented young biology professor, played an instrumental role in propagating "mendelism" in the United States. He was quick to apply the theory to problems of human heredity and about 1905 he secured a large gift from Mrs. E.H. Harriman (the wife of the railroad magnate) to develop and sustain a eugenics research facility at Cold Spring Harbor on Long Island, an entity which operated independently of the Station for Experimental Evolution that he had founded there a couple of years earlier. One of his most important decisions was to recruit a midwestern high school teacher named Harry Hamilton Laughlin to direct the Eugenics Record Office. In so doing, Davenport, a highly respected biologist who would become a member of the National Academy of Sciences, tied human genetics to eugenics and provided eugenics with a cloak of scientific legitimacy that it wore for more than three decades (5).

The indefatigable Laughlin became an ardent eugenicist who, from 1918 to 1939, played the premier role as a strategist for the eugenics movement in the United States. One of his early projects was to train cadres of young women as eugenic field-workers. Armed with knowledge on how to prepare detailed family pedigrees, many of these workers reviewed the records of thousands of institutionalized persons and interviewed their relatives, thereby gathering the raw material that became the basis for what might be considered the nation's first foray into sociobiology (3). Between 1910 and 1920 eugenicists working in association with the Eugenics Record Office and sometimes assisted by Laughlin's field-workers published a number of lengthy monographs with colorful names, such as *The Hill Folk: Report of a Rural Community of Hereditary Defectives*. These monographs reinforced the eugenic ideas propounded in *The Jukes* and the equally famous *The Kallikaks* (1912), written by Henry Herbert Goddard, a prominent psychologist who worked at the Vineland Training School in New Jersey and who imported IQ testing from France about 1905 (6). These and similar works caught the attention of American journalists. About 1910 articles on eugenics constituted the second most popular topic in the print media according to the Index to Periodical Literature (3).

Laughlin played a critically important role in the effort to secure enactment of federal laws to limit the immigration of persons that he and many others believed were of inferior racial stock. He conducted and published surveys that purported to show that immigrants from southern Europe and Russia were much more likely than immigrants from northwestern Europe to wind up in charity hospitals or need public assistance. In 1922 he served as the official expert on eugenics to the committee in the United States Congress charged with immigration matters. In that role he provided testimony that offered an apparent scientific basis to rationalize a legislative quota system that favored the immigration of some ethnic groups over others from 1924 to 1968 (7).

During the 1920s Laughlin also spent an immense amount of time drafting and lobbying for the enactment of laws to permit state officials to sterilize institutionalized retarded persons without their consent. He helped propagate the second wave of such laws that swept through America in the 1920s, and he provided an important deposition in the lower court proceedings that led to *Buck v. Bell* (8), in which the United States Supreme Court ultimately upheld the constitutionality of a law he helped to craft. This opinion removed lingering doubts in many state legislatures and made possible the enactment of about a dozen new laws (discussed below) in the ensuing five years.

Impact of Vasectomy

Prior to 1910 in the United States proponents of eugenics tried a variety of methods to reduce child-bearing by persons thought to be unfit. In some states, notably California, the focus was on the insane; in others such as New York and New Jersey, there was special concern for "protecting" mentally retarded young women who, it was feared, were especially vulnerable to unscrupulous men. The typical proposals were to segregate the sexes in institutions and prohibit trysts. In society at large the problem was dealt with by the enactment of marriage restriction laws that forbade the insane, the retarded, the epileptic, public drunkards, those with tuberculosis, and even, in some states, the poor, from marrying. Unlike anti-miscegenation laws which were vigorously prosecuted, marriage restriction laws were not much enforced. Those few that were legally challenged in the years before World War I did not survive constitutional scrutiny.

A clinical advance, the development of the vasectomy, about 1897 had an obvious, material impact on the rise of sterilization laws. A.J. Ochsner, chief surgeon at St. Mary's Hospital in Chicago, published his surgical experience with several cases in the *Journal of the American Medical Association* in 1899 (9). The paper carried the remarkable title: "Surgical Treatment of Habitual Criminals." After describing the new surgical technique. Dr. Ochsner asserted: "if it were possible to eliminate all habitual criminals from the possibility of having children, there would soon be a very marked decrease in this class." Further, the same treatment "could reasonably be suggested for chronic inebriates, imbeciles, perverts and paupers." Although there was no comparable, safe operation for women, Ochsner opined that since most female criminals were also prostitutes and were highly likely to become infertile due to the impact of untreated gonorrhea, their class would produce relatively few children.

Ocshner's proposal received a large boost in 1902 when Dr. Harry C. Sharp, the surgeon for the Indiana Reformatory in Jeffersonville, reported on the follow-up of 42 prisoners who had agreed to undergo vasectomy, claiming that the patients "feel that they are stronger, sleep better, their memory improves, the will becomes stronger, and they do better in school" (10). He urged his fellow physicians to lobby the leaders of state institutions of all kinds "to render every male sterile who passes its portals, whether it be almshouse, insane asylum, institute for the feeble-minded, reformatory or prison." Policy makers, who had viewed earlier, sporadic proposals to castrate criminals with distaste, greeted the suggestion to use the much less mutilating vasectomy with enthusiasm (3).

In the years 1902 to 1912, Sharp was an outspoken advocate for vasectomy as a social tool. He spoke on the subject frequently at regional and national medical meetings (often exaggerating the salubrious effects of sterilization), wrote political pamphlets on eugenics, and button-holed state legislators. It is no accident that in 1907 Indiana became the first state (indeed, the first political jurisdiction in the world) to enact an involuntary sterilization law that had a demonstrably eugenic underpinning (3).

First Sterilization Laws

The nation's first sterilization bill was introduced in the Michigan legislature in 1897, but did not come to floor vote. In 1905 the Pennsylvania House of Representatives became the first legislative body to pass a bill proposing involuntary sterilization of certain institutionalized persons, but the governor vetoed it. On April 9, 1907, Indiana governor J. Frank Hanly, a month after a sizable majority of both houses had voted favorably, signed the nation's first eugenic sterilization law. The statute authorized the compulsory sterilization of "confirmed criminals, idiots, imbeciles and rapists" residing in a state institution if, after appropriate review, a panel of one physician and two surgeons concluded that it was "inadvisable" that the individual procreate and that there was "no probability of improvement." The new law was crafted to legitimize the program that Dr. Sharp was already vigorously pursuing in Jeffersonville, except it eliminated the pretense of obtaining consent (3).

In 1909 the legislatures in California, Connecticut, Oregon, and Washington passed similar laws. Despite overwhelming support in the legislature, the governor of Oregon vetoed the bill sent to him; the other three governors promptly signed their bills into law. In the ensuing four years (1910–1913), ten states (Iowa, New Jersey, Nevada, New York, North Dakota, Michigan, Kansas, Wisconsin, Vermont, and Nebraska) passed sterilization laws. In general, there was little opposition and most votes were lopsided. Only in the state where the vote was close (96–82 in the House), Vermont, did a governor cast a veto (3).

The California law, which launched the most active eugenical sterilization program in the United States until well into the 1920s, was slightly more sophisticated. Section 1 of the law covered institutionalized persons who had been diagnosed with "hereditary insanity," "incurable chronic manic," and "dementia," requiring that their discharge must be premised on "asexualization." Used in the early days as a synonym for vasectomy, this term was often confused with castration. Section 2 targeted recidivists in state prison. It identified three persons (the resident prison physician, the general superintendent of state hospitals, and the secretary of the state board of health) to review the cases of persons who had been convicted twice of rape or sexual assault or three times of having committed other crimes, and who while in prison continued to show evidence that they were moral or sexual degenerates. If two of the three reviewers concluded that there was no hope of "moral recovery," they could order that the prisoner be sterilized without his consent. The third section of the law directed the state to pay for the sterilization of institutionalized retarded children or adults so long as their parents or guardians consented. This relatively enlightened section may explain why the constitutionality of this law was not challenged (11).

By 1913 involuntary sterilization programs were active in 14 states. There were significant differences in the their scope and pace, partly because in a number of the states opponents of eugenics attacked the constitutionality of the enabling laws. In every instance (Indiana, Iowa, Michigan, Nevada, New Jersey, New York, and Washington) in which the constitutionality was put at issue, the courts invalidated the laws, usually on the grounds that they violated the requirements of the Due Process Clause of the Fourteenth Amendment (12). Laws that targeted prisoners were held to violate the Eight Amendment prohibiting cruel and unusual punishments (13). In Oregon the sterilization law, which also was challenged on constitutional grounds, was repealed by public referendum only months after the governor signed it (3).

From 1907 to 1922 activities in California account for the vast majority of reported involuntary sterilizations conducted pursuant to state law. Of a national total of 3,233 operations, 2,558 were performed there, most on institutionalized, mentally ill persons. During that era a substantial fraction of the men and women who were discharged from a California state hospital were sterilized. The women usually were subjected to oophorectomy (removal of the ovaries), in those days a risky operation that resulted in several deaths. Due to the vigorous commitment of its medical superintendent, Dr. John Reily, during this era the South California State hospital in Patton sterilized 1,009 of its residents. In March 1918, Dr. Reily reported that he had sterilized 43 persons (3).

The constitutional challenge brought against the New York sterilization statute is of special interest because of the involvement at trial of some of the leading strategists of the national eugenic movement. The case arose because of disagreements among public officials about the scientific rationale that purported to justify sterilizing retarded persons to prevent the birth of children who too would be retarded. To resolve the dispute, the Board of Examiners of the Custodial Asylum in Rome, New York, ordered that a 22-year-old man named Frank Osborne be sterilized, knowing that his attorney would challenge the law. At trial Dr. Francis Bernstein, the superintendent of the facility, vigorously opposed the law, claiming correctly that if feeblemindedness was a recessive genetic disorder (a commonly held view in that time) than it was unlikely that Osborne would father a genetically retarded child. Charles Davenport also testified, no doubt disappointing many eugenicists by favoring a policy of segregation of the sexes over sterilization. Another prominent eugenicist, Bleeker Van Wagenen, favored sterilization but acknowledged it would be preferable to obtain the consent of the retarded person's guardian. The court, in part because of the weakness of the scientific arguments offered in support of sterilization, struck down the law (14).

During the years just before and during World War I, there was a hiatus in the enactment of sterilization laws. Almost certainly this was because of the failure of existing statutes to survive constitutional challenge. But the hiatus was probably also influenced by the sharp drop in immigration during that era. In the first two decades of the twentieth century American eugenicists were far more vexed by the massive influx of what they were convinced were racially inferior people than by the slowly growing number of retarded persons who were housed in state institutions. As the tide of immigration rapidly subsided during 1914 to 1919, the sense of urgency among eugenicists may have relaxed (15).

RESURGENCE OF STERILIZATION LAWS

Eugenic Thinking in the 1920s

During the 1920s the eugenics movement, which prior to the World War had begun to decline, grew and prospered. August institutions, such as Yale University, the Cold Spring Harbor Laboratory, and the New York Museum of Natural History numbered intellectual leaders of eugenics on their faculties. The Second International Congress of Eugenics met in New York City in 1921. By 1924 the New York based American Eugenics Society was lobbying in Albany against bills that its members thought were dysgenic (e.g., including those intended to provide financial assistance to poor women with school age children) (3). Across the nation, local eugenic societies flowered. In 1926 the Human Betterment Foundation, the pet project of an eccentric California millionaire named Ezra S. Gosney, emerged as a major voice for eugenics on the West Coast (3). In Cleveland in 1928 Charles F. Brush, a successful inventor, launched the Brush Foundation for the Betterment of the Human Race with one stated goal being the propagation of eugenic goals. In Topeka, J.H. Pile, a self-made millionaire who had studied at Yale, founded the Eugenic Babies Foundation. Especially in the midwest, interest in positive eugenics (the search for methods to have genetically superior children) captured the imagination. County fairs sponsored "fitter family" contests in which people, not unlike the prize hogs or cattle they showed, competed for blue ribbons based on their pedigree, physical examinations, and their children's report cards (16).

At the Eugenics Record Office, Laughlin, stung by the constitutional defeats suffered by the involuntary sterilization laws between 1913 and 1918, produced a massive tome on the societal benefits of eugenic sterilization (11). He carefully analyzed the laws that the courts had found flawed, and then drafted and circulated a model sterilization law that he hoped would satisfy the constitutional concerns. In the early 1920s his work was widely used by legislators who wanted to sponsor such bills. His polemics on sterilization found their way to Nazi Germany where Laughlin was held in such high regard that he was awarded an honorary degree by the University of Heidleberg in 1934 (7).

Beginning in 1923 there was a major resurgence of sterilization laws. After five years of legislative inactivity, new laws were enacted in Delaware, Michigan, Montana, and Oregon. Virginia adopted a law in 1924, and governors signed seven of the nine bills that were passed in 1925. By January 1926 eugenic sterilization laws were on the books of 17 states and small bands of pro-sterilization lobbyists were urging many others to follow suit. In some states directors of state institutions permitted involuntary eugenical sterilizations to occur despite the absence of enabling laws (3).

In mid-1925 the Michigan Supreme Court upheld the constitutionality of the new law which it held was "justified by the findings of Biological Science," and was a "proper and reasonable exercise of the police power of the state" (17). This greatly encouraged other legislatures, but many still wondered how such laws would fare before the United States Supreme Court. In a bold move, pro-sterilization forces in Virginia decided to find out. The resulting decision, *Buck v. Bell* (8), was the single most important event in the history of the sterilization laws in the United States.

Buck v. Bell

In 1924 the Virginia legislature had by a nearly unanimous vote (30–0 in the Senate and 75–2 in the House) passed a law that authorized the superintendents of the five state institutions for the retarded to petition a special board for permission to sterilize their wards. A few months later, Dr. A.S. Priddy, the superintendent of the State Colony for Epileptics and Feeble-Minded, who had decided to test the constitutionality of the law (which he supported) selected an 18-year-old woman named Carrie Buck as his first candidate for surgery. By 1925 surgeons had become fairly adept at performing tubal ligations, although such surgery was certainly much more risky than was vasectomy. One reason why Priddy chose Carrie for the test case was that she was allegedly the daughter of a retarded woman and she had recently given birth to a child who was also reported to be retarded. The superintendent thought he had a clear case of hereditary mental retardation (18).

To bolster his case, Dr. Priddy asked a number of experts to examine Carrie or review her records. Among them was Harry Laughlin, who after reviewing her case file, concluded that Carrie was part of the "shiftless, ignorant, and worthless class of anti-social whites of the South." He opined that the possibility that her feeblemindedness arose from nonhereditary causes was "exceptionally remote" (18). On September 10, 1924, the review board approved the sterilization petition, and as planned, R.G. Shelton, her court-appointed attorney, immediately filed an appeal in the local circuit. The constitutional challenge was heard in November. The appellees assembled a formidable array of eugenics experts who testified on behalf of the scientific validity of sterilization programs. Attorney Shelton did not offer a single expert to rebut those claims. In April 1925 Judge Bennet Gordon upheld the law and ordered the operation to take place within 90 days. On appeal the Virginia Supreme Court unanimously upheld the law, finding that it was intended to benefit the persons who would be sterilized (19).

Shelton appealed to the United States Supreme Court. On May 2, 1927, by a vote of 8–1 the high court held the Virginia law to be a constitutionally valid exercise of the police power that did not run afoul of the Equal Protection Clause. Writing for the court, Judge Oliver Wendell Holmes, Jr. asserted: "It is better for all

the world, if instead of waiting to execute degenerate offspring for crime, or to let them starve for their imbecility, society can prevent those who are manifestly unfit from continuing their kind. The principle that sustains compulsory vaccination is broad enough to cover cutting the Fallopian tubes" (8). Holmes, who in private correspondence revealed that he shared many commonly held eugenic tenants, wrote to a colleague that he was particularly proud of the social contribution he made by upholding the Virginia law (3).

Sterilization Data

Buck v. Bell put the constitutional questions to rest. The legislative response was swift. In 1929 nine states enacted sterilization laws, and by the end of 1931 laws had been enacted in 28 states. At the Eugenics Record Office in Cold Spring Harbor, it was Laughlin's pleasant task to monitor the spread of sterilization programs and keep tabs on the victories. His work and the surveys conducted annually by the Human Betterment Foundation (to which the state agencies dutifully reported their activities) (20) provide substantial documentation of the number of persons sterilized for eugenic reasons. The next 15 years were unquestionably the heyday of eugenical sterilization in the United States.

In California the number of reported sterilizations rose from 322 in 1925 to 2,362 in the period from January 1928 through December of 1929, about a 300 percent annual increase. In Minnesota in the decade 1916 to 1925 only 21 eugenical sterilizations were performed. In the ensuing decade nearly 1300 were performed pursuant to the same law. In 1938 the Minnesota School for the Feeble-Minded in Faribault admitted 452 new patients and sterilized 151. Nationally there were about 2,500 to 3,000 operations each year. The highest reported annual figure was 3,921 in 1932, but in 1940 the nation's institutions still reported in excess of 2,800 operations (3). These are well-documented, minimum estimates. The archives of the Human Betterment Foundation contains correspondence in which state officials and judges acknowledge that illegal sterilizations were being performed that for obvious reasons could not be officially reported. A Michigan probate judge wrote that he knew of 71 illegal sterilizations; the Maine assistant attorney general reported that "many more" operations were being performed than were being reported (3).

Despite the relative uniformity of the state laws, the programs differed substantially among states, among institutions within states, and from year to year. In most states the single most critical factor was the attitude of the superintendent of each institution. If the superintendent opposed eugenic sterilization, as was frequently the case in the northeastern states, few operations were performed with or without enabling laws. If the superintendent vigorously supported sterilization, the programs could be ensured of adequate budgets and political protection and could flourish.

During the late 1920s and 1930s, even as the pace of eugenical sterilization picked up, the rationales and the goals shifted. Perhaps the most significant change was that with each year young women accounted for a higher percentage of those who were sterilized. In 1924 the cumulative data showed that 47 percent of those who had been sterilized were women. By the end of 1940 the data indicated that women accounted for 58 percent of the all time totals. Between 1927 and the end of 1941 about 18,300 women were sterilized, while 11,200 men underwent the procedure (3). Several factors contributed to this pattern. By 1930 American surgeons had substantial experience with tubal ligation which, being a much less risky procedure than hysterectomy and oophorectomy, they were more willing to perform. Also by this period early convictions that mental retardation was explained by a few dominant or recessive genes that obeyed Mendel's laws had dissolved, and few good physicians were willing to predict the risk that a person would parent a "defective" child. But, they were willing to predict who would be a "defective" parent.

The impact of the Depression Era economy was almost certainly very significant. Strapped by limited resources, some state officials abhorred the thought of poor, young women bearing children out of wedlock many of whom would become wards of the state. In many states the profile of the individual most likely to be admitted to the local institution was a mildly retarded teenage girl who had either born one child out of wedlock or was thought very likely to become pregnant by an unscrupulous man. In many instances these young women were admitted mainly so that could be sterilized, and rapidly thereafter returned to the community. A 1928 Wisconsin report concluded: "Many mentally deficient persons by consenting to the operation are permitted to return, under supervision, to society where they become self-supporting social units and acceptable citizens. Those inmates unwilling to consent to the operation remain segregated for social protection as well as individual welfare" (21, p. 28). Wisconsin prided itself on not performing sterilization's without consent, but seemed to overlook the fact that when the alternative is incarceration, consent cannot be voluntary.

The Wisconsin report was echoed in many other states where again and again sterilization was tied to release. Many agencies generated follow-up studies that glowingly reported that, once sterilized, midly retarded women were easily maintained at home or made wonderful live-in domestic servants. At the same time there was less interest in sterilizing retarded men because they were considered unable to have access to partners outside of the institution. Also sentiment in favor of sterilizing habitual criminals and rapists had faded as the eugenical rationale had been found scientifically wanting. At the turn of the century leading criminologists accepted a biological basis for many crimes (22), but by 1930 criminological thinking as to cause greatly favored socioeconomic forces as being of paramount importance. A genetic rationale was fading fast, but the idea that certain persons were not fit to be parents remained strong.

Another curious aspect of the history of eugenic sterilization in the United States is that the state programs seemed to flourish in different regions in different decades. During the first quarter of the century, California far outpaced the rest of the nation in number of persons sterilized. In the late 1920s and 1930s, programs in several midwestern states were the most active. In the 1950s and 1960s, three southern states accounted for more than half of the sterilizations performed on institutionalized persons in the nation (23).

Critics

Despite its legislative successes and the level of its programmatic activity in the United States during the late 1920s and 1930s, eugenical sterilization met with sustained criticism in some quarters. They included biologists and physicians, social scientists and lawyers, and, most notably, the Catholic Church. Although many scientists were sharply critical of eugenic theorists for having built social policy upon shoddy scientific data, few academic biologists were willing to take on the public role that was needed to testify against proposed bills. Two notable exceptions were the distinguished zoologist, Herbert Jennings, and Raymond Pearl, director of the Institute for Biological Research, both of whom worked at Johns Hopkins (24). The brazen manner in which Nazi Germany made its perverse eugenic ideology ever more public during the 1930s does not seem to have had much impact on academic geneticists in the United States. A significant minority of sociologists, social workers, and psychologists resisted eugenic sterilization programs from their origins, but with only modest effect. During the 1940s prominent geneticists like L.C. Dunn and Theodozius Dobzhansky argued vociferously against the eugenic proposals (25) and their influence was palpable.

Although many physicians expressed doubt about sterilization programs, only a few took up the cause. No one was more effective than Dr. Abraham Myerson, a Tufts neurologist who was especially troubled by trends in Germany. In 1936 he coauthored a report that advocated that sterilization programs should only proceed if it was based on a freely given consent, that enabling laws should apply equally to all citizens, and that advice about human sterilization should be sought only from recognized experts (26).

In many states the Catholic Church provided the only major opposition to sterilization bills. The decision by the governor of Colorado in 1927 to veto a sterilization bill can in part be traced to the strong opposition to the program from organized groups of Catholics. Eugenicists reported that Catholic groups had constituted the main (and, ultimately, effective) opposition to bills in New York and Connecticut. In 1930 Pope Pius XI issued *Casti Connubi* (On Christian Marriage), an encyclical that harshly criticized eugenics (27). During the 1940s the National Catholic Welfare Conference lobbied hard against eugenics (23). Well-placed eugenic strategists asserted that the organized Catholics were instrumental in defeating a sterilization bill in Wyoming (3).

EUROPE AND CANADA

England, the birth place of Social Darwinism and of eugenics, never enacted an involuntary sterilization law, nor came close to implementing eugenic social programs. Certainly it had its share of unabashed advocates, such as Robert Reid Rentoul, a Liverpool physician who published a book entitled, *Proposed Sterilization of Certain Mental and Physical Degenerates* in 1903. During the decade before World War I the Eugenics Education Society was fashionable at Oxford and Cambridge, and its rolls numbered many scientists. But efforts to include sterilization programs in the Mental Deficiency Act of 1913 failed, and the issue was rarely, if ever again, debated in Parliament. This may have been in part because the academic geneticists in England argued that there were insufficient data to draw any inferences concerning the genetic influence on mental retardation, much less the likelihood that a particular person or couple would parent a retarded child (28).

Canada seems to have been influenced by events in the United States. Beginning in 1928 the Province of Alberta operated a sterilization program remarkably similar to the model advocated by Laughlin which resulted in the sterilization of several thousand persons. It continued to operate out of the spotlight until 1960 when government officials ended it. During the 1990s Alberta was the defendant in a class action lawsuit filed by some of those who had been sterilized (29). On November 2, 1999, the Government of Alberta reached an out of court settlement awarding $82 million to a group of 246 persons who had been sterilized pursuant to its law.

Involuntary sterilization in the name of eugenics reached its apotheosis in Nazi Germany from 1934 to 1945. Popular interest in eugenics swept through Germany early in the century. Interest was high enough that a sterilization bill was introduced in the Reichstag in 1907, but it was soundly rejected. World War I and its devastation curtailed interest, but in Germany, as elsewhere, there was a resurgence during the 1920s. The first university professorship in eugenics was established in Bavaria in 1923. In 1921 the German Society for Race Hygiene adopted a 41 point manifesto on eugenics that favored the right of defective persons "to be sterilized by their own wish" (30). About this time Adolph Hitler was writing *Mein Kampf*, in which he urged that "to prevent defective persons from reproducing equally defective offspring is an act dictated by the clearest light of reason. Its carrying out is the most humane act of mankind. It would prevent the unmerited suffering of millions of persons, and, above all, would, in the end, result in a steady increase in human welfare" (3). This language is strikingly similar to the opinion in *Buck v. Bell*.

A comprehensive German eugenic sterilization law was enacted on July 14, 1933. Pursuant to it, the nation set up a network of Hereditary Health Courts empowered to sterilize persons about whom, in "the experience of medical science, it may be expected with great probability that their offspring may suffer severe physical damage." At first persons with any one of nine conditions were targeted: inborn feeblemindedness, schizophrenia, manicdepressive insanity, hereditary epilepsy, Huntington's chorea, hereditary blindness, hereditary deafness, severe hereditary physical deformity, and severe habitual drunkenness." Each special court had three members: a district judge, a local public health official, and a physician deemed to be expert in making the evaluations of the individuals thought to be at risk (31).

The scale of the eugenic sterilization program in Nazi Germany dwarfed those in all other nations, including the United States. In 1934 more than 200 courts received 84,500 petitions to sterilize. These were sometimes filed by doctors or local public health officials, but often they were filed by one family member about another. In one of the more extreme examples of patriotism, substantial numbers of deaf persons volunteered to be sterilized as a show of support for the "fatherland." Of the 64,499 petitions that were heard, the courts decided for sterilization in 56,244 for a eugenic conviction rate of 87 percent. By 1935 more than 150,000 sterilizations had been approved, many based on judicial proceedings that must have taken under an hour (32).

Over the ensuing years the scope of the law was broadened. For example, in 1934 it was amended to apply to non-Germans living in Germany. During the 1940s people were often sterilized on the weakest of pretenses, such as being half-Jewish. In 1951 the Central Association of Sterilized People in West Germany estimated that the Nazi programs had sterilized 3,500,000 persons, although it is not possible to document the claim (3).

Nazi Germany also operated a positive eugenics program known as Lebensborn that fostered reproduction by ideal Aryan couples selected for that purpose by public health officials (33). Such couples and their offspring received a variety of extra public benefits (tax breaks, funds for child support). At about the same time in the United States a private organization, The Pioneer Fund, with the approval of federal officials, operated a similar program that offered financial aid to any officer in the U.S. Air Corps with three children who had another one during the year 1940. About a dozen children eventually received such scholarships (34).

Eugenical sterilization laws were enacted in other European nations as well. Norway and Sweden adopted programs in 1934 and Finland did so in 1935. The program in Sweden, which remained active into the 1970s, was not anchored to unsupported genetic assumptions concerning the transmission of genes that caused mental retardation. Instead, it targeted individuals who were not thought fit for parenthood, arguably a much larger cohort (3). During the early 1990s Sweden repudiated its sterilization program. In November 1999 Swedish officials disclosed that more than 500 persons who alleged that they had been involuntarily sterilized had filed for compensation, and that it had awarded and would continue to award 175,000 crones ($21,250) to each person who it determined had been sterilized without consent.

STERILIZATION IN THE UNITED STATES AFTER WORLD WAR II

Although many state sterilization programs went into sharp decline or ceased during World War II, it is inaccurate to assert, as has frequently occurred, that revulsion over the crimes committed by the Nazis was the major impetus for this change. The large reduction in sterilizations from 1942 to 1945 was mainly due to the unavailability of trained nurses and surgeons to serve the state institutions. The urgency of the war effort put the programs on hold. The war almost certainly speeded a decline that was inevitable. Also, during this hiatus and into the late 1940s and 1950s, genetics continued to mature. Early notions that a phenotype as complex as mental retardation could be explained by a few purely mendelian alleles were now regarded as at best quaint and at worst dangerous. Justice Holmes dictum that "three generations of imbeciles are enough" rang scientifically hollow and was embarrassing.

Nevertheless, in some quarters sterilization programs flourished. Of particular interest is the work of the Human Betterment Leagues, the brainchild of Dr. Clarence Gamble, a physician and an heir to the Proctor & Gamble soap fortune (35). In 1944 Gamble joined Birthright, a successor to the Sterilization League of New Jersey, an unabashedly proeugenics group run by a former social worker, Marion Norton. From 1941 to 1942 this group had tried, but repeatedly failed, to secure a sterilization law for New Jersey. Gamble arranged for the Cincinnati-based Gamble Trust Fund to grant Birthright a gift of $10,000 for educational work. These funds were used, in effect, to decentralize Birthright by founding a number or local programs in other states (3). These Human Betterment Leagues sprung up in several states in the midwest and south. The effort had its greatest impact in North Carolina from 1945 to 1963 (36).

In 1945 Gamble funded a study of IQ among rural residents of North Carolina and reported the results to public health officials. The disturbing findings led the state officials to permit a pilot voluntary sterilization program in Orange county in which trained sociologists identified women whom they felt were incompetent to be mothers and offered them the option of sterilization. After two years officials considered the program a great success, and by 1948 social workers throughout rural North Carolina were searching for appropriate candidates. From 1948 to 1955 about 186 women (mostly from rural areas), half of whom never had resided in a state institution, were sterilized (36). In Iowa a similar effort undertaken from 1944 to 1947 resulted in about 50 allegedly voluntary sterilizations each year.

During the years from 1952 to 1958 from 50 to 75 percent of all eugenical sterilizations conducted pursuant to state law were performed in just three states—Georgia, North Carolina, and Virginia. In 1958 those states reported sterilizing 574 persons, 76 percent of the nation's total.

There is, with one exception, no evidence that programs in the southern states were racially motivated. In fact state sterilization programs in the deep south (Alabama, Georgia, and South Carolina) originally targeted whites (23). This was because during the 1930s when eugenic sterilization programs in the deep south commenced, most state institutions for the retarded and the mentally ill provided comparatively few or no beds for blacks. It also reflected the overriding concern among southern eugenicists to purify the Caucasian race. In the end, as segregated state facilities were made available to blacks, they and whites were sterilized in numbers that roughly mirrored the composition of the general population. Of course they were sterilized in racially segregated facilities.

South Carolina is the only state in which there is evidence that sterilization programs were aimed directly at black persons residing in state institutions. Although the state facility for the mentally retarded did not admit blacks, the facility that housed the incurably mentally ill did. During the years from 1949 to 1960, 104 inmates of the later institution were sterilized, of whom 102 were black (3).

Georgia offers a good example of the idiosyncratic manner in which state sterilization laws were often implemented. Although covering all state institutions for the retarded and/or insane, the vast majority of sterilizations were performed on the residents of only one, the huge (12,000 residents) Milledgeville State Hospital for the mentally ill. Among residents at Milledgeville, schizophrenia was the most prevalent single diagnosis. During the heyday of sterilization, most Georgia physicians and alienists (psychiatrists) held the thesis that schizophrenia was largely hereditary. For well over a decade physicians at Milledgeville sterilized more than 200 persons each year. Unlike officials at other state institutions, authorities at Milledgeville kept the sterilization effort quiet. This was one reason it continued into the early 1960s (23).

During the 1960s the number of reported eugenical sterilizations, already much below the numbers reached in 1930s, declined to very low levels. There is some evidence that inside the gates of a few state institutions sterilizations continued to occur, but these were usually done at the behest of relatives who feared that a retarded young woman would be seduced or raped and become pregnant. From June 1970 through April 1974, at least 23 sterilizations were performed on residents of North Carolina institutions pursuant to the eugenics statute (37).

Although one cannot point to a moment in which state-sanctioned eugenical sterilization in the United States ended, a satisfactory date is 1983 when a class-action lawsuit brought by women in Virginia who had been sterilized without their consent while in state facilities was settled. The case, *Poe v. Lynchburg*, was filed by the American Civil Liberties Union (ACLU) in December 1980 on behalf of five plaintiffs and their class. The allegations of one plaintiff are illustrative. She claimed that she was not retarded (38). She had been admitted to the Lynchburg Training School in 1949 after giving birth at age 14 to a child whom she said was conceived when she was raped by her stepfather. At the facility she was told she needed an appendectomy; shortly after undergoing the operation she was discharged. A few years later she married but did not discover that she was infertile because she had undergone a tubal ligation until more than two decades later. The records of many state institutions include summary surgical statistics that support this, for they indicate that a incomprehensibly large number of appendectomies were performed on young women.

The lawsuit, which among other things, challenged the constitutionality of the Virginia sterilization law, had little chance of success, for it was taking on the very statute that had been upheld in *Buck v. Bell*. However, under the 1983 settlement the state of Virginia agreed to attempt to locate all living persons who had been sterilized pursuant to the law, and to provide them with modest compensation. Relatively few came forward (3).

Sterilization in the Courts

Given the legitimacy conferred upon eugenical steriliza-tion programs by the United States Supreme Court in *Buck v. Bell*, courts played a relatively unimportant role in the demise of state sterilization laws in the United States after World War II. In 1942 the high court had an opportunity in *Skinner v. Oklahoma* to revisit eugenical sterilization when it considered a constitutional challenge to the Oklahoma Habitual Criminal Sterilization Act, a 1935 statute that authorized the state to sterilize any person upon his or her third felony conviction of crimes involving "moral turpitude" (39). The statute exempted certain crimes that today we might call "white collar" (e.g., income tax evasion, embezzlement) from the reach of the law, a feature that the Supreme Court found to vio-late the Equal Protection Clause. No doubt influenced by practices in Nazi Germany, Justice William O. Douglas, writing for the Court, warned that such laws "in evil or reckless hands" ... can cause races or types that are inim-ical to the dominant group to wither and disappear" (39). But the Court decided the matter narrowly, declining to review the broader issues raised in *Buck v. Bell*. Thus the constitutional status of laws targeting institutionalized retarded persons was unaffected by the 1942 holding.

The Supreme Court has only considered the power of the state to sterilize persons without their consent on one other occasion—growing out of a judicial proceeding in a county court in Indiana in 1971. At issue in *Stump v. Sparkman* was whether an Indiana judge could be sued for granting a petition brought by a woman to have her mildly retarded 15-year-old daughter sterilized because the mother feared that the daughter would become pregnant (40). The central issue was the scope of judicial immunity for actions taken on the bench. Finding that no Indiana statute forbade the judge to consider such petitions and noting in passing that Indiana had enacted an eugenical sterilization law, the high court ruled that the judge could not be sued by the woman who had been sterilized once she discovered that fact (40). As was so often the case, she had been told that she would undergo an appendectomy, and only learned the true nature of the surgery after she married and realized she was infertile.

Perhaps the most significant decision against steril-ization practices was handed down by the federal district court for Washington, DC, in 1976, resolving a lengthy bat-tle over procedures to be followed in permitting federally funded medical clinics to sterilize individuals. Growing out of a widely reported incident in which several poor, black teenage girls were sterilized without having first given informed consent, the case, *Relf v. Weinberger*, forced the U.S. Department of Health and Human Services to for-mulate strict guidelines concerning procedures (including informed consent for adults and a prohibition of the steril-ization of minors) that must be adhered to if a sterilization was requested (41).

During the 1970s and 1980s, a period in which there were virtually no statutorily based sterilizations (although only a few of the enabling laws were repealed),

a new question was presented to the supreme courts of more than a dozen states. Under what circumstances, if any, may a noninstitutionalized retarded person be sterilized? Typically such cases arose as sterilization petitions brought by the mothers of mildly to moderately retarded teenage daughters whom they feared would become sexually active or raped. Faced with such questions state courts had two options: to hold that, absent express legislative authorization, they lacked power to decide the matter or to rule on the matter. Most of these cases wound up in the state supreme courts where in a series of cases in the 1970s the courts embraced a "best interests" test. That is, they would approve the sterilization petitions if they were convinced that the sterilizations were intended primarily to benefit the young woman (the cases never involved men) who were the subject of the petitions. The courts almost invariably found a judicial path to approve sterilization petitions brought by caring parents. During this era, a period in which society was struggling to integrate persons with disabilities into everyday life, the courts essentially held that, just as did persons of normal intelligence, persons with developmental disabilities had a right to be sterilized if it was in their best interests. In the 1970s nine state supreme courts articulated this right, followed by several more in the 1980s. A New Jersey decision, *In the Matter of Grady*, is among the most thorough discussions of the matter (42).

EUGENIC LAWS TODAY

Globally speaking, government-sponsored sterilization programs that were premised on the value of negative eugenic policies (e.g., sterilization and immigration restriction laws) declined rapidly after World War II. In the United States about five states have repealed their laws, but in most cases the laws remain on the books, although no programs are active. The law in Alberta, Canada was repealed in 1972 (43) and Sweden repealed its law in the early 1990s. Japan enacted a "Eugenic Protection Law" in 1948 which permitted persons affected with or at risk for a litany of disorders, some of which were correctly identified as genetic, to obtain sterilization. The law was not compulsory. The statute was recently amended, and the term eugenic was dropped (44).

No modern European state has a law authorizing authorities to sterilize persons without their consent. Indeed, many have enacted laws that expressly or implicitly forbid state-supported eugenic sterilization. These include France, Germany, Norway, the United Kingdom, Spain, and Switzerland. The Council of Europe's Convention for the Protection of Human Rights and Dignity of the Human Being with regard to the Application of Biology and Medicine: Convention on Human Rights and Medicine (1997) opposes eugenic programs. The United Nations has recently endorsed a Universal Declaration on the Human Genome and Human Rights (1997) which a eugenic sterilization law would clearly contravene (45).

In the modern era practices in India and China have raised concerns that eugenic thinking is alive and well (46). Although now officially forbidden by the governments, there are states in India and provinces in China where it is relatively common practice to use medical technology and selective abortion to avoid the births of girls. This, together with the once not uncommon practice in China of denying lifesaving treatments to infant girls who are ill, has led to claims that as many as 100 million girls are missing from the Asian continent.

In 1983 Lee Kuan Yew, the autocratic Prime Minister of Singapore, launched a social program based on unscientific assumptions concerning positive eugenics. Essentially he authorized programs to encourage well-to-do, highly educated persons to marry and have large families. The unspoken, but clear, message was that such persons would be more likely to produce bright talented children than would persons in the lower classes (47).

China enacted a Maternal and Infant Health Care Law in 1994 that contains many laudable elements (48). However, it also includes language that many Western observers have interpreted as revitalizing long discredited eugenic notions (49). The law requires medical counseling before marriage for people whose families have a relative with one of a listed group of conditions (including mental retardation, epilepsy, and mental illness) that the law seems to presume are hereditary. Related language has been interpreted to require sterilization or the monitored use of long-term contraception as a precondition of marriage if a person is determined by a doctor to be at risk for parenting such children. However, the law includes no penalty for noncompliance, and some have interpreted it as expressing more an ethical obligation than a legal requirement. Although the law is not generally used to restrict child-bearing, it appears to be a public health policy in China to discourage reproduction by mentally retarded persons (49).

CONCLUSION

Ever greater understanding of the human genome has led to ever greater certainty that complex phenotypes such as intelligence emerge from countless interactions between genes and the environment in which a person develops and lives. The naive application of mendelism to complex human conditions that was common in the first three decades of the twentieth century is no longer scientifically accepted. For this reason social programs built on unsupported quasi-genetic tenets have virtually no adherents among biologists and physicians.

The late twentieth century has witnessed in the West at least a remarkable surge of concern for the well-being and rights of persons with developmental disabilities. Persons who once were housed inside the walls of state institutions for the retarded are today living and working in the community. The majority of citizens applaud this change. The Americans with Disabilities Act, arguably the most important civil rights legislation enacted in the United States since the 1960s, reifies a national commitment to treat disabled persons as equals.

Today the notion of involuntary eugenical sterilization of a person to prevent the infinitesimally small contribution to the gene pool that would be caused by his or her reproducing is scientifically ludicrous. But one cannot discount the possibility that misinformed and prejudiced persons and political entities will choose to rationalize their acts with eugenic arguments. The twentieth century drew the millennial curtain on a world in which "ethnic cleansing," a political goal frighteningly similar to the Nazi ideology of the 1930s, was being attempted in several nations on several continents.

While it is highly unlikely that state-supported eugenic sterilization programs will reassert themselves in Western nations, it is likely that eugenic thinking will manifest itself in other ways. In 1971 Nobel laureate Williams Shockley suggested to the American Psychological Association that persons of low intelligence (as measured by IQ scores) should be offered financial incentives to be sterilized with the incentive growing as the IQ score dropped (16). Comparable ideas have been floated with some regularity in every decade.

Much more important to consider is the impact of prenatal diagnosis. As this technology includes an ever larger array of tests that ever more women will use, an ever larger number of fetuses that would in earlier times have been born with disabilities will be aborted. This trend is already well underway in respect to the fate of fetuses ascertained through screening programs designed to warn women about the risk of bearing children with spina bifida (50). Similarly widespread use of prenatal screening coupled with selective abortion is causing a significant decline in the number of children born with Down syndrome. Such outcomes are the result of free choices made by thousands of women when confronted with knowledge delivered to them by the application of new tools. The results are not the product of a state law, yet they may represent a new form of eugenics. The tools are used and the choices are made in a climate that seems to accept those born with disabilities while promoting efforts to avoid such births.

The challenge before us is to mobilize advances in genetics to maximize benefits for individuals while blocking the efforts of malevolent or ignorant persons to misuse those tools in the name of a false science. In this regard there is no weapon as powerful as education. Only when we all understand that humans result from an ultimately unfathomable complex of gene-environmental interactions and that it makes (with rare exceptions) no sense to attempt to predict human phenotypes will we truly be confident that another sad episode in the history of eugenics does not lurk in our future.

BIBLIOGRAPHY

1. D.W. Forrest, *Francis Galton: The Life and Work of a Victorian Genius*, Taplinger, New York, 1974.

2. F. Galton, *Inquiries into the Human Faculty and its Development*, Macmillan, London, 1883.

3. P.R. Reilly, *The Surgical Solution: A History of Involuntary Sterilization in the United States*, Johns Hopkins University Press, Baltimore, MD, 1991.

4. R.L. Dugdale, *The Jukes: A Study in Crime, Pauperism, and Heredity*, 4th ed., Putnam, New York, 1910.

5. C.E. Rosenberg, *Bull. Hist. Med.* **35**, 266–276 (1961).

6. H.H. Goddard, *The Kallikak Family*, Macmillan, New York, 1912.

7. F.J. Hassencahl, *Harry H. Laughlin, Expert Eugenics Agent for the House Committee on Immigration and Naturalization, 1921 to 1931*, Ph.D. Dissertation, Case Western Reserve University, 1970, University Microfilms, Ann Arbor, No. 50.

8. *Buck v. Bell*, 274 U.S. 200 (1927).

9. A.J. Ochsner, *J. Am. Med. Assoc.* **53**, 867–868 (1899).

10. H.C. Sharp, *N.Y. Med. J.* 411–414 (1902).

11. H.H. Laughlin, *Eugenical Sterilization in the United States*, Chicago Psychopathic Laboratory, Chicago, 1922.

12. *Smith v. Board of Medical Examiners of Feebleminded*, 88 Atl. 963 (1918).

13. *Mickle v. Henrichs*, 262 F. 2D 687 (1918).

14. *In re Thompson*, 103 Misc Rep. 23, 169 N.Y.S. 538 (1918).

15. R.L. Garis, *Immigration Restriction*, Macmillan, New York, 1927.

16. D.J. Kevles, *In the Name of Eugenics: Genetics and the Uses of Human Heredity*, Knopf, New York, 1985.

17. *Smith v. Probate*, 231 Mich. 4409 (1925).

18. R.J. Cynkar, *Columbia Law Rev.* **881**, 1418–1461 (1981).

19. *Buck v. Bell*, 143 Va. 310 (1925).

20. E. Gosney and P. Popenoe, *Sterilization for Human Betterment*, Macmillan, New York, 1929.

21. State Board of Control of Wisconsin, *Nineteenth Biennial Report: Period Ending June 30, 1928*, Democratic Printing, Madison, WI, 1928.

22. H.M. Boies, *Prisoners and Paupers*, Putnam, New York, 1893.

23. E.J. Larson, *Sex, Race, and Science: Eugenics in the Deep South*, Johns Hopkins University Press, Baltimore, MD, 1995.

24. H.S. Jennings, *The Biological Basis of Human Nature*, Norton, New York, 1930.

25. L.C. Dunn and T. Dobzhansky, *Heredity, Race, and Society*, New American Library, New York, 1952.

26. A. Myerson, *Eugenical Sterilization: A Reorientation of the Problem*, Macmillan, New York, 1936.

27. *Five Great Encyclicals*, Paulist Press, New York, 1939.

28. G.R. Searle, *Eugenics and Politics in Great Britain, 1900–1914*, Noordhoof International, Leyden, The Netherlands, 1976.

29. *Muir v. Alberta*, Atla. L.R. 3d 305 Alt. Q.B. (1996).

30. F. Lenz, Eugenics in Germany, *J. Heredity* **15**, 223–231 (1924).

31. P. Popenoe, *J. Heredity* **26**, 257–260 (1935).

32. R. Cook, *J. Heredity* **26**, 485–489 (1935).

33. M. Hillel and C. Henry, *Of Pure Blood*, McGraw-Hill, New York, 1976.

34. D.A. Blackmon, A Breed Apart. *Wall Street J.*, August 17, p. 1 (1999).

35. C. Gamble, *J. Am. Med. Assoc.* **141**, 773 (1949).

36. M. Woodside, *Sterilization in North Carolina*, University of North Carolina Press, Chapel Hill, NC, 1950.

37. *North Carolina ARC v. North Carolina*, 430 F. Supp. 451 (1976).

38. *Poe v. Lynchburg*, 518 F. Supp. 789 (1978).

39. *Skinner v. Oklahoma*, 316 U.S. 535 (1942).

40. R. Macklin and W. Gaylin, eds., *Mental Retardation and Sterilization: A Problem of Competency and Paternalism*, Plenum Press, New York, 1981.

41. *Relf v. Weinberger*, 565 F. 2d 722 (1977).

42. *In the Matter of Grady*, 426 N.E. 2d 467 (1981).

43. T. Caulfield and J. Robertson, *Alberta Law Rev.* **35**, 59–80 (1996).

44. T. Tsuchiya, *Newslet. Network Ethics Intellectual Disabil.* **3**(1), 1–4 (1997).

45. B. Knoppers, *Nature Genet.* **22**, 23–26 (1999).

46. D.C. Wertz and J. Fletcher, *Soc. Sci. Med.* **46**(2), 255–273 (1998).

47. Chan, *Int. J. Health Serv.* **15**(4), 707–712 (1985).

48. Law of the People's Republic of China on maternal and infant health care (official translation) 1994, Legislative Affairs Commission of the Standing Committee of the National People's Congress of the People's Republic of China, Beijing.

49. Board of Directors of the American Society of Human Genetics, *Am. J. Hum. Gen.* **64**, 335–338 (1999).

50. H.S. Cuckle and N.J. Wald, *Prenat. Diag.* **7**, 91–99 (1987).

See other entries DISABILITY AND BIOTECHNOLOGY; EUGENICS, ETHICS.

F

FDA REGULATION OF BIOTECHNOLOGY PRODUCTS FOR HUMAN USE

Michael J. Malinowski
Widener University School of Law
Wilmington, Delaware

OUTLINE

INTRODUCTION

The 1990s transitioned life science from a century of profound discovery and commercial development into a new millennium that holds unprecedented potential to improve health care through technology. The Food and Drug Administration (FDA), which approved more new drugs in 1996 to 1999 than in any three-year period since 1962 (1) continues to implement a comprehensive modernization mandate from Congress to accelerate the accessibility of health care innovation (2). The pipeline of applied life science innovation, from laboratory to market, never has been filled with so much promise to alleviate suffering from debilitating diseases and generally to improve human health. "The industry is now delivering its second generation of products, a generation that includes humanized monoclonal antibodies, protease inhibitors, traditional small molecules, proteins that serve as drugs, and combinations of delivery devices/drugs that uniquely link diagnostics and therapeutics" (3).

The first generation of biotech products has reached the clinic and market. Close to 100 biotech drugs now are commercially available, and FDA approved 24 biotech products in 1997 alone—a 12-fold increase since 1994

(4). "In fact, of the 350 new biotech drugs moving through clinical trials toward FDA approval, some 30% are in the late stages of testing. More than a third of those target various types of cancer; the rest target AIDS-related diseases, autoimmune disorders, diabetes, infectious diseases, and other ailments, according to a 1998 survey published by the Pharmaceutical Research and Manufacturers of America" (5).

These accomplishments tower high above the expectations of most health care providers at the commencement of the Human Genome Project (HGP) in 1990. Nevertheless, biotechnology's achievements are merely the beginning of markedly more rapid scientific progress that will change health care fundamentally and comprehensively during the early decades of the twenty-first Century (6). Notably scientists are coupling biotechnology and informatics to identify the intricacies of protein interactions and their impact on cell function and disease pathways, a field known as proteomics. People's individual genotypes are now being taken into account in pharmaceutical clinical trial design. Increasingly, the delivery of health care, including the prescription of pharmaceuticals, will become tailored to personal genotypes through pharmacogenetic profiling.

Change is the theme of this article. The first part summarizes the FDA's official approach to the regulation of biotechnology and presents a primer on how the FDA generally regulates the major groupings of products developed through biotechnology. Parts III and IV offer discussion of present changes to, within, and around the agency, and changes the agency will face as biotechnology and health care fully integrate over the next several years.

BACKGROUND: FDA REGULATION OF PRODUCTS DEVELOPED WITH BIOTECHNOLOGY

As a matter of federal policy, the United States evaluates and regulates products, including products derived through biotechnology, based on what they are rather than according to the processes used to make them (7,8). This official approach to the regulation of biotechnology, known as the Coordinated Framework for Regulation of Biotechnology and adopted in the mid-1980s (9), has distinguished the United States:

> Where other countries have tried to write entire new bodies of jurisprudence in response to recent medical advances, American lawmakers have said that questions raised by biotechnology can all be answered within the body of existing law. As a result, while other nations' biotech industries have become mired down in legal wrangles, the industry in America is booming, with 1997 sales of $13 billion ... (10).

Although agency compliance with the Coordinated Framework policy has not been uniform, the FDA generally has adhered to the policy. In fact the FDA was instrumental in the policy's formation and adoption: "During the biotech regulation formation process, the FDA determined that its regulatory infrastructure could handle biotechnology while EPA and USDA concluded that rDNA techniques introduce, per se, an incremental risk in new products" (7,9,11).

As discussed later, during the 1990s the FDA went through a period of public questioning fueled by collaborations among the research-driven life science industries, a Republican Congress, and patient groups. This questioning inspired new legislation, coupled with self-assessment and reform from within the agency. The Prescription Drug User Fee Act of 1992 (PDUFA) (12), which was proposed and heavily supported by industry, significantly expanded the staff of FDA through drug sponsor fees, and thereby accelerated review and approval times. Although PDUFA did not directly address clinical development requirements, the Food and Drug Agency Modernization Act of 1997 (FDAMA), which reauthorized the collection of user fees, includes several provisions intended to reduce drug development times (13). The net effect of these reforms has been documented by the Tufts Center for Drug Development: FDA approved 108 new chemical entities in 1996 to 1998, the largest total number of approvals in a three-year period since 1962 (1). "The 1990s value represents a 47% increase over that of the 1980s" (13).

The following is a primer on FDA review of major biotech product groupings. The groupings addressed include therapeutics, diagnostics and medical devices, vaccines, tissue products, and food.

Therapeutics

FDA and sponsors of therapeutic products are drawn together when new chemical entities advance from animal studies into human clinical trials (14). The U.S. Food, Drug, and Cosmetic Act (FDCA), which prohibits the introduction of drugs into commerce in the absence of data sufficient to establish safety and efficacy, includes an express exception for clinical experimentation that complies with regulatory precautions to protect human subjects (15). To exercise this exception, sponsors must submit an investigational new drug application (IND), and FDA must approve that application (14). FDA authority over clinical trials also is bolstered by their expense and the fact that sponsors are proceeding with the objective that FDA ultimately will accept the resulting data in conjunction with an application and find it persuasive of both safety and efficacy. Human clinical trial highlights are set forth in Table 1 (14).

Table 1. Human Clinical Trial Highlights

Characteristics	Descriptions
Phase I	Closely monitored studies
Primary objectives	Determine *toxicity* and whether the drug generally is safe for human use
	Determine the *preferred route of administration*
	Determine the *safe dosage range*
Secondary objective	Make a preliminary determination of effectiveness
Subjects	Small number of subjects (less than 100) and, in the U.S., usually healthy volunteers
Time frame (U.S.)	From six months to one year
Prerequisites (U.S.)	Approval of an investigational new drug application (IND)
	Protocol approval by an institutional review board (IRB) and
	For gene therapies, special protocol approval by FDA and perhaps also by the National Institutes of Health (NIH)
Prerequisites (EU)	Ethics committee approval and/or approval of the equivalent of an IND
Phase II	Placebo-controlled and double-blind
Primary objectives	Develop *dosage and toxicity data*
	Assess the *risks of administration*
	Obtain preliminary evidence of *effectiveness*
Subjects	Several hundred subjects (usually 100 to 300) who are patient volunteers with the target condition
Time frame (U.S.)	Two years
Phase III	Randomized, double-blind studies
Primary objectives	Verify *effectiveness*
	Determine the incidence of *adverse reactions over time*
	Overall, gather enough data to make a *meaningful risk-based assessment*
Secondary objectives	*Refine dosage* and administration ranges
	Determine appropriate *labeling*
	Perhaps address *pharmacoeconomic considerations* (cost-benefit analysis of the drug for targeted consumers)
Subjects	Approximately 1000 patient volunteers
Time frame (U.S.)	Three years (subject to acceleration, especially for fatal conditions without alternative treatments)

Historically FDA has drawn a regulatory distinction between new drugs and new biologics. Although all new drugs, referred to as new chemical entities (NCEs), are regulated under FDCA (15), biologics have been subject to sometimes onerous additional requirements under the Public Health Services Act (PHSA) (14,16). The primary objective of PHSA is to control manufacturing processes, meaning that, relative to sponsors of traditional drug products, sponsors of biologics have been subjected to additional licensing requirements for manufacturing (14). FDA's definition of biologics is broad enough to encompass virtually all biotech therapeutics: "A biologic drug is a virus, therapeutic serum, toxin, antitoxin, vaccine, blood component or derivative, allergenic product or analogous product, or arsphenamine or its derivatives (or any other trivalent organic arsenic compound) used for a therapeutic purpose" (17).

FDA bureaucracy reflects this historic division in the regulation of drugs and biologics for, administratively, there are two regulatory pathways to approval—the Center for Drug Evaluation and Research (CDER) and the Center for Biologics Evaluation and Research (CBER). Sponsors petition for the classification—CBER or CDER—they desire, and FDA has 60 days to respond in writing (2). If FDA fails to respond on time, the applicants' recommendation becomes binding. FDA may modify the classification only with the consent of the application or based upon public health reasons (2).

During the 1990s the historic distinction between drugs and biologics has been blurring in favor of harmonization between CBER and CDER. The introduction of the biologics license application (BLA) constitutes a major bridge between the Centers. Prior to the BLA, drug sponsors filed new drug applications (NDAs) with CDER while biologics sponsors had to file product license applications (PLAs) coupled with establishment license applications (ELAs) and other license applications, and those had to be filed with CBER. BLA is a single application covering all biologics (14,17). Today drug sponsors still file NDAs with CDER, but biologics sponsors may petition to file BLAs with either CBER or CDER (2,17). The advent of the BLA also marks a much more fundamental harmonization between CBER and CDER through the FDA reform movement and modernization:

> The combined impact of the Food and Drug Administration Modernization Act (FDAMA) of 1997, reinventing government (REGO) initiatives, improved bioanalytical methods, FDA's increased familiarity with recombinant DNA product safety, and good manufacturing practices (GMPs) have allowed for more harmonized review between the Center for Drug Evaluation and Research (CDER) and the Center for Biologic Evaluation and Research (CBER) for specified biologics (18).

The net effect of harmonization between CBER and CDER is the lessening of incentives on the part of NCE sponsors to attempt to steer products to one division or another and force them into product classifications to achieve this purpose. Ideally the end result of the new consistency will be less gaming of the system by sponsors, and more honesty, transparency, consistency, and predictability throughout the process, which in turn should improve the reliability of the review process and lower transaction costs for industry. Implementation of FDAMA is still underway, so time will tell.

Diagnostics and Medical Devices

"Medical device" has served as a regulatory catch-all for health care products for human use and consumption other than pharmaceuticals, including components, parts and accessories of devices; diagnostic aids, including reagents, probes, and antibiotic sensitivity discs; and test kits for use in laboratories. In fact, under FDCA, the term "device" encompasses all health-care products that do not achieve their primary intended purpose through chemical action in or on the body, or by being metabolized (15).

FDA Classification and Review of Devices. The FDA has established a comprehensive system to regulate the safety and effectiveness of medical devices under the FDCA as amended by the Safe Medical Devices Act of 1990 and Medical Device Amendments of 1992 (15). The most fundamental difference between the regulation of therapeutics and devices is that devices are subject to classification at the outset, and the classification is correlated with varying degrees of scrutiny and regulation (14). Virtually all new and changed devices are classified at the outset pursuant to FDA's premarket notification (PMN) or 510(k) notification process (15). Devices are classified I to III based on the concerns about safety and effectiveness, with Class I devices subject to only general controls, Class II devices subject to special controls, and Class III devices generally subject to full premarket review and approval to ensure safety and effectiveness (15).

Before developers make medical devices available to the public even for research, they must apply for an investigational device exemption (IDE)—the device analogue to an IND for new drugs (15). However, developers may be able to circumvent the IDE requirement by establishing that there is an independent means to confirm the validity of their tests—such as by establishing that the product is the substantial equivalent of a previously marketed product (meaning the 510(k) clearance process discussed below) or by just directly obtaining a premarket approval (PMA).

Devices, depending on their classification and other considerations, may reach the commercial market through any of the following: (1) a PMA, (2) a 510(k) exemption with a PMN, or (3) a 510(k) exemption and a PMN exemption under Section 510(m) of FDAMA (2,14). The 510(k) exemption is for diagnostics that are the substantial equivalent of others already approved or otherwise exempted under section 510(k) of the FDCA (15,19). Under FDAMA (2), the FDA has been downgrading device classifications, and many devices are entering the market even without PMN reporting. FDA now publishes lists of Class I and Class II devices that qualify for this double (510(k) and PMN) exemption.

Any change in a device's design triggers additional review (20), and FDA also regulates device construction and manufacturing pursuant to good manufacturing process (GMP) requirements. These requirements, which are tailored to all stages of the manufacturing process, can be detailed. FDA monitors compliance through factory inspections (at least once every two years for Class IIII products) (15) and postmarketing reports, such as Medical Device Reports on adverse incidents (21).

FDA Review of Biotech-Based Diagnostics and Predictive Tests. FDA has adopted an ad hoc approach to biotechnology-based products, which has made classification of biotechnology-based diagnostics and other tests, including predictive genetic tests, extremely unpredictable (14). Also, due to a number of ongoing federal regulatory initiatives and multitude of state legislation, the manufacturers of gene-based diagnostics and other tests must expect additional requirements, including mandatory counseling requirements and specified informed consent requirements.

As suggested at the outset of this article, in addition to breakthrough therapeutics, understanding the function of genes offers the promise of tailoring health care to individual genotypes and radically increasing the efficacy and capabilities of medicine through genetic profiling and pharmacogenetic testing to predict individual reactions (positive or adverse) to drug interventions (22). FDA regulation of this technology depends on whether genetic testing is offered as a kit for others to perform or as a testing service performed internally by the manufacturer. While kits are regulated by FDA as devices, testing services do not fit squarely within the FDA regulatory infrastructure. Companies and laboratories that perform testing services, sometimes by accepting samples through the mail, are not subject to FDA regulation to ensure safety, effectiveness, and market responsibility. Federal regulation is limited to the Clinical Laboratory Improvement Act and Amendments (CLIA) (23), which requires only that laboratories performing these tests meet standards for technical competence (24). However, laboratories also are licensed on the state level, and laboratories associated with academic and research institutions usually are subjected to oversight by Institutional Review Boards (IRBs) and adhere to prescribed human subject protocols. Similarly, some private laboratories have voluntarily subjected themselves to IRB oversight (24).

At this time, most genetic testing takes place in academic settings. According to a study reported in September 1997 by the Task Force on Genetic Testing, in comparison with biotech companies, twice as many nonprofit organizations are engaged in genetic testing (25). Beyond the context of newborn screening programs and diagnostic use for (monotype) conditions, regulatory uncertainty has raised reservations among health care providers and the general public, and thereby impeded the business of commercial predictive genetic testing. Notably, *predictive* genetic testing has raised a multitude of issues, such as the potential for employment and insurance discrimination based upon resulting information even when that information is obtained in an authorized manner.

National regulatory infrastructure for commercial predictive genetic testing may be introduced soon, but perhaps not by FDA, and certainly not in a voluntary manner. Although FDA regulates test kits and the quality of analyte-specific reagents used by clinical laboratories (26), clinical testing services too closely approximate the practice of medicine. FDA holds broad authority to regulate under the Medical Device Amendments to FDCA (15), but the Agency is expressly prohibited from interfering with the practice of medicine (24,27–29). Moreover FDA is preoccupied with pressing demands associated with the implementation of FDAMA. If needed regulatory infrastructure is introduced, it is likely to rise out of the work of the Secretary's Advisory Committee on Genetic Testing (SACGT) established in 1998 by the Secretary of the Department of Health and Human Services (HHS), or in conjunction with the mandate for federal medical privacy protection under the Health Insurance Portability and Accountability Act of 1996 (HIPAA) (30). Pursuant to HIPAA, on October 29, 1999, HHS released Proposed Standards for Privacy of Individually Identifiable Health Information (31) for public comment. Response to the Secretary's proposal also could move Congress to craft a legislative solution.

In contrast to the limited presence of genetic testing in the commercial consumer market, the technology is notably present in commercially sponsored research, including the design of clinical trials. With the increased role of genomics, informatics, and proteomics in drug development, the technologies are rapidly advancing in the research context. Presumably these genetic testing technologies will enter commerce in conjunction with the therapeutics they are instrumental in developing. Herceptin, a drug to treat advanced breast cancer that was introduced commercially by Genentech, Inc. in 1998, is representative of the next generation of pharmaceuticals and the first full generation of commercially viable genetic tests (22). Herceptin is tied to expression of the HER2 gene associated with an aggressive form of breast cancer, which is found in 25 to 30 percent of breast cancer patients. Genentech, in conjunction with another company, Dako, developed a genetic test to identify carriers of the HER2 gene. Although the HER2 precision of Herceptin has narrowed the drug's labeling and limited the size of the market for this drug, that same precision is enabling Genentech to sell the drug at a premium price of approximately $19,000 per treatment, roughly twice the cost of Taxol, another innovative cancer treatment (4).

Vaccines

Vaccines are reviewed and regulated within the context of therapeutics regulation as set forth above. However, traditionally the methodology for vaccine development has been to introduce weakened strains of the target viruses. The risk associated with this approach has necessitated extensive study of the animal and human cells in which the weakened viruses are grown, followed by animal trials that are generally more comprehensive than drug trials (32). FDA regulation of vaccines postapproval also is more intense than for drugs. FDA has established a comprehensive adverse event reporting system for vaccines, known as the Vaccine Adverse Event Report System (VAERS), which was introduced in response to the National Childhood Vaccine Injury Act of 1986 and

Table 2. Representative Vaccines Developed with Biotechnology

Name	Indication	Approval	Developer
Engerix-B	Prevention of hepatitis B in individuals suffering from chronic hepatitis C	August 1998 for the U.S. market	SmithKline Beecham
LYMErix	Preventions of Lyme disease	December 1998 for the U.S. market	SmithKline Beecham
Primavax	Active immunization (primary vaccination and booster) against hepatitis B, dipththeria, and tetanus in infants	February 1998 for Europe	Pasteur Merieux

is managed by both FDA and the national Centers for Disease Control and Prevention (33).

Only a handful of biotech-based vaccines have reached the commercial market. Some of those most recently approved are identified in Table 2 (34).

Nevertheless, biotechnology is revolutionizing vaccine development. The technology is being used to develop vaccines for myriad infectious diseases, including herpes, tuberculosis, and meningitis, and illnesses such as cancer (5). Biotechnology's primary contribution is to enable vaccines to be developed that do not expose patients to actual viruses. "Scientists can now produce a single viral protein that tricks the immune system, getting it ready to defeat the virus should it show up.... Strictly speaking, these are not 'vaccines' but 'therapies' because they are given after a patient comes down with the disease; but the process is the same" (5). Primary examples are AIDSVAX, an AIDS vaccine being developed by VaxGen (Brisbane, California) that has been reported to be close to FDA approval (5).

Unfortunately, advancement of these innovative vaccines now may be impeded by some recent events. Notably in October 1999 a Federal health advisory panel, the Advisory Committee on Immunization Practices to the Centers for Disease Control, withdrew a recommendation that all infants be immunized with RotaShield against rotavirus, a virus that causes a severe form of diarrhea. FDA had approved Rotashield for the U.S. market in August 1998 for oral administration to infants in a three-dose series at 2, 4, and 6 months of age (34). Subsequently Rotashield was linked to a painful and potentially fatal bowel obstruction known as intussusception (35). The federal government had licensed the vaccine from the manufacturer, American Home Products, a year earlier, and much of the work to develop the vaccine over 23 years was done at NIH (35). This incident closely followed other adverse incidents, including vaccine-associated polio and the death of an infant soon after receiving a dose of a pertussis (whooping cough) vaccine (32).

Tissue Products

The multidisciplinary field of tissue engineering couples cellular biology and chemical engineering, and much of the potential of this field is attributable to the ability of genetically engineered cells to assimilate into the environment in which they are placed (14). Cells are modified and cultivated to create body parts, including both implants and replacements. Potential and actual commercial applications include artificial skin, tendons, bone, corneas, bioartificial organs, blood and blood vessel substitutes, heart valves, expansion of bone marrow stem cells, neurological implants, tissue-engineered vascular grafts, various orthopedic devices such as tissue-engineered cartilage, and the use of artificial skin and its equivalents for toxicity testing (14). Scientists also are studying the use of modified human cells to treat viral infections, Parkinson's disease, and diabetes, as well as other disease conditions. The ultimate in tissue engineering may be gene therapy and xenotransplantation, or animal-to-human transplants (36).

Under the prevalent existing technology, for most of these applications, either cells are genetically modified to achieve a very specific genetic assimilation that will achieve a targeted therapeutic impact (e.g., gene therapies) and/or living tissue cells are applied to biodegradable plastic scaffolds. Apligraf, engineered by Organogenesis of Canton, Massachusetts, is one of the first commercially available products to contain human cells. Apligraf is a substitute skin approved by FDA in May 1998. Beyond applications such as Apligraf, the pipeline from this technology includes the possibility of tissue implants that deliver therapeutic drugs and hormonal secretions, and myriad gene therapies. At this time more than 3000 patients are in gene therapy trials (5). "From a health care perspective, the need for tissue-engineering products is unquestionable: 'In the United States each year there are 20,000 transplants [and] there are 2,000,000 implants.... The need for implantable parts and devices is staggering, and this need cannot be met through [traditional] organ and tissue transplantation'" (14,36).

FDA announced on September 30, 1999, that the Agency intends to establish a comprehensive new system for the regulation of human cellular and tissue-based products (37). The agency's proposals include amending current GMP regulations that apply to human cellular and tissue-based products — whether regulated as drugs, medical devices, and/or biologics — to incorporate into existing GMP regulations new (1) donor-suitability procedures (i.e., improved screening against the epidemiological dangers associated with conditions such as HIV and hepatitis) and (2) procedures for the proper handling, storage, and processing of these products (37). According to FDA, (1) the regulations will embody appreciation for the fact that there is a wide spectrum of products in this category that carry varying degrees of risk, (2) the regulations will be responsive to the specific level of risk of communicable disease involved, and (3) this approach will avoid unnecessary and duplicative regulations. As stated by FDA (37).

> [T]he agency has tailored the proposed testing and screening requirements to the degree of communicable disease risk associated with the various types of human cellular and tissue-based products. The testing and screening for donors of cells and tissues that pose a high degree of communicable disease risk will be more extensive than for donors of cells and tissues with lesser risk. Where the risk is quite low (e.g., cells or tissues used autologously), FDA will recommend testing and screening, but will not require them; however, certain labeling will be required.

Food

Through the Center for Food Safety and Applied Nutrition, the FDA regulates $240 billion worth of domestic foods, $15 billion worth of imported foods, and $15 billion worth of cosmetics sold across state lines (38). FDA coordinates regulation of food with the United States Department of Agriculture, which regulates field testing (39), and the Environmental Protection Agency, which has jurisdiction over genetically modified plant and microbial pesticides under the Federal Insecticide, Fungicide, and Rodenticide Act (FIFRA) (14,40).

FDA has required notification and consultation from manufacturers before they market bioengineered foods, but FDA has required an official approval and labeling only in certain circumstances — such as when the foods contain (1) a known toxic substance, (2) nutrients different from those found in the unmodified version, (3) other new substances, (4) a known food allergen, or (5) antibiotic-resistant genes (41). However, FDA regulation of bioengineered foods presently is the subject of public and political scrutiny and debate, and is susceptible to change. Domestic pressure is mounting in response to media coverage of consumer concerns in Europe, which have fueled a trade war. The influence of consumer concern in the European Union (EU) is augmented by the EU's lack of a counterpart to FDA for food regulation (42,43). In response to these pressures, in October 1999 the FDA announced public meetings on bioengineered foods and an openness to reevaluation based on new scientific information (44).

Moreover the United States is expected to comply with the Biosafety Protocol (BP) agreed to by 120 to 30 nations in February 2000, following the November to December 1999 World trade Organization (WTO) meeting debacle in Seattle, Washington (45). The BP (1) requires exporters to label shipments that "may contain" bioengineered commodities, and (2) allows countries to block imports of genetically modified organisms (GMOs) on a precautionary basis in the absence of sufficient scientific evidence about their safety (45).

CHANGES TO, WITHIN, AND AROUND THE FDA

The accomplishments of contemporary life science place extraordinary external demands on FDA for change at a time when the agency is moving into a new era under the leadership of Commissioner Henney. Foremost, the Agency is working to implement a comprehensive collection of reforms required under FDAMA that are the ongoing focus of intense industry and patient group lobbying. Under FDAMA, the FDA's longstanding role of protector and promoter of public health now includes increased recognition of a responsibility to make innovative products available as expeditiously as possible, albeit without sacrificing safety and efficacy (2).

> "[P]ublic confidence in safety and effectiveness of the drug supply depends on a system of extensive premarket testing, but business competition, the expense of drug development, and, increasingly, the vocal advocacy of patient groups, create countervailing pressures on manufacturers and regulators to shorten the investigational period and to speed new drugs to market" (46).

Many FDA changes reflect responsiveness to break-throughs associated with biotechnology, and there is sincere enthusiasm at the agency for this technology. However, concern that speedier agency approvals may be leading to a higher rate of adverse drug events (ADEs) and safety-related drug withdrawals (47) has increased tension between the goals of patient accessibility and safety. In June 1999, an FDA Task Force on Risk Management reported that both premarket and postmarketing safety-review programs would be bolstered (47).

The challenge before FDA to meet its restated mission is exacerbated by the fact that information technology now is integrating with all stages of drug development and health care to increase the pace of innovation and patient access to experimental treatments. The College of American Pathologists (CAP), a national medical specialty society with a membership of board-certified physicians that has significant interface with FDA on clinical laboratory issues, supports increased communication with all interested parties (47). CAP has recommended (47):

- Greater communication with and inclusion of stakeholders in all FDA activities — including development of policy and regulations and advisory committees;
- Better internal communication among FDA offices and between FDA centers;
- Clearly defined policy on the authority of guidance documents and the processes and circumstances surrounding their issuance; and
- More efficient use of the FDA's Web site for distribution of information to the general public.

The maturation of biotechnology will continue to have a profound impact on FDA well into this century. Moreover the agency continues to change in ways that directly affect biotech research, development, and commercialization. The primary forces of change presently bearing upon FDA and the agency's regulation of biotechnology are highlighted below.

Commissioner Henney

Dr. Jane E. Henney, an oncologist who served as FDA Commissioner Kessler's Deputy Commissioner for

Operations from January 1992 to March 1994, was appointed Commissioner of FDA on November 30, 1998. Prior to her appointment, Commissioner Henney served as Deputy Director of the National Cancer Institute and was Vice President of Health Sciences at the University of New Mexico.

Commissioner Henney succeeds Dr. Kessler, who left the agency in 1997. During the interim, Deputy Commissioner Michael Friedman negotiated PDUFA and FDAMA. Now Dr. Henney must fully implement the reforms associated with this legislation. In fact she has identified her priorities to include implementing the "letter and spirit" of FDAMA and strengthening the FDA's science base (49). Internally, Commissioner Henney must lead the transformation of the agency. During the period of self-reform preceding FDAMA, the need for extreme leadership, especially in the absence of a commissioner, necessitated heavy staffing in the Office of the Commissioner. Commissioner Henney now must shift FDA staff away from her office and refocus the Agency. Rather than self-assessment and self-reform, the agency must implement Congressionally mandated reforms while maintaining (even increasing) the speed of product review and approval, which is closely monitored by industry and patient groups. In addition, despite all these pressures and the challenges associated with escalating scientific innovation, Commissioner Henney must ensure that there is no catastrophic, widespread adverse event under her watch.

Modernization of the Agency

The passage of FDAMA in 1997 with virtually unanimous bipartisan support marked the culmination of years of intense lobbying by a coalition comprised of the biotechnology industry, pharmaceutical industry, patient groups, and the academic life science establishment. Industry and its allies used FDA's dependence on user fees under the Prescription Drug User Fees Act and the public's enthusiasm for accomplishments in life science to modernize the agency through codification of comprehensive changes designed to increase the agency's predictability, speed, accountability, and constructive communication with drug sponsors (4,14,42).

Implementation of FDAMA is an ongoing process, the outcome of which will be determined largely by the commitment of the interests responsible for its passage. FDAMA-mandated reforms focus on five general areas:

1. Reauthorization of the Prescription Drug User Fee Act of 1992 ("PDUFA II") with new performance goals for the FDA
2. Enhanced collaboration between manufacturers and the FDA throughout the approval process
3. Expansion of expedited drug and device approval tracks
4. Improvements in the economy and efficiency of manufacturing
5. Increased access to information for practitioners, health care organizations and consumers (14,50)

Moreover, through FDAMA, Congress has given FDA a mission that encompasses an obligation to promote public health by making new products available as quickly as possible, albeit without sacrificing safety and efficacy. Under FDAMA (2),

> The Administration shall—
>
> (1) promote the pubic health by promptly and efficiently reviewing clinical research and taking appropriate action on the marketing of regulated products in a timely manner;
> (2) with respect to such products, protect the public health by ensuring that—
>   (A) foods are safe, wholesome, sanitary and properly labeled;
>   (B) human and veterinary drugs are safe and effective;
>   (C) there is reasonable assurance of the safety and effectiveness of devices intended for human use;
>   (D) cosmetics are safe and properly labeled; and
>   (E) public health and safety are protected from electronic product radiation;
> (3) participate through appropriate processes with representatives of other countries to reduce the burden of regulation, harmonize regulatory requirements, and achieve appropriate reciprocal arrangements; and
> (4) as determined to be appropriate by the Secretary, carry out paragraphs (1) through (3) in consultation with experts in science medicine and public health, and in cooperation with consumers, users, manufacturers, importers, packers, distributors, and retailers of regulated products.

The fees collected under PDUFA II provide an ongoing incentive for FDA to implement FDAMA while bestowing on the agency the financial means to maximize its performance; the FDA will collect an estimated $740 million in fees under PDUFA II (12,51). In accordance with PDUFA II and its FDAMA-prescribed mission, FDA has demonstrated initiative to engage in constructive interaction with industry by introducing regulatory infrastructure that would make such interaction standard operating procedure. For example, the FDA published a draft guidance in 1999 to introduce procedures for requesting, scheduling and conducting meetings between the FDA and sponsors (52). This draft guidance suggests that the agency is receptive to increased collaboration in the design of clinical trials, especially for making outcome determinations and establishing effectiveness.

FDA also has been responsive to the call under FDAMA and PDUFA II for the agency to establish performance goals that expedite the review processes. Several of the implementation regulations promulgated by the agency introduce response time lines (42). For example, FDA has issued a final rule that requires the agency to respond to a Humanitarian Device Exemption application (explained below) within 75 days of receipt (53). The agency also has issued a final rule stating that it will respond within 30 calendar days to a sponsor's written requests to remove a hold placed on the sponsor's clinical trails of a drug or biologic product (54).

Several key FDAMA provisions are especially beneficial to biotechnology. For example, FDAMA harmonized BLA application procedures with those used for NDAs to introduce more uniform evaluation of products going to market. Now sponsors may file a single BLA, rather than a PLA and ELA (55). Also Section 112 of FDAMA introduced a "fast track" for new drugs that are "intended for the treatment of a serious or life-threatening condition and [demonstrate] the potential to address unmet medical needs for such condition" (2). A disproportionate number of biotech products qualify for fast-track designation, for biotech products are more inclined to be innovative and address unmet health care needs and serious health conditions as defined by FDA (7).

Orphan Drug Act and Humanitarian Devices

The biotechnology industry has been the commercial beneficiary of the Orphan Drug Act (56), which is subject to ongoing interpretation and modification (57). This act is applicable to drugs that treat rare diseases when the target patient population is not significant enough to make development of the drug economically feasible, hence the "orphan" label. "Before 1983, the development of these drugs was left to the benevolence of the drug companies, who would occasionally develop them as a public service" (56,58).

The Orphan Drug Act bestows direct economic incentives and incentives that facilitate the development process, many of which are highlighted in Table 3. Notably, the first applicant to obtain such designation and product application approval in the United States is entitled to market exclusivity for a period of seven years — meaning that no other company can market a molecularly identical drug for the FDA-approved use for seven years following approval of the original orphan (7). The period of exclusivity, which begins once the drug is approved, offers sponsors an opportunity for maximum pricing.

The Orphan Drug Act has been accomplishing its policy objective to the extent that it has resulted in a significant increase in the number of drugs available to treat rare diseases, and the sponsors of those drugs are largely biotechnology companies (58). However, the act is being criticized on the grounds that, in light of the benefits bestowed to sponsors, experimental treatments are not being made readily available to patients (59).

FDA approved nine orphan drugs in 1998, six of which were sponsored by biotech companies (7). The biotechnology industry's utilization of the act reflects that (1) many of the products being developed by the biotechnology industry are for genetic diseases that also happen to be rare diseases, and (2) the scope of protected markets under the Orphan Drug Act, although insignificant by most pharmaceutical company standards, often constitute a meaningful incentive for small biotech companies. Despite "orphan" designation, several of these drugs have reaped enormous profits for their sponsors — such as AZT (HIV infection /AIDS); pentamidine isethionate (pneumonia associated with AIDS); human growth hormone (hGH) (improper growth in children lacking the enzyme); erythropoietin (EPO) (anemia associated with end-stage renal disease); and CeredaseTM (Gaucher's disease) (58).

Sponsors are increasingly applying for orphan drug status, which means ongoing interpretation and regulatory modification, and the possibility of conditions being placed on orphan drug status where there is unexpected profitability. Moreover the EU now is close to fully implementing a counterpart to the act, and FDA has introduced a device counterpart in the form of analogous benefits for "humanitarian devices." Specifically, FDA has discretion to grant an exemption from the effectiveness requirements of Sections 514 (performance standards) and 515 (premarket approval) of the FDCA (15) after finding that a device:

- is designed to benefit a target disease population of not more than 4,000 individuals in the U.S.;
- would not be available otherwise;

Table 3. Benefits Under the Orphan Drug Act

Incentive	Description
Marketing exclusivity	First drug sponsor to have its application approved by FDA receives a 7-year period of market exclusivity against all other sponsors of the same drug approved for the same condition.
	Exclusivity is still subject to revocation based on whether (1) the sponsor fails to produce enough of the drug to meet demand, or (2) a competitor demonstrates clinical superiority — e.g., based on improved efficacy or diminution in adverse reactions.
Tax credits	Sponsors receive a tax credit for qualified clinical testing expenses, meaning a tax credit for 50% of the amounts spent conducting clinical trials.
Protocol assistance	FDA assists to distinguish exactly what tests and experiments sponsors need to complete in order to secure drug marketing approval.
	Objective is to enable sponsors to overcome the increased difficulty in structuring trials associated with the smallness of the sizes of orphan drug patient groups.
Grants	Sponsors may receive grants to defray the costs of qualified testing; this provision covers all testing after a drug is designated an orphan drug.
Open protocols	Sponsors may make drugs available to people not participating in their clinical trials while those trials are still ongoing; this creates an opportunity for sponsors to recoup some costs through charges to users before full approval.

- has no alternative available to treat or diagnose the disease or condition;
- will not expose patients to unreasonable or significant risk of illness or injury; and
- presents benefits to health from its use that outweigh associated risks.

However, there are some conditions. Devices granted this exemption may only be used at facilities that have an established institutional review board. Also, the humanitarian use must be approved by the board before studies begin (60).

Judicial Developments

In 1998, the pharmaceutical industry increased direct-to-consumer (DTC) advertising 23 percent to reach $1.32 billion (61). During that same year, the United States District Court for the District of Columbia enjoined FDA from enforcing pre-FDAMA guidance documents for off-label drug use (i.e., uses for purposes not reflected in FDA-approved labeling), concluding that FDAMA's labeling/marketing restrictions unduly burdened commercial free speech in violation of the First Amendment (62). Although the guidance documents permitted some dissemination of off-label information, they also restricted manufacturer use of textbooks, journal reprints and other educational materials that promote off-label use of drugs. FDA allowed distribution of printed and graphic materials addressing the safety and effectiveness of off-label drug use, provided that the manufacturer applied for FDA approval of the new use within six months of initial distribution. FDA was attempting to strictly limit dissemination through these media to avoid companies telling physicians about benefits but not risks. FDA wanted authority to review published studies before companies could give them to doctors. The agency also desired the power to require drug makers to give doctors additional studies of the drug, and the authority to limit distribution of studies until firms entered the federal review process for the additional uses.

The court deemed that the proposed FDA regulation was more extensive than necessary given less burdensome alternatives — for example, full, complete, and unambiguous disclosure by the drug's manufacture — and unduly burdened commercial free speech (42,62). The court granted summary judgment and issued a permanent injunction against the FDA restricting dissemination of information on off-label use of drug and medical devices. The court subsequently issued an amended order staying final determination of the legality of off-label drug use regulations implementing FDAMA, pending the filing of further submissions by FDA and the Washington Legal Foundation (63). On July 28, 1999, the District Court ruled that the FDAMA provisions requiring a supplemental application for the approval of dissemination of information concerning off-label drug use violate First Amendment commercial speech protections (42,64). The court has enjoined FDA and U.S. Department of Health and Human Services from prohibiting or restricting the following services:

- Dissemination of articles to doctors and medical professionals about off-label uses for drugs or medical devices when those articles are published in a bona fide peer-reviewed professional journal, regardless of whether the article focuses on non-FDA-approved uses.
- Dissemination of reference textbooks published by a bona fide independent publisher, regardless of whether they discuss nonapproved uses.
- Suggestion of content or speakers to independent continuing medical education (CME) program providers regardless of whether they discuss non-approved uses.

The net effect of these rulings is that pharmaceutical companies have greater leeway in the promotion of drugs for unapproved uses. Companies may give doctors copies of published medical studies that highlight the uses of drugs not approved by the FDA. However, the studies companies provide to doctors cannot be false or misleading, drug company sales representatives must disclose any association between the company and researcher, and the company must disclose whether the treatments detailed in the studies are FDA-approved.

From a policy perspective, these ruling have brought DTC marketing issues to the forefront and raised a call for empirical data that can provide guidance. An FDA survey by the Division of Drug Marketing, Advertising, and Communications (DDMAC) released in 1998 found that DTC advertisements for prescription drugs increased patient compliance with drug therapies. According to this survey, a solid description of the benefits and risks of a drug, compared to just name identification, induced greater consumer confidence (65).

CHALLENGES IN THE NEW MILLENNIUM

Biotechnology has raised both capabilities and expectations for life science to an unprecedented height, especially for medicinal applications. For biotechnology to realize its potential, both regulators and developers of this technology will have to overcome a multitude of emerging obstacles. FDA and those engaged in product development share the challenge of meeting expectations. They also are mutually dependent on each other's success in overcoming their respective challenges.

FDA

In the absence of a tragic mistake of a magnitude that shifts public opinion (Phen-Fen criticism has been mostly sponsor-specific), FDA will be under continued pressure to make new technology available. Moreover, as we enter an era of increased transparency and industry interaction, technology will continue to deluge the agency and demand greater scientific expertise. Given that these pressures will be coupled with accelerated patient access through mechanisms such as the fast track, more adverse events are probable. The FDA has identified the following as key challenges in its immediate and near future (66):

- Research and development (R&D) fueled pressures on regulatory responsibilities
- Greater product complexity driven by breakthroughs in technology
- Growth in recognized adverse effects associated with product use
- Unpredictable, new health and safety threats
- More targeted needs and awareness of citizen-shareholders
- Emerging regulatory challenges in the international arena
- Increased volume and diversity of imports
- Federal budget constraints

Product Developers

After decades of chemistry-based pharmaceuticals approved for broad market use that drew tremendous revenues throughout the duration of their sponsors' patents, biotechnology is introducing precision and lessening the hit-or-miss nature of therapeutics. Genentech's Herceptin for advanced breast cancer in women with the HER2 gene (approximately 35 percent of the women who have the disease, and the most aggressive form of the disease) marks the beginning of the forthcoming generation of pharmaceuticals (67).

Approved use and product labeling will be tied to genetic profiling, meaning that therapeutics will be coupled with genetic tests to determine their likely effectiveness and also susceptibility to adverse events. In terms of development, this precision will dramatically streamline clinical trials and accelerate FDA approval. FDA's standard for demonstrating "substantial effectiveness" under Section 355(d) of Title 21 of the United States Code, FDCA §505(d), no longer will demand multiple, independent clinical trials.

However, unless genetic profiling is utilized to create reimbursable, preventive care use of these therapeutics under managed care and such use is extensive enough to offset the streamlining of therapeutic use, pharmaceutical markets will fraction dramatically. Increasingly the combination of genomics, proteomics, and informatics will result in pharmaceuticals tailored to individual genetic profiles, thereby dramatically streamlining markets by historic standards. The net effect will be the introduction of a bounty of highly safe and effective therapeutics approved for tailored patient populations. In many ways today's orphan drugs could foreshadow tomorrow's mainstream pharmaceuticals.

The transition into this new era in life science will involve a series of pressing challenges. Arguably, the first collection of challenges are the most overarching and demand some of the more difficult and fundamental changes. Foremost, pharmaceutical innovation is shifting health care expenditure from hospital and provider care, including expensive surgeries and other specialized care, to drugs. Consequently the pharmaceutical industry increasingly is being blamed for the social and economic costs of contemporary health care, and the inadequacies and inequities of the U.S. health care system (68). Moreover there now is awareness that the pharmaceutical and biotechnology industries have been integrated through alliances. The biotechnology industry, once perceived as clusters of small, entrepreneurial companies with the mission of curing cancers and embodying many ideals of corporate America, is becoming recognized as big business, including, for example, commercial agriculture. Marketplace criticisms of commercial life science — such as pricing, aggressive direct-to-consumer (DTC) marketing, and limited labeling information — threaten to counterbalance public and political support for life science research, to the extent that many supporters of research are demanding a quid pro quo in the form of pricing controls.

These public pressures are rising at a time when the industry is about to shift from a period of record profitability to one of inability to meet sales forecasts despite extraordinary R&D expenditures. In addition to the factioning of consumer markets as addressed above, some 50 major patents held by pharmaceutical companies will expire by 2005, and each pharmaceutical company must produce 45 new drugs annually just to meet standard shareholder expectations and maintain market shares (7 percent annual sales growth) (69). "This will be a formidable task. The top seven pharmaceutical companies produce 45 drugs per year combined. Accelerating this development pace will move from being a competitive advantage to a competitive necessity" (69).

To remain competitive, pharmaceutical and biotechnology companies now are investing approximately $39 billion annually on R&D, $21.1 billion in the United States and $14.1 billion in the EU (70). The U.S.-based companies have more than doubled their R&D expenditure since 1990, when they spent $8.42 billion collectively (59). Moreover investment in R&D is expected to continue to rise — 11 percent this year alone (70). Development pressures are posing an extraordinary strain on pharmaceutical R&D operations. The focus of these operations is shifting from developing new drug targets to selecting among targets and significantly increasing the pace of advanced product development through better data management, clinical trial design, and regulatory submissions. To cut costs and accelerate the pace of clinical research, many pharmaceutical companies have shifted clinical research from academic medical centers and teaching hospitals to contract research organizations such as Quintiles and Parexel. At this time, Quintiles is reputed to be engaged in more clinical research than any other entity in the world.

The pharmaceutical industry also is attempting to lessen its dependency on outside clinical investigators and to reduce competition for patients. The latter has risen significantly in recent years within an overall increase in the amount of clinical drug development associated with accomplishments in biotechnology. Although patients have become more receptive to experimental treatments, they also are identifying these treatments on the Internet and gaining access to them via compassionate use and equivalent approvals rather than subjecting themselves to clinical trials that hold the risk of receiving placebos (71). In addition to turning to global research entities such as Quintiles, pharmaceutical companies are creating an alternative through simulated data. "Companies are beginning to create populations of 'software people' designed to behave like the real thing and computer-based organs that can be used to test potential therapies before involving humans.... According to a report released by Pricewaterhouse Coopers LLP, New York, virtual trials will probably reduce the amount of clinical resources required in the short term by 10%" (70).

The regulatory accomplishment of FDAMA also poses some longer-term challenges to the developers of innovative commercial life science products. For example, by lessening the distinction between CBER and CDER, the sponsors of biologics are making themselves susceptible to market competition from generic drug manufacturers faced by traditional drug sponsors. Biologic drug products, meaning those approved under 42 U.S.C §262, are excluded from Title I of the Waxman-Hatch Act (72), meaning that they are not susceptible to competition from generics. But the traditional definition of biologic, as set forth under Section 351(a) of the Public Health Service Act and reiterated under section 123(e) of FDAMA, is extremely broad and is being diluted and blurred (18):

> The review and approval of biotech-derived products within the classic drug domain challenges us to define a true biologic, and in doing so brings entire classes of products closer to the realm of generic challenge....With so many scientific, technical, financial, legal, and political impediments, will there ever be a generic biological process? Yes. Maybe it won't appear as familiar as the ANDA [abbreviated new drug application], but the framework for a generic biologics process is taking shape.

CONCLUSION

Today is a glorious but challenging time for commercial life science. FDA now is being modernized and realigned with public and political enthusiasm for the accomplishments of biotechnology. The combination of genomics and informatics establishes a foundation for seemingly infinite scientific possibilities. In the absence of discernible diminishing returns in the laboratory, time and the limitations of imagination are the only firm checks on biotechnology's potential.

However, the accomplishments and potential of biotechnology to improve human health rest upon commercial and regulatory uncertainty, especially given weaknesses within the U.S. health care system. Although the inadequacies of this system are long-standing, they are being exacerbated by advancements in life science. A generation of breakthrough technology now is entering the market and beginning to dramatically expand the ability to treat. Yet this technology also, by introducing costly treatments where there were previously none and turning once fatal conditions into chronic conditions, is testing the limits of health care finance and increasing payer resistance to new technologies. This paradox is sobering, especially given that the primary opportunity cost is improvements to human health.

Life science is highly susceptible to regulation by its very nature, and the FDA is a political entity. The pharmaceutical industry faces extraordinary pressures, some of which will continue to be redirected toward FDA by industry and the people awaiting industry's products. Similarly FDA, positioned within the Department of Health and Human Services, will experience the tremors of a health care system realigning. How the economic realities of contemporary health care will affect the FDA's role in commercial life science and, in particular, the agency's regulation of biotechnology, is an open question. However, the combination of FDAMA and ongoing accomplishments in biotechnology are certain to provide an ongoing incentive for policy makers and public officials inside and outside the agency to work towards a resolution.

ACKNOWLEDGMENT

Associate Professor of Law, Widener University School of Law. The author wrote this article while a SmithKline-Beecham-sponsored Distinguished Fellow in Law and Genetics, Center for the Study of Law, Science and Technology, Arizona State University, School of Law. The background information relayed in this article summarizes a comprehensive, technical treatment presented in Michael J. Malinowski, *Biotechnology: Law, Business, and Regulation* (Aspen Law and Business/Little Brown, 1999). The author thanks Erica Rose for her thoughtful corrections and suggestions.

BIBLIOGRAPHY

1. Tufts Center for the Study of Drug Development, *Impact Report: Analysis and Insight into Critical Drug Development Issues*, July 1999.

2. Food and Drug Administration Modernization Act of 1997, Pub. L. No. 105-115, 111 Stat. 2296 (codified throughout 21 U.S.C.).

3. *Med. Ad. News* (July), 20–23 (1999).

4. Ernst & Young, *Biotech 1999: Bridging the Gap*, No. 48 (December 1998).

5. *Forbes*, May 31 (1999), 1999 WL 19911831.

6. J.M. Nardone and G. Poste, in M.J. Malinowski, ed., *Biotechnology: Law, Business and Regulation*, Aspen/Little Brown, Boston, MA, 1999.

7. M.J. Malinowski and N. Littlefield, in T. Caulfield and B. Wiliams-Jones, eds., *The Commercialization of Genetic Research: Ethical, Legal and Policy Issues*, Plenum, New York, 1999.

8. H. Miller, *Policy Controversy in Biotechnology: An Insider's View*, R.G. Landes, Austin, TX, 1997, p. 198.

9. 51 Fed. Reg., 23,302-93 (1986).

10. A. Katz-Stone, *Wash. Bus. J.* **P21**, (Special Report) (1998).

11. S. Krimsky and R.P. Wrubel, in *Agricultural Biotechnology and the Environment*, University of Illinois Press, Urbana, IL, 1996, p. 251.

12. Reauthorization of the Prescription Drug User Fee Act of 1992, Pub. L. No. 102-571, 106 Stat. 4491 (1992).

13. K.I. Kaitin and E.M. Healy, Drug Development Trends in the User Fee Era. Paper presented at the Annual Meeting of the American Society for Clinical Pharmacology and Therapeutics, San Antonio, TX, March 19, 1999.

14. M.J. Malinowski, in *Biotechnology: Law, Business and Regulation*, Aspen Law & Business/Little Brown, Boston, MA, 1999, ch. 11.

15. Federal Food, Drug, and Cosmetic Act, as amended by the Safe Medical Devices Act of 1990, Pub. L. No. 101-629, 104 Stat. 4511 (codified in scattered sections of 21 U.S.C. §301, 42 U.S.C. §§263b-263n (1968, as amended), and Medical Device Amendments of 1992, Pub. L. No. 102-300, 106 Stat. 239.

16. Public Health Services Act, 42 U.S.C. §262.

17. 21 CFR 600.3(h) (1994).

18. B. Zeid, *BioPharm.*, WL 14995071, March 1, 1999.

19. 21 CFR pt. 807.

20. 21 CFR 807.81(a)(3)(i).

21. 21 C.F.R. Pt. 803 (post-marketing reports, such as Medical Device Reports, on adverse incidents).

22. Boston Consulting Group, *The Pharmaceutical Industry into Its Second Century: From Serendipity to Strategy* **63**, 64 (January 1999).

23. Clinical Laboratory Improvement Act of 1967 and Clinical Laboratory Improvement Amendments of 1988, Pub. L. No. 100-578, 102 Stat. 2903 (codified at 42 U.S.C. §201 note, 263a, 263a note, 263a).

24. A. Huang, *Food Drug Law J.* **53**, 555, 558, 574 (1998).

25. Task Force on Genetic Testing, National Institutes of Health–Department of Energy, *Final Report: Promoting Safe and Effective Genetic Testing in the United States*, U.S. Government Printing Agency, Washington, DC, September 1997.

26. 62 Fed. Reg., 62,243 (November 21, 1997).

27. 37 Fed. Reg., 16,503 (August 15, 1972).

28. *Linder v. United States*, 268 U.S. 5, 18, 45 S.Ct. 446, 449, 69 L.Ed. 819 (1925).

29. *United States v. Evers*, 453 F. Supp. 1141 (M.D. Ala. 1978).

30. Health Insurance Portability and Accountability Act of 1996, §264, Pub. L. 104-191 (August 21, 1996).

31. U.S. Department of Health and Human Services, *Proposed Standards for Privacy of Individually Identifiable Health Information*, Available at: *www.aspe.hhs.gov/admnsimp/pvcsumm.htm*

32. Isadora B. Stehlin, *How FDA Works to Ensure Vaccine Safety*, Available at: *www.verity.fda.gov/search97*

33. U.S. Food and Drug Administration, *Adverse Event Reporting System*, Available at: *www.fda.gov/cber/vaers/vaers.htm*

34. S. Engel, *Med. Ad. News* 96–108, July (1999).

35. L.K. Altman, *N.Y. Times*, A10, October 23 (1999).

36. R.M. Nerem, in L.V. McIntire and F.B. Rudolph, eds., *Biotechnology: Science, Engineering, and Ethical Challenges for the twenty-first Century*, Joseph Henry Press, Washington, DC, 1996, p. 97.

37. 64 Fed. Reg., 52696 (September 30, 1999).

38. U.S. Food and Drug Administration, *Inside FDA: Center for Food Safety and Applied Nutrition*, Available at: *vm.cfsan.fda.gov*

39. U.S. Department of Agriculture, *Biotechnology: An Information Resource*, Available at: *www.nal.usda.gov/bic/*

40. Federal Insecticide, Fungicide and Rodenticide Act, Pub. L. No. 102-300, 106 Stat. §239, codified as amended, 7 U.S.C. §§136 et seq. (1984).

41. U.S. Food and Drug Administration, *FDA's Policy for Foods Developed by Biotechnology, FDA/CFSAN*, Available at: *http://vm.cfsan.fda.gov/lrd/biopolicy.html*

42. N. Littlefield and N. Hadas, *Food Drug Law J.* **55**, (Spring 2000).

43. *Economist*, 15, June 19 (1999).

44. U.S. Food and Drug Administration, *FDA Announces Public Meetings on Bioengineered Foods*, Available at: *vm.cfsan.fda.gov/lrd/hhbioeng.html*

45. *Economist*, February 5 (2000), 2000 WL 8140683.

46. J. Kulynych, *Food Drug Law J.* **54**, 127–128 (1999).

47. *Drug Topics*, 36, June 7 (1999), 1999 WL 10021762.

48. Thomas P. Wood, MD, President, College of American Pathologists, Correspondence, Dockets Management Branch, Food and Drug Administration (dated September 10, 1998).

49. J.E. Henney, *Food Drug Law J.* **54**, 1–4 (1999).

50. N. Littlefield and M.S. Webb, *Intellectual Property Rights News* 3 (Summer 1998).

51. U.S. Food and Drug Administration, *PDUFA II Five-Year Plan*, Available at: *www.fda.gov/oc/pdufa2/5yrplan/pdufa2.pdf*

52. CDER and CBER, *Draft Guidance for Industry on Formal Meetings with Sponsors and Applicants for PDUFA Products*, FDA, Washington, DC, February 1999.

53. 63 Fed. Reg., 59217 (1998) (codified at 21 C.F.R. pt. 814).

54. 63 Fed. Reg., 68676 (1998) (codified at 21 C.R.F. pt. 312).

55. 63 Fed. Reg., 40858 (1998). 63 Fed. Reg. 40858 (1998) (to be codified at 21 C.F.R. pts. 3, 5, 10, 20, 207, 310, 312, 316, 600, 601, 607, 610, 640, and 660) (proposed July 31, 1998).

56. Orphan Drug Act, Pub. L. No. 97-414, 96 Stat. 2049 (1982), codified as amended at 21 U.S.C. §§360aa–360ee (1994), 26 U.S.C. §45C (Supp. II 1994), 42 U.S.C. 236 (1994).

57. U.S. Food and Drug Administration, *The Orphan Drug Regulations*, Available at: *www.verify.fda.gov*

58. G.A. Pulsinelli, *Santa Clara Computer High Tech. Law J.* **15**, 299, 301, 316–324 (May 1999).

59. D. Grady, *N.Y. Times* November 16 (1999).

60. U.S. Food and Drug Administration, *Requirements of Laws and Regulations Enforced by the U.S. Food and Drug Administration*, Available at: *www.fda.gov/opacom/morechoices/smallbusiness/blubook.htm*

61. *F-D-C Reports, The Pink Sheet*, No. 36 (April 26, 1999).

62. *Washington Legal Foundation v. Friedman*, 13 F. Supp.2d 51 (D.D.C. 1998).

63. *Washington Legal Foundation v. Friedman*, 36 F. Supp.2d 418 (D.D.C. 1999).

64. *Washington Legal Foundation v. Henney*, No. 94-1306 (D.D.C. 1999).

65. *F-D-C Reports, The Pink Sheet*, No. 6 (August 17, 1998).

66. FDA, *Food and Drug Administration Modernization Act of 1997: FDA Plan for Statutory Compliance* 5, FDA, Washington, DC, November 1998.

67. R. Bazell, *Her2: The Making of Herceptin, a Revolutionary Treatment for Breast Cancer*, Random House, New York, 1998.

68. D. Rosenblum, *N.Y. Times*, Sect. 4, November 14, pp. 1, 16 (1999).

69. *Med. Ad. News* (July), 3–13 (1999).

70. A. Humphreys, *Med. Ad. News* **64**(July), 1 (1999).

71. G. Kolata and K. Eichenwald, *N.Y. Times*, A1, 30, October 3 (1999).

72. Waxman-Hatch Act, Title I, Pub. L. No. 98-417, 98 Stat 1585 (1984).

See other entries ANIMAL AND ANIMAL MEDICAL BIOTECHNOLOGY; FEDERAL REGULATION OF BIOTECHNOLOGY PRODUCTS FOR HUMAN USE, FDA, ORPHAN DRUG ACT; GENE THERAPY, LAW AND FDA ROLE IN REGULATION; GENETIC INFORMATION, LEGAL, FDA REGULATION OF GENETIC TESTING; HUMAN SUBJECTS RESEARCH, LAW, FDA RULES; RESEARCH ON ANIMALS, LAW, LEGISLATIVE, AND WELFARE ISSUES IN THE USE OF ANIMALS FOR GENETIC ENGINEERING AND XENOTRANSPLANTATION; TRANSGENIC ANIMALS: AN OVERVIEW.

FEDERAL POLICY MAKING FOR BIOTECHNOLOGY, CONGRESS, EPA

WENDY E. WAGNER
Case Western Reserve University
Cleveland, Ohio

OUTLINE

INTRODUCTION

The Environmental Protection Agency's (EPA's) regulation of biotechnology products provides a fascinating case study of the challenges that this important, but still rapidly evolving technology presents to a large federal bureaucracy. In the absence of specific directions from Congress, EPA has used existing regulatory programs originally intended for conventional chemicals to address the new and often very different regulatory challenges posed by biotechnology. The results of EPA's regulatory effort are often disappointing, both in creating clear and effective rules to guide the biotechnology industry and in ensuring adequate public input into important choices about the future. The specific roles that EPA currently plays both as regulator and promoter of biotechnology and the problems that continue to plague EPA's biotechnology programs are the subject of this article.

THE HISTORY OF EPA'S INVOLVEMENT IN BIOTECHNOLOGY

In many areas of public health and environmental protection, Congress has developed legislation that is quite specific with regard to the methods and targets of regulation of environmental and public health risks (1). Yet Congress still has not provided specific legislative direction on the regulation of biotechnology products, despite the several decades over which the public has expressed concern about biotechnology's long-term safety. EPA's regulation of biotechnology thus does not derive from Congress' grand biotechnology plan. Rather, the regulation comes from authorities EPA and other executive branch authorities have read into existing statutes intended for the related, but still quite different purpose of regulating industrial chemicals in food, drugs, pesticides, and consumer products (2).

Congress' silence may be attributed in part to its general pattern of avoiding public controversies until a crisis forces legislative action (3). But Congress may have also been forestalled by aggressive Executive Branch initiatives to fend off such legislation because of White House concerns that Congress might overreact to the public's fear of biotechnology products and restrict the industry in damaging ways (4). In 1974 the Office of Science and Technology Policy (OSTP)—a White House agency—began developing a framework for coordinating the various federal agencies' regulation of biotechnology, taking as its premise that the products of biotechnology do not demand new statutory attention and that the agencies have sufficient existing statutory authorities to regulate these products (5,6). In 1986 the OSTP published its *Coordinated Framework for Regulation of Biotechnology*, which directs agencies to use existing statutory authorities to regulate biotechnology and "to operate their programs in an integrated and coordinated fashion that together should cover the full range of plants, animals and microorganisms derived by the new genetic engineering techniques" (6). Moreover, "[w]here regulatory oversight or review for a particular product is to be performed by more than one agency, the policy establishes a lead agency,..." (6). In 1992 the OSTP provided still further direction by instructing agencies to utilize a risk–benefit test for determining whether and how to regulatory biotechnology products (7).

Because of its statutory mandates to regulate pesticides and other commercial chemical substances not regulated elsewhere, EPA acts as the lead regulatory agency under the OSTP framework for overseeing the safety of these two biotechnology products (8,9). EPA has not always been an enthusiastic participant in the OSTP's coordinated framework, however; it has periodically expressed concern about this "entirely new area of responsibility" in biotechnology regulation that extends well beyond its traditional regulatory mission of regulating only conventional chemical products and wastes (10). Ironically, in fact, even when EPA has passed biotechnology regulations in accordance with OSTP policies, the regulations have occasionally met White House resistance because of the costs they impose on the important and rapidly growing biotechnology industry (11,12).

In considering EPA's programs, it is important to keep this larger Executive Branch context in mind. Professor William Rodgers has characterized the federal government's regulation of biotechnology as consisting of a number of "regulatory agencies with truncated responsibilities ... spread across the landscape like a minefield with its full potential untested" (8,p. 503). Even more charitable characterizations of existing regulatory programs cannot avoid recognition of the adverse consequences that emerge from the current legislatively deficient approach to biotechnology. Because its programs are so rudderless, the EPA's regulations seem both slow to develop and lacking in coherence and comprehensiveness.

EPA AS REGULATOR OF BIOTECHNOLOGY

Most of EPA's regulatory authority over biotechnology targets two categories of products: pesticide products, and other products of biotechnology that are not pesticides, drugs, food additives, or strains of plants. In far more limited circumstances, EPA may also have the authority to regulate the release of biotechnology products into specific environmental media. These different types of regulatory programs and authorities are discussed in turn.

Pesticides

In the Federal Insecticide and Rodenticide Act (FIFRA), passed initially in 1947 and amended significantly throughout the 1970s, 1980s, and 1990s (13), Congress tasks the EPA with the daunting responsibility of establishing a testing and approval process for new pesticide products, reevaluating and reregistering existing pesticides, and overseeing the safety of the distribution, sale, and use of pesticides (14). In these responsibilities Congress has instructed EPA to ensure that registered pesticides will not present "unreasonable adverse effects on the environment" (15). Since a number of pesticide products are now developed with biotechnology, such as genetically engineered microbes and genetically altered plants that produce their own pesticidal substances ("plant pesticides"), EPA has included these biotechnological pesticides within its regulatory purview. Expectedly, however, biotechnological pesticides present a host of unique characteristics that raise new, and to some extent more difficult challenges to regulators. In response to these challenges, EPA is gradually creating additional subprograms within its pesticide unit that exclusively address the unique products of biotechnology.

Traditional Pesticide Program. The most significant regulatory task for EPA under FIFRA is the registration of all new pesticides. This registration functions predominantly as a "preclearance regulatory regime where adequate study is supposed to be a precondition to commercial use" (8, p. 506). Accordingly, before marketing a pesticide, a manufacturer of a new pesticide must submit data in support of registration that meet EPA requirements for the particular type of pesticide product. These requirements are less onerous than the pre-approval requirements for new drugs under the Food, Drug, and Cosmetic Act described elsewhere in this encyclopedia, but they are nevertheless resource-intensive and time-consuming (16).

For conventional chemical pesticides, a manufacturer follows a standard testing regimen that usually occurs without EPA oversight. The testing and data requirements are standardized in the Code of Federal Regulations (17). EPA oversight begins only when the requisite large-scale testing commences (on at least 10 acres of land or one surface-acre of water). At this point, the manufacturer must secure an Experimental Use Permit (EUP) from EPA (18).

After the necessary testing has been completed, the manufacturer submits a final registration application to the EPA. Based on the application, EPA makes an ultimate determination to approve, reject, or condition approval of the pesticide based on evidence of its safety. EPA's final decisions can be appealed by adversely affected parties in the federal courts (19). After receiving EPA's approval, the manufacturer generally can commence marketing.

Biotechnology Pesticides. Although conventional chemical pesticides continue to dominate its regulatory agenda, EPA has interpreted FIFRA to provide it with regulatory authority over all pesticides, which also include pesticidal substances produced by living organisms (11). But regulating these biological pesticides, especially those biologics produced using biotechnology, requires more specialized programs that adjust to the different scientific uncertainties, as well as to the different potential benefits and risks posed by a genetically engineered organism. EPA is in the process of developing two specialty programs—one for genetically engineered microbial pesticides and one for plant pesticides.

Genetically Engineered Microbial Pesticides. Microbial pesticides are those that employ microorganisms (e.g., bacteria, fungi, algae, and protozoa) and viruses for pesticidal purposes. Although EPA often allows microbial pesticides to undergo less rigorous testing for FIFRA registration, microbial pesticides that are the product of genetic engineering are determined to require slightly higher regulatory intervention (21) because of concerns that these nonnative microbes could spread into the environment in unexpected and even catastrophic ways (22).

As might be expected, EPA's first hurdle in regulating genetically engineered microbial pesticides is defining what they are in terms that are clear but sufficiently flexible to accommodate the often case-by-case safety issues posed by these organisms. After several years of effort, EPA promulgated a definition in 1994 that defines biotechnological microbial pesticides as those with "pesticidal properties [that] have been imparted or enhanced by the introduction of genetic material that has been deliberately modified" (23). Considerable debate surrounds this definition. Scientists criticize EPA's process-based definition as overinclusive, including within its regulatory reach biotechnology products that are almost riskless but yet which must undergo extended regulatory scrutiny simply because of the process by which they are produced (24). Others, including the EPA, have acknowledged the rough cut of this definition but have concluded after surveying alternatives that it provides the greatest clarity to regulated entities, a value that offsets the costs of overinclusiveness (25). Indeed, a survey of most environmental and public health environmental programs reveals similar, bright-line jurisdictional determinations for what will be included within a regulatory program (26).

The second step of EPA's pesticide program for biotechnology products concerns clarifying the requirements for registration. There appear to be two points of significant divergence in the registration of genetically engineered microbes as compared with EPA's treatment of conventional chemical pesticides. The first difference occurs with EPA's oversight of testing activities. In contrast to EPA's oversight of conventional chemical and natural biological pesticide testing, manufacturers of genetically engineered microbial pesticides must notify the EPA before conducting even small-scale testing (less than 10 acres). Manufacturers of biotechnology microbes can avoid this regulatory oversight only if the testing facility meets prescribed containment conditions or if the engineered pesticides are believed to be low risk (27). Despite criticism for this early regulatory intervention during the testing process (8), EPA has consistently justified it as necessary because of the unusual and unpredictable risks associated with accidental release of genetically engineered microbes into the environment (28).

Manufacturers intending to conduct this testing must provide EPA not only with notification, but with data that will provide EPA sufficient information upon which to judge the application for testing (29). After notification, EPA has 90 days to conduct an assessment to determine whether or how to permit the testing (30). Testing may not commence until after EPA has issued its determination (31). If the manufacturer is dissatisfied with either the conditions for testing or EPA's rejection of their application, it has available both regulatory and ultimately judicial avenues for challenging the agency's determination, although great deference is given to the agency during these layers of review (32).

The second difference from conventional chemical pesticides are the regulatory requirements themselves. Data requirements and registration reviews of genetically engineered microbes are often individually tailored to the risks and benefits presented by a particular product. Individual manufacturers are advised to contact the agency directly, rather than commence testing in accordance with EPA's FIFRA regulations (33). EPA's review of the applications, including its use of risk assessment protocols, also appear to be undertaken on a case-by-case basis (34).

Plant Pesticides. EPA has also been working on a specialty program for plant substances that are genetically engineered (using recombinant DNA techniques, not through traditional plant breeding) to produce more or additional natural toxins within the plant that repel specific pests—termed "plant pesticides" (35). Although the program is not yet final, considerable effort and discussion has been dedicated to determining whether and the extent to which EPA should regulate genetically engineered plant pesticides.

Under its proposed policy, EPA considers the active ingredient of a plant pesticide (which is EPA's target of regulation under FIFRA) to be both the substance produced in the plant having the pesticidal effect as well as the genetic material necessary for the production of the substance (36). This EPA policy has been the subject of scientific controversy since it involves regulating the "inherited traits of plants" that "cannot be separated for regulatory purposes from the plant itself," which is quite different from the traditional external pesticides EPA typically regulates under FIFRA (37). Industry has criticized the program as well, both because it includes within its regulatory reach low- to no-risk plant pesticides, and because it may ultimately impose "substantial costs on plant breeding" and discourage the development and use of genetically engineered crops (34,38).

Because certain genetically engineered plant pesticides are considered to be virtually no risk, EPA will likely develop a list of "exempted" plant pesticides in its final plant pesticide policy (25,39). EPA is also considering whether pest resistance management plans should be required as part of the approval process for plant pesticides, although industry maintains that it is already voluntarily dedicating considerable resources into pest resistance management (40). For example, in its approval of Bt corn products (genetically modified corn that contains a bacterium to resist pests), EPA did condition registration on adherence to specified refuge requirements, although the pesticide resistance plans were produced initially by the Bt corn industry (41).

Under EPA's proposed program the nonexempted plant pesticides will still be regulated under the traditional FIFRA process. Differences from EPA's regulation of conventional chemical pesticides may occur, however, with regard to the requirements for testing plant pesticides, data requirements for registration, and conditions for approval (42). In fact, conditions for approval—such as warnings—may be particularly problematic for plant pesticides (43). For example, warning users of a plant pesticide to avoid using the plant within danger areas (e.g., near a water body) may be insufficient to keep the plant from spreading to that area naturally. Thus EPA ultimately may reject applications for some plant pesticides because of the realistic obstacles to restrictive labeling, even though in theory the plant could be used safely if warnings could be provided.

Pesticide Residues

EPA has also been charged with regulating pesticide residues in raw agricultural commodities and processed foods under the Food, Drug, and Cosmetic Act (FDCA) (44). This responsibility generally requires EPA to set tolerances for pesticide residues based on risks posed by human dietary consumption. Thus, unlike FIFRA which requires EPA to consider *all* environmental and human health risks associated with release of a pesticide into the environment, FDCA requires EPA only to consider how much pesticide residue can remain on food products to ensure protection of the public health.

With regard to products of biotechnology, it is expected that the FDCA will pose few if any additional requirements. EPA is developing exemptions for certain plant pesticides because of the extremely low risks presented from consumption (45). These exemptions target various plant pesticides that do not result in significantly different dietary exposures, that are derived from nucleic acids, or that consist of plant virus protein coats (46).

"Toxic Substances" Not Otherwise Regulated

Under the Toxic Substances Control Act (TSCA), EPA is authorized to regulate chemicals substances to ensure their safety. This includes the authority to screen new chemicals and to impose controls on those that pose "unreasonable risk of injury to health or the environment" (47). EPA's TSCA program, like its FIFRA program, has been stretched to include the regulation of biotechnology products. The future uncertainties regarding biotechnology products, coupled with the constraints imposed by the TSCA statute that was passed with conventional chemicals in mind, continue to reveal gaps in EPA's authority and appear to pose impediments to a more coherent and comprehensive regulatory program.

The major features of EPA's TSCA program mirror FIFRA in the sense that EPA must first define those products of biotechnology that fall under the jurisdiction of TSCA, since the statute itself provides very little guidance on the matter. Once the jurisdictional bounds have been defined, EPA then reviews biotechnology products in accordance with its recently conceived and evolving specialty rules.

General. The primary emphasis of TSCA is to require all "new chemical substances" to go through a registration and approval process that evaluates, however cursorily, their safety (48). These registration requirements are considerably less onerous than those of FIFRA or the FDCA. Before marketing a new substance, the manufacturer must submit a pre-manufacture notice (PMN) that provides all available data regarding the health and environmental safety of the substance (49). The most notable contrast with FIFRA and FDCA is that if the data are incomplete, there is *no* affirmative obligation on the manufacturer to conduct the research itself unless EPA specifically requires additional testing: The manufacturer must include in its notice only the health or environmental test data in the manufacturer's possession or control (50). EPA's ability to require additional testing is read to be justified only after EPA is able to show that the substance presents the possibility of an unreasonable risk or that it will be produced in substantial quantities. Therefore EPA's authority to require added testing, as a practical matter, is generally constrained for substances where the scientific uncertainties are particularly high or the safety testing scant (51). Indeed, some have suggested that the design of TSCA may tend to lead to less safety testing rather than more (52).

After filing the PMN with all available information, EPA has 90 days within which to consider the application (53). If it believes that the information leads to a conclusion that the substance might be risky, it can require additional testing or deny the application (54). Perhaps because of the time, resource, and burden of proof and related statutory problems that the EPA encounters in administering TSCA, the denial of an application is a rare event for conventional chemical substances (55). Since the newly emerging products of biotechnology are only now moving through the requirements of TSCA, it is difficult to tell how rigorously TSCA will be applied to this industry.

Specific. Recurrent biotechnology-related problems continue to arise under TSCA that have led EPA to create specialty regulations to streamline the process for products of biotechnology. Most notable are added requirements for research and development, a tailor-made review process for biotechnology microbial products, and bright-line exemptions for small quantity users and biotechnology products that are presumptively low risk. Before detailing these specialty programs and requirements, however, the threshold jurisdictional question must be resolved: Is the manufacturer seeking to market a biotechnology product that EPA considers within the jurisdictional reach of TSCA?

Jurisdictional Reach. TSCA is a "gap-filler" statute that applies largely to new chemical substances in commercial use that are not regulated elsewhere as food additives, drugs, or pesticides (8,56). In accordance with this statutory definition, there are particularly thorny jurisdictional issues that must be resolved before EPA will assert its regulatory authority over a biotechnology product, all of which surround the definition of "chemical substance." In interpreting what constitutes a "chemical substance" (57), EPA has determined that living organisms may be included within the definition. In order to limit the consequences of this generous reading of its regulatory jurisdiction under the statute, EPA currently has restricted its regulatory program to microorganisms "used in conversion of biomass for energy, pollutant degradation, enhanced oil recovery, metal extraction and concentration, and certain non-food and non-pesticidal agricultural applications, such as nitrogen fixation" (58).

EPA's regulatory jurisdiction over products of biotechnology is still further limited by its definition of when a microorganism is genetically engineered—only when it results from the "deliberate combination of genetic material originally isolated from organisms of different taxonomic genera" (59). Like FIFRA, this definition results from the EPA's attempt to develop a clear, yet generally inclusive definition based on perceived risks of current products of biotechnology (60). The TSCA regulatory coverage of only commercial research further constrains EPA's regulation of emerging biotechnologies, although to a much more limited extent (12).

Once within the jurisdiction of TSCA, the manufacturer must make itself aware of additional requirements and exemptions that govern EPA's regulation of biotechnology.

Added Considerations and Requirements. The approval process for genetically engineered microbes on its face appears similar to the process for conventional chemicals. In order to commercialize new genetically engineered microbes, the manufacturer, importer, or processor must submit to EPA the biotechnology equivalent of PMNs, termed "Microbial Commercial Activity Notices" (MCANs) (61). EPA then has 90 days to determine whether the organism "may present an unreasonable risk to human health or the environment" (62,63). But several noteworthy differences in biotechnology products present new questions for this established regulatory program. None of these questions appear to have been anticipated by Congress when it passed TSCA in 1976. Some of these questions also appear to remain unresolved by EPA (11).

First, EPA has focused almost exclusively on genetically engineered microbes as the covered set of biotechnology products. While these appear to be the largest category of biotechnology products not otherwise regulated under other statutes, it is not the only set of products that could in theory fit within the statute. One commentator, for example, has suggested that genetically engineered fish could be included within the jurisdictional reach of TSCA (64). Thus the classification of biotechnology products that EPA will regulate under TSCA remains an open issue for debate and possibly the subject of future regulatory expansions.

A second regulatory problem that arises is similar to that occurring for genetically engineered microbial pesticides—the uncertain risks associated with release into the environment and the resulting perceived need for earlier regulatory intervention during the testing stage (65). Thus, in contrast to conventional chemical substances where "small quantity" research and development is exempted from regulatory oversight, EPA maintains an active oversight role in research and development of genetically engineered microbes and requires manufacturers of these microorganisms to submit abbreviated applications (called "TSCA Experimental Release Applications" or TERAs) 60 days before testing intergeneric organisms in the environment (66). In an effort to minimize the adverse impact of this oversight on low-risk genetically engineered microbes, EPA is also promulgating a rule that will exempt from regulatory oversight certain low-risk biotechnology products used in research and development activities (67).

Third, and relatedly, standard reporting requirements and assessment procedures used for conventional chemicals often fall short of providing helpful guidance in overseeing the safety of genetically engineered microbes. For example, the TSCA section 8(e) reporting requirement that manufacturers report "substantial risks" arising from their chemical substances focuses largely on potential health risks such as cancer, rather than on the release and spreading concerns posed by the use of genetically engineered products (68). A similar complaint has been made regarding the use of EPA's risk assessment protocols that were originally developed for traditional TSCA chemical substances. Although EPA has responded to these discontinuities by providing guidance on the types of information to submit for MCANs (69), and in relying heavily on ecological risk assessment protocols for reviewing these applications (70), these adjustments are far from complete or streamlined (70). As a result, EPA appears to determine reporting requirements for biotechnology projects on a case-by-case basis, with the resulting requirements being much less predictable than the more standardized reporting requirements that apply to conventional chemical products (71). In fact, to date, EPA's sole decision under TSCA to approve the commercialization of a genetically engineered microorganism for release into the environment (RMBPC-2, a soil bacterium that fixates nitrogen for alfalfa plants) involved nine years of data (including five years of test data) on the product, generated some controversy within the EPA's scientific advisory panel, and ultimately attained approval conditioned on limited production and monitoring requirements (72).

EPA has also recognized that in some cases complete exemptions from TSCA oversight are justified for genetically engineered microbes for which the risks can be reliably predicted in advance to be minimal to nonexistent. Thus, with the assistance of several years of deliberations, input from its Science Advisory Board, and the public comment process, EPA has also promulgated a variety of general and research-specific exemptions for certain types of genetically engineered microbes (73).

Miscellaneous Other Statutory Authorities

EPA may also have authority under other statutes to regulate the release of a wide variety of genetically engineered organisms into the environment. Under the Clean Water Act, for example, EPA may be able to justify regulation or restriction of the introduction of bioengineered fish into U.S. rivers and streams through its authority to restrict the discharge of "pollutants" into navigable waters (64,74,75). Both the Resource Conservation and Recovery Act (RCRA) and the Comprehensive Environmental Response, Compensation, and Liability Act (CERCLA) may provide EPA with the ability to regulate or require cleanup of genetically engineered organisms that can be characterized as "hazardous wastes" or "hazardous substances" respectively (75). Finally, the Endangered Species Act, with its prohibition against impairing the critical habits of endangered species, may also provide mechanisms for regulating the introduction of biologically engineered organisms into these often narrowly bounded environments (64,76). None of these authorities appear to have been used by EPA to restrict the release of genetically engineered organisms, although their potential has been recognized by commentators.

EPA AS PROMOTER OF BIOTECHNOLOGY

Although EPA's relationship to the biotechnology industry is primarily one of regulator, EPA also acts in a much more limited capacity as a promoter of biotechnology through its effort to carve out administrative exemptions to existing regulatory programs and through its small grants program for research. EPA's activities in streamlining regulatory programs and creating exemptions to regulatory oversight when the risks of a biotechnology product are minimal are motivated in part by the Executive Branch's relatively consistent support of biotechnology. The agency's efforts to minimize regulatory impediments to biotechnology innovation also result from its conviction that many genetically engineered organisms—such as plant pesticides—are safer and more effective than conventional chemical products.

EPA's role as a promoter of biotechnology not only consists of its effort to minimize the regulation of biotechnology products where justified, but to encourage research in areas like bioremediation and related fields (77). Grants administered through EPA provide some support for biotechnology research activities (78). Most notable is EPA's continued support of genetically engineered bioremediation technologies for hazardous waste cleanup (79). It seems likely that EPA's encouragement of these developments will only increase over time.

PROBLEMS

Biotechnology regulatory programs that are shoe-horned into regulatory programs originally designed for conventional chemicals run the risk of not only inheriting the existing weaknesses of these programs, but exacerbating the weaknesses by stretching the statutes to include the new and very different regulatory challenges posed by products of biotechnology (80). EPA's regulatory programs appear to confirm these dismal predictions.

Incoherent and Incomplete Regulatory Programs

National policy regarding the regulation of biotechnology remains muddled in the separate and often quite different regulatory programs of the various agencies, such as FDA, EPA, and the U.S. Department of Agriculture (USDA). The resulting national regulatory approach has been condemned by some critics as "develop[ed] in response to jurisdictional arguments that can be made to protect and project agency authority instead of in response to national policy considerations" (2, p. 241). EPA has in fact conceded that some important gaps remain in its ability to fit biotechnology products into existing regulatory programs. Regulatory oversight of genetically engineered wildlife and fish remain problematic with regard to existing statutory authorities (64). Existing gaps in regulatory authority will likely grow only more serious as the biotechnology industry expands. Yet in some circumstances, parties injured by inadequately regulated releases of biotechnology products may be left with disappointing legal remedies precisely because existing federal programs could be read to preempt their common law claims (81).

Perhaps equally serious is the complex and often unpredictable nature of EPA's regulatory programs for the biotechnology industry (82). EPA's regulatory programs are not only confused by the fact that its statutory authority never mentions or even alludes to biotechnology, but also because EPA's programs must comply with the Executive Branch biotechnology policy statement that at times appears quite remote or even at odds with the purported original grant of legislative authority (2). Although EPA has dedicated considerable effort to defining its jurisdiction for biotechnology and developing specialty programs to address some of the more categorically different problems that biotechnology presents to existing programs, considerable regulatory uncertainties remain. For example, to determine what sorts of pre-approval testing requirements apply to biotechnology pesticides, a manufacturer must generally *meet* with EPA to hammer out the case-specific tests. The resultant "uncertainty costs" are of continuing concern to the biotechnology industry (83). In contrast, manufacturers of conventional chemical pesticides can generally determine their testing requirements by reference to the Code of Federal Regulations. To make matters worse, the regulatory review process for biotechnology products also tends to be considerably more unpredictable with regard to its costs, delays, and ultimate outcomes.

The absence of any solid legislative grounding to EPA's biotechnology regulatory programs also subjects these programs to seemingly endless waves of revisions, exemptions, and other regulatory changes without a clear, overarching regulatory plan. Rather than falling ahead of the curve and engaging in important regulatory planning for the inevitable biotechnological future, EPA seems often to be endeavoring to catch up to last year's problems. This "regulatory gap" then produces a damaging "false start" for an industry that invests in innovation only to find that their ultimate marketing is severely restricted (84). The resulting outdated, or at least constantly shifting, nature of EPA's programs likely also takes its toll on public understanding and support (85).

Scientific Uncertainties that Complicate Regulation

EPA's effort to fit biotechnology regulation within existing regulatory structures also exacerbates preexisting weaknesses in these statutory programs (2,75). Both TSCA and FIFRA regulatory programs have been criticized for their unjustified over-reliance on quantitative risk assessment, even with regard to assessing the risks of conventional chemicals (2,63,86). In the highly uncertain area of biotechnology, the obstacles to reliable quantitative risk assessments and cumulative impacts are still more daunting. Because the uncertainties often overwhelm the information that science is able to provide, EPA is left with analytical tools that are sorely inadequate, but upon which Congress, the public, and the courts consistently demand that EPA's regulatory judgments be based (2,86). In order to meet these external expectations and demands, EPA may overstate the scientific grounding of its risk assessments for biotechnology products, a tendency that likely improves the short-term credibility of its regulatory actions, but that over time may undercut public understanding, participation, and even support for some of the agency's underlying policy judgments (2). As one author has argued, "[m]aking quantitative risk assessment the sole basis for biotechnology regulation ... is likely to increase public anxiety, mask important political choices in purportedly neutral, scientific terms and ultimately, fail to consider many of the potential hazards presented by biotechnology" (2, pp. 249–250).

In addition, existing statutory programs appear to assume that added testing will resolve remaining, material uncertainties, an assumption that necessarily tends to overemphasize those risks that can be quantified (2). The statutory policies of FIFRA, for instance, tend to presume that a battery of tests will adequately assess the health and environmental risks of pesticides, an assumption much more appropriate for pesticides posing primarily cancer risks than for biopesticides that may behave in unpredictable ways once released into the environment in large numbers (11,87). Even more inappropriate are the policies of TSCA that require the EPA to demonstrate that a product presents a likelihood of health or environmental risk before requiring the manufacturer to conduct additional testing. Since there is almost no information on how these products will react in the environment (and the manufacturer, absent a command from EPA, is not required to conduct the testing on products not otherwise covered under FIFRA or the FDCA), the burden placed on EPA undercuts its ability to effectively encourage adequate safety research on these new products (63,88).

Finally, the unfortunate consequences of these uncertainties seem to be multiplied by the current regulatory approach that attempts to ignore or cover them up. Current incremental and ad hoc approaches do not make headway in gaining public confidence. Indeed, growing fears of the safety and acceptability of genetically engineered foods by some of the public may threaten the plant pesticide industry more than tentative and still incomplete regulatory oversight by EPA (89).

CONCLUSION

EPA's federal policy making on the products of biotechnology is diverse and complex, due at least in part to the Executive Branch's historic efforts to fend off legislative intervention. Although EPA seems to be endeavoring mightily toward the incremental creation of subprograms for different types of biotechnological pesticides and products, skepticism remains with regard to whether this administrative approach will ultimately meet the escalating needs of both the industry and the concerned public.

BIBLIOGRAPHY

1. W.H. Rodgers, Jr., *Environmental Law*, vols. 1–5, West Publishing, St. Paul, MN, 1988.
2. P. Mostow, *Pace Environ. Law Rev.* **10**, 227,240–242 (1992).
3. J.P. Dwyer, *Ecol. Law Q.* **17**, 233 (1990).
4. Streamlining Federal Regulation of Biotechnology Products, Press Release from Council on Competitiveness, February 23, 1992.
5. 1992 National Biotechnology Policy Report at E-4, reprinted in *Biotechnol. Law Rep.* **12**, 127 (1993). Several documents were produced as a result of this effort, the most recent being a 1992 Federal Oversight document that clarifies the scope of federal regulation of biotechnology. *Fed. Regist.* **57**, 6753 (1992).
6. *Fed. Regist.* **51**, 23302–23303 (1986).
7. *Fed. Regist.* **57**, 6753 (1992). For a critique of this policy, see Mostow (2).
8. W.H. Rodgers, Jr., *Environmental Law: Pesticides and Toxic Substances*, vol. 3, West Publishing, St. Paul, MN, 1988, §6.12, pp. 504–505, Chart I.
9. *Fed. Regist.* **9**, 50859–50861 (1984).
10. *Hearings Before the Subcommittee on Investigation and Oversight and the Subcommittee on Science, Research and Technology of the House Committee on Science and Technology, 98th Cong., 1st sess.*, pp. 131–132 (1983). (Statement of Dan Clay, Acting Asst. Admin., Office of Pesticides and Toxic Substances, EPA).
11. L. Maher, *J. Environ. Law Litig.* **8**, 133,195 (1993).
12. C.C. Vito, *Me. Law Rev.* **45**, 329,341–346 (1993).
13. W.H. Rodgers, Jr., *Environmental Law: Pesticides and Toxic Substances*, vol. 3, West Publishing, St. Paul, MN, 1998, §5.3, pp. 32–54.
14. 7 U.S.C. §§136, *et seq.*
15. 7 U.S.C. §§136a(c)(5)(D), 136d(b).
16. W.H. Rodgers, Jr., *Environmental Law: Pesticides and Toxic Substances*, vol. 3, West Publishing, St. Paul, MN, 1998, §5.10, pp. 127–145.
17. 40 CFR, pt. 158.
18. 7 U.S.C. §136c; 40 CFR, pt. 172.
19. 7 U.S.C. §136n.
20. 7 U.S.C. §136(u).
21. *Fed. Regist.* **59**, 45605 (1994).
22. *Fed. Regist.* **59**, 45608–45609,45611–45612 (1994).
23. *Fed. Regist.* **59**, 45606–45607,45614 (1994).
24. H.J. Miller, *Science* **266**, 1815 (1994).
25. M.J. Angelo, *U. Fla. J. Law Public Policy* **7**, 257,281 (1996).
26. R.V. Percival, A.S. Miller, C.H. Schroeder, and J.P. Leape, *Environmental Regulation: Law, Science, and Policy*, Little, Brown, New York, 1996, pp. 176–179.
27. *Fed. Regist.* **59**, 45602,45606–45607,45614 (1994).
28. *Fed. Regist.* **49**, 40660 (1984).
29. 40 CFR §172.48.
30. 40 CFR §§172.50, 172.57, 172.59.
31. 40 CFR §172.50(a).
32. 40 CFR §172.10.
33. J.R. Holmstead, *ALI-ABA* **C948**, 709,712 (1994); *Daily Environ. Rep. News (BNA)* **138**, A-7 (1997).
34. *Daily Environ. Rep. News (BNA)* **143**, A-3 (1997).
35. *Fed. Regist.* **59**, 60519 (1994); but see *Chem. Regul. Rep. (BNA)* **23**, 184 (1999).
36. *Fed. Regist.* **59**, 60496 (1994); *Fed. Regist.* **59**, 65500 (1994).
37. *Chem. Regul. Rep. (BNA)* **20**, 657 (1996) (quoting report issued by the Institute of Food Technologists).
38. J.E. Beach, *Food Drug Law J.* **53**, 181,191 (1998); *Chem. Regul. (BNA)* **22**, 1930 (1999).
39. *Fed. Regist.* **59**, 60519,60522,60525–60528 (1994).
40. *Chem. Regul. Rep. (BNA)* **21**, 711,1220 (1997).
41. *Chem. Regul. Rep. (BNA)* **23**, 857 (1999); see also *Daily Environ. Rep. News (BNA)* **161**, A-5 (1999) (reporting on research detecting adverse effects of Bt corn on monarch butterflies). EPA has also sought the assistance of its science advisors in determining appropriate pest resistance plans for particular biotechnology products. *Chem. Regul. Rep. (BNA)* **21**, 1414 (1998).
42. 40 CFR §172.3.
43. *Fed. Regist.* **59**, 60510 (1994).
44. 21 U.S.C. §§346a and 348.
45. *Daily Environ. Rep. News (BNA)* **95**, A-2 (1997).
46. *Fed. Regist.* **59**, 60535,60537,60538 (1994).
47. 15 U.S.C. §§2603(a) and 2605(a).
48. 15 U.S.C. §2604(a).
49. 15 U.S.C. §2604(a)(1).
50. 15 U.S.C. §2604(d)(1).
51. 15 U.S.C. §2603(a); D.J. Hayes and A. Claassen, *ALI-ABA* **C864**, 191,208 (1993).
52. M.L. Lyndon, *Mich. Law Rev.* **87**, 1795,1825–1827 (1989).
53. 15 U.S.C. §2604(a)(1).
54. 15 U.S.C. §§2603(a) (require testing); 15 U.S.C. 2604(e)(1)(A) (block manufacture pending testing); 15 U.S.C. 2605 (ban manufacture).
55. General Accounting Office, *Toxic Substances: EPA's Chemical Testing Program Has Made Little Progress*, U.S. Government Printing Office, Washington, DC, 1990.

56. 15 U.S.C. §2602(2)(B) (defining "chemical substance" in relation to other statutes).
57. 15 U.S.C. §2602(2).
58. *Fed. Regist.* **51**, 23324 (1986).
59. 40 CFR §725.3.
60. *Fed. Regist.* **51**, 23325–23326 (1986). EPA can also use other authorities to reach existing products that are used in significantly new ways under its "significant new use rule" (SNUR). 15 U.S.C. §2604(a)(2).
61. 40 CFR §§725.100–190.
62. 40 CFR §725.170.
63. R. Harlow, *Yale Law J.* **95**, 553,565,570–574 (1986).
64. A.G. Gorski, *Univ. Balt. J. Environ. Law* 3(1), 24–25 (1993).
65. Jaffe, *Harv. Environ. Law Rev.* **11**, 491,540 (1987).
66. 40 CFR §§725.200–288; *Chem. Regul. Rep. (BNA)* **22**, 7 (1998).
67. *Fed. Regist.* **61**, 63143 (1996); *Daily Environ. Rep. News (BNA)* **13**, G-2 (1997).
68. *Daily Environ. Rep. News (BNA)* **206**, A-5 (1997) (quoting EPA biotech attorney, David Giamporcaro).
69. 40 CFR §§725.155–725.160; *Chem. Regul. Rep. (BNA)* **21**, 245 (1997); *Fed. Regist.* **62**, 17910 (1997).
70. *Daily Environ. Rep. News (BNA)* **169**, A-4 (1997).
71. *Daily Environ. Rep. News (BNA)* **206**, A-5 (1997).
72. *Daily Environ. Rep. News (BNA)* **187**, A-2 (1997); *Chem. Regul. Rep. (BNA)* **18**, 1749 (1995); *Chem. Regul. Rep. (BNA)* **21**, 1610 (1998).
73. 40 CFR Subparts E,F, and G.
74. 33 U.S.C. §§1251–1387 (1994).
75. D.E. Hoffmann, *Drake Law Rev.* **38**, 471,498 (1988). (Considering similar regulations under the Clean Air Act).
76. 16 U.S.C. §§1536 and 1538.
77. *Environ. Rep. (BNA)* **29**, 192 (1998).
78. *Fed. Regist.* **59**, 348,359,401 (1994); *Chem. Regul. Rep. (BNA)* **15**, 1646 (1992).
79. *Chem. Regul. Rep. (BNA)* **20**, 279 (1996); *Daily Environ. Rep. News (BNA)* **124**, d21 (1993).
80. C.R. Manor, *Rutgers Comput. Tech. Law J.* **3**, 409 (1987).
81. C.M. Steen, *Ariz. Law Rev.* **38**, 763,791 (1996).
82. *Chem. Regul. Rep. (BNA)* **18**, 1549 (1995).
83. *Daily Environ. Rep. News (BNA)* **143**, A-3 (1997).
84. S. Goldberg, *Culture Clash: Law and Science in America*, New York University Press, 1994, pp. 84–111.
    M. Malinowski and M. O'Rourke, *Yale J. Regul.* **13**, 163 (1996).
85. *Chem. Regul. Rep. (BNA)* **18**, 1549 (1995).
86. W.E. Wagner, *Columbia Law Rev.* **95**, 1613 (1995).
87. *Chem. Regul. Rep. (BNA)* **2**, 566 (1997) (reporting criticisms of EPA's risk assessment protocols for genetically engineered microbes).
88. J.S. Applegate, *Columbia Law Rev.* **261**, 319–330 (1991).
    R.A. Chadwick, *Hofstra Law Rev.* **24**, 223 (1995).
89. *Daily Environ. Rep. News (BNA)* **143**, A-3 (1997).

See other entries AGRICULTURAL BIOTECHNOLOGY, LAW, AND EPA REGULATION; see also FEDERAL POLICY MAKING FOR BIOTECHNOLOGY entries.

FEDERAL POLICY MAKING FOR BIOTECHNOLOGY, EXECUTIVE BRANCH, DEPARTMENT OF ENERGY.
See INFORMING FEDERAL POLICY ON BIOTECHNOLOGY: EXECUTIVE BRANCH, DEPARTMENT OF ENERGY.

FEDERAL POLICY MAKING FOR BIOTECHNOLOGY, EXECUTIVE BRANCH, ELSI

AUDRA WOLFE
University of Pennsylvania
Philadelphia, Pennsylvania

OUTLINE

Introduction
Organizational Structure
    Origins of ELSI
    Form and Function
ELSI in Action: Projects and Policy
    Topics and Questions
    Projects
ELSI's Critics and Federal Science Policy
    Policy or Research? Restructuring ELSI
    A New Role for Government?
    Ethics and the Role of History
Bibliography

INTRODUCTION

The Ethical, Legal, and Social Implications (ELSI) Program, part of the Human Genome Project (HGP), is a unique federal policy and research program that attempts to anticipate potential negative consequences of the very project of which it is a part. Between 3 and 5 percent of the budget for the HGP is dedicated to projects that investigate various ethical, legal, and social aspects of genomic research, genetic testing, and genetic information. The ELSI Program refers to a set of administrative and grant-making bodies in both the National Institutes of Health (NIH) and the Department of Energy (DOE) that make policy recommendations, determine the ELSI agenda, and sponsor conferences and research. As a funding agency, ELSI administers extramural and intramural contracts and grants to historians, philosophers, legal scholars, sociologists, bioethicists, and policy analysts on a wide variety of issues. As an advisory board, ELSI has managed task forces and formulated policy recommendations for topics including genetic testing, genetic discrimination in insurance and employment, and the protection of human subjects in genetic research. As the HGP moves toward the completion of human DNA sequencing, ELSI has more recently emphasized questions of human genetic variation and diversity, clinical and nonclinical integration of genetic information, multicultural interactions with genetic information, and fairness in access to genetic testing and treatments. As the first federal scientific research program to dedicate portions of its own budget to anticipating and analyzing its potential social impact, ELSI represents a landmark in federal science and bioethics policy.

ORGANIZATIONAL STRUCTURE

Origins of ELSI

The government's commitment to funding research into ethical, legal, and social implications of the HGP began immediately with Dr. James Watson's appointment as director of the Office of Human Genome Research of the National Institutes of Health. On October 1, 1988, at the press conference announcing his appointment, Watson suggested that 3 to 5 percent of the NIH HGP funds should support work on the ethical, legal, and social implications of new knowledge about human genetics. Recognizing that these issues were not necessarily new, Watson noted that they would nonetheless be associated with the genome project and deserved serious attention within the project (1). By the end of October, a Joint NIH-DOE ELSI Working Group had been established by the Program Advisory Committee of the HGP with psychologist Nancy Wexler (former director of the NIH's Huntington's Commission) serving as chair (2). Although initiated by Watson and originally housed in the NIH, the DOE began to co-sponsor ELSI programs in December 1989 under congressional pressure.

The original Working Group core included (in addition to Wexler) Thomas Murray, bioethics; Jonathan Beckwith, molecular genetics; Robert Murray, clinical genetics; Patricia King, law and policy; Victor McKusick, human genetics; and Robert Cook-Deegen, policy. In addition to these Working Group members, the individual programs at NIH and DOE were overseen by separate administrators. Eric Juengst, with a background in philosophy and bioethics, served as chief of the NIH ELSI program from 1990 to 1994; Michael Yesley, with experience in the law, originally conducted the DOE ELSI program (2).

Although Congress had not suggested an ELSI component in the HGP, members of Congress were nevertheless concerned about the possible consequences of new genetic information. Stressing questions of commercialization, genetic discrimination, and the choices influencing decisions to screen populations, both the National Research Council and the Office of Technology Assessment discussed social and ethical implications in their preliminary studies of the HGP. In addition Thomas Murray expressed his concern during the HGP hearings before Congress in 1988. The idea to set aside part of HGP's funds for ethical, legal, and social research, however, was Watson's and was only instituted through the NIH's budget recommendations (1).

Form and Function

While the Working Group, as a subcommittee of HGP, received funds from both NIH and DOE, the ELSI Program has both historically and currently functioned through separate NIH and DOE programs. DOE research funds are provided through an ELSI Program in the Office of Health and Environmental Research. Beginning in 1990,

NIH extramural grants were administered through the National Human Genome Research Initiative (NHGRI) by the ELSI Branch (later renamed the ELSI Research Program). In 1995 a NIH Office of Policy Coordination (OPC), housed in the Office of the NIH director, was created to sponsor conferences on and analysis of ELSI-type policy issues. In addition the NHGRI's Division of Intramural Research (DIR) maintains an Office of Genome Ethics and Policy Analysis to explore ethical and legal issues arising from applied genetics. While the majority of ELSI projects have been sponsored by either of the grant programs, various task forces and conferences have been supervised by the Joint NIH-DOE ELSI Working Group.

Though unofficial, a division of labor has existed in some of the issues addressed by the NIH and DOE programs. DOE ELSI projects have been much more prominent in the areas of genetic privacy, public education, and intellectual property issues. The most visible NIH ELSI projects, on the other hand, have involved genetic discrimination and genetic testing issues on diseases ranging from cystic fibrosis to breast cancer. Both DOE and NIH, however, have administered a broad range of topics spanning the issues posed by HGP. Both agencies maintain Web sites with full descriptions of their current programs and grants, publications resulting from ELSI funds, and full-text press releases, task force reports, and workshop reports (3,4).

ELSI IN ACTION: PROJECTS AND POLICY

Topics and Questions

In its initial 1990 NIH grant announcement, ELSI stressed that projects should focus on the possible impacts of disease-related genetic information. While those projects that offered policy solutions would be granted the highest priority, the Working Group expressed a commitment to the traditional methods of the social sciences and humanities, bridging perspectives from morality, ethics, the law, policy, history, and public understanding of science. ELSI grant applicants were encouraged to propose projects in nine topics areas: fairness in insurance, employment, the criminal justice system, education, adoption, and the military; psychological and societal responses to individual genetic information; privacy and confidentiality; genetic counseling, including prenatal and presymptomatic testing, testing in the absence of therapeutic options and population screening versus testing; issues of reproductive choice; medical practice, including standards of care, training, the doctor–patient relationship, and patient education; historical abuses of genetic information, especially eugenics; commercialization, including property and intellectual property rights; and philosophical issues such as the meaning of identity, definitions of health and disease, and questions of determinism and reductionism. Clearly, the working group had established a broad agenda for the ELSI experiment (5).

The Working Group prioritized its topics in two sessions in February and September 1990. Without downplaying the research-driven humanities topics included in the original grant announcement, the working group outlined four areas as high priority for policy in the first five years. Policy recommendations for clinical implementation of new genetic tests, genetic privacy concerns, genetic discrimination in insurance and employment, and professional and public education were seen as especially pressing issues. Although the Working Group's approach to each of these priorities differed, in each area they went beyond extramural grant administration to coordinate the production of policy options.

Projects

Within a year, funding was underway for 16 projects, including several interdisciplinary conferences. By September 1991 that number had increased to 25 external grants and 10 national conferences. ELSI began with 3 percent of the HGP budget; in 1992 NIH increased its commitment to 5 percent (2). Reflecting ELSI's wide range of topics and goals, these projects ranged from specific policy measures to academic conferences designed to define what might constitute ELSI-type issues.

Cystic Fibrosis Consortium. One of ELSI's first high-profile projects implemented the policy goal of determining appropriate uses of genetic testing. At the same time that the Working Group had been formulating its goals, medical scientists had developed a test for heterozygotes (carriers) of cystic fibrosis (CF). The ELSI Working Group emerged as a voice advocating caution before widespread use of a test of not only unknown reliability but also unexplored psychosocial consequences. What did a subject need to know to use the test's results to his or her advantage? What sorts of counseling measures were necessary? Would the public at large be interested in testing for CF? To help answer some of these questions, the Working Group solicited proposals for clinical trials evaluating social as well as medical aspects of CF testing and then created a coordinated project among several applicants. The American Society for Human Genetics bolstered ELSI's power and legitimacy in CF testing by urging restraint until the results of the trials were made clear. The recognition of the importance of "client-centered criteria" was one of the most significant results of the trials. Any understanding of the "success" of genetic tests must take into account individual families' reactions to and uses of the information. Empowering the patient had become at least a nominal goal of CF trials. ELSI sponsored a similar program for evaluating genetic tests for those at risk of cancer, and these programs have set precedents for similar investigations sponsored by the Heart Institute, the National Institute of Mental Health, and the National Child Health Institute (1,2).

Genetic Testing. By 1994 the possibilities for genetic testing had grown far beyond CF. To the rising concern of the ELSI Working Group, a number of tests of variable quality had been introduced to the public without much discussion of patient reception. It was clear that a more systematic approach to the problem of genetic testing had become necessary. As a first step toward addressing this problem, the Working Group commissioned a study by the National Academy of Science's Institute of Medicine on assessing genetic risks. This report raised special concerns about the lack of treatment—preventive

or otherwise—for diagnosed diseases and insufficient genetic expertise among those who developed or performed genetic tests. The Working Group established the Task Force on Genetic Testing in 1994 to further explore these issues and to formulate policy recommendations. The voting portion of the Task Force consisted of 15 members nominated by diverse genetic testing interest groups; several nonvoting liaison members from agencies such as the Department of Health and Human Services (HHS) and the Food and Drug Administration (FDA) completed the committee.

The Task Force began their ambitious agenda by surveying the quality of genetic tests, the quality of the laboratories providing the tests, and the competence of testing personnel. They commissioned a series of background papers on the state of genetic testing and clinical laboratories and conducted a survey of all organizations likely to be involved in genetic testing. As a follow-up to the survey, in-depth interviews were conducted at 29 of the almost 500 organizations either currently conducing genetic testing or considering it in the future. The Task Force met in seven different sessions; halfway through the process, they invited public comment on preliminary conclusions and recommendations. In 1997 the Task Force issued its final report and policy recommendations.

The Final Report established several overarching principles before moving on to their specific studies. Informed consent, consideration of an individual family's values in regard to prenatal or carrier testing, confidentiality, and the prevention of discrimination were considered crucial goals regardless of specific policy recommendations. Policy measures focused on questions of oversight, licensure, and expertise. The Task Force expressed concern that genetic tests did not fall under FDA supervision, so the report strongly encouraged interactions with IRBs and the Office of Protection of Human Subjects to determine appropriate guidelines for conducting clinical trials on genetic tests. Tests should be subject to external peer review to ensure their efficacy and scientific merit. The report further forcefully stated that laboratories conducting genetic tests should have Clinical Laboratory Improvement Amendments (CLIA) certification. To demonstrate expertise in medical genetics, laboratory directors or technical supervisors should have American Board of Medical Genetics certification. To improve health care providers' understanding of genetic testing and genetic disease, the Task Force strongly recommended that medical, public health, and nursing education include more human genetics instruction and suggested that the inclusion of human genetics questions on licensing exams might spur the development of medical residencies in genetics. Finally, the Task Force recommended that measures were required to ensure the continued development of tests for rare genetic diseases as well as those possibly effecting a large portion of the population (6).

To implement all of these proposals, the Task Force advised the creation of a formally chartered advisory committee on genetic testing in the Office of the Secretary of Health and Human Services. Such an office was created in June 1998 by Secretary Donna Shalala. While not all of the recommendations have been implemented, the Task

Force's Final Report has been received as an important document in determining the future of genetic test development, and the creation of the Advisory Committee represents a key step in maintaining a formal discussion on issues raised by genetic testing.

Genetic Discrimination. Especially in ELSI's early years, questions of genetic discrimination generally concerned either employment or insurance. In dealing with matters of employment, the Working Group appealed to the Americans with Disabilities Act (ADA) as a possible source of guidelines. For insurance, the Working Group created a special task force that participated in national debates on the American health care system.

The ADA ensures that employers cannot consider disabilities as a factor in hiring decisions. The Equal Employment Opportunity Commission (EEOC) interprets and enforces the ADA. The possibility of pre-symptomatic genetic testing introduced the potential for a new kind of disability categorization and subsequent discrimination. Would the EEOC offer protection to those denied employment on the basis of genetic test results that suggested that a person was at risk for a disease, or of parenting a child with a disease, in the future? Could heterozygotes for recessive illnesses be considered disabled for the purposes of the ADA?

The Working Group responded to these concerns by submitting policy recommendations to the EEOC in April 1991. While NIH and DOE themselves cannot make recommendations concerning legislation and regulations, their subcommittees can. ELSI's policy statement on ADA was therefore incorporated into a recommendation by the NIH-DOE Joint Committee on the Human Genome. This action—though not immediately effective—demonstrated that ELSI might be able to exercise policy options not available to the NIH itself.

ELSI's recommendations included three main points for strengthening ADA. First, heterozygotes for recessive and X-linked disorders should be considered "impaired" for the purposes of protection by ADA. Second, employment entrance exams administered after a job offer should either be voluntary or limited to job-related physical or mental impairments. This amendment would protect against discrimination on the basis of HIV status as well as genetic disorders. Third, EEOC should consider ways to protect employees from employer access to personal medical information unrelated to specific insurance claims (7). EEOC did not initially incorporate ELSI's recommendations into their interpretation of ADA; after several years of negotiations, however, EEOC altered their guidelines to explicitly offer protection from employment discrimination based on either genetic disease or the results of tests predicting the development of genetic disorders in the future.

In the area of health insurance, ELSI created a special Task Force on Genetic Information and Insurance consisting of two Working Group members, representatives from the health insurance industry, public-interest groups, government officials, and ELSI grantees. Operating on the assumption that the advance of genomic science would blur the distinction between genetic

and nongenetic information, the task force concluded that specific recommendations based solely on genetic information offered too narrow a solution. Instead, the task force concluded that only the development of a new system of American insurance—one not calculated on individual risk underwriting—could adequately and equitably address the issues raised by genetic testing. The task force's bold recommendations included seven points. First, health status—past, current, or future—should not be used to deny health care coverage. Second, access to the health care system should be expanded so that all Americans would have access to basic medical care. Third, genetic services, including counseling, testing, and treatment, should be included as basic aspects of medical care. Fourth, individual insurance costs should not be based on information concerning health status—past, present, or future. Fifth and sixth, access to health care or insurance programs should neither be based on employment nor disclosure of genetic or other medical information. Finally, additional steps should be taken to protect individuals until the time that universal health coverage should arrive (8).

The preliminary ELSI recommendations were passed along to the White House Task Force on Health Care Reform in 1992 and were included in the Clinton health care reform bill, the Health Care Security Act of 1993. Having made its recommendations, the Task Force then dissolved, and their recommendations were defeated along with the rest of the Clinton health care plan (1). While ELSI measures to protect individuals from genetic discrimination in insurance largely failed, the Health Insurance Portability and Accountability Act of 1996 now provides some measure of protection by excluding genetic tests, in the absence of disease, from preexisting condition clauses. In other words, a patient does not have a preexisting condition until he or she develops a genetic disease regardless of the indications of a genetic test (9).

Genetic Privacy. ELSI's approach to genetic privacy offered legislation instead of task force recommendations. The DOE's ELSI program commissioned several studies to investigate various philosophical, sociological, and legal aspects of genetic privacy. Concerns ranged from the protection of medical records to an individual subject's control over the fate of his or her own genetic samples. Under the direction of this program, George J. Annas, Leonard H. Glantz, and Patricia A. Roche (collaborating at the Health Law Department of the Boston University School of Public Health) drafted ELSI's first legislative document (9). Their "Genetic Privacy Act" was released in 1994. The central tenet of the legislation was that because of both its predictive value and potential implications for other family members, genetic information warrants a higher level of protection than other medical information.

Though a complex bill, the Genetic Privacy Act stressed two issues. First, under no circumstances, except legal order or forensic investigation, should genetic analysis be conducted without the express written consent of individuals (special provisions were included to deal with pregnant women, embryos, minors, and incompetent persons). A subject should be informed of the nature of testing or research being conducted, and medical personnel and research scientists should abide by any restrictions a subject might make. Second, the act restricted research access to unidentifiable DNA samples. Any sample that could be traced to an individual patient through a name, address, social security number, health insurance number, or any other identifying information should be excluded from genetic research. The Act defined DNA samples very broadly, including banked blood, tissue samples, cell lines, and saliva as well as samples taken expressly for genetic analysis. Their guidelines placed the dignity of the research subject at the center of biomedical ethics — if a subject refused to permit his or her DNA to be used for research or commercial purposes, researchers would be required to respect his or her wishes regardless of the scientific importance of the sample (10). Not surprisingly, some members of the research community and the pharmaceutical industry have responded to the proposed Act with alarm, charging that such measures impede the process of medical research. Although Oregon passed a bill modeled on the Genetic Privacy Act in 1995, New Jersey Governor Christine Whitman vetoed a similar measure in 1996 under pressure from the pharmaceutical industry. Debates about the merits and drawbacks of such a bill continue at both the state and national level (9).

Public Education. Though ELSI placed high priority on increasing public and professional literacy about genetics, genetic testing, and HGP, the Working Group itself has sponsored few public forums on this matter. The DOE ELSI program, on the other hand, has been quite active in creating educational materials for the general public, teachers and schoolchildren, and professionals who encounter genetic information. Though not directly a part of the ELSI program, DOE's Human Genome Management Information System (HGMIS) has also been essential in disseminating information about HGP to the public through press releases, published documents, and an extensive website.

DOE ELSI education projects vary widely in form, content and audience. General public education programs have included exhibits at the American Museum of Natural History and the San Francisco Exploratorium, a radio program (*The DNA Files*) broadcast over National Public Radio, video documentaries such as *A Question of Genes* and *The Secret of Life* for public television and classroom use, forums to inform minority groups about HGP and to spark discussions about its meanings, and Spanish-language radio programming on HGP. To reach a student audience, DOE ELSI has sponsored training programs in genomic techniques for high school and community college teachers, and a pilot program in Seattle, Washington, allows students to perform DNA sequencing experiments in their classrooms. Workshops, conferences, and seminars have also been conducted for medical professionals and judges who increasingly interact with genetic evidence in the courtroom. Another important program at Cold Spring Harbor educates policy makers, legislators, journalists, and other opinion makers

on HGP and human genetics in an attempt to influence "downstream" popular perceptions of HGP (11).

As is clear from this partial list of programs, much ELSI education has focused on increasing scientific literacy. While parties on all sides of HGP tend to agree that education is theoretically important, the question of what constitutes education has sparked much controversy. Will Seattle students who understand how to sequence DNA be more capable of dealing with the social implications of HGP through their exposure to scientific techniques? For the architects of the DOE ELSI program, improving scientific literacy is the first step in preparing the public to deal with genetic issues and is an important part of dispelling damaging misconceptions of the goals and methods of HPG. From another perspective, however, these sorts of educational programs divert attention from the real ELSI concerns arising from the genome project.

ELSI'S CRITICS AND FEDERAL SCIENCE POLICY

Policy or Research? Restructuring ELSI

Although ELSI was not the first federally sponsored bioethics board, it did mark a unique path in science policy. As both its critics and supporters have noted, HGP was the first federal science project to fund self-criticism as well scientific research. ELSI's grant structure, however, was more suited to stimulate discussion across a wide spectrum of scholars than to implement policy decisions. Over the past 10 years, the purpose of ELSI as either a policy board or a research enterprise has been a matter of contention.

As discussed in the previous section, the ELSI Working Group actively participated in policy discussions about genetic testing, privacy, discrimination, and education. To its critics, however, the Working Group's extramural grant program appeared to lack coordination and seemed to bear little relation to actual policy concerns. ELSI's most visible early products, such as the Genetic Privacy Act or the Insurance Task Force Report, were spearheaded by individual grant recipients or particular members of the Working Group rather than the Working Group as a whole. In 1992 the Committee on Governmental Operations issued a report stating that ELSI lacked mechanisms for issuing policy measures and recommended the formation of a formal advisory commission to address policy issues (12). Under further criticisms that ELSI projects tended to address "contextual" issues surrounding the HGP rather than specific issues stemming from genomic work, the new director of NGHRI, Francis Collins, strongly suggested that ELSI focus its work on instrumental projects. In 1996, the Working Group came under review by an 11-member Joint NIH-DOE ELSI Evaluation Committee to determine its future role in the HGP.

The committee of legal scholars, scientists, health care professionals, and one science studies scholar released their final report on the future of ELSI in December 1996. The Evaluation Committee pointed to the increasing pace of genomics research, the turn toward applied genetics, and the increasing integration of genetic research into medical practice as indications of a crucial need for effective policy mechanisms on ELSI issues. Although the report contained several specific criticisms of the effectiveness of the ELSI Working Group, the crux of the Evaluation Committee's criticisms suggested that ELSI's mix of research programs and policy development was incompatible with efficient policy formulation. While stressing the importance of an independent ELSI grant program for public acceptance of HGP, the committee concluded that the ELSI Working Group, as it was then configured, occupied too Byzantine a niche and pursued too many concurrent goals to ever be effective. To resolve these issues, the Evaluation Committee recommended three changes in the structure of ELSI programs: (1) the redesign of the ELSI Working Group as the ELSI Research Evaluation Committee, which would coordinate and follow-up on ELSI grants and prioritize research agendas; (2) the creation of a NIH-wide policy office, housed in the Office of the Director, to formulate policy issues and monitor compliance with NIH guidelines on genetic research; (3) and the establishment of a federally chartered Advisory Committee on Genetics and Public Policy to be housed in the Office of the Secretary of HHS (13).

While the recommendations of the Evaluation Committee were not implemented exactly as outlined in their report, its assessment of ELSI as an impotent program has had repercussions in ELSI's form and function. The tasks of the Working Group, the grants programs, and additional ELSI programs within NIH and HHS have been realigned to give stronger voices to scientists and policy experts. NIH and DOE now collaborate through an ELSI Research Planning and Evaluation Group (ERPEG) similar to that suggested by the Evaluation Committee. NIH's OPC performs many of the NIH-wide functions advocated by the Evaluation Committee. Finally, the Secretary's Advisory Committee on Genetic Testing, while perhaps addressing a more narrow set of policy questions than those envisioned by the Evaluation Committee, has been established as a high-profile policy entity. As conceived by both the Evaluation Committee and the Task Force on Genetic Testing, this Advisory Committee interacts with other federal agencies such as FDA, the Environmental Protection Agency (EPA), the Department of Justice, and the National Bioethics Advisory Committee when addressing issues of broad public and policy interest. As an Advisory Committee on Genetic Testing, however, this HHS committee cannot be expected to issue policy on the complete spectrum of ELSI issues.

A New Role for Government?

ELSI was not the federal government's first foray into biomedical ethics. Its novelty lie instead in its source of funding in the very project it was meant to critique. Over the past 30 years, however, various government agencies have acted as ethical and social commentators on scientific research. Housed in the Department of Health, Education, and Welfare, the National Commission for the Protection of Human Subjects of Biomedical and Behavioral Research (1974–1978) participated in debates over fetal tissue research, set guidelines for experiments involving human subjects, and established the guiding questions of the discipline of bioethics in its landmark Belmont Report (1978). The National Commission was followed by the President's Commission for the Study of Ethical Problems in Medicine and Biomedical and Behavioral Research (1980–1983), which addressed concerns on genetic screening, genetic counseling, and human gene therapy as well as other issues in bioethics. In 1984 this President's Commission was replaced with the less successful Congressional Biomedical Ethics Board and its expert arm, the congressionally appointed Biomedical Ethics Advisory Committee (BEAC, 1988–1989). While the BEAC attempted to address issues of genetic screening and genetic discrimination in insurance, its potential as a force in federal biomedical ethics policy was lost in a political battle over abortion in 1989 (2). More recently the President established a National Bioethics Advisory Committee (NBAC) with a broad mandate to issue policy recommendations on any bioethical issues arising from research (or clinical applications of such research) on human biology and behavior. When established in 1995, the highest priorities of the NBAC were the protection of the rights of human research subjects and questions of the uses of genetic information. The NBAC, then, offers another policy arena in which ELSI-type issues enter public discussion without the involvement of actual ELSI programs.

While federally chartered bioethics commissions oversee biomedical research from outside the scientific research establishment, a self-critical federal science program could be created in numerous ways. In one configuration, scientists regulate their own work to protect the public from potential harm. The NIH guidelines for working with recombinant DNA provide an example of this sort of program. Responding to public concerns about the safety of genetic engineering techniques, scientists created technical guidelines for permissible levels of risk. Individual proposals for research surpassing a certain level of risk require approval from a scientist-run NIH committee. Many of the scientists involved in HGP had experienced this sort of scientific policing either through reviews of their own recombinant work or participation on a review committee. In a different sort of self-critical science, uniquely represented by ELSI, scientific agencies provide funds for the public discussions of medical ethics and social concerns resulting from scientific advances. While the first evaluates individual research proposals in light of current research guidelines determined by scientists, the second type may allow scientific outsiders to determine what those guidelines should be in the first place. By focusing on the ethical, legal, and social *implications* of genome work, however, the ELSI program has largely reacted to consequences of genome work rather than formulation of research guidelines.

Ethics and the Role of History

From its inception, ELSI has garnered serious criticism. The original NIH guidelines for ELSI programs implied that ELSI would fund projects criticizing approaches and consequences of the HGP. Some scientists questioned whether this was a wise use of HGP funds. In addition, the provision of funding for an ELSI program within the HGP—and not other federal research projects—necessarily implied that the HGP, by definition, required a higher level of public scrutiny than most federal science projects. Those scientists who saw such fears as unwarranted were especially alarmed at the attention ELSI might bring to the HGP. Many critics of science, on the other hand, wondered whether those receiving ELSI funds would have the intellectual freedom to criticize the HGP. In other words, the very idea of ELSI seemed to represent a conflict of interest. Another criticism was whether introducing such relatively large sums of money into the pool of grants available to historians, sociologists, science policy analysts, and bioethicists might unduly skew research topics away from other pressing issues in contemporary biomedicine. Finally, it has been suggested that ELSI serves nothing better than the interests of science. By encouraging limited criticism, scientists at the HGP are freed from the claim that they have ignored public concerns over the uses of genetic technologies such as those publicly voiced during the recombinant DNA controversy of the 1970s.

The strongest criticism has been a difficult one for ELSI to answer. By focusing on the *implications* of the HGP, ELSI leaves little room for discussions of *whether* the HGP should proceed. Questions of resource allocation in federal science research funding, for example, are not appropriate uses of ELSI funds as originally defined. Furthermore this issue highlights the potential for conflict of interest. Is it realistic to expect scholars working with ELSI funds to recommend a reduction in the HGP's budget? The history of ELSI thus far suggests that scholars working with ELSI funds have been able to maintain a reasonable amount of academic freedom. The extramural grant structure of ELSI projects offers a first level of protection since projects are evaluated by peer review instead of HGP (though now reviewed as well by ERPEG), and the very nature of ELSI projects in exploring the social, cultural, and legal context of HGP has encouraged researchers to question HGP itself. Historians and philosophers have been especially interested in the elements of reductionism inherent in HGP and exploring the limitations of such a methodological approach (1).

Underlying most of these criticisms is a question of the singularity of the problems caused by HGP. Regardless of the medical possibilities or "therapeutic gaps" presented by genetic testing technologies, cultural notions associated with genetics do present special concerns. Justifiably or not, Americans have attributed special powers to genetics—over education, socioeconomic status, or psychology—in controlling individual destiny. This attitude was most clearly represented, of course, in the involuntary sterilization eugenic campaigns of the early twentieth century. XXY screening programs represent a more recent example. The historical precedent for misuses of genetic information not only suggests HGP warrants an extra level of caution above that granted other medical technologies but also imparts a particular importance to understanding the past. Discussions of "past abuses of genetics" almost always arise in discussing

the function of HGP. In this context the various ELSI programs have been essential in securing public support for HGP and will remain so throughout the completion of the project. ELSI-type issues, however, will not disappear with the completion of genomic sequencing. It remains to be seen how concerns over the ethical, legal, and social implications of HGP will be handled after the project itself ends and leaves only its "implications."

BIBLIOGRAPHY

1. E.T. Juengst, *Soc. Policy Philos.* **13**(2), 63–95 (1996). Available at NIH ELSI Homepage: *http://www.nhgir.nih.gov/ELSI/*

2. R. Cook-Deegan, *The Gene Wars: Science, Politics, and the Human Genome*, Norton, New York, 1994.

3. DOE ELSI Homepage: *http://www.ornl.gov/hgmis/resource.elsi.html* Maintained by the Human Genome Management Information System. Accessed 12/13/99.

4. NIH ELSI Homepage: *http://www.nhrgi.nih.gov/ELSI/* Maintained by the National Human Genome Research Initiative. Accessed 12/13/99.

5. National Center for Human Genome Research, *NIH Guide Contracts Grants* **19**(4), 12–15 (1990).

6. N. Holtzman and M.S. Watson, eds., *Promoting Safe and Effective Genetic Testing in the United States: Final Report of the Task Force on Genetic Testing*, 1997. Available at: *http://www.nhgri.nih.gov/elsi/tfgt_final* Site maintained by the NHGRI, accessed 12/14/99.

7. M.S. Yesley, *Hum. Genome News* **3**(3), 12–14 (1991).

8. NIH-DOE Working Group on Ethical, Legal, and Social Implications of Human Genome Research, *Genetic Information and Health Insurance: Report of the Task Force on Genetic Information and Insurance* NIH-DOE, Washington, DC, 1993. Available at: *http://www.nhgri.nih.gov:80/About_NHGRI/Der/Elsi/itf.html* Site maintained by the NHGRI. Accessed December 14, 1999.

9. P.R. Reilly, *Hum. Genome News* **8**(3 and 4), 1–3 (1997).

10. G.J. Annas, L.H. Glantz, and P.A. Roche, *The Genetic Privacy Act and Commentary*, 1995. Available at: *http://www.ornl.gov/hgmis/resource/privacy/privacy1.html* Site maintained by the Human Genome Management Information System, Accessed 12/13/99.

11. D. Drell and A. Adamson, *DOE ELSI Program Emphasizes Education, Privacy: A Retrospective, 1900–1999*, 1999. Available at: *http://www.ornl.gov/hgmis/resource/elsiprog.html* Site maintained by the Human Genome Management Information System, Accessed 12/13/99.

12. Committee on Governmental Operations, House of Representatives, U.S. Congress, *Designing Genetic Information Policy: The Need for an Independent Policy Review of the Ethical, Legal, and Social Implications of the Human Genome Project*, U.S. Government Printing Office, Washington, DC, 1992.

13. Committee to Evaluate the Ethical, Legal, and Social Implications Program of the Human Genome Project, *Report of the Joint NIH/DOE Committee to Evaluate the Ethical, Legal, and Social Implications Program of the Human Genome Project*, 1996. Available at: *http://www.nhgri.nih.gov:80/Policy_and_public_affairs/Elsi/elsi_recs.html*

See other entries Education and training, public education about genetic technology; see also Federal policy making for biotechnology entries.

FEDERAL POLICY MAKING FOR BIOTECHNOLOGY, EXECUTIVE BRANCH, NATIONAL BIOETHICS ADVISORY COMMISSION

HAROLD T. SHAPIRO
Princeton University
Princeton, New Jersey

OUTLINE

INTRODUCTION

The establishment of national bioethics commissions with rather broad responsibilities to advise the federal government regarding public policy matters in the bioethical arena, possibly by recommending new regulations and/or legislation, is, like the discipline of bioethics itself, a relatively new phenomena. In the United States such commissions have been established for a wide variety of complex reasons, under the auspices of various federal entities, within somewhat different venues and with more or less enthusiasm, resources and authority. Overwhelmingly, however, these commissions have been only advisory in nature. Moreover these commissions, unlike their counterparts abroad, have not been created as standing bodies but have been appointed for relatively short terms, presumably at those moments in time where the federal government felt the need for such assistance. In any case, these commissions have been appointed principally to advise the federal government on its ethical responsibilities within biomedical science and clinical medicine, especially in morally controversial areas where the imperatives of advances in the biomedical sciences and associated clinical applications stand in some tension to what some may feel are important ethical obligations, or where our evolving moral sensibilities require a change in traditional practices. In this respect they are, very broadly speaking, expected to clarify the relationship between the new opportunities that biomedical science research practices and clinical applications continue to offer to human welfare and the limitations that might result on these activities because of other important ethical obligations.

America is a morally pluralistic society (i.e., a society dedicated to incorporating a diverse set of ideas regarding those values that will allow for the greatest human flourishing). As such, there is a special need for some mechanism to articulate a national consensus in bioethical matters where that is possible, or at least to delineate the points of disagreement among us and the common ground (if any) where mutual empathy and understanding are the most that can be expected. This is especially so given the rapid advances in biomedical science that inevitably produce new ethical questions. At times these commissions may have additional roles such as identifying emerging issues, defusing controversy, delaying action, or giving the illusion of acting, endorsing a decision that is already made, reviewing the effectiveness of existing rules, giving the impression of open-mindedness on the part of policy makers, and educating the public and professionals on bioethical issues. Many critics have pointed out that the appointment of broad-based commissions with their inevitably process-orientated approach that often focuses on mid-level principles is not an obvious way to resolve controversial moral issues. Indeed, some claim that such a deeply sociopolitical process is unlikely to generate morally adequate recommendations, since commissions rely a great deal on a certain level of consensus in generating their recommendations.

On the other hand, the experience of recent decades provides strong evidence that under the right conditions such commissions not only can become important agents of change in long-established practices, but they also can enhance public discourse on bioethical issues. While commissions are unlikely to generate new moral theories, they can be effective in bringing clarity to an area of moral confusion and/or controversy, providing a forum for ideas about the appropriate use of new biomedical technologies and in finding sufficient common ground to foster consensus and empathetic understanding among groups with different moral perspectives. That is to say, commissions can serve to identify emerging issues, defuse controversy, and monitor on-going compliance with existing regulations and professional standards.

Advances in science and medicine often call for difficult choices, in which benefits must be weighed against risks or in which individual needs may conflict with social norms or laws, or decisions must be made under conditions of great uncertainty. These dilemmas surround us, sometimes engulfing individuals and families, and often entering the political arena. Examples include the debate over doctor-assisted suicide, the use of fetal tissue for therapeutic transplantation, protection of human subjects in medical research, advances in medically assisted reproduction, and the trend to cut the costs of medical care through rationing of resources. As a result often a new scientific advance simultaneously is considered both a blessing and a challenge to certain existing moral commitments. Underlying the debate is a struggle to understand better how to define the moral status of human life at its various steps of development, the respect owed to human life, and the relative importance of an individual life when weighed against the importance of many lives.

These issues are at the heart of the burgeoning field of bioethics, once the sole province of philosophers and theologians. Increasingly, however, American society finds itself turning to its government for resolution of thorny ethical issues generated by advances in biomedical science and our evolving moral sensibilities, perhaps an inevitable consequence of living at a time of startling scientific advances and in a pluralistic society where no single religious or other moral authority dominates.

As already noted, the creation of federal commissions and the use of federal funds to deliberate on the nature of the ethical constraints under which biomedical science should proceed arises from the need for some public mechanisms to articulate common values and foster consensus in the face of growing cultural and religious heterogeneity and rapidly advancing science. The establishment of these deliberative bodies also signals the increasing importance of medical and biological technologies in national life and the pressing need for reasonable groups of diverse individuals to consider these issues in a public forum. The work of some of these commissions has been widely used in courts to decide cases, in federal and state legislatures to devise statutes, in the formation of professional standards, and as intellectual and policy landmarks.

HISTORICAL BACKGROUND

Over the past quarter century, government forays into bioethics have had significant impacts on the conduct of biomedical research and the delivery of health care. In the United States, four major bioethics bodies have been established by Congress since 1968: the National Commission for the Protection of Human Subjects of Biomedical and Behavioral Research (National Commission), the President's Commission for the Study of Ethical Problems in Medicine and Biomedical and Behavioral Research (President's Commission), the Biomedical Ethics Advisory Committee (BEAC), and the National Bioethics Advisory Commission (NBAC). A fifth federal initiative, the Ethics Advisory Board (EAB) was created in response to a recommendation of the National Commission.

Other public ethics bodies have been ad hoc advisory panels of the Department of Health and Human Services. The Human Fetal Tissue Transplantation Research Panel was formed in 1988 to consider the ethical issues in the potential use of human fetal tissue for the treatment of diseases such as Parkinson's disease. In 1994 the National Institutes of Health Human Embryo Research Panel was formed to consider various areas of research involving the ex utero preimplantation human embryo and to provide advice as to those areas of research that are acceptable for federal funding, warrant additional review, and are unacceptable for federal support.

Over the same period of time, bioethics has been addressed sporadically by the Institute of Medicine, National Academy of Sciences, and the now defunct congressional Office of Technology Assessment. In addition

several states have launched bioethics commissions to advise state executives and legislatures on a range of concerns, including surrogate parenting, determination of death, the use of advance directives in medical care, and organ transplantation (1).

More recently, the President established in 1995 NBAC, the first standing committee charged to address bioethical issues since 1983. We consider the work of each of these.

THE NATIONAL COMMISSION FOR THE PROTECTION OF HUMAN SUBJECTS OF BIOMEDICAL AND BEHAVIORAL RESEARCH (THE NATIONAL COMMISSION)

In the 1960s a series of scandals involving research with human subjects (i.e., the inappropriate treatment of human subjects) signaled to Congress that some biomedical and behavioral scientists were not adequately self-policing themselves and that some sort of independent oversight was necessary. Most notable among these scandals were the Tuskegee syphilis trials, the Willowbrook hepatitis experiments, the use of prisoners to test drugs, and whole body radiation experiments sponsored by the Department of Defense. Although the scientific community was initially resistant to a formal system of oversight, fearing that it might unnecessarily stymie important research, the powerful message of events like Tuskegee convinced most investigators that important research might actually be stalled for lack of public support if adequate protections were not in place.

The National Commission was established through Title II of the National Research Act of 1974 (Public Law 93-348). The bill, sponsored by Senator Edward Kennedy, reflected congressional concern about reported abuse of human research participants and the moral status of research using biological materials from aborted fetuses.

Congress created the National Commission as part of the Department of Health and Human Services (DHHS), then Department of Health, Education, and Welfare (DHEW). Eleven members were appointed by the Secretary of DHEW: five scientists, three lawyers, two ethicists, and one person in public affairs. In establishing the National Commission, Congress gave it the task of articulating the ethical principles of protecting human subjects in research. It was instructed to then employ those principles to recommend actions by the federal government. Congress also asked the National Commission to address fetal research, which it did in the four-month period allotted.

The Commission issued its first report, "Research on the Fetus," in May 1975 (2). By late 1975 the National Commission's recommendations had been translated into regulations. This first report presaged many more reports that laid a formal foundation for human subject protection in the United States. Much of the ensuing work of the National Commission focused on broader issues in human subjects protection. In essence, the National Commission codified existing DHEW policies on research with human subjects into formal federal regulations and established the requirement of Institutional Review Board (IRB) review and approval of all federally funded research involving human subjects. The resulting regulations,

45 CFR, established IRB review procedures and detailed the required elements of informed consent. In 1978, based on the National Commission's recommendations, DHEW revised its human subjects regulations and added regulations covering pregnant women, fetuses, in vitro fertilization, and prisoners. In its work on the protection of human subjects in research, the National Commission articulated three basic principles: respect for persons, beneficence, and justice. It laid great emphasis on autonomy, elaborated and extended the notion of informed consent, and recognized the special vulnerability of specific populations (e.g., children, prisoners, those institutionalized as mentally infirm).

The National Commission operated from 1974 until 1978, issuing 10 reports (2–11). As noted, many of these reports were translated, often directly, into the now-familiar federal regulations for research involving human subjects (45 CFR 46). Other aspects of its work, for example on psychosurgery, were largely ignored, and its recommendations regarding research on the "institutionally mentally infirm" were never implemented.

One of the last acts of the National Commission was its recommendation that there be a broad-based on-going federal entity to review controversial areas of research, the Ethics Advisory Board or EAB. DHEW incorporated this recommendation into its regulatory framework (45 CFR 46.204) and established the EAB (1). The National Commission also recommended that a successor body be created, but with broader authority to address issues beyond protection of human participants in research. The time was ripe for such a recommendation, since the nation was again confronted with new ethical issues created by the accelerated development of new biological and medical technologies. At the same time, issues regarding the safety of recombinant DNA were being debated on Capitol Hill, and termination of treatment was rapidly becoming a national issue in the wake of the Karen Ann Quinlan case and other court challenges to medical authority. Finding the work of the National Commission generally useful, Congress concurred that a more general mandate for a national bioethics organization was in order and created the President's Commission for the Study of Ethical Problems in Medicine and Biomedical and Behavioral Research (the President's Commission).

THE ETHICS ADVISORY BOARD

Following the recommendation of the National Commission, the Ethics Advisory Board (EAB) was established in 1978 as an 11-member board that included lawyers, a theologian, a philosopher, clinicians, researchers, and a member of the public. It operated for two years. Although federal regulations define EAB's purview as research involving the fetus, pregnant women, and in vitro fertilization, the EAB charter grants it a broader role. It was intended as an ongoing, standing body charged to review specific proposed protocols or controversial areas of research.

EAB had some of its marching orders from day one. In 1975 DHEW had announced it would fund no proposal for research on human embryos or on the external fertilization

of human eggs unless it was reviewed and approved by an independent ethics advisory board. Specifically, federal regulation required an EAB review prior to funding any research on human in vitro fertilization (45 CFR 46.204d). In vitro fertilization was the first topic addressed by EAB and its 1979 report stipulated several criteria for approval of such experiments (12). However, DHHS never implemented any of the general policy recommendations of the EAB, and DHHS disbanded the EAB in 1980 at the direction of the White House's Office of Science and Technology Policy (OSTP). As a result the Board never approved a single proposal before its dissolution, and the moratorium on human in vitro research remained in effect until lifted by President Clinton in 1993. Before it closed its doors, the EAB accomplished a few other tasks. It recommended granting a waiver to permit fetoscopy to diagnose hemoglobinopathies and handled several issues related to Freedom of Information Act inquiries in DHEW.

The budget for the EAB was diverted to the President's Commission in 1979. Since that time, few efforts have been made to reestablish the Board (1). In fact, in 1988 DHHS proposed reestablishment of the EAB and published a proposed charter for a new EAB (53 FR 35232), which expanded membership to 21 individuals. The revised charter was never signed by President Ronald Reagan, and no efforts have been made to revitalize the EAB despite recommendations to do so by various bioethics organizations.

THE PRESIDENT'S COMMISSION

The President's Commission was established by Section III of Public Law 95-622. The enabling legislation specified several tasks, as it had for the National Commission, but it also gave the President's Commission the authority to undertake studies at the request of the President or upon its own initiative. The new Commission was elevated to independent presidential status, in comparison with the National Commission which had been housed within DHHS. The range of the President's Commission work was broadened to encompass activities of the entire federal government and was extended beyond human subjects research to include medical practice.

The eleven members of the President's Commission, drawn from specific areas of expertise by law, were sworn in by President Jimmy Carter in January 1980. It operated until March 1983 and issued eleven reports including a summary of its work (13–23). As a matter of explicit policy, the Commission made few specific recommendations, instead producing consensus reports (24,25).

A focus of much of the President's Commission work concerned protection of human subjects in research, including health research regulations and compensation for research injuries (14,15,20). In 1980 the President's Commission began its congressionally mandated investigation into the adequacy and uniformity of the federal laws governing human subjects research, a topic on which it was supposed to report every two years. In its first biennial report, Protecting Human Subjects (14), the Commission reported that it was "satisfied that the basic regulations of the Department (DHHS) were adequate if not above improvement." Its first recommendation was that all federal agencies adopt the HHS regulations set forth in 45 CFR 46. The Commission then focused its attention on determining uniformity among federal agencies as measured by the extent to which their rules confirmed the basic regulations of DHHS. Its second biennial report on this issue, Implementing Human Research Regulations (20), recommended that a program of routine site visits to Institutional Review Boards (IRBs) be implemented by relevant federal agencies on a coordinated basis and that all relevant agencies keep a record of the IRBs subject to their jurisdiction. The aim of the report was to increase the adequacy and uniformity of the implementation of existing regulations. In 1991, 16 federal agencies and departments eventually adopted a single set of regulations, known as the "Common Rule."

Its report Defining Death (13) became the foundation for statutory changes adopted in many states. This report helped to formulate and explain the Uniform Determination of Death Act. The Commission also confronted controversies about termination of treatment at the end of life in its reports on making health care decisions (16), and more specifically in deciding to forgo life-sustaining treatment (19). This report, undertaken at the Commission's own initiative, addressed highly contentious issues that continue to be debated in the courts and some legislatures, particularly in Oregon. The Commission also directly confronted the arguments for and against the use of life-sustaining treatments (26). The Commission encouraged patient- and family-centered decision making, and made recommendations regarding appointment of surrogate decision makers. It also concluded that nutrition and hydration were not fundamentally different from other medical treatments, a source of great controversy in several nationally prominent cases (e.g., Karen Ann Quinlan, Nancy Cruzan). This conclusion immersed the Commission in a controversy that led some Senate conservatives to argue that even federal bioethics committees could not be trusted on matters of great social import (26).

The President's Commission also took on the difficult issues of equitable access to health care. Its report, Securing Access to Health Care (22), was the only report that drew a dissenting vote from a commissioner. For a variety of reasons it has been more highly criticized than other Commission reports, perhaps because of the complexity and intensity of the issue, and because of its importance to broader sets of interests (27–30).

The President's Commission also addressed issues not yet fully debated at the national level concerning applications of human genetics research. It issued reports on genetic screening and counseling (21) and on human gene therapy (17). The report "Splicing Life" emphasized the distinction between genetically altering somatic cells, which would not lead to inherited changes, and germ cells (sperm, egg cells, and their precursors), which would induce inherited changes. This distinction illustrated that there could be some forms of gene therapy that would not be morally different from other forms of treatment.

The Commission suggested that there be in place policies of research protocol review and ongoing means to evaluate new developments in this area of research. The Commission recommended that the National Institutes of Health review gene therapy through its Recombinant DNA Advisory Committee (RAC). In response, a RAC working group on human gene therapy was established and drafted the "Points to Consider in the Design and Submission of Human Somatic Cell Gene Therapy Protocols" (31,32). In the end, the report served to broaden the discussion about a controversial area of research and kept open the possibility for some forms of gene therapy to be considered in a political climate that was quickly moving toward unnecessarily restrictive legislation.

The Commission's term expired at the end of 1982. Extension of its term was debated in the Senate where conservative interests argued that bioethics should be brought under more direct congressional scrutiny. Ironically, a conclusion of "Splicing Life" would lay the groundwork for the establishment of two subsequent federal efforts, the BEAC and later NBAC. The report noted that there was a need for public debate, which could be mediated by an ad hoc commission on genetics or by a standing federal bioethics commission. Then-Congressman Albert Gore subsequently introduced legislation to create a President's Commission on Human Genetic Engineering, favoring permanent oversight of advances in human genetics and reproduction. This became the seed for legislation to create BEAC with a broader mandate than human genetics, as then-Congressman Gore became convinced that a broader mandate would be more useful (33). During this same period there were several alternate congressional proposals including initiatives to extend the life of the President's Commission, to give the Institute of Medicine (IOM) a mandate to do studies in bioethics, and to have the congressional Office of Technology Assessment (OTA) also do such work (26).

THE BIOMEDICAL ETHICS ADVISORY COMMITTEE

Congress took bioethics into its own hands in 1985 when it passed the Health Research Extension Act (Public Law 99-158), despite President Ronald Reagan's veto of the measure. The BEAC was a 14-member group whose multidisciplinary membership was appointed by the Biomedical Ethics Board (BEB), comprised of 12 Members of Congress, three each from the majority and minority parties of the House and Senate. BEAC was to be directly responsible for carrying out the studies of topics in biomedical ethics mandated by legislation or specified by the congressional Board. It took almost a year for the party leaders of the House and Senate to appoint the 12 members of the congressional Board, which then took on the responsibility of appointing the 14 members of BEAC, the operational arm. The appointment process took nearly two and a half years and resulted in deepened mistrust among members of the congressional Board, particularly around issues concerning abortion.

BEAC was required to prepare at least three reports on specified topics, as well as to provide annual reports. The first mandated report, on implications of human genetic engineering, stemmed from Representative Gore's bill proposing an extension of the President's Commission (H.R. 98-2788). The deadline for the second report, on fetal research, expired before BEAC members were appointed. The fetal research mandate was reinstated in the Omnibus Health Extension Act of 1988 (Public Law 100-607) with the deadline delayed until November 1990. The third mandate stemmed from Senator William Armstrong's proposed amendment to the 1988 AIDS bill. BEAC finally met in September 1988, less than a week before its authorization expired (33). Congressional haggling over appointment to BEAC of anti-abortion members essentially closed BEAC down before it ever began its work; the office was closed at the end of September 1989.

THE NIH HUMAN FETAL TISSUE TRANSPLANTATION RESEARCH PANEL

Another effort to address bioethics at the national level, this time on an ad hoc basis, began in May 1988, when Health and Human Services Assistant Secretary Windom initiated a moratorium on the use of fetal tissue in transplantation research funded by the federal government. Fetal tissue had long been used in research, and the National Commission had previously developed the guidelines that were incorporated into federal regulations for use of fetuses in research (45 CFR 46). When scientists proposed using fetal tissue for neural grafting as an experimental treatment for Parkinson's disease, questions were raised as to whether the existing guidelines adequately covered therapeutic intent.

Lacking an Ethics Advisory Board, DHHS directed the National Institutes of Health (NIH) to convene a panel to advise DHHS about the ethical implications of such research, specifically whether the moral issues surrounding the source of such tissue (elective abortions) could ethically be separated from the therapeutic use to which such tissue is put. The issues proved to be complex and divisive (34). The panel heard testimony from disease groups, researchers, and those opposed to the research and voted on a set of specific recommendations. A majority were in favor of permitting such research as long as three conditions were met in addition to IRB approval, namely (1) the decision to donate tissue was kept separate from and made only after the decision to abort, (2) the process for abortion was not altered in any way, and (3) the informed consent of both parents was obtained in cases when the fathers could be contacted. The majority of the panel argued that they did not have to directly engage in questions about the morality of abortion, since the practice was legal (34). Two separate dissenting statements argued that the abortion issue should not be sidestepped. The report was approved by the Advisory Committee to the Director, NIH, urging acceptance of the recommendations (35). No action was ever taken on the panel's report.

THE NIH HUMAN EMBRYO RESEARCH PANEL

In 1994 research use of fetal tissues would again be the focus of bioethics debates, this time ex utero

preimplantation human embryos produced by in vitro fertilization or other sources. Because there was no EAB, NIH was sitting on several protocols that could not be funded because they involved the research use of early-stage human embryos. The status of these protocols, in terms of need for EAB review, was unclear because they concerned only research that involves extracorporeal human embryos or parthenogenetically activated oocytes. Research involving in utero human embryos, or fetuses, was not at issue, since guidelines for such research were already embodied in federal laws and regulations governing human subjects research. Research involving human germ-line gene modification also was not within the Panel's scope. In addition therapeutic human fetal tissue transplantation research, the topic taken up by this Panel's predecessor, was also not a part of the Panel's mandate because guidelines were already in place to govern such research (36).

In 1994 the NIH Director appointed the Human Embryo Research Panel to consider various areas of research involving the ex utero preimplantation human embryo and to provide advice as to those areas that (1) are acceptable for federal funding, (2) warrant additional review, and (3) are unacceptable for federal support. For those areas of research considered acceptable for federal funding, the Panel recommended specific guidelines for the review and conduct of this research. In addition to developing guidance for research deemed acceptable, the Panel addressed issues surrounding the sources of gametes and embryos for research, transfer of embryos to a uterus, parthenogenesis, and systems for review and oversight.

One of the most difficult issues the Panel had to consider was whether it is ethically permissible to fertilize donated oocytes expressly for research purposes or whether researchers should be restricted to the use of embryos remaining from infertility treatments that are donated by women or couples. The Panel concluded that studies that require the fertilization of oocytes are needed to answer crucial questions in reproductive medicine, and that it would therefore not be wise to prohibit altogether the fertilization and study of oocytes for research purposes. It concluded that the use of oocytes fertilized expressly for research should be allowed only under two stringent conditions: when the research by its very nature cannot otherwise be validly conducted or when a compelling case can be made that it is necessary for the validity of a study that is potentially of outstanding scientific and therapeutic value. One member of the Panel dissented from the Panel conclusion that under this condition oocytes may be fertilized expressly for research purposes.

It was this one of several recommendations that brought yet another federal bioethics panel into great controversy. Before the Panel had even finished presenting its findings to the NIH Director and his Advisory Panel, President William Clinton issued an Executive Order prohibiting the use of federal funds for research in which oocytes were fertilized expressly for research purposes. A subsequent congressional ban extended the prohibition to include any research that involves exposing embryos to risk of destruction for nontherapeutic research (P.L. 104-91 and P.L. 104-208).

BIOETHICAL ANALYSES IN OTHER FEDERAL SETTINGS

During the past three decades, bioethics debates have found a place in several federal settings besides focused national commissions and panels. For example, the OTA was established in 1972 as an analytical arm of Congress, to anticipate how science and technology would raise issues for policy makers, and to advise Congress on federal policies affecting science and technology development. It rendered technical advice about how to promote or regulate science and technology, and gave early warning about the impacts of emerging technologies. Eliminated by Congress in 1995, OTA contributed markedly to bioethics debates in its 24-year history, especially during the interludes when there was no standing federal commission. Its 1983 report on genetic testing in the workplace (37) explicitly incorporated bioethics analysis. A 1984 report on "Human Gene Therapy" (38) addressed the ethical issues of the use of recombinant DNA techniques in therapeutic treatment. A succession of reports included chapters on ethical considerations or had extensive discussion of ethical issues (37–50). Although bioethics was becoming an important component of the OTA analyses, OTA was not a bioethics commission and held a much broader mandate from Congress.

Matters of bioethics at the IOM, National Academy of Sciences (NAS), grew out of concerns with medical practice and public health. In its early years IOM fostered a small bioethics program, issuing a 1974 report, "The Ethics of Health Care," an inspection of the ethical underpinnings of medical practice (51). Since then IOM has continued to incorporate sections or chapters on bioethics into many reports. In 1994 IOM completed a study of issues in genetic testing (52) and in 1995 completed a systematic review of past and ongoing bioethics commissions at the federal, state, and international levels (53).

The National Human Genome Research Institute at the NIH and a parallel program under the Department of Energy (DOE) since their inception have devoted a fraction of their genome research budgets to analysis of the social and ethical implications. The Ethical, Legal, and Social Implications (ELSI) Program, established in 1990, is a grant-making and policy-making body within the NIH. It is currently the largest federal supporter of bioethics research, with an annual budget of approximately $7 million. Over the years ELSI research projects have focused on a wide range of issues including discrimination in insurance and employment based on genetic information, when and how new genetic tests should be integrated into mainstream health care services, informed consent in genetic research protocols, and public and professional education about genetics research and bioethics. An ELSI Working Group has an advisory role in overseeing the research portfolio of NIH and the DOE and helps formulate the requests for grant proposals and program announcements. Its mandate also includes the formulation of policy options. It has issued policy statements on the need for pilot studies of cystic fibrosis screening (54), on protection from genetic discrimination under the Americans with Disabilities Act (55), and on genetic discrimination and breast cancer.

PRESIDENT'S ADVISORY COMMITTEE ON HUMAN RADIATION EXPERIMENTS

In response to allegations that the United States had performed human radiation experiments and exposed unwitting human subjects to dangerous levels of radiation during the cold war, President Clinton in 1994 established the Advisory Committee on Human Radiation Experiments (ACHRE). The committee was charged with identifying the ethical and scientific standards by which experiments conducted by the federal government between 1944 and 1974 should be judged, and whether those experiments met those standards. Furthermore President Clinton asked the committee to consider whether there were identifiable medical or scientific purposes for the experiments, whether there was follow-up care for the subjects, and whether the experiments met the pre-1974 or current standards for informed consent and other ethical principles for research involving human subjects.

The fourteen member committee, comprised of one citizen, two lawyers, two ethicists, four scientists, three physicians, one statistician, and one professor of humanities, delivered its report a year and a half later, in October 1995. In accordance with its mandate, the report starts with an overview of the ethics of human subjects research from 1944 to 1974, which details the ethical principles adopted by the DOE and the then Department of Health, Education and Welfare, as well as the then Atomic Energy Commission. As additional background, common practices involving humans subjects research in medical research are detailed. The second part of the report focuses on specific experiments, such as plutonium injections and total-body irradiation, examining in depth the protocols used, the level of consent obtained by the researchers, as well as the risks and benefits to which the subjects were exposed. The third part then addresses the current picture in human subjects research, focusing on the existing federal system of human subjects protection. As part of its efforts to gain a complete picture of current practices, ACHRE also conducted an independent study of current research protocols as well as the perceptions of subjects involved in research.

In October 1995 ACHRE reported its findings and recommendations, finding "evidence of serious deficiencies in some parts of the current system for the protection of rights and interests of human subjects." In particular, following a survey of research proposal documents, the ACHRE concluded that these materials provided insufficient evidence with which to make judgments about the voluntariness of the subjects' participation and about the justification for involving individual specific subjects in the research. Review of consent forms also evidenced deficiencies, according to the committee, and patient-subjects interviewed in a separate study seemed to be confused about the difference between research and therapy (56). In addition there were several intentional releases of radioactive substances into the environment without the knowledge or consent of the surrounding community and hundreds of uranium miners died as a result of exposure to radon and other radioactive materials at levels in excess of those known to be hazardous, even though they were being monitored by the federal government.

While ACHRE found that most of the research conducted between 1944 and 1974 involved only minimal risk, they also identified experiments that violated accepted norms of informed consent and placed subjects at risk for cancer and other illnesses without appropriate consent processes. The committee also concluded that although protections mandated by federal regulations for human subjects research were by and large in place, these rules were adequate only when applied to healthy and independent human subjects.

The committee recommended that the government personally apologize to subjects who had involuntarily been exposed to substantial radiation and, where appropriate, offer them financial compensation. In addition the committee took note of both deficiencies in current protections for children and the mentally ill who were serving as subjects in medical experiments and with research subjects' perceptions that research is primarily therapeutic. Finally, ACHRE made a series of recommendations focused on removing current difficulties in interpreting current federal regulations, and, where appropriate, expanding them.

ESTABLISHMENT OF THE NATIONAL BIOETHICS ADVISORY COMMISSION

In the fall of 1993, the White House Office of Science and Technology Policy (OSTP) was approached by NIH, the DOE, and other agencies to consider establishing a standing expert commission on bioethics. The proposal stemmed in part from a congressional request that NIH and DOE establish an advisory committee on genetic privacy. OSTP, however, expressed a need for a high-level group to serve as a shared resource to address a broad set of ethical issues, including genetic privacy, and to complement specialized committees and boards already supported by the various mission agencies.

In August 1994 OSTP published in the Federal Register a draft charter for a National Bioethics Advisory Commission (NBAC). The resulting NBAC charter reflects public comments received as well as bipartisan input from Congress. NBAC provides advice and makes recommendations to the National Science and Technology Council and to other appropriate government entities on relevant bioethical issues.

In addition to chartering NBAC, the President also charged the executive branch agencies that conduct, support, or regulate research involving human subjects to review their policies and procedures for protection of research subjects. This directive was a response to the recommendations contained in the report of the President's Advisory Committee on Human Radiation Experiments, which had just concluded its review of protections (or lack thereof) of U.S. citizens exposed to radiation experiments several decades ago, including American soldiers who were purposely exposed to radiation during atmospheric nuclear tests. Three issues raised by the Advisory Committee on Human Radiation Experiments provided some of the impetus to create NBAC: (1) the need for a continuing public forum on the interpretation and application of ethics rules and principles for the conduct of human subject research, (2) the need to maintain consistency in ethical standards for human subjects research across the 19 federal agencies and departments that support such efforts, and (3) the need to review the current Institutional Review Board (IRB) system.

According to the President's Executive Order, federal agencies were required to report the results of their review to NBAC, which was to pursue, as its first priority, protection of the rights and welfare of human research subjects. The charter also requires that NBAC consider "issues in the management and use of genetic information, including but not limited to human gene patenting." The Commission also may consider additional issues suggested by executive branch agencies, Congress, and the public, or that originate within the Commission itself. NBAC is not a regulatory committee and does not review or approve individual projects. Rather, it defines and identifies broad overarching principles to govern the ethical conduct of research. The 18 members of NBAC are presidentially appointed and represent science, medicine, law, ethics, theology, and public policy. The Commission held its inaugural meeting on October 4, 1996.

In February 1997 the work of the Commission was diverted toward an unexpected development. Within days of the published report of the apparently successful cloning of a sheep using a technique called somatic cell nuclear transfer, President Clinton instituted a ban on federal funding for research related to cloning of human beings. In addition the President asked NBAC to address within 90 days the ethical and legal issues that surround the subject of cloning human beings. This provided an opportunity for initiating a thoughtful analysis of the many dimensions of the issue, including a careful consideration of the potential risks and benefits. It also presented an occasion to review the current legal status of cloning and the potential constitutional challenges that might be raised if new legislation were enacted to restrict the creation of a child through somatic cell nuclear transfer. The Commission quickly commissioned eight papers on the scientific, legal, ethical, religious, and policy aspects of the prospect of human cloning and met five times over the following 90 days. It delivered its report, "Cloning Human Beings," to the President at a White House ceremony on June 9, 1997.

The second topic undertaken by the Commission was the review of the federal agency reports on human subjects protections required by the Executive Order creating NBAC. This is an ongoing project with a comprehensive report due in the fiscal year 1999–2000.

The second topic also concerned an aspect of the protection of the rights and welfare of human research subjects; namely how ethically acceptable research can be conducted with human subjects who suffer from mental disorders that may affect their decision-making capacity. The Commission's report, which was delivered to the President in January 1999, made a number of recommendations to strengthen regulations in this area.

A third topic was an inquiry into the appropriate use of human biological materials, particularly those materials that have been collected in tissue banks over the last century. The analysis was carried out with a focus on the fact that developments in biomedical technologies now enabled investigators to gather much more personal information about those who donated these samples than had ever been anticipated when most of these samples were collected. The Commission therefore addressed the increasing concern that the use of genetic information found in these materials might infringe upon an individual's privacy, and if misused could result in discrimination. In particular, NBAC assessed the adequacy in this new context of existing federal regulations for the protection of human subjects that are incorporated in the so-called Common Rule. The report with its recommendations was delivered to the President in July 1999.

Late in 1998 the President again made a special request to the Commission as a result of the announcement that scientists had been able to isolate and culture human embryonic stem cells. This immediately raised a number of long-debated ethical issues, since the only current sources of these materials were early embryos or fetal tissue. The President asked the Commission to advise on how to best take advantage of the great promise of these materials while also giving consideration to a broad range of the ethical issues involved. The Commission delivered its report and recommendations to the President in August 1999.

To date, NBAC has scheduled the release of two further reports, both in fiscal 1999–2000. The first of these will take up the ethical issues involved in biomedical research protocols involving sponsors and/or investigators and/or research sites within a number of different countries that may not all share the same ethical concerns in the biomedical area. They may differ, for example, on how human research subjects should be protected, or the conditions under which drug trials ought to be allowed to proceed. The second report will be the comprehensive assessment of the current efforts of federal departments and agencies and their grantees to implement existing regulations.

BIBLIOGRAPHY

1. U.S. Congress, Office of Technology Assessment, *Biomedical Ethics in U.S. Public Policy*, OTA-BP-BBS-105, U.S. Government Printing Office, Washington, DC, 1993.

2. National Commission for the Protection of Human Subjects of Biomedical and Behavioral Research, U.S. Department of Health, Education and Welfare, *Research on the Fetus*, U.S. Government Printing Office, Washington, DC, 1975.

3. National Commission for the Protection of Human Subjects of Biomedical and Behavioral Research, U.S. Department of Health, Education and Welfare, *Research Involving Prisoners*, U.S. Government Printing Office, Washington, DC, 1976.

4. National Commission for the Protection of Human Subjects of Biomedical and Behavioral Research, U.S. Department of Health, Education and Welfare, *Research Involving Children*, U.S. Government Printing Office, Washington, DC, 1977.

5. National Commission for the Protection of Human Subjects of Biomedical and Behavioral Research, U.S. Department of Health, Education and Welfare, *Psychosurgery*, U.S. Government Printing Office, Washington, DC, 1977.

6. National Commission for the Protection of Human Subjects of Biomedical and Behavioral Research, U.S. Department of Health, Education and Welfare, *Disclosure of Research Information under the Freedom of Information Act*, U.S. Government Printing Office, Washington, DC, 1977.

7. National Commission for the Protection of Human Subjects of Biomedical and Behavioral Research, U.S. Department of Health, Education and Welfare, *Special Study: Implications of Advances in Biomedical and Behavioral Research*, U.S. Government Printing Office, Washington, DC, 1978.

8. National Commission for the Protection of Human Subjects of Biomedical and Behavioral Research, U.S. Department of Health, Education and Welfare, *Research Involving Those Institutionalized as Mentally Infirm*, U.S. Government Printing Office, Washington, DC, 1978.

9. National Commission for the Protection of Human Subjects of Biomedical and Behavioral Research, U.S. Department of Health, Education and Welfare, *Institutional Review Boards*, U.S. Government Printing Office, Washington, DC, 1978.

10. National Commission for the Protection of Human Subjects of Biomedical and Behavioral Research, U.S. Department of Health, Education and Welfare, *Ethical Guidelines for the Delivery of Health Services by DHEW*, U.S. Government Printing Office, Washington, DC, 1978.

11. National Commission for the Protection of Human Subjects of Biomedical and Behavioral Research, U.S. Department of Health, Education, and Welfare, *The Belmont Report: Ethical Principles and Guidelines for the Protection of Human Subjects of Research*, U.S. Government Printing Office, Washington, DC, 1978.

12. Ethics Advisory Board, U.S. Department of Health, Education, and Welfare, *Report and Conclusions: Support of Research Involving Human In vitro Fertilization and Embryo Transfer*, U.S. Government Printing Office, Washington, DC, 1979.

13. President's Commission for the Study of Ethical Problems in Medicine and Biomedical and Behavioral Research, *Defining Death*, U.S. Government Printing Office, Washington, DC, 1981.

14. President's Commission for the Study of Ethical Problems in Medicine and Biomedical and Behavioral Research, *Protecting Human Subjects*, U.S. Government Printing Office, Washington, DC, 1981.

15. President's Commission for the Study of Ethical Problems in Medicine and Biomedical and Behavioral Research, *Whistleblowing in Biomedical Research*, U.S. Government Printing Office, Washington, DC, 1982.

16. President's Commission for the Study of Ethical Problems in Medicine and Behavioral Research, *Making Health Care Decisions* (with Vols. 2 and 3 appendixes), U.S. Government Printing Office, Washington, DC, 1982.

17. President's Commission for the Study of Ethical Problems in Medicine and Biomedical and Behavioral Research, *Splicing Life*, U.S. Government Printing Office, Washington, DC, 1982.

18. President's Commission for the Study of Ethical Problems in Medicine and Biomedical and Behavioral Research, *Compensating Research Injury*, U.S. Government Printing Office, Washington, DC, 1982.

19. President's Commission for the Study of Ethical Problems in Medicine and Biomedical and Behavioral Research, *Deciding to Forgo Life-Sustaining Treatment*, U.S. Government Printing Office, Washington, DC, 1983.

20. President's Commission for the Study of Ethical Problems in Medicine and Biomedical and Behavioral Research, *Implementing Human Research Regulations*, U.S. Government Printing Office, Washington, DC, 1983.

21. President's Commission for the Study of Ethical Problems in Medicine and Biomedical and Behavioral Research, *Screening and Counseling for Genetic Conditions*, U.S. Government Printing Office, Washington, DC, 1983.

22. President's Commission for the Study of Ethical Problems in Medicine and Biomedical and Behavioral Research, *Securing Access to Health Care*, U.S. Government Printing Office, Washington, DC, 1983.

23. President's Commission for the Study of Ethical Problems in Medicine and Biomedical and Behavioral Research, *Summing Up*, U.S. Government Printing Office, Washington, DC, 1983.

24. M.B. Abram and S.M. Wolf, *N. Engl. J. Med.* **310**(10), 627–632 (1984).

25. A.J. Weisbard, *Ethics* **97**, 776–785 (1987).

26. K.E. Hanna, R.M. Cook-Deegan, and R.Y. Nishimi, *Politics Life Sci.* **12**(2), 205–219 (1993).

27. J.D. Arras, *Cardozo Law Rev.* **6**(2), 321–346 (1984).

28. R. Bayer, *Cardozo Law Rev.* **6**(2), 303–320 (1984).

29. N. Daniels, *Just Health Care*, Cambridge University Press, New York, 1985, pp. 81–83.

30. N. Daniels, *Am I My Parents Keeper? An Essay on Justice between the Young and the Old*, Oxford University Press, New York, 1988.

31. J.C. Fletcher, *Hum. Gene Ther.* **1**(1), 55–68 (1990).

32. T.H. Murray, *Hum. Gene Ther.* **1**(1), 49–54 (1990).

33. R.M. Cook-Deegan, *The Gene Wars*, Norton, New York, 1994.

34. J. Childress, in K.E. Hanna, ed., *Biomedical Politics*, National Academy Press, Washington, DC, 1991, pp. 215–248.

35. Advisory Committee to the Director, *Human Fetal Tissue Transplantation Research*, National Institutes of Health, Bethesda, MD, 1988.

36. National Institutes of Health (NIH), *Report of the Human Embryo Research Panel*, NIH, Bethesda, MD, 1994.

37. U.S. Congress, Office of Technology Assessment, *The Role of Genetic Testing in the Prevention of Occupational Disease*, OTA-BA-194, U.S. Government Printing Office, Washington, DC, 1983.

38. U.S. Congress, Office of Technology Assessment, *Human Gene Therapy—A Background Paper*, OTA-BP-BA-32, U.S. Government Printing Office, Washington, DC, 1984.

39. U.S. Congress, Office of Technology Assessment, *Impacts of Neuroscience*, OTA-BP-BA-24, U.S. Government Printing Office, Washington, DC, 1984.

40. U.S. Congress, Office of Technology Assessment, *Reproductive Health Hazards in the Workplace*, OTA-BA-266, U.S. Government Printing Office, Washington, DC, 1985; reprinted by Lippincott, Philadelphia, PA.

41. U.S. Congress, Office of Technology Assessment, *Alternatives to Animal Use in Research, Testing, and Education*, OTA-BA-273, U.S. Government Printing Office, Washington, DC, 1986.

42. U.S. Congress, Office of Technology Assessment, *New Developments in Biotechnology, 1: Ownership of Human Tissues and Cells—Special Report*. OTA-BA-337, U.S. Government Printing Office, Washington, DC, 1987; reprinted by Lippincott, Philadelphia, PA.

43. U.S. Congress, Office of Technology Assessment, *New Developments in Biotechnology, 2: Public Perceptions of Biotechnology—Background Paper*, OTA-BP-BA-350, U.S. Government Printing Office, Washington, DC, 1987.

44. U.S. Congress, Office of Technology Assessment, *Life-Sustaining Technologies and the Elderly*, OTA-BA-306, U.S. Government Printing Office, Washington, DC, 1987; reprinted by Lippincott, Philadelphia, PA.

45. U.S. Congress, Office of Technology Assessment, *Losing a Million Minds: Confronting the Tragedy of Alzheimer's Disease and Other Dementias*, OTA-BA-323, U.S. Government Printing Office, Washington, DC, 1987; also available through Lippincott, Philadelphia, PA and Human Sciences Press, New York.

46. U.S. Congress, Office of Technology Assessment, *Infertility: Medical and Social Choices*, OTA-BA-358, U.S. Government Printing Office, Washington, DC, 1988.

47. U.S. Congress, Office of Technology Assessment, *Mapping Our Genes—Genome Projects: How Big? How Fast?* OTA-BA-373, U.S. Government Printing Office, Washington, DC, 1988; reprinted by Johns Hopkins University Press, Baltimore, MD.

48. U.S. Congress, Office of Technology Assessment, *Biotechnology 4. U.S. Investment in Biotechnology*, OTA-BA-360, U.S. Government Printing Office, Washington, DC, 1988.

49. U.S. Congress, Office of Technology Assessment, *New Developments in Biotechnology 5. Patenting Life*, OTA-BA-370, U.S. Government Printing Office, Washington, DC, 1989; reprinted by Dekker, New York.

50. U.S. Congress, Office of Technology Assessment, *Neural Grafting: Repairing the Brain and Spinal Cord*, OTA-BA-462, U.S. Government Printing Office, Washington, DC, 1990.

51. L.R. Tancredi, ed., *Ethics of Health Care. Papers of the Conference on Health Care and Changing Values, November 27–29, 1973*, National Academy Press, Washington, DC, 1974.

52. Institute of Medicine, *Assessing Genetic Risks*, National Academy Press, Washington, DC, 1994.

53. Institute of Medicine, *Society's Choices: Social and Ethical Decision Making in Biomedicine*, National Academy Press, Washington, DC, 1995.

54. Working Group on Ethical, Legal, and Social Issues in Human Genome Research, *Workshop on the Introduction of New Genetic Tests*, National Institutes of Health and Department of Energy, Washington, DC, 1990.

55. Joint Working Group on Ethical, Legal, and Social Issues, *Genetic Discrimination and the Americas with Disabilities Act*, Submitted to the Equal Employment Opportunity Commission, Washington, DC, 1991.

56. Advisory Committee on Human Radiation Experiments, *Final Report*, U.S. Government Printing Office, Washington, DC, 1995.

See other entries CLONING, ETHICS; CLONING, OVERVIEW OF HUMAN CLONING; EDUCATION AND TRAINING, PUBLIC EDUCATION ABOUT GENETIC TECHNOLOGY; see also FEDERAL POLICY MAKING FOR BIOTECHNOLOGY entries.

FEDERAL POLICY MAKING FOR BIOTECHNOLOGY, EXECUTIVE BRANCH, NATIONAL HUMAN GENOME RESEARCH INSTITUTE

KATHI E. HANNA
Science and Health Policy Consultant
Prince Frederick, Maryland

OUTLINE

Introduction

An Early Commitment to Ethical Analyses

Ethical, Legal and Social Implications Program

INTRODUCTION

The Human Genome Project of the National Institutes of
Health (NIH) and the Department of Energy (DOE) was
initiated in fiscal year 1988 as a line item in the federal
budget to map and sequence the entire complement of
genetic information in the human genome. The project,
the first major, federally funded biology initiative, was
originally scheduled to take 15 years at a cost of
approximately $3 billion. As of early 2000 the project
is ahead of schedule and under cost.

The project is headquarted in the National Human
Genome Research Institute (NHGRI), originally called the
National Center for Human Genome Research (NCHGR).
NHGRI is one of 24 institutes, centers, or divisions that
make up NIH, the federal government's primary agency for
the support of biomedical research. The collective research
components of NIH make up the largest biomedical
research facility in the world. NIH is part of the U.S.
Department of Health and Human Services (DHHS).

Although much of the genome research effort takes
place in the United States, the Human Genome Project
is a worldwide scientific effort with the goal of analyzing
the structure of human DNA by determining the location
of the estimated 100,000 genes in the human genome
and identifying the sequence of its 3 billion base pairs.
The four major goals of the Human Genome Project
include (1) mapping and sequencing the human genome;
(2) mapping and sequencing the DNA of model organisms;
(3) computerized data collection, storage, and handling
of this information; and (4) examining and addressing
related ethical, legal, and social implications of such a
research effort. The information generated by the Human
Genome Project will be a major resource for the areas of
basic and applied biomedical and behavioral research in
the twenty-first century.

The ambitious nature of the scientific goals of the
project was itself a source of initial controversy. Simultaneously hailed as the search for the biological "holy
grail" (1), and big science at its worst (2–4), the Human
Genome Project is unprecedented in many ways. Besides
being "big biology," the research alliance between NIH
and DOE was also a unique first (5–8), as was the allocation of 3 percent of the research budget for the study

of ethical, legal, and social implications of the application
of knowledge gained from the mapping and sequencing
research enterprise. Never before had the federal government rushed headlong into such an ambitious research
program, while at the same time supporting efforts that
would raise questions about the wisdom, pace, and potential social consequences of its actions.

Although these ethical and social concerns were
not new when the Human Genome Project was first
conceived — they were previously raised in concert with
early genetic diagnostic capabilities such as sickle cell
carrier screening and the use of prenatal diagnosis for
selective abortion — the debate about the Human Genome
Project brought many of these issues to the surface once
again because of the scale and magnitude of the mapping
effort. Whereas ethical, legal, and social concerns raised
by the application of genetic technology to human health
were previously addressed on a case-by-case basis, the
accelerated pace of new discoveries from the Human
Genome Project was likely to exponentially increase the
volume and intensity of concerns, rendering a casuistic
approach dangerously obsolete. The genome project is
leading to genetic tests that will be faster, cheaper, more
accurate, and more applicable to a multitude of diseases.
Thus it was believed by the first leadership of the project
that a more broad-stroked policy approach was required.

This scientific effort is already producing information
that is leading to the detection and diagnosis of
genetic disorders. The long-range goal, however, is to go
beyond diagnostics and to provide improved treatment,
prevention, and ultimately cures. The interim phase,
the phase in which gene detection is possible but
understanding of gene function is limited (and therefore
treatment is unavailable), has been thought by some to be
the period in which the most significant deleterious social,
ethical, and legal consequences might arise, particularly
concerning the use and potential abuse of such information
in terms of discrimination, stigmatization, and potential
medical harm (9,10).

An Early Commitment to Ethical Analyses

James D. Watson, co-discoverer of the molecular structure
of DNA and an early proponent of a federal effort to
map the human genome, recognized the need to confront
the policy issues raised by the possible applications of
genetic information early in the project. He reiterated his
commitment to addressing concerns about the risks posed
by misuse of genetic information at a press conference
in October 1988 announcing his appointment as the first
head of the NIH Office of Human Genome Research:

> Some very real dilemmas exist already about the privacy of
> DNA. The problems are with us now, independent of the
> genome program, but they will be associated with it. We
> should devote real money to discussing these issues. People
> are afraid of genetic knowledge instead of seeing it as an
> opportunity (11).

Watson felt that the NIH program should spend some of
its genome money on pursuing the social, legal, and ethical
issues raised by rapid advances in genetic knowledge. His

advocacy led to the creation of the Ethical, Legal, and Social Implications (ELSI) Program, a grant-making and policy-making body within NIH. In recent years NHGRI has committed 5 percent of its annual research budget to the ELSI program. The DOE Office of Energy Research, NHGRI's partner in the Human Genome Project, also reserves a portion of its funding for ethical and legal research and education.

Never before had a leading scientist taken such a strong position regarding the need to commit federal funds (which otherwise might have gone to funding research) to the study of the ethical implications of research. Watson continued to defend his surprising and somewhat controversial proposal as the debate about federal support for the project went on (11). Because concerns about the social and ethical implications of genetic research were not new in Washington—and, in fact, the subject of several congressional hearings as well as the work of the National Academy of Sciences (NAS), the congressional Office of Technology Assessment (OTA), and the President's Commission for the Study of Ethical Problems in Medicine and Biomedical and Behavioral Research (President's Commission) (12,13)—some argue that Watson was wise to take the bull by the horns and preempt any attempt by policy makers to prematurely and perhaps unnecessarily inhibit progress through overzealous regulation or legislation.

The fact that an icon in modern American molecular genetics would argue so strongly for public funding for social studies of science was welcome news to some observers and suspect to others, who viewed the diversion of funds from science to social research as, at best, an "unavoidable political tax" that the shrewd Watson was willing to pay to accomplish scientific goals. An ethics tax, like any tax, is not without controversy. While it is encouraging that the Human Genome Project has an ethics component, the value of such an organization in affecting health care decisions, research agendas, and policy remains to be seen, even today. And some initial observers were downright skeptical. In the words of Judith Swazey:

> ELSI—an imagistically unfortunate acronym—certainly is being taken seriously by the social scientists, ethicists, lawyers, and assorted other scholars, who have seldom had such financial largesse available to them, and their studies should yield a body of interesting and in some case practically useful findings and recommendations. But in both the short term and the long run, the significance of the ELSI component will be greatly diminished if the concerns that generated it, and its work and results, are seen by scientists and clinicians as politically necessary but basically irrelevant appendages to the "real work" of the Genome Project (14).

ELSI is not the only component of the NHGRI involved in policy issues. The director's office provides overall leadership to the institute, sets policies, and develops scientific, fiscal and management strategies for both the extramural and intramural programs. The office oversees intramural, collaborative, and field research to study human genetic diseases and formulates research goals and long-range plans to accomplish the mission of the

Human Genome Project, including the study of the ethical, legal, and social implications of genome research. The office coordinates the NIH human genome program with those of other federal and private agencies and with other international programs, and fosters support for international meetings, workshops, and other activities to promote efficient international coordination and data exchange.

ETHICAL, LEGAL AND SOCIAL IMPLICATIONS PROGRAM

The planners of the U.S. Human Genome Project recognized that the information gained from mapping and sequencing the human genome would have profound implications for individuals, families, and society. When the original Genome Center was approved and funded by Congress in 1990, its Advisory Committee and NIH and DOE staff had already developed a five-year scientific plan (15,16). Part of the plan addressed ethical, legal, and social considerations, specifically: (1) develop programs addressing the understanding of the ethical, legal, and social implications of the human genome project, and (2) identify and define the major issues and develop initial policy options to address them (15,16).

The Genome Center Advisory Committee, in its initial deliberations, decided to spin off working groups to address specific areas of the project, such as genetic testing and insurance discrimination and genetic testing and policies related to disability. Federal rules concerning working groups are intended to make them temporary. Advisory Committee members must chair each working group; the other members are actually ad hoc technical consultants, serving at the pleasure of the director of NHGRI.

Development of the ELSI Agenda

The first ELSI Working Group met in 1989 to define and develop a plan of activities. The Working Group operationalized its mission by agreeing to the following activities (16):

- Stimulate research on issues through grant making
- Refine the research agenda through workshops, commissioned papers, and invited lectures
- Solicit public input through town meetings and public testimony
- Support the development of educational materials
- Encourage international collaboration in this area.

Thus, at the operational level, the ELSI Working Group developed realistic and practical goals following the model of data gathering and dissemination. In a sense, its early mission was to study what policy makers and the public should study. In terms of policy making, the group developed the following objectives (16):

- Clarify the ethical, legal, and social consequences of mapping and sequencing the human genome through a program of targeted research
- Develop policy options at professional, institutional, governmental, and societal levels to ensure that

genetic information is used to maximize the benefit to individuals and society

- Improve understanding of the issues and policy options through educational initiatives at public, professional, and policy-making levels
- Stimulate public discussion of the issues and policy options.

In addition specific topics were recommended for research support, including fairness in the use of genetic information, the impact of knowledge of genetic variation on the individual, privacy and confidentiality, the impact of the Human Genome Project on genetic counseling, reproductive decisions influenced by genetic information, issues raised by the introduction of genetics into mainstream medical practice, uses and misuses of genetics in the past and the relevance to the present, and commercialization of the products of the Human Genome Project (16). At the time this agenda was set, much policy research had already been conducted or was underway on some of the topics, such as the use of genetic information by employers (17,18), in the criminal justice system (19,20), commercialization (21–23) and genetic testing when no therapy is available. One wonders whether the Working Group found existing work to be so inconclusive as to warrant repeat attention. Nevertheless, the development of a laundry list for topics to be addressed by future grantees is an expansive, if inefficient, method for setting priorities.

Eventually three sets of issues were identified as particularly important initial considerations: (1) privacy of genetic information, (2) safety and efficacy of new genetic testing options, and (3) fairness in the use of genetic information (16). While all critical issues, the issues were narrowly confined to what could be considered a civil liberties orientation. The narrow agenda was likely due to the lack of diversity in perspectives and membership among the Working Group membership, which was largely constituted of academicians and policy researchers (24). Were the membership of the Working Group more diverse, other equally important issues might have been placed on the agenda, such as the effects of commercial interests on the research agenda, intellectual property rights, conflicts of interest for genome scientists, and quality assurance and control beyond issues of safety and efficacy.

Beyond setting a research agenda, the ELSI program was initially assigned the broad goal to "develop the safeguards required as new genetic information is put to practical purposes" (15). The language of the ELSI documents is filled with ambitious verbiage such as "develop sound policy recommendations that will govern the confidentiality of genetic test results, insure equal access to adequate education and counseling for patients, establish minimum qualifications for clinicians, assure quality control for genetic tests, establish guidelines for genetic testing programs, and define ethical and legal responsibilities of clinicians who perform tests" (15).

The basic flaw in the original design of the ELSI program was that it failed to consider the fact that it has no authority to affect policy and no clear route for communicating the information it gathers to the policy arena. These flaws would be corrected later, when

the Genome Center received NIH Institute status, and through the creation of its policy and ethics office, which could approach thorny issues with a more pragmatic and policy-oriented perspective.

In the Beginning: Policy Making through Extramural Research

What distinguishes ELSI from other national ethics bodies is its mandate to administer a grants program. The ELSI Research Program, established in 1990, is responsible for funding and managing research grants and education projects that examine ELSI issues at institutions throughout the United States. It also supports workshops, research consortia, and policy conferences related to funded research and education projects. The ELSI grants programs solicit proposals through program announcements (PAs) and requests for applications (RFAs). At NIH, the Center for Scientific Review (previously the Division of Research Grants) reviews all grants applications and assigns them to the appropriate study sections for peer review for scientific merit. The multidisciplinary review groups consist of bioethicists, educators, genetic counselors, lawyers, theologians, philosophers, psychologists, and geneticists.

Although the development of PAs and RFAs is an iterative process that ideally accounts with addressing critical needs identified by the sponsoring agency and the scientific community, the peer-review method of selecting grants cannot guarantee proper and appropriate attention is being given to important issues. It is not the best way to set a policy agenda because the only citizens with access to the process are those schooled in an academic or professional discipline and capable of responding to the requirements of grant writing. In many ways it is a reductionist process that runs the risk of ignoring the most pressing policy issues. Academicians are not representative of society and can be dangerously naive when it comes to public policy. On the other hand, setting a policy agenda through a bottom-up approach provides the potential for more long-term analytical approaches to issues that might otherwise be subjected to political winds (24,25).

ELSI grantees are hardly representative of the general population or a broad array of disciplines. They are all specialists in genetics and ethics having written numerous publications on the topics they propose to study. It is a small universe that directs and benefits from the ELSI grants program.

On the other hand, the ELSI program is principally designed to support academic research, and this it does well. In fact, one of the major products of the ELSI program has been articles published by the investigators it has funded. As of 1996 the extramural research effort had funded over 125 research and education projects and related activities. These projects have resulted in the publication of over 150 journal articles and books, the development of education programs aimed at health professionals and the general public, and the establishment of policy recommendations on issues ranging from the use of genetic tests to preventing discrimination based on genetic information (26).

Surely such productivity enhances the scholarly writings in the field of bioethics, and is consistent with the traditional output of federally funded research, but is it not yet clear whether these efforts have reached the public in the most effective manner possible. The majority of the writings that have arisen from the grant funds appear in peer-reviewed journals and the academic press, hardly accessible to most policy makers and much of the public.

Until recently there was no mechanism for ensuring that the results of these scholarly pursuits made their way back to the policy arena unless one relied, in the words of one grantee's abstract, on absorption of the facts by "a general audience of intelligent readers." This lack of feedback from its extramural program into the policy process was perhaps the most important barrier to the early ELSI program's efforts as a policy-making body.

Formation of the NHGRI Offices of Policy Coordination and Genome Ethics

Recognizing the need to analyze and coordinate policy issues within NHGRI and the broader community, NHGRI established the Office of Policy Coordination in the Office of the Director in 1995 (intramural program). This Office provides information and analysis on ELSI policy and legislative issues and sponsors workshops and conferences, which assist in the development of policy options and recommendations related to ELSI issues. In addition, the Office has established collaborative relationships with a number of other NIH Institutes and has formed cooperative relationships with a number of other federal agencies, such as DOE, the Centers for Disease Control and Prevention (CDC), the Health Resources Services Administration (HRSA), the National Science Foundation (NSF), and the Food and Drug Administration (FDA).

In 1996 NHGRI established the Office of Genome Ethics (OGE) in the Division of Intramural Research to assist genome researchers in the NHGRI intramural program in identifying and addressing ethical issues arising from genome research.

The addition of these two offices moved NHGRI to a new level in policy making. Staff with the authority to act on behalf of the Institute are able to enter into policy discussions with congressional committees, other agencies, the scientific community, health advocacy groups, and the public.

SPECIFIC AREAS IN WHICH NHGRI HAS CONTRIBUTED TO POLICY DEBATES

The ELSI program and its bureaucratic counterparts in policy (mentioned above) have contributed to policy debates through support of extramural research, publications, and more recently, participation in legislative debates.

Privacy and Fairness in the Use and Interpretation of Genetic Information

As of 1996, the ELSI program had funded 31 research projects designed to examine the use (or misuse)

and interpretation (or misinterpretation) of genetic information, including use by insurers.

Current projects examine the suitability of using genetic technologies for forensic and other law enforcement purposes; the need for standards; the development and application of supporting technology and instrumentation; the current understanding of the statistics and population genetics required in the interpretation of the data; and the social, legal, and ethical issues surrounding the use of these technologies (26).

Another project is developing opportunities for the public to comment on emerging genetic technologies. This project seeks input from lay and professional communities about how each feels about genetic technologies. The investigators are working with these communities to formulate model laws, institutional policies, professional standards of practice, and approaches to clinical decision making.

A current ELSI-funded study is designed to examine the assumptions made in medicine about health, normality, disease causation, and disease susceptibility, and another study is designed to explore the meaning of human genetics in popular culture. Gaining insight into both medical and popular ideas about the interpretation and understanding of genetic information can help in understanding the impact of this information on health care decisions, human relationships, and social policies.

The ELSI Working Group undertook one intensive effort to influence policy directly, rather than through discussion and grant making. As far back as 1990, members of the ELSI Working Group and the ELSI program staff had interactions with the U.S. Equal Employment Opportunity Commission (EEOC), expressing concerns about the lack of employment protections in place for individuals who might be identified as having genetic predisposition to disease. There were fears that although individuals with disabilities would be protected from discrimination under the Americans with Disabilities Act (ADA), individuals who were suspected or known to have a genetic predisposition to develop a disability or have children with disabilities would not be protected. As a result of discussions with the ELSI Working Group, ELSI program staff, and ELSI grantees, and following Commissioners' deliberations, the EEOC provided guidance on March 15, 1995, that clarifies that protection under ADA extends to individuals who may be discriminated against in employment decisions based on genetic information (27).

The ELSI Working Group had long been concerned about the fair use of genetic information, particularly as it relates to health insurance. In response to this concern, the ELSI Working Group spun off its first task force in 1991 on genetic information and health insurance. The Task Force released its recommendations in 1993, which included among others, recommendations that would prohibit the use of genetic information in denying or limiting health care coverage or services and ensure universal access to and participation in a program of basic health services, including genetic health care services (10).

In 1995 the ELSI Working Group developed and published the following recommendations for state and federal policy makers to protect against genetic discrimination (10):

- Insurance providers should be prohibited from using genetic information, or an individual's request for genetic services, to deny or limit any coverage or establish eligibility, continuation, enrollment or contribution requirements.

- Insurance providers should be prohibited from establishing differential rates or premium payments based on genetic information, or an individual's request for genetic services.

- Insurance providers should be prohibited from requesting or requiring collection or disclosure of genetic information. Insurance providers and other holders of genetic information should be prohibited from releasing genetic information without prior written authorization of the individual. Written authorization should be required for each disclosure and include to whom the disclosure would be made.

Genetic discrimination was also a priority for the National Action Plan on Breast Cancer (NAPBC), a public–private partnership established to address the research, education, and policy issues in breast cancer. Building on their shared concerns, the ELSI Working Group and the NAPBC co-sponsored a workshop on July 11, 1995, to address the issue of genetic discrimination and health insurance. Based on the information presented at the workshop and subsequent discussions, the ELSI Working Group, NAPBC, and NHGRI staff developed and published recommendations designed to protect against genetic discrimination.

The group recommended that employment organizations should be prohibited from using genetic information to affect the hiring of an individual or to affect the terms, conditions, privileges, benefits, or termination of employment unless the employment organization can prove this information is job related and consistent with business necessity. In addition employment organizations should be prohibited from requesting or requiring collection or disclosure of genetic information prior to a conditional offer of employment, and under all other circumstances employment organizations should be prohibited from requesting or requiring collection or disclosure of genetic information unless the employment organization can prove this information is job related and consistent with business necessity, or otherwise mandated by law. Furthermore written informed consent should be required for each request, collection, or disclosure. Employment organizations should be restricted from access to genetic information contained in medical records released by individuals as a condition of employment, in claims filed for reimbursement of health care costs, and other sources. Employment organizations should be prohibited from releasing genetic information without prior written authorization of the individual. Written authorization should be required for each disclosure and include to whom the disclosure will be made. Violators of these provisions should be subject to strong enforcement mechanisms, including a private right of action (29).

Since the publication of these recommendations, several bills have been introduced in state legislatures and the U.S. Congress to address the issue of genetic discrimination in health insurance The NAPBC/ELSI Working Group recommendations were considered during the development of a number of these bills. They were also taken into consideration in the deliberations about broader health insurance reform (26).

Clinical Integration of New Genetic Technologies

NHGRI has funded nearly 50 research projects to examine the impact of integrating genetic technologies into health care practice, to establish a better understanding of the current state of knowledge by health professionals, and to develop recommendations about how best to improve knowledge and incorporate these technologies into health care practice (26). The focus of work in this area has been on basic research in clinical ethical issues, professional issues and standards, and applied research designed to examine the impact of genetic testing and counseling. The ELSI program has sponsored two special initiatives in this area. The first, in 1991, was a Request for Applications (RFA) that solicited applications to study issues surrounding genetic testing and counseling for cystic fibrosis mutations. The second RFA, released in 1994, was designed to stimulate the study of issues surrounding genetic testing and counseling for heritable risk of breast, ovarian, and colon cancer.

In 1989 the cystic fibrosis gene was discovered, and mutations, which resulted in disease, were identified. Shortly after the discovery, concerns were expressed that there would likely be an increasing demand for such testing and that inadequate numbers of health professionals were prepared to provide such testing. Further concerns were expressed that not enough was known about how such testing could best be carried out safely and with appropriate pre-test education and post-test counseling. As a result NHGRI, along with the National Institute of Child Health and Human Development, and the National Institute of Nursing Research released an RFA, "Studies of Testing and Counseling for Cystic Fibrosis Mutations," to solicit applications in order to help establish practices that improve professional interpretation and patient understanding of CF testing, and more generally to examine alternative approaches to genetics education, testing, and counseling.

As a result of this initiative, eight research projects were funded. The findings of these studies suggested that interest in testing for cystic fibrosis mutations was much lower than had been expected in the general population. Investigators also discovered that although there was limited knowledge about genetics and cystic fibrosis in the population, it was possible to develop a variety of satisfactory alternative education strategies (e.g., videos and brochures) about testing. Furthermore the investigators saw no evidence of undue anxiety in most individuals tested (30,31).

A second major initiative was undertaken in 1994 by NHGRI in anticipation of the discovery of a number of cancer predisposing genes. This initiative (also an RFA) was co-sponsored by the National Cancer Institute, the National Institute of Mental Health and the National Institute of Nursing Research (32). It solicited applications for studies designed to examine the psychosocial and clinical impact of using gene-based diagnostic tests in families with heritable forms of breast, ovarian, and colon cancer to identify those individuals who have an increased risk of developing cancer. Knowledge and attitudes about genetic testing for cancer risks are also being assessed, and information is being gathered to establish clinical protocols for the optimum use of these risk assessment technologies in the future. Once completed, these projects will provide valuable experience-based guidance for genetic testing for cancer susceptibility genes.

In 1996 the ELSI Working Group formed a Task Force on Genetic Testing. The Task Force was charged with examining the current state of genetic testing in the United States and (if needed) making recommendations to ensure the development and delivery of safe and effective genetic tests. This group specifically examined the scientific validation of new genetic tests; laboratory quality assurance; and education, counseling, and delivery of genetic tests. It sought broad participation by federal agencies, professional societies, the private biotechnology industry, insurers and consumers. A set of principles was published in 1997 (33).

Issues Surrounding Genetic Research

The ELSI program has supported research projects aimed at examining ethical, legal, philosophical, and ethnocultural issues surrounding genetics research. Research has been or is being conducted to examine issues surrounding informed consent in genetics research, explore how the research agenda was set for the Human Genome Project, study academe–industry relationships in genetics research, develop a legal research agenda, examine the impact of the Human Genome Project on women, and identify strategies for documenting the history of the Human Genome Project as it occurs. A research project designed to gather information about the status of informed consent for genetics research resulted in the development of recommendations regarding the components of informed consent for genetics research using stored samples (34).

To address concerns about informed consent in genetics research, the NIH Office for Protection from Research Risks (OPRR) and the ELSI program collaborated to convene a workshop to develop guidance for investigators and Institutional Review Boards (IRBs) who were increasingly being asked to approve genetics research protocols. The deliberations of this group resulted in the publication of a chapter on "Human Genetics Research" in the most recent version of OPRR's *IRB Guidebook*, which is distributed to IRBs. This is the first time that guidance on human subjects protections for genetics research has been provided in the *Guidebook* (35).

Stored tissue samples are valuable resources for genetics research. Due to increasing concerns about the adequacy of informed consent and privacy protections when stored tissue samples are used in genetics research, the CDC and the ELSI program supported a meeting to explore these issues. After intensive deliberations, recommendations were developed and published in December 1995 in the *Journal of the American Medical Association* (34). As a direct result of these deliberations, a number of other groups took up this issue, including

the American College of Medical Genetics, the American Society of Human Genetics (36), the College of American Pathologists (37), and the National Bioethics Advisory Commission (38).

Public and Professional Education

ELSI supports projects designed to educate health and other professionals about genetics and genetic technologies, to develop formal curriculum materials for kindergarten through college-age students, and to educate consumers and the public about these issues.

For example, one study found that knowledge of genetics and genetic tests among physicians is increasing, but deficiencies in knowledge still exist (39). The study also revealed that primary care physicians are more likely to be directive when providing genetic tests rather than providing options from which patients may choose. Another survey revealed the limited amount of education in genetics of a wide variety of health professionals in university-affiliated programs (40). These health professionals reported that they deal on a daily basis with individuals with genetic disorders and their families and that they participate in providing genetic information and counseling to those families (41). They further recognize the need for more education in this area. This survey also revealed that consumers were more likely to have heard about the Human Genome Project than were health professionals. Information obtained through such surveys has been valuable in the ELSI program's efforts to examine its educational priorities and has led to the designation of health professional education as a high-priority area (25).

Another project was designed to educate state policy makers about the Human Genome Project and increase their knowledge about the social, legal, and ethical issues surrounding the research. Regional meetings were held around the country and a publication was developed and widely distributed to state lawmakers and other interested policy makers. A related project was designed to educate appellate judges and journalists about the Human Genome Project and its implications for the future. During the course of this project, an integrated textbook, a casebook, and a teaching manual appropriate for each group was developed and educational workshops provided.

CONCLUSIONS

A 1996 review of the ELSI program concluded that the establishment of an ELSI program at NHGRI was a "novel departure and an experiment" (25). The initial goals of ELSI, to raise the level of awareness of the ethical, legal, and social issues surrounding genetics research, have not been fully met, although tremendous progress has been made. The establishment of the Offices of Policy Coordination and Genome Ethics within NHGRI has increased the chances of advancing these goals. These offices have the authority and the capability to take the findings of ELSI-supported studies and communicate them to the communities making policy. Their creation has elevated the level of policy discourse about these issues.

Although NHGRI and its ethics and policy programs will continue to play a central role in addressing the ethical issues presented by the Human Genome Project, NHGRI does not stand alone in this effort. The U.S. Congress, other agencies of the federal government, professional organizations, universities and other research institutions, regulatory agencies, and industrial enterprises also have a vested interest in how these issues are debated and resolved. The ELSI program must work with all these parties, building on the experience it has gained and the relationships it has established in the first decade of the program.

BIBLIOGRAPHY

1. Hall, *Smithsonian* February (1990).

2. R. Lewin, *Science* **232**, 1598–1600 (1986a).

3. R. Lewin, *Science* **233**, 620–621 (1986b).

4. J. Walsh and J. Marks, *Nature* **322**, 590 (1990).

5. R.M. Cook-Deegan, in K.E. Hanna, ed., *Biomedical Politics*, National Academy Press, Washington, DC, 1991.

6. National Research Council, *Mapping and Sequencing the Human Genome*, National Academy Press, Washington, DC, 1988.

7. U.S. Congress, Office of Technology Assessment, *Mapping Our Genes — Genome Projects: How Big? How Fast?* No. OTA-BA-373. U.S. Government Printing Office, Washington, DC, 1988.

8. L. Roberts, *Science* **245**, 1438–1440 (1989).

9. N.A. Holtzman, *Proceed with Caution: Predicting Genetic Risks in the Recombinant DNA Era*, Johns Hopkins University Press, Baltimore, MD, 1989.

10. K.L. Hudson et al., *Science* **270**, 391–393 (1995).

11. J.D. Watson, NIH Press Conference, Videotape, Available at: National Reference Center for Bioethics Literature, Georgetown University, Washington, DC, 1988.

12. President's Commission for the Study of Ethical Problems in Medicine and Biomedical and Behavioral Research, *Splicing Life*, U.S. Government Printing Office, Washington, DC, 1982.

13. President's Commission for the Study of Ethical Problems in Medicine and Biomedical and Behavioral Research, *Screening and Counseling for Genetic Conditions*, U.S. Government Printing Office, Washington, DC, 1983.

14. J. Swazey, in G.J. Annas and S. Elias, eds., *Gene Mapping: Using Law and Ethics as Guides*, Oxford University Press, New York, 1992.

15. National Center for Human Genome Research, *Annual Report I — FY 1990*, National Institutes of Health, Bethesda, MD, 1990.

16. U.S. Department of Health and Human Services and U.S. Department of Energy, *Understanding Our Genetic Inheritance; The U.S. Human Genome Project: The First Five Years FY 1991–1995*, No. 90-1580, National Institutes of Health, Bethesda, MD, 1990.

17. U.S. Congress, Office of Technology Assessment, *The Role of Genetic Testing in the Prevention of Occupational Disease*, No. OTA-BA-194, U.S. Government Printing Office, Washington, DC, 1983.

18. U.S. Congress, Office of Technology Assessment, *Genetic Monitoring and Screening in the Workplace*, No. OTA-BA-455. U.S. Government Printing Office, Washington, DC, 1991.

19. U.S. Congress, Office Technology Assessment, *Genetic Witness: Forensic Uses of DNA Tests*, U.S. Government Printing Office, Washington, DC, 1990.

20. National Research Council, Committee on DNA Typing, *DNA Technology in Forensic Science*, National Academy Press, Washington, DC, 1992.

21. U.S. Congress, Office of Technology Assessment, *New Developments in Biotechnology, 1: Ownership of Human Tissues and Cells — Special Report*, No. OTA-BA-337, U.S. Government Printing Office, Washington, DC, 1987.

22. U.S. Congress, Office of Technology Assessment, *New Developments in Biotechnology, 4. US Investment in Biotechnology*, No. OTA-BA-360, U.S. Government Printing Office, Washington, DC, 1988.

23. U.S. Congress, Office of Technology Assessment, *New Developments in Biotechnology. 5. Patenting Life*, No. OTA-BA-370, U.S. Government Printing Office, Washington, DC, 1989.

24. K.E. Hanna, in R.E. Bulger et al., eds., *Society's Choices: Social and Ethical Decision Making in Biomedicine*, National Academy Press, Washington, DC, 1995.

25. K.E. Hanna, R.M. Cook-Deegan, and R. Nishimi, *Politics Life Sci.* **12**(2), 205–219 (1993).

26. Review of the Ethical, Legal and Social Implications Research Program and Related Activities (1990–1995), Available on *www.nhgri.nih.gov*

27. American Council of Life Insurance and Health Insurance Association of America, *Report of the ACLI-HIAA Task Force on Genetic Testing*, American Council of Life Insurance, Washington, DC, 1991.

28. NIH-DOE Working Group on Ethical, Legal, and Social Implications of Human Genome Research, *Genetic Information and Health Insurance: Report of the Task Force on Genetic Information and Insurance*, No. 93-3686, National Institutes of Health, Bethesda, MD, 1993.

29. K. Rothenberg et al., *Science* **275** (1997).

30. S. Loader et al., *Am. J. Hum. Genet.* **59**, 234–247 (1996).

31. P.T. Rowley, S. Loader, and J.C. Levenkron, *Gene. Test.* **1**(1), 53–59 (1997).

32. National Human Genome Research Institute, National Institute of Mental Health and National Institute of Nursing Research, *NIH Guide to Grants and Contracts.* **25**(13), April 26 (1996).

33. NIH-DOE Working Group on Ethical, Legal, and Social Implications of Human Genome Research, *Promoting Safe and Effective Genetic Testing in the United States: Final Report of the Task Force on Genetic Testing*, National Institutes of Health, Bethesda, MD, September 1997.

34. E.W. Clayton et al., *J. Am. Med. Assoc.* **274**(22), 1786–1792 (1995).

35. Office of Protection from Research Risks, *Protecting Human Research Subjects: Institutional Review Board Guidebook*, U.S. Government Printing Office, Washington, DC, 1993.

36. American College of Medical Genetics Storage of Genetic Materials Committee, *Am. J. Hum. Gen.* **57**, 1499–1500 (1995).

37. American Society for Investigative Pathology Position Statement, *Balancing Research Progress and Informed Consent*, October 1995.

38. National Bioethics Advisory Commission, *Research Involving Human Biological Materials: Ethical Issues and Policy Guidance*, U.S. Government Printing Office, Rockville, MD, 1999.

39. F.S. Collins, *J. Am. Med. Assoc.* **278**(15), 1285–1286 (1997).

40. E.V. Lapham and J.O. Weiss, in *Human Genome Education Model Project*, Georgetown University, Washington, DC, 1996.

41. D. Eunpu and J. Weiss, *J. Genet. Couns.* **2**, 93–113 (1993).

See other entries FEDERAL POLICY MAKING FOR BIOTECHNOLOGY.

FEDERAL POLICY MAKING FOR BIOTECHNOLOGY EXECUTIVE BRANCH OFFICE OF MANAGEMENT AND BUDGET

GILBERT S. OMENN
University of Michigan
Ann Arbor, Michigan

OUTLINE

Introduction
OMB Role in Biotechnology
Budget Initiative and Analysis
Budget Follow-on to the Budget Initiative
Bibliography

INTRODUCTION

The Office of Management and Budget (OMB) is the nerve center of all Executive Branch agencies. OMB, with the President's directive and/or concurrence, issues budget guidance for the upcoming year, monitors budget commitments and outlays in the current year, and negotiates with agency heads the agency budget and policy and programmatic initiatives. Major disagreements are taken to the President by Cabinet secretaries or agency heads and the OMB director for resolution. OMB staff also coordinates all agency responses to congressional inquiries by reviewing the substance of these written documents and helps to anticipate the kinds of questions Senators, Congress members, and their staffs might raise. The aim of the annual Budget Request to Congress, and of all the attendant negotiations and testimony, is to try to put the President's stamp on the direction and priorities of the departments and agencies.

The budget process involves overlapping work on three fiscal years. During 1999, for example, the Congress enacted, with the President's signature, 13 appropriations bills covering the various departments' and agencies' budgets for fiscal year 2000 (FY00), beginning October 1, 1999. Sometimes Congress is unable to complete that process on time, leading to Continuing Budget Resolutions to keep the government functioning and able to pay its bills until final appropriations are enacted. Meanwhile the agencies function between October 1 of the preceding year and September 30 of the current year under the budget for the current fiscal year. OMB monitors the budget authority committed and the outlays actually made during the current fiscal year. In the example of the budget process during 1999, the agencies were already deeply engaged in budget preparations for fiscal year 2001 to be submitted to OMB in the fall of 1999 and by the President to Congress in February 2000.

The OMB began as the Bureau of the Budget in 1921, pursuant to an Act of Congress. In 1969 industrialist Roy Ash led a commission that recommended to President Nixon substantial changes in OMB, creating a management division and renaming the Bureau the Office Management and Budget. Considerable emphasis was laid on management improvement goals.

In *The Prune Book: The 100 Toughest Management and Policy-making Jobs in Washington*, the first of an invaluable series of books from the nonpartisan Council for Excellence in Government, the deputy director position in OMB was highlighted as focused on the management side of OMB. Government wide efforts to enhance the administrative operations of the Executive Branch drew more attention during the 1980s due to rapidly growing budget deficits and became a high-profile activity of Vice-President Gore during the Clinton administration.

The legendary OMB official Paul H. O'Neill, who rose from entry budget examiner in the Kennedy administration to deputy director of OMB in the Ford administration, told Prune Book author John Trattner that "OMB is a wonderful place to be. Every important issue of government goes through it" (1, p. 75) The title "Prune Book" is a knockoff of the title of the quadrennial "Plum Book," listing the approximately 4000 federal government positions to which the newly elected or re-elected president can make political appointments. Among the plums, the prunes require really able, well-prepared appointees!

The work of OMB was managed in the 1980s by program associate directors responsible for the areas of economics and government (e.g., Departments of the Treasury, Commerce, Justice); human resources, veterans, and labor (Departments of Health and Human Services, Veterans Affairs, Labor, Education, Social Security Administration); national security and international affairs (Departments of Defense and State, Agency for International Development, CIA, and other security agencies); and natural resources, energy, and science (Departments of Energy, Interior, Agriculture, National Science Foundation, NASA, Environmental Protection Agency). Subsequently, in the Clinton administration, the OMB directorate was divided to address Social Security and social welfare agencies in one division and health care financing and health research agencies in the other division. Before the 1970 reorganization, there were no appointed program associate directors in the Bureau of the Budget; the only political appointee was the director.

Another section of OMB, called the Office of Information and Regulatory Affairs, has grown substantially since the late 1970s and early 1980s. Efforts to characterize the full impact on the economy of federal regulations, especially health, safety, and environmental regulations, included concepts of a "regulatory impact," the costs and benefits to the economy and to society of certain major regulations; a "regulatory budget," the idea of limiting the aggregate of such costs or net costs in any year to some politically chosen maximum; "cost-effectiveness analysis," comparing alternative paths to achieve similar protection of health or the environment; and "cost–benefit analysis," putting all health, environmental, and social benefits into dollar terms, then estimating the costs of compliance and of goods forgone, and calculating the ratio or net benefit. The Reagan administration and, later, the Republican Congress of 1995 to 1996 put a high priority on regulatory reform, which was interpreted by environmentalists and many Democrats as a dismantling of the regulatory agencies and regulatory protections so hard-won, rather than just a drive for more cost-effective, efficient regulatory priorities and programs, which all administrations have aimed to achieve. Currently agencies and OMB are working to interpret and implement the Government Performance and Results Act of 1993.

A certain tension exists in OMB. Only the several top officials, including the program associate directors, are political appointees who change with each administration. The career staff represent the ongoing collective institutional history of the Executive Branch. They have a long tradition of pride in their professionalism. Former OMB political appointees commonly describe the career staff as "the best in government." In recent administrations, the career staff have been tapped for important budget and administrative assistant secretary positions in various agencies by shrewd agency heads and Cabinet secretaries. Outsiders, like congressional appropriations staff members, have described the internal process at OMB as requiring for a program associate director position "someone who is very politically sensitive and knows how to deal with the Hill [Congress] and the agencies, or someone who is very analytic. The best is a combination of both. You have a pretty high-powered staff throwing a lot of information and data at you and trying to push you into a position usually against the agencies, and you have to be able to look at that critically" (1, p. 81).

OMB ROLE IN BIOTECHNOLOGY

OMB has central responsibility for the review of budget requests from all agencies conducting or funding biotechnology research, including the National Institutes of Health (NIH), the National Science Foundation (NSF), the Department of Agriculture (USDA), and the Environmental Protection Agency (EPA). Similarly OMB can become involved with agencies whose policy interests are in the development of this sector of the economy (Department of Commerce); in safety for workers, the public, and the environment (Occupational Safety and Health Administration in the Department of Labor, Food and Drug Administration, Environmental Protection Agency); in potential for military applications (DOD, CIA and other security agencies, Department of State); and in international cooperation in all aspects of R&D, applications, and regulation (EPA, Justice, State, NIH, NSF, Defense, Agriculture, AID, Commerce). OMB also has responsibility for extensive vetting within the administration and coordination of views and statements before Executive Branch officials testify before congressional committees or submit written responses to such committees.

As long as the agency activities are continuing on a generally approved course consistent with administration policy and there is good cooperation among agencies without conflicts or turf battles requiring mediation by OMB, any topical area like biotechnology gets little attention in a given year. Periodically, however, a department, agency, or the President may decide to highlight an agency initiative or research and development (R&D), as has recently happened with biotechnology. Alternatively, external events — like congressional fights over fetal research, the reported cloning of the sheep Dolly in Scotland, business sector demands for greater access to government biotechnology research, disputes over what is patentable, or reports of biological warfare agents in other countries — bring this topic to the fore.

In addition OMB often cooperates with other Executive Office agencies, such as the Office of Science and Technology Policy, the Office of Environmental Policy, and the Domestic Policy staff, in reviews of R&D initiatives and performance and coordination of R&D agencies. Thus OMB was an active participant in the late 1970s as recombinant DNA research and biotechnology start-up companies were emerging, and again in the 1980s as biotechnology faced challenges from environmentalists and the Congress about potential hazards of environmental, agricultural, medical, and chemical industry applications. OMB staff participated in an initiative that led to a 1986 report from the White House Domestic Policy Council's Working Group on Biotechnology; that Working Group was co-chaired by Bernadine Healy, then associate director of the Office of Science and Technology Policy (OSTP) and later director of the NIH, and David Kingsbury, associate director of the NSF. The Working Group report laid out a plan for coordination of agency roles and for a balance in the regulation of biotechnology research and the stimulation of product development. Biotechnology applications were again highlighted by the Clinton administration's National Science and Technology Council, a broad interagency effort coordinated by the OSTP, in 1993 and 1994.

A major and continuing investment is the National Human Genome Project, the effort to sequence the human genome. Its scale has attracted the interest of the OMB and the President, and there is sufficient turf competition to require OMB mediation. Discourse in the scientific community during 1985 and 1986 led to first a competitive and then a cooperative initiative by both NIH and the Department of Energy's (DOE) Office of Energy Research and National Laboratories with line items in the FY88 budget. Interestingly the budget request for FY88 from the Department of Health and Human Services (DHHS) for the NIH effort on the Human Genome Project was $25 million. According to OMB sources, it was OMB that increased the request to the nice, round figure of $100 million suitable for emphasis in the President's budget request. For excellent scientific reasons, the mandate was expanded to sequence genomes of other organisms — both infectious agents and "model organisms," like yeast and earthworms and mice, for studying underlying biology and comparing and inferring gene relationships and evolutionary changes with humans.

The NIH program matured into a distinct NIH organ, the National Human Genome Research Institute, in the mid-1990s. Both the NIH and DOE programs have included components of special funding on ethical, legal, and social implications of the new genetics, research and conferences that generate lots of policy issues requiring the administration's attention and OMB coordination.

BUDGET INITIATIVE AND ANALYSIS

Fiscal year 1991 stands out as the year in which the President's Budget Request to Congress included a major section on R&D, titled "Expanding the Human Frontier." Another section dealt with "Improving Productivity and the Quality of Life through Biotechnology" (2, pp. 59–63). Detailed presentation of that budget document is revealing about the role of OMB and of the government more generally in this important area of R&D and social policy.

Using data from 1987 published by the Congressional Office of Technology Assessment in 1988, the document included an introductory figure showing that the sources of U.S. biotechnology investment for 1987 were 59.4 percent federal government, 38.3 percent industry, and 2.3 percent state governments. Of course, "biotechnology" had to be defined and be put in context. Thus the document begins, "Biotechnology is an ancient practice that includes such familiar applications as the use of yeast in baking bread and cultures in making cheese. Recent breakthroughs in biotechnology, such as recombinant DNA techniques, cell fusion, and gene therapy, offer unprecedented opportunities for improving the nation's productivity, health, and well-being. Uncertainties in the returns on biotechnology investment, however, stemming from market barriers and unnecessary regulation, have retarded progress. Increasing Federal investment in basic biotechnology research will spur further advances, as will initiatives that improve the payoffs on investments" (3, pp. 72–75).

The characterization of regulation, an editorial insertion, reflects policy views of the Bush administration. It was then felt that the long struggle from the beginning of the recombinant DNA era had overcome fears of magical and mystical powers of the technology, as capable of creating wholly new living things with unpredictable properties. This was 14 years after science advisers in 1977 had explained to senior staff in EPA, OMB, and the Domestic Policy office what recombinant DNA techniques were. Many of these individuals were attorneys, as were some of the congressional staff addressing these issues in light of the scary, though wise, decision of the scientific community to issue a moratorium on recombinant DNA research three years earlier at the famed Asilomar Conference, in order to develop biological containment techniques. The 1977 discussions helped explain the use of standard techniques of preparation of DNA and proteins, and actions of enzymes; use of chromatography or electrophoresis to separate and identify molecules; special and predictable roles of the remarkable restriction and ligase enzymes; and specific experiments and proposed experiments. A particularly memorable event was the November 8, 1977, hearing of the Senate Committee on

Science and Technology, chaired by Senator Adlai Stevenson of Illinois, at which prominent scientists, NIH Director Donald Frederickson, and Presidential Science Adviser Frank Press testified (4). In their exuberance about the potential for recombinant DNA applications and their disdain for the emerging NIH Guidelines for Recombinant DNA Research, some testifying scientists nearly brought the wrath of the Congress down upon the research community in the form of prohibitive regulatory requirements. Such a result was averted through cooperative review of the NIH process and the scientific methods themselves (5).

The FY91 Budget claimed that "Advances in biotechnology hold much promise. They can help improve the availability and quality of the food supply; prevent, identify, and cure disease; and reduce the hazards of industrial waste. Cell fusion, the merging of the genetic material of two cells of different species, can accelerate the selective breeding process for producing hardier and more fruitful crops and livestock. Gene therapy, replacing defective genetic material with normal DNA, may enable doctors to attack directly the source of major diseases, including cancer. In drugs, foods, agriculture, waste management, and energy, biotechnological advances offer the possibility of improvements that will make a real difference in people's lives. In this sense, biotechnology is an "enabling" technology: we may be able to make products safer or more cheaply, and we may be able to produce goods that we could not produce at all using traditional methods" (2, p. 59).

Clearly, the Bush administration, led by the OMB Director, Richard Darman, was hitching itself to the biotechnology revolution. "Biotechnology is a classic case of investing for the future. U.S. industry is spending at least $2 billion a year on biotechnology research and development, even though sales of products manufactured using biotechnology only reached the $1 billion mark for the first time in 1989. It is clear that the private sector believes the return on this investment will be great. The budget reflects a similar belief for the Federal investment" (2, p. 60).

The FY91 Budget proposed $8.6 billion, an increase of $218 million (6 percent) over the FY90 Budget, for biotechnology research and development. As all readers might imagine, the "budget" reported for such R&D depends mightily on the criteria for inclusion or exclusion. In fact it is common for experienced officials who must respond to requests from policy makers, Congress, or journalists for the amount of spending on a given need or opportunity to decide first whether it is in one's policy interest to generate a "big number," indicating a lot going on, or a "small number," indicating a need for much more investment, given the presumed opportunities. A smaller number offers the attractive feature of making any increment a bigger percentage increase, of course.

The estimate was built up through a host of not necessarily consistent definitions in 12 specified agencies: the DHHS (primarily the NIH, but also the Food and Drug Administration, FDA); the DOE, Commerce (including the National Institute for Standards and Technology, NIST, and the National Oceanic and Atmospheric Administration, NOAA), Defense, Veterans Affairs, and

Agriculture; the NSF, and the EPA. In fact NIH has struggled for decades with the definitions devised for characterization of research as "basic" or "applied" for governmentwide summaries. A close reading of the FY91 narrative several years later reveals some inconsistencies even in the document. There was no tabulation of the sources, by agency, of the $8.6 billion total. Only two agencies' budgets were highlighted. NIH was described as having a proposed increase of $280 million (more than the $218 million total for all agencies combined, noted above), of which $48 million was in the Human Genome Project. The National Research Initiative for Agriculture was described as having a doubling of the competitive research grants program, from $48 to $100 million, of which half was expected to be related to plant and animal biotechnology research. The description of DOE roles had no budget information.

The Budget did describe the emergence of 400 start-up firms and diversification of an estimated 200 existing companies into biotechnology and 200 companies supporting biotechnology with materials, instruments, equipment, and services. Data from 1987, probably the OTA report (6) (see the discussion above) were cited as showing $110 million in biotechnology-related expenditures by states, of which 38 claimed to be active in such investments in a 1986 survey through centers of excellence, university initiatives, incubator facilities for new firms, or grants for research projects.

On the management side, the administration highlighted the use of Cooperative Research and Development Agreements (CRADAs) by the NIH. The DOE and its national laboratories similarly utilized this mechanism to stimulate collaborations with industry and spin-offs of federal research into companies. Federal scientists were permitted and encouraged to begin filing patent applications. The Budget expanded funding for the Patent and Trademark Office (PTO) of the Department of Commerce, which in 1988 had instituted a 13 point plan to accelerate the review and award of biotechnology patents. PTO joined forces with the biotechnology industry to create a Biotechnology Institute to enhance training and technical expertise in the Patent Examining Corps. The FY91 Budget's process also introduced the concept of a system of user fees at FDA to provide staffing and other resources that would speed up the review of drugs and medical devices, including products that use techniques developed through biotechnology.

With regard to regulation, the FY91 Budget document emphasized the Coordinated Framework for Regulation of Biotechnology, the 1986 report from the Domestic Policy Council published in the Federal Register and mentioned above, and the ongoing Biotechnology Science Coordinating Committee, under the aegis of the Council on Competitiveness. Citing scientific assessments from the National Academy of Sciences and the Congressional Office of Technology Assessment, the Framework concluded that regulation should focus on the characteristics and risks of an organism or product, not on the process by which it was produced (i.e., by recombinant DNA, or cell fusion, or other biotechnology methods versus chemical synthesis). At the agency level the FDA concluded that

no new procedures are required for genetically engineered products. In contrast, the USDA developed a new rule to review genetically engineered organisms that could possibly pose a high risk as a plant pest, and the EPA developed new regulations to tailor its rules and review procedures for microorganisms proposed as pest control agents or as bioremediation agents for soil and water contamination.

BUDGET FOLLOW-ON TO THE BUDGET INITIATIVE

The Budget for fiscal year 1992 contained a similar section (3) entitled "Expanding the Human Frontier through Biotechnology." Some of the budget numbers are surprising, revealing inconsistencies year-to-year in the Budget analysis. This time there was a tabulation of agency budgets for "biotechnology R&D," as shown in Table 1.

As shown in the table the narrower definition did not change the percentage increase, but it certainly did make the total effort seem smaller than did the figures highlighted in the FY91 document. OMB staff confirmed that the narrative was thematic, not tied analytically to the actual Budget. OSTP and OMB worked together, and with NIH and the Office of the Assistant Secretary for Health, DHHS, to seek standardized definitions and made big changes from year to year. As documented here, these shifts revealed one of the fundamental problems with biotechnology as a budget priority: The research was so hard to define even for descriptive purposes within and across agencies that no common understanding emerged and the budget effort was abandoned.

Otherwise, the narrative closely tracked the FY91 document, with highlights of achievements and trends in the diverse technologies being applied in human health, agriculture, foods and animal husbandry, and bioremediation and waste management for the environment. University/government/industry cooperation and investment in research training were emphasized, as was the ongoing influence of the 1986 Coordinated Framework for Federal Regulation of Biotechnology.

Current OMB senior staff have confirmed that no similar Budget highlights have focused on biotechnology

Table 1. Agency Budgets for Biotechnology

Department or Agency	Budget Authority in ($millions)	
	FY91 enacted	FY92 proposed
DHHS	$3296	$3557
USDA	119	139
DOE	110	140
NSF	130	132
DOD	118	123
VA, EPA, and other agencies	17	17
Total, all agencies	3788	4107 (+8%)
In addition:		
Directly-related activities	1663	1810
Broader science-based activities	1998	2144
Scale-up activities	25	32
Grand total (not given)	7474	8093 (+8%)

and no across-the-government estimates of budgets for biotechnology R&D have been generated by OMB since 1992. (OSTP did make an estimate of $4.3 billion in FY94.) In part, this observation reflects emphasis on other initiatives, and a kind of stay-the-course investment in the Human Genome Project and related areas. More important, it reflects the fact that biotechnology has become embedded in all areas of cell and molecular biology research and applications to human and animal diseases and industrial and environmental needs.

BIBLIOGRAPHY

1. J.H. Trattner, *The Prune Book: The 100 Toughest Management and Policy-Making Jobs in Washington, DC*, Madison Books, the Center for Excellence in Government, Washington, DC, 1988; 6th ed., 2000.

2. Office of the President of the United States, *The Budget for Fiscal Year 1991*, Office of Management and Budget, Government Printing Office, Washington, DC, 1990.

3. Office of the President of the United States, *The Budget for Fiscal Year 1992*, Office of Management and Budget, Government Printing Office, Washington, DC, 1991.

4. F. Press, *Science* **211**, 139–145 (1981).

5. G.S. Omenn, A.N. Schechter, R.F. Goldberger, and A. Dean, eds., *The Impact of Protein Chemistry on the Biomedical Sciences, an International Symposium in Honor of Christian B. Anfinsen*, Academic Press, New York, 1984.

6. Office of Technology Assessment, *U.S. Investment in Technology*, vol. 2, Government Printing Office, Washington, DC, July, 1988.

See other entries FEDERAL POLICY MAKING FOR BIOTECHNOLOGY.

FEDERAL POLICY MAKING FOR BIOTECHNOLOGY, EXECUTIVE BRANCH, RAC

LEROY WALTERS
Kennedy Institute of Ethics, Georgetown University
Washington, District of Columbia

OUTLINE

EARLY HISTORY AND WORK OF RAC, 1974 TO 1983 AND 1984 TO 1990

The Recombinant DNA Advisory Committee (RAC) of the National Institutes of Health (NIH) has had a long and distinguished history. RAC was established in fall 1974, shortly before the Asilomar meeting on research with recombinant DNA. Credit for the creation of the advisory committee and for the idea of devising guidelines for the safe conduct of recombinant DNA research is shared by a committee of scientists chaired by Paul Berg, who suggested the approach, and by NIH Director Donald S. Fredrickson, who implemented the plan and shepherded the field through its early years (1,2, pp. 80–96, 154–263; 3, pp. 272–274; 4,5). The committee met for the first time in February 1975, immediately after the Asilomar meeting (2, pp. 99–153; 3, pp. 274–278). From that moment until the early 1980s the RAC set the safety standards for all recombinant DNA research being conducted in the United States. These standards became known as the NIH "Guidelines for Research Involving Recombinant DNA Molecules" (6). The NIH Guidelines were adopted, in whole or in part, by many other industrialized countries.

In the early years most recombinant DNA research was funded by NIH and the National Science Foundation (NSF), so academic researchers had little choice but to follow the Guidelines. However, private companies also voluntarily complied with the RAC's Guidelines, in part to avoid regulation by their states or municipalities. While Congress considered numerous bills that would have regulated recombinant DNA research, especially in 1977, in the end the Congress deferred to the NIH and the RAC (2, pp. 312–337).

Historical research has demonstrated quite conclusively that a major violation of the Guidelines occurred in January 1977. A plasmid that had not yet been certified by NIH (pBR322) was used in the laboratory of Herbert Boyer of the University of California at San Francisco (UCSF) in an experiment that successfully cloned the rat insulin gene (7, pp. 112–180). NIH was informed of the violation in March 1977, but because of upcoming congressional hearings and pending legislation, the incident was kept quiet (7, p. 136). There is disagreement among witnesses about whether the prohibited clones were completely destroyed during March 1977; it is at least possible that the rat insulin genes were removed from the clones and re-used in an experiment with a certified vector in April (7, pp. 137–139, 167–173). In May 1977 the UCSF researchers announced their success in cloning the rat insulin gene at a press conference, and their paper documenting the experiment appeared in the June 17th issue of *Science*. However, a September 30th article by Nicholas Wade in *Science* raised questions about the possible use of an uncertified vector by the UCSF researchers (8). This article led to an intense confrontation between Senators Adlai E. Stevenson III and Harrison Schmitt, on the one hand, and researchers Herbert Boyer and William Rutter, on the other, at a November 1977 hearing (7, pp. 169–173). Despite the criticism directed at the researchers by the senators, no legislation followed.

The first major revision of the Guidelines occurred in December 1978. The Secretary of Health, Education, and Welfare, Joseph Califano, became deeply involved in the revision process. While the substantive provisions of the Guidelines were being substantially relaxed, Califano expanded the RAC to 25 members, appointed several additional nonscientist members, and insisted that NIH promptly prepare a plan to assess the risks of recombinant DNA research to human health and the environment (2, pp. 247–248).

By late 1978 and 1979 it was becoming clear that most kinds of laboratory research with recombinant DNA were safe for both laboratory workers and the environment. New issues arose, however, such as the use of recombinant DNA techniques for large-scale production of human insulin (2, p. 355; 9) and the deliberate release of recombinant DNA into the environment, for example, to lower the temperature at which strawberry plants freeze. These new technologies, while initially overseen by the RAC, gradually moved to the appropriate regulatory agencies, the Food and Drug Administration (FDA) and the Environmental Protection Agency (EPA).

As NIH and the RAC ceded oversight authority regarding these areas of biotechnology to other agencies in the early 1980s, it appeared that RAC's advisory role would gradually diminish and might, in fact, disappear. Novel host-vector systems were seldom submitted to RAC, and most problems of physical and biological containment seemed to have been solved. Through a strange and perhaps fortuitous quirk of history, the research base for a new technique called "human gene therapy" was gradually being developed in the early 1980s. This technique had certain continuities with the laboratory research that had been RAC's central focus in the 1970s. From one perspective, gene therapy was the introduction of recombinant DNA (or products derived from recombinant DNA) into human beings. However, gene-therapy research was clearly a hybrid field. On the one hand, it was highly technical and required the expertise of molecular biologists, virologists, and human geneticists. On the other hand, gene-therapy research was human-subjects research, which was governed by its own set of rules and which was, at least in its broad outlines, quite comprehensible to laypeople.

Two events in 1980 had sparked public interest in the topic of human gene therapy — as well as in genetic engineering more broadly (10,11; 12, pp. 146–150). The first was a letter sent in June to President Carter from leaders of the Jewish, Catholic, and Protestant religious communities, expressing concern about the potentially deleterious effects of genetic engineering (13, pp. 95–96). A few months later the *Los Angeles Times* broke the story of Martin Cline's unapproved attempts to perform gene therapy on two patients afflicted with β-thalassemia, one patient in Israel and the other in Italy (14, 15, pp. 189–267). From these two pivotal events one can draw a direct line to subsequent ethical and public-policy discussions of gene therapy in the United States.

In 1982 a report by a presidential commission on bioethics, *Splicing Life* (13), and a congressional hearing on human genetic engineering (16) framed the major ethical issues in gene-therapy research. In response to those hearings, NIH and RAC began in 1983 to consider whether the committee should volunteer to review gene-therapy research protocols on a case-by-case basis. Over the course of a year NIH and RAC moved step-by-step toward accepting the oversight of gene-therapy research, in part because its other work was essentially finished and in part because no other agency or committee was prepared at that time to review this emerging field of research. A working group on human gene therapy was established during the summer of 1984 as a subcommittee to RAC, and this working group began developing Guidelines for gene-therapy research, which became known as the "Points to Consider," in the fall of that year (10,11). Once again, Congress deferred to the Executive Branch and to its public advisory committee, RAC. It did not pass legislation regulating gene-therapy research, nor did it establish a presidential advisory committee on the Human Applications of Genetic Engineering, as recommended by Congressman Albert Gore, Jr., in H.R. 2788 (April 27, 1983). The Congressional Office of Technology Assessment also published a report in late 1984, *Human Gene Therapy: A Background Paper* (17) that seemed to accept the merits of the approach being taken by NIH and RAC.

What were the central ethical questions to be asked about any proposed gene-therapy research protocol? The many questions asked in the RAC's Guidelines — the Points to Consider document — can be reduced to four rather simple and straightforward questions:

1. What are the potential harms and benefits of the research to the research subjects who will participate in a planned study?

2. How will these potential harms and benefits be communicated to prospective research subjects so that they can make voluntary and informed decisions about whether to participate in the research?

3. How will the selection among potential research subjects be made in a fair and equitable way, especially in cases where more people want to participate than can be enrolled in a study?

4. How will the privacy of research subjects be protected and the confidentiality of their medical information preserved?

If it is possible to develop guidelines for an emerging field of biomedical research too early, RAC and its working group did so. Working group members hurried to finish polishing the Points to Consider document in the spring and summer of 1985, then had to wait for almost two years for even a "preclinical" gene-therapy protocol. In the summer and fall of 1988, the first gene-marking study was reviewed and approved by the working group (now called the Human Gene Therapy Subcommittee) and the parent committee, RAC. Finally, in 1990, two gene-therapy studies were reviewed and approved. On September 14, 1990, the first officially sanctioned gene-therapy study began when W. French Anderson, R. Michael Blaese, and their colleagues administered genetically modified T-cells to a four-year-old girl named Ashanti DeSilva (15, pp. 13–52, 326–349; 18, pp. 227–240).

In its guideline-writing efforts and its review of the earliest preclinical and clinical protocols, the RAC was supported by a series of excellent NIH staff people in an office called the Office of Recombinant DNA Activities (ORDA). The professionalism of this staff, its commitment to the public health and the protection of human subjects, and the long tenure of many of its members all contributed significantly to any success that RAC has had in its oversight responsibilities over the years.

PARALLEL EFFORTS BY NIH AND FDA, 1991 TO 1995

Gene-therapy research gradually expanded during the early 1990s. The number of new gene-therapy and gene-marking protocols submitted year by year to RAC can be summarized as follows:

1990: 2	1993: 31
1991: 9	1994: 33
1992: 23	1995: 44

An important innovation in the monitoring of clinical protocols adopted by RAC deserves special mention. Pursuant to a suggestion by the late Brigid Leventhal, a pediatric cancer researcher from Johns Hopkins University, RAC created a system that asked researchers to submit for the annual reports on adverse events and biological (as opposed to clinical) efficacy in ongoing gene-therapy studies (19). This safety monitoring activity of RAC eventually became known as "data management" (20,21). Dr. Leventhal presented the initial data-monitoring report to RAC in December 1992 and covered 40 patients (24).

The data management activities of the years 1992 through 1994 laid the groundwork for one of RAC's finest achievements in its oversight of gene-therapy research. In preparation for the June 1995 RAC meeting, RAC members and the ORDA staff undertook a comprehensive review of gene-therapy and gene-marking studies that had been reviewed and approved to date. This review, which was published in *Human Gene Therapy* on September 10, 1996 (23), revealed that during the first four years of intensive gene-therapy research there were hints of benefit in several studies but that in no case had a patient been cured of his or her disease by this new experimental approach.

As RAC continued to review clinical protocols in the mid-1990s, several widely acknowledged problems in the conduct and oversight of clinical research began to become apparent for gene-therapy research as well. The first cluster of problems surrounded the role and work of local Institutional Review Boards (IRBs). IRBs frequently lacked the necessary expertise to evaluate the technical aspects of HGT protocols. In addition, there were sometimes conflicts of interest within the local institution, especially when gene-therapy programs were considered to be showpieces for academic medical centers. Further IRBs' work involved primarily the front-end approval of paper protocols, with little monitoring of the actual conduct of the trials. RAC encountered a second set of problems as it examined the consent forms submitted with gene-therapy research protocols. Consent forms were often incomplete, omitting important information about the proposed research. In addition the forms were often

misleading, especially in their descriptions of Phase I studies. The questions of what the local institution would do in case a research subject were injured and who would pay for experimental procedures were frequently evaded by means of artful circumlocution. Retrospective analysis also revealed that consent forms were sometimes not updated to report, for example, toxicities encountered during the course of a trial.

In the early 1990s FDA had also greatly enhanced its capability to review Investigational New Drug (IND) applications that employed gene-therapy techniques (24,25). FDA officials and reviewers regularly attended RAC meetings and increasingly participated in RAC discussions. Researchers began to note differences in the kinds of information being sought by RAC and FDA, and some researchers also complained that they had to jump over two regulatory hurdles rather than one.

In response to these complaints and similar complaints by some AIDS activists and biotechnology companies, NIH and FDA sought, in 1994, to work out a system for dual submission of protocols and coordinated review. In retrospect, it seems quite clear that this well-meaning effort did not go far enough and that serious differences in emphasis and approach remained between NIH and its advisory committee, RAC, on the one hand, and FDA, on the other. The two agencies also failed to agree on how to develop a data-management system for gene-therapy research.

VERMA COMMITTEE REPORT AND ORKIN-MOTULSKY COMMITTEE REPORT, SEPTEMBER 1995 AND DECEMBER 1995

In September 1995 a committee chaired by Inder Verma submitted recommendations to NIH Director Harold Varmus regarding the appropriate role of RAC in the review of gene-therapy research. The committee concluded that the RAC had an important ongoing role in the review of such research but recommended that RAC publicly review only research protocols that raise novel questions, for example, protocols that employ a new vector or seek to treat a new disease. For all other protocols, those that do not raise novel questions, the Verma Committee recommended that the review be conducted solely by FDA (26,27).

Three months later, in December 1995, a committee chaired by Stuart Orkin and Arno Motulsky delivered a somber verdict on the first five years of publicly reviewed and approved gene-therapy research: Not a single study had demonstrated clinical benefit to patients from gene therapy alone. The committee recommended that more attention be paid to the infrastructure for gene-therapy research, including the development of better vectors and of a better understanding of human immunology (28,29).

EIGHTEEN MONTHS OF UNCERTAINTY, MAY 1996 TO OCTOBER 1997

In May 1996 NIH Director Harold Varmus announced his intention to abolish the RAC in a speech delivered in Hilton Head, South Carolina (30). This proposal was formulated more precisely in a *Federal Register* notice published in July 1996 (31). Over the next year and a quarter RAC's future role was debated by academic people, patient advocacy groups, biotechnology companies, several members of Congress, and RAC members themselves. Two general revisions of the original plan were published in the *Federal Register*, the first in November 1996 and the second in February 1997 (32,33). Finally, on October 31, 1997, a new oversight system for gene-therapy research was formally announced in the *Federal Register* (34). According to this final plan, RAC and NIH would no longer approve or disapprove gene-therapy research protocols. Instead, the RAC would discuss protocols that raised novel issues and make suggestions to the authors of the protocols. It was understood by all that RAC discussions would also inform FDA reviewers in their confidential negotiations with the sponsors of gene-therapy research who had submitted the same protocols as part of the IND review process.

There are five other features of the October 1997 plan that are especially worthy of note. First, the Office of Recombinant DNA Activities accepted responsibility for developing a data-management system to assist RAC in its review of adverse events and its annual audit of gene-therapy research. Second, gene-therapy researchers had a clearly articulated duty to inform ORDA and RAC of any changes in RAC-reviewed protocols that occurred between the time of RAC review and time that the researchers received permission from FDA to proceed with their proposed research (under an IND). Third, gene-therapy researchers also had a clearly articulated duty immediately to report to ORDA the occurrence of "any serious adverse event" in a gene-therapy research protocol. Fourth, researchers were required to submit annual data reports to ORDA for inclusion in the data-management system and analysis by RAC. Finally, ORDA and RAC would plan Gene Therapy Policy Conferences to look at broad themes surrounding gene-therapy research, both in the present and in the future.

FROM OCTOBER 1997 TO THE PRESENT: HOW IS THE NEW SYSTEM WORKING?

There is some good news to report from the early years of the new oversight arrangement. The Gene Therapy Policy Conferences have been highly successful in promoting interdisciplinary discussion of three important topics: genetic enhancement, in utero gene therapy, and the use of lentiviruses as vectors. RAC members continue to be deeply committed to their public roles and have been quite forthright in expressing concern about being asked to treat adverse-event reports as proprietary information. Similarly the staff people at ORDA (recently made a part of the NIH Office of Biotechnology Activities, OBA), have devoted long hours to fulfilling the roles assigned to them under the October 1997 agreement.

However, in September 1999 the world of gene-therapy research was shaken to its foundations by the unexpected death of an eighteen-year-old patient, Jesse Gelsinger. Mr. Gelsinger, whose ornithine transcarbamylase (OTC) deficiency was under reasonable control through a combination of drugs and diet, was a participant in a study being conducted at the University of Pennsylvania. He died four days after receiving an infusion of an adenoviral vector and the OTC gene (35,36). Months of intensive review and a full day of discussion at the December 1999 RAC meeting failed to clarify the precise cause of Mr. Gelsinger's death.

In December 1999 inspectors from FDA charged that the researchers conducting the OTC deficiency protocol in which Mr. Gelsinger died had violated several FDA regulations (37,38). Further inspections led to a publicly released set of "investigational observations" and FDA's placing a clinical hold on all gene-therapy protocols currently being conducted by the University of Pennsylvania research group (39,40). The Penn researchers replied to the observations in February 2000 (41,42), and a final resolution of this tragic incident is awaited in the near future.

In response to the death of Mr. Gelsinger, the NIH Office of Biotechnology Activities (OBA) and FDA's Center for Biologics Evaluation and Review (CBER) initiated an intensive joint review of all gene-therapy protocols using adenoviral vectors. This review, conducted in the last three months of 1999, sought to gather and analyze all adverse events that had occurred in gene-therapy studies using adenoviral vectors — approximately 25 percent of all U.S. protocols. In addition, OBA reminded gene-therapy researchers whose protocols had been registered with NIH of their duty to report all serious events to NIH. At the December 1999 RAC meeting, a working group on adenoviral vectors reported a pervasive lack of standardization in the characterization and use of such vectors (43).

Taken together, the events and discussions that occurred from September 1999 through February 2000 revealed that there are serious problems in the current oversight system for gene-therapy research in the United States. First, an online database for the data-management system, discussed and planned for since 1994, is still not available in the year 2000 (44). Initially delays occurred because of FDA's 1995 decision not to collaborate in the development of the database. In recent years ORDA has not had sufficient staff or resources to complete the development of the database.

Second, many gene-therapy researchers who are covered by the NIH "Guidelines for Research Involving Recombinant DNA Molecules" have neglected to file immediate reports with ORDA of serious adverse events that have occurred in the trials they are conducting. According to a December 21, 1999, letter from former NIH Director Harold Varmus, to Congressman Henry Waxman, only 39 (or 5.6 percent) of 691 serious adverse events in gene-therapy research using adenoviral vectors had been reported to ORDA before October 1999, when NIH and FDA began their vigorous joint effort to gather and analyze those events (45).

Third, the lack of coordination between NIH and RAC, on the one hand, and FDA, on the other, has continued in certain arenas. The two parent agencies have had different histories and sometimes reflect those histories in divergent approaches to the same question. Important issues remain unclarified, for example, whether RAC is advisory to

FDA or not. Some modes of FDA-NIH cooperation that could have been initiated by October 1997, at the latest, commenced only in late 1999, in response to a crisis. In December 1999 FDA changed its standard operating procedure on two important points and began providing weekly summaries to OBA of amendments to gene-therapy research protocols and adverse-event reports that it has received during the preceding week.

CONCLUSION

At the beginning of 2000 both RAC and the field of gene-therapy research face uncertain futures. It is still the case that no published report has demonstrated clear efficacy for gene-therapy procedures alone—that is, without adjunctive therapy. Now one relatively healthy patient has died in a gene-therapy trial, and several additional patients seem to have experienced laboratory or clinical toxicities. The RAC is frequently cited as a model for the oversight of emerging biomedical technologies, for example, the field of xenotransplantation research. However, several important questions about the RAC's role vis-à-vis gene therapy and the role of future committees vis-à-vis emerging biomedical technologies remain to be resolved. Among these questions, some of the most important are the following:

- What should be the oversight body's relationship to the major federal funding agency for biomedical research—NIH?
- How should the oversight body relate to the major federal regulatory agency for the approval and licensing of new biotechnology products—FDA?
- What should be the relationship of the oversight body to the major federal regulatory office for the protection of human subjects—now called the Office for Human Research Protections?
- Should different guidelines apply to human-subjects research conducted with public support and research conducted with private funds?
- What would be the best mechanism for independent safety monitoring of ongoing clinical trials in important new fields of biomedical research?
- And how can the consent process be structured, and consent forms be written, in a way that discloses all pertinent information to prospective subjects, and does so in a setting where they can freely decide whether to become volunteers in the war on disease?

ACKNOWLEDGMENT

I wish to thank my colleague, Linda A. Powell, for helpful editorial assistance.

BIBLIOGRAPHY

1. P. Berg et al., *Science* **185**, 3034 (1974).
2. S. Krimsky, *Genetic Alchemy: The Social History of the Recombinant DNA Controversy*, MIT Press, Cambridge, MA, 1982.
3. D.S. Fredrickson, in K.E. Hanna, ed., *Biomedical Politics*, National Academy Press, Washington, DC, 1991, pp. 258–307.
4. J. Lear, *Recombinant DNA: The Untold Story*, Crown Publishers, New York, 1978.
5. N. Wade, *The Ultimate Experiment: Man-Made Evolution*, Walker and Company, New York, 1977.
6. Office of Recombinant DNA Activities, *Recombinant DNA Research: Documents Relating to "NIH Guidelines for Research Involving Recombinant DNA Molecules,"* vols. 1–20, National Institutes of Health, Bethesda, MD, August 1976–December 1995.
7. S.S. Hall, *Invisible Frontier: The Race to Synthesize a Human Gene*, Atlantic Monthly Press, New York, 1987.
8. N. Wade, *Science* **197**, 1342–1345 (1977).
9. *Fed. Regist.* **45**, 24968–24971 (1980).
10. L. Walters, *Hum. Gene Ther.* **2**, 115–122 (1991).
11. E. Milewski, *Recomb. DNA Techn. Bull.* **8**, 176–180 (1985).
12. L. Walters and J.G. Palmer, *The Ethics of Human Gene Therapy*, Oxford University Press, New York, 1997.
13. United States, President's Commission for the Study of Ethical Problems in Health Care and Biomedical and Behavioral Research, *Splicing Life*, U.S. Government Printing Office, Washington, DC, November 1982.
14. P. Jacobs, *L. A. Times*, October 8, 1, 26 (1980).
15. L. Thompson, *Correcting the Code: Inventing the Genetic Cure for the Human Body*, Simon & Schuster, New York, 1994.
16. United States, Congress, House, Committee on Science and Technology, Subcommittee on Investigations and Oversight, *Human Genetic Engineering*, 97th Congr., 2nd Sess., November 16–18, 1982.
17. United States, Congress, Office of Technology Assessment, *Human Gene Therapy: A Background Paper*, OTA, Washington, DC, December 1984.
18. J. Lyon and P. Gorner, *Altered Fates: Gene Therapy and the Retooling of Human Life*, Norton, New York, 1995.
19. Minutes of the Human Gene Therapy Subcommittee Meeting, July 29–30, 1991, *Recomb. DNA Res.* **14**, 907, 910 (1994).
20. Minutes of the Human Gene Therapy Subcommittee Meeting, November 21–22, 1991, *Recomb. DNA Res.* **15**, 289–293 (1994).
21. Minutes of the RAC Meeting, June 1–2, 1992, *Recomb. DNA Res.* **15**, 707–708 (1994).
22. Minutes of the RAC Meeting, December 3–4, 1992, *Recomb. DNA Res.* **16**, 582–585 (1994).
23. G. Ross et al., *Hum. Gene Ther.* **7**, 1781–1790 (1996).
24. Center for Biologics Evaluation and Research, *Points to Consider in Human Somatic Cell Therapy and Gene Therapy*, FDA, Rockville, MD, 1991.
25. D.A. Kessler et al., *N. Engl. J. Med.* **329**, 1169–1173 (1993).
26. Ad Hoc Review Committee for the Recombinant DNA Advisory Committee (Inder Verma, Chair), "Executive Summary of Findings and Recommendations," September 8, 1995, Unpublished document submitted to NIH Director Harold Varmus.
27. E. Marshall, *Science* **267**, 1588 (1995).
28. Office of the Director, National Institutes of Health, "Report and Recommendations of the Panel to Assess the NIH Investment in Research on Gene Therapy," Stuart H. Orkin and Arno G. Motulsky, Co-Chairs, December 7, 1995. Unpublished report.
29. E. Marshall, *Science* **270**, 1751 (1995).

30. E. Marshall, *Science* **272**, 945 (1996).
31. *Fed. Regist.* **61**, 35774 (1996).
32. *Fed. Regist.* **61**, 59725 (1996).
33. *Fed. Regist.* **62**, 7108 (1997).
34. *Fed. Regist.* **62**, 59032 (1997).
35. R. Weiss and D. Nelson, *Washington Post*, September 29, A1, A21 (1999).
36. N. Wade, *N.Y. Times*, September 29, A20 (1999).
37. R. Weiss and D. Nelson, *Washington Post*, December 8, A1, A10 (1999).
38. S.G. Stolberg, *N.Y. Times*, December 9, A22 (1999).
39. R. Weiss and D. Nelson, *Washington Post*, January 22, A1, A12 (2000).
40. S.G. Stolberg, *N.Y. Times*, January 22, A1, A10 (2000).
41. R. Weiss and D. Nelson, *Washington Post*, February 15, A3 (2000).
42. S.G. Stolberg, *N.Y. Times*, February 15, A16 (2000).
43. P. Smaglik, *Nature* **402**, 707 (1999).
44. Minutes of the RAC Meeting, June 8–9, 1995, *Hum. Gene Ther.* **7**, 417–422 (1996).
45. H. Varmus to H. Waxman, Letter dated December 21, 1999, p. 7.

See other FEDERAL POLICY MAKING FOR BIOTECHNOLOGY entries.

FEDERAL REGULATION OF BIOTECHNOLOGY PRODUCTS FOR HUMAN USE, FDA, ORPHAN DRUG ACT

ROBERT A. BOHRER
California Western School of Law
San Diego, California

OUTLINE

The Orphan Drug Act was enacted to provide sufficient additional incentives to spur the development of therapeutics for smaller patient populations (1). One of the key incentives is market exclusivity for orphan drugs. The Federal Drug Administration (FDA) may not approve a second application for the same drug for the same orphan indication for seven years. The incentives available under the Orphan Drug Act have proved attractive to biotechnology companies, since several of the first and most important products of biotechnology, including human growth hormone and erythropoietin, were approved as orphan drugs. Both of these products were the cause of controversy over the interpretation of the "same drug" market exclusivity. FDA's regulations in response to those early problems have failed to eliminate the "same drug" interpretation problem, as can be seen from the most recent biotechnology orphan drug approvals of two forms of interferon-β for multiple sclerosis.

OVERVIEW OF THE ORPHAN DRUG ACT

Orphan Drug Act's Purpose and Incentives

A drug can be an "orphan drug" if it is for a rare disease or condition that affects fewer than 200,000 patients in the United States or for which there is no reasonable expectation of recovering the cost of developing the drug from sales in the United States (§360bb). Ten to 20 million Americans (about 9 percent of the population) suffer from more than 5000 rare diseases and that number will likely increase as the Human Genome Project uncovers more genetic causes of human diseases.

The purpose of the Orphan Drug Act (ODA) is to provide incentives for the research and development of new drugs for these smaller patient populations. ODA is intended to assist drug companies with two of the key drug development constraints, the cost and duration of the FDA approval process and the issue of intellectual property protection. ODA can reduce the development costs of orphan drugs before FDA approval by facilitating and expediting the FDA review process, and it can increase the financial returns from the development of orphan drugs after approval by providing additional market exclusivity. The development costs required to obtain FDA approval are reduced through a number of ODA's provisions. FDA provides help to pharmaceutical companies regarding the FDA's drug approval process, and FDA advice on the planning of clinical trials may be of particular value to small biotechnology companies with little prior experience in drug development. The Internal Revenue Code (2) provides tax breaks for expenses related to orphan drug development, and the FDA may help fund the clinical testing necessary for approval of an orphan drug (§360ee). In addition the Orphan Products Board coordinates the federal agencies involved in drug research and regulation. After approval, the intellectual property protection provisions of ODA (§360cc) provide a seven-year term of exclusive marketing rights for the drug in the orphan disease population to increase the financial returns for the orphan drug sponsor.

For manufacturers the market exclusivity provision is likely to be the most important incentive offered by ODA because of the potential for very large profits during the period of exclusivity. ODA market protection is narrow, since only the use of that drug for treating the designated rare disease is protected. However, if the drug is not approved for any other medical indication, then orphan drug exclusivity is intended to be more or less as effective as patent protection. A second pharmaceutical manufacturer may seek FDA approval of a different drug for the same disease (or the same orphan drug for different orphan diseases or non-orphan diseases), but the sponsor of a subsequent drug for the same disease bears the burden of proof to demonstrate that its drug is different.

In placing the burden of distinguishing its drug on the second orphan drug sponsor, ODA parallels patent law, an older and more comprehensive body of law for providing exclusive rights as an incentive for innovation. Every patent applicant must disclose all prior relevant inventions and public knowledge ("prior art") to the Patent Office and explain how her invention differs from the prior art to a significant degree (the difference, to be significant, must be nonobvious) (3). The simple premise for market exclusivity, whether under ODA or through a patent, is that awarding a monopoly to an innovative product is generally economically justified (despite the monopoly output restrictions and the correspondingly higher prices) when the investment in innovation would be unlikely without market protection. Indeed, when enacting the ODA, Congress determined that the pharmaceutical industry needed these special economic incentives to undertake research and development for diseases affecting fewer people.

Procedures of the Orphan Drug Act

Qualification for orphan drug benefits is a two-step process: (1) designation and (2) drug approval. After drug approval the pharmaceutical manufacturer obtains a product license to sell the drug with ODA's seven-year market exclusivity.

Designation. A pharmaceutical manufacturer (sponsor) seeks orphan status designation for a drug by (1) certifying that the product is for a rare condition, (2) providing a scientific rationale for using the drug for that rare condition, and (3) providing supporting epidemiological data. The designation process gives the sponsor an early opportunity to interact with the FDA and learn of any significant issues that might arise later in the course of development and approval. The required scientific rationale for the orphan drug's usefulness need only be an explanation of the sponsor's hypothesis and some experimental evidence from animal model or laboratory studies. However, a sponsor may seek orphan drug designation at any point in the research and development process, even after the final stages of human clinical

testing, before submitting an application for marketing approval.

ODA does not limit the number of drugs that may be designated for a particular rare disease. If a first orphan drug has obtained market approval, however, then FDA must not approve an application for designation by a second sponsor of the same drug until seven years have passed. FDA can grant orphan drug designation status to new versions of an already marketed drug, but the second, similar drug will not be approved unless FDA determines that the second applicant's is clinically superior to the already marketed drug.

Approval. While any sponsor of an orphan drug may receive the development-phase benefits of ODA, only the first manufacturer to receive full FDA drug approval receives the exclusive marketing rights for any one drug. Although the FDA is liberal in awarding orphan drug designation, the standard for approval is consistently high.

The FDA drug approval process consists of preclinical studies supporting the safety and possible efficacy of the drug in animals followed by three phases of clinical investigation. The pharmaceutical manufacturer must submit extensive scientific and medical data including chemical, pharmacological, and clinical studies. The marketing approval applications for both drugs and biologics must show that the products are "safe and effective" for their intended use.

Two major statutory exceptions to the seven-year orphan drug market exclusivity are (1) where the market exclusivity holder cannot provide sufficient quantities of the drug to patients who need it and (2) where the market exclusivity holder consents to subsequent approvals. The FDA must, before making a finding of nonavailability, give the market exclusivity holder notice and the opportunity to comment. Orphan drugs have generally spent less time in development than non-orphan drugs. This may be because an orphan drug application usually involves fewer patients and fewer clinical trials (which is inevitable for diseases with fewer sufferers). Orphan drugs are often the only available treatment for the rare disease or condition, which is why orphan drug protection is necessary to make the research and development commercially viable. Furthermore an orphan drug sponsor can seek FDA approval to allow patients access to the drug even before marketing approval, either through a Treatment IND (4) or an orphan drug open protocol (§360dd).

Orphan Drug Market Exclusivity in Biotechnology Business Strategy

One of the greatest challenges for the emerging biotechnology company and its legal counsel is to integrate its intellectual property and regulatory strategies with its financial plan. After discovering a new compound that may have beneficial medical properties, a company typically applies for a patent, since patent protection precludes other companies from selling or obtaining the patent on the compound. However, there are many compounds that have entered into the public domain by sitting on the shelf for a number of years after a first attempt to prove it effective against a particular disease. AZT was originally tried against cancer and abandoned for that use. AZT was later given orphan drug exclusivity because at the time of its designation less than 200,000 persons in the United States were identified as having progressed from (HIV) positive status to (AIDS). It has brought its sponsor billions of dollars in revenues as an AIDS drug.

Congress had expected that ODA would be primarily used by sponsors of orphan drugs that did not qualify for product, or composition of matter, patents. Indeed, one of the most powerful and widely used applications of biotechnology is to enable the production of commercially viable quantities of otherwise rare or very difficult to produce compounds, the prior knowledge of which may preclude composition of matter patent protection. Recombinant Human Growth Hormone, the subject of the first ODA dispute, was precisely such a drug. However, orphan drug exclusivity can still be of significant value even where a compound is also the subject of a patent. The statutory patent term of 20 years from the filing date of the patent application is running during the more than seven years it takes to obtain perform the required preclinical and clinical tests required to obtain FDA approval for the average new drug application. The Drug Price Competition and Patent Term Restoration Act of 1984 (5) can partially restore a portion of the years spent in clinical development. However, the patent term remaining after FDA approval can be less than seven years, raising the value of the orphan drug period of market exclusivity, which does not begin until FDA approval.

ODA thus provides incentives for biotechnology companies in the intersecting concerns of intellectual property protection and rapid drug approval. In a biopharmaceutical development strategy, it is important to get into the marketplace sooner, with a longer patent term remaining, even to the extent of influencing, if not dictating, the choice of initial target indication for a biotechnology company's lead compound. As long as the time period necessary for gaining product approval effectively consumes a substantial portion of a drug's patent-life or biotechnology companies use genetic engineering to produce recombinant versions of known proteins, ODA will continue to be a central part of many biotechnology companies product development strategy.

Effect of the Orphan Drug Act on Drug Innovation

Pharmaceutical manufacturers have long argued that ODA procedures do not provide sufficient certainty to guide their research and development investments. One source of such uncertainty is the potential of competition among sponsors of the same or similar orphan drug. Although two sponsors of the same drug may receive orphan drug designation during the development phase, the seven-year marketing exclusivity is awarded solely to the first company to achieve market approval for the drug. Thus a company may be the first to conceive of an orphan drug program, such as interferon-β in the treatment for multiple sclerosis, invest resources into preclinical and clinical research, and then lose that investment if another company is the first to receive drug approval for that orphan drug. Applications for orphan designation, which are usually made when clinical trials are ready to begin,

are kept secret until the FDA approves the designation and because companies need not apply for designation until late in the development process, a company may not realize that it is in a race with another company for orphan drug approval until it has completed its preclinical development and readied its lead compound for human trials. At that point a company may well feel pressure from its investors to reach the next milestone, with the competition continuing and the stakes growing ever higher. The initial uncertainty and the potential for an expensive, winner-take-all race can substantially undermine the incentives of the orphan drug.

Although such uncertainty undermines the goal of encouraging innovation, the uncertainty of a race to develop a product may be unavoidable. Of course, the risk of being beaten to the marketplace is quite different than the risk of being first to market only to find that winning the race provided no real victory at all. That may result when what were initially believed to be minor differences between two versions of a drug result in the orphan drug approval of both (as was the case with different forms of interferon-β, i.e., Avonex and Betaseron) for multiple sclerosis. The ODA has been successful in bring many orphan drugs to market, with more new drugs undergoing research and development despite the uncertainty of the scope of its market exclusivity. Nevertheless, the new issues raised by the FDA's recent decision to approve Avonex without closely examining its degree of identity to Betaseron may substantially impair the future effectiveness of ODA.

Small biotechnology companies have developed most orphan drugs. For small biotechnology companies, demonstrating to investors the ability to successfully develop any product is a significant corporate milestone. Also diseases affecting 200,000 Americans are "rare" under the law but may be sizable, even hugely profitable, markets for small companies. The orphan drug designation, based on U.S. disease populations, does not account for the additional potential profits on international sales, a key target for both large pharmaceutical companies and biotechnology companies. Furthermore drugs developed for a rare disorder may also work on more common diseases. For example, the biotechnology company Genentech originally developed human growth hormone (hGH) to treat children with hypopituitary dwarfism, but hGH is also useful in treating other growth deficiencies "off-label." Such extremely successful orphans, like human growth hormone and erythropoietin (EPO), are "blockbuster drugs," and both were the subject of major battles over the scope of ODA protection.

Critics of ODA, spurred by such instances of enormously profitable orphan drugs, are concerned that market exclusivity leads to higher prices that limit access to the drug. Critics also assert that profitable orphan drugs violate the spirit behind ODA, apparently believing that only minimal commercial incentives are justified. These criticisms led, in October 1990, to an amendment to ODA, passed by both houses of Congress, to permit simultaneous licensing of the same orphan drug for the same rare disease under some circumstances. President George Bush "pocket-vetoed" the amendment, however, because he believed that the bill would weaken "the marketing incentives provided by the Act." It would be difficult to legislatively fine-tune the financial returns available under ODA to provide just enough to encourage development without going so low as to negatively impact the number of future orphan drugs. Unless the important unmet needs of orphan disease suffers can be met in some other way, the potential lucrative incentives of the ODA will continue to be essential to providing new therapies for those patients.

SCIENTIFIC CONTEXT OF THE PROBLEM OF ORPHAN DRUG MARKET EXCLUSIVITY

Since the 1985 to 1986 battle over two versions of recombinant Human Growth Hormone, FDA has been repeatedly confronted with a continuing problem with ODA, the "same or different" problem. If FDA considers the two structurally very similar drug variants to be "different," then FDA approves both drugs. When "same or different" is thus narrowly construed and an orphan drug's market exclusivity thereby narrowed, the incentives to develop such orphan drugs are substantially diminished by the risk that a second manufacturer could enter the market with a similar, almost copycat, variation. This is especially problematic for biotechnology companies, because while the research and development costs to bring a drug to market are high, it is relatively easy to make minor changes in genes and proteins that are unlikely to have pharmacological significance. On the other hand, some seemingly minor sequence changes may result in significant differences in protein activity.

The problem for FDA in defining sameness under ODA is to distinguish between significant and insignificant changes in a way that provides as much guidance and clarity as possible for companies considering the development of orphan drugs, while preserving the incentives to develop orphan drugs. A very clear, very narrow definition would substantially reduce the potential profit of an orphan drug, while a very clear, very broad definition might well deprive patients of the benefit of improved versions of a drug. Prior to the development of recombinant proteins by the biotechnology industry, when smaller, simpler chemical structures provided the basis for most drugs, two drugs were considered the same if they have the same active moiety. This strict structural approach worked well because, for small molecules produced by chemical syntheses, even slight changes in the chemical structure of the active moiety are reasonably likely to result in significant pharmacological differences. For large biological molecules, however, slight structural variations often do not result in pharmacological differences. With the growth of the biotechnology industry, an increasingly large percentage of designated orphan drugs are proteins or other large biological molecules.

All of the biotechnology orphan drugs have been proteins and glycoproteins. Proteins consist of strings of up to several hundred amino acids. It is usually possible to substitute similar amino acids for other amino acids in the protein without any noticeable change in biological activity. The problem of insignificant variation is even

greater for glycoproteins. Glycoproteins are proteins with attached carbohydrate (sugars or saccharides) groups. The saccharide portion is attached to the protein enzymatically after the amino acid primary structure is produced, and depends on the particular glycosylation enzymes of the producing cell, rather than being genetically determined like the primary amino acid structure of the protein. Thus different species of animals add different saccharides to the same primary-structure protein. Even different cell lines of the same species and different cells in the same body can add different saccharides to the same primary-structure protein. Thus, when recombinant proteins are made by inserting the same gene sequence in different organisms, such as *Escherichia coli*, yeast, and Chinese hamster ovary (CHO) cells, the resulting proteins will certainly be glycosylated differently (*E. coli* and other bacteria do not glycosylate at all) and may nevertheless have the same biological activity. Whether or not such differences in glycosylation will result in differences in biological activity is not currently predictable.

For a company pursuing the very costly development of a gene-based, protein-based, or glycoprotein-based drug, the problem of predicting intellectual property protection against variant competitor drugs is a major concern.

HISTORY OF "SAME DRUG" UNDER THE ORPHAN DRUG ACT

Human Growth Hormone

Almost from the start ODA was plagued by controversy over whether two competing products were the same or were sufficiently different that the second product could be approved for the same indication. The ODA definition of a "different" drug was the issue in litigation over the orphan drug designation for human growth hormone (6). Human growth hormone is a protein secreted by the human pituitary gland that can strongly affect the growth rate and adult height of children. Administration of this hormone increases growth in children with hypopituitary dwarfism, for which indication Genentech received FDA approval. The protein has been known for some time, and was originally purified from the pituitary glands of human cadavers, a source that ultimately proved to be prone to serious contamination.

Genentech developed, with the assistance of orphan designation funding, a genetically engineered human growth hormone, and was granted market exclusivity for this hormone (Protropin). Patent protection was unavailable on the product itself because the natural protein was previously known. Eight months later FDA granted Eli Lilly & Co. orphan drug market exclusivity for a human growth hormone that differed by only a single amino acid. Genentech's Protropin had an additional amino acid that Lilly's human growth hormone (Humatrope) lacked. From a medical and clinical standpoint, there was no difference in safety and efficacy between Genentech's Protropin and Lilly's Humatrope.

Genentech first filed a citizen petition with FDA claiming that Lilly's drug was, for the purposes of ODA, the same as Protropin and therefore ineligible for marketing approval. When FDA granted Lilly approval and market exclusivity for Humatrope, Genentech went to court to ask for judicial review of FDA's action.

The U.S. District Court for the District of Columbia held that the two human growth hormones were not the "same" drug under ODA, but also explicitly declined to provide any "universal rule for determining whether two drugs are different.... That responsibility is statutorily imposed on the FDA. Until the FDA endeavors to meet that obligation, the courts will be forced to make case-by-case determinations based on the broad polices embodied in the Act." (6, p. 306).

Erythropoietin

The protein erythropoietin was also the subject of a "same or different" orphan drug controversy, in an astronomically high-stakes race to clone the gene for the human hormone erythropoietin. Erythropoietin is a glycoprotein that stimulates red blood cell production, which is useful in the treatment of anemia. While the protein was known, no method of making commercially practicable quantities was available prior to the application of genetic engineering. Erythropoietin remains the key product in Amgen's success as a biotechnology company, accounting for over a billion dollars a year in annual revenues.

In 1989 Amgen received orphan drug marketing exclusivity for its erythropoietin product Epogen, for the treatment of the chronic anemia associated with end-stage renal disease, a "rare disease." Seeking access to the market, Chugai and its marketing partner Genetics Institute tried to obtain orphan drug status for their erythropoeitin product, Marogen, arguing that because the Marogen glycosylation pattern differed from Amgen's Epogen, Marogen should not be blocked from the market. Eventually FDA denied the Chugai/Genetics Institute application. The battle over erythropoietin continued, however, until Chugai and Genetics Institute lost a patent dispute with Amgen. Amgen gained exclusive control of the protein by virtue of their patent on the gene sequence used to produce the protein (7).

FDA Orphan Drug Regulations

The problems for orphan drug developers raised by the growth hormone and erythropoetin cases caused FDA to issue regulations under ODA to refine the definition of "same" drug by defining "different." FDA orphan drug regulations tried "to ensure both that improved therapies will always be marketable and that orphan drug exclusivity does not preclude significant improvements in treating rare diseases" (8). To be considered different under ODA, small molecule drugs cannot have the same "active moiety." Large biological drugs cannot contain the "same principal molecular structure" (9). FDA orphan drug regulations provide a "presumption of sameness" even when differences occur in protein structure. FDA believed this broader protection was consistent with congressional intent.

When it proposed its ODA regulations, FDA also considered a definition of same that would have held similar macromolecules to be the same unless the structural differences could be reasonably expected to have

pharmacological significance. However, FDA realized that the regulations did not and could not specify the kinds of structural differences likely to be related to differences in pharmacological activity and therefore the regulation would create a new area of uncertainty that might also have undermined the incentives of ODA.

The final regulation, adopted in 1993, clearly rejected the notion that a macromolecule's chemical structure should be the ultimate determinant of whether or not a second orphan drug application was for the same drug as a prior approved drug. Rather, FDA decided that a second sponsor should always be able to overcome the presumption of sameness for macromolecules with the same principal structure by demonstrating significant clinical differences. "With regard to macromolecular drugs, clinical superiority by itself will render a subsequent drug different." This "clinical differences" standard was based on the principle that the market exclusivity should not create a barrier to needed patient therapies. Thus clinical superiority by itself will always lead to approval of a subsequent drug despite its substantial similarity to a prior approved orphan drug. In its ODA regulations, FDA essentially changed the focus from the underlying question of same and different to the question of under what circumstances the FDA will determine that a second drug is clinically superior to the first.

The FDA regulations define a "clinically superior" drug as one that "is shown to provide a significant therapeutic advantage over and above that provided by an approved orphan drug." Therapeutic advantage, or clinical superiority, can be show one of three ways: (1) greater effectiveness, (2) greater safety, or (3) in "unusual cases," demonstration that the drug makes a major contribution to patient care (10). To demonstrate greater effectiveness, the same kind of evidence is needed as that generally required to support a comparative effectiveness claim for two different drugs, that is, an improvement as assessed by the drug's "effect on a clinically meaningful endpoint in adequate and well controlled clinical trials." To support a claim of superior safety, the company seeking approval of the second product must establish that its product provides "greater safety in a substantial portion of the target populations, for example, by the elimination of an ingredient or contaminant that is associated with relatively frequent adverse effects." FDA interpretation is that even "a small demonstrated ... diminution in adverse reactions may be sufficient to allow a finding of clinical superiority" (11). Finally, a second drug can be considered "clinically superior" if it makes some other "major contribution to patient care." FDA intends this to be "a narrow category." such as the development of an oral dosage form for which there had been only a parental form (12).

The FDA's decision to rely on clinical superiority as the ultimate determinant of whether a second orphan drug was different than a prior, similar drug, left the FDA with considerable discretion to decide when head-to-head comparisons of the two drugs would be required to prove clinical superiority. One comment in response to FDA orphan drug regulations suggested that as proof of clinical superiority, FDA should always require a demonstration in head-to-head, double-blind studies, just as for comparative efficacy claims. FDA acknowledged the value of head-to-head studies but chose nevertheless to retain discretion to make a finding of clinical superiority on the basis of other well-designed studies that it finds demonstrate a benefit on a clinically significant endpoint.

To summarize the FDA's regulations on the same and different problem, the general principal is that structurally similar drugs will be considered the same unless the second drug is shown to be clinically superior to the first. Where the issue is efficacy, the situation is quite clear: The only way to show that drug B is more effective than drug A is to directly compare their performance on an important efficacy endpoint. Where the difference is safety, or adverse effects, or "contribution to patient care," the guidelines become significantly less clear, as became obvious in the most recent case involving the variant forms of Interferon-β. If the safety of two drugs is judged on a case-by-case basis, without head-to-head data (because the second drug has a different dosage schedule or route of administration), a second drug sponsor can quite literally go to school on the first drug's data in an effort to produce fewer adverse effects.

The first sponsor of a drug for a difficult clinical indication such as multiple sclerosis would ordinarily design clinical trials to demonstrate safety and efficacy at the maximum tolerated dose. The general course of development for new compounds follows that pattern because, in most cases, a higher dose is more likely to have a statistically significant impact on the primary clinical endpoints. Following approval, clinicians often find that lower doses are effective in actual practice (13). However, a second sponsor would then have every incentive to do its studies at a lower dose or different dosing regimen, in an effort to look for efficacy under conditions that decreased adverse reactions. In defining the Orphan Drug Act's "such drug" protection in terms of clinical superiority, the FDA shifted the focus from molecular structure to clinical significance without substantially reducing the real uncertainty that Orphan Drug sponsors face.

BETASERON AND AVONEX: THE ORPHAN DRUG ACT'S LATEST CONTROVERSY

The FDA's decision that interferon-β produced in CHO cells and interferon-β produced in *E. coli* are sufficiently different to approve both for the treatment of multiple sclerosis once again raises substantial questions about ODA's market exclusivity incentive. Berlex Laboratories, Inc. had received FDA approval to market an *E. coli*-produced drug, interferon-β-1b, trade-named Betaseron, as an orphan drug for the treatment of relapsing-remitting multiple sclerosis. Berlex felt that FDA clearly erred when it decided that Biogen's drug, interferon-β-1a, trade-named Avonex, was sufficiently different to receive approval for the same patient population. Accordingly Berlex filed a suit against FDA seeking to have its determination reversed. On October 7, 1996, the U.S. District Court for the District of Columbia dismissed the lawsuit, (*Berlex Laboratories, Inc. v. FDA*), (14). The court held FDA acted lawfully when it determined that Avonex

was "clinically superior" to Berlex's interferon-β-1b drug Betaseron and that "FDA's determination that Avonex is safe, pure and potent is amply supported by the record."

The question raised by the controversy over variant forms of interferon-β is not the correctness of the District Court's decision to uphold FDA. The District Court's opinion was properly based on considerations of administrative law, particularly the issue of the proper relationship between a court and an agency on a technical matter within the primary competence of the administrative agency. The Court did not attempt to decide whether or not Avonex and Betaseron are sufficiently different that Avonex is entitled to be a second entrant into the market for relapsing-remitting multiple sclerosis. Rather, the Court reviewed the regularity and sufficiency of FDA's administrative process. As a decision about administrative law and judicial deference to agency expertise, the decision in Berlex Labs is unexceptionable. What is needed is an analysis of whether FDA's regulations, in light of the interferon-β controversy, need further clarification to avoid negatively affecting the development of future orphan drugs.

Interferon-β provides an important context for examining FDA's interpretation of ODA because it is in fact a paradigmatic example of the Act's real-world effects. The potential therapeutic benefits of interferon-β had spurred both companies' efforts to produce human interferon-β by recombinant DNA procedures. While the protein is also being fought over in complex patent litigation, the ODA incentives were very likely to have been significant in the decision of the competing sponsors to attempt to prove its value in the treatment of multiple sclerosis (MS). The competing drugs each sell for about $7000 for an annual supply; both are taken by injection. Betaseron is injected under the skin every second day at a dose of 250 micrograms per injection, while Avonex is injected into muscle once a week at a dose of 30 micrograms per injection, a difference that may have played a key role in FDA's approval.

Betaseron is produced from a human interferon-β-1b gene that has been cloned and expressed in the bacterium *E. coli*. The gene and protein sequence of Betaseron varies from that of the natural molecule by one codon and its corresponding amino acid, a difference that is related to the bacterial expression of proteins. Because it is produced in bacteria, Betaseron is not glycosylated but does have an antiviral activity similar to that of native human interferon-β, thus indicating that glycosylation is probably not essential for full biological activity.

Because FDA's approval of Betaseron was based on data from Berlex' clinical trial involving patients with relapsing-remitting MS, the results pertain only to the relapsing-remitting patient group. The trial showed that injection of Betaseron every other day under the skin (subcutaneously) decreased the frequency of flare-ups and kept more patients free of flare-ups over a two-year treatment period. Adverse reactions to Betaseron included inflammation and pain at the injection site and flu-like symptoms.

About 1991 Biogen began to manufacture Avonex. On May 17, 1996, FDA approved Avonex for the treatment of active relapsing forms of MS to slow the deterioration of physical ability and decrease the frequency of attacks. The definition of "active relapsing" MS included patients with both relapsing-remitting and relapsing-progressive forms of the disease, a more diverse population than the one studied by Berlex and approved for Betaseron.

When approving the licensing of Avonex, FDA relied on a randomized, double-blind, multicenter trial of active relapsing MS patients. In that trial, patients receiving a weekly injection of Avonex into their muscles (not subcutaneously) had a 37 percent reduction in the risk of clinically significant disability progression within the period of the study, compared with patients who received placebo. Furthermore 32 percent of placebo-treated patients had three or more exacerbations over the course of two years, compared with only 14 percent of Avonex-treated patients. Patients in the Avonex group also had a statistically significant reduction in active brain lesions seen on magnetic resonance imaging scans. The overall Avonex treatment was well-tolerated. Only 9 percent of the patients receiving the drug stopped treatment, half of them due to side effects (flu-like symptoms, muscle aches, fever, chills, and asthenia). Injection-site reactions associated with Avonex treatment occurred in only 4 percent of patients, not significantly different from patients in the placebo group. There were no reports of tissue death at the injection site, possibly because of the much lower dose, the much less frequent injections, and the intramuscular, rather than subcutaneous, injection route.

FDA's decision was based on its ODA regulations, providing that a new drug will not be considered the same as the previously approved drug if the new drug is "clinically superior" because it offers "greater safety in a substantial portion of the target populations" (15). Based upon these clinical results of the trial results, the FDA concluded that Avonex was clinically superior to Betaseron and therefore sufficiently different for ODA approval. The FDA decision rested on the substantially less frequent occurrece of injection site necrosis associated with Avonex (0 percent) compared with Betaseron (5 percent) and lower incidence of even lesser injection site reactions (4 percent of Avonex compared to 85 percent of Betaseron patients). FDA's decision was apparently based solely on the clinical data in support of the two manufacturer's applications. FDA did not attempt to determine whether the difference in the adverse effects of the two drugs was due to the dosing and administration differences or to the actual biochemical properties of the two drugs. Nor did FDA determine whether Avonex was more effective than Betaseron in treating the underlying disease.

SAME AND DIFFERENT FOR DRUGS IN OTHER FDA DETERMINATIONS

FDA Guidance Document Concerning Demonstration of Comparability of Human Biological Products

Just weeks before FDA approved Avonex, it published a Guidance Document regarding changes in the manufacturing processes of "well-characterized" biological drugs (16).

FDA issued this Guidance Document to provide pharmaceutical manufacturers with more flexibility in bringing biological products to market more efficiently and expeditiously. Until that time, companies developing biotechnology drugs such as recombinant proteins or monoclonal antibodies faced a considerably more complicated regulatory pathway through FDA's Center for Biologics (CBER), than did companies developing small molecules through FDA's Center for Drugs (CDER).

Prior to the Guidance Document, CBER's approval of a biologic required two separate applications and approvals: (1) the approval of the product as safe and effective, through the submission of a product license application (PLA) and (2) the submission of an establishment license application (ELA) for approval of the manufacturing facility that produced the actual material that was used to generate the data in the PLA. The reason for the two-part, interrelated approval process for biologics was that the manufacture of biological molecules was such a highly variable process that a change from one facility to another could produce a change in the product itself. In addition the manufacturer was required to verify the product's identity for each lot produced. The FDA refused to approve applications unless clinical trials were conducted with the specific product to be licensed because any change in the manufacturing process could mean that the clinical data was in fact generated by different macromolecules and could not support the marketing approval for just one variant, or worse, an even newer and untested variant.

The Guidance Document was part of a greater policy to harmonize the requirements across the FDA for pharmaceutical manufacturers to produce "well-characterized" biotechnology drugs. The policy of more flexibility is possible because recent advances have provided the scientific ability to control the manufacture of biotechnology drugs and to determine the consistency of the identity of macromolecules produced by different processes. The Guidance Document thus recognizes this increasing technical ability of manufacturers to show that the protein produced in one facility and that was used in clinical trials is the same as the protein produced in another facility.

In allowing the clinical data for a "precursor" product be used for a later product, the FDA is may rely on the results of analytical testing, biological assays (in vitro or in vivo), assessment of pharmacokinetics and/or pharmacodynamics and toxicity in animals, in place of clinical testing, to determine whether or not the later product is the same as a prior composition. In other words, the FDA now feels that the science of genetic engineering of proteins has advanced to the point where a variety of tests can determine whether a change in manufacturing a protein yields the same protein or a different one for product identity purposes. Ironically, although the FDA relied on such principles to determine that Avonex could use clinical data generated by a prior compound, it ignored them completely in making its ODA determination that Avonex and Betaseron were different drugs based on its finding of clinical superiority.

When it proposed the FDA orphan drug regulations, FDA was concerned about determinations that involved too much judgment and discretion from FDA officials. At that time FDA rejected any approach where the kinds of structural differences likely to be related to differences in pharmacological activity were not specified. With the Guidance Document now in place, FDA officials would appear to have more of the tools required to make judgments about when structural differences are likely to be related to differences in pharmacological activity for purposes of deciding the similarity of well-characterized biological drugs. FDA officials also have the discretion to decide whether further clinical trials are required. The Guidance Document provides a major science policy basis for a further clarification of ODA's "same or different" problem.

Generic Drugs

FDA has had a great deal of experience in determining when two drugs are identical, because that function is required during the approval of generic versions of drugs for which patent protection is expiring. Manufacturers of the generic versions need only submit abbreviated new drug approvals containing bioavailability and bioequivalence data. The process of generic drug approval is much simpler and less costly than the approval of new drugs. A manufacturer of an orphan drug that is also protected by patent exclusivity is unlikely to open itself to generic competition by seeking approval for a non-orphan indication. Instead, the manufacturer can build off-label usage for its orphan drug by distributing journal articles that support the additional uses. If the non-orphan indication were instead added to the label, a generic competitor could quickly gain approval for the non-orphan indication by submitting bioavailability and bioequivalence studies and, after approval, sell the cheaper generic off-label into the orphan market.

A generic version of a drug is chemically identical to the original pioneer drug and therefore can rely on the safety and efficacy studies done for the pioneer drug, submitting only manufacturing data and bioavailability and bioequivalence data to establish chemical identity (5). The tests for bioavailability and bioequivalence measure important aspects of the pharmacokinetics of a drug, the overall way in which a drug is processed by the body. If two drugs that are very similar in structure have nearly identical pharmacokinetics, one could presume that whatever differences in structure exist are not pharmacologically important or clinically relevant. That is because such important pharmacologically or clinically relevant differences, for example, in the two drugs' affinity for their target molecule (ligand), or their relative antigenicity, would cause marked differences in pharmacokinetics for most, if not all drugs.

CONCLUSION

The scope of orphan drug exclusivity for biotechnology drugs remains uncertain. In patent law the inquiry as to whether a second item that is more or less closely related to an earlier invention, is entitled to its own protection or infringes the earlier patentees rights, takes place under the doctrine of equivalence, the doctrine of reverse equivalents, and the doctrine of unexpected results. Each of these doctrines addresses a common, core concern: Does the second item make a substantial and independent contribution, or is it merely an attempt to profit from the exploitation of the earlier inventor's contribution? ODA, enacted in 1983, created an analogous problem in prohibiting FDA from approving the same drug for the same indication for a period of seven years. The obviously analogous problem for the administration of ODA is when a second, similar drug is the same and therefore barred from FDA approval and when a second, similar drug is sufficiently different to be entitled to its own market exclusivity.

Soon after the enactment of ODA, in the absence of any administrative interpretation of the scope of the "such drug" protection, the courts applied a much narrower interpretation to ODA protection than would have been the case under patent law. The damage that such narrow readings would do to the incentives of the Act was obvious. FDA ultimately responded with regulations that defined the scope of ODA exclusivity in a way that is much closer to the way that patent law resolves the issue. When FDA concluded that the ultimate basis for determining whether a second, similar drug could be approved would rest on its demonstration of clinical superiority over the prior drug, the agency was apparently searching for a test that essentially measured whether the second applicant made a substantial and independent contribution to the treatment of the orphan disease, or whether it simply sought to profit from the same work as the first applicant.

FDA's reliance on clinical superiority serves two purposes. One purpose is clearly furthering the interests of the patients in receiving the benefits of any significant advance in therapy. At the same time it serves a purpose much like that of the doctrine of unexpected results or the reverse doctrine of equivalents in patent law. Clinical superiority is intended by FDA to be sufficient evidence to support a determination that the difference between the two drugs is in fact significant.

Unfortunately, the Berlex Labs case reveals the shortcomings in FDA's attempt to solve the "same or different" drug problem by looking to clinical superiority. If clinical superiority can be demonstrated by the frequency of adverse effects where direct, head-to-head comparisons are not possible, it is difficult to know whether the second drug's "superiority" is due to the substantial independent contribution of the second applicant or the advantage of learning from the first applicant's data. At the same time the FDA has clearly indicated that it would not favor a policy that required head-to-head comparisons for all structurally similar second drugs. What remains is the question of when head-to-head comparisons are most necessary and whether the cases in which head-to-head comparisons should be required can be sufficiently clarified by further guidelines.

ACKNOWLEDGMENT

This entry is based in part of the article by R.A. Bohrer and J. Prince, *Harv. J. Law Technol.* **12**, 365 (1999).

BIBLIOGRAPHY

1. Orphan Drug Act, Pub. L. No. 97-414, 96 Stat. 2049 (1982), codified as amended at 21 U.S.C. §§360aa–360ee (1994), 26 U.S.C. §45C (Supp. II 1994), 42 U.S.C. §236 (1994).

2. 26 U.S.C. §28.

3. 35 U.S.C. §103.

4. 21 CFR §§312.80–88.

5. Drug Price Competition and Patent Term Restoration Act, codified at 35 U.S.C. §§156 (c)(2), 156 (d)(5)(F)(i), 21 CFR §314.55.

6. *Genentech, Inc. v. Bowen*, 676 F Supp. §§301, 306 (D.D.C. 1987).

7. *Amgen, Inc. v. Chugai Pharmaceutical Co.*, 927 F.2d 1200, 18 USPQ2d 1016 (1991) ("Amgen II") cert. den., *Genetics Inst. v. Amgen, Inc.*, 502 U.S. 856 (1991).

8. 56 Fed. Reg. 3338.

9. DHHS, Food and Drug Administration Agency, 21 CFR §316.3, Orphan Drug Reg., No. 85N-0483, RIN 0905-AB55, 57 FR 62076 (December 29, 1992).

10. 21 CFR §316.3(b)3(i–iii).

11. 57 Fed. Reg. 62076 at comment 18.

12. 56 Fed. Reg. 3338–3342, 3343.

13. *N.Y. Times*, October 12, F1 (1999).

14. *Berlex Laboratories, Inc. v. FDA*, 942 F. Supp. 19 (D.D.C. 1996).

15. 21 CFR §316.3(b)(ii).

16. FDA, *Guidance Concerning Demonstration of Comparability of Human Biological Products, Including Therapeutic Biotechnology-Derived Products*, U.S. Government Printing Office, Washington, DC, April 1996, Available at: *http://www.fda.gov/cber/guidelines.htm* and *http://www.fda.gov/cber/gdlns/comptest.txt*

(a) Bioavailability means the rate and extent to which the active ingredient or active moiety is absorbed from a drug product and becomes available at the site of action.

(e) Bioequivalence means the absence of a significant difference in the rate and extent to which the active ingredient or active moiety in pharmaceutical equivalents or pharmaceutical alternatives becomes available at the site of drug action when administered at the same molar dose under similar conditions in an appropriately designed study.

ADDITIONAL READINGS

R.A. Bohrer, *Symposium issue, intellectual property and the FDA*: Forward: *Biotechnology Business Strategy: A Lawyer's Perspective*, 33 CAL. W.L. REV. 1, 6 (1996).

Brief of Amicus Curiae Chiron Corporation in Support of Respondent, Warner-Jenkinson Co., Inc. v. Hilton Davis Chemical Co., 117 S. Ct. 1040 (1997).

G.M. Cavalier, *Law Policy Int. Business* **27**(2), 447, 450 (1996).

P. Ducor, *Rutgers Comput. Tech. Law. J.* **22**, 369, 473 (1996).

V. Henry, *J. Legal Med.* **14**, 617 (1993).

Li-Hsien Rin-Laures and D. Janofsky, Note, *Harv. J. Law Technol.* **4**, 269, 271–272 (1991).

E. Knight, Jr., *Proc. Natl. Acad. Sci. U.S.A.* **73**(2), 520, 522 (1976).

J.A. Levitt and J.V. Kelsey, *Food Drug Law J.* **48**, 525, 527 (1993).

J.T. O'Reilly, Food and Drug Administration, 2nd ed., §13.22 (1995).

E. Weck, in *From Test Tube to Patient: New Drug Development in the United States: an FDA Consumer Special Report 1988*, PLI Patents, Copyrights, Trademarks & Literary Property Course Handbook Ser. No. G4-3819, 1988.

See other entries FDA REGULATION OF BIOTECHNOLOGY PRODUCTS FOR HUMAN USE; GENE THERAPY, ETHICS, SOMATIC CELL GENE THERAPY; MEDICAL BIOTECHNOLOGY, UNITED STATES POLICIES INFLUENCING ITS DEVELOPMENT; PATENTS AND LICENSING, ETHICS, INTERNATIONAL CONTROVERSIES; SCIENTIFIC RESEARCH, POLICY, TAX TREATMENT OF RESEARCH AND DEVELOPMENT.

G

GENE THERAPY, ETHICS, AND INTERNATIONAL PERSPECTIVES

ANDREA L. BONNICKSEN
Northern Illinois University
DeKalb, Illinois

OUTLINE

INTRODUCTION

An active debate has been ongoing about the ethics of human gene therapy since the early 1970s, when scientists announced innovations in recombinant DNA research. This debate intensified in the 1980s, when fertilization outside the human body brought closer the prospect of genetically manipulating human eggs, spermatozoa, and embryos. In the 1980s various nations set up ethics commissions to weigh ethical issues associated with embryo research, reproductive technologies, and genetic manipulations. From the recommendations of these commissions, officials in the United Kingdom, Germany, and other nations enacted laws that included provisions for gene therapy. Public attention broadened in the 1980s and 1990s, when regional and global international organizations began also to deliberate about the ethical implications of human genetic knowledge and applications. Faced with the need to achieve a consensus that crossed cultures, these organizations crafted more principle-oriented statements than did national governments, which tended to enact laws related to named reproductive techniques.

Gene therapy involves both somatic cell gene therapy (SCGT) and germ-line gene therapy (GLGT). In SCGT genes are modified and used to treat a patient without introducing changes that will be passed to subsequent generations through the germ cells. In germ-line gene therapy, genetic alterations affect the germ cells (oocytes, spermatozoa) or embryos in a way that presumably would be passed to the next generations. Research on somatic cell gene therapy, which is practiced in only a few countries and with mixed success, is generally regarded as ethical if standard safety protections and provisions for informed consent are followed. Because genetic changes are limited to the patient and are not passed to the next generation, SCGT approximates other forms of medical therapy and does not raise the same degree of ethical concern as the inheritable GLGT. One exception would be the use of SCGT on fetuses, which could inadvertently affect the fetal germ cells and become a germ-line intervention.

The still-hypothetical prospect of GLGT, in contrast, is not associated with the same degree of consensus as SCGT. According to one perspective, GLGT is a logical extension of medical research that differs by degree rather than by a clear line. If GLGT is shown to be safe through animal research and laboratory studies, according to this point of view, germ-line alterations will more efficiently eliminate genetic disease than will somatic cell technologies. Faced with this promising method of combatting human suffering, it is argued, medical personnel are obligated to develop new treatments for genetic diseases. In the process, scientific inquiry into germ-line manipulations will yield knowledge useful in its own right (1,2). According to an opposing perspective, GLGT is a labor-intensive and expensive alternative to other treatments that would be available primarily to wealthy people and would divert scarce resources from therapies that would benefit greater numbers of people. Those wary of GLGT have particular concern about the technique's inheritability, which means mistakes from this unusually risky treatment would pass from one generation to the next. Germ-line interventions are also thought to generate the potential for genetic enhancement, in which genes would be manipulated to promote socially desirable attributes that are not medically necessary. Germ-line enhancement, in turn, according to this argument, would lead to a differential evaluation of individuals, with those who have been enhanced for traits deemed socially important valued more highly than those who have not received a genetic enhancement. Germ-line interventions would also, according to the detractors of GLGT, undermine respect for human diversity, divert societal resources from ill to healthy people and impose unacceptable risks on individuals (3). For these and other reasons it has been argued there are simpler and safer ways of avoiding the birth of children with serious genetic disease, including preimplantation genetic diagnosis to identify and not transfer embryos with disease-linked genes (3).

National and international governmental and non-governmental organizations have reached a range of conclusions about the ethical acceptability of GLGT, and the deliberations remain active as genetic knowledge advances. More consistency is apparent in the matter of germ-line enhancement, with consensus that nonmedical genetic interventions on the germ-line are not ethically acceptable (4). The proactive nature of the debate over GLGT, started well before human germ-line interventions were remotely possible, underscores the sensitivity of the issue and the seriousness with which citizens regard the ethics of GLGT. Because a broader and more problematic array of ethical issues is associated with GLGT than with SCGT, this entry will focus on national and international responses to GLGT.

GENE THERAPY IN INDIVIDUAL NATIONS

Nations with laws in effect or in process relating to GLGT are primarily European and Anglo-American, and these nations have developed stable yet diverse approaches to assisted reproductive technologies ARTs in general and to GLGT in particular. National policies on GLGT tend to be primarily restrictive or permissive. Restrictive policies bar GLGT either explicitly by targeted provisions in national laws or implicitly by prohibitions on embryo research. Permissive policies leave a door open for GLGT through flexibly worded laws or through the absence of laws. Nations with laws governing assisted reproduction have been called "framework nations"; nations that rely on principles and rules developed by commissions and professional associations are known as "guideline nations" (5,6).

Restrictive Policies

Germany's Embryo Protection Act of 1990 (7), called the "world's most restrictive law" in reproductive medicine (8), was enacted to protect human embryos from nontherapeutic research and to draw lines for ethically contentious applications in assisted reproduction. The Act makes the misuse of human embryos a criminal act, where "misuse" is defined as buying, passing on, or acquiring a human embryo "for purposes other than preserving its development." Under this law, to cause the further development of an embryo "for purposes other than causing a pregnancy" is to commit a criminal act. An embryo is defined as a fertilized human oocyte that includes "any totipotent parts" that could develop into an individual being. According to this definition, the embryo is a human being and so are the totipotent (undifferentiated) cells in the early embryo.

The German law bars GLGT in two ways. First, it forbids preimplantation genetic diagnosis (PGD), in which a cell from the embryo of a couple at risk for passing a genetic disease to their offspring is removed and tested for the presence of the disease-linked gene. Not only is it illegal under the law to discard an affected embryo, but it is also a criminal act to discard the biopsied cell even if the embryo were found to be without the disease gene and transferred to the patient's uterus. By barring PGD, which would be a precondition of gene correction, the law precludes GLGT. Second, Germany's law explicitly makes it a criminal offense to "manipulate the genetic information of a human germ cell" or "use genetically manipulated germ cells for the purpose of fertilization."

Germany's embryo protection law was designed to exclude practices that might lead to eugenic applications. Particular features of German history, particularly the eugenic experiments and goals of Nazi Germany before and during World War II, make gene therapy a highly emotional issue in Germany. According to Mauron and Thevoz, the German approach to reproductive medicine, which is shared by other Germanic nations, reflects a cultural view that assumes negative consequences will follow from genetic applications. This view anticipates a technological imperative in which possessing a new capability is the same as using it (9).

Denmark also has a prohibitory assisted reproduction law that extends to gene therapy. Its Law No. 460 (1997) forbids the genetic modification of germ cells and fertilized eggs. The law provides that assisted conception will take place only for the purpose of "uniting a genetically unchanged (unmodified) oocyte with a genetically unchanged (unmodified) spermatozoan" (10). It also states that fertilized eggs used in therapeutic or nontherapeutic research will not be transferred "unless the fertilized oocytes are genetically unchanged." The Danish law allows PGD if there is "a known and considerable risk that the child will be affected by a serious hereditary disease." This underscores a value in Danish law that limits genetic procedures to serious medical situations only and that eschews selective embryo transfer for eugenic reasons. The Danish law was passed with significant involvement of the Danish public. It places fewer restrictions on ARTs and reveals less distrust of the procedures than the German law.

France's restrictive embryo research law forbids experimentation on the human embryo unless the embryo is to be transferred (11). It also forbids interventions that would adversely affect the embryo's "developmental capacities." Although French law forbids embryo research, it pairs PGD with prenatal diagnosis rather than with embryo research (12). Consequently, to remove cells for PGD is not regarded as a form of embryo experimentation. Physicians may perform PGD under limited conditions and if the couple is at demonstrated risk for bearing a child affected with a serious and incurable genetic disease (12). Ethical debates on PGD in France have revealed particular concern about eugenic uses, in which embryos would be selectively transferred for their socially preferred traits rather than selectively nontransferred for disease-related conditions (13). The French Senate's version of the 1994 embryo research law would have forbidden PGD altogether because of this concern, but the law's final version allowed PGD in limited circumstances (14). Despite the permissibility of PGD under French law, the technique is not practiced in France because the government only recently published the requisite decree rendering the embryo research law applicable. French law is more forthright about GLGT than about PGD. No study may be carried out if "its object is to change, or if it is likely to change, the genetic heritage of the embryo."

Other nations with restrictive GLGT policies include Austria, Switzerland, and Spain. Austria's embryo research law states that "interventions involving the germline shall not be permitted" (15). In Switzerland, a 1992 amendment to the Federal Constitution forbids "genetic manipulation" and imposes other restrictions on assisted reproduction (16). The referendum passed with a 71 percent approval vote in a national referendum and set up a committee of experts appointed by Parliament to interpret the amendment (17). Spain's law on assisted conception and embryo research deems GLGT a "very serious offense" if the genetic manipulation is for "nontherapeutic purposes or for therapeutic purposes that are not authorized" (18). This wording leaves the door open for eventual authorized therapeutic GLGT.

Other nations limit GLGT indirectly in their restrictions on embryo research. Norway's law, for example, prohibits "research on fertilized eggs," which would include GLGT in its experimental stage (19). Such laws would not necessarily forbid therapeutic GLGT if it were effective and safe. In fact nonexperimental therapeutic GLGT might be more acceptable than PGD in some nations because it would aim to treat rather than discard embryos that have the disease gene in question.

At the subnational level, the Infertility Treatment Act of 1995 in Victoria, Australia, forbids specified techniques such as cloning and GLGT and regulates other ART procedures. It set up a Standing Review and Advisory Committee to advise the health minister on, among other things, "the use of treatment procedures or related procedures to avoid genetic abnormalities or disease" (20).

Permissive Policies

Nations are permissive regarding GLGT either by flexible laws or by default due to the absence of proscriptive laws. The United Kingdom exemplifies the former in that it has a national law that accepts embryo research and could be congenial to GLGT. British policy derives from the 1984 report of the Warnock Commission, one of the first national ethics commissions set up world wide to evaluate the ethics of ARTs. Its report recommended against specified interventions on the embryo but left GLGT open. Commission members pointed out that public anxiety over genetic therapy focused not on therapy but on "the deliberate creation of human beings with specific characteristics" premised on eugenics (21). Commission members presumed such uses would be covered by the controls recommended in the report, including a licensing body to review and approve embryo research.

Although pointedly critical of other techniques that were still speculative in 1984, the Warnock Commission was relatively permissive in the matter of gene therapy. Its report identified selective breeding but not therapy as onerous, and it invited the licensing body to determine what might not be ethical. Coming at the end of the report, the section on gene therapy was less restrictive than other sections recommending that certain activities, such as "the placing of a human embryo in the uterus of another species for gestation," should be criminal offenses. The Warnock Commission recommended that a licensing body be responsible for identifying research that "would be unlikely to be considered ethically acceptable in any circumstances and therefore would not be licensed."

From the Warnock Report emerged the Human Fertilization and Embryology Act of 1990 (HFEA) (22). This Act created a statutory licensing authority (Human Fertilization and Embryology Authority) to license laboratories working with human embryos and to monitor and set the conditions for embryo research. Under the law, researchers need a license to create, keep, or use embryos. The United Kingdom's policy on embryo research has been characterized as "highly permissive" (23). The Act lists categories of research for which licenses can and cannot be granted, and it adds a category of uncertain status into which gene therapy fits. Under the Code of Practice of the Human Fertilization and Embryology Authority, however, research licenses at present may not be granted for "altering the genetic structure of any cell" while the cell becomes part of an embryo. This also precludes gene therapy for nonmedical reasons. According to the 1992 report of the Committee on the Ethics of Gene Therapy, "in the present state of knowledge any attempt by gene modification to change human traits not associated with disease would not be acceptable" (24).

The British national policy on embryo research and assisted reproduction reflects a broader cultural optimism about medicine and the ability of humans to draw lines at unacceptable applications than appears in policies in Germanic-speaking nations. According to Mauron and Thevoz, the British policy reflects a utilitarian mode of thinking in which observers "seem as a rule more confident [than countries holding a pessimistic view] in their ethico-legal capability to promote good and prevent evil" (9). As policy the British HFEA is perhaps the most thorough on assisted reproduction in the world. It is flexible in the way it relies on a licensing system that has room for clinical judgments and gives government officials discretion in deciding categories of research for which licenses can be granted (25). It also encourages continued review of emerging techniques.

Other European nations have flexible policy climates for gene therapy by default in the absence of explicit embryo research or ART laws. These guideline nations rely on existing laws, norms, professional guidelines, and other protections to oversee innovations in ARTs. Some, such as Belgium and Canada, are in the process of formulating laws relating to bioethics. The process of developing legislation has been ongoing for over a decade in Canada. In 1989 the Canadian government appointed the Royal Commission on the New Reproductive Technologies to examine and report on the developments in and implications of the new reproductive technologies. Members of the Commission met over a period of 14 months, solicited written and oral testimony from interested citizens, and issued a two-volume report, *Proceed with Care* (3).

In its report, the Royal Commission concluded that somatic cell gene therapy was ethically acceptable provided principles governing informed consent and other protections were followed. Noting the risks of GLGT and the presence of PGD and selective nontransfer of embryos as a less risky alternative, however, it concluded that GLGT was "inconsistent with the Commission's guiding principles" and that it should be forbidden with no federal funds used to support human GLGT research. In 1995 Canada's Minister of Health formed a Discussion Group on Embryo Research. Several months later, she called for a voluntary moratorium on nine embryo research applications, including GLGT (26). Recommendations from the Royal Commission were used to draft Bill C-47 in 1996, which among other things would have forbidden

GLGT. The Bill died when the 1997 federal election was called, leaving the voluntary moratorium to limit GLGT. While it imposes no penalty, the moratorium is respected as a pronouncement of the federal government. If the Royal Commission report is an indication of public opinion, it reveals the desire of most Canadians to proscribe GLGT and enhancement interventions.

The U.S. policy on GLGT is de facto permissive except for several states with embryo research laws that might preclude GLGT. Concern about newly discovered abilities genetically to modify organisms through recombinant DNA research led the government to establish a Recombinant DNA Advisory Committee (RAC) in 1976 to review proposals from researchers seeking federal funding to conduct experiments involving genetic recombinations of nonhuman organisms. In 1984 the RAC formed a Working Group on Human Gene Therapy to consider the possibilities of human gene therapy. In the same year the Working Group disseminated for public review a Points to Consider document that posed questions about what would be necessary to establish ethical and safe therapy for SCGT (24). The final document was approved and implemented in 1985. One section excludes funding for GLGT research protocols:

> The RAC and its Subcommittee will not at present entertain proposals for germ line alterations ... in germ line alterations, a specific attempt is made to introduce genetic changes into the germ (reproductive) cells of an individual, with the aim of changing the set of genes passed on to the individual's offspring (27).

This sole mention of GLGT in the U.S. federal regulations precludes federal funding for human germ-line research but does not forbid the research. At present, no central forum has been mobilized to weigh ethical issues associated with GLGT, although the National Bioethics Advisory Commission, situated in the Executive Branch, could be asked to issue a statement on gene therapy, as could a Gene Therapy Policy Conference under the auspices of the RAC. In 1998 the RAC announced it would review its SCGT submission procedures, which would present an opportunity to revisit the provision in those procedures that the RAC would not "at present" entertain proposals for germ-line alterations. A geneticist's proposal to conduct in utero SCGT, which could inadvertently affect fetal germ cells, was submitted to the government in 1998 to provoke a revisiting of the RAC funding guidelines (28). The absence of a proscriptive GLGT policy in the United States makes this guideline nation congenial for GLGT research relative to other countries.

In the mid-1990s Jones surveyed practitioners in twenty five nations in which in vitro fertilization is practiced, and he determined that nine had voluntary guidelines but no national regulations or laws on assisted reproduction and another four did not have either laws or clear guidelines (5). Schenker and Shushan surveyed practitioners in 16 Asian and Middle-Eastern nations and found that only Taiwan had legislation directed to ARTs (29). These surveys suggest that a guideline approach is common among nations in the area of ARTs and that a sizable number of nations lack even clear guidelines. Because policies on GLGT are often appended to ART laws, this suggests that when one takes numbers of nations into account, silence is a prevailing international perspective on GLGT policy.

Summary

A number of European and Anglo-American nations have debated the ethics of GLGT in national forums and have developed formal or informal policies related to gene therapy. Nations with GLGT policies cover a range of approaches, with some permissive and others restrictive. These nations, in which scientists engage in genetics research, make up a minority of the world's nations. Nations in which genetics research is nearly absent or peripheral generally have no laws directly related to GLGT and often have conducted no extensive debate on a national GLGT policy, although they may have set up review committees to review gene protocols (24). Although firm generalizations cannot be drawn, certain features of national responses to GLGT identify directions for future ethical inquiry and policy formation.

First, among nations in which gene therapy debates have ensued, it is generally the case that SCGT is regarded as ethically acceptable, provided safety and informed consent procedures are followed. Germ-line interventions for medical reasons, on the other hand, are generally proscribed but not necessarily permanently. Genetic interventions for enhancement, nonmedical, reasons are nearly universally held to be ethically unacceptable, although this may be by convention rather than through explicit mention in the law.

Second, national positions on GLGT reflect stable ideological perspectives related to the political culture and history of individual nations. Germany's restrictive embryo research law reflects fears of eugenic applications traced to reactions against the eugenic premises of the Nazi ideology during the World War II period and a notion of justice designed to protect vulnerable groups that include embryos and generations of humans to come. There is also in Germany a distrust of genetic alterations in general that extends to the genetic modification of plants and animals. The United Kingdom's policy, in contrast, reflects a greater trust in medical genetics. Not having directly experienced a national ideology that embraced the political strategy of selective breeding to bring in those with socially desirable traits and to select out those with traits deemed unworthy, British researchers conduct genetics research in a different milieu from that in Germany. In fact the United Kingdom, as the home of two significant events in the last quarter of the twentieth century—the first successful in vitro fertilization and the first cloning of a mammal, a lamb, from the body cell of an adult sheep—reveals an ethic of discovery and innovation in assisted conception. Distrust of plant and animal genetic engineering is, however, pronounced.

Third, legislation in European nations tends to emphasize the bioethical principle of justice. According to Knoppers and Chadwick, the ethical debate on germ-line interventions gains momentum from a desire for justice toward future generations, where future generations are deemed to be vulnerable populations. The debate also stresses the importance of the principles of autonomy, privacy, quality, and equity (16). A European theme that regards human genes as a common heritage (see the discussion below) also orients thinking toward principles of distributive justice more than in the United States and the United Kingdom, where protection of the autonomous choices of patients is a primary value. Within Europe, however, GLGT policies differ according to national decision-making styles. According to Byk, bioethical decision-making processes are in place in Norway, Denmark, Sweden, and the United Kingdom as a result of shared moral values and a strong constitutional orientation (30). These nations also share a pragmatic style of thinking that tolerates the lack of a clear resolution for certain ethical issues. Other nations, such as Italy and Belgium, lack the same decision-making process for bioethics, argues Byk, and have fundamental divisions that can make consensus elusive.

Fourth, ethical debates and policies about GLGT in many European nations are anticipatory. On the one hand, this protects individual rights by addressing potential injuries before the science is imminent and before violations of human dignity occur, and it amounts to a preventive rather than curative ethics (31). On the other hand, restrictive anticipatory policies may cut off debate prematurely and before a context develops for reasoned, experienced-based inquiry (32). Moreover they may quickly become dated or difficult to interpret as technologies change. For example, GLGT policies revolve around definitions that consider GLGT to be the splicing of genes into nuclear DNA. Recently, however, researchers have suggested manipulations that might modify the definition of a germ-line intervention. Some genetic diseases are linked to mutations in mitochondrial DNA, a form of DNA found in the cell's cytoplasm. These diseases might be circumvented by transferring the nucleus from the egg of a woman with a mitochondrial disease to a donor egg from which the nucleus has been removed and discarded. This would result in an inheritable change in which the new mitochondrial DNA would appear in each cell of the child to be, including egg cells, which would make it a germ-line innovation. National laws cover interventions on nuclear DNA, but it is not clear in any nation except Germany whether they would apply to a procedure in which cytoplasms and hence mitochondrial DNA, is substituted (33). This suggests that anticipatory laws can produce formulas that are not sufficiently supple to oversee unexpected innovations in ARTs. A similar problem will conceivably arise when policy makers try to figure out what is meant by nonmedical (or enhancement) germ-line interventions. Nations with ART licensing procedures can arguably respond to definitional ambiguities more easily than those without such mechanisms.

Fifth, ethical debate on GLGT across nations is intertwined with views about the genetic modifications of plants and animals, which indicates the powerful symbolic content of genetics and genetic engineering. It is not coincidental that citizens in Germany and Switzerland, nations that restrict human genetic manipulations, also are distrustful of genetic modifications of plants and animals. The United States, in contrast, holds a relatively more trustful view of genetic interventions in humans as well as in agriculture and commerce, and many of the world's most active biotechnology companies are in the United States.

GENE THERAPY AND INTERNATIONAL ORGANIZATIONS

Variations in regulations across countries, especially within the close borders of European nations, generate concern that a "procreative tourism", will arise in which patients will travel from one nation to the next in order to circumvent restrictive ART laws in their home countries (34). A persistent sense that interventions on the human germ line are so important that protections must span national borders has caused policy makers to identify principles that ought to prevail in all nations. Efforts to identify cross-national guidelines for genetics research have taken place regionally in Europe and globally in governmental and nongovernmental decision-making bodies.

European Regional Organizations

Europeans, primarily through the Council of Europe, have taken a leadership role in identifying common principles to span variations in laws within European nations and in publishing the results of their deliberations (35). Europeans are well poised to do this in that they have been actively involved in protecting human rights in medical and scientific settings. From this tradition, as seen in the Nuremberg Code for medical research and the European Convention on Human Rights, emerged recommendations protecting the rights of dying persons and the mentally ill (36). In addition, in Europe, an historical collectivist ethic has undergirded the idea of the human genome as a collective heritage.

The Council of Europe has actively sought principles to govern new reproductive and genetic technologies. Established in 1949 to promote cooperation among nations, the Council currently has 40 member states. The Council's "leading conscience" is its Parliamentary Assembly, which is made up of appointed representatives of national parliaments of member states and geared to formulating resolutions and recommendations (36). In 1982, in response to widespread concerns about genetic engineering and inspired by a parliamentary hearing on genetic engineering and human rights held in Copenhagen in 1981, the Parliamentary Assembly delivered Recommendation 934 to the Committee of Ministers (37). This early public policy statement on genetics was positive in tone, and its scope was broad in defining genetic engineering as "artificially recombining genetic material from living organisms," which covered plants as well as humans.

Recommendation 934 affirmed the promise and responsibilities of scientific inquiry in general and gene therapy in particular by noting that gene therapy "holds great promise for the treatment and eradication of certain diseases which are genetically transmitted." Members of the Parliamentary Assembly developed Recommendation 934 with an eye to Articles 2 and 3 of the European Convention on Human Rights, which dealt with rights to life and human dignity. The members interpreted these articles to "imply the right to inherit a genetic pattern which has not been artificially changed." This right need not be absolute; Recommendation 934 stated that gene therapy on embryos, fetuses, and minors should not be conducted without the "free and informed consent of the parent(s) or legal guardian(s)," which leaves the door open for therapy on embryos. In its Recommendation 934, the Assembly encouraged a European agreement to be drawn up regarding genetic engineering so that national legislation could be aligned accordingly, and efforts could be made to "work towards similar agreements at world level." It called for an "explicit recognition in the European Convention on Human Rights of a right to a genetic inheritance which has not been artificially interfered with, except in accordance with certain principles which are recognised as being fully compatible with respect for human rights (as, for example, in the field of therapeutic applications)." Such a statement would leave the door open to GLGT if the intervention conformed with therapeutic and research principles and was not conducted for nontherapeutic ends. In 1989 the Parliamentary Assembly issued Recommendation 1100, which would have limited embryo research and would have forbidden "any form of therapy on the human germinal line" (38). The Recommendation did not pass the Committee of Ministers, however, so its broad proscription of any form of GLGT was not officially approved (39).

In 1991 the Council of Europe acted on the Parliamentary Assembly's recommendation to recognize a right to a genetic inheritance that has not been artificially changed and to seek a transnational harmonization of genetic principles. In that year the Council convened a bioethics convention. The convention was designed to create a legally binding instrument based on a "limited set of general principles founded on human rights values" (36). It was a prospective agreement designed to guide national legislation in a document flexible enough to reflect differing cultures and legal systems. The guiding assumption was that the convention would reflect values that would apply across different cultures and legal systems. It also was designed to accommodate rapid technological change and to address public fears about genetics (39). In 1996 the convention, known as the Convention for the Protection of Human Rights and Dignity with Regard to Biology and Medicine ("Bioethics Convention"), was endorsed by the Parliamentary Assembly. It was opened for signature of the Council's member states in 1997. Nations signing the convention were legally obligated to bring their own laws into conformity with the principles, unless they signed with a reservation that they would not conform with individual articles covered by their existing laws (40).

The Bioethics Convention uses human rights themes and is based on protections of human dignity inherent in the Universal Declaration of Human Rights (41). The Preamble to the Bioethics Convention reiterates the idea that individuals have the right to a gene line that has not been deliberately altered; "an intervention seeking to modify the human genome may be undertaken for preventive, diagnostic, or therapeutic purposes and only if its aim is not to introduce any modification in the genome of any descendants" (42). This provision bars GLGT if the genomes of descendants would be modified, which effectively precludes the most commonly envisioned forms of GLGT. It is more restrictive than Recommendation 934 because it adds a stipulation against introducing inheritable changes. It leaves the door open to noninheritable GLGT, which scientists suggest may be possible if, for example, genetic changes to the germ cell could be introduced via a dispensable extra chromosome that would be engineered not to pass to the next generation (43).

Twenty-eight of the 40 member states of the Council of Europe signed the Bioethics Convention within three years of its completion, which obligated them to harmonize their national laws with the principles of the convention. Fourteen others did not indicate whether they would sign. Germany was reluctant because the Convention was less restrictive than its own practices (e.g., it allowed embryo research) (44), and the United Kingdom was reluctant because certain provisions were more restrictive than its own legislation. Nonmember observer states that participated in the convention's development (Australia, Canada, Japan, the Holy See, and, the United States) were also invited to sign but none did (40).

European discussions reveal a wariness of GLGT and an intense distrust of enhancement interventions. As early as the 1980s the European Community set up a crossdisciplinary working party to examine ethical issues associated with the new reproductive technologies. With Jonathan Glover as chair, the group issued the Glover Report in 1989, which concluded that GLGT should be rejected for humans because of its "serious risks" and also because of the ethical concerns associated with changing the genomes of the patient's offspring. The authors argued that given the current state of knowledge, germ-line interventions should be rejected and that enhancement genetic engineering is unethical "at least until policies have been worked out to cope with the huge problems it raises" (45). In 1988 the European Medical Research Councils, representing the Medical Research Councils of Austria, Denmark, Finland, France, the Netherlands, Norway, Spain, Sweden, Switzerland, the United Kingdom, and West Germany, concluded that genetic enhancement "should not be contemplated" in light of the ethical problems it would raise (31). In 1989 the Council of Europe's Ad hoc Committee of Experts on Bioethics published a report on assisted conception and embryo research. Its nonbinding recommendations intimated that procedures on embryos were acceptable only to benefit or observe the embryo, but that states could allow other "investigative and experimental procedures" for a "preventive, diagnostic or therapeutic purpose for grave diseases of embryos." While this left the door open for GLGT, it proscribed enhancement interventions: assisted reproduction should not be used "for obtaining particular characteristics in the future child" (46).

A recurring concept in European organizations is the notion of a shared genetic heritage or genetic patrimony (*patrimoine genetique*). This notion, which is not generally embraced in Anglo-American nations, extends beyond the genetic endowment of any one individual to regard the human germ-line as a shared, collective resource. Under this concept the genome represents "the collective assets of a community (or of mankind)" that is "both irreplaceable and of enduring worth, and therefore subject to specific forms of social protections" (9). As defined by Agius, "the collective human gene pool knows no national or temporal boundary, but is the biological heritage of the entire human species." Originating with discussions of the law of the sea in 1967, the concept of a common heritage embraces access to common resources. A common heritage cannot be owned; it is instead passed from one generation to the next as a set of openly accessible goods. As a consequence no generation has the right to use GLGT to "alter the genetic constitution of the human species" (47). Genes are part of the common heritage because they are inherited; moreover a sharing of genes across the generations unites humans as a species. The knowledge that humans share a common genetic structure helps humans see themselves as a "collectivity of rights and responsibilities." Thus humans as a species have the right "to inherit a healthy and diversified genetic heritage" that has not been appropriated by patenting or other actions that benefit the individual (48). Genetic patrimony implies an international sharing of genes as a resource and an international policy that involves broad participation on behalf of all members of the human species (48).

This concept brings GLGT into a framework of human rights and responsibilities. Germ-line interventions are more than mere science or medicine; they touch upon something with a nearly mystical aura—a collective genetic heritage. While the intensity of the concept varies among countries, at root it embraces notions that genetic interventions are societal rather than individual and that they pose harms as well as benefits. It also includes the idea, influenced by the German philosopher Hans Jonas, that to possess technological knowledge is eventually to use it in a kind of technological imperative (9).

Organizations and Associations Beyond Europe

International commissions and organizations beyond Europe have also weighed the ethics of GLGT. The Council of International Organizations of Medical Sciences (CIOMS), which holds international conferences to discuss medical and scientific topics, convened a meeting in 1990 on Genetics, Ethics and Human Values that was co-sponsored by the World Health Organization (WHO) and the United Nations Educational, Scientific, and Cultural Organization (UNESCO). The 102 participants from 24 nations concluded in the CIOMS Declaration of Inuyama that GLGT is unique because of the possibility for "permanent genetic change," but that it "might be the only means of treating certain conditions," and "must not be prematurely foreclosed," (24). The statement limited GLGT to medical conditions that "cause significant disability," and it did not condone interventions to "enhance or suppress cosmetic, behavioural or cognitive characteristics unrelated to any recognized human disease."

The World Medical Association (WMA), founded in 1947 to promote high international standards in medicine and health care, advised physicians in 1987 to respect international ethical codes when engaging in research on genetic diagnosis and treatment. In response to what it perceived to be "uncompromising opposition" to the human genome project, the WMA in 1992 issued a Declaration on the Human Genome Project that urged a rational assessment of ethics using the same guiding principles to evaluate risks and benefits as used for any diagnostic or therapeutic innovations—respect for the patient as a human being and for his or her autonomy and privacy (49).

The Human Genome Organization (HUGO) was established in 1989 to promote cooperation among scientists engaged in human genome research and to encourage public debate on the multiple implications of genome projects (50). This independent body now has over 1000 members from over 50 nations. The Ethical, Legal and Social Issues Committee of HUGO, chaired by Bartha M. Knoppers, developed and released in 1995 an aspirational Statement on the Principled Conduct of Genetics Research. This Statement issued ten recommendations in response to concerns about genome research, none of which called for a ban on GLGT. It also identified four principles as essential in genetics research: regarding the human genome as "part of the common heritage of humanity"; respecting "international norms of human rights"; recognizing the "values, traditions, culture, and integrity of participants"; and upholding "human dignity and freedom" (51). The reiteration of the concept of the common heritage reflects in the HUGO document a European voice.

In early 1999 the Executive Board of the WHO also approved draft bioethics guidelines that repeat the idea that the genome is the "common heritage of humankind." The guidelines do not preclude GLGT, bar enhancement, or cover embryo research, but they do warn against "hurried and premature legislation in the rapidly evolving field of genetics" (52). The nonbinding draft guidelines were presented to the WHO General Assembly in mid-1999.

An aspirational document of particular importance is the Universal Declaration on the Human Genome and Human Rights, signed in 1997 by the 186 member states of UNESCO (53). In light of expanding genetic knowledge occasioned by the human genome project and increasingly voiced concerns about the impact of genetic knowledge on human rights, UNESCO charged its International Bioethics Committee (IBC), chaired by Nicole Lenoir, a member of the French Constitutional Council, to consider an international framework for protecting the human genome (50). IBC proposed a nonbinding, aspirational declaration rather than a binding treaty. The UNESCO General Conference accepted this proposal and requested a draft Declaration for its next meeting in 1997. It was hoped the Declaration could be ready for signatures of its member states to coincide with the 50th anniversary of the United Nations Declaration of Human Rights in 1998. Over the next several years, IBC broadly consulted parties and documents from around the world. IBC was an independent body made up of 55 individuals from 40 nations who were selected for their "competence and personal attributes," and not to represent individual nations. Overall, committee members drafted and examined eight versions of the Declaration before handing its final version to the General Conference in mid-1997 (50).

IBC faced numerous challenges in its deliberations. Dealing with a scientific area marked by rapid change, it aimed to create an enduring document that would not be dated within a short time. It also worked for a document that would be given to the nearly 200 nations of the world, each of which had different values, cultures, and national perspectives. In only a few of these nations were scientists engaged in genetics research, yet it was an aim of the document that research holding great promise for humans should not unduly be restricted (51).

Lenoir regarded UNESCO as an "ideal forum" for producing an international document linking genetics and human rights. For one thing, UNESCO is the only agency of the United Nations with jurisdiction over science. For another, its constitution, which underscores the ideals of "dignity, equality, and respect for human rights," links it with ethical ideals. The UNESCO Declaration was meant to balance individual rights with the promise generated for all humans through genetic advances. It was, according to Lenoir, meant to be a reminder of the solidarity of richer and poorer nations in enjoying the benefits of science. As a document embracing universal human rights, the Declaration was based on "the unity of mankind and the equal dignity of individuals, as upheld by the principle of the universality of human rights" (50).

The IBC draft Declaration was given to the General Conference in the summer of 1997, where it took its ninth and final draft. Differing national perspectives were expressed in these deliberations. Germany wanted more direction in the document, with lists of technologies to be banned with germ-line interventions as a priority. The United States, which acted as an observer with three representatives participating in the IBC meetings, favored a more open framework for genome research (54). The United Kingdom and Singapore also acted as observers even though they, along with the United States, were not UNESCO members.

By late 1997 the Declaration was presented for signature, whereupon it was signed by UNESCO's 186 member states. The document does not explicitly forbid GLGT or any other specific technique except cloning. It does, however, echo the European sentiment that the genome is a shared resource. The final wording of Article 1 states the following:

> The human genome underlies the fundamental unity of all members of the human family, as well as the recognition of their inherent dignity and diversity. In a symbolic sense, it is the heritage of humanity.

Other articles establish principles for the ethical conduct of genetic inquiry. For example, the document regards genetics research as a way to "offer relief from suffering and improve the health of individuals and humankind as a whole." This embraces the principle of justice in that it places a value on the equitable sharing of the benefits of genetic medicine among all individuals and not just the wealthy few.

The Declaration also places a high value on individual dignity and respect for diversity. Article 2, for example, states that individual dignity "makes it imperative not to reduce individuals to their genetic characteristics and to respect their uniqueness and diversity." The status placed on diversity echoes the preamble of UNESCO's constitution (41) and parallels other United Nations documents such as the Convention of Biological Diversity, which respects cross-species diversity. The concern about genetics as a common heritage, although modified in the final version, echoes UNESCO's tradition of protecting other entities of universal interest such as the moon and cultural achievements. It also reveals the influence of the European tradition (55).

Summary

International documents related to gene therapy generally rely more on principles than technique-driven rules. This strategy is designed to help ensure consensus among participants who represent a variety of political and cultural systems and to allow the documents to withstand the test of time better than were they geared to particular techniques. The documents present a general framework around which nations can harmonize their legislation; they do not impose rules on the nations, which are left to devise their own regulations (56). International statements also differ from national laws in that they avoid reaching conclusions on the moral status of the embryo or making detailed rules about embryo research. This contrasts with national laws that are intimately tied with presumptions about the embryo's moral status and the ethics of embryo research. The documents also provide protection to citizens in signatory nations that do not have germaine policies and are unlikely to craft their own legislation.

Differences exist between the regional European-based and internationally based positions on gene therapy, as seen by contrasting provisions in the European Bioethics Convention and the worldwide UNESCO Declaration. In the main, the UNESCO Declaration, which is oriented solely to genetics issues, has a broader reach that recognizes the global import of genetic interventions, addresses genetic reductionism, and warns of the dangers of eugenics by acknowledging the range of factors that contribute to individual personalities (41). The Bioethics Convention, in contrast, covers bioethical issues other than genetics and has a more procedure-oriented focus on genetics. Byk suggests a similar regional approach would be suitable for Asia in that a regional approach, which harmonizes rules when fundamental human rights are at stake, more easily accommodates regional cultural factors than a global approach (30).

One similarity among international documents is the concept of the genome as the common heritage of humanity, which arises in the Bioethics Convention, the UNESCO Declaration, HUGO's ethics statement and policies in Switzerland, Canada, Germany, France, Australia, and other countries (47). This reminds nations of the shared benefits as well as costs that can emerge from medical genetics.

THE UNITED STATES AND INTERNATIONAL PERSPECTIVES

International perspectives set a broad context for examining GLGT, but it is not clear how influential those perspectives will be for U.S. policy. First, the United States has no specific GLGT policy and does not appear to be in the process of developing one. Second, while the United States was involved in drafting major international documents, it has not ratified them. It played an active role in developing the UNESCO Declaration, but it did not ratify the document even though it could have signed it as a nonmember. It could also have ratified the Bioethics Convention by virtue of its status as an observer state, but it did not (40). As Annas points out, the United States ratified the United Nations International Covenant on Civil and Political Rights but not the International Covenant on Economic, Social, and Cultural Rights, possibly because the latter was perceived to be inconsistent with the emphasis on private property in the U.S. economic system (57). The United States also failed to support the negotiated version of the 1998 Biosafety Protocol, which would have been the first global treaty to regulate genetically altered products, in part to protect its economic interests. The Bioethics Convention, which limits GLGT, may be inconsistent with U.S. economic interests even though GLGT is hypothetical and of uncertain ethical acceptability, and it may not turn out to be economically feasible.

Third, the U.S. and European perspectives differ in significant ways. The European perspective tends to be collectivist and infused with a worldview ethic. Justice, including respect for future generations, is a primary concern. The U.S. perspective, on the other hand, is more individualistic, with the autonomous choices of potential parents and freedom of scientific inquiries as primary values. Thus the idea of the genome as a common heritage of humankind is less compatible with American beliefs than with beliefs in Europe. Scholars on both sides of the Atlantic have voiced criticism of the concept. Juengst, for example, questions its validity as a scientific concept when he argues that the genome is not a thread connecting all humans from one generation to the next. On the contrary, "each organism's germ-line terminates in its gametes" and the embryo begins a new germ line that is the product of the mother's and father's germ lines, which now end (47). According to Juengst, the idea of a common gene pool is uncomfortably close to old and inaccurate ideas about blood lines. Moreover genetic modification is continually ongoing in the form of reproductive decisions, which opens to question why GLGT is singled out as particularly threatening. Cook-Deegan similarly asks why the genome should be considered a public resource in societies in which reproductive privacy is highly valued (58). In Europe, Sass recommends viewing the genetic heritage as an individual as well as familial concept characterized by an "ever-modifying and mutating pool of human DNA," (55) and Mauron and Thevoz question whether the human genome is a "resource" (9).

Although the genome as a human resource has not informed U.S. policy making, the interest in the ethical implications of germ-line interventions provides international forums for discussing GLGT issues. Reports such as those from the Royal Commission on the New Reproductive Technologies in Canada and the Warnock Commission in the United Kingdom and policies such as the UNESCO Declaration enlighten citizens and policy makers about emerging international norms. They also serve as an eloquent reminder that the repercussions of GLGT are global. This reminds policy makers of the need to craft policies aimed at avoiding potential injuries to future generations. Perhaps more important, if GLGT proceeds in the United States and other industrial nations and is shown to be safe and effective, this reminds us of the need to ensure that the benefits of germ-line alterations are shared equitably among citizens in nations of differing economic levels. If the genome is a resource, fellow humans deserve access to beneficial applications as well as protections from preventable harm.

CONCLUSION

In 1991 Walters examined 20 policy statements on human gene therapy issued by governmental and nongovernmental bodies worldwide between 1980 and 1990 (4). He found that most, but not all, opposed GLGT and none supported germ-line enhancement. Differing conclusions about the ethical acceptability of GLGT continue to permeate international perspectives today, and they have produced an array of policies at the national and international levels. Individual nations can be thought of having either frameworks or guidelines, where the former means having laws directly monitoring assisted reproduction and gene therapy and the latter means reliance on voluntary guidelines and tangential policies to protect human research subjects.

International policies echo this framework/guideline distinction as well. A framework international policy that is legally binding on its signatories is the Bioethics Convention of the Council of Europe, which states that interventions on the genome may be undertaken only if the aim is not to produce a modification in descendants (42). The UNESCO Universal Declaration on the Human Genome and Human Right, in contrast, in an aspirational, nonlegally binding guideline that sets forth principles for the just and equitable use of genetic knowledge (53).

Apart from a few notable national laws and the Bioethics Convention, which has not been signed by key nations in which scientists are engaged in innovative genetics research, the overall perspective on GLGT can be said to be cautionary but not absolutely prohibitory. It is not uncommon for warnings about GLGT to be paired with caveats about the state of knowledge at this time. Such utilitarian warnings are based on the assumption that the risks of correcting genes at the germinal level now far outweigh any benefits, given the primitive knowledge about human GLGT. Were animal and human laboratory experiments to reveal the requisite heightened level of safety needed to proceed, however, the cost–benefit calculation would be open to revision. Risk-based reasons for not proceeding with gene therapy at this time, in other words, leave the door open to eventual application.

A differing perspective regards the intentional manipulation of the germ line as fundamentally illicit because it would interfere with a resource that is shared across humanity. Still the idea of the genome as a collective human heritage is not autonomatically paired with a prohibition on GLGT. Instead, the notion of a collective heritage, which has been moderated in meaning from the earlier, more rigid concept of genetic patrimony, can serve to guide eventual research. The Ethics Committee of HUGO, for example, used the collective heritage concept to guide scientists but not prohibit them from human gene therapy (51).

Many facets of human gene therapy remain to be explored in the international arena. For example, it is taken for granted that enhancement genetic interventions ought not be pursued, yet policies are bereft of guidance about what a nonmedical germ-line intervention might be and where and how a line would be drawn between medical and nonmedical applications. In addition definitions of GLGT are being challenged by research advances that suggest inheritable manipulations can occur in ways other than through nuclear DNA splicing or that germ-line interventions need not be inherited.

Ethical inquiry into GLGT remains fertile and provocative. As organizations, agencies, associations, and governments continue the discussion, the UNESCO Universal Declaration provides a backdrop, signed by most of the world's nations, that sets forth durable principles to advise the judicious use of genetic knowledge. International attitudes have modulated in the last quarter of the twentieth century to embrace concerns about whether benefits can be equitably shared as well as whether individual and societal harms can be prevented.

BIBLIOGRAPHY

1. E.T. Juengst, *J. Med. Philos.* **16**(6), 587–592 (1991).

2. N.A. Wivel and L. Walters, *Science* **262**, 533–538 (1993).

3. Royal Commission on New Reproductive Technologies, *Proceed With Care: Final Report of the Royal Commission on New Reproductive Technologies*, Minister of Government Services of Canada, Ottawa, 1993.

4. L. Walters, *Hum. Gene Ther.* **2**(2), 115–122 (1991).

5. H.W. Jones, Jr., *11th World Congr. In vitro Fert. Alternative Assist. Reprod.*, Vienna, 1995, pp. 121–125.

6. R. Scott, *Hum. Reprod.* **12**(11), 2342–2343 (1997).

7. A German Embryo Protection Act (October 24, 1990), *Gesetz zum Schutz von Embryonen*; reprinted in *Hum. Reprod.* **6**(4), 605–606 (1991).

8. H.M. Beier and J.O. Beckman, *Hum. Reprod.* **6**(4), 607–608 (1991).

9. A. Mauron and J.-M. Thevoz, *J. Med. Philos.* **16**(6), 649–666 (1991).

10. Denmark Law No. 460 of 10 June 1997 on artificial fertilization in connection with medical treatment, diagnosis, and research, etc. (Lortindende, 1997, Part A, 11 June 1997, No. 84, pp. 2195–2198), Den. 97.25; reprinted in *Int. Dig. Health Legis.* **48**(3/4), 231–233 (1997).

11. Decree No. 97-613 of 27 May 1997, on studies conducted on human embryos in vitro, and amending the Public Health Code (Second Part: Decrees made after consulting the Conseil d'Etat). (Journal officiel de la Republique Francaise, Lois et Decrets, 1 June 1997, No. 126, pp. 8623–8624), Fr.97.60; reprinted in *Int. Dig. Health Legis.* **48**(3/4), 355–357 (1997).

12. S. Viville and I. Nisand, *Hum. Reprod.* **12**(11), 2341–2345 (1997).

13. J. Testart and B. Sele, *Hum. Reprod.* **10**(12), 3086–3390 (1995).

14. *Science* **264**, 1655 (1994).

15. Austria, Federal Law of 1992 (Serial No. 275) regulating medically assisted procreation (the Reproductive Medicine Law), and amending the General Civil Code, the Marriage Law, and the Rules of Jurisdiction. Date of entry into force: 1 July 1992. (Bundesgesetzlatt fur die Republik Osterreich, 4 June 1992, No. 105, pp. 1299–1304) Austr. 93.1; reprinted in *Int. Dig. Health Legis.* **44**(2), 247–248 (1993).

16. B.M. Knoppers and R. Chadwick, *Science* **265**, 2035–2036 (1994).

17. A. McGregor, *Lancet* **339**, 1345 (1992).

18. Law No. 35/1988 of 22 November 1988 on Assisted Reproduction Procedures. Boletin Oficial del Estado: 24 November 1988, No. 282, pp. 33373–33378; reprinted in *Int. Dig. Health Legis.* **38**(4), 782–785 (1989).

19. Law No. 68 of 12 June 1987 on artificial fertilization. Norsk Lovtidend, Section I, 26 June 1987, No. 13, pp. 502–503, Norway; reprinted in *Int. Dig. Health Legis.* **38**(4), 782–785 (1987).

20. Australia (Victoria). The Infertility Treatment Act 1995. No. 63 of 1995. Date of Assent: 27 June 1995 (151 pp.), Ausl. (Vic) 97.1; reprinted in *Int. Dig. Health Legis.* **48**(1), 24–33 (1997).

21. M. Warnock, *A Question of Life: The Warnock Report on Human Fertilization and Embryology*, Basil Blackwell, Oxford, 1985.

22. Human Fertilization and Embryology Act 1990, reprinted in *Int. Dig. Health Legis.* **42**(1), 69–85 (1991).

23. D. Morgan and L. Nielsen, *J. Law, Med. Ethics* **21**, 30–42 (1993).

24. L. Walters and J.G. Palmer, *The Ethics of Human Gene Therapy*, Oxford University Press, New York, 1997.

25. S. McCarthy, Presentation at Conference, Approaches to ART. Oversight: What's Best in the U.S.? Sponsored by The National Advisory Board on Ethics in Reproduction (NABER), RESOLVE and the U.S. Centers for Disease Control and Prevention, Washington, DC, 1998.

26. P. Baird, *Hum. Reprod.* **12**(11), 2343–2345 (1997).

27. NIH Guidelines for Research Involving Recombinant Molecules (NIH Guidelines), *Fed. Regist.* **55**, 7438–7448 (1990).

28. J. Couzin, *Science* **282**, 27 (1998).

29. J.G. Schenker and A. Shushan, *Hum. Reprod.* **11**(4), 908–911 (1996).

30. J.C. Byk, *Eubois J. Asian Int. Bioethics* **5**, 59–61 (1995).

31. B.A. Brody, *The Ethics of Biomedical Research: An International Perspective*, Oxford University Press, New York, 1998.

32. J.C. Fletcher, *Politics Life Sci.* **13**(2), 225–227 (1994).

33. A.L. Bonnicksen, *Politics Life Sci.* **17**(1), 3–10 (1998).

34. B.M. Knoppers and S. LeBris, *Am. J. Law Med.* **27**(4), 329–361 (1991).

35. European Commission, *Ethical, Legal and Social Aspects of the Life Sciences and Technology Programmes of the Fourth Framework Programme*, EUR 17767 EN, European Commission, 1998.

36. J.C. Byk, in D.R.J. Macer, ed., *Bioethics for the People by the People*, Eubois Ethics Institute, 1994, pp. 68–73.

37. Document from the Parliamentary Assembly of the Council of Europe, 1982. Thirty-Third Ordinary Session Recommendation 934 (1982) on genetic engineering; reprinted in *Hum. Gene Ther.* **2**(4), 327–328 (1991).

38. Parliamentary Assembly of Council of Europe Adopts Recommendation on Use of Human Embryos and Foetuses for Research Purposes; reprinted in *Int. Dig. Health Legis.* **49**(2), 485–491 (1989).

39. M.A.M. de Wachter, *Hastings Cent. Rep.* **27**(1), 13–23 (1997).

40. F.W. Dommel, Jr. and D. Alexander, *Kennedy Inst. Ethics J.* **7**(3), 259–276 (1997).

41. M.F. Munayyer, *Am. Univ. J. Int. Law Policy* **12**, 687–731 (1997).

42. Convention for the Protection of Human Rights and Dignity of the Human Being with Regard to the Application of Biology and Medicine: Convention on Human Rights and Biomedicine, Oviedo, 04.IV.1997. Available at: *http://www.coe.fr/eng/legaltext/160e.htm*

43. *Engineering the Human Germline Symposium*, Summary Report, University of California at Los Angeles (UCLA), 1998.

44. A. Abbott, *Nature (London)* **389**, 660 (1997).

45. J. Glover, *Ethics of New Reproductive Technologies: The Glover Report to the European Commission*, Northern Illinois University Press, DeKalb, 1989.

46. Council of Europe Publishes Principles in the Field of Human Artificial Procreation, *Int. Dig. Health Legis.* **40**(4), 907–912 (1989).

47. Quoted in E. Juengst, in E. Agius and S. Busuttil, eds., *Germ-Line Intervention and Our Responsibilities to Future Generations*, Kluwer Academic, Dordrecht, The Netherlands, 1998, pp. 85–102.

48. E. Agius, in E. Agius and S. Busuttil, eds., *Germ-Line Intervention and Our Responsibilities to Future Generations*, Kluwer Academic, Dordrecht, The Netherlands, 1998, pp. 67–83.

49. *Declaration on the Human Genome Project*, 44th World Med. Assembly, Marbella Spain, 1992.

50. N. Lenoir, *Kennedy Inst. Ethics J.* **7**(1), 31–42 (1997).

51. Statement on Research, HUGO Ethical, Legal and Social Issues Committee Report to HUGO Council. Available at: *http://www.gene.ucl.ac.uk/hugo/conduct.ht*

52. D. Butler, *Nature (London)* **398**, 179 (1999).

53. Universal Declaration on the Human Genome and Human Rights. Available at: *http://www.unesco.org/ibc/uk/genome/projet/index.htm*

54. Memo from E.M. Meslin to L. Staheli et al., *Report from UNESCO Meeting of Government Experts to Finalize the Draft Declaration on the Human Genome and Human Rights*, Paris, 1997.

55. H.-M. Sass, *J. Med. Philos.* **23**(3), 227–233 (1998).

56. D. Shapiro, *Politics Life Sci.* **13**(2), 233–234 (1994).

57. G.J. Annas, *N. Engl. J. Med.* **339**(24), 1778–1781 (1998).

58. R.M. Cook-Deegan, *Politics Life Sci.* **13**(2), 217–220 (1994).

See other Gene therapy and International aspects entries.

GENE THERAPY, ETHICS, GENE THERAPY FOR FETUSES AND EMBRYOS

Carol A. Tauer
College of St. Catherine
St. Paul, Minnesota

OUTLINE

INTRODUCTION

Trials of gene therapy for human fetuses and embryos have not yet been conducted. Discussions of in utero gene transfer have led to a consensus that it is too early to begin such trials. Gene transfer into early embryos has been debated as the most likely method for germ-line gene modification, and is similarly considered inappropriate for human trials at this time. However, animal research is making progress with both fetal and embryo gene transfer, indicating the importance of public discussion and education on possible human applications of these technologies.

TWO TYPES OF GENE THERAPY

Gene therapy in fetuses and gene therapy in embryos may appear to be quite similar, but the ethical and social issues raised by the two technologies are actually quite different. Gene therapy in fetuses, or in utero gene therapy, is technically and ethically similar to somatic cell gene therapy in born infants, children, and adults. Some additional ethical issues arise because interventions toward the fetus necessarily involve the pregnant woman. But these issues are similar to those that arise in any intervention or therapy directed toward the fetus, and are not specific to genetic interventions.

On the other hand, gene therapy in embryos is ordinarily understood to refer to genetic manipulation of in vitro fertilized embryos that have not yet been transferred to a woman. With the increasing availability of preimplantation diagnosis to identify genetic defects in early embryos, gene therapy offers a hope for correcting or ameliorating these defects rather than discarding the affected embryos. Since genetic intervention in embryos would occur at the developmental stage when all cells of the embryo are totipotent, or able to differentiate into all the cell and tissue types of the human organism, it is expected that genetic modifications of early embryos would affect germ cells as well as somatic cells. Thus genetic therapy in embryos is usually considered to be germ-line gene therapy, hence raising all the ethical and social issues identified in the germ-line intervention debate. In addition the development of embryo gene therapy presupposes the ethical acceptability of research involving early embryos, an issue that is unresolved in American public policy. As of May 2000, federal funding of human embryo research is prohibited by act of Congress, although there is no prohibition or regulation of privately funded embryo research.

Because they involve significantly different ethical and social issues, gene therapy for fetuses and gene therapy for embryos will be discussed separately in this article.

GENE THERAPY FOR FETUSES

Prenatal diagnosis through amniocentesis became a possibility in the late 1960s, providing prospective parents the option of abortion if a genetically compromised fetus was identified. Even before induced abortion was legalized throughout the United States, such abortions were generally permitted under the heading of therapeutic abortion. While many opponents of abortion regarded prenatal diagnosis negatively, even describing it as a "search and destroy" mission, the hope was that eventually it would be possible to correct the genetic anomalies that were diagnosed (1).

Two streams of clinical research converge in fetal gene therapy. The first type is research on fetal treatment in general. Trials of pharmacologic and surgical interventions in fetuses have shown some promise in treating hereditary and other congenital conditions (1,2). The second type is research on somatic cell gene therapy, particularly gene transfer in infants and children, or transfer directed toward conditions where early intervention is preferable or even essential (3). Fetal gene therapy would combine these two types of innovative interventions by utilizing the procedures of somatic cell gene therapy to treat the fetus in utero.

One specific type of fetal treatment, the in utero transfer of hematopoietic stem cells, is particularly pertinent to the development of fetal gene therapy. Similar to a bone marrow transplant, the transfer of pluripotent hematapoietic stem cells is believed to hold great promise for the treatment of congenital blood disorders. Although trials reported thus far indicate only a handful of successes, all with fetuses having immunodeficiency disorders, proponents believe that interest and application are likely to increase. The advantages of stem cell transfer in utero rather than after birth are (1) the absence of an immune response to foreign cells in early gestation, (2) the possibility of developing "tolerance" to foreign cells that would allow for further treatment after birth, and (3) intervention that is early enough to correct a disorder before clinical and uncorrectable manifestations develop (4,5).

The RAC and In Utero Gene Transfer

When the Food and Drug Administration (FDA) began to receive applications for approval of investigational bone marrow and stem cell transplants into fetuses, it recognized that in adults, these procedures have been forerunners to trials of gene therapy. Thus in late 1994 it urged the Recombinant DNA Advisory Committee (RAC) of the National Institutes of Health (NIH) to begin to examine gene therapy in fetuses (6,7). In addition to ethical issues related to experimental gene transfer procedures and issues related to the involvement of the pregnant woman when treatment in utero is contemplated, scientists recognized a further issue. It is believed that in utero gene transfer, particularly at an early gestational stage, carries a risk of unintentionally affecting germ cells, a risk not incurred in trials of somatic cell gene therapy in children and adults. Thus in utero gene therapy could result in the first germ-line effects from gene transfer trials, brought about as unintentional side effects of a procedure classified as somatic cell gene therapy. Since the RAC has decided not to consider protocols that involve germ-line transfer at this time, such an unintended outcome would circumvent one of the ethical and social barriers currently believed to be prudent (6).

Also in 1994 NIH began a reassessment of the regulatory role of RAC regarding the approval of gene therapy propocols. Prior to 1994 all protocols had to be individually approved by RAC as well as FDA, resulting in what many perceived as duplication of review and unnecessary delays. A series of compromise proposals were considered by NIH administration, FDA, and RAC, and in 1996 Harold Varmus, director of the NIH, initially decided to eliminate RAC and to transfer its role in developing public policy to a new body within NIH. However, public comments weighed heavily against this plan, and Varmus decided to retain RAC, to continue requiring simultaneous submission of gene therapy proposals to RAC and FDA, but to require only FDA approval of individual protocols (8).

NIH's final policy decision was announced on October 31, 1997. It stated RAC's new functions as identifying specific human gene transfer proposals that raised novel issues deserving of public discussion, transmitting to the NIH director its recommendations on such proposals, and initiating consideration of forthcoming gene transfer procedures raising new ethical and social issues (8). Such consideration could be initiated in the absence of specific submitted protocols, with the intention of raising public awareness and obtaining public input prior to actual implementation of the novel procedures. Besides maintaining public access to its meetings, RAC together with NIH's Office of Recombinant DNA Activities (ORDA) would be expected to sponsor regular conferences on new developments in order "to serve as a unique public forum for the discussion of science, safety, and ethics of recombinant DNA research" (9).

These changes in the role of RAC are particularly pertinent to consideration of gene transfer in fetuses. Although no protocols for trials of in utero gene transfer had yet been submitted, in January 1999 RAC and ORDA sponsored a conference on "Prenatal Gene Transfer: Scientific, Medical, and Ethical Issues." This conference continued the discussion initiated when FDA referred the matter to RAC in 1994. The topic became even more timely as a result of the submission of two "preprotocols" for in utero gene transfer brought to RAC in July 1998 by

W. French Anderson and Esmail Zanjani. Their intention was not to seek approval for research trials but to stimulate discussion of the issues raised by this new application of gene transfer. While such preliminary discussion would be similar to discussions that took place before the first somatic cell gene therapy protocol was approved in 1990, it was particularly appropriate to the new role and function of RAC.

The two preprotocols on in utero gene transfer were the major topic of discussion for the RAC meeting of September 24, 25, 1998. Besides reviews submitted by individual RAC members, eight ad hoc consultants contributed reviews and six of them participated in the discussion at the meeting (10). Issues from this meeting were brought to the January 1999 conference on "Prenatal Gene Therapy," which was organized around questions assigned to three working groups: Preclinical Research Issues, Clinical Research Issues, and Ethical, Legal, and Societal Issues. By the end of the conference, areas of consensus and of disagreement had been identified. The main conclusion of the conference was stated as follows:

> At present, there is insufficient clinical data to support the initiation of clinical trials involving prenatal gene transfer. A substantial number of critical scientific, ethical, legal, and social issues must be addressed before clinical trials proceed in this arena (9).

This statement was followed by a listing of 26 specific areas in which more data were needed before clinical trials could be considered.

At the RAC meeting of March 11, 12, 1999, chairs of the three working groups presented their reports, responded to questions, and led discussion of the issues. While there were many areas of consensus, there were also some points of disagreement (11). The committee decided, however, that it was prepared to make a public statement in order to clarify its position:

> The RAC continues to explore the issues raised by the potential of *in utero* gene transfer research. However, at present, the members unanimously agree that it is premature to undertake any human *in utero* gene transfer experiment (11).

The RAC returned to the topic of in utero gene transfer at its June 14, 1999, meeting. At this meeting it reviewed a document that would eventually be published as a more detailed report of RAC's findings regarding prenatal gene transfer and that would be incorporated into the NIH *Guidelines for Research Involving Recombinant DNA Molecules* (12,13).

Summaries and minutes of RAC meetings, available through the NIH Web site and eventually through publication in the journal *Human Gene Therapy*, are an indispensable source of information regarding current issues in prenatal gene transfer. While committee discussion attempts to separate scientific and clinical questions from ethical and social issues, these areas overlap in many respects. For example, questions about the safety of gene transfer procedures may be regarded as preclinical or clinical questions, but the presence of significant risk to the fetus or pregnant woman also raises an ethical problem. In the following discussion, the focus on ethical and social issues will incorporate preclinical and clinical questions.

Areas of Ethical Consensus

Both in the ethics literature and in the discussions at RAC meetings, some areas of broad consensus have been identified, though consensus does not preclude the possibility of dissenting voices (1,10,11). Some consensus points relate to whether and when fetal gene transfer trials should begin, while others relate to requirements or conditions for protocols at a time when such trials are implemented.

Terminology. In the literature on research ethics there is continuing concern that research subjects may expect therapeutic benefit from their participation in research, even though there is no evidence for such benefit. This expectation, called the "therapeutic misconception," may lead subjects to accept risks that they would not accept if they realized that the procedures to which they are asked to consent are intended to gain knowledge for the treatment of future patients but are unlikely to benefit the research subjects (14). The therapeutic misconception may be reinforced by describing research procedures as "therapies" and by identifying research subjects as "patients."

Historically the research on gene transfer procedures aimed at therapeutic goals has been characterized as gene therapy research. However, nearly a decade of this research "has quite understandably failed to produce results swiftly leading to viable gene therapies" (15). In this situation, continuing to describe protocols as "gene therapy" contributes to a therapeutic misconception by potential research subjects as well as unwarranted expectations in the public. Because it is important that subjects have a realistic grasp of risks relative to expected benefits in order to give fully informed consent, misleading terminology should be avoided.

Several commentators have urged that the terms "gene therapy," "treatment," and "patient," be replaced by "gene transfer," "research," and "subject" in all gene transfer protocols (15). This proposal was supported by LeRoy Walters, former chair of RAC, at the meeting of September 24, 25, 1998, when he urged the committee to adopt the revised language (10). Examination of recent RAC documents indicates compliance with this proposal.

Genetic Counseling. In utero gene transfer would be performed on fetuses who are diagnosed as having a genetic condition. In turn, such prenatal diagnosis would be offered to couples who are at risk for transmitting a genetic disease. Thorough and unbiased genetic counseling must be available to all couples who have such risk factors. All options, including the option of abortion of an affected fetus, must be presented to them. After a genetic condition is diagnosed prenatally, the time frame for decision making is often brief. Yet time must be taken for the necessary information to be provided, for all options to be explained, and for reflection to occur.

Informed Consent. The consent process and form must clearly state that fetal gene transfer is an experimental procedure or intervention, that it may not benefit the fetus and will not medically benefit the woman, and that it carries specific (enumerated) risks to the fetus and to the pregnant woman. The pregnant woman provides consent for all interventions toward the fetus and herself. Though it is highly recommended that her partner be involved in discussions, he can neither consent to nor refuse interventions against her wishes. However, with regard to experimental interventions such as in utero gene transfer, it would be inadvisable to proceed if there were disagreement between the two partners.

Prenatal gene transfer cannot be made conditional on the woman's agreement not to seek a later abortion, nor on her agreement to have an abortion should the gene transfer not prove successful (1). (Note, however, that it may not be possible to test the effectiveness of the transfer until after birth.) Because an autopsy may be important in assessing the results of a clinical trial, the woman may be asked to agree to autopsy of the fetus or herself if either of them should die. However, if she changes her mind, an autopsy cannot be forced on her. The same thing is true of a requirement for long-term follow-up for herself and the prospective child. She may be asked to consent to follow-up assessments, but as with any research protocol, she cannot be forced to continue should she choose to withdraw.

Efficacy. As indicated in the RAC consensus statement, there is agreement that not enough is known about the potential effectiveness of in utero gene transfer to support clinical trials at this time (11). There is some promising research with animals, for example, the apparently successful reversal of cystic fibrosis in mice by gene therapy in utero (16). Trials of somatic cell gene therapy in born humans, however, have been disappointing. Since the first clinical trial in 1990, over 300 human gene transfer protocols have been registered with NIH, but none has been a clear therapeutic success. In the words of W. French Anderson, pioneer and proponent of gene therapy:

> The efficiency of gene transfer and expression in human patients is … still disappointingly low. Except for anecdotal reports of individual patients being helped, there is still no conclusive evidence that a gene-therapy protocol has been successful in the treatment of a human disease (17).

Before subjecting pregnant women and their fetuses to experimental gene transfer procedures, researchers must have adequate evidence of potential efficacy in order to balance the risks involved. Many preclinical and clinical questions remain unanswered as of 1999. What is the most efficient way to transfer genes to the fetus? How can gene transfer be targeted to particular cell types without inadvertently exposing other cells, especially germ cells, to modification? When gene expression requires regulation to be successful, how would this be accomplished? The RAC working group on clinical issues agreed that clinical trials should not be undertaken until animal studies indicated that expression of transferred genes occurred at a level "conducive to correction of the phenotype rather than

merely a slight change," or in other words, at a high level of effectiveness (11).

Safety. Gene transfer studies indicate that "the procedure appears to carry a very low risk of adverse reactions" (17). On September 17, 1999, the death of an 18-year-old man four days after a gene transfer into his liver marked the first fatality attributed to an experimental gene transfer procedure (18). One could argue that one fatality in over 300 protocols is a small number, but the event serves as a warning that gene transfer can involve serious risks.

In the case of treatment in utero, safety to the pregnant woman as well as to the developing fetus must be considered, and risks should be weighed separately for woman and fetus. The pregnant woman may consent to accept some additional risks for the sake of her fetus, but risks to her health should be minimized as they would be in any research involving a healthy subject. In most cases risks to the pregnant woman should be limited to risks arising from the method of gene delivery, for example, the risks of a surgical procedure.

In relation to the fetus, some of the safety issues are parallel to issues recognized in adult gene transfer procedures: the potential risk of mutagenesis caused by insertion of the transfer vectors, the possibility of vectors being replication-competent, and the risk of harmful unregulated gene expression. Additional questions raised by fetal gene transfer include: How might the process of fetal development be affected by introducing a vector? Is transplacental migration of virus vectors a possibility (thus potentially affecting the pregnant woman also)? Though the fetus may not have the same immune response to foreign DNA that is seen in adults, what other immune response problems might arise (11)?

Areas of Ethical Disagreement

Points of disagreement largely focus on which diseases, or types of diseases, would be the best candidates for the first trials of human in utero gene transfer. Should the first trial be for a disease where there is an animal model, and where effectiveness and safety have been clearly demonstrated in that model? Some diseases do not have animal models but nonetheless might be good candidates for human trials.

Should the first trial be for a disease where postnatal gene therapy of infants, children, or adults has been successful? For example, trials provide some evidence for the effectiveness of gene therapy in children with the immunodeficiency adenosine deaminase-severe combined immunodeficiency disorder (ADA-SCID) (1). However, in these trials an insufficient number of cells have been transduced to produce adequate ADA, so children in the gene transfer research have continued to receive supplementary administration of polyethyleneglycol (PEG)-ADA by shots that are both painful and expensive. It is possible that in utero gene transfer would improve transduction efficiency and hence eliminate the need for supplementary PEG-ADA. For this reason one of the preprotocols submitted to RAC in 1998 by Anderson and Zanjani related to ADA-deficient SCID (10). This preprotocol suggested a trial of in utero gene transfer for a condition where there is some evidence for the effectiveness of postnatal transfer efficacy, and where there is a belief that prenatal treatment would be more effective.

While there is agreement that the first trials of fetal gene transfer should be for serious diseases, and certainly not for mere improvement of desirable traits, disagreement remains on these questions: Should in utero gene transfer be limited to diseases for which there is no effective nongenetic treatment? An affirmative answer would eliminate ADA-SCID, since PEG-ADA is an effective but burdensome treatment. Should fetal gene transfer be limited to diseases in which irreversible damage will occur in utero if the disease is not corrected at an early gestational stage?

A further disagreement among RAC members relates to the treatment of diseases that are ordinarily fatal in utero. This issue arose in connection with the second of the Anderson-Zanjani preprotocols, gene transfer for α-thalessemia. Besides being lethal to the fetus, this disease also produces toxic symptoms in the pregnant woman, and is thus an indication for therapeutic abortion. Although various in utero therapies for α-thalessemia have been attempted, none has been therapeutically effective, although there have been cases of partial correction resulting in the live birth of a severely affected infant (10).

Comments on the α-thalessemia gene transfer preprotocol expressed two contrasting points of view. Some RAC members and consultants argued that the disease was a good choice for a gene transfer experiment because it would otherwise be fatal in utero, and because there were no existing treatments for the disease. Others maintained that a disease that was fatal in utero was not a good choice for gene transfer, but that it would be better to let nature take its course. The risk of contributing to the birth of a severely ill newborn for whom there was no effective treatment, plus the serious risk to the woman from carrying a potentially toxic pregnancy to term, both argue against the selection of α-thalessemia for prenatal intervention.

The controversial issues involved in determining which types of diseases would be most appropriate for the first trials of fetal gene transfer are typical of the kinds of questions on which RAC and NIH seek public input. Given the RAC position that clinical trials of any in utero gene transfer would be premature at this time, there is a window of opportunity for public education and discussion to occur.

GENE THERAPY FOR EMBRYOS

Embryo Gene Transfer as Germ-Line Modification

Discussions of germ-line gene therapy agree that genetic modification of the one-celled zygote or early embryo is probably the most feasible way to make germ-line alterations. Because early embryonic cells are totipotent, or able to differentiate into any of the cell or tissue types of the organism, a genetic modification introduced at this stage has the potential to affect the developing germ cells are the sperm and the eggs. The only way to make the zygote or early embryo available for gene transfer procedures is through in vitro fertilization (IVF),

thus requiring use of this reproductive technology as preparation for gene transfer. It may also be possible to achieve germ-line changes through modification of sperm stem cells. However, since females possess all their egg cells at birth, direct modification of egg cells is highly unlikely (19).

Animal research involving the genetic modification of mouse embryos shows that gene transfer before cell differentiation can produce changes that are transmitted to offspring. For example, Leroy Hood's experiments with "shiverer mice" demonstrate that genetic alterations of affected embryos can correct expressions of this phenotype, not only in the mice that develop from the treated embryos but also in their descendants. Other research groups have prevented the transmission of serious diseases in generations of mice by genetically altering mouse embryos (20,21). According to LeRoy Walters and Julie Palmer, there has been more success with germ-line genetic intervention than with somatic cell gene therapy in laboratory animals (19).

The ethical and social issues involved in gene therapy for embryos encompass all the controversies raised by germ-line transfer in general. In fact RAC has not explicitly considered the topic of gene therapy in embryos, while it has taken a position regarding germ-line gene therapy. Its position is stated in the NIH *Guidelines*: "[The] RAC will not at present entertain proposals for germ-line alterations" (13). The RAC sees its role as providing a forum for discussion of germ-line transfer, both at its meetings and through sponsored conferences, as long as it is clear that such discussion does not imply RAC endorsement of germ-line transfer (22).

Advantages of Gene Therapy for Embryos

The most obvious benefit from genetic modification of early embryos would be the correction of lethal and other serious diseases so that they will not be transmitted to offspring and later descendants. While somatic cell gene therapy may cure or help an individual who suffers from a genetic disease, it will not prevent transmission of a heritable disease to the next generation, since somatic cell gene transfer will not alter germ cells. When a disease is undesirable for an individual, it is equally undesirable for that individual's descendants, and its elimination may appear to have only good consequences.

Not only does embryo gene therapy offer benefits to the individual and family involved, but it promises long-term benefits related to improving the human gene pool. Somatic cell gene therapy may cure or ameliorate a disease in an existing individual, but in extending that person's life and enabling him or her to reproduce, it could have the effect of increasing the total number of persons (offspring) suffering from the disease. In preventing this outcome, germ-line gene therapy may be preferable to somatic cell therapy.

For some genetic conditions, somatic cell gene transfer would not work in principle. Diseases expressed in nondividing cells, like the cells of the neural system, are intractable to many somatic cell gene transfer techniques. Thus a disease like Lesch-Nyhan syndrome might respond to gene transfer into affected embryos, while somatic cell gene transfer would be ineffective. Similarly somatic cell techniques that require removal of cells before their modification would not work with diseases expressed in nonremovable cells (19).

Some genetic diseases, for example, cystic fibrosis, affect many different organs and cell types in the body. Gene transfer at the early embryo stage would allow delivery of a corrective gene to all affected cell types, while later somatic cell transfer could require repetition of a variety of gene transfer procedures. This argument also applies to cancers that result from inborn genetic factors and subsequent mutations. Such cancers have the propensity to affect more than one organ or system, and hence might be prevented most effectively through correcting the genetic defect at the early embryonic stage (19).

Finally, some genetic conditions result in irreversible damage, often to the brain, during the first trimester of pregnancy. Corrective gene transfer into the fetus or after birth could not ameliorate this damage. Since the embryo in utero is not accessible to treatment until some weeks after implantation, early intervention requires that the gene transfer be done in the laboratory after IVF. Gene transfer into the zygote or early embryo may be the only way to prevent the intrauterine harms that are foreseen.

Arguments Against Gene Therapy for Embryos

Standard arguments against gene therapy for embryos are summarized by Walters and Palmer (19). These arguments generally view embryo gene transfer as a form of germ-line gene therapy, and hence focus on the issue of modifying the genetic heritage of individuals and of the human race.

Safety. A germ-line gene transfer procedure may have negative effects, including some that do not show themselves until later generations. In the case of somatic cell gene transfer, negative effects harm only the individual who is treated (or in the case of a fetus in utero, possibly also the pregnant woman). But when a genetic modification affects the germ cells, as is the case with modification of early embryos, then all descendants of the transfer recipient could be negatively affected. There is speculation that techniques will be developed to remove or render inoperable an inserted gene that is causing trouble for later generations (19). However, there are no guarantees that such reversal will be available.

Safety issues must be resolved through animal studies before attempts are made to modify human embryos. Even when animal studies demonstrate that gene transfer techniques are safe and effective, there is some risk in moving to human applications because of possible species differences. For successful gene transfer into embryos, the added genes must integrate without disrupting normal development of the resulting fetuses. The genes must integrate into all cells of the early embryo rather than merely some of them, a situation that could produce a genetic chimera. And the integrated genes must be properly expressed later in the born human being (19).

Use of Resources. Gene transfer into embryos would be a complex, multi-step procedure. First, the couple at risk for transmission of a genetic defect would have to conceive through IVF. While this procedure is ordinarily chosen by couples who are infertile, here it would be used in order to make embryos available for diagnosis and manipulation. Second, it would be necessary to identify the specific embryos that have the genetic defect by using preimplantation genetic diagnosis (PGD). This procedure is itself experimental, and its application has involved some erroneous diagnoses (23). Third, some type of gene transfer would be utilized in order to correct the genetic defect in the embryos that were identified.

Each of the three steps in this treatment protocol is expensive and technology intensive. Most likely the protocol could be made available only to affluent families, unless alternate funding were available during the research phase. In a society where millions of people lack access to basic health care, it is questionable whether such a procedure is a prudent use of medical, scientific, and technological resources.

Inviolability of Genetic Heritage. Some opponents of germ-line gene transfer argue that making genetic modifications that will be passed to future generations involves an improper tampering with the future of evolution and the human gene pool. These critics may rely on religious arguments, claiming that scientists would be "playing God," or they may hold that there are natural limitations built into the universe that humans ought not to exceed (19).

A related argument claims that human beings have a right to receive a genetic heritage that has not been tampered with. Stated as a type of human right, this prerogative has been enunciated by the Parliamentary Assembly of the Council of Europe. In its strongest form, the presumed right has led to adoption of a constitutional ban on germ-line intervention in Switzerland (24).

Positions like these are deontological, or based on a theory of rightness and wrongness that is independent of the consequences that are actually produced. Hence they cannot be refuted by arguments that point to the potential elimination of lethal or other serious genetic diseases. They might be countered, however, by making analogies to other medical procedures that seem to interfere with the ordinary course of nature, or to other ways in which we alter genetic heritage, for example, by selective breeding of animals. Opponents might still argue that human genetic heritage is different and should be regarded as inviolable.

Alternatives to Gene Therapy in Embryos

One particular form of opposition to embryo gene therapy focuses on alternatives that are regarded as preferable. Two options are offered. Since embryo gene therapy must be preceded by preimplantation genetic diagnosis, the genetic disease under consideration could be avoided by simply discarding affected embryos. Alternatively, in order to avoid the complex procedures of IVF and PGD, prenatal diagnosis during an established pregnancy followed by abortion of an affected fetus would achieve the same goal (25).

Either of these options may be viewed as safer, both in terms of achieving the desired result and in order to avoid harmful side effects or possible negative outcomes in a later generation. However, there may be situations in which the options are not available; for example, when both partners have two copies of the same malfunctioning gene, that is, both are afflicted with the same recessive genetic disorder. In this situation all genetic offspring of the couple would necessarily have the same disease as their parents (19).

Additionally some couples may have moral objections to aborting an affected fetus, or to discarding embryos that are identified as affected. Walters and Palmer suggest that attempting to treat a genetic disease, rather than discarding embryos or eliminating fetuses that have the disease, fits more closely the mission of the health sciences and shows greater respect for born persons who have a genetic disease or disability (19). Munson and Davis argue that medicine by its very nature has a therapeutic obligation to pursue the development of genetic therapy as a way of curing and eliminating disease (26).

The close relationship between preimplantation genetic diagnosis and gene therapy in embryos suggests that the ethical and social issues posed by these procedures should be considered together. Yet, as Pergament and Bonnicksen note, the form that public policy discussions have taken has led to their separation, with PGD seen as an issue related to research on reproductive technologies, and embryo gene therapy viewed as germ-line gene transfer and thus coming under the purview of RAC (27).

Research Involving Early Human Embryos

Public policy in the United States does not allow federal funding of research that involves early human embryos. Thus research on preimplantation genetic diagnosis, a necessary forerunner to gene therapy in embryos, may not be funded or sponsored by the National Institutes of Health.

In 1994, as a result of a change in the congressional language on appropriations for NIH, it appeared that some research involving IVF and early embryonic development would be fundable. At that time the director of NIH appointed a panel, the Human Embryo Research Panel (HERP), to develop guidelines for such funding. In its report HERP recommended federal funding for a variety of types of research related to infertility. It also gave its approval to funding for research on preimplantation genetic diagnosis (28).

The Panel was explicitly directed not to consider the issue of research on germ-line gene transfer. On the assumption that such research fell within the purview of RAC, the charge to the Human Embryo Research Panel stated that "Research involving human germ-line gene modification . . . is not within the Panel's scope" (28, p. ix). This somewhat artificial separation of tasks eliminated the possibility of joint consideration of two closely related procedures, preimplantation genetic diagnosis and embryo gene transfer (27). Because of the limitations stated in its charge, HERP did not discuss embryo gene transfer in any way.

Similarly RAC has given essentially no attention to the fact that research on germ-line gene transfer would most likely involve research on early human embryos. If germ-line transfer is ever to be seriously considered as a public policy issue, this aspect of the issue would have to be debated.

Debate would be further complicated by the fact that shortly after the HERP report was approved by the Advisory Committee to the Director of NIH, Congress rescinded its approval of federal funding for IVF research. Beginning with fiscal year 1996 the appropriations bill for NIH has specifically prohibited funding of any research in which early human embryos are harmed or destroyed. It is unlikely that the congressional prohibition will be removed in the foreseeable future, with the possible exception of allowing research on embryonic stem cells that may have great therapeutic potential for treating diseases in already-born people.

Embryo Gene Transfer for Enhancement Purposes

While many advocates of gene transfer, whether germ-line or somatic cell, have stressed that it ought to be used only for therapeutic goals, others have recognized the likelihood that its use will eventually be extended to the enhancement of desirable human traits. Wivel and Walters believe that "Germ-line gene modification for serious disease will inevitably lead to the next step, genetic enhancement" (29). RAC has raised this concern about germ-line transfer, acknowledging that once the technique is successful as a medical procedure, it has the potential for "off-label" use for enhancement purposes. The issue of genetic enhancement is thus an important topic for debate by RAC, one where the committee recognizes its role in facilitating public discussion, education, and input (30).

A number of examples illustrate the tendency to extend use of FDA-approved therapies to enhancement purposes. When recombinant DNA procedures made human growth hormone (HGH) available in large quantities, its therapeutic use to treat dwarfism resulting from HGH deficiency was quickly expanded to the treatment of children who were short but had no medical condition (31). Breast implants that were developed for women who had radical surgery because of cancer became cosmetic prostheses for women who desired larger breasts. In the gene transfer area, recent reports of mice whose hair growth was stimulated by insertion of the Sonic hedgehog gene note that therapeutic use in humans would be directed to persons with hair loss due to chemotherapy. However, cosmetic uses of gene transfer to reverse baldness are anticipated, either with enthusiasm or with hesitation, depending on the perspective of the commentator (32).

While early discussions contrasting gene transfer for therapy with gene transfer for enhancement suggested that the distinction is clear, this is by no means the case (19,33). The preceding examples suggest that the line between a supposed therapeutic application and an enhancement or cosmetic application may be somewhat murky. Walters and Palmer distinguish between enhancement goals that are health related and those that are not. An enhancement that is health-related alters one's physical condition so that one is more resistant to disease. For example, immunization is widely accepted as a physical enhancement that renders the body immune to certain infectious diseases. Similarly a gene transfer procedure might provide immunity to acquired immunodeficiency syndrome (AIDS), or enable the body to fight cancer more effectively (19).

Arguments by analogy with current practices thus seem to give some support to genetic modification for health-related enhancements like disease immunity. But once again, it is not always easy to distinguish health-related enhancements from non-health-related enhancements. Cognitive improvements such as improved memory might initially seem non-health-related. Yet, if one were able to use gene transfer to improve the cognitive functioning of a mentally retarded person, this enhancement might be seen as health related, or even therapeutic.

Some authors have viewed the prospect for genetic enhancement as the most perplexing problem in the germ-line gene transfer debate (25). For those who view enhancement uses of genetics as undesirable, the potential for these applications provides a strong argument against pursuing germ-line gene transfer. The fears of opponents are reinforced by assertions that "the only plausible reason to insert genetic material into embryos would be for genetic enhancement" (34). Given the availability of less complex means to avoid the birth of genetically compromised children (prenatal or preimplantation diagnosis), enhancement may eventually become the main reason for germ-line gene transfer, or for gene transfer in early embryos.

BIBLIOGRAPHY

1. J.C. Fletcher and G. Richter, *Hum. Gene Ther.* **7**, 1605–1614 (1996).

2. M.I. Evans et al., in A. Milunsky, ed., *Genetic Disorders and the Fetus: Diagnosis, Prevention, and Treatment*, 3rd ed., Johns Hopkins University Press, Baltimore, MD, 1992, pp. 771–797.

3. C. Coutelle et al., *Nature Med.* **1**(9), 864–866 (1995).

4. A.W. Flake and E.D. Zanjani, *J. Am. Med. Assoc.* **278**(11), 932–937 (1997).

5. Gene Therapy Advisory Committee, Health Departments of the United Kingdom, *Hum. Gene Ther.* **10**, 689–692 (1999).

6. H. Gavaghan, *Nature Med.* **1**(3), 186–187 (1995).

7. P.D. Noguchi, *Food Drug Law J.* **51**, 367–373 (1996).

8. R.A. Merrill and G.H. Javitt, in T.J. Murray and M.J. Mehlman, eds., *Encyclopedia of Ethical, Legal & Policy Issues in Biotechnology*, Wiley, New York, 2000.

9. National Institutes of Health, *Gene Therapy Policy Conference: Prenatal Gene Transfer*, Available at: *http://www.nih.gov/od/orda/gfpcconc.htm* (1999).

10. National Institutes of Health, Recombinant DNA Advisory Committee, *Minutes of meeting, September 24, 25, 1998*, Available at: *http://www.nih.gov/od/orda/9-98rac.htm#IV* (1999).

11. National Institutes of Health, Recombinant DNA Advisory Committee, *Minutes of meeting, March 11, 12, 1999*, Available at: *http://www.nih.gov/od/orda/3-99rac.htm* (1999).

12. National Institutes of Health, Recombinant DNA Advisory Committee, *Minutes of meeting, June 14, 1999*, Available at: *http://www.nih.gov/od/orda/6-99rac.htm* (1999).

13. National Institutes of Health, *Guidelines for Research Involving Recombinant DNA Molecules*, Appendix M., Available at: *http://www.nih.gov/od/orda/apndxm.htm* (1999).

14. P.S. Appelbaum, L.H. Roth, and C.W. Lidz, *Int. J. Law Psych.* **5**, 319–329 (1982).

15. N.M.P. King, *Hum. Gene Ther.* **10**, 133–139 (1999).

16. J.E. Larson et al., *Lancet* **349**, 619–620 (1997).

17. W.F. Anderson, *Nature* **392**(Supp. April 30), S25–S30 (1998).

18. R. Weiss and D. Nelson, *Washington Post*, September 29, p. 1 (1999).

19. L. Walters and J.G. Palmer, *The Ethics of Human Gene Therapy*, Oxford University Press, New York, 1997.

20. J.W. Gordon, *Int. Rev. Cyt.* **115**, 171–229 (1989).

21. R.D. Palmiter and R.L. Brinster, *Ann. Rev. Gen.* **20**, 465–499 (1986).

22. National Institutes of Health, Recombinant DNA Advisory Committee, *Hum. Gene Ther.* **10**, 489–533 (1999).

23. A.L. Trounson, *Hum. Reprod.* **7**(5), 583–584 (1992).

24. A. Mauron, *Politics Life Sci.* **13**, 230–232 (1994).

25. N. Holtug, *Cambridge Q. Healthcare Ethics* **6**, 157–174 (1997).

26. R. Munson and L.H. Davis, *Kennedy Inst. Ethics J.* **2**(2), 137–158 (1992).

27. E. Pergament and A. Bonnicksen, *Am. J. Med. Genet.* **52**, 151–157 (1994).

28. National Institutes of Health, Human Embryo Research Panel, *Report of the Human Embryo Research Panel*, National Institutes of Health, Bethesda, MD, 1994.

29. N.A. Wivel and L. Walters, *Science* **262**, 533–538 (1993).

30. National Institutes of Health, Recombinant DNA Advisory Committee, *Hum. Gene Ther.* **9**, 911–932 (1998).

31. C.A. Tauer, *IRB* **16**(3), 1–9 (1994).

32. N. Wade, *N.Y. Times*, October 5, pp. C1, C9 (1999).

33. E. Juengst, in T.J. Murray and M.J. Mehlman, eds., *Encyclopedia of Ethical, Legal & Policy Issues in Biotechnology*, Wiley, New York, 2000.

34. J.R. Botkin, *J. Law Med. Ethics* **26**, 17–28 (1998).

See other GENE THERAPY entries; HUMAN SUBJECTS RESEARCH, ETHICS, RESEARCH ON HUMAN EMBRYOS; REPRODUCTION, ETHICS, MORAL STATUS OF THE FETUS; REPRODUCTION, LAW, WRONGFUL BIRTH, AND WRONGFUL LIFE ACTIONS.

GENE THERAPY, ETHICS, GERM CELL GENE TRANSFER

ROBERT NELSON
University of Pennsylvania
Philadelphia, Pennsylvania

OUTLINE

Introduction
Discussion Overview
Technical Issues with GCGT

INTRODUCTION

Germ cell gene transfer (GCGT) is distinguished from somatic cell gene transfer by the fact that the intervention alters the DNA of the germ cells (with or without altering the DNA of somatic cells) and the alteration is transmitted to the individual's progeny. Thus GCGT can be directed to the germ cells of a differentiated organism (sperm or egg) through gametocyte modification, or to the undifferentiated organism at an early embryonic stage (prior to the cellular distinction between somatic and germ cell lines), hereafter referred to as "pre-embryo modification." Somatic cell gene transfer, whether for therapeutic or enhancement purposes, is limited in effect to the individual who is the recipient of the technique. Gametocyte modification is directed solely at the transmission to an individual's progeny of either undesired or desired genes, independent of whether the parent has undergone the same modification. Pre-embryo modification would likely result in the genetic alteration of both somatic and germ cell lines so that the resulting individual and his or her progeny would benefit from the gene transfer. In addition to altering intergenerational transmission of genes, pre-embryo modification may emerge as an effective technique for somatic cell gene transfer. As such, we need to consider whether GCGT as a result of pre-embryo modification can be considered an acceptable albeit indirect effect (1). Nevertheless, any technique of GCGT would be precluded if there were compelling moral reasons why we should never directly or indirectly intend to alter the human genome so as to impact on our future progeny.

DISCUSSION OVERVIEW

This article begins with a brief review of some technical issues in GCGT to set the stage for considering the ethics of GCGT research. It is inappropriate to discuss the issue of GCGT as therapy before considering GCGT as research in light of current federal research guidelines. Four conditions will be identified which must apply for initial GCGT research to proceed: (1) no alternative treatments exist (including somatic cell gene transfer), (2) the phenotypic injury occurs early in fetal development, (3) the outcome is uniformly fatal, and (4) the condition is caused by a single gene defect. With this as a foundation, we then explore the major objections to GCGT based on an illegitimate "tampering" with our genetic endowment balanced by an obligation to heal those afflicted with a genetic disease. This obligation to treat an individual, however, may be interpreted as an obligation to prevent genetic disease in a population, raising concerns about the eugenic use of GCGT. Nevertheless, from an individual perspective, a parent may appropriately choose to enhance a child's opportunities through the use of genetic technology. If so, the challenge will be to distinguish between legitimate and illegitimate eugenic uses of GCGT—a challenge that raises the issue of the social and political control of GCGT technology. Before addressing the control of genetic technology, we need to address the similarities and differences between pre-embryo discard as part of GCGT procedures and the controversial issue of abortion. In addition GCGT needs to be considered in the context of the alternatives—most of which currently involve either embryo discard or abortion.

TECHNICAL ISSUES WITH GCGT

Gene transfer techniques that may be applied to germ cells fit into three broad categories: gene augmentation, gene modification, and gene excision and splicing. Gene augmentation in which a functional gene is inserted into a cell to direct the synthesis of an otherwise missing or defective gene product, or gene modification in which a functional gene is inserted into the nuclear DNA, have been used for somatic cell gene transfer. Gametocyte modification using gene augmentation probably will not be an effective technique given the need for the gene to be distributed predictably to subsequent cells (such as gametes) (2). Gene modification through the insertion of a replacement gene into nuclear DNA would allow for the transmission of the inserted gene to the subject's gametes; however, the insertion may disrupt otherwise functional genes, uncover or create proto-oncogenes, or lead to gene expression in inappropriate tissues (2). Given the need for precise timing and expression of gene products during embryonic and fetal development, it is likely that only gene excision and splicing targeted to particular missing or nonfunctional genes will meet the necessary standards of safety, accuracy and cellular integrity (3). However, accurate gene excision and splicing techniques do not exist at the present time.

Gene transfer techniques targeted at either somatic or germ cells may be effective only with single gene disorders given the necessary causal assumption that the simple absence of a functional gene results in a diseased phenotype. The correction of a dominant disorder may require the removal of the offending gene, whereas the simpler technique of gene augmentation or modification may correct a recessive disorder. Pre-embryo modification also depends on the ability to diagnose the targeted genetic disease at an early stage of in vitro development in order to determine the need for or success of treatment. Finally, there are a number of diseases involving mitochondrial DNA that may be amenable to either the insertion of a functional gene or the complete exchange of functional for dysfunctional mitochondria (3,4). GCGT techniques that target mitochondrial DNA can be considered a special case of either gene augmentation or modification and will be discussed separately.

ETHICS OF GCGT RESEARCH

The ethical application of GCGT to either the treatment of human disease or the enhancement of human characteristics would first require that the techniques undergo a thorough research evaluation. Before the development and implementation of a human protocol, the gene transfer techniques to be used would need to be thoroughly tested in animals, including trials in nonhuman primates (5). The lack of an animal model for many human diseases that may be appropriate candidates for GCGT inevitably gives rise to uncertainty in moving from the nonhuman to the human subject. However, given that the human applications of GCGT require that the procedures first undergo research evaluation, it is useful to reflect on GCGT in the context of current federal research regulations (6).

The first question to consider is who is the subject of the research. The adult is clearly a participant in any research involving gametocyte modifications; however, one could argue that a child who is the anticipated product of the research should also be considered a research subject. To eliminate this possibility, it is conceivable that an adult who is otherwise at risk for passing a genetic defect onto his or her progeny, yet does not want children, may volunteer for early testing of gametocyte modification. The safety of the technique for the adult subject, as well as the efficacy in altering the genome of the resulting gametes, then could be studied without concern for a future child. With voluntary and informed consent, we currently allow competent adults to participate in nontherapeutic research which places them at some personal risk. However, the ethical acceptability of this early testing of gametocyte modification assumes that the adult subjects will not procreate — a stipulation that cannot be guaranteed if the subjects remain fertile after the research. A requirement that adult subjects be sterilized as the last step of a research protocol would prevent harm to any resulting progeny; however, such a requirement would be impossible to enforce as current research guidelines permit an individual to exit voluntarily from a research study at any point.

It is problematic to impute the right to be free from harm to an individual who does not exist (7); however, if a fetus is injured during research involving gene transfer, the ensuing child both exists and suffers as a result of the research. Since an adult participant may have a child after undergoing gametocyte modification, the risks and benefits of GCGT research protocols involving either gametocyte or pre-embryo modification should be evaluated from the perspective of the fetus/child who is both the product and the subject of the research (8). Although the definition of a "human subject" would need to be expanded beyond "a living individual," GCGT then would be judged according to the federal guidelines governing research using fetuses and children (6). Finally, the woman who becomes pregnant after pre-embryo modification and implantation could not be considered the research subject, for this would preclude the research as not being directed toward the "health needs of the mother" and the risk to the fetus is clearly greater than minimal (6).

The general requirements for approval of a research protocol involving an adult stipulate that the risks of the procedure are minimized and are reasonable in relation to any anticipated benefits and/or the importance of the knowledge (6). Unless a technique for targeting gametocytes can be developed that does not alter the subject's somatic cells, the subject will be at risk for the complications of somatic cell gene transfer such as the inappropriate expression of gene products in different tissues and the disruption of otherwise functional gene products or the unmasking of oncogenes through insertional mutations (2). Precise gene excision and splicing in which the defective gene is removed and replaced with a functional gene theoretically would reduce or eliminate these risks; however, this technique is not possible at this time. In effect, in vivo GCGT techniques involving the adult gametocyte should not be approved until the development of selective targeting or until these same techniques are deemed safe for somatic cell gene transfer. These difficult technical requirements for an appropriate research protocol along with the availability of in vitro fertilization after either in vitro gametocyte or pre-embryo modification make it unlikely that an in vivo GCGT protocol will be developed or approved.

Any GCGT protocol involving in vitro gametocyte or pre-embryo modification necessarily involves an adult as either a gamete donor or a pre-embryo recipient. A GCGT protocol may present no greater than minimal risk to an adult participant provided that (1) gamete procurement uses standard nonresearch collection techniques, (2) any in vitro genetic modification takes place prior to implantation, and (3) the pre-embryo modification does not alter the biology of pregnancy, so that the woman who is the pre-embryo recipient is not placed at any additional risk beyond that associated with standard in vitro fertilization procedures. Appropriate nonhuman primate studies may be necessary to establish this third condition, for otherwise the pregnant woman would need to be considered a research subject and the research would be disallowed under current federal guidelines governing research with pregnant women (6).

Currently, neither the Recombinant DNA Advisory Committee of the National Institutes of Health (NIH) in the United States nor the Joint Medical Research Council in the United Kingdom will consider a GCGT protocol (5). Although there is no direct ban on GCGT research, there is a ban on the use of federal funds either for the creation of a human embryo for research purposes or for research in which a human embryo is "destroyed, discarded, or knowingly subjected to risk of injury or death" beyond that allowed by existing federal guidelines (9,10). Unless the research is privately funded, this ban would need to be modified to allow for the creation of a human embryo even if the intent was to implant and not discard all created embryos. Once the pre-embryo is implanted, federal guidelines require that the research is designed to meet the "health needs of the particular fetus" and restrict any risk to the minimum necessary to meet such needs (6). In effect, the fetal guidelines are similar to those for "greater than minimal risk" research with children. The risk of the research must be justified by the "anticipated benefit" for the child. The balance of risk and benefit must also compare favorably with alternative approaches. Treating the future child as a research subject does not preclude the development of a pre-embryo modification protocol due to the inability to obtain consent, for the assent of the child may be waived if the benefit is not available outside of the research (6).

The development and approval of a GCGT protocol, as discussed above, would need to build on prior experience using somatic cell gene transfer for the same or at least similar conditions (5). Consequently somatic cell gene transfer may serve as a viable alternative for certain conditions. The benefit of germ cell modification would be to eliminate the need for both the individual person and his or her future progeny to undergo somatic cell modification. The risks of germ cell modification would need to be balanced against those of somatic cell modification — a balancing that likely would favor somatic cell modification given the complexity and uncertainty of embryonic, fetal, and child development. The claims of the adult suffering from the genetic condition under consideration to want to free his or her progeny from the putative guilt of defective reproduction or the burden of somatic cell treatment may not be compelling if a safe and effective somatic cell treatment is available. Nevertheless, for conditions that require a more extensive and uniform distribution of the transferred gene, or that will have an impact on the future child at an early stage of in utero development, GCGT may be the only effective method for somatic cell modification. Thus concern about scientific uncertainty is met by a specification of the conditions under which such uncertainty is worth the risk. The conditions then that may be suitable for initial testing of GCGT techniques are those for which there are no available alternative treatments (including somatic cell gene transfer), the impact of the genetic defect occurs early in fetal development, the outcome is uniformly dismal or fatal, and the phenotypic condition is caused by a single gene defect (11). Once GCGT is shown to be safe and effective for this limited range of genetic disorders, the techniques could be extended (again within a research protocol) to other less ominous diseases, or to diseases for which safe and effective somatic cell therapies may exist. GCGT aimed at the enhancement of human characteristics would not be approved, if ever, until GCGT has been shown safe and effective for a variety of diseases.

MAJOR OBJECTIONS TO GCGT

The primary motivation leading to the development of GCGT technology is therapeutic; that is, it begins with a parent's desire to bear a child who is free from the burden of an otherwise untreatable genetic disease. In the absence of strong counterobjections, it can be argued that medicine has a prima facie duty to pursue research on the therapeutic use of GCGT. Some argue the strong moral claim that the character of medicine

as seeking knowledge for healing purposes mandates the exploration of GCGT in the absence of compelling objections (2,5,12,13). If objections to the development and implementation of GCGT rests primarily in the fear of unknown and potentially disastrous consequences, the cautious and gradual implementation of germ cell protocols guided and controlled by an already established research review process may provide for an early recognition and mitigation of untoward consequences (13). Nevertheless, we would prevent the transmission of our defective gene(s) to the children of our children. Is this "cleansing of our family line" an unacceptable outcome, even if it is not our primary intent? Are GCGT techniques fundamentally wrong under all circumstances such that we can never choose it as a means to an otherwise acceptable end?

Munson and Davis interpret principled objections to GCGT as involving a basic claim of illegitimate "tampering" with either individual rights, social order, or nature itself. First, concerning individual rights, the Council of Europe asserts the right of the unconceived and unborn to a genetic inheritance which has not been artificially altered. An exception is made for therapeutic interventions based on a distinction between pathological and nonpathological conditions, that is, between treatment of disease and enhancement. The Council bases such a right on appeals to human dignity, integrity, the "normal" or "natural," genetic divergence or an appeal to the preservation of being human — concepts that neither entail an "alleged right to an untouched genome" nor escape the definitional ambiguities inherent in the concept of disease. Even if we reject the position that our future progeny have the right to be left alone, we may still hold that they have the right not to be knowingly harmed. The limits of this right would involve our ability to predict the results of our otherwise well-intentioned interventions, relieving us of the unreasonable burden of knowing in advance all of the potentially negative consequences of our gene transfer technology. This balancing of the risks and benefits of intervention is involved when GCGT is considered as a research protocol aimed at freeing an individual of genetic disease (12).

Second, GCGT may give rise to social disorder or conflict. Will a parent be required to submit to germ cell modification in order to prevent the transmission of a genetic disease? The apparent conflict between community interests and individual freedom is not new. The availability of genetic testing for such conditions as Huntington's disease, and the knowledge of the fetal impact of maternal drug or alcohol abuse, already have led some to advocate for restrictions on individual behavior. If GCGT is used for the enhancement of desirable biological traits, existing socioeconomic inequalities may be exacerbated. However, current inequalities in access to and distribution of health care resources already reinforce existing socioeconomic differences. These are important problems that we must face in the design, implementation and control of health care technology; however, they are not unique to GCGT (12).

The third version of "tampering" concerns "playing God" or altering the "very order of nature." Munson and Davis identify three arguments against such tampering. First, the concern that GCGT will inevitably lead from treatment of disease ("negative eugenics") to enhancement ("positive eugenics") is met with an empirical claim that eugenic practices such as selective breeding have existed for centuries yet rarely been used. Acknowledging that the distinction between positive and negative eugenics may be difficult to sustain, Munson and Davis prefer to question the assumption that our desire to improve ourselves through genetic technology is wrong. Munson and Davis also take a cautious empirical approach to the second concern that unforeseen hazards may exist, for example, in the loss of biological diversity as genes are eliminated. They see no reason to fear a disaster from the development and application of GCGT more than from other applications of genetic technology (13,14). Finally, they address the concern that GCGT threatens our "humanity" by creating the possibility of human/nonhuman hybrids or an evolution into a superior yet nonhuman species. As an empirical concern, it is unlikely that the elimination or addition of genes through GCGT will corrupt or eliminate a genetic structure that is somehow essential to our humanity. By linking our sense of self-worth to the inviolability and integrity of our genetic structure, the opponents of GCGT appear to adopt the same reductionist assumption of genetic determinism that proponents of genetic technology are assumed to hold. Munson and Davis propose that behind all three objections is a belief in either the "wisdom of evolution" or the "design of a good and wise Being." However, not only is it impossible to establish the "sanctity" or "special moral standing" of human nature apart from specific theological or moral commitments, the very existence of genetic disease belies the wisdom of natural selection and the coherence of providential design (12,15). It is as likely that our ingenuity in developing and applying GCGT technology is either an evolutionary adaptation that enhances our chances for human survival or an affirmation of our God-given stewardship over creation (13).

Practical objections to the development of GCGT involve scientific uncertainty and the unpredictability of long-term risk, both addressed through an incremental process of research as outlined above. Some assert that GCGT will never be sufficiently cost-effective to merit the allocation of necessary social and economic resources. This objection, however, is true of all unproved technologies prior to our attaining sufficient knowledge to evaluate the costs and benefits. For some, the strongest pragmatic argument against GCGT is the existence of pre-embryo screening and selective implantation (16,17).

EUGENICS AND THE DESIRE TO PREVENT GENETIC DISEASE

GCGT has the potential to be more effective than somatic cell gene transfer in preventing the onset of a genetic disease in an otherwise affected individual through providing for broader cellular coverage. Putative alternatives such as in vitro fertilization and pre-implantation embryo screening reduce the risk of disease through avoiding the birth of an affected individual, rather than in preventing the onset of disease. Juengst

is critical of the ease with which the therapeutic use of GCGT to prevent the expression of a genetic disease in an individual (so-called phenotypic prevention) is equated with the use of GCGT to prevent the transmission of the genetic disease to future generations (so-called genotypic prevention). Juengst believes that it is this confusion over the purpose of germ cell gene transfer that fuels concerns about the eugenic use of this technology (18).

Juengst identifies four problems that geneticists face in adopting as a goal the prevention of certain genotypes (18). First, genotypic prevention, Juengst argues, understands the diseases it prevents as being caused by the associated gene abnormality rather than at the level of pathophysiological expression. He implies that this is a limited metaphor by which to understand genetic disease; however, those disorders that are amenable to GCGT techniques necessarily may fit within this model of causality. Nevertheless, the deterministic causal assumptions on which genetic therapy relies may be overly simplistic and subject to the risk of both false negative and false positive predictions of clinical disease.

Second, the decision to prevent the birth of an individual affected by a certain disease, Juengst points out, assumes that the burden of living with the disease outweighs any other value that the individual may experience or bring to the life of the family or community (18). However, the decision to risk GCGT in order to avoid the burden of genetic disease relies on the same calculus. Furthermore decisions to limit or withdraw life-sustaining treatment may be predicated on the value judgment that death (or nonexistence) is preferable to life under certain conditions. Our ambivalence in applying this same calculus to pre-implantation embryo selection may rest in the conceptual difficulties associated with so-called wrongful life and the active nature of the selective intervention. Beyond the stigmatization of those affected with a genetic disease is the prejudicial impact on carriers, either on the parents of those who undergo correction or on those whose parents either could not afford or who chose not to undergo carrier correction.

Third, Juengst is concerned that the traditional commitment to an individual's voluntary reproductive choice may give way to the economic and public health interests of society if medicine endorses "genotypic prevention." Finally, Juengst is concerned that the definition of "pathological genotypes" will inevitably be influenced by "larger cultural ideologies and social values" such as contemporary concerns to prevent either "reproductive anxiety and interpersonal aggression" (18). However, the need to define the domain of genetic conditions deemed suitable for the use of GCGT techniques remains a problem, even if geneticists eschew the professional goal of genotypic prevention. Unless individual requests are given unfettered access to genetic technology, some social and political definition of warranted requests will need to be established. The conceptual difficulties in defining a disease so as to exclude inappropriate requests for, say, genetic enhancement remain salient. Phenotypic prevention can be for the purpose of treating disease or enhancing human characteristics. Genotypic prevention can be aimed at the elimination of diseases or the selection of desirable characteristics in the population as a whole. Thus the distinction between treatment and enhancement remains a problem independent of whether we accept or reject the goal of genotypic prevention.

TREATMENT AND ENHANCEMENT: A ROLE FOR POSITIVE EUGENICS

With the acceptance of somatic cell gene therapy and the continued discussion of germ cell gene therapy, Fletcher and Anderson observe that the moral differences between these two forms of gene transfer technology appear less significant than the distinction between the treatment of disease and the enhancement of human characteristics. Reflecting the general condemnation of the use of genetic technology for enhancement, they argue for drawing a moral line, not between somatic cell and GCGT techniques, but between treatment of disease and enhancement of human characteristics (5). However, this distinction is difficult to maintain as concepts of disease and illness involve complex and subtle evaluations regarding the scope of medical interventions. Any attempt to distinguish between human needs (treatment) and desires (enhancement) also falters on the historical and cultural diversity of conditions that medicine in fact has treated. Finally, the concepts of disease and health incorporate moral and nonmoral values and goals, making it difficult if not impossible to discover an objective yet morally significant line between disease and various competing, positive notions of health (15). Still, the concept of malady has been put forward in an attempt to draw such a line.

A malady is defined as an existing condition that causes a person to suffer or risk suffering an evil such as death or disability in the absence of a distinct sustaining cause. A genetic condition that fulfills the definitional criteria of malady can be treated through gene transfer as negative, not positive, eugenics. Berger and Gert claim that the definition involves the avoidance of universal (and thus objective) evils, such as death, pain, disability, loss of freedom or pleasure, which are not dependent on particular cultures. The notion of a "sustaining cause" seems to hinge on a distinction between an internal or physiologic cause and external or environmental cause — whose removal leaves the individual at increased risk of injury or evil. A full discussion is beyond the scope of this entry. Suffice it to say that the concept of malady will not escape the problems associated with distinguishing between the treatment of disease and the enhancement of health. Notions of disease and disability are notoriously culture dependent (19); the notion of causality itself deeply intentional (20). The recent controversy over the use of recombinant human growth hormone to treat children with short stature who are not growth hormone deficient illustrates the difficulty in defining a disease so as to draw a line between treatment and enhancement (21).

To render this distinction more problematic, Torres presents an example of a somatic cell gene transfer treatment protocol which, strictly speaking, qualifies as an enhancement technique. Gene transfer techniques are used to enhance the natural resistance of hematopoetic stem cells to the effects of anticancer drugs in patients receiving chemotherapy for solid tumors. Given that a clinical intervention may or may not be justified given its purpose, Torres proposes that gene transfer techniques may be used for enhancement provided that "such enhancement constitutes a necessary condition for the success of treatment designed to suppress the causes, symptoms or effects of severe pathology." The enhancement is thus a means and not an end. Torres cites two reasons that should dissuade us from the use of gene transfer techniques for enhancement: the risks of toxicity, and the discrimination involved in unequal access to genetic technology and the devaluing of those left with the unenhanced trait (22).

It may be inevitable that the introduction of GCGT for the treatment of disease would, after further research and development, be applied to the enhancement of human traits. If the purpose of the enhancement is an accepted social goal, such as providing for minimal functionality given the design of social space (i.e., short stature), or if it's purpose is an individual goal (i.e., enhancement of height in order to procure a lucrative sports contract) that is equally available to all — is enhancement necessarily precluded? It would clearly be unjust and a violation of parental autonomy for governments to perform genetic enhancement without parental consent. However, parents have the prima facie right both to attempt to prevent disease and to offer any advantage or benefit to their children through human GCGT. Although this right could be overridden by the risk of harm to future generations or by grave social, political, and economic injustices, Resnik proposes that the goods achieved by enhancement can be regulated according to accepted principles of justice so as to not exacerbate social inequities. The challenge then will be to distinguish between legitimate and illegitimate enhancement, both conceptually and through appropriate social policy and regulation (2,23).

THE ANALOGY BETWEEN PRE-EMBRYO DISCARD AND ABORTION

Mauron and Thevoz suggest that society can be protected from the use of GCGT for illegitimate enhancement through reinforcing the patient or client-centered ethics of genetics (16). Similarly Juengst wants to preserve the traditional allegiance of geneticists to respecting the freedom of parental reproductive choices — an allegiance that is threatened by the acceptance of genotypic prevention as a social and professional goal. The professional obligation to support the personal goals of the patient, Juengst claims, has little to do with the content of those goals and can peacefully co-exist with the parental motivation to eliminate specific genetic diseases and thus genotypes from an individual family heritage (18). Whereas Munson and Thevoz assert that this traditional focus enables medicine to "keep its sphere of action technically and morally manageable," Juengst recognizes that this traditional allegiance to reproductive freedom does not resolve a number of ethical problems (16,18). For example, as genetic testing increasingly becomes a prelude to prenatal interventions to improve the health of the fetus, the fetus emerges as a "patient" with its own associated moral claims that may be in tension with parental choice. In addition, as mentioned above, a simple allegiance to freedom of reproductive choice may result in a laissez faire genetic economy in the absence of social or professional limits to the range of offered genetic services — limits which may reintroduce the problem of defining genetic conditions which are deemed pathological (18). Finally, if Juengst wants to insulate GCGT from antiabortion arguments against embryo discard, an appeal to reproductive freedom is less than reassuring given the importance of this argument to establishing access to abortion in the first place. The argument for parental reproductive freedom and autonomy may obviate the need to benefit any particular embryo and undermine appeals against pre-embryo discard to the extent that parental autonomy implies ownership of gametes and pre-embryos (11).

The analogy, though disputable, between the discard of defective pre-embryos and the controversial issue of abortion may impede the development of GCGT. During the development phase of pre-embryo modification, even after extensive animal testing, it is unlikely that the procedures would be perfected such that all modified pre-embryos are appropriate for implantation. The inevitable presence of a defective pre-embryo, even if created with the intent to implant, raises the question of discard given the woman's voluntary and informed consent (or dissent) to have such a pre-embryo implanted. This dilemma does not appear to be different from current in vitro fertilization practices, or from the decision to abort a fetus that is determined to be genetically defective based on prenatal testing. Thus GCGT does not present any new or different problems with respect to embryo discard (or abortion) than currently exists. In vitro gametocyte modification, if technically feasible, would allow for in vitro fertilization procedures to proceed as currently supported, given that only modified gametes could be used to create a pre-embryo suitable for implantation. Our experience with the abortion debate over the past two decades would suggest that if moral and political opposition to GCGT is based on the potential destruction of living pre-embryos, a simple principled or political solution may not be forthcoming (2).

Juengst's proposal would allow GCGT techniques to be developed for the purpose of more effectively preventing the "onset of a genetic disease in a patient." It undermines the practical argument that GCGT is unnecessary given the existence of pre-implantation embryo selection by highlighting an essential conceptual difference between the discard of defective embryos and the treatment of affected embryos (18). However, although individuals who are against the discarding of the products of human conception may support this distinction, the discarding of human embryos may be a necessary component of any research program seeking to establish the safety and efficacy of such techniques (11).

Juengst also suggests that rejecting the goal of genotypic prevention would allow for the correction of an adult's carrier status through GCGT techniques involving gametocytes, while precluding pre-embryo modifications directed toward the same goal of eliminating an offspring's carrier status (18). Although this option is rejected

as a form of genotypic prevention through limiting the reproductive freedom of a child to transmit a deleterious gene, it is unclear that such a decision is outside of the range of discretionary choices that parents may make on behalf of their children. It may be difficult to insulate GCGT techniques from supporting the goal of preventing the transmission of deleterious genes. Juengst's arguments depend on two questionable assumptions: first, that the professional morality of geneticists can be divorced from an analysis and critique of the consequences of parental action; second, that personal moral choices can be supported while rejecting the social, cultural and political context and implications of such choices. A more productive approach may be to address directly the fundamental issue underlying the affirmation of individual reproductive freedom, that is, the social and political control of genetic technology.

ALTERNATIVES TO GCGT

The alternatives available for individuals who are at risk for transmitting a genetic disease to their progeny include (1) selection of a reproductive partner who reduces or eliminates the risk, (2) selection of a reproductive mechanism that reduces risk such as donor gametes, artificial insemination, or pre-embryo selection and in vitro fertilization, (3) avoid procreation altogether, (4) prenatal testing and selective abortion, (5) somatic cell gene therapy for affected offspring, and, finally, (6) GCGT techniques.

Restricting access to GCGT technology in light of the alternatives of avoiding procreation or selective termination of pregnancy generally is believed to be an unacceptable infringement on reproductive choice (2,13). The ability to conceive, nurture a pregnancy, and give birth to a healthy baby is accepted as an aspect of reproductive health appropriately addressed by medical technology. Individual screening and then selection of appropriate reproductive partners is technically feasible. However, rarely do affairs of the heart submit to such a rational and premeditated approach. For many individuals, the acceptability of alternative approaches using surrogates and donor gametes founders on the desire to produce a genetically related offspring. For such individuals who knowingly bear a genetic defect, the only options, other than the identification and termination of defective pre-embryos or fetuses, are germ cell and somatic cell gene transfer techniques.

Current in vitro fertilization techniques involve the discard of pre-embryos that are not suitable (for whatever reason) for implantation. Once we are able to identify pre-embryos that carry a certain genetic defect, we will be able to selectively eliminate these pre-embryos prior to implantation. Such a procedure may be technically and morally preferable to prenatal diagnosis and selective abortion, primarily due to lower maternal risk; however, it is conceptually no different from the point of view of a pre-embryo or fetus in the absence of any ontological distinction between the two. The development of gamete selection coupled with in vitro fertilization and implantation may eliminate the conceptual link with abortion. However, there would remain rare instances such as couples who are homozygous for a genetic disorder where their desire for a genetically related yet healthy offspring could not be met in the absence of GCGT technology.

GCGT IN THE TREATMENT OF MITOCHONDRIAL DISEASE

Recently Rubenstein and colleagues proposed a protocol involving the treatment of genetic diseases associated with defects in mitochondrial DNA. The inheritance of mitochondrial DNA is strictly maternal, given that the cytoplasm of the ovum contributes most if not all of the mitochondria incorporated into the developing embryo. In effect, the procedure they propose transfers the nucleus of the carrier's ovum into the enucleated cytoplasm of a donor ovum, followed by standard in vitro fertilization and implantation. The procedure thus involves the transfer of genetic material into a germ cell for the purpose of correcting both the phenotypic expression and the vertical transmission of an otherwise debilitating and potentially fatal disease (4).

Although Rubenstein and colleagues present a complex and less than compelling classification of GCGT based on the level of cellular penetration, the essence of their argument hinges on an ethical distinction between manipulation of mitochondrial DNA and nuclear DNA (3). The protocol appears to satisfy the arguments in favor of GCGT, that is, medical utility and necessity, prophylactic efficiency, respect for parental autonomy and the pursuit of scientific knowledge within the bounds of ethical research. The protocol also renders the arguments against GCGT less compelling (24). Preserving the integrity of nuclear DNA may reduce the scientific uncertainty and risks to future generations inherent in other forms of GCGT, while avoiding concern about the putative right of an individual to an unmodified genetic endowment. The technology appears feasible and likely not to generate significant costs beyond those currently associated with some in vitro fertilization techniques. Finally, it is unlikely that the mitochondrial DNA will be amenable to enhancement. The development and application of GCGT techniques involving mitochondrial DNA may lead to a greater acceptance of germ cell manipulations for the treatment of inherited disease. However, given the special nature of mitochondrial DNA, it is unlikely that such acceptance will impact significantly on the debate concerning other forms of GCGT.

THE SOCIAL AND POLITICAL CONTROL OF GENETIC TECHNOLOGY

The ability to prevent the illegitimate use of GCGT technology will depend, not on the strength or weakness of such conceptual distinctions as the treatment of disease versus the enhancement of human traits, but on how we conceive of and establish the social and political control of genetic technology. The right to be free from harm, though not to be left alone, is the right of children with respect to their parents, not with respect to the state. The reproductive freedom of parents is asserted as a limit

to state intervention. However, an appeal to individual freedom in the application of genetic technology neglects issues in the social control over development of this same technology. For example, the choices of parents who are at risk for transmitting either cystic fibrosis or sickle cell disease will be shaped by the resources committed to the investigation and development of these same alternatives — resources committed through a complex process of political advocacy, community activism, and private marketing.

The therapeutic application of somatic cell gene transfer, when combined with genetic diseases that may remain resistant to somatic cell approaches, likely will create moral and political pressure for the incremental development of GCGT techniques to address these diseases. As we gain experience with genetic technology, the development of GCGT protocols could proceed within the context of federal research guidelines assuming appropriate modifications to allow for the involvement of pre-embryos. The ethics of GCGT technology reduce to the ethical issues of the creation and use of pre-embryos. In the development phase, the discard of pre-embryos is likely despite attempts to conceptually link GCGT with the treatment of individual pre-embryos and thus divorce it from the issue of pre-embryo discard and abortion. However, in the application phase, it is likely that the availability of GCGT technology would greatly reduce the use of either pre-embryo selection or selective abortion for the prevention of genetic disease.

BIBLIOGRAPHY

1. M. Lappe, *J. Med. Philos.* **16**(6), 621–639 (1991).

2. B.K. Zimmerman, *J. Med. Philos.* **16**(6), 593–612 (1991).

3. M.D. Bacchetta and G. Richter, *Cambridge Q. Healthcare Ethics* **5**(3), 450–457 (1996).

4. D.S. Rubenstein, D.C. Thomasma, E.A. Schon, and M.J. Zinaman, *Cambridge Q. Healthcare Ethics* **4**(3), 316–339 (1995).

5. J.C. Fletcher and W.F. Anderson, *Law, Med. Health Care* **20**(1–2), 26–39 (1992).

6. 45 *Code of Federal Regulations* 46 (1983).

7. B. Steinbock, *Life before Birth: The Moral and Legal Status of Embryos and Fetuses*, Oxford University Press, New York, 1992.

8. T.H. Murray, *The Worth of a Child*, University of California Press, Berkeley, 1996, pp. 96–114.

9. Section 512 of the FY 1997 Appropriations Act, Public Law 104–208 (September 30, 1996).

10. J.C. Fletcher and G. Richter, *Hum. Gene Ther.* **7**(13), 1605–1614 (1996).

11. K. Nolan, *J. Med. Philos.* **16**(6), 613–619 (1991).

12. R. Munson and L.H. Davis, *Kennedy Inst. Ethics J.* **2**(2), 137–158 (1992).

13. President's Commission for the Study of Ethical Problems in Medicine and Biomedical and Behavioral Research, *Splicing Life: The Social and Ethical Issues of Genetic Engineering with Human Beings*, U.S. Government Printing Office, Washington, DC, 1982.

14. E.M. Berger and B.M. Gert, *J. Med. Philos.* **16**(6), 667–683 (1991).

15. H.T. Engelhardt, *Soc. Philos. Policy* **13**(2), 47–62 (1996).

16. A. Mauron and J.-M. Thevoz, *J. Med. Philos.* **16**(6), 649–666 (1991).

17. W. Glannon, *Bioethics* **12**(3), 187–211 (1998).

18. E.T. Juengst, *Hum. Gene Ther.* **6**(12), 1595–1596 (1995).

19. A. Kleinman, *Writing at the Margins: Discourse between Anthropology and Medicine*, University of California Press, Berkeley, 1995, pp. 19–92.

20. H. Putnam, *The Many Faces of Realism*, Open Court, LaSalle, IL, 1987, pp. 23–40.

21. R.M. Nelson, in J.P. Kahn, A.C. Mastroianni, and J. Sugarman, eds., *Beyond Consent: Seeking Justice in Research*, Oxford University Press, New York, 1998, pp. 56–57.

22. J.M. Torres, *J. Med. Philos.* **22**(1), 43–53 (1997).

23. D. Resnik, *J. Med. Philos.* **19**(1), 23–40 (1994).

24. E.T. Juengst, *J. Med. Philos.* **16**(6), 587–592 (1991).

See other GENE THERAPY entries; HUMAN ENHANCEMENT USES OF BIOTECHNOLOGY: OVERVIEW.

GENE THERAPY, ETHICS, RELIGIOUS PERSPECTIVES

GERALD P. MCKENNY
Rice University
Houston, Texas

OUTLINE

INTRODUCTION

Theologians were among the first to address ethical issues in genetic science, and religious bodies and individual theologians have contributed to debates over the ethics of gene therapy from the beginning (1). Their contributions are important for several reasons. First, religious ideas inevitably play a role in shaping public attitudes to gene therapy. It is therefore important for those who have research, commercial, or policy interests in gene therapy to be informed of religious responses to the latter. Second, there is a common perception that religious traditions are hostile to, or suspicious of, genetic technology. This perception is fueled by a few instances in which religious leaders have taken high-profile stances against research on germ-line gene therapy or the patenting of genes, and by the use of religious or quasi-religious language by certain opponents of genetic technologies. However,

as the following survey shows, the index of support for somatic cell gene therapy (SCT) among the vast majority of religious groups and writers is very high, while the range of responses to germ-line gene therapy (GLT) generally tracks that of the informed public as a whole. Third, many people who work in gene therapy or related areas adhere to a religious tradition and may consider it important to know what that tradition says about their work. Finally, many religious responses to gene therapy contain arguments or insights that are missing in secular debates.

CHRISTIANITY

Ecumenical

The World Council of Churches (WCC) and the National Council of the Churches of Christ in the USA (NCC) each produced multiple documents during the 1980s that addressed gene therapy along with other issues in genetics and biotechnology. With one exception (2), none of these documents represent the official position of the body in question, though all of them received official recognition at some level.

World Council of Churches. The WCC published two reports on genetic technology during the 1980s (3,4). The reports share a focus on science and technology as forms of power as well as knowledge, and emphasize social, political, economic, and ideological factors in biotechnology as a global enterprise. While the specific conclusions regarding gene therapy are conventional, attention to these factors affects the analysis of the broader context in which gene therapy is or will be carried out.

SCT is regarded as no different from other forms of experimental therapy; accordingly, it should be undertaken only in the absence of adequate alternative treatments and carried out under the usual protections governing research with human subjects. GLT is more problematic insofar as germ-line interventions involve alterations that will persist over many generations. Of course, the effect on future generations can be a strong argument in favor of GLT: "By overcoming a deleterious gene in future beings, the beneficial effect of such changes may actually be magnified" (3). However, weighing these potential benefits against risks would require extensive knowledge of long-term consequences that we do not yet possess; research on GLT should therefore be banned at present. But this ban need not be permanent; the report that advocates it goes on to call for ethical reflection leading to future guidelines (4).

The earlier of the two reports exhibits a tension found in other religious statements. On the one hand, the justifiability of gene therapy seems to be connected with its use for "recognized diseases of genetic etiology." On the other hand, there is considerable skepticism about whether reliable lines can be drawn between therapy and enhancement ("Correction of mental deficiency can move imperceptibly into enhancement of intelligence, and remedies of severe physical disabilities into enhancement of prowess") or between negative and positive eugenics ("There is no absolute distinction between eliminating 'defects' and 'improving' heredity"). There may be agreement that certain conditions "are so deleterious as to deserve first call" on gene therapy resources, but beyond these cases choice "becomes a matter of subjective judgment which will differ according to personal values."

In place of drawing lines between negative and positive eugenics or therapy and enhancement, both WCC reports raise other ethical issues. These issues all involve power and ideology as factors that arise in the social, political, economic, clinical, and research contexts in which gene therapy is or will be carried out. One ethical issue is the role of social and cultural prejudices in identifying certain genes as "defective" or, more generally, the potential for gene therapy to result in discrimination or in eugenic policies that institutionalize prejudices (4). When genetically transmitted characteristics become societal liabilities, gene therapy may be used to alter those characteristics rather than society altering its values and prejudices (3). A second worry at the societal level is that resources devoted to gene therapy will divert attention and resources from (1) nongenetic diseases, (2) protection of genes from avoidable damage (from mutagens, carcinogens, and man-made radiation), and (3) providing each person with opportunities to develop their existing capacities (3,4). At the political level, genetic interventions should remain options and not requirements (legal or otherwise) that parents are obligated to fulfil in their offspring (3,4). A similar worry pertains to economic pressures to secure certain characteristics for one's children (4). Also in the economic realm, one report calls for legal safeguards to protect individuals and their potential descendents with regard to quality control of materials and methods used in gene therapy, and misrepresentation of possible benefits by commercial advertisers or by scientists (3). In the clinical context, there is concern that desperately ill patients might try unproved techniques of doubtful efficacy. Patient-subjects should therefore be fully informed of possible negative effects. The report does not address the current question of whether desperately ill people who indicate awareness of possible negative effects should be allowed to enroll in early phases of experimental trials.

A different set of criticisms denounces the mechanistic worldview of contemporary science and technology that, the report claims, objectifies life for utilitarian and instrumental ends and whose primary goal is "the maximizing of material advantages for those few most able to appropriate and profit from the extraction of the earth's resources" (4). This conviction appears to undergird a concern that genetic engineering reduces persons to interchangeable parts. This reduction threatens the "inalienable dignity" of persons, which is the basis of mutual respect (3). However, it is not clear what exactly is under condemnation. At one point the target is the transformation "of offspring into interchangeable parts to be selected at will"—the concern, echoed in many religious statements, that genetic knowledge and interventions will turn children into products—but this is followed by a more general criticism that genetic engineering in and of itself, as a form of knowledge and practice, "converts the human subject into a composite object of interchangeable

elements." This latter concern would seem to render all genetic interventions suspect, including those directed at conditions the WCC itself would consider serious diseases. The report goes on to ask, "In what ways do we, by manipulating our genes in other than simple ways, change ourselves to something less than human?" Unable to draw lines between acceptable and unacceptable *kinds* of intervention, the WCC latches on instead to a concern with the *degree* of intervention.

Finally, both WCC documents call for a prohibition of research on zygotes and embryos, with exceptions for therapeutic purposes under well-defined conditions (4). Both the prohibition and the exception derive from the status of zygotes and embryos as potential but not actual persons.

National Council of Churches. The NCC produced three reports during the 1980s. While not ignoring the warnings highlighted by the WCC, the NCC urges churches to take a positive stance toward genetic science and technology and to participate in public debates, and asks scientists to accept public scrutiny of genetic research. The reports applaud the role of genetics in fostering awareness of the interconnection of human life with other forms of life and the responsibility of humans for other life forms, though these convictions are in some tension with theological views that emphasize the special status of human life (5).

The reports are primarily designed to foster discussion of genetic technology; they therefore identify advantages and disadvantages of gene therapy rather than issuing prescriptions or proscriptions. SCT is no different in principle from other experimental therapies, but it poses potential dangers: The host cell into which the transferred DNA integrates may produce too little or too much of the desired product; the transferred DNA may disrupt the functioning of existing cells (5). GLT departs from standard medical therapy insofar as alterations are passed on to future generations. This holds out the promise of reducing the frequency of deleterious genes in the population but also raises ethical concerns that accompany eugenics (5). These concerns apparently include the possibility of involuntary participation in eugenic research, compulsory treatment in the name of eugenics, elitism in the determination of desirable and undesirable characteristics, and encouragement of an illusory quest for human perfectability (2,5). GLT also has the advantage of offering a possible treatment for genetic diseases that affect multiple tissues, but the technical difficulties with gene expression and the risk of disrupting cellular functioning remain (5). Future use of GLT is not ruled out, but the unknown and uncontrollable risks require extreme caution, and the interests of future descendents may have to be represented by a guardian ad litem (though the report is silent on the question of what those interests are) (2,5).

The reports emphasize the legitimacy, in principle, of intervening into genetic processes for the betterment of human life (2,5,6). They refute, on theological grounds, the objection sometimes attributed (usually wrongly) to Christianity that genetic interventions ipso facto exceed proper human limits or violate a normative natural order. However, the reports are consistently skeptical of efforts to identify and prioritize genetic conditions for possible intervention: Some so-called bad genes may serve beneficial purposes. Social and environmental factors, which often can and should be altered, make some conditions liabilities. Efforts to prioritize genetic conditions by placing them on a spectrum from trivial to serious are relative to social, economic, medical, and value variables (6). Rather than attempting to prioritize interventions, the NCC invokes human dignity and distributive justice to impose certain limits and requirements on any such interventions. However, the conception of human dignity is vague and inconsistent; its grounding is unclear [is it an *alien* dignity "conferred by God's love" or is it "related to human powers and to human transcendence over the rest of nature" along with "human reverence and human relations with the rest of nature"? (6)]. Human dignity functions primarily as a placeholder for four ethical concerns: the sanctity of life that prohibits deliberate distortion or destruction of human beings in genetic research, rejection of the notion that genetic health or normality is a criterion of human worth, affirmation of the possibility that some kinds of suffering can serve a purpose, and recognition of the limits of some kinds of control over nature. The report concedes that no firm prescriptions follow directly from these concerns, but insist that the latter "establish a context of awareness" (6).

The claims on behalf of distributive justice are more specific, though they are short on argument and attention to practical implementation. The reports question the development of procedures and products that, because of demand and cost, will benefit only a few; pose the problem of how to balance the current treatment needs of afflicted individuals with research that might someday cure these diseases; and express a concern that basic health care not suffer neglect due to the pursuit of "exotic techniques of genetic control" (5,6). The reports also call for the benefits of genetic technology to be made available to all "regardless of geographic location, economic ability, or racial lines," especially when the products of genetic research result from public funding, and strongly oppose the disparity in standards for the protection of human subjects in the case of products used in the United States but initially tested elsewhere (2,5,6).

Orthodox

Few Orthodox individuals or groups have directly addressed issues of gene therapy. Two exceptions are John Breck (7) and Demetrios Demopulos (8). The lack of official statements by patriarchs or bishops and the scarcity of work by individual thinkers makes it difficult to determine how representative these commentators are, though both approach genetic technology with characteristic Orthodox themes and concerns.

Both Breck and Demopulos welcome SCT but oppose GLT. Breck, citing the "unacceptable risk" of transmitting irreversible consequences of errors to future generations, calls for a moratorium on GLT research. Demopulos points to the unknown consequences of eliminating genes from the gene pool and the likelihood that the development

of gene therapy techniques would involve the discarding of embryos (a concern that other Orthodox are likely to share). These objections do not seem to rule out the justifiability of GLT in principle or in perpetuity; they address risks that may someday fall within acceptable levels and moral wrongs (in the case of discarded embryos) that may eventually be avoidable. However, Breck and Demopulos set the knowledge conditions and the moral strictures on the processes by which GLT would be developed very high. Even if GLT is eventually acceptable on their terms, its process of development will almost certainly have been unethical.

Breck and Demopulos both begin with the characteristic Orthodox view of humanity as the "icon of God." "Human nature in this sense is a process of moving toward the Archetype which is Christ incarnate" (8). As such, humanity is also a microcosm of creation and the link between God and the rest of creation. The purpose of humanity is "to proceed toward union with God and achieve ontological actualization, and to bring the rest of creation with it" (8). These convictions might at first seem to support ambitious efforts at genetic enhancement and eugenics, which might be understood in terms of ontological actualization or cooperating with God's "intent to transfigure the cosmos" (7). However, Breck and Demopulos sharply reject this interpretation. Breck supplies the theological reason: *theosis*, or union with God, is not achievable through genetic means but only "through a process of continual repentance and the free exercise of moral choice" that permits the practice of virtue. Breck therefore distinguishes "therapeutic" from "innovational" interventions. The latter (which appear to encompass enhancements and positive eugenics) are unable to produce the characteristics (repentance and moral choice) that truly matter while, if ever successfully developed, they would likely be used for traits associated with enhancing competitiveness, which Orthodox Christianity would consider suspect. In addition their development would almost certainly violate moral norms, including moral limits on the treatment of embryos. On the other hand, Breck is surprisingly supportive of negative eugenics (though he would not support the use of GLT for this purpose or, presumably, involuntary measures). Demopulos also restricts the role of genetics in ontological actualization to the reduction of sickness, which would enable persons to live longer, giving them more time to pursue union with God by nongenetic means.

These analyses raise serious questions about the role of genetics in various characteristics and the distinction between therapy and enhancement. Regarding the latter, Breck admits that the line is unclear, but argues that since enhancement of character traits is, at best, far in the future, more pressing issues deserve primary attention. These issues include whether access to beneficial technologies will be limited to those who can pay for them, and how standards for research using human subjects will be set. To these concerns, Demopulos adds the question of the priority of gene therapy research relative to other medical needs. However, beyond identifying these issues as priorities, Breck and Demopulos do little in the way of analyzing or resolving them.

Roman Catholic

Catholic approaches to bioethical issues traditionally rely on natural law theories that claim validity apart from appeals to revelation or distinctively Christian theological claims. However, debates over the role of human experience and culture in interpreting natural law, whether natural law issues in absolute prohibitions, and the relation between natural law and virtue divide Catholic thinkers. Some of these debates are reflected in the following treatments of gene therapy.

Pope John Paul II. In the early 1980s Pope John Paul II issued two declarations on genetic technology (9,10). The declarations establish moral norms that no genetic interventions may violate and that cut across distinctions between SCT and GLT, between therapy and enhancement, and the like. These norms are derived from a broad notion of human dignity ultimately grounded in the biblical notion of humanity as "created in God's image, redeemed by Christ and called to an immortal destiny" (10). However, in accordance with Catholic natural law theory, these norms are knowable apart from biblical revelation. For clarity's sake they may be grouped under three headings: respect for life, human dignity in a narrower sense, and liberty. Respect for life entails the right to life "from the moment of conception to death" and status as an end and not a mere means to the collective good (10). This rules out genetic interventions that destroy embryos or subject them to experimentation (9). And, should they ever become possible, it rules out "manipulations tending to modify the genetic store and to create groups of different people, at the risk of provoking fresh marginalizations in society" (presumably because of their inferior status) (9,10). Dignity in the narrow sense refers to the integral unity of humanity as one in body and soul (9,10). This rules out genetic interventions that might make use of forms of reproduction that separate the procreative act from the biological and spiritual union of husband and wife (e.g., artificial insemination and in vitro fertilization). It also rules out interventions that would distort or destroy this integral unity, though it is unclear what kinds of interventions the pope has in mind (10). Finally, liberty is violated when a genetic intervention "reduces life to an object, when it forgets that it has to do with a human subject, capable of intelligence and liberty ..." (10). Again, it is not clear what kinds of interventions or procedures would violate liberty in this sense.

These considerations lead to more specific conclusions about gene therapy, some explicitly drawn by the pope, others that may be inferred. The primary distinction is between therapeutic and nontherapeutic interventions, a distinction the pope assumes without elaboration. Both declarations express strong support for therapeutic interventions so long as they tend to improve one's overall condition. In principle, this could apply to both SCT and GLT. But the restrictions on treatment of embryos and forms of human reproduction would make most GLT research currently envisioned morally unjustifiable. Nontherapeutic interventions, whether somatic or germ line, must avoid the violations of human dignity and liberty identified above. They also must avoid racist assumptions and a materialist view of human happiness (10). This certainly stops well short of ruling out enhancements or positive eugenics; one can imagine circumstances in which such interventions would not create inferior beings, distort or destroy body-soul unity, objectify persons, or carry out racist and materialist attitudes. But the pope's final remark, contrasting "adventurous attempts aimed at promoting I know not what superman" with "salutary efforts aimed at correcting maladies, such as certain hereditary maladies," seems to cast a general suspicion on enhancements and eugenic efforts (10).

Catholic Bishops' Joint Committee on Bioethical Issues (U.K.). In 1996 a working party of this committee, consisting mostly of physicians and fellows of the Linacre Centre, published what is probably the most comprehensive statement on gene therapy from a religious perspective (11). The report analyzes gene therapy in light of fundamental questions about human nature and fulfillment, the role of medicine in promoting human fulfillment, and responsibilities for the genetic health of oneself and one's children.

Human nature, as in the papal declarations, is a unity of soul and body. Since the soul is the body's "life principle," a living human body is never without a human soul; personhood therefore coincides with being a distinct living organism, which usually begins at conception. In accordance with a recent natural law theory, human fulfillment consists in the pursuit of certain basic goods such as life, knowledge, and sociability (12). (Health is one of these basic goods but is also a condition for pursuing the other goods.) These goods are never to be deliberately attacked (though they need not be promoted in all situations). Hence "[t]he life and health of some may not be promoted by means of an attack on innocent others; for example, by means of destructive experimentation on human subjects." Since personhood begins at conception, this principle extends to research using human embryos. The role of medicine is to promote the basic good of health, defined as "the complex of *functional, goal-directed, psychophysical systems* ... in the contribution they make to the good of the whole." Health is best promoted "through the normal channels of human *activity*, whether conscious or non-conscious"; medicine should intervene, in the form of prevention, cure, or palliation, only when a functional defect renders the normal means to fulfillment unavailable or unsatisfactory. Finally, while persons have some responsibility for their genetic health and that of their children, there are natural and environmental limits to the elimination of genetic disease, and often the most appropriate response to a genetic disorder will be social or environmental rather than medical.

These general considerations enshrine theories of natural law and of medicine that are controversial in themselves and not fully integrated with each other. However, they lead to specific judgments about SCT, GLT, and genetic enhancements. SCT is no different in principle from other forms of medical treatment. As such, it should be evaluated according to principles that govern other experimental therapies, including consent, independent review, proportionality between risks and burdens of treatment and degree and likelihood of benefit, consideration of the risks to others (e.g., the risk to the mother from SCT performed on a fetus), and restricting initial use of SCT to cases of serious disease for which there is no satisfactory alternative treatment. There is a risk that SCT will have inadvertent germ-line effects, but many other therapies pose the same risk.

GLT is problematic for several reasons aside from current technical difficulties. First, its development and use, at least in early stages, would involve in vitro fertilization (which violates the respect owed to life in its transmission) and the destruction of embryos. However, these objections would not apply to GLT performed on ova or spermatagonia followed by normal marital intercourse [a conclusion also drawn by the Catholic Health Association (13)], to treatment of the pre- or postimplantation embryo in situ, or to the removal and treatment of the pre-implantation embryo followed by implantation. None of these germ-line interventions need involve IVF or the discarding of embryos. A second problem with GLT is the risks it poses to various existing and future persons. At the present stage of technology, the risk to the embryo would be considerable, as recent animal experiments demonstrate. GLT procedures involving IVF and/or therapy on ova pose risks to the mother (as would treatment of the embryo in situ or its removal and replacement, though these are not mentioned in this context). Finally, there are long-term risks of adverse effects on the germ line. These risks are still too great to consider GLT even in cases involving serious diseases. But what if technological advance reduces these risks sufficiently to justify GLT in cases of extremely serious conditions? Three problems would remain. First, the risks of GLT would still be significant and would apply to descendants—assuming they would have existed at all—who are not affected by the condition, and who therefore would be subjected to risks without having the condition that justified taking those risks. Second, GLT would be costly and would compete against other pressing medical needs. Since GLT would mostly affect those individuals who will exist only because GLT will have been available, it is harder to justify the cost (or the risks) over against the needs of individuals and their descendants who would have existed whether or not GLT had been available. Both of these objections raise significant questions about future persons that the report does not address. Third, GLT would almost certainly become safe and effective due to immoral research on human embryos. While it would not necessarily be immoral to make use of it at that point—parents could always request that no embryos be harmed in their case—appearing to condone the means by which a technique was developed and the witness to the sanctity of life entailed in refusing it would be significant factors to take into account. However, the report explicitly rejects the common European claim of a right to inherit an unaltered genome (14). Changes to the genome do not affect the uniqueness of the person any more than do changes to other parts of the body. The genome "like other parts of the body . . . may *in principle* be altered, to cure some defect of the body." Indeed, assuming that the objections noted above could all be met, the possibility of

eliminating a devastating disease from a family would in many cases be not simply a right but an obligation.

The report addresses nontherapeutic genetic interventions (enhancements) by distinguishing between "environmental" and "mechanical" interventions. The former involve "a mere *response* to *selected existing potential* of the child" and are "open-ended" in that they do not specify the exact characteristics or the degree to which the intervention will prove favorable. The latter, which include genetic interventions, involve "an *amendment* of existing potential" and "are something that *happens* to the child rather than something the child *does* in a certain environment." There are two arguments in favor of environmental over mechanical interventions. First, because health should, whenever possible, be promoted through the normal channels of conscious or nonconscious human activity (see above), there is a presumption in favor of the former. Second, while both types of intervention run the risk that parents will consider the child as a product or something they control, this is more likely in the case of mechanical interventions. Mechanical interventions, then, "would at least sometimes be unjustified, and conducive to further acts of parental manipulation." This stops short of a prohibition, but the report does not address questions of what circumstances would justify overriding the presumption against mechanical interventions, whether the latter are justifiable for conditions (e.g., short stature) that do not generally admit of environmental solutions, and what conditions should be candidates for enhancement at all. However, the report notes that if the common belief that strangers are not entitled to perform nontherapeutic mechanical interventions on children is justified, then germ-line enhancements would be immoral, since we are strangers with regard to our future descendants in a way that we are not with regard to our children.

Other. In place of pre–Vatican II natural law theories and the new natural law theory adopted by the Catholic Bishops' Joint Committee, Richard McCormick proposes a criterion for genetic interventions that relies heavily on human experience, asking of each proposed intervention whether it will "promote or undermine human persons 'integrally and adequately considered'" (15). This criterion "is necessarily inductive, involving experience and reflection upon it." Experience and reflection alone, however, could be appealed to in support of almost any conclusion; fortunately, McCormick identifies certain values that are meant to supplement, or perhaps specify, the criterion. One is the sacredness of human life, which opposes undue risks and especially discriminatory distribution of risks, and requires informed consent. McCormick does not discuss embryo research here, but elsewhere he argues for a presumption against the latter with exceptions approved "by an appropriate authority" (16). A second value refers to the interconnection of life systems, on which grounds McCormick rejects genetic interventions that accomplish short-term benefits at the risk of long-term harms. Since GLT risks eliminating deleterious genes that may have long-term beneficial effects, it is suspect on these grounds. Third, human diversity and individual uniqueness are important aspects of the human condition that are threatened by some eugenic interventions. A related concern is that genetic enhancements could lead us to evaluate persons "not for the *whole* that they are . . . but for the *part* that we select." Finally, social responsibility requires distributive justice in both research priorities and access to medical benefits. These values, however, are still quite general; it is not clear how one moves from them to judgments regarding specific interventions and policies.

James Keenan criticizes the dominance of concerns about rights in discussions of the ethics of gene therapy (17). "We must ask not whether we have a right to enter these areas or not. We must ask what type of people we could become by entering into any of these areas." Keenan's major concern is the potential of gene therapy for objectifying persons. This potential apparently resides in a combination of reductionism and the mode in which genetic technology intervenes into evolution. While humans have always intervened into evolution, genetic technology does so from within human nature rather than from without; just as the process of directing nature has objectified external nature, so genetic interventions will objectify the human subject. Keenan does not show how this kind of objectification is more significant than more familiar interventions that also work "from within," ranging from psychotherapy to ascetic practices, nor does he question the viability of the idea of a free subject that underlies his analysis. Instead, he describes a progression of objectification: SCT objectifies the disease, GLT collapses the distinction between person and disease, enhancement takes the genotype itself—not simply correction of a disorder—as its object, while eugenics aims precisely at the objectification of the genome so that the person is an object before being a subject. GLT also risks the objectification of parenting (because it is difficult to see a gamete or zygote as more than an object) and research (because consent cannot be obtained). Keenan does not argue that the threat of objectification renders gene therapy unjustifiable; rather, the moral challenge "will be the creation of conditions in which the person, though objectified, is not solely treated as an object." Despite the problems with the category of objectification and its applications, Keenan's central question—what kind of people could we become through genetic interventions — opens up a promising line of inquiry that too few commentators, whether religious or secular, have followed.

Protestant

Anglican. The Episcopal Church offically adopted a brief resolution on gene therapy in 1991 stating that there is no theological or ethical objection to gene therapy if proved effective without undue risk, if aimed at "prevention or alleviation of serious suffering," and if benefits are available to all who need them for these purposes (18). None of these conditions are elaborated in any detail, and "serious suffering" is not defined. In 1992 the Governing Body of the Church of Wales commended to the church a report on genetic screening and therapy written by the church's Division of Social Responsibility (19). The report finds no objection to the use of SCT to correct serious

genetic defects for which there are no alternative cures, but supports a 1988 statement by a group of European medical research councils (20) that GLT should never be carried out (presumably because of its unknown consequences to future generations, though this is not clear). Finally, the report rejects any "attempt to manipulate the human genome for other than therapeutic reasons. Any proposal to engineer particular traits or characteristics in human beings should be rejected as frivolous and regarded as unethical." No effort is made to distinguish enhancement from therapy or to address hard cases that may not be readily assignable to either category. No reason is given for the opposition to enhancements, though the first of three concerns the report raises about genetics in general, namely the concern that parents will choose characteristics of their children based on individual whim, could be one such reason. The other two concerns — that the state will require that parents carry out genetic interventions on their children, and that discrimination against or diminishment of respect for those who will continue to be born with congenital anomalies despite genetic interventions will occur — would remain even if, as the report recommends, gene therapy were restricted to serious diseases.

Evangelical/Holiness/Adventist. Ethical issues of gene therapy have been addressed chiefly by individual evangelicals rather than by churches or organizations. If those individuals are representative, evangelicals may be less interested than some others in drawing lines between different kinds of genetic intervention. In place of distinctions between SCT and GLT or between therapy and eugenics, John Feinberg proposes a line between gene therapy "to fight something in human beings that is clearly a result of the consequences of sin and living in a fallen world" and gene therapy to alter what is simply part of the diversity of creation (21). Feinberg classifies cystic fibrosis, Huntington's chorea, Parkinson's disease, and other physical and psychological conditions under the former category, while hair color, skin color and left-handedness fall under the latter. In effect, Feinberg simply redefines the distinction between diseases and traits as a distinction between the effects of sin and genetic diversity. Genetic interventions designed to alleviate the former are in principle permissible, though the standard moral and scientific preconditions for performing any medical intervention may supply reasons not to perform them in a given case or in general (as is the case with germ-line interventions until they are proved to be safe and effective). What about the latter—is it permissible to change one's hair color, for example, even though this is a matter of human diversity rather than sin? It depends on one's motive. Belief that certain of these traits are inferior to others, that changing them will increase one's own value as a human being, that those who possess other such traits are less valuable, or that everyone should possess a certain trait — all of these motives are immoral according to Feinberg. But there are problems with this approach. First, as Feinberg concedes, it is difficult to determine whether some traits, such as aggressive behavior, are the results of sin or diversity.

Second, while certain characteristics — skin color or body shape, for example — are from Feinberg's standpoint due to genetic diversity and not sin, the discrimination some people face due to these traits is the effect of sin. Would Feinberg permit one to fight this consequence of sin by changing the trait? Third, one may question on theological grounds whether it is always justifiable in principle to fight the effects of sin. If, as Feinberg himself believes, death is a consequence of human sin, should Christians support the use of genetic technology to attain immortality? Like Feinberg, James Peterson questions the line between SCT and GLT (22). In practice, some somatic cell interventions (e.g., keeping those with deleterious genes alive long enough to reproduce) could have germline effects, while ethically, the generational factor in GLT brings not only greater risks but also greater potential benefits. Peterson also questions the line between therapy and enhancement, arguing that if one proposes a normal range of human functioning as normative, there is no reason for excluding some characteristics and functions from being brought into that range and, indeed, no reason for not trying to increase the normal range itself. Instead Peterson proposes five criteria any genetic intervention should meet. They should be (1) incremental in order to minimize the degree and extent of unanticipated harms, (2) choice-expanding in the sense of not limiting one to a particular kind of life, (3) parent-directed in order to decentralize choice and thus avoid large-scale eugenic programs, (4) kept within societal boundaries that set minimum conditions that parents must meet with regard to their children but allow flexibility within those conditions, and (5) carried out by acceptable means, namely with limited risks and as noninvasive as possible. Peterson recognizes the limits of this approach: Parents, in seeking advantages for their children, and corporations, in promoting the use of genetic technologies, will ignore these limits. Nevertheless, because of their potential benefits, humans are responsible to God to pursue genetic interventions — "[e]ven some instances of germline enhancement" — within the five conditions.

The Church of the Nazarene, in the holiness tradition of American evangelicalism, included in its Manual from 1993 to 1997 a statement approving of gene therapy for the prevention and cure of disease but opposing "any use of genetic engineering that promotes social injustice, disregards the dignity of persons, or that attempts to achieve racial, intellectual, or social superiority over others (Eugenics)" (23). The statement does not articulate what is meant by social injustice or human dignity.

The Seventh-Day Adventist Church adopted a document on genetic interventions in 1995 (24). The general justification for gene therapy is theological: The genetic endowment of Adam and Eve was perfect; hence gene therapy is welcomed as a form of cooperation with God in recovering more of the original condition of creation and alleviating the results of sin. However, gene therapy must be carried out in accordance with Christian principles. These include (1) the rejection, given current knowledge, of GLT on the ground that it could affect the image of God in future generations; (2) the exercise of "great caution," in light of human sinfulness, the possibility of abuse, and

unknown biological risks, in attempts to modify physical or mental characteristics in persons free of genetic disorders; and (3) the availability of the benefits of genetic research to all who need them. The document does not tell how to distinguish therapy from enhancements, nor does it explain why genetic interventions affecting future generations, but not those affecting present individuals, threaten the image of God.

Lutheran. Lutheran theologians in the United States and theologians and church groups in Europe together constitute a spectrum of theological and ethical evaluations of GLT. Among Americans, Ted Peters has called for "keeping the door open" to GLT and eugenics (25). Peters argues that if God's creative activity is understood as giving the world a future and humanity is understood as a "created co-creator," then "ethics begins with envisioning a better future." We should therefore keep open the possibility of improving the genetic makeup of the species. Peters rejects the various arguments against GLT and eugenics: Unforeseen consequences to future generations is no reason to prohibit GLT but rather to proceed in accordance with growing knowledge of those effects; eugenics can be dissociated from historical abuses; prejudice and discrimination against those who fall short of standards of genetic perfection already occur without GLT and need not accompany GLT. Peters addresses another criticism made by some European Lutheran theologians and by nonreligious writers in the United States such as Jeremy Rifkin (26) and Robert Sinsheimer (27): neither nature in general nor DNA in particular is sacred or represents, in its present form, God's final plan for humanity; rather, because nature is created ex nihilo, it has no ultimacy or sanctity in itself, while because God continues to create (*creatio continua*) nature as it is has no normative status (28). By describing creation entirely in terms of future and novelty and human co-creating entirely in terms of technology, Peters gives the impression that his theology simply reinscribes the modern narrative of progress. Also Peters offers no guidance regarding what future possibilities, aside from the treatment of fatal diseases, are and are not worth pursuing. With one notable exception, he also says little about the means by which they may be pursued. The exception concerns embryos. The latter may not be persons in the full sense, but they possess a moral status that makes genetic manipulation of gametes prior to fertilization morally preferable to genetic manipulation of zygotes (29).

In contrast to Peters, Gilbert Meilaender opposes GLT (30). Meilaender concedes that GLT could be therapeutic in intent and effect, and that it could spare future generations serious problems. This last feature, however, is precisely the problem; GLT exercises control over future generations. "Such interventions would aim ... at shaping the nature of others still to come. Not only a human being but humankind is then the object of our intervention." Medicine, apparently, should focus on the person with the disease, neither eliminating the person (as with selective abortion) nor eliminating the disease from humanity as a whole. However, Meilaender does not indicate why medicine should be judged differently from other interventions (e.g., those involving public health and the environment) that aim at humanity as a whole. In any case, if SCT but not GLT is justifiable, can we distinguish between treating diseases and enhancing traits? Meilaender's definition of diseases as "disorders that bring pain or hinder an individual in carrying out the biological functions necessary for personal or species survival" is admittedly narrow; more promising is his suggestion that in place of such a line we cultivate "a renewed sense of the mystery of the human person and the limits to our own efforts at shaping and transforming character" together with the virtue of love "that in its open-hearted acceptance of an other disciplines and restrains the urge to transform and remake." Meilaender worries that without these attitudes and virtues children will become products — made rather than begotten; with these attitudes and virtues, and the discernment that comes with them, there is little to fear from SCT. But why, if these attitudes and virtues are capable of guiding the use of SCT, are they not also capable of guiding GLT? If GLT is ruled out because it exercises control over others (and not simply because it aims at humanity as a whole), then why not also rule out SCT performed on one's children? And if unqualified love and a sense of mystery are sufficient to prevent abuses of enhancements in the case of SCT, why are they not sufficient to prevent the same abuses in the case of GLT?

European Lutherans tend to side with Meilaender against Peters (31,32). A few isolated individuals view SCT as the first step toward breeding human beings or argue that placing the human genome at human disposal nullifies human dignity, but the vast majority support SCT provided that standard ethical conditions governing experimental therapies are met. GLT, however, is almost universally rejected, either permanently or in light of present knowledge and current (European) moral conventions. The reasons vary. Principled objections refer to the illicitness of embryo research and to the claim that GLT violates the genetic integrity of humanity (as humanistically inclined theologians argue) or that humanity precisely as it is, is created in the image of God (as more biblically oriented theologians argue). These latter sorts of claim seem to presuppose a genetic essentialism that most theologians and church groups attempt to avoid; moreover they raise the question again of why the germ line, and not somatic cells, is the locus of human integrity or the image of God. A second set of objections rules out GLT because of certain problems inevitably connected with it, namely the impossibility of drawing a line between GLT and eugenics or the impossibility of determining, in the final sense GLT implies, what are healthy and diseased genes. A final set of objections could be overcome by future developments. These refer to the unreliable results of animal research at present, to the consequences errors in GLT may have for descendants, and to the relative risks and cost of GLT.

Methodist. In 1992 the United Methodist Church adopted a comprehensive report on genetic science that endorsed SCT for the alleviation of suffering caused by disease, opposed GLT until its safety and certainty of its effects can be demonstrated and its risks to human life shown to be minimal, and opposed the use of gene therapy for eugenic purposes and for enhancements designed only for cosmetic purposes or social advantage (33). These conclusions are listed without supporting arguments, but the report as a whole provides reasons that at least partially support most of them. Human beings are understood as stewards of God's creation. This role emphasizes the sustaining of creation, which allows for enhancing creation but also requires acknowledgement of limits to human creativity and power. Genetic diversity reflects the goodness of creation and therefore must be preserved, while the unity of humanity in creation and in Christ rules out discrimination based on biological factors and requires recognition of the worth of the most defenseless. These convictions could serve as arguments against ambitious eugenic programs. The report cites several specific concerns regarding gene therapy. Three of these — the danger to individuals from experimental procedures, unanticipated adverse effects of combining genes from different species, and the larger numbers of people who are helped by gene therapy but who may be carriers of genetic diseases that are difficult or expensive to treat — apply to SCT as much as, or more than, to GLT. The other three — the long-term effects on the species, the unanticipated long-term health and genetic consequences of genetic enhancement, and the vision or goal that governs efforts to control evolution — apply primarily to GLT and/or efforts at eugenics and enhancement. Nevertheless, the report does not indicate why these concerns outweigh the potential benefits of GLT, enhancements and eugenics, but permit SCT.

J. Robert Nelson, a United Methodist theologian, supports SCT but proposes a present ban on GLT (34). Nelson is not opposed to GLT in principle — he recognizes its therapeutic potential, first for the organism itself, and then for its progeny — but only because of our insufficient knowledge, at present, of long-term generational consequences. Like most Christians, Nelson rejects any notion that the genetically unmodified person is normative; gene therapy cannot be rejected on grounds that it is unnatural.

Reformed. Both the Presbyterian Church, U.S.A. and the United Church of Christ have approved statements on genetics that include brief treatments of gene therapy. In 1990 the Presbyterian Church's General Assembly resolved to "[s]upport the discovery of new genetic knowledge that can improve the treatment and eradication of disease ..." (35). In 1983 the General Assembly approved a report that supported the potential of genetic research for "relieving suffering and enhancing life" but warned against the "threat of idolatry in the search for the 'perfect human being' ..." and concluded that "[t]he pursuit of 'superior' human beings through genetic manipulation should be explored only with great caution, if at all" (36). No elaboration or arguments clarify or support these declarations. The United Church of Christ statement, approved in 1989, is marginally more substantive, approving of SCT and noting that GLT may have unforeseen consequences that preclude it now, though future developments may alleviate this problem (37).

Support for gene therapy in principle is grounded in human covenantal responsibility to participate in God's creative and redemptive work, as made known in the healing ministry of Jesus.

In 1995 the General Assembly of the Church of Scotland welcomed a report written by a study group of the church's Board of Social Responsibility (38). The report, which focuses on genetic diagnosis, screening and therapy, is perhaps the most restrained of any religious treatment of gene therapy. Its authors believe most genetic interventions, including SCT beyond very limited uses and any use of GLT, are highly unlikely; rather than analyzing ethical issues they consider wildly hypothetical, they give reasons why they believe the techniques that pose such issues are so improbable. These reasons refer to technical obstacles, moral and regulatory requirements regarding research using human subjects, costs, and the presence of less risky (and less costly) alternative interventions (e.g., in vitro fertilization followed by selective implantation or fetal testing followed by abortion as alternatives to GLT). Not all of these reasons are convincing (e.g., GLT would be a superior alternative for those opposed to selective implantation or abortion on moral grounds), but all of them have been raised by other individuals and groups as well. In general, the report seems to endorse the conclusions of the Clothier Report of 1992 (39), which supported SCT under the current regulations governing experimental therapies and opposed GLT for the present. However, it also includes sociologist Margaret Stacey's criticism of the Clothier Report for ignoring the cultural context in which "genetic manipulation will itself inevitably change perceptions and beliefs about what it is proper for individuals to ask others to do to them or their children. ..." Stacey's point, that technology in its cultural context generates new obligations and ideals, offers a promising line of inquiry which religious analyses of gene therapy have so far ignored.

Ronald Cole-Turner takes issue with most theological responses to genetic technology for leaving unclear the moral status of genetic disease and its cure (40). First, they do not resolve the question of whether genetic illness is natural or a defect of nature; they therefore have no basis for determining whether God wills illness or its cure. Second, they emphasize creation rather than redemption. As (created) co-creators, human beings are authorized to explore novel genetic combinations, but the question is whether they are permitted to identify and correct genetic defects. The latter requires a view of redemption as the restoration and reordering of creation. Cole-Turner seeks to rectify these problems. Nature is good, but also disordered: "A gene is identified for research and possible therapy because it causes human suffering. But it is regarded as a genetic defect because it is taken as a manifestation of the moral disorder of nature in reference to the intentions of the Creator. ..." This gives gene therapy a moral ground: "That which is defective is that which may be changed or altered. Indeed, altering it would be seen as an act of participation in the redemptive work of God." What genetic interventions, then, may be welcomed as redemptive? Cole-Turner has no satisfactory answers to this question. He does,

however, refer to the healing ministry of Jesus as paradigmatic of redemption. The conditions into which Jesus intervened — Cole-Turner identifies skin diseases, neurological conditions, and mental disorders — indicate what kinds of conditions are contrary to the purposes of God. And Jesus's special concern with the weak, the sick, and the poor, reversing natural selection by favoring the retention of their genes, indicates what research priorities and marketing arrangements should, from a Christian perspective, govern the development of gene therapy. But these claims are open to criticism. First, why is the ministry of Jesus the sole paradigm of redemption? Second, how strictly does Cole-Turner want to interpret this ministry? If Jesus healed only these sorts of diseases and restricted his healing to individuals whom he encountered more or less directly, does this rule out enhancements and GLT, respectively?

Judaism

Nearly all practicing Jews accept the normative status of the law, or *halakhah*, as given in the Talmud and in the commentaries, codes, and responsa that constitute the rabbinic tradition. However, there are important areas of disagreement. One disagreement is over how strictly to interpret the law. Another concerns whether there is an ethic outside the *halakhah*, and if so, what is its content, and how is it related to the law. Thus far the most detailed treatments of gene therapy have come from Orthodox commentators such as Barry Freundel, Azriel Rosenfeld, and Fred Rosner. Because they largely agree with one another and their views overlap, they will be discussed together, along with other Jewish thinkers whose less direct treatments of gene therapy supplement the work of the three Orthodox commentators. However, this should not give the impression that there is a single Jewish position on gene therapy or that, as other Jewish thinkers begin to address this issue, the consensus among these commentators will necessarily hold up.

Judaism brings to gene therapy a tradition of strong encouragment of therapeutic interventions, based on the importance of saving life, that amounts to a justification of gene therapy in principle. Arguments that genetic engineering falls under a class of illicit alterations of nature (along with sowing diverse seeds, mating different kinds of animals, mixing certain fabrics) are not unknown in Judaism, but Rosner's counterargument — that genetic engineering is a permissible alteration because it falls within the physician's divine license to heal — reflects a nearly universal view (41). Indeed, "even the most conservative Orthodox thinking provides no support for the view that such genetic manipulation would be an unallowable 'tampering' with nature" (42). Like other interventions that save or improve human life, gene therapy falls under the divinely ordained task of *tikkun olam*, namely healing or repairing and perfecting the world (43). In principle, then, gene therapy is permitted at all stages, including preconception, in utero, or following birth. SCT is judged by the same risk–benefit criteria used to judge other medical procedures, though a slightly higher risk factor may be tolerated for the fetus (who is not fully a person in Jewish law) or the infant prior to thirty days (who is considered not yet fully viable). GLT also falls under the permissibility given to therapeutic interventions. What distinguishes Judaism from other perspectives, religious and secular, is its attitude toward the risk of unanticipated ill consequences posed by gene therapy (especially GLT). Freundel argues that Jewish law deals with what exists in the present: "A person who is ill today is to be helped to the extent possible. What results in later generations will be dealt with then." Freundel grounds this in "a Talmudic principle that enables us to assume that when we do our best G_d will take care of what we could not foresee or anticipate. If things do not work out, the theological question is G_d's to answer, not ours" (43).

Given this strong endorsement of gene therapy in principle, discussions have focused on whether or not Jewish law forbids certain means of accomplishing it. In an early article Rosenfeld considers possible objections to gene surgery on human ova (their removal, genetic modification, and replacement) and transplantation of genes from a donor into germinal cells (44). As for gene surgery, Rosenfeld appeals to an "indisputable principle" that any surgery permitted on a person is permitted on germinal cells, which are at most potential persons, prior to conception. If there were a surgical cure for hemophilia, it would be permissible; hence gene surgery to cure hemophilia is also permitted. Similarly, if (as many authorities agree) cosmetic surgery is permitted to relieve psychological distress, then gene surgery to achieve cosmetic effects should also be permitted. These permissions assume, however, that the procedures are safe enough that the ova are (almost) never destroyed; otherwise, they violate the prohibition against "destruction of the seed." Gene transplants raise the questions of whether they involve a possible illicit sex act and whether the child whose birth followed the procedure would be considered related to the donor, with implications for inheritance and restrictions on marriage into that family. Rosenfeld argues that no sex act is involved, since the donor genes need not come from the reproductive cells of the donor and the transplantation is carried out outside the body of the recipient. Nor would the child be related to the donor: If ovaries or testicles were transplanted, a child conceived after the transplantation would not be regarded as related to the donor; why, then, would he or she be in the case of genes? It is impossible that transplanting submicroscopic parts of sex cells—which, as invisible to the human eye, are generally excluded from consideration in Jewish law—would have more effect on the status of a child than whole sex organs. Since later commentators tend to repeat Rosenfeld's analysis, it is difficult to know what halakhic difficulties, if any, alternative methods of gene therapy would present.

Rosenfeld's reference to cosmetic surgery raises the question of enhancements more generally. Freundel and Rosenfeld both refer to a Talmudic story according to which Rabbi Yohanan sat on the road from the ritual bath so that women would see him before resuming sexual relations with their husbands. Yohanan's purpose was that women would think of him and thereby produce offspring as handsome as he and as accomplished a scholar. For both Freundel and Rosenfeld, the story

authorizes the use of genetics for intellectual-ethical and aesthetic purposes. According to Freundel, genetic interventions for such purposes are no different from psychological or behavioristic interventions. Are there any limits to what characteristics can be modified? Freundel argues that certain characteristics—he mentions speech, intellect, love, and creativity—are "manifestations of the presence of the soul, but not the same as the soul." To discover the genetic sites of these characteristics is permissible. "Great concern, however, would exist about tampering with such sites either in terms of damaging something fundamentally human or in terms of potentially diminishing free will and individuality." Freundel's concern, however, stops short of a permanent prohibition; a judgment on such interventions would have to be made in accordance with the nature and impact of the interventions. On the basis of the stories about the creation of golem in Jewish lore, Jewish commentators tend to rule out the use of genetics to create humanlike beings lacking fundamental human qualities, though it is not clear what qualities would have to be absent or distorted (the golem often lacked speech).

Jewish attitudes to eugenics are shaped by the horrendous suffering of Jews at the hands of Nazi eugenic policies. Laurie Zoloth-Dorfman locates this history within a longer history of discrimination and violence in which the difference of the Jewish body from the male gentile body served to mark Jews as dangerous or less than fully human (45). Zoloth-Dorfman also notes the readiness of some Jewish physicians to use medicine in order to make the bodies of Jews conform to gentile "normality." Her point is that Jews and gentiles must remember this history as a warning not use genetic medicine to underwrite a suspect "normality" to which others are made or expected to conform. Freundel argues that Jews would not accept any eugenic program that kept those who wanted to procreate from doing so or that defined certain people as undesirable.

Rosner mentions two other (extra-halakhic) ethical issues. First, gene therapy is subject to common ethical principles governing novel therapies and research on human subjects. Second, gene therapy runs the risks of furthering the mechanization of human life. As with the extra-halakhic concerns regarding artificial beings and eugenics, however, it is not clear whether or how these concerns qualify the acceptance of gene therapy. Also missing from these accounts, but prevalent in analyses of other bioethical issues by Jewish commentators, Orthodox and other, is attention to questions of allocation of and access to gene therapy.

Islam

By almost all accounts, Islam places no limitations on the pursuit of scientific knowledge, including genetic knowledge. However, Hassan Hathout and B. Andrew Lustig note that applications of scientific knowledge are subject to five Islamic governing rules (46). The third of these rules refers to "changing God's creation," a phrase uttered by Satan in connection with his plans for leading humankind astray. Hathout and Lustig claim that there is a consensus among Islamic scholars that this phrase does not support a ban on genetic engineering; otherwise, it would rule out many forms of life-saving and life-promoting surgery (appendectomy, tonsillectomy, cholecystectomy, and others) that involve a change in God's creation. They conclude that genetic engineering is permissible, but that the fourth rule — "Wherever the welfare exists, there stands the statute of God" — requires juridical sanctions to ensure that applications of genetic research will be used for human benefit. However, Gamal Serour, who has written widely on genetic and reproductive issues in Islam, restricts the justifiability of gene therapy to its therapeutic uses (47). Use of genetic technology for enhancement or eugenic purposes "would involve change in the creation of God" that could lead to imbalance in the universe as a whole or in humanity.

Other

The survey thus far presents an incomplete picture of religious attitudes to gene therapy. It excludes the many religious traditions, large and small (including the major traditions of South or East Asian origin), which have not yet addressed gene therapy in a substantive way. Among those it surveys, it concentrates on official or quasi-official statements and leading theologians, which may only approximately reflect the attitudes of large numbers of adherents. Finally, it ignores the views of the growing number of people who do not identify with an official religious body but who have spiritual commitments, often drawn from a wide range of sources. Popular culture often expresses possibilities or fears related to genetics in religious language (48), while opponents of genetic technologies, including Jeremy Rifkin (26) and Robert Sinsheimer (27), describe DNA as sacred or quasi-sacred. Is this latter view, rejected by most public spokespersons of mainstream religious traditions, widely shared, and does it indicate a principled stance against many genetic interventions or only an expression of ambivalence toward the latter? The importance public representatives of mainstream religious traditions place on criticizing such views indicates their precarious hold on the language of the sacred and, by extension, the religious response to gene therapy.

BIBLIOGRAPHY

1. J.R. Nelson, Hum. Gene Ther. 1, 43–48 (1990).
2. National Council of the Churches of Christ in the U.S.A., Genetic Science for Human Benefit, National Council of Churches of Christ in the U.S.A., New York, 1986.
3. World Council of Churches, Manipulating Life, World Council of Churches, Geneva, 1982.
4. World Council of Churches, Biotechnology: Its Challenges to the Churches and the World, World Council of Churches, Geneva, 1989.
5. F. Harron, ed., Genetic Engineering: Social and Ethical Consequences, Pilgrim Press, New York, 1984.
6. National Council of the Churches of Christ in the U.S.A., Human Life and the New Genetics, National Council of Churches of Christ in the U.S.A., New York, 1980.
7. J. Breck, St. Vladimir's Theol. Q. 32(1), 5–26 (1988).
8. D. Demopulos, in J.T. Chirban, ed., Ethical Dilemmas: Crises in Faith and Modern Medicine, University Press of America, Lanham, MD, 1994, pp. 91–99.
9. Pope John Paul II, Origins 12(21), 342–343 (1982).
10. Pope John Paul II, Origins 13(23), 385–89 (1983).
11. The Catholic Bishops' Joint Committee on Bioethical Issues, Genetic Intervention on Human Subjects: The Report of a Working Party, Linacre Centre, London, 1995.
12. G. Grisez, J. Boyle, and J. Finnis, Am. J. Juris. 32, 99–151 (1986).
13. Catholic Health Association of the United States, Human Genetics: Ethical Issues in Genetic Testing, Counseling, and Therapy, Catholic Health Association of the United States, St. Louis, MO, 1990.
14. Europaishes Parlament, Betr. die Gen-Manipulation, Empfehlung 934, Bundestagsdrucksache 9/1373, 1982, pp. 11–13.
15. R.A. McCormick, The Critical Calling: Reflections on Moral Dilemmas since Vatican II, Georgetown University Press, Washington, DC, 1989, pp. 261–272.
16. R.A. McCormick, The Critical Calling: Reflections on Moral Dilemmas since Vatican II, Georgetown University Press, Washington, DC, 1989, p. 345.
17. J.F. Keenan, Hum. Gene Ther. 1, 289–298 (1990).
18. Episcopal Church, Resolution 1991-A095, Archives of the Episcopal Church, New York, 1991.
19. Division for Social Responsibility, Church in Wales Board of Mission, Human Genetic Screening and Therapy: Some Moral and Pastoral Issues, Church in Wales Publications, Penarth, 1992.
20. Lancet (1988).
21. J.S. Feinberg, in J.F. Kilner, R.D. Pentz, and F.E. Young, eds., Genetic Ethics: Do the Ends Justify the Genes? Eerdmans, Grand Rapids, MI, 1997, pp. 183–192.
22. J.C. Peterson, in J.F. Kilner, R.D. Pentz, and F.E. Young, eds., Genetic Ethics: Do the Ends Justify the Genes? Eerdmans, Grand Rapids, MI, 1997, pp. 193–202.
23. Church of the Nazarene, Manual/1993-97, Nazarene Publishing House, Kansas City, MO, 1993, 904.1.
24. Seventh-Day Adventist Church, Statements, Guidelines and Other Documents, Seventh-Day Adventist Church, Silver Spring, MD, 1995, pp. 91–98.
25. T. Peters, J. Med. Philos. 20, 365–386 (1995).
26. J. Rifkin, Algeny, Penguin, New York, 1983.
27. R. Sinsheimer, Technol. Rev. 86, 14–17, 70 (1983).
28. T. Peters, in T. Peters, ed., Genetics: Issues of Social Justice, Pilgrim Press, Cleveland, OH, 1998, pp. 1–45.
29. T. Peters, For the Love of Children: Genetic Technology and the Future of the Family, Westminster John Knox Press, Louisville, KY.
30. G. Meilaender, Christian Century 107, 872–875 (1990).
31. H. von Schubert, Evangelische Ethik und Biotechnologie, Campus Verlag, Frankfurt, 1991.
32. K. von Kooten Niekerk, in J. Hubner and H. von Schubert, eds., Biotechnologie und Evangelische Ethik, Campus Verlag, Frankfurt, 1992, pp. 324–347.
33. United Methodist Church, New Developments in Genetic Science, United Methodist Publishing House, Nashville, TN, 1992.
34. J.R. Nelson, Human Life: A Biblical Perspective on Bioethics, Fortress Press, Philadelphia, PA, 1984.

35. General Assembly, Presbyterian Church, U.S.A., *Social Policy Compilation*, Presbyterian Church, U.S.A., Louisville, KY, 1990.

36. Presbyterian Church, U.S.A., *The Covenant of Life and the Caring Community*, Presbyterian Church, U.S.A., Louisville, KY, 1983.

37. United Church of Christ, *The Church and Genetic Engineering*, United Church of Christ, Cleveland, OH, 1989.

38. W. Storer and I. Torrance, eds., *Human Genetics: A Christian Perspective*, The Church of Scotland Board of Social Responsibility, Edinburgh, 1995.

39. C. Clothier, *Report of the Committee on the Ethics of Gene Therapy*, H.M. Stationery Office, London, 1992.

40. R. Cole-Turner, *The New Genesis: Theology and the Genetic Revolution*, Westminster John Knox Press, Louisville, KY, 1993.

41. F. Rosner, *Modern Medicine and Jewish Ethics*, 2nd ed., Yeshiva University Press, New York, 1991.

42. R. Green, *Judaism* **34**(3), 263–277 (1985).

43. B. Freundel, in J.R. Nelson, ed., *On the New Frontiers of Genetics and Religion*, Eerdmans, Grand Rapids, MI, 1994, pp. 120–136.

44. A. Rosenfeld, in F. Rosner and J.D. Bleich, ed., *Jewish Bioethics*, Hebrew Publishing Co., Brooklyn, NY, 1979, pp. 402–408.

45. L. Zoloth-Dorfman, in T. Peters, ed., *Genetics: Issues of Social Justice*, Pilgrim Press, Cleveland, OH, 1998, pp. 180–202.

46. H. Hathout and B.A. Lustig, in B.A. Lustig et al., eds., *Bioethics Yearbook*, vol. 3, Kluwer Academics, Dordrecht, The Netherlands, 1993, pp. 133–148.

47. G.I. Serour, in B.A. Lustig et al., eds., *Bioethics Yearbook*, vol. 5, Kluwer Academics, Dordrecht, The Netherlands, 1997, pp. 171–188.

48. D. Nelkin and M.S. Lindee, *The DNA Mystique: The Gene as Cultural Icon*, Freeman, New York, 1995.

See other GENE THERAPY and RELIGIOUS VIEWS ON BIOTECHNOLOGY entries.

GENE THERAPY, ETHICS, SOMATIC CELL GENE THERAPY

NELSON A. WIVEL
University of Pennsylvania School of Medicine
Philadelphia, Pennsylvania

OUTLINE

Introduction

Development of Science and Concepts Leading to Gene Therapy

Origins and Evolution of Public Oversight of Human Gene Therapy

Principles of Review for Human Gene Therapy

Human Gene Therapy and the Industry Model for Drug Development

Challenges for the Future

Conclusion

Bibliography

INTRODUCTION

With the initiation of any new area of scientific research, a whole host of unknowns must be addressed. In the preclinical stages of development, moral restraints do not play a major role, but once human experimentation is contemplated, a significant number of ethical parameters have to be considered. This has been particularly true for human gene therapy, an application of recombinant DNA technology that involves the insertion of functioning genes into the somatic cells of a patient, either to correct an inborn genetic error or to impart a new function to the cell.

In that interim period between medical experimentation and actual development of medical treatment, researchers continually must confront the necessity of creating a positive risk–benefit ratio. This requires careful analysis with regard to the choice of disease target, the prognosis for a given disease, the cohort of research subjects to be studied, and the types of therapy currently available. This matrix of issues must be addressed both for gene therapy and all other types of human subjects research. In this particular sphere, gene therapy is no different than other types of experimental intervention.

One of the major ethical debates surrounding gene therapy has focused on the question of whether or not this particular application of biotechnology is qualitatively different from preceding types of medical therapies. A superficial analysis might suggest this to be the case, but a more careful consideration will lead to the conclusion that gene therapy is simply an extension of a therapeutic continuum (1). Since somatic cell gene therapy targets only nonreproductive cells, the genetic changes are limited to the patient, and there is little or no chance of affecting future offspring of that patient. Further the products of gene transfer are proteins that function in a manner analogous to drugs. In addition the use of gene transfer has the potential to produce many of the same results as allogeneic organ transplantation. For example, diabetes mellitus can be treated pharmacologically with insulin, or one can attempt to treat this disease with islet cell transplantation, or one could postulate treatment with gene therapy in which the gene encoding for insulin is given to the patient and the control of gene expression is physiologically regulated. In the foregoing context, gene therapy is not so intrinsically different from other treatments, and it takes advantage of the knowledge derived from more standard therapies.

A principal purpose of this discussion is to examine some of the early findings that led to the development of gene therapy, to look at origins of public oversight of human gene therapy, to review the major ethical questions concerning this type of research, and to look at the challenges for the future that are posed by this particular form of molecular medicine.

DEVELOPMENT OF SCIENCE AND CONCEPTS LEADING TO GENE THERAPY

If one looks at the history of molecular biology and molecular genetics, it encompasses a period of approximately

50 years, and the developments in this area of biological science have provided the necessary infrastructure for the establishment of human gene therapy. A critical discovery was made by Avery, MacLeod, and McCarty at Rockefeller University when they demonstrated that a gene-inducing transformation in bacteria could be transferred in nucleic acids (2). This satisfied a longstanding desire to identify chemical means by which hereditary traits could be transferred from one generation to the next. A few years later Watson and Crick were able to complete the model for the helical structure of DNA (3); this was followed by discovery of mRNA, and the subsequent development of the central dogma of molecular biology that postulates the flow of genetic information from DNA to RNA to protein (4). Another major advance occurred with the discovery of the restriction endonucleases, enzymes that could cut DNA at specific recognition sites, thus paving the way for creating DNA molecules in which sequences that are not naturally contiguous can be placed next to each other (5). Cohen and Boyer and their colleagues were able to construct functionally active recombinant DNA molecules (6), and it soon became possible to move genes from one species to another without loss of function. Several years later the development of gene delivery vehicles became a reality with the report that retroviruses could be modified to insert genes into cells (7). Although the first human gene therapy protocol was not approved until 1990, the techniques that made it possible had their origins in the many fundamental discoveries emanating from the continued interest in genetics.

As is so often true in science, its practitioners are able to frame concepts well in advance of the actual experiments. For a particularly intriguing history of gene therapy one is referred to the article of Wolff and Lederberg (8). They have pointed out that Edward Tatum, in 1966, was the first to suggest that viruses could be used to insert genes into cells and that modification and regulation of gene activities ultimately might be used to treat cancer (9). Lederberg proposed a potential gene therapy for hemophilia in 1968 when he stated that fractionated DNA containing the normal alleles of the hemophilia gene could be introduced into the liver in experiments analogous to the attempts at transforming bacteria (10). Arthur Kornberg, who successfully used DNA polymerase to synthesize DNA in vitro, predicted that hereditary defects might be cured by attaching a therapeutic gene to a harmless virus that would serve to infect the cell and deliver the gene (11). It is now popular to refer to DNA as a designer drug, but in 1970 Aposhian advanced the idea that if the purpose of a drug was to restore the normal function to a physiological process, then the time might arrive when DNA would become the ultimate drug (12).

If the basic understanding of molecular genetics suggested ideas for gene therapy itself, there were also discussions of the ethical implications of gene therapy and one of the first was presented by Marshall Nirenberg who played a key role in deciphering the genetic code. In 1967 he predicted that cells would be programmed with synthetic messages within 25 years, but that the technology might surpass our ability to assess the long-term consequences of such alterations (13). He expressed the particular concern that this knowledge might not benefit humankind unless it was applied with sufficient wisdom. Robert Sinsheimer accelerated his misgivings by suggesting that designed genetic change was simply a new form of eugenics that could escape the boundaries of the selective processes that occur in nature (14). It was his contention that these new types of choices imposed an extreme need for responsibility.

Although the first human gene therapy trial was not approved by local and national oversight bodies until 1990, there was an earlier attempt at gene transfer that occurred in the 1960s. Stanfield Rogers had done fairly extensive work with the Shope papilloma virus and observed that it apparently had arginase activity. Animals infected with this virus exhibited reduced blood levels of arginine, and laboratory personnel who studied this virus also had reduced blood arginine but no apparent side effects. Three siblings who suffered from arginemia because of arginase deficiency were injected with the Shope virus, but there was no reduction in blood arginine; fortunately there was no evidence of toxicity (15,16). There was some concern expressed about the ethics of these experiments, but the principal investigator defended his actions on the grounds that this type of intervention offered the only reasonable chance to prevent progressive deterioration in these children (17). Another pioneer in the field of gene therapy, French Anderson, supported the experiment on the basis that there were several decades of experience documenting the safety of the Shope virus, and that there was an absolute certainty of suffering and death associated with arginemia (18). Still other investigators expressed the concern that the experiment lacked sufficient preclinical data and that it would serve as stimulus for other groups to proceed in an unprepared fashion (19). There were no further human experiments until 1980 (these will be described at a later point in this discussion).

In attempting to characterize the confluence of events that supported the actual development of human gene therapy, one is properly forced to consider the analogies to pharmacology and surgery. While the administration of drugs preceded our current understanding of molecular genetics, it is our present-day knowledge of genetic principles that allows us to consider the use of DNA fragments in a pharmacotherapeutic context. Similarly gene therapy might be considered as molecular surgery since the incorporation of a therapeutic gene into cell chromosomes has the potential to modify tissues or organs for the life span of the treated patient (8). Thus human gene transfer for the treatment of disease does not require entirely new ethical paradigms, but rather it requires careful attention to those issues that bear on any type of human experimentation, the choice of disease, the choice of patients, the risk–benefit ratio, the need for informed consent, and the right of patient privacy.

ORIGINS AND EVOLUTION OF PUBLIC OVERSIGHT OF HUMAN GENE THERAPY

In order to develop the proper perspective for the review processes that were created explicitly for human gene therapy clinical trials, one has to look at the history of recombinant DNA research. Some of the initial concerns that were applied to recombinant DNA technology abated over the course of five to six years when it became readily apparent that many of the postulated hazards were not going to occur, but there was a recrudescence of concern as the possibilities for human gene therapy became concrete.

Marshall Nirenberg's caution about genetic engineering was translated to the public sector and by 1968, Senator Mondale introduced a resolution for the specific purpose of establishing a Commission on Health, Science, and Society. One of the proposed tasks for this commission was the study of the moral and ethical questions surrounding genetic intervention. Several prominent geneticists testified at hearings held during the spring of 1968 and emphasized that there were few societal hazards associated with the "new" genetics. In the period between 1968 and 1982, Congress did not concern itself with human gene therapy, but in the mid-1970s there was a spirited debate about the potential hazards of recombinant DNA research.

As early as 1971 Paul Berg was having success in developing the first recombinant viral vector, using the simian virus SV40. His experiments provoked a spirited discussion at a meeting at the Cold Spring Harbor Laboratory. Because of the legitimate differences of opinion, the first Asilomar Conference was convened in 1973. There was a fairly systematic discussion of some of the potential hazards associated with certain kinds of experiments such as the insertion of antibiotic resistance genes into bacteria that do not normally possess such properties, or moving part or all of the genes of one animal virus into a plasmid or another virus. Because of the cumulative uncertainties, a voluntary moratorium was suggested and was communicated in the form of a letter to the journal, *Science*, with many of the most active investigators as signatories (20). In February 1975 the second Asilomar Conference was held, and the results were suggestive of a new order for the scientific community. Despite internal disagreements the participants in this meeting ultimately reached a consensus that there should be a scheme for control of recombinant DNA experiments that would at least minimize, or preferably eliminate, potential biohazards. Two kinds of containment were proposed: One was physical containment that would require appropriate facilities, and the second was biological containment that would require the engineering of microorganisms so that they would have a selective disadvantage for survival in the laboratory environment.

Immediately following this conference, the first meeting of the National Institutes of Health (NIH) Recombinant DNA Advisory Committee (RAC) was convened. The primary task for this group was to create the "NIH Guidelines for Research Involving Recombinant DNA Molecules"; the task required about 16 months, and this document was published on June 23, 1976, in the *Federal Register* (21). As a part of its charge, RAC reviewed all recombinant DNA research that was conducted in institutions receiving NIH support.

In the period from 1976 to 1980, it became apparent that many of the predicted hazards of recombinant DNA

research did not materialize, and the NIH Guidelines were revised into a less stringent format. However, 1980 became a sentinel year when attention toward human gene therapy reached a new level. Shortly after the Supreme Court ruled that recombinant microorganisms could be patented, a letter was sent to President Carter. It was signed by the General Secretary of the National Council of Churches (Protestant), the General Secretary of the Synagogue of America, and the General Secretary of the United States Catholic Conference. The thesis of this document was that questions about the proper use of genetic engineering were moral, ethical, and religious questions. Misuse of this technology was seen as a threat to the fundamental nature of human life and the dignity and worth of the individual human being. There was a specific request for the formation of a body of wide-ranging interests and expertise that could advise the government in its necessary oversight role. A Presidential Commission was formed and its initial meeting took place in July 1980; it accepted the task of studying the ramifications of genetic engineering.

In the meantime, Dr. Martin Cline of the University of California, Los Angeles (UCLA) School of Medicine discovered that DNA from a methotrexate-resistant Swiss 3T6 cell line could be successfully transfected into mouse bone marrow cells (22). By using functional markers, he was able to establish that the transfected bone marrow cells could be successfully transplanted into irradiated mice, and that resistance to methotrexate was maintained. Based on this experimental evidence, he attempted to transfect the β-globin gene into human bone marrow cells that were then transplanted into two patients with thalassemia (one in Israel and one in Italy). Because this protocol had not received approval by the local safety committees at UCLA, an investigation was conducted by NIH with the result that Dr. Cline was censured and lost research funding (23,24).

By 1982 the Presidential Commission had published its report entitled, *Splicing Life*, and it concluded that there were no fundamentally new social and ethical questions raised by somatic cell gene therapy (25). In response to this report, Congressman Albert Gore convened hearings on human genetic engineering. Scientists, clinicians, ethicists, and lawyers all testified, and the recurring question from Mr. Gore alluded to the need for some kind of government body to oversee the development of human gene therapy.

In the final chapter of the *Splicing Life* report, it was suggested that a reconstituted RAC might be an appropriate oversight body for human gene therapy studies. Given the fact that RAC had been functioning as a recombinant DNA review body since 1976, it clearly possessed the most in-depth experience and expertise in this particular technology. Although the original RAC was composed only of scientists, the composition of its membership was changed in 1978 by Joseph Califano (then Secretary of Health, Education, and Welfare) so that two-thirds were scientists and one-third were "public" members.

In April 1983 the chairman of RAC asked the members of this committee if they wished to respond to the *Splicing*

Life report. The response was positive and addressed two issues, the establishment of a Working Group to develop guidelines and review procedures, and acceptance of the responsibility for review of actual protocols at such time that it would become necessary. This Working Group had two meetings in 1983 and then recommended that a larger interdisciplinary group be convened; both RAC and the NIH director concurred with this suggestion and a 15-member Working Group on Human Gene Therapy was created. Members represented clinical medicine, laboratory science, ethics, and law. Throughout 1984 a document entitled "Points to Consider in the Design of Human Gene Therapy Protocols" was prepared, and a first edition was published in the *Federal Register* in 1985. At its September 1985 meeting the full membership of the RAC accepted the revised version of the Points to Consider. By February 1986, the executive secretary of RAC sent a letter to all potential investigators, asking for the submission of preclinical data pertaining to the development of human gene therapy protocols. During 1986 and 1987 the Working Group was known as the Human Gene Therapy Subcommittee (HGTS), and a number of essential discussions were held that addressed such topics as retroviral vectors, the use of trangenic animals as disease models, and the FDA process for regulation of investigational new drugs (INDs). By 1987, a group of investigators, headed by French Anderson, submitted to HGTS a compendium entitled, "Human Gene Therapy: Preclinical Data Document." This was reviewed in the context of being an actual protocol but was actually a prelude to the first proposal for human gene transfer.

In July 1988, the first request for a protocol was submitted to HGTS by Steven Rosenberg, French Anderson, and others; this was not a true "gene therapy" protocol, but rather a "gene-marking" trial to determine if retroviral vectors containing the transgene encoding for neomycin resistance could be given to human subjects without untoward effects. Following an initial discussion, approval of the protocol was deferred, based on the need for additional data. In September 1988, HGTS again requested more information, but in October, the parent body, RAC, approved this protocol with considerably less than a unanimous vote. On October 18, 1988, the director of NIH, Dr. James Wyngaarden, did not approve the protocol and sent it back to the HGTS with the request for additional data. By December 1988, HGTS had approved the protocol and members of the RAC gave their approval via a telephone conference call.

In January 1989, the NIH director publicly announced the approval of this "gene-marking" protocol. Almost immediately a lawsuit was filed by the Foundation on Economic Trends, litigation that was designed to prevent patients from enrolling in the trial. It was argued that a telephone conference call among RAC members was not equivalent to a public meeting and thus a violation of the NIH Guidelines. Following several months of legal interactions, the matter was brought to a successful conclusion. It was during December 1988 that the two-stage process for national review was established. By joint committee agreement it was decided that initial review of protocols would be conducted by HGTS, and once full

approval was granted, said protocols would be forwarded to RAC. At this stage of the oversight process, most of the expertise for gene therapy review was concentrated in HGTS, although there were some members who served both on HGTS and RAC.

In March 1990, French Anderson and Michael Blaese submitted a protocol for the study of adenosine deaminase (ADA) deficiency, a form of severe combined immune deficiency that often caused the death of patients within the first two years of life. At its June meeting, HGTS agreed to provisional approval of this protocol, but requested additional data relating to proof of a selective survival advantage for lymphocytes transduced with the normal ADA gene. On July 30 and 31, 1990, several groundbreaking events occurred. On July 30, HGTS approved two protocols, the ADA protocol and a cancer protocol designed to use tumor-infiltrating lymphocytes as the delivery vehicle for tumor necrosis factor, a means of treating melanoma by adoptive immunotherapy. On July 31, the RAC convened its meeting and approved both protocols.

Although the results of the "gene-marking" protocol were a necessary prologue, the first actual gene therapy trial was conducted on September 14, 1990, when Ashanti De Silva received approximately one billion of her own peripheral blood T lymphocytes that had been transduced with the normal ADA gene. Although the chronology of these early events has been compressed for the purposes of this discussion, a detailed history can be abstracted from the published minutes of the RAC meetings (26–30).

At the outset, there were two independent and parallel processes for the review of human gene therapy protocols, the entirely public review process conducted by RAC and the FDA review process, a mandatory regulatory exercise dictated by federal statute. There was no option for open review by FDA because of legal requirements that all information pertaining to the development of drugs or cell therapies or gene therapy be treated as proprietary and therefore confidential. In retrospect, it was probably fortuitous that NIH RAC accepted the responsibility for public review, since the absence of such an activity could have negatively influenced public acceptance of this experimental form of molecular medicine.

Two public opinion surveys, taken in 1986 and 1992, yielded results that are interesting, but also reflective of the fact that the general public had very little knowledge of genetic engineering and gene therapy. In 1986, before the first protocol was approved, 52 percent of a random sample of polled individuals felt that it was not morally wrong to genetically alter human cells to treat disease, while 47 percent strongly approved of gene therapy to treat genetic diseases and 41 percent somewhat approved of this approach (31). In 1992, after a number of gene therapy protocols had been approved, and several initiated, 30 percent of respondents were very willing to undergo gene therapy to correct a serious or fatal genetic disease before symptoms appeared in late life, and 49 percent were somewhat willing. When asked about the willingness to have a child undergo gene therapy for a usually fatal genetic disease, 52 percent were very willing and 36 percent were somewhat willing (32). Despite a lack of in-depth understanding, these new technologies were seen as acceptable in the treatment of genetic disease, and this is an important finding since the majority of research monies supporting gene therapy, and all the basic research that augments gene therapy, represent tax dollars, and not private investment.

As gene therapy clinical research has developed, there has been a parallel with recombinant DNA research itself in that many of the fears about safety have failed to materialize. Experience has been an important factor in the rather constant modification of the oversight process. By 1991, it became apparent that the two-stage national review of gene therapy was becoming redundant, unnecessarily time-consuming, and of questionable value. In October 1991, RAC decided to consider the disbanding of the HGTS and transferring its membership to the parent body. It was correctly assumed that the major responsibility of RAC, at this point in its history, was human gene therapy and not bacterial genetics. By February 1992, HGTS was formally disbanded, and the sole responsibility for public national review was vested in RAC, whose meetings were increased from three times per year to four. A one-year transition period was established for the purpose of transferring the members of HGTS, not already on the RAC, to the RAC (33).

While the review process was time-consuming, investigators took their responsibilities seriously, and while the open dialogues often served to highlight legitimate differences of opinion, the end result was an improvement over the initial submission. However, in 1992, Drs. Ivor Royston and Robert Sobol requested that the director of NIH and the commissioner of the FDA grant them a compassionate plea exemption so that they could use gene therapy to treat a patient with a brain tumor. This request was particularly challenging because it would bypass all the usual oversight procedures that had been put into place. Although FDA had set a precedent for compassionate plea exemptions for single patients, no analogous mechanism was in place at NIH.

At the December 1992 meeting of RAC, members raised serious concerns about this request. Since the entire field of research was still so new, there was no unequivocal evidence of efficacy in any of trials. These particular investigators had a paucity of preclinical data, and therefore it was difficult to assess the risks associated with the proposed treatment. If the request had represented a minor variation on previously approved protocols, it might have been possible to approve it, but that was not the case. After vigorous discussion, the RAC declined to approve the protocol.

Subsequently the NIH director and the FDA commissioner approved the request on a compassionate plea basis, an action that was entirely within their prerogatives. Committees such as the RAC are simply advisory to the NIH director and their actions constitute recommendations that can be accepted or rejected. It is of note that this particular circumstance was never repeated, but RAC responded to the episode by creating procedures for expedited review, with provisions for using ad hoc reviewers who could be available on short notice (34). A further revision occurred in that protocol categories were established

and some of these categories were exempted from full RAC review.

A further change in oversight was initiated in 1994 as a result of the formation of the National AIDS Task Force on Drug Development. This task force had an unusual charter in that the government officials on the task force were required to be at the level of agency head. Thus the Assistant Secretary of Health, the NIH director, and the commissioner of FDA were members of this working group. Other members included physicians, pharmacologists, AIDS clinical researchers, senior pharmaceutical company executives, and members of several AIDS activist groups. This latter contingent had a single element agenda and that was a change in the approval procedures for human gene therapy protocols. It was the contention of the AIDS activists that gene therapy offered one of the best hopes for the cure of this disease, and that the dual agency review of gene therapy protocols was unnecessarily inhibiting progress. They proposed that RAC be abolished and that sole review of human gene therapy protocols be confined to FDA. This clearly was not acceptable to the entire task force and a compromise was reached.

The new plan was crafted on the idea of consolidated review in which both NIH and FDA would review all new protocols simultaneously. Designated staff members from the two agencies would consult; if the protocol represented a marked departure in concept from previous protocols, it would be fully reviewed by both NIH RAC and FDA. If the protocol lacked any notable differences from previous protocols, it would receive only one review by FDA. First, the AIDS Task Force on Drug Development accepted this new scheme for protocol review, and RAC, at its September 1994 meeting, voted to make the appropriate changes in the NIH Guidelines (35). At this point, consolidated review became the standard operative procedure.

Another significant event occurred in 1995 when NIH Director Dr. Harold Varmus appointed an ad hoc review committee to assess the function of RAC, to develop recommendations about its future role, and to identify ways in which this committee could best support gene therapy research. This committee met several times and issued a summary of its findings in September 1995 (36). It affirmed the basic tenets of the consolidated review process, it recommended that RAC should continue to provide advice on gene therapy policy matters, and it recommended that a data management system should be devised to enable RAC and the Office of Recombinant DNA Activities to monitor all gene therapy clinical trials, even in the absence of case-by-case review.

In 1996, the NIH director announced that he planned to abolish RAC and to replace it with a small number of scientists and ethicists who would meet on an ad hoc basis to render advice on public policy issues relevant to human gene therapy research. After publishing a Notice of Intent in the *Federal Register*, a large number of comments was received and the vast majority were in favor of retaining RAC as the principal public advisory committee for human gene therapy research. In September 1996, the NIH director announced that RAC would be retained but would operate under the following conditions. Membership would be reduced from 25 to 15, the approval process for gene therapy protocols would be relinquished, and a major new assignment would be the responsibility for organizing gene therapy policy conferences to cover such topics as lentivirus vectors, in utero gene therapy, and the use of gene transfer for the purposes of enhancement.

In 1996, the RAC began operating under its new mandate. Several gene therapy policy conferences have been held and selected protocols have been reviewed even though approval is no longer required. Thus RAC continues to provide a venue for public discussion of new scientific and ethical issues in the area of gene therapy clinical trials and can compliment the privately conducted review activities of FDA.

PRINCIPLES OF REVIEW FOR HUMAN GENE THERAPY

From the standpoint of process, there were two principal elements in the NIH oversight of human gene therapy protocols. At the local level, two separate committees were involved, the Institutional Review Board (IRB) or the ethics board responsible for protecting patients from unnecessary research risks, and the Institutional Biosafety Committee (IBC), a product of the requirements detailed in the NIH Guidelines (37). In looking at the division of labor, IRB placed a primary emphasis on the informed consent document, and IBC made its principal focus the science of the pertinent recombinant DNA technology. At the national level, both HGTS and RAC were initially involved, and later RAC itself assumed sole review responsibility. With the exception of the IRB, all of the review bodies relied on the Points to Consider or what is now Appendix M of the NIH Guidelines.

In a previous publication, Walters and Palmer have reduced more than 100 specific questions in Appendix M to seven central ethical questions. They are as follows:

1. What is the disease to be treated, and why is it a good candidate for gene therapy?
2. What alternative treatments are available for this disease?
3. What is the potential harm associated with the genetic intervention?
4. What is the potential benefit associated with the intervention?
5. What steps will be taken to ensure that participants in the study are selected in a manner that is fair to everyone who wants to take part in the study?
6. What steps will be taken to ensure that the consent of study participants is both informed and voluntary?
7. What steps will be taken to protect the privacy of participants and the confidentiality of medical information about them (38)?

A thorough perusal of Appendix M of the NIH Guidelines emphasizes that the great multitude of specific questions are directed toward the seven aforementioned general questions. In developing a background and rationale for a trial, the principal investigator must provide information concerning the disease to be studied (General Question 1). The following subset of questions must be addressed. Why is the disease selected for treatment by means of gene therapy a good candidate for such treatment? What is the natural history and range of expression of the disease selected for study? What objective and/or quantitative measures of disease activity are available? Are the usual effects of the disease predictable enough to allow for meaningful assessment of the results of gene therapy? Is the protocol designed to prevent all manifestations of the disease, to halt the progression of the disease after symptoms have begun to appear, or to reverse manifestations of the disease in seriously ill patients?

Under General Question 2 (alternative therapies), the principal investigator is required to describe what alternative therapies exist and characterize the groups of patients for whom these therapies are effective. It is also necessary to indicate the relative advantages and disadvantages as compared with the proposed gene therapy. This particular matter assumed great importance in the very first gene therapy trial that was approved. Because of the very prolonged review process for the ADA trial, two alternative therapies came into existence, partially matched bone marrow transplants and the use of conjugated enzyme therapy (polyethylene-glycol-ADA, or PEG-ADA). Both showed modest degrees of efficacy. In the final analysis, the first two patients enrolled in the trial were maintained on PEG-ADA, and this action had sound ethical origins. It would have been impossible to justify the withdrawal of PEG-ADA and substitute an entirely experimental gene therapy of unknown therapeutic potential.

Under General Questions 3 and 4 (potential harms and potential benefits), it is necessary to provide a wealth of information about research design, including appropriate preclinical studies. In an important sense, high-quality science carefully applied to the study of patients is a necessary precursor to establishing a positive risk–benefit ratio. Thus it is essential to provide a complete description of the methods and reagents to be employed for gene therapy. It is necessary to describe the structure of the cloned DNA and to completely characterize the gene vector or gene delivery vehicle. This means that one has to perform a complete nucleotide sequence analysis or a detailed restriction enzyme map of the total vector–gene construct. All the regulatory elements in the construct must be identified, including the promoters, enhancers, polyadenylation sites, and origins of replication. In essence, one has to describe the molecular structure of the material that will be administered to the patient. It is also incumbent on the principal investigator to demonstrate the safety, efficacy, and feasibility of the proposed procedures using animal and/or cell culture model systems and to justify why the chosen models are the most appropriate. Very early in this history of human gene therapy research, it became apparent that there would be a number of diseases for which there was no adequate animal model system; it was necessary to introduce some flexibility into the review system and data from cell culture models were allowed to stand in lieu of animal models, when necessary.

In order to accurately assess the risks and benefits the vector or delivery system is subjected to the following kinds of questions. What cells are the intended target cells of recombinant DNA? Is the delivery system efficient? What percentage of the target cells contain the added DNA? Is the added DNA extrachromosomal or integrated? How many DNA copies are present per cell? What is the minimal level of gene transfer and/or expression that is estimated to be necessary for the protocol to be successful in humans? Is the gene expressed in cells other than target cells?

As a further adjunct to assessing risks and benefits, the clinical elements of the study have to be described in detail, including procedures for patient monitoring. In answering the following questions, many of the key elements related to the success or failure of the study will be characterized. Will cells be removed from patients and treated ex vivo? Will patients be treated to eliminate or reduce the number of cells containing malfunctioning genes (radiation or chemotherapy)? How will the treated cells be administered? How will it be determined that new gene sequences have been inserted into the patient's cells and if these sequences are being expressed? What studies will be conducted to assess the presence and effects of contaminants? What are the clinical endpoints of the study? How will patients be monitored to assess specific effects of the treatment on the disease? What is the sensitivity of the analyses? What are the major beneficial and adverse effects of treatment anticipated? What measures will be taken in an attempt to control or reverse adverse effects if they occur?

When one looks at the levels of risk, there are two types of risk to take into consideration, the direct risk to the patient and the risks to health care workers or family, namely the public health considerations. To address these issues, the following kinds of questions must be answered. Is there a significant possibility that the added DNA will spread from the patient to other persons or to the environment? What precautions will be taken against such spread? In light of possible risks to offspring, including vertical transmission, will birth control measures be recommended to patients?

In summary, there are many component concerns that must be addressed before one can establish a risk–benefit ratio. In effect, this ratio will be altered by the type of disease being studied and the principal aims of the study itself. Yet for all the seeming complexities associated with this particular type of clinical research, the approval of individual protocols rests on a positive risk–benefit ratio, and with almost no exceptions, protocols that have failed approval have lacked critical information concerning this issue.

The fair and equitable selection of patients (General Question 5) poses a problem for most emerging new technologies as they reach the stage of clinical trials. Patient interest often exceeds the resources available for a Phase I trial, which is really a pilot study. Very often the diseases are extremely serious, and not infrequently, they are uniformly fatal. Given what are often dire circumstances, it is not surprising that patients and their families are seeking any possible solution.

A significant problem occurred rather early in the development of human gene therapy protocols. Investigators at

NIH were given approval to conduct a Phase I trial for the study of glioblastoma multiforme, a type of brain tumor that has a particularly relentless course. Most of the patients die within less than a year after diagnosis. As initially approved, a maximum of 20 patients could be studied. Because of the intense interest and the publicity given to the study, approximately 2000 patients made inquiries about participation. Because of this overwhelming response, a screening committee was organized, and by choice, the principal investigators did not participate in this screening exercise in order to avoid any potential conflict of interest.

General Question 6 addresses one of the most complex issues pertaining to clinical research, that of informed consent. Informed consent from adults presents one particular set of challenges, but obtaining informed consent from minors means that parents or guardians will be the actual signers of the document. Interestingly enough, there are two differing ethical perspectives with regard to entering children into clinical trials. Almost 30 years ago, a dominant position was that clinical trials should be done in adults first, and then children could be studied secondarily (39). More recently the prevailing opinion is that persons who participate in clinical trials actually have the primary access to potential benefits that are not available to the public at large. This being the case, no class of individuals whether they be members of ethnic minorities, or women, or children, should be denied access to timely participation in clinical trials (40).

A principal challenge for investigators in gene therapy involves conveying the critical information about the disease, the major alternative treatments and the procedures to be followed in the clinical trial. If one adheres to the dictum that the information has to be understandable to someone with an eighth-grade education, this requires that such elements as recombinant DNA technology, the nature of vectors, the insertion of genes into cells, and the possible complications of the trial, have to be simplified in very imaginative ways.

Throughout its period of gene therapy protocol review, RAC spent considerable time revising the portion of Appendix M of the NIH Guidelines pertaining to informed consent. In essence, it became more detailed with time. Investigators were asked to provide information about the manner in which the study would be communicated to potential research subjects, the personnel involved in the process, the measures taken to avoid conflict of interest, particularly if the researcher has responsibility for the patient's medical care, the length of time for decision making, any special arrangements in place for pediatric or mentally handicapped subjects, the nature of reproductive risks or the need for reproductive restrictions, the need for long-term follow-up, and the indication that permission to perform an autopsy will be sought from the family, whatever the cause of the patient's death (41).

RAC's review of gene therapy protocols represents one of the few forums for public discussion of informed consent, and not infrequently, the discussions were both lively and marked by some disagreements between investigators and reviewers. If there was a recurring theme, it was focused on unbridled optimism on the part of some investigators that resulted in optimistic statements about potential therapeutic benefit in Phase I trials, an inappropriate inclusion for a study that is designed to answer questions about safety and not efficacy. Many of the informed consent documents lacked clarity when it came to a discussion about the charges associated with certain research procedures. Obviously, the third-party payers or insurance companies have little interest in paying for anything but standard medical procedures that are relevant to the treatment of a given patient.

Another complicating element in the RAC review of informed consent was derived from the fact that the primary responsibility for government oversight of human subjects research was posited in the Office of Protection from Research Risks (OPRR) of the Department of Health and Human Services (DHHS). Its mandate was derived from a Code of Federal Regulations [45CFR46] (42). Thus RAC was reduced to functioning in a purely advisory role because the final control of informed consent documents resides with IRB, or the local ethics committee of the institution where the research is being conducted. If the investigators chose to ignore RAC's requests or if the local IRB chose to ignore them, there was no means for appeal. However, there has been a positive outcome to all these discussions in that there is documentable evidence that RAC had an influential role in shaping both the process and the text of the informed consent.

Major Ethical Question 7 addresses the issue of patient privacy and much attention was directed to procedures for maintaining such privacy when the original Points to Consider document was being drafted. Since gene therapy primarily was designed for the treatment of the so-called single gene deficiency diseases, and since many of these diseases affect children, the need for patient and family privacy was given appropriate recognition. It is generally accepted that it is the responsibility of the principal investigator and the members of the research and medical teams to preserve privacy if that is the request of the patient and his/her family. What has happened since the advent of the first approved protocol is that there have been notable exceptions to the general request for privacy. While the first two girls to be studied in the initial ADA protocol remained anonymous for several years, they were later featured, including names and pictures, in a popular magazine (43). During the second phase of the ADA trial, in which a number of technical changes were made, the parents of the newborns who were treated, specifically requested that the pertinent information be disclosed (44). Within the last year there has been a rather intense public interest in gene therapy trials in Boston that are designed to treat severe peripheral artery disease in adults. One of the first patients to be admitted to that trial consented to a public interview, and later on, the Public Broadcasting System television network ran a special program describing this trial and conducting patient interviews (45). While it is impossible to predict the wishes of a particular patient or family, the principal investigator and his/her team have a primary obligation to protect the family's intentions if privacy is requested; conversely, there is no choice but to step aside if the decision is to make the trial participation public.

HUMAN GENE THERAPY AND THE INDUSTRY MODEL FOR DRUG DEVELOPMENT

Strangely enough, the failure of human gene therapy to demonstrate unequivocal evidence of efficacy in any of the clinical trials, has delayed the onset of another significant problem. At such time that gene therapy is successful in the treatment of one of the rare monogenic deficiency diseases, the question of industry funding will be a point of active debate. It is particularly instructive to look at the pattern of clinical trials over the past eight years. While the rare genetic diseases provide the best conceptual models for the use of gene therapy, they have not been a major focus of investigation thus far. Of the 244 trials approved up to this point, 73 percent have been devoted to the study of cancer, 14 percent have involved monogenic deficiency diseases, and 9 percent are for the study of AIDS (46). Since many of these early trials have received significant funding from industry, it suggests that large or commercially promising markets are a key element in product development.

The rare disease issue is not new to the pharmaceutical industry, and indeed, Congress passed the Orphan Drug Act so that companies could be granted financial incentives to develop treatments for uncommon disorders. This legislation has been useful, but it is not known if it could be adapted to cover gene therapy. It is possible that new statues would be required in order to stimulate companies to develop treatments when there is a limited potential for payback.

There are a number of theoretical circumstances in which gene therapy could ameliorate a disease on the basis of a single treatment, and it is probable that the theoretical will shift to the actual within the next 15 to 20 years. In order for this to occur, it will be necessary to successfully insert a therapeutic gene into a cell such as a bone marrow stem cell and for that gene to be permanently expressed; further the level of expression will have to be sufficient to correct the genetic disorder. When this series of events occurs, there will now be in place a therapeutic model that is the exact opposite of the drug model in which a patient takes a pill one or more times a day. How would one price a gene therapy treatment in this context? Already there are several genetic diseases that can be treated with recombinant protein products; recombinant glucocerebrosidase is available for patients with Gaucher's disease, and in some instances it may cost as much as $300,000 a year to treat a single patient. Assuming that gene therapy could be used to treat this disease and do it effectively with one intervention, how much would one charge? Even if one charged $300,000 for gene therapy that might be only 5 or 10 percent of the income derived from using the recombinant protein. There is little in this scenario to tweak the interest of the marketing department of a biotechnology company.

In recognition of this problem, FDA and NIH have begun considering alternative ways of product development for gene therapy of rare genetic diseases. This might involve altering the standard paradigm for drug development in that clinical trials could be compressed into fewer phases and would require fewer patients (47). When the patient base is very small, the amount of gene product needed will be relatively small and it may be feasible to have nonprofit distribution centers, located in academic medical institutions, serve the role of traditional drug and biotechnology companies. Naturally all the standards of quality control and quality assessment would have to be met in precisely the same way that is required of the for-profit sector. It will be interesting to watch what models actually develop as gene therapy for rare diseases becomes a reality.

CHALLENGES FOR THE FUTURE

This discussion has focused on ethical issues when gene transfer is used in the context of treating disease, but another significant challenge may develop in the not-too-distant future when this technology may be adapted for purposes other than the treatment of disease. This alternative use has been designated as enhancement. Between 1980 and 1993, there have been 28 international policy statements concerning gene therapy, but there are no specific references to genetic enhancement in these documents (38). There are two exceptions to the aforementioned statements; these emanate from advisory committees in Canada and the United Kingdom and find genetic enhancement to be ethically unacceptable (48,49). There have been two polls in the United States that have attempted to define public attitudes toward genetic enhancement, one taken in 1986 and one taken in 1992. In 1986, respondents were asked about their attitude toward genetic manipulation to improve physical characteristics in children, and 16 percent strongly approved, 27 percent somewhat approved, 21 percent somewhat disapproved, and 33 percent strongly disapproved; in 1992, the responses to the same question revealed that 16 percent strongly approved, 28 percent somewhat approved, 23 percent somewhat disapproved, and 31 percent strongly disapproved (31,32). When asked about their attitude toward genetic manipulation to improve intelligence in children, respondents in 1986 indicated that 17 percent strongly approved, 25 percent somewhat approved, 20 percent somewhat disapproved, and 35 percent strongly disapproved. In the 1992 survey, 18 percent strongly approved, 26 percent somewhat approved, 22 percent somewhat disapproved, and 31 percent strongly disapproved (31,32). Given the limits of error in this kind of poll, it appears that a slight majority of Americans are opposed to genetic engineering for enhancement, but a significant minority are in favor of it.

In 1997, NIH sponsored a Gene Therapy Policy Conference that was organized by the Office of Recombinant DNA Activities. Two issues were defined by the discussions that took place. First, there was the prediction that procedures for enhancement might be available in the near future, and second, it was agreed that the gradation from treatment of disease to enhancement might be subtler than once perceived. With regard to the second issue, it was pointed out that a biotechnology company had been successful in inserting the tyrosinase gene into the cells of the hair follicle (50). Current research is directed toward genes that promote hair growth. One of the primary objectives is to develop a product to treat alopecia

or hair loss associated with cancer chemotherapy, just as erythropoietin is used to treat the anemia associated with cancer drugs. Beyond the primary indication for treating hair loss in cancer patients, there is a large audience of men and some women who would not object to using a product to counteract baldness. The latter use is perfectly legal and would constitute a so-called off-label use of the product, a practice commonly allowed by FDA. As an example, it is well known that certain anticonvulsant drugs are regularly used to effectively treat patients with a variety of affective disorders, although the original drug approval was solely for the treatment of seizures.

What about other scenarios? Recombinant proteins have already been used in settings that could be described as enhancement. Some young children of very small stature have been treated with the human growth hormone (HGH) protein, even though there were no data to support HGH deficiency. Admittedly, short stature in males could be a serious social problem, but it does not constitute a disease. One could envision the use of the gene encoding HGH instead of using the protein. Another area that is vulnerable to enhancement is the world of athletics where the participants are always seeking to gain a competitive advantage. An athlete wishing to increase his/her muscle mass might be more than willing to use the gene encoding for vascular endothelial growth factor (VEGF) to increase the blood supply to this tissue or perhaps to use a combination of VEGF and a gene encoding for one of the dystrophins to further augment bulk and muscle strength.

Undoubtedly, there are a number of risks in using gene transfer either for disease treatment or for enhancement, but it is much easier to justify said risks when one is attempting to treat a serious disease. The fact that people will take risks for a purely cosmetic result attests only to personal interests. Should this technology be made available for enhancement purposes with a warning label attached? It would be difficult to find a consensus on this point. However, the matter of control could turn out to be a very elusive one. One nation or series of nations could elect to ban gene transfer for enhancement, but another sovereign state could take advantage of a commercial opportunity, knowing that there would be a ready supply of customers for whom personal expense and travel are negligible barriers.

Various writers have commented on the matter of gene transfer for various types of intellectual enhancement including efficiency of memory, general cognitive ability or intelligence, or other behavioral traits such as antisocial behavior (38). While such treatises provide fascinating reading and address such issues as the fundamental importance of individual liberty, they fail to take cognizance of the scientific infrastructure. If one considers that our ability to treat single gene deficiency diseases is in a most rudimentary state, then it stands to reason that alteration of polygenic diseases or personality traits must be consigned to the rather distant future. To achieve these types of changes will require an intrinsic knowledge of the interactions among multiple sets of genes plus the ability to control gene expression, another area in which current knowledge is extremely lacking. There is no reason to assume that some of the critical scientific knowledge will never be available, but gene transfer for various types of intellectual enhancement awaits quantum leaps in our knowledge of molecular genetics.

CONCLUSION

In viewing the end of the twentieth century, it is apparent that the past 50 years have represented the flowering of molecular genetics. There have been many important applications of the basic research in this area, and none is more potentially exciting than somatic cell gene therapy. While the early gains have been modest, there is every reason to assume that gene therapy will have a major impact on the practice of medicine within the next 20 to 25 years. To the credit of the scientific community and the public, there have been extensive public discussions of this technology, and the first national oversight program was a completely public process. When Marshall Nirenberg expressed his concerns in 1967, he alluded to the necessity of an informed society to make appropriate decisions about biochemical genetics. There have been systematic attempts to create that informed society and until now, there has been a pattern of responsibility in place. The challenges of the future are to protect the rights of the patient as additional disease states are subject to intervention, and to carefully assess the societal consequences of using gene transfer for nonmedical enhancement.

BIBLIOGRAPHY

1. H.I. Miller, *Hum. Gene Ther.* **1**, 3–4 (1990).
2. O.T. Avery, C.M. MacLeod, and M. McCarty, *J. Exp. Med.* **70**, 137–158 (1944).
3. J.D. Watson and F.H.C. Crick, *Nature (London)* **171**, 964–969 (1953).
4. H.F. Judson, *The Eighth Day of Creation: The Makers of the Revolution in Biology*, Simon & Schuster, New York, 1979.
5. D. Nathans and H.O. Smith, *Ann. Rev. Biochem.* **44**, 273–293 (1975).
6. S.N. Cohen et al., *Proc. Natl. Acad. Sci. U.S.A.* **70**, 3240–3244 (1973).
7. D.J. Jolly, J.K. Yu, and T. Friedmann, *Methods Enzymol.* **149**, 10–25 (1987).
8. J.A. Wolff and J. Lederberg, *Hum. Gene Ther.* **5**, 469–480 (1994).
9. E.L. Tatum, *Perspect. Biol. Med.* **10**, 19–32 (1966).
10. J. Lederberg, *Proc. World Cong. Fertil. Steril., 6th, 1968*, 18–23 (1970).
11. A. Kornberg, in M. Burnet, ed., *Genes, Dreams, and Reality*, Basic Books, New York, 1971, p. 71.
12. H.V. Aposhian, *Perspect. Biol. Med.* **14**, 987–1008 (1970).
13. M.W. Nirenberg, *Science* **157**, 633 (1967).
14. R. Sinsheimer, *Eng. Sci.* **32**, 8–13 (1969).
15. S. Rogers and M. Moore, *J. Exp. Med.* **117**, 521–542 (1963).
16. H.G. Terheggen et al., *J. Exp. Med.* **119**, 1–3 (1975).
17. S. Rogers, in M. Lappe and R.S. Morison, eds., *Ethical and Scientific Issues Posed by Human Uses of Molecular Genetics*, New York Academy of Sciences, New York, 1976, pp. 66–70.
18. W.F. Anderson, in M. Hamilton, ed., *The New Genetics and the Future of Man*, Eerdmans, Grand Rapids, MI, 1972, pp. 109–124.
19. T. Friedmann and R. Roblin, *Science* **175**, 949–955 (1972).
20. P. Berg et al., *Science* **185**, 303 (1974).
21. Guidelines for Research Involving Recombinant DNA Molecules (NIH Guidelines), *Fed. Regist.* **41**, 27902 (1976).
22. M.J. Cline et al., *Nature (London)* **284**, 422–425 (1980).
23. N. Wade, *Science* **210**, 509–511 (1980).
24. N. Wade, *Science* **212**, 1253 (1981).
25. President's Commission for the Study of Ethical Problems in Medicine and Biomedical and Behavioral Research, *Splicing Life: The Social and Ethical Issues of Genetic Engineering with Human Beings*, U.S. Government Printing Office, Washington, DC, 1982.
26. National Institutes of Health (NIH), *Minutes of the Recombinant DNA Advisory Committee*, January 30, NIH, Washington, DC, 1989, pp. 6–14.
27. National Institutes of Health (NIH), *Minutes of the Recombinant DNA Advisory Committee*, October 6, NIH, Washington, DC, 1989, pp. 25–27.
28. National Institutes of Health (NIH), *Minutes of the Recombinant DNA Advisory Committee*, February 5, NIH, Washington, DC, 1990, pp. 5–14.
29. National Institutes of Health (NIH), *Minutes of the Recombinant DNA Advisory Committee*, March 30, NIH, Washington, DC, 1990, pp. 4–7.
30. National Institutes of Health (NIH), *Minutes of the Recombinant DNA Advisory Committee*, July 31, NIH, Washington, DC, 1990, pp. 3–17.
31. U.S. Congress, Office of Technology Assessment, *New Developments in Biotechnology — Background Paper: Public Perceptions of Biotechnology*, U.S. Government Printing Office, Washington, DC, 1987.
32. March of Dimes Birth Defects Foundation, *Genetic Testing and Gene Therapy Survey: Questionnaire and Responses*, March of Dimes Birth Defects Found., White Plains, NY, 1992.
33. National Institutes of Health (NIH), *Minutes of the Recombinant DNA Advisory Committee*, February 10, NIH, Washington, DC, 1992, pp. 37–41.
34. National Institutes of Health (NIH), *Minutes of the Recombinant DNA Advisory Committee*, December 2, NIH, Washington, DC, 1993, pp. 51–53.
35. National Institutes of Health (NIH), *Minutes of the Recombinant DNA Advisory Committee*, September 12, NIH, Washington, DC, 1994, pp. 43–48.
36. Ad hoc Review Committee of the Recombinant DNA Advisory Committee, *Executive Summary of Findings and Recommendations*, 1995.
37. Guidelines for Research Involving Recombinant DNA Molecules (NIH Guidelines), *Fed. Regist.* **62**, 59032 (1997).
38. L. Walters and J.G. Palmer, *The Ethics of Human Gene Therapy*, Oxford University Press, New York, 1997.
39. J. Katz, A.M. Capron, and E.S. Glass, *Experimentation with Human Beings: The Authority of the Investigator, Subject, Professions, and State in the Human Experimentation Process*, Russell Sage Foundation, New York, 1972.
40. A.C. Mastroianni, R. Faden, and D. Federman, *Women and Health Research: Ethical and Legal Issues of Including Women in Clinical Studies*, National Academy Press, Washington, DC, 1994.

41. N. Wivel and W.F. Anderson, in T. Friedmann, ed., *The Development of Human Gene Therapy*, Cold Spring Harbor Lab. Press, Cold Spring Harbor, NY, 1999, pp. 1–19, 671–689.

42. Protection of Human Subjects (45 CFR 46), *Fed. Regist.* **48**, 9269 (1983).

43. L. Thompson, *Time*, June 7, 1993, pp. 50–53.

44. L. Jaroff, *Time*, May 31, 1993, pp. 56–57.

45. G. Kolata, *N.Y. Times*, December 13, 1994, p. C1.

46. Office of Recombinant DNA Activities, ORDA Rep., *Human Gene Therapy Protocols*, National Institutes of Health, Bethesda, MD, 1998.

47. A. Pollack, *N.Y. Times*, August 4, 1998, p. B1.

48. Royal Commission on New Reproductive Technologies, *Proceed with Care*, vol. 2, Minister of Government Services, Ottawa, Canada, 1993, p. 945.

49. United Kingdom, Committee on the Ethics of Gene Therapy, Report. HM Stationery Office, London, 1992, p. 22.

50. L. Li and R.M. Hoffman, *Nat. Med.* **1**, 705–706 (1995).

See other GENE THERAPY entries.

GENE THERAPY, LAW AND FDA ROLE IN REGULATION

RICHARD A. MERRILL
University of Virginia
Charlottesville, Virginia

GAIL H. JAVITT
Johns Hopkins School of Hygiene and
Public Health
Baltimore, Maryland

OUTLINE

INTRODUCTION

The discovery of methods to "recombine" DNA from different species, that is to transfer genes from one organism to another, has opened a new frontier for therapeutic medicine. It offers the prospect that some diseases and disorders — those known to be caused by specific and identifiable genetic defects — can be corrected and even prevented through skillful intervention in the body's genetic instructions. But scientists and investors who perceive the potential of this new technology have to confront not only daunting technical challenges and enormous biological uncertainties; they have to take account of societal regulation of new technology.

As this article describes, primary responsibility for such regulation at the federal level has fallen to the U.S. Food and Drug Administration (FDA), a unit of the Department of Health and Human Services (HHS). Congress first enacted national food and drug legislation nearly a century ago (1) to "protect consumers from dangerous products" by providing uniform federal regulation of therapeutic drugs (2). The FDA derives its current authority from several dozen laws passed by Congress over the ensuing 90 years. The most prominent of these is the Federal Food, Drug, and Cosmetic Act (FD&C Act), enacted in 1938 (3) and substantially revised in 1962 (4), 1976 (5), 1990 (6), and 1997 (7). This statute provides the basic framework for FDA regulation of medical products defined as "drugs" or "medical devices."

FDA also administers the Biologics Act of 1902 (8), now codified as part of the Public Health Service (PHS) Act (9). The Biologics Act gave federal officials authority over "biological products," which include "any virus, therapeutic serum, toxin, antitoxin, vaccine, blood, blood component or derivative, allergenic product, or analogous product" (10). Responsibility for administering this and another provision of the PHS Act, aimed at preventing the transmission of communicable disease, was originally vested in the National Institutes of Health (NIH) but it was transferred to the part of FDA now known as the Center for Biologic Evaluation and Research (CBER) in 1972 (11).

Thus FDA has for several decades been the federal agency exclusively responsible for regulating virtually all medical products. It usually exercises this authority by requiring the sponsor of a new treatment modality to demonstrate, through carefully controlled clinical trials, that it is safe and reliably produces beneficial effects before it can be made available to the medical profession generally. Most of the laws that FDA administers, however, were enacted before the era of biotechnology, and none were enacted with gene therapy specifically in mind. At the time researchers first began exploring the potential of gene therapy, it was therefore by no means clear that FDA would or should be the agency responsible for its oversight or, if the agency were to have a role, that it possessed the appropriate tools for regulation.

At the outset another agency, also part of HHS, claimed a dominant interest in the subject. This was the NIH, which in 1974 established the Recombinant DNA Advisory Committee (the "RAC"), an interdisciplinary body, to review and at least tacitly approve federally funded research using the techniques of recombinant DNA technology (12). Because the early experiments did not involve human subjects, FDA did not initially seek to play a significant oversight role. When research began to move into the clinic, however, the RAC lost its initial regulatory monopoly.

FDA saw an important role for itself in regulating gene therapy, derived from its statutory responsibility for assuring the safety and effectiveness of commercially distributed therapeutic products. If the materials used in gene therapy fit the statutory definition of drug, device, or biological product—and, given the breadth of those definitions, it would be hard to assert that they did not—FDA was obligated to address them. And, because FDA has long overseen the clinical investigation of the medical products whose marketing it regulates, this has meant that the agency has major responsibility for monitoring the experimental applications of gene therapy as well. Reasoning in this way, FDA eventually made clear its intention to exert regulatory oversight and equipped itself, administratively and in personnel, to perform that role.

To many, FDA's assertion of responsibility appeared to duplicate, if not compete with, the role of the NIH RAC, and, more specifically, with the RAC's Gene Therapy Working Group. The first part of this article describes the uneasy relationship between the two bodies, a relationship that some saw as "rivalry" and others as natural evolution as research moved from the laboratory into the clinic. By the early 1990s FDA's claim to the primary oversight role had been established, while the RAC continued to conduct what many regarded as duplicative review of individual protocols. By 1997, however, FDA and NIH had agreed that FDA would be exclusively responsible for approving individual gene therapy protocols, and that the RAC would no longer review protocols for their technical merit. At the same time the RAC would continue its recognized role as the forum for public discussion of the social and ethical issues raised by novel applications of, or approaches to, gene therapy.

The second part of the article describes FDA's evolving regulatory policy and analyzes the requirements it has imposed. It is noteworthy that FDA has never promulgated formal regulations specifically addressing gene therapy products; all of its policy pronouncements have taken a less formal guise. This approach is consistent with FDA's position that gene therapy products are simply another type of drug, biologic, or device. But it also reflects a regulatory regime that is still a work in progress. The broad framework of FDA regulation of gene therapy will be familiar to any student of FDA drug law, but the decisions made within that framework betray the complexity and novelty of the technology involved.

EVOLUTION OF FDA REGULATION OF GENE THERAPY

1974–1984: Prehistory

In 1974 scientists announced that they had developed a method to recombine DNA from different species (e.g., two different types of microbes or viruses) to form new biological entities (13). Concerns about the potential impact of this new capability led to an event unprecedented in the history of science: Researchers themselves called for a moratorium on recombinant DNA experiments pending public review of its risks (13,14). The same year, the National Academy of Sciences established a Committee on Recombinant DNA Molecules to examine the risks associated with recombinant DNA research and to recommend specific precautionary measures (15). The NAS Committee's recommendations were published in the journal *Science* in July 1974 (16). The Committee recommended, among other things, that certain experiments be voluntarily deferred, and that the director of the NIH establish a committee to evaluate hypothetical risks, to develop procedures to minimize the spread of recombinant DNA molecules, and to recommend guidelines to be followed by investigators (16). Within four months the Department of Health Education and Welfare (now HHS) chartered the RAC and directed it to establish appropriate biological and physical containment practices and procedures for recombinant DNA research (12). In 1976 the RAC codified these practices as "guidelines," which had to be followed in all research conducted using NIH funds (17).

The scientists who in 1974 had called for a moratorium on recombinant DNA research had done so in part because of fears that a genetically modified organism inadvertently released into the environment could become an "Andromeda Strain" (13). The RAC's early preoccupation with containment methods accordingly reflected these concerns. While several authors called attention to the therapeutic, as well as the social and ethical, implications of genetic manipulation of the human genome (18), it was not until 1980 that gene therapy drew public attention. In that year an American physician, Dr. Martin Cline, chief of the division of hematology/oncology at the University of California at Los Angeles (UCLA), conducted unauthorized gene therapy experiments in Israel and Italy on two women suffering from beta thalassemia, a rare but often fatal genetic defect affecting the red blood cells (19). Cline had submitted his protocol to UCLA's Institutional Review Board (IRB) but neglected to wait for a decision. He later maintained that at the time he carried out the experiments he fully expected to receive approval (20). The UCLA IRB, however, ultimately rejected the protocol, citing insufficient animal studies (19). In addition Cline failed to inform authorities in Italy and Israel or the patients that the protocol entailed the first-ever purposeful insertion of recombinant DNA into humans (19).

Dr. Cline's experimental treatments did not improve or exacerbate the patients' condition but his own professional career was severely compromised. After investigating the incident, NIH issued a report censuring Cline (21), whose research it had supported. NIH canceled Cline's NIH funding and attached a copy of its report to his subsequent applications for grant support (19).

Following this episode three major religious organizations — the U.S. Catholic Conference, the Synagogue Council of America, and the National Council of Churches — wrote to President Carter expressing concern about rapid advances in genetics and the absence of a federal oversight mechanism (22). The letter was referred to the recently formed President's Commission for the Study of Ethical Problems in Medicine and Biomedical and Behavioral Research. In November 1982, before a congressional hearing chaired by then-Representative Albert Gore, Jr., the President's Commission released its assessment, "Splicing Life" (23). The report saw a need for an oversight body to review gene therapy experiments, and also recommended the formation of a permanent federal bioethics commission (23). Speaking to the merits, the report took the position that somatic cell gene therapy — which does not affect the genetic material of reproductive cells — should be permitted to proceed, whereas germ-line gene therapy should not be undertaken without prior public debate (22).

1984–1989: Early Rivalry Between FDA and NIH

In 1983 the RAC formed the Working Group on Human Gene Therapy to study and respond to the report of the President's Commission. The working group comprised three laboratory scientists, three clinicians, three ethicists, three lawyers, two public policy specialists, and a representative from the public. It recommended that the RAC add to its purview experiments involving the "Deliberate transfer of recombinant DNA or DNA derived from recombinant DNA into human subjects" (24). It also drafted a document that it titled "Points to Consider," a nearly 4000 word instruction manual detailing the information that researchers seeking the RAC approval for gene therapy experiments in humans would have to submit (24).

The RAC's "Points to Consider" required researchers to address the scientific aspects of their protocol, including the research design, the anticipated methods of gene delivery, and the results of animal experimentation. In addition researchers were instructed to address a broad range of social concerns pertaining to gene therapy, signaling the RAC's intent to judge not only the scientific validity of human gene therapy protocols but also their social acceptability. The RAC's commitment to public review of all protocols was underscored by its expectation that the first proposals submitted for RAC review would contain no proprietary information or trade secrets, which would enable all aspects of the review to be open to the public (24). An initial draft of the "Points to Consider" document was unveiled on January 22, 1985, and was published in the Federal Register for public comment (24).

Although Congress enacted no legislation directing the formation of the RAC Working Group or authorizing NIH oversight of gene therapy protocols, the RAC assumed this role with substantial congressional as well as public support, and most notably the support of then-Senator Albert Gore, Jr. As one commentator noted, "In the absence of any other duly constituted body, the Working Group on

Human Gene Therapy has become the locus for broad social discussion of [gene therapy] issues" (25).

FDA officials, however, were not sanguine about NIH's determination to take the lead regulatory role nor content with the approach it outlined for reviewing gene therapy protocols. NIH officials may have believed that FDA would not seek to participate in the regulation of gene therapy protocols, except for the individual contributions of Dr. Henry Miller as FDA liaison to the RAC, but if so, they proved to be mistaken (26).

On December 31, 1984, three weeks before the RAC's draft "Points to Consider" was published for public comment, FDA issued a policy statement that both dispelled doubt about its intention to regulate clinical trials of gene therapy products and that implied dissatisfaction with NIH's contemplated regulatory approach (27). The policy statement set forth FDA's position that existing laws conferred sufficient authority to regulate all the commercial applications of biotechnology within its jurisdiction. Furthermore, the statement announced FDA's view that gene therapy was not a fundamentally different therapeutic modality that required special scrutiny or new oversight mechanisms. According to the policy statement, "Nucleic acids used for human gene therapy trials will be subject to the same requirements as other biological drugs." The policy statement cautioned against the adoption of "[i]nconsistent or duplicative domestic regulation" of biotechnology that could "put U.S. producers at a competitive disadvantage." Yet, in its sole reference to NIH's oversight of gene therapy protocols, the statement acknowledged that "[i]t is possible that there will be some redundancy between the scientific reviews of these products performed by the National Institutes of Health and FDA" (27,28).

During 1985 it became apparent that a jurisdictional rivalry was brewing between NIH and FDA (26,29). The first gene therapy protocols were expected to be ready for clinical trials within the year (30). Perhaps in anticipation of their arrival, Dr. Henry Miller, Commissioner Frank Young's adviser on biotechnology issues and later head of FDA's Office of Biotechnology, publicly criticized the composition of the RAC Working Group, which he thought gave undue prominence to ethicists and lawyers at the expense of scientists and clinicians (26). Dr. Miller also took the position that since non-germ-line gene therapy was not a qualitatively unique form of therapy, physicians planning to initiate clinical trials with human gene therapy should simply file an investigational new drug (IND) application, permitting FDA to review the experimental gene and vector to be used (26). Submission of an IND to FDA is a standard prerequisite for clinical trials of new drugs.

In what some viewed as another FDA effort to undercut the RAC's hegemony, Commissioner Young strongly supported a proposal prepared in the Office of the Assistant Secretary for Health to create a federal Biotechnology Science Board (BSB) within HHS (31). The Board, which would have reported to the Assistant Secretary, would have had broad authority over research and development in genetic engineering. It would also have diminished NIH's authority over gene therapy (26,31). Specifically, the proposal would have (*1*) barred NIH from publishing a final version of its "Points to Consider" document without BSB review and (*2*) precluded RAC review of clinical trials of genetically engineered products that were also under the jurisdiction of another regulatory agency (31). The proposal was submitted to the White House Office of Science and Technology Policy (OSTP) on August 1, 1985.

This early attempt by FDA to exert control over the regulation of gene therapy research was unsuccessful. The deputy director of OSTP, Bernadine Healy, favored maintaining a strong RAC (31). Critics viewed the BSB proposal as an attempt by FDA Commissioner Young — who was a supporter of research in the field and a leading candidate to become Assistant Secretary of Health — to become the Reagan administration's "biotech 'czar' " (31). Moreover on August 30, 1984, Senator Gore wrote letters to both Healy and Secretary of HHS Heckler, insisting that they immediately stop FDA's effort to "usurp" RAC's rope in overseeing human gene therapy experiments. Gore's endorsement of the RAC was clear, albeit lukewarm. It had, he wrote, "done an adequate job of addressing the scientific issues to date, and it appears capable of continuing to do so for the immediate future" (32). Gore also forecast the imminent passage of his legislation to establish a national commission on bioethics, which would "undoubtedly consider whether a formalized regulatory structure, like the BSB, for gene therapy will be needed" (32).

The Reagan administration ultimately rejected the BSB proposal in favor of a new interagency committee within the OSTP's Federal Coordinating Council for Science, Engineering, and Technology (FCCSET) (32). The committee would develop scientific policy recommendations for the agencies involved in recombinant DNA research, including NIH and FDA. Thereafter NIH and FDA reached an uneasy truce. The RAC continued to require researchers who contemplated gene therapy experiments to submit their protocols for review addressing the areas identified in the "Points to Consider," while FDA insisted that researchers seeking to conduct gene therapy trials using viral vectors must comply with FDA's IND regulations (31). NIH agreed to modify its "Points to Consider" to state explicitly that they applied only to "institutions receiving support for recombinant DNA research from the NIH" (33). While this limitation had been tacitly acknowledged previously, fear of public criticism and possible liability led unfunded academic and industrial sponsors of biotechnology research to voluntarily submit their projects for the RAC review (31).

The revised "Points to Consider" also included a footnote acknowledging that FDA "has jurisdiction over drug products intended for use in clinical trials for human somatic-cell gene therapy," and directing applicants to review FDA's Policy Statement. This version was published in the Federal Register on August 19, 1985 (33).

In February 1987 the RAC called attention to the continuing rivalry by modifying its "Points to Consider" explicitly to prohibit NIH-funded researchers from undertaking gene therapy clinical trials absent prior RAC review and approval, even if the experiment had been approved by another agency. FDA's Miller opposed

this amendment on the ground that demanding separate RAC review would delay expeditious use of gene therapy in patients who might urgently need such treatment (34).

1989–1991: First Gene Therapy Protocols Authorized

The debate over which agency or agencies would regulate gene therapy experiments — and the appropriate parameters of such oversight — had preceded by several years the first authorized clinical trial of gene therapy in the United States. The first clinical experiments were expected as early as 1985, but technical difficulties delayed the first attempt to introduce foreign genes into patients (35). Finally, on May 22, 1989, NIH researchers did so (36). The protocol entailed removing tumor-infiltrating lymphocytes (TIL) from a cancer cell, inserting a bacterial gene, and giving the TILs back to the patient. The experiment was not intended to be therapeutic; rather, the bacterial genes were intended to serve as a "marker" that would permit scientists to trace the path of the tumor-fighting cells (37). Prior to conducting the experiment, the protocol had undergone 26 hours of formal review hearings, including review by 7 advisory committees (36,37). At the eleventh hour, Jeremy Rifkin — fierce opponent of biotechnology in all applications — further delayed commencement of the experiment by filing a lawsuit claiming that NIH had failed to follow proper procedures in approving the protocol (38).

Less than a year later NIH researchers Michael R. Blaese, W. French Anderson, and Kenneth Culver received approval from both the RAC and FDA to conduct the first clinical trial using gene therapy to treat a genetic disorder. On September 14, 1990, genetically modified white blood cells containing a gene for adenosine deaminase (ADA) were introduced into a four-year-old girl with severe combined immunodeficiency disorder (SCID), which is caused by a defect in the ADA gene (13). Children lacking a functional ADA gene cannot mount immune responses and usually die in early childhood. The researchers hoped that the new gene would begin to manufacture ADA in the patient's body and thereby restore immune function.

1991–1993: FDA Steps Up Regulatory Efforts

The first experiments helped quiet public fear and rekindled enthusiasm over the seemingly limitless possible applications of gene therapy to treat, for example, cancer, heart disease, high cholesterol, and AIDS (39). While the initial clinical trials did not unambiguously show effectiveness, the harms to patients that had been feared, such as illness caused by the viral vectors, did not occur. The regulatory environment for manufacturers of the viral vectors used in gene therapy, however, remained very uncertain. Maryland-based Genetic Therapy Inc. (GTI), which had supplied the viral vectors used in the initial gene therapy experiments under cooperative agreements with NIH, acknowledged in its May 1991 prospectus that "the precise regulatory requirements with which the company will have to comply are uncertain at this time due to the novelty of the human gene therapies currently under development" (40).

Anticipating a flurry of INDs from industry, FDA's CBER made available a draft Guidance Document addressing somatic cell therapy and gene therapy. FDA emphasized that the document was not a binding regulation. Rather, it was intended to inform manufacturers engaged in the production and testing of products for these therapies about the issues that the agency believed should be considered and that, by implication, should be addressed in IND submissions (41). The guidance document recommended that sponsors of experiments address, among other issues, (1) quality control procedures, (2) procedures to prevent cell culture contamination by adventitious agents, (3) proper characterization of gene constructs, and (4) vector insertion methods. It also detailed the types of studies FDA believed should be performed to establish product safety.

In December 1992, FDA took another step to formalize its regulatory authority over anticipated gene therapy products. CBER announced its plan to establish, as of January 1993, a new Office of Therapeutics Research and Review (OTRR) that would oversee four new "lab-based" divisions, including a Division of Cellular and Gene Therapies (42). Creation of OTRR would allow CBER to remain "on the cutting edge" of "many very novel, very innovative types of therapy," according to then-Deputy Director of CBER, Janet Woodcock, M.D. (43). Woodcock acknowledged that there were then no marketed products under the purview of the new Division, but explained that FDA reviewers would be preparing "to deal with these burgeoning new types of therapeutic approaches" as they became ready for submission. Dr. Philip Noguchi was designated to head the division.

In October 1993, FDA issued another guidance document explaining how the agency's statutory authorities applied to human somatic cell therapy and gene therapy products. FDA stated that it was publishing the guidance document "in response to requests that the agency clarify its regulatory approach and provide guidance to manufacturers of products intended to be used in somatic cell therapy or gene therapy" (44). As the trade press observed, the document "essentially codifies current FDA practice to regulate ... all gene therapies" (45).

Contemporaneously FDA Commissioner David Kessler took another step to explain the agency's evolving policy. Deploying a tactic that became familiar during his tenure, Kessler and several FDA officials within OTRR published an article in the *New England Journal of Medicine*. Their professed objectives were to "examine the regulation of somatic-cell and gene therapy by the Food and Drug Administration (FDA) in the context of the agency's traditional role in the development of biologic products and to stimulate discussion in areas in which policy is still being formulated." The tone and scope of the article suggested that another goal was to underscore FDA's centrality in the gene therapy approval process. The authors, however, professed no desire to supplant the RAC. Rather, they acknowledged that FDA and NIH had "important, complementary functions," and said that RAC review "ensures broad public discussion of the scientific evaluation of this new technology, particularly with regard to social and ethical concerns." FDA, in contrast, "focuses on the development of safe and effective biological products, from their first use in humans through

their commercial distribution." Therefore "[p]roducts used in protocols subject to review by the Recombinant DNA Advisory Committee must also undergo FDA review" (46). This endorsement of the RAC's unique role as a forum for public debate of the societal and ethical issues raised by gene therapy proved prophetic.

1991–1994: The RAC Move Toward Streamlined Review

As FDA elaborated its regulatory approach, the RAC saw signs that its primary role in overseeing gene therapy research might be in jeopardy. At a July 31, 1991, meeting of the RAC, researcher Dr. W. French Anderson urged RAC members to combine the functions of the Human Gene Therapy Subcommittee with that of the full RAC. Lamenting that the lengthy and redundant review process for gene transfer protocols was a "difficult experience" for investigators, Anderson predicted that investigators with private sources of funding would seek to evade the RAC review and submit their protocols only to FDA for review to avoid the associated regulatory burdens (47). Anderson expressed the concern, however, that the loss of the public forum provided by the RAC might jeopardize public confidence in gene therapy research. He emphasized that, unlike the RAC, FDA conducted review of all submissions in private (48).

Soon after Anderson's cautionary warning about "redundancy," Viagene, a biotech company, seemed poised to prove his point. In May 1992, Viagene filed in IND with FDA for a Phase I (preliminary) gene transfer protocol aimed at preventing the onset of AIDS in HIV positive patients. The protocol had not been submitted to the RAC. Nevertheless, in June 1992, an FDA advisory committee recommended that FDA approve Viagene's clinical trial, representing the "first time that such an experiment has been treated as a routine drug trial rather than a foray into unknown territory that requires extraordinary safety and ethics review" (49). Commenting on this apparent sea change in the U.S. regulatory process, the British journal *Nature* opined that, "in the face of increasing acceptance of gene therapy, those who still wish for special reviews of everything from the basic biology to the theology of simple gene-transfer experiments appear to be losing ground to those who argue for business as usual" (49). *Nature* was premature in predicting the demise of the RAC review: The following year Viagene voluntarily submitted a related HIV gene therapy protocol to the RAC. The company explained that it expected that future testing of the gene therapy product would involve NIH-funded institutions, and it wished to have the RAC involved from the outset (50). Criticism from within the industry for its attempt to circumvent the RAC review process may also have played a role in its about-face (51).

In the face of (1) FDA's expanding role in overseeing gene therapy protocols, (2) the increasing number of protocols being submitted by both research institutions and industry, and (3) diminishing public concern over gene-based therapies, members of the RAC debated what the committee's oversight role should be. Some felt that parallel review by FDA and NIH was unnecessary, while others asserted that the RAC should focus on cross-cutting policy issues relating to gene therapy and on public education.

The committee members agreed that review of individual research protocols needed to be streamlined (52).

A July 1994 proposal by the National Task Force on AIDS Drug Development forced both the RAC and FDA to reconcile their overlapping regulatory efforts. Chaired by HHS Assistant Secretary for Health Philip Lee and including both FDA Commissioner Kessler and NIH Director Harold Varmus, the Task Force unanimously approved a proposal to streamline gene therapy protocol reviews. Under the proposed review process, all gene transfer protocols would be submitted directly *to* FDA and in the format required *by* FDA. If the RAC review were also deemed necessary, it would accept the same format (53).

Upon receipt of an application, FDA would begin its review. Simultaneously NIH's Office of Recombinant DNA Research (ORDA) would evaluate the protocol to decide whether RAC review was necessary. The Task Force proposal identified several factors that should be considered in making this determination, including whether the protocol (*1*) employed novel approaches, (*2*) involved a new disease, (*3*) involved unique applications of gene transfer, or (*4*) involved other issues requiring public review (53). Finally, the proposal recommended eliminating the RAC's Points to Consider document.

1994–1996: FDA Assumes Lead Role in Review of Gene Therapy Research

While both Kessler and Varmus agreed to the Task Force proposal, support for it was not universal. Criticism came from both extremes. Andrew Kimbrell, attorney for Jeremy Rifkin, declared that "[t]his is not the time for decreased reviews," and argued that the government should be doing more to follow up early experiments to discover possible late-developing harms from gene therapy (54). Kimbrell threatened a lawsuit if the RAC's regulatory role were diminished. A more eloquent defense of the RAC's continued relevance was offered by gene therapy pioneer W. French Anderson in his journal *Human Gene Therapy*. Anderson asserted that the "public's trust in this new experimental treatment is in large part the result of, and maintained by, the RAC's resolute role as a watchdog" (55).

Taking an opposing view, Henry Miller, now the former FDA Director of Biotechnology, maintained that the RAC review was an anachronism now that gene therapy experiments involved human subjects and potential products. He disdained the HHS effort to streamline reviews as "purely cosmetic" (54). Leonard Post, a former member of the RAC and an industry spokesman, supported Miller's view that RAC's review of gene therapy protocols unnecessarily duplicated FDA regulation (56).

Nor did members of the RAC favor the compromise. Although they reluctantly voted to approve in principle the consolidation of FDA and RAC reviews (57), they vehemently opposed deleting the "Points to Consider" from the RAC guidelines. In response to their criticism of certain aspects of the Task Force proposal, FDA's Noguchi offered a compromise, under which the RAC's "Points to Consider" would be retained and FDA, ORDA, and RAC would together determine the need for the RAC review of individual clinical protocols (58). NIH Director

Varmus emphasized that he did "not intend to see the demise of RAC" but concluded that the increased number and diversity of gene therapy protocols requiring review necessitated coordination with FDA (58). He noted that numerous investigators had complained to the Task Force on AIDS Drug Development that the review process was too slow and was delaying the initiation of new approaches to the treatment of HIV. The RAC met only episodically while FDA received and reviewed INDs every day. Varmus also acknowledged receiving "a long series of complaints … from my colleagues in the field who claim that there [have] been undue delays in reviews of protocols in RAC." Admitting that the RAC's criteria for approval were unclear, Varmus directed the committee to establish an ad hoc review panel to examine its criteria for approving gene therapy protocols (59). This committee, which was headed by the Salk Institute's Inder Verma, Ph.D., a future RAC member, and became known as the "Verma Committee" (60).

Into 1995 the RAC and FDA wrestled with the details of coordinating review of experimental protocols, as advocates for industry and patient groups demanded quicker review (61). The irony of the situation was not lost on W. French Anderson, who wrote "this was the first time in history that anyone had wanted to go to the FDA because another federal review process was too slow!" (55). RAC member Alexander Capron found it "puzzling" that "AIDS activists want streamlining of gene therapy protocols, but others say we're going too fast" (61).

In March 1995 the members of the RAC unanimously approved a proposal for consolidated protocol review, which allowed simultaneous submission of protocols to FDA and NIH and incorporated elements of both the original Task Force proposal and the Noguchi compromise. Under the new process the RAC and FDA would receive simultaneous submissions of all protocols, which would include information required by RAC's "Points to Consider" document. FDA and ORDA would then decide which protocols merited RAC review. These would include protocols using new vectors or methods of gene delivery, targeting new diseases, employing unique applications of gene transfer, or raising ethical issues warranting public review (62). However, the committee members rejected a proposal by Viagene that would have further streamlined review. In a February 24, 1995, letter, Viagene Regulatory Affairs Director Sheryl Osborne recommended that the RAC not review expansions of previously approved Phase I gene therapy trials in Phase II and III, that sponsors be permitted to seek concurrent IRB and FDA/RAC approval, and that the NIH director be required to complete protocol review within 15 calendar days (62).

FDA's Noguchi initially praised the compromise, stating that it could cut industry's waiting time for approval from a minimum 8 months to as little as 45 days. He also announced FDA's plan to establish a registry to track every American who received gene therapy treatment (63). Within a few months, however, Noguchi was complaining that the RAC had failed to reduce review times and warning that continued FDA funding of the patient registry was in jeopardy because of the RAC's lack of cooperation. In a memorandum to the

RAC, Noguchi asserted that a "year after the mandate to streamline the ... process ... I am obligated to say that the RAC has not appeared to be as accommodating or committed to sharing tasks," such as assisting in development of the database. In addition he noted that "modest proposals to further streamline the RAC process, such as allowing concurrent review by the local IRBs have been unanimously disdained." Noguchi stated that he had "taken a lot of heat from my superiors for their impressions that FDA funding of the gene therapy registry has been for the NIH benefit, rather than for the FDA. Sadly, I reluctantly concur with that impression" (64).

The RAC's protocol review role was soon to be further curtailed, for a number of reasons. First, notwithstanding the official compromise between the RAC and FDA, industry spokespersons remained confused over when RAC review was and was not required (65). In addition the RAC's Verma Committee recommended that the RAC devote more attention to policy issues while delegating scientific review of most protocols to FDA (66). Furthermore proponents of biotechnology, like Henry Miller, continued to rail against the RAC's inefficiency and lack of technical expertise, and to argue that overregulation of gene therapy would inevitably delay development of new therapies (67). Miller criticized NIH Director Varmus for failing to eliminate the RAC (67). Pharmaceutical industry representatives concurred that the RAC review of gene therapy protocols was unnecessary. For example, Washington, DC, attorney Bruce Mackler opined that "[i]t is hard to identify in a quantifiable fashion a uniqueness of gene therapy risks and benefits that would call out for a particular regulation that differs from other biological products." Mackler contended that "the RAC's domain should be ... assuaging public fears and advancing scientific quality of knowledge" (68).

1996–1997: The RAC's Role in the Aftermath of Consolidation

During 1996 NIH Director Varmus considered three options: (1) terminate the RAC, (2) maintain consolidated FDA and NIH review, or (3) maintain the RAC as a vehicle for public accountability and access to information, while terminating its role in the review and approval of individual protocols. After much internal debate Varmus initially chose to pursue elimination of the RAC and transfer of its public policy functions to a new body to be developed within NIH. Following public protestations against elimination of the RAC, however, Varmus decided to retain the RAC and also to maintain simultaneous submission of gene therapy protocols to the RAC and FDA, even though only FDA would actually review individual protocols under most circumstances. Significantly, FDA would have exclusive authority to approve gene therapy protocols (after review and approval by local IRBs).

On July 8, 1996, NIH published a Notice of Intent announcing that NIH was considering amending its Guidelines to eliminate the RAC's review of most gene therapy protocols and to transfer *all* approval responsibility for such protocols to FDA (69). At the same time NIH stressed that its oversight of gene therapy would be "enhanced" through three new mechanisms: (1) the establishment of

the Office of Recombinant DNA Activities Advisory Committee (OAC) — a 6 to 10 member interdisciplinary body that would "ensure public accountability for recombinant DNA research and relevant data"; (2) implementation of Gene Therapy Policy Conferences to "augment the quality and efficiency of public discussion of the scientific merit and the ethical issues relevant to gene therapy clinical trials"; and (3) "continuation of the publicly available, comprehensive NIH database of human gene transfer clinical trials," including adverse event reporting.

NIH explained its plan as simply another step in the normal "devolution" of NIH authority as scientists gained experience with gene therapy protocols and the concerns about the technology therefore shifted. Specifically, NIH characterized the current plan as a continuation of the 1995 decision to consolidate FDA and NIH protocol review:

> In 1995, a similar devolution of NIH oversight of human gene therapy occurred. By this time, the RAC had reviewed and approved 113 gene therapy protocols and over 1,000 patients had been enrolled in worldwide trials. The RAC, the scientific community, and the public had a substantial base of information regarding the use and safety of many of the vectors employed in, and target diseases addressed by, human gene therapy. Subsequent analyses revealed that the human health and environmental safety concerns expressed at the inception of gene therapy clinical trials had not materialized (70).

NIH described its further curtailment of RAC's protocol review function as resulting from the very success of the 1995 consolidation:

> Since the implementation of consolidated review in July 1995, only six of the 36 protocols submitted to ORDA required RAC review and approval; and five of those six protocols were already in the system before consolidated review. The consolidated review process proved to be so successful in eliminating the need for RAC review and approval, that NIH canceled both the March and June 1996 RAC meetings due to the lack of novel protocols requiring RAC attention (70).

Two features of the NIH notice are particularly noteworthy. First, NIH repeatedly stressed that the proposal should not be viewed as simply an "elimination" of the RAC but rather as a reallocation of NIH resources to avoid regulatory redundancy and permit NIH to carry out its role in leading public discussion: "Eliminating RAC protocol approval reduces duplication of effort with the FDA while enhancing the time and effort devoted to both ongoing anticipated gene therapy policy issues deserving of substantial public discussion" (70). In addition, refocusing NIH's efforts would assure NIH's ability to maintain public accountability over gene therapy research:

> NIH concludes that it is not the RAC per se that is critical for public accountability, but the system by which NIH continues to provide public discussion of the scientific, safety, and ethical/legal issues related to human gene therapy (70).

Second, NIH for the first time publicly acknowledged that only FDA possessed the statutory authority to review and approve gene therapy protocols: "The NIH Director has concluded that the current proposal ... is timely and appropriate based on ... the duplication of review and approval by the NIH while the FDA holds the statutory authority" (70). Similarly, NIH acknowledged that, while the contemplated Gene Therapy Policy Conferences would enhance NIH's ability to provide advice on policy matters pertaining to gene therapy, "[t]he NIH cannot ... give the RAC, or any other NIH standing or ad hoc body, the authority to give policy advice or make recommendations to the FDA" (71).

In response to its Notice of Intent, NIH received 71 comments from individuals or groups reflecting the interests and concerns of academe, industry, patient advocacy groups, consumer advocacy organizations, professional scientific societies, ethicists, other federal agencies, NIH-funded investigators, past and present RAC members, and private citizens (72). According to NIH, of the 61 comments that addressed the proposal to terminate the RAC, 20 expressed support and 41 opposition (73).

The Biotechnology Industry Organization (BIO) whole-heartedly embraced the NIH proposal, agreeing that NIH review of individual gene therapy protocols had become redundant, and that NIH resources would be "better spent addressing truly novel gene therapies or those raising significant ethical issues such as in utero or germ line gene therapies" (74). Additionally BIO concurred with NIH's proposal to replace the RAC with the smaller OAC: "Since the number of protocols warranting full NIH-RAC review has dropped to close to zero, maintenance of a large panel is no longer necessary" (74). Similar support for the NIH proposal was conveyed in the comments of numerous industry members—ranging from established pharmaceutical companies such as Merck (75), Glaxo Wellcome (76), Parke Davis (77), and Baxter Healthcare Corporation (78) to entrepreneurial biotechnology outfits such as Cell Genesys (79), Genzyme (80), Auragen (81), and IntraImmune Therapies, Inc. (82). Henry Miller applauded the proposal to eliminate NIH oversight of gene therapy protocols as "long overdue" but criticized NIH's proposal to establish the OAC and to maintain a database of gene therapy clinical trials as a step in the wrong direction:

> In the future, NIH should approach gene therapy in a way no different from other kinds of techniques and treatments, except as medical or scientific considerations dictate. Circumstances do not now require retention of ORDA, the creation of a new advisory committee, or the maintenance of a gene therapy database (83).

Opponents of eliminating the RAC—including current and former RAC members, bioethicists, academe, consumer and patient advocacy groups, and some members of the public—were equally fervent in their insistence that the RAC continue to review individual research protocols. They argued, in essence, that the RAC had developed significant expertise in considering the social and ethical consequences of gene therapy experiments and that it had gained the public's trust in carrying out this role. Elimination of the RAC, they feared, would undermine that trust and jeopardize public accountability for or acceptance of gene therapy experiments. They rejected the OAC as a poor substitute that would have neither the experience

nor reputation to lead public debate of the social and ethical issues surrounding gene therapy (84). Their concerns were eloquently expressed by a science teacher from Illinois:

> Gene therapy intimately affects the future of the human race and the technology must not be allowed to proceed only in the name of good science. We must work hard to prevent the technology from driving the ethical, moral, and legal issues. While the advisory panel is no guarantee that "what's best for humanity" will drive genetics research and therapy, it does provide a necessary oversight function and should not be disbanded (85).

In response to the number and fervor of comments advocating the retention of the RAC, Varmus retreated, stating that he had "underestimated the historical purpose and significance" of the RAC (86). Varmus offered a "compromise" proposal under which the RAC's size would be reduced but several of its functions maintained (86). However, Varmus contemplated that under the compromise the RAC would no longer participate in reviewing or approving individual gene therapy protocols (86).

On October 31, 1997, NIH announced its final decision. In its revised Guidelines, NIH reduced the RAC from 25 to 15 members while retaining its interdisciplinary composition. The RAC's functions were amended to include (1) identifying novel human gene transfer experiments deserving of public discussion by the full RAC, (2) transmitting to the NIH director specific comments or recommendations concerning specific experiments or categories of experiments, and (3) identifying novel social and ethical issues relevant to specific human applications of gene transfer and recommending appropriate modifications to NIH Guidelines to provide guidance in preparing relevant informed consent documents and in designing and submitting human gene transfer clinical trials (72,87).

The revised Guidelines provided that NIH would relinquish all approval responsibilities for gene therapy experiments to FDA. However, no human gene therapy experiment could proceed if the protocol had not been simultaneously submitted to NIH and FDA. Submissions to NIH would need to comply with NIH's "Points to Consider" (Appendix M), while submissions to FDA would be required to comport with FDA's regulations pertaining to the content and format of IND submissions (88). In addition investigators who had received FDA approval would still be required to report any adverse events to both agencies (89).

The Guidelines provided that submissions to NIH would be for "registration purposes," and would "ensure continued public access to relevant human gene transfer information conducted in compliance with the NIH Guidelines" (72,90). If NIH/ORDA determined that a protocol possessed novel features or raised novel concerns that should be discussed by the full RAC, the principal investigator would be notified. In determining whether the full RAC discussion was warranted, NIH/ORDA reviewers would examine "the scientific rationale, scientific content (relative to other proposals reviewed by RAC), whether the preliminary in vitro and in vivo safety data were obtained in appropriate models and are sufficient, and whether questions related to relevant social and ethical issues have been resolved" (90). Otherwise, FDA alone would review the protocol. Whether or not the RAC reviewed the protocol, FDA would be solely responsible for determining whether or not to grant approval.

NIH acknowledged that the Guidelines were obligatory for those institutions and investigators receiving NIH funding or entities collaborating with NIH-funded institutions (90). However, it encouraged continued voluntary compliance by all other entities conducting gene therapy research (87).

ANALYSIS OF FDA'S EVOLVING APPROACH TO GENE THERAPY REGULATION

With the October 31, 1997, announcement, NIH formally and finally acknowledged FDA's exclusive statutory authority to approve gene therapy protocols and, by implication, to regulate the development and marketing of gene therapy-derived products, whether privately supported or federally funded. This had been FDA's position from the mid-1980s. FDA's position was that gene therapy protocols represented "business as usual" and that FDA would therefore review gene therapy research protocols — and eventually marketing applications — within the same framework it applied to other medical products.

This part reviews the formal pronouncements that FDA issued describing and justifying this position. It concludes with a brief discussion of the agency's experience applying its "off the shelf" principles of regulation to individual research protocols. The reader should be cautioned, however, that our text does not attempt to provide a full account of what the sponsor of a research protocol will experience in dealing with the agency. This experience varies so much with the technology, the evidence, the agency's current workload, and the personalities of both sponsor and reviewer that no broad generalizations can be reliable. The discussion does not even attempt to describe how FDA will review applications for approval to commercialize gene therapy products. No such applications have come before the agency, and it would be foolish to speculate how it will process the first of these inherently precedent-setting technologies.

Between 1984 and 1998, FDA issued five separate policy documents outlining the agency's plans for regulating gene therapy. These documents provided progressively more explicit guidance for sponsors of gene therapy research regarding the statutory authorities on which FDA relied to regulate their activities, as well as the procedures that should be followed and the clinical data that must be assembled to demonstrate the safety and effectiveness of gene therapy products. From the outset these documents reflected a consistent regulatory approach, one that affirmed FDA's existing legal authorities were broad enough to encompass and flexible enough to fit this emerging technology. There has, in FDA's view, been no need to develop a new regulatory paradigm.

1984 Policy Statement

FDA first described its approach to regulating products derived from biotechnology generally, including gene therapy, on December 31, 1984. The agency's "Statement of Policy for Regulating Biotechnology Products" (27) (Policy Statement) was published for public comment in the Federal Register two months after the RAC Working Group reviewed a draft "Points to Consider in the Design and Submission of Human Somatic-Cell Gene Therapy Protocols," and three weeks before a draft of the RAC "Points to Consider" was published in the Federal Register for public comment. FDA's Policy Statement can therefore be viewed as the agency's response to — and at least mild dissent from — the NIH approach.

The very title of the Policy Statement is revealing; it signaled FDA's (and the Reagan administration's) view, which would be reiterated in subsequent policy documents, that gene therapy should be viewed as merely one example of a much larger class of therapeutic modalities derived from biotechnology, for which FDA already possessed adequate statutory authority and scientific expertise. The introduction to the statement explained:

> A small but important and expanding fraction of the products the Food and Drug Administration (FDA) regulates represents the fruits of new technological achievements. ... It is also noteworthy that technological advancement in a given area may give rise to very diverse product classes, some or all of which may be under FDA's regulatory jurisdiction. For example, new developments in recombinant DNA research can yield products as divergent as food additives, drugs, biologics, and medical devices (27).

The Policy Statement acknowledged that "there are no statutory provisions or regulations that address biotechnology directly" (27), but asserted that the agency's existing regulatory authorities were broad enough to extend to these products and, moreover, that such extension was appropriate:

> The Agency possesses extensive experience with the administrative and regulatory regimens described as applied to the products of biotechnological processes, new and old, and proposes no new procedures or requirements for regulated industry or individuals (27).

Consistent with FDA's existing approach to regulating other medical products, the Policy Statement announced that "the administrative review of products using biotechnology is based on the intended use of each product on a case-by-case basis" (27). FDA thus preserved its discretion to choose among available statutory authorities and to assign administrative responsibility as it judged appropriate.

The statement then proceeded to summarize the basic Act requirements applicable to manufacturers of human drugs:

- Under the FD&C Act, a "drug" is defined to include articles (1) "intended for use in the diagnosis, cure, mitigation, treatment, or prevention of disease in man," or (2) "intended to affect the structure or any function of the body of man" (91). The Act defines a "new drug" as a drug that is not "generally recognized by qualified scientific experts as safe and effective for its proposed use" (92). Thus the manufacturer of a new drug must establish its safety and effectiveness before marketing. And, to do this, the manufacturer must submit to FDA clinical data from investigations of the drug in human subjects.

- The "sponsor" of a clinical investigation — the entity responsible for the clinical trials — must first file an IND with FDA and obtain the agency's approval to conduct the study or studies described in it (27,93). (This submission follows approval by a research facility's IRB, whose review typically encompasses both ethical and scientific merit.) The IND must contain information sufficient to demonstrate the propriety of testing the drug in human subjects. This would include, for example, drug composition, manufacturing and controls data; results of animal testing, training, and experience of the investigators; and a plan for clinical investigation (27,94). Furthermore the IND sponsor must ensure that informed consent will be obtained from the human subjects who participate in the studies and that the rights and safety of the human subjects will be protected (27,95).

- At the conclusion of three phases of clinical investigation, the manufacturer of a new drug must, if it wishes to distribute the product commercially, submit a New Drug Application (NDA) to FDA. An NDA must contain information including (1) full reports of investigations, as well as the results of clinical investigations, demonstrating the drug's safety and effectiveness, (2) a list of components of the drug and a statement of the drug's quantitative composition, and (3) a description of the methods used in, and the facilities and controls used for, the manufacturing, processing, and packaging of the drug (27,96).

After outlining the framework for regulation of new human drugs, the FDA Policy Statement compared the NDA process to that required for biological products (27,97). FDA's authority to regulate biological products derives chiefly from section 351 of the PHS Act, which defines a "biological product" as:

> any virus, therapeutic serum, toxin, antitoxin, vaccine, blood, blood component or derivative, allergenic product, or analogous product ... applicable to the prevention, treatment, or cure of diseases or injuries of man. ... (98)

The Policy Statement explained that biological products "are regulated similarly to new drugs during the IND phase" (97). Approval for marketing biological products, however, "is granted by license, which is only issued upon demonstration that both the manufacturing establishment and the product meet standards designed to ensure safety, purity, potency, and efficacy" (97). To obtain approval, the manufacturer must submit a Product License Application (PLA) and include information demonstrating that both the manufacturing facility and the product meet FDA requirements (97). The facility must also pass a prelicensing inspection and be separately licensed (97).

The Policy Statement emphasized that manufacturers of both new drugs and biological products must comply

with "good manufacturing practice" (GMP) regulations, which specify requirements for, among other areas, (1) manufacturing facilities, (2) personnel training, and (3) processing methods (97). The Policy Statement also reviewed FDA's quite different statutory authority to regulate medical devices, which do not achieve their intended effect through chemical action in or on the body or through metabolism (99).

After completing its survey of FDA's general requirements for drugs, biological products, and devices, the Policy Statement addressed their application to specific products, including products that are "genetically engineered." Here the agency spoke in vague terms. It reiterated that its approach to regulating these products was "product specific" rather than "technology specific" (27,97,100). Taking genetically altered viruses used as vaccines as an example, the agency declared:

> The composition, concentration, subtype, immunogenicity, reactivity, and nonpathogenicity of the vaccine preparation are all considerations in the final review, whatever the techniques employed in "engineering" the virus (100).

The 1984 Policy Statement addressed gene therapy in just a single sentence: "Nucleic acids used for gene therapy trials will be subject to the same requirements as other biological drugs" (100). This brevity was consistent with agency's position that FDA regulation "must be based on the rational and scientific evaluation of products, and not on *a priori* assumptions about certain process" (100). In other words, that gene therapy might or would entail the use of genetically engineered products to achieve a therapeutic effect would not alter FDA's regulatory approach. The agency would place such a product into the statutory framework that best fit the product's intended use and mode of action.

The Policy Statement was similarly terse in describing the role of the RAC Working Group in reviewing gene therapy protocols: "It is possible that there will be some redundancy between the scientific reviews of these products performed by the National Institutes of Health and FDA" (100).

Certain features of FDA's initial Policy Statement bear emphasis. The document surely cannot be described as a "how to" manual for developers of biotechnology products generally, much less for sponsors of gene therapy protocols. It provided no concrete guidance for sponsors of gene therapy products, save that they would be required to submit and gain approval for an IND before undertaking clinical experiments. Indeed, the Policy Statement did not specify whether products used in gene therapy would be regulated as "drugs," "devices," or "biological products," but instead left open the option to regulate them under any one — and perhaps some combination — of these categories. Of course, given FDA's premise — that gene therapies fell within the agency's customary jurisdiction — this lack of specific guidance was not surprising.

The Policy Statement did not specify which biological component or components used in gene therapy would be considered to be the "drug" or "biological product." The agency thus left open the possibility that it would regulate the genetic material itself, the vector used to deliver the genetic material, some other component used in the gene transfer process, or all of these.

Also notable was the FDA's oblique criticism of the NIH role in regulating gene therapy. To be sure, the regulatory issues addressed in the Policy Statement overlapped in several respects with those identified by the RAC's "Points to Consider," although these points specified the relevant considerations in substantially more detail. Like FDA's Policy Statement, the RAC's "Points to Consider" emphasized the need for sponsors of clinical trials to address the study's design, results of animal studies, investigator and personnel qualifications, and the adequacy of the laboratory facilities. However, the issues FDA addressed were a subset of a substantially larger range of issues that NIH required sponsors to address — some of which FDA's Policy Statement implied were not necessary and perhaps not even appropriate for its consideration.

For example, under "Description of Proposal," NIH asked investigators to address (1) why the disease in question was appropriate for gene therapy and what alternative therapies existed, (2) what "equity issues" would likely arise in the selection of patients and how they would be addressed, and (3) whether the "innovative character" of gene therapy would be discussed with patients (24,101). In addition, under the heading "Social Issues," NIH asked investigators to address (1) the steps that would be taken to ensure that accurate information was made available to the public regarding concerns raised by the study and (2) whether the investigator intended to protect the products or procedures used in the proposed study under patent or trade secret laws and, if so, what steps would be taken to permit full communication among investigators and clinicians concerning research methods and results (101). The initial draft of the "Points to Consider" stated that the RAC and its working group would also consider (1) whether the proposed study would likely affect the reproductive cells of the patient, (2) whether the proposed study was an "extension of existing methods of health care" or represented a "distinct departure from present treatments of disease," and (3) whether somatic cell gene therapy would likely lead to germ-line therapy (i.e., therapy affecting reproductive cells and therefore future generations, enhancement of human capabilities through genetic means, or government-sponsored eugenic programs (101). However, the final version, adopted on September 29, 1986, deleted the section addressing germ-line and eugenic implications of gene therapy (102).

FDA's Policy Statement reflected the agency's position that, from its perspective, medical applications of biotechnology generally and gene therapy specifically represented "business and usual." By contrast, NIH's "Points to Consider" signaled the RAC Working Group's belief that gene therapy was a fundamentally new approach to therapy, which posed unique social and ethical questions as well as complex scientific challenges.

Following review of public comments, FDA in June 1986 published a substantially unchanged final version of the 1984 Policy Statement. For the next five years FDA issued no other formal statements regarding gene therapy or its regulation.

FDA's 1991 "Points to Consider"

On August 27, 1991, almost a year after the first clinical experiment employing gene therapy, FDA issued a document entitled *Points to Consider in Human Somatic Cell Therapy and Gene Therapy* ("Points to Consider") (44,103). Unlike its 1984 Policy Statement, which addressed a wide range of applications of biotechnology that could be subject to regulation, FDA's 1991 document focused specifically on gene therapy products (as well as somatic cell therapy products). FDA stated that the purpose of issuing its "Points to Consider" was to provide information to "manufacturers engaged in the production and testing of products for these therapies (103, p. 3). Compared to the 1984 Policy Statement, the FDA "Points to Consider" provided concrete guidance to manufacturers regarding the procedures they should be using to develop their products and the type of supporting data the agency would expect. FDA stressed that its "Points to Consider" did not constitute binding regulations but merely "represent issues that the Center for Biologic Evaluation and Research (CBER) staff believes should be considered at this time" (103, p. 3).

FDA's "Points to Consider" set forth two definitions. Somatic cell therapy was defined as "the administration to humans of . . . living cells which have been manipulated or processed ex vivo" (103, p. 3). Gene therapy was defined as "a medical intervention based on modification of the genetic material of living cells." The agency continued:

Cells may be modified ex vivo for subsequent administration to humans, or may be altered in vivo by gene therapy given directly to the subject. When the genetic manipulation is performed ex vivo on cells which are then administered to the patient, this is also a form of somatic cell therapy. The genetic manipulation may be intended to have a therapeutic or prophylactic effect, or may provide a way of marking cells for later identification (103, p. 3).

The agency observed that initial approaches to gene therapy "have involved the alteration and administration of somatic cells," but it forecast that "future techniques may include approaches such as the direct administration to patients of retroviral vectors or other forms of genetic material" (103, pp. 3–4).

The FDA document then identified several subjects that sponsors of gene therapy and somatic cell therapy procedures should consider during product development, as well as information that should be presented to the agency. Sponsors are expected to address (1) quality control, (2) development and characterization of cell populations for administration, (3) preclinical testing, (4) lot-to-lot manufacturing control and release testing, and (5) clinical trials (103). FDA emphasized that it is essential to characterize the gene sequence being introduced into cells and the vector used to do this, and specified that "each distinct vector is considered a different product and should be fully characterized and tested for safety" (103, p. 12). The agency also stated that manufacturers should address the methods used to insert the vector into cells and the implications of the method used. Specifically, manufacturers should be able to demonstrate that the introduced genes were integrating at the correct location in the chromosome, and that they were functioning appropriately once integrated (i.e., were expressing the correct gene product in the correct quantity) (103).

FDA noted that clinical trials using gene therapy raise "some novel concerns due to the nature of the therapeutic agents" (103, p. 19). For example, use of viral vectors "may in special cases require testing of clinical personnel or household contacts to confirm lack of infectious spread," (103, pp. 19–20). In addition "the product of the inserted gene must be considered as a potential source of immune reactions" (103, p. 20); thus sponsors might in some cases need to document whether such reactions were observed and whether they altered the safety or therapeutic effectiveness of the product (103, p. 20).

While FDA's "Points to Consider" provided details not contained in the 1984 Policy Statement, both documents reflected the same regulatory approach. The "Points to Consider" concluded with the observation that somatic cell and gene therapy raise issues common to all biological products as well as issues unique to these types of therapies. It reflected FDA's confidence in its ability to address these issues, and any new issues that the future might reveal, within the framework of existing legislation (103).

1993 Guidance

Shortly after establishing the new Division of Cellular and Gene Therapy, FDA published in the Federal Register a notice entitled "Application of Current Statutory Authorities to Human Somatic Cell Therapy Products and Gene Therapy Products" (44). This document, though it largely reiterated information contained in the 1984 and 1991 documents, said it was being published "in response to requests that the agency clarify its regulatory approach and provide guidance to manufacturers of products intended to be used in somatic cell therapy or gene therapy" (44).

Like its 1984 Policy Statement, FDA's 1993 self-described "guidance" identified the FD&C Act and the PHS Act as the legal sources of its authority to regulate gene therapy. It also reiterated that clinical trials of biological products—including gene therapy—to gather data on safety and effectiveness must be conducted under an IND. In addition the document stated that a product could be subject to regulation under both statutes concurrently: "Products considered to be biological products subject to the provisions of section 351 of the PHS Act are simultaneously also drugs or devices subject to the applicable provisions under the Act" (44). While FDA asserted that products regulated as biological products must also meet requirements applicable to drugs or devices, it also emphasized that sponsors were not required to submit multiple applications for marketing approval; for any product the agency would require only a biologic application (PLA), new drug application (NDA), or a device application (PMA) (104).

The 1993 guidance offered additional but still general information about FDA's statutory authorities over drugs, medical devices, and biological products—including, for example, its authority to regulate labeling, its requirements for licensing and inspection of manufacturing facilities, and its enforcement powers.

The 1993 guidance for the first time provided concrete examples of treatments that the agency would regulate as gene therapy products, either by the mechanism of a PLA or an NDA:

> Final products containing the genetic material intended for gene therapy are regulated as biological products requiring PLA's (e.g., viral vectors containing genetic material to be transferred, ex vivo transduced cells and analogous products) or as drugs requiring NDA's (e.g., synthetic products) regardless of whether they are intended for use in vivo or ex vivo. Gene therapy products that are licensed biological products will be approved as biological products intended for further manufacture if they are intended to be used ex vivo during the manufacture of genetically altered cells.
>
> Examples include the following: (1) A synthetic polynucleotide sequence intended to alter a specific genetic sequence in human somatic cells after systemic administration is regulated as a drug requiring an NDA; (2) a retroviral vector containing the adenosine deaminase (ADA) gene, intended to be administered intravenously to the patient, is regulated as a biological product requiring a PLA, and (3) a retroviral vector containing the ADA gene and intended to modify cells ex vivo is regulated as a biological product intended for further manufacture requiring a PLA (105).

1996 Addendum

In 1996 FDA issued a draft Addendum to its 1991 "Points to Consider" document (106,107). The Addendum focused on the regulatory requirements for viral vectors used in gene therapy. FDA explained that since the 1991 document was issued, "the range of proposals has expanded to include additional classes of vectors and also the in vivo use of vectors (direct vector administration to patients)" (106). Accordingly it was issuing the Addendum to "provide manufacturers with current information regarding current regulatory concerns for production, testing, and administration of recombinant vectors for gene therapy" (107, p. 2). The Addendum instructed manufacturers regarding proper characterization of the gene sequence of the vector, proper maintenance of the cells used to produce the vectors, and appropriate methods for testing vectors for purity, potency, and safety. The Addendum also addressed issues applicable to specific types of vectors.

All three of FDA's previous publications discussed the authority to regulate gene therapy products generally, offering virtually no information about its views on any particular facet or application of the technology. The 1996 Addendum was the first to address the requirements pertaining to a specific component of gene therapy, the gene delivery system. Apparently FDA no longer considered it necessary to repeat its claims to regulatory authority, assuming perhaps that any doubts on that point had been resolved by acquiescence and the passage of time. Now the agency could focus exclusively on what sponsors must do to satisfy their legal obligations.

1998 Guidance

In March 1998 FDA issued a guidance document concerning human somatic cell therapy and gene therapy (108). FDA stated that the guidance document was intended to update and replace the 1991 "Points to Consider" document "with new information intended to provide manufacturers with current information regarding regulatory concerns for production, quality control testing, and administration of recombinant vectors for gene therapy; and of preclinical testing for both cellular therapies and vectors" (108, p. 1).

STATUS OF PROTOCOLS

A recent compilation from NIH's registry of gene therapy protocols reveals that, between 1989 and 1999, a total of 280 gene transfer protocols were submitted to NIH and FDA for review (109–111). These include both protocols intended to provide therapeutic benefit and those in which the protocol's intent is not therapeutic (e.g., "marker" gene experiments). Of the 280, 107 were subject to review by the full RAC, approval by the NIH director, and IND approval by FDA—this review process is no longer in effect (109–111). An additional seven protocols were subject to the accelerated RAC review, NIH/ORDA approval, and FDA IND approval—this review process also is no longer in effect (109–111). Ten protocols were subject to FDA IND approval and full RAC discussion—this form of review is currently in effect for some protocols (109–111). The NIH list does not reflect whether protocols approved by NIH were subsequently approved by FDA.

An additional 151 protocols have been screened by NIH/ORDA and referred for exclusive FDA review (109–111). Submission to NIH/ORDA has been required for the purpose of data monitoring and adverse event reporting. As of February 10, 1999, four protocols are awaiting NIH/ORDA's determination of their review requirements (109–111). Finally, one protocol was submitted for sole FDA review and was also voluntarily submitted to NIH/ORDA, presumably by an individual or entity not funded by NIH (109–111).

In early 1997 FDA and NIH approved the first gene therapy trial to be conducted in healthy volunteers. The protocol provides for the injection of healthy individuals with a viral vector in order to measure their immune response (112). The data can be used to establish a baseline that will help evaluate the vector's therapeutic effect in sick people. Use of healthy volunteers is standard in early phase drug development studies (112).

The early sponsors of gene therapy protocols were NIH researchers or investigators at medical research institutions. Between 1989 and 1991, a total of 11 protocols were approved by NIH, all of which were sponsored by researchers at NIH or in academe (109). Beginning in 1992, the caseload began to include more industry sponsors and the pace of submissions rose dramatically. In 1992 and 1993, 44 protocols were approved by NIH, 8 of which were identified as having industry sponsors (109). Between 1994 and 1996, 27 out of 115

protocols received, reviewed, or approved by NIH were sponsored by industry. (During this time period NIH increasingly began referring protocols for sole FDA review, and NIH's publicly available list of protocols does not reflect the fate of the protocols thereafter) (109). Of 110 protocols received by NIH/ORDA between 1997 and 1999, 44 were sponsored by industry (109). It should also be noted that private firms may, and often do, participate in and contribute financially to government or university-sponsored protocols without being listed as sponsors.

As the private sector has assumed a growing role in financial sponsorship of gene therapy research, the types of protocols submitted for review have changed. Although gene therapy was originally conceived as a means of treating or preventing rare diseases caused by "monogenic" defects, namely defects in a single gene—ADA being a prime example—only 36 experimental protocols for such diseases have been approved (110). Most of the industry-sponsored protocols have been aimed at widespread diseases such as cancer and heart disease, or at AIDS. While these diseases may have a genetic component or be susceptible of treatment through gene therapy techniques, they certainly are not the "classical" genetic diseases for which researchers first contemplated gene therapy (51).

As both researchers and FDA become more expert in and comfortable with the techniques of gene therapy, new, and perhaps more difficult questions are emerging. Initially, FDA's concerns focused on the safety of research subjects, such as how would cells be removed from patients? would these cells become contaminated while outside the patients? how would reintroduction of gene-altered cells be achieved (51)? Increasingly, however, FDA's concerns have shifted from "can you" to "should you." NIH recently sponsored its first Gene Therapy Policy Conference to discuss the use of gene therapy for "enhancement," namely for use in non-life-threatening diseases or conditions, such as baldness (113). Public discussion of this issue, FDA's Noguchi emphasizes, is warranted because "the concept of gene therapy is being pushed to lots of different diseases, some of which are more social than physiological" (51).

According to Dr. Noguchi, the RAC is seen as having a crucial role in ensuring that the social and ethical issues raised by particular applications of gene therapy are fully debated and debated in public. Unlike FDA, which must conduct most of its review of gene therapy protocols in private to protect the trade secret information they embody or contain, the RAC is "well constituted for public discussion" (51).

CONCLUSION

The past two decades have seen remarkable advancements in gene therapy research. Given the pace with which experimental protocols are being pursued, it would not be surprising to see the first approved gene therapy product before the official close of the twentieth century. The rapid pace of research has also posed profound challenges for those federal agencies charged with the task of fashioning social controls on scientific technology. As the story of FDA's regulation of gene therapy reveals, the very first challenge was to determine which agency — NIH or FDA — would take the lead in designing and implementing these social controls. More recently the NIH has formally recognized FDA's statutory authority to review and approve individual gene therapy protocols, and FDA has in turn acknowledged the important role played by the NIH RAC in leading the public discussion on the social and ethical implications of gene therapy. FDA and NIH, working together, thus seem well prepared to confront the challenges that will soon emerge when gene therapy makes the leap from the laboratory to the pharmacy.

ACKNOWLEDGMENTS

The authors are grateful to Marilynn Whitney for her assistance in preparing this article and to Peter Barton Hutt for helpful comments on an early draft.

BIBLIOGRAPHY

1. Federal Food and Drugs Act of 1906, *U.S. Statutes at Large* **34**, 768 (1906).
2. *United States v Sullivan*, 332 US 689, 696 (1948).
3. *U.S. Statutes at Large* **52**, 1040 (1938), codified at *U.S. Code*, vol. 21, secs. 301 et seq.
4. Drug Amendments of 1962, *U.S. Statutes at Large* **76**, 780 (1962).
5. Medical Device Amendments of 1976, *U.S. Statutes at Large* **90**, 539 (1976).
6. Safe Medical Devices Act of 1990, *U.S. Statutes at Large* **104**, 4511 (1990).
7. Food and Drug Administration Modernization Act of 1997, *U.S. Statutes at Large* **111**, 2296 (1997).
8. *U.S. Statutes at Large* **32**, 728 (1902).
9. *U.S. Statutes at Large* **58**, 682 (1944), codified at *U.S. Code*, vol. 42, secs. 201 et seq.
10. *U.S. Statutes at Large* **58**, 702 (1944), as amended, codified at *U.S. Code*, vol. 42, sec. 262(a).
11. Public Health Service and Food and Drug Administration, *Fed. Regist.* **37**, 12865 (1972).
12. National Institutes of Health, *Fed. Regist.* **39**, 39306 (1974).
13. J. Lyon and P. Gorner, *Altered Fates*, Norton, New York, 1996, p. 61.
14. S.G. Michaud, *Business Week*, September 19, 1977, p. 18.
15. National Institutes of Health, *Fed. Regist.* **61**, 35774 (1996).
16. Committee on Recombinant DNA Molecules, *Science* **185**, 303 (1974).
17. National Institutes of Health, *Fed. Regist.* **41**, 27902 (1976).
18. J. McWethy, *U.S. News and World Report*, July 12, 1976.
    J. Goodfield, *Playing God: Genetic Engineering and the Manipulation of Life*, Random House, New York, 1977.
19. J.C. Fletcher, *Hum. Gene Ther.* **1**, 55, 60–61 (1990).
20. W. Froelich, *San Diego Union-Tribune*, January 1, 1985.
21. National Institutes of Health, *Report of the NIH Ad Hoc Committee on the UCLA Report Concerning Certain Research Activities of Dr. Martin J. Cline* (memo. May 21), NIH, Washington, DC, 1981, p. 8.
22. R.M. Cook-Deegan, *Hum. Gene Ther.* **1**, 163–170 (1990).
23. President's Commission for the Study of Ethical Problems in Medicine and Biomedical And Behavioral Research, *Splicing Life: The Social and Ethical Issues of Genetic Engineering with Human Beings*, U.S. Government Printing Office, Washington, DC, 1982.
24. National Institutes of Health, *Fed. Regist.* **50**, 2940 (1985).
25. B.J. Culliton, *Science* **227**, 493 (1985).
26. B.J. Culliton, *Science* **229**, 736 (1985).
27. Food and Drug Administration, *Fed. Regist.* **49**, 50878 (1984).
28. Food and Drug Administration, *Fed. Regist.* **49**, 50880 (1984).
29. B.J. Culliton, *Sci. News*, August 31, 1985.
30. S. Squires, *Washington Post Health*, October 30, 1985.
31. R. Rhein, *Chem. Week*, October 27, 1985.
32. R. Rhein, *Chem. Eng.*, November 11, 1985.
33. National Institutes of Health, *Fed. Regist.* **50**, 33462–33463 (1985).
34. Human gene therapy research should have NIH review, even if funded by another federal agency, RAC decides. *F-D-C Rep. — The Blue Sheet*, February 4, 1987.
35. R. Steinbrook, *Los Angeles Times*, April 13, 1986.
36. S. Squires, Sally, *Washington Post*, May 30, 1989.
37. P. Gorner, *Chicago Tribune*, May 21, 1989.
38. L. Roberts, *Science* **243**, 734 (1989).
39. R.M. Henig, *N.Y. Times Mag.*, March 31 (1991).
40. Genetic Therapy, Inc. goes public as fourth gene therapy collaborative trial with NIH is set to begin. *F-D-C Rep. — The Blue Sheet*, July 3, 1991.
41. Food and Drug Administration, *Fed. Regist.* **56**, 61022 (1991).
42. Food and Drug Administration, *Health News Daily*, December 11, 1992.
43. Food and Drug Administration, *F-D-C Rep. — The Pink Sheet*, December 14, 1992.
44. Food and Drug Administration, *Fed. Regist.* **58**, 53248 (1993).
45. Synthesized gene therapy products will fall under FDA drug review regulations, agency states. *F-D-C Rep. — The Blue Sheet*, October 20, 1993.
46. D.A. Kessler et al., *N. Eng. J. Med.* **329**, 1169–1173 (1993).
47. Cancer vaccine/TIL gene therapy protocols proposed by NCI's Steven Rosenberg okayed by RAC subcmte. *F-D-C Rep. — The Blue Sheet*, July 31, 1991.
48. J. Palca, *Science* **253**, 624 (1991).
49. C. Anderson, *Nature (London)* **357**, 615 (1992).
50. Viagene's Phase I in vivo HIV gene transfer protocol receives unanimous RAC approval June 7. *F-D-C Rep. — The Blue Sheet*, June 9, 1993.
51. Interview by authors with Dr. Philip Noguchi, November 24, 1997.
52. S. Jenks, *J. Nat. Cancer Inst.* **85**, 1544–1546 (1993).
53. FDA and NIH will consolidate gene transfer protocol review by December; HHS' Lee calls for progress report at October AIDS Task Force Meeting. *F-D-C Rep. — The Pink Sheet*, July 25, 1994.
54. E. Marshall, *Science* **265**, 599 (1994).
55. W. French Anderson, *Hum. Gene Ther.* **5**, 1309–1310 (1994).
56. L.E. Post, *Hum. Gene Ther.* **5**, 1311–1312 (1994).
57. S. Sternberg, *BioWorld Today*, September 14, 1994.

58. NIH 'Points-to-consider' retained in new gene therapy NIH/FDA consolidated review proposal from FDA's Noguchi; RAC will stay intact, NIH's varmus assures. *F-D-C Rep. — The Pink Sheet*, September 19, 1994.

59. RAC protocol approval criteria should be assessed by ad hoc review group, NIH Director Varmus recommends, noting complaints about confusing procedures. *F-D-C Rep. — The Pink Sheet*, September 19, 1994.

60. RAC termination called 'Timely and appropriate' in notice submitted by NIH; Sen. Pryor concerned. *F-D-C Rep. — The Blue Sheet*, July 3, 1996.

61. NIH and FDA RAC balks at regulatory reforms. *Cancer Res. Wkly.*, December 19, 1994.

62. National Institutes of Health, *Fed. Regist.* **60**, 20726 (1995).

NIH committee approves combined FDA/NIH submissions of gene therapy protocols. *Gene Ther. Wkly.*, April 17, 1995.

63. L. Neergaard, *The Ledger (Lakeland, FL)*, May 25, 1995.

64. FDA/NIH gene therapy database future funding may be threatened, FDA's Noguchi warns at RAC meeting. *F-D-C Rep. — The Blue Sheet*, June 14, 1995.

65. So you want to develop gene therapy products? What lies ahead in regulatory hurdles. *Bio Venture View*, July–August 1995.

66. FDA should conduct most reviews of gene therapy protocols, while NIH RAC should continue as policy body — Ad hoc groups; Database development supported. *F-D-C Rep. — The Pink Sheet*, September 4, 1995.

67. H.I. Miller, *Hum. Gene Ther.* **6**, 1361–1362 (1995).

68. Gene therapy should be regulated within traditional biological framework, asserts Fenwick & West's Mackler; Calls current RAC role counterproductive. *F-D-C Rep. — The Pink Sheet*, January 22, 1996.

69. National Institutes of Health, *Fed. Regist.* **61**, 35774 (1996).

70. National Institutes of Health, *Fed. Regist.* **61**, 35775 (1996).

71. National Institutes of Health, *Fed. Regist.* **61**, 35776 (1996).

72. National Institutes of Health, *Fed. Regist.* **62**, 59032 (1997).

73. National Institutes of Health, *Fed. Regist.* **61**, 59726–59727 (1996).

74. Letter from Alan Goldhammer, Ph.D., BIO, to Office of Recombinant DNA Activities, August 2, 1996.

75. Letter from Bonnie J. Goldmann, M.D., Merck Research Laboratories, to ORDA, August 7, 1996.

76. Letter from M. Lynn Pritchard, Ph.D., Glaxo Wellcome Inc., to Dr. Harold Varmus, NIH, August 6, 1996.

77. Letter from Leonard E. Post, Parke-Davis, to Office of Recombinant DNA Activities, July 31, 1996.

78. Letter from Robert C. Moen, M.D., Ph.D., Baxter Healthcare Corporation, to Office of Recombinant DNA Activities, August 5, 1996.

79. Letter from Bridget P. Binko, Cell Genesys, to Office of Recombinant DNA Activities, August 5, 1996.

80. Letter from Alexander E. Kuta, Ph.D., Genzyme, to Office of Recombinant DNA Activities, August 7, 1996.

81. Letter from James G. Timmins, Auragen, to Office of Recombinant DNA Activities, July 29, 1996.

82. Letter from Susan D. Jones, Ph.D., IntraImmune Therapies, Inc., to Office of Recombinant DNA Activities, July 25, 1996.

83. Letter from Henry I. Miller, M.D., Hoover Institution, to Office of Recombinant DNA Activities, August 7, 1996.

84. Comments of Charles R. McCarthy, LeRoy Walters, John C. Fletcher, Foundation on Economic Trends, Council for Responsible Genetics, Alliance of Genetic Support Groups, Friends of the Earth, Genetic Therapy, Inc.

85. Letter from David K. Vanderberg to Dr. Harold Varmus, NIH, June 25, 1996.

86. J.L. Fox, *Nat. Biotechnol.* **15**, 11–12 (1997).

87. National Institutes of Health, *Fed. Regist.* **62**, 59042 (1997).

88. *Code of Federal Regulations*, vol. 21, p. 312.

89. National Institutes of Health, *Fed. Regist.* **62**, 59046 (1997).

90. National Institutes of Health, *Fed. Regist.* **62**, 59034–59035 (1997).

91. *U.S. Code*, vol. 21, secs. 321(g)(B) and (C).

92. *U.S. Code*, vol. 21, sec. 321(p)(1).

93. *Code of Federal Regulations*, vol. 21, sec. 312.20.

94. *Code of Federal Regulations*, vol. 21, secs. 312.23–312.31.

95. *Code of Federal Regulations*, vol. 21, sec. 312.23(a)(iv).

96. *Code of Federal Regulations*, vol. 21, sec. 314.50.

97. National Institutes of Health, *Fed. Regist.* **49**, 50878–50879 (1984).

98. *U.S. Code*, vol. 42, sec. 262(a).

99. *U.S. Code*, vol. 21, sec. 321(h)(3).

100. Food and Drug Administration, *Fed. Regist.* **49**, 50880 (1984).

101. National Institutes of Health, *Fed. Regist.* **50**, 2942–2944 (1985).

102. National Institutes of Health, *Fed. Regist.* **54**, 10956 (1989).

103. CBER, *Points to Consider in Human Somatic Cell Therapy and Gene Therapy*, Center for Biologic Evaluation and Research, 1991.

104. Food and Drug Administration, *Fed. Regist.* **58**, 53249 (1993).

105. Food and Drug Administration. *Fed. Regist.* **58**, 53251 (1993).

106. Food and Drug Administration, *Fed. Regist.* **61**, 5786 (1996).

107. Food and Drug Administration, *Fed. Regist.* (1991).

108. CBER, *Guidance for Industry: FDA Guidance for Human Somatic Cell Therapy and Gene Therapy*, CBER, 1998.

109. Office of Recombinant DNA Activities, *Human Gene Therapy Protocols* (last modified February 10, 1999). Available at: <http://www.nih.gov/od/orda/protocol.pdf>.

110. Office of Recombinant DNA Activities, *Protocol List Table*. Available at: <http://www.nih.gov/od/orda/prottab.pdf> The protocol list contains 284 protocols, however, only 280 are included in the accompanying table. According to information received from NIH/ORDA, the table does not include the four protocols because they are incomplete submissions.

111. Telephone call with Gene Rosenthal, May 6, 1999.

112. J.L. Fox, *Nat. Biotechnol.* **15**, 314 (1997).

113. National Institutes of Health, *Fed. Regist.* **62**, 44386 (1997).

See other GENE THERAPY entries; see also HUMAN SUBJECTS RESEARCH, LAW, FDA RULES.

GENE THERAPY, LAW, RECOMBINANT DNA ADVISORY COMMITTEE (RAC). See RECOMBINANT DNA, POLICY, ASILOMAR CONFERENCE.

GENE THERAPY: OVERVIEW

LeRoy Walters
Kennedy Institute of Ethics, Georgetown University
Washington, District of Columbia

OUTLINE

TERMINOLOGY

The phrase "gene therapy" was used for the first time in a published article in 1970 (1,2). However, functional equivalents to this phrase had been used in academic discussions from the early 1960s on. Among the synonymous phrases were "genetic surgery," "nanosurgery," "euphenics," "genetic engineering," "gene replacements," "directed mutation," "directed genetic change," "designed genetic change," "algeny," "programming cells with genetic messages," and "genetic therapy" (3–14). The phrase "genetic(al) engineering" had been employed at least as early as the Sixth International Congress of Genetics held in 1932 in Ithaca, New York (15), and some authors continued to use the phrase "genetic engineering" even after the advent of molecular biology in the early 1950s.

As the preceding paragraph suggests, there was during the 1960s a striking diversity in the terminology used by commentators on the new techniques that molecular biology might make possible. This diversity is reflected in several publications and reminiscences from that important decade. At a Ciba Foundation symposium held in London on November 26–30, 1962, Hermann J. Muller suggested that it would be preferable to secure gametes from genetically fit individuals rather than to use what he called "nano-needles" to "cause prespecified changes in [those individuals]" (16,17). In opposition to Muller, Joshua Lederberg advocated the use of molecular techniques in the manipulation of germ cells to achieve "the direct control of nucleotide sequences in human chromosomes, coupled with recognition, selection and integration of the desired gene" (18, p. 265). At an April 1963 symposium sponsored by Ohio Wesleyan University, Salvador Luria, Edward Tatum, and Muller favored the phrase "genetic surgery," although Luria also discussed the "removal, addition, and replacement of genes" (19,

p. 10), while Tatum at times employed the alternative terminology of "directed gene mutation" and "genetic engineering" (20, p. 22).

As the decade progressed, additional academic disciplines became involved in the discussion of genetic intervention, and more precise descriptions of the actual techniques that might be employed in gene transfer began to appear. At a symposium held at Gustavus Adolphus College in January 1965, theologian Paul Ramsey and biologist Tatum both considered the theme, "Genetics and the Future of Man" (21). Ramsey described one of two possible approaches to genetic control in the following terms:

> The first [method] is some direct attack upon the deleterious mutated gene, either by what is called "genetic surgery," "micro-surgery" or "nano-surgery" [see Ref. 22] or by the introduction of some anti-mutagent chemical that will cause the gene to mutate back or will eliminate it from among the causes of genetic effects (23, pp. 9–10).

In his presentation at the same symposium Tatum described three methods for achieving what he called "*manipulation* [his italics] of genetic change." The second method identified by Tatum sounds like a blueprint for what in the 1990s came to be called ex vivo gene therapy.

> Another potential future approach to directed mutation is via the synthesis in the laboratory of a desired molecule of DNA. This tailored molecule, or any desired DNA molecule if it can be isolated from an organism or cell, can probably be amplified by already known enzymatic replication processes to any needed quantity. This new or isolated gene can then hopefully be introduced into mammalian cells in culture, as in bacterial transformation.
>
> If the rare desired transformed cell can be selected and cultured, the new cells so derived could conceivably be transplanted into a living organism, there to correct a defective function of the original host cells (24, p. 58).

By the late 1960s biologists were describing what would later be called somatic-cell gene therapy in slightly different terms. For example, in a 1968 essay, Lederberg suggested that

> an attempt could then be made to transform liver cells of male offspring of haemophilic ancestry by the introduction of carefully fractionated DNA carrying the normal alleles of the mutant haemophilia gene. This experiment would appear to be entirely analogous to the typical attempts at transforming bacterial forms. However, it is not clear whether one should regard this as a pure example of genetic engineering, since the practical outcome would probably be best achieved by influencing the nuclear constitution of somatic tissues rather than by direct tackling of the germ line (26).

In the same year, 1968, Robert Sinsheimer gave a lecture at the Fordham Chapter of the Society of the Sigma Xi. This lecture, published in the spring of 1969, drew a clear distinction between somatic-cell and germ-line approaches to what he called "designed genetic change" (27). Sinsheimer's specific suggestion for somatic-cell genetic change was that the almost-available technique could be used for the treatment of diabetes.

If we could obtain a virus analogous to simian virus 40, able to persist within altered cells and, let us say, carrying an expressable gene for proinsulin [the precursor to insulin] in lieu of a normal viral gene, we might indeed be able to provide a genetic alternative to the daily injection of insulin (27, p. 140).

The most widely cited essay during this era was almost certainly Bernard Davis's paper on "Threat and Promise in Genetic Engineering," first given at a December 1969 symposium, then revised and published a year later in *Science* under the title, "Prospects for Genetic Intervention in Man" (28). Davis drew important distinctions between polygenic (including many behavioral) and monogenic traits and between somatic-cell and germ-cell alteration. He also pointed to the technical obstacles standing in the way of even the simplest types of somatic-cell gene alterations. However, Davis also noted a potential advantage of a genetic approach to therapy.

Such a one-shot cure of a hereditary disease, if possible, would clearly be a major improvement over the current practice of continually supplying a missing gene product, such as insulin (28, p. 1280).

By 1970 there was a consensus among biologist-commentators like Lederberg, Davis, and H. Vasken Aposhian that the phrase "genetic engineering" was unduly alarming to the public (28–30). As noted above, the phrase "gene therapy" began to be used in the published literature during the same year. By the time of two major symposia on this subject in May 1971 the alternative formulations were virtually always "genetic therapy" and "gene therapy" (31,32). Beginning with two articles by Aposhian in 1970 (1,2), one finds an additional three articles that use the phrase "gene therapy" in their titles in 1971 (33–35), and two additional articles plus the proceedings of an NIH symposium in 1972 (36–38). Perhaps the most decisive publication in this series was Theodore Friedmann and Richard Roblin's widely cited *Science* article from March 1972 entitled "Gene Therapy for Human Genetic Disease" (39). (The phrase "gene therapy" was added to the *Bioethics Thesaurus of Bioethicsline* in 1980, to the Library of Congress Subject Headings in 1985, to the National Library of Medicine's Medical Subject Headings in 1989, and to the Dewey Decimal Classification with the 21st edition of 1996.)

It is worth pausing to note that the transition from "genetic engineering," often associated with the evolution of the human species and with voluntary programs of positive eugenics, to the more modest goals of "gene therapy" paralleled a shift in emphasis from classical genetics to microbiology and molecular biology. H.J. Muller and those who sympathized with his views were concerned about an increasing load of mutations in the human gene pool and accented the difficulty of modifying polygenic traits through molecular techniques. Muller himself argued that the basic unit of genetic improvement was the human sperm or egg cell, derived from a willing donor who possessed many desirable characteristics. In contrast, the microbiologists and molecular biologists who wrote on human genetic intervention in the 1960s and early 1970s extrapolated from laboratory research involving bacteria and bacterial and animal viruses. Some commentators, for example, J. Lederberg, were willing to consider targeted genetic changes in both germ-line and somatic cells, but the general trend of the discussion in the 1960s was clearly toward a focus on somatic cells and on the effort to alleviate diseases like hemophilia or diabetes. By the early 1970s the language of "genetic engineering" had been left behind by most biomedical scientists, and germ-line effects were taking second place to the conception of "gene therapy" as a new kind of drug, or biologic, or transplant. This paradigm shift occurred even before the exquisite techniques of recombinant DNA research were widely available.

The consensus in favor of the phrase "gene therapy" (with the word "human" sometimes included as a prefix) remained intact from 1972 through approximately 1998. Journals called *Human Gene Therapy* and *Gene Therapy* were inaugurated to track developments in the field, and the conceptual distinction between somatic-cell gene therapy and more ambitious programs of genetic intervention became firmly established among both academics and members of the general public. However, in a searching article published in 1998, Larry Churchill and his colleagues criticized the use of the phrase "gene therapy" for implying benefit to patients when, at least until that point, little therapeutic success had been achieved in gene therapy trials (40). Churchill and associates argued that the terms "gene therapy" and "gene therapy research" should be deleted from all federal documents describing this new technique and should be replaced by more neutral and modest terminology like "gene transfer research" (40). In their view, this alternative language "more accurately conveys the experimental practice that is currently at issue" (40, p. 45). In the remainder of this article the phrase "gene transfer" will be employed instead of "gene therapy" whenever possible.

MAJOR MODES OF HUMAN GENETIC INTERVENTION

The foregoing discussion has highlighted one of the central distinctions in scientific and ethical discussions of genetic intervention—the distinction between genetic alterations that affect the recipient of the gene transfer alone and alterations that will be passed on to the recipients' descendants through the germ line. Implicit in the early discussions and in the use of the word "therapy" was a second distinction, namely the distinction between the treatment or prevention of disease, on the one hand, and the enhancement of human capabilities or characteristics, on the other. The possible types of enhancements have been classically categorized by Muller and subsequent commentators as physical, intellectual, and moral enhancements (41,42). Thus one can conceive of two-by-two matrix (Table 1) that depicts these two distinctions (43,44).

More recent commentators, especially Matthew Bacchetta and Gerd Richter, have argued that this simple two-dimensional framework is inadequate to deal with another type of potential genetic intervention, namely a deliberate alteration of the mitochondrial DNA that is present in the cytoplasm of mammalian cells (45). In the

Table 1. Gene Therapy Techniques

	Somatic	Germ Line
Prevention, treatment, or cure of disease	1	1
Enhancement of capabilities or characteristics	3	4

Source: Ref. 44, p. xvii.

view of Bacchetta and Richter, the inclusion of genetic interventions to treat or prevent mitochondrial disease yields a three-dimensional matrix, involving the six possibilities shown in Figure 1 (45).

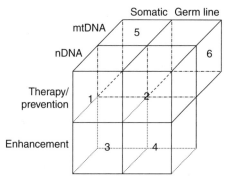

Figure 1. Dimensions of genetic interventions in the human genome: cube 1, somatic gene therapy within nDNA; cube 2, germ-line gene therapy within nDNA; cube 3, enhancement in somatic cells within nDNA; cube 4, enhancement in germ-line cells within nDNA; cube 5, somatic mitochondrial gene therapy; and cube 6, mitochondrial germ-line therapy. *Source*: Ref. 45, p. 456.

EARLY ATTEMPTS AT HUMAN GENE TRANSFER

In 1980 there was clearly an experimental attempt made to perform gene transfer in two subjects who were afflicted with beta-thalassemia. However, three leading accounts of the history of this topic also discuss an event that had occurred a decade earlier, the effort by Stanfield Rogers to infect three German sisters aged 5, 2, and a few months old with the Shope papilloma virus. Rogers, a biochemist and physician at Oak Ridge National Laboratory, hoped that the virus would carry a gene into the sisters' bodies that would counteract the effects of an inherited disease that resulted in toxic levels of an amino acid, arginine, in the children's livers (25,46,47). Rogers later acknowledged that the intervention did not help any of the three girls, although a delay in administration of the virus to the youngest of the three may have rendered the preparation inactive (48–50). The experiment was reported in the *New York Times* on September 21, 1970, by Harold Schmeck (51). At a May 1971 symposium (52) and in a 1972 article published in *Science* (53) Rogers was severely criticized for having undertaken this clinical trial with insufficient preclinical research.

A second early study of gene transfer bore an uncanny resemblance to Rogers's experiment. On May 30, 1979,

Martin Cline, a hematologist at the UCLA Medical Center, and three colleagues submitted a clinical protocol to the local institutional review board (IRB)—called the Human Subject Protection Committee (47). What Cline and his colleagues proposed to do was to perform gene transfer experiments in subjects who had hemoglobin disorders like sickle cell anemia or thalassemia. The precise strategy that the researchers planned to follow was to introduce functioning beta-globin genes into bone marrow cells that had been removed from the subject's body. Their hope was to be able to confer a selective advantage on the genetically modified cells so that, when the cells were reintroduced into the body of the subject, they would outgrow the native, deficient bone marrow cells. After modifying the protocol to exclude the use of recombinant DNA and after having waited for more than a year for approval by the local IRB, Cline took matters into his own hands. In the summer of 1980 he flew to Israel and Italy; in each country he performed a gene transfer experiment on a subject who was afflicted with β-thalassemia (47). Once again, the press broke the story about the experiment. On October 8, 1980, Paul Jacobs published a story in the *Los Angeles Times* entitled "Pioneer Genetic Implants Revealed" (54). In response, National Institutes of Health (NIH) Director Donald Fredrickson appointed an ad hoc committee to review Cline's action. After the committee's final report in May 1981, Cline was punished by both UCLA and NIH for having violated both federal regulations that protect human subjects and the NIH "Guidelines for Research with Recombinant DNA Molecules" (47).

DEVELOPMENT OF A REVIEW PROCESS FOR GENE TRANSFER RESEARCH IN THE UNITED STATES

The lively discussion of human genetic intervention that had occurred between 1962 and 1972 came to an almost-abrupt halt in 1972. During the remainder of the 1970s two other topics in genetics and molecular dominated ethical discussion — genetic testing and screening and recombinant DNA research (44). Laboratory research with recombinant DNA, in particular, was brought to the attention of the public by scientists concerned that their work could harm either laboratory workers or the public's health, and the Asilomar meeting held in February 1975 marks a decisive moment in the recombinant DNA debate. Out of the discussion process that led to Asilomar there emerged a public-oversight body for this important field of research, the NIH Recombinant DNA Advisory Committee (RAC). By the late 1970s the concerns of scientists, policy makers, and the public about the potential biohazards had been allayed by additional research and further discussion.

In 1980 public attention began to return to the topic of "genetic engineering," which had so abruptly disappeared from sight in the early 1970s. The resuscitation of the topic began in Europe, where in January of 1980 Mr. B. Elmquist, a Danish member of the Legal Affairs Committee for the Parliamentary Assembly of the Council of Europe, requested that the assembly pass a "recommendation on the protection of humanity against genetic engineering" (44, p. 146). Mr. Elmquist's request initiated a two-year discussion process that led to a parliamentary hearing in May 1981 and to Recommendation 934 (1982) in January 1982. This recommendation, framed in terms of the postwar human-rights tradition, asserted the right of every human being to "inherit a genetic pattern that has not been artificially changed" (55). On the other side of the Atlantic, a June 1980 letter from three religious leaders to President Carter asserted that "We are moving into a new era of fundamental danger triggered by the rapid growth of genetic engineering" (56, p. 95). In response, an already-existing presidential commission on bioethics decided to conduct a formal study of genetic engineering as applied to human beings. News of Cline's unauthorized experiments, which broke later in the same year, and two widely cited essays on "gene therapy" in the *New England Journal of Medicine* (57,58), also contributed to a renewed awareness of a topic that had been actively debated in the 1960s but almost forgotten during most of the 1970s.

The report of the presidential commission, entitled *Splicing Life*, and a congressional hearing on "Human Genetic Engineering" held in November 1982 led gradually but almost inexorably to the establishment of a public-oversight system for human gene transfer research in the United States (44). The NIH advisory committee that had nurtured recombinant DNA research through its early years, the Recombinant DNA Advisory Committee (RAC), now turned its attention to human gene transfer. Through an interdisciplinary working group that was later renamed the Human Gene Therapy Subcommittee, RAC devised guidelines called the "Points to Consider" for a new arena of human-subjects research. These guidelines were reactive in the sense that Martin Cline had already performed a human gene transfer experiment in 1980. However, they were also proactive: the "Points to Consider" were essentially complete in 1985, yet the first gene-marking proposal was not formally submitted to the RAC until 1988. Two years later the first gene-transfer experiments that aimed to be therapeutic were proposed and publicly debated (46,47).

EARLY YEARS OF PUBLICLY APPROVED GENE TRANSFER RESEARCH

On September 14, 1990, the era of publicly approved gene transfer research with a therapeutic aim began. NIH researchers W. French Anderson, R. Michael Blaese, Kenneth Culver, and their colleagues administered genetically modified T-lymphocytes to a girl named Ashanti DeSilva, who had just turned four. Ashanti was suffering from a rare genetic disease called severe combined immune deficiency, which was caused by a defect in a single gene, the gene that produces an enzyme called adenosine deaminase (ADA). A synthetic enzyme had initially helped strengthen Ashanti's weakened immune system, but the beneficial effects of the enzyme therapy had gradually diminished. The researchers, RAC members, members of the press covering the story, and the general public all hoped that this new mode of somatic-cell genetic intervention would prove to be beneficial (46,47).

During the next four-and-a-half years the field of human gene transfer research grew at a rather steady

Table 2. Number of Gene Transfer Protocols Submitted by Year, 1990 to 1995

1990	1991	1992	1993	1994	1995
2	9	23	31	33	44

pace in the United States and commenced in several additional countries. By early 1995 the number of gene therapy protocols approved by RAC stood at 82, and the number of gene-marking studies had reached 25 (59). The number of subjects involved in these 107 early trials was small: Only 597 participants took part in the studies. Of the early gene therapy protocols, all but three were Phase I studies, aimed primarily at determining whether the gene transfer procedure would be toxic to patients. Two studies were categorized as Phase I/II studies, and only one study, approved in March 1995, was a Phase II study—a preliminary evaluation of efficacy (59). The gradual growth of activity in human gene transfer research is indicated in Table 2.

In the early studies of gene transfer in the United States several types of diseases were targeted, and multiple vectors were employed. More than 60 percent of the gene therapy studies (51/82) sought to combat cancers of various kinds. Twenty studies were directed toward monogenic diseases, including cystic fibrosis (11), Gaucher disease (3), and severe combined immune deficiency (1). The remaining 11 protocols sought to combat HIV infection or AIDS (9), peripheral artery disease (1), and rheumatoid arthritis (1). The vast majority of the initial 107 gene transfer studies in the United States employed retroviral vectors (76/107). Adenovirus vectors began to be employed in 1993 and were used in 15 of the initial 107 studies. In 1994 the first adeno-associated viral vector was proposed and approved. Non-viral delivery vehicles included liposomes (12/107), plasmid DNA (2), and particle mediation (1) (59).

GENE-TRANSFER RESEARCH AND OVERSIGHT POLICIES IN OTHER NATIONS

For gene transfer research conducted outside the United States the earliest published information began to appear in the journal *Human Gene Therapy* in December of 1992. In the earliest annual summary of gene transfer research the following three non-U.S. protocols were listed (60):

Fudan University and Changhai Hospital, Shanghai, China: Hemophilia B

Centre Leon Berard, Lyon, France: cancer (a gene-marking study)

San Raffaele Scientific Institute, Milan, Italy: severe combined immune deficiency

By the end of 1993 two additional protocols had been initiated in the Netherlands (61). At the end of 1994, the number of non-U.S. gene transfer protocols had increased to 13. New countries added in the 1994 summary included the United Kingdom, Germany, and Sweden (62). In 1995, or approximately the same time as the U.S. audits discussed below, the number of non-U.S. protocols stood at 17, with Canada and Poland joining the list as new sites in 1995 (63). At mid-year in 1996, a more comprehensive report compiled by Tony Marcel and J. David Grausz cited the following additional countries in which gene transfer studies either were being planned or had already been undertaken: Switzerland, Egypt, Spain, Australia, Finland, Japan, and Israel (64).

Oversight bodies similar to RAC or FDA regulate the conduct of human gene transfer research in several European countries and in Japan. In the United Kingdom the Gene Therapy Advisory Committee (GTAC) reviews all gene transfer protocols on the basis of guidance that the committee has prepared for researchers (65). In 1996 France passed a special law for cell and gene therapies that requires these biologics to be reviewed by the French Medicines Agency (66). Japan's oversight system for human gene transfer includes two national review bodies, one of which meets publicly and the other of which convenes in private (66).

AUDITS OF 1995 AND THE CRISIS OF 1999 IN THE UNITED STATES

The June 1995 RAC report from which the information in the preceding section was derived reached three primary conclusions about the field of human gene transfer. First, gene-marking studies had advanced the science of bone marrow transplantation by allowing researchers to distinguish between cells that had been removed from a patient's body and purged and cells that had not been removed. Second, there was little evidence of toxicity in the early gene transfer studies. And third, "It is clearly too early ... to assess the therapeutic efficacy of gene therapy or even to predict its promise" (60, p. 1789).

In December 1995 a committee co-chaired by RAC member Arno Motulsky and Stuart Orkin delivered a more somber verdict to NIH Director Harold Varmus and the NIH Director's Advisory Committee. According to the Orkin-Motulsky committee,

> While the expectations and the promise of gene therapy are great, clinical efficacy has not been definitively established at this time in any gene therapy protocol, despite anecdotal claims of success and the initiation of more than 100 Recombinant DNA Advisory Committee (RAC) approved protocols (67).

The Orkin-Motulsky committee stopped short of recommending a moratorium on clinical trials of human gene transfer, but the committee clearly thought that the balance between preclinical and clinical research should be radically shifted.

In retrospect, one can discern a clear, though gradual downward trajectory in human gene transfer research in the U.S. from the end of 1995 through late 1999. There were several elements in this decline. The early hope for therapeutic success continued to be frustrated, and the 1995 verdicts of the RAC and the Orkin-Motulsky committee remained valid through 1996, 1997, and 1998. During this same period gene transfer research became much less visible in the United States because

of changes in the oversight system. The existing system of review and approval or disapproval of all human gene transfer protocols by both RAC and the Food and Drug Administration (FDA) gave way first to a system of more selective review and approval or disapproval by RAC, then to sole regulatory authority by FDA, with informal advice from RAC on protocols that raised novel issues. Researchers were asked to continue submitting reports of serious adverse events both to RAC and FDA, as well as to provide an annual report to the RAC staff on the progress of their research. However, the RAC staff lacked both the number of people and the database capabilities required to tabulate and analyze the reports that were submitted. There was, in addition, increasing evidence that not all of the required information was being provided to the RAC and its staff by researchers and companies. As a result of these multiple factors, no annual reports on progress (or the lack of progress) in human gene transfer research were produced by the RAC and its staff between 1995 and 1999.

The crisis of 1999 for human gene transfer research in the United States began when RAC noted at its September meeting that one researcher and one company had labeled adverse event reports as "proprietary information" and had asked RAC not to discuss the events publicly. Even as RAC responded by insisting that adverse event reports be a matter of public record, a second event occurred that would change the field of gene transfer research for the foreseeable future. On September 17, 1999, Jesse Gelsinger, an eighteen-year-old young man who had ornithine transcarbamylase (OTC) deficiency, died as a direct result of having received gene transfer with an adenoviral vector. The tragedy was compounded by the fact that Mr. Gelsinger's disease had been relatively well controlled through all of 1999 by a combination of drugs and diet (68,69). To their credit, the team of researchers conducting this study at the University of Pennsylvania promptly reported Mr. Gelsinger's death to both NIH and FDA.

The crisis of 1999 continued as FDA placed several gene transfer studies on clinical hold. In addition a vigorous attempt by the RAC staff and FDA to gather all serious adverse events that had occurred in human gene transfer trials, especially those using adenoviral vectors, revealed that less than 10 percent of these events had been reported to RAC and its staff in a timely fashion. At the December 1999 RAC meeting, a working group on adenoviral vectors criticized gene transfer researchers for their lack of standardization in calculating doses of vector and in assuring that the properties of the vector had not changed before it was administered to human subjects. At the same meeting FDA alleged that the University of Pennsylvania research group had committed several violations of FDA regulations in its conduct of the OTC deficiency gene transfer trial. These oral criticisms were followed in January 2000 by FDA's release of "inspectional observations" detailing FDA's charges. The Penn researchers responded to the charges in February, acknowledging some technical errors but denying that those errors were causally related to Mr. Gelsinger's death (70–73).

A February 2000 Senate hearing chaired by Senator William First explored the current oversight system for gene transfer research in the United States. At the hearing Mr. Gelsinger's father, Paul Gelsinger, asserted that he and his son had not been told important information from the preclinical studies that preceded the start of the human gene transfer study for OTC deficiency. More specifically, the deaths of monkeys in preclinical studies employing earlier generations of adenoviral vectors had not been disclosed. Mr. Gelsinger also reported that he and his son had been led to believe that the son's participation in the gene transfer study was likely to be clinically beneficial, even though the study was a Phase I trial. Other witnesses, including the present author, commented on weaknesses in the current oversight system for gene transfer research in the United States and on steps that are being undertaken by NIH and FDA in an effort to remedy those weaknesses.

PROSPECTS FOR THE FUTURE

As of late 1999, approximately 320 U.S. gene therapy (as distinct from gene-marking) trials had been registered with NIH and RAC. For studies being conducted in other countries, the most recent figures available suggest that at least 36 gene transfer trials have been initiated or completed in 11 countries (74). This latter number is almost surely a gross underestimate. Many companies seem to prefer reporting gene transfer trials only on a confidential basis to the regulatory agencies that oversee those trials. The earlier generalizations about the lack of demonstrated clinical efficacy, at least in published articles, continue to hold, although there are rumors of initial success with severe combined immune deficiency in several children and in hemophilia with adults. The 1995 conclusion about the lack of toxicity in gene transfer trials has now had to be reversed, in the light of Mr. Gelsinger's death and several other reported laboratory and clinical toxicities.

It is fair to say that in late 1999 and early 2000 the field of human gene transfer research was undergoing an agonizing reappraisal. Success in this use of these techniques may lie ahead, but it will not be as easy as all hoped and many believed in 1990, when the first approved study was initiated. The precise role for gene transfer, cell transfer (often called "cell therapy"), organ and tissue transplantation (including xenotransplantation), biologics, and drugs in the armamentarium of the future remains to be clarified. The future success of gene transfer in the treatment of disease cannot be guaranteed. What can definitely be achieved, however, is the creation of a transparent, accountable oversight system that assures that the human subjects who make this research possible will be dealt with honestly and with the highest measure of respect.

ACKNOWLEDGMENTS

I wish to thank my colleague, Linda A. Powell, for helpful editorial assistance. I am indebted to my colleague, Martina Darragh, for information on the first recorded usage of the phrase "gene therapy."

BIBLIOGRAPHY

1. H.V. Aposhian, *Perspect. Biol. Med.* **14**, 98–108 (1970).
2. J.V. Osterman, A. Waddell, and H.V. Aposhian, *Proc. Nat. Acad. Sci.* **67**, 37–40 (1970).
3. H.J. Muller and J. Lederberg, in G. Wolstenholme, ed., *Man and His Future*, J. & A. Churchill, London, 1963, pp. 255, 265.
4. S.E. Luria, E.L. Tatum, and H.J. Muller, in T.M. Sonnenborn, ed., *The Control of Human Heredity and Evolution*, Macmillan, New York, 1965, pp. 1, 9–11, 17, 29, 40, 42–43, 101, 104, 108–109, 119, 121, 123, 125–127.
5. E.L. Tatum and P. Ramsey, in J.D. Roslansky, ed., *Genetics and the Future of Man*, Appleton-Century-Crofts, New York, 1966, pp. 53–54, 57–58, 118–119, 162–153, 155.
6. R.D. Hotchkiss, *J. Hered.* **56**, 197–202 (1965).
7. J. Lederberg, *Bull. Atomic Scientists* **22**, 4–11 (1966).
8. E.L. Tatum, *Perspect. Biol. Med.* **10**, 19–32 (1966).
9. M.W. Nirenberg, *Science* **157**, 633 (1967).
10. J. Lederberg, *Science* **158**, 313 (1967).
11. R.L. Sinsheimer, *Am. Sci.* **57**, 134–142 (1969).
12. B.D. Davis, in P.N. Williams, ed., *Ethical Issues in Biology and Medicine*, Schenkman Publishing, Cambridge, MA, 1973, pp. 17–32.
13. B.D. Davis, *Science* **170**, 1279–1283 (1970).
14. J.A. Wolff and J. Lederberg, *Hum. Gene Ther.* **5**, 469–480 (1994).
15. H.D. Goodale, in D.F. Jones, ed., *Proc. Sixth International Congress of Genetics*, vol. 2, Brooklyn Botanic Gardens, Ithaca, NY, 1932, pp. 65–66.
16. H.J. Muller, in G. Wolstenholme, ed., *Man and His Future*, J. & A. Churchill, London, 1963, p. 255.
17. H.J. Muller, *Perspect. Biol. Med.* **3**, 1–43 (1959).
18. J. Lederberg, in G. Wolstenholme, ed., *Man and His Future*, J. & A. Churchill, London, 1963, p. 265.
19. S.E. Luria, in T.M. Sonnenborn, ed., *The Control of Human Heredity and Evolution*, Macmillan, New York, 1965, p. 10.
20. E.L. Tatum, in T.M. Sonnenborn, ed., *The Control of Human Heredity and Evolutions*, Macmillan, New York, 1965, p. 22.
21. J.D. Roslansky, ed., *Genetics and the Future of Man*, Appleton-Century-Crofts, New York, 1966.
22. H.J. Muller, in T.M. Sonnenborn, ed., *The Control of Human Heredity and Evolution*, Macmillan, New York, 1965.
23. P. Ramsey, in J.D. Roslansky, ed., *Genetics and the Future of Man*, Appleton-Century-Crofts, New York, 1966, p. 118. Reprinted in P. Ramsey, *Fabricated Man: The Ethics of Genetic Control*, Yale University Press, New Haven, CT, 1970, pp. 9–10.
24. E.L. Tatum, in J.D. Roslansky, ed., *Genetics and the Future of Man*, Appleton-Century-Crofts, New York, p. 58.
25. J.A. Wolff and J. Lederberg, *Hum. Gene Ther.* **5**, 471 (1994).
26. J. Lederberg, in *Proc. Sixth World Congress on Fertility and Sterility*, May 20–27, 1968, Israel Academy of Sciences and Humanities, Jerusalem, 1970, pp. 18–23.
27. R.L. Sinsheimer, *Am. Scientist* **57**, 134–142 (1969).
28. B.D. Davis, *Science* **170**, 1279–1283 (1970).
29. J. Lederberg, *N.Y. Times*, September 26, 28 (1970).
30. H.V. Aposhian, *Perspect. Biol. Med.* **14**, 106–107 (1970).

31. E. Freese, ed., *The Prospects of Gene Therapy*, Fogarty International Center, National Institutes of Health, Bethesda, MD, 1972.

32. M.P. Hamilton, ed., *The New Genetics and the Future of Man*, Eerdmans, Grand Rapids, MI, 1972, pp. 107–175.

33. S. Rogers, *Res. Commun. Chem. Pathol. Pharmacol.* **2**, 587–600 (1971).

34. J.V. Osterman et al., *An. N.Y. Acad. Sci.* **179**, 514–519 (1971).

35. P.K. Qasba et al., *Proc. Nat. Acad. Sci.* **68**, 2345–2349 (1971).

36. T. Friedmann and R. Roblin, *Science* **175**, 949–955 (1972).

37. E. Freese, *Science* **175**, 1024–1025 (1972).

38. H.V. Aposhian et al., *Fed. Proc.* **31**, 1310–1314 (1972).

39. T. Friedmann and R. Roblin, *Science* **175**, 949–955 (1972).

40. L.R. Churchill et al., *J. Law, Med. Ethics* **26**, 38–47 (1998).

41. *Nature* **144**, 521–522 (1939).

42. H.J. Muller, *Perspect. Biol. Med.* **3**, 22, 37–38 (1959).

43. L. Walters, *Nature* **320**, 225–227 (1986).

44. L. Walters and J.G. Palmer, *The Ethics of Human Gene Therapy*, Oxford University Press, New York, 1997.

45. M.D. Bacchetta and G. Richter, *Camb. Q. Healthcare Ethics* **5**, 450–457 (1996).

46. J. Lyon and P. Gorner, *Altered Fates: Gene Therapy and the Retooling of Human Life*, W.W. Norton, New York, 1995.

47. L. Thompson, *Correcting the Code: Inventing the Genetic Cure for the Human Body*, Simon & Schuster, New York, 1994.

48. S. Rogers, in M. Lappé and R.S. Morrison, eds., *An. N.Y. Acad. Sci.* **265**, 66–71 (1976).

49. S. Rogers, in S. de Grouchy, ed., *Proc. 4th Int. Congr. Hum. Genet.* (Paris, September 6–11, 1971), Excerpta Medica, Amsterdam, 1971, pp. 36–40.

50. H.G. Terheggen et al., *Z. Kinderheilkunde* **119**, 1–3 (1975).

51. H.M. Schmeck, *N.Y. Times*, September 21, 28 (1970).

52. P. Ramsey, in M.P. Hamilton, ed., *The New Genetics and the Future of Man*, Eerdmans, Grand Rapids, MI, 1972, pp. 162–163.

53. T. Friedmann and R. Roblin, *Science* **175**, 949–955 (1972).

54. *L.A. Times*, October 8, 1, 26 (1980).

55. Council of Europe, Parliamentary Assembly, *Texts Adopted by the Assembly*, 33rd Ordinary Session, Third Part, January 25–29 the Council, Strasbourg (1982).

56. United States, President's Commission for the Study of Ethical Problems in Medicine and Biomedical and Behavioral Research, *Splicing Life*, U.S. Government Printing Office, Washington, DC, November 1982.

57. W.F. Anderson and J.C. Fletcher, *N. Engl. J. Med.* **303**, 1293–1297 (1980).

58. K.E. Mercola and M.J. Cline, *N. Engl. J. Med.* **303**, 1297–1300 (1980).

59. G. Ross et al., *Hum. Gene Ther.* **7**, 1781–1790 (1996).

60. *Hum. Gene Ther.* **3**, 729 (1992).

61. *Hum. Gene Ther.* **4**, 856 (1993).

62. *Hum. Gene Ther.* **5**, 1550–1551 (1994).

63. *Hum. Gene Ther.* **6**, 1676–1678 (1995).

64. T. Marcel and J.D. Grausz, *Hum. Gene Ther.* **7**, 2025–2046 (1996).

65. Gene Therapy Advisory Committee, U.K. Department of Health, *Gene Ther.* **1**, 1–20 (1994).

66. O. Cohen-Haguenauer, *Current Opinion Biotechnol.* **8**, 361–369 (1997).

67. Stuart H. Orkin and Arno G. Motulsky, Co-Chairs, "Report and Recommendations of the Panel to Assess the NIH Investment in Research on Gene Therapy," December 7, 1995. Unpublished report available from the Office of the Director, National Institutes of Health, Bethesda, MD.

68. R. Weiss and D. Nelson, *Washington Post*, September 29, A1, A21 (1999).

69. N. Wade, *N.Y. Times*, September 29, A20 (1999).

70. R. Weiss and D. Nelson, *Washington Post*, January 22, A1, A12 (2000).

71. S.G. Stolberg, *N.Y. Times*, January 22, A1, A10 (2000).

72. R. Weiss and D. Nelson, *Washington Post*, February 15, A3 (2000).

73. S.G. Stolberg, *N.Y. Times*, February 15, A16 (2000).

74. *Hum. Gene Ther.* **10**, 3119–3123 (1999).

See other GENE THERAPY entries.

GENETIC COUNSELING

ANNE L. MATTHEWS
Case Western Reserve University
Cleveland, Ohio

OUTLINE

Introduction
  History
The Genetic Counseling Process
  Data Gathering
  Counseling
  Follow-up
Genetic Counseling Settings
  Prenatal Setting
  General Genetics Setting
  Specialty and Management Clinic Settings
Genetic Counselors
  Genetic Counseling Training Programs
  Curriculum
  National Society of Genetic Counselors
  Certification and Accreditation
  Employment
  Issues in Genetic Counseling
Other Health Professionals
Summary
Bibliography

INTRODUCTION

Genetic counseling is relatively new as a recognized health profession. The need for genetic counseling services as a discipline unto itself arose out of technological advances in human genetics and their applications to clinical medicine and health care. The profession's continued growth and maturity is a reflection of the continuing explosion of new knowledge in the discipline of human and medical genetics and of its impact on health care.

History

The concept of genetic counseling can be considered eons old. Genetic advice has been provided to families by their relatives, friends, and neighbors since humans began communicating with one another. The Bible, other religious writings, and civil laws have provided families for centuries with rules and laws regulating reproductive behavior based on heredity. However, the term "genetic counseling" as currently utilized is attributed to Sheldon Reed, a Ph.D. geneticist, who published his philosophy of genetic counseling in his text entitled *Counseling In Medical Genetics* (1,2). Reed defined the concept of genetic counseling as "a kind of social work done for the benefit of the whole family entirely without eugenic connotations" (3, p. 335). He stated that the primary function of counseling was to provide people with an understanding of the genetic problems they had in their families. For Reed, in order for one to provide genetic counseling the individual had to have some knowledge of human genetics, have a "deep respect for the sensitivities, attitudes, and reactions of the client," and have a "desire to teach, and to teach the truth to the full extent that was known" (1, pp. 11–12).

Providers of genetic information during the first 40 years of the twentieth century were primarily doctorally trained geneticists in academic institutions. Their major interests focused on laboratory approaches and population genetics to answer questions regarding evolution and how to decrease the presence of disease genes in the gene pool. When consulted, genetics experts provided information dealing with the genetic contribution of heredity and the prevention of genetic diseases and birth defects. Their counseling strategy was one of advice and recommendations. This fit well with the basic tenets of eugenics that were so prevalent in academic and scientific circles during this time period (4). It was not until the later part of the 1940s and 1950s that medicine began to demonstrate an interest in human genetics. Clinics dealing with genetic questions began to appear, two of the first were the Hereditary Clinic at the University of Michigan and the Dight Institute of Human Genetics at the University of Minnesota (5). However, much of this information continued to be presented with the overall intent of decreasing the incidence of genetic disease and birth defects (6).

By the 1960s technological advances in a number of areas — including cytogenetics, biochemical, and molecular genetics, population screening of newborns for phenylketonuira, and the technological advances making prenatal diagnosis a reality (7) — had a tremendous impact on physicians' interests in the field of human genetics (5). Medical genetics was becoming a recognized area of practice as physicians began to replace the laboratory-based geneticist in providing genetic information to patients and families. And while the emphasis of counseling remained preventative in nature, by assisting families in a process of making informed, rational decisions regarding reproductive choices, the need to respond to the psychosocial aspects of genetic disorders was beginning to emerge as an appropriate goal of genetic counseling (8).

In the 1970s the movement in medicine and health care as a whole shifted toward greater patient autonomy (9). In the genetic counseling arena, effectiveness of counseling was no longer solely based on whether or not reproductive plans were altered or information and recurrence risks recollected. It was also reflected by the importance of patients and families making reproductive decisions based on personal values and family circumstances. This philosophical shift, from prevention of genetic disease to concern for the patient's or counselee's total well-being, brought medical geneticists closer to Reed's definition of genetic counseling. Moreover this shift was coincident with the initiation of the first master's level genetic counseling program, a team approach to providing genetic services, and the general public's request for access to such services. This philosophy was also legitimized when the Ad hoc Committee on Genetic Counseling of the American Society of Human Genetics published their definition of genetic counseling that acknowledged the importance of the psychological dimensions of the counseling process and the role of master's level trained personnel who were advocates of this approach (10). The definition that was adopted by the American Society of Human Genetics in 1975 continues to be accepted as the definitive definition of genetic counseling:

> Genetic counseling is a communication process which deals with the human problems associated with the occurrence, or the risk of occurrence, of a genetic disorder in a family. This process involves an attempt by one or more appropriately trained persons to help the individual or family (*1*) comprehend the medical facts, including the diagnosis, the probable course of the disorder, and the available management; (*2*) appreciate the way heredity contributes to the disorder, and the risk of recurrence in specified relatives; (*3*) understand the options for dealing with the risk of recurrence; (*4*) choose the course of action which seems appropriate to them in view of their risk and the family goals and act in accordance with that decision; and (*5*) make the best possible adjustment to the disorder in an affected family member and/or to the risk of recurrence of that disorder (10, p. 240).

THE GENETIC COUNSELING PROCESS

To understand the role of the genetics team, and more specifically genetic counselors, it is helpful to understand what the process of genetic counseling encompasses. In simple terms, genetic counseling can be thought of as an information exchange provided in a team approach milieu (11). Based on information gleamed from the patient or counselee (terms used interchangeably) and working in concert with team members, the genetic counselor provides information and psychological support regarding specific concerns or questions of the family. The process often begins with the identification of individuals within the family affected with birth defects or genetic disorders or with the identification of those who may be at increased risk for a variety of inherited conditions. The process consists of a number of different steps, often posed as questions by families. The first such question is usually: "What is it?" or "Am I at increased risk because of my genetic make-up?" The next step in the process is to attempt to answer the questions of "What caused it?" and "What can be done about it?" Ideally this aspect of the process involves establishing an accurate medical diagnosis, which in turn forms the basis for genetic counseling. While the genetic evaluation consists of many of the same components found in any medical evaluation, the emphasis is often quite different. In particular, medical information regarding extended family members is often a prerequisite to providing accurate information and genetic counseling. After the establishment of a specific diagnosis, a discussion of the prognosis, treatment, and management options based on the most current information available can ensue. Counseling next addresses the question of "Will it happen again?" Recurrence risks can be discussed and families provided with a variety of options regarding future reproductive choices.

Equally important, the genetic counseling process assists individuals and families in coping with the emotional burdens and adjustments required where the person or a family member is at risk for or affected with a genetic condition.

Data Gathering

In a genetics evaluation the information obtained from families provides important and often defining data needed to reach a specific diagnosis. Major categories of information elicited include prenatal, perinatal, medical, developmental, family, and psychosocial histories. The prenatal and perinatal histories provide an overview of fetal and newborn well-being. Documentation of such information as a potential teratogenic exposure, maternal disease and acute illness, or fetal growth and behavior, such as reflected in fetal movement, may provide important clues to identifying a specific diagnosis and etiology. These specific histories are often helpful in differentiating between prenatal etiologies of abnormalities and those resulting from birth injury (12).

Medical and developmental histories provide the genetic counselor with information regarding the natural history of the disorder in the affected individual. They also provide important information regarding variability, namely how a specific etiology affects what is seen in a specific patient (the phenotype). The medical and developmental histories can also provide the genetics team with direction in delineating a list of differential diagnoses that could help establish a specific diagnosis.

Obtaining a reliable family history can be extremely helpful in clarifying an etiology, diagnosis, and/or risk of recurrence in a family. The genetic counselor constructs a detailed and extensive pedigree that outlines in pictorial format three to four generations in a family. The genetic counselor obtains specific information about family members based on his/her knowledge of the variability of birth defects and the often multiple expressions of a genetic disorder. Families may not always be aware of how variable the expression of a genetic disorder can be, and thus may not be aware that other family members are also affected. For example, a child with a cleft palate, a heart defect, and learning disabilities is diagnosed with velocardiofacial syndrome (a genetic disorder that can have numerous effects including abnormalities of the palate, heart, and other organ systems as well as a characteristic facial appearance) (13). By obtaining a detailed family history and asking questions that would guide the family to describe specific findings, two other family members are identified (and the diagnosis is confirmed on physical examination and/or genetic testing as also having features of velocardiofacial syndrome). Based on this information, the genetic counselor can provide the family with specific information regarding this diagnosis including prognosis, treatment and management options, and specific recurrence risks to future offspring or other family members.

The physical examination part of the genetic evaluation differs somewhat from a routine medical examination. Performed by the physician geneticist, it is aimed not only at detecting major and minor malformations (dysmorphic features) but also at describing a pattern or constellation of findings that may provide clues in determining a diagnosis (14). For example, a congenital heart defect may be an isolated finding. However, if other features are noted such as poor muscle tone, upslanting eyes, and single palmer creases (simian creases), one might suspect a diagnosis of Down syndrome. Moreover such detailed descriptions are helpful in determining whether or not a physical finding is of significance or represents a normal variation or familial finding. Laboratory data collection is another important aspect of the data gathering. If a genetic disorder is suspected, diagnostic studies may be useful in delineating a specific diagnosis. Such laboratory tests as chromosome analysis, molecular DNA or biochemical studies, radiographs, or organ imaging may be appropriate. Additionally the patient may need to be referred to other specialties such as neurology or ophthalmology or for developmental studies.

Lastly, families may be asked for permission to take photographs. In many cases the adage "A picture is worth a thousand words" is well deserved, as photographs provide accurate descriptions of what has been noted on the physical examination. This also allows the genetic counselor to share the patient's findings more accurately with other clinicians in order to get assistance in reaching a specific diagnosis.

Counseling

Once all of the information has been gathered, and the genetics team has had the opportunity to consider possible diagnoses and review pertinent literature and databases, discussion with the client and/or family can begin. If a diagnosis has been identified, the geneticist or genetic counselor discusses the etiology and its genetic basis (what caused it), the medical and developmental implications of the diagnosis (what is it); the prognosis, treatment, and/or management options (what can be done about it); as well as the recurrence risks and availability of prenatal testing (will it happen again). The counselor also identifies and discusses the psychosocial impact of the disorder on the affected individual as well as on other family members, what needs the family may have, and what resources and support groups may be available to help the family or patient to take the next step forward in the process of learning and coping with new information.

If more testing is needed before a specific diagnosis can be made, the team provides the family with their impressions and/or an explanation of the diagnoses being considered. The genetic counselor discusses the specific testing or other specialty evaluations that the genetics team has recommended to confirm or rule out a diagnosis(es) and provide the family with information regarding costs, referrals, insurance issues, and possible date(s) when results may be expected. The team then discusses any interventions that may be useful such as a referral for physical and occupational therapy. And finally, the counselor sets up a plan with the family regarding how best to relay results of recommended tests as well as how best to provide follow-up information to the family and their health care providers.

Finally, if no diagnosis can be made, the ensuing discussion with the family includes the level of suspicion that the condition is genetic and the possibility of further evaluation and new testing in the future. The counselor provides the family with a range of recurrence risks depending on the possible genetic etiologies for similar findings, and with options for monitoring future pregnancies, if appropriate. Discussion also includes any available options regarding management or therapy for symptoms. When no specific diagnosis can be reached, a frustrating experience for both counselor and counselee, the emphasis shifts to helping the family address the impact of the lack of a diagnosis with supportive follow-up counseling for the family as needed.

Follow-up

Perhaps one of the most important, although often unacknowledged, aspects of the genetic counseling process is that of follow-up. As mentioned above, a major aspect of the genetic counseling process is to help the patient and/or family incorporate the information component of the session with the psychosocial, emotional impact of that information. Kessler notes that the "psychological responses of counselees are not only a normal, but often a necessary step in comprehending, integrating and coping with a medical diagnosis or the content material of genetic counseling" (15, p. 19). Following any counseling session, a written summary of the information provided and recommendations made is sent to the family. This written summary provides the family with an opportunity to review and refresh their memories of all the information that they received, as well as allowing them to share the information with other family members if they so desire. Specific recommendations can be reviewed at a later date, or the importance of additional counseling can be reemphasized if reproductive plans or circumstances change. The summary may also identify any misinformation or misconceptions that need to be addressed (9).

The genetic counselor will often follow up with phone contact to answer remaining questions or clarify issues for the family. If further testing or evaluations are pending, the counselor will arrange for the family to return to the clinic to discuss any additional information that has been obtained. These contacts also allow the counselor some ongoing assessment of how the family is assimilating information and coping and adjusting to the condition. In many situations the counselor is able to provide the family with anticipatory guidance and support regarding a number of different issues. However, when situations arise that are beyond their therapeutic expertise, the counselor will refer the client or family to the appropriate health care provider. The counselor may also help families to identify support groups or other resources, such as financial or educational resources, that would be appropriate for their particular situation. In many cases the genetic counselor may have an ongoing relationship with the client and family as new information or family circumstances arise.

GENETIC COUNSELING SETTINGS

Genetic counseling services are provided in a number of different settings. These include prenatal diagnostic clinics, diagnostic or general genetics clinics, and specialty and management clinics. Genetic services may also be available as part of a consultative service in some hospitals. In each setting, while specific aspects may vary and different components of the genetic counseling process may be emphasized, the overall approach and goals of the genetic counseling encounter remain constant.

Prenatal Setting

In the prenatal setting, pregnant women are usually referred for genetic counseling services based on one of two scenarios: the ideal situation when counseling occurs prior to prenatal screening or procedures, and the more common scenario when an abnormality is noted on ultrasound or maternal serum screening results change a woman's risk for abnormalities. In either scenario, genetic counseling can assist pregnant couples by providing a forum for discussion and detailed information on which informed decisions about further procedures can be made.

In the ideal scenario, one in which a referral is made prior to screening or testing because of a known risk factor (i.e., pregnant woman over the age of 35 or a positive history of a genetic disorder), there is usually a two-step process involving a counseling session followed by the actual prenatal testing procedure. Patients meet with a genetic counselor who obtains information regarding their indication for referral, their pregnancy, and their extended family history in the form of a pedigree. The counselor reviews the couple's risk for having a fetus affected with a birth defect or genetic disorder based on all of this information. At this juncture the couple may decide to have further diagnostic testing initiated or may decide not to pursue further evaluation. It is incumbent on the genetic counselor to provide the appropriate support in this decision-making process in a nondirective (i.e., noncoercive) manner (16) so that couples feel comfortable with the process and counseling does not become a mandate for subsequent procedures. Walker notes that "there is still a widespread perception that the sole purpose of prenatal diagnosis is to identify anomalous fetuses so that their birth can be prevented by pregnancy termination" (9, p. 604). There are a number of good reasons beyond termination why prenatal testing may be appropriate for couples, including reassurance about the well-being of the fetus, anticipatory management based on a fetal diagnosis such as where or how the delivery should occur (in a tertiary hospital setting vs. local facility), the opportunity to plan for treatment in the neonatal setting, and, equally important to many patients, the opportunity to adjust, plan ahead, and gather support and resources prior to the baby's birth (17). If the patient elects to have one of the diagnostic procedures, such as chorionic villus sampling, amniocentesis, or chordocentesis, the procedure is reviewed and discussed in detail. Associated risks of the procedure and follow-up are also described. After the test results are received, which are usually available within 10 days to 3 weeks depending on what studies are being conducted, the referring physician is notified. If the results are abnormal, the patient often returns to the genetic counselor to discuss the test results, what the meaning of such results may have for the present pregnancy, and what options may be available to the family such as termination or continuation of pregnancy and management at delivery. Whatever the decision, the counselor continually acknowledges the impact of such findings and subsequent actions on the patient and the family. Continued acknowledgment and validation regarding couples' decisions should remain a primary concern for the counselor (17). Even if the pregnancy is terminated after a diagnosis of a disorder in the fetus, the genetic counselor provides continued support and referral resources.

The more common prenatal diagnosis scenario occurs when the patient or couple is referred for genetic counseling following ultrasound of the pregnancy where an abnormality is noted or the maternal serum screening test is abnormal. In this all too common situation, the genetic counselor must provide information, counseling, and support in a milieu of anxiety and fear to a couple or patient that did not previously know of any increased risk and is now facing unanticipated decisions (9). The counselor will need to focus the discussion on what is known and not known and what other procedures may be helpful in better delineating the abnormality or problem. This process is conducted within a time-limited framework, where decisions regarding possible options available to the couple must be made quickly amid confusion, stress, and high anxiety.

General Genetics Setting

The general genetics clinic, once thought of as the realm of pediatrics, now often includes a growing number of adult patients as more knowledge is gained regarding late-onset genetic disorders or as affected individuals live longer with excellent medical care. In general, the initial visit to the genetics clinic is preceded by some form of intake process, during which the genetic counselor or clinic staff contacts the family to briefly discuss their concerns, gain some preliminary information, and provide the family with some overview of what they can expect during the visit. Additionally medical records and past testing results are requested prior to the visit.

At the appointment, families are again asked about their understanding of why they were referred, what their concerns are, or what they hope to gain from the visit. Previously obtained information is confirmed and additional prenatal, perinatal, developmental, medical, and psychosocial histories are elicited as needed. A detailed four-generation pedigree is also constructed. If appropriate, the medical geneticist then performs a physical examination, paying particular attention to minor findings and variations. At this point, testing and other evaluations may be ordered or recommended to confirm or rule out a suspected diagnosis. If a specific diagnosis is made, the genetics team provides information regarding the diagnosis, prognosis, treatment, and management, risk of recurrence, and available reproductive options or testing. Available resources and referrals to support groups may be made. Other family members at risk for developing the disorder or for having an affected offspring are identified. It may also be suggested that other family members may wish to seek genetic counseling to discuss the same or similar issues. In most clinic settings, a written summary of the counseling session is then sent to the family and health care providers for their records.

Sometimes, despite the best of efforts and for a number of different reasons, no diagnosis can be established or test results are inconclusive. If the patient is a child, a family may be asked to return to the clinic at some point in the future as, not rarely, a child will "grow into" a diagnosis (12). This also allows the genetics team to observe growth and development over time, since they may provide clues to a diagnosis. In any case, the only options available to the genetic counselor may be reviewing the information that is available regarding possible categories of diagnosis, recurrence risks, and future options. This unfortunate scenario is as frustrating for the genetic counselor as it is for the families. Nevertheless, the genetic counselor will help the family address the issue and impact of the lack of a diagnosis and provide support and follow-up for the family as needed.

Specialty and Management Clinic Settings

Specialty and management clinics, in which the medical geneticist and genetic counselor are integral members of a multidisciplinary team, are designed to assist families and affected individuals deal with a myriad of problems associated with a specific disorder or constellation of findings that can arise over a lifetime. Such specialty or management clinics may deal with a specific genetic disorder such as hemophilia or Marfan syndrome, or with a grouping of disorders or diagnoses such as craniofacial clinic or neuromuscular clinic. The clinic team works closely with the families and their primary care professionals to provide specialized health care services and treatments as well as address such issues as financial concerns and special needs such as equipment. The genetic counselor can provide information and counseling to parents of affected children regarding the genetics of the diagnosis as well as discuss recurrence risks and future reproductive options. These same issues can be discussed with affected adults and their spouses and can continue into the next generation with the children born to these families.

Two of the fastest growing areas of specialty clinics are the hereditary cancer clinics and those that deal with late-onset disorders, such as Huntington disease or Alzheimer's disease. These clinics have seen tremendous growth in the last few years due to the technological advances made in human genetics and the discovery of specific genes responsible for some inherited cancers and certain neurological disorders. In both cancer and late-onset disorders, families are actively seeking information regarding susceptibility or presymptomatic testing (18). In the case of cancer, genetic testing for specific inherited cancers has sparked great interest from individuals and families although only 15 percent of cancer cases are known to have an inherited susceptibility gene (19). For many patients, test results may influence medical management. In late-onset disorders such as Huntington disease, presymptomatic testing (i.e., the ability to identify the gene causing the disorder years before any symptoms appear) has led people to seek testing in order to provide themselves with information on which to make choices about career paths, lifestyle changes, or reproductive decisions. These genetic counseling sessions may be extremely complex as they need to cover a wide range of concerns, not only about the disorder in question, but about the impact of testing and test results on the patient and family, including such issues as confidentiality, discrimination, and possible loss of life or health insurance (20).

In summary, genetic counseling is a complex and multifaceted process that draws on a field of professional expertise that involves diagnosis, provision of information, and counseling of individuals about their genetic makeup and their chances of being affected with a genetic disorder or having offspring with birth defects or genetic disorders. It is usually best accomplished as a team approach by genetics professionals including genetic counselors, medical geneticists, Ph.D. trained geneticists, and other specialized health care professionals such as genetic nurse specialists and social workers.

GENETIC COUNSELORS

Having defined the specific roles a genetic counselor carries out, we turn our attention to the process of acquisition and official recognition of these skills, given the genetic counselor's unique relationship between the counselee and their family. Genetic counselors are health professionals with specialized graduate training and experience in the areas of medical genetics and counseling (21). They work as members of the health care team, providing information and support to families with or at risk for genetic disorders and birth defects. In addition to providing expertise regarding genetic disorders, genetic counselors also provide supportive counseling, serve as patient advocates, and act as educational and resource professionals for their patients, other health care providers, and the general public. Many counselors are also actively engaged in research activities related to the field of medical genetics and genetic counseling. Bartells and colleagues state that "genetic counselors work at the intersection between the information produced by scientists and the hopes, dreams,

and fears of clients whose lives could dramatically change as a result of receiving that information" (22, p. ix).

Genetic Counseling Training Programs

Sarah Lawrence College developed the first master's level genetic counseling program in 1969. Program developers, drawing on the expertise of several well-known geneticists, developed a genetic counseling program that balanced theoretical coursework in human and medical genetics with counseling theory and clinical experience in genetics centers providing genetic counseling services (23). This program based its counseling curriculum on a client-centered approach in which students learned interviewing skills, the concept of nondirective counseling, and "employing empathic responses against the background of unconditional positive regard" (23, p. 20).

Curriculum

In continuing efforts to define the role of the genetic counselor as a major provider of genetic services and to standardize the minimum educational requirements for a genetic counselor, directors of genetic counseling programs and other genetics professionals met three times between the years of 1974 and 1979 to address standards of genetic counseling education. The conclusions and recommendations were outlined by Dumars and colleagues in 1979 in *Genetic Associates: Their Training, Role and Function* (24). It was the consensus of the program directors that genetic counseling programs needed to maintain a dialogue among themselves and with other genetics professionals to assure the quality and effectiveness of the professionals being trained (25). Currently, there are 26 genetic counseling programs recognized by the American Board of Genetic Counseling, three international programs, and three graduate nursing programs that have a clinical nurse specialist tract in genetic counseling (26).

Most genetic counseling programs are two year curricula that include theoretical coursework, laboratory exposure, research experience, and extensive clinical training. The curricula have expanded considerably in both depth and complexity from the initial recommendations outlined in 1979. Coursework usually includes human and medical genetics, cytogenetics, biochemical genetics and molecular genetics as well as principles of quantitative and population genetics. Most programs balance the basic science and medical genetics courses with courses that stress the psychosocial aspects of genetic counseling, including principles of genetic counseling and the counseling process, interviewing techniques, ethical, legal, social, and ethnocultural issues pertaining to genetic evaluation, screening, testing, and counseling. Clinical training covers all areas provided by genetic services. Students usually receive extensive exposure in prenatal diagnosis, general genetics and metabolic clinics including both children and adults, specialty clinics such as cystic fibrosis or craniofacial clinic, and such clinics as hereditary cancer and neurodegenerative disorders. Students are expected to take an active role in most counseling sessions under the supervision of a certified genetic counselor and/or medical geneticist in order for the counseling session to be used toward formulating a logbook of cases (27). Today's genetic counseling student may have well over 1000 hours of clinical practicum (28), far greater than the 1979 "optimum curricula" recommendations that suggested 400 hours of supervised clinical placement in a minimum of two settings (25).

National Society of Genetic Counselors

The genetic counseling profession continued to establish its presence as a knowledgeable member of the genetics health care team with the formation and incorporation of its professional society, the National Society of Genetic Counselors (NSGC) in 1979 (21). NSGC instilled the sense of professionalism needed by this emerging group of health care professionals. Today NSGC has over 1700 members. Its mission is promoting "the genetic counseling profession as a recognized and integral part of health care delivery, education and public policy" (21). NSGC promotes the professional interests of genetic counselors and provides a network for professional communications, local and national continuing education opportunities, and the discussion of issues relevant to human genetics and the genetic counseling profession (21). In 1991 NSGC adopted and then published its Code of Ethics (29). This publication provided NSGC with a framework within which its members could function and practice. Moreover as NSGC continues to grow, its presence and influence are becoming more widespread regarding programs and policy in medical genetics and genetic counseling. NSGC has representatives on a number of policy-making groups and works closely with such groups as the American Society of Human Genetics, the American College of Medical Genetics, the American Board of Genetic Counseling, the American Board of Medical Genetics, the International Society of Nurses in Genetics, and the National Coalition for Health Professionals Education in Genetics.

Certification and Accreditation

Another important milestone in the professional development of genetic counselors occurred with the initiation of a certification examination process in 1981 by the American Board of Medical Genetics (25). Initially physicians, Ph.D. trained geneticists, and genetic counselors who met specific credential requirements could apply to sit for the certification examinations within five categories. Certification, or being eligible to sit for certification examinations, gave employers some degree of confidence that health professionals hired to provide genetic services possess a minimum level of knowledge in human and medical genetics and genetic counseling. Most employers require that genetic counselors be certified or eligible to sit for the certification examinations as prerequisites for employment.

In 1990 the genetic counseling subspecialty section of the certification examinations came under the auspices of its own board, the American Board of Genetic Counseling (ABGC). Current practice requires that applicants applying to sit for certification examinations have graduated from an accredited genetic counseling program. The applicant must also demonstrate that they have acquired a wide range of clinical experiences

documented by a logbook of cases supervised and signed by a board certified genetic counselor or medical geneticist. Currently over 1300 genetic counselors have achieved board certification.

The most recent achievement in the professional evolution and development of the genetic counseling field has been the establishment of an accreditation process for master's programs in genetic counseling. Following a consensus meeting that included genetic counseling program directors, members of ABGC and experts in a number of different fields, the ABGC published its criteria for accreditation including the practice-based competencies as defined by the consensus meeting (30). Programs must demonstrate adequate institutional support and facilities, adequate leadership and management, including board certified genetics professionals, and a curriculum that provides educational experiences, including theoretical courses, clinical training, and supplementary activities that would provide graduates with the necessary knowledge and skills to perform, accurately and reliably, the functions of a genetic counselor (27).

Continuing Education. With the explosion of new information in human genetics, genetic counselors are finding it necessary to continue to expand their knowledge base and keep abreast of new developments in the field. This continuing education effort has taken the form of requiring certified genetic counselors to obtain documentation of continuing educational activities through continuing education units (CEUs). While NSGC developed a continuing education model in the early 1980s to ensure quality educational programs, it was not until 1996 that courses and conferences sponsored by NSGC had to meet established continuing education criteria (31). Certified genetic counselors may demonstrate their continuing education by either sitting for a re-certification examination or obtaining a minimum of 250 approved contact hours during a period of 10 years. NSGC is continuing its efforts to broaden the types of activities by which counselors can obtain the appropriate educational experiences.

Employment

As noted previously, genetic counselors usually work as members of the health care team in a number of settings. Once every two years, NSGC conducts a status survey of its membership to obtain information about employment, professional roles, and activities (32). In the survey conducted in spring 1998, almost half (47 percent) of the counselors reported working in a university medical center and 24 percent reported working in a private hospital or medical facility. The remaining counselors were scattered between other categories such as HMO's (7 percent), diagnostic laboratories (6 percent), federal/state/county offices (5 percent), and self-employed (1 percent). Almost three-quarters of counselors (73 percent) stated they worked in two or more specialty areas. These areas included prenatal genetic counseling (70 percent), followed by pediatric genetics (45 percent), and cancer genetics (34 percent). Of potential interest to new graduates of genetic counseling training programs, the 1998 survey noted that it took less than two months for members to obtain genetic counseling positions following graduation (32).

Issues in Genetic Counseling

While there are a number of issues that genetic counselors encounter such as dealing with nonpaternity or ambiguous findings on genetic test results (covered elsewhere in this text), two issues that continue to be important to the profession are nondirective counseling and financial reimbursement of genetic counseling services.

Nondirective Counseling. From Sheldon Reed's first publication defining genetic counseling as a "a kind of social work done for the benefit of the whole family entirely without eugenic connotations" (3, p. 335), genetic counseling has been equated with the concept of nondirectiveness. Nondirectiveness appealed to the genetics community as a way to distance itself from the eugenics movement associated with Nazi Germany. Genetic counseling embraced Carl Rogers's client-centered counseling approach. Rogers felt that counselors need to provide a warm, accepting environment free from pressure or coercion for clients to reach a successful self-acceptance and self-understanding (33). Nondirectiveness is understood to mean nonprescriptive. Fine defines nondirectiveness as a genetic counseling strategy that supports autonomous decision making by clients (34). NSGC Code of Ethics states, "Therefore, genetic counselors strive to: ... Enable their clients to make informed independent decisions, free of coercion, by providing or illuminating the necessary facts and clarifying the alternatives and anticipated consequences" (35). Genetic counselors therefore are facilitators and advocates of informed decision making, with the goal of having the counselee make a decision based solely on his or her own values and beliefs.

However, consensus regarding the terms directiveness and nondirectiveness is difficult to find among genetics professionals and in the literature (36). Kessler notes that depending on how one defines the term nondirective will determine whether or not it can be achieved (37). White notes that nondirectiveness is often equated with value-neutrality, which may "either imply that the counseling approach as a whole does not represent any values or moral positions, or it may refer to value-free communication, representing an ideal in which concepts and facts are expressed in impartial terms" (38). A number of authors have argued that counseling is never value-neutral. The types of information provided or not provided, the tone of voice, body language, all convey counselor values (39,40,37). Singer elaborates on this theme by noting that many of the decisions that patients make take place in an atmosphere of crisis and that the issues are often highly emotional (41). Counselors relay information that is often highly technical, while most counselees are likely to have limited knowledge of the biological and statistical issues that arise; they are a vulnerable population. Thus a counselor, who has a duty to provide all the information that clients need in order to make informed decisions, must decide what information the counselee needs and how to present the information. In this sense the genetic counselor

utilizes her expertise to decide what and how much information the counselee needs to make the best possible decision for her. Brunger and Lippman (42) would agree. They conclude that genetic counseling is not a "one-size-fits all" endeavor; rather, it is information that is tailored to specific counselees in specific situations. For some, this would be considered a directive approach. However, authors such as Kessler or Singer would suggest that the counselor is facilitating the goal of genetic counseling by providing information upon which counselees can make autonomous, independent, and informed decisions (41).

One of the often quoted mechanisms for deciding whether or not a counselor is being directive is to ask whether or not counselors answer questions such as "What would you do in my situation?" Michie and colleagues (43) analyzed 131 transcripts of genetic counseling sessions and quantified directiveness based on how often counselors provided advice (what was best for the client), evaluation (provided views about an aspect of the counselee's situation), and reinforcement (provided statements that affirmed or rejected a counselee's behavior, thoughts or emotions). The authors found a mean of 5.7 advice statements per counseling session. Moreover none of the genetic counselors in these sessions rated their counseling as "not at all" directive. Interestingly there was no significant association found between any of the measures of directiveness and such outcome measures as satisfaction with information, nor any information on whether the counselees' expectations were met or on the amount of anxiety and concern (43). Bernhardt concluded that the study found that the practice of genetic counseling "is not characterized ... as uniformly nondirective" (43, p. 40), and that it provided "data to substantiate the long held impression that nondirective genetic counseling is impossible to achieve" (44, p. 17). Kessler interpreted Michie and colleague's data differently based on a definition of directiveness that emphasizes coercion. He noted that if there is an attempt by the counselor, through deception, threat, or coercion, to undermine the individual's autonomy and compromise his or her ability to make an autonomous decision, that can be defined as directiveness (37). Moreover Kessler points out that one does not have to answer the "What would you do" question to be directive.

In a study by Bartels et al. (45), 781 members of NSGC were surveyed to assess how they defined nondirectiveness and what they actually did in practice. Of the 383 respondents, 96 percent reported viewing nondirectiveness as very important; however, 72 percent stated they were sometimes directive. Bartels concluded that although nondirectiveness was a goal of genetic counseling, it was not the only goal. They found that counselors made important distinctions between concerns for directiveness about the decision-making process and directiveness about decision outcomes. Counselors noted that they should take responsibility for directing a counseling session such as clarifying counselee expectations and questions, sharing genetic information, and facilitating understanding and communication. Counselors felt that recommending genetic or medical testing was consistent with informed consent, and that principle outweighed being nondirective (45). However, counselors were more conflicted regarding decision outcomes such as having counselors share personal biases or values with their patients. The study participants described the need to be very careful in what and how these values would be presented. However, once counselees made a decision, genetic counselors felt that the decision should reflect the counselee's values and not the counselor's values.

While the literature regarding nondirective counseling remains unsettled, research suggests that nondirectiveness is not the only guiding principle employed by genetic counselors. In their efforts to provide and facilitate autonomous, independent, and informed decision making by counselees, genetic counselors strive to maintain a delicate balance between a nondirective stance and enhanced counselee understanding.

Reimbursement. The issues of financial reimbursement of genetic counseling services is also of major concern to genetic counselors. Currently there is a paucity of literature that speaks to reimbursement of services provided by counselors. In most instances, reimbursement for genetic counseling services provided by a genetic counselor is under the control of the physician provider. While nonphysicians can bill insurers for their services, they may or may not be reimbursed as some insurers do not recognize nonphysicians as health care providers. Moreover Medicare and Medicaid require that professional services of a nonphysician must be rendered under the physician's direct supervision (Medicare Part B Carriers Manual). Currently the Economics of Genetic Services Committee of the American College of Medical Genetics is working with the American Medical Association CPT Editorial Committee in an effort to obtain billing codes that are more representative of genetic services (46).

NSGC has also actively pursued licensure as a possible avenue to improving the reimbursement of services provided by genetic counselors. Currently the only state considering licensure of genetic counselors is California (47). A continuing and frustrating challenge for genetic counselors is to seek avenues of reimbursement within the context of a managed care environment. Another challenge remains how to educate other health care providers, hospital administrators, insurers, and the public to the skills and expertise that genetic counselors have to offer.

OTHER HEALTH PROFESSIONALS

Medical genetics is fast becoming central to the delivery of health care and preventive services. This process is in large part due to the progress in the Human Genome Project which promises, and in some cases has brought to fruition, wide-ranging applications in the diagnosis, treatment, and prevention of human diseases (48). Gene testing for a number of different disorders is now available to large segments of the population. For example, gene testing for breast and colon cancer susceptibility is available for individuals with a positive family history (49), an NIH Consensus Development Conference (50) has advocated offering DNA-based carrier testing for cystic fibrosis

to pregnant couples, and population screening is being considered for hemochromatosis (51). These advances, coupled with inquiries regarding genetic testing from patients, have made it increasingly necessary for primary care providers to be informed about genetic information and testing. Ideally genetic services are best provided by geneticists and genetic counselors who have had extensive training and exposure to the complex genetic principles and issues regarding genetic testing, screening, and counseling. Although genetic counselors remain the "gold standard" for the provision of genetic services, there is a growing demand that other health care providers be able to provide some of these services. Nevertheless, it is evident from a number of studies that neither physicians nor nurses, the primary sources of health care, have received the training and knowledge needed to deal with the proliferation of new genetic information. Two recent surveys, one of physicians and one of nurses, demonstrated that while the majority of health care professionals stated they were providing some genetic related services (at least occasionally), provider knowledge of genetics was fragmented and uneven (52,53). Giardiello (54), in a study of the use and interpretation of gene testing for Familial Adenomatous Polyposis (FAP), found that one-third of physicians ordering the tests were misinterpreting the results and did not fully understand their meaning. The authors concluded that the "use of genetic counseling before testing would be expected to eliminate many ... errors" (54, p. 826).

With the move toward managed health care and a focus on primary care, it is the physicians and nurses who will now be ordering and interpreting many genetic tests, providing genetic counseling, and assuming responsibility for obtaining informed consent and protecting client privacy. In an effort to rectify some of the knowledge deficit, a number of organizations have come together to ensure that health professionals are prepared for this era of genetic technologies and testing. The National Coalition for Health Professional Education in Genetics (NCHPEG), a group of professionals spearheaded by the National Institute for Human Genome Research, the American Medical Association, and the American Nurses Association, has brought together genetics and other health care professionals to ensure that physicians and nurses have the knowledge, skills, and resources to integrate new genetic knowledge and technologies into the prevention, diagnosis, and management of disease (55). NCHPEG has established a number of goals for developing and implementing a comprehensive genetics education initiative. Its goals include persuading health care organizations to establish genetics education as a top priority; creating mechanisms for collaboration between genetics professionals, other health care professionals, consumers, industry, and government; and identifying existing and future genetics education activities.

Primary health care providers may assume a number of different roles in the care they give to individuals who have or are at risk for a genetic disorder. There are major areas and roles identified by both physician (56) and nursing genetics professionals (53), in which all primary care providers should be knowledgeable. These include

(*1*) identifying individuals and families who would benefit from genetic services and counseling; (*2*) knowing how to obtain genetic services once a risk factor for a genetic disorder has been identified, (*3*) being able to interpret risk factors and information in genetic test results and explaining the information to patients and their families, and (*4*) having a foundation to read and interpret the medical literature in order to provide competent guidance to patients and families regarding their questions about genetic testing, gene therapy, and genetic disorders. Additionally primary care providers can help families learn about what to expect in terms of a genetic evaluation and the genetic counseling; they can coordinate care for the family by assisting in the arrangements for diagnostic testing, referrals for further evaluations, and appointments with other specialists if appropriate; and they can reinforce, interpret, and clarify information obtained during a genetic counseling evaluation. Certainly the primary care provider could continue to assess family dynamics, coping strategies, and other psychosocial responses and serve as a resource person and educator to his/her patients and families regarding birth defects and genetic disorders.

These roles and responsibilities, however, should not be construed as a substitute for referring patients and families for genetic services and counseling. Ideally the primary care provider will work in partnership with the genetic counselor and recognize that a major component in the genetic counseling process is the opportunity to generate a dialogue that will result in a patient being able to make an informed decision. This is particularly necessary in presymptomatic or susceptibility genetic testing, such as Huntington disease and BRCA1 and BRCA2 testing for breast cancer, or in screening programs such as cystic fibrosis (50). The dialogue becomes part of an informed decision-making process rather than merely the transmission of information (57). Counseling skills required for such interactions must combine respect for a patient's right to make an autonomous decision with an appropriate level of support to facilitate the decision-making process. Genetic counseling provides this component and blends the informational, educational aspect of the counseling session with a dialogue on the benefits and risks of a genetic test. Rarely does a primary care provider have the time or specific training to be responsible for this process.

SUMMARY

The advent of recombinant DNA research and the initiation of the Human Genome Project has produced a revolution in human genetics. Every week scientists announce the identification and chromosomal location of another gene associated with a human genetic disorder. With these discoveries comes the ability to test for these disorders, both diagnostically and predictively. For some people it may mean the development of new treatment strategies. It is imperative that health professionals, in particular primary care providers, be informed about human genetics, genetic testing, and the ethical, legal, and social issues that accompany these areas of research

and practice. Primary care providers are on the front line of providing health care services to families. Thus they need to be able to identify, assess, counsel, and refer patients, clients, and families with or at risk for genetic disorders and birth defects.

Finally, as new tests for carriers are developed and the diagnosis of genetic conditions and genetic susceptibility to disease continues to grow, the number of individuals utilizing genetic services will also increase. As the number of individuals and families seeking services increases, the need for genetic counselors to provide quality genetic counseling will also grow. Genetic counselors, utilizing their expertise and skill within the genetic counseling process, will be needed to explain facts, educate patients regarding inheritance and recurrence, provide counseling about benefits and risk, aid in decision making, and help set public policy. Well-trained, board-certified genetic counselors will continue to expand their presence in a number of settings: in hospitals and clinics to counsel families who are affected or may be at risk for genetic disorders; in diagnostic labs as resources for physicians and their patients; and in government agencies to design genetics education programs for health care providers, shape public health policy, and develop more effective ways of communicating the many new findings to employers, insurers, and the general public.

BIBLIOGRAPHY

1. S.C. Reed, *Counseling in Medical Genetics*, W.B. Saunders, Philadelphia, PA, 1955.

2. S.C. Reed, in H.G. Hammons, ed., *Hereditary Counseling: A Symposium sponsored by the American Eugenics Society*, Hoeber & Harper, New York, 1959.

3. S.C. Reed, *Soc. Biol.* **21**, 332–339 (1975).

4. R.H. Kenen, *Soc. Sci. Med.* **18**, 541–549 (1984).

5. F.C. Fraser, *Birth Defects Original Article Series, National Foundation March of Dimes* **15**, 5–15 (1979).

6. C. Stern, *Principles of Human Genetics*, 3rd ed., W.H. Freeman, San Francisco, CA, 1973.

7. N.L. Nadler, *Pediatrics* **42**, 912–918 (1968).

8. J.R. Sorenson and A.J. Culbert, *Birth Defects Original Article Series, National Foundation March of Dimes* **15**, 85–102 (1979).

9. A.P. Walker, in D.L. Rimoin, J.M. Connor, and R.E. Pyeritz, eds., *Emery and Rimoin's Principles and Practice of Medical Genetics*, 3rd ed., Churchill Livingstone, New York, 1996, pp. 595–618.

10. C.J. Epstein et al., *Am. J. Hum. Genet.* **27**, 240–242 (1975).

11. J.G. Hall, in R.E. Stevenson, J.G. Hall, and R.M. Goodman, eds., *Human Malformations and Related Anomalies*, vol. 1, Oxford University Press, New York, 1993.

12. E. Sujansky and A.L. Matthews, in G.B. Merenstein and S.L. Gardner, eds., *Handbook of Neonatal Intensive Care*, 4th ed., Mosby-Year Book, St. Louis, MO, 1998, pp. 604–624.

13. J. Leana-Cox et al., *Am. J. Med. Genet.* **65**, 309–316 (1996).

14. J.M. Aase, *Diagnostic Dysmorphology*, Plenum Medical Book Company, New York, 1990.

15. S. Kessler, in S. Kessler, ed., *Genetic Counseling: Psychological Dimensions*, Academic Press, New York, 1979.

16. S. Kessler, *Am. J. Med. Genet.* **72**, 164–171 (1997).

17. A.L. Matthews, *Birth Defects Original Article Series, National Foundation March of Dimes* **26**, 168–175 (1990).

18. J.L. Benkendorf et al., *Am. J. Med. Genet.* **73**, 296–303 (1997).

19. K.A. Schneider, *Counseling about Cancer: Strategies for Genetic Counselors*, Graphic Illusions, Dennisport, MA, 1994.

20. E. Kodish, G.L. Weisner, M.J. Mehlman, and T.H. Murray, *J. Am. Med. Assoc.* **279**, 179–181 (1998).

21. National Society of Genetic Counselors, *Genetic Counseling as a Profession*, NSGC Executive Office, Wallingford, PA, 1999.

22. D.M. Bartels, in D.M. Bartels, B.S. LeRoy, and A.L. Caplan, eds., *Prescribing Our Future: Ethical Challenges in Genetic Counseling*, Aldine de Gruyter, New York, 1993.

23. J.H. Marks, in D.M. Bartels, B.S. LeRoy, and A.C. Caplan, eds., *Prescribing Our Future: Ethical Challenges in Genetic Counseling*, Aldine de Gruyter, New York, 1993.

24. K.W. Dumars et al., eds., *Genetic Associates: Their Training, Role and Function: A Conference Report*, U.S. Department of Health, Education and Welfare, Washington, DC, 1979.

25. J.A. Scott, A.P. Walker, D.L. Eunpu, and L. Djurdjinovic, *Am. J. Med. Genet.* **42**, 191–199 (1987).

26. ABGC News Corner, American Board of Genetic Counseling, Spring 2000, <http://www.faseb.org/genetics/abgc/abgc menu.htm>

27. American Board of Genetic Counseling, Requirements for graduate programs in genetic counseling seeking accreditation by the American Board of Genetic Counseling, American Board of Genetic Counseling, Inc. Bethesda, MD, 1996.

28. A.C.M. Smith, *J. Genet. Counsel.* **2**, 197–211 (1993).

29. J.L. Bendendorf et al., *J. Genet. Counsel.* **1**, 31–43 (1992).

30. B.A. Fine, D.L. Baker, M.B. Fiddler, and ABGC Consensus Development Consortium, *J. Genet. Counsel.* **5**, 113–121 (1996).

31. D.L. Baker, J.L. Schuette, and W.R. Uhlmann, *A Guide to Genetic Counseling*, Wiley, New York, 1998.

32. K.A. Schneider and K.J. Kalkbrenner, *Perspect. Genet. Counsel.* **20**(3), S1–S8 (1998).

33. C. Rogers, *Client-Centered Therapy: Its Current Practice, Implications, and Theory*, Houghton Mifflin, Boston, MA, 1951.

34. B.A. Fine, in D.M. Bartels, B.S. LeRoy, and A.C. Caplan, eds., *Prescribing Our Future: Ethical Challenges in Genetic Counseling*, Aldine de Gruyter, New York, 1993.

35. National Society of Genetic Counselors, *J. Genet. Counsel.* **1**, 41–43 (1991).

36. S. Kessler, *J. Genet. Counsel.* **1**, 9–17 (1992).

37. S. Kessler, *Am. J. Hum. Genet.* **61**, 466–467 (1997).

38. M.T. White, *J. Genet. Counsel.* **6**, 297–313 (1997).

39. A. Clarke, *Lancet* **338**, 998–1001 (1991).

40. A. Lippman and B.S. Wilfond, *Am. J. Hum. Genet.* **51**, 936–937 (1992).

41. G.H.S. Singer, in B. Gert et al., eds., *Morality and the New Genetics: A Guide for Students and Health Care Professionals*, Jones and Bartlett, Boston, MA, 1996.

42. F. Brunger and A. Lippman, *J. Genet. Counsel.* **4**, 151–167 (1995).

43. S. Michie, F. Bron, M. Bobrow, and T.M. Marteau, *Am. J. Hum. Genet.* **60**, 40–47 (1997).

44. B.A. Bernhardt, *Am. J. Hum. Genet.* **60**, 17–20 (1997).

45. D.M. Bartels, B.S. LeRoy, P. McCarthy, and A.L. Caplan, *Am. J. Med. Genet.* **72**, 172–179 (1997).

46. D.L. Doyle, Personal communication, September 1999.

47. A.P. Walker, Personal communication, October 1999.

48. E.P. Hoffman, *Am. J. Hum. Genet.* **54**, 129–136 (1994).

49. S.J. Laken, G.M. Petersen, and S.B. Gruber, *Nat. Genet.* **17**, 79–83 (1997).

50. Genetic Testing for Cystic Fibrosis, NIH consensus statement, No. 15, April 14–16, 1997, pp. 1–37.

51. T. Cox, *Nat. Genet.* **13**, 386–388 (1996).

52. K.A. Hofman et al., *Acad. Med.* **68**, 625–632 (1993).

53. C. Scanlon and W. Fibison, *Managing Genetic Information: Implications for Nursing Practice*, American Nurses Publishing, Washington, DC, 1995.

54. F.M. Giardiello et al., *N. Engl. J. Med.* **336**, 823–827 (1997).

55. National Coalition for Health Professional Education in Genetics, Planning meeting, July 23, 1996.

56. J. Stephenson, *J. Am. Med. Assoc.* **279**, 735–736 (1998).

57. G. Geller et al., *J. Am. Med. Assoc.* **277**, 1467–1474 (1997).

See other entries GENETIC INFORMATION, ETHICS, FAMILY ISSUES; see also REPRODUCTION, ETHICS entries.

GENETIC DETERMINISM, GENETIC REDUCTIONISM, AND GENETIC ESSENTIALISM

ROBERT WACHBROIT
University of Maryland
College Park, Maryland

OUTLINE

Introduction

Determinism: Genetic and Otherwise

Reductionism: Genetic and Otherwise

Essentialism: Genetic and Otherwise

Conclusion

Bibliography

INTRODUCTION

In the public debate over the significance of genetic discoveries, the terms "genetic determinism," "genetic reductionism," and "genetic essentialism" are often used interchangeably. This is unfortunate because these terms refer to quite distinct claims about the importance of genes, claims that vary greatly in their plausibility, their implications, and their popular acceptance. Thus, for example, few people would accept genetic determinism, but many would subscribe to doctrines that are characteristic of some version of genetic reductionism or essentialism. The purpose of this article is to clarify what each of these terms mean, and how they differ from each other. The first, I will show, is a claim about causation; the second, about explanation; and the third, about identity.

Being clear about these terms is not just a matter of lexographic tidiness; it is critical for avoiding serious misunderstanding. The significance of this misunderstanding can be illustrated by contrast with another common confusion in the area of genetics; between genes and alleles.

Articles and discussions about genetics often confuse genes and alleles, using "genes" when the proper term should be "alleles." But this muddle is usually harmless. The knowledgeable and attentive reader can usually understand which term the writer intends even if they happen to be misused at the time. No significant issue is raised by this common conflation.

The situation is quite different with regard to the above "isms." Many writers reject what they regard as an inflated importance attached to genetics by rejecting genetic determinism, genetic reductionism, or genetic essentialism. These three positions are, however, quite different ways of articulating what the importance of genetics consists in. Not only does one not entail either of the other two, but also, as we will see below, each raises different significant issues.

The underlying assumption of this article is that the semantic structure of these "isms" should be taken seriously: Genetic determinism is a kind of determinism. Its implications, significance, considerations for and against, should be shaped by the history of the dispute over determinism itself. Otherwise, the expression is misleading. Similar assumptions hold for genetic reductionism and genetic essentialism. If they are not a kind of reductionism and a kind of essentialism, respectively, then we need to be told what they mean.

Finally, let me repeat, the purpose of this article is to clarify succinctly the differences between these "isms." The aim is not to provide a comprehensive survey of the many issues and the extensive literature surrounding these terms. If interested, the reader is advised to look to the bibliography.

DETERMINISM: GENETIC AND OTHERWISE

Determinism is a thesis about universal causation — every event has a cause sufficient for its occurrence. Thus identical states of the universe would have identical outcomes (1). The phrase "genetic determinism" would, strictly speaking, mean that every event has a genetic cause that is sufficient for that event's occurring. No one takes genetic determinism in this way. It is usually understood to be restricted to a specific class of events or properties — such as organism's physical and mental traits. Thus genetic determinism is the thesis that an organism's physical and mental traits are entirely the causal result of its genes.

So understood, the general consensus is that genetic determinism is false. This view is embedded in the standard distinction in genetics between an organism's genotype — the combination of the organism's genes — and its phenotype — its set of mental and physical traits. Genes alone do not yield any traits; various environmental factors must be present for the trait to arise. Regardless of how close the tie is between a certain gene and trait, there is always an environment where the presence of the gene does not result in the presence of the trait, including hostile environments in which the organism does not survive long enough to develop the trait.

Nevertheless, noting that genetic determinism is false does not mean determinism is false as well. One could reject genetic determinism and still hold that the combination of genes and environment determines traits. Or one could reject determinism even for the genes plus environment combination, noting the role of various stochastic processes in the development of traits (2, chap. 2). As we will see, this difference affects the significance of claiming that genetic determinism is false.

The significance of genetic determinism lies in its implications for predictability and manipulability. If genetic determinism is true, then it should be theoretically possible to predict people's traits solely from a knowledge of their particular genes. Furthermore the only way to alter an individual's particular trait would be to manipulate the relevant genes. If the rejection of genetic determinism is a rejection of determinism, then predictability and manipulability are also rejected. But if genetic determinism is rejected in favor of a "genes plus environment" determinism, then the rejection is much less significant. Prediction is still theoretically possible, though practically much more difficult given the complex variety of environmental factors that need to be considered. And manipulability becomes theoretically easier since there are now more options available.

Despite the widespread rejection of genetic determinism, we should acknowledge the occasions when researchers will talk about a gene "determining" a trait — for example, they might say that having a certain structure on chromosome 4 determines that the individual will develop Huntington's disease. This expression of genetic determinism rests on a special restriction on the events being considered. All circumstances in which the individual dies before he manifests the disease are being excluded. Furthermore we are confining attention to only known circumstances — no one is denying that researchers might some day discover that currently unknown circumstance in which we are able to prevent or cure that disease. Once these restrictions are understood, such talk does not really amount to an endorsement of genetic determinism.

REDUCTIONISM: GENETIC AND OTHERWISE

Reductionism is one of the main topics of modern philosophy of science, and during that time it has undergone considerable sophistication and complexity in its formulation and arguments. For a standard introduction, see Ref. 3. For an advanced discussion of this topic as applied to genetics, see Refs. 4–6. Fortunately, for our purposes, many of these details can be ignored. The core claim of reductionism — a thesis about the relation between two kinds of things (X and Y) — is that X's are nothing but Y's. Different ways of specifying the meaning of the relation "is nothing but" as well as the different sorts of things at issue yield different types of reductionism. For example, taking the "is nothing but" relation to be that of composition, we get a class — sometimes called "ontological" — of reductionist positions. Illustrations from the physical sciences provide some of the least controversial examples, such as "physical objects are nothing but swarms of atoms" and "water is nothing but hydrogen and oxygen."

One of the most familiar reductionist positions in this class is variously called "materialism" or "physicalism." It is roughly the view that everything is nothing but atoms, parts of atoms, and their interactions. This reductionism is a central principle of modern biology. Biologists reject vitalism and instead believe that biological objects are composed entirely of the stuff of physical objects—molecules, free-floating atoms, electrons and the like.

Genetic reductionism, understood ontologically, is the position that organisms consist of nothing but genes. It should be clear that no one is a genetic reductionist in this sense since the position is so obviously false. We are clearly composed of more than just DNA molecules. Indeed, the occasional expression such as "we are nothing but our genes" cannot be seriously meant literally. It must be understood as figurative way of referring to a different sense of reductionism, to be described below. Thus, while all biologists, including geneticists, are reductionists, in the sense of being materialists, no one is a genetic reductionist, understood ontologically. There is therefore no need to discuss it further.

The other major class of reductionist positions are those where the "is nothing but" relation is understood in terms of explanation and Y is understood as a theory. X can itself be a theory, where Y is a more fundamental or basic theory than X. Or X can be a representation or description of one type of phenomena, and Y is a theory typically about phenomena of a different, more fundamental realm. As opposed to ontological, this class of reductionist positions is sometimes called "epistemological" or "methodological." Fairly uncontroversial examples from the physical sciences would be "chemistry is nothing but physics" and "heat is nothing but molecular motion." In the first case, the thesis is that chemical theories can be reduced to physical theories: For example, the correctness of explanations in chemistry can be explained by physics. In the second case, the thesis is that heat, a phenomenon characterized by thermodynamics, can be explained by the more fundamental theory of statistical mechanics. In the end, both types of cases are similar: Chemical explanations can be replaced, at least in principle, by physical explanations; thermodynamic explanation can be replaced, at least in principle, by statistical mechanical explanations. Their differences are more in the methods of achieving reduction. In the first case, reduction is achieved by translating or associating the language of one theory into the language of a more fundamental theory; in the second case, reduction is achieved by directly explaining the phenomenon by using a fundamental theory.

Genetic reductionism, understood in this sense, can refer to either the thesis that biological explanations can be translated into or reformulated as genetic explanations or the thesis that biological phenomena can be explained by genetics. As we will see, unlike genetic determinism, genetic reductionism is an open question. It depends on how the science of genetics develops. Given the present state of the science, genetic reductionism is at best a working hypothesis, although some scientists claim that it is not a particularly promising hypothesis. In any case, whether genetic reductionism is true or false is an empirical matter.

Assessing the truth of genetic reductionism plainly turns on our understanding of what constitutes a genetic explanation. A precise characterization of genetic explanation is challenging, especially since there is considerable dispute over what constitutes an explanation (particularly explanations in biology!), but the intuitive idea is fairly clear. An explanation of a phenomenon is genetic just in case all the nongenetic activities or interactions that are part of the phenomenon are relegated to the background or are deemed to have minor importance. Thus a genetic explanation does not need to deny the existence of any nongenetic activity. Consider, for example, the genetic explanation for the probability of inheriting a certain (i.e., Mendelian) trait. One begins by determining whether to associate the trait with a dominant or recessive gene, and then one calculates the probability that the gene will be transmitted to offspring. The role of environment in the development of the trait is not denied; it is relegated to a background condition, deemed not a salient factor in the cases to be explained.

Similarly nongenetic explanations, such as environmental explanations, need not deny the existence of genes or genetic processes. A sociological explanation of family dynamics need not deny that genetic processes typically play a role in the existence of families. Such processes are deemed background conditions.

Not surprisingly, just as there is a controversy over what is an explanation, there is a controversy over what constitutes a background condition. Some have suggested objective criteria—for example, if it makes a comparatively small contribution to the phenomena, it is a background condition. An illustration of this can be found in explanations of differences. Suppose that we wanted to explain the differences in health between monozygotic twins. One might argue that since there is little genetic difference between the twins, genetic factors will be background conditions. So the explanation of differences in health will not be a genetic explanation.

Others have suggested subjective or pragmatic criteria—for example, factors that are not salient for the context of inquiry are background conditions. An explanation is here seen as primarily a response to a particular question, which characterizes, explicitly or implicitly, certain factors as salient. For example, a traffic accident is the result of how the car functions and how the driver functions. Whether we explain the accident in terms of the driver's behavior or the car's functioning can depend on which is, from the context of inquiry, deemed normal. Normality—either in behavior or in functioning—is usually relegated to a background condition.

We need not pursue any further the details of the concept of a background condition—or of explanation or of theory—in order to discuss the significance of genetic reductionism. As a scientific hypothesis about the power and scope of genetic explanations, it is plainly important for scientific research. But, as we will see, it also has social and ethical significance.

From the standpoint of science, genetic reductionism is a thesis about the direction scientific research should take. If genetic reductionism is true, investigations that focus on the genetic aspects of biological phenomena are more likely to come up with interesting or important results. The patterns that emerge are more likely to reflect fundamental principles of biology than accidental or derivative correlations. Moreover such research is more likely to discover unifying themes in that divers biological phenomena might be explicable by the same type of (genetic) explanation. Indeed, genetic reductionism indicates what constitutes progress in biology—being able to account for more and more biological phenomena in terms of genetics.

The social significance of genetic reductionism is closely connected to these scientific implications. Genetic reductionism suggests that the "real" explanation of a biological phenomenon is a genetic explanation. Even if a nongenetic explanation of a particular biological phenomenon were available, it would not get to the heart of the matter. Nongenetic factors play a relatively minor role. It is a short step from this to holding that the nongenetic is of little importance or value. The significance of this becomes especially clear in the case of explaining human behavior. If many of our standard ways of explaining human behavior—explanations in terms of character or in terms of motivation or in terms of intentions or in terms of beliefs, desires, and circumstances—are held to be not "real" explanations or are held to be genetic explanations in disguise, then it can seem that character, intention, circumstances, and so forth, are unimportant or are themselves reducible to genetics. Indeed, some writers reject genetic reductionism because they maintain that nongenetic factors can be important.

At this point it might be worth summarizing some of the differences between genetic determinism and genetic reductionism. Their clearest divergence is with inter-actionism—the view that both genetic and nongenetic factors play a causal role in most biological phenomena. Interactionism is incompatible with genetic determinism but can be compatible with genetic reductionism. Recall that a genetic reductionist can acknowledge nongenetic causal factors as long as they amount to no more than background conditions of proper scientific explanations of biological phenomena. Thus the falsity of genetic determinism does not entail the falsity of genetic reductionism. (Whether the converse is true—whether genetic determinism entails genetic reductionism—turns on issues concerning the relation between causation and explanation, which we cannot pursue here.)

ESSENTIALISM: GENETIC AND OTHERWISE

The general doctrine of essentialism is linked to a particular conception of what change and identity consists in. It begins with the question, What is the difference between an object changing and an object ceasing to exist and being replaced by a different object? According to essentialism, an object has two kinds of properties—essential properties and accidental properties. An alteration in an accidental property results in the object being changed; an alteration in an essential property results in the object ceasing to exist. For example, the temperature of gold is an accidental property; change the temperature, and the result is still gold, only warmer

or colder. In contrast, the atomic number of gold is an essential property; change it by even one digit and the result is no longer gold.

Genetic essentialism is the view that the genetic properties of an organism are essential. Given the multitude of genes, genetic essentialism fans out into a spectrum of specific views. At the extreme is the view that every gene—and hence every genetic property—is essential; if even one gene in the organism changes, say, from some environmental damage, the result is a different organism. Indeed, there is no such thing as a harmless error in gene replication on this view. An error means that the organism has ceased to exist and a new one, albeit quite similar, is now in its place. Moreover gene therapy must be seen as logically impossible. A procedure that alters the gene of an individual could not be therapy for that individual, since the alteration would destroy the individual and replace him with another. Hardly anyone is a genetic essentialist of this sort.

The situation becomes more complicated once we consider changes in certain genes or in many genes. How many and which phenotypic traits of an organism would have to be different—and to what degree—in order for us to regard the result as no longer the same organism? Some might hold that changing the genes that determine the organism's sex results in a different organism. Others might hold that in the case of persons, altering the genes associated with mental capacities and abilities can result in a different person. And others might hold that only those genetic changes that constitute a different species result in a different organism. Where we draw the line—when (genetic) change becomes replacement—is the central controversy regarding genetic essentialism. It also indicates the conceptual limits of gene therapy since altering an essential property can never be therapeutic (7).

This last point has been especially important in discussions of the alleged paradox of "genetic harm," which refers to a harm that is the result of an abnormal gene. The paradox arises if the abnormal gene is an essential gene. In that case, removing the harm, altering the essential gene, would not be altering the organism—presumably for the better—but rather replacing that organism with a different one. It would seem then that treating genetic harms when they involve essential genes could never be beneficial to the individual.

Genetic essentialism makes no claim about causation and so is distinct from genetic determinism, as we would expect from general discussions in philosophy regarding essentialism and determinism. Essential properties need not be deterministic (i.e., sufficient causes) and deterministic properties need not be essential. But see Ref. 8 which treats genetic essentialism as incorporating theses of determinism and reductionism.

Nor need genetic essentialism have anything to do with genetic reductionism of either type. A genetic reductionist of the ontological type—one who believes that organisms are composed of nothing but genes—need not be committed to any view regarding which or how many genes are essential to the identity of the organism. This claim is in line with the larger view that materialism doesn't entail any view about which material is essential. A genetic reductionist of the epistemological type — one who believes that biological phenomena can be entirely captured in genetic explanations — need not be committed to any particular genetic essentialism. Essential properties do play a critical role in the explanation of change — how an organism can remain the same even when some of its properties are different — but it plays no role in scientific explanations. In short, a genetic reductionist need not have any view about which genes, if any, are essential.

CONCLUSION

Advances in genetics have highlighted the role genes play in various biological phenomena. This increased attention to genetics has led to various assertions and denials of the importance of genes.

Genetic determinism, genetic reductionism, and genetic essentialism are three different ways of stating what that importance consists in. The primary purpose of this article has been to articulate these differences. Any useful discussion of where genes are important and where they are not, if it invokes any of these "isms," must be clear about what exactly is meant.

BIBLIOGRAPHY

1. John Earman, *A Primer on Determinism*, Reidel Publishing Dordrecht, The Netherlands, 1986.

2. R. Lewontin, *Biology as Ideology: The Doctrine of DNA*, Harper Perennial, New York, 1992.

3. E. Nagel, *The Structure of Science*, Chapter 11, Hackett Publishing Company, Indianapolis, IN, 1979.

4. P. Kitcher, *Philosophical Review* **93**, 335–373 (1984).

5. S. Sarkar, *Genetics and Reductionism*, Cambridge University Press, New York, 1998.

6. K. Schaffner, *Discovery and Explanation in Biology and Medicine*, University of Chicago Press, Chicago, 1993.

7. D. Wasserman, Personal identity and the moral appraisal of prenatal therapy, in L. Parker and R. Arkeny, eds., *Evolving Disciplines: Genetics, Medicine, and Society*, Reidel Publishing Company, Holland, in press.

8. R. Dreyfuss and D. Nelkin, The Jurisprudence of Genetics, *Vanderbilt Law Review* **45**, 313–348 (1992).

See other entries BEHAVIORAL GENETICS, HUMAN; HUMAN ENHANCEMENT USES OF BIOTECHNOLOGY: OVERVIEW; MEDIA COVERAGE OF BIOTECHNOLOGY; PROFESSIONAL POWER AND THE CULTURAL MEANINGS OF BIOTECHNOLOGY; PUBLIC PERCEPTIONS: SURVEYS OF ATTITUDES TOWARD BIOTECHNOLOGY.

GENETIC INFORMATION, ETHICS, AND INFORMATION RELATING TO BIOLOGICAL PARENTHOOD

JEAN E. MCEWEN
National Human Genome Research Institute
Bethesda, Maryland

INTRODUCTION

Genetic testing now makes it possible to determine parental and other familial relationships with a remarkable degree of accuracy. For this reason genetic testing in disputed paternity cases (at least in cases of the traditional type, where the paternity of a child born to unmarried parents is at issue) has gained increasing acceptance, and the ethical and legal concerns raised by such testing are lessening. However, new ethical and legal concerns are being raised in the growing number of cases in which either putative or presumed fathers seek genetic paternity testing to rebut the longstanding legal presumption that a child born to a mother during the course of a marriage is the biological offspring of the mother's husband. The practice of artificial insemination by donor is also creating new challenges in the area of forensic paternity determination.

As techniques such as surrogacy, egg donation, and embryo donation come to be used more widely, forensic challenges regarding *maternity* determinations can also be expected to arise with greater frequency. In addition the increased use of all types of alternative reproductive methods (coupled with advances in genetic testing more generally) is presenting courts with complicated questions regarding the posthumous determination of parentage or other familial relationships. Should human cloning eventually become feasible, the task of sorting through the myriad potentially recognizable familial relationships will become even more complex.

Unanticipated findings regarding parentage (or regarding other familial relationships) can also sometimes occur as a consequence of genetic testing undertaken for nonidentification purposes, raising challenging ethical and legal dilemmas. Incidental unexpected findings of misattributed paternity, in particular, occur quite frequently in the course of genetic testing. Several possible strategies are available for dealing with findings of this type, ranging from full disclosure, nondisclosure, partial disclosure, or disclosure only to the woman, to handling the issue through the informed consent process. However, each approach carries with it a separate set of potential problems.

Previously undisclosed adoption, artificial insemination, or an incestuous mating within the family can also be inadvertently brought to light in the course of genetic testing. In rare cases, genetic testing may also reveal prior mix-ups, such as that two babies were switched in a hospital nursery, that sperm samples were switched in the course of artificial insemination, or that frozen embryos were switched during in vitro fertilization (IVF) procedures. These situations raise problems similar, although not identical, to those that arise when misattributed paternity is discovered, and they will require novel approaches.

GENETIC DETERMINATION OF PARENTAGE FOR FORENSIC PURPOSES

General Background

Objective scientific methods for the determination of biological parentage are a relatively recent development. Only a few decades ago the methods available for determining disputed paternity were limited and relatively primitive. Courts were typically required to calculate the date of the child's conception to determine whether or not the putative father could have conceived the child (1) and to make highly subjective assessments of physical resemblances (2).

Modern scientific tests provide far more reliable evidence of biological parentage by analyzing inherited characteristics: either the physical expression of the DNA in the child (phenotype) or the DNA itself (genotype) (3). The earliest phenotypic tests for paternity involved the ABO blood grouping system, but ABO testing, while capable of excluding an individual (i.e., ruling him out) as a child's father, could not make a definitive positive determination of paternity (4). However, ABO testing can be combined with other kinds of phenotypic testing, such as the analysis of red blood cell antigens, serum proteins, enzymes, and human leukocyte antigens (HLA) (5). When this is done, the cumulative impact of the test results including an individual as the child's father (i.e., placing him within a population of men who could be the father) becomes considerably stronger.

Direct genotypic testing of the DNA, which first became available in the late 1980s, provides an even more highly discriminating form of parentage testing. It can establish paternity (or maternity, if disputed) to degree that experts today agree is nearly conclusive (6). Every state in the United States now has a statute that provides for the admissibility of genetic paternity test results (7). Although the language of these laws varies, most states have patterned their laws on one of two model uniform paternity statutes: the Uniform Parentage Act (UPA) (8) or the Uniform Act on Paternity (UAP) (9).

State statutes vary in the statistical analysis methods (if any) that can be invoked in court to help describe the probative value of a test result including a man as the possible father. For example, some statutes require use of the "paternity index," which is the probability that a child born to the mother and the alleged father would have the observed phenotypes or genotypes divided by the chance of such types appearing in the child of the mother and a man randomly selected from the general population of men. Others require use of the "probability of paternity," which is obtained by multiplying the paternity index by the prior odds of paternity and converting the resulting "posterior odds" to probability form. Still others mandate the "probability of inclusion" (or exclusion) approach, which simply asks which proportion of the male population would be included (or excluded) as the possible father (10).

Regardless of the particular statistical analysis authorized, most state statutes mandate that when paternity test results reach a specified level (typically in the range of 95 to 99 percent), they create a rebuttable presumption of paternity; some statutes go even further and mandate a *conclusive* presumption of paternity when the results exceed a certain threshold (11). The growing confidence of legislators and courts in the reliability of genetic paternity testing, and federal child support enforcement legislation enacted in 1993 requiring all states to enact laws authorizing simple voluntary procedures for the establishment of paternity (12), have resulted in most disputed paternity cases now being settled by agreement of the parties without the need for trial (although statistical issues and questions regarding the adequacy of laboratory procedures are still often raised) (13). As the subjectivity and unreliability of the methods associated with earlier methods of determining paternity have been replaced by modern genetic testing, the ethical and legal issues associated with the determination of paternity (at least in the traditional situation involving unmarried parents and child support) have also gradually lessened.

Emerging Issues in Genetic Determination of Parentage

Rebutting the Presumption of Paternity. Although most paternity disputes continue to arise out of proceedings for the support of children born to unmarried parents, courts are increasingly being called upon to adjudicate paternity in other contexts. For example, notwithstanding the longstanding legal presumption that the parent of a child born during the course of a marriage is the mother's husband (unless the husband was sterile, impotent, or geographically distant at the time of conception), a growing number of putative fathers of children born to married mothers (i.e., men who are not married to the children's mothers but who believe they may have fathered the children) are filing lawsuits to rebut that presumption and assert various parental rights. Although the United States Supreme Court has held that such putative fathers have no right to challenge the presumption of paternity as a matter of constitutional law (14), the continued usefulness of the presumption might well be questioned in cases where undisputed evidence shows that the mother's husband could not possibly be the biological father.

In a related development, some former husbands of women who gave birth to children the former husbands believe were fathered by someone else have begun to challenge their own legally presumed paternity, typically in an effort to relieve themselves of postdivorce child support obligations. At the time of this writing, a petition for *certiorari* in a case raising this issue was pending before the United States Supreme Court (15). On the one hand, it can be argued that no man should be compelled to support a child for whose birth he is not actually responsible, and that no court should be complicit in aiding such a fundamental deception. On the other hand, overfocus on biological parenthood in such cases, and underfocus on the role of social parenting, could harm the children involved, some of whom could be left altogether fatherless and with a great sense of betrayal should their presumed fathers be permitted so easily to relinquish responsibility. While a wrong may clearly have been perpetrated against the husbands in these cases, the question becomes whether the law should create the potential for an even greater wrong to be visited against the innocent children involved. Still, where a family situation has already deteriorated to the point that a man who has previously held himself out as a child's father announces publicly his belief that the child is not biologically his and states that he no longer wishes to be responsible, there may be little to be gained by perpetuating a legal fiction.

Paternity and Artificial Insemination by Donor. Continuing increases in the numbers of children conceived through alternative methods of reproduction, such as through artificial insemination by donor (AID), will further expand the range of cases in which courts will be asked to make forensic parentage determinations. For example, currently state statutes modeled either on the UPA or on the Uniform Status of Children of Assisted Conception Act (USCACA) provide that in a case of AID of a married woman, the woman's husband—not the sperm donor—is treated as the natural father of the child conceived (so long as the husband has consented to the procedure) (16). Likewise statutes patterned on the Uniform Putative and Unknown Fathers Act (UPUFA) accord no legal recognition to men who donate sperm under circumstances indicating that the donor did not anticipate having an interest in the resulting child (17). Moreover the current practice in AID is to keep the identity of sperm donors strictly anonymous, thus making it virtually impossible for anyone (including their resulting children) ever to trace their identity (18). For this reason genetic testing is unlikely to play a role in the establishment of biological paternity in the context of AID, at least for the immediate future. This could change, however, should the standard of practice for AID move toward the adoption of enhanced recordkeeping and liberalized disclosure policies. Significantly the anonymity that currently pervades AID is quite analogous to the secrecy and sealed records practices that for many years pervaded traditional adoption. The justification for such practices is increasingly being called into question as the public becomes more and more aware of the importance for every child of having access to a complete and accurate family history (19).

Genetic Determination of Maternity. Before the advent of modern reproductive technology, determining the *maternity* of a child was essentially never at issue. The woman who gave birth to the child was considered the child's mother as a matter of biological necessity. Increasingly, however, techniques such as surrogacy, egg donation, and embryo donation are raising questions regarding who should be recognized as the legal mother of a child so conceived when agreements involving the use of these technologies break down (20). The UPA expressly creates an action to declare the mother-child relationship but contains no specific provisions regarding the adjudication of such cases.

Surrogacy agreements are presenting especially novel challenges for the courts. There are two forms of surrogacy. In a traditional surrogacy arrangement, a woman agrees to be artificially inseminated with the sperm of the intended father (a man other than her husband), to carry to term the child thereby conceived, and to relinquish the child after birth to the intended father (and presumably also to his wife or partner). In a pure gestational surrogacy arrangement, by contrast, a woman agrees to carry an embryo created not through her own genetic material but through in vitro fertilization of the egg and sperm of the intended parents (or of a donated egg and/or sperm), and later to relinquish the child to the intended parents. The distinction between the two forms of surrogacy is thus that whereas the "traditional" surrogate mother both provides the egg for the pregnancy and gestates the baby to term, the pure gestational surrogate bears no actual genetic relationship to the child.

Judicial resolution of the question of who should be considered the child's legal mother in surrogacy cases has differed depending whether the underlying agreement was one for traditional surrogacy or gestational surrogacy. In the leading U.S. case, the New Jersey Supreme Court held that the woman who gestates (and provides the egg for) the child—not the partner of the man who contracted with her to bear the child—is to be treated as the child's mother (21). By contrast, in the leading case involving a purely gestational surrogacy agreement, the court held that while both genetic consanguinity and giving birth are recognized means of establishing a mother–child relationship, in cases where the two means do not coincide in one woman, the woman who intended to raise the child at the time the agreement was entered into—not the woman who gave birth—should be treated as the mother (22). The court distinguished this situation from a true "egg donation" situation, in which a woman gestates and gives birth to a child formed from the egg of another woman with the intent to raise the child as her own.

The USCACA is drafted so as to give states the option either to accord legal recognition to preapproved surrogacy contracts that meet specified statutory requirements, or to make surrogacy contracts unenforceable, and it specifically provides that except in those cases involving preapproved contracts, a woman who gives birth to a child is to be considered the child's mother. However, in states that have not adopted the USCACA or that have not otherwise clarified this issue legislatively, considerable uncertainty remains regarding the establishment of maternity in surrogacy cases, as well as in cases involving egg and embryo donation. Moreover, outside the United States, approaches to surrogacy have been quite different. For example, in most countries, in a contest between a genetic mother and a pure gestational mother, the gestational mother generally prevails (23).

Posthumous Determination of Parentage or Other Familial Relationships. Because DNA is present in almost all human cells and remains unchanged long after a person has died, DNA testing technology now makes it possible to make an accurate determination of paternity (or of grand-paternity, great-grand-paternity, or even more distant relationships) long after the putative father (or other more distant relative) has died (24). For this reason the advent of DNA testing has brought with it an increase in requests for exhumation of remains to conduct such testing. Persons may be interested in establishing paternity (or other family relationships) posthumously for a variety of reasons, ranging from establishing entitlement to inherit, immigrate, or receive government benefits, to satisfying concerns of purely genealogical interest.

In the 1990s DNA testing was used for the first time to support the claims of several persons long claiming to be descendants of deceased celebrities or historical figures, and there is reason to expect that the number of such cases will increase in the future (25). Issues may also arise posthumously concerning the parentage of children conceived through alternative reproductive methods. For example, the sperm of a man may be frozen and then used after his death to artificially inseminate a woman, resulting in the conception of a child months or even years later. In such a situation, resort to posthumous genetic testing may become necessary to establish the biological relationship between the resulting child and the deceased father. At least one such case, involving an application by Social Security survivor's benefits on behalf of a child, who was conceived by gamete intrafallopian transfer three months after the death of her biological father, has already been litigated (26). The Social Security Administration initially denied the claim, reasoning that because the child was born 13 months after her biological father's death, she could not have been his legal heir. However, the agency subsequently reversed its position, reasoning that conclusive proof of the child's biological paternity could provide a constitutional nexus for securing her entitlement (27).

The technology for the freezing of eggs has not yet developed sufficiently to allow for eggs to be fertilized and successfully implanted following the death of the woman from whom they were derived, but should this technology improve, analogous issues will arise as the resulting children seek posthumously to establish their genetic maternity. Related issues will also emerge as a growing number of children are created from embryos that have been cryopreserved for use in IVF procedures and that are not implanted until after (perhaps even many years after) the death of both biological parents. For example, in Australia, in the early 1980s controversy erupted when a married couple died simultaneously in an airplane crash after having frozen embryos for use in IVF. Because the couple was exceptionally wealthy, a question arose over whether the not-yet-implanted embryos should be provided to a third party for implantation—and if so, whether any resulting children would be legally those of the deceased couple (and thus eligible to inherit their estate) (28).

The USCACA would resolve some of the legal uncertainty regarding the posthumous determination of genetic parentage by providing that an individual who dies before the implantation of an embryo, or before a child is conceived other than through sexual intercourse, using the individual's egg or sperm, is not considered the parent of the resulting child. Once again, however, in states that have not patterned their laws on the USCACA, the resolution of these issues remains unclear. In fact, separate and apart from resolving issues of the *parentage* of children resulting from the use of such techniques, courts are still struggling with preliminary questions regarding the legal status of frozen sperm (29) and the frozen embryos (30) *themselves*—that is, whether they should be treated as property, as human life, or as something "in between."

Human Cloning and the Determination of Familial Relationships. Should human cloning (the creation of a human being through somatic cell nuclear transfer technology or a similar technology) someday become feasible, the task of sorting through the myriad potentially recognizable familial relationships will become even more complex. Although the National Bioethics Advisory Commission concluded in its 1997 report on human cloning that it would at this time be morally unacceptable for anyone to attempt to create a child using somatic cell nuclear transfer technology (31), it is nonetheless likely that successful attempts at human cloning will eventually occur, and when they do, the question will arise who is the "parent" of the clone. The process of cloning will result in a child having genetic material from as many as four individuals: the person from whom the cell nucleus was derived, that individual's two biological parents, and the woman contributing the enucleated egg cell (which contains a small fraction of mitochondrial DNA). In addition, if the egg with the transferred nucleic material is implanted in a surrogate gestational mother, the child will have two other potential parents: the gestational mother and (if she is married) her husband. There may also be intended parents unrelated to the person who is cloned, such as in cases where the cloned person is deceased or a celebrity (32). In such cases, not only will it be necessary to decide whether the child's genetic parent(s) should be given precedence over the biological (but nongenetic parent) or over the purely social parent(s), but it will also be necessary to determine who should be recognized as the child's genetic parent(s) in the first place.

GENETIC DETERMINATION OF PARENTAGE AS A CONSEQUENCE OF GENETIC TESTING FOR NONIDENTIFICATION PURPOSES

Findings of Misattributed Paternity

While genetic testing in the forensics context is the common way in which information regarding biological parentage or other familial relationships is brought to light, such information can also be uncovered inadvertently in the course of genetic testing undertaken for completely unrelated purposes, such as in clinical medicine. The common incidental finding that occurs in the context of nonforensic genetic testing is the finding of misattributed paternity (or sometimes grand-paternity). The true incidence of misattributed paternity is unknown, and undoubtedly varies widely depending on geographical region, age group, and cultural or ethnic group, among other factors. However, 10 percent is the figure most commonly cited, and estimates as high as 30 percent have been proposed (33). While more recent studies suggest that both of these figures may be substantial exaggerations (34,35), the accumulated experience of large-scale genetic screening programs (e.g., newborn screening programs) shows that the aggregate number of children born each year whose paternity is misattributed is by no means insignificant.

The incidental finding that the presumed father of a child is not the biological father can arise in a number of situations, such as when several family members are being tested to locate a suitable donor for a bone marrow or organ transplant, to take part in genetic linkage testing, or to participate in other types of genetic risk assessment that require samples from multiple family members. In cases where testing for bone marrow or donor organ compatibility yields evidence of misattributed paternity, it is often possible to communicate the fact that the individual tested and the intended recipient (e.g., two half-siblings) would not be a good match without mentioning anything about misconceptions regarding the degree of their biological relatedness. This is because it is easy to explain the *fact* of a mismatch without going into the apparent *reason* for the mismatch; there may be many reasons other than misattributed paternity why two people might not be suitably matched for transplantation purposes. The nondisclosure approach in the transplantation situation can also probably be justified ethically and legally because at least in most cases the nondisclosure is unlikely to have any direct, potentially adverse, effect on the parties' future personal medical or reproductive decision making.

On the other hand, when a finding of misattributed paternity surfaces in the context of genetic risk assessment, the stakes are higher because genetic risk estimates are based on the assumption that the biological relationships assumed to exist within a family are correct. A person's misunderstanding about his or her biological relationship to other family members can confound the clinical determination of whether he or she is at increased risk for an inherited disorder or for passing on an inherited disorder—with crucial ramifications for health and reproductive planning.

A common situation regarding misattributed paternity occurs when genetic testing is sought to determine recurrence risk following the birth of a child affected with an autosomal recessive genetic disorder, for which both parents must be obligate carriers. In some cases the woman may already suspect that another man fathered her child, and may thus seek out counseling on her own without involving her husband or partner. However, where the woman does not realize (or is in denial of the possibility that) the child has a different father, the entire family may become involved. If carrier testing in such a case reveals the presumed father not to be a carrier, this means that he cannot be the biological father. The genetic counselor or other provider then faces a dilemma: how to convey to the couple the reason *why* they are not at increased

genetic risk for bearing another affected child without simultaneously disclosing the fact that the child they already have must have been fathered by someone other than the husband (or other presumed father).

Reconciling the competing interests in cases like these can be very difficult, and no one strategy for resolving the issue is likely to be entirely satisfactory. In fact international surveys of genetic service providers performed as recently as the 1990s revealed a marked lack of consensus regarding the appropriate resolution of this dilemma, even though it is one that genetics professionals have been wrestling with for years (36,37).

Alternative Approaches to Handling Findings of Misattributed Paternity

Full Disclosure. One approach to the problem of misattributed paternity—the approach recommended in 1983 by the President's Commission for the Study of Ethical Problems in Medicine and Medical and Behavioral Research—is for the genetic counselor or other provider frankly to disclose the finding, including the conclusion that the recurrence risk in any future pregnancy of the couple is virtually zero because the existing affected child is not biologically the husband's (38). This approach accords maximal weight to the principles of autonomy and beneficence. It also reflects the practical consideration that deception regarding a child's paternity is likely eventually to be discovered in any event, and that in the long run, greater disruption to the family may result from this than from the frank revelation of misattributed paternity made when the information first surfaces in the clinical setting.

However, this approach can been criticized (at least in many cases) as placing form over substance, giving insufficient allegiance to the integrity of the family unit, and naive in its failure to recognize that many women—especially those who are in abusive relationships or who are economically disempowered—may suffer tangible detriment, in the form of physical, psychological, social, or economic harm, from the disclosure to their husbands or partners of misattributed paternity. In fact, in some cultures, the social environment may be such that an almost certain consequence of such a revelation would be clear harm to both the woman and the child. Indeed, it was this concern that led the Hereditary Diseases Programme of the World Health Organization to conclude that there is probably never a justification for a provider to reveal such a finding to a husband (39). Thus, at the very least, in cases where a decision is made to reveal information regarding misattributed paternity to both partners simultaneously, the provider should be prepared to offer appropriate psychological and other support.

Nondisclosure, Partial Disclosure, or Disclosure Only to the Woman. Another possible approach to dealing with findings of misattributed paternity is either to misrepresent the finding (or the basis for the finding) or skirt the issue in some other way, either through some form of partial disclosure or by telling the woman alone. The justification for this approach is that the genetic counselor owes greater loyalty to the integrity of the family unit than to any one family member, and that for the reasons mentioned above, revealing the complete truth simultaneously to both the husband and wife could do more harm than good.

An approach that misrepresents the facts, however, is also problematic, both from an ethical and a legal standpoint. First, to the extent that overt deception is involved (e.g., explaining away the child's disorder, and thus the reason for the lack of recurrence risk, as merely a spontaneous mutation or some other anomaly, or explaining away the test results as having been confounded by a mix-up in the testing laboratory), it risks jeopardizing the provider's professional integrity and lowers the standards of practice (40). Moreover, if the explanations given are viewed by the couple as implausible, the approach is likely to engender suspicion and mistrust. In fact, should the deception eventually be discovered, the provider could conceivably be liable for medical malpractice.

A particularly risky practice from the standpoint of legal liability is for the provider to lie outright if asked by the husband whether he is in fact the biological father (41). Nevertheless, surveys indicate that many providers follow this approach, sometimes justifying it on the basis that because the genetic testing is not being done for the purpose of discovering paternity, they have no obligation to reveal the finding, even when asked. In fact two-thirds of all United States geneticists in one survey stated that they would not tell a man that he is not the father of a child, even if he asked (42). On the one hand, in cases where a genuine and serious risk of harm to the woman or to the family appears likely if the information were to be disclosed, this approach may have considerable justification. On the other hand, as earlier discussed, it can be argued that making a genetic counselor complicit in the woman's intentional deception is always unethical, that secrets of this type are in any event unlikely to remain buried forever, and that when the truth does come out, the well-being of the woman and the family may be even more seriously jeopardized than if the deception had never been perpetuated in the first place.

A form of partial disclosure that may be less risky legally but that still raises significant ethical concerns is simply to avoid any discussion of the specifics regarding actual recurrence risk (e.g., by characterizing the results as inconclusive), thus obviating the need to discuss the husband's noncarrier status. This approach, however, may lead the couple to make inappropriate future reproductive decisions based on the erroneous belief that they are both in fact carriers (and thus have a 25 percent chance of having another affected child), when the actual risk is close to zero. Based on this inaccurate assumption, the couple may later resort unnecessarily to artificial insemination by donor, forgo future pregnancies altogether, or even divorce and seek new noncarrier mates. If they do decide to conceive another child together based on the misapprehension that such a child is at increased genetic risk, they may suffer needless anxiety and incur needless risk and expense associated with amniocentesis or other prenatal testing that is not in fact medically indicated.

Another approach is to convey to the couple the actual risk (close to zero, in the example discussed) while withholding the information about genetic transmission that would explain the reason for the risk and raise suspicions regarding nonpaternity (the fact that in order for a child to be born with an autosomal recessive disorder, both parents must carry the gene). This approach, however, may leave the couple feeling anxious and confused and lead them to suspect that something important is being withheld.

Yet another approach is to relate the finding only to the woman (who is likely in many cases to suspect anyway that another man fathered her child) and leave with her the choice as to whether or how to tell her husband or partner. This approach—the approach recommended by the Institute of Medicine's Committee on Assessing Genetic Risks in its 1993 report (43)—avoids the above-described difficulties that may arise when such a finding is revealed by an outsider, and at the same time, requires no overt misrepresentation or skirting of the issues in the provider's conversations with the couple. However, the approach seems difficult to reconcile with the notion that the ethical and legal obligations of genetic counselors run equally to both partners (44). Thus, if this approach is followed, the potential psychological benefits of disclosure, including relief from the burden of keeping a secret and greater honesty in family relationships, should be stressed with the woman. However, the potential for adverse consequences should also be raised, and once again, the provider should stand ready to provide other necessary support.

Informed Consent Approach. An emerging approach to dealing with unexpected findings of misattributed paternity is to try to avoid many of the above-described problems by addressing the issue before the testing takes place, as part of the informed consent process. Under this approach the woman (or in some cases, both partners) are informed, prior to taking the test, of the possibility that misattributed paternity will be discovered. The woman (or the couple) can then (at least in theory) agree *in advance* on the way such information, if discovered, will be handled. This approach has the advantage of making the persons most likely to be affected active participants in any decision about disclosure. However, it too has limitations, due the practical realities of the context in which genetic testing typically occurs. The very inclusion of the subject of paternity among the subjects treated in an informed consent document may provoke anxiety, and where the woman is aware that misattributed paternity may be an issue, she may well "panic" in the situation, having never seriously thought before about the ramifications of that possibility. Even if the pre-test counseling is done separately, the woman may feel confused about how to now "get out of" a test she had previously seemed to agree to (before the possibility was called to her attention). In the end this approach could discourage some women who would like to obtain genetic information from participating in genetic testing—perhaps, in some cases, to the future detriment of themselves and their families.

Some genetic testing centers include in their standard informed consent form a reference to the possibility of an incidental finding of misattributed paternity, but simply state what the center's policy is regarding the communication of such findings without giving the woman (or the couple) an opportunity to communicate her (or their) preference in this regard. This approach is problematic for many of the same reasons. In addition it is based on the (typically erroneous) assumption that a couple who is uncomfortable with a center's policy can simply "go elsewhere" for testing. Testing for some genetic disorders (particularly those that are relatively rare) may, as a practical matter, only be available at a single location. Insurance and other practical constraints may also limit a couple's ability to "shop around" for a center with a more favorable disclosure policy.

Other Possible Unanticipated Findings Regarding Familial Relationships

Misattributed paternity is not the only type of unanticipated finding regarding biological parentage that may surface when genetic testing is sought for nonidentification purposes. Genetic testing may also bring to light the fact that a child has been adopted, was conceived through donor insemination or another alternative reproductive method, or is the product of an incestuous mating (perhaps within the extended family). Situations like these raise ethical concerns similar to, but slightly different from, those present in cases where paternity has simply been misattributed by the mother to her husband or partner. Unlike in the misattributed paternity situation where disclosure of the finding is likely to come as a surprise to the presumed father, the basic facts surrounding an adoption, donor insemination, or incest will already (presumably) be known to both the husband and the wife. For this reason the risks associated with disclosure of the child's biological status are unlikely to interfere directly with the couple's relationship. The decision whether to disclose could, however, have crucial ramifications for the *parent–child* relationship, as well as for the child's own sense of psychological stability, because in such cases it is the child who has not yet been made aware of the family secret. Thus, once again, in these situations the genetic counselor is faced with a dilemma: Disclosing to the child his or her biological status may upset both the parents (who may arguably have sound reasons for opposing such disclosure) and the child (who may be upset by the revelation). On the other hand, *not* disclosing the information may make the counselor complicit in a deception that could ultimately be viewed by the child as a betrayal should the facts later come to light.

In general, where the child has not yet reached majority, it would generally seem appropriate for the counselor to respect the parents' wishes not reveal the information (although the counselor should advocate strongly that the child be told). Once the child has reached adulthood, however, the situation becomes more complex. In general, if the adult individual (perhaps suspecting that certain information has long been withheld) specifically *asks* whether the genetic test results reveal anything unusual regarding his or her parentage, it would seem ethically required for the counselor to reveal the information, even if doing so may upset the parents. If the adult individual does *not* ask, but if it appears

that the information could be highly relevant to his or her own health or reproductive planning, the provider must balance the possible medical risks associated with nondisclosure against the possible psychological risks associated with a disclosure that turns out not to have been desired by anyone in the family (including the individual adult most directly affected) (45). The optimal approach is to try to anticipate such eventualities before the testing takes place, by raising the issue directly during the informed consent process.

Genetic testing performed in the context of clinical medicine may occasionally reveal that babies were switched (whether inadvertently or deliberately) in a hospital nursery. It may also bring to light the finding that sperm samples were mixed up in the course of artificial insemination (e.g., that the husband's sperm was confused with that of an anonymous donor, resulting in the birth of a child whom the parents had mistakenly believed was biologically related to both of them). Or, it may reveal that embryos used in the process of IVF were switched (e.g., where a woman gives birth to a baby whose phenotype suggests strongly that it had been fathered by a man of another race). In these cases any deception or other fault is unlikely to lie with either party to the marriage, but rather with the physician or other hospital personnel who were responsible for the mistake. For this reason the primary considerations underlying the decision whether to disclose will involve not so much potential risks to the mother's well-being as the potential for disruption to the family as a whole. The decision whether to disclose will also have important ramifications for the legal liability of the persons responsible for the original error. There are no easy resolutions in such extremely sensitive cases, and such cases are likely to arise even more frequently in the future as the number of children created through alternative reproductive methods continues to grow.

ACKNOWLEDGMENT

Note: The views expressed are solely those of the author and do not represent the views of NHGRI, NIH, or DHHS.

BIBLIOGRAPHY

1. *State v. Van Guilder*, 271 N.W. 473 (Minn. 1937).
2. *Andrade v. Newhouse*, 128 P.2d 927 (Cal. App. 1942).
3. J.W. Morris and D.W. Giertson, in D.L. Faigman, ed., *Modern Scientific Evidence: The Law and Science of Expert Testimony*, 1997.
4. 1 McCormick on Evidence, 4th ed., §205, at 895 (1992).
5. D.H. Kaye and R. Kanwischer, *Fam. Law. Q.* 22, 109–116 (1988).
6. D.H. Kaye, *Fam. Law. Q.* 24, 279–304 (1990).
7. A.Z. Litovsky and K. Schultz, *Jurimetrics* 39, 79–94 (1998).
8. *Uniform Parentage Act, Uniform Laws Annotated*, vol. 9B, West Publishing, St. Paul, MN, 1973, pp. 287–345.
9. *Uniform Act on Paternity, Uniform Laws Annotated*, vol. 9B, West Publishing, St. Paul, MN, 1960, pp. 347–368.
10. R.H. Walker, ed., *Inclusion Probabilities in Parentage Testing*, 1983.

11. N.M. Vitek, *Disputed Paternity Proceedings*, 5th ed., Matthew Bender, New York, 1997.

12. Omnibus Reconciliation Act of 1993, Pub. L. 103-66, 42 U.S.C. §666(5).

13. L.W. Morgan, *Am. J. Fam. Law.* **12**, 204–211 (1998).

14. *Michael H. v. Gerald D.*, 491 U.S. 110 (1989).

15. *Miscovich v. Miscovich*, 688 A.2d 726 (Pa. 1997).

16. *Uniform Status of Children of Assisted Conception Act, Uniform Laws Annotated*, vol. 9B, West Publishing, St. Paul, MN, 1988, pp. 191–206.

17. *Uniform Putative and Unknown Fathers Act, Uniform Laws Annotated*, vol. 9B, Supp., West Publishing, St. Paul, MN, 1988, pp. 91–105.

18. A.T. Lamport, *Am. J. Law Med.* **14**(1), 109–124 (1988).

19. American Society of Human Genetics, Social Issues Committee, *Am. J. Hum. Genet.* **48**(5), 1009–1010 (1991).

20. M. Coleman, *Cardozo Law Rev.* **17**, 497–530 (1996).

21. In Re Baby M, 537 A.2d 1227 (N.J. 1988).

22. *Johnson v. Calvert*, 851 P.2d 776 (Cal. 1993).

23. Office of Technology Assessment, U.S. Congress, *Infertility: Medical and Social Choices*, OTA-BA-358, U.S. Government Printing Office, Washington, DC, 1988.

24. C.N. LeRay, *BC Law Rev.* **35**, 747–798 (1994).

25. H.B. Jenkins, *Alabama Law Rev.* **50**, 39–61 (1998).

26. *Hart v. Shalala*, Case No. 94-3944 (E.D. La. 1994).

27. E. Garside, *Loy. Law Rev.* **41**, 713–734 (1996).

28. *Syracuse Law Rev.* **36**, 1021–1053 (1985).

29. *Hecht v. Superior Court*, 20 Cal. Rptr. 2d 225 (Cal. App. 1993).

30. *Davis v. Davis*, 842 S.W.2d 588 (Tenn. 1992).

31. National Bioethics Advisory Commission, *Cloning Human Beings: Report and Recommendations of the National Bioethics Advisory Commission*, 1997.

32. N. Elster, *Hofstra Law Rev.* **27**, 533–555 (1999).

33. S. MacIntyre and A. Soosman, *Lancet* **338**, 869–871 (1991).

34. M.-G. LeRoux et al., *Lancet* **340**, 607 (1992).

35. P.J.H. Brock and A.E. Shrimpton, *Lancet* **338**, 1151 (1991).

36. D.C. Wertz, *Health Law J.* **3**, 59 (1995).

37. D.C. Wertz, *Clin. Genet.* **54**, 321–329 (1988).

38. United States President's Commission for the Study of Ethical Problems in Medicine and Biomedical and Behavioral Research, *Screening and Counseling for Genetic Conditions: A Report on the Ethical, Social, and Legal Implications of Genetic Screening, Counseling, and Education Programs*, United States Government Printing Office, Washington, DC, 1983.

39. D.C. Wertz, J.C. Fletcher, and K. Berg, *Guidelines on Ethical Issues in Medical Genetics and the Provision of Genetic Services*, World Health Organization, Hereditary Diseases Programme, Geneva, 1995.

40. B. Biesecker, in M.A. Rothstein, ed., *Genetic Secrets: Protecting Privacy and Confidentiality in the Genetic Era*, Yale University Press, New Haven, CT, 1997, pp. 108–125.

41. P.R. Reilly, *Suffolk U. Law Rev.* **27**, 1327–1357 (1993).

42. D.C. Wertz, *J. Contemp. Health Law Policy* **13**, 299–346 (1997).

43. Institute of Medicine, Committee on Assessing Genetic Risks, *Assessing Genetic Risks: Implications for Health and Social Policy*, National Academy Press, Washington, DC, 1994.

44. G.S. Omenn, J.H. Hall, and K.D. Hansen, *Am. J. Med. Genet.* **5**(2), 157–164 (1980).

45. S.M. Suter, *Mich. Law Rev.* **91**(7), 1854–1908 (1993).

See other entries GENETIC INFORMATION, ETHICS, FAMILY ISSUES; GENETIC INFORMATION, ETHICS, PRIVACY AND CONFIDENTIALITY: OVERVIEW; GENETIC INFORMATION, LEGAL, FDA REGULATION OF GENETIC TESTING; GENETIC INFORMATION, LEGAL, GENETIC PRIVACY LAWS; GENETIC INFORMATION, LEGAL, REGULATING GENETIC SERVICES.

GENETIC INFORMATION, ETHICS, ETHICAL ISSUES IN TISSUE BANKING AND HUMAN SUBJECT RESEARCH IN STORED TISSUES

CURTIS NASER
Fairfield University
Fairfield, Connecticut

SHERI ALPERT
University of Notre Dame
Notre Dame, Indiana

OUTLINE

Introduction

Human Tissues: What They Are, and How They Are Used

Current Federal Regulations for Protecting Human Subjects in Biomedical Research

    Exemption Implied by the Definition of "Human Subject"

    Exemption for Existing Tissues and Medical Records

    Conditions Necessary for the Waiver of Informed Consent

    General Requirements for Informed Consent and Special Provisions for Genetics Research

Autonomy, Privacy, and the Social Good

    How the Regulations Encode and Interpret Autonomy and Beneficence

    Defining Privacy

    Consequentialist Interpretation of Autonomy

    Deontological Approach to Autonomy

Conclusion

Bibliography

INTRODUCTION

Research using human tissues and cells has contributed immensely to progress in the medical and basic biological sciences. These materials have been essential in developing and testing new drugs and vaccines, investigating infectious diseases, and exploring the mechanisms of virtually all disease processes. The use of human tissues and cells is also the foundation upon which much of the current biotechnological revolution has been based. The project to identify, map, and ultimately sequence the human genome would not be possible without the thousands of human tissue specimens from which DNA is routinely extracted and analyzed. Human tissues are also used for a variety of other medical purposes, including the transplantation of whole organs (kidney, heart, liver, eye), the transfusion of blood and blood products, and diagnostic testing (blood chemistry, pathological identification of diseased tissues). In addition human tissues may be used in nonmedical contexts, such as the forensic identification of suspects based on trace tissues left at crime scenes, or the identification of the remains of soldiers.

As the collection, storage, and use of human tissues in biomedical research has increased, and the power of the scientific methods to analyze and unlock the secrets held within these tissues has grown, a number of ethical and public policy issues have been identified and debated in the research community. Traditionally excess tissues removed at the time of surgery or in other diagnostic procedures have been viewed as "waste" abandoned by patients and left to the disposal of the hospital or clinical lab. It has been presumed that patients would have no further interests in the disposition of their tissues and that most if not all would be glad to have medical researchers putting them to productive use. However, with recent advances in biotechnology, especially in the area of genetics, suddenly these "waste" tissues have become what one commentator has called a "coded future diary" of the individual and his/her family (1). Intimate knowledge about a person's medical condition (both their current status and potential future status) may be gleaned from the tiniest samples of human tissue.

Since many human tissues are used in research without the knowledge or consent of the persons from whom they are derived (tissue sources), the new-found powers of molecular and genetic analysis raise many difficult questions: What, if anything, should tissue sources be told about the results of research findings? Since information gleaned from tissue specimens, especially genetic information, can adversely affect an individual's employability and insurability, how can sources of research tissues best be protected from these social risks? Who *owns* these tissues and what rights, if any, do tissue sources have to financial gains derived from the use of their tissues in research? And, indeed, is "ownership" even an appropriate concept in this context? How do we protect individuals whose religious faiths or cultural practices impose special restrictions upon the disposition (or burial) of body parts? And, most important, should the informed consent of the tissue source be required for the research use of their tissues, and in what circumstances and how much information should be provided in the process?

HUMAN TISSUES: WHAT THEY ARE, AND HOW THEY ARE USED

Tissues (also referred to as "human biological material") can include everything from organs and parts of organs, cells and tissues (like bone, muscle, connective tissues and skin), to subcellular structures (e.g., DNA) and cell products, blood, gametes (sperm and ova), embryos and fetal tissue, and waste (urine, feces, sweat, hair, nail clippings, epithelial cells, placenta (2,3). Tissue specimens can be stored in many forms depending on both the reason for their collection, and their intended use: in paraffin blocks, slides, formalin fixed, frozen, tissue culture,

extracted DNA, or dried blood spots (e.g., on Guthrie cards). Cryogenic storage is generally used for cord blood, gametes, and embryos.

Human tissues are collected and stored (or banked) through a variety of means and for a variety of purposes. They are most commonly collected in conjunction with diagnostic procedures or surgical treatment. For instance, at the time an individual has a surgical procedure to remove diseased tissue, that tissue may be examined to determine whether it contains malignant cells. The diseased tissue is generally preserved and maintained by pathology laboratories for several years (indeed, some tissue archives retain samples that were collected over 100 years ago). This storage of clinically derived specimens may be legally required in some states since they are regarded as part of the patient's clinical record. If questions arise at a later date about the adequacy of the laboratory testing of the specimen, the preserved specimen may be used to confirm (or refute) the original diagnosis.

Additionally excess diagnostic specimens may be used for follow-up clinical care, but most commonly they are used for educational and research purposes, as well as laboratory quality control. For instance, tissues can be used to ensure that equipment in diagnostic and pathological laboratories is functioning properly. Pathology slides may be used for the education of medical students and other specimens may be used to train technicians in testing procedures. Excess specimens are often made available to researchers for a variety of purposes. Fresh specimens may be cultured into immortalized cell lines which provide a perpetual source of DNA and may be used for a variety of other research purposes. Live tissues may also be cultured for use in pharmaceutical research. For instance, the first anti-viral drug for the treatment of AIDS—AZT—was initially demonstrated to be effective by testing it on HIV-infected cell cultures (4,5). Preserved pathological specimens may be used for a variety of research purposes, from studies of enzymes, proteins, and cell physiology to genetic analyses.

Many tissues enter into biomedical research through explicit research protocols. In gene mapping studies, for instance, DNA samples will be collected from entire families suspected of harboring a specific disease gene. Often, immortalized cell lines will be cultured from their blood samples to provide a constant source of DNA without having to return to the family members for further blood samples. Tissue specimens collected for one research purpose may also be used for secondary research investigations, either related to the initial research purpose or for completely unrelated research purposes.

Tissues may be stored in a variety of locations, depending upon the reason for their collection and their ultimate use. The storage facilities include military facilities, forensic DNA banks, government laboratories, diagnostic pathology and cytology laboratories, university- and hospital-based research laboratories, commercial enterprises, and nonprofit organizations (6). Tissue collections can range in size from fewer than 200 specimens to over 92 million, with a conservative estimated total of at least 282 million specimens (from over 176 million cases) (6). The two largest collections of human tissue in the world (National Pathology Repository and the DNA Specimen Repository for Remains Identification) both reside within the U.S. Armed Forces Institute of Pathology (AFIP), which stores over 94 million specimens (3). An additional 13.5 million specimens are accounted for by newborn screenings, and an estimated 160 million or more specimens can be found throughout the various U.S. graduate medical education teaching institutions (3). These figures do not capture the additional information and tissues that may be found in cancer registries in many states, nor do they contemplate specimens that may be collected as part of the Human Genome Diversity Project.

In all these applications and uses, the tissues themselves can retain varying levels of identifiability — namely the ease with which the identity of the tissue source can be established. The protections afforded human research subjects in the United States are closely tied to how easily the identity of the tissue sources can be discerned. It is thus useful to lay out a general taxonomy of tissue identifiability:

- *Identified*. The tissue source is known and the individual's identity is tied to the sample. (This would be the case with specimens being analyzed for diagnostic and treatment purposes.)
- *Identifiable*. The tissue source is tied to the specimen through the use of a link (e.g., a code number), but the identity of the source is not directly known without tracing the link. Many pathology specimens, for instance, are archived according to a pathology record number.
- *Anonymized*. The tissue source's identity is irrevocably unlinked from the specimen, so that the individual's identity cannot be discerned (i.e., the tissue is unidentifiable).
- *Anonymous*. The tissue source's identity is never known, since the specimen is collected with no identifiers at all (i.e., the sample is unidentified).

It should be noted that as biomedical research progresses, becoming increasingly sophisticated in its ability to tease apart the molecular components of tissues and cells, the notion of having an "anonymous" or even "anonymized" sample will likely diminish, if not disappear altogether. In the future, determining the sequence of a tissue sample's (and tissue source's) DNA could become a routine procedure. If this were to happen in the clinical context, which is quite likely given the different analytical systems now under development (7), this information will invariably end up in a patient's medical record. Since these records are increasingly being stored and processed electronically, it could require only a small endeavor to match sequences from a medical record against the sequences from stored tissues. When that happens, the protections to safeguard tissue source identity currently in place, discussed in the next section, could become wholly inadequate to the task.

CURRENT FEDERAL REGULATIONS FOR PROTECTING HUMAN SUBJECTS IN BIOMEDICAL RESEARCH

The U.S. federal regulations (8,9) governing the use of human subjects in biomedical research establish three primary requirements intended to protect human subjects. First and foremost is the requirement for informed consent, which includes among other things a statement of the purpose of the research and its probable risks and benefits to the subject. Second is the requirement that all research involving human subjects be reviewed by an Institutional Review Board, (IRB) that is composed of other scientists and physicians, at least one nonscientist, and at least one member not affiliated with the institution. This committee is charged with evaluating the adequacy of informed consent, establishing that the research poses a favorable risk benefit ratio to the subject, and ensuring that the research design will yield useful results.

The third requirement is that institutions conducting federally funded research file an "assurance" with the Department of Health and Human Services (DHHS) Office for the Protection from Research Risks (OPRR), which administers and enforces the regulations. This assurance contains a statement of the ethical principles to be followed in conducting the research as well as the constitution of the IRB and its operating procedures. Institutions which conduct multiple research protocols may apply for a "multiple project assurance" which permits a single IRB to review each protocol. Institutions holding an MPA are required to ensure that *all* research involving human subjects conducted by their employees and affiliates complies with the regulations, including projects not funded by the federal government. Thus a wide array of research is subject to these regulations. There are approximately 450 institutions in the United States currently holding an MPA (10). Institutions which do not regularly conduct federally funded research may not be covered by these regulations. Privately funded research institutions, such as pharmaceutical firms, may engage in research involving human tissues outside of the regulatory context, including review by an IRB and requirements for informed consent. The Food and Drug Administration maintained until 1991 separate regulations governing research involving human subjects that was undertaken by drug and medical device manufacturers. In 1991, regulations governing research on human subjects conducted by the many agencies of the Federal government were consolidated into what is known as the Common Rule. However, the FDA maintains a separate system of enforcement (largely by audit) and its regulations do not cover the use of human tissues or medical records, but apply only to research conducted under the auspices of an Investigational New Drug application (IND) or an Investigational Device Exemption (IDE) as part of the FDA approval process prior to marketing. Thus, basic research conducted by private corporations and research labs prior to an application to the FDA for marketing approval is not regulated, unless these labs are otherwise regulated through their receipt of Federal funds.

The current regulations and the IRB system of peer review are recent phenomena that grew out of revelations in the 1960s and 1970s of unethical research practices, including the infamous Tuskegee syphilis study. These abuses led to the first formal regulations issued in 1974 by the U.S. Department of Health, Education, and Welfare (HEW) (11). That same year, Congress passed the National Research Act, which established a National Commission for the Protection of Human Subjects of Biomedical and Behavioral Research, which was charged with reviewing all aspects of the involvement of human subjects in research and making recommendations to improve the system of protections. The National Commission issued a series of reports, including the *Belmont Report* (12) which set out the broad ethical guidelines that research involving human subjects should follow. The Commission also issued specific recommendations regarding the IRB review process. HEW was required by law to codify these recommendations into its regulations. Proposed regulations were published in 1979 by HEW (13) and after public comment, were finally adopted in January 1981 by what had then become DHHS (14).

The research use of tissues (and medical records) was explicitly considered by the National Commission and as part of its recommendations to HEW, the Commission suggested that these research projects need not obtain informed consent provided that "the importance of the research justifies the invasion of privacy" (15). The National Commission recognized that the use of human tissues and medical records constituted an "invasion of privacy," but argued that if adequate safeguards to protect individual's confidentiality were provided, this research was of minimal or no risk to subjects and could proceed in the absence of informed consent. Nevertheless, the Commission's recommendation would have required IRB review of all such research protocols. The National Commission also recommended that institutions that anticipate using tissues and records in research should in lieu of informed consent, notify all patients and provide a mechanism (i.e., a blanket consent) by which they could opt out, if they so desired.

However, in codifying these recommendations, HEW opted not to follow completely the National Commission's recommendations. Concerned about IRB workloads and convinced that much of this research was of minimal or no risk, HEW instead proposed two exemptions for such research from the IRB regulations. In addition, in these cases, HEW provided that IRB's could waive or modify informed consent.

The first exemption from the requirements of IRB review and informed consent is implied by the definition of "human subject" and the statement [at 45 CFR 46.102(f)] that the regulations apply only to research involving *human subjects*. It is possible that some research protocols can claim that their use of human tissues, under narrowly defined circumstances, do not involve human subjects as defined, and therefore the regulations do not apply. The second exemption is an explicit exemption governing the use of existing specimens and records, provided only that investigators do not record any patient identifiers or links to identifiers. Protocols that meet the requirements of either exemption may proceed in the absence of IRB review and in the absence of any informed consent, unless state laws or institutional policies provide otherwise.

For protocols that do not meet the requirements for either exemption, it is possible that the IRB nonetheless may permit a waiver or modification of informed consent. There are four conditions to be met before an IRB may waive or modify consent. Because of their ambiguity and because they point to important ethical issues, the discussion of these conditions below will extend to some of the important ethical and policy questions surrounding the use of human tissues in research, including the recontact of tissue sources with research findings, the ownership and commercial use of tissues, as well as the potential risks to which new biotechnologies may give rise.

Research protocols involving the use of human tissue that do not meet any of the above conditions for exemption or waiver of consent are required to obtain the informed consent of the subjects. Many tissue protocols may follow standard informed consent requirements used at the investigator's institution, but some protocols, especially those involved in genetics research, may require special considerations.

Exemption Implied by the Definition of "Human Subject"

This first exemption from IRB review arises out of the regulation's definition of "human subject." If a research protocol can demonstrate that it does not involve human subjects, then the regulations do not apply. The regulations define "human subject" as a "living individual about whom an investigator (whether professional or student) conducting research obtains" either of the following: (1) data through intervention or interaction with the individual, or (2) identifiable private information (§46.102(f)).

Interventions and Interactions. Criterion 1 is an important consideration because obtaining human tissues may often require some intervention or interaction with a person, usually an invasive procedure such as a needle stick to draw blood, or a tissue biopsy. Since these procedures may involve pain or risk harm to subjects, the regulations require that the interventions be subject to the oversight of an IRB, though it is possible that the IRB may waive the requirement for informed consent (see below).

One circumstance that easily gives rise to confusion in the application of this criterion is when tissues are removed from a patient for clinical reasons. If the attending physician is also a researcher who hopes to use some of the tissues for research purposes, and if the diagnostic procedure is genuinely clinically indicated, then the physician might assume that there is no interaction or intervention which would not have taken place in the absence of the research use of the tissue. However, the regulations do not distinguish between such dual roles of clinician and investigator. If the clinician is also an investigator, then, by definition, there is interaction, if not intervention, by the clinician qua investigator—whether or not the procedure is ordered solely for clinical purposes. The duality of roles in such cases presents the possibility that individuals may be subjected to research procedures under the guise of clinical care, or at the very least that a clinician's judgment may be influenced by his or her research interests, and thus would modify the patient's care accordingly. Such modifications of patient care for research purposes, however, present the possibility of further risks to these subjects, and as such, the regulations require at least the oversight of an IRB to evaluate those risks, and to advise the investigator on appropriate procedures and safeguards.

Identifiable Private Information. Many protocols that seek to collect, store, and experiment on human tissues use excess tissues from clinical procedures with which the investigators are not associated. These tissues may be gathered from blood banks, blood chemistry labs, pathology labs, cytogenetics labs, or any other clinical diagnostic laboratory. Since more diagnostic specimens are routinely collected than are strictly necessary to perform the clinical tests and analyses, there is often excess tissue which may be made available to investigators. In these cases, research protocols can clearly meet the first criterion in the definition of human subject, namely that the protocol does not involve any intervention or interaction with an individual.

In order to meet the second criterion of the exemption, however, the investigators must not obtain any "identifiable private information." OPRR has interpreted this phrase to mean that if investigators have access, *however briefly*, to individual identifiers (such as names or social security numbers) or to links to identifiers (such as a medical record number or pathology record number or other such code) then the research involves human subjects as defined and the protocol would fail this exemption (16).

While some research protocols involving human tissue may be able to use anonymized tissue specimens, more powerful research can be conducted on identified or linked specimens. By maintaining a link back to the originating laboratory and/or the medical record, it is possible to update a tissue bank's database with subsequent morbidity and mortality data from the sources of the tissue specimens. Several tissue banks and tissue procurement services exist (17) or have been proposed in the literature (18) that are designed to maintain a one-way flow of information from medical record to tissue bank and on to individual investigators, while ensuring through the use of codes that no research information finds its way back to individual patients or subjects. Since some research, especially in genetics, can generate sensitive information about an individual, it is important to maintain a clear distinction between research results and clinical information in order to protect the source from the harmful consequences of research as well as from the uncertainty of results that have not been further validated. While such one-way tissue banks make very powerful research tools, they would fail to meet the exemption implied by the definition of "human subject," since they must maintain linking codes back to the originating laboratories and ultimately back to patient identifiers.

Exemption for Existing Tissues and Medical Records

The second exemption for research involving human tissues is in Section 46.101(b) of the regulations and it lists a number of specific exempt categories of research. Most of these categories pertain to educational, social or psychological research which involve the use of surveys and questionnaires. However, §46.101(b)(4) explicitly addresses the use of tissues and medical records in research. It exempts:

> [r]esearch involving the collection or study of existing data, documents, records, pathological specimens, or diagnostic specimens, if these sources are publicly available or if the information is recorded by the investigator in such a manner that subjects cannot be identified, directly or through identifiers linked to the subjects.

The key determination in applying this exemption is that the research materials must be *existing*. OPRR has clarified the interpretation of "existing" to mean that the materials (specimens, medical records, etc.) must be already existing *at the time the research is proposed* (19). Thus this exemption only applies to tissues and records that have already been produced and are being stored when the protocol solicits exemption status from the IRB. Such protocols are typically referred to as *retrospective*. Although the historical record of the promulgation of these regulations does not discuss the reasoning behind this limitation, it can be easily seen that it serves two purposes. First, because the specimens must already be in existence and therefore will already have been removed from individual patients, it would be impossible for investigators to influence clinicians—diagnostic technicians, surgeons, and so forth—to alter their procedures to obtain more or different kinds of tissue than they would otherwise remove in the course of clinical care and diagnosis.

Second, it would be much more difficult and in many cases impossible to recontact former patients to inform them of the intentions of the investigators to use their tissue specimens in research and to ask their consent. Many patients will have moved or died and recontact could only be made at great expense; expense that would be prohibitive for many research protocols. It is also possible that sampling bias might result if investigators were limited to only those tissues for which consent could be obtained.

It should be noted that this exemption does permit investigator access to patient identifiers, though it requires that investigators *record* no identifiers or links to identifiers. The purpose of this limitation is the same as that expressed in the limitation under the definition of "human subject" that investigators obtain no "identifiable private information." Because the exemption for existing specimens is more liberal in permitting investigators access to identifiers, the first exemption based upon the definition of "human subject" applies primarily to *prospective* protocols, that is, protocols that seek to collect specimens produced or procured after the protocol is proposed to the IRB.

While tissue banking and collection protocols that meet either of these exemptions are therefore exempt from the federal requirement of informed consent, the regulations do not preempt state laws. While some states laws defer to the federal regulations in matters concerning the protection of research subjects, other states may have independent laws and regulations governing human subjects research. Before investigators proceed on the basis of exemption from the regulations they, or their IRBs, would be prudent to confirm their compliance with local law.

Who Decides What is Exempt? One problem in applying the exemptions that is not addressed in the regulations is deciding *who* determines what is exempt in the first place. Applying the exemption categories can be complex and confusing. Investigators may easily misinterpret them and fail to submit to the IRB research that in fact would not be exempt and may in fact even require informed consent. The MPAs at some institutions may require that the IRB or IRB chair make this determination. In other cases, it may be institutional policy that only the IRB may make exemption determinations. This is the safest policy, provides for a uniform application of the regulations, and limits the possibility that institutional funding could be jeopardized by the failure of an investigator to properly interpret the regulations.

Conditions Necessary for the Waiver of Informed Consent

For protocols that do not meet the criteria for exemption from the regulations, as described above, the possibility remains that the IRB could waive or modify the requirements for informed consent, as provided for in §46.116(d). In order to do so, the IRB must find and document that a protocol meets all four of the following conditions:

1. The research involves no more than minimal risk to the subjects.
2. The waiver or alteration will not adversely affect the rights and welfare of the subjects.
3. The research could not practicably be carried out without the waiver or alteration.
4. Whenever appropriate, the subjects will be provided with additional pertinent information after participation.

The discussion that follows will look at the ethical issues that arise in the application of each of these four conditions, since many of the issues that arise in tissue research will do so within the context of one or more of them. The more general ethical question of whether informed consent should or should not be required for these protocols (and this includes the exemptions) will be taken up later in this article.

Minimal Risk. Minimal risk is defined in the regulations at §46.102(i):

> Minimal risk means that the probability and magnitude of harm or discomfort anticipated in the research are not greater in and of themselves than those ordinarily encountered in daily life or during the performance of routine physical or psychological examinations or tests.

This definition presents two pairs of issues that need to be clarified. First, the scope of the term "harm" needs to be established, and second, the distinction between the *probability* and the *magnitude* of harm that may result from participation in a research protocol needs to be clarified.

Physical Harms. Many research protocols involving the collection, use, and storage of human tissues involve no contact with the tissue sources, the tissues being procured from excess diagnostic or pathological specimens which have been generated for clinical (or other) purposes. In such protocols, there is virtually no risk of physical harm to these sources that may result from the research use of their tissues. Thus the use of genuinely excess tissue specimens, procured by whatever means, is generally of minimal *physical* risk.

One exception to this may arise when the investigators explicitly solicit tissues from the clinicians who order the diagnostic or surgical procedures, or when these clinicians are the investigators themselves. In such cases, it is possible that clinical procedures may be altered to accommodate the research need for more tissue. Whether or not the extraction of additional tissue constitutes a greater than minimal risk to the subject depends on the kind of procedure the patient will undergo. If it is simply a matter of drawing an additional vial of blood, then the additional physical risk is minimal if not non-existent for most subjects. The same may be true of procuring additional bone marrow during a biopsy procedure. But risks may be heightened, for instance, if additional spinal fluid is procured during a spinal tap or additional liver tissue is procured during a liver biopsy. The risks of taking additional tissue will depend on the specific circumstances of the procedures being used. See Holder and Levine (20), who argue that in many cases, small amounts of additional tissues taken during surgery for research are of minimal or no risk and do not require informed consent. There are no hard and fast guidelines in this area and ultimately IRBs must make their own judgment of whether the procurement of additional tissue for research purposes during a clinical procedure is of minimal risk or not.

It is important to note, however, that the risks that result from the procurement of additional tissue for research purposes do not include the risks of the procedure itself, which is being conducted for clinical purposes. Thus the risks of infection from a spinal tap procedure are not a result of the research use of the additional spinal fluid taken, but result from the procedure itself which is ordered for clinical purposes. Likewise the risks of surgery are not a *research risk* if the surgery is undertaken for clinical purposes, while a small amount of additional tissue is taken for research. On the other hand, if an invasive diagnostic procedure is undertaken strictly for research purposes, then the full risks of the procedure must be considered in determining whether it is of minimal risk. A blood draw, for instance, is a routine diagnostic procedure and may be judged to be of minimal risk. A bone marrow biopsy, however, entails greater risks of infection and greater possibility of pain and discomfort and therefore may not be of minimal physical risk.

Psychosocial Harms. In addition to the physical risks of harm and discomfort, tissue sources may be at risk of social, psychological, economic, and legal harms if research results and any medical record information collected from the tissue source are not kept confidential. Such risks will depend on the specific experiments to be conducted upon the tissue specimens. Analyses of protein structure may have little impact on one's self-image, insurability, or employability, but some genetics research results can have a devastating impact on the source if the results were to find their way back to the individual's medical record or into the hands of insurers and employers. Although the definition of minimal risk adopted in the current regulations does not clearly identify these risks as relevant considerations, the National Commission in the *Belmont Report* clearly indicated that IRBs should be concerned with these possible consequences:

> Many kinds of possible harms and benefits need be taken into account. There are, for example, risks of psychological harm, physical harm, legal harm, social harm and economic harm and the corresponding benefits. While the most likely types of harms to research subjects are those of psychological or physical pain or injury, other possible kinds should not be overlooked (12, p. 15).

A number of commentators have argued that, depending upon the specific circumstances, genetic information developed in research protocols may entail more than minimal risks to subjects (21–24). Others have argued that the excellent record of researchers in the United States in maintaining the confidentiality of research data is a sign that the risks of a breach of confidentiality are so low that we may judge all research involving tissues to be of minimal risk (25–28). Part of this judgment will depend on what, if any, information is communicated back to the tissue sources. If the sources are identified and research results either are or can be communicated back to them, then clearly these persons may be at risk of learning things about themselves which they never consented to finding out. If sources are not identified, or there are no intentions to provide sources with research results, then the risks that these results may have an adverse impact on them are greatly minimized.

However, despite the intentions of investigators not to inform sources of research findings, there may be occasions when research records are used in public health or even criminal investigations. If codes are maintained that would link to individual identifiers, it is possible that individuals will become involved in such activities as a result of the use of their tissue specimens or medical records. For instance, tissue specimens from the Navaho Health and Nutrition Survey maintained by the Centers for Disease Control were used in the investigation of the hanta virus outbreak in the four corners region of the southwest, even though the sources of these tissues never consented to such use (29). Public health investigations are not governed by these regulations and because they often involve infectious diseases and seek to minimize imminent risks to the public, they may override the usual considerations for the protection of human subjects. The same justifications can be made in criminal investigations, though for certain classes of research, especially research involving illicit drug use, psychiatric problems and even genetics, investigators may obtain a "certificate of confidentiality" from DHHS (30,31) which protects research data and

materials from most court ordered subpoenas. Certificates of Confidentiality are issued under the Public Health Service Act §301(d), 42 U.S.C. §241(d). Categories of research information for which such certificates are issued include what "would normally be recorded in a patient's medical record, if the disclosure ... could reasonably lead to social stigmitization or discrimination," "information ... damaging to an individuals financial standing, employability, or reputation," and "genetic information."

Distinguishing between Probability and Magnitude of Harm. Whether the risks are physical, psychological, or social, it is essential to clearly distinguish the difference between the *probability* of harm and the *magnitude* of harm that may result from a subject's participation in a research protocol (12, p. 15). Some harms may be of very low probability. For instance, the record of researchers maintaining the confidentiality of private information is excellent. Thus, in general, the probability of a breach of confidentiality may be quite low in many research projects, especially when investigators take steps to protect confidentiality by the use of codes and locked file cabinets. However, the magnitude of the harm such a breach would cause depends on the sensitivity of the information itself and the context in which the breach occurs.

For instance, disclosure of the histological tissue type of an identified specimen may be of little consequence to the tissue source, since this information has little or no clinical or social relevance, except in organ transplantation. However, disclosure that a given identified tissue contains the genetic mutation for Huntington's disease may have a profound impact upon the source, both psychologically as well as socially and economically. Such an individual may find it difficult or impossible to obtain health insurance, or even find a job. While the risk of a breach of confidentiality may be low, the magnitude of social, psychological, or economic harms that may result if confidentiality is breached may be quite high. Again, the definition of minimal risk does not provide a great deal of guidance on how IRBs are to weigh the risks and magnitudes of harms and the current literature is divided on this question as well. Under the current regulatory framework it remains to the individual IRBs to make this judgment as best they can.

Rights and Welfare. The second criterion to be met is that the waiver or modification of informed consent does "not adversely affect the rights and welfare of the subjects." This has been a difficult phrase to interpret. The phrase "rights and welfare" has served as a catch-all idea at least from the earliest days of federal involvement in the protection of human subjects of biomedical research (32,33). It is hard for investigators and IRB members alike to know precisely what is meant by this wording (34). On the one hand, if informed consent is a "right" which the regulations bestow (or recognize), then clearly the waiver of informed consent violates such a right (35). On this interpretation, requiring that the waiver not violate a subject's rights would be self-contradictory. Furthermore, since "minimal risk" is a requirement for the waiver of informed consent, it is difficult to understand how this waiver could be construed as adversely affecting the welfare of a subject, since a protocol would have to already minimize any such adverse effects.

Nevertheless, it is necessary for IRBs to interpret this clause and document that the waiver of consent does "not adversely affect the rights and welfare of the subjects." Since the promulgation of these regulations in 1981, biological science and technology have advanced at an ever-increasing pace, and issues which at that time were perhaps just beyond the horizon have come to the fore with increasing frequency. In particular, two issues that may fall within the scope of "rights and welfare" have been debated and are relevant to the waiver of informed consent for the research involving human tissues: (*1*) the ownership and/or commercial exploitation of human tissues and cells and the patenting of gene sequences; (*2*) the adverse impact some genetic research may have upon ethnic, racial, or other groups.

Ownership and Commercial Use of Human Biological Specimens. A potentially contentious issue is that of "ownership" of tissues, for it may juxtapose the rights and welfare of tissue sources against the interests of researchers in freely pursuing scientific knowledge. From this pursuit of scientific knowledge comes most of the breakthroughs that allow new drugs and therapies to be developed, and any concomitant financial rewards. Part of this tension arises because the issues of "ownership" are enmeshed in the language of "property rights," and the attendant legal lexicon. Additionally "ownership" implies that all interests associated with the tissues can be couched in economic terms.

Probably the most famous court case involving these issues is that of John Moore. In its 1990 decision in *Moore v. Regents of Univ. of Cal* (36), the California Supreme Court ruled that Mr. Moore did not have property rights in his removed tissues/cells (which had been transformed into a profitable product by his physician, a colleague of the physician, and a pharmaceutical company, without his knowledge or consent). However, the court also ruled that his physician had breached his fiduciary duty to Mr. Moore by not disclosing his financial interest in treating and extracting tissue specimens from Mr. Moore. This case has become the touchstone for how many institutions deal with issues of tissue "ownership," even though, as legal precedent, it applies only in California.

The Moore case does not exhaust the possible ways to address "ownership" of tissues. Indeed, there are at least four different ways in which to view these issues:

1. Tissue sources have no "ownership rights" in their tissues; researchers do (e.g., the Moore paradigm).

2. Tissue sources share "ownership rights;" the question becomes how best to compensate them and when.

3. Tissue sources do not have "ownership rights" but researchers owe them recompense when their tissues become profitable.

4. "Ownership" is not an appropriate construct in this context, either for researchers or sources.

1. *Tissue sources have no "ownership rights," researchers do*. On this view, tissue sources are presumed to either abandon tissues (if provided in a clinical encounter) or donate their tissues for research use. Consent forms for clinical encounters (e.g., where surgery is involved) may have merely stated that removed tissues would be disposed of by the institution. Most patients would probably interpret this to mean that their excess tissues would be thrown out, even though this has generally not been the case. "In many — perhaps most — cases, individuals were not aware that their specimens were being stored or had no knowledge that they might be used for various research purposes by a number of investigators" (6, p. 41). Furthermore, according to a 1995 study of informed consent forms for genetics research, of the 23 documents reviewed, 4 explicitly mentioned that the investigator or institution were the sole owners of any tissue samples or transformed cell lines. The other 19 were silent on the issues of ownership or recompense to tissue sources in the case of profits realized (37).

In the language of John Locke, the sort of property right in tissues that would be claimed by researchers would be a natural right claim — namely that they have a right to any amount of wealth the fruits of their labor have produced. On this moral theory, whenever researchers come into possession of tissues they presume are abandoned, which they then modify in a way that renders the materials commercially valuable, the right to the financial rewards inheres to the researcher — the tissue source does not share in the profits, for s/he did not contribute to the labor that produced the value from the tissue.

2. *Tissue sources share "ownership rights," and the question is one of compensation*. According to this view, tissue sources and researchers share the profits realized from the transformation of tissues. There would be several challenges in trying to implement this sort of model, many of them logistical in nature. First, there is the problem of identifying at the outset which tissues might, through transformation, produce a marketable product. For instance, the case of John Moore is highly exceptional in that his doctor saw the commercial potential in the unique characteristics of his cells at the outset of his clinical treatment. This has rarely been the case, not only in terms of the early recognition of potential value, but also in the exclusive existence of the cells within one person. In most cases, any individual's "raw" tissues are of low economic value (leaving aside any questions of organ donation).

Even if one were able to predict at the outset which extracted tissues could produce value, the next challenge would come in instituting a scheme that could compensate the tissue sources. The issues here include keeping track of individuals whose tissues eventually do lead to a valuable commodity, often years after the tissues are obtained, which could eventually cost more than the actual compensation realized from any commercial product made from those tissues. In addition it is infrequent that any *one* person's tissues will lead to a viable product. That means that an additional logistical hurdle would have to be overcome, that is, calculating proportionality. In other words, researchers would have to determine what proportion of each person's tissues led to the product (assuming there was a desire to achieve compensation based on the value of each contribution of the unique tissues), and distribute compensation accordingly. This presumes that researchers would have a methodology to keep track of whose tissues were transformed in what ways, and that this "inventory" carried forward to other researchers who might also get the tissues for related or unrelated research. Finally, most profitable discoveries are made on the basis of a variety of tissues, and the proportional compensation of each tissue source would create a logistical ordeal to administer.

In a different context, a similar issue has led some scholars to argue that individuals should have property rights in the personal information about them that is bought and sold. The sale of personal information forms the basis of and supports a multi-billion dollar industry. In this context, these scholars propose forming a clearinghouse (similar to ASCAP and BMI in the music industry) that would receive royalties on behalf of individuals whose personal information is bartered (38,39). While their schemes have not been adopted by the information industries, the idea is still one worth considering for compensating tissue sources for the contribution their tissues make to science.

3. *Tissue sources do not have "ownership rights" but they are owed recompense when their tissues become profitable.* Because the body and its parts are generally not held by most people to be commodities, some have suggested that as an alternative to "ownership" in their tissues, when tissues do form the basis of a profitable product, some recompense should be provided to those individuals. As Thomas Murray has stated,

> There is something very special about human organs and tissues, even when removed from the body. We do retain moral interests in them, so that at the least they are not misused or treated in an undignified manner. And we have certainly recognized that body parts, whatever their dignity, can also have a price. But, on balance, we have rejected the idea that they should be bartered on the market (40).

Researchers finding a way to compensate tissue sources for the contribution their tissues have made to a profitable product would demonstrate respect for the people from whom the tissues came, and an appreciation for their contribution. While it may still not be easy to determine precisely whose tissues led to the profitable product, the issue does not have to be determined definitively, since "ownership" by tissue sources is not involved. For instance, for groups of related individuals (see the discussion that immediately follows), products that are derived from their tissues can be fairly easily compensated, since the group's collective contribution makes the product possible. Consider some of the new genetic tests that are now available. Research done on tissues collected from Ashkenazi Jews looking for a genetic basis for some types of breast cancers led not only to the discovery of BRCA1 and BRCA2 but also led to the marketing of genetic tests for the mutation. One possible way for researchers and pharmaceutical companies marketing these genetic tests to show their respect for the research subjects would be to donate some of their profits to Jewish organizations

or synagogues that serve the communities from which the research subjects came. Alternatively, the groups that would most likely be tested (if they formed the pool from which researched tissues came) could receive a discounted price on the testing. In the former case, the compensation is spread over more of the community, while in the latter, it is concentrated on those potentially most likely to directly "benefit" from the testing.

4. *"Ownership" is not an appropriate construct in this context, either for researchers or sources.* Finally, there is a case to be made for the outright rejection of the concept of ownership in this context. It may be that it is more productive to think of the holding of tissues for research purposes as a "custodial" relationship between the tissue source and the researcher. Certainly the legal discourse surrounding property "rights" could be abandoned. Moreover this approach can also accommodate more readily the noneconomic interests people have in their tissues. Some have suggested that an alternative is an independent trust model, where a disinterested nonprofit organization could hold cell lines (or other tissues) in a custodial arrangement, and grant licenses for use of the cell lines or other tissues to researchers and others (41). The licensing agreements could conceivably contain conditions under which tissue sources do not want their tissues used (e.g., genetic research). Any such alternative model will require great care and thought to ensure that the shortcomings of the current "property paradigm" are actually accommodated.

There is little regulatory guidance for researchers and IRBs alike in this controversial area, and it thus remains to IRBs to individually determine how they will handle questions of ownership and the commercial exploitation of human tissues in biomedical research. Some tissue banking protocols, such as the Cooperative Human Tissue Network (17), have stipulated that their tissues may not be used for commercial purposes nor patented by the researchers who use them. But it must also be recognized that for better or worse, private enterprise is increasingly becoming the economic engine that drives scientific progress. How these economic interests fit into the other ethical considerations — particularly the presumed altruism which has historically justified researcher access to tissues and records — which must be considered when weighing the waiver of informed consent remains an open question.

Risks to Ethnic and Racial Groups. The rights and welfare of groups of related individuals are becoming more of a concern in the context of biomedical (and, for now, genetic) research. The concept of "groups" used here is identical to that which was used in a paper written for the National Bioethics Advisory Commission on privacy issues in analyzing tissues (42):

> ...a collective of individuals who are culturally or ethnically related, where shared genetic characteristics are either likely or possible (or perhaps simply inferred). "Groups" can usually be characterized by a demographic label; e.g., African-American; Pacific Islander; American Indian; Scandinavian American; Ashkenazi Jew; etc. This notion of "group" does not necessarily extend to nuclear families as the unit of analysis, although certainly nuclear families may be members of larger

cultural/ethnic groups. Culturally or ethnically related people are, for the *present time*, the most easily recognizable as members of particular groups, within social contexts, and therefore, potentially the most readily stigmatized by genetic characteristics predominantly associated with that group. The distinction amongst different types of groups may become less relevant in the future, as we accumulate more knowledge about the genetic makeup of the entire population and all its constituent groups.

With the increase in molecular genetic research, groups of genetically related individuals are increasingly becoming a desirable "unit of analysis," particularly where there is thought to be a "group" component to the genetic trait — namely that the trait is more prevalent in certain genetically-related individuals. That members of these groups might have concerns about how their group is understood and characterized should not be too surprising, since much of one's self-identity comes from their interactions with others like them (i.e., other members of their group). This is particularly the case where some groups historically have been the targets of discrimination and stigmatization. Many individuals find that their associations with groups make important contributions to their self-development, self-discovery, and even their self-image. The often mutually supportive nature of groups and collectivities plays a key role in making these contributions. This may be even more the case in groups in which the members have an ethnic, racial, or cultural commonality. In other words, group identification, particularly in these latter cases, can be as important to the development of an individual's self-definition and self-respect as it is to the group's self-definition and continuity. When an encroachment on an individual (as a member of a group) or on the group itself occurs, the violation may be felt as being an affront to both the individual and the group (42). As Larry Gostin writes:

Derogatory information associated with a group can result in real harms such as discrimination against members of the group in employment, housing, or insurance. Derogatory information can also cause intangible hurt to groups such as lowering their self-esteem or racial or cultural pride. Derogatory information about a sub-population can stigmatize and wound its people as much as breaches of confidentiality can affect an individual. The information collected from groups, just as information about individuals, need not be blatant or intentional to cause harm or hurt. Even the best intentioned and careful research can trigger concerns about privacy (43).

As discussed above, the federal regulations exempt both the retrospective and prospective collection of some tissues and records from IRB review, provided investigators meet the specific requirements for access to identifiers. From the standpoint of public policy, these exemptions from the regulations may need to be rethought, where the tissues are known to originate *within a particular group*. The current policies make the presumption that *individual* identity is the only form of identity that is relevant to the research being conducted — and to ensure the protection of human subjects. Of particular concern is the possible impact upon various ethnic and racial groups of genes associated with personality, behavior and intelligence, though other less socially charged genetic traits may give rise to economic discrimination based on ethnicity or race. Such concerns have led several commentators to argue that the current exemptions from IRB review should be abandoned (23,42,44,45) in order that there be some "independent, social mechanism to ensure that research is ethically acceptable and that the rights and welfare of subjects will be protected" (13, p. 47692).

Where a group or identifiable community is the "unit of analysis" for the genetic research, researchers should involve members of the affected community in the research process, from recruitment through to the publication of results. The Human Genome Diversity Project (HGDP) provides much guidance in this regard. Henry Greely states that:

Research inevitably provides information about a group, as well as the individuals who constitute it. The group . . . is really the research subject. It is the group's collective autonomy that is challenged if researchers, with the informed consent of only a few individuals in the group, can probe for information about the whole group (46).

Indeed, it is partly for this reason that the model protocol for HGDP requires that researchers obtain the informed consent of the population, "through its culturally appropriate authorities where such authorities exist" (47) prior to sampling. Furthermore, if the population's authorities choose not to participate, HGDP would not accept any samples from any member of that population. "We believe . . . that the population-based nature of this research requires population-based consent, and we will insist on it" (47, p. 1444).

Within the United States, finding the "culturally appropriate authority" can be a difficult, if not impossible task for many groups. While it may be possible to find people who can facilitate discussions within the community (e.g., religious leaders), many groups are simply too populous and dispersed (e.g., Scandinavian-Americans or African-Americans) to have an authority with the power to make decisions for the entire group. In these cases, it is still important to hold frank discussions within the community to facilitate trust in the process (42).

Moreover, even though the regulations to protect human subjects address only the protection of individuals and not the protection of identifiable groups, when researchers propose additional retrospective research on tissues belonging to such groups, they should utilize the full IRB process (e.g., not expedited review) to justify that additional research. Part of the justification should be an indication of how the researcher can mitigate the harm that can be caused by the information obtained as a result of the research. In other words, even though the regulations treat these tissues as anonymous, researchers and IRBs would be prudent to do more than the regulations require in this case, and treat protocols using these tissues as using identified samples.

"Practicality" of Research without the Waiver. Of the four conditions this requirement is the easiest to interpret, though it may be the condition upon which most protocols founder. The waiver or modification of informed consent cannot be granted unless it can be demonstrated that "the research could not practicably be carried out without the waiver or alteration." The key term to interpret is "practicably." While the regulations provide no further guidance, this criterion was explicitly commented upon by the National Commission in the *Belmont Report*:

> In all cases of research involving incomplete disclosure, such research is justified only if it is clear that (1) incomplete disclosure is truly necessary to accomplish the goals of the research ... Care should be taken to distinguish cases in which disclosure would destroy or invalidate the research from cases in which disclosure would simply inconvenience the investigator (12, p. 12).

The primary concern in considering the "practicality" of the research without the waiver is the *scientific validity* of the study. The National Commission had in mind primarily social and psychological research which required some element of deception or incomplete disclosure, and the Commission expressed its concern that such alterations and deceptions not be taken lightly, but be justified by scientific necessity. This concern was underscored in the Commission's recommendations to HEW in 1978:

> *Nondisclosure must be essential to the methodological soundness of the research* and must be justified by the importance or scientific merit of the research (48).

The Commission was also clear to distinguish such methodological issues from the question of whether obtaining informed consent was *inconvenient* to the investigator (see also Ref. 35, p. 62). Accordingly, then, if we are to base our interpretation of this clause on the National Commission's own reflections, "practicality" refers to scientific necessity and not to the extra work an informed consent requirement might entail.

Although the National Commission had in mind primarily deception and incomplete disclosure in social and psychological research in its comments on this criterion, their application to tissue collection, use, and storage protocols is relatively straightforward and is best analyzed in light of those categories of protocols that fail to meet the exemption criteria discussed above: (*1*) retrospective and identified and (*2*) prospective and identified. Following this discussion, we will (*3*) analyze the role of prior "blanket consent" in the waiver of consent for specific protocols.

Retrospective and Identified. Many studies involving tissue specimens and/or medical records benefit by the inclusion of identifiers, or indeed may methodologically require such identifiers. Such studies are not therefore candidates for exemption under §46.101(b)(4). Especially in epidemiological studies based on medical record reviews, a number of commentators have argued that waivers of consent are methodologically necessary to the statistical validity of the studies (27,49). The same considerations arise in some epidemiological research involving the collection of tissue samples that already exist (50). Such protocols can involve thousands of tissues

and medical records, making the attempt to obtain consent not only exceedingly expensive but impossible in many cases where the patients have moved and no forwarding address is available (51). The impossibility of obtaining consent in these cases undermines the validity of the study by introducing selection bias in the data.

On the other hand, some studies may involve smaller numbers of tissues, and in other cases the issue of selection bias will not be relevant. In those cases recontact of individual patients to obtain their consent may be practical, though an inconvenience to the investigators. It will remain for individual IRBs to judge what number of subjects is too many to make obtaining informed consent impractical, as well as when the scientific validity of the study depends on the waiver or modification of informed consent.

Prospective and Identified. There are two issues to be analyzed here. First, there is again the problem that obtaining consent will lead to selection bias, since some subjects will inevitably refuse. As in the case with retrospective protocols, investigators would have to demonstrate that selection bias is genuinely a problem for the study they are conducting, that there are no alternative methods that would provide equally valid results without the waiver, and that the value of the research to society justifies the violation of subjects' privacy and autonomy. Since the validity of some studies will not suffer as a result of selection bias, it is disingenuous to propose that *all* tissue protocols should be exempt from requirements for informed consent.

The second problem concerns the time and expense of obtaining informed consent. Although these protocols are *prospective* and hence do not face the obstacle of contacting subjects that *retrospective* studies face, still investigators may argue that because they have no professional relationship with the subjects, who may be surgical patients or patients undergoing routine or invasive diagnostic procedures, contacting these patients is *impractical*. Some have argued, for instance, that even if contact were practical, the consent process itself is more burdensome to the subject than the minimal risks involved in the research (20).

Again, however, the validity of these arguments will depend on the specific nature of the protocols and where the patients interact with the health care system. When the investigators are directly involved in ordering or performing the clinical procedures from which the tissues are collected, obtaining informed consent for the research use of tissues and review of medical records is convenient and feasible. Other studies, however, may involve the collection of tissues from a large number of satellite hospitals and clinics and enormously increase the burdens on the investigators if consent is required. One solution to this problem is to name local personnel at these institutions as co-investigators, who can then make the necessary arrangements to obtain informed consent.

We should be careful though, to distinguish between *impractical* in the sense that a study would be methodologically impossible to perform and *impractical* in the sense that the consent process is simply burdensome to the investigators. Would it be just an inconvenience to investigators in a pathology lab to contact surgical patients to inform them about the nature of a tissue bank to which they hope to send samples of the patient's tissues, or would such a requirement make the collection impossible or impractical? How burdensome must the consent requirement be in order to find it impractical? Is the financial cost of obtaining informed consent a relevant consideration in these cases? Several commentators have argued that the increased costs to research of obtaining informed consent wastes limited financial resources because the informed consent process provides virtually no further protection from harm to subjects since the protocols under consideration for the waiver are by definition "minimal risk" (27,49,50,52). Melton has argued that since well over 95 percent of patients surveyed in Minnesota would gladly consent to the research use of their medical records (27), the requirement for informed consent is an added burden to patients at a time of stress, and the high costs of requiring consent are not justified by consideration for the abstract right of patient autonomy. As Phillip Reilly puts it:

> The cost of such research would be indirectly increased by the invocation of rules that, to me, only abstractly protect individual autonomy. So few people are likely to forbid their samples to be used for anonymous research that the expense attached to asking the question and tracking the few samples that are not available for study seems a poor use of resources (53).

However, as Veatch has argued, since disclosure of the purposes of research is an essential element of informed consent (35, p. 46), the federal regulations take seriously the right of subjects to decide for themselves what research purposes they wish to contribute to, a point not discussed by those who argue that consent should not be required for the use of tissues in research. Individuals may have a variety of reasons for wishing not to have their tissues used for different research purposes, from religious convictions regarding the disposition of body parts, to concerns about specific types of research and the commercial exploitation of their tissues or the patenting of genes (see Ref. 54 for a full discussion of these issues). The question of increased costs to research of requiring informed consent, and at what point those costs make the research *impractical*, leads to a direct confrontation with the respect for individual autonomy. We will discuss this dilemma later, but note here that judgments regarding the practicality of research without a waiver of informed consent are ethically complex, and there is little guidance in the regulations for IRBs and investigators alike. At the very least, it is prudent for IRBs to judge the impracticality of the consent requirement for each protocol on a case-by-case basis.

Blanket Consent. One important modification of informed consent should not be overlooked in applying this criterion to tissue collection protocols. Some hospitals contain either in their admissions literature or in their surgical and diagnostic consent forms provisions notifying the patient that tissues and medical record information may be used for research purposes. A few institutions provide the patient with the option to dissent from this use of these materials. These "blanket" consents

or notifications simply tell patients in advance that their records and specimens may be used for research purposes. The National Commission had recommended that this type of notification or blanket consent be used in institutions that anticipated using tissues and records for research, and indeed, the National Commission tied this recommendation to their recommendation that explicit informed consent may then generally be waived for specific research protocols. Unfortunately, HEW/DHHS opted not to include this recommendation in the regulations. Nevertheless, institutions which employ blanket consents give researchers and IRBs the opportunity to use notification as a *modification* of informed consent that would not violate individuals' autonomy. Explicit informed consent may be waived since the subject has already consented to the use of their tissues and records for any research purpose.

The adequacy of these blanket consents and notifications, typically included in the small print of surgical consent forms, has been questioned however. In the context of genetics research, some commentators have argued that broad statements that tissues and medical records may be used for research purposes are inadequate to the complexities of genetics research (22, p. 1791; 35, p. 173; 55–58). Others, including the National Commission and the Privacy Protection Study Commission, concluded that the benefits of relatively unrestricted access to tissues and records, coupled with the minimal risks to subjects involved in their use, justify the use of blanket consent or notification measures (15,25, pp. 111–112; 59–61). Lacking further regulatory guidance, IRBs must rely upon the collective judgments of their members in determining the adequacies of blanket consent or notification for the specific protocols that they review.

Providing Subjects with Additional Pertinent Information after their Participation. The history of the promulgation of these regulations makes it clear that this requirement was intended to be applied primarily to psychological and social research which for methodological reasons involved deception or incomplete disclosure. In these studies, debriefing subjects afterward can help allay anxieties and stress that may have arisen through the deception, and in general speaks to the respect of the subjects as persons. In research involving solely the use of human tissues and associated medical records such issues do not arise, and in the absence of informed consent prior to the use of the tissues, recontacting subjects would place researchers in the difficult position of explaining to them that — through their tissues and medical records — they had been involved unawares in a research protocol.

Nevertheless, in some research involving identified or identifiable human tissues, researchers may discover clinically relevant information either as a direct result of the research or by happenstance. If informed consent is required at the outset, then it is possible to state up front under what conditions, if any, subjects will receive research results, and when necessary, be provided with adequate pre- and post-test counseling. A difficult problem arises when *unanticipated* research findings are discovered, as happened for instance, when a strong correlation was discovered between the apo-E4 gene and Alzheimer's disease upon analysis of data gathered to study the relationship between apo-E family of genes and hypercholesterolemia and heart disease (62). These cases present the same dilemma that arises in studies for which the requirement for informed consent has been waived by the IRB. Such studies will typically involve identified or identifiable tissues (studies involving anonymous and anonymized tissues for the most part being exempt from the regulations), and therefore researchers may find that their results could be of clinical relevance to the tissue sources, who in these cases, will never have been informed of, much less consented to, the research use of their tissues.

This requirement for the waiver of informed consent stipulates that *"where appropriate, the subjects will be provided with additional pertinent information after participation."* The central question then is whether it is *appropriate* to provide research findings to subjects who have not given prior consent to the use of their tissues. The literature on this question, which has focused primarily on genetics research, is split. Those who argue for a duty to contact tissue sources do so based on two arguments: (1) a legal argument drawn on analogy to the clinical duty to recontact past patients with new information regarding their treatment, and (2) the obligation of researchers to *benefit* subjects whenever possible. Arguments against the duty to contact cite (1) the increased financial and administrative burdens that contact would place on research protocols, (2) the uncertainty of research generated results which may not be fully understood or validated by subsequent investigations, and (3) the psychological, social, and economic harms to subjects that may result if research findings are disclosed without adequate prior consent and counseling.

Arguments for a Duty to Contact

1. *Do Researchers Using Tissues Have Clinical Obligations?* To date there have been no cases litigated that would establish an investigator duty to contact tissue sources. Legal arguments suggesting such a duty have been based on analogy to cases in which physicians and health care institutions have been held liable for not re-contacting former patients with important findings regarding treatments previously provided. One of the most famous cases is that of the drug DES given to women in the 1950s. The University of Chicago was held liable for delaying notification to women who received this drug at their hospital four to five years after its toxic side effects became known in 1971. Pelias comments that the court found that the doctor–patient relationship is "on-going, especially when future injury to a client can be attributed to the relationship" (63). Pelias suggests that courts may also view the relationship of investigator to subject along analogous lines.

This point is also made by Hannig et al. who argue that the obligations of physician to patient are recognized, both legally and morally, to spread over the entire health care team, including consultants whom some state courts have found to have a legal duty of care to patients they may not have even seen in person (64). Research is then understood as just another part of a complex health care system, and the obligations placed on clinical health care providers should apply to researchers as well:

Yet the law imposes a host of requirements on the practice of medicine regardless of the individual physician's actual motives. One cannot avoid these obligations simply by asserting that research is somehow different. Where, as here, research requires the assistance of certain individuals because they or their relatives have a problem that is the object of study and where the research is directed toward the diagnosis or treatment of this condition, research assumes the mantle of health care. In that setting, the law should not hesitate to impose on the researchers some duties of care toward those subjects as well, at least in the absence of explicit agreements to the contrary (64, p. 259; 65).

There are three responses to these arguments. First is the claim that their position risks confusing the separate roles of researcher and clinician. A number of commentators (25, p. 2; 66–67), including the National Commission, have argued that a clear separation between clinical and investigator roles must be maintained, even (and especially) when they are borne by the same person. Though this argument has historically been addressed to research involving the evaluation of therapeutic interventions, it applies equally well to the use of *information* derived from tissue specimens that may be of clinical relevance.

Ultimately the ground for clearly distinguishing research from clinical practice is based on the second argument, which concerns whether research findings are of sufficient validity on which to base clinical decisions. We will take this argument up below, but in short, the claim is that research findings are preliminary and may not be fully validated. The use of such research findings in the clinical context risks basing clinical decisions on incomplete or possibly false information that may lead to substantial harms to the patient.

The third argument seeks to dissolve the entire dilemma by ensuring that subjects are *informed* and counseled up front about what, if any, research findings they may expect to receive. Adequate planning at the start coupled with an explicit informed consent process will head off most dilemmas that may arise out of the research findings, anticipated or not (56, p. 87; 57). But this requires informed consent, and so the argument here is really against the waiver of informed consent entirely, if it is possible that research findings may be of clinical value to the subject. This question will be analyzed later in the general discussion of the ethics of the waiver of informed consent.

2. *The Obligation of Researchers to Benefit Subjects.* Several commentators have argued that researchers always have an obligation to serve the best interests of their subjects. Robert Veatch in particular has argued that human subjects should be viewed as *partners* in the research enterprise (35), a view also expressed by Jonas in a famous essay of 1969 (68). Veatch argues subjects of research are entitled to any benefits that the research may produce, including being informed of research results. If the results are merely preliminary, then subjects may be counseled regarding their uncertain status. Veatch does not, however, address the problems of contacting subjects who have never consented to the use of their tissues in research, since he argues that the waiver of consent

(and the exemptions) are inconsistent with the respect for individual autonomy that underpins the regulation and ethics of human subjects research.

Arguments against a Duty to Contact

1. *Increased Administrative and Financial Burdens.* The burdens to a research protocol of contacting subjects whose tissues have been used in research without their consent will depend on the individual protocol itself and the origin of the tissues used. Some protocols may involve relatively small numbers of tissues. In these cases the added record keeping, time, and resources spent in contacting the tissue sources would be small. Other protocols may involve larger numbers of tissue specimens, anywhere from hundreds to the tens of thousands. For instance, the NHANES III study conducted by the CDC (69) contains over 17,000 blood and DNA specimens (55,70). Many genetic epidemiological protocols similarly will involve very large numbers of tissue specimens. Although these research protocols may generate genetic or other information that would be of value to the tissue sources (and may even be of high quality and certainty), the costs of keeping track of and contacting large numbers of tissue sources could often exceed the entire budget of the research protocol in the first place.

Second, and in rejoinder to the argument that researchers owe subjects the benefit of research results, some commentators (55,71) have argued that the goal of biomedical research is to provide benefits to society at large and that insisting on providing benefits to research subjects whose participation is of minimal risk diverts resources that otherwise would contribute to medical and scientific progress and ultimately to the public good. If and when research results find their way into clinical practice, individuals whose tissues are used for research will have the same opportunity as the rest of the population to benefit from this new knowledge.

2. *The Uncertainty of Research Generated Results.* Data and conclusions derived from individual research protocols are often tentative and uncertain. The path from an intriguing research result to a validated clinical test, treatment, or procedure is long and arduous, requiring a series of clinical trials with increasing numbers of subjects. Many results that flow directly from research protocols are unsuitable for use in the clinical context without further testing and evaluation. If the conclusions derived from a research protocol are flawed and yet the results are passed along to the tissue source and their physician, it is possible that they will decide on an inappropriate course of action that may end up producing more harm than good for the patient. Here again, commentators on this problem have pointed to the necessity of maintaining clear distinctions between the often dual roles of researcher and clinician. Merz et al. in particular have recommended that tissue procurement and banking protocols be constructed so as to permit only a one way flow of information from patient to tissue bank to investigators so as not to confuse research and clinical information (18; see also Ref. 28).

In any event, the clinical validity of research results must be evaluated by investigators and IRBs on a case-by-case basis. The possibility that some research results are

relevant to clinical care should not be ruled out, especially when there are interventions available that may reduce or minimize harm to the patient. When contact is made with an unconsenting subject, however, a number of problems may arise which are discussed in the next section.

3. *Disclosing Research Results without Adequate Prior Counseling.* When informed consent has not been required for researchers to obtain and use human tissues, contacting subjects presents a dilemma. In standard clinical practice, diagnostic tests typically are preceded by either the tacit consent of the patient whose presence in the physician's office indicates their willingness to investigate a particular health problem, or by the explicit and sometimes written authorization of the patient if the test is particularly invasive or will produce sensitive information. For instance, most genetic testing clinics require extensive pre-testing and post-testing counseling of their patients (72–77). Such counseling includes information about the nature of the test itself, its relation to symptoms and disease, including the penetrance of the gene and limits of the predictive value of the test, and the social consequences of the genetic information, including the possibility of various forms of economic and social discrimination. When the test results come back, patients are more psychologically and emotionally prepared to cope with the consequences of bad news and often undergo further counseling.

It is a well documented fact that depending on the disease gene in question (and especially whether any treatment for it exists), many patients who contemplate a genetic test opt not to perform the test after initial counseling (78,79). Some patients decide that the burdens and risks of knowing are not justified by the benefits. However, if individuals have never consented to the research testing of their tissues, there is no way for investigators to know whether they would want to know the results or not. It is largely a value judgment that individual patients make for themselves when they decide for or against having a given diagnostic test. If unconsenting subjects are recontacted, some subjects may be forced to learn things about themselves and their potential medical future to which they never would have consented in the first place had they been given the opportunity.

One solution to this dilemma advocated by a number of commentators is the use of a general newsletter from researchers describing aggregate results to subjects (56). This presupposes prior informed consent in order to justify this form of contact with the subjects, but publication of study results to the general population from which the tissue sources were derived may fulfill the same role. It should also not be overlooked, as pointed out above, that as research knowledge finds its way into clinical practice in the form of validated procedures, the unwitting tissue sources will have the same opportunity to benefit from these procedures as the general population and in a manner in which their autonomy and values are respected.

The duty to contact is a complex issue, but despite the ongoing debates, the research community has generally tended to limit contacting unconsented tissue sources. There may be instances where this has not been the case, and IRBs and investigators should look at each case carefully, but in general it is more prudent not to permit contact of unconsented tissue sources with research findings. Investigators who find themselves with information that they feel is compelling enough to warrant the risks of contact should certainly consult their IRB before proceeding.

General Requirements for Informed Consent and Special Provisions for Genetics Research

Protocols that fail to meet the criteria for either exemption or the waiver of informed consent are required by the regulations to be reviewed by the IRB and must conform to the requirements for informed consent. Since these protocols may be eligible for expedited review we will in this section examine (1) the conditions for such review, and then turn to the general discussion of (2) informed consent and (3) the special provisions for genetics research.

Expedited Review. The review of tissue protocols, if they present no more than minimal risk to subjects, may be done on an expedited basis according to the conditions outlined in §46.110 of the regulations. These regulations refer to the Expedited Review List published by DHHS in 1981 (80). A large number of the items contained in this list refer to the collection of human biological specimens, including the collection of hair and nail clippings, external secretions, amniotic fluid and placenta (at the time of birth), excreta, blood draws of less that 450 ml, dental plaque and saliva, and the study of existing pathological and diagnostic specimens. This latter category would involve only studies in which identifiers or links are recorded by investigators, since otherwise these protocols would be exempt.

The expedited review may be carried out by the IRB chair person or an experienced member of the IRB. This reviewer may exercise all the powers of the IRB, including the waiver of informed consent, but protocols under expedited review must meet all the conditions for IRB approval stipulated in the regulations. The expedited review process simply permits a more timely review of protocols and was designed to minimize the work of the full IRB.

DHHS recently expanded the list of expedited review categories (81). The list increases the amount and frequency of blood drawn and most notably includes research which collects identifiable records and pathological and diagnostic specimens, whether they are collected retrospectively or prospectively, provided that they are excess or have been produced for nonresearch purposes. However, a report by the DHHS Inspector General's Office on IRB performance noted plans to give IRBs "added responsibility in the areas of genetics and confidentiality" (10). Permitting more of these and related protocols to be reviewed on an expedited basis would seem to work against the need for greater IRB vigilance and expertise in these areas.

General Requirements for Informed Consent. Section 46.116 lists eight elements required for informed consent:

1. A statement that the study involves research, an explanation of the purposes of the research and the expected duration of the subject's participation, a description of the procedures to be followed, and identification of any procedures that are experimental.

2. A description of any reasonably foreseeable risks or discomforts to the subject.

3. A description of any benefits to the subject or to others that may reasonably be expected from the research.

4. A disclosure of appropriate alternative procedures or courses of treatment, if any, that might be advantageous to the subject.

5. A statement describing the extent, if any, to which confidentiality of records identifying the subject will be maintained.

6. For research involving more than minimal risk, an explanation as to whether any compensation and an explanation as to whether any medical treatments are available if injury occurs and, if so, what they consist of, or where further information may be obtained.

7. An explanation of whom to contact for answers to pertinent questions about the research and research subjects' rights, and whom to contact in the event of a research-related injury to the subject.

8. A statement that participation is voluntary, refusal to participate will involve no penalty or loss of benefits to which the subject is otherwise entitled, and the subject may discontinue participation at any time without penalty or loss of benefits to which the subject is otherwise entitled.

Not all of these requirements are appropriate for tissue protocols: Specific consent requirements depend on whether the intervention is research related or for clinical purposes, or whether only excess tissue specimens are used or the collection of additional tissue is required. Many research institutions maintain standard informed consent templates that investigators may then customize to fit the individual needs of their protocol. It is important to remember that the risks of procedures ordered for clinical purposes need not be detailed in the research consent process. Only those risks that arise from the specific research activities need to be detailed, since the clinical consent process should cover the risks attendant to the diagnostic or surgical intervention.

Special Provisions for Genetics Research. Some research protocols using human tissues requiring informed consent will be of little risk to the subject. For instance, research that uses normal blood specimens to study the clotting process is relatively straightforward, and the informed consent process need not be lengthy or complex. However, the use of tissues, especially for genetics studies, can present an array of problems that must be dealt with in the informed consent process. The purposes of the research, the communication of research results to the sources, the psychosocial risks of such disclosure, questions of ownership of tissues and rights to commercial profits, the use of specimens for other research purposes or the disposal of specimens after the research is complete, may all be relevant considerations.

While the clinical genetics community has reached a consensus on the elements of clinical informed consent, good counseling practices, and the protection of patients from psychosocial harms arising from the dissemination of genetic testing results, the research community continues to debate both the need for, as well as the extent of, informed consent for genetic studies. On the one hand are those like George Annas, who advocate for a robust informed consent requirement. Annas has proposed that:

> No collection of DNA samples destined for storage is permissible without prior written authorization of the individual that a) sets forth the purpose of the storage; b) sets forth all uses, including any and all commercial uses, that will be permitted of the DNA sample; c) guarantees the individual (i) continued access to the sample and all records about the sample and (ii) the absolute right to order the identifiable sample destroyed at any time; and d) guarantees the destruction of the sample or its return to the individual should the DNA bank significantly change its identity or cease operation (56).

Annas would further require that DNA samples be used only for the purposes for which they were originally collected and would prohibit the use of open-ended consents that permit the use of the sample for any other research purpose. He would also prohibit third-party access to research results but would require notification and adequate counseling of sources concerning research results that may have a significant health impact.

Annas's proposals as embodied in the Genetic Privacy Act (GPA) (82), a proposed model law, have been influential in a variety of legislative proposals around the country, but they have met with opposition from the research community. Some commentators have argued that singling out genetic information from other types of medical information ("genetic exceptionalism") is impossible to apply in practice, since a variety of medical information can be considered "genetic" (28,83). It is also suggested that elaborate informed consent, more consistent with interventional research, is inappropriately applied to the collection and use of tissues and medical records, since little or no risk is involved in the research use of tissues. Furthermore an elaborate consent may frighten subjects and raise the spectre of significant harms that are of extremely low probability, leading to what Korn has called "uninformed denial" (28, p. 25). Most of the objections to Annas's position concern the question of whether informed consent should be required at all for such protocols, and we will discuss this question in detail below.

An alternative idea — a tissue "advance directive" — has been proposed by Robert Weir (84). This document would be filled out by patients as they enter a health care institution and could limit the research use of their tissues to specific types of diseases, allow no research use at all, or permit any research use. This detailed blanket consent could then be entered into the patient's computerized medical record and easily tracked by pathologists and researchers. This proposal would have

the virtue of avoiding a detailed consent process each time a researcher wanted to use an individual's tissue. Individuals could update their "tissue directive" as well as specify whether they desired to be informed of any research results. Whether such a system is feasible warrants further study, for it would obviate the need for specific informed consent for many research uses of tissues and records, while giving individuals more control over the purposes for which these materials are employed.

AUTONOMY, PRIVACY, AND THE SOCIAL GOOD

Although some controversy still surrounds the extent and detail of informed consent for the use of human tissue in research, the primary debate has centered on the question of whether informed consent should be required at all for these protocols. As seen above, the regulations support several exemptions from both informed consent and IRB review in addition to the waiver of informed consent. Those who support these measures and would seek to expand them make six general arguments: (1) requiring informed consent is administratively and financially burdensome; (2) requiring informed consent may introduce selection bias into research; (3) the informed consent process is itself a burden to patients and subjects and as such outweighs any benefits the consideration for autonomy and privacy would produce through informed consent; (4) there is a long tradition in medicine of free and unfettered access to tissues and medical records for research purposes, and requirements for informed consent and/or IRB review for currently exempt categories of research would infringe upon academic freedom; (5) the social benefits of minimal regulation of research involving human tissues and medical records, including the omission of informed consent, greatly outweighs any violation of individual autonomy; and (6) there is no violation of individual privacy in the use of tissues and records, since confidentiality is scrupulously protected by researchers.

In response, those who advocate for greater individual control over their tissues and records as well as regulatory and IRB oversight of this research argue that (1) the ethical foundation of informed consent is a respect for individual autonomy, and permitting research on tissues and records without informed consent violates a growing tradition in research ethics that has served patients and subjects very well; (2) social benefits are not a sufficient ground to undermine the respect for autonomy by not requiring informed consent, and indeed, social benefit is incommensurate with respect for autonomy and cannot simply be weighed against it; (3) as a result of the successes of the Human Genome Project, human tissues contain a vast amount of medical information that easily can increasingly be tapped, and subjects have a right to determine whether and how this highly personal information is used; (4) public trust in the health care industry and in biomedical research is undermined when researchers use individuals' tissues without their consent or knowledge for purposes they may not share, and especially when researchers might profit financially from them; and (5) our current health care system does not distribute its benefits equally across the population, and many minorities and the poor are excluded or underserved. Justifing the violation of individual autonomy by the claim that research on tissues serves the public good fails to recognize that this public good does not return equally to all segments of the population.

With the exception of this last argument which is predicated upon the principle of *justice*, it can readily be seen that the core of the ethical debate over whether to require informed consent pits those for whom social benefit — a concept derived from the general principle of *beneficence* — is held as the highest moral principle, against those for whom the principles of individual *autonomy* and *privacy* take precedence over competing principles and values. Beneficence, according to the National Commission, implies two separate obligations: "do not harm and maximize possible benefits and minimize possible harms" (12). These benefits and harms may serve to advance or hinder the interests of either individuals or society in general. Those advocating less regulation and no informed consent typically will view autonomy as a species of this more general category of beneficence: autonomy and privacy simply represent *interests* that are important to individuals but that must be weighed against competing *interests* such as general social benefits. On the other hand, those who advocate for the requirement of informed consent typically will understand autonomy as a principle independent of beneficence and either conceive of it as a moral *right* or as entailing a moral *duty*.

How the Regulations Encode and Interpret Autonomy and Beneficence

The involvement of the U.S. government in the regulation of human subjects research evolved out of revelations in the 1960s and 1970s of various abuses that risked or caused grave harm to research subjects. It was recognized that such abuses would be less likely to occur if subjects were fully informed of the purposes of, and risks (and benefits) of the research, since subjects would be reluctant to submit to risks without the corresponding possibility of benefits. But informed consent also came to serve another purpose: respect for individual autonomy, both moral and legal. Legally, those who practiced medical interventions upon unconsenting individuals were liable for *battery*, defined as unwanted or unconsented touching of one's body. In the legal domain, autonomy is expressed as a right to control one's body and what is done to it. This legal right entails an obligation or duty on the part of others not to violate this right. Morally, the requirement of informed consent is an expression of the respect for the independent will or freedom of the individual to associate and participate in activities of their own choosing. In the moral domain, autonomy is expressed first as a duty, not as a right, though some have inferred a corresponding moral right of autonomy. This duty is one that demands that in all our actions we respect the free will and self-determination of others. As a moral right, autonomy has come to mean, in our society, the freedom to choose.

In most cases the principles of autonomy and beneficence work hand in hand for the protection of human subjects in biomedical research. However, research involving human tissues and medical records, insofar as

it is genuinely of minimal or no risk requires that we evaluate the justification for informed consent solely on the basis of autonomy, since the beneficence role informed consent plays in minimizing harm is, by definition, out of play. The role that the principle of beneficence then plays is to ground the research project itself as contributing to the social good. Thus the stage is set for the conflict between autonomy (*informed consent*) and beneficence (*social good*).

In its seminal *Belmont Report*, the National Commission identified three basic ethical principles that apply to research involving human subjects: justice, beneficence, and autonomy. The National Commission recognized that informed consent served two ethical masters (beneficence and autonomy), but constructed its recommendations to HEW in such a manner that neither came into conflict with the other. The commission did indeed recognize that research involving human tissues and medical records involved an "invasion of privacy," but this invasion, in the absence of informed consent, was justified, on the one hand, by the social value of the research and, on the other hand, by the recommended requirement that institutions conducting such research institute a blanket consent at the time of admission. The commission also recognized that IRB review was necessary to evaluate the social value of the research, which, in the absence of informed consent, would entail a judgment that the research is of a socially acceptable nature — that is, that the research would not likely be objected to by those whose tissues and records are conscripted without their informed consent.

In its reasoning, the National Commission was guided by the conclusions of the Privacy Protection Study Commission (PPSC), which had published its report in 1974. The PPSC had not considered the use of tissues but did examine carefully the research use of medical records and recommended that this use is legitimate provided that individuals are *notified* in advance (59). The National Commission took a stronger stance with regard to tissues and records, recommending blanket *consent* rather than mere notification. The debate over the adequacy of blanket consent aside, this provision represents the recognition by the National Commission of the independence of autonomy as a ground for informed consent. IRB review would stand as proxy for the *informed* consent of individual subjects, what Veatch has called "constructed consent" (35, p. 63; see also Ref. 25, p. 150, and Ref. 85), while blanket *consent* for the research use of tissues and records would permit those who might object, for whatever reasons, to opt out of any research use of their tissues and records.

As we have seen in the discussion above, HEW did not follow the National Commission's recommendations for a standard waiver of consent for tissue and medical record research coupled with a blanket consent requirement at institutions conducting such research. Instead, HEW proposed, and DHHS adopted, the two exemption categories as well as a general provision for the waiver of informed consent. But neither the proposed nor the final regulations contained any provision for blanket consent. There is no comment on this omission in the public record, but HEW/DHHS may have decided that it would have been an unwarranted intrusion of government regulation into *clinical* affairs to require blanket consent at all hospitals and health care facilities that might supply tissues and records for research. But blanket consent was an essential element in the National Commission's recommendations that preserved respect for autonomy while greatly simplifying research access to tissues and records. And IRB review was essential to ensure that the purposes of the research were generally unobjectionable in the public eye. There are two ways, then, to interpret this omission by HEW/DHHS. Either the compromise reached in the final regulations does not recognize the independent validity of autonomy as a ground for informed consent, or the compromise reached simply distorted the ethical ground so carefully crafted by the National Commission.

There is good reason to believe that the latter is the correct interpretation. In his *Patient as Partner*, Veatch has argued that the informed consent requirement (§46.116(a)(1)), that subjects be informed of the *purposes* of the research protocol, can only be justified by appeal to subject autonomy and cannot be based on any beneficence considerations to prevent or minimize harm (35, p. 99). By requiring that the purposes of the research be stated clearly in the consent form, subjects are given the opportunity to accept and adopt such purposes as something to which they are willing to contribute. Informing a research subject that a protocol will contribute to the treatment of a particular disease or, perhaps, make abortion less risky to women, plays no role in the subject's evaluation of the physical risks of participating in the protocol but speaks instead to the values they wish to promote or not promote.

Veatch argues that the exemption categories as well as the provision for the waiver of informed consent are fundamentally at odds with the independent role of autonomy in the justification and constitution of the informed consent requirement. By exempting a large class of research protocols as well as permitting the waiver of informed consent, the regulations usurp the ability of individual subjects to make value judgments regarding the goals of the research protocols to which their genomes, tissues, and medical records contribute. It was perhaps the view of the authors of the regulations that biomedical research was of such a high and uncontested value that no one would reasonably object to contributing their tissues and records to the common good. But this notion would apply equally well to invasive and physically risky research protocols as it does to protocols involving only the use of tissues and records. There is no sufficient ground for explaining to subjects the purposes of research in the one case but not in the other. It therefore follows that the regulations must recognize the independent validity of autonomy as an ethical justification for informed consent (86).

Defining Privacy

We noted above that the National Commission viewed researcher access to tissues and records in the absence of informed consent to be an "invasion of privacy." It is useful at this point to clarify the concept of privacy, since it is intimately related to the concept of autonomy and suffers from the same ambiguity of interpretation.

One of the most influential articulations of what is meant by the legal "right to privacy" (as distinct from "privacy" as a moral interest or value) appeared in a now-famous *Harvard Law Review* article in 1890 by Samuel Warren and Louis Brandeis. The right to privacy, they said, is "the right to be let alone" (87). It was from this simple articulation that common law began to recognize a right to privacy in certain circumstances. This recognition, however, has not made defining the concept of "privacy" any easier. Indeed, privacy is a "notoriously vague, ambiguous, and controversial term that embraces a confusing knot of problems, tensions, rights, and duties" (88).

Privacy is usually described as being related to notions of solitude, autonomy, anonymity, self-determination, and individuality: It is experienced on a personal level. Within socially and culturally defined limits, privacy allows us the freedom to be who and what we are as individuals. By embracing privacy, we exercise discretion in deciding how much of our personhood and personality to share with others. Moreover we generally feel less vulnerable when we can decide for ourselves how much of our personal sphere we will allow others to observe or scrutinize (89). Complicating the process of defining what privacy "is" is the fact that it often means something different to nearly everyone, and the experience with and perception of what invades privacy will likely differ significantly from person to person (42).

In trying to "break apart" the notion of privacy, much of the literature focuses on the following elements:

1. *Autonomy*. Respecting the dignity of each individual to make decisions for themselves, free from coercive influences. It also encompasses our need for solitude and intimacy. A National Research Council report stated that the protection of individual autonomy is a fundamental attribute of a democracy (90). Autonomy is also addressed in other analyses as "decisional privacy" (91).

2. *Informational privacy*. Defined by how much personal information is available from sources other than the individual to whom it pertains. Informational privacy encompasses the ability to limit access to one's personal information (which also supports the autonomy aspects of privacy), from both a quantitative and qualitative perspective — namely the amount and type of information one surrenders, either voluntarily or by coercion. It also involves when such information should be communicated or obtained, and what uses of it will be made by others. It includes the collection, storage, use, maintenance, dissemination/disclosure, and disposition of personal information.

3. *Freedom from intrusion/surveillance*. Encompasses, in part, an individual's desire to preserve his or her anonymity and solitude (both physical and emotional solitude). This notion includes not only the individual's desire to limit access to information about him/herself but also to be free from physical intrusion and observational surveillance by others. Surveillance can have a chilling effect on individuals,

as noted by many sociologists and studies of electronic monitoring. Individuals often change their behavior to conform to what they believe those monitoring their movements/actions will find "acceptable" or "normal" (92,93). Freedom from intrusion is addressed in other analyses as "physical privacy" (91).

The concept of privacy is often confused with or treated as synonymous with two other distinct concepts: confidentiality and security. Confidentiality

> refers broadly to a quality or condition accorded to information as an obligation not to transmit that information to an unauthorized party....Confidentiality has meaning only when the promises made to a data provider can be delivered, that is, the data gatherer must have the will, technical ability, and moral and legal authority to protect the data (90).

The following is a simple way to differentiate between these three concepts: *security* measures provide the technical (and sometimes physical) means to safeguard the *confidentiality* of personal information, which in turn protects the *privacy* of individuals. Within the doctor–patient relationship, confidentiality is used to describe the relationship of trust that must exist for appropriate clinical care to be rendered. In its essence, confidentiality advances the protection of personal information that is exchanged or generated between doctor and patient (whether through verbal exchanges of information or information generated through physical examinations). This is the most fundamental way in which the patient's privacy is preserved (42).

One of the arguments made against the unconsented use of human tissues and medical records is that it violates this fundamental trust between patient and physician. Even if health care is delivered by a team of nurses, physicians, and other personnel, there is a fundamental expectation on the part of patients that their medical information will be kept confidential. What this means to most patients is that this information will not be divulged to other persons or institutions for purposes other than what the patient initially provided it for, namely their own medical benefit and, correlatively, for purposes of payment. The Privacy Protection Study Commission (PPSC) concurred with this understanding in its 1974 report when it recommended that medical records could be used for:

> ...conducting a biomedical or epidemiological research project, provided that the medical-care provider maintaining the medical record: (i) determines that such use or disclosure does not violate any limitations under which the record or information was collected (59, p. 306).

Of course, virtually all standard medical record information is collected with the expectation of confidentiality. It was thus necessary for the PPSC to insist in its recommendation 12:

> ...that each medical-care provider be required to notify an individual on whom it maintains a medical record of the

disclosure that may be made of information in the record without the individual's express authorization (59, p. 313).

The National Commission, as we have seen, took a stronger stance and recommended blanket consent rather than mere notification. But regardless, only through such notification or blanket consent procedure could private medical information be disclosed for research purposes in a manner that did not violate the *expectation* upon which the information was first gathered.

We may therefore understand and interpret "privacy" according to the intentions and expectations of patients when they provide specimens or medical information to their providers. Confidentiality is the *respect* providers pay to these intentions of their patients by not disclosing medical information. Or, confidentiality is a respect for the autonomy of the individual to determine the purposes to which information (and their diagnostic specimens) are put. In the medical context, privacy becomes a species of autonomy and confidentiality a species of the respect for this autonomy, or more generally, a respect for persons. Thus privacy is not violated if the patient knowingly *consents* to the disclosure of information (or the use of tissue). It should therefore be clear that privacy and autonomy with respect to one's tissues and records is of equal concern whether the records are identified or used anonymously. Although the use of identifiers or links to identifiers may raise the risks of harm to the source and correlatively the use of anonymous tissues protects *confidentiality*, it is the expectations placed upon the information and tissues by patients which determines them as private in the first place. Only through *consent* can their use respect the privacy and autonomy of sources (for a contrary view, see Ref. 85, p. 176).

As thus interpreted, either privacy may then be evaluated as an *interest* patients have in controlling their medical information and therefore may be evaluated alongside other such interests and goods within a consequentialist framework, or privacy may be understood deontologically as a duty on the part of others to respect what is private, that is, not to violate the intentions according to which one may receive private information. The ambiguity that attends the interpretation of autonomy — whether it may be subsumed by beneficence or is independent of beneficence concerns — also attends the interpretation of privacy. Indeed, autonomy and privacy are so intertwined that it may well be the same ambiguity in either case.

Consequentialist Interpretation of Autonomy

As we have seen the exemptions and waiver provisions of the regulations appear to simply disregard the independent role of autonomy as an ethical justification for the requirement of informed consent. Because the regulations permit research access to tissues and records in the absence of specific informed consent and yet do not require any form of blanket consent, the regulations appear to authorize an "invasion of privacy" that two federal commissions recognized as violating patient autonomy.

There remains one route out of this dilemma that would save the consistency and coherence of the regulations. It is the argument offered by some proponents of greater research access to tissues and records without informed consent. The only way in which autonomy can be traded off against beneficence concerns (informed consent balanced against, or sacrificed for, the social good) is if autonomy itself is understood as the expression of certain *interests* that individuals and society regard as especially important. Once autonomy is understood as the expression of interests, the task is either to minimize the harm to these interests that research may cause or to demonstrate that such harms to them as do occur are minimal and do not *outweigh* the benefits to society of the research. In a paper commissioned by the National Bioethics Advisory Commission (NBAC), Allen Buchanan has made precisely this argument. Since this argument has formed the basis of many of the recommendations the NBAC report (6) on the use of human tissues in research, it is worth while analyzing this argument in detail.

This argument begins with the assertion commonly made by those who would require informed consent for tissue research that individual autonomy is a *right* that permits patients to control private information about them, including their tissues. The language of rights is an appealing framework for moral argument, since rights often stand as moral trump cards overriding other interests.

If we begin with the assertion of individuals' autonomy rights, or privacy rights, against the interest researchers have in freer access to tissues and medical records, the outcome of the argument will be determined by how these putative moral rights are justified or grounded. Buchanan suggests, in line with some traditions of moral and political thought, that rights are "protectors of morally important interests" (94). Specifically:

> ...rights-statements are assertions that certain interests are of such importance from a moral point of view that they deserve especially strong protections (94).

The claim that such interests are protected by "rights" simply asserts that the interest is of such high priority as to overrule other competing interests and rights. As Buchanan points out, such assertions are the conclusions to moral arguments that must be provided in order for the assertion to be justified.

Having established that a right is simply a proxy for an important interest, the task is to demonstrate that the interest that justifies the right is of greater or lesser priority relative to competing interests. Interests are defined as "an ingredient in someone's well-being" (94, p. B-5). Well-being in turn relates to the goods and benefits that individuals pursue, either as a matter of necessity or to satisfy their desires. Interests thus represent goods or benefits. Insofar as a right is violated, the interest in that benefit is set back; that is, the individual is harmed. This argument has succeeded in converting the principle of autonomy into a species of the more general principle of beneficence, and the deontological force of autonomy, that is, the *duty* to respect another's autonomy, is measured according to the consequences of respecting or violating the interest expressed by this putative right.

One simply reckons up the benefits and harms according to a utilitarian calculus.

In the case of the research use of human tissues and medical records, one need only reckon up the harms (and benefits) to individuals alongside the harms and benefits to the social good that the use entails with and without consent. The argument from this point is straightforward. Since only in special circumstances does society permit the unconsented risk of harm to individuals for the sake of social benefit (e.g., military conscription), the task of research interests is to demonstrate that harms to individuals are minimal or *inconsequential*. By protecting confidentiality, adverse psychosocial consequences to subjects are minimized. We are then left with what Reilly has called the abstract protection of individual autonomy (53) weighed against the obvious social benefits of biomedical research. Since little of *consequence* hinges upon the exercise of autonomy in this instance, and the benefits of biomedical research to health care and the economy in general are so high, overriding the "right" of autonomy is easily justified. Thus Buchanan concludes that even blanket consent may not be necessary:

> But it would be hyperbole to say that a system that does not include the requirement of blanket consent violates anyone's "right to autonomy." For one thing, ... not all choices warrant the stringent protections that talk about a right to autonomy implies; some choices are relatively insignificant because they are largely irrelevant to a person's well-being and value (94 p. B-18).

If we interpret autonomy along these lines, the exemptions and waiver of informed consent provisions in the regulations in the absence of a blanket consent requirement no longer appear inconsistent with the otherwise independent validity of the principle of autonomy. On this interpretation, stating the purpose of research when informed consent *is* required is justified by the respect for the autonomy *interests* of the subjects. It costs research little to include this. The requirement for informed consent itself is justified by the beneficence concern to avoid harm by affording individual subjects the opportunity to evaluate the risks and benefits, as well as the overriding legal concern not to put individuals at risk without their prior informed consent. In the case of the research use of tissues and medical records, however, since there are few or no risks that cannot be managed through the maintenance of confidentiality, the beneficence ground of informed consent does not apply. Lacking any further ground for informed consent, it is morally justified, according to this argument, not to require it.

This argument forms the foundation for the six arguments cited at the beginning of this section supporting researcher access to tissues and records without the informed consent of sources.

1. The requirement for informed consent, in addition to the requirement for IRB review, adds administrative and financial burdens to research protocols. It requires the filing of paperwork and the tracking of subject consent documents in addition to the time and personnel resources to contact patients

and go through the informed consent process with them. When protocols involve multiple centers for the collection of tissues and records, these burdens are multiplied many times over. In addition some epidemiological studies include such large numbers of subjects that the costs of obtaining consent would far exceed the protocol budgets, thus making the studies impossible to perform (27,49). If the risks of confidentiality are minimized through appropriate protections, the social benefits of research can be maximized by eliminating the costly burdens of informed consent.

2. Selection bias may be introduced into research by requiring informed consent. Many protocols examine trends and small statistical differences. Permitting some subjects to opt out of the use of their tissues and records can introduce uncontrollable statistical bias. Furthermore, in some areas of research, because of the sensitivity of the topic (e.g., psychiatric or sexual dysfunction), informed consent may make it difficult to obtain sufficient specimens and records, making such research impossible. As a result some areas of research could be abandoned that could produce important benefits for these same populations (27,49,50,53).

3. Requiring informed consent for the use of excess surgical and diagnostic tissues would place an added burden upon patients, often at times of emotional stress. Approaching patients prior to surgery may confuse them and raise unnecessary fears in their mind because they may fail to understand the complexities and subtleties of the research use of their tissues. Furthermore, since the risks to subjects are minimal if not nonexistent, there is little benefit to be gained by informing subjects and asking their consent. The informed consent process itself is a burden to the subjects with little or no corresponding benefit (20, p. 76).

4. Researchers have long enjoyed unfettered access to human tissues and medical records. The benefits of this access are innumerable, and the record of researchers protecting the few risks that might arise is impeccable. This argument is partially predicated upon the claim that sources simply do not have any rights to control the use of their tissues, rights which they gave up when the tissues were removed (53). Researchers and hospitals consider tissues not used for direct patient benefit in diagnostic testing or treatment to be "waste," which the institution may dispose of as it sees fit, provided it does not directly harm the patient in the process (51, p. 6).

5. The social benefits of research outweigh any putative violation of individual autonomy (53, p. 380). Indeed, since autonomy has no other meaning in this argument than to promote certain benefits and prevent certain harms, and since researchers control for these harms by maintaining confidentiality, autonomy has no role to play in such research. This is the essence of the consequentialist argument presented above.

6. There is no violation of any putative "right" to privacy as researchers scrupulously protect the confidentiality of the information and tissues they receive. Patients in the contemporary health care system must consent to a wide array of individuals and institutions gaining access to their medical records. Research is an integral part of the health care system and provided protections are in place to maintain confidentiality, subject privacy is maintained (95).

It is worth noting here that there is also one argument that employs the idea of the social good as a justification for the requirement of informed consent. Clayton et al. argue that respect for autonomy is an important social good that the informed consent requirement promotes. We should not sacrifice this good for the lessor goods of scientific progress, which could undermine public trust in the research establishment (22, p. 1787). In a similar vein Hans Jonas has argued that scientific progress is an *optional* good, one which brings benefits but not benefits that are so *essential* to the public welfare that they justify overriding individual autonomy (68, p. 230). Although a majority of individuals would gladly consent to such uses, they also find benefit in the asking. For instance, in a 1993 Harris Poll survey, 64 percent of patients surveyed indicated that permission should be obtained before medical records are used for research, even if individuals are not identified in any publication (96). Whether the *benefits* of respecting the autonomy of patients and subjects by requiring informed consent outweigh the social benefits of research conducted in the absence of this requirement is fundamentally a value judgment, which in the regulatory context must be decided through political process.

Deontological Approach to Autonomy

The consequentialist interpretation of autonomy is predicated upon two premises: (*1*) moral duties derive from moral rights and (*2*) moral rights are nothing more than the expression of strong moral interests. Accepting these two premises leads inevitably to the conclusions analyzed above. The fateful move, however, is to accept the first premise. Although the use of the language of rights in moral discourse may be quite common today, it is a relatively recent phenomenon. In the history of ethics, ethical *duties* have rarely been interpreted according to rights language, and even more rarely actually derived from rights.

The derivation of duties in a deontological system may be seen by way of contrast to the consequentialist approach. We may evaluate actions according to their *consequences*, that is, the value of the results of particular actions. This is the core of utilitarian ethics in its many different forms. It depends, then, on how we define the *good* we seek to promote by our actions before we can evaluate these consequences. Alternatively, we may evaluate the actions according to the motivations or reasons for the action. This is the way of a *deontological* ethics. In the former case, it is the ends that justify the means, whereas in a deontological system, the means must be justified independent of the ends, that is, the action itself must have moral worth independent of any consequences which flow from it. The method according to which we define the *good* in action will depend upon the ethical system we employ. In a deontological moral system, agents have a *duty* to perform those actions which are good, regardless of the consequences.

There have been several articles in the literature that employ deontological reasoning to evaluate the question of requiring informed consent in general and in particular for the research use of tissues and records. In a general analysis of consent, both Lebacqz and Levine (97) and Marshall (98) cite autonomy as defined by Immanuel Kant as the foundation of informed consent alongside the beneficence concern to minimize harm. Clayton et al. cite a deontological interpretation of autonomy as the ground for requiring informed consent for research involving tissues and records (22, p. 1788). And Faden et al. follow the historical rise of autonomy in the regulation of research (and in the practice of clinical medicine) as an increasingly important ethical foundation for informed consent (33).

Capron provides an intuitively clear argument that highlights the difference between a deontological and consequentialist ethics by distinguishing between harms and wrongs (99). Suppose someone enters your house without permission and looks through your belongings without taking or damaging any of them. Capron argues one may not be *harmed* by this invasion of privacy, but one is certainly *wronged* by it, since one's private life has been exposed without one's permission. This analogy is then extended to the medical record that contains a great deal of intimate information about the individual, which is also the case with human tissues. The use of medical records and tissues *without consent* may not *harm* the individual (though harm is possible), but it does violate the individual's privacy, or *wrongs* them. In the consequentialist domain, wrongs are simply measured by the severity of harms. In a deontological framework, the wrong is constituted not by any harm to the individual but rather by the violation of their own self-determination. The intruder into the home and the medical researcher who obtains private medical information without consent, both fail to respect the self-determination, or autonomy, of the individual, regardless of whether harm is suffered or not.

The concepts of *respect for persons* and *autonomy* have long been used interchangeably. The National Commission interpreted respect for persons in the *Belmont Report* as the requirement that individuals "should be treated as autonomous agents" (12, p. 4), and defined such an agent thus:

> An autonomous person is an individual capable of deliberation about personal goals and of acting under the direction of such deliberation. To respect autonomy is to give weight to autonomous persons' considered opinions and choices while refraining from obstructing their actions unless they are clearly detrimental to others (12, p. 5).

The key to this interpretation of autonomy is the idea of self-determination, that individuals may set their own goals and act accordingly, within the limits of not acting in a detrimental fashion toward others, that is, either harming them or violating their autonomy. Autonomy is thus related to the concept of *freedom*, and is typically understood as a matter of *choice*.

Our contemporary concept of autonomy derives for the most part from the moral philosophy of Immanuel Kant, and is intimately related to the concept of freedom (100). For Kant, freedom has two aspects: negative and positive. Negative freedom is the absence of constraint. One is free to the extent that one is not constrained in one's actions by external constraints (chains, iron bars, poverty), or by internal constraints, such as the influence of desire upon one's decision making. But negative freedom is incapable of steering one toward any particular action. For this, we need *positive* freedom — a ground for the determination of our actions. This positive freedom Kant argued is *autonomy*, which he understands in its etymological meaning of *self* (auto) *law* (nomos). Far from being the freedom to do anything that one wishes or wants, autonomy for Kant is the freedom to legislate the moral law *for oneself* and in this sense is genuinely self-determining. Positive freedom is thus the ability to self-legislate the moral law. For Kant, autonomy is a much richer concept than the mere "freedom to choose" by which it is typically understood.

In Kant's famous categorical imperative, the moral worth of an action (and the moral law) may be determined by whether its maxim — a simple statement of the principle of the action — can be willed as a universal law. If the maxim can stand as a universal law for all to follow and does not lead to logical contradiction in the process of universalization, then the maxim is lawful and it becomes a duty of the person to act in accordance with it. For Kant, moral worth, or the *good*, is determined simply and solely by the *lawfulness* of the maxim of the action, that is, its ability to be applied universally and without contradiction, and duty is the necessity to act in conformity to this law which one has legislated for oneself. Thus, autonomy derives from each person's ability to formulate for themselves the moral law that applies to everyone, since a law that applies only to some is not universal. It is by virtue of the light of reason in each person that they have moral freedom and moral responsibility, since it is through our reason that we are able to conceive and indeed legislate the moral law.

This capacity — autonomy — is the *highest good* according to Kant, for it is the condition of all other goods, which are of relative worth. Only *persons* may be accorded this *respect*, all other goods and values being merely *things*. "Things," according to Kant, have a relative price and can be exchanged with one another, but persons are accorded a *dignity*, that is, a worth that is beyond all measure and comparison. Therefore persons cannot be bartered or traded for something of equivalent value, for each person has the capacity to legislate the moral law. Consequently the *value* of autonomy cannot be exchanged, balanced, weighed against, or superceded by other values, for instance, social benefit.

Insofar as individuals determine the moral good through their self-legislation of the moral law, and thereby determine themselves to action, it follows that each individual sets his or her own moral *ends* or purposes. Kant thus offered an alternative formulation of the categorical imperative which is most helpful in this context: "Act in such a way that you always treat humanity, whether in your own person or in the person of any other, never simply as a means, but always at the same time as an end." "Rational nature," writes Kant, "exists as an end in itself" (100, p. 96). Respect for persons is the necessity or duty to always treat others as ends in themselves, that is, as persons who set their own goals and purposes and who in their actions determine the moral law.

Kant recognized that in everyday interactions we are constantly involved in relations with others through which we are reciprocally means to each others' ends. The force of this imperative is to stipulate that in these relationships we always *also* respect each other as self-determining persons—autonomous—that is, as persons who set our own goals. Thus in our everyday activities it is wrong to use other persons for purposes (ends) that they do not share. To do so is to treat other persons merely as *means*, that is, as mere *things*, and thus to violate their dignity as persons (101).

There are three conclusions to be drawn from Kant's argument. First, there is no claim here to a moral *right* of autonomy. Rather, duty derives from the rational capacity of each agent. Respect for autonomy is a *duty* of the researcher who seeks to use human subjects as a means to the production of scientific knowledge, just as it is a duty of every individual to always treat others as ends, even as we use each other as means. Only through such respect can actions have moral worth. This duty is the necessity of the researcher to submit their purposes to the dignity and sovereignty of the person who is the research subject, which is to say, it is their submission to the moral law itself. Moral duty thus does not derive from a "right," the right of the other to determine their own ends, but rather duty derives from the moral law itself. By starting with the premise that autonomy is a right that simply represents important moral *interests*, the consequentialist interpretation of autonomy fundamentally loses sight of the deontological dimension of the concept, that duty depends on the moral law. Autonomy may indeed represent certain "interest," but these are not the basis of the deontological concept of autonomy and the duty to respect others as ends in themselves.

Second, although other goods may be achieved through failure to respect the dignity of persons and the moral law, it is the goodwill of each rational agent that is the highest good and prerequisite of all other goods. To sacrifice this highest good, from which all other relative goods derive their value, for the sake of a dependent and relative good is to undermine the very good that one would hope to achieve in the action. It is self-defeating. From this deontological point of view, the respect for autonomy has nothing to do with the consequences of action, good or bad, but is rather the condition for achieving any good at all. Social good is not an end in itself but rather is a relative good, or means, to the realization of the highest good, the achievement of morally good action. The fundamental claim Kant is making is that there can be no good at all, including social good, without moral freedom, or autonomy.

In usurping this freedom by failing to inform and ask the consent of patients for the use of their tissues and records, we limit the moral freedom of individuals and thereby diminish the possibility of realizing the good to be achieved through the social benefits of the research. It is the problem of putting the cart (social good) before the horse (individual moral freedom). Individuals whose privacy has been violated are indeed *wronged* as Capron argues. The wrong is a violation of their moral freedom and consequently a violation of their *dignity*. The pejorative *abstract autonomy* to which Reilly refers, and which the consequentialist interpretation produces in this context, fails to recognize that the very possibility of moral freedom and responsibility is at stake when we fail to respect the autonomy of others.

Third, it is necessary that research subjects, in order that they not be treated *merely* as means, be informed of the purposes of the research and freely accept these purposes as their own in their participation in the research protocol. This is the insight for which Veatch argues so strenuously. It is also the basis of the National Commission's definition of autonomy. It follows, as Veatch argued, that the exemption from, and waiver of, informed consent is immoral, since only through informed consent is it possible to respect the autonomy of the persons who are subjects of research. Failure to inform subjects and ask their consent for participation is to treat them merely as *means* to ends which they may or may not share. It is to treat persons as *things* which lack dignity.

Autonomy, understood in this light, conflicts with the consequentialist interpretation which would subsume autonomy as a species of beneficence. On the contrary, the deontological interpretation of autonomy places consequentialist reasoning as subordinate to and derivative from the concept of good which autonomy determines. This good is incommensurate with the relative goods to be achieved through the consequences of particular actions. For the deontologist, arguments concerning the increased social benefits to be derived from research simply miss the point or end up subverting the concept of autonomy. Beneficence concerns are legitimate but not overriding.

It can readily be seen that the absoluteness of the duty to respect autonomy in the deontological interpretation is behind some of the arguments for requiring the informed consent which we cited at the beginning of this section.

1. The ethical foundation of informed consent is a respect for individual autonomy, and permitting research on tissues and records without informed consent violates a growing tradition in research ethics that has served patients and subjects very well. We have demonstrated the independent validity of autonomy as an ethical underpinning of the requirement for informed consent, and the National Commission gave a deontological interpretation of autonomy. Although the independence of autonomy has emerged through the historical rise of government regulation in human subjects research, it has served as an important and powerful check against researchers who would privilege the social benefits of scientific knowledge over the free will of research

subjects, which in the past has led to disasterous consequences.

2. Social benefits are not a sufficient ground to undermine the respect for autonomy by not requiring informed consent. This is the heart of the deontological argument which finds that benefits and harms in general are incommensurate with deontological principles.

3. As a result of the successes of the Human Genome Project, human tissues contain a vast amount of medical information that can be tapped with increasing ease. There was a time when tissues represented very little of consequence, either directly to patients or indirectly by virtue of the purposes to which researchers might put them. But the production of scientific knowledge is less of an unequivocal good now. Not all research subjects may share the research purposes of some or even many protocols. Thus the moral freedom of individual subjects is at stake when the research community argues for less stringent requirements for informed consent.

4. Public trust in the health care industry and in biomedical research is undermined when researchers use individuals' tissues without their consent or knowledge for purposes they may not share, and especially when researchers profit financially from them. This argument, as we saw above, may be grounded on a consequentialist analysis, but it also may be grounded on the deontological requirement not to involve other persons in purposes that they do not share or have not accepted. Especially when research is increasingly motivated by profit, the presumed altruism that has played a role in justifying researcher access to tissues and records without informed consent is undermined. This presumed altruism is really the presumption that individuals *share* the goals of biomedical research. But the ulterior motives that also animate research to an increasing degree cannot so easily be presumed to be shared by the general public.

The idea of patient altruism has a correlate in the idea that patients who benefit from biomedical advancement owe a *debt* to past research subjects and thus have an *obligation* to contribute to the future advancement of biomedical science. One relatively easy way for patients to discharge this obligation is to provide easy access to tissues and medical records for research purposes. However, both Caplan (102) and Jonas (68) have argued that while we may owe some debt of gratitude to past subjects of biomedical research, those subjects cannot possibly have considered that this debt would be discharged through violating the autonomy of subjects that follow, though they may well hope that others would make similar contributions to the common good. Furthermore, were this debt one that the research community could *exact* from the population of patients, there would be no reason why the requirement to participate in research would not extend to all biomedical research, including the testing of invasive

procedures and drugs. Contributing to biomedical progress is like the general moral obligation to be charitable. We do indeed have an obligation to be charitable, but charity must be given freely and according to the individual's own view of the good to be achieved thereby. To require that patients contribute their tissues and records to biomedical research amounts to a *tax* on patients, in much the same way that welfare programs for the poor are funded by a tax on the public. It would be a serious mistake to refer to one's tax contributions to welfare programs as charitable, and it is no less a mistake to presume that the health care industry can collect on patient debts to medical progress by collecting their tissues without consent.

The last argument for requiring informed consent derives from the fact that our current health care system does not distribute its benefits equally across the population. Many minorities and the poor are excluded or underserved. The justification that the violation of individual autonomy by the claim that research on tissues serves the public good fails to recognize that these benefits do not return equally to the individual sources of tissues and records. This argument is founded on neither beneficence nor autonomy but the independent principle of justice. Justice in this context concerns the *fair* distribution of benefits and burdens and relates to the previous argument concerning the obligations of patients to participate in research. Clearly, those who have not shared in the benefits of biomedical research as fully as others cannot be said to have an equal obligation to that system, especially when the inequities of our health care system are largely the result of a conscious political choice not to remedy them. Autonomy does play a role here insofar as those who have received less than equal share in the benefits of biomedical science and treatment may no longer share the goals of this industry, at least, until they and others are included as full partners.

CONCLUSION

Having formulated the contrast between the consequentialist and deontological interpretations of autonomy so starkly, or what amounts to the same thing, the contrast between beneficence and autonomy, we must be careful how we proceed. It would be tempting, perhaps, to simply *weigh* these two principles against each other. But this would beg the question, for we must have some criteria by which to evaluate these competing principles, a common scale as it were, and by definition there can be no such common scale. Marshall (98) has argued that the National Commission fell prey to this error by viewing autonomy, beneficence, and justice as principles to be *weighed* or *balanced* against each other when they lead to contradictory conclusions. The metaphor of *balancing* implies that the contradictory terms are commensurate. But as Marshall concludes "…the point of the Kantian principles is precisely to say that certain things cannot be 'balanced out'" (98, p. 6). The metaphor of "weighing" and "balancing" is primarily a utilitarian strategy of reckoning up the goods and ills of consequences. To suggest that a

deontological principle may be balanced against a consequentialist principle is to eviscerate the deontological core
of the concept, and thus begs the question.

We are left therefore with incommensurate principles
which, when applied to research involving tissues
and records, lead to contradictory conclusions. Each
principle appears to assert its priority over the other.
Either we must sacrifice autonomy as an independent
deontological principle if we wish to privilege beneficence
in justifying not requiring informed consent, or we
must accept the deontological principle of autonomy as
taking precedence over beneficence considerations, and
thus sacrifice some measure of the common good to
be achieved through the use of tissues and records
without informed consent. Notice that in this dilemma,
privileging beneficence over autonomy does violence to
its deontological interpretation, while recognizing the
priority of autonomy does not in any way corrupt the
principle of beneficence. Consequentialist reasoning is
inherently blind to its deontological counterpart, while
from the deontological point of view, there is no inherent
contradiction in reckoning up consequences, provided only
that in doing so we conform to our moral duty. If we are
to accept the deontological interpretation, the inevitable
conclusion is that autonomy is an independent moral
principle that cannot be subsumed under beneficence.

If public policy is then to recognize the independent
validity of autonomy, we must reconsider the exemptions
and waiver of informed consent that the current
regulations permit. This does not necessarily mean that
we must default to always requiring the full informed
consent of subjects for the use of their tissues and records
in research. As we have seen, the National Commission
was reluctant to give up the social benefits of scientific
research that would be lost if specific informed consent
were required for all uses of tissues and records. But
recognizing that individual autonomy could not simply be
sacrificed at the altar of social expediency, the commission
recommended two provisions that would preserve the
respect for autonomy. First, it required IRB review of
all research involving human tissues and records, and
second, it required that institutions notify all patients
that these materials may be used for research and provide
patients with a blanket consent to opt out if they so
chose. The requirement of IRB review ensures the social
acceptability and value of the research and thus stands in
as proxy, or "constructed" consent, for subjects who do not
specifically consent to these protocols. The requirement
for blanket consent recognizes, however, the sovereignty
of the autonomous subject to choose whether to participate
in such research activities. Clearly, this compromise
crafted by the National Commission limits the exercise
of individual autonomy, but at the time it was perhaps
much less likely that subjects would object to the research
use of their tissues.

Research involving tissues, however, has changed
dramatically since the late 1970s when the National
Commission issued its recommendations. The revolution
in genetics and biotechnology in general raises a host of

new risks to subjects, and the erosion of privacy in our
evolving information society and within health care itself
has made the public more aware and more cautious in
guarding privacy. It may be that blanket consent is no
longer an adequate expression of respect for the autonomy
of patients. Robert Weir's concept of an advance directive
for tissues would go a long way toward returning to
patients and subjects the control that is an expression of
their self-determination. It is at least worthwhile exploring
the feasibility of instituting such a system.

If we are to pay more than lip service to autonomy, it
will be necessary for IRBs to pay close attention to the
scientific necessity of the waiver of informed consent. The
current categories of exemption should be abandoned in
favor of IRB review in which IRBs may decide that the
waiver of the informed consent requirement is necessary
to accomplish the goals of the research protocol, and
not just a matter of inconvenience to the investigators.
The current exemption categories permit investigators to
use human tissues and records for virtually any purpose
without the knowledge and consent of the sources. While
most such purposes are generally laudable and serve the
public good, this may not always be the case. If the waiver
of informed consent is genuinely necessary to conduct
the research, then the research community owes it to
the individual sources of tissues and records to submit
the research to an independent evaluation to ensure that
the research is not foreseeably objectionable. This does
not provide for those who may have particular objections,
but those who are concerned not to have their tissues
and records used for purposes to which they might object
would have the opportunity to opt out of participation
ahead of time through a blanket refusal of consent or an
advance directive prohibiting the research use of tissues
and records.

In order for IRBs adequately to take on this increased
responsibility, the conditions for waiver of informed
consent need to be tailored specifically to research
involving tissues and records. The ambiguities in the
waiver criteria analyzed in the second part of this article
should be clarified, and specific criteria relating to tissues
and records should be made explicit. The questions
concerning risks, "rights and welfare," ownership and
commercial use, recontact of subjects and especially the
"practicality" of research, that is, the scientific necessity
of the waiver, need to be specifically tailored to research
involving tissues and records.

Second, IRB membership, as noted in the Inspector
General's report on the reform of IRBs (10), needs to be
broadened with additional noninstitutional and nonscien-
tific members representative of community interests so
that evaluations of the social value or objectionability of
research conducted without specific informed consent can
be more representative of the local populations. IRBs also
need greater expertise in the areas of genetics and biotechnology in order to be in a position to recognize and analyze
the many issues these protocols may raise.

These modifications to the current system of protections
would ensure that both autonomy is respected and that

the benefits of scientific research may accrue through as minimal interference with the research enterprise as is consistent with this respect for autonomy.

BIBLIOGRAPHY

1. G.J. Annas, *J. Am. Med. Assoc.* **270**(19), 2346–2350 (1993).

2. Nuffield Council on Bioethics, *Human Tissue: Ethical and Legal Issues*, Nuffield Council on Bioethics, London, 1995, p. 19.

3. E. Eiseman, *Stored Tissue Samples: An Inventory of Sources in the United States*, in Research Involving Human Biological Materials: Ethical Issues and Policy Guidance; Volume II, Commissioned Papers; National Bioethics Advisory Commission, Rockville, MD, 2000, p. D-5.

4. U.S. Pat. 5,409,810 (December 1, 1992), B.A. Lardar and S.D. Symons (to the Burroughs Wellcome Co.).

5. B.A. Lardar, K.E. Coates, and S.D. Kemp, *J. Virol.* **65**(10), 5232–5236 (1991).

6. National Bioethics Advisory Commission, *Research Involving Human Biological Materials: Ethical Issues and Policy Guidance; Volume I, Report and Recommendations of the National Bioethics Advisory Commission*, National Bioethics Advisory Comm., Rockville, MD, August 1999, p. 1.

7. C.R. Naser, in A. Thompson and R. Chadwick, eds., *Genetic Information: Acquisition, Access and Control*, Kluwer Academic, Plenum, New York, 1999, pp. 105–120.

8. U.S. Department of Health and Human Services, Office of Protection from Research Risks, *Federal Policy for the Protection of Human Subjects*, Title 45, Code of Federal Regulations, Part 46, Subpart A. All citations to this policy will appear in the text by section number, for example: "§46.101(b)(4)."

9. U.S. Department of Health and Human Services, Office of Protection from Research Risks, *Fed. Regist.* **56**(117), 28003–28032 (1991).

10. U.S. Department of Health and Human Services, Office of Inspector General, *Institutional Review Boards: A Time for Reform*, OEI-01-97-00190; USDHHS, Washington, DC, 1998, p. B-1.

11. U.S. Department of Health, Education, and Welfare, *Fed. Regist.* **39**(105), 18914–18920 (1974).

12. National Commission for the Protection of Human Subjects of Biomedical and Behavioral Research, *The Belmont Report: Ethical Principles and Guidelines for the Protection of Human Subjects of Research*, U.S. Government Printing Office, U.S. Department of Health, Education and Welfare, Washington, DC, 1978.

13. U.S. Department of Health, Education and Welfare, *Fed. Regist.* **44**(158), 47688–47734 (1979).

14. U.S. Department of Health and Human Services, *Fed. Regist.* **46**(16), 8366–8392 (1981).

15. National Commission for the Protection of Human Subjects of Biomedical and Behavioral Research, *Institutional Review Boards: Report and Recommendations*, U.S. Government Printing Office, Washington, DC, 1978, p. 21.

16. Personal communication from J.T. Puglisi, Ph.D., Director, Division of Human Subjects Protections, Office for Protection from Research Risks, March 28, 1997.

17. V.A. LiVolsi et al., *Cancer (Philadelphia)* **71**(4), 1391–1394 (1993).

18. J.F. Merz et al., *J. Invest. Med.* **45**(5), 252–257 (1997).

19. G.B. Ellis, Director, Office of Protection from Research Risks, *Memorandum: Response to Frequently Asked Questions*, April 17, 1996, Chart 2.

20. A.R. Holder and R.J. Levine, *Clin. Res.* **24**(2), 68–77 (1976).

21. J. Merz, *IRB* **18**(6), 7–8 (1996).

22. E.W. Clayton et al., *J. Am. Med. Assoc.* **274**(22), 1786–1792 (1995).

23. E.W. Clayton, *J. Law, Med. Ethics* **23**(4), 375–377 (1995).

24. K.T. Kelsey et al., *J. Am. Med. Assoc.* **275**(14), 1085–1086 (1996).

25. R.J. Levine, *Ethics and Regulation of Clinical Research*, Urban & Schwarzenberg, Baltimore, MD, 1981.

26. P.R. Reilly, M.F. Boshar, and S.H. Holtzman, *Nat. Genet.* **15**(1), 16–20 (1997).

27. L.J. Melton, *N. Engl. J. Med.* **337**(20), 1466–1470 (1997).

28. D. Korn, *Genetic Privacy and the Use of Human Tissues in Research, Presented at Risk, Regulation, and Responsibility: Genetic Testing and the Use of Information*, American Enterprise Institute, Washington, DC, 1997. Available at: *http://www.aei.org*

29. Personal communication from J. Cheek, Ph.D., Principle Epidemiologist for the Indian Health Service, June 1998.

30. Privacy and Protection of Research Subjects: *Certificates of Confidentiality*, Information sheet available from PHS through the NIH Office of Resource Management. (Investigators may obtain a Certificate of Confidentiality by applying to the NIH. To receive a complete information packet, call 301-443-3877.)

31. C.L. Earley and L.C. Strong, *Am. J. Hum. Gene.* **57**(3), 727–731 (1995).

32. W.J. Curran, in P.A. Freund, ed., *Experimentation with Human Subjects*, George Braziller, New York, 1970, pp. 402–454.

33. R.R. Faden, T.L. Beauchamp, and N.P. King, *A History and Theory of Informed Consent*, Oxford University Press, New York, 1986, p. 208.

34. R.S. Dresser, *IRB* **3**(4), 3–4 (1981).

35. R.M. Veatch, *The Patient as Partner: A Theory of Human-Experimentation Ethics*, Indiana University Press, Bloomington and Indianapolis, 1987, p. 43.

36. Moore v. The Regents of the University of California, 215 Cal. App. 3d 709, 249 Cal. Reptr. 494 (1988), *aff'd in part, rev'd in part*, 51 Cal. 3d 120, 793, 271 Cal. Reptr. 146 (1990).

37. R.F. Weir and J.R. Horton, *IRB* **17**(5,6), 1–8 (1995).

38. K.C. Laudon, *Dossier Society*, Columbia University Press, New York, 1986.

39. J.B. Rule, *Newsday*, July 30, p. A40 (1997).

40. T.H. Murray, *IRB* **8**(1), 1–5 (1986).

41. E.R. Gold, *Body Parts: Property Rights and the Ownership of Human Biological Materials*, Georgetown University Press, Washington, DC, 1996, pp. 161–162.

42. S. Alpert, *Privacy and the Analysis of Stored Tissues*, in Research Involving Human Biological Materials; Ethical Issues and Policy Guidance; Volume II, Commissioned Papers. National Bioethics Advisory Commission, Rockville, MD, 2000.

43. L. Gostin, *Law, Med. Health Care* **19**(3–4), 191–204 (1991).

44. L.M. Kopelman, *J. Med. Philos.* **19**(6), 525–552 (1994).

45. L.H. Glantz, *IRB* **1**(6), 5–6 (1979).

46. H. Greely, *Houston Law Rev.* **33**, 1397–1430 (1997).

47. North American Regional Committee of the Human Genome Diversity Project, *Houston Law Rev.* **33**, 1431–1473 (1997).

48. U.S. Department of Health Education and Welfare, *Fed. Regist.* **43**(231), 56174–56198 (1978).

49. K.J. Rothman, *N. Engl. J. Med.* **304**(10), 600–602 (1981).

50. W.W. Grody, *Diagn. Mol. Pathol.* **4**(3), 155–157 (1995).

51. O. O'Neill, *J. Med. Ethics* **22**, 5–7 (1996).

52. E.G. Knox, *Br. Med. J.* **304**, 727–728 (1992).

53. P.R. Reilly, *J. Law, Med. Ethics* **23**, 378–381 (1995).

54. L. Andrews and D. Nelkin, *Lancet* **351**, 53–57 (1998).

55. E. Marshall, *Science* **271**, 440 (1996).

56. G.J. Annas, in T.F. Murphy and M.A. Lappé, eds., *Justice and the Human Genome Project*, University of California Press, Berkeley, 1994, pp. 75–90.

57. P.S. Harper, *Br. Med. J.* **306**, 1391–1394 (1993).

58. J. Benbassat and M. Levy, in P. Allebeck and B. Jansson, eds., *Ethics in Medicine: Individual Integrity Versus Demands of Society*, Raven Press, New York, 1990, pp. 159–165.

59. Privacy Protection Study Commission, *Personal Privacy in an Information Society*, U.S. Government Printing Office, Washington, DC, 1977, pp. 596–598.

60. LeR. Walters et al., *IRB* **3**(3), 1–3 (1981).

61. A.R. Holder, *IRB* **3**(3), 4–5 (1981).

62. G. Kolata, *N.Y. Times*, October 24, pp. A1, C6 (1995).

63. M.Z. Pelias, *Am. J. Med. Genet.* **39**, 347–354 (1991).

64. V.L. Hannig, E.W. Clayton, and K.M. Edwards, *Am. J. Med. Genet.* **47**, 257–260 (1993).

65. M. Feinleib, *J. Clin. Epidemiol.* **44**(Suppl. I), 73s–79s (1991).

66. E.D. Pellegrino, in H.T.J. Engelhardt, S.F. Spicker, and B. Towers, eds., *Clinical Judgment: A Critical Reappraisal*, Reidel Publ., Boston, MA, and Dordrecht, The Netherlands, 1979, pp. 169–194.

67. K. Lebacqz, *Villanova Law Rev.* **22**, 357–366 (1977).

68. H. Jonas, *Daedalus* **98**(2), 219–247 (1969).

69. National Health and Nutrition Examination and Survey (NHANES), conducted by the Centers for Disease Control, Atlanta, GA, July 1994.

70. K. Steinberg, in R. Weir, ed., *Stored Tissue Samples: Ethical, Legal, and Public Policy Implications*, University of Iowa Press, Iowa City, 1998, pp. 82–88.

71. C.R. Naser, in R.F. Weir, ed., *Stored Tissue Samples: Ethical, Legal, and Public Policy Implications*, University of Iowa Press, Iowa City, 1998, pp. 160–181.

72. President's Commission for the Study of Ethical Problems in Medicine and Biomedical and Behavioral Research, *Screening and Couseling for Genetic Conditions*, U.S. Government Printing Office, Washington, DC, 1983.

73. D.L. Breo, *J. Am. Med. Assoc.* **269**(15), 2017–2022 (1993).

74. K.F. Hoskins et al., *J. Am. Med. Assoc.* **273**(7), 577–585 (1995).

75. A. Clarke, *Lancet* **338**, 998–1001 (1991).

76. F.M. Giardiello et al., *N. Engl. J. Med.* **336**(12), 823–827 (1997).

77. G.G.S. Singer, in B. Gert et al., eds., *Morality and the New Genetics: A Guide for Students and Health Care Providers*, Jones & Bartlett International, London, 1996, pp. 125–146.

78. M. Decruyenaere, G. Evers-Kiebooms, and H. Van den Berghe, *J. Med. Genet.* **30**, 557–561 (1993).

79. M. Huggins et al., *Am. J. Hum. Genet.* **47**, 4–12 (1990).

80. U.S. Department of Health and Human Services, *Fed. Regist.* **46**(16), 8392 (1981).

81. U.S. Department of Health and Human Services, *Fed. Regist.* **63**(216), 60364–60367 (1998).

82. G.J. Annas, L.H. Glantz, and P.A. Roche, *The Genetic Privacy Act and Commentary*, Boston University School of Public Health, Boston, MA, 1995.

83. T. Murray, in M. Rothstein, ed., *Genetic Secrets: Protecting Privacy and Confidentiality in the Genetic Era*, Yale University Press, New Haven, CT, and London, 1997, pp. 60–73.

84. R.F. Weir, in R.F. Weir, ed., *Stored Tissue Samples: Ethical, Legal, and Public Policy Implications*, University of Iowa Press, Iowa City, 1998, pp. 236–266.

85. B.M. Dickens, *Law, Med. Health Care* **19**(3–4), 175–183 (1991).

86. R.J. Levine, *Hastings Cent. Rep.* **9**(3), 21–26 (1979).

87. S.D. Warren and L.D. Brandeis, *Harv. Law Rev.* **4**, 193 (1890); reprinted in F.D. Schoeman, ed., *Philosophical Dimensions of Privacy: An Anthology*, Cambridge University Press, Cambridge, MA, 1984, pp. 75–103.

88. C.J. Bennett, *Regulating Privacy: Data Protection and Public Policy in Europe and the United States*, Cornell University Press, Ithaca, NY, 1992, p. 13.

89. S.A. Alpert, *Santa Clara Comput. High Technol. Law J.* **11**(1), 97–118 (1995).

90. G. Duncan, T. Jabine, and V. de Wolf, *Private Lives and Public Policies: Confidentiality and Accessibility of Government Statistics*, National Academy Press, Washington, DC, 1993, p. 27.

91. A. Allen, in M. Rothstein, ed., *Genetic Secrets: Protecting Privacy and Confidentiality in the Genetic Era*, Yale University Press, New Haven, CT, 1997, pp. 31–59.

92. A.F. Westin, *Privacy and Freedom*, Atheneum, New York, 1967.

93. V.M. Brannigan and B. Beier, *Datenschutz und Datensicherung*, April 1985.

94. A. Buchanan, *An Ethical Framework for Biological Samples Policy*, in Research Involving Human Biological Materials: Ethical Issues and Policy Guidance; Volume II, Commissioned Papers, National Bioethics Advisory Commission, Rockville, MD, 2000, p. B-4.

95. M.B. Visscher, *Mod. Med.*, pp. 62–64 (Feb. 15, 1975).

96. Harris-Equifax, *Health Information Privacy Survey 1993*, Study No. 934000009, Harris-Equifax, New York, 1993, p. 45.

97. K. Lebacqz and R.J. Levine, *Clin. Res.* **25**(3), 101–107 (1977).

98. E. Marshall, *IRB* **8**(6), 5–6 (1986).

99. A.M. Capron, *J. Clin. Epidemiol.* **44**(Suppl. I), 81s–89s (1991).

100. I. Kant, *Groundwork of the Metaphysics of Morals* (H.J. Paton, trans.), Harper & Row, New York, 1956.

101. G. Dworkin, in T.L. Beauchamp et al., eds., *Ethical Issues in Social Science Research*, Johns Hopkins University Press, Baltimore, MD, 1982, pp. 246–254.

102. A.L. Caplan, in S.F. Spicker et al., eds., *The Use of Human Beings in Research*, Kluwer Academic Publishers, Boston, MA, 1988, pp. 229–248.

See other entries GENETIC INFORMATION, ETHICS, PRIVACY AND CONFIDENTIALITY: OVERVIEW; GENETIC INFORMATION, LAW, LEGAL ISSUES IN LAW ENFORCEMENT DNA DATABANKS; GENETIC INFORMATION, LEGAL, GENETIC PRIVACY LAWS; PATENTS AND LICENSING, ETHICS, MORAL STATUS OF HUMAN TISSUES: SALE, ABANDONMENT OR GIFT.

GENETIC INFORMATION, ETHICS, FAMILY ISSUES

Eric T. Juengst
Case Western Reserve University
Cleveland, Ohio

OUTLINE

Introduction
Family Virtues and Genetic Testing
    Loyalty
    Intimacy
    Security
Family Styles
    Organic Families
    Social Families
    Virtual Families
Conclusion
Bibliography

INTRODUCTION

Of all the many "publics" that are affected by advances in genetics, families that use genetic services are the closest to our society's grass roots: They cut across all other sociological categories and lie behind all the usual interest groups that contend over our society's health policies and practices. Moreover these families and their members have always been portrayed by the biomedical community as the focus of the genetic services it generates. In the absence of effective therapies, the promise of accessible genetic information lies almost entirely in its ability to allow families to identify, understand, and sometimes control their inherited health risks. Against the excesses and abuses of the eugenicists' population-oriented concerns, contemporary geneticists are firm in their conviction that "the fundamental value of genetic screening and counseling is their ability to enhance the opportunities for individuals to obtain information about their personal health and child-bearing risks, and to make autonomous and non-coerced choices based on that information" (1). This puts families at the moral fulcrum of the enterprise: If genetic services are to be judged a success, it must be from the recipients' point of view, in terms of their ability to use the results to support their flourishing as individuals and families. That, in turn, gives a special urgency to getting clear about the impact of genetic services on family life, and the ethical issues they can raise for family members. This article summarizes what is known about these "family matters" in genetics and suggests an agenda for futher research in this crucial area.

FAMILY VIRTUES AND GENETIC TESTING

Over the last decade a large literature has evolved anticipating and addressing the ethical, legal, and social implications of advances in human genetics. For overviews, see Juengst (2) and Thomson (3). One of the weakest spots in that literature, however, is work addressing the ethical issues faced by the individuals and families who might avail themselves of the fruit of those advances. This relative neglect is not entirely surprising, despite the centrality of the issues to the success of the genetic enterprise. First of all, most of our efforts have been directed toward the development of policy capable of optimizing our uses of genetic advances: rules, guidelines, agreements, and positions that can be generalized across all the situations that raise the issues they address. That is much easier to accomplish for entities like states, institutions, and professions than it is for "the family," since our many different families subscribe to no unified process for making universally binding "family policy" on ethical issues. Moreover there is no well-received generic account of the moral dynamics of family life to draw upon in even attempting to develop such policies. With some recent exceptions (4,5), the ethics of family interactions has been a black box for contemporary applied ethics, protected from intellectual scrutiny as well as state intervention by our liberal traditions. In part, this is because there is such a rich pluralism of strongly held specific theories on the subject, which reflect the convictions and experiences of different family histories and traditions (6). Since these specific theories are usually intertwined tightly with our most important beliefs and values, contemporary moralists tend to give them the deference that we give to other transrational matters of conscience, like personal religious commitments. Against that variegated and sensitive background, the thinking goes, one would be foolish to try to generalize about what ethical issues any new genetic service might raise for families outside of some parochial cultural perspective (7).

Be that as it may, the existing literature does suggest three ways in which advances in genetics will challenge the ethics of at least many American families. Despite our differences, there are three familial virtues that most socialized Americans would not be surprised to find listed among the qualities that have traditionally been ascribed to the "good family" in our culture. I will call these qualities the familial virtues of loyalty, intimacy, and security. While these three virtues are neither necessary nor sufficient to an adequate ethical theory of family life, they are keys to the familial ethics of genetic testing, because they are value commitments that do seem particularly challenged by our emerging genetic testing practices. To the extent that these virtues are at least widely intelligible as ingredients in the moral dynamics of family life, their analysis provides a starting point for further discussion. New advances in genetics will challenge the ability of the families who endorse them to live up to the ideals that they represent.

Loyalty

The members of good families accept special obligations to serve their kin. Whether this means grown children joining their parents' businesses, siblings helping each other out of debt, or cousins hosting visiting cousins, members of a good family are expected to aid and assist one another without having made any of the explicit promises, offers, or agreements to do so that would govern such service between strangers. Moreover this familial loyalty can supersede the interests of individual family members in significant ways: when mothers delay careers to raise unexpected children, for example, or when children's educational interests are sacrificed in order to care for their infirm grandparents, or when siblings come home for the holidays at the cost of their holidays. Of course, like any virtue, familial loyalty can become a vice *in extremis*, as when excessive concern for "family honor" generates vendettas between families, or self-sacrifice becomes unnecessarily self-destructive. Identifying the proper demands of family loyalty and balancing them against our other interests is one of the perennial challenges of moral life within a family. However, the way we approach that balancing problem is itself oblique evidence for the value we give to the virtue. As Nelson and Nelson (5, p. 76) point out, "Moral relationships among family members can certainly be strained by betrayal or violence, but it takes a catastrophe to dissolve them." ... In our culture we understand the vices of extreme loyalty precisely as a problem of overdoing a good thing: we can sympathize with their perpetrators in ways that we cannot with those who abandon or betray their kin. On the whole, we applaud the families that stick together because we see a reciprocal concern for each other's best interests as a constitutive ingredient of what it means to be a good family. For example, Post (4, pp. 81–108) argues that the balance between giving and receiving love in the context of the family is a common moral expectation that is a truly universal intuition.

New advances in human genetics challenge the virtue of familial loyalty in two ways. They simultaneously illuminate both the connections among and the differences between family members. First of all, genetics is, by definition, a science of family connections. Clinically useful genetic information about individuals often requires knowing the background against which the individual's genome presents itself: the pattern of inheritance of the traits and markers in question within the larger family. This means that in order for a genetic test to be useful to an individual family member, other members of the family have to be willing to provide that background and, in the process, discover their own status within the pattern.

Moreover, like most medical interventions, genetic testing usually motivated by a crisis — someone diagnosed with breast cancer or genetic disease — which creates a sense of urgency to "get the family in for testing." For those other family members, the decision to participate in a testing program raises a basic moral question: What are the demands of my loyalty-based obligation to help my kin learn their genetic risks? In particular, must I sacrifice my own "right not to know" in order to help my relative enjoy the "right to know," and join him or her in braving the psychosocial risks of having that personal information known about me? When family members decide to protect their own interests and decline to participate, the same question is passed "downstream" to their children and grandchildren: If those downstream kin should decide to be tested, the status of the declining member could be revealed as a simple matter of deduction. What interests

must they sacrifice, then, in order to give the decliner the filial respect that they deserve (8)?

Finally, if a decliner's kin do become interested in learning their own genetic risks, but cannot do so without involving the reluctant relative, to what lengths may the family go to persuade the unwilling to do their familial duty? Split decisions about genetic testing have already been observed by genetic counselors to lead to familial discord in some cases (9), while unanimous decisions in other cases have raised suspicions of undue familial pressure to participate (10).

The moral bite of these questions within families can be seen by the ways in which they seem to be provoking health professionals involved genetic testing to clarify their own allegiances. Despite the long-standing reluctance of clinical geneticists to interfere in the personal choices that their clients make about genetic testing and the use of test results, some now argue that clinicians should help their clients persuade reluctant family members to participate, by reminding them of their familial obligations (11). Others stand by the conviction that each family member has equal standing as a client, and thus, "it is important that individuals within families are supported to make their own decisions about testing and are not coerced by eager relatives" (9, p. 25). Still others now argue that, if the clinical geneticist's "patient" is best understood as the family as a whole, perhaps clinicians have no more business attempting to regulate the influences on the family's collective decision than they would in second-guessing an individual's informed consent. From that point of view, whatever internal dynamics produce the decision are protected from the clinician's interference by the same sphere of familial privacy that grounds their commitment to nondirective counseling at the individual level (12). Just like the family members involved, the professionals are forced by genetic testing to think about the moral significance of the genetic connections that link up the individuals with whom they interact.

Of course, genetic information is as much about our differences as it is about our shared traits, and illuminating those differences is another way in which genetics can challenge the familial virtue of mutual loyalty. As we are able to sort out which lineages within families, and which individuals within a lineage, carry a family's risk-conferring mutations, tension will be created between the divergent interests of the two groups. Whatever their commitment to family solidarity, the family will have to face the fact that it will be in its "normal" "members" interests to reveal their noncarrier status in some circumstances and in the interests of the carriers to conceal theirs. Family members free of the mutations in question, for example, will find it in their interests to use that information to counter their family history of a disease in applying for insurance (13). In doing so, however, they will inevitable raise questions about their kin who do not volunteer their test results in turn. Should families be expected to stick together "in sickness and in health" as we ask of married couples, or do the "limited sympathies" of human nature give us leave to concentrate on the welfare of our own threads within the familial patterns of inheritance?

For families in our society, market institutions like risk-rated health and life insurance that create serious competitive advantages out of genetic differences make it increasingly difficult for even conscientious families to present a united front on the matter of disclosing test results. As a spokesman for the insurance industry recently wrote, for high risk families faced with gaining access to adequate health and life insurance, "Harsh as it may sound to ears of a society that subscribes to egalitarian principles, solidarity ends with a negative genetic test" (14). Since, short of cloning, human families will always weave together a combination of different genetic threads, new abilities to identify those differences will continue to expose families to this kind of external pressure, as long as we live in a society that uses such differences to allocate its opportunities.

Intimacy

A second traditional virtue of family life is intimacy. As our concept of "family secrets" suggests, we expect family members to communicate with each other about private matters more readily than they do with their neighbors, coworkers, or even friends. Family members are allowed to know more about each other, and expected to share important personal information with each other (births, deaths, marriages) before they tell the rest of the world. Again, it is possible to see this virtue reflected in the way we treat its vicious distortions. Thus excessive communication (i.e., gossip) within families is a recognizable moral problem for family life, but we are usually more concerned about the keeping of secrets between close family members, particularly when the secret bears on their own relationship. In theory, disclosures of behavior that would be scandalous in public can be safely made within families not only because the secret is safe there but also because the expected reaction is not a moral judgment; it is rather an attempt at understanding. Intimacy in family life is not only about exposing our vulnerabilities to one another; it also requires taking no offense at the disclosures.

Unfortunately, despite our commitments to intimacy, information about shared genetic health risks seems particularly hard to share with our kin. For families that are already aware that they have a hereditary history with a particular disease, revealing that one is a mutation carrier can invite stigmatization as a bearer of the "family curse," and premature assignment to the sick role. On the other hand, "survivor guilt" seems to be relatively common among those testing negative in these families, and it also inhibits disclosure (15)? In these "at risk" families, genetic test results place individual members on one side or the other of the "watchful waiting" that characterizes their family's corporate identity and, in the process, segregates them from one of the central dramas of their family's life. For example, Alice Wexler explains her own decision to remain uncertain about whether she carries the gene for Huntington disease this way:

> One man who tested negative felt as if he were missing a limb, a part of his identity. "I had lost my creative terror" he said. . . . Choosing not to take the test is a decision one can easily revoke, unlike the situation after testing. A Geri Harville said, "You

and your family will be affected by this information forever. Once you have the information, you cannot give it back." I have made my peace, more or less, with uncertainty. . . . Perhaps I even enjoy the ambiguity, resisting sharp categories and binary definitions, the border guards insisting that we place ourselves in one camp or the other. (16, p. 238)

Clearly, genetic test results can fall in with mis-identified paternity as among the hardest kinds of personal secrets to share with kin: secrets that seem to deny the very kinship that licenses their intimate disclosure.

On the other hand, genetic test results are also almost always "about" more than the individual who undertakes them. For some, this overlap serves to strengthen the obligation to share genetic test information, because they understand themselves to be privy to a secret about their at risk kin (17). For members of families that do not have a recognized history of the genetic risk, their test reveals this poses another kind of challenge to familial intimacy. In these situations the disclosure does not segregate the source from kin but rather exposes a new and troubling connection between them. Here, feelings of shame and the urge to shield the family from harm seem to combine to encourage keeping genetic risk information secret from kin (18).

These pressures raise for family members the question: What are the limits of my intimacy-based obligations to share my genetic test results? Can my personal sphere of privacy extend to information that is also about my relatives without becoming secrecy? Of course, the problem with keeping specific secrets from other family members is that families are particularly hard organizations to keep individual secrets in without leaving the family altogether. Yet, when secrets have been kept and do emerge, it immediately calls into question the true intimacy of the relationships involved, with potentially devastating effects. Thus Alice Wexler writes that, in the wake of her mother's diagnosis with Huntington disease, and its exposure of her long-held secret about her family history,

> [W]hat my sister and I thought we knew about our family suddenly shifted and everything had to be rethought, reinterpreted. Who we were had suddenly been called into question, and everything had to be reconfigured taking into account the presence of the disease. It was as if we had been experiencing fallout from some unseen bomb for all these years, and suddenly the great mushroom cloud had come into view and we could see the source of all that radiation (16, p. 75).

Looking to law and social policy for indications of social consensus on these questions is only partially helpful. For example, the fact that there are laws that ensure adoptees access to genetic information about their birth parents suggest that we do feel strongly about the obligations of parents, like the Wexlers, to share their genetic health secrets with their children (19). On the other hand, most policy statements on the genetic testing of children stress the need to protect their genetic privacy even in the face of parental requests, suggesting that these intimacy obligations are not always reciprocal (20). Moreover, as Lori Andrews points out in her review of the issue, the law is particularly silent on familial duties outside the parent-child relationship: "With respect to an individual's duties towards other relatives (such as siblings or cousins), there might be a moral duty to disclose a person's genetic information, but there is probably would not be a legal one" (19, p. 273).

Again, the importance of these ethical questions for families can be seen by the extent to which they bleed over into professional ethical questions for clinicians. Thus clinical experience with patients' reluctant to warn their families of their risks has spawned a renewed discussion of the limits of the professional's commitment to confidentiality in the genetic testing context, with some arguing that the patient's familial obligation to disclose shared risks should translate into a professional duty to warn third parties (21), others standing by the individual proband's authority to make disclosure decisions, and still others arguing that if clinicians take the family seriously as a collective client, they should never make promises of privacy to any particular piece of it but interact with the family as a whole throughout the testing process (22).

Security

Good families are safe places for their members, as the virtues of loyalty and intimacy both suggest. In part, that is because good families accept special obligations to protect their vulnerable members from harm, even when those members cannot ask for help. Thus we expect adult family members to practice a benevolent paternalism with their immature and infirm kin, and we expect our kin to defend us against calumny and slander even when we are unaware of it. Of the three familial virtues, this is the one whose applications and dynamics have been most discussed in biomedical ethics, because it is the one that is called into question in health care when parents are faced with making medical decisions about the welfare of their children (23), and adult children are called to make medical decisions for their aged parents (24). At both ends of life, our bioethical discussions are often struggling to define the limits of people's duties to stand guard over the bodies of their loved ones in particular circumstances, but they almost always start from the premise that they do have those duties.

Like the others, this familial virtue is also thrown into relief by our responses to the vices on its borders. We can sympathize with overly protective parents even as we criticize them as excessive, but we find cultural practices that seem to deny special protections for the vulnerable (e.g., prenatal sex selection) morally callous and hard to understand when they occur in our society. Moreover the strength of our allegiance to this virtue is reflected again in professional ethics and law, with the recognition that "In general, the appropriate presumption is that the family of the incompetent individual is to be the principal decision-maker." Thus it is received wisdom in medical ethics that:

The chief reasons in support of the presumption that the family is to serve as surrogate decision-maker are both obvious and compelling. The family is generally both knowledgeable about the incompetent individual's good [e.g., intimacy] and his or her previous values and preferences [e.g., loyalty], and most concerned about the patient's welfare. [e.g., security]" (25).

With genetic testing, the limits of this virtue are pressed most vividly by the questions raised by the prospect of testing our offspring. How far should parents go in seeking to protect their vulnerable children from genetic harms? Does prenatal testing and selective abortion count as a "preventive" or "protective" intervention for the children tested, or simply a means of choosing which children to bear (26)? Studies already show that declining to have prenatal testing is perceived as causing the bad outcome when a disabled child is subsequently born, which suggests that many families do think that prenatal screening falls within the protections that parents should afford their offspring (27). This perception, in turn, can put the pressure on couples to be "responsible reproducers" in their own eyes and the eyes of other family members in ways that can seriously compromise their own reproductive autonomy (28). Clinicians, however, have traditionally viewed prenatal testing as enhancing familial rather than fetal security: It is understood as a service to the prospective parents that allows them to reduce the risk of burdening themselves and their family with more than they can handle. This perception of prenatal testing simplifies the professional ethical challenges that the practice poses, but challenges parents to clarify the limits of their ability to nurture children with disabilities, and to define for themselves what will count as an acceptable quality of life for their children (29).

Similar challenges accompany decision making about pediatric genetic testing. Should parents include screening for diseases of late adult onset amongst their obligations to protect the welfare of their children? There has been a lively debate in the literature on this topic, aimed at establishing clinical policies for practitioners (30). Interestingly, however, much of the argument is parental ethics, not professional ethics. Critics of such practices see in them the vice of over-protectiveness and "pre-emptive paternalism," while defenders argue that:

Children are not discrete monads who develop in isolation until they reach adulthood when they can seize autonomy and begin to make significant decisions about their lives. They are shaped within the context of their family and community as they make their way to adulthood. It is the responsibility of parents to provide for their children's nurture and education. ... While such socialization can affect the choices children will make as adults, it is considered so much a part of parental responsibility that parents who fail to teach their children a value system can be said to have failed in their duty to them. ... Yet responsible parenthood would have it no other way. Moreover, refraining from predictive testing is, in effect, teaching children that ignorance is a good way of life. That is at best a controversial message (31).

FAMILY STYLES

Of course, it would be naive to believe that the families in our society that still subscribe to the three familial virtues of loyalty, intimacy, and security all resemble one another. The institution of the family has undergone tremendous change in our society, and it continues to evolve in several directions at once (32). Unsurprisingly, the challenge that genetic testing will pose to a family's ability to uphold the three virtues will vary with the kind of family involved.

For convenience, consider three of the many kinds of American families that sociologists describe: The sessile organic family, the blended social family, and the diasporadic virtual family. Each of these will feel the challenge of genetic testing differently. Interestingly, however, it is the first, most traditional form of the family, presumably the source of these virtues, in which the challenge may be most severe.

Organic Families

The organic family is the kind of family that best exemplifies what some authors call the "biological" concept of family (33): a multigenerational clan that lives together in geographic proximity, like the Walton's on Walton's Mountain, for example, or the Venezuelan "Huntington disease families" of Lake Maricaibo. In some ways this kind of family might be best situated to benefit from a genetic testing program. Experience with presymptomatic testing for Huntington disease, for example, suggests that in families of "blood relatives" of any particular proband, intrafamilial communication is efficient, and the familial nature of any recurrent health problem will be readily understood. Clinical case studies show that familiarity with the disease in relatives may also improve family members' motivation for testing and diminish the stigmatization of those affected (34).

On the other hand, this kind of family is also more likely to be hierarchically organized, with clear lines of internal authority over decisions that affect the family as a whole. Under this structure the same features that make it possible to support strong family loyalties, intimacies, and protections can also exacerbate ethical problems for individual family subunits. For example, consider the following case report, described in a recent review of the experience of Indiana University's testing program for Huntington disease:

> Mr. Crawford's son married a woman at risk for HD. The couple, now divorced, were married for eight years and had a son and a daughter. These children are at 25 percent risk for HD. The former daughter-in-law has repeatedly said that she is not interested in testing for herself. Mr. Crawford is contemplating setting up a trust fund for his son's children to ensure adequate care for them if they should develop HD. In order to know whether he should make these arrangements, Mr. Crawford wants to have his grandchildren tested (35).

Cases of this sort pose ethical problems for genetic counselors to manage because of the way they challenge the individualistic, "nondirective" ethos that characterizes the profession's ideal. From the counselor's point of view, the principal issue in this case is the appropriateness of testing the children for a late onset disorder like Huntington disease before they are able to give their own autonomous consent to the procedure, and that is how the case is discussed in the review that reports it. But for the family, the scenario raises more fundamental moral questions than the children's ability to consent. Rather, they are more likely to be preoccupied with the relative weight of the various familial loyalties, protections, and

intimacies involved. In seeking to improve the security of his grandchildren, has Mr. Crawford overstepped the proper limits of his role? Is it fair to ask the ex-daughter-in-law to risk discovering her own status in order to help Mr. Crawford in his attempt to improve her children's welfare? What role should the history of her divorce from his son play in the deliberations? As this case suggests, families that are organized in ways that facilitate power differentials between members may be most at risk for becoming coercive in pressing members to be tested on behalf of their relatives, and for holding individuals to familial standards of "responsible parenting" that curtail their parental autonomy and procreative liberty. This is also the family structure in which it is most difficult to keep a secret, raising concerns about how to protect the interests of those who choose not to learn their own genetic risks and what information is fair to use in deciding to override them. Moreover all these challenges can emerge insidiously within these families, because they all begin in positive affirmations of the familial virtues. Unlike abandonment, neglect, and silence, the risks of coercion, paternalism, and gossip will be hard to detect and difficult to demonstrate, especially by the less powerful family members who, by definition, will bear the burden of their harms.

Social Families

At the other extreme are families that are the products of serial monogamy, adoption, and new reproductive technologies: the "Brady Bunch," for example. These families exemplify the "household" concept of family: "an aggregate or group of actual (living) members, who are closely associated by living arrangement or by commitment, for better or worse (36). For these families the medical value of genetic testing is lowered by the fact that fewer family members are expected to share the same genetic risks. For these families genetic testing may be less threatening, if only because they are likely to be less attractive. Families seeking to build their identities on ties of love and commitment rather than blood will tend to downplay the importance of shared health problems, which will lower pressure to test "loyalty" through genetic testing. On the other hand, by underlining the differences between the lineages that make up the family, genetic testing does pose another kind of risk for social families: the potential divisiveness that testing could create within the blended family structure between at-risk and low-risk members. For example, consider this case report:

> Angela Smith, a 30-year-old Caucasian woman, came to an outreach clinic for genetic counseling because her husband, David, had two half siblings who died from cystic fibrosis. She was interested in finding out her risk to have an affected child. ... Angela was tested first; the results showed that she carried the F508 deletion. David had blood drawn for testing. No mutations were detected. Based on this information and family history, the lab calculated his carrier risk to be 1 in 8. His carrier status could potentially be clarified by testing his father, an obligate carrier. Mr. Smith Sr. was tested but no mutation was identified. For further clarification, studies were proposed for David's stepmother, Sue, and his surviving half siblings, Robert, Dale and Karen. If Sue had an identifiable

mutation which was also present in one of the half siblings, linkage analysis could potentially establish whether David had inherited a chromosome with a high risk of carrying as mutation. Sue, Karen and Robert lived in another state, but agreed to be tested. Sue and Karen had the F508 mutation. Robert was negative for any mutation. Because Robert fell into the same category as David, his sample could not be used to clarify risk. An additional complication arose when Ruth learned that linkage studies indicated that Karen was not Mr. Smith Sr.'s daughter. The family was told that linkage studies were uninformative. If they wished to proceed, a sample would be needed from Dale. The family has made no further efforts to pursue testing (37).

In discussing this case report, its authors focus on issues raised for the genetic counselor by the inadvertent finding of Karen's mis-identified paternity. But behind the counselor's dilemma, the outlines of this blended family's own moral struggle can also be seen. How far should their sense of familial loyalty oblige Sue and her children to go to help her husband's son's wife clarify her reproductive risks? Most of them are willing to be tested, and to endure the inconvenience of out-of-state testing. By agreeing to be tested, Sue even risked exposing the secret of Karen's paternity and the familial harm that might have resulted. Should Dale now also agree to be tested in turn, despite his initial reluctance? How hard should Angela and David press to convince him to help them? Familial fractures along risk status lines are already observed in other disease contexts, in which at-risk members become overly zealous in protecting their branches from the disease of the afflicted. In the analogous context of HIV testing, see levine (38). Genetic testing in Blended families will impose the additional tension of tracing health risks along lines already marked by parentage.

Virtual Families

Somewhere between the extended and the blended family are families that retain the identity of a biological extended family but no longer live together. Here is the incarnation of what some call the "abstract" concept of families: "an idea or ideal that refers to a family name or genetic line, the extended family in the largest sense, whose boundaries or members extend over both space and time" (36, p. 47). They are "virtual families" in that except for high holidays, most of their interactions are mediated by communication technology. Like organic families, these families may be motivated to seek testing, since they define themselves in terms of their biological connections. However, both distance and increased control over communication serve to make extended family members more inaccessible and more autonomous. As sociologists point out, for today's virtual families, "contact with relatives outside the nuclear family depends not only on geographical proximity — not to be taken for granted in our mobile society — but also on personal preference. Even relations between parents and children are matters of individual negotiation once children have left home" (39).

Clinicians involved in genetic testing already describe cases that suggest how the familial virtues might be challenged in families that fit this description as well.

For example, consider the following case from the Indiana study:

Paul, a healthy twenty-five year old man, contacts a testing center to request genetic testing for Huntington disease. Paul's father has HD. The father and his family live in another part of the country and are not known to the center. Paul has an identical twin brother, Michael, who does not wish to have the predictive test. Paul has moved away from his family for work reasons, but he maintains regular telephone contact with his parents, twin brother, and older sister. . . . (35, p. 21).

In discussing the management of this case, the authors frame its professional ethical challenge as a conflict between Paul's "right to know" his HD risk status, and Michael's "right not to know" his own identical risk. Interestingly, however, Paul apparently uses no "rights" language at all in framing the issue he faces. The case notes simply report that:

Paul is planning to get married; he and his partner Linda are determined to find out whether he will develop HD before they start a family. Michael is single, and, according to Paul, does not think he could cope with the knowledge if he knew that he carried the HD mutation. Paul understands that his test result will reveal Michael's gene status. He states that he will not tell Michael or other family members that he is undergoing the predictive test. If he receives a positive result, he says that he will not tell Michael. However, he might tell Michael if he were to receive a negative result.

Against the background of his loyalty to his future family, Paul is wrestling with the tension between his existing obligation to protect his vulnerable brother and the costs to his natal family's intimacy such secrecy will exact. While he thinks he could succeed in keeping the test a secret from his natal family in order to protect Michael from bad news, how could he not share good news with Michael if he got it?

It is interesting that the authors of this case study concur with Paul's reasoning in this case and recommend accepting him for testing, despite their strong defense of obligate carriers' "right not to know" in more organic familial circumstances. As a result of the structural loosening of family ties in virtual families, it is also often easier to keep secrets within these families, increasing local control over both decisions to be tested and the privacy of test results. On the other hand, virtual families also face a structural challenge in preserving their intimacy for the same reasons, and that challenge poses its own risks. Paul and his wife will find themselves having to conceal increasingly more important aspects of their life if his test is negative, or else attempt to reveal their secret to Michael without the benefit of the mental preparation their own pre-test counseling provided them. Moreover efforts to recruit dispersed families for studies of hereditary mental health problems have shown that as familial communication becomes attenuated, it is possible for the dispersed branches to lose familiarity with the health problems to which they are prone, making it potentially more stigmatizing to reveal genetic risks across the family tree and reducing the motivation for distant members to contribute to clinical studies of the extended family (40).

CONCLUSION

From the current discussions of professional ethical problems in clinical genetics, it appears that at least three traditional virtues of family life are being severely tested by the increasing availability of predictive genetic risks assessment tools: loyalty, intimacy, and security. For blended and virtual families, the prospect of genetic testing can threaten our commitment to the familial virtues by focusing us on what separates us from each other. Within organic families, however, genetic testing can serve to reinforce our sense of solidarity. In that context the risk is not that the traditional virtues will be undermined but that we will be encouraged to go too far in pursuit of what it means to be a good family and our black sheep will suffer for it. To that extent, ironically, it may be that it is the most traditional form of the family, the sessile organic family, in which the challenge to our traditional familial virtues will be most difficult to anticipate and address.

Clearly, these claims can only be suggestions for further empirical research at this point: They assume that the categories of families employed here are useful within the context of genetic testing and that the associated familial virtues are the values that are at stake. To the extent that these ideas are worth exploring further, however, they do serve to suggest one important conclusion: as important as the professional ethical and public policy challenges of human genome research are, they may ultimately be eclipsed by the impact on how we, the public, understand our relations with our closest kin and the obligations those relations entail.

BIBLIOGRAPHY

1. President's Commission for the Study of the Ethical Problems in Medicine and Biomedical and Behavioral Research, *Screening and Counseling for Genetic Conditions*, U.S. Government Printing Office, Washington, DC, 1983, p. 55.
2. E.T. Juengst, *Am. J. Hum. Genet.* **54**, 121–128 (1994).
3. E. Thomson, *Ethical, Legal and Social Implications Program: Grants, Contracts, and Related Activities*, National Human Genome Research Institute, Bethesda, MD, 1996.
4. S. Post, *Spheres of Love: Toward a New Ethics of the Family*, Southern Methodist University Press, Dallas, TX, 1994.
5. H. Lindemann Nelson and J. Lindeman Nelson, *The Patient in the Family: An Ethics of Medicine and Families*, Routledge, New York, 1995.
6. S. Mintz and S. Kellogg, *Domestic Revolutions: A Social History of American Family Life*, Free Press, New York, 1988.
7. R. Shinn, in M. Frankel and A. Teich, eds., *The Genetic Frontier: Ethics, Law and Policy*, AAAS, Washington, DC, 1993, pp. 9–25.
8. R.M. Green and A.M. Thomas, *J. Genet. Couns.* **6**, 245–254 (1997).
9. B. Biesecker and L. Brody, *J. Am. Med. Women's Assoc.* **52**, 22–27 (1997).
10. B. Biesecker et al., *J. Am. Med. Assoc.* **269**, 1970–1974 (1993).
11. M. Yarborough, J. Scott, and L. Dixon, *Theor. Med.* **10**, 139–149 (1989).
12. L. Parker and C. Lidz, *IRB: Rev. Hum. Subj. Res.* **16**, 6–12 (1994).
13. T.M. Powell, D. Cox, S. Jeffery, and D. Cua, *Am. J. Hum. Genet.* **63**(4), A204 (1998).
14. R.J. Pokorski, *Am. J. Hum. Genet.* **60**, 205–216 (1997).
15. M. Huggins et al., *Am. J. Hum. Genet.* **47**, 4–12 (1990).
16. A. Wexler, *Mapping Fate: A Memoir of Family, Risk and Genetic Research*, Random House, New York, 1995.
17. J.K.M. Gevers, *Med. Law* **7**, 161–165 (1988).
18. P.R. Winter et al., *Am. J. Med. Genet.* **66**, 1–6 (1996).
19. L. Andrews, in M. Rothstein, ed., *Genetic Secrets: Protecting Privacy and Confidentiality in the Genetic Era*, Yale University Press, New Haven, CT, pp. 255–281.
20. E.W. Clayton, *Am. J. Med. Genet.* **57**, 630 (1995).
21. M. Shaw, *Am. J. Hum. Genet.* **26**, 243–246 (1987).
22. R. Wachbroit, *Suffolk Univ. Law Rev.* **27**, 1391–1410 (1993).
23. T.H. Murray, *The Worth of a Child*, University of California Press, Berkeley, 1996.
24. D. Callahan, *Hastings Cent. Rep.* **15**, 32–37 (1985).
25. A. Buchanan and D. Brock, *Deciding for Others: The Ethics of Surrogate Decision-making*, Cambridge University Press, Cambridge, UK, 1989, p. 136.
26. A. Buchanan, *Soc. Philo. Policy* **13**, 18–47 (1996).
27. T. Marteau and H. Drake, *Soc. Sci. Med.* **40**, 1127–1132 (1995).
28. A. Charo and K. Rothenberg, in E. Thomson and K. Rothenberg, eds., *Women and Prenatal Testing: Facing the Challenges of Genetic Technology*, Ohio State University Press, Columbus, 1994, pp. 105–131.
29. R. Faden, in E. Thomson and K. Rothenberg, eds., *Women and Prenatal Testing: Facing the Challenges of Genetic Technology*, Ohio State University Press, Columbus, 1994, pp. 88–98.
30. E. Clayton, *J. Med. Philos.* **22**, 233–251 (1997).
    A. Fryer, *Arch. Dis. Child.* **73**, 97–99 (1995).
    ASHG/ACMG, *Am. J. Hum. Genet.* **57**, 1233–1241 (1995).
31. C. Cohen, *Kennedy Inst. Ethics J.* **8**, 111–130 (1998).
32. B. Gottleib, *The Family in the Western World*, Oxford University Press, Oxford, UK, 1993.
33. R. Macklin, *Hastings Cent. Rep.* **21**, 5–12 (1991).
34. G.M. Petersen and P.A. Boyd, *J. Natl. Cancer Inst.* **17**, 67–71 (1995).
35. D. Smith et al., *Early Warning: Cases and Ethical Guidance for Presymptomatic Testing in Genetic Diseases*, Indiana University Press, Bloomington, 1998, p. 72.
36. P. Smith, in D. Meyers, K. Kipnis, and C. Murphy, eds., *Kindred Matters: Rethinking the Philosophy of the Family*, Cornell University Press, Ithaca, NY, 1993, p. 47.
37. J. Maley, *An Ethics Casebook for Genetic Counselors*, Center for Biomedical Ethics and Division of Medical Genetics, University of Virginia, Charlottesville, 1994, pp. 36–41.
38. C. Levine, *Milbank Q.* **68**(Suppl. 1), 37–50 (1990).
39. R. Bellah et al., *Habits of the Heart: Individualism and Commitment in American Life*, University of California Press, Berkeley, 1985, p. 89.
40. D. Shore et al., *Am. J. Med. Genet.* **48**, 17–22 (1993).

See other entries GENETIC COUNSELING; GENETIC INFORMATION, ETHICS, AND INFORMATION RELATING TO BIOLOGICAL PARENTHOOD; GENETIC INFORMATION, ETHICS, PRIVACY AND CONFIDENTIALITY: OVERVIEW; GENETIC INFORMATION, LEGAL, FDA REGULATION OF GENETIC TESTING; GENETIC INFORMATION, LEGAL, GENETIC PRIVACY LAWS.

GENETIC INFORMATION, ETHICS, INFORMED CONSENT TO TESTING AND SCREENING

KIMBERLY A. QUAID
Indiana University School of Medicine
Indianapolis, Indiana

OUTLINE

INTRODUCTION

The rapid advance of genetic knowledge in the past decade has led to dramatic increases in our ability both to diagnose individuals with a variety of genetic disorders as well as to identify individuals at increased risk for developing genetic disorders at some time in the future. The relative newness of this technology, as well as its rapid dissemination into general medical practice, has raised numerous issues regarding the role of genetic testing in public health, in medical care, and in the lives of individuals at risk for or affected by diseases of genetic origin. Debates on these issues are clouded by the fact that our ability to treat many of these disorders lags far behind our ability to identify those at high risk, the treatments that are available are often expensive and less than entirely satisfactory, and the cost of testing remains fairly high. In the absence of general societal consensus on the larger issues surrounding genetic testing such as who should be tested, for what indications, what type of treatment and counseling should be available prior to testing, who pays and who should have access the information, the issue of informed consent has gradually emerged as a central focus in the testing debate. This article will discuss the general issue of informed consent, delineate the differences between genetic screening and genetic testing, and present the salient informed consent issues to be considered.

DEFINITION OF INFORMED CONSENT

An in-depth analysis of the issue of informed consent is beyond the scope of this article, and the reader is referred to other articles that address this topic at length. For our purposes it is important simply to understand the basic concept of informed consent. Informed consent has been defined as an *autonomous authorization* by a subject or patient that allows a professional either to involve the subject in research or to initiate a medical plan for the patient, or both (1). Informed consent occurs if and only if a patient or subject with a substantial understanding and in a substantial absence of control by others intentionally authorizes a professional to do something (2).

Traditionally medical ethics has been dominated by a commitment to the principle of medical beneficence—the principle that obligates physicians to further the medical best interests of their patients (1). Within this tradition and over time, the general consensus has emerged that "informed consents should be obtained for research or experimental interventions (therapeutic as well as non-therapeutic), for innovative interventions, for interventions that carry with them substantial or unknown risks and in situations of choice between substantially different medical plans. Also, all other things being equal, informed consents need not be obtained for routine interventions that pose little or no risk to the patient and that are not part of a research protocol" (1, p. 39). This level of general agreement as to when informed consent is required provides little guidance for specific medical treatments, interventions, or innovations such as genetic testing, although one author has advanced the idea that most applied genetic technology—screening, diagnosis, counseling, or treatment—should be characterized as therapeutic clinical research (3) requiring informed consent. The history and development of the concept of informed consent for genetic testing and screening is relatively recent, and it has been affected by myriad historical events, public policy and economic shifts, and technological innovations. These variables have resulted in two divergent approaches to informed consent such that consent is usually required for genetic testing but not always for genetic screening. The development of these two approaches to consent will be discussed below.

GENETIC SCREENING AND THE PUBLIC HEALTH MODEL

Definition of Genetic Screening

The term "genetic screening" is generally reserved for the public health context in which interventions that can detect disease or risk of disease are employed on a populationwide basis without regard to family history. A committee of the National Academy of Sciences (NAS) has defined genetic screening as "a search in a population of apparently healthy individuals for those genotypes which place them or their offspring at high risk for disease" (4). With genetic screening, carriers of deleterious genes who have never had an affected child can be identified as can fetuses, newborns, or adults who do not have an affected relative (5). For this reason, in most of the populations targeted for genetic screening—African-Americans for sickle cell, Jews of Ashkenazi descent for Tay-Sachs, virtually all newborns for PKU—parents or individuals will have no first-hand knowledge of the disorder, its symptoms, severity, course, or risk for recurrence due to a lack of family history.

Public health is concerned with the prevention and reduction of morbidity and mortality in a given community. Within the public health framework, the principle of beneficence, which focuses on considerations of human welfare or well-being, is the guiding ethical principle. The principle of beneficence asserts a duty to confer benefits and to work actively to prevent and remove harms (2). Equally important is the duty to balance possible benefits that might accrue as a result of screening against possible harms. In the public health context, the benefit to be sought is the health or welfare of the community as a whole rather than that of the individual. The goal of genetic screening therefore is to reduce the incidence of morbidity and mortality related to genetic diseases in the community (6). If the benefit to be sought is the welfare of the community, what might some of the harms be? In general, harms that may occur during screening programs will occur to individuals and include labeling (7), stigmatization (8), misunderstanding (9), and false reassurance (10).

The degree to which a screening program, once implemented, is successful in reducing morbidity and mortality depends on several variables. These include the prevalence of the condition in the population to be screened, the sensitivity and specificity of the screening tool, the availability of a treatment or intervention for the condition, and the follow-up plans for those identified as positive.

Criteria for Screening

Principles to be followed with regard to genetic screening have been suggested by several authors (3,11,12) These principles include the following:

1. The disorder should be of high burden to the affected individual.
2. The inheritance and pathogenesis of the disorder should be understood.
3. The disorder should be preventable and practical therapy available, including genetic counseling and reproductive alternatives.
4. Patient's right to informed consent, voluntary participation, and confidentiality should be protected.
5. The benefit-to-cost ratio to the patient (public) should be greater than one.
6. The laboratory screening method should minimize false positive and exclude all false negative results.
7. A diagnostic test should be available.
8. Both screening and diagnostic tests should be available to all who require it.

In practice, however, these recommendations are rarely followed. A short history of our experience with newborn, prenatal, and carrier screening may prove instructive.

Experience with Screening

Newborn Screening for PKU. The first widespread experience with newborn screening emerged in the early 1960s with the invention by Robert Guthrie of the Guthrie assay to detect phenylketonuria (PKU). PKU is a hereditary metabolic disorder in which a deficiency of an intracellular enzyme results in the accumulation of the amino acid phenylalanine. PKU has an incidence in the United States and Europe of approximately 1 in 10,000 live births with the primary manifestation of the disorder being severe mental retardation (13). The Guthrie assay could be performed on apparently healthy newborns using a blood sample obtained by heel prick and was simple, inexpensive, and relatively painless. These characteristics were seen to weigh in favor of testing as many individuals as possible relative to the cost of not screening. At the time, however, the ability of a low-phenylalanine diet to prevent retardation had not been proved. Despite this lack of a clear medical benefit to screening, lobbying by Guthrie and the National Association for Retarded Children soon resulted in laws in most states that mandated the screening of newborns for increased concentrations of phenylalanine in the blood (14) even in the face of opposition to mandatory screening on the part of prominent groups including the American Academy of Pediatrics (15). In the rush to initiate widespread screening, some normal children with transient hyperphynylalaninemia were incorrectly labeled as having PKU and prescribed diets deficient in phenylalanine with deleterious outcomes including death (4,16).

Today, in most states, parental consent for PKU screening is not sought (17). In fact genetic screening, especially for newborns, is often carried out without informed consent or parental permission (18). Current common newborn screening tests performed without express informed consent include tests for hypothyroidism, PKU, sickle cell, and/or other hemoglobinopathies, although parents may object to screening on religious grounds (17). This exception to mandatory screening, however, may be seen as meaningless because parents seldom learn about the test until after it has been performed (19).

Some have concluded that informed consent ought not to be necessary for a procedure that offers great benefit and little risk (20). According to this argument, parental autonomy in decision making about newborn screening is grounded in the principle of beneficence, which holds that parents are the people best suited to act in the best interests of the infant. If it is generally agreed that it is in the best interest of the child to be screened, parental consent is superfluous and need not be sought. While this argument has clear appeal in the case where treatment is available (i.e., PKU), it is less compelling in situations in which the benefits of screening are less clear-cut. A second justification for not seeking informed consent in the context of newborn screening has been lodged in the notion of *therapeutic privilege* according to which a physician may legitimately withhold information based on sound medical judgment that to divulge the information would be potentially harmful. Harmful outcomes that would qualify for consideration of therapeutic privilege include endangering life, causing irrational decisions, and producing anxiety or stress (2). Physicians have justified the lack of informed consent by "maintaining that they do not wish to alarm women unnecessarily, since the tests will be normal in the vast majority of cases" (5, p. 184).

Finally, a third reason underlying the sentiment favoring lack of informed consent is that women might actually refuse to have the test done (5). Empirical investigation of this question for PKU, however, has shown that parents can be educated about this test at little cost (21) and that, if asked, less than 0.05 percent of parents would refuse testing (19). Some commentators feel that the scope of therapeutic privilege should be severely circumscribed and that at the very least, the privilege should not apply in situations when the potential harm to the patient from full disclosure would result not from the disclosure itself but from the decision that the practitioner fears that the patient might make as a result of the information disclosed (22).

Prenatal Screening. In addition to screening tests performed on newborns, screening tests are also often performed on pregnant women without providing any explanation or obtaining informed consent (23). This screening includes conditions that are primarily of interest because of the risk they pose to the developing fetus, such as rubella and Rh factor, as well as diseases with medical implications for both mother and fetus, such as tuberculosis, gonorrhea, gestational diabetes, and herpes simplex. Justification for this practice lies in the fact that for each of these disorders, effective interventions are available to prevent or markedly reduce harmful consequences (20).

Carrier Screening. The history of carrier screening in the United States might be characterized as one of mixed success. One notable failure is the experience with screening for sickle cell disease in the 1970s (23). Sickle cell disease is an autosomal recessive hemolytic anemia occurring most frequently in blacks but also in persons of Mediterranean, Asian, Caribbean, Middle Eastern, and South and Central American origin. Sickle cell disease is estimated to occur in as many as 1 in 400 African-American newborns (24). The public health implications of sickle cell disease, especially for African-Americans were brought to national attention in the early 1970s. By April 1973, 10 states had laws requiring mandatory screening for sickle cell disease despite the fact that no effective treatment was available. This rush to mandatory screening was not accompanied by provisions or funding for either genetic counseling or education of those screened (14). As a result there was great confusion between sickle cell trait or being a carrier for sickle cell and actually having the disease, on both the part of the general public and those who were screened. Identification of having sickle cell trait or sickle cell disease resulted in documented cases of job discrimination most notably in the military where for years the U.S. Air Force did not train black recruits with sickle cell trait to become pilots (25). Life insurance companies charged higher premiums for individuals with sickle cell trait or refused to insure them at all (26). Other reported hazards of screening for sickle cell included inappropriate medication and treatment of individuals whose symptoms were falsely attributed to sickle cell disease, delays in the adoption of children suspected to have the disease or trait, and the exposure of nonpaternity (27).

This experience made clear to what extent hastily planned, poorly executed, and underfunded screening programs mounted in the name of public health had the potential to cause great harm to individuals. When coupled with the growing importance of the principle of autonomy which advanced the right of those approached for testing to be fully informed as to the profound effects that testing may have on their lives as well as the lives of other family members, this experience led some to question the wisdom of the old approach. The genetics community, if not the wider medical community, also became sensitized to the concept that psychosocial risks, in contrast to medical/physical risks, were most likely to occur as a consequence of these programs and others like them. More recently our experience with screening for the evidence of human immunodeficiency virus (HIV) infection has further dramatized the psychosocial risks of diagnostic or predictive testing (coerced diagnosis, anxiety, loss of privacy, stigmatization, and discrimination). For the individuals most intimately involved, these risks came to be seen as important as physical risks and highlighted the potential downside of being identified as having or being at risk for developing a disorder that was poorly understood or considered highly undesirable by the society at large (14). These lessons have become apparent in the relatively recent handling of genetic screening for another autosomal recessive disorder, cystic fibrosis.

As early as 1983 the President's Commission for the Study of Ethical Problems in Medicine and Biomedical and Behavioral Research saw cystic fibrosis (CF) as a likely candidate for widespread genetic screening (12). CF is the most common autosomal recessive disorder among northern Europeans, with 1 in 25 being known carriers (10). The President's Commission came out in favor of explicit informed consent for CF screening and emphasized the fact that in their view, the "fundamental value of genetic screening and counseling lies in its potential for providing individuals with information they consider beneficial for autonomous decision-making" (12, p. 97).

In 1985 identification of DNA probes linked to the CF gene allowed prenatal diagnosis for families who had a previous child with CF and for whom specimens for the affected child and both parents could be analyzed. In 1989 the DNA sequence of the gene associated with CF was published (28) making widespread screening for carrier status possible. Mass screening was widely advocated for several reasons, with the most compelling being the ability to counsel heterozygote carriers of reproductive age. Arguments against mass screening were equally compelling. The initial test was able to detect only 75 percent of carriers (the frequency of the first detectable mutation ΔF_{508}). In addition the cost of screening was estimated to be as high as 2.2 million for each cystic fibrosis birth avoided (29). Shortly after the announcement of the CF gene, the board of directors of the American Society of Human Genetics (ASHG) issued a statement that "Routine CF carrier testing of pregnant women and other individuals is not yet the standard of care in medical practice" (30).

Continuing research showed that CF was genetically heterogeneous with over 100 different mutations giving rise to the same symptoms. As more mutations were found and the ability of the test to detect carriers improved to over 90 percent, CF came to be seen as a test case for the application of future genetic technologies. Funding for pilot projects was made available by the National Center for Human Genome Research with the express research goals of gathering information to identify clinical practices that best increase patient understanding of disease–gene carrier testing and test results, and best protect individuals and families from test-related psychological harm, stigmatization, and discrimination (31).

In contrast to the previous experience with sickle cell screening, two important developments had taken place. The first was that changes in technology had spawned the development of numerous biotechnology companies eager to capitalize on the potential market for genetic tests and poised to market tests directly to physicians and consumers. The prospect of a test that could potentially be marketed to every Caucasian of child-bearing age was an enormous incentive to tout the benefits of testing. The second development was a major shift in perception of what the proper goals of routine carrier screening ought to be. Rather than promoting the public health model, wherein the goal of a carrier screening program would be a reduction in the incidence of genetic disease, there now appeared to be some level of agreement that the appropriate goal of carrier screening is to help clients make more informed decisions (32). With this change in emphasis came a concomitant change in the definition of what would count as the mark of a successful program. If the public health context dictated that a successful program should result in the reduction of morbidity and mortality in a population, the new emphasis on informed consent suggested that the effectiveness of screening programs should be assessed by whether the participants in the screening program have become informed (29) regardless of the ultimate decision made and regardless of that decision's ultimate effect on morbidity and mortality.

The CF pilot studies established an important precedent by incorporating the assessment of psychosocial impact into clinical studies usually dominated by concerns of medical safety, reliability and efficacy (31). This change reflected the now-common assumption that genetic "screening is a medical intervention with serious psychosocial risk" (29). The feeling was now that mass genetic screening programs should be considered experimental public health programs, implemented only after a favorable assessment of a program design that effectively achieves its goal while minimizing the potential medical, ethical, legal, and social problems (29). For that reason informed consent should be obtained to ensure that the client understands the risks of taking the test and is willing to accept them (33). Pre-test counseling or education has become the most accepted means to promote patient autonomy by providing accurate and understandable information designed to allow patients to decide for themselves whether the benefits outweigh the risks. In the context of screening for CF, the information to be presented should include information about CF, including its prognosis, treatment, and costs. The fact of screening

raises the possibility that both members of a couple may be identified as carriers of the CF gene. In that case such a couple would have a one in four chance of having a child affected with CF. This possibility raises the issue of alternative reproductive options. Because one or more of these options, such as abortion, may not be acceptable to some families, these options need to be discussed in detail. Finally, the risks of testing, including testing errors and the possibility of stigmatization or discrimination, should be disclosed (29). This advancement of patient autonomy as the proper goal of screening programs may serve as a potent counterweight to commercial and other pressures in favor of mass screening.

Future of Informed Consent for Genetic Screening

There is a clear precedent for performing prenatal, newborn, and carrier screening for a variety of disorders without obtaining informed consent, although recent experience with screening for cystic fibrosis has placed informed consent clearly on the table as desirable if not required. As the Human Genome Project proceeds apace, the number of screening tests that can be performed on a single blood sample will increase rapidly. The proliferation of genetic screening tests coupled with a perceived shortage of personnel trained in genetics (29) has raised the question of whether obtaining informed consent, or even attempting educational efforts about the tests and their potential findings, is practical (34). George Annas and Sherman Elias have suggested that "generic consent" may be a more useful approach because it may not be feasible to explain the risks and benefits of each test. This approach is based on the presumption that "it will soon be impossible to do meaningful prescreening counseling about all available carrier tests" (34, p. 1611). The authors warn against the dangers of "information overload" and envision a situation in which patients would be informed of broad concepts and common denominator issues in genetic screening rather than specific information about each test and each condition that may be detected. The authors stress that the concept of generic consent is not a waiver of the individual patient's right to information. Rather, "it would reflect a decision by the genetics community that the most reasonable way to conduct a panel of screening tests to identify carriers of serious conditions is to provide basic, general information to obtain consent for the screening and much more detailed information on specific conditions only after they have been detected" (34, p. 1612). This view raises the issue of who will decide for what disease, symptoms, or conditions the screening will be done. Ultimately a single blood test may determine not only the risk to infants for Down syndrome, trisomy 18 and 13, but also sex chromosome abnormalities such as Turner's syndrome and Klinefelter's syndrome and conditions not present at birth but manifest in middle age including breast cancer, Huntington's disease, and Alzheimer's disease. This approach needs to be evaluated in view of the fact that now, and in the foreseeable future, we have few cures or treatments to offer for these genetic disorders. Individuals identified as affected, at risk, or carriers face difficult choices involving mating, reproduction, and abortion. The continued subtle but steady erosion of a woman's right to abortion should be factored into the prospect of potentially identifying hundreds of thousands of individuals as affected, or at risk for, serious untreatable disorders.

This model has been criticized as driven primarily by technological capabilities rather than one based on established ethical principles (35). However, this model is perfectly in keeping with the historical framing of genetic screening as a public health issue whereby a group of influential experts decides what is in the best interest of the health of the population. A more sober note regarding this practice, and one that we might do well to ponder, is the observation that public health, rather than individual therapy, was the driving force behind the Nazi medicalization of eugenics that brought about the horrors of the holocaust (36).

GENETIC TESTING AND THE PSYCHOSOCIAL MODEL

Definition of Genetic Testing

The term genetic testing encompasses a number of technologies used to detect genetic traits, changes in chromosomes, or changes in DNA. Genetic testing is normally performed for two purposes; (1) to confirm a diagnosis in a person with symptoms of a particular disease, or (2) to clarify risk in an asymptomatic person who is known to be at risk for a specific disorder. This risk clarification in asymptomatic individuals can be further divided into presymptomatic or predictive testing in those individuals where disease genes have been identified that have high sensitivity and specificity and penetrance, and so-called susceptibility testing where the presence of a particular disease gene may be a necessary but not sufficient precondition for the development of the disorder. Examples of diseases for which predictive genetic testing are available would include Huntington disease, caused by an expanded trinucleotide repeat on chromosome 4 (37), and early-onset Alzheimer's disease associated with mutations on chromosomes 1 (38), 14 (39), and 21 (40). Examples of diseases for which susceptibility testing is available (although not necessarily recommended) include colon cancer (41) and breast cancer associated with mutations in BRCA1 (42) or BRCA2 (43).

In contrast to genetic screening, individuals, whether affected or at risk, for whom genetic testing might be appropriate usually have first-hand knowledge of the disease in question by virtue of having an affected relative. One advantage with regard to informed consent is that these individuals will often have a deep and personal familiarity with the disease, its symptoms and its course. They may not, however, have a clear understanding of the risk to their own offspring or of their options for avoiding the conception or birth of an affected child (5).

Genetic testing is typically initiated by a patient and/or his or her family and is performed in consultation with his or her medical care provider. Until very recently almost all genetic testing was performed in specialized medical centers primarily by professionals trained in the various subdisciplines of genetics including clinical genetics, genetic counseling, molecular genetics, and cytogenetics.

The issue of context is important in this discussion. Although medical genetics may be viewed as a subspecialty of medicine as evidenced by the recent formation of the American College of Medical Genetics and the recognition of medical genetics as a primary subspecialty by the American Board of Medical Subspecialties in 1991, medical genetics has evolved with a very different moral tradition. Because of a previous history of eugenics, and for many years its almost exclusive involvement in highly personal and difficult issues of human reproduction, medical genetics almost from its inception has adopted as its primary moral principle respect for autonomy of the client or patient (44). "Respect for autonomy obligates professionals to disclose information, to probe for and to insure understanding and voluntariness, and to foster adequate decision-making" (2, p. 127).

In part, this commitment to respect for autonomy in genetics can be explained by the fact that most of what medical genetics has had to offer is information regarding recurrence risks for specific disorders rather than the more traditional medical benefits such as treatment or cure. What has evolved in genetics is a commitment to the idea of nondirective counseling in the provision of genetic information (45) and in the convention of obtaining informed consent prior to all diagnostic genetic testing (46). The goal of genetic testing is seen as to improve the ability of people affected or at risk to make informed personal and reproductive decisions in light of their genetic status (47). It has been argued therefore that before an individual agrees to be tested for a genetic condition, pre-test education and informed consent are necessary, and post-test counseling must be provided (29).

Genetic Testing for Huntington's Disease

This approach to genetic testing is most clearly seen in the experience with predictive testing for Hunting's disease (HD). In 1983 a linked marker to the HD gene was found (48). Ensuing discussions among genetic researchers, clinicians working with HD patients and their families as well as interested laypersons resulted in preliminary guidelines for predictive testing including eligibility criteria and testing protocols. These protocols included neurological examination, psychiatric and psychological screening, pretest counseling and follow-up (49–51). National and international guidelines for testing were soon issued by the Huntington's Disease Society of America (HDSA) (52–53) and the World Federation of Neurology Research Group on Huntington's chorea (54–55).

These protocols were guided by two main ethical principles, nonmaleficence and autonomy. Nonmaleficence is termed a negative duty to avoid doing harm. Many argue that predictive genetic testing is a complex endeavor and despite ten years of experience the jury is still out on the overall benefits of testing (56). Many of those most intimately involved in the testing process, either as persons at risk or experienced clinicians continue to urge caution (57–59). Mandatory pre-test counseling was, and remains to this day, a major element of testing for HD. The purpose is to ensure, to the extent possible, that individuals considering tested have clearly understood the risks and benefits of testing to them, and have made a decision regarding testing that is consistent with their own personal goals and values. In other words, mandatory pre-test counseling was seen as a means by which to promote autonomous decision-making regarding this predictive test. While this approach has been criticized as being overly paternalistic (60), there is some evidence that this approach is useful. A recent survey of 12 testing centers following recommended testing procedures in Canada has indicated a small number of catastrophic events, defined as a completed suicide, suicide attempt, or psychiatric hospitalization, have occurred as a result of testing (61). A recent survey of all testing centers in the United States has indicated that one advantage to the recommended approach to testing is the ability to identify the estimated 3 percent of individuals for whom testing is inappropriate or who would benefit from further counseling prior to testing (62). Debate continues regarding how much a person needs to know prior to testing and whether the standard approach to informed consent (33) can be maintained as the number and complexity of potential genetic tests increase.

The experience with HD and other disorders, as well as thoughtful commentary by a number of people, have alerted us to a number of other ethical issues that need to be considered in the context of informed consent for genetic testing. One of these is the issue of confidentiality with regard to genetic information. In truth, the results of a genetic test often have important ramifications beyond the individual patient including children or other blood relatives. For example, if an individual at risk for HD by virtue of having an affected parent tests positive for the mutation that causes HD, each of his or her children now has a 50 percent of having inherited the HD mutation from his or her parent. While confidentiality is basic to all health care relationships, limits to confidentiality have been accepted in the case of communicable diseases or intent to harm another person. In the HD testing, a common example is the situation in which an individual who has tested positive refuses to inform his or her children about their risk for HD. Situations such as these have led to discussions that at least consider the justification of breaching confidentiality, although the issue remains a thorny one (63).

A different, but related issue is that of privacy of genetic information. Should results of genetic tests be placed in medical records whereby they might be easily accessed by employers or insurance companies. Concern for confidentiality and privacy of genetic information is highest in those situations were the results of a genetic test may affect an individuals ability to obtain or to keep health insurance. This concern has been heightened by reports of individuals denied insurance, especially those stories based on misunderstanding of the genetic condition, confusions between being affected with a particular disease and being a carrier, or between having a positive genetic test and actually being affected with a particular condition (64). These concerns also extend into the arena of employment where individuals fear being asked to undergo genetic testing as a prerequisite for employment.

Genetic Testing for Cancer

As discussed above, genetic testing has traditionally been performed for well-defined, monogenetic syndromes that are inherited in a strict Mendelian fashion such as cystic fibrosis and HD. For these diseases, inheritance of the disease gene leads to the expression of the disease phenotype. While raising interesting issues in their own right, these diseases are not the best models for the future of genetic testing and the challenges that are likely to arise. For a better view of the future of genetic testing, we must look to the inherited cancers to understand the full complexity of genetic testing and counseling.

Although chromosomal locations for several cancer-predisposing genes, including those for hereditary retinoblastoma (Rb) gene, the WT1 gene for Wilms's tumor, neurofibromatosis type 1 gene, the APC gene for familial polyposis coli, and the p53 gene for Li-Faumenia syndrome, have been mapped, cancers due to these mutations are rare or are usually preceded by distinctive clinical manifestations. For these reasons, discussions concerning genetic testing for these disorders are rare (65). With the discovery of a major breast–ovarian cancer susceptibility gene on chromosome 17, the situation has changed dramatically.

In 1997 alone, approximately 180,000 American women developed breast cancer and 44,000 of these women died of this disease (66). The first breast cancer predisposition gene (BRCA1) was located to chromosome 17q in 1990 (67) and cloned in 1994 (42). BRCA1 is a large tumor suppressor gene with 22 exons, and more than 100 mutations have been identified. A second breast cancer predisposition gene on chromosome 13 was identified in 1995 (43). While mutation analysis is possible in high risk families, it is not yet feasible in the general population (68). It is estimated that as many as 1 in 300 women may carry germ-line mutations in one or more breast cancer susceptibility genes (69). In contrast to the Mendelian disorders, studies of inherited breast cancer susceptibility provide evidence of incomplete penetrance suggesting that the inheritance of an altered breast cancer gene is not sufficient to produce the disease. Other factors, including additional genetic and/or environmental factors may be necessary. The fact of incomplete penetrance presents the potential for prevention, a factor that may weigh heavily in favor of testing unaffected individuals from families with known hereditary breast cancer (70).

The prospect of widespread screening for cancer susceptibility provides the opportunity to take a close look at the process of informed consent, not only for cancer but for any clinical context in which there is inherent uncertainty about the benefits and risks of a specific test (71). From the beginning, those involved in genetic testing for breast cancer took the approach used in HD as their model and structured testing protocols based on the principles of autonomy, beneficence, confidentiality, and equity or justice (71). The research protocols through which testing for breast cancer was first offered included pretest counseling and education and follow-up (73). In virtually all discussions of genetic testing for breast cancer susceptibility, the issue of informed consent is considered of crucial importance. Reasons for this emphasis may be the influence of the HD model, the fact that most testing involves adults and/or the fact that the benefits of testing are unclear while the risks and limitations of testing are becoming better known.

Three key limitations of genetic testing for breast cancer susceptibility have been identified (74). First, genetic testing for breast cancer susceptibility may not be informative. Second, the genetic information is probablistic in nature indicating an increased or decreased risk but not certainty. A negative genetic test result for BRCA1 leaves an individual with the population lifetime risk of breast cancer due to environmental and other factors and which is estimated at approximately 11 percent (75). Third, there are very real limitations to the current methods for cancer prevention and early detection. It is unclear whether current prevention methods of treatment (surgery, radiation, chemotherapy) or surveillance practices (mammography, clinical breast examination, and self-examination) actually reduce morbidity and mortality or what role genetic testing should play. In addition individuals must consider the risk of genetic discrimination in employment or in obtaining or maintaining insurance (64) and the potential for negative psychological consequences of learning one's genetic status (74,76–77).

The major benefit of genetic testing for susceptibility is the possibility of reassuring information if one is found not to carry a particular genetic mutation associated with increased risk for cancer (78). For those who test negative, one risk may include survivor guilt and shame (77). For those who test positive, testing may prove beneficial if this information facilitates actions to help prevent cancer or to aid in its early detection. This benefit will not be realized, however, if increased worry about cancer interferes with adherence to surveillance measures such as breast self examination, clinical breast examinations, and mammograms (79). Data suggest that a substantial proportion of women identified as gene carriers experience some level of psychological distress including persistent worries, depression, confusion and sleep disturbance (76).

Several studies have been designed specifically to determine what people want to know about susceptibility testing for cancer. In one study the authors concluded that at a minimum, disclosure statements needed to include information on (1) risk factors for cancer, (2) practical details of testing, (3) current limitations of testing, (4) available follow-up options, (5) known benefits of testing, and (6) known risks of testing (80). Other groups have recommended that informed consent be obtained through full disclosure of all the risks and benefits of testing including information about test accuracy, importance of correct cancer diagnosis, laboratory error rate, physical risks of blood drawing, potential problems with insurance or employment, psychological risks, and benefits of health care planning (68). A third study focused on two essential goals of informed consent, assuring that patients have substantial understanding and that their decisions whether to accept or to reject interventions are substantially voluntary. Two specific concerns arose with regard to these two goals, the relationship between patients' backgrounds, beliefs, and their understanding and the role of provider recommendations in voluntary

decision making (71). One conclusion reached was that patients often expect and request that their providers play an active role in decision making. This approach is somewhat at odds with the traditional stance of nondirective counseling as held by genetic counselors and is time intensive, presenting a major challenge to our current system of health care delivery.

Genetic Testing for Children and Adolescents

The genetic testing of minors, especially for late-onset disorders, has long been a subject of much controversy due to questions about the ability of minors to give informed consent and because of the sometimes conflicting views of lawyers and ethicists. Anglo-American common law has long granted almost absolute authority to parents (81), and in the context of medical treatment, parents are allowed to do almost anything that is not harmful in and of itself or not clearly against the best interests of the child (82). From the beginning, however, general testing protocols and guidelines for HD prohibited the testing of those under 18 years of age at their parents' request (52–55). This prohibition was based on a desire to preserve the autonomy of the child and the ability to make his or her own decision about testing, and the desire to avoid causing harm. Several professional societies have published statements upholding the presumption against testing asymptomatic children for late-onset disorders for which there is no treatment or cure (83–85). Research has shown, however, that many health professionals would test children at their parents' request including 53 percent who would test for HD (86), and the issue remains controversial (87–88).

Potential harms of testing are seen to include misconceptions about the future, stigmatization, feelings of unworthiness, fear of intimacy or interpersonal relationships, harm to self-concept, guilt, and blame. Depending on the condition, potential benefits may include increased medical surveillance in those at high risk, relief in those found to be at low risk, the opportunity to obtain accurate information from a trained professional rather than reliance on parental knowledge, and allowing more time for adaptation to the possibility of future illness (89).

In general, genetic testing of minors can be seen to fall into four general categories based on utility where (1) testing offers immediate medical benefits for the minor, (2) there are no medical benefits but testing may be useful in making reproductive decisions, (3) the parents or the minor requests testing, and (4) testing is done solely for the benefit of another family member (90). The general consensus on the acceptability of testing varies widely among these categories as does the discussion of what might consitute proper counseling, how to assess minors' ability to give informed consent, and what are the actual outcomes of such testing. While the debate goes on, evidence suggests that more and more children are being tested (91).

Future of Informed Consent for Genetic Testing

The traditional role of informed consent for genetic testing is facing many challenges. One major challenge to the role of informed consent is the rapid movement of genetic testing out of specialized genetics centers and into general medical practice. The increasing use of heavily marketed, commercially available tests by physicians who may not have the necessary training or background in genetics to either inform patients properly about the risks and benefits of testing or to interpret the results of testing accurately may seriously undermine the current practice of obtaining informed consent. One recent study examining genetic testing for the germ-line mutation of the adenomatous polyposis coli (APC) gene that causes colon cancer found that patients who underwent APC testing often received inadequate counseling, did not provide informed consent, and would have been given incorrectly interpreted results (92).

Evidence exists that biotechnology companies are also offering tests directly to consumers raising questions of whether patients and families are adequately informed about the tests, their limitations, and the risks and benefits of choosing to be tested (91). As the number of tests increases, especially for disorders that are relatively common in the population, the pressure on the part of biotechnology companies to market these tests will increase as well as the pressure on health care professionals to make them available. It remains to be seen whether and to what extent the fragile adherence to the notion of informed consent will survive.

BIBLIOGRAPHY

1. R.R. Faden, in J. Rodin and A. Collins, eds., *Women and New Reproductive Technologies: Medical, Psychosocial, Legal and Ethical Dilemmas.* Lawrence Erlbaum Associates, Hillsdale, NJ, 1991, p. 39.

2. T.L. Beauchamp and J.A. Childress, *Principles of Biomedical Ethics*, 4th edn., Oxford University Press, New York, 1994.

3. L.J. Elsas, *Emory Law J.* **39**, 811–853 (1990).

4. Committee on Inborn Errors of Metabolism, *Genetic Screening: Programs, Principles and Research*, National Academy of Sciences, Washington, DC, 1975.

5. N.A. Holtzman, *Proceed with Caution: Predicting Genetic Risks in the Recombinant DNA Era*, Johns Hopkins University Press, Baltimore, MD, 1989.

6. R.R. Faden, N.E. Kass, and M. Powers, in R.R. Faden, G. Geller, and M. Powers, eds., *AIDS, Women and the Next Generation: Towards a Morally Acceptable Public Policy for HIV Testing of Pregnant Women and Newborns*, Oxford University Press, New York, 1991, pp. 3–26.

7. A.B. Bergman and S.J. Stamm, *N. Engl. J. Med.* **176**, 1008–1013 (1967).

8. G. Stamatoyannopoulos, in S.G. Motulsky and W. Lenz, eds., *Birth Defects*, Excerpta Medica, Amsterdam, The Netherlands, 1974, pp. 268–276.

9. M.T. Pugilese et al., *Pediatrics* **80**, 175–182 (1987).

10. H. Bekker et al., *J. Med. Genet.* **31**, 364–368 (1994).

11. L.J. Elsas, in A. Rudolph and J. Hoffman, eds., *Pediatrics*, 18th ed., Appleton and Lange, Norwalk, CT, 1987, pp. 222–225.

12. President's Commission for the Study of Ethical Problems in Medicine and Biomedical and Behavioral Research, *Genetic Screening and Counseling: The Ethical, Social and Legal Implications of Genetic Screening and Education Programs*, U.S. Government Printing Office, Washington, DC, 1983.

13. F.D. Ledley, A.G. DiLella, and S.L.C. Woo, *Trends Genet.* **I**, 309–313 (1985).

14. B.S. Wilfond and K. Nolan, *J. Am. Med. Assoc.* **270**(24), 2948–2954 (1993).

15. American Academy of Pediatrics, *Pediatrics* **39**, 623–624 (1967).

16. S.P. Bessman and J.P. Swazey, in E. Mendelsohn, J.P. Swazey, and I. Traviss, eds., *Human Aspects of Biomedical Innovation*, Harvard University Press, Cambridge, MA, 1971, pp. 49–76.

17. K.L. Acuff, in R.R. Faden, G. Geller, and M. Powers, eds., *AIDS, Women and the Next Generation: Towards a Morally Acceptable Public Policy for HIV Testing of Pregnant Women and Newborns*, Oxford University Press, New York, 1991, pp. 121–165.

18. L.B. Andrews, *Medical Genetics: A Legal Frontier*, American Bar Foundation, Chicago, IL, 1987.

19. R.R. Faden et al., *Am. J. Pub. Health* **72**, 1347–1352 (1982).

20. R.R. Faden, N.A. Holtzman, and A.J. Chwalow, *Am. J. Pub. Health* **72**, 1396–1400 (1982).

21. N.A. Holtzman et al., *Pediatrics* **72**, 807–812 (1983).

22. A.M. Capron, *Univ. of Pennsylvania Law Rev.* **123**, 364–376 (1974).

23. K.L. Acuff and R.R. Faden, in R.R. Faden, G. Geller, and M. Powers, eds., *AIDS, Women and the Next Generation: Towards a Morally Acceptable Public Policy for HIV Testing of Pregnant Women and Newborns*, Oxford University Press, New York, 1991, pp. 59–93.

24. D.L. Rucknagle, *Arch. Int. Med.* **133**, 595–606 (1974).

25. L. Andrews, *State Laws and Regulations Governing Newborn Screening*, American Bar Foundation, Chicago, IL, 1985.

26. P. Reilly, *Genetics, Law and Social Policy*, Harvard University Press, Cambridge, MA, 1977.

27. E. Beutler et al., *N. Engl. J. Med.* **285**(26), 1486–1487 (1971).

28. J.R. Riordan et al., *Science* **245**, 1066–1073 (1989).

29. B. Wilfond and N. Fost, *J. Am. Med. Assoc.* **263**, 2777–2783 (1990).

30. C.T. Caskey, M.M. Kaback, and A.L. Beaudet, *Am. J. Hum. Genet.* **46**, 393 (1990).

31. K.E. Hanna, in *Society's Choice: Social and Ethical Decision-Making in Biomedicine*, Institute of Medicine, National Academy Press, Washington, DC, 1995, pp. 432–457.

32. M. Lappe, J. Gustafson, and R. Roblin, *N. Engl. J. Med.* **286**, 1129–1132 (1972).

33. R.R. Faden and T.L. Beauchamp, *A History and Theory of Informed Consent*, Oxford University Press, New York, 1986.

34. S. Elias and G.J. Annas, *N. Engl. J. Med.* **330**, 1611–1613 (1994).

35. L.G. Biesecker and B.S. Wilfond, *N. Engl. J. Med.* **331**, 1024 (1994).

36. A.L. Caplan, in G.J. Annas and S. Elias, eds., *Gene Mapping: Using Law and Ethics as Guides*, Oxford University Press, New York, 1992, pp. 128–141.

37. Huntington's Disease Collaborative Research Group, *Cell* **72**, 971–983 (1993).

38. E. Levy-Lehad et al., *Science* **269**, 970–973 (1995).

39. R. Sherrington et al., *Nature* **375**, 754–760 (1995).

40. A.M. Goate et al., *Nature* **349**, 704–706 (1991).

41. I. Nishisho et al., *Science* **253**, 665–669 (1991).

42. Y. Miki et al., *Science* **266**, 66–71 (1994).

43. R. Wooster et al., *Nature* **378**, 789–792 (1995).

44. K.L. Garver and B. Garver, *Am. J. Med. Genet.* **49**, 1109–1118 (1991).

45. B.A. Fine, in D.N. Bartels, B.S. LeRoy, and A.L. Caplan, eds., *Prescribing Our Future: Ethical Challenges in Genetic Counseling*, Aldine de Gruyter, New York, 1993, pp. 101–117.

46. B.S. LeRoy, in D.N. Bartels, B.S. LeRoy, and A.L. Caplan, eds., *Prescribing Our Future: Ethical Challenges in Genetic Counseling*, Aldine de Gruyter, New York, 1993, pp. 101–117.

47. K. Nolan and S. Swenson, *Hastings Center Rep.* **18**(5), 40–46 (1988).

48. J.F. Gusella et al., *Nature* **306**, 234–238 (1983).

49. G.J. Meissen et al., *N. Engl. J. Med.* **318**, 535–542 (1988).

50. S. Fox et al., *Am. J. Med. Genet.* **32**, 211–216 (1989).

51. J. Brandt et al., *J. Am. Med. Assoc.* **261**, 3108–3114 (1989).

52. Huntington's Disease Society of America, *Guidelines for Predictive Testing for Huntington Disease*, Huntington's Disease Society of America, New York, 1989.

53. Huntington's Disease Society of America, *Guidelines for Geneitc Testing for Huntington's Disease*, Huntington's Disease Society of America, New York, 1994.

54. L. Went, *J. Med. Genet.* **27**, 34–38 (1990).

55. L. Went, *Neurol.* **44**, 1533–1536 (1994).

56. A.M. Codori and J. Brandt, *Am. J. Med. Genet.* **54**, 174–184 (1994).

57. K.A. Quaid, *J. Genet. Counsel.* **1**, 277–302 (1992).

58. C.V. Hays, *N. Engl. J. Med.* **327**, 1449–1451 (1992).

59. M.A. Chapman, *Am. J. Med. Genet.* **42**, 491–498 (1992).

60. D. DeGrazia, *J. Clinic Ethics* **2**, 219–228 (1991).

61. M. Bloch et al., *Am. J. Hum. Genet.* **49**(4), A33, #1939 (1996).

62. M.A. Nance et al., *Am. J. Hum. Genet.* **61**(Supp.), A225 (1997).

63. D. Smith et al., *Early Warning: Cases and Ethical Guidance for Presymptomatic Testing in Genetic Diseases*, Indiana University Press, Bloomington, IN, 1998.

64. P.R. Billings et al., *Am. J. Hum. Genet.* **50**, 476–482 (1992).

65. F.P. Li et al., *J. Nat. Cancer Inst.* **84**, 1156–1160 (1992).

66. American Cancer Society, *Cancer Facts and Figures — 1997*, American Cancer Society, Atlanta, GA, 1997.

67. J.M. Hall et al., *Science* **250**, 1684–1689 (1990).

68. B.J. Baty et al., *J. Genet. Counsel.* **6**, 223–244 (1997).

69. E.B. Claus, N. Risch, and W.D. Thompson, *Am. J. Hum. Genet.* **48**, 232–242 (1991).

70. J.E. Garber and A.F. Patenaude, *Cancer Surveys* **25**, 381–397 (1995).

71. G. Geller et al., *Hastings Center Rep.* (March–April) **27**, 28–33 (1997).

72. K.M. Kash, *Ann. N.Y. Acad. of Sci.* 41–52 (1995).

73. J.R. Boukin et al., *J. Nat. Cancer Inst.* **88**, 872–882 (1996).

74. C. Lerman and R. Croyle, *Arch. Intern. Med.* **154**, 609–616 (1994).

75. K. Offit, *Clinical Cancer Genetics: Risk Counseling and Management*, Wiley-Liss, New York, 1998.

76. H.T. Lynch et al., *Arch. Intern. Med.* **53**, 1979–1987 (1993).

77. B.B. Biesecker et al., *J. Am. Med. Assoc.* **269**, 1970–1974 (1993).

78. K.G. MacDonald et al., *Can. Med. Assoc. J.* **154**, 457–464 (1996).

79. C. Lerman et al., *Health Psychol.* **10**, 259–267 (1991).

80. B.A. Bernhardt et al., *J. Genet. Counsel.* **6**, 207–222 (1997).

81. P. Aries, *Centuries of Childhood: A Social History of Family Life*, Vintage Books, New York, 1965.

82. R.H. Nicholson, *Medical Research with Children: Ethics, Law and Practice*, Oxford University Press, Oxford, UK, 1986.

83. Clinical Genetics Society, *J. Med. Genet.* **31**, 785–797 (1994).

84. American Medical Association, Council on Ethical and Judicial Affairs, *Code of Medical Ethics Reports*, vol. 6, No. 2, Rep. 66. American Medical Association, Chicago, IL, 1995.

85. American Society of Human Genetics Board of Directors and the American College of Medical Genetics Board of Directors, *Am. J. Hum. Genet.* **57**, 1233–1241 (1995).

86. A. Clarke, *The Genetic Testing of Children: Report of a Working Party of the Clinical Genetics Society*, Birmingham Maternity Hospital, Edgbaston and Birmingham, UK, 1993.

87. N.F. Sharpe, *Am. J. Med. Genet.* **46**, 250–253 (1993).

88. K.A. Quaid, *Am. J. Med. Genet.* **49**, 354 (1994).

89. J.H. Fanos, *Am. J. Med. Genet.* **71**, 22–28 (1997).

90. D.C. Wertz, J.H. Fanos, and P.R. Reilly, *J. Am. Med. Assoc.* **272**, 875–881 (1994).

91. D.C. Wertz and P.R. Reilly, *Am. J. Hum. Genet.* **61**, 1163–1168 (1997).

92. F.M. Giardiello et al., *N. Engl. J. Med.* **226**, 823–827 (1997).

See other entries GENETIC INFORMATION, ETHICS, PRIVACY AND CONFIDENTIALITY: OVERVIEW; GENETIC INFORMATION, LEGAL, ERISA PREEMPTION, AND HIPAA PROTECTION; GENETIC INFORMATION, LEGAL, FDA REGULATION OF GENETIC TESTING; GENETIC INFORMATION, LEGAL, GENETIC PRIVACY LAWS; GENETIC INFORMATION, LEGAL, REGULATING GENETIC SERVICES.

GENETIC INFORMATION, ETHICS, PRIVACY AND CONFIDENTIALITY: OVERVIEW

MADISON POWERS
Georgetown University
Washington, District of Columbia

OUTLINE

Introduction

Privacy, Confidentiality, and Anonymity

Uniqueness of Genetic Information

Research Contexts

Confidentiality in Clinical Contexts

Nonmedical Uses of Genetic Information

Conclusion

Bibliography

INTRODUCTION

Research into the genetic contribution to disease holds out the promise of scientific discoveries that will revolutionize the diagnosis, prevention, and treatment of numerous medical conditions, thereby reducing premature mortality and excess morbidity. However, the expected benefits from genetic research and the incorporation of its techniques into clinical practice are accompanied by risks to individual privacy. The importance of protecting individual privacy, especially in circumstances involving highly personal, often sensitive information such as that revealed by the analysis of an individual's genome, finds justification in a number of closely related moral arguments.

A major defense of stringent privacy protection is the claim that the ability to limit the access others have to personal, highly sensitive information is an essential element in any social policy committed to the preservation of individual autonomy (1). The ability to limit third party access to personal information also is necessary for establishing trust and intimacy within personal and professional relationships, for making reproductive and other medical decisions without the undue influence and interference of others, and for preserving valuable social and economic opportunities for persons whose life prospects may be diminished unfairly by the disclosure of information to persons with competing interests (2,3).

Genetic testing reveals just the sort of information that those concerned with the preservation of individual privacy are most anxious to protect. It may reveal a person's current medical condition, increased susceptibility to particular illnesses, or status as a carrier of genes that affect offspring. Such information can be used in ways highly detrimental to many of an individual's most fundamental opportunities and greatly limit his or her range of available economic and social choices. The potential adverse consequences of unwanted disclosures include limitations on access to health, life and disability insurance, lowering of medical treatment priority within public or private health care programs, loss of employment and educational opportunities, restriction of or added burdens placed upon available reproductive options, and social stigma and discrimination (3).

Although the present or future impact of the proliferation of genetic information regarding individuals remains uncertain and not easily quantifiable, persons having first hand experience with genetic testing within their immediate families view the threats to privacy as substantial, and many within the scientific community are concerned that public policies and clinical practices should be designed to mitigate or prevent the kinds of harms that can result from unwanted disclosure of personal genetic data (4). Indeed, many favoring increased genetic research and the rapid translation of genetic knowledge into clinical and public health benefit also favor enhanced efforts to protect the privacy, confidentiality, and anonymity interests of patients and research subjects.

PRIVACY, CONFIDENTIALITY, AND ANONYMITY

Although the precise contours of privacy and related concepts remain the subject of ongoing debate within both legal and philosophical circles a few basic definitions are needed to better understand the issues associated with the collection and use of genetic information. Privacy is often regarded as a state or condition in which personal information about an individual is inaccessible to others (3,5). A person can therefore be said to enjoy genetic privacy when there are limitations or constraints on the access third parties have to information about a person's individual genetic makeup. The most stringent genetic privacy protection policies, accordingly, are those

that allow individuals to decide for themselves what if any individualized genetic information will be collected and analyzed.

Complete privacy, however, is both impossible and often undesirable for a variety of reasons. Genetic privacy, or the inaccessibility of others to genetic or any other type of information about individuals, is often a matter of degree. In some instances, what matters most to individuals is not complete privacy — namely where no one has access to personal information — but control over the access particular individuals or institutions have to information regarding one's genetic makeup. Often the central moral concern focuses on the identities of those who have access to their genetic information and the purposes for which that information will be used. For example, individuals often have prudential medical or reproductive reasons to make information about genetic susceptibility to some disease available to their spouses, children, physicians, or genetic counselors, but they may not want to reveal that information to their employers, insurers, or specific family members. In short, the paramount concern in many circumstances is for confidentiality: persons may not want information disclosed in confidence to health professionals to be shared with third parties for whom subsequent disclosure is not authorized.

Complete genetic privacy is impossible to assure for more basic scientific reasons as well. One reason is that sophisticated techniques for collecting and analyzing DNA make it impossible to guarantee that individual genetic information will not be obtained without his or her consent and subsequently used to his or her detriment. Blood, hair and other bodily materials from which DNA can be extracted are available to many third parties in numerous medical settings, non-medical institutional settings such as prisons and the military, as well as the conduct of criminal investigations (6,7). The opportunities third parties have for gaining unwanted and unconsented access to individual genetic information thus extend well beyond the clinic and research institutions.

Another reason that complete genetic privacy is impossible to assure is that many conclusions about genotype can be inferred from phenotype. For example, medically knowledgeable persons can make reliable inferences about the existence of genetic conditions such as Marfan's syndrome from visual observation of physical attributes. Moreover genetic privacy is more difficult to assure because often it is not information over which any individual always can retain exclusive control. Third-party knowledge of the genetic makeup of some individuals can be obtained from testing conducted on an individual's relatives. Researchers and health care providers will therefore learn about the genetic makeup of persons who are neither research subjects nor patients and will have access to genetic information about those who had no opportunity to refuse disclosure.

In other instances, a central concern is the prevention of unwanted third party access to genetic information through the preservation of anonymity. For example, individuals may be motivated by altruistic reasons to make DNA samples available to researchers studying inheritance of a particular disease within a family or population geneticists studying genetic variation. Apart from any concern about subsequent disclosures of confidential information there are, in addition, risks that others may inadvertently obtain genetic knowledge about an individual who provided DNA samples. In such cases those persons providing the sample reasonably may insist on complete anonymity, or the assurance that once DNA is made available for research the researchers will not retain any personally identifying links (8,9).

Even the term anonymity can be misleading in what it implies about protection from loss of privacy. Genuinely anonymous collection of genetic information occurs when samples are not linked to specific persons. Studies using genetic information where specific personal identifiers have been collected but later discarded have been "anonymized" and thus direct identification of individuals is impossible (9). In instances in which personal identifiers have been stripped but a researcher retains the possibility of linking individuals to samples through a code persons who contributed the samples are not directly identifiable. The possibility of loss of privacy, however, is not eliminated entirely for anonymized (or even anonymous) samples. Under some circumstances (discussed below) researchers or members of the general public may be able to infer the identity of individual participants in studies, and a prudent privacy protection policy may include additional precautions to prevent publicly available or published information about the population from whom DNA samples were collected to be used to gain access to information about specific persons (10).

Privacy thus appears to be the most central conceptual category of moral analysis. Confidentiality, anonymity, and other policies may be largely strategic in achieving the goal of limiting access others have to personal information.

UNIQUENESS OF GENETIC INFORMATION

Those responsible for public policies, research protocols, or clinical practices designed to ensure the privacy, confidentiality, or anonymity of individuals must address the question of whether genetic information is uniquely sensitive and subject to a greater degree of abuse than other types of information. Such claims are often referred to under the rubric of "genetic exceptionalism," or the view that because the risk of privacy loss and its potential for adverse consequences may be greater for genetic information than for other types of medical information, separate and more stringent genetic privacy standards and policies are sometimes warranted (11).

There are three distinguishable questions within the genetic exceptionalist debate. First, is genetic information sufficiently distinct from other types of medical information? Second, does the intimate nature and sensitivity of such information represent a greater threat to individual privacy interests, thus justifying a heightened degree of protection? Third, are there feasible mechanisms for treating genetic information differently than other medical information?

The issue of distinctiveness is a threshold concern. Unless some workable definition of genetic information can be found, neither the question of its special sensitivity

nor the issue of appropriate policies for its protection can be resolved. Genetic information clearly includes knowledge that may be gained from genetic testing of individual DNA samples. However, knowledge of an individual's genome also can be obtained by biochemical tests designed to detect gene products such as the production of enzymes or proteins that result from a particular genetic mutation (11,12). Moreover, inferences about an individual's genetic makeup can be made from family histories. For example, hemophilia A is an X-linked, recessive condition, and a diagnosis of a male child with hemophilia A reveals that it was inherited from the mother, and that the mother's female siblings also have a 50 percent probability of being carriers of the disease-causing gene (13).

To the extent that the aim is restriction of access third parties have to genetic information about specific individuals, a focus on limiting access to the results of genetic tests may be too narrow. However, the case for employing a broader conception of genetic information presents a dilemma; the fact that matters of family history and the results of biochemical tests are integral parts of patients's medical records undermines efforts to afford more stringent protection for genetic information than other types of medical information. Moreover the goal of improved clinical care will necessitate increased availability of genetic information, including the incorporation of results of individual tests into the patient record. Thus effective protection of genetic privacy often will be dependent upon the development of effective medical privacy protection policies generally (14).

The special privacy concerns about genetic information mirror similar concerns for other highly sensitive medical information that, if made available to third parties, may undermine individual autonomy and result in arguably unfair economic and social burdens on those whose genes have been identified. There is already a fairly well-established social consensus regarding the special sensitivity of medical information bearing on sexual behavior, mental health, sexually transmitted diseases, alcoholism, and drug use. These matters are deemed especially sensitive because they reveal intimate details of an individual's life and often carry a degree of social stigma or may result in discrimination in employment or insurance. Even without stigma or loss of social opportunities, many persons will view public knowledge of intimate aspects of an individual's life as an assault on human dignity and the ability to exercise autonomous control over matters at the heart of individual personality (3).

Genetic information can be seen as analogous to these already established categories of sensitive medical information because of similar risks associated with their disclosure to others, and accordingly it is not unreasonable to argue that genetic information should be added to the list of medical information for which added privacy precautions are warranted. Moreover, to the extent that stigmatized medical conditions such as alcoholism and mental illness have a genetic component, the case for more stringent privacy protection of these types of genetic information is strengthened.

There are other aspects of genetic information that enhance the arguments for the view that the collection of genetic information poses even greater privacy risks than other similarly sensitive medical information. One such argument is that genetic information, unlike cholesterol screening or testing for treatable sexually transmitted diseases, has more significant familial and reproductive implications. Because testing of one person can reveal information about an individual's siblings, parents, and children the privacy interests at stake go well beyond the risks any unwanted disclosure poses for a single individual (14). Hence, even if the genetic information arguably is no more sensitive or intimately revealing than some other types of nongenetic information, the privacy interests of many more persons may be implicated with each instance of privacy breach.

Another argument for the special moral importance of genetic privacy is the fact that much genetic information is merely probabilistic, and that learning the existence of some genetic susceptibility to disease often does not predict a particular individual's risk of acquiring a disease, its severity, or its date of onset. Thus the disclosure of genetic information is said to carry an inherent risk of being used to falsely label an individual on the basis of an increased statistical risk of having a mental or physical disease that has not and indeed may not result in any symptoms or disability (5,15). Of course, many other kinds of medical tests are merely probabilistic indicators of increased risk of disease as well, and genetic information may not be unique in this respect (11). However, the deeper concern seems to be that genetic predictors often are mistakenly seen as deterministic, immutable guides to assessing an individual's mental and physical capabilities (15).

A further argument for the uniqueness of genetic information and the need for more stringent genetic privacy protection lies in the observation that our DNA contains hidden information that may in the future reveal more about us than either we or those who collect and analyze DNA samples now realize. DNA has been called a "future diary," "a code not yet cracked," holding currently undecipherable but highly personal, sensitive information bearing on one's medical and social prospects (6,7,16). Two points are of special concern in this claim. First, the availability of DNA samples that have been analyzed for one gene may be subject to later analysis for another gene whose existence and function are not known at the time of consent to testing. It may not be reasonable to suppose that those who have consented to tests revealing one type of genetic information would have consented to the disclosure of genetic information of another type.

A second worry is that even testing for an allele with a known function can in the future reveal knowledge about an individual that no one could have anticipated at the time of testing. Genes typically have many functions, only some of which are known. A single gene may contribute to some disease burden such as sickle cell anemia but that same gene also may play a protective role such as improving one's resistence to malarial infection. Because there are no inherently "good" or "bad" genes, even testing for a "good" gene today may reveal the same gene as a "bad" gene tomorrow. Because persons who consent to

testing may not be able to anticipate all that is revealed by a single test, adherence to the duty of confidentiality is a crucial moral component of any comprehensive genetic privacy protection approach.

It remains an important empirical question whether genetic information is inherently more sensitive than other types of information, and whether there is likely to be any consensus on what types of information are more important to protect than others. Certainly, all types of genetic information are not on par in their sensitivity, given the differences in what genetic tests may reveal about individuals and the potential consequences of their disclosure. Some persons may be more concerned with limiting the information others have about their sexual behavior and drug use than their increased genetic susceptibility to beryllium disease. Or a well-known trial lawyer may have great concern about the public disclosure of a heart condition for fear that potential clients will doubt his capability for aggressive advocacy, while being indifferent to what others know about his sexual conduct.

Individual differences in privacy attitudes are of tremendous moral significance. To the extent that the moral arguments for privacy protection are grounded chiefly in the protection of individual autonomy and dignity, matters that ought to be accorded the greatest protection are largely for the individual to decide for herself. However, to the extent that privacy interests are grounded in more general concerns for the protection of valuable social opportunities, then the issue of whether genetic information is more sensitive and therefore deserving of greater protection will depend on the extent to which the configuration of the major social and economic institutions leave persons more vulnerable to unfair disadvantage based on individual genotype.

Given uncertainties of whether genetic information is uniquely sensitive, its increasing integration with other types of medical information, and the competing moral rationales offered for its protection, debate about how best to deal with the protection of genetic privacy is likely to continue for some time.

RESEARCH CONTEXTS

Many who have considered the moral demands of privacy within research contexts argue for individual control over the creation of individually identifiable genetic information. Concerns about individual privacy take on great significance when there is a more than minimal risk of privacy loss for the subject and the contribution of DNA samples by research subjects offers little or no realistic expectation of any personal benefit. Such protections are embodied in current guidelines for federally funded medical research in the United States, and others also argue that this should be the norm in clinical practice, especially when collection of samples involves subsequent research uses of genetic information (9). These policy guidelines and suggested moral norms recognize the fundamental importance of obtaining informed voluntary consent to the collection of blood or other DNA samples for a variety of moral reasons, including but not limited to privacy concerns. Thus there is broad consensus that individually identified stored blood or tissue samples ought not be made available to others for subsequent analysis without the explicit consent of the persons providing the samples, and that in instances where additional testing not contemplated at the time of obtaining the initial informed consent is proposed researchers should recontact subjects to obtain appropriate consent for these new uses (8,9).

Moreover many argue that in addition to the routine disclosure of medical risks associated with research, risks of privacy loss should be included among the core elements of the informed consent process (17,18). When links to identifiable persons who contributed genetic material to a study are retained, researchers cannot guarantee that confidentiality will not be compromised, but they can and should provide subjects with information about the steps they plan to take to minimize the risks to confidentiality.

One issue that continues to divide knowledgeable commentators is when informed consent for analysis of DNA samples is necessary. Some have concluded that identified or identifiable subjects should be given the option of restricting the use of their samples to the study for which samples were initially collected (8). Others also recommend that since many samples are obtained in nonresearch settings, all individuals should be given the opportunity to decide whether their samples may be used in research and whether they are willing to have those samples used in identifiable or linked research (9).

An alternative approach is to ask research subjects to give a blanket consent to additional testing for other genes or other genetic conditions not examined initially and perhaps not possible for researchers to contemplate at the time of obtaining informed consent (6–8,17,18). The main benefit of such a proposal would be one of efficiency. Researchers would have ready at hand a supply of DNA samples, perhaps in the form of immortalized cell lines, which can be used as a resource for many research studies, and in the process save both time and money associated with collecting and preserving DNA samples.

The major drawback in such a proposal is a concern over whether is it plausible to suppose that genuinely informed consent is possible for types of research not within the contemplation of the research community. In particular, informed consent to subsequent research uses may pose privacy risks (as well as other moral concerns about use or purpose of the research itself) that differ in kind or in magnitude from those that might be recognized initially. Even if the probability of privacy loss is not increased with re-testing for a different genetic condition, the magnitude of harm from an unwanted identification of any individual may be much greater than the harm the subject would have had in mind at the time of contributing the DNA samples. For example, agreeing to be tested for a remote risk of unwanted disclosure of some genetic mutation with no obvious adverse implications for a subject's ability to obtain insurance is quite different from testing for a different condition that does have insurance implications. The upshot of any policy of obtaining broad, open-ended consent for genetic testing for research purposes is that the subject may end up waiving any rights to protect his or her

privacy interests without any feasible basis for knowing how significant those privacy interests might be.

Another area of continuing debate concerns the use of anonymous and anonymized samples and how best to deal with the risk of inferential identification of individual subjects mentioned previously. The recommendations of some commentators restrict the requirement of informed consent for research to instances in which actual personal identifiers are linked to the samples and are retained by the researchers performing the genetic tests. They therefore conclude that no informed consent is necessary either for prospective or retrospective tests when samples are collected anonymously or later are anonymized (8). Where genetic analysis is performed and the identities of subjects are removed from the samples but individual subjects remain directly identifiable through means of a code maintained by researchers, informed consent for the use of such samples is claimed to be necessary for prospective studies and usually necessary for retrospective studies, unless there is no more than minimal risk to privacy and the research could not be carried out practicably otherwise.

Others, however, suggest that in some instances, even anonymized samples may pose nonnegligible risks to individual privacy when it is possible for researchers or others to identify individual persons having a genetic mutation putting them at increased risk for some specific disorder. In such cases they conclude that there may be a need for recontacting individuals to obtain informed consent for subsequent use of samples originally contributed for a different purpose, and that only genuinely anonymous samples—where it is impossible under any circumstances to identify the individual source—clearly qualifies as exempt from any need for informed consent or other institutional efforts to safeguard individual privacy (9). Critics, however, point out that conducting research without specific informed consent has been the norm for most retrospective studies using patient medical records not involving genetic analysis, and while the critics do not reject categorically the stricter recommendations, they do worry that the proper balance between the legitimate aims of public health research and individual privacy requires further reflection and public discussion (19).

Inferential identification of individual subjects may occur in a variety of ways. Suppose that a group of researchers obtains DNA samples for a longitudinal study of several hundred members of families with a high prevalence of some type of cancer (e.g., cervical cancer) and that as a part of that study researchers also collect detailed medical histories of each study participant. The researchers therefore need to retain identifying links to each individual that are necessary for follow-up and for purposes of contacting the subjects with new information that might be of medical benefit to those subjects. Researchers therefore promise only confidentiality because the needs of the study as well as the medical needs of the subjects make anonymous collection of DNA samples undesirable. Suppose then that a later date a second group of researchers propose to conduct a different study of another medical condition also suspected of greater than average prevalence in those families, and they therefore seek permission to re-analyze the DNA samples for a second gene mutation. Even if the second group of researchers receive the samples without individual identifiers—an anonymized study—publication or other dissemination of study results may enable family members to infer the identities of specific individuals who were tested and thereby reveal previously undisclosed information about who among those families had been affected either by cervical cancer or the second medical condition. Moreover publication of detailed family pedigrees showing family lineages and various characteristics—including additional medical data such as miscarriages or abortions potentially relevant to findings regarding cervical cancer risk—can magnify both the likelihood of inferential identification of specific individuals and the harm flowing from inadvertent disclosure of other sensitive medical information in the process (10).

The risk of inferential identification may be increased, for example, when the initial pool of subjects is small, subjects are drawn from a patient population publicly associated with some particular medical or research institution, the second medical condition tested for is rare, or some discernible phenotypic characteristics are described in the publication of the study findings (9). The publication of study results may allow individual researchers, and perhaps more significantly, other family members, to more easily infer the identities of individual subjects.

Genetic research poses other privacy risks not ordinarily raised by many other types of medical research. It is not only patients whose medical records may be examined by researchers and research subjects who voluntarily contribute DNA samples for analysis that risk having others learn information about mutations predisposing them to genetic disease or their status as carriers of mutations that may affect their offspring. Personal genetic information about individuals with no personal relationship to researchers or medical caregivers may appear in another family member's medical records used by researchers or become part of a data bank or genetic registry developed and maintained by researchers from information obtained from probands (5,14,17). Detailed, comprehensive, and accurate family histories may be essential to genetic linkage studies, and thus sensitive and perhaps inaccurate information about family members not participating in the research may be obtained without their permission or knowledge.

Some risks of breach of confidentiality can be minimized through efforts to build a strong wall of separation between these specialized research records and other medical data banks and individual patient records. While the risks may be small that persons will be harmed through third-party access to these records, individuals concerned about privacy on the ground that respect for human dignity requires an ability to control the kinds of sensitive personal information others can maintain in data banks will not agree that confidentiality protections are sufficient. As the boundaries between research and clinical practice erode the feasibility of more stringent privacy protections for

genetic information is diminished and the burdens of privacy protection will fall increasingly on the shoulders of medical practitioners, genetic counselors, and researchers alike (20).

CONFIDENTIALITY IN CLINICAL CONTEXTS

The fact that genetic testing of one individual can reveal information relevant to the health status and medical care of other family members is the source of one of the most hotly debated moral issues in clinical genetics. When patients refuse to disclose genetic information that would benefit other family members, health care professionals are presented with a dilemma: They must choose whether to preserve patient confidentiality or to breach that duty of confidentiality for the benefit of third parties. The dilemma is often especially keenly felt when information regarding increased familial risk for some disease can be used by other family members to seek medical interventions that may prevent or delay the onset of disease or reduce its likely severity (e.g., familial breast or colon cancer), make informed reproductive decisions (e.g., CF, thalassemia, or Tay-Sachs), or make social and financial plans to cope with an anticipated physical or mental impairment when no treatment is available (e.g., Huntington's disease).

The dilemma, of course, is not unique to cases involving physicians, genetic counselors, and the family members of patients; researchers without any clinical connection to the person tested may obtain information bearing on the well-being of others, and interests other than those of family members may be at stake (21). For example, a widely discussed case involves researchers who learn that an airline pilot has the gene for Huntington's disease, which ultimately will impair his ability to perform his job safely.

Nonetheless, the dilemma is often felt more acutely in contexts involving clinical caregivers and those who are genetically related to their patients. The prospect of breaching patient confidentiality runs against the grain of the clinician's deepest professional commitments (22). That commitment in its strongest form is reflected is reflected in the fifth-century B.C. Hippocratic Oath: "What I may see or hear in the course of the treatment or even outside the treatment in regard to the life of men, which on no account one must spread abroad, I will keep to myself holding such things shameful to be spoken about."

More recent formulations of a physician's duties of confidentiality are more permissive. A 1980 statement of the American Medical Association (AMA), for example, states that a physician shall "safeguard patient confidences within the constraints of the law." The implication of the AMA statement is that legal requirements may override moral duties of confidentiality, and in fact court decisions in the famous *Tarasoff* case and in similar cases subsequently have provided much of the grist for moral discussions of the morally permissible exceptions to the ordinary duties of confidentiality (22,23). Although the *Tarasoff* case involved the issue of whether a psychologist must breach confidentiality in order to protect a third party from impending threats of bodily harm learned in the course of a therapeutic relation, both courts and the numerous legal and ethical commentators have extended the reasoning to other contexts as well. For example, analogous duties to disclose information to persons at risk of harm have been argued for in the context of contagious diseases such as HIV. Predictably lawyers and ethicists have pondered the extension of these duties to cases involving genetic information.

The legal and ethical issues are, of course, distinct, but the reasoning relied upon in both realms is strikingly similar and the answers given by lawyers rely heavily on ethical concerns (22–25). Both raise the fundamental question, Under what conditions, if at all, a health care professional's duty to protect third parties from harm outweighs the duty to maintain the confidentiality of genetic information obtained from a patient? A secondary question is, To what extent ought these moral duties be reflected in the law?

Several positions on these questions can be found in the literature. Some, such as the members of the President's Commission on Biomedical Issues, the Nuffield Council on Bioethics, and the authors of the recent Institute of Medicine report on genetic testing counsel against disclosure of any medical information, including genetic test results, except in very narrowly defined circumstances (3,23,26). The analytic framework of the President's Commission provides the starting point for most of the discussions the exceptions to confidentiality for genetic information. They recommended four necessary conditions that should be met before disclosure is justified:

1. There must have been an unsuccessful attempt to persuade the patient to disclose the information to the relevant party.
2. There must be a high probability of harm without disclosure and a high probability that the disclosure itself will avert the anticipated harm.
3. The potential harm must be serious.
4. Only the degree of informational detail necessary to avert harm should be disclosed.

These criteria restrict significantly the instances in which confidentiality may be overridden. For example, in cases where no effective treatment is available, no disclosure would be warranted. Similarly, where third parties have other opportunities through which they may learn of their increased genetic risk for some specific disease, the duty of confidentiality would not be overridden.

Other commentators appear sympathetic to a more liberal attitude toward disclosure in a broader array of cases. They appear more concerned, for example, about the role of health professionals in promoting the health of the entire family and often speak of the family unit rather than any individual as the patient. When conceptualized in this fashion, strictly speaking no duty of confidentiality is broken. If, as Dorothy Wertz and John Fletcher claim, "hereditary information is a family possession rather than simply a personal one" the dilemma of confidentiality versus the duty to protect others is dissolved, and the issue is recast as a question about how best to discharge the health care professional's duty to make decisions for the best interests of their patients collectively (27).

The shift to viewing the family as patient nonetheless fails to resolve what may be the central issues lurking behind claims on behalf of patient confidentiality in the context of these familial disagreements. Important questions include whether health care professionals should be presumed to possess superior knowledge and insight about what is best in these circumstances, and whether they should be seen as having the moral authority to act as enforcers of a view of family communication or of an ideal of familial loyalty and connectedness which everyone may not share. Indeed, part of the moral justification given by the defenders of patient privacy is the claim that individuals, not governments or members of professional elites, ought to control highly personal, sensitive information, even in cases where failure to disclose such information may make it harder for others to take steps to prevent harm to themselves.

Others cite a second reason for their reluctance to leave it to health care professionals to decide when to breach patient's confidentiality. They argue that genetic harm is different from some other cases in which the duty to protect from harm is thought to outweigh the duty of confidentiality. They note that the patient, whose interest in confidentiality is sacrificed, is not normally the direct causal vector of potential harm to third parties (24). Disclosure of patient information in the genetic context imposes burdens on persons who simply fail to assist others, not on persons who, for example, through their own violent or risky behaviors impose risks on innocent third parties. Thus, unlike the psychiatric or HIV cases, where it can be claimed that individuals, through their own behavior, may have forfeited their privacy rights, no such claims can be made for genetic information.

Whether a particular factual situation meets the criteria set out by the President's Commission will vary considerably. To the extent that there is any agreement on these thorny issues, most concede that case for breach of patient's confidentiality is strongest when it involves a disclosure of information obtained from a patient who is the direct and morally culpable cause of another's harm, and the harm to third parties is grave, not otherwise preventable, and immediate.

NONMEDICAL USES OF GENETIC INFORMATION

The range of potential nonmedical uses of genetic information is lengthy and limited only by available technology and imagination. Moreover, the loss of genetic privacy may result from the generation or disclosure of genetic information gathered in nonmedical contexts, many of which are not governed by established ethical and legal norms of privacy governing conduct in medical and research arenas. A few widely discussed examples illustrate the extent to which genetic privacy concerns arise outside of traditional clinical and research contexts.

Two of the nonmedical uses of genetic information receiving the most critical attention arise within employment and insurance contexts. The use of such information to deprive a person of employment or health, life or disability insurance, are among the most economically consequential implications of a loss of genetic privacy (28,29).

The connection between genetic privacy issues related to employment and various forms of insurance is especially close in the United States, since a great majority of working persons obtain their insurance as a benefit of employment (30). The consequence, for example, of an employee having some genetic predisposition for increased susceptibility to some substance found in the industrial workplace are powerful economic incentives for employers to avoid hiring persons who may be both expensive to maintain as a member of its group health insurance plan and may in the future expose the employer to some additional risk of greater tort liability for employee illnesses developed after exposure to workplace hazards.

Many other potential nonmedical uses of genetic information also may arise as a consequence of their relevance in judicial proceedings. One such use now in the public spotlight is the collection of DNA samples for criminal forensic use (31). While the use of what is popularly known as DNA fingerprinting may aid in the identification of criminals and the exoneration of persons falsely accused of crimes, the widespread and systematic collection of DNA samples from convicted felons has some attendant privacy risks. The use of DNA in forensic analysis relies on a pattern of DNA sequences often referred to as junk DNA because they do not reveal specific portions of the genome believed to be associated with any medical condition, increased susceptibility to disease, or a person's carrier status. However, state and federal law enforcement agencies that collect DNA samples for identification purposes often retain the original DNA sample, and it can be analyzed and sensitive genetic information about individual persons may be revealed. In addition to the routine ways genetic information may be used to the detriment of any individual, such information could further be used as a basis for subsequent judicial decisions about sentencing, parole, or postconviction confinement to mental health or sex offender treatment programs (32).

The potential uses of genetic information in civil litigation are limitless as well. In addition to the defensive use employers might make of such information in exclusionary hiring and other workplace decisions, employers may also assert genetic predisposition as a partial defense against lawsuit by employees, consumers, and members of the general public who are exposed to substances in their products found to contribute to illnesses that also have a genetic component in their development. The scenario some envision is one in which genetic privacy may be routinely sacrificed for the sake of apportioning legal liability for harm suffered by workers and consumers (33). Beyond the strictly legal aspects of such uses of genetic information the use of genetic information in judicial contexts raises profound ethical worries about the fairness or justice of how risks and benefits are distributed within society's major social institutions.

Other morally problematic uses of genetic information in judicial contexts raise related issues of how individual responsibility for conduct and wrongdoing are assessed and the fitness of individuals to occupy professional roles involving a high degree of trust and fiduciary integrity. For example, in a California disbarment proceeding,

DNA evidence was used to show that a lawyer who had embezzled client funds was genetically disposed to alcoholism (34). Although the California case resulted in the lawyer being placed on probation on the grounds that his genetic predisposition should be viewed as a mitigating factor in assessing his responsibility for his ethical lapses, other equally controversial uses of genetic information can be imagined as well. For example, some might argue that genetic information should be relevant in the assessment of an individual's fitness for entry in to professions such as law, medicine, or nursing. The moral objection this prospect raises — apart from the obvious worries that statistical associations between certain genes and behavior may not survive subsequent scrutiny by researchers — is that such information may be both a poor predictor of any specific individual's behavior and too intrusive into the private lives of individuals.

A few final examples reflect the potential uses of genetic information by courts, schools, and parents in making decisions affecting the custody and education of children. If detailed genetic information is added to the existing body of health information prospective adoptive parents can receive about children, problems of hard-to-place children will be exacerbated. If couples come to view genetic information as relevant to their decisions about adding an abandoned child to their family the risk is that more of the already unfortunate children will be put in an even greater disadvantage (35).

Similarly a parent, grandparent, or other party to a divorce or custody battle may make information regarding an opposing litigant's genetic illness or predisposition to illness a factor in determining the child's best interest, which is the touchstone of how judges make these difficult decisions (34). In addition schools and day care centers may assert a need to know genetic information about a child in order to make assessments of potential learning disabilities or judgments regarding the behavioral traits bearing on their suitability for enrollment (36).

The use of genetic information by those who resort to judicial proceedings to resolve disputes related to family life is already a fact of modern life, and the extent that role such information may hasten greater governmental intrusion into family life remains an open question.

CONCLUSION

In each context discussed in this article, we have seen how the increased collection and use of genetic information adds layers of complexity to old and familiar ethical problems, takes us into novel and poorly charted moral terrain in balancing privacy claims against other legitimate social goals, and often requires a fresh look at the moral foundations and value assumptions underlying the practices of many of our most basic social, medical, legal, and economic institutions.

BIBLIOGRAPHY

1. J. Rachels, *Philos. Public Affairs* **4**, 323–333 (1975).
2. J. Reiman, *Philos. Public Affairs* **6**, 27–44 (1976).

3. L. Andrews et al., *Assessing Genetic Risks: Implications for Health and Social Policy*, National Academy Press, Washington, DC, 1994.

4. J. Benkendorf et al., *Am. J. Med. Genet.* **73**, 296–303 (1997).

5. M. Powers, in M.S. Frankel and A. Teich, eds., *The Genetic Frontier: Ethics, Law and Policy*, AAAS Press, Washington, DC, 1994, pp. 77–100.

6. G.J. Annas, *JAMA, J. Am. Med. Assoc.* **270**, 2346–2350 (1993).

7. G.J. Annas, in T.F. Murphy and M.A. Lappé, eds., *Justice and the Human Genome Project*, University of California Press, Berkeley, 1994, pp. 75–90.

8. American Society of Human Genetics, *Am. J. Hum. Genet.* **59**, 471–474 (1996).

9. E.W. Clayton et al., *JAMA, J. Am. Med. Assoc.* **274**, 1786–1792 (1995).

10. M. Powers, *IRB: Rev. Hum. Subj. Res.* **15**, 7–11 (1993).

11. T. Murray, in M. Rothstein, ed., *Genetic Secrets: Protecting Privacy and Confidentiality in the Genetic Era*, Yale University Press, New Haven, CT, 1997, pp. 60–73.

12. M. Yesley, *Microb. Comp. Genom.* **2**(1), 9–35 (1997).

13. R. Wachbroit, *Rep. Inst. Philos. Public Policy* **9**, 9–11 (1989).

14. L.O. Gostin, *J. Law, Med. Ethics* **23**, 320–330 (1995).

15. E. Juenst, *Genome Sci. Technol.* **1**(1), 21–30 (1995).

16. G.J. Annas, *Am. J. Public Health* **85**, 1196–1197 (1995).

17. P.R. Reilly et al., *Nat. Genet.* **15**, 16–20 (1997).

18. R.F. Weir and J.R. Horton, *IRB: Rev. Hum. Subj. Res.* **17**, 1–4 (1995).

19. B.M. Knoppers and C. Laberge, *JAMA, J. Am. Med. Assoc.* **274**, 1806–1807 (1995).

20. G. Geller et al., *J. Law, Med. Ethics* **21**, 238–240 (1993).

21. P. Harper, *Br. Med. J.* **306**, 1391–1394 (1993).

22. D. Orentlicher, in M. Rothstein, ed., *Genetic Secrets: Protecting Privacy and Confidentiality in the Genetic Era*, Yale University Press, New Haven, CT, 1997, pp. 77–91.

23. President's Commission for the Study of Ethical Problems in Medicine and Biomedical and Behavioral Research, *Screening and Counseling for Genetic Conditions*, U.S. Government Printing Office, Washington, DC, 1992.

24. S. Suter, *Mich. Law Rev.* **91**, 1854–1908 (1993).

25. M.Z. Pelias, *Am. J. Med. Genet.* **39**, 347–354 (1991).

26. Nuffield Council on Bioethics, *Genetic Screening: Ethical Issues*, Nuffield Council on Bioethics, London, 1993.

27. D.C. Wertz and J.C. Fletcher, *Bioethics* **5**, 212–232 (1991).

28. *Genetic Information and Insurance*, Task Force Report, Genetic Information and Health Insurance, National Institutes of Health, National Center for Human Genome Research, Bethesda, MD, 1993.

29. N. Kass, in M. Rothstein, ed., *Genetic Secrets: Protecting Privacy and Confidentiality in the Genetic Era*, Yale University Press, New Haven, CT, 1997, pp. 299–316.

30. M.A. Rothstein and B.M. Knoppers, *Eur. J. Health Law* **3**, 143–161 (1996).

31. J.E. McEwen, *Am. J. Hum. Genet.* **56**, 1487–1492 (1995).

32. B. Scheck, *Am. J. Hum. Genet.* **54**, 931–933 (1994).

33. M.A. Rothstein, *J. Law Health* **9**, 109–120 (1994–1995).

34. R.C. Dreyfuss and D. Nelkin, *Vanderbilt Univ. Law Rev.* **45**, 313–348 (1992).

35. L. Andrews, in M. Rothstein, ed., *Genetic Secrets: Protecting Privacy and Confidentiality in the Genetic Era*, Yale University Press, New Haven, CT, 1997, pp. 255–280.

36. L. Rothstein, in M. Rothstein, ed., *Genetic Secrets: Protecting Privacy and Confidentiality in the ·Genetic Era*, Yale University Press, New Haven, CT, 1997, pp. 317–331.

See other entries GENETIC INFORMATION, ETHICS, AND INFORMATION RELATING TO BIOLOGICAL PARENTHOOD; GENETIC INFORMATION, ETHICS, ETHICAL ISSUES IN TISSUE BANKING AND HUMAN SUBJECT RESEARCH IN STORED TISSUES; GENETIC INFORMATION, ETHICS, FAMILY ISSUES; GENETIC INFORMATION, ETHICS, INFORMED CONSENT TO TESTING AND SCREENING; GENETIC INFORMATION, LAW, LEGAL ISSUES IN LAW ENFORCEMENT DNA DATABANKS.

GENETIC INFORMATION, LAW, LEGAL ISSUES IN LAW ENFORCEMENT DNA DATABANKS

PAUL GIANNELLI
SHARONA HOFFMAN
WENDY WAGNER
Case Western Reserve University, School of Law
Cleveland, Ohio

OUTLINE

INTRODUCTION

Jean Ann Broderick was sexually assaulted and murdered on November 17, 1991, in Minneapolis. There were no suspects, and the possibility of another unsolved crime loomed large. The police, however, discovered semen at the crime scene, extracted a DNA profile from this evidence, and entered the profile into the state DNA databank.

The computer responded with what is known as a "cold hit"—a match that in an electronic second transformed a "no suspect" case to one with overwhelming prosecutorial merit. It was the "first case in American history in which the new tool of DNA databanking was used to solve a rape or murder case" (1). The prosecutor would later remark, "Without a DNA pool, there is no way we would have been able to identify the suspect. And we certainly would not have been able to get the conviction" (1).

As the Broderick case illustrates, DNA databanks are a significant advancement in crime solution. As of January 1999, over 400 "hits" have been recorded. These databanks are similar to the computerized fingerprint databank, called AFIS (Automated Fingerprint Identification System), which has been operational during the last decade. In some ways DNA databanks may be of greater utility than AFIS. While the wearing of gloves prevents the leaving of fingerprints, it is more difficult to prevent the deposition of some type of evidence that contains the perpetrator's DNA—especially in rape cases. Indeed, Virginia officials claim that "material susceptible to DNA analysis, including blood, skin tissue, hair follicles, and semen, may be found at thirty percent of all [violent] crime scenes" (2). Saliva and sweat should be added to this list (3). Furthermore, fingerprints cannot be dated; they can place a suspect at a specific location but cannot, by themselves, establish when the suspect was there, a significant limitation in cases in which the suspect has innocent access to the crime scene location. In contrast, semen found in a rape victim eliminates the "dating" problem in cases where the suspect claims mistaken identification.

An understanding of the role of DNA databanks in the criminal justice system requires some appreciation of the impact of DNA evidence in criminal prosecutions.

DNA EVIDENCE

In 1985 Dr. Alec Jeffreys of the University of Leicester, England, recognized the utility of DNA profiling in criminal cases. Its first use in American courts came the following year. The initial appellate case, *Andrews v. State* (4), was reported in 1988. By January 1990 forensic DNA evidence had been admitted in at least 185 cases by 38 states and the U.S. military (5). Today DNA evidence, in one form or another, is admissible in every state and federal circuit (6). These developments are remarkable. No other scientific technique had gained such widespread acceptance so quickly. No other technique had been as complex or evolved so rapidly. DNA profiling raised issues at the cutting edge of modern science (7). New DNA technologies were introduced even as cases litigating the older procedures worked their way through the court system; there have already been three generations of tests — Restriction Fragment Length Polymorphism (RFLP), Polymerase Chain Reaction (PCR) for discrete alleles, and the current state of the art, Short Tandem Repeats (STRs). In addition, testimony based on mitochondrial DNA has been admitted in evidence (8). Moreover, nonhuman DNA has proved useful in litigation, "ranging from homicide prosecutions to patent infringement litigation, with organisms as diverse

as household pets, livestock, wild animals, insects, plants, bacteria, and viruses" (9).

Finally, no other technique has been as potentially valuable. One court commented that DNA evidence may be the "single greatest advance in the 'search for truth' ... since the advent of cross-examination" (10). Even its critics acknowledge that "[a]ppropriately carried out and correctly interpreted, DNA typing is possibly the most powerful innovation in forensics since the development of fingerprinting in the last part of the 19th Century" (11). For instance, in the World Trade Center bombing prosecution, an FBI expert matched saliva on an envelope sent by the terrorists to the *New York Times* with the DNA of one of the defendants (12). The next crime tool on the horizon is a credit-card-size device that would permit the analysis of DNA at a crime scene (3).

DNA evidence's power to convict is matched by its power to exculpate. This was underscored by a Department of Justice report that discussed the exoneration of 28 convicts through the use of DNA technology — some of whom had been sentenced to death (13). By mid-1999, more than 70 convicts had obtained postconviction relief based on exculpatory DNA test results. This development has already resulted in a change in some legal procedures. For example, the basis for motions for new trials has historically been quite limited; after a trial with the full panoply of constitutional protections, "finality" becomes a significant, if not paramount, interest. Thus, courts are skeptical of witnesses who subsequently "change" their testimony or post-trial "confessions" by unavailable third parties. Due to its reliability DNA evidence alters the calculus between "finality" and justice. Consequently, New York and Illinois have statutorily extended the time period for post-trial challenges to convictions based on DNA evidence, and the Department of Justice's Commission on the Future of DNA has advocated adoption of similar provisions in other jurisdictions (14).

Similarly, legislatures are reconsidering time limits on statutes of limitation in criminal cases (15). Ohio, for instance, has increased its statute of limitations for felonies from 6 to 20 years. DNA evidence reduces, to some extent, the danger of conviction based on evidence that is unreliable because it is stale. Indeed, to toll the Wisconsin statute of limitations, one creative prosecutor indicted a "John Doe" rapist based solely on his DNA code (16).

DNA DATABANKS

There are a variety of organizations that collect samples for DNA analysis — the Department of Defense collects blood and tissue samples from every U.S. service member, states authorize laboratories to secure dried blood samples (and often other tissue samples) from newborns, and reproductive laboratories and blood banks store samples from patrons. Each of these DNA systems raises a number of important privacy concerns (17). This discussion will focus, however, only on DNA databanks created by state or federal law for criminal enforcement purposes. These databanks present the full range of privacy and related issues in the much more dramatic context of criminal enforcement, with the government, rather than a private employer or insurer, poised as the entity that threatens an individual's right to genetic privacy.

The first DNA databank used for criminal enforcement purposes was established by the Virginia legislature in 1989 (18). Today every state has enacted databanking legislation (19). The DNA Identification Act of 1994 provides federal funds to assist in this endeavor (20). Although each state legislates the conditions under which DNA samples are taken, the FBI has established a national databank system, called CODIS (Combined DNA Index System), into which the state profiles can be entered (21,22). Now states can search the databases of other states (23).

The state databank statutes vary widely with respect to their coverage. Some states require only sex offenders to provide samples for databank use (24). Other states also include different crimes of violence (25). Still others reach all convicted felons (26). One statute extends to those arrested for felony sex offense crimes or other specified offenses (27). Some states include juvenile offenders (28,29) and others cover probationers as well as parolees (30). Several databases also contain DNA profiles of missing persons and victims of mass disasters (31). The method of collection differs; some statutes require the collection of blood (sometimes a finger prick) (32) while others collect cheek swabs (33). Some statutes contain expungement procedures, under which a person's profile may be removed from the database if that person's conviction is reversed on appeal (19,34).

The state databanks also vary in other respects. Some states have legislated the circumstances under which their database can be searched. For example, two states allow their databases to be searched only for the investigation of sex-related or violent crimes (19). States also vary in the resources dedicated to DNA collection and analysis. Some jurisdictions have made considerable headway entering samples into their databases, while others face a tremendous backlog of samples yet to be analyzed (19,35,36). One report notes: "So while a new national FBI databank and state databanks now hold a total of 270,000 DNA profiles, there is also a backlog of roughly 500,000 unanalyzed DNA specimens. And the DNA of an estimated 1 million more people is supposed to be added by law, but some jurisdictions are already so far behind they're not even bothering to collect new samples" (37).

While variations in the coverage and procedures for state databases produce inconsistencies, state databanks do share important similarities as well. First, DNA profiles are generally kept in a database that identifies them by a coded identification number. To determine the identity of the person, a separate database must be accessed that decodes the identification number and links the profile to a specific individual. These security measures help to ensure that the DNA profile does not provide readily usable information about the identity of a particular individual (38). Second, databases generally contain one set of DNA profiles that have been taken from identified individuals, and a second set of profiles, usually taken from crime scenes, for which a match is sought (39). If a crime

scene profile does not result in a match, it remains in the system. Some time in the future it may be matched with the profile of a subsequently convicted offender (39). Or, it may be matched with another crime scene profile, alerting the police that they are looking for a serial offender (40).

The present success and future potential of these databases for determining the identity of criminal perpetrators is clear. Commentators have suggested that dramatic deterrence and criminal enforcement benefits will be gleaned from the enhanced enforcement potential (19,36,39). Much of the criticism of these databases, in fact, focuses not on the lack of benefits to criminal enforcement but rather on deficiencies in confidentiality assurances that protect individuals' remaining rights to privacy.

PRIVACY AND RELATED CONCERNS

The privacy issues associated with DNA profiling were recognized from the beginning. In 1990 Congress's Office of Technology Assessment highlighted this issue: "Citing the inherent intimacy of genetic information, the current and developing ability to test for personal information other than unique identity, and the difficulties of maintaining the confidentially in a computer network, experts raise concerns that genetic information could be used unfairly to deny future benefits to persons with criminal records, and that genetic profiling within the criminal justice sphere could lead to wider testing and broader threats to privacy" (5, p. 35).

The National Academy of Science's 1992 DNA report also took note of privacy concerns, citing developments in both molecular biology and computer technology. "Molecular geneticists are rapidly developing the ability to diagnose a wide variety of inherited traits and medical conditions. The list already includes simply inherited traits, such as cystic fibrosis, Huntington's disease, and some inherited cancers. In the future, the list might grow to include more common medical conditions, such as heart disease, diabetes, hypertension, and Alzheimer's disease. Some observers even suggest that the list could include such traits as predispositions to alcoholism, learning disabilities, and other behavioral traits (although the degree of genetic influence on these traits remains uncertain)" (41). The report goes on to state: "Even simple information about identity requires confidentiality. Just as fingerprint files can be misused, DNA profile identification information could be misused to search and correlate criminal-record databanks or medical record databanks. Computer storage of information increases the possibilities for misuse. For example, addresses, telephone numbers, social security numbers, credit ratings, range of incomes, demographic categories, and information on hobbies are currently available for many of the citizens in our society from various distributed computerized data sources" (41).

Privacy concerns also arise from inadequate confidentiality protections for the DNA profiles, and more importantly, for stored samples in many state DNA databanks (17,42). Only some of the states provide a meaningful penalty for the unauthorized use of DNA samples or profiles by private parties (43–45). There also appear to be few, if any, common law remedies available to deter this avenue for abuse, although there has been some discussion about legislative reforms that would provide individuals with protected privacy rights in their genetic information (45). It is not clear whether the unauthorized use of DNA profiles will ultimately present serious privacy intrusions, since the profiles are done with "junk DNA" that is currently believed to reveal little if any information about personal traits (46). Inadequate protection of the original biological samples is a completely different matter, however, and there is unanimity that these samples contain very private information about an individual (38,47,48). Yet in some state databank laws, the original biological samples receive less protection than the DNA profiles (19,48). The danger that DNA information contained in original samples may be disseminated in unauthorized ways becomes even more worrisome, since states may retain samples indefinitely in order to adapt to possible future changes in the profiling system (48).

In sum, DNA databanking will be a powerful tool in solving, and perhaps deterring, crime, but the possibility for misuse exists — probably in ways that cannot be anticipated. Some prior precedents demonstrate the possibility of abuse. For example, a former assistant U.S. Attorney "recalls an incident from his days as a prosecutor in the 1970s in which police officers were caught selling confidential police records to private investigators" (39). The harm that can result from unauthorized release of genetic information could be still greater. Because of privacy concerns, bioethicist Eric Juengst argues that "any DNA taken for identification purposes should only be typed for information-free markers" and "no physical DNA samples should be banked" (44, p. 64).

LEGAL CHALLENGES

Databanks have been challenged on a wide range of constitutional grounds — for example, freedom of religion (49) and the right to privacy (50). Also statutory attacks under the Religious Freedom Restoration Act have been advanced (51). Several attacks have been quite creative. For example, the Tenth Circuit has rejected arguments that taking DNA samples violates the Ninth Amendment and deprives offenders of a property interest in their blood without due process (52). None of these challenges has prevailed; often well-accepted legal principles foreclosed many of these attacks (53).

Six constitutional grounds are discussed in this article: (1) self-incrimination, (2) ex post facto, (3) equal protection, (4) due process, (5) cruel and unusual punishment, and (6) unreasonable search and seizure. In addition, states may provide greater protection under state constitutions or statutes than the U.S. Supreme Court has recognized under the federal constitution; independent state grounds have been raised, but no challenge has yet prevailed on this basis (54–57).

SELF-INCRIMINATION CLAUSE

Challenges to the collection of blood or saliva grounded in the Self-incrimination Clause of the Fifth Amendment

have been quickly dismissed based on well-established precedent. The leading case is *Schmerber v. California* (58). While being treated at a hospital for injuries sustained in an automobile collision, Schmerber was arrested for driving under the influence of alcohol. At the direction of the investigating police officer, a physician obtained a blood sample from Schmerber. Although the defendant objected to this procedure on the advice of counsel, his blood was extracted and analyzed for alcoholic content. Before the Supreme Court, Schmerber argued that the extraction of blood violated the privilege against self-incrimination. Rejecting this argument, the Court held that the privilege covers only *communicative or testimonial evidence*, not *physical or real evidence*. According to the Court:

> It is clear that the protection of the privilege reaches an accused's communications, whatever form they might take.... On the other hand, both federal and state courts have usually held that it offers no protection against compulsion to submit to fingerprinting, photographing, or measurements, to write or speak for identification, to appear in court, to stand, to assume a stance, to walk, or to make a particular gesture. The distinction which has emerged, often expressed in different ways, is that the privilege is a bar against compelling "communications" or "testimony," but that compulsion which makes a suspect or accused the source of "real or physical evidence" does not violate it (58).

Subsequent Supreme Court cases reaffirmed the testimonial-physical evidence distinction recognized in *Schmerber*. In *United States v. Wade* (59), the Court held that compelling an accused to exhibit his person for observation was compulsion "to exhibit his physical characteristics, not compulsion to disclose any knowledge he might have" (59) and thus not proscribed by the privilege. In *Gilbert v. California* (60), the Court concluded that the compelled production of a "mere handwriting exemplar, in contrast to the content of what is written, like the voice or body itself, is an identifying physical characteristic outside [the Fifth Amendment's] protection" (60). Similarly, in *United States v. Dionisio* (61), the Court ruled that compelling a defendant to speak for the purpose of voice analysis did not violate the Fifth Amendment because the "voice recordings were to be used solely to measure the physical properties of the witnesses' voices, not for the testimonial or communicative content of what was to be said" (61,62). Cheek swabbing falls into the same category.

Courts addressing the Fifth Amendment arguments in the databank context have applied these precedents when rejecting such arguments (63–65).

EX POST FACTO CLAUSE

The United States Constitution prohibits the retroactive application of criminal laws. Article I provides that neither Congress nor any state shall pass an "ex post facto Law" (66). According to the Supreme Court, this prohibition means that "[l]egislatures may not retroactively alter the definition of crimes or increase the punishment of criminal acts" (67). The ex post facto argument is limited to convicts who were already incarcerated at the time the databank legislation took effect; prospective application does raise this issue (68). Nevertheless, the efficacy of the databanking program would be severely undercut if only the profiles of persons convicted of sex offenses in the future were in the databank; many of these new defendants would not be released for years, while previously convicted inmates would be released into the community without inclusion in the databank.

Some courts have ruled that the ex post facto prohibition does not apply because databanking statutes are not *penal* in nature (69–71). For example, the Ninth Circuit rejected such a challenge to the Oregon statute because its "obvious purpose is to create a DNA data bank to assist in the identification, arrest, and prosecution of criminals, not to punish convicted murderers and sexual offenders" (72).

The ex post facto issue, however, does not necessarily disappear merely because a statute is labeled "non-penal." Ex post facto principles apply when punishment is retroactively increased (73), and that may occur if a sanction for refusal to provide a DNA sample is the denial of parole or the forfeiture of good time credits (credits awarded for a period of good behavior in prison). Much depends on how a parole or good time statute is written. Of course, many states have eliminated both parole and good-time credit.

Parole Release

If parole is purely discretionary, a parole board may consider a refusal to comply with a valid prison regulation, such as one requiring a DNA sample, in determining the appropriateness of parole. In contrast, an increase in the length of a sentence caused by new conditions in a mandatory parole jurisdiction is suspect. For example, the Virginia parole statute mandated parole six months before the sentence release date, and the Fourth Circuit ruled that withholding release for failure to provide DNA samples would be unconstitutional (74). This does not necessarily mean that these inmates can escape providing a sample; a state may make it a new crime to refuse to provide a sample (75).

Good-Time Credit

Reduction of good-time credit raises somewhat different issues. In *Weaver v. Graham* (76), the Supreme Court ruled that the elimination of good-time credit constituted an increase in punishment because "a prisoner's eligibility for reduced imprisonment is a significant factor entering into both the defendant's decision to plea bargain and the judge's calculation of the sentence to be imposed." *Weaver*, however, involved inmates whose good-time credit was legislatively reduced across the board, even if they had *not* violated any prison regulation. Several courts have distinguished databank statutes on this basis, finding that at the time of sentencing good-time credits were known to be contingent on compliance with legitimate prison regulations and the nature of those regulations may be amended while the prisoner is serving penitentiary time (77–79).

EQUAL PROTECTION CLAUSE

The Fourteenth Amendment establishes that no state may "deny any person within its jurisdiction the equal protection of the laws." Several inmates have asserted equal protection grounds as a basis for striking down databank statutes. They claim, for example, that sex offenders are treated differently from other offenders in violation of the equal protection mandate.

The Supreme Court has developed a multi-tiered classification for reviewing equal protection claims. A state statute is subjected to "strict" scrutiny if it adversely affects a suspect class. Utilizing strict scrutiny analysis, a court will require the state to prove that it has a *compelling* governmental interest and that it is employing the *least restrictive means* to achieve its compelling goal. Suspect classifications that warrant strict scrutiny under the Equal Protection Clause are race, alienage, and national origin (80).

A statute can be challenged under the Equal Protection Clause even if it does not adversely affect a suspect classification. Thus, a databank statute that requires DNA sampling only of sex offenders and violent felons may be attacked on the ground that it treats those particular criminals unequally in violation of the equal protection requirement. If no suspect classification is involved, however, courts use a lower level of scrutiny, namely, what is known as the "rational basis" test.

The rational basis test is derived from a long line of Supreme Court decisions (81). Under this type of judicial review, a "statute is presumed to be valid and will be sustained if the classification is rationally related to a legitimate state interest" (82). In *Boling v. Romer* (83), the Tenth Circuit rejected the argument that taking DNA samples only from sex offenders violated the Equal Protection Clause. The court held that there was a "rational relationship" between the "government's decision to classify inmates as convicted sex offenders and the government's stated objective to investigate and prosecute unsolved and future sex crimes" (83, p. 1341).

In *State v. Olivas* (84), the Washington Supreme Court considered a challenge to the state statute that required a DNA sample from anyone convicted of a sexual or violent offense. The court held that "[t]here is a rational relationship between the interest of the government in law enforcement and the application of the statute to this class of persons" (84, p. 1087). The statute's purpose of facilitating the investigation and prosecution of sex offenses and violent crimes was sufficiently important to defeat the equal protection challenge.

DUE PROCESS

Both the Fifth and the Fourteenth Amendments forbid the denial of life, liberty, or property "without due process of law." Inmates have asserted two different due process arguments: substantive due process and procedural due process.

Substantive Due Process

The Supreme Court has stated that "[d]ue process of law is a summarized constitutional guarantee of respect for those personal immunities which ... are 'so rooted in the traditions and conscience of our people as to be ranked as fundamental' ... or are 'implicit in the concept of ordered liberty'" (85). Recognized fundamental rights include the right to bodily integrity (86), which is arguably violated when the state conducts a medical procedure over an individual's objection. State action that infringes a fundamental right protected by the Constitution is subject to the strict scrutiny test (87).

In the 1952 case of *Rochin v. California* (85), the Supreme Court held that the forcible stomach pumping of a suspect to recover narcotic pills "shock[ed] the conscience" and did not comport with traditional ideas of fair play and decency, thereby violating due process. By contrast, the Court, faced with a due process challenge in the 1957 case of *Breithaupt v. Abram* (88), upheld the involuntary extraction of blood from an unconscious suspect after an automobile accident in order to determine whether he was intoxicated. In distinguishing *Rochin*, the Court emphasized that unlike the extraction of stomach contents, the extraction of blood was performed "under the protective eye of a physician" and was a routine and scientifically accurate method that did not involve the "brutality" and "offensiveness" present in *Rochin* (88, pp. 435–437).

The *Rochin* and *Breithaupt* decisions predated the applicability of the Fourth Amendment to the states through the Due Process Clause of the Fourteenth Amendment in 1961 (89,90), and thus the continued validity of an independent substantive due process analysis in these cases is questionable. Challenges to databank statutes no longer need be addressed in terms of due process, but rather as possible violations of specific constitutional guarantees enumerated in the Bill of Rights, such as the right to be free of unreasonable searches and seizures (91). The Supreme Court specifically held in *Schmerber v. California* (92) that the manner in which evidence is obtained from a suspect is subject to the reasonableness requirement of the Fourth Amendment (92, p. 771), and in *Winston v. Lee* (93), the Court applied the Fourth Amendment to the surgical removal of a bullet from a suspect. Thus, virtually all DNA databank and other cases that are potentially subject to attack on substantive due process grounds are better analyzed under the Fourth Amendment (94–96).

Procedural Due Process

Procedural due process mandates that a person cannot be deprived of "life, liberty, or property" without a hearing and attendant procedural safeguards, although the nature of the safeguards differs depending on the interest involved (97). Some inmates have challenged DNA databank statutes on the ground that the taking of a DNA sample without a hearing deprives them of a liberty or a property interest in their genetic material without due process of law. These challenges have uniformly failed.

In *Rise v. Oregon* (98), the plaintiffs argued that the Due Process Clause required prison officials to provide an opportunity for a hearing before requiring felons to submit a blood sample in accordance with Oregon's databank statute. The court held that "[t]he extraction of blood from an individual in a simple, medically acceptable manner, despite the individual's lack of an opportunity to object to the procedure, does not implicate the Due Process Clause" (98, pp. 1562–1563; 99). Consequently, the felons did not have a liberty or property interest at stake.

Similarly, in *Boling v. Romer* (100), the plaintiff challenged a Colorado statute that required inmates convicted of sexual assault offenses to submit a DNA sample as a condition of release on parole. Without providing the sample, inmates could not regain their liberty. The court nevertheless found that plaintiff's argument that the state "unconstitutionally deprived him of a property interest in his blood without due process" was "unpersuasive" (100, p. 1340). The court explained that parole in Colorado was discretionary and that convicts have no constitutional right to be conditionally released before the expiration of their valid sentences.

CRUEL AND UNUSUAL PUNISHMENT

Several challenges to DNA databanks focused on the Eighth Amendment, which proscribes cruel and unusual punishment (101). In *Sanders v. Coman* (102), inmates argued that the use of force to obtain blood samples violated the amendment; they alleged: "The uses of force have included instances of several officers surrounding an inmate while one held his arm still, the spraying of mace, and bending inmates' wrists in a painful manner to induce compliance." An Eighth Amendment violation, however, occurs only if force is applied for the purpose of causing harm (103), or if the force is excessive (104). Neither theory, in the district court's view, applied in this context. Here, force was used to compel compliance with a valid prison regulation (105,106).

Courts have also held that placement in solitary confinement for failing to comply with an order to provide a blood sample does not violate the Clause (107). In *Boling v. Romer* (108), the plaintiff argued that DNA sampling in a prison constituted cruel and unusual punishment because it exposed him to potential abuse from fellow inmates. He claimed that when prison authorities indicated in front of other prisoners that he was required to submit to a DNA test, they identified him as a sex offender and thus made him vulnerable to possible physical harm from inmates who were apparently particularly hostile toward sex offenders (108, p. 1341). The Tenth Circuit rejected this argument, finding it insufficient to support an Eighth Amendment claim.

Another prisoner asserted that DNA testing violated the Eighth Amendment because the blood test itself was painful (109). Not surprisingly, the district court found that the argument lacked merit, noting that the blood was withdrawn by a trained technician in accordance with medically accepted procedures.

The Cruel and Unusual Punishment Clause also prohibits deliberate indifference to an inmate's serious medical needs (110). Inmates asserting that they were injured because of the blood test, however, must show more than mere negligence in withdrawing blood (111).

Consequently, the Eighth Amendment does not present an obstacle to databanking. Moreover, DNA samples need not be blood. Profiles can be created from cheek swabs, which inflict no pain and are extremely unlikely to cause injury.

SEARCH AND SEIZURE

The most significant legal challenge to databanks is based on the Fourth Amendment's prohibition of unreasonable searches and seizures. Although the U.S. Supreme Court has yet to address the issue, its decisions in other areas provide a framework for analysis.

The Fourth Amendment is intended to ensure "privacy, dignity, and security of persons against certain arbitrary and invasive acts by officers of the Government or those acting at their direction" (112). There are three distinct Fourth Amendment issues raised in this context. First, there is a "seizure" of the person, which brings that person under the control of the government agents. Second, there is a subsequent search for and seizure of a biological sample or trace evidence from this person (113). Third, the use to which the genetic information in the sample is put raises a final Fourth Amendment issue.

A finding that the Fourth Amendment applies does not mean that a procedure is unconstitutional. That is merely the first step in the analysis. As the Supreme Court has often remarked: "[T]he Fourth Amendment does not proscribe all searches and seizures, but only those that are unreasonable" (114).

Applicability of Fourth Amendment

Seizure of the Person. In the databanking context the first issue — seizure of the person — is not problematic because convicts are already incarcerated. The seizure would be an issue for parolees, probationers, or previously released convicts. Nevertheless, notifying such persons to report and provide DNA samples would be a reasonable seizure. Indeed, it is probably not a "seizure" within the meaning of the Fourth Amendment (113). As for arrestees, probable cause is required, but an arrest warrant is not mandated if the arrest takes place in a public place (115).

Search to Obtain Samples. The leading case on defining which governmental activities are "searches" within the meaning of the Fourth Amendment is *Katz v. United States* (116). *Katz* substituted a privacy approach for the traditional property approach to this issue. According to the Supreme Court: "[T]he Fourth Amendment protects people, not places. What a person knowingly exposes to the public, even in his own home or office, is not a subject of Fourth Amendment protection. ...But what he seeks to preserve as private, even in an area accessible to the public, may be constitutionally protected" (116, p. 351).

There is little dispute that taking blood samples is a search. In *Schmerber* the Supreme Court held that the extraction of blood for the purpose of scientific (blood/alcohol) analysis "plainly constitutes searches of the

'persons'" within the meaning of the Fourth Amendment. In *Skinner v. Railway Labor Executives' Ass'n* (117), which involved a drug testing program, the Court wrote that "it is obvious that this physical intrusion, penetrating beneath the skin, infringes an expectation of privacy that society is prepared to recognize as reasonable" (117, p. 616). In addition to blood samples, lower courts have generally treated the taking of hair (118–120) and saliva (121) samples as searches.

In contrast, the taking of fingerprints (122), voice exemplars (113), or handwriting samples (123) do not constitute searches because such physical characteristics are constantly exposed to the public. (Note the difference between fingerprints and blood or cheek swabbings; it will be important in discussing arrestees later in this article.)

Use of Genetic Information. In *Skinner* the Supreme Court also ruled that the subsequent chemical analysis of the blood sample to obtain physiological data "is a further invasion" of privacy interests — informational privacy (124). This point was further refined when the Court considered the collection of urine samples. Even though this procedure did not involve a bodily intrusion, the Court held that it was a search. Like blood, the chemical analysis of urine can "reveal a host of private medical facts," including whether a person is epileptic, pregnant, or diabetic (124, p. 617).

The courts addressing the constitutionality of databank statutes have acknowledged the applicability of the Fourth Amendment to the taking of a sample (125–127) as well as its subsequent analysis (127). Consequently, the databanking litigation has focused on the second step in Fourth Amendment analysis — the reasonableness of these programs.

Reasonableness of Search

As noted above, the Fourth Amendment does not prohibit all searches, only unreasonable ones. Traditionally, reasonable searches are those conducted pursuant to a warrant issued by a neutral and detached magistrate and based on probable cause. Moreover, search warrants must describe the place to be searched and the items to be seized with "particularity." The particularity requirement circumscribes the police's discretion in executing a search warrant. Nevertheless, exceptions to these traditional requirements have been recognized, and courts have cited several in upholding DNA databank statutes.

The databank cases can be grouped around three lines of Supreme Court precedents: (*1*) administrative searches, (*2*) "special needs" searches, and (*3*) prisoner searches (127). These categories, however, are not mutually exclusive — and they all involve a balancing of interests in determining the reasonableness of the procedure. The next sections focus on sex offenders, the most common category in DNA databank statutes. Later sections discuss persons convicted of other crimes and arrestees.

Administrative Searches. Originally, the phrase "administrative search" was used to describe non-law enforcement searches. For example, the landmark case, *Camara v. Municipal Court* (128), involved housing inspections. The purpose of these inspections was not to gather evidence of criminal conduct but rather to ensure compliance with health and safety standards. Housing inspectors rather than police officers conducted these searches, although violation of the regulations could result in criminal prosecution.

In *Camara*, the Court held that the reasonableness of an administrative search is determined by balancing the governmental interest against the nature and extent of the intrusion on privacy.

> The ... argument is in effect an assertion that the area inspection is an unreasonable search. Unfortunately, there can be no ready test for determining reasonableness other than by balancing the need to search against the invasion which the search entails. But we think that a number of persuasive factors combine to support the reasonableness of area code-enforcement inspections. First, such programs have a long history of judicial and public acceptance. Second, the public interest demands that all dangerous conditions be prevented or abated, yet it is doubtful that any other canvassing technique would achieve acceptable results. Many such conditions — faulty wiring is an obvious example — are not observable from outside the building and indeed may not be apparent to the inexpert occupant himself. Finally, because inspections are neither personal in nature nor aimed at the discovery of evidence of crime, they involve a relatively limited invasion of the urban citizen's privacy (128, pp. 536–537).

The Court found the inspection system "of indispensable importance to the maintenance of community health" (128, p. 537). Thus, in *Camara*, the Court concluded that housing inspection programs were supported by the compelling government interest of avoiding dangerous living conditions and maintaining housing stock and that the inspection programs were a reasonable means for achieving these societal interests.

Later cases involved the inspection of gun dealerships (129), mines (130), and the workplace pursuant to the Occupational Safety and Health Act (OSHA) (131). Perhaps the most familiar administrative search is the metal detector procedures at airports (132).

New York v. Burger (133), decided in 1987, is a transitional case. It involved a New York statute authorizing warrantless administrative searches of automobile junkyards, which the Supreme Court upheld. The *key* point is that the statute was aimed specifically at finding evidence of *crime*. In contrast, prior administrative searches had focused on governmental interests such as health and safety. Moreover, the junkyard inspections were conducted by the *police*. In a later case the Court employed the balancing test to uphold sobriety roadblock checkpoints (134).

While the balancing approach provides flexibility in achieving significant government objectives, such as airline passenger safety, the danger exists that this approach will result in the "balancing" away of constitutional rights. Therefore, this analysis demands rigor. For example, while the Supreme Court upheld the drug testing of railroad employees after an accident and custom's officers involved in drug interdiction operations, it has struck down the drug testing of political candidates as mandated by a Georgia statute (135). The Court found that the justification for the latter procedure was simply

not compelling. In one case, which involved jail searches, the Supreme Court explained the "balancing" analysis as follows:

> The test of reasonableness under the Fourth Amendment is not capable of precise definition or mechanical application. In each case it requires a balancing of the need for the particular search against the invasion of personal rights that the search entails. Courts must consider the scope of the particular intrusion, the manner in which it is conducted, the justification for initiating it, and the place in which it is conducted (136).

Roe v. Marcotte (137), a Second Circuit case decided in 1999, can be used to illustrate this approach. In this case the court reviewed the Connecticut databank statute, which is limited to sex offenders. First, the court correctly found the government interest — solving past and future violent sex crimes — both legitimate and significant. Moreover, the databank system "may" deter future crimes by those whose profile is in the system. Second, the means selected to accomplish these objectives were reasonable. The state cited studies showing a high rate of recidivism for sexual offenders and DNA evidence is "particularly useful" in investigating these crimes "because of the nature of the evidence left at the scenes of these crimes and the demonstrated reliability of DNA testing" (138). Third, the blanket testing of all sex offenders eliminated the need for discretionary decisions, an historical concern in Fourth Amendment jurisprudence. Fourth, the intrusion — the extraction of blood — was slight ["minimal" in the Supreme Court's view (124)] and did not raise a health risk. In these circumstances, the court held that the balance tipped in favor of the databanking statute.

Three other aspects of the Connecticut scheme are noteworthy. First, trained medical personnel are required to take the blood sample. Second, the identifying information associated with the DNA profile remains anonymous until a match is made. Third, procedures limiting access to and dissemination of information in the system are specified.

"Special Needs" Searches. Over time, the rationale underlying administrative searches was extended to other procedures, commonly called "special needs" searches. The Supreme Court in *New Jersey v. T.L.O.* (139) applied this rationale to searches of public school children by teachers; the "special need" was the maintenance of a safe, orderly, and contraband-free school environment in order to create a healthy learning atmosphere. To achieve the desired environment, the Court recognized that "the school setting requires some easing of the restrictions to which searches by public authorities are ordinarily subject" (139, p. 340).

Similarly, in *Griffin v. Wisconsin* (140), the Supreme Court upheld a Wisconsin regulation that permitted a warrantless search of a probationer's home if there existed "reasonable grounds" to believe that the probationer possessed contraband. The Court observed that "[a] State's operation of a probation system, like its operation of a school, government office or prison, or its supervision of a regulated industry, likewise presents 'special needs' beyond normal law enforcement that may justify departures from the usual warrant and probable-cause requirements" and that "in certain circumstances government investigators conducting searches pursuant to a regulatory scheme need not adhere to the usual warrant or probable-cause requirements as long as their searches meet 'reasonable legislative or administrative standards'" (140, pp. 873–874).

Subsequently, the Court applied this rationale in cases involving government-required alcohol and drug testing for railroad employees (124) and customs agents involved in drug interdiction (141). In the school and probationer cases, the special need resulted in a lesser standard (reasonable suspicion instead of probable cause) to justify an invasion of privacy (142), while the drug testing cases upheld regulatory schemes that did not require any quantum of proof.

A number of courts have used the "special needs" rationale to uphold databank statutes (143–145). In contrast, other courts have balked at applying the "special needs" rationale in this context, noting that this category is limited to governmental objectives "beyond normal law enforcement" (146,147). These courts note that DNA databanks are intended *only* for law enforcement purposes. Other courts point out, however, that "special needs" searches, such as probationer searches, are also associated with law enforcement but do not involve the investigation of a specific crime (148).

More important, as noted above, the administrative search and "special needs" categories are not mutually exclusive — indeed, they often overlap. This is because the "special need" beyond normal law enforcement is typically some administrative objective. For example, an inventory search of the personal belongings of arrestees prior to placement in a jail cell is reasonable, whether classified as a "special need" or an administrative search (149). Similarly, this procedure could also be considered a prisoner search, the next category to be considered. The important point is the "balancing" rationale employed in determining reasonableness. There may, however, be a tendency in some opinions to use "special needs" as a talismanic incantation, curtailing further inquiry.

Fourth Amendment Rights of Prisoners. In *Jones v. Murray* (150), the Fourth Circuit adopted a third type of analysis. In upholding the Virginia statute, the Fourth Circuit relied on several Supreme Court decisions that had held that prisoners had reduced expectations of privacy under the Fourth Amendment. In *Bell v. Wolfish* (151), for example, the Supreme Court upheld the constitutionality of body cavity inspections of pretrial inmates following "contact visits," even in the absence of probable cause. The Court's rationale in determining the reasonableness of the procedure focused on the significant security dangers inherent in this environment: "A detention facility is a unique place fraught with serious security dangers. Smuggling of money, drugs, weapons, and other contraband is all too common an occurrence. And inmate attempts to secrete these items into the facility by concealing them in body cavities are documented in this record" (151, pp. 558–560).

In a later case, *Hudson v. Palmer* (152), the Supreme Court upheld cell searches ("shakedown" inspections) for the purpose of discovering contraband in a prison. The Court, in a 5–4 decision, ruled that a prisoner did not have a reasonable expectation of privacy in a cell. Yet, this holding (like *Wolfish*) was justified on institutional security needs. The Court wrote: "The recognition of privacy rights for prisoners in their individual cells simply cannot be reconciled with the concept of incarceration and the needs and objectives of penal institutions" (152, p. 526).

There are no institutional security needs in the databanking context, and thus this rationale is simply inapplicable. Indeed, some statutes apply even in the absence of incarceration (146). Moreover, both *Wolfish* and *Hudson* acknowledged that the Court's jurisprudence in prisoner rights cases recognizes the applicability of constitutional protections: "There is no iron curtain drawn between the Constitution and the prisons of this country" (153,154). Similarly, in another case the Court recognized that although lawful incarceration "brings about the necessary withdrawal or limitation of many privileges and rights," a retraction had to be "justified by the considerations underlying our penal system" (155).

Finally, these cases are, in the words of one court, "nothing more than special needs cases" (146). In sum, the administrative search rationale provides the best approach; it does not torture the "special needs" rationale, nor misapply the Fourth Amendment prison cases.

Inadequate Remedies. Under any rationale the most troublesome aspect of the databank statutes is the lack of meaningful remedies. As discussed previously, in many states there is little to no deterrent for unauthorized dissemination of DNA profiles and samples. The Virginia statute makes unauthorized dissemination of databank information a misdemeanor, but other statutes do not. By contrast, unauthorized disclosure of information in the federal databank system is punished by a $100,000 fine. While significant criminal penalties should be enacted, criminal prosecution may be insufficient. It often requires proof of intentional conduct, a standard that may be difficult to establish beyond a reasonable doubt. More important, prosecutors have enormous discretion in charging crimes, including the power not to charge at all — a distinct possibility considering the close relationship between prosecutors and the police.

Civil remedies should thus also be included in all databank statutes. Because databank statutes involve constitutional rights, civil rights suits under Section 1983 are possible, although not without impediments. Appropriate remedial models are not hard to find, however. For example, the federal eavesdropping and wiretap act provides for civil damages and injunctive relief in addition to felony sanctions (156). For some violations a plaintiff may recover actual damages or statutory damages of $100 a day or $10,000, whichever is greater. In addition, punitive damages are permitted as well as reasonable attorney's fees and other litigation costs (157). The Privacy Protection Act of 1980 provides for civil (actual) damages but not less than liquidated damages of $1000, reasonable attorney's fees, and other litigation costs (158).

Comparable provisions should be added to the databanking statutes. Proponents of databanking could not object to such provisions because they assert that violations will be few.

Expansion of Coverage Beyond Sex Offenders

Most databank statutes are limited to sex offenders. These provisions are supported by empirical research on recidivism; more than 50 percent of brutal and violent crimes, e.g., rape and murder, are carried out by repeat offenders (3). Recidivism is noted by several courts in upholding databank schemes (159). The nature of these offenses — their brutality and their often serial nature (3, p. 48) — is a critical point. However, some statutes also encompass homicides and other crimes of violence. Still others include all felons. The justification for including prisoners who have been convicted of white-collar felonies is difficult to discern. Even the sex offender category is problematic if it includes prostitution and public indecency as some statutes do.

Jones v. Murray (160), the Fourth Circuit case discussed above, addressed this issue because all felons are included in the Virginia DNA databank system. To buttress its position, the court cited recidivism studies encompassing all felons (161). The inmates, however, argued that the statistics on nonviolent felons undercut the state's position. The inmates' "statistics indicate[d] that 97% of the cases in which DNA evidence was used to link a defendant with a crime involved murder or rape, and further, less than 1% of all nonviolent offenders are later arrested on murder and rape charges" (162). In response, the majority merely noted that the percentages need not be high where the objective is significant and the privacy intrusion is limited.

The dissent in *Jones* believed that the distinction between violent and nonviolent felons was critical: "The only state interest offered by the Commonwealth for including non-violent felons is administrative ease," but such an interest does not suffice "to outweigh a prisoner's expectation of privacy in not having blood withdrawn from his body when that prisoner is not significantly more likely to commit a violent crime in the future than a member of the general population" (162, pp. 313–314). Indeed, the Virginia senate report concluded that the recidivism data *only* "supported the inclusion of plaintiffs convicted for felony sex offenses, assault, capital murder, first and second degree murder, voluntary manslaughter, larceny and burglary" (162, p. 314). All felons were added to make the databank "more efficient and cost effective." The dissent also pointed to other statistics in the record: "United States Justice Department statistics provided in the record show that only %0.4 of non-violent felons are later arrested on rape charges, and only %0.8 are later arrested on murder charges. One might assume non-violent drug offenders would be more likely to commit violent crime subsequent to release than other non-violent felons; yet, only %0.4 of them are later arrested for rape, and %0.3 for murder." The dissenting judge concluded: The lack of justification "leads me to a deep, disturbing, and overriding concern that, without a proper and compelling justification, the Commonwealth may be successful in taking significant strides toward the establishment of a future police state, in which broad and vague concerns for

administrative efficiency will serve to support substantial intrusions into the privacy of citizens" (162, p. 314).

The British experience, which commenced earlier than that of the United States, may be instructive. The British initially focused on sex offenses but later included burglaries and car theft because of the high number of matches. They found cross-over among offenses. According to one official, "People who commit serious crime very often have convictions for petty crime in their history" (163). While the cross-over concept is significant, the scope of the British system is breathtaking; they expect to "eventually include a third of all English men between 16 and 30, the principal ages for committing crimes."

The category of crimes subject to databanking should be supported by empirical data or persuasive reasons. There is apparently some support for including some nonviolent crimes (23), such as burglary, but each offense should be specified. For example, historically, burglary was not considered a "property" crime; it was a crime against habitation, intended to protect people in their dwellings (164). The crime that is the objective of the burglary need not be larceny or theft; it could be any felony including murder or rape. Burglars must anticipate what action they will take if surprised by an occupant, including the use of force. Therefore an argument to include burglary could be made, but felony tax evasion would be a different issue.

Expansion of Coverage to Arrestees

The Louisiana statute applies to sex offender and other specified *arrestees*. New York Police Commissioner Howard Safir has proposed that DNA be collected from *all* arrestees (23). Not satisfied with that proposal, New York Mayor Rudy Giuliani suggested that all newborns be tested (39). These proposals raise significant legal and policy problems.

Unlike a conviction, which is either based on a jury verdict and the "beyond a reasonable doubt" standard or a guilty plea with its attendant constitutional safeguards (including the right to counsel), *one* police officer can make an arrest based on his or her own view of probable cause, which is not a high standard. If the arrest occurs in a public place, an arrest warrant is not required (165). There is a requirement for judicial review of the probable cause determination within 48 hours (166,167), but the DNA sample will have been taken by then. In any event, this judicial review occurs in an ex parte procedure — that is, without the presence of the arrestee or defense counsel. The difference between an arrest and conviction is immense. In 1994, 65.3 percent of murder arrests resulted in conviction, but the conviction rates for some other crimes were much lower: robbery (39.3 percent), aggravated assault (14.1 percent), and burglary (38.8 percent) (168). Moreover, the FBI has reported that one-third of the initial rape suspects identified by the police are exonerated by DNA profiling (169); this statistic further underscores the difference between arrest and conviction. In short, the expansion of databank coverage to arrestees raises significant constitutional issues. Several theories that may be used to justify such expansion are discussed next and rejected.

Search Incident to Arrest. One possible theory for obtaining blood samples or cheek swabs would be a search incident to apprehension, a well-established exception to the warrant requirement (170,171). Under this exception, once a suspect has been arrested based on probable cause, a search of the arrestee's person and the area within her immediate control is permitted. In *Chimel v. California* (172), the Supreme Court set forth a twofold justification for this exception: (*1*) protection of the arresting officer and (*2*) prevention of the destruction of evidence. The search is automatic once there is an arrest; no additional showing is required.

The Supreme Court, however, has shown a greater concern about searches involving bodily intrusions than about other types of incident searches. In *Schmerber v. California* (173), the Court considered the constitutionality of extracting blood for the purpose of blood-alcohol analysis. The Court rejected the notion that the extraction of blood would automatically be encompassed by the search incident to arrest doctrine. According to the Court, the justifications underlying the search incident to arrest rule

> have little applicability with respect to searches involving intrusions beyond the body's surface. The interests in human dignity and privacy which the Fourth Amendment protects forbid any such intrusions on the mere chance that desired evidence might be obtained. In the absence of a *clear indication* that in fact such evidence will be found, these fundamental human interests require law officers to suffer the risk that such evidence may disappear unless there is an immediate search (173, pp. 769–770).

The Court further considered the necessity of securing a warrant based on probable cause as a prerequisite to the extraction of blood. It found the purpose underlying the warrant requirement—the intervention of a neutral detached magistrate between the police and the citizen—applicable to bodily intrusions: "The importance of the informed, detached and deliberate determinations of the issue whether or not to invade another's body in search of evidence of guilt is indisputable and great" (174). Nevertheless, because the alcohol content of blood diminishes with the passage of time, the Court recognized an "emergency" exception to the warrant requirement, which was necessary to preclude the destruction of evidence. This emergency exception, however, does not apply in other contexts—for example, when blood is sought for the purpose of genetic testing, including DNA profiling, a physical characteristic that remains constant (175,176).

Search for Blood. There may be grounds to take a DNA sample for testing *outside* the databank context. For example, if the crime that is the basis for an arrest involves blood, semen, or other evidence, there may be probable cause to issue a search warrant in that *specific case*. Often the probable cause to arrest also provides probable cause to search the arrestee for a DNA profile—for example, to compare the suspect's DNA with that from a semen stain in a rape case. A search warrant, as distinguished from an arrest warrant, requires probable cause that (*1*) the subject committed a crime and (*2*) blood analysis results would be evidence of that crime (177). In other words, the search is not automatic upon arrest, and this type of search differs from an administrative search (databanking), which is based on future or past crimes. Once a sample is obtained for this purpose, a search of a databank (for past offenses) would involve only a slight incremental privacy invasion (178).

Identification Rationale. Another rationale that has been suggested is an "identification" exception, applicable at a stationhouse "booking" after arrest. *Jones v. Murray* (179) alluded to this rationale: "[W]hen a suspect is arrested upon probable cause, his identification becomes a matter of legitimate state interest and he can hardly claim privacy in it. ...[T]he identification of suspects is relevant not only to solving the crime for which the suspect is arrested, but also for maintaining a permanent record to solve other past and future crimes" (179, p. 306). While *Jones* did not involve an arrestee, it went on to cite an analogy to fingerprinting—as have other databank cases (180,181).

Unquestionably, the proper identification of a person arrested is a legitimate governmental objective (182); it is not unusual for fugitives to use an alias (183,184). Moreover, fingerprinting arrestees is a reasonable method to accomplish this goal, as a number of courts (but not yet the U.S. Supreme Court) have recognized (185–187). Prior to the time that fingerprinting became routine, photographing (188) and the Bertillion system (based on physical measurements) were used for this purpose (189).

Nevertheless, the "identification" rationale is problematic for several reasons. First, there are significant differences between fingerprinting and DNA sampling. The former does not involve the kind of privacy issues raised by DNA samples (as opposed to profiles). As the Supreme Court has noted, "[f]ingerprinting involves none of the probing into an individual's private life and thought that marks an interrogation or search" (190). The fingerprint system is less intrusive and not as subject to abuse as some present methods of DNA sample collection. Second, the availability of fingerprinting undercuts the need to use DNA for the purpose of identification; every person whose DNA profile is in the database has fingerprints in AFIS. There are 226 million fingerprint cards in the FBI's Criminal Justice Information Services Division.

Third, fingerprinting arrestees as a means of identification developed before computers automated the process in AFIS. Prior to AFIS, the FBI used the Henry classification system, which was based on friction ridge patterns (e.g., arches, loops, whorls, ridge counting) and required prints from all *ten* fingers to identify arrestees. This took weeks if not months. In contrast to the classification of fingerprints, the identification of a partial crime scene print was based on ridge detail (e.g., ridge endings, bifurcations, enclosures) and required *suspects* because prints of all ten fingers are rarely left at a crime scene (191). In short, a single crime scene print could not be matched to the FBI's central depository. For example, a serial rapist, known as the "Westside Rapist" terrorized Cleveland in the 1980s. He was eventually convicted of raping 29 women (192). The police had partial fingerprints from several crime scenes, and the perpetrator had a record of prior convictions, but there was no way at that time to connect the prints without a suspect or suspects. Stated another way, there were no "cold hits" before AFIS. Accordingly, when the identification "exception" was judicially recognized, stationhouse fingerprinting did not solve past or future crimes. Consequently, the fingerprint "precedent" cannot be cited without further analysis.

Other Uses: Medical and Administrative Purposes

Some databank statutes do not limit use of DNA profiles or stored samples to criminal identification. For example, one statute authorizes databank samples to be used in medical research, even though informed consent has not been obtained from any of the subjects (48, p. 1491). Somewhat similarly, the Massachusetts statute permits disclosure for "advancing other humanitarian purposes" (193). Although many research uses of criminal databanks may ultimately prove to be ethically and legally acceptable, added safeguards such as mandatory approval by medical review boards are essential (17,19,44,47). The greater possibility that individual genetic information will be identified or released in the course of research also underscores the need for enhanced security for DNA samples and profiles described above, particularly civil remedies.

Military and Medical Records

In the future, one can expect criminal enforcement officials to turn to other repositories of DNA, such as military and medical databanks, to search for matches with crime scene profiles (47,195). Commentators have only begun to analyze the privacy issues raised by government use of DNA for law enforcement when the original biological samples were collected by other entities for entirely different purposes. While such a dramatic expansion of criminal DNA databanks seems susceptible to the legal challenges outlined above, there is currently neither judicial guidance nor academic consensus on whether this type of dramatic expansion to the criminal enforcement artillery will survive constitutional challenge (17,19,21,45,163). Additional legislative prohibitions or limitations seem inevitable if DNA databanks expand in this way (195).

CONCLUSION

DNA databanking offers a powerful tool for crime solution, especially in violent crime cases such as rape. Unfortunately, possible infringements of essential rights to privacy may also be made possible by the collection and centralization of individuals' genetic information. While the courts have begun to consider the constitutionality of DNA databank programs, other privacy concerns presented by the databank statues are likely to evade judicial scrutiny. There is thus need for more anticipatory legislation to ensure that the best of DNA databanks is harnessed and used for the good, without eroding the privacy rights that remain for those whose genetic

material is stored in state DNA databanks. As bioethicist Eric Juengist has remarked, "It is ... individuals' 'informational privacy' that is at stake in the prospect of widespread [state DNA databank laws], and it is in those terms that the policy challenge ... should be framed. What should society be allowed to learn about its citizens in the course of attempting to identify them?" (44, p. 63). While important steps have been taken over the past few years to improve state databank laws, work remains to be done.

BIBLIOGRAPHY

1. H. Levy, *And the Blood Cried Out: A Prosecutor's Spellbinding Account of the Power of DNA*, Basic Books, New York, 1996, p. 128.
2. *Jones v. Murray*, 962 F.2d 302, 304 (4th Cir.), *cert. denied*, 506 U.S. 977 (1992).
3. *Popular Science* **48**, 49 (1999).
4. 533 So. 2d 841 (Fla. Dist. Ct. App. 1988), *rev. denied*, 542 So. 2d 1332 (Fla. 1989).
5. *Office of Technology Assessment, U.S. Congress* 99, *Genetic Witness: Forensic Uses of DNA Tests*, U.S. Government Printing Office, Washington, DC, 1990.
6. P.C. Giannelli and E.J. Imwinkelried, *Scientific Evidence*, 3rd ed., Lexis Law Publishing, Charlottesville, VA, 1999.
7. P.C. Giannelli, *J. Crim. L. Criminol.* **88**, 380 (1997).
8. *State v. Council*, 515 S.E.2d 508, 518 (S.C. 1999).
9. G. Sensabaugh and D.H. Kaye, *Jurimetrics J.* **39**, 2–3 (1998).
10. *People v. Wesley*, 533 N.Y.S.2d 643, 644 (Co. Ct. 1988), *aff'd*, 633 N.E.2d 451 (N.Y. 1994).
11. R.C. Lewontin and D.L. Hartl, *Science* **254**, 1745–1746 (1991).
12. *United States v. Salameh*, 152 F.3d 88 (2d Cir. 1998); A. Blum, *Nat. Law J.* **8** (October 25, 1993).
13. E. Connors et al., *Convicted by Juries, Exonerated by Science: Case Studies in the Use of DNA Evidence to Establish Innocence after Trial*, National Institute of Justice, U.S. Department of Justice, Washington, DC, 1996.
14. G.W. O'Reilly, *Judicature* **81**, 114 (1997).
15. J.W. Diehl, *Jurimetrics J.* **39**, 431 (1999).
16. B. Dedman, *N.Y. Times*, October 7, 1 (1999).
17. G.J. Annas, *J. Am. Med. Assoc.* **270**, 2346 (1993).
18. Va. Code Ann. §19.2–310.2.
19. M. Hibbert, *Wake Forest Law Rev.* **34**, 767, 775 (1999).
20. 42 U.S.C. §1370.
21. J.S. Deck, *Vermont Law Rev.* **20**, 1057, 1065–1067 (1996).
22. R. Hoyle, *Nat. Biotechnol.* **16**, 987 (1998).
23. M. Hansen, *Am. Bar Assoc. J.* **85**, 26 (1999).
24. Colo. Rev. Stat. §17-2-201(5)(g).
25. Wash. Rev. Code §43.43.754; Mo. Stat. §650.055.
26. Ala. Code §36-18-24; N.M. Stat. Ann. §29-16-6; Va. Code Ann. §19.2-310.2; Wyo. Stat. Ann. §7-19-403.
27. La. Rev. Stat. Ann. §15-609.
28. Maricopa County Juvenile Action, 930 P.2d 496 (Ariz. Ct. App. 1996).
29. In re Mitchell, 880 P.2d 958 (Or. Ct. App. 1994), *cert. denied*, 522 U.S. 1004 (1997).
30. *Landry v. Attorney General*, 709 N.E.2d 1085 (Mass. 1999).
31. N.C. Gen. Stat. §§15A-266.1.

32. Code Mass. Regs. 55, §1.03-1.05.

33. *Shelton v. Gudmanson*, 934 F. Supp. 1048, 1050 (W.D. Wisconsin 1996).

34. Mass. Gen. Laws ch. 22E, §15.

35. J.E. McCain, *Am. J. Hum. Genet.* **56**, 1487–1489 (1995).

36. R.W. Schumacher, II, *Fordham Urb. Law J.* **26**, 1635, 1667–1668 (1999).

37. *U.S. News & World Rep.*, October 25, 33 (1999).

38. Y.H. Yee, *Am. J. Crim. Law* **22**, 461, 483 (1995).

39. M. Higgins, *Am. Bar Assoc. J.* **85**, 64, 66 (1999).

40. FBI, *Crime Lab. Dig.* **20**, 51 (1993).

41. National Research Council, *DNA Technology in Forensic Science*, National Academy Press, Washington, DC, 1992, p. 114.

42. R. Murch and B. Budowle, in M. Rothstein, ed., *Genetic Secrets: Protecting Privacy and Confidentiality*, Yale University Press, New Haven, CT, 1997, pp. 212–231.

43. U.S. General Accounting Office, *National Crime Information Center: Legislation Needed to Deter Misuse of Criminal Justice Information*, GAO/T-GGD-9341, U.S. Government Printing Office, Washington, DC, 1993.

44. E.T. Juengst, *Chicago-Kent Law Rev.* **75**, 64–67 (1999).

45. M.J. Markett, *Suffolk Univ. Law Rev.* **30**, 185, 208–215 (1996).

46. D.L. Burk, *Univ Tol. Law Rev.* **24**, 87, 91–92 (1992).

47. D. Kaye and E. Imwinkelried, *Forensic DNA Typing: Selected Legal Issues*, Report to the Working Group on Legal Issues, National Commission on the Future of DNA Evidence, Washington, DC, 2000, pp. 53–54.

48. J.E. McEwen and P.R. Reilly, *Am. J. Hum. Genet.* **54**, 941, 955–956 (1994).

49. *Ryncarz v. Eikenberry*, 824 F. Supp. 1493, 1502 (E.D. Wash. 1993).

50. *Jones v. Murray*, 962 F.2d 302 (4th Cir. 1992); *State v. Olivas*, 856 P.2d 1076 (Wash. 1993).

51. *Shaffer v. Saffle*, 148 F.3d 1180 (10th Cir. 1998).

52. *Boling v. Romer*, 101 F.3d 1336, 1340 (10th Cir. 1996).

53. *State v. Olivas*, 856 P.2d 1076, 1086–1067 (Wash. 1993).

54. *People v. Calahan*, 649 N.E.2d 588, 592 (Ill. App. 1995).

55. *People v. Wealer*, 636 N.E.2d 1129, 1137 (Ill. App. 1994).

56. State *ex rel. Juv. Dept. v. Orozco*, 878 P.2d 432 (Oregon 1994).

57. *State v. Olivas*, 856 P.2d 1076, 1080 (Wash. 1993).

58. 384 U.S. 757 (1966).

59. 388 U.S. 218 (1967).

60. 388 U.S. 263 (1967).

61. 410 U.S. 1 (1973).

62. *Fisher v. United States*, 425 U.S. 391, 408 (1976).

63. *Shaffer v. Saffle*, 148 F.3d 1180, 1181 (10th Cir.)(Oklahoma statute), *cert. denied*, 119 S.Ct. 520 (1998).

64. *Boling v. Romer*, 101 F.3d 1336, 1340 (10th Cir. 1996)(Colorado statute).

65. *Cooper v. Gammon*, 943 S.W.2d 699, 705 (Mo. App. 1997).

66. U.S. Const. art. I, §9, CL. 3 and §10.

67. *Collins v. Youngblood*, 497 U.S. 37, 42 (1990).

68. *Weaver v. Graham*, 450 U.S. 24, 29 (1981).

69. *Shaffer v. Saffle*, 148 F.3d 1180, 1182 (10th Cir. 1998), *cert. denied*, 119 S.Ct. 520 (1998).

70. *Gilbert v. Peters*, 55 F.3d 237, 239 (7th Cir. 1995).

71. *Jones v. Murray*, 962 F.2d 302, 309 (4th Cir. 1992).

72. *Rise v. Oregon*, 59 F.3d 1556, 1562 (9th Cir. 1995), *cert. denied*, 517 U.S. 1160 (1996).

73. *Calder v. Bull*, 3 U.S. Dall. 386, 390 (1798).

74. *Jones v. Murray*, 962 F.2d 302, 310 (4th Cir. 1992).

75. Mass. Gen. L. ch. 22E, §11.

76. 450 U.S. 24, 32 (1981).

77. *Gilbert v. Peters*, 55 F.3d 237, 239 (7th Cir. 1995).

78. *Ewell v. Murray*, 11 F.3d 482, 486 (4th Cir. 1993), *cert. denied*, 511 U.S. 1111 (1994).

79. *Jones v. Murray*, 962 F.2d 302, 309–310 (4th Cir. 1992).

80. *Cleburne v. Cleburne Living Ctr., Inc.*, 473 U.S. 432, 440 (1985).

81. *Chapman v. United States*, 500 U.S. 453, 465 (1991).

82. *Bankers Life and Cas. Co. v. Crenshaw*, 486 U.S. 71, 81 (1988).

83. 101 F.3d 1336 (10th Cir. 1996) (Colorado statute).

84. 856 P.2d 1076 (Wash. 1993).

85. *Rochin v. California*, 342 U.S. 165, 169 (1951).

86. *Washington v. Glucksberg*, 117 S.Ct. 2258, 2267 (1997).

87. *San Antonio Independent School District v. Rodriguez*, 411 U.S. 1, 17 (1973).

88. 352 U.S. 432 (1957).

89. *Mapp v. Ohio*, 367 U.S. 643 (1961) (exclusionary rule).

90. *Wolf v. Colorado*, 388 U.S. 25 (1949) (core value).

91. *Yanez v. Romero*, 619 F.2d 851, 854 (10th Cir.), *cert. denied*, 449 U.S. 876 (1980).

92. 384 U.S. 757 (1966).

93. 470 U.S. 753 (1985).

94. *County of Sacramento v. Lewis*, 523 U.S. 833, 849 n. 9 (1998).

95. *People v. Bracamonte*, 540 P.2d 624, 631 (Cal. 1975) (en banc).

96. *Yanez v. Romero*, 619 F.2d 851, 853 (10th Cir 1980).

97. Compare *Gagnon v. Scarpelli*, 411 U.S. 778 (1973)(probation revocation), with *Goss v. Lopez*, 415 U.S. 912 (1974) (school disciplinary hearing).

98. 59 F.3d 1556 (9th Cir. 1994).

99. *Cooper v. Gammon*, 943 S.W.2d 699, 706 (Mo. App. 1997).

100. 101 F.3d 1336 (10th Cir. 1997).

101. U.S. Const. amend. VIII.

102. 864 F.Supp. 496, 498 (E.D. N.C. 1994).

103. *Hudson v. McMillian*, 503 U.S. 1 (1992).

104. *Whitley v. Albers*, 475 U.S. 312, 319 (1986).

105. 864 F. Supp. at 500.

106. W.R. LaFave and A. Scott, *Substantive Criminal Law*, §2.14(f), West Publishing, St. Paul, MN, 1986.

107. *Cooper v. Gammon*, 943 S.W.2d 699, 707 (Mo. App. 1997).

108. 101 F.3d 1336 (10th Cir. 1996).

109. *Kruger v. Erickson*, 875 F. Supp. 583, 588 (D. Minn. 1995).

110. *Estelle v. Gamble*, 429 U.S. 97, 104–105 (1976).

111. *Ryncarz v. Eikenberry*, 824 F. Supp. 1493, 1502 (E.D. Wash. 1993).

112. *Skinner v. Ry. Labor Executives' Association*, 489 U.S. 602, 613–614 (1989).

113. *United States v. Dionisio*, 410 U.S. 1, 8 (1973) (grand jury subpoena for voice exemplar not a seizure).

114. *Skinner v. Ry. Labor Executives' Ass'n*, 489 U.S. 602, 619 (1989).

115. *United States v. Watson*, 423 U.S. 411 (1979).

116. 389 U.S. 347 (1967).

117. 489 U.S. 602 (1989).

118. *Bouse v. Bussey*, 573 F.2d 548, 550 (9th Cir. 1977).

119. *United States v. D'Amico*, 408 F.2d 331, 333 (2d Cir. 1969).

120. *State v. Sharpe*, 200 S.E.2d 44, 47–48 (N.C. 1973).

121. *United States v. Nicolosi*, 885 F. Supp. 50, 55 (E.D. N.Y. 1995).

122. *Cupp v. Murphy*, 412 U.S. 291, 294 (1973).

123. *United States v. Mara*, 410 U.S. 19, 21 (1973).

124. Skinner, 489 U.S. at 616.

125. Schlicher (NFN) Peters, I and II, 103 F.3d 940, 942 (10th Cir. 1996).

126. *Landry v. Attorney General*, 709 N.E.2d 1085, 1090 (Mass. 1999).

127. *People v. Wealer*, 636 N.E.2d 1129, 1132 (Ill. App. 1994).

128. 387 U.S. 523, 538 (1967).

129. *United States v. Biswell*, 406 U.S. 311 (1972).

130. *Donovan v. Dewey*, 452 U.S. 594 (1981).

131. *Marshall v. Barlow's Inc.*, 436 U.S. 307 (1978).

132. W.R. LaFave, *Search and Seizure: A Treatise on the Fourth Amendment*, §9.6(c), 3rd ed., West Publishing, St. Paul, MN, 1996.

133. 482 U.S. 691 (1987).

134. *Michigan Dept. of State Police v. Sitz*, 496 U.S. 444, 455 (1990).

135. *Miller v. Chandler*, 502 U.S. 305 (1997).

136. *Bell v. Wolfish*, 441 U.S. 520, 559 (1979).

137. 193 F.3d 72 (2d Cir. 1999).

138. 193 F.3d at 79.

139. 469 U.S. 325 (1985).

140. 483 U.S. 868 (1987).

141. *National Treasury Employees Union v. Von Raab*, 489 U.S. 656, 665–666 (1989).

142. *Terry v. Ohio*, 392 U.S. 1 (1968).

143. *Roe v. Marcotte*, 193 F.3d 72 (2d Cir. 1999).

144. *Shelton v. Gudmanson*, 934 F. Supp. 1048, 1051 (W.D. Wisconsin 1996).

145. *State v. Olivas*, 856 P.2d 1076, 1086 (Wash. 1993).

146. *People v. Wealer*, 636 N.E.2d 1129, 1135 (Ill. App. 1994).

147. *State v. Olivas*, 856 P.2d 1076, 1092 (Wash. 1993).

148. *Shelton v. Gudmanson*, 934 F. Supp. 1048, 1050–1051 (W.D. Wis. 1996)(Wisconsin statute).

149. *Illinois v. LaFayette*, 462 U.S. 640, 643 (1983).

150. 962 F.2d 302 (4th Cir. 1992).

151. 441 U.S. 520 (1979).

152. 468 U.S. 517 (1984).

153. *Wolff v. McDonnell*, 418 U.S. 539, 555–556 (1974).

154. P.C. Giannelli and F.A. Gilligan, *Va. Law Rev.* **62**, 1045 (1976).

155. *Price v. Johnston*, 334 U.S. 266, 285 (1948).

156. 18 U.S.C. §2521.

157. 18 U.S.C. §2520(b)(2) & (3).

158. 42 U.S.C. §2000aa.

159. *Shelton v. Gudmanson*, 934 F. Supp. 1048, 1051 (W.D. Wis. 1996).

160. 962 F.2d 302 (4th Cir. 1992).

161. A.J. Beck and B.E. Shipely, *Recidivism of Prisoners Released in 1983*, U.S. Dept. of Justice, Office of Justice Programs, Bureau of Justice Statistics, Washington, DC, 1989, p. 1.

162. 962 F.2d at 308.

163. N. Wade, *N.Y. Times*, October 12, A1 (1998).

164. W.R. LaFave and A. Scott, *Substantive Criminal Law*, West Publishing, St. Paul, MN, 1986.

165. *United States v. Watson*, 423 U.S. 411 (1976).

166. *County of Riverside v. McLaughlin*, 500 U.S.44 (1991).

167. *Gerstein v. Pugh*, 420 U.S. 103 (1975).

168. *N.Y. Times, 1998 Almanac* 323 (1997)(citing Bureau of Justice Statistic Bull. 1994).

169. FBI, *The Application of DNA Testing to Solve Violent Crimes* (1993).

170. W. LaFave, *Search and Seizure: A Treatise on the Fourth Amendment*, West Publishing, St. Paul, MN, 1996.

171. E. Imwinkelried et al., *Courtroom Criminal Evidence*, 3rd ed., Michie, Charlottesville, VA, 1998.

172. 395 U.S. 752 (1969).

173. 384 U.S. 757 (1966).

174. 384 U.S. at 770.

175. *Graves v. Beto*, 301 F. Supp. 264, 265 (E.D. Tex. 1969), *aff'd*, 424 F.2d 524 (5th Cir.), *cert. denied*, 400 U.S. 960 (1970).

176. *Mills v. State*, 345 A.2d 127, 132 (Md. App. 1975), *aff'd*, 28 Md. 262, 363 A.2d 491 (1976).

177. In re Lavigne, 641 N.E.2d 1328, 1331 (Mass. 1994).

178. H.J. Krent, *Tex. Law Rev.* **74**, 49, 53 (1995) (for a different view).

179. 962 F.2d 302 (4th Cir. 1992).

180. *Landry v. Attorney General*, 709 N.E.2d 1085, 1092 (Mass. 1999).

181. *Olivas*, 856 P.2d 1076, 1093 (Wash. 1993).

182. *United States v. Laub Baking Co.*, 283 F. Supp. 217, 221 (N.D. Ohio 1968).

183. *United States v. Valencia-Lucena*, 925 F.2d 506, 513 (1st Cir. 1991).

184. *United States v. Boyle*, 675 F.2d 430, 432 (1st Cir. 1982).

185. *Napolitano v. United States*, 340 F.2d 313, 314 (1st Cir. 1965).

186. *Smith v. United States*, 324 F.2d 879, 882 (D.C. Cir. 1963).

187. *United States v. Krapf*, 285 F.2d 647 (3d Cir. 1961).

188. *People v. Sallow*, 165 N.Y.S. 915 (Ct. Gen. Sess. of Peace 1917).

189. *Bartletta v. McFeeley*, 152 A. 17 (N.J. 1930).

190. *Davis v. Mississippi*, 394 U.S. 721, 727 (1969).

191. *United States v. Laub Baking Co.*, 283 F. Supp. 217, 223 (N.D. Ohio 1968).

192. J. Neff, *Unfinished Murder: The Capture of Serial Rapist*, Pocket Books, New York, 1995.

193. Mass. Gen. ch. 22E, §10(d)(4).

194. A. Lippman, *Am. J. Pub. Health* **86**, 1030 (1996).

195. R.C. Scherer, *Geo. Law J.* **85**, 2007, 2016 (1997).

See other entries GENETIC INFORMATION, ETHICS, ETHICAL ISSUES IN TISSUE BANKING AND HUMAN SUBJECT RESEARCH IN STORED TISSUES; GENETIC INFORMATION, ETHICS, PRIVACY AND CONFIDENTIALITY: OVERVIEW; GENETIC INFORMATION, LEGAL, ERISA PREEMPTION, AND HIPAA PROTECTION; GENETIC INFORMATION, LEGAL, FDA REGULATION OF GENETIC TESTING; GENETIC INFORMATION, LEGAL, GENETIC PRIVACY LAWS.

GENETIC INFORMATION, LEGAL, ERISA PREEMPTION, AND HIPAA PROTECTION

MARY ANNE BOBINSKI
University of Houston Law Center
Houston, Texas

OUTLINE

INTRODUCTION

Researchers are constantly discovering new linkages between human traits, diseases, or conditions and human genes. The information obtained through the Human Genome Project is expected to be used for human benefit. But should the use of this information be restricted in any way, and if so, by whom? People might be reluctant to gather useful information about their genetic heritage if there is a chance this information will be used to their detriment by others. "Genetic discrimination" has a long history in the United States and is defined as "discrimination against an individual or a member of an individual's family solely on the basis of that individual's genotype" (1). Fears about genetic discrimination could limit the utility of new genetic discoveries.

A number of states have attempted to protect individuals by enacting legislation restricting the use of genetic information by employers, insurance companies, and others (Tables 1 and 2, Appendix). At the same time the federal government has enacted statutes that affect the necessity for, and utility of, state intervention. One federal statute in particular, the Employee Retirement Income Security Act of 1974 (ERISA), has created difficulties for states that wish to regulate the use of genetic information (2). The federal government recognized the difficulties faced by the states and enacted the Health Insurance Portability and Accountability Act (HIPAA) (3). This Act directly regulates the permissible use of genetic information by employment-based plans and authorizes states to engage in additional regulatory action in this area. This article is devoted to examining the roles of ERISA, HIPAA, and state statutes in regulating the use of genetic information. The discussion concludes that federal legislation has provided an important supplement to state regulation of the use of genetic information.

GENETIC INFORMATION

Although the technology for genetic testing has only recently become accessible, genetic information has been available for a long time. People have long understood the general concept of heredity and have been aware that some conditions, such as hemophilia, tend to be inherited within families. The general concept of hereditable conditions gained a new scientific foundation with Watson and Crick's description of the structure of DNA in the 1950s. Since that time scientists have been able to make an ever-increasing number of connections between the structure of an individual's DNA and that individual's traits or the expression of a variety of genetic conditions. What this means, first of all, is that the problems presented by the availability of genetic information are not new. Some genetic information has been available for use, properly or improperly, and with good or bad intent, for much of this century. Individuals with a family history of a known genetic condition have used this information to make important decisions, such as decisions about whether to reproduce or whether to provide health or life insurance for a member of the family.

States have also used this information to sanction the sterilization of individuals based on their genotype. For example, in 1927 the United States Supreme Court upheld the constitutionality of a Virginia statute that allowed the involuntary surgical sterilization of Carrie Buck, a female resident in a Virginia institution for the "feebleminded." In this now infamous case, *Buck v. Bell*, Justice Holmes concluded that "it is better for all the world, if instead of waiting to execute degenerate offspring for crime, or to let them starve for their imbecility, society can prevent those who are manifestly unfit from continuing their kind.... Three generations of imbeciles are enough" (4,5). Although modern courts are not likely to uphold state-mandated sterilization programs, this decision vividly demonstrates past misunderstandings about the nature of the hereditability of various traits as well as the misuse of that information to infringe individual rights.

While the availability of genetic information has a long history, the technology used to obtain this information has undergone rapid change. Advances in genetic research have improved the accuracy of attempts to relate the content of one's genetic code with its real life effects. In addition research has created an exponential growth in the number of conditions with an identified genetic correlate. This has enabled scientists to test for the presence of the particular gene or gene combination. Within the next several decades we are likely to have access to over 100 different tests for genetic variations that predispose persons to common diseases such as cancer, cardiovascular diseases, autoimmune diseases, and so on (6).

Table 1. Selected Recent State Legislative Genetic Information Initiatives

State (year enacted)	Type of Genetic Information	Area of Protection	Type of Protection
Alabama (1997)	Genetic test for predisposition to cancer	Health benefit plans (insured, self-insured)	No genetic testing or use of result for coverage or rates
Alaska (1997)	HIPAA-related provisions; genetic information	Health insurers	No discrimination based on health status; health status included genetic information; no preexisting condition unless genetic condition diagnosed
Arizona (1997)	Genetic condition	Life or disability insurance contracts	No coverage or rate discrimination unfair unless actuarial guidelines met
Arkansas (1997)	HIPAA-related provisions; genetic information	Group health plans	Genetic information not a preexisting condition without diagnosis of condition related to the information; no discrimination based on health status; health status includes genetic information
California (1994, 1995, 1996, 1998)	Asymptomatic genetic condition or predisposition	Health benefit plans (insured, self-insured, and MEWA) Disability insurance for medical expenses also regulated	No nontherapeutic use of information; confidentiality protections
Colorado (1994)	Direct tests for presence or absence of alterations in genetic material associated with disease	Health, group disability, long-term care insurance Does not cover life insurance or individual disability insurance	Therapeutic use permitted; no underwriting use in covered insurance
Connecticut (1997)	Information about genes or inherited characteristics derived from individual or family member	Individual or group health insurance	No use for coverage or rate determinations Can refuse to cover or can apply preexisting clause to person with symptomatic genetic disease.
Delaware (1998)	Broad definition of genetic information "about inherited genes or chromosomes, and of alterations thereof, whether obtained from an individual or family member, that is scientifically or medically believed to predispose an individual to disease, disorder or syndrome or believed to be associated with a statistically significant increased risk of development of a disease, disorder or syndrome"	Informed consent and confidentiality Health Insurance	Restricts access to genetic information, except where person consents or otherwise permitted by law Insurers are permitted to have access under some circumstances Prohibits discrimination in access or rates of health insurance based on genetic characteristics
Florida (1997)	Information from genetic testing for asymptomatic persons DNA analyses	Health benefit insurers (insured, self-insured plans) No protections for coverage and rate determinations of life insurance, disability income policies, long-term care policies, or certain other insurance policies Public or private entities performing DNA analyses	Health insurers cannot use genetic information in coverage or rate determination unless diagnosis of disease Other than in statutorily limited situations, public or private entities performing DNA analysis must obtain informed consent of person and result is property of individual

Table 1. *Continued*

State (year enacted)	Type of Genetic Information	Area of Protection	Type of Protection
Georgia (1995)	Genetic testing for asymptomatic genetic conditions.	Health and sickness plan payers	Therapeutic use only
		Does not cover self-insured plans subject only to ERISA	Confidentiality protection
		Statute does not protect against use in life insurance, disability income policies, long-term care, Medigap, and other policies	
Hawaii (1997)	Genetic information of individual or family member	Health Insurance. Does not apply to life insurance, disability income insurance, and long-term care insurance	May not use genetic info for coverage or rate determinations
	HIPAA provisions		Confidentiality protection
		Group and individual health insurance	Implements HIPAA
Idaho (1997)	Small employer reforms	Small employers	Limits on use of health status
Illinois (1997)	Genetic testing for abnormalities or deficiencies linked to current disorders or susceptibility. No confidentiality protection for determination that person suffers from disease, "whether or not currently symptomatic."	Accident, health insurance policies	Accident and health insurers may not seek or use for nontherapeutic purposes
		Employers	Confidentiality protection
			Individual can release favorable results to insurers for consideration
			Employers may use if consistent w/ADA
Indiana (1997)	Results of genetic tests	Insurers other than life insurers	Insurance companies not entitled to access to genetic test results unless individual gives specific written consent
	Direct genetic screening or testing of individual's genes for defects linked to disorder or susceptibility or damage	Health care services coverage	Non-life insurer may not use individual or family members genetic testing results to determine coverage or rates
	Does not cover detection of genetic disorder through its manifestation		Insurer may consider favorable results released by individual
			Confidentiality protection
Iowa (1992, 1998)	Genetic Testing in Employment	Employers, labor organizations, licensing authorities	No mandatory pre-employment genetic testing; employee can volunteer and give informed consent for genetic testing related to occupational risks
	Small group reforms	Small groups	
			Restrictions on use of health status, including genetic information
Kansas (1997)	Genetic screening or tests	Health insurers, HMOs	Covered entities should not request test or condition insurance on testing; covered entities should not establish rates based on results
	HIPAA -related	Group and individual policies must be renewed without consideration of health status, which includes genetic information	Life insurance, disability, and long term care coverage must set rates based on reasonable risks
			Life insurers not covered

(continued)

Table 1. *Continued*

State (year enacted)	Type of Genetic Information	Area of Protection	Type of Protection
			Group and individual policies must be renewed without consideration of health status, which includes genetic information
Kentucky (1998)	Genetic testing or information	Group or individual health plan; group insurers; disability income insurers	"A group or individual health benefit plan or insurer offering health insurance in connection with a health benefit plan or an insurer offering a disability income plan may not request or require an applicant, participant, or beneficiary to disclose to the plan or insurer any genetic test about the participant, beneficiary, or applicant"
Louisiana (1997)	Genetic information is all information about genes, inherited characteristics, or family history/pedigree expressed in common language	Health insurers, including employee benefit plans Life disability income, and long-term care insurance policies excluded	Health insurers may not use genetic information of individual or family member in coverage or rate determinations Confidentiality protection
Maine (1997)	Genetic tests or results	Employers Health insurers Life, disability, long term care insurers	Employers prohibited from discriminating based on refusal to take genetic test or results of genetic test unless bona fide occupational qualification Health insurers cannot use genetic test results to discriminate Life, disability, long-term care insurers may discriminate only if reasonably related to expected claims experience
Maryland (1997)	Genetic test used to identify alterations in genetic material associated with disease or illness	Health insurance policies or contracts Section does not apply to life insurance policies, annuity contracts or disability insurance policies	May not use genetic test results for coverage or rate determinations Confidentiality protections
Minnesota (1995)	Presymptomatic test of genes associated with genetic conditions or predispositions	Health plan companies Life insurance policies	Health plan companies may not require or use genetic tests in coverage and rate determinations Life insurance companies may require use of genetic test but informed consent and confidentiality provisions apply
Missouri (1998)	Genetic testing and information	Insurers, employers	Insurers cannot require tests or consider results Excludes disability income insurance and long-term care insurance Employers can use genetic information when directly related to job responsibilities Confidentiality protections

Table 1. *Continued*

State (year enacted)	Type of Genetic Information	Area of Protection	Type of Protection
Montana (1991, 1999)	Genetic condition Genetic information and genetic tests	Life and disability insurance Individual or group insurers	Coverage and rate discrimination permitted only where "substantial" differences in claims likely Insurers may not require genetic testing Insurers may not discriminate in coverage or rates Provisions do not apply to life insurance, disability insurance, or long-term care insurance
Nebraska (1997)	Genetic information	Individual and group insurers	Prohibits discrimination based on health status, which includes genetic information
Nevada (1997)	Genetic information is that obtained from a genetic test to determine abnormalities linked to disorder or susceptibility to disease	Health insurance Disability income and long-term care coverage excluded	Health insurers cannot require test for individual or family members and cannot use results in coverage and rate determination
Nebraska (1997, 1996)	Genetic information or a typical hereditary cellular or blood trait; genetic testing Genetic characteristics are inherited gene or chromosome or alteration thereof scientifically or medically believed to predispose individual to disorder or to predispose to disorder	Employers Health insurance; life insurance and annuities	Employers cannot use genetic information, etc., as basis for discrimination Confidentiality protections Genetic characteristics cannot be used in coverage and rate determinations for health insurance Discrimination in life insurance must be reasonably related to expected claims experience
Nevada (1997)	Genetic information and testing HIPAA	Group and individual health insurers HIPAA Implementation	Health insurers prohibited from requiring genetic tests or from using the results of such tests to discriminate in coverage or rates Long-term care or disability insurance not covered HIPAA Implementation
New Hampshire (1995)	Genetic Testing	Employers, labor organizations	Cannot require genetic testing or use to affect terms and conditions of employment Individual can consent to tests for susceptibility to workplace chemicals if employer takes no adverse actions based on results Genetic tests can be used to determine insurability for life, disability income, or long-term care insurance as part of employee benefit plan

(continued)

Table 1. *Continued*

State (year enacted)	Type of Genetic Information	Area of Protection	Type of Protection
	Genetic information	Health insurers Life, disability income, and long-term care insurance	Cannot require individual or family member to undergo testing and cannot use information in coverage or rate determinations Life, disability income, and long-term care insurers can use information
New Jersey (1996, 1997)	Genetic Information and Testing Employers not permitted to require genetic tests Genetic characteristics HIPAA implementation	Restrictions on testing apply to all "persons" Special rules for life insurance or disability income insurance. Employers Group and individual insurers Group and individual coverage	Informed consent required for most genetic testing Commissioner on Banking and Insurance to establish regulations Limits ability of employers to require testing or require access to test results Prohibits discrimination in coverage or rates based on genetic characteristics HIPAA implementation
New Mexico (1998)	Genetic Information Privacy Act	Applies to most "persons"	Genetic analysis prohibited without informed consent Life, disability, long- term care entities are exempted Prohibits discrimination based on genetic analysis, genetic information, or genetic propensity, except that life, disability income, and long-term care insurers are permitted to make actuarily reasonable adjustments
New York (1996)	Genetic predisposition Genetic tests, genetic predispositions	Employers All persons; insurance companies	Unlawful for employers to discriminate against individuals based on their disability, genetic predisposition, or carrier status Informed consent and confidentiality required; special rules for consent and confidentiality for insurers
North Carolina (1997)	Genetic information: from individual or family member, about genes, gene products or inherited characteristics HIPAA amendments	Health benefit plans, including all those where regulation permitted by ERISA Excluded types of plans include disability income, long-term care, and Medigap coverage Persons, corporations, etc. Group and individual plans	Insurers may not make coverage or rate determinations based on genetic information No person corporations, etc., can engage in employment discrimination based on genetic information about the individual or family members HIPAA conforming amendments
North Dakota (1997)	Genetic information	Hospital and medical insurance	Genetic information is not a preexisting condition absent a diagnosis

Table 1. *Continued*

State (year enacted)	Type of Genetic Information	Area of Protection	Type of Protection
Ohio (1997, repealed effective 2004; replacement provision)	Genetic screening or testing	Health insurers; government self-insurers	Health insurers cannot require testing or use results unless favorable results volunteered by applicant
			Provision repealed effective 2004; protections for genetic information obtained before 2004 remain
Oklahoma (1998)	Genetic Nondiscrimination Insurance Act; genetic information means result of a genetic test and does not include family history	Health and accident insurance, not disability income or long-term care	Prohibits discrimination based on genetic information "except to the extent and in the same fashion as an insurer limits coverage, or increases premiums for loss caused or contributed to by other medical conditions presenting an increased degree of risk"
			Insurers may discriminate based on manifestations of conditions
			Weak confidentiality protections
Oregon (1995, 1997)	Genetic characteristic: gene or chromosome or alteration thereof believed to cause or predispose to disease	Insurance providers include all those subject to state regulation; health providers	Informed consent and confidentiality protections
			Insurers cannot use favorable genetic tests as inducement to purchase insurance
	Genetic information can be about individual or family		Genetic information cannot be used negatively in hospital and medical expense insurance
Pennsylvania (1996)	PKU and Insurance	Insurance companies	Insurance must cover PKU-related formula
Rhode Island (1997)	Genetic testing	Individual or group coverage, HMOs, nonprofit health corps and insurers	No use of genetic tests or results to affect coverage or rates for health coverage
			Disability income and long-term care policies not covered
South Carolina (1998)	HIPAA Implementation	MEWAs	Impermissible "health status" discrimination includes use of genetic information
	Genetic privacy act protecting genetic characteristics and genetic information	Health coverage	Informed consent and confidentiality provisions; prohibits discrimination in coverage or rates based on genetic information
			Excludes disability income, long-term care coverage, and other nongeneral health insurance types of policies
South Dakota (1997)	Genetic information and preexisting medical conditions	Individual, group, and small employer policies	Preexisting condition cannot include genetic information unless related condition has been diagnosed

(continued)

Table 1. *Continued*

State (year enacted)	Type of Genetic Information	Area of Protection	Type of Protection
Tennessee (1997)	Genetic information; carrier status, genes that cause or predispose to disease Questions about family history excluded Insurers are permitted to ask questions about health of applicant and family HIPAA implementation	Insurance providers, but focused on medical or health insurance; excluding life insurance, disability income, long-term care policy and other types of insurance Small group market, individual and group coverage	No coverage or rate discrimination. Confidentiality protection Restrictions on ability of insurer or plan to consider health status; health status includes genetic information
Texas (1997)	Genetic information: that derived from genetic test for genes associated with predisposition to disorder	Employers, licensing authorities; group health benefit plans as permitted by ERISA Excluded insurance plans: specific disease plans; accidental death or dismemberment; Medigap, works compensation, long-term care, etc.	Discrimination based on genetic information or refusal of testing prohibited by employers, group health benefit plans, and licensing authorities Confidentiality protection. Insurers may not coerce abortion in pregnant women carrying children with genetic conditions
Vermont (1997)	Results of genetic testing Genetic testing in employment, licensure, and insurance	Business of insurance Employers, licensing authorities, insurers	Genetic test results can be used where there is a reasonable relationship between the information and anticipated claims experience Employers or licensing authorities cannot require or use genetic test results Health insurers cannot require testing or use results in underwriting Disability or long-term care coverage excluded
Virginia (1996)	Genetic Information HIPAA-related reforms on genetic information and health status	Health insurers Insurer issuing group or individual coverage	No discrimination in coverage or rates based on genetic information Confidentiality protections Disability income insurance excluded Genetic information cannot constitute preexisting condition unless diagnosed condition
Washington (1988)	PKU-related provisions	Insurers, HMOs	Coverage of PKU-related formula
West Virginia (1997)	HIPAA implementing provisions	Various health plans	Limitation on use of health status; health status includes genetic information; preexisting medical condition only where genetic condition diagnosed
Wisconsin (1991, 1997)	Genetic test for disease or predisposition Genetic tests or information HIPAA-related amendments	Insurer or self-insured governmental programs Employment Various insurers or plans	Cannot require testing or revelation of results and cannot use in coverage and rate determinations Use in life insurance and income continuation insurance must be reasonably related to risks

Table 1. *Continued*

State (year enacted)	Type of Genetic Information	Area of Protection	Type of Protection
			Employers may not require or use genetic tests; employees may request and provide informed consent for genetic tests related to occupational safety issues
			Genetic information not a preexisting condition without diagnosis of condition
Wyoming (1997, 1998)	HIPAA-related amendments	Group and small employer health plans	Insurers cannot discriminate based on health status; health status includes genetic information; no preexisting condition unless genetic condition diagnosed

Note: The table was first created using a Westlaw search of the state legislative database in Fall 1997. The table was updated with an additional Westlaw search in Fall 1999. The search results were compared with a separate survey published in Ref. (25).

There are at least two sources of genetic information: the general medical records of the individual or his or her family (which may reveal information about genetic disease), and the results of specific genetic tests performed on the individual or his or her family. Policy makers interested in protecting genetic information must regulate both sources. Advances in genetic testing have further complicated the issue by providing information about an individual's susceptibility to particular genetic conditions. An individual may be determined to be a "carrier" of a gene, that is, able to pass the trait on through reproduction but at no risk for expression of a genetic condition. For example, individuals can carry the gene associated with Tay-Sachs without personality experiencing the disease. Alternatively, genetic testing can reveal that a person has a genetic condition or disease ("diagnostic testing"), such as hemochromatosis. It can also reveal that a person will develop a genetic condition at some future point ("predictive testing"), such as Huntington's disease, or that the person is at greater-than-average risk for developing the condition or disease being tested, such as certain types of colon or breast cancer (7).

USE OF GENETIC INFORMATION BY EMPLOYERS, INSURERS, AND OTHERS

Buck v. Bell involved the use of genetic information by the state, which mandated sterilization as a method of protecting the public welfare. But private entities, such as insurance companies and employers, have also been interested in obtaining genetic information. These private parties have an interest in using the genetic information to make decisions about employment, insurance, and other issues. Employers might want to exclude potential employees who present higher health insurance costs or who might be susceptible to workplace injury or illness (8). In the 1970s, for example, employers began screening workers for sickle cell anemia, which led to stigmatization and discrimination against sickle cell anemia carriers in employment (1). Insurance companies selling health,

disability, life, long-term care, or other types of policies might want to exclude applicants or raise premiums for those who have higher rates of illness, disability, or premature death (7,8).

Individuals who anticipate these forms of discrimination and social stigmatization will have an interest in restricting access to genetic information. There are some circumstances, however, where an individual might want to release genetic information to employers or others in order to gain more favorable treatment than they would otherwise receive. This discussion is concerned with the state and federal regulation of the use of genetic information by private parties in the context of health, life, and other types of insurance. The central issue in this complex web of regulation is control. Who will control the decision about whether to undergo genetic testing, the individual or some third party such as an employer or an insurer? Who will control access to and use of genetic information: the individual or a third party?

A number of different commissions, working groups, and other organizations have concluded that it is important to restrict access to genetic information. The view of the Task Force on Genetic Testing created by the National Institutes of Health (NIH) and Department of Energy (DOE) Working Group on Ethical, Legal, and Social Implications of Human Genome Research is typical:

Protecting the confidentiality of information is essential for all uses of genetic tests.... Results should be released only to those individuals for whom the test recipient has given consent for information release.... Under no circumstances should results ... be provided to any outside parties, including employers, insurers, or government agencies, without the test recipient's written consent....No individual should be subjected to unfair discrimination by a third party on the basis of having a genetic test or receiving an abnormal test result. (9, pp. 14–15)

The common belief that the privacy and confidentiality of genetic information must be protected to encourage individuals to undergo testing does not resolve all questions.

Table 2. Legal Documents Used for Table 1

Alabama Stat. §§27-53-1 to 27-53-4 (Alabama 1999) (effective 1997)

Alaska Stat. §§21.54.100, 21.54.110 (Alaska & Mathew Bender 1999) (HIPAA-related provisions enacted 1997)

Arizona Revised Statues Annotated §20-448 et seq. (West 1999) (main genetic provisions enacted 1997). See also, 2000 Ariz. S.B. 1330 (amendments enacted during publication process).

Ark. Code Ann. 23-86-subch. 3 (Arkansas 1999) (HIPAA-related provisions enacted 1997)

Cal.Civ.Code Ann. §56.17 (West 1999) (main provisions enacted in 1995 and 1996)

Cal. Health & Safety Code 1374.7 (West 1999) (multiple enactments and amendments, including 1995, 1996, 1998)

Cal.Health & Safety Code Ann. §124975 et seq. (West 1999) (enacted 1995)

Cal. Ins. Code §742.405, 10123.3, 10140, 10143, 10146 et seq. (West 1999) (multiple amendments, including 1994, 1995, 1996, 1998)

Colorado Rev. Stat.Ann. §10-3-1104.7 (West 1999) (enacted 1994)

Connecticut General Statutes Annotated §§38a-476, 38a-476a (West 1999) (enacted 1996 for HIPAA conformity)

Delaware Code Annotated, Title 16, §1220-1227 (Del. 1998) (informed consent and confidentiality for genetic information) (effective 1998)

Delaware Code Annotated, Title 18, §2317 (Del. 1998) (prohibiting discrimination in health insurance) (effective 1998)

Florida Statutes Annotated §§627.4301, 760.40 (West 1999) (enacted 1997)

Official Code of Georgia Annotated §33-54-1 to -8 (1997) (enacted 1995)

Haw. Rev. Stat. Ann. §§431:10A-118 (individual health insurance); 432:1-507 (group health coverage); 432:1-607 (mutual benefit societies); 431:2-201.5 (HIPAA) (West 1999) (enacted 1997)

Id. Code §§41-4708 (Lexis 1999) (amendments including 1997)

Illinois Compiled Stats. Ann. 215, §5/356v (West 1999) (effective 1998)

Illinois Compiled Stats. Ann. 410, §§513/5, 15, 20, 30 (West 1999) (effective 1998)

Burns Indiana Code Ann. §16-39-5-2 (West 1999) (amended 1997)

Burns Indiana Code Ann. §27-8-26-1 to 11 (West 1999) (enacted 1997)

Iowa Code §§5132B.9A, 513B.10 (small group reforms, 1997); §729.6 (genetic testing in employment, 1992)

Kansas Statutes Annotated §40-2209, -2257, -2259 (Revisor, Kansas 1998) (sections amended or enacted 1997, 1998)

Kentucky Rev. Stat. Ann. §§304.12-085 (West 1999) (enacted 1998)

Louisiana Rev. Stat. Ann. §§22:213.7, 22:250.1 to 22:250.16, 22:1214 (West 1999) (provisions enacted 1997)

Maine Rev. Stat. Ann. 5:§19302, 24:§2159-C (West 1999) (enacted 1997)

Maryland Ins. Code Annotated §§27-208; 27-909 (1997)

Minnesota Stat. §72A.139 (West 1999) (enacted 1995)

Vernon's Missouri Statutes Ann., 24:375.1300 to .1312 (West 1999) (enacted 1998)

Montana Code Anno. §33-18-206 (enacted 1991), 33-18-901 to 904 (enacted 1999) (West 1999)

Neb.Rev.St. §§44-787, 44-6910, 44-6915,44-6916 (Neb. 1999) (enacted and amended, 1997, 1998, 1999)

Nevada Rev. Stat. §689A, §689B.420, §695B.069, §695.317 (enacted 1997) (West 1999)

New Hampshire Revised Statutes Annotated 141-H:1 to H:6 (N.H. and Lexis 1999) (enacted 1995)

Table 2. *Continued*

New Jersey Statutes Ann. 10:5-12 (non-discrimination in employment); 10:5-43 to 10:5-49 (genetic privacy act); §17:48-6.18 (individual or group health policies); 17:48A-6.11 (medical service corporation contracts); 17:48E-15.2 (health service contracts); 17B:26-3.2 (individual health policies); 17B:27-36.2 (group health insurance); 17B:27-54 (HIPAA implementation); 17B:30-12 (insurance trade practices) (West 1999) (most provisions enacted in 1996, HIPAA implementation in 1997)

New Mexico Statutes, Ch. 24, Art. 21 (N.M. 1999) (enacted 1998)

McKinney's New York Civ. R.Law §79-1 (consent and confidentiality); McKinney's New York Ins. Law §2612, 3221 (insurance consent and confidentiality); McKinney's New York Exec. Law §292,296 (employers; discrimination) (West 1999) (many provisions enacted in 1996)

North Carolina Gen. Stat. §58-3-215 (genetic information in health insurance); §95-28.1A (prohibiting discrimination in employment); Ch. 58, Art. 68 (HIPAA amendments) (Lexis 1999) (most enactments 1997)

North Dakota Century Code 26.1-36.4-03.1 (Lexis 1999) (enacted 1997)

Ohio Rev. Code Annotated §1751.64-65; 3901.49 to 3901.50 (West 1999) (major portions enacted 1996)

Oklahoma Stat., Title 36, §§3614.1 to 3614.4 (West 1999) (enacted 1998)

Oregon Revised Statutes §659.036 (employers use of genetic information limited to bona fide occupational qualifications); §§659.700-.720 (genetic privacy), §746.135 (discrimination in health insurance) (1995, 1996)

Pa. Cons. Stat. §3902 (West 1999) (enacted 1996) (insurance must cover PKU-related formula)

R.I. Gen. Laws §27-18-52 (genetic testing and insurance), §§27-20-39 & 27-19-44 (non-profit hospital services/non-profit insurance), §27-41-53 (HMOs) (R.I. and Lexis 1998) (enacted 1998)

S.C. Code Ann. §38-41-45 (MEWAs), §38-71-670 (HIPAA and individual coverage), §38-71-840 (HIPAA and group coverage), §38-93-10 to -60 (S.C. 1999) (HIPAA implementation 1997, Genetic Privacy enacted 1998)

S.D. Codified Laws §§58-17-84, 58-18-45, 58-18B-27 (S.D. 1999) (pre-existing condition cannot include genetic information unless related condition has been diagnosed) (enacted 1997)

Tennessee Code Ann. §56-7-2701 to 2708 (genetic information non-discrimination in health insurance act), §56-7-pt 28 (HIPAA implementation) (Tenn. 1999) (enacted 1997)

Texas Labor Code Ann. §21.401 to 21.405 (West 1999) (discriminatory use of genetic information prohibited enacted 1997); Texas Insurance Code Ann. §21.73 (West 1999) (non-discrimination in insurance, enacted 1997; Texas Civ. St. Ann. Art. 9031 (limitation on use by licensing authority) (West 1999) (enacted 1997)

Vermont Stat. Ann. §§8:4724, 8:9331-9335, 18:9332-9333 (Vt. 1999) (enacted 1997)

Va. Code Ann. §§38.2-508.4 (genetic information privacy), 38.2-613 (confidentiality, enacted 1996) (Lexis 1999); §§38.2-3431, 38.2-3432.3 (HIPAA-related amendments in 1997, 1998, 1999) (Lexis 1999)

Washington Rev. Code Ann. §§48.20.520, 48.21.300, 48.44.440, 48.46.510 (West 1999) (PKU provisions, enacted 1988)

W.Va. §§33-15-2a, 33-16-1a, 33-16-3k (Lexis 1999) (enacted 1997)

Wisc. Stats. Ann. §§111.372, 111.39 (employment, 1991), §631.89 (genetic testing and insurance, 1991), §§632.746 (HIPAA implementation, 1997) (West 1999)

Wyo. Stat. Ann. §§26-19-107, 26-19-306 (Wyo. 1999) (HIPAA amendments enacted 1997, 1998)

Which level of government should become involved in regulating genetic information, the federal government or the states? What uses of genetic information are permissible? When is the use of genetic information "unfair"?

OVERVIEW OF STATE AND FEDERAL REGULATION OF GENETIC INFORMATION

It is important and useful to understand the distinctions between the regulatory authority of the federal and state governments regarding the control of genetic information. The federal government has power over interstate commerce which theoretically could include the regulation of insurance companies and the provision of health and other benefits by employers. Nevertheless, the federal government for many years failed to enact much substantive regulation touching on the problems created by genetic information. Indeed, under the McCarran-Ferguson Act, the federal government has long ceded primary authority over the regulation of the business of insurance to the states (10). Where federal law existed, it tended not to impose substantive standards of conduct regarding the use of genetic information.

Under the Employee Retirement Income Security Act of 1974 (ERISA), the federal government established various requirements for the maintenance and operation of employee benefit plans (2). These federally regulated benefit plans include those in which the employer undertakes to provide medical or health coverage for its employees. But ERISA did not originally regulate the substantive content of those employee benefit plans in any way that impinged on the ability of the employee benefit plan to seek out and make use of genetic information. Other federal statutes, such as the Rehabilitation Act of 1973 or the Americans with Disabilities Act (ADA), prohibited certain types of discrimination by employers and others, but these statutes were not drafted to make clear that they included protections for all types of genetic information or discrimination (11,12).

Thus, for several years, states were alone in regulating access to and use of genetic information in insurance or employment-related areas. Many states enacted specific statutory protections for genetic information as a matter of self-initiated state policy (see Table 1). State legislation in this area was sparked by the knowledge that advances in genetic technology were creating conflicts between individuals, insurers, and employers. Individuals often wanted to know relevant genetic information but then realistically or unrealistically feared discrimination by others (11,13).

The federal ERISA statute governing employee benefit plans created some significant barriers to state regulation (2). ERISA contains a vigorous "preemption" clause. The clause invalidates state attempts to regulate many types of employee benefits, including state efforts to regulate the way in which employers use genetic information in certain employee benefit plans such as in health benefits coverage. Congress appeared to recognize the regulatory vacuum created by the ERISA preemption clause when it enacted the Health Insurance Portability and Accountability Act (HIPAA) (3). This federal statute established

substantive guidelines for the use of genetic information in health plans and also explicitly authorized consistent state regulation. HIPAA sparked a new wave of state legislative activity so that now nearly all states have enacted measures governing the use of genetic information in at least some contexts (Table 1).

STATE REGULATION AND ERISA PREEMPTION BEFORE THE HIPAA AMENDMENTS

State Regulation Before HIPAA

States began to be concerned with the use and misuse of genetic information over 20 years ago because of the problems associated with sickle cell anemia testing. These concerns were augmented over time as geneticists began to identify a host of conditions thought to be related to underlying genetic traits. States sought to encourage individuals to use genetic testing services by protecting the voluntary and confidential nature of genetic testing. On the other hand, states recognized that providing access to genetic information for individuals but not insurers could create problems. Is it "fair," for example, to permit an individual who knows she is susceptible to early disability and death, but does not share this information with her insurer, to purchase large amounts of disability and life insurance coverage (7,14,15)? In an effort to deal with some of these problems, states attempted to establish when insurers, or others, should be permitted to gain access to an individual's genetic information and to establish certain domains within which discrimination based on genetic characteristics would be permissible.

As Table 1 indicates, a number of states responded to these conflicts by enacting strong privacy protections for genetic information in the early to mid-1990s. State genetic privacy acts, such as those enacted in California, Georgia, Nebraska, and New York, generally provided strict protections for the confidentiality of genetic information. Some states even provided that a person's genetic information would be the "property" of the individual. Typically an individual could not be tested without giving specific informed consent, and genetic test results could not be released without the individual's consent.

Some states prohibited employment discrimination based on genetic factors. New Hampshire, Nebraska, and New York, for example, prohibit employers from using genetic test results to affect the terms and conditions of employment. These state employment discrimination statutes supplement the sometimes weak protections offered by existing federal statutes.

Some states also sought to regulate the use of genetic information by insurers, particularly health insurers before HIPAA's enactment. California, Colorado, Georgia, Minnesota, Montana, Oregon, and Wisconsin adopted restrictions on the use of genetic testing or genetic information by insurers. Each of these states focused on health insurance, leaving life insurers and other types of insurers relatively free from state regulation.

The early state efforts to regulate the use of genetic information by insurers appeared to be driven by three principles: (1) genetic information is a "good" which individuals should be encouraged to obtain, (2) access to medical and health insurance should not be restricted based on genetic information, and (3) genetic discrimination in other types of insurance, such as disability income or life insurance, should not be prohibited. The fact that so many states attempted to regulate in this area suggests a high level of state concern about the impact of genetic information on the health insurance market.

State efforts to regulate health insurance were complicated by a complex federal and state regulatory structure. Congress largely delegated the regulation of the business of insurance to the states in the McCarran-Ferguson Act (10). Yet two important federal statutes affected the ability of states to regulate in this area. The first, and most important, impediment to state regulation of the insurance market was ERISA (2). The second relevant federal statute was HIPAA, which amended, ERISA and prohibital certain types of genetic discrimination (3).

Employee Retirement Income Security Act of 1974

ERISA regulates employee benefit plans, including health coverage, disability, and life insurance provided as a benefit of employment (2). The benefit plans covered by ERISA include those in which an employer enters into a contract with another entity such as an insurance company. ERISA also applies where an employer "self-insures," or bears the risk of paying the benefits directly. (Governmental plans, church plans, and a few others are excluded from ERISA regulation.) (2).

About 70 percent of the people who have private health insurance coverage obtain ERISA benefit plans through employment (19). However, most employees who have health insurance through employment are covered by plans that are "self-insured," that is, their employer bears the risk of medical expenses (19).

Until recently ERISA was silent about the employer's ability to take genetic conditions into account in the terms and conditions of his or her employee benefit plans. Nevertheless, the ERISA preemption clause presented a substantial barrier to effective state regulation of genetic information.

Federal Preemption of State Regulation of Genetic Information

Unlike many federal laws, ERISA was drafted with an explicit and broad preemption clause that prevents states from regulating employee benefit plans. The statute states that ERISA "shall supersede any and all state laws insofar as they may now or hereafter relate to any employee benefit plan" (2). The major exception, for our purposes, is the "savings clause," which explains that "nothing in this subchapter shall be construed to exempt or relieve any person from any law of any state which regulates insurance" (2). This provision "saves" state laws regulating insurance from ERISA preemption. However, ERISA makes clear, that in regard to employer self-insured plans, "an employee benefit plan shall [not] be deemed an insurance company ... or to be engaged in the business of insurance ... for purposes of any law of

any state purporting to regulate insurance companies" (2). Self-insured employee benefit plans are thus completely protected from state regulation.

The Supreme Court has often been called upon to interpret this complex and confusing provision. In order to decide whether ERISA preempts a state law, the Court must start with the presumption that Congress does not intend to supplant state law (20,21). In evaluating whether the normal presumption against preemption has been overcome, the Court considers whether the state law "relates to" an employee benefit plan. The Supreme Court has said that a state law "relates to" an employee benefit plan if it (1) has a "connection with" or (2) "reference to" such a plan (22). State laws that "relate to" an employee benefit plan in this fashion are preempted and given no effect because of the ERISA preemption clause, unless an exception to preemption can be found.

Once the Court rules that a state law "relates to" an employee benefit plan and thereby falls under ERISA preemption, it must determine if the law "regulates insurance" and thus escapes preemption under the "savings clause." To determine if a law regulates the "business of insurance," the Court must first consider whether, from a commonsense view, the contested prescription regulates insurance (23). The Court must also consider three factors to determine whether the regulation fits within the "business of insurance," the phrase used in both the McCarran-Ferguson Act and the ERISA: (1) whether the practice has the effect of transferring or spreading a policyholder's risk, (2) whether the practice is an integral part of the policy relationship between the insurer and the insured, and (3) whether the practice is limited to entities within the insurance industry (23).

Under the "deemer" clause, a state law that relates to an employee benefit plan but is "saved" from preemption because it constitutes the state regulation of insurance can still not be applied to self-insured employee benefit plans. A "self-insured" plan—one in which the employer bears the risk of his or her employee's health care costs directly—cannot be treated as an insurance plan and thereby be subjected to state regulation of insurance (20).

ERISA's broad preemption clause has presented a significant barrier to state regulation of the use of genetic information. A large number of states have sought to prevent entities from using genetic information to determine whether to provide health benefits for individuals. California law provides, for example, that genetic information should not be used in making coverage or rate determinations for medical benefits (see Table 1). Since most persons who have private insurance obtain it as a benefit of employment, California's statute is only effective if it is applied to employee benefit plans. Under the ERISA preemption clause, California's statute may be considered state "regulation of insurance" and may be "saved" from preemption (2). This means that the state statute can be applied to insurance companies that sell health insurance to employers for the benefit of their employees. The statute cannot be applied, however, to employers who self-insure their health benefits plans. Employers who self-insure are immune from state regulation under the "deemer" clause of ERISA. It is for this reason, to avoid state regulation, that many employers provide employee benefits, such as health care coverage, through self-insurance.

The same principles apply to state attempts to regulate the use of genetic information in other areas, such as disability income and life insurance. Under ERISA, states can regulate insurance companies selling these types of benefits and may prohibit or permit the use of genetic information in determining the terms and conditions of the insurance plan. However, states are barred from attempting to regulate the use of genetic information by employers who self-insure a disability or death benefit plan. Consequently most states permit the use of genetic information as an underwriting consideration in life and disability income insurance (24). Indeed, ERISA preemption has little impact on employers who self-insure because state law often permits use of genetic information in coverage or rate determinations for disability and life insurance.

EFFECT OF THE HEALTH INSURANCE PORTABILITY AND ACCOUNTABILITY ACT OF 1996

HIPAA Amendments to ERISA

Although by 1996 many states had attempted to regulate the use of genetic information, the ERISA preemption clause limited the effectiveness of state regulation, since it shielded self-insured employee benefit plans, such as those providing health coverage, from state regulation. In addition, more than half the states had not enacted *any* legislation governing the use of genetic information. Critics were quick to note that state regulation had not been an effective method of restricting improper use of genetic information in health care coverage determinations.

Congress responded to these critics with amendments to ERISA that govern the use of genetic information in many types of health insurance coverage. While recognizing that state regulation is still important, particularly in areas outside employer provision of health coverage, in 1996 Congress amended ERISA to provide protection from genetic discrimination in some contexts.

HIPAA amends ERISA to directly regulate the use of genetic information in several important ways (3). The most serious restrictions are imposed on group health plans. HIPAA prohibits the use of "genetic information" in making eligibility determinations for group health plans, whether the health plan is insured or self-insured. Group plans are also prohibited from charging individual members higher premiums because of genetic information about themselves or their dependents. HIPAA provides that these plans may not consider an individual's genetic characteristics to be a preexisting condition unless the genetic characteristic has given rise to a diagnosis of an actual condition related to the genetic information (3).

Other sections of HIPAA, while not specifically focused on the issue of genetic information, provide important benefits for individuals who have a genetic condition. The statute establishes a "credit" for prior insurance coverage that may be applied to the preexisting limitations period of a new policy. This means that an individual with a diagnosed genetic condition will actually have greater job mobility, since he or she will be able to achieve continuous coverage for medical treatments even after switching jobs (3). The statute also contains some protections that apply to health insurer issuers (including HMOs) and individual health plans. The statute establishes standards for guaranteed availability and renewability, for example, that could prove helpful to individuals who have genetic conditions (3).

Finally, HIPAA does not amend the general ERISA preemption clause but does provide that states may continue to regulate certain aspects of how health insurance companies use genetic information (3). States will monitor the implementation of the federal rules governing insurers and HMOs, for example. States can also enact alternate protections, so long as their efforts do not prevent the application of federal law. Thus states may enact certain measures that provide greater protection to individual enrollees than that under federal law. Most states already have passed state legislation implementing HIPAA's protections in the state insurance market.

Even with HIPAA, there are still big gaps in protection for those concerned about loss of confidentiality and/or genetic discrimination (17). First, HIPAA regulates the "use" of genetic information but does not otherwise protect the creation or privacy of the information. Second, HIPAA focuses federal regulation on group plans; individual health insurance policies are offered lesser protection. Third, even for group plans, HIPAA does not prohibit an insurer from charging higher overall rates for a plan based on genetic information about plan participants.

Current Status of State Regulation

As Table 1 shows, most of the states that now regulate the use of genetic information enacted their legislation during or after 1996, the year of HIPAA's enactment. Most state regulation now focuses on insurance issues. A large number of states attempt to control the use of genetic information by various types of insurers. There are three basic variables that characterize state legislation: the definition of "genetic information," the types of insurers or entities regulated, and the substance of the insurance regulation (25).

The first important variable is the type of genetic information covered under the state statute. States have taken a variety of approaches to defining the types of genetic information subject to statutory protection. Most states focus on the use of "genetic testing" and protect test "results." Several take into consideration the need to protect both the test results of the "individual" and his or her "family members." Louisiana recognizes the need to protect other types of genetic information, such as information about genetic conditions that might be found in the family history or medical records of an individual.

A significant number of jurisdictions, such as California, protect only genetic information about "predispositions" or "susceptibility," while permitting disclosure of information about actual genetic conditions or diseases. Some states, such as North Carolina, seem to protect the full range of genetic information, up to and including a determination that an individual in fact is currently exhibiting the effects of a genetic condition.

The second variable in state regulation is the identity of the entities subject to regulation. As shown in Table 1, states typically have sought to regulate a wide range of insurers and insurance-like entities. Thus state laws focus on health benefit plans, whether insured or self-insured (e.g., see Alabama, California, and Florida). The statutes also attempt to restrict the use of genetic information by health maintenance organizations, preferred provider organizations, and other health insurer-hybrids. State regulatory schemes mention other types of insurance, such as life insurance, disability income insurance, Medigap coverage, and other types of policies.

The third variation in state regulation is the differential treatment of genetic information based on the insurance coverage type. Each state regulating the use of genetic information by insurers has enacted strong prohibitions against use by health or medical insurance entities. A large number of states, however, permit discrimination — or at least "actuarially sound" discrimination — by life, disability income, and other types of insurers. Most of states regulating the use of genetic information in insurance thus have established that discrimination in health or medical benefits is "unfair" while discrimination in other types of insurance is often "fair" and should be permitted (5).

Some of the legislation would doubtlessly have been enacted even without HIPAA. But many of the post-1996 statutes have clear objectives on the implementation of HIPAA-type provisions within the state. Post-HIPAA, state legislation that relates to employee benefit plans will consequently not be preempted so long as it does not prevent the application of federal law (3). Indeed, states can and have considered provisions that would establish standards more protective than those offered under HIPAA.

CONCLUSION

Genetic information is becoming an increasingly important area for state and federal regulation. Most states have already specifically regulated the confidentiality and use of genetic information. There are significant barriers, however, to effective state regulation, including federal preemption under ERISA.

HIPAA's enactment and the subsequent flurry of state legislative activity have been helpful to those concerned about the confidentiality of genetic information and the risk of genetic discrimination. However, commentators have identified the gaps that still remain. It is clear that further federal action must address at least two of these gaps: (1) the risk of discrimination in group premiums or rates based on genetic information, and (2) the risk of the loss of privacy for genetic information (17). Before the Congress are versions of the Patients Bill of Rights that would protect individuals in group plans from discrimination in the provision of services based on genetic information (26,27). Congress is also considering bills specifically designed to protect the confidentiality of genetic information and to restrict the imposition of genetic testing (28).

BIBLIOGRAPHY

1. M.B. Kaufmann, *Loyola Univ. Chicago Law J.* **30**, 393–438, 400–401 (1999).
2. Employee Retirement Income Security Act of 1974, 29 U.S.C.A. §§1001–1169 (West 1997).
3. Health Insurance Portability and Accountability Act, Public Law 104–191.
4. *Buck v. Bell*, 274 U.S. 200 (1927).
5. P.A. Lombardo, *N.Y.U. Law Rev.* **60**, 30–62 (1985).
6. L. Hood and L. Rowen, in Rothstein, ed., *Genetic Privacy*, 1997.
7. J. Gaulding, *Cornell Law Rev.* **80**, 1646–1694 (1995).
8. K. Rothenberg et al., *Science* **275**, 1755–1757 (1997).
9. Task Force on Genetic Testing, NIH-DOE Working Group on Ethical, Legal, and Social Implications of Human Genome Research, *Final Report: Promoting Safe and Effective Genetic Testing in the United States*, N.A. Holtzman and M.S. Watson, eds., U.S. Government Printing Office, Washington, DC, 1997.
10. McCarran-Ferguson Act, 15 U.S.C.A. §1011–1014 (West 1999).
11. Rehabilitation Act of 1973, 29 U.S.C.A. §§701–796 (West 1999).
12. Americans with Disabilities Act, 42 U.S.C.A. §12101–12213 (West 1999).
13. *Norman-Bloodsaw v. Lawrence Berkeley Lab.*, 135 F.3d 1260 (9th Cir. 1998).
14. E.M. Holmes, *Kentucky Law J.* **85**, 503–664 (1996, 1997).
15. R.A. Bornestein, *J. Law Policy* **4**, 551–610 (1996).
16. L.F. Rothstein, Disabilities and the Law, §5.02 (Shepard's/McGraw-Hill, 1992 and 1997 Supplement).
17. M. Rothstein and S. Hoffman, *Wake Forest Law Rev.* **34**, 849–888 (1999).
18. Civil Rights Act of 1964 (Title VII), 42 U.S.C.A. §2000e-2 (West 1999).
19. M.A. Rothstein, *J. Law Health* **9**, 109, 113 (1994, 1995).
20. M.A. Bobinski, *University of California at Davis Law Rev.* **24**, 277–278 (1990).
21. *De Buono v. NYSA-ILA Med. and Clinical Serv. Fund*, 520 U.S. 806 (1997).
22. *California Div. of Labor Standards Enforcement v. Dillingham Const.*, 519 U.S. 316 (1997).
23. *UNUM Life Ins. Co. v. Ward*, 119 S. Ct. 1380 (1999).
24. C.M. Keefer, *Indiana Law J.* **74**, 1375–1395 (1999).
25. W.F. Mulholland II and A.S. Jaeger, *Jurimetrics* **39**, 317–320 (1999).
26. Patients Bill of Rights Act of 1999, 1999 U.S. Congr. Senate Bill 1344 [introduced by Senator Lott (R)].
27. Patients Bill of Rights Act of 1999, 1999 U.S. Congr. Senate Bill 1256 [introduced by Senator Daschle (D)].
28. Genetic Information Disclosures, 1999 U.S. Congress House Bill 2555, 1st Sess. 106th Congr.

See other entries GENETIC INFORMATION, ETHICS, INFORMED CONSENT TO TESTING AND SCREENING; GENETIC INFORMATION, LAW, LEGAL ISSUES IN LAW ENFORCEMENT DNA DATABANKS; GENETIC INFORMATION, LEGAL, FDA REGULATION OF GENETIC TESTING; GENETIC INFORMATION, LEGAL, GENETIC PRIVACY LAWS.

GENETIC INFORMATION, LEGAL, FDA REGULATION OF GENETIC TESTING

Anny Huang
Wachtell, Lipton, Rosen, and Katz
New York, New York

OUTLINE

INTRODUCTION

Genetic tests are the means by which individuals may unlock the wealth of information contained in their genetic constitution, and as the understanding and significance of genetic information advances, the scope of genetic testing expands accordingly. However, growth in genetic testing has opened new avenues of commercial exploitation that raise problems with ensuring adequate evaluation of the scientific legitimacy of genetic tests being offered to the public. The Food and Drug Administration (FDA) subjects genetic tests sold as kits to full regulatory review but leaves genetic tests provided as services entirely free from FDA scrutiny. The federal government does not review genetic testing services for the soundness of the scientific claims made for the tests, and there is very little substantive review of genetic testing services at the state level. The issue of whether the current state of regulation and the distinction between test kits and testing services is reasonable becomes more pressing as the genetic testing industry gains momentum.

Like many achievements in biotechnology, developments in genetic testing have been shadowed by concerns about their use and implications: scientists worry how their research will be used, physicians wonder whether the benefits outweigh the harms, and ordinary people face philosophically uneasy choices about whether knowledge leads one to the garden or away from it. The nuanced and weighty problems surrounding the use of genetic technology have been struggling for years toward resolutions that might suggest a basis for practical action. While numerous states have passed laws restricting the use of genetic information (1,2) and a consensus appears to

be growing regarding the privacy of genetic information, basic questions about regulating genetic tests themselves remain mired in complex theoretical and scientific controversies. For many years the Task Force on Genetic Testing, an entity organized by the National Institutes of Health (NIH) and the U.S. Department of Energy, provided leadership in the development of national policy on regulating genetic tests. The Task Force produced its pivotal Final Report in 1997 (3) which contained extensive analysis of the regulatory and scientific issues related to genetic testing and provided groundbreaking detail on the economics and practice of the genetic testing industry (3). However, the role of FDA, a natural candidate to lead regulation of the safety and efficacy of genetic tests, was left surprisingly murky in the Report's recommendations. In fact the Final Report did not specify an appropriate candidate to monitor the scientific validity of genetic testing services. This result is remarkable in light of the fact that the Task Force's creation derived in part from an Institute of Medicine report criticizing the disparity between test kits regulated by the FDA and unregulated testing services (4,5). Panel members were clearly attentive to the issue and had questioned publicly the business strategy of marketing tests as services rather than kits specifically for the purpose of evading FDA oversight. Dr. Neil A. Holtzman, chairman of the Task Force, observed, "Companies don't create kits so they can circumvent the FDA regulatory process" (6). Yet the Final Report refrained from proposing an extension of FDA oversight to genetic testing services.

Such reserve, however, may have derived more from the Task Force's consensus style of decision making than from a fundamental objection to FDA involvement in genetic testing regulation. Part of what made the Task Force's conclusions so forceful and well-informed was the diversity of interests represented on the panel, including from the biotechnology industry (3). The Final Report of the Task Force had the distinction of being unanimously approval by its members with no abstentions. However, several Task Force members had openly doubted whether the FDA was the right entity to regulate genetic testing services (3). In fact, as early as April 1997, the Task Force on Genetic Testing "tabled any efforts to come up with a recommendation for the role that the FDA should play in regulating genetic testing" (3). In addition FDA itself showed considerable reluctance in assuming greater responsibility. In her testimony before the House Subcommittee on Technology, the FDA Deputy Commissioner Mary K. Pendergast stated that, "[t]o date, the FDA has minimal involvement with genetic testing," and cautioned that careful weighing on the relevance of further FDA efforts would be necessary:

> If the FDA is to do any of the additional regulating, we would have to evaluate how these concerns fit with other concerns facing the Agency, e.g., product approval and regulation, infectious disease transmission through foods, blood, and tissues, and examine how the harms from inaccurate genetic testing stack up against those other priorities. If additional oversight is mandated, there is the question of resources and how to pay for the oversight (7).

The agency has repeatedly taken the position that it will not exercise jurisdiction over tests marketed as services (8). Another full-length study, which supported the imposition by "federal regulators" (9) of minimum standards for the positive predictive value (PPV) of genetic testing services, came to the following conclusion: "In light of the public and political pressures on the FDA, such regulation might best be introduced through the [Centers for Disease Control], [Federal Trade Commission], [Health Care Financing Administration], or [Department of Health and Human Services] by, for example, modifying [the Clinical Laboratories Improvement Act]" (9, p. 1299).

While the new Advisory Committee on Genetic Testing in the Department of Health and Human Services (DHHS), formed at the recommendation of the Task Force, closely involves FDA in its activities, it is proceeding with the collection and analysis of data on the analytical validity, clinical validity and clinical utility of genetic tests leaving aside, as premature, the issue of the appropriate source and level of oversight (10). The question that remains unanswered is why the distinction between commercial services, which FDA believes it has authority to regulate, and kits makes sense. A Task Force member affiliated with OncorMed expressed the view that FDA did not have the appropriate level of expertise (6), but the criticism applies equally well to the agency's competence in regulating genetic testing kits, whose ranks can be expected to swell in the next few years. According to an FDA official: "At present we estimate that there are, or soon will be, dozens of companies or laboratories offering hundreds of different genetic tests to the public and it is projected that this number will grow substantially" (7). Currently 5,000 genes associated with genetic disorders have been cataloged for the Human Genome Initiative by a group at Johns Hopkins University (11). It appears that amid herculean efforts to collect information, resolve grave social issues, and achieve consensus among the various constituencies in genetic testing, the proposals refrained from charging any particular regulatory body with assuring consumers that the genetic tests being performed as a service, which represents the vast majority of emerging tests, are scientifically valid, despite agreement as to the necessity of such measures (9,12). That is not to say that detailed and fact sensitive proposals were not made to secure the validity and analytical sensitivity of these tests through improvements at the provider or clinical level. With regard to some of the principles, however, it cannot be enough to presume the clinics will regulate themselves, and, perhaps due to political constraints, the Final Report did not specify which regulatory body should enforce the recommendations.

The regulatory standards were introduced presumably to ensure that genetic testing is made available in clinical laboratories whose clinical validity has been established, that is, their positive predictive value (PPV), unless it is collecting data on clinical validity under either an IRB-approved protocol or conditional premarket approval agreement with the FDA.

The task of developing an understanding of how FDA *should* and, from a legal perspective, *could* fit into a coherent and flexible regulatory plan remains open. The efforts of the Task Force and other institutions and professionals have laid the difficult ground work for the development of standards and protocols in the genetic testing community, and a slight shift in perspective could suggest a path to greater clarity in regulations and liability. By focusing on the problem from a test recipient's perspective, the issue becomes remarkably simple. Despite some empirical differences in delivery, the information dynamics from the patient's perspective produce as great a need for protection in the area of commercial genetic testing services as for a test packaged as a kit. Indeed, apparent factual distinctions that may seem important from a regulatory standpoint turn out to be irrelevant when considered from the view of a genetic test recipient. Emphasizing these considerations is quite appropriate in determining the applicability of FDA oversight, furthermore, because the national protection of consumers is the animating purpose behind food and drug legislation even when it encroaches upon the prerogatives of medical practice. Subject to the boundary issues involved in legal jurisdiction (14), FDA is the most logical and efficient choice as the regulatory actor. Nothing in the Final Report or in available information defeats this view or suggests a better alternative. While it is important to recognize the political and institutional constraints that may dissuade the agency from undertaking regulation of genetic testing services, it is equally important to be aware that these external considerations may be charting a course in the wrong direction.

Ensuring public health safety in an age of new biological technologies will undoubtedly subject FDA to evolutionary pressures, but the novelty and complexity of genetic technology should not deflect the agency from its traditional goals and areas of oversight. FDA already has begun to review the early products of genetic technology, and it will doubtless face many more in this millennium (16–18). The use of innovative forms to deliver essentially commercial products should not confound regulation. It is important, however, to make a distinction between commercial testing services that have the character of products and traditional laboratory analysis. A bright-line rule not only serves regulatory goals and eases compliance, but is advisable to stay well within the bounds of federal authority. The dangers of genetic information are profound and warrant the public delegating the resources and mandate to the FDA to ensure that the troubling issues and agonizing choices occasioned by genetic testing are not compounded by poorly developed or even misleading information.

BACKGROUND ON GENETIC TESTING

The potential scope of genetic testing as a commercial enterprise has become increasingly clear over the last decade (19). In 1986 the conductors of the survey found only 118 companies likely to be offering or researching genetic tests, and of those 85 companies responded to the survey. Only 22 were performing or developing tests. In contrast, the 1996 survey found a target audience of 594 biotechnology companies of which 461 responded. About a third of respondents were engaged in genetic testing activity (3). The once largely academic activity has moved into the marketplace, where more and more companies are realizing that it can be big business. A survey conducted in 1986 found 22 biotechnology companies offering or developing genetic tests, whereas a similar survey conducted ten years later found 147 companies. Tests have been developed for hundreds of conditions (20), including Alzheimer's, colorectal cancer, and melanoma with others such as asthma and even deafness on the horizon (21–24) and according to NIH, more than 450 research programs are working to develop more (21). It should be noted that the press and literature on genetic tests quote a large variety of figures for how many genetic tests have been developed deriving largely from differences in definition. Because thousands of genes linked to disorders have been identified, it would be possible to say that thousands of tests have been developed. In many of these cases, however, the disorder is extremely rare or the linkages have not been established with sufficient generality to justify use in a clinical setting. Nonetheless, the commercial potential of viable tests is considerable. The total DNA diagnostic market is estimated to exceed $6 billion by 2005 (25). Depending on the complexity involved and the number of genes to be screened, a single test may cost anywhere from a few hundred to several thousand dollars (21,26,27) and might be used by millions every year. For example, the most commonly used genetic screening test for phenylketonuria is used on millions of newborns annually (20). The proceeds from even one test could prove very lucrative. Consider, for example, the osteoporosis test being developed by Medical Science Systems, Inc. (28). The company estimated that if 2 percent of all affected Americans, approximately 500,000 people, are tested at $200 each, the market for this one test alone is worth $100,000,000 in revenues (28). MSSI focuses on the national market in estimating the commercial potential of a test. (Also companies such as Myriad, Salt Lake City, Utah, and Genetics and I.V.F. Institute, Fairfax, Virginia, comply with the regulations of jurisdictions as far away as New York in their quest for a national audience.)

Like the rest of the biotechnology industry, however, these endeavors are new and are dependent on recent breakthroughs in science (3). It is a young and dynamic industry (29). According to the Task Force Report, "The companies engaged in testing activities operate in an extremely dynamic environment, frequently undergoing restructuring, forming new partnerships, embarking on new initiatives, or dropping projects" (3, app. III). The various obstacles involved in entering the genetic testing market at this time — regulatory uncertainties, ethical issues, rapid technological evolution, the inherent limit on the utility of a test to once per patient — serve to dampen rapid maturation in this sector (3). Also most single-gene disorders are quite rare and have less market potential, so to earn high profits the tests must involve more common disorders, which often turn out to be more genetically complex (3,30). Investing in research for more common multiple-gene disorders such as cancers and heart ailments is expensive and risky (3), but promises the biggest payoffs in terms of patents and first to market profits. Consider, for example,

the unseemly race to patent the BRCA1 breast cancer gene test, which generated a considerable amount of controversy in 1996 (31). Alternatively, a company could commercialize technology licensed from an academic or research institution (3). In either situation an attractive way to reap the full benefits of developing the product and to leverage research and development and production costs is promotion to a national market (28,32). For example, the fee for a single breast cancer genetic test is $2000, for the two genes involved in here ditary nonpolyposis colorectal cancer, $870. The international competition to identify the first breast cancer gene was probably the most publicized scientific "race" of the 1990s. And when the winners — Myriad Genetics, a Utah-based biotechnology company, working with the University of Utah — promptly sought a very broad patent over BRCA1 in 1994, there was considerable disquiet. Opposition came not only from "genetic interest groups," arguing that genes are natural human blueprints which should not be patented, but also from other scientists who had co-operated with the Utah group during earlier stages of the research (31).

This commercial model for test development, however, is of fairly recent origin (26). Most genetic testing still takes place in research and academic settings (33). In the survey of genetic testing discussed above, twice as many nonprofit organizations were engaged in genetic testing activities as biotechnology companies (3). Perhaps more significantly, most of the basic research underlying the science of testing continues to be generated in public and nonprofit institutions and universities through sequencing under the Human Genome Project (HGP) and investigational studies of particular disorders (34–37). A comparison of recent headlines is instructive. Several research centers organized by the National Institute for Alcohol Abuse have found links between specific chromosomal areas and a propensity for alcohol abuse (34). The development of private sector initiatives is part of a general burgeoning of genetic testing research and practice rather than representative of a transition from public to private leadership. Symbolic of the growth of both research and commercial endeavors, a new publication known as the *Journal of Genetic Testing* was recently introduced (38).

What is a Genetic Test?

The landscape for the policy debate on genetic testing takes place against a complex and rapidly evolving scientific background of genetic testing technology. The details and methods used in genetics are complicated, but the general concept of a genetic test is fairly simple. By now the image of DNA as a double helix should be quite familiar (20). DNA stands for deoxyribose nucleic acid. The deoxyribose, in conjunction with a phosphate group, alternately link together to form the backbone of the helix similar to the sides of a ladder. The nucleic acid, of which there are four kinds, projects off the deoxyribose molecule to form the step of the ladder known as the base. Each step in the winding ladder of the helix is called a base-pair, and a person's genetic code contains billions of base-pairs (20). Imagine if the two sides of the ladder were pulled apart, splitting all the steps in half (20). (More specifically 23 human chromosomes contain about 3 billion base-pairs. Each side of the ladder is a reverse of the other because of the base-pairing.) Each half of a base-pair is called a *nucleotide*. The sequence of nucleotides provides specific directions for the production of all the parts of a living organism. Originally a *gene* was defined as a set of nucleotides that codes for a protein (20), although the term now commonly is used to identify sets of nucleotides that are associated with more general traits. Although variation in human genes is as natural as variation in fingerprints, there are specific variations in some genes that cause or are linked to a genetic disease (20). For example, the common inherited disorder of cystic fibrosis results from mutations or deletions in a transmembrane protein known as cystic fibrosis transmembrane conductance regulator. Thus, locating and cloning a disease gene can be important in determining the molecular pathology of inherited disorders. A genetic test determines whether an individual has a harmful genetic variation or any other specified genetic sequence. At present directly sequencing a gene (20) is a method reserved primarily for finding the gene originally. It is uncommon for a genetic test to simply sequence each person's gene at the relevant location because sequencing has until recently been a time-consuming and expensive process. Increasingly powerful engines for the sequencing of DNA, however, make the prospects of such an approach not only possible, but likely in the future. For example, the Perkin-Elmer's 3700 DNA analyzer, which gained so much notoriety as the engine behind a bid to take HGP private, can analyze thousands of base-pairs a day and sells for $300,000 (39–42).

Causation in Genetic Diseases. There are several relatively high incidence single locus gene disorders such as sickle cell anemia, thalassemia, and cystic fibrosis where genetic diagnosis is now common (4). Of course, identifying a disease-related genetic variation is not always conclusive, nor is failing to identify one. There may be an extremely large number of disease-inducing variations amid many benign ones, requiring extensive clinical follow-up and research. Expression of a gene or genetic predisposition is subject to complex laws of inheritance and the influence of environmental factors (3,29). In addition common disorders such as heart and Alzheimer's diseases are usually multifactorial, which greatly increases the complexity of analysis (3,20,29). Tracing the relative significance of variable genetic and environmental factors is extremely troublesome from a research perspective and impedes conclusive determinations of risk and probability. The complexity of the underlying genetics requires flexibility and sensitivity to situation-specific information for effective genetic testing generally. Indeed, the pace of innovation subjects testing practices to continual change. These conditions, and the technical difficulty level involved, encourages testing to be performed as a service rather than packaged as kits at least in the near and medium term (19).

Future Technology. In the relatively near future it may become possible to screen a larger number of patients more routinely for a larger number of genetic diseases, and to do so for diseases with more allelic variation. For example, the "gene chip," a device with space for up to 409,000 probes, has been developed (44). This new technology is in the final stages of gaining approval for its first commercial use detecting mutations in the P53 gene, which may lead to development of certain cancers. This mutation usually is acquired during the course of a person's life, however, and there currently is no chip-based test commercially available for inherited diseases (44). There is no reason, however, that this technology could not be used to test more cheaply and rapidly for a large number of genetic diseases at once. Additionally the DNA chip may make it more feasible to test for diseases with a large number of different disease alleles such as cystic fibrosis (a three base-pair deletion detectable by a single probe accounts for 68 percent of cystic fibrosis mutations, but the other 30 percent arise from over 60 different kinds of mutations (20,45). Although there are no commercial genetic tests for inherited disorders based on this technology, the chip may be a view into the future of such testing. Given the chip's probable status as a device, it will be a future likely overseen by FDA.

Potential Harms in Genetic Testing

The technical aspects behind genetic testing indicate that use of these technologies may involve considerable uncertainty. Tests for multifactorial inheritance disorders such as Alzheimer's and breast cancer have generated the most concern because the clinical utility of a positive test in a healthy individual is often still unclear at the time the tests become commercially available. For example, a panel of experts has recently concluded that testing for ApoE4, the Alzheimer's "susceptibility" gene, does not provide sufficient predictive value to justify its use outside of research labs (46). While the test improves the clinical diagnosis of symptomatic patients, it is not appropriate for testing at large. These ambiguities and uncertainties, however, probably are not significantly greater than in other medically complex areas of diagnosis. What distinguishes genetic diagnosis from other kinds of complex or controversial forms of testing? In truth, there are many similarities, especially from a medical perspective. Misdiagnosis could lead to unnecessary, painful, and risky treatments (4), and the failure to diagnose might discourage diligence in taking care of the body and seeking medical attention. Imperfect tests in general create difficult choices, suffering, and uncertainty. In fact, when used to diagnose a symptomatic individual, there are probably few differences between genetic tests and other methods of assessing disease. Predictive genetic testing, however, is performed on healthy individuals and even on fetuses in utero. The tests frequently offer only estimates and probabilities of risk, and may become available before the validity and reliability of its predictive value have been established. The genetic diseases for which tests are being developed are generally more dangerous and severe within the range of ailments (2,30), and unfortunately there are rarely effective treatments, much less cures or means of prevention (26). The fundamental and basic nature of genetic abnormality changes the entire dynamic of disease. One's genetic heritage, unlike a virus, bacterium, or a mutant cell, simply cannot be destroyed or abated (39,47). Genetic conditions, in a sense, are inseparable from the existence and identity of a person, and they can confer potential disease status upon someone with perfect health and no symptoms. The medical realities of genetic diseases form a background for a much broader, and sometimes more troubling, range of social and psychological concerns. The basic character of genetic information brings it to bear on fundamental life choices and understandings in a way that raises unique possibilities of harm to the individual, and perhaps, even to society.

Inherited Breast Cancer. The paradigmatic case and original source of much of the controversy and concern over the commercialization of genetic testing was the discovery and marketing of the BRCA breast cancer gene tests (31). The case provides a good illustration of the issues involved in the commercialization of genetic testing. Normal or wild-type BRCA genes apparently inhibit the growth of tumor cells in breast and ovarian tissue (48). Mutations in the genes may confer a lifetime risk of 80 percent for developing breast cancer and 50 percent for developing ovarian cancer (48). There are many deletions and variations, however, that may appear in a woman's BRCA gene of "unknown significance" (49), and in fact the high correlations to risk cited above are based on specific mutations found in women that were previously determined to be high risk (48,50). The possibility of selection bias compromises the probabilities (51–53). It has been speculated that the risk posed by the BRCA mutation for women with no family history of breast cancer might only be 40 percent or lower (54). Amid this background of uncertainty, and with the drastic measure of a prophylactic mastectomy as the only potentially preventative option, several companies moved to provide BRCA tests commercially (55). The Genetics and IVF. In vitro Fertilization Institute in Fairfax, Virginia, began offering the test to the general public on the theory that women had a right to know about their genetic status (56). Myriad Genetics, in its promotional literature to physicians, suggested that almost everyone could benefit from taking the test (57). Since then a University of Washington study has concluded that women without a strong family history of breast cancer should not worry about getting tested (58). Another study at Leiden University in the Netherlands found that the polymerase chain reaction (PCR) process used in the gene tests was missing about a third of the disease-causing mutations because they were too large (59).

At the center of the confusion and controversy, women were making private, painful decisions about whether to undergo testing despite few medical remedies except for surgically removing both breasts. Parents had to decide for daughters and weigh the benefits of alertness in life to symptoms of the disease against the possibility of ruining her healthy years with depression and apprehension. A woman told that she has a significantly increased risk would have to decide how to react to news reports that having no children or having them late increases the risk of breast cancer (60) and that smoking seems to reduce the risk (61). She would have to contemplate the possibility of passing on the gene to her children and the guilt that might entail. Some of the healthy women who tested positive chose the difficult course of prophylactic double mastectomies and oophorectomies (removal of the ovaries), although breast cancer could still occur in remaining tissue or even in the abdomen or colon (62). Questions remain whether the estimated 2.5 million other American women with BRCA mutations should undergo similar procedures (62). Today new options such as tamoxifen offer women alternative means of prevention and hold promise for future developments (51–53,63). An at-risk woman, however, would still have the burden of figuring out how reducing the risks posed by her particular mutation compares to side effects like increasing the risk of uterine cancer and life-threatening blood clots (64). The result is a biological version of Russian roulette that many women may simply avoid by not playing at all (65).

Dangerous Information. The discussion of breast cancer gives some brief indication of the hard choices and painful consequences that genetic testing can create. Only a superficial sense of these issues can be conveyed here, but it is important to realize the scope of the effects of genetic information. Perhaps the most obvious effect, as well as the most difficult to describe, is that testing positive may be "psychologically devastating" (26). It can cause deep depression or anxiety, and disrupt family relations. Even cases of suicide of otherwise healthy individuals have been reported (4,66). The problems seem most acute with respect to incurable late on-set disorders, such as Huntington's disease, where the utility of testing in light of the lack of treatment has been the subject of considerable controversy (67). Genetic information also substantially affects important life decisions such as whether to have children and even whom to marry (68). Such information may even define personal identity and status within one's community (69).

The broader social implications of genetic testing also are profound. Genetic testing already has begun to enable genetic discrimination by employers and insurers (70,71). The government has started creating databases for the DNA of employees, all military personnel, and criminals (70,71). Prenatal diagnosis provides a means of extending subtle social discriminations to selective termination. In California, the only state keeping track of prenatal genetic testing, 70 percent of women receive prenatal diagnosis (72). It is estimated that at least half the pregnant women in the country do so, and that about half the women who receive a positive test for a serious illness choose selective termination (72). There is little room to be sanguine in these matters. Consider, for example, the recent research indicating the genetic basis of Lou Gehrig's disease (73), and imagine that parents began terminating births of afflicted fetuses. An advanced society might think that the private activity of parents has resulted in a net benefit, but in fact, could the world have done without the likes of Stephen Hawking, or for that matter, Lou Gehrig (74)? An Institute of Medicine study commented:

> The development and widespread use of genetic tests ... raises issues about discrimination and privacy ... that

people found to possess certain genetic characteristics will lose opportunities for employment, insurance and education.... [G]enetic testing raises worries about inequities and intolerance ... that not everyone will share equitably in the benefits of genetic testing, that some will be stigmatized, and that the beauty of human diversity will be denigrated due to a narrowed definition of what is acceptable (4, p. 30).

In Germany genetic testing already is being used in immigration policies (75), and in the United States suits for "wrongful birth" have been brought against doctors who botched genetic tests indicating disability (76).

The stakes in genetic testing are not only the traumatic personal ones discussed above, but include far-ranging social ones involving democracy, privacy, and delicate notions of human worth. As one commentator has noted, "Although their proponents invariably proclaim that new technologies will bring unprecedented prosperity and freedom, they can also threaten our civic values. What Thomas Jefferson called 'cherished liberty' is not determined by our genes. It is determined by our eternal vigilance" (70). The difficult private and public decisions that genetic information requires, about the power of genetic selection, eugenics, and so on, seem impossible to render. Before individuals and society are asked to go where angels fear to tread (77–80); however, it seems reasonable to ensure at least that the momentous consequences provoked will flow from information sufficient and reliable enough to make a basic scientific determination. It is not easy to advise a parent whether to terminate selectively a fetus that has an 80 percent (or perhaps only 40 percent) chance of developing breast cancer (72) or for society to decide whether to intervene. In fact the only easy thing to say about these troubling and complex issues is that people should not be asked to resolve them based on information with no established clinical validity. It is perhaps the only easy decision to make in minimizing the harms that can result from genetic information. The question remains, however, over who should be responsible for doing so.

HISTORY AND JURISDICTION OF THE FOOD AND DRUG ADMINISTRATION

Though FDA regulates genetic testing kits, it has historically had a policy of not regulating clinical services (82). Before turning to the original reasons behind adoption of this policy, it might be appropriate to explain why the option of restraint might have particular currency with respect to genetic testing. The political climate in which FDA now operates greatly heightens the risks and barriers to undertaking new forms of regulatory oversight. Since 1994 there has been a remarkable shift in public, and especially congressional, attitudes toward FDA. While critical campaigns in the past traditionally centered on the agency's failure to protect sufficiently the public health or to meet fully its statutory obligations with respect to imposing regulations and review (83,84) recent criticisms have focused on excessive bureaucracy and oversight that prevented medical benefits from reaching the public with adequate speed (85,86). Comparisons were made to the availability of new products in

European markets (87), and, particularly in the case of HIV treatments, the social cost of standards and requirements for testing and proof of efficacy became the basis for public outrage and demonstrations (88). A newly elected Republican Congress, sensing an opening, concentrated much of its energy on FDA in its campaign to deregulate American business (89–92). The fierce cost-cutting budgetary environment put further restrictive pressure on the scope of the agency's regulatory mandate.

The FDA also has not had positive experiences taking strong regulatory positions in the testing area. Its last few attempts to protect the public from too much self-knowledge proved to be an institutional bellyache. In the late 1980s FDA adopted a complete ban on HIV home testing kits, which came under considerable derision (93–97). Accusations of paternalism were leveled and public health experts argued that without encouraging testing of populations not apparently at-risk, efforts to control spread of the disease would not be effective (98). FDA retreated from its position in 1995 (99). Similar efforts to block the distribution of drug-testing kits also suffered considerable unpopularity (100).

Finally, the agency's lack of enthusiasm may stem from the intuition that regulation of genetic testing is a thankless task. The extraordinary volume of scholarly writings on genetic testing exploring the social and moral implications often include critiques of potential government policies, which writers prophesize will lack nuance and sensitivity to the exploding issues related to genetic testing (101–105). Almost any guidelines that can be produced by the limited resources available will prompt passionate objections and painstaking critiques. While physicians, scientists, consumer advocates and other thoughtful individuals repeatedly declare that some effort to establish validity should be undertaken (2,8,11,12), the general public may still question attempts to limit their access to information, however uncertain the information may be. People might prefer to judge for themselves.

In addition to these present considerations, the policies and possibilities of FDA regulation of genetic testing are shaped by its institutional and jurisdictional history. Its range of administration is formulated according to a legislatively constructed and evolving societal role that is defined by specific purposes and by its position in relation to other guardians of the public health.

Historical Account of the FDA's Mandate

FDA's creation and evolution have resulted largely from successive periods of public crisis (106). Its charter began at the turn of the century when industrialization created an urban workforce dependent on produce and packaged goods transported from rural areas (107). The market proved to be a poor regulator of the quality and content of these goods, and the horrors of mislabeled and adulterated foods became the subject of public outcries and criticism (108,109). For example, food labeled "potted chicken" or "potted turkey" in North Dakota was found by state government officials to contain no discernible amounts of chicken or turkey (108). The first food and drug law was passed in 1906 and was aimed solely at punishing transportation of food and drugs that were adulterated or misbranded (110). There was no provision for direct regulation of safety and efficacy.

Congress began to entertain the possibility of more expansive regulation during the New Deal, but could not come to an agreement on the issue for many years (107). From the outset there was concern about administrative overreaching into the discretion and business decisions of the industry and fear that regulation would hinder growth (107). The trade-off between preservation of the public health and economic and technological development was immediately apparent. While Congress debated, in 1937 the elixir sulfanilamide disaster struck (111). The product, which had been tested for appearance and taste but not for safety, caused the deaths of almost 100 people. The only federal statutory violation was labeling the product as an elixir, which technically can apply only to alcohol solutions (111). The product was actually a sulfa compound dissolved in diethylene glycol with unfortunately fatal side effects. The manufacturer was not required to test for safety, or even to disclose the fatal ingredients. It was a tragic introduction to the inadequacies of existing legislation, and within a year the basis of modern food and drug law, the Federal Food, Drug, and Cosmetic Act of 1938 (FDCA), was passed (107,112).

In 1962, following the development of powerful new pharmaceutical products, regulation for effectiveness was added to the goals of FDA (107). The amendment was an important step in blending the work of the medical profession with the obligations of federal regulators. While the thrust of the laws continued to be the protection of individual consumers of mass-produced and marketed products, comprehensive regulation of all drugs (not merely for whether people could be harmed but for whether they would be helped) represented a major inroad on medical discretion. Extending FDA guardianship over both the safety and effectiveness of medical devices through the Medical Device Amendments of 1976 continued the assault (108). Congress was aware of the overlap, and specifically provided that effectiveness need only be reasonably substantiated (113). The statute made clear that no interference in medical practice, particularly with respect to the physician–patient relationship, was intended (114). The reluctance to regulate clinical services derives from this historical deference to medical practice.

With each successive wave of technological advancement in various health-related fields, the agency's purview has expanded and adapted to encompass novel issues and risks. Despite political attacks and criticisms of FDA, it will rarely be heard that the agency should be abolished. FDA's necessity has been demonstrated by experience, and despite complaints about the bureaucracy there is a certain efficiency to its existence. By having one national gatekeeper to monitor for potentially harmful drugs or devices, the independent review necessary to secure the public health takes place only once and is rendered in a methodical and dependable manner rather than duplicatively in offices and hospitals with inconsistent results depending on variations in time, access to information, and interest. The public benefits from consistency, standardization of review, avoidance of redundancy, and accumulation of institutional expertise generated by the existence of a central monitoring agency.

Boundaries of FDA Jurisdiction

While a fairly uncontroversial argument can be made that the agency has some authority to regulate the activities of services under the Medical Device Amendments to the FDCA (116), it seems equally clear that its discretion to do so is not unlimited. The question of where to draw the line is a matter of some novelty and ambiguity. There are in fact two separate but related inquiries involved: first, whether the regulation of an apparently intrastate service falls within the constitutional boundaries of federal power, and, if so, whether such power has been delegated by Congress to FDA. The extent to which a historical deference to the medical profession may bear upon proposed efforts to intervene in auxiliaries to clinical practice also must be considered.

Interstate Commerce Clause. FDCA derives its authority ultimately from the interstate commerce clause of the U.S. Constitution (117), which grants Congress the power, "[t]o regulate Commerce with foreign Nations, and among the several States, and with the Indian Tribes" (118). The significance of the clause has undergone considerable evolution as the legitimacy of national power has grown, the most memorable point of departure taking place during the New Deal. During the 1930s President Franklin D. Roosevelt oversaw enactment of a series of laws designed to help the country out of the Depression including the National Industrial Recovery Act. The Supreme Court invalidated the Act and other efforts as unconstitutional in several famous cases including *ALA. Shechter Poultry Corp. v. United States*, 295 U.S. 495 (1935) and *Carter v. Carter Coal Co.*, 298 U.S. 238 (1936). Roosevelt's response was a proposal to change the structure of the court to include one additional justice for each justice over the age of 70 who did not retire. In 1937 there were six such justices. Several key opinions on the Court changed in what has come to be known as "the switch in time that saved Nine." After that time the New Deal vision of national power was vindicated completely, and almost no judicial restraints on the federal commerce power seemed to remain:

> By 1945 the supreme court had come to the position that the primary and perhaps exclusive federalism-based constraints on Congress were imposed by the political process. Although Congress's regulation of the national economy has continued to grow, nearly all of its work falls well within the boundaries set by cases such as *Wickard* and *Darby* (119).

Present jurisprudence interprets the commerce power to permit Congress (*1*) to regulate the use of the channels of interstate commerce, (*2*) to regulate and protect the instrumentalities of interstate commerce, or persons and things in interstate commerce, even though the threat may come only from intrastate activities, and (*3*) to regulate those activities having a substantial relation to interstate commerce (120). FDCA already has been sustained as a constitutional exercise of the interstate commerce power (121–123). In determining the permissible extent of federal regulation of genetic testing, however, the constitutional question warrants preliminary consideration, because under the doctrine of "constitutional doubt" statutes are construed, if fairly possible, so as to avoid raising a serious question as to its constitutionality (124).

The substantial relation to commerce theory is probably the best basis for regulating genetic testing services. A classic case illustrative of the generous boundaries of that test is *Wickard v. Filburn* (125), the authority of which recently was reiterated by the Supreme Court (120). *Wickard* sustained an implementation of the Agricultural Adjustment Act of 1938 that imposed a wheat quota on a farmer whose wheat was intended entirely for consumption on the farm (125). Not only was the wheat not intended for marketing or distribution out of the state, it was not intended for commercial sale at all. Determining that even the home consumption of wheat bore on the demand for it in the national market, the Court concluded, "This record leaves us in no doubt that Congress may properly have considered that wheat consumed on the farm where grown if wholly outside the scheme of regulation would have a substantial effect in defeating and obstructing its purpose to stimulate trade therein at increased prices" (125). The Court's holding, however, was premised on a specific finding that home consumption substantially affected the economics of the wheat industry. In *United States v. Lopez* the Supreme Court rejected the claim that bearing weapons near schools substantially affected interstate commerce" (120). There clearly are limits to the effects that will support jurisdiction. With respect to commercial genetic testing where samples or patients cross state lines to obtain service for fees or where the company itself has clinics in several states, regulation is clearly well within the bounds of Congress' constitutional power. The activity clearly bears on commerce that is interstate in nature. Even if a test is "home brewed" entirely in state and performed on local residents, it would probably fall within the rule enunciated in *Wickard* and subsequent cases that intrastate activity affecting interstate commerce may be regulated. The Supreme Court repeatedly has sustained regulation of intrastate activity that bears a substantial relation to interstate commerce (120). A more troublesome question arises with respect to programs offered by academic and research centers that are nonprofit. The findings of the Task Force suggest that despite the volume of testing in which such institutions engage, they do not compete significantly with commercial producers and providers (2). If the programs attract patients from out of state then regulation might be supported under the second prong of the commerce power, namely protecting persons in interstate commerce (120). Regulation of solely regional test centers and nonprofit organizations therefore present increasingly marginal constitutional cases for regulation.

Federal Food, Drug, and Cosmetic Act. The more substantial inquiry focuses on the breadth of Congress' delegation of regulatory power to the FDA as it bears on genetic testing. The relevant operating provision in the FDCA is codified at 21 U.S.C. §331, which prohibits:

(a) The introduction or delivery for introduction into interstate commerce of any food, drug, device, or cosmetic that is adulterated or misbranded.

(b) The adulteration or misbranding of any food, drug, device, or cosmetic in interstate commerce.

(c) The receipt in interstate commerce of any food, drug, device, or cosmetic that is adulterated or misbranded, and the delivery or proffered delivery thereof for pay or otherwise....

(k) The alteration, mutilation, destruction, obliteration, or removal of the whole or any part of the labeling of, or the doing of any other act with respect to, a food, drug, device, or cosmetic, if such act is done while such article is held for sale (whether or not the first sale) after shipment in interstate commerce and results in such article being adulterated or misbranded.

Pursuant to its authority to regulate devices in this section and under other MDA provisions, FDA has asserted repeatedly that it has the power to regulate genetic testing services (126). To support its authority, FDA must identify both a device and the presence of interstate commerce. The language of FDCA is therefore more limited than the constitutional boundaries, regulating activity in interstate commerce rather than activity merely bearing a substantial relation to interstate commerce. The agency has several alternatives to establish jurisdiction: It may premise jurisdiction either upon the materials shipped in interstate commerce from which the genetic test is assembled, on the genetic test itself, or some blend of the two.

Materials Received in Interstate Commerce. FDA's regulation of analyte-specific reagents is an example of regulating ingredients or materials received in interstate commerce (116). Extending regulation to the entire genetic test assembled after the components are shipped in interstate commerce, however, adds a twist. Section 331(k) of title 21 of the *U.S. Code* prohibits adulteration or misbranding of a "device" after shipment in interstate commerce, and the statutory definition of "device" includes "any component, part, or accessory ... intended for use in the diagnosis of disease or other conditions" (127) clearly encompassing the materials used to assemble a test. A logical and persuasive argument can be made that the components previously shipped in interstate commerce can be "adulterated" within the meaning of Section 331(k) by assembling them into a genetic test that does not conform to regulations prescribed by FDA. A similar reasoning was adopted by the U.S. Court of Appeals for the Ninth Circuit in *Baker v. United States* with respect to the sale of misbranded drugs (128). The court held that Section 331(k) applied where the ingredients were shipped in interstate commerce even though the final manufacture and sale of the drug took place in California (128). The U.S. Court of Appeals for the Eighth Circuit adopted a similar approach to extend regulation to animal biologics that were only distributed intrastate (129). In its argument for jurisdiction under this theory the FDA would be assisted by the long-standing principle in food and drug law jurisprudence that the "high purpose of the Act to protect consumers who under present conditions are largely unable to protect themselves ... [should not] be easily defeated" (130). Ensuring the public health is a prime example of national power and prerogative, especially where the circumstances are buttressed by other commercial and interstate elements. The weakness in the approach is that a testing service could theoretically evade regulation by assembling the test from materials which never crossed state lines, even if the service openly and aggressively engaged in national consumer-oriented marketing and served residents from outside the state.

Commercial Testing as Interstate Commerce. Alternatively, jurisdiction may be premised on the genetic test itself if it can be deemed to be an object in interstate commerce. There is some question whether a product technically manufactured and consumed locally may nevertheless, in light of the surrounding circumstances, be an object in interstate commerce. In the past courts have found interstate commerce where the consumer traveled from another state (131) or where information crossed state lines (132), both of which often are involved in providing commercial testing services by test producers. There appear to be no cases, however, where jurisdiction was premised on a device that did not itself at some point cross state lines (133). By providing testing as a service, companies inadvertently have found a means for circumventing the regulation of diagnostics. It is an exception not limited to genetic testing, and generates some disturbing dynamics. Beyond the lack of FDA regulation, it also makes manufacturers no longer dependent on the approval of the medical establishment and allows them to advertise and sell directly to the consumer. Consider, for example, the recent controversy over an Alzheimer's test marketed as a service by Nymox Pharmaceutical Corporation (134). The company ran an advertisement in *TV Guide* implying a conclusive test despite considerable doubt as to the scientific validity of the $400 test (134). A nationally advertised single-test "service," such as the Nymox Alzheimer's test, bears a rather suspicious resemblance to a nationally distributed product, the very object of historical FDA regulation. Adopting the theory that interstate commerce could be created by out-of-state advertising and servicing would require some revision of existing interpretations (135). Courts might be willing to view the test as taking place in interstate commerce under Section 331(a) or (b) if the regulation was restricted specifically to commercial providers who marketed their tests outside the state. Regional services that advertised only to local medical practitioners would provide little basis for regulation. The more interstate elements that are added in defining the scope of those regulated, the more compelling the case for applying Section 331(a) or (b). Using Section 331(a) or (b) seems to better reflect the substance of the activities' interstate nature. However, premising jurisdiction on the test components shipped in interstate commerce and proceeding under Section 331(k) seems a more certain approach under present case law.

Unwarranted Interference with the Practice of Medicine. Although there are dependable grounds for asserting statutory jurisdiction, the FDA must be wary of encroaching on medical practice. While many of the FDA's activities constrain the practice of medicine, such as regulating drugs and screening medical devices (136), there are limits to the permissibility of interference, perhaps even of a constitutional nature. In a very early opinion the Supreme Court stated, "Obviously, direct control of medical practice in the States is beyond the power of the federal government..." (139). Although it should be noted that this statement was made prior to the changes in the federal commerce power wrought by the New Deal (119), there is probably still some active residue of the principle remaining today. In *United States v. Evers*, the FDA tried to regulate a physician's use of a chelation drug in treating arteriosclerosis (138). The district court concluded that it was an unwarranted agency intrusion, stating:

> The courts have rather uniformly recognized the patients' rights to receive medical care in accordance with their licensed physician's best judgment and the physician's rights to administer it as it may be derived therefrom (138).

On appeal the U.S. Court of Appeals for the Fifth Circuit declined to endorse this reasoning and affirmed on entirely different grounds (139). The situation involved in *Evers* is quite different than the setting of genetic testing involving commercial labs and regulation of substantive changes in the device rather than merely regulating its medical use. In addition *Evers* reflects somewhat dated, paternalistic notions of the privileges of the medical profession that may have little currency in the modern era of patient's rights. Despite doubts about its authority, however, the reasoning of *Evers* strongly suggests that some care should be taken to avoid encroaching on the practice of medicine in fashioning regulation for genetic testing (140).

THE CASE FOR FDA REGULATION

FDCA arose partly out of a recognition that drug companies and medical products manufacturers, although they assist doctors in saving lives and relieving suffering, are not parties to the Hippocratic Oath. Their motive is profit. Regulation represents a determination that the dangers posed to public health are not sufficiently accounted for in the risk calculus of companies when they introduce products. As revealed in the historical discussion of FDA, the nation learned that lesson through painful experience and since that time has attempted to anticipate potential problems by delegating broad authority that can evolve to cover novel technologies (141,142). There seems to be little disagreement that some authority should determine that genetic tests being offered as services have positive predictive value for the disorder being tested (9,12). In tension with this simple purpose is a broad and varied range of settings in which genetic testing takes place. Any DNA analysis performed at the request of a physician or researcher, whether in a chemical or hospital laboratory including individualized linkage analyses used for diagnosis, can constitute a genetic test. Over 500 laboratories in universities, hospitals, public health departments, and commercial centers perform analysis utilizing innumerable target genetic variations and rapidly evolving test strategies (143). If that is the objective of regulation, it is quite apparent that FDA is not exceptionally suited to such a task. It seems, however, that the dangers generating unique concerns do not arise in all settings where genetic testing services are provided. Some testing takes place in a research environment or as an extension of medical practice similar to other forms of laboratory analytical services where there has been no general call for regulation. Describing these settings is a helpful prelude to distinguishing precisely the objective of regulating genetic tests. When the goal is carefully identified, the strengths and expertise necessary to its implementation will also become apparent.

Information Dynamics in Genetic Testing

Genetic testing providers may engage in a wide range of activities: researching genetic linkages (144), developing new diagnostic products (28), performing genetic analysis, and serving people directly through clinics (145). To clarify the information dynamics and make the realities of the industry's structure simpler and more immediate, the discussion proceeds by tracing the progress of a hypothetical consumer of genetic testing services through the process. An individual may obtain a genetic test either as part of a clinical research program, through a physician, or directly from a commercial testing service. The fourth option of purchasing a home test kit is not currently available in the United States, although an over-the-counter genetic test for cystic fibrosis was recently introduced in the United Kingdom (146–148).

Patient Participating in Research. Individuals asked to participate in research programs often already have developed the condition or are related to someone who has the condition (4,9). The linkage studies used to identify disease genes are most productive using a familial database and comparing the genes of those who have the disorder with those who do not have it (144). The researchers are necessarily genetic specialists and academics who must comply with human subject requirements and other investigative protocols (9). The potential harms of poor information in this scenario are minimal since the individuals tested are at-risk to begin with, the information is often accompanied by education and counseling (26), and the results are delivered by professionals with the specialized knowledge necessary to explain the significance and experimental nature of the information and who have motive other than commercial profit. Sometimes the results are not even given to the test recipient (4). In addition to taking place in a "protected" setting (26), testing for research purposes yields important social benefits in increased scientific understanding of the human body and produces precisely the kind of information necessary to make medical determinations more reliable.

Requesting Testing through a Physician. There are basically three purposes for which genetic testing might be ordered through a physician. First, the physician might be using it to further the diagnosis and treatment of a symptomatic individual, which in the case of genetic disorders is usually a child (a related case involves presymptomatic testing of a close relative). Second, an expectant mother might seek prenatal diagnosis, or her obstetrician might recommend it based on a family

history of previous birth defects. Finally, a physician may recommend it to an individual with a general family history of a condition, or the individual might raise the possibility on his or her own after having heard of the test on the news or even through advertisements.

Diagnosis of a Symptomatic Individual. As discussed above, when genetic tests are used to make or improve diagnosis of a symptomatic individual, the dangers implicated are not significantly different from those occasioned in other types of testing. Because the test recipient presently is endangered by illness, any medically useful information may facilitate treatment and prevent imminent physical harm. The ordering physician, furthermore, is usually a specialist in the particular disorder being tested for and has a special capacity to determine the presence of a valid indication for testing (149) and evaluate the medical implications of a positive result. The case will involve a patient under continuing care whose medical and family history will have been explored in conjunction with the disorder. For family disorders, close relatives who are tested will be subject to similar conditions. Although a high rate of misinterpretation may persist even in these circumstances because of the complexity of genetics (149), these tests generally offer a greater degree of medical certainty as a starting point (149) and physicians who work with genetic disorders are more likely to be able to understand and evaluate genetic information and research (2).

Prenatal Diagnosis. Prenatal diagnosis involves somewhat special considerations, and also blends aspects of the other two cases. In a survey of physicians, obstetricians earned some of the highest marks for genetic knowledge (2). Obstetricians' work exposes them to more genetics, and they are one of the main users of clinical genetic services, including genetic counseling (29). Furthermore, until recently, most prenatal diagnosis utilized the same tests employed by pediatricians and specialists to diagnose symptomatic individuals. To have the test performed, the obstetrician may send a sample to one of the national commercial laboratory services, a regional service or the laboratory at a local hospital, perhaps one with which she is affiliated. Because of the risks involved in prenatal diagnosis (48), the tests remain conservative and focus largely on severe, predominantly single-gene disorders with well-established genetic tests. Commercialization of tests like the one for breast cancer susceptibility, however, may not leave prenatal diagnosis immune. An article noted that "some couples already are asking for [testing] the genes that confer a fifty to eighty five percent breast cancer risk" (72). The Genetics and IVF Institute has responded that they might even do it "after counseling and careful consideration" (72).

General Physician Referral. Primary care providers are likely to be on the frontlines of both demands for genetic tests by patients (150) and advertising by biotechnology companies and national laboratories who develop tests. As one journalist noted, "It's almost a daily occurrence to pick up the morning newspaper and read that scientists have found another gene that causes a medical malady" (26). Furthermore, "Once we find the gene, the stories imply, a genetic cure may be just around the corner"

(26). Most Americans are confident in the ability of their primary care physician to tell them if they are at risk and believe that the family practitioner can correctly interpret a genetic test (151). Unless their doctor has had some previous experience with genetic testing, however, their reliance may be misplaced (150). A 1993 study found that primary care physicians earned low scores on average in a test of genetic knowledge, although there was marked improvement for more recent years of graduation (152). Unfamiliarity with the essentials of genetics makes the primary care provider a somewhat unlikely candidate for ensuring independent review of the clinical utility of a new genetic test. In fact, the media and even advertising may take the decision out of the hands of primary care providers as consumers are persuaded to pursue the test based on these extrinsic sources, and physicians feel pressured not to interfere in the patient's decision making (149).

At present there are only a handful of companies developing new, advanced tests (2). Most are young, aggressive companies "hoping to take the lead in a potentially lucrative field, unconstrained by the complex FDA regulations that beset drug approvals (28,153,154). The rest of the industry is comprised of major diagnostic companies, such as Boehringer Mannheim, Ciba-Geigy, and Johnson & Johnson, and national behemoths in genetic testing services, such as Genzyme, Lab-Corp., Abbot Laboratories, and SmithKline-Beecham, which have been slower to commit resources (153). The existence of the small companies revolves around their venture capital structure, which requires quick turnarounds on profit and creates intense pressure to get a product quickly to market, perhaps even before the test's safety and effectiveness can be established (153). The smaller regional commercial testing centers, such as Clinical Diagnostic Services serving the New York metro area and Laboratories for Genetic Services in Texas, traditionally have offered a more conservative array of genetic analysis but seem to have been expanding in recent years (155). Their advertising is almost exclusively local and is directed at physicians rather than the broader public because of expense considerations. Laboratories affiliated with universities and hospitals generally offer a smaller range of testing, specialize in particular kinds of tests, and rarely advertise their services.

Getting a Test on One's Own. The last consumer is one who has not been invited to be part of a research study and has not been referred for a test by a physician. Perhaps, the individual does not wish to divulge family disorders for fear of insurance discrimination, or he or she is motivated more by curiosity and apprehension than medical necessity. The laboratories at universities, hospitals, and regional commercial services are largely unavailable without a physician referral (155). A few testing services that cater to a nationwide audience, however, have given consumers direct access to genetic testing. A prominent and controversial example is the Genetics and IVF Institute, which permits walk-in testing for BRCA mutations (145). Third-party screening of the test's validity in this scenario obviously would be nonexistent.

Need for Regulation. The need for objective information on the soundness of tests being purveyed is greatest with respect to commercial services marketing tests directly to the public. In these instances, a genetic testing service bears the greatest resemblance to a conventional commercial product. Prior to commercialization of genetic testing, there were established, informal networks for obtaining a test through specialists and other experienced providers. If a test was well-proven, knowledge about the test filtered through the professional establishment to the appropriate physicians. Now, producers promote novel tests directly to all physicians and the public (153). The consuming physician and patient need not have any independent information to obtain access. The problems of misinformation and defective information seem to be most critical where the elements of commercial interest, puffery, ready access by the public, and lack of independent review converge. In fact, the essence of these elements lies in the flow of information, that is, who has it and who depends on it. A negative change in the source of information away from the supervising physician to a financially self-interested company or service may be summarized in one word: marketing. Relying for information on the very parties who have the most to benefit from praising the test, however, obviously leaves a great deal to be desired.

Current State of Regulation

However, the existing framework of regulation focuses not on where the greatest harm may exist, but on whether a test is packaged or performed on site. Providers and developers of genetic testing are currently subject to a patchwork of federal and state regulation based on the form of the activity engaged in rather than the substance. Biotechnology companies who package the test as a kit to be distributed to laboratories and physicians are subject to FDA regulation. Companies and laboratories that deliver the testing as a service, sometimes by accepting samples through the mail, are not subject to any regulations to ensure safety and effectiveness. Instead, only the technical competence of a laboratory's testing is regulated under the Clinical Laboratory Improvement Act. Laboratories also are licensed by the state and voluntarily may undertake limited FDA compliance by setting up Institutional Review Boards (IRBs) (8). Laboratories associated with academic and research institutions are regulated similarly, and usually have IRBs and prescribed human subjects protocols (8).

FDA Regulation of Kits. FDCA as amended by MDA (107) and the Safe Medical Devices Act of 1990 (156) established a comprehensive system to regulate the safety and effectiveness of medical devices. The level of review depends on a device's classification (159). Class I devices are devices that pose no unreasonable risk to health and whose safety and effectiveness can be ensured by general controls (158). Class II devices require specialized controls to ensure their safety and effectiveness (158). Most genetic tests, however, would probably fall into Class III because of their complexity (160,161). Class III devices require submission of a premarket approval application for FDA review unless "substantial equivalence" can be established with a previously approved device through Section 510(k) notification (162). When an application is received, FDA assigns the product to an examiner who possesses the relevant scientific background to review the application. Although no more than a handful of genetic testing products have been submitted for review (163), the agency already has designated an examiner to specialize in genetic tests (164). Oncor's Inform Her-2/Neu breast cancer genetic diagnostic was one of the first genetic tests to pass FDA review (165). When Oncor first submitted a version of the gene diagnostic kit, FDA rejected the application (166). It was not until after the company "expanded reproducibility studies, set forth manufacturing criteria, and ... develop[ed] a training program for the physicians and technicians using the test" that it was approved for use in predicting the likelihood of recurrence in previously symptomatic patients (166). FDA also requires compliance with good manufacturing practice (GMP) standards, monitors advertising, and expedites clearance by utilizing postmarket control (83). The agency further might condition approval on the company developing a system to ensure counseling or physician consultation (167).

Clinical Laboratory Improvement Amendments. The Clinical Laboratory Improvement Amendments of 1988 (CLIA) were passed to establish minimum quality standards for laboratory testing (168). The Health Care Financing Administration (HCFA), which implements CLIA, reviews some 158,000 laboratories across the country pursuant to the act (169). CLIA sets forth quality standards for proficiency testing, patient test management, quality control, personnel qualifications and quality assurance (170). There is no effort under CLIA's regulatory administration to establish the utility of or the validity of using any particular test. Under CLIA, the laboratory need only demonstrate that its procedures result in accurate and reliable identification of the target, namely, technical competence in test performance (2). Furthermore, much of the review process is conducted through third-party accreditation organizations and programs (171). The scope of review is limited to laboratory procedures and its purpose, to certify labs for payment under federal health plans, is largely administrative.

State Regulation. The states of Washington, Oregon, and New York have laboratory review regulations rigorous enough to qualify them for exemption from CLIA regulation by the HCFA (172). However, only New York undertakes any significant assessment of genetic testing. If a company or laboratory located in New York or providing testing to New York residents wishes to offer a genetic test, it must become certified for the particular test, a process that involves consideration of the quality and effectiveness of the test (28,173).

Choosing a Regulatory Actor

There is thus a sharp regulatory divide between genetic testing kits, which receive the full intensity of FDA review of safety and effectiveness, and widely marketed genetic tests offered as a service, which generally receive no review of safety and effectiveness. It is a regulatory divide not warranted by any significant factual distinctions. A genetic testing recipient requires as much protection from the harms of a proprietary genetic test provided as a service, as a genetic test sold as a kit. If a company promotes a genetic test in interstate marketing directly to the public, the test, even if performed as a service, loses the characteristics of medical practice and carries the hallmarks of commercialization. It is a product being purveyed to the public, and as such, should be subject to the scrutiny of the nation's gatekeeper of medical devices, the FDA. In fact FDA is the only agency whose existing legislative authorization might cover regulation of genetic testing services.

The lack of convincing alternatives emphasizes the appropriateness of FDA oversight. Besides FDA, there is a fairly short list of other serious potential regulators of genetic testing: the Centers for Disease Control (CDC), the HCFA, or the Department of Health and Human Services (DHHS) (8). Both CDC and HCFA are agencies within DHHS (as is FDA). CDC "is responsible for promoting health and quality of life by preventing and controlling disease, injury, and disability" (174). CDC is the nation's prevention center; its efforts are directed primarily at monitoring, researching, and preventing diseases, and cover a broad range of social activities including collaborative projects with either HCFA or FDA (175,176). HCFA administers the Medicare and Medicaid programs and CLIA (177). HCFA's duties in certifying laboratories and reviewing testing proficiency are related to its administration of Medicare and Medicaid and are intended to ensure that laboratories are qualified for the work for which they are being reimbursed, which is similar to its authority to prescribe nursing home standards (178). Finally, DHHS is mentioned not so much to suggest that the department itself undertake regulation of genetic testing, but that a new body within DHHS be created to do so. As an initial matter, it is not apparent that these alternative regulatory bodies have any existing statutory authorization to regulate the clinical validity of genetic testing, which creates a substantial hurdle to effective action and suggests a lack of traditional jurisdiction. Under CLIA, the CDC determines the categorization for tests performed by laboratories, and HCFA reviews their competence. As discussed, however, CLIA does not include the authority to review the quality or scientific validity of a genetic test and focuses instead exclusively on proficiency. However, the public policy question remains as to whether the CDC or HCFA have more expertise and a more appropriate structure than FDA to substantively regulate genetic tests, or whether all the agencies are so poorly suited to the task that a new body should be invented.

HCFA does not appear to have developed particular genetics expertise in its regulation of clinical laboratories (179,180). While it may have a better infrastructure to undertake widespread regulation of many different sites, the clinical validity of a type of genetic test only needs to be established once. Nor does HCFA seem to have familiarity in conducting a substantive scientific review. CDC could be a more promising candidate because of its scientific expertise and prominent historical role in disease control. Setting up a new process for conducting safety and effectiveness review for genetic testing services under CDC, however, seems redundant because FDA already is implementing similar measures to regulate genetic testing kits. The science and protocols for evaluating a genetic test for clinical utility should not change based on how the test is delivered. Furthermore the issues raised in regulating genetic testing conform to those routinely raised for FDA such as identifying a professional standard, expediting clearance through postmarket controls and tracking, monitoring advertising, and determining equivalence with a previously approved produce.

In fact, the only benefit of choosing another agency or charting a new one comes not from any advantage in scientific qualifications or institutional expertise, but in avoidance of perceived excessive bureaucracy at FDA. This concern, however, may be misjudging the matter. Government review and approval of a diagnostic is inherently bureaucratic. Creating a new organization to undertake the task or assigning it to one with no experience is nothing more than reinventing the wheel and hoping it turns out better this time. Not only is such a course somewhat doubtful in rationale, it is duplicative of existing FDA regulation of genetic testing kits and therefore wasteful of public resources. From the perspective of expertise and experience, it is clear that FDA may most readily provide the regulation called for in genetic testing services.

CONCLUSION

Perhaps the most disturbing aspect of the issues discussed is the immediate need for oversight. FDA is the best positioned agency from the perspective of jurisdiction and institutional expertise to undertake the task, and what it requires is not so much legislative authorization as a social and political mandate. Promptness in action is not only essential in light of the particular circumstances in genetic testing but foreshadows the proficiency of society to address these problems in the future. Commercialization of genetic testing is one of the first challenges presented by the wondrous and rapid developments in genetic technology, and it only can be hoped that the government will prove itself capable of managing the dangers with the energy and foresight necessary to ensure that the public safely receives the benefits of a new era in genetics.

ACKNOWLEDGMENT

Originally published as A. Huang, *Food Drug Law J.* **53**, 555–591 (1998). Reprinted with the permission of The Food and Drug Law Institute.

BIBLIOGRAPHY

1. President W.J. Clinton, *U.S. Newswire*, February 8 (2000).
2. S.N. Hurd, *Law Polic Rep.*, October, 161 (1997).
3. Task Force on Genetic Testing, NIH–DOE, *Final Report: Promoting Safe and Effective Genetic Testing in the United States*, U.S. Government Printing Office, Washington, DC, September 1997.
4. L. Andrews et al., *Institute of Medicine Assessing Genetic Risks: Implications for Health and Social Policy*, 1993.
5. E. Lane, *Newsday*, November 5, 17 (1993).
6. L. Seachrist, *Bioworld Today*, April 7, 1 (1997).
7. M.K. Pendergast, FDA Deputy Comm'r. & Special Advisor to the Comm'r. Hearing before House Comm. on Science, 104th Cong. 2nd Sess. (1996).
8. B. Burlington, Director, Center for Devices and Radiological Health, to FDA, Task Force on Genetic Testing, April 3 (1997). Contents available at: FDA Freedom of Information Office, 5600 Fisher's Lane #12A16, Rockville, MD 20857.
9. M.J. Malinowski and R.J.R. Blatt, *Tul. Law Rev.* **71**, 1211, 1296 (1997).
10. Department of Health and Human Services, Advisory Committee on Genetic Testing, *The Gray Sheet*, November 1 (1999).
11. Human Genome Initiative *Genome Database* (visited July 21, 1998), Available at: <http://www.gdb.org/>
12. *Nat. Biotech.*, December, 1627 (1996).
13. L. Wasowicz, *Assoc. Press, BC Cycle*, November 23, 1996, *Available in:* Westlaw, Allnews Database.
14. Medical Device Amendments (MDA), Pub. L. No. 94-295, 90 Stat. 539 (codified at 15 U.S.C. §55 (1994), scattered sections of 21 U.S.C.).
15. J.W. Hulse et al., *Food Drug Law J.* **48**, 285–287 (1993).
16. L.M. Fisher, *N.Y. Times*, January 1, A1 (1998).
17. *Bus. Wire*, May 25, 1998, *Available in:* Westlaw, Allnews Database.
18. *PR Newswire*, August 22, 1989, *Available in:* Westlaw, Allnews Database.
19. R. Weiss, *Washington Post*, May 26, A1 (1996).
20. J.D. Watson et al., *Recombinant DNA*, 2nd ed., 539, 1992.
21. M. Guttman, *USA Today (Weekend)*, February 9 (1997).
22. S. Seshadri et al., *Archi. Neurol.* 1074 (1995).
23. T.M. Maugh II, *L.A. Times, reprinted in The Record* (N. NJ), March 16, 1998, at H04.
24. *Pulse*, February 28, 32 (1998).
25. *Chem. Bus. NewsBase*, February 16 (1999).
26. N.J. Nelson, *J. Natl. Cancer Inst.*, January 17, 70 (1996).
27. W.R. Wineke, *Wis. St. J.*, March 9, 1C (1998).
28. R. Osborne, *Bioworld Today*, February 24, 1 (1998).
29. *Standard & Poor's*, August 28, 2 (1997).
30. H.M. Kingston, *ABC Clin. Genet.* 31 (1997).
31. C. Cookson, *Fin. Times (Tech.)*, June 27, 12 (1996). *Available in:* Westlaw 1996 WL 10598276.
32. Genetic Testing Quality Assurance Office, New York State Department of Public Health, New York State Genetic Testing Laboratories F12, F13 (1998).
33. Congressional testimony of Francis S. Collins, 1998 WL 12760444 (May 21, 1988).
34. S. Okie, *Washington Post*, May 26, Z11 (1998).
35. *Irish Times*, May 21, 15 (1998).
36. *Chicago. Trib.*, May 6, 7 (1998).
37. *Gene Therapy Wkly*, April 27 (1998).
38. P.S. Harper, in P.S. Harper and A.J. Clarke, eds., *Genetics, Society and Clinical Practice*, vol. 7, 1997.
39. *Marketletter*, May 18 (1998), *Available in* 1998 WL 11622549.
40. Z. Moukheiber, *Forbes*, June 1, 119 (1998).
41. R. Kotulak, *Chicago. Trib.*, May 14, 1 (1998).

42. J. Siegel-Itzkovich, *Jerusalem Post*, November 23, 10 (1997).

43. P. Recer, *Assoc'd. Press*, May 17 (1998).

44. D. Brown, *Washington Post*, February 20, E1 (1998).

45. D.L. Wheeler, *Chron. Higher Educ.*, November 29, A24 (1999).

46. M. Elias, *Red Flag Raised on Gene Test, USA Today*, October 27, 1D (1997).

47. N. Peterson, *Breast Cancer Action* (Breast Cancer Action, San Francisco, CA), June 1996.

48. P.A. Hoffee, *Med. Mol. Genet.* 279–280 (1998).

49. D. Shattuck-Eidens et al., *J. Am. Med. Assoc.* **278**, 1284 (1997).

50. P. Kahn, *Science* **274**, 496–498 (1996).

51. J. Gillis, *Washington Post*, April 7, C2 (1998).

52. Stanford University Program in Genomics, Ethics, and Society, Report of the Working Group on Genetic Testing for Breast Cancer Susceptibility (November 23, 1996), Available at: <http://www-leland.stanford.edu/dept/scbe/brcaexec.htm>

53. B.A. Koenig et al., *J. Women's Health* **7**, 531–545 (1998).

54. A. Marshall, *Nat. Biotech.* **14**, 1642 (1995).

55. B.A. Brenner, *Breast Cancer Action* (Breast Cancer Action, San Francisco, CA), June 1996.

56. G. Kolata, *N.Y. Times*, April 1, A1 (1996).

57. Myriad Genetics Laboratories, Inc., BRCA1 Genetic Susceptibility for Breast and Ovarian Cancer: A Reference for Healthcare Professionals in Anticipation of BRCA1 Genetic Susceptibility Testing (January 1996).

58. B. Newman et al., *J. Am. Med. Assoc.* **279**(12), 915–921 (1998).

59. A. Petrij-Bosch et al., *Nat. Genet.* **17**(3), 341 (1997).

60. L. Jennings, *Comm. Appeal*, April 26, F1 (1998).

61. T.H. Maugh II, *L.A. Times*, May 25, S2 (1998).

62. P. Lynden, *Am. Health for Women*, June 1, 29 (1997).

63. C. Wallis et al., *Time*, May 18, 48 (1998).

64. L. Seachrist, *Bioworld Today*, April 13 (1998).

65. B. Kuska, *J. Nat. Cancer Inst.* **87**, 1578 (1995).

66. L. Sowers, *Houston Chron.*, August 17, 1 (1997).

67. J. Brandt et al., *J. Am. Med. Assoc.* **261**, 3108 (1989).

68. N.M. Resnick, *Pittsburgh Post-Gazette*, March 17, G1 (1998).

69. D. Grady, *N.Y. Times*, January 7, C3 (1997).

70. P. Bereano and R. Sclove, *Washington Post*, March 22, C5 (1998).

71. *Smith v. Olin Chem.* Corp., 555 F.2d 1283 (5th Cir. 1977).

72. K. Painter, *USA Today*, August 15, 1A, 2A (1997).

73. J. Talan, *Newsday*, March 21 (1998), at 1998 WL 2663370.

74. Mayo Clinic, *Lou Gehrig and Stephen Hawking: Stars Despite ALS* (visited September 4, 1998), Available at: <http://www.mayohealth.org/mayo/9604/htm/als_sb.htm>

75. *Wkly J.*, January 30, 12 (1998).

76. C. Francke, *Baltimore Sun*, October 30, 1A (1998).

77. J. Price, *Washington Times*, September 2, A1 (1997).

78. C. Russell, *Washington Post*, February 24, Z08 (1998).

79. R. Kotulak, *Chicago. Trib.*, October 17 (1997).

80. M. Downey, *Atlanta J. & Atlanta Const.*, October 19, 1 (1997).

81. S. Roan, *L.A. Times*, April 29, 13 (1997).

82. *Fed. Regist.* **61**, 10,484 (1996) (to be codified at 21 CFR §809 & 864).

83. R.A. Merrill, *Virginia A. Law Rev.* **82**, 1753 (1996).

84. Regulation of New Drug R&D by the Food and Drug Administration, 1974: Joint Hearings before the Subcomm. on Health of the Senate Comm. on Labor and Public Welfare and the Subcomm. on Administrative Practice and Procedure of the Senate Comm. on the Judiciary, 93d Cong., 2d Sess. 207 (1974) (statement of Dr. Alexander M. Schmidt, the FDA Comm'r).

85. H. Burkholz, *The FDA Follies* 105-41 (1994). J. Schwartz and R. Marcus, *Washington Post*, July 13, A4 (1996).

86. J. Schwartz, *Washington Post*, July 15, A17 (1996).

87. M. Kraus, *Cal. W. Law Rev.* **33**, 101 (1996).

88. P. Duggan, *Washington Post*, October 12, B1 (1988).

89. J. Schwartz, *Conservative Foes of Government Regulation Focus on the FDA, Washington Post*, January 21, A7 (1995). 142 CONG. REC. H5631-38 (daily ed. May 29, 1996) (statements of Reps. Greenwood, Barton, Klug, Burr, and Fox).

90. M. Simons, *L.A. Times*, January 2, A1 (1995).

91. *World News Tonight* (ABC television broadcast, February 21, 1996).

92. *Med. Health* **49**, (Feb. 20), 1 (1995).

93. W.E. Leary, *N.Y. Times*, June 23, A18 (1994).

94. J. Mann, *Washington Post*, April 22, E3 (1994).

95. E.J. Millenson, *Washington Post*, August 27, C5 (1989).

96. *Buffalo News*, June 24, C2 (1994).

97. H.H. Harris, *Washington Post*, April 9, G1, G5 (1994).

98. C.T. Fang et al., *Patient Care* **23**, October 30, 19 (1989).

99. L. Garrett, *Newsday*, June 23, A7 (1994).

100. D.J. Murphy, *Investors Bus. Daily*, October 1, A4 (1994).

101. J.C. Fletcher and D.C. Wertz, *Emory Law J.* **39**, 747 (1990).

102. T. Duster, *Backdoor to Eugenics*, 1990.

103. D.B. Paul, *Soc. Res.* **59**, 663–683 (1992).

104. D.J. Kevles, *In the Name of Eugenics: Genetics and the Uses of Human Heredity*, 1985.

105. P. Reilly, *Genetics, Law, and Social Policy*, 1977.

106. J.L.J. Nuzzo, *Food Drug Law J.* **53**, 35, 40 (1998).

107. J.E. Hoffman, *Food and Drug Law*, 1–2 (1991).

108. Regier, *Law Contemp. Probs.* **1** (1933).

109. I. Scott Bass, *Food Drug Law* **61** (1991).

110. Pure Food and Drugs Act, Pub. L. No. 59–384, 34 Stat. 768 (1906) (codified at 21 U.S.C. §§1–15 (1934)) (repealed in 1938 by 21 U.S.C. §392(a)).

111. *Report of the Secretary of Agriculture on Deaths Due to Elixir Sulfanilamide-Massengill*, S. Doc. No. 124, 75th Cong., 1st Sess. (1937).

112. Pub. L. No. 75-717, 52 Stat. 1040 (codified at 21 U.S.C. §§301 et seq.).

113. 21 U.S.C. §355(d) (Supp. 1998).

114. S. Rep. No. 1744, 87th Cong., 2d Sess. 16 (1962).

115. *Fed. Regist.* **37**, 16,503 (August 15, 1972).

116. *Fed. Regist.* **62**, 62,243 (November 21, 1997).

117. *United States v. Apex Distributing Co.*, 148 F. Supp. 365 (D. R.I. 1957).

118. U.S. Const. art. I, §8, cl. 3.

119. G.R. Stone et al., *Constitutional Law* 194 (1991).

120. *United States v. Lopez*, 514 U.S. 549, 558 (1995).

121. *United States v. Sullivan*, 332 U.S. 689 (1948).

122. *United States v. Funk*, 412 F. 2d 452 (8th Cir. 1969).

123. *Dean Rubber Mfg. Co. v. United States*, 356 F. 2d 161 (8th Cir. 1966).

124. *Almendarez-Torres v. United States*, 118 S. Ct. 1219, 1227 (1998); *United States ex rel. Attorney General v. Delaware & Hudson Co.*, 213 U.S. 366, 408 (1909).

125. 317 U.S. 111 (1942).

126. N.A. Holtzman et al., *Science*, October 24, 602 (1997).

127. 21 U.S.C. §331(k) (FDCA §301(a)). 21 U.S.C. §321 (h) (FDCA §201(h))

128. 932 F.2d 813 (9th Cir. 1991).

129. *Grand Labs. v. Harris*, 660 F.2d 1288 (8th Cir. 1981) *cert. denied sub nom. Grand Labs. v. Schweiker*, 456 U.S. 927, 102 S. Ct. 1972, 72 L. Ed. 2d 442 (1982).

130. *Kordel v. United States*, 335 U.S. 345, 349 (1948).

131. *Drown v. United States*, 198 F.2d 999 (9th Cir. 1952).

132. *Western Union Tel. Co. v. Foster*, 247 U.S. 105 (1918).

133. *Barnes v. United States*, 142 F.2d 648 (9th Cir. 1944).

134. J. Weber, *Bus. Week*, March 16, 108 (1998).

135. *United States v. An Undetermined Number of Unlabeled Cases*, 21 F.3d 1026 (10th Cir. 1994).

136. C.J. Walsh and A. Pyrich, *Rutgers Law Rev.* **48**, 883, 890 (1996).

137. *Linder v. United States*, 268 U.S. 5, 18, 45 S. Ct. 446, 449, 69 L. Ed. 819 (1925).

138. 453 F. Supp. 1141 (M.D. Ala. 1978).

139. *United States v. Evers*, 643 F.2d 1043 (5th Cir. 1981).

140. Food and Drug Administration Modernization Act of 1997 (the FDAMA), Pub. L. No. 105–115, 111 Stat. 2296 (1997).

141. FDCA §201(h), (codified it 21 U.S.C. §321).

142. J.M. Zitter, *A.L.R. Fed.* **129**, 343 (1996).

143. University of California, San Diego, Biochemical Genetics, *Biochemical Genetics Test List* (UCSDW3BG) (visited July 22, 1998), Available at: <http:www.biochemgen.ucsd.edu>

144. S. Gottlieb, *Times Union*, March 12, A11 (1998).

145. Telephone Interview with Genetics and IVF Institute customer service representative (Apr. 8, 1998) (tape on file with author).

146. C. Cookson and C. Adams, *Fin. Times*, September 24, 12 (1997).

147. E. Chang, *Washington Post* (Magazine), September 7, W19 (1997).

148. A. Heller, *Drug Store News*, October 6, 21 (1997).

149. F.M. Giardiello et al., *N. Engl. J. Med.* **12**, 823, 825 (1997).

150. L. Seachrist, *Bio World Today*, March 12, 1 (1998).

151. *U.S. Newswire*, March 11 (1998).

152. K.J. Hoffman et al., *Acad. Med.* **68**, 625 (1993).

153. P.H. Silverman, *Hastings Cent. Rep.*, May 1, S16 (1997).

154. M. Guidera, *Baltimore Sun*, April 2, 1D (1998).

155. Telephone Interview with Joe Robinson, Coordinator, New England Regional Genetics Group (May 22, 1998).

156. Pub. L. No. 101–629, 104 Stat. 4511 (codified in scattered sections of 21 U.S.C. §§301 note-383note, 42 U.S.C. §263b-263n).

157. 21 U.S.C. §360c (FDCA §513).

158. 21 U.S.C. §360c(a)(1)(A) (FDCA §513(a)(1)(A)).

159. 21 U.S.C. §360c(a)(1)(B) (FDCA §513(a)(1)(B)).

160. M.J. Malinowski and M.A. O'Rourke, *Yale J. Reg.* **13**, 163, 206 (1996).

161. 21 U.S.C. §360c(a)(1)(C) (FDCA §513(a)(1)(C)).

162. Food and Drug Administration Modernization Act of 1997, Pub. L. No. 105–115, 111 Stat. at 2296.

163. Telephone Interview with Dr. Steven Gutman, Dir., Div. of Clinical Laboratory Devices, FDA (May 1, 1998).

164. Food and Drug Administration, Center for Devices and Radiological Health, Monitoring the Human Genome Project for Impact on Developments in Medical Devices [contents available at the FDA Freedom of Information Office, 5600 Fisher's Lane #12A 16, Rockville, MD 20857].

165. B. Berselli, *Washington Post*, January 26, F5 (1998).

166. *Daily Record*, June 11, 2 (1997).

167. *Chicago Trib.*, May 15, 3 (1996).

168. Pub. L. No. 100–578, 102 Stat. 2903 (codified at 42 U.S.C. 201 note, 263a, 263a note).

169. Health Care Financing Administration, *Clinical Laboratory Improvement Amendments* (visited July 28, 1998), Available at: <http://www.hcfa.gov/medicare/hsqb/clia1.htm>

170. 42 CFR 493 *et seq.*

171. Regulatory notices HCFA-2246-N (published 4/9/98) (approving the Joint Commission on Accreditation of Healthcare Organizations, the American Association of Blood Banks, and the American Osteopathic Association) and HSQ-242-N (published 5/19/97) (approving the Commission on Office Laboratory Accreditation).

172. Regulatory notices HSQ-243-N (published 4/9/98) (exempting Washington), HSQ-231-N (published 6/31/96) (exempting Oregon) and HSQ-230-N (published 8/28/95) (exempting New York).

173. Telephone Interview with Nanette P. Healy, Assistant Administrator, New York Genetic Testing Quality Assurance Office (January 20, 1998).

174. Centers for Disease Control and Prevention, *CDC FY 1999 President's Budget Request, General Statement from the FY99 Congressional Justification* (visited Aug. 11, 1998), Available at: <http://www.cdc.gov/fy99bdgt/99genstm.htm>

175. CLIA Program; Cytology Proficiency Testing, *Fed. Regist.* **60**, 61,509 (November 30, 1995) (to be codified at 42 C.F.R. pt. 493), 1995 WL 701433.

176. *Med. Malpractice Law Strategy* **14**, 1 (January 1997).

177. Office of the Inspector General, Health Care Financing Administration Projects, 995 PLI/Corp. 585, 639 (June 1997).

178. Medicaid Program; Standards for Intermediate Care Facilities for the Mentally Retarded, Proposed Rules, DHHS, HCFA, *Fed. Regist.* **51**, 7520 (March 4, 1986) (to be codified at 42 C.F.R. pts. 435, 442) 1986 WL 90426.

179. *CLIA & Genetic Testing*, F-D-C REP. ("The Blue Sheet"), (February 11, 1998), Available in Westlaw, 1998 WL 9283076.

180. Summaries of board meetings are available at HCFA, Office of Information Services, Division of Freedom of Information and Privacy, 7500 Security Boulevard, Room C2-01-11, Baltimore, MD 21244-1850.

See other entries FDA REGULATION OF BIOTECHNOLOGY PRODUCTS FOR HUMAN USE; see also GENETIC INFORMATION, ETHICS and GENETIC INFORMATION, LEGAL entries.

GENETIC INFORMATION, LEGAL, GENETIC PRIVACY LAWS

Mary R. Anderlik
University of Houston
Houston, Texas

Rebecca D. Pentz
M.D. Anderson Cancer Center
Houston, Texas

OUTLINE

INTRODUCTION

Privacy is a broad concept. In the legal context it includes at least four categories of concern: (1) access to persons and personal spaces; (2) access to information by third parties, and also any subsequent disclosure of this information by third parties (the category of concern best captured by the term "confidentiality"); (3) third-party interference with personal choices, especially in intimate spheres such as procreation; and (4) ownership of materials and information derived from persons (1). Typically statutes characterized as "genetic privacy acts" address a number of these concerns, regulating genetic testing and other means of generating genetic information, limiting access to genetic information, and ruling out certain uses of genetic information by third parties. However, some states have laws that are much narrower. Further, even laws that are broad in terms of the range of privacy concerns addressed may be narrow in scope, targeting a particular industry or type of information keeper. For example, many genetic

privacy laws were passed in response to abuses, actual or perceived or potential, in the insurance industry and by employers. Some laws are tailored to problems that arise in the areas of research or law enforcement or family relations.

For organizational purposes, this article is divided into four sections. Each section covers major developments at the state and federal levels, and common (or judge-made) law as well as statutory law, to the extent relevant. Given the range and complexity of the issues, this entry is intended as an overview. For more in-depth treatment, the reader may wish to consult the sources cited in the references.

LAWS REGULATING THE GENERATION, DISCLOSURE, AND USE OF GENETIC INFORMATION

State Laws

Genetic Privacy Laws. At present, the only laws that comprehensively address genetic privacy are at the state level. At least 35 states have some statutory law relating to genetic privacy (excluding laws authorizing creation of DNA banks and databases for law enforcement purposes, which may include privacy protections). See Table 1. The most comprehensive of these laws include general provisions covering genetic testing and the handling of genetic information, accompanied by more focused provisions addressing special concerns that arise in connection with insurance and employment, and in Texas and Wisconsin, occupational licensing.

Most genetic privacy laws cover at least two aspects of privacy, access to persons (for testing) and access to information. See Table 2. About half prohibit genetic testing of persons or samples without prior informed consent, subject to certain exceptions, for example, law enforcement, paternity determination, court order, and anonymous research. In many states the elements of the consent form are specified; standard elements include a description of the test, a statement of the purposes of testing, and the names of the persons or entities to whom results may be released. It is common for laws to contain a statement that genetic information is confidential, or "confidential and privileged," meaning that it is protected from subpoena in a civil proceeding, although production can still be compelled by a specific court order. Disclosure of genetic information to a third party without written authorization or consent is generally prohibited unless an exception applies. The standard list of exceptions parallels the list of exceptions to the consent requirement. Some states also permit disclosure of genetic information for the benefit of blood relatives if the subject is dead (2). Usually the authorization for disclosure must be specific rather than general. Many state laws provide that these legal protections only apply to genetic material or information that can be identified as belonging to an individual or family.

In the area of insurance, a major issue is breadth of application of genetic privacy laws. Many states limit special privacy protections for genetic testing and information to *health* insurance, leaving consumers with few or no safeguards in their dealings with life, disability

Table 1. State Genetic Privacy Laws (as of November 30, 1999)

State	Statute
Alabama	Ala. Code §27-53-1 et seq.
Arizona	Ariz. Rev. Stat. Ann. §20-448.02
California	Cal. Civ. Code §56.17; Cal. Health & Safety Code §1374.7; Cal. Ins. Code §§742.405, 742.407, 10123.3, 10123.35, 10140, 10140.1, 10146 et seq.
Colorado	Colo. Rev. Stat. §10-3-1104.7
Connecticut	Conn. Gen. Stat. Ann. §46a-60
Delaware	Del. Code Ann. tit. 16, §1220 et seq.; Del. Code Ann. tit. 19, §711
Florida	Fla. Stat. Ann. §§627.4301, 636.0201, 760.40
Georgia	Ga. Code Ann. §33-54-1 et seq.
Hawaii	Haw. Rev. Stat. §§431:10A-118, 432:1-607, 432D-26
Illinois	Ill. Comp. Stat. Ann. §513/1 et seq.
Indiana	Ind. Code Ann. §27-8-26-1 et seq.
Kansas	Kan. Stat. Ann. §44-1002 et seq.
Kentucky	Ky. Rev. Stat. Ann. §304.12-085
Louisiana	La. Rev. Stat. Ann. §22:213:7
Maine	Me. Rev. Stat. Ann. tit. 5, §19301-02; Me. Rev. Stat. Ann. tit. 24-A, §2159-C
Maryland	Md. Ins. Code Ann. §27-909
Minnesota	Minn. Stat. Ann. §72A.139
Missouri	Mo. Ann. Stat. §375.1300 et seq. (Vernon's)
Montana	Mont. Code Ann. §33-18-901 et seq.
Nevada	Nev. Rev. Stat. Ann. §§629.111 et seq., 689A.417, 689B.069, 689C.198, 695C.207, 695B.317
New Hampshire	N.H. Rev. Stat. Ann. §141-H:1 et seq.
New Jersey	N.J. Stat. Ann. §§10:5-12, 10:5-43 et seq., 17B:30-12
New Mexico	N.M. Stat. Ann. §24-21-1 et seq.
New York	N.Y. Civ. Rights Law §79.1; N.Y. Ins. Law §2612 (McKinney)
North Carolina	N.C. Gen. Stat. §58-3-215
Ohio	Ohio Rev. Code Ann. §§1751.64, 1751.65, 3729.46, 3901.49, 3901.491, 3901.50, 3901.501 (Baldwin)
Oklahoma	Okla. Stat. Ann. §tit. 36, §3614.1 et seq.
Oregon	Or. Rev. Stat. §§659.036, 659.700 et seq.
Rhode Island	R.I. Gen. Laws §§27-18-52, 27-19-44, 27-20-39, 27-41-53, 28-6.7-1 et seq.
South Carolina	S.C. Code Ann. §38-93-10 et seq.
Tennessee	Tenn. Code Ann. §56-7-2701 et seq.
Texas	Tex. Labor Code Ann. §21.401 et seq.; Tex. Ins. Code Ann. §21.73; Tex. Rev. Civ. Stat. Ann. §9031 (Vernon's)
Vermont	Vt. Stat. Ann. tit. 18, §9331 et seq.
Virginia	Va. Code Ann. §§38.2-508.4, 38.2-613
Wisconsin	Wis. Stat. Ann. §942.07

income, and long-term care insurers, among others. Texas is unusual in limiting protections affecting insurance to group health benefit plans (3). Legislation specifically directed at life insurers has been enacted by a number of states, but the protections afforded applicants for life insurance are minimal. Typically these laws require only that informed consent be obtained prior to performance

of any genetic testing and/or that any use of genetic information in medical underwriting meet standards of actuarial fairness. Note that state protections are generally considered to be inapplicable to self-funded employer-sponsored benefit plans, including employer-sponsored health and life insurance, due to the operation of the federal Employee Retirement Income Security Act of 1974, commonly known as ERISA (4).

Many genetic privacy laws are silent on the issue of retention of samples (i.e., biological specimens obtained or retained for the purposes of genetic testing). A few states require destruction of samples upon specific request, or after the purpose for which the sample was obtained has been accomplished. The New York law requires that the sample be destroyed at the end of the testing process or not more than 60 days after the sample is taken, unless a longer period of retention is expressly authorized (5). Laws that require destruction of samples typically include exceptions related to research and law enforcement, areas discussed in more detail below.

States also vary in the sanctions imposed for violations of privacy protections. In most states, a violation is a misdemeanor punishable by fine or jail time or both. (A willful violation may be a felony.) Further, a number of states allow individuals to sue for equitable relief, such as an order to stop a violation, and damages, costs and attorney fees (6). The Louisiana law relating to health insurance allows recovery of the greater of actual damages or $50,000 ($100,000 in cases of willful violation) against persons who violate the law by negligently collecting or disclosing genetic information, and the law authorizes treble damages where a violation resulted in monetary gain (7).

Definitional Issues. Whether a law relates to insurance, employment, or has a more general orientation, the choices legislators make in two key areas have significant implications: (1) how to define the category or categories of protected material or information, and (2) whether and how to address the problem of compelled consent.

State genetic privacy laws typically focus on the generation, disclosure and use of "genetic information," a term that may be linked to a definition of "genetic test" and/or "genetic characteristic." Many states that have genetic privacy laws use one or more of these definitions to limit protections to persons who are presymptomatic or asymptomatic for disease. (The protections in an Alabama law are even more limited, only applying to presymptomatic testing for a predisposition to cancer [8].) From a policy perspective, this limitation may be defensible; it is unclear why persons with diseases that are genetic in origin should be favored vis-à-vis persons with diseases that arise in some other fashion. On the other hand, the "genetic revolution" has blurred many lines that formerly appeared sharp, even creating questions about what counts as a disease. For example, the Illinois Genetic Information Privacy Act provides that results of genetic testing that indicate the person is already afflicted with a disease, "whether or not currently symptomatic," are not subject to the confidentiality requirements of the act (9). This provision suggests that a disease may exist

Table 2. Summary of Key Provisions in State Genetic Privacy Laws (as of November 30, 1999)

Provision	State
General provisions	
Genetic testing generally prohibited without prior informed consent	AZ, DE, FL, GA, NV, NH, NJ, NM, NY, OR, SC, VT
Standard exceptions include law enforcement, paternity determination, court order, anonymous research	
Standard elements of consent form include description of test, statement of purpose(s), who receives results	
Person requesting genetic testing must advise of risk of discrimination	VT
Genetic information confidential and/or privileged	AZ, CA, CO, GA, IL, MO, NY, OK, OR, SC, TX
No release of genetic information without specific authorization	AZ, CA, CO, DE, FL, GA,
Standard exceptions include law enforcement, paternity determination, court order, anonymous research, benefit of relatives (subject deceased)	HI, IL, LA, MD,[a] MO, NV, NJ, NM, NY, OR, SC, TN,[a] TX,[b] VT, VA,[a] WI,[c]
Requests for genetic services expressly protected	DE, HI, LA, KY, MD, MN, MO, NV, NH, NM, RI, VT, VA
Destruction of (identifiable) sample generally required upon request and/or accomplishment of purpose	DE, LA, NV, NJ, NM, NY, OR, TX
Definitions	
Narrow definition of genetic test/characteristic/information:	
• Test (direct) for alterations in genes	CO, IN, KS, MO, OH
• Test for predisposition and/or environmental damage and/or carrier status	AZ, DE, IN, ME, NJ, NM, NY, TX, WI
• Person must be a- or presymptomatic	AL, CA, FL, GA, IL,[d] KY, ME, MD, MN, MT, OK, SC, TN, VT, VA
Genetic information expressly excludes family history	FL, MO, MT, OK, TN
Insurance and employment	
Health insurers prohibited from requiring genetic testing[e]	AL,[f] CA, FL, GA, HI, IL, IN, KY, LA, ME, MD, MN, MO, MT, NV, NH, NJ, OH, OK, RI, SC, TN, TX,[g]
Health insurers prohibited from requesting genetic testing and/or information (some states add "for nontherapeutic purposes")[e]	CA, CO, FL, GA, HI, IL,[h] IN,[h] KY, LA, MD, MN, MO, MT, NH, OH, OK, RI, TN
Employers prohibited from requiring genetic testing (some states add "unless job related" or similar language)	CT, DE, KS, ME, NH, NJ, OK, OR, RI, TX, VT
Employers prohibited from requesting genetic testing and/or information (some states and "unless job related" or similar language)	CT, DE, KS, NH, OK, OR, VT
Sanctions	
Individual right to sue	CA, CO, DE, GA, IL, LA, NV, NH, NJ, NM, RI, SC

[a] Insurance context.

[b] Employment, group health insurance, and occupational licensing contexts.

[c] Employment context.

[d] Provision in Act states that confidentiality protections do not apply if genetic testing indicates that the individual is at the time of the test afflicted with a disease, whether or not currently symptomatic.

[e] Self-funded plans covered by ERISA may be exempt.

[f] Genetic tests for predisposition to cancer.

[g] Group health insurers.

[h] Insurers may consider favorable test results voluntarily submitted by an individual.

purely on the basis of genotype, without any phenotypic manifestation. Another area of variation is whether "genetic information" includes biological materials such as tissue samples.

Some laws specify that only testing or information relating to inherited genes or genetic characteristics is covered, pointing up an ambiguity in the adjective "genetic." Many diseases, including all cancers, can be described as genetic in the sense that they are triggered by altered genes. However, only 5 to 10 percent of cancers are thought to be closely linked to a particular set of inherited

genetic defects. Another area of variation is the stringency of the definition of genetic test. Some states limit the term to direct tests for alterations in genetic materials, while others include tests of proteins and other gene products. It is also common to exclude routine medical tests, such as cholesterol tests, human immuno-deficiency virus (HIV) tests, and drug tests from the definition of genetic test.

Genetic tests are not the only source of information about predisposition to disease or disease risk. Some of the first documented cases of genetic discrimination involved inferences from family history (10). Several states use a broad definition of genetic information that appears to encompass information in a family history. For example, the Connecticut law defines genetic information as "information about genes, gene products or inherited characteristics that may derive from an individual or family member" (11). Louisiana expressly includes family history in its definition of genetic information, while Florida, Missouri, Montana, Oklahoma, and Tennessee expressly *exclude* family histories from protection (12). Finally, a number of states expressly extend privacy protections to requests for genetic services (13).

Compelled Consent. Genetic privacy laws commonly prohibit health insurers and employers from requiring genetic testing as a condition of insurance or employment and from using any genetic information acquired in a discriminatory fashion. Privacy advocates have long argued that these protections are fairly meaningless if insurers and employers can persuade or pressure unsuspecting individuals into submitting to genetic testing or sharing genetic information, or obtain genetic information from other sources. Once a third party has possession of information, it is very difficult to police its use (14).

To address these problems, some states prohibit covered insurers and/or employers from even requesting genetic testing or genetic information (15). Vermont, while not prohibiting requests, requires that the person requesting genetic testing advise the individual to be tested that the results may become part of the individual's permanent medical record and may be material to his or her ability to obtain insurance benefits (16). At least four states prohibit covered insurers from seeking genetic information for any nontherapeutic purpose (17). A Delaware law passed in 1998 makes it unlawful for an employer to "intentionally collect" genetic information unless it can be demonstrated that the information is job related and consistent with business necessity or is sought in connection with a bona fide employee welfare or benefit plan, and laws in Oklahoma and Oregon provide that an employer may not "seek to obtain" genetic information concerning an employee or prospective employee (18). It is unclear whether these laws will have any effect on employer access to health insurance claims data, a major area of concern for privacy advocates.

Generic Privacy Laws. In some states, either by default or design, privacy protections are "generic" rather than "genetic," that is, such protections as exist apply to the general category of medical record or health information rather than to genetic information per se. Generic approaches avoid many of the definitional or line-drawing problems that arise in statutes that seek to single out genetic tests and information for protection. They also avoid stigmatizing genetic conditions by treating them differently (19). However, general laws tend to be fairly weak. Laws providing for the confidentiality of physician–patient communications limit disclosure by providers of health care, but they typically permit blanket releases of information to insurance companies and other third parties. Oregon is one of the exceptions. The state-mandated form for medical record release authorizations requires that sensitive information (HIV/AIDs-related records, mental health information, genetic testing information, and drug/alcohol-related information) be initialed by the patient or legal surrogate in order to be included in the release (20).

The first-generation Insurance Information and Privacy Protection Model Act, released by the National Association of Insurance Commissioners in 1981 and adopted in 10 states, covers areas such as pretext interviews (i.e., attempts to gain information involving a misrepresentation of or refusal to provide identity), disclosure of information practices to applicants and policyholders, the content of authorization forms, access to recorded personal information including medical record information, opportunities for correction of recorded personal information, limitations and conditions affecting disclosure of personal information, and penalties and remedies for violations. It does not block exchanges of sensitive information or describe specific steps insurers must take to minimize the risk of unauthorized disclosure. It also bars any action for defamation, invasion of privacy, or negligence for disclosure of personal information in a manner permitted under the state insurance code, even if the information is false, absent malice or willful intent to injure, and a provision of this nature is common even in states that have not adopted the model law.

A Connecticut law passed in 1999 to prohibit the sale of medical record information and restrict disclosure for marketing purposes, and a similar Maine law passed in 1998, are examples of the next generation of generic privacy laws (21). The Connecticut law has a broad definition of "medical-record information," which expressly includes information obtained from a pharmacy or pharmacist. (However, the definition excludes information that lacks personal identifiers *or* has been encrypted or encoded.) The law requires insurance-related entities that regularly collect, use or disclose medical record information to develop and implement written policies, standards and procedures for its management, transfer and security. These must include limiting access to persons who need the information in order to do their jobs, employee training, institution of disciplinary measures, periodic monitoring of employee compliance, and an additional layer of protection of "sensitive health information," which includes information regarding genetic testing such as the fact that an individual has undergone a test.

The National Association of Insurance Commissioners' Health Information Privacy Model Act, released in 1998 and enacted in at least one state (Montana) in 1999, also protects health information generally rather than singling out genetic information. In the employment arena, an innovative Minnesota law imposes a "job-relatedness" condition on all medical evaluation by employers, and laws in a few other states offer similar protections (22). Massachusetts has a general Privacy Act that authorizes the award of damages as well as equitable remedies for "unreasonable, substantial, or serious" interventions with a person's privacy (23).

Constitutional and Common Law Protections. Genetic privacy may also be protected, generically, under state constitutions. Several state constitutions recognize a right to privacy (24). The case of *Norman-Bloodsaw v. Lawrence Berkeley Laboratory* concerned unauthorized testing of clinical and administrative workers for conditions including sickle cell trait. Because the employer, a research laboratory, was operated by a California state agency and a federal agency, the workers had claims under the state and federal constitutions, as well as federal antidiscrimination laws. The Ninth Circuit Court of Appeals found that constitutionally protected privacy interests encompass medical information and extend beyond unauthorized disclosure to reach collection of information by illicit means. According to the court, "the most basic violation possible involves the performance of unauthorized tests—that is, the nonconsensual retrieval of previously unrevealed medical information" that may be unknown even to the test subjects (25). The court concluded that genetic conditions are among the conditions that should enjoy the greatest level of protection. Mere consent to a general medical examination and the taking of samples for routine testing would not constitute authorization for these highly sensitive tests.

Common law may be another source of protection and redress. Invasion of privacy is a blanket term for a number of separate claims under common law. The two most relevant to the genetic context are (*1*) public disclosure of private facts, and (*2*) intrusion upon seclusion. In *Doe v. High-Tech Institute*, a case involving unauthorized testing for HIV/AIDS, a Colorado court concluded that a person has a privacy interest in a blood sample and in the medical information obtained from it, and that an additional, unauthorized test can be sufficient to state a claim for relief for intrusion upon seclusion (26). The elements of the claim are intentional intrusion (which need not be physical) upon the solitude or seclusion of another or his private affairs or concerns, and evidence that the intrusion would be highly offensive to a reasonable person. The court in *Doe* commented that in a case of unauthorized testing the intrusion is the interference with autonomy, namely the right to control important health decisions such as whether to undergo testing for a particular disease, condition, or genetic trait. Offensiveness depends on the nature of the specific test and the circumstances under which the test was performed. A law prohibiting a particular kind of testing without consent can serve as evidence of a significant privacy interest.

Ownership of Genetic Information. A few states have statutes that at least obliquely address the question of ownership, the fourth category of concern subsumed under privacy. A section of the Oregon genetic privacy act states that "an individual's genetic information and DNA sample are the property of the individual except when the information or sample is used in anonymous research," but the law also states that this provision "does not apply to any law, contract or other arrangement that determines a person's rights to compensation relating to substances or information derived from a sample of an individual from which genetic information has been obtained" (27). (The Delaware and New Jersey laws contain similar disclaimers.) A 1999 amendment to the Oregon law,

effective August 2, 1999, to January 1, 2002, declares any research conducted in accordance with federal regulations for the protection of human subjects "anonymous" (28). Several states have laws that contain general declarations that genetic information is the "unique" or "exclusive" property of the person tested (29). In a much-discussed case, *Moore v. Regents of University of California*, a court rejected a patient's argument that commercial development of a cell line derived from his excised cells amounted to theft, while recognizing that patients have a right to information concerning their physicians' economic and research interests as part of the consent process (30).

Federal Laws

Certain members of Congress have taken an interest in genetic privacy, and bills addressing genetic privacy have been circulating for several years. Most of the genetic privacy bills introduced recently are limited to the health insurance context, although one addresses health insurance and employment. As of May 15, 2000, none of these bills had been enacted. Hence state law remains the primary source of protection for genetic information, although a diverse array of federal laws that are broader in scope may have some bearing on conduct in this area. For example, laboratories that perform genetic testing may be subject to the Clinical Laboratory Improvement Act, and attention is currently being given to confidentiality standards. In the employment context, the Americans with Disabilities Act (ADA) of 1990 regulates the timing and scope of medical examinations conducted by employers. The ADA also requires that employers keep information obtained from medical examinations confidential.

A number of federal laws, as well as bills or regulations under consideration, address the privacy of medical records or health information generally. The Privacy Act of 1974 regulates the handling of health-related information by federal agencies. The Health Insurance Portability and Accountability Act (HIPAA) of 1996 provides protection against discrimination based on genetic information and specifies how genetic information is to be handled for purposes of preexisting condition clauses in group health insurance. HIPAA also mandates further attention to privacy concerns. On August 12, 1998, pursuant to a HIPAA requirement, the Department of Health and Human Services (HHS) published a proposed rule establishing security standards for all electronic transactions involving health information. A proposed rule intended to protect the privacy of individually identifiable health information was published on February 3, 1999. The rule does not single out genetic information for heightened protection. It does provide a framework for regulation of the use and disclosure of individually identifiable health information by health plans, health care clearinghouses, and health care providers that engage in electronic transactions. The rule would not preempt stronger privacy protections at the state level (31). If Congress does not take action, the rule will be issued in final form sometime in 2000 or 2001.

Several bills have been introduced in Congress that would introduce more comprehensive privacy regulation at the federal level, including the "Health Care Personal

Information Nondisclosure Act of 1999" (S. 578), the "Medical Information Protection Act of 1999" (S. 881), the "Medical Information Privacy and Security Act" (S. 573), the "Medical Information Protection and Research Enhancement Act of 1999" (H.R. 2470), and the "Health Information Act" (H.R. 1941). As of November 30, 1999, none had been enacted. Major points of contention include the scope of protected health information (the Medical Information Privacy and Security Act appears to be unique in explicitly including tissue samples) and the level of anonymity required for exclusion, the acceptability of blanket authorizations, rights of minors, preemption of state laws offering greater protections, and the creation of an individual right to sue for violations. Interestingly all provide that protections continue after death.

The U.S. Constitution is yet another source of privacy protection for health-related information. In *Whalen v. Roe*, a case involving a New York law requiring physicians to send copies of prescriptions for certain classes of drugs to the state health department, the U.S. Supreme Court ruled that the privacy interests secured under the U.S. Constitution extend to medical record information (32). However, the court found that the steps taken by the state to protect confidentiality were sufficient to meet constitutional requirements. In *Norman-Bloodsaw*, discussed above, a federal appeals court ruled that the right to privacy extends to genetic testing.

LAWS REGULATING RESEARCH

Basic Regulatory Framework

The Federal Policy for the Protection of Human Subjects (often referred to as the "Common Rule") and other regulations governing human subject research are codified at Title 45 Part 46 of the Code of Federal Regulations. The Office for Protection from Research Risks (OPRR) within the National Institutes of Health has primary responsibility for implementation. Significantly, these regulations only apply to research conducted, supported, or subject to regulation by a department or agency of the federal government. Further, since the term "human subject" is limited to living individuals, biological materials and data derived from deceased persons are not covered. Also, to be covered, research must involve (*1*) intervention or interaction with the individual or (*2*) *identifiable* private information, meaning that the identity of the subject is or may readily be ascertained by the researcher or associated with the information. If neither condition is met, research would not be considered human subject research. OPRR has taken the position that information is identifiable where codes can be broken with the cooperation of others (33). Some research, while covered, may be eligible for an exemption from regulatory requirements. The categories of exempt research include research involving the collection or study of existing data or specimens, if these sources are publicly available or if the information is recorded by the investigator in such a manner that subjects cannot be identified directly or indirectly through identifiers.

For research that falls within the scope of the regulations and does not qualify for an exemption, there are two basic requirements: approval must be given by an Institutional Review Board (IRB) and proper informed consent must be obtained from participants. One requirement for IRB approval is that "when appropriate" adequate provisions are made to protect the privacy of subjects and maintain the confidentiality of data. The consent requirement has aroused more controversy. Traditionally tissues, blood, and other biological materials used for research have been stored and studied without explicit consent based on a waiver provision in the Common Rule. Section 46.116(d) states that informed consent for research can be altered or waived if an IRB finds and documents that (*1*) the research involves no more than minimal risk to the subjects, (*2*) the waiver or alteration will not adversely affect the rights and welfare of the subjects, (*3*) the research could not practicably be carried out without the waiver or alteration, and (*4*) whenever appropriate, the subjects will be provided with additional pertinent information after participation. With the advent of genetic research, some have argued that the risk associated with the use of identified or identifiable data or specimens is no longer minimal and consent should be obtained, particularly where protocols propose sharing information with the source or a third party (34).

In August 1999 the National Bioethics Advisory Commission issued recommendations for the interpretation or modification of existing federal regulations relating to the use of human biological materials in research. The Commission found that use of specimens for purposes other than the purpose for which they were originally collected raises the strongest privacy concerns. Pursuant to the recommendations, even research limited to manipulation of existing specimens, if identified or coded (identifiable), would be judged greater than minimal risk and therefore ineligible for a consent waiver unless an IRB were to find that a particular study adequately protects the confidentiality of personal information and incorporates an appropriate plan for whether and how to reveal findings to the donors or their physicians. Consent forms for research would provide a menu of options; future research uses involving unidentified and unlinked samples would generally be accorded similar treatment, as would identified and coded samples (33).

Turning to the state level, states that have genetic privacy laws often exempt research activities, but only if there is some level of anonymity. Requirements vary, ranging from a requirement that the genetic information used for research be free of all identifiers, to a requirement that the only identifier be a code, to a requirement that the identities of donors not be released to researchers. States may parse the research exemption as an exemption from the mandate to obtain informed consent before performing any genetic testing or as an exemption to the mandate to obtain authorization for disclosure of genetic information to a third party. The Missouri law exemplifies the latter approach: genetic information is confidential and cannot be disclosed without written authorization except for health research conducted in accordance with the Common Rule or health research "using medical archives or databases in which the identity of individuals is protected from disclosure by coding or encryption, or by removing all identities" (35). Florida's genetic privacy law contains no exemption for research (36).

Like much biological materials research, medical records research has traditionally been conducted without the knowledge of patients, let alone their informed consent. State medical record confidentiality laws frequently permit access to records for research purposes, so long as researchers maintain the confidentiality of identifiable information (37). Minnesota generated much controversy when it passed a law imposing unusually strict requirements on external researchers seeking access to medical records. An executive of a biotechnology company that conducts a large proportion of its clinical trials at the Mayo Clinic reports that this state law has dramatically reduced the medical records available for research, from 97 percent when the law was passed to 70 percent in 1999 (38).

The privacy bills currently under consideration by Congress would permit disclosure of protected health information for health research that satisfies certain requirements, typically compliance with the Common Rule (or the equivalent for the Food and Drug Administration, FDA) or, in the case of existing information, review by an IRB or IRB-like entity. The Medical Information Privacy and Security Act would make all health research subject to Part 46 of Title 45 of the Code of Federal Regulations, whether federally funded or not. The proposed rule on health information privacy, referred to above, would subject all medical records research to scrutiny by an IRB or privacy board, but only with respect to certain privacy criteria (31).

The federal and state laws and regulations that protect research subjects either by requiring consent or by requiring anonymity for health information and biological materials used in research do not address the potential for abuse of information about specific populations. Research involving biological materials that have been stripped of individual identifiers should not produce information that will result in direct harm to particular donors, but it may reveal sensitive information about the groups to which the donors belong.

DNA Banking for Research Purposes

Banking of biological materials for research, and storage of related information, is largely unregulated. DNA banks range from the collections of individual researchers to the extensive stores of the U.S. Centers for Disease Control and Prevention. State public health agencies may also have extensive collections of biological materials and health information (e.g., cancer registries and "Guthrie cards," specimens of infant's blood on filter paper), subject to varying levels of privacy protection. At least four states require that genetic information be destroyed at the completion of a research study or upon withdrawal of the person from the study, unless retention is explicitly authorized. In a controversial move, the Michigan Commission on Genetic Privacy has recommended that the state permanently preserve blood samples of newborns, which were originally collected to screen for rare congenital disorders, for use in research

conducted in accordance with the Common Rule (unless parents elect to opt-out of research) (39).

Research and Commercial Development

The potential for commercial development based on genetic research has added another layer to research informed consent. While the court in *Moore* refused to recognize a patient's property rights in a cell line derived from his tissue, it did find that physicians must disclose personal interests unrelated to the patient's health that may affect their judgment and failure to do so may give rise to a cause of action for performing a medical procedure without informed consent or breach of fiduciary duty. In response to the *Moore* ruling, consent forms have been expanded to include explicit permission to retain cells for development with potential commercial value. It is not clear that current laws adequately address issues of commercialization. The Uniform Anatomical Gift Act, adopted in some form by the majority of states, gives the donor the authority to decide particular use of body tissue, while the National Organ Transplantation Act prohibits the sale of human tissue and organs for transplantation. Neither law was designed to address the commercialization of genetic material (40).

Law and Chilling Effects on Research

Some have argued that the new emphasis on informed consent for research use of biological materials and medical records will have a chilling effect on research (41,42). The major concern is that the ethical benefits of obtaining informed consent (e.g., demonstrating respect for persons and for privacy as personal choice) are far outweighed by the burdens to research. Consent requirements mean added time, expense, and paperwork, and selective participation may introduce significant bias into research, detracting from its value. A further concern is the lack of uniformity in state laws. Major research projects frequently cross state lines in order to obtain adequate accrual of subjects.

Research may also suffer due to uncertainty concerning confidentiality protections for data in the possession of researchers. A number of celebrated products liability cases from the 1980s showed the willingness of courts to compel production of research data. For example, in *Deitchman v. E.R. Squibb*, Dr. Arthur L. Herbst was required to produce research data linking adenocarcinoma of the genital tract and exposure in utero to diethylstilbestrol (DES); Herbst was not a party to the case nor was he even being called to testify (43). A provision of the Omnibus Consolidated and Emergency Supplemental Appropriations Act (P.L. 105-277) passed in October 1998 requires that federal agencies make data produced under awards to nonprofit organizations available to the public through Freedom of Information Act procedures. To be sure, individual identifiers are removed before data are released to litigants or requestors, but this does not allay the concerns of researchers and privacy advocates. With expanded capacities to manipulate data, redaction to eliminate obvious identifies may not adequately protect the privacy of individuals and groups. Under the

Public Health Service Act, Certificates of Confidentiality are available for certain types of sensitive research including genetic research. Certificates of Confidentiality, however, only protect against forced disclosure of identifying information about individuals, not forced disclosure of aggregate data.

Other Emerging Issues

The fecundity of biological materials raises two further concerns. Particular samples may have potential for usefulness in future research unanticipated at the time of collection. Should the individual be recontacted for permission to do further research on the sample or is a generic consent to use of the sample in future research sufficient? More pressing, with advancements in genetics, the amount of useful medical information that a sample can yield continues to expand. What is the responsibility of a researcher to inform the donor of the sample if medically significant information is discovered? Since most state laws do not require total anonymization of samples used in research, it is theoretically possible to identify the donor. With the possibility of identification may come a responsibility to inform, but the potential psychological harms and risk of discrimination that accompany unsolicited disclosure or "inflicted insight" must be carefully evaluated. Cases addressing the obligations of treating physicians to disclose information to potentially affected family members (discussed below) appear to be the closest analogue. Subjects might be queried concerning their desire for recontact for further studies and for contact with medically significant information, yet questions of interpretation will remain.

LAWS REGULATING DNA BANKS AND DNA DATABASES

Large collections of DNA samples, and records reflecting the results of analysis of those samples, are assembled primarily for one of three purposes: law enforcement, identification of human remains (chiefly in the military context), and medical treatment and research. This section addresses collections of DNA samples (or "banks") and DNA records (or "databases") related to the first two purposes. Interest in this area has been focused on the first two categories of privacy concern, bodily integrity and confidentiality. In particular, questions have arisen concerning the constitutionality of coercive extraction of DNA samples and subsequent testing, and the adequacy of measures to ensure confidentiality and protect samples and information from misuse by third parties. Although their primary goal is identification of perpetrators of crimes, DNA banks and databases maintained by law enforcement agencies may also contain DNA records of unidentified persons, and, in a few states, relatives of missing persons. Texas also authorizes inclusion of DNA records of "a person at risk of becoming lost" (44).

DNA Collection by Law Enforcement

The collection and analysis of DNA samples for law enforcement purposes is concentrated at the state level, with some coordination through the Federal Bureau of Investigations. All 50 states have passed authorizing legislation. States must meet certain federal guidelines in order to participate in data exchange through the Combined DNA Index System or CODIS, established under the DNA Identification Act of 1994 (45). Recently, controversy has centered on the scope of state collection efforts. State laws usually authorize collection of a sample (blood or a buccal swab) upon conviction of a crime and/or as a condition of parole or release from custody. Many state laws cover juvenile offenders (or children adjudicated delinquent) as well as adult offenders, and they contain no special provision for the expungement of DNA records or samples upon attainment of majority.

Initially the focus was on individuals convicted of sex crimes and other serious felonies, offenses associated with high rates of recidivism and biological evidence. However, at least 14 states have added burglary, a nonviolent property crime, to the list of covered offenses. Further, in 1999, several states considered legislation that would permit the creation of DNA records for felony suspects at the time of arrest, a practice contemplated under a Louisiana law that has been enacted but is not yet effective (46). (Significantly the DNA Identification Act limits entries into CODIS to DNA identification records of persons convicted of crimes, analysis of DNA samples recovered from crime scenes, and analysis of DNA samples from unidentified human remains.) Such an expansion of scope may implicate group as well as individual privacy interests. Given that certain minority groups are disproportionately represented in the prison population, and therefore in DNA databases, these technologies may be more effective in "fingering" members of those groups. This problem is likely to be compounded if all arrestees are sampled. The Attorney General has asked the National Commission on the Future of DNA Evidence to study the legality of taking samples upon arrest.

Inmates challenging state laws on Fourth Amendment unreasonable search and seizure grounds have lost with a fair degree of consistency. For example, in *Boling v. Romer*, the Tenth Circuit Court of Appeals found that obtaining and analyzing the DNA of an inmate convicted of a sex offense is a search and seizure raising Fourth Amendment concerns, but that the search and seizure was a reasonable one in light of an inmate's diminished privacy rights, the minimal intrusion of saliva or blood tests, and the legitimate government interest in the investigation and prosecution of unsolved and future crimes by the use of DNA in a manner not significantly different from the use of fingerprints, that is, for purposes of identification (47). First Amendment free exercise, Fifth Amendment self-incrimination and due process, Eighth Amendment cruel and unusual punishment, Fourteenth Amendment equal protection, and ex post facto clause challenges have also failed. The balancing tests may come out differently, however, when the subjects of collection and testing have yet to be convicted of any crime. It is significant that most state laws provide for expungement of samples and information when a conviction is overturned. The Rhode Island law also provides that all identifiable information and samples be destroyed upon official proof that the subject has been deceased for at least three years (48).

At present the biggest deterrent to expansion of DNA sampling by law enforcement agencies is a lack of capacity to analyze samples, rather than any fear of lawsuits. As of June 1999 the backlog of samples awaiting analysis exceeded 500,000 nationwide (49). In response, states are sending samples to private laboratories, a practice that heightens concerns about security and potential for misuse of data.

The DNA profiles entered in CODIS databases currently consist of 13 DNA loci or markers. These DNA markers were selected for their variability, rather than their intrinsic information value (e.g., links to disease or personality traits). Hence advocates for the use of DNA databases in law enforcement believe the potential for abuse is limited. In this regard it is important to distinguish between the samples stored in DNA banks, which can be used to generate other, more sensitive information so long as they are in existence, and the profiles stored in DNA databases. Given the interest in behavioral genetics, one can imagine the eagerness with which some researchers might pursue access to genetic information concerning this particular subset of the population.

State laws address confidentiality concerns in a number of ways. They typically contain a statement that DNA profiles are not public records. Some laws require custodians of samples or records to implement security measures. The Rhode Island law is one of the more detailed in this respect, requiring that an encryption code be used for access to records, that DNA samples be securely locked with access only by director of health department and head of laboratory, and that all identifiers be removed when samples are provided to third parties for creation of DNA records (50). Some state statutes require the relevant agency to develop privacy standards for laboratories under contract. The California law declares that the computer software and database structures, as well as the data, are confidential (51).

Most state laws mimic the language of the DNA Identification Act limiting disclosure of DNA analysis to criminal justice agencies for law enforcement purposes, in judicial proceedings, and, to defendants for criminal defense purposes. Where personally identifiable information is removed, disclosure is also permitted for population statistics databases, for identification research and protocol development purposes, and for quality control purposes. While the reference to population statistics databases might be interpreted broadly, in this context the term refers to analysis of the frequency of occurrence of genetic characteristics in local populations, a type of research closely related to the identification objective for which the DNA banks and databases have been established. Even so, a few states seem to take a fairly permissive view. A number allow dissemination of statistical or research information if there are no identifiers, and the Alabama law permits the state's DNA population statistical database to be used to provide data relative to the causation, detection and prevention of disease or disability and "to assist in other humanitarian endeavors including, but not limited to, educational research or medical research or development" (52). The Massachusetts law allows release of DNA

records for the purpose of assisting in the identification of human remains from mass disasters, assisting the identification and recovery of missing persons, and "advancing other humanitarian purposes" (53). At the other end of the spectrum, Indiana, Rhode Island, and Wyoming expressly prohibit the use of the state collections for the purpose of obtaining information about human physical traits or predisposition for disease (54).

DNA Collection by the Military

The Department of Defense is compiling the world's largest employer-held DNA bank; all inductees and all active duty and reserve personnel are required to provide samples to the military's DNA Specimen Repository. The purpose of the repository is to permit identification of remains, with DNA extracted from samples only as needed. However, concerns have been expressed about possible diversion of the repository to other uses, such as criminal investigation (55). In 1995 a federal district court upheld the sampling program against a challenge on constitutional and other grounds, although the judgment was vacated on mootness grounds after the plaintiffs had been honorably discharged (56).

LAWS REGULATING FAMILY RELATIONS

Disclosure to Potentially Affected Family Members

In the clinical context, a major question has been whether health care providers have a legal and/or ethical duty to disclose genetic information to family members for whom the information may have medical or reproductive implications. There is little direct statutory guidance in this area, and where statutes exist, they tend to permit disclosure rather than mandate it.

The Uniform Health Care Information Act (National Conference of Commissioners on Uniform State Laws, 1985) recognizes several exceptions to the general obligation to keep information confidential. One exception permits, but does not require, a health care provider to disclose information from a patient's record where the health care provider has reason to believe that disclosure will avoid or minimize imminent danger to the health or safety of the patient or another individual. The exception developed out of common law precedents permitting or requiring disclosure in situations where a patient had an infectious disease and the facts were such that the patient could be expected to infect others absent disclosure, or where a patient threatened harm to self or others. The Act has been adopted in only two states, Montana and Washington, and Montana declined to include this provision (57). Other statutes that limit disclosure of health care information fail to address this issue directly, and it has fallen to the courts to define the scope of a legal duty to disclose genetic information.

Judges have grappled with this issue in two cases, *Pate v. Threlkel* and *Safer v. Estate of Pack*, with somewhat different results (58). In interpreting the results, it is important to recognize that both cases were decided at a stage in the proceedings where the court was required to accept as correct the plaintiffs' assertions concerning the standard of care required of physicians during the relevant time period (and other factual matters). Also the decisions in these cases are binding only in the respective jurisdictions, Florida and New Jersey. Other courts will accord them weight based only on the persuasiveness of the reasoning relative to the law in their home states.

Pate was decided first, with the basic facts as follows: Heidi Pate discovered that she had medullary thyroid carcinoma, an autosomal dominant disorder, three years after her mother received treatment for the same disease. She sued her mother's physicians and their employers, arguing that the physicians had a duty to warn her mother of the risk of genetic transmission and to recommend testing of any children. The Florida Supreme Court ruled that if the standard of care was to warn a patient of the genetically transferable nature of a condition, as Pate alleged, then the intended beneficiaries of the standard would include the patient's children as well as the patient. In other words, the patient's children would be entitled to recover for a breach of the standard of care. However, in light of state laws protecting the confidentiality of medical information, the court found no requirement that a physician warn a patient's children. Rather, the court found that in any circumstances in which the physician has a duty to warn of a genetically transferable disease, that duty will be satisfied by warning the patient. The court believed a more expansive standard would be burdensome and unmanageable.

Safer involved similar facts. Donna Safer's father was treated over an extended period of time for colon cancer associated with adenomatous polyposis coli, an autosomal dominant disorder. Almost two decades after his death, Safer was diagnosed with metastatic colon cancer associated with adenomatous polyposis coli. Safer then sued the estate of George Pack, her father's physician, for Pack's failure to warn of the risk to her health. Two additional facts, taken as true for purposes of the court's decision, were significant. First, Safer's mother testified that on at least one occasion she asked Pack whether what he referred to as an "infection" would affect her children and was told not to worry. Second, Safer contended that careful monitoring of her condition would have provided an opportunity to "avoid the most baneful consequences of the condition," a consideration giving force to her argument that the standard of care in the circumstances was to warn any children at risk.

Examining the precedents, the New Jersey Appeals Court found no essential difference between this case with its "genetic threat" and traditional duty-to-warn cases involving the menace of infection or threat of physical harm. The court concluded that a duty to warn in the genetics context would be quite manageable, commenting that those at risk are easily identified. The court noted the potential to avert or minimize substantial future harm by a timely and effective warning to potentially affected persons. The court failed to state how the duty to warn might be discharged, especially in cases involving small children. The relation of this ruling to confidentiality protections was considered only in the court's concluding speculations about possible complications, such as the existence of an instruction from the patient not to disclose

information to family members. The court offered no resolution of this dilemma.

Some states do have statutes that address the question of disclosure for the benefit of family members in very narrow circumstances. For example, a section of the California Welfare and Institutions Code concerning involuntary commitment provides that information pertaining to the existence of a genetically handicapping condition may be released to a qualified professional for purposes of genetic counseling for a blood relative upon the request of the blood relative (59). This can happen either with the consent of the patient or after reasonable attempts have been made over a two-week period to get a response from the patient.

It is important to remember that broad dissemination of genetic information is not an unalloyed benefit to potentially affected family members. In addition to possible psychological harms, family members may face discrimination if genetic information finds its way into their medical records or becomes part of their knowledge base and so must be disclosed on applications for insurance. This could happen without their cooperation where a health care provider shares information without first verifying that family members with to receive it, or where an insurer or other third party stores information concerning more than one family member and fails to prevent information flow. An awareness of this problem appears to explain a New York law that prohibits any person in possession of information derived from a genetic test from incorporating that information into the records of a nonconsenting individual who may be genetically related to the tested individual (60).

Testing of Children

The right of parents to order genetic testing of their children has also provoked some discussion. The issue is particularly vexing where the testing is for an adult-onset condition that cannot be prevented, ameliorated or cured by any action taken during childhood. In such cases it is hard to argue that testing confers any benefit on the child, or any benefit on the parents such as the ability to plan for burdens likely to affect the family unit (as might be the case with an incurable condition manifesting in childhood). The general rule is that parents control medical decision making for their children. The Delaware genetic privacy act expressly states that it "does not alter any right of parents or guardians to order medical and/or genetic tests of their children" (61). An Illinois law addresses a situation in which testing is requested by or ordered for a child and a question arises concerning the proper recipient of the results. The statute requires a health care provider who orders a genetic test for a minor to notify the minor's parent or legal guardian of the results of the test, but only if the provider determines that notification would be in the best interest of the minor and has first sought unsuccessfully to persuade the minor to give the notice (62).

Paternity Determination

The use of genetic testing in the determination of paternity is most analogous to the use of genetic testing in law enforcement. As in the law enforcement context, testing aims at identification rather than production of information about health or behavior, and similar privacy interests are implicated, namely the interest in being free of unwanted bodily invasion, and the interest in preventing access to or disclosure of personal information. If an individual refuses to submit to paternity testing, the result may be a contempt citation or a presumption that the results would be adverse to the individual. Laws governing genetic testing to determine paternity typically provide that the results of testing are confidential and cannot be disclosed to third parties absent exceptional circumstances. Sample retention is not always addressed. Some laws do provide that if a man is found not to be the father of a child in a paternity proceeding, the man's genetic material must be destroyed (63).

Adoption

Adoption is yet another context in which genetics, privacy concerns, and the law come together. A critical issue here is the extent of adoptive parent and adoptee access to genetic information. State adoption laws typically require preparation of a report of the complete family medical and social history of a candidate for adoption (so long as parental identity is not disclosed), including identification of any known genetic disorders. Ten states require information concerning extended family if available (64). Generally, the adoptive parents, and the adopted child upon attainment of majority, are allowed access to this information. In a few states, where an adopted child has died, genetic information (and other medical information) is available to the spouse of the adopted child if he or she is the legal parent of the adopted child's progeny, and also to any progeny of the adopted child age 18 or older.

Some states create voluntary registries that allow nonidentifying genetic information to flow back *to* birth parents (e.g., notice that the child has been diagnosed with sickle cell anemia, meaning that the birth parents are carriers of sickle cell trait), as well as permitting transfer of additional information *from* birth parents as it becomes available. In Arizona, other biological children of the birth parent can also access nonidentifying information in adoption records upon request (65). In several states, a certified statement from a physician explaining why genetic or other critical medical information should be communicated to adoptive parents or an adopted child, or an adopted child's genetic parent or sibling, is the trigger for an effort by the registry to notify the affected person(s) that the nonidentifying information is available from the registry (66). North Dakota places information exchange within the discretion of the child-placing agency (67).

Opening of sealed adoption records that include identifying information is usually permitted only by court order upon a showing of "good cause." Proceedings to open records have been brought by adopted children seeking to contact genetic relatives as potential donors of tissue or bone marrow for life-saving transplants, or for other health-related purposes. In such cases courts typically apply a balancing test, weighing the interests of the adopted child, the adoptive parents, the biological parents, and society. In *Golan v. Louise Wise Services*, New York's highest court reversed a lower court order that would have permitted an adoptee, seeking genetic information for evaluation of his heart condition, access to the identities of his biological parents. The court found that consideration of the adoptee's well-being alone in cases involving medical problems with genetic implications would "swallow" the state's strong policy against disclosure. The court suggested correspondence through a guardian ad litem as an alternative to unsealing adoption records (68).

Other scenarios put forward by commentators, in which privacy interests may be at odds with other interests, are more speculative, and appear not to be directly addressed under current law. For example, there are concerns about excessive testing of children prior to adoption. Genetic testing or genetic information might also be demanded of prospective adoptive parents, as bearing on their suitability for parenthood. Concerns have also been expressed about the use of genetic information as ammunition in child custody disputes, as affecting the likelihood that a parent will "be there" for a child in the future (10,69).

Postmortem Testing for the Benefit of Family Members

Several states authorize sharing of genetic information with family members once a person is dead. The next logical step is genetic testing of the dead to obtain information desired by family members. Under normal circumstances the next-of-kin usually controls disposition of the body and has the right to consent to or refuse an autopsy (or the use of samples in research). An argument could be made that by extension family members have the right to order genetic testing, although laws prohibiting genetic testing without consent with only limited exceptions might be cited as evidence against such a right. In at least one case, reported in the media, a daughter was given access to a sample of her deceased father's blood for paternity testing. Her mother sued the hospital to try to block testing and lost (70). There appear to be few laws directly addressing this question. A notable exception is a New York law that expressly authorizes genetic testing on specimens from deceased persons if informed consent is provided by the next of kin (71). (Autopsies may also raise the unsolicited disclosure problem, where a pathologist obtains genetic information that may be material to family members in the course of an autopsy and must determine whether to disclose this information to family members.) The Native American Graves Protection and Repatriation Act addresses the issue by stipulating that Native American remains in the possession or control of federal agencies are the property of lineal descendants or tribes and must be repatriated upon request (72).

MISCELLANEOUS

Genetic information is increasingly sought in the context of civil litigation. Defendants in personal injury lawsuits may be eager to prove that injuries resulted from the plaintiffs' genetic defects rather than their own negligent conduct. As noted above, state laws may declare that genetic information is privileged and hence protected from routine discovery in the investigational phase of a civil proceeding. However, a judge may order testing or disclosure of information if persuaded of its relevance. For example, a defendant in a lawsuit arising out of an automobile accident sought to compel genetic testing of the plaintiff for Huntington's disease, as a possible causal factor, and the court ordered the testing over the plaintiff's objections (73). Colorado law creates a barrier to recovery for injury arising from genetic counseling and screening, prenatal care, labor, delivery, or postnatal care where it can be established that the injury was the result of a genetic disease or disorder (74). A provision of the law expressly permits discovery of medical information concerning the plaintiff and makes this information admissable as evidence at trial. In addition the law allows discovery of medical information relating to genetic siblings, parents, and grandparents of the plaintiff, if the defendant cannot secure voluntary releases and persuades the court of the possible relevancy of the information.

BIBLIOGRAPHY

1. A.L. Allen, in M. Rothstein, ed., *Genetic Secrets: Protecting Privacy and Confidentiality in the Genetic Era*, Yale University Press, New Haven, CT, 1997, pp. 31–59.
2. E.g., Del. Code Ann. tit. 16, §1224; Nev. Rev. Stat. §629.171; N.J. Stat. Ann. §10:5-47; Or. Rev. Stat. §659.720; S.C. Code Ann. §38-93-30; Tex. Labor Code Ann. §21.403, Tex. Ins. Code art. 21.73, §4, and Tex. Rev. Civ. Stat. Ann. art. 9031, §3 (Vernon's).
3. Tex. Ins. Code, art. 21.73.
4. 29 U.S.C. §§1140 et. seq.
5. N.Y. Civ. Rights Law §79.1.
6. E.g., Colo. Rev. Stat. Ann. §10-3-1104.7(12); Nev. Rev. Stat. Ann. §629.201; N.M. Stat. Ann. §24-21-6.
7. La. Rev. Stat. Ann. tit. 22, §213.7(F).
8. Ala. Code §27-53-1 et seq.
9. Ill. Stat. Ann. ch. 410, para. 513/15.
10. L.N. Geller et al., *Sci. Eng. Ethics* **2**(1), 71–88 (1996).
11. Conn. Gen. Stat. Ann. §46a-60(a)(11) (West).
12. La. Rev. Stat. Ann. tit. 22, §213.7(A)(8); Fla. Stat. Ann. §627.4301; Mo. Ann. Stat. §375.1300 (Vernon's); Mont. Code Ann. §33-18-901; Okla. Stat. Ann. tit. 36 §3614.1; Tenn. Code Ann. §56-7-2706.
13. E.g., N.M. Stat. Ann. §24-21-2.
14. M.A. Rothstein, *Am. J. Law, Med. Ethics* **26**, 198–204 (1998).
15. E.g., Conn. Gen. Stat. Ann. §46a-60(a)(11); Fla. Stat. Ann. §627.4301(2)(b); Ga. Code Ann. §33-54-3; Haw. Rev. Stat. Ann. §431:10A-118; Ill. Stat. Ann. ch. 410 §513.20); Ind. Code Ann., §27-8-26-5; 1999 Kansas Sess. Laws S.B. 22; Ky. Rev. Stat. Ann. §304.12-085; 1999 Maryland Laws ch. 50 (S.B. 774); Minn. Stat. Ann. §72A.139; Mo. Ann. Stat. §375.1303; Ohio Rev. Code Ann. §1751.65; R.I. Gen. Laws §27-18-52; Tenn. Code Ann. §56-7-2704.
16. Vt. Stat. Ann. tit. 18, §9332.
17. E.g., Cal. Health & Safety Code §1374.7(b); Ga. Code Ann. §33-54-4; Mont. Code Ann. §33-18-904; Colo. Rev. Stat. Ann. §10-3-1104.7(3)(b).
18. Del. Code Ann. tit. 19, §711; Okla. Stat. Ann. tit. 36, §3614.2; Or. Rev. Stat. §659.036.
19. M.A. Rothstein, *Am. J. Law Med.* **24**, 399–416 (1998).
20. Or. Rev. Stat. §192.525.
21. 1999 Conn. Pub. Acts No. 99-284; Me. Rev. Stat. Ann. tit. 22, §1711-C.
22. Minn. Stat. Ann. §363.01-.20.
23. Mass. Gen. Laws Ann. ch. 214 §1B.
24. E.g., Constitution of the State of California, Art. I, §1; Florida Constitution, Art. I, §23; Constitution of the State of Illinois, Art. I, §§6, 12; Constitution of the State of Louisiana, Art. I, §5; Constitution of the State of Montana, Art. II, §10.
25. 135 F.3d 1260 (9th Cir. 1998).
26. 972 P.2d 1060 (Colo. Ct. App. 1998).
27. Or. Rev. Stat. §659.715; Del. Code Ann. §1225; N.J. Stat. Ann. §10:5-46.
28. 1999 Oregon Laws Ch. 921 (S.B. 937).
29. E.g., Colo. Rev. Stat. Ann. §10-3-1104.7(1)(a); Fla. Stat. Ann. §760.40; Ga. Code Ann. §33-54-1; La. Rev. Stat. Ann., tit. 22, §213.7(E).
30. 793 P.2d 479 (Cal. 1989), *cert. denied*, 499 U.S. 936 (1991).
31. Standards for Privacy of Individually Identifiable Health Information, Proposed Rule, 64 *Fed. Reg.* 59,918 (1999) (to be codified at 45 CFR Parts 160 through 164).
32. 429 U.S. 589 (1977).
33. Report and Recommendations of the National Bioethics Advisory Commission, *Research Involving Human Biological Materials: Ethical Issues and Policy Guidance*, vol. 1, National Bioethics Advisory Commission, Rockville, MD, 1999.
34. E.W. Clayton et al., *J. Am. Med. Assoc.* **274**, 1786–1792 (1995).
35. Mo. Ann. Stat. §375.1309 (Vernon's).
36. Fla. Stat. Ann. §760.40.
37. *The State of Health Privacy: An Uneven Terrain*, Health Privacy Project, Georgetown University Institute for Health Care Research and Policy, Washington, DC, 1999.
38. *Bluesheet* **42**, 18 (1999).
39. Michigan Commission on Genetic Privacy and Progress, *Final Report and Recommendations*, February 1999.
40. R.G. Harman, *J. Legal Med.* **14**, 463–477 (1993).
41. L.J. Melton III, *N. Engl. J. Med.* **337**, 1466–1470 (1997).
42. W. Grizzle et al., *Archiv. Pathol. Lab. Med.* **123**(4), 296–300 (1999).
43. 740 F.2d 556 (7th Cir. 1984).
44. Tex. Gov. Code §411.142.
45. 42 U.S.C. §13701 et seq.
46. *DNA Detection of Sexual and Violent Offenders Law*, La. Rev. Stat. Ann. §15:601 et seq., suspended by H.C.R. 40 (1999).
47. 101 F.3d 1336 (10th Cir. 1996).
48. R.I. Stat. Ann. §12-1.5-13.
49. R.S. Wilson, L. Forman, and C.H. Asplen, *Corrections Today* **61**(3), 20–22 (1999).
50. R.I. Gen. Laws §12-1.5-10.
51. Cal. Penal Code §299.5.
52. Ala. Code §36-18-31.
53. Mass. Gen. Laws Ann. ch. 22E, §10.
54. Ind. Code Ann. §10-1-9-18; R.I. Gen. Laws §12-1.5-10; Wyo. Stat. §7-19-404.
55. J.E. McEwen, in M. Rothstein, ed., *Genetic Secrets: Protecting Privacy and Confidentiality in the Genetic Era*, Yale University Press, New Haven, CT, 1997, pp. 231–251.

56. *Mayfield v. Dalton*, 901 F.Supp. 300 (D. Haw. 1995), *vacated*, 109 F.3d 1423 (9th Cir. 1997).

57. Mont. Code Ann. §50-16-529; Wash. Rev. Code Ann. §70.02.050.

58. 661 So.2d 278 (Fla. 1995); 291 N.J. Super. 619, 677 A.2d 1188 (N.J. Super. Ct. App. Div. 1996), *cert. denied*, 683 A.2d 1163 (N.J. 1996).

59. Cal. Welf. & Inst. Code §5328.

60. N.Y. Civ. Rights Law §79.1.

61. Del. Code Ann. §1226.

62. Ill. Stat. Ann. ch. 410, para. 513/30.

63. E.g., Mich. Stat. Ann. §722.716a.

64. L.B. Andrews, in M. Rothstein, ed., *Genetic Secrets: Protecting Privacy and Confidentiality in the Genetic Era*, Yale University Press, New Haven, CT, 1997, pp. 255–280.

65. Ariz. Rev. Stat. Ann. §8-129.

66. E.g., Minn. Stat. Ann. §259.83; Miss. Code Ann. §93-17-205; Va. Code Ann. §63.1-236.02; Vt. Stat. Ann. §tit. 15A, §6-104.

67. N.D. Cent. Code §14-15-16.

68. *Golan v. Louise Wise Services*, 69 N.Y.2d 343, 507 N.E.2d 275 (N.Y. Ct. of Appeals 1987).

69. C. Lerman et al., *J. Health Care Law Policy* **1**, 353–372 (1998).

70. D. Nelkin and L.B. Andrews, *Chron. Higher Educ.* **45**, B6 (1999).

71. N.Y. Civ. Rights Law §79.1. Or. Rev. Stat. §659.715.

72. Native American Graves Protection and Repatriation Act, 25 U.S.C. §3001–30013.

73. M.A. Rothstein, in M. Rothstein, ed., *Genetic Secrets: Protecting Privacy and Confidentiality in the Genetic Era*, Yale University Press, New Haven, CT, 1997, pp. 451–495.

74. Colo. Rev. Stat. Ann. §13-64-502.

See other GENETIC INFORMATION, ETHICS and GENETIC INFORMATION, LEGAL entries.

GENETIC INFORMATION, LEGAL, GENETICS AND THE AMERICANS WITH DISABILITIES ACT

MARY R. ANDERLIK
University of Houston
Houston, Texas

OUTLINE

INTRODUCTION

The Americans with Disabilities Act (ADA) is one of the landmark pieces of legislation of the twentieth century (1). Its purpose is to "provide a clear and comprehensive national mandate for the elimination of discrimination against individuals with disabilities" (2). Many people affected by genetic conditions fear discrimination, especially in the areas of employment and insurance. The ADA contains provisions that address disability discrimination affecting the conduct of employers and insurers. However, the protections of the ADA may be of little help to those who believe they have suffered discrimination owing to a genetic characteristic. This is so owing to the restrictive definition of disability in the ADA and other features such as a provision that shelters traditional insurance underwriting practices from challenge under the ADA.

Structural Reasons for Discrimination

It may be helpful to review some of the reasons for genetic discrimination, and the available evidence concerning its prevalence, before proceeding to a discussion of the statute itself. Discussions of genetic discrimination tend to focus on the employment and insurance contexts. This is so because given the current arrangements for financing health care, both employers and insurers have strong financial incentives to discriminate on the basis of genetic information (and other information concerning present or future health status) in order to control costs.

Employment

In the United States, employers frequently play a role in obtaining health insurance for employees. Where an employer assumes responsibility for paying some part of the premiums for experience-rated health insurance or chooses to self-insure, employee health problems have a direct effect on the employer's financial performance. Experience rating means that premiums are set based on the claims history of persons covered under the employer's group policy. For small employers, in particular, one employee with an illness that is expensive to treat can send premiums sky-rocketing, increasing costs to the employer and to other employees. Where an employer chooses to self-insure, the employer bears the burden of any expenditures for covered medical care and captures the benefit from any reduction in such expenditures. In 1997, 13 percent of all employers had self-funded health plans, and 56 percent of firms with 500 or more employees had self-funded health plans (3). Strategies that self-insured employers might be tempted to pursue to hold down health care related costs include cutting salaries or increasing employee cost-sharing to offset any increases in health care costs, weeding out workers likely to have health problems through hiring and firing processes, identifying conditions that are likely to prove costly and excluding them from coverage (a strategy that is more effective if the health risks of actual employees are known), and dropping health coverage entirely (4).

In addition employers may seek to exclude individuals with above-average susceptibility to a toxin from certain jobs out of paternalistic concern or fear of liability. Another response would be to increase protections (e.g., reduce exposures) for all workers. In the employment context, it is important to distinguish between screening and monitoring. Screening refers to efforts to test applicants or employees for conditions that may render them more susceptible to harm from workplace substances than the average person. The target is the individual, and the goal is to improve productivity and lower workers' compensation and health insurance costs. Monitoring refers to the performance of periodic examinations of employees to identify and assess changes. The target is the active workforce and the goal is to identify workplace risks that can be reduced through implementation of prevention programs (5). Finally, employers may wish to exclude persons with genetic conditions from the workforce based on concerns about attendance and productivity.

Insurance

Insurers, such as issuers of health, disability, long-term care, and life insurance policies, discriminate among insureds and applicants for insurance based on health status due to bottom-line concerns and philosophical commitments. The bottom-line concern may be solvency or profitability. In addition many insurers have a commitment to "actuarial fairness," meaning that policies are priced to accurately reflect risk or expected losses. The forms of health status related discrimination operative in insurance are captured by the term "medical underwriting." Insurers obtain health information in order to decide as a threshold matter whether an applicant is an acceptable insurance risk, and also for use in setting premiums. Insurers can also design benefit packages strategically to limit potential losses. For example, health insurers can exclude certain conditions or procedures from coverage, or they can use "preexisting condition" clauses to eliminate coverage for any health care needs that relate to a condition diagnosed or treated prior to enrollment.

As a justification for these practices, insurers cite the problem of adverse selection. Adverse selection is the disproportionately heavy purchase of insurance by individuals at higher risk for claims than their insurers are aware. When individuals learn that genetic or other factors put them at high risk for disease, disability, and/or early death, they may load up on insurance. If insurers are ignorant of the information concerning risk, they cannot incorporate it into the process of underwriting. As the proportion of higher-risk individuals in an insurance pool increases, payouts will increase, and as payouts increase, premiums for all policyholders will go up. In a voluntary system of insurance, premium increases can be expected to drive out lower-risk individuals, resulting in a further increase in the proportion of higher-risk individuals, and so on. At least for health insurance, the problem of adverse selection could be addressed through universal coverage (6). The preferred solution for U.S. insurers has been to ensure that they have access to any health-related information that may be available to applicants for insurance.

Anecdotal and Survey Evidence of Genetic Discrimination

Although incentives to engage in genetic discrimination are present, the limited evidence available suggests that few employers or insurers systematically collect or use information derived from genetic testing. The expense involved in testing is surely a factor. A government-sponsored survey of Fortune 500 companies in 1989 found that of the 330 respondents, 12 reported conducting current biochemical genetic screening of employees. None reported conducting direct-DNA screening (5). Significantly more employers may solicit family histories containing genetic information.

In testimony before a congressional committee in 1998, a spokesperson for the Health Insurance Association of America stated that a survey of association members found none required genetic testing or solicited information regarding genetic testing as part of the application process (7). Authors of a recent study of genetic discrimination in health insurance concluded that a person with a serious genetic condition who is presymptomatic currently faces little or no difficulty in obtaining health insurance (8). Further over 75 percent of Americans obtain health insurance through their employers or the government and would not be subject to individual medical underwriting (9).

By way of contrast, almost 75 percent of life insurance policies are individual policies (10). As a result the majority of Americans with life insurance coverage are subject to medical underwriting. An individual may be denied life insurance if a family member suffers from a heritable disorder associated with premature death. For example, the practice among life insurers has been to decline to write individual policies for children of people with Huntington's disease until they are over 50 years old (11). Underwriting standards for individual policies of disability-income insurance may be even stricter. A number of researchers have surveyed currently healthy people with known genetic predispositions to disease or family members affected by genetic disorders. These researchers consistently find that a number of respondents report personal knowledge of instances of genetic discrimination in insurance and employment, although the magnitude of the problem is difficult to establish from these kinds of surveys (12).

Investigative reports and lawsuits are another source of information on genetic discrimination. A *New York Times* investigative report on genetic testing in the workplace in 1980 found that DuPont de Nemours & Co. tested black job applicants for sickle cell trait. DuPont also tested job applicants for two enzyme deficiencies correlated with ethnicity. The purpose, according to the company's medical director, was to determine whether these tests would be of value in protecting the health of susceptible employees. (Another employer, Dow Chemical Company, engaged in genetic monitoring to detect any changes in chromosomal structure that might be attributable to workplace exposures.) Many years later, Lawrence Berkeley Laboratory, one of the national laboratories involved in the Human Genome Project, was sued for testing clerical and administrative employees for sickle cell trait, allegedly without employee consent. In each of these cases, it was unclear how the genetic information was used in decision making. DuPont apparently reassigned a small number of employees based on results of one of the enzyme tests (13).

Although the evidence concerning the actual extent of genetic discrimination is sparse, there is overwhelming evidence that people fear genetic discrimination. Nearly two-thirds of respondents in a 1997 survey reported that they would not undergo genetic testing if employers and health insurers would have access to the results. A 1995 survey found that over 85 percent of respondents were very or somewhat concerned about access to and use of genetic information by employers and insurers (12).

ADA: OVERVIEW

The ADA represents a blending of civil rights law and disability law. Key provisions were influenced by Title VII of the Civil Rights Act of 1964 and by the Rehabilitation Act of 1973 and its implementing regulations. From civil rights law the ADA derives its broad ambition of emancipating people with disabilities from a history of discrimination. From disability law it derives its demand for accommodation rather than simple equality, its mandate of community integration to combat isolation and segregation, and its emphasis on individualized or case-by-case decision making, taking account of the reasonableness of any accommodations requested and the burdens imposed on public and private actors. In addition the ADA reflects the belief that the denial of opportunity translates into dependency and nonproductivity, giving rise to substantial social costs.

The ADA was signed into law on July 26, 1990; most provisions became effective on or before July 26, 1992. Its structure is fairly neat. The law begins with a recital of findings and purposes and a number of general definitions, including a definition of the key term "disability." Each of the first three titles addresses a particular arena in which discrimination may occur. Title I governs employment. It applies to employers with 15 or more employees, excluding the United States and certain private clubs. Title II governs public services and applies to state and local governments and specified transportation agencies. Title III governs public accommodations and services and applies to all private entities with operations affecting interstate commerce. Each of these titles contains a prohibition of discrimination—with some variation

in phrasing—and an enforcement provision. A section labeled "miscellaneous provisions" offers guidance on interpretation of the statute generally, and Title IV concerns telecommunications.

The U.S. government is subject to the antidiscrimination mandates of Section 501 of the Rehabilitation Act (14). The Rehabilitation Act also applies to federal government contractors (Section 503) and any entity receiving federal financial assistance (Section 504). Differences between the ADA and the Rehabilitation Act include the language in the prohibitions on discrimination. The Rehabilitation Act prohibits discrimination "solely" by reason of a person's disability (15). Titles I through III of ADA use "because of" or "on the basis of" disability, suggesting that the disability need not be the only factor motivating the adverse treatment (16). Enforcement is another area of difference. The Equal Employment Opportunity Commission (EEOC) is charged with issuing regulations under Title I of ADA and enforcing its provisions, and the Department of Justice is responsible for Titles II and III of ADA. The Rehabilitation Act does not concentrate authority in this manner. For example, the Department of Health and Human Services (DHHS) issues implementing regulations governing its grantees and enforces the law as applied to its own activities and those of its grantees through its Office for Civil Rights. The remedies available under the two statutes also differ. For example, the courts have established that money damages may be awarded under the Rehabilitation Act, whereas this remedy is sometimes unavailable under ADA.

The sections of ADA that are most relevant to the area of genetic discrimination are the definition of disability, the provisions in Titles I and III affecting employment and insurance, and the miscellaneous provisions that address the relationship of ADA to other laws and to insurance. While there are similarities, Titles I, II, and III are not perfectly symmetrical. In particular, a provision of Title I that regulates the collection of medical information by employers has no equivalent elsewhere in ADA. This important prophylactic provision will be discussed first, before consideration of whether and how the ADA might affect genetic discrimination in employment and other areas.

THRESHOLD QUESTION: IS A GENETIC CONDITION A "DISABILITY"?

Statutory Language

The definition of disability is important because the antidiscrimination provisions of ADA protect individuals with disabilities from discrimination based on disability. Under ADA, disability can be established in one of three ways. First, an individual can show that he or she has "a physical or mental impairment that substantially limits one or more of the major life activities of such individual." Second, an individual can show that he or she has "a record of such an impairment." Third and finally, an individual can show that he or she is "being regarded as having such an impairment" (17). Although other provisions of the statute categorically exclude certain conditions from the definition of disability (e.g., transvestitism, pedophilia, compulsive gambling), the determination of whether a particular condition satisfies one of the three prongs of the definition of disability is generally made on a case-by-case basis.

Under the first prong of the definition, an affected individual might argue that a genetic condition is a physical or mental impairment that presently substantially limits the major life activity of reproduction, or will in the future substantially limit one or more major life activities of the individual. Under prong three, an affected individual might argue that he or she is being regarded as having a physical or mental impairment that presently substantially limits one or more major life activities, or that the anticipated future impairment, whether certain or merely more likely than for the average person, is being imputed to the present as evidenced by the discriminatory conduct of a third party. The strength of the argument will likely depend on the nature of the condition. The adjective "genetic" alone is fairly uninformative for ADA purposes; generally, ADA is concerned with function rather than causation. Commentators have identified at least seven categories of genetic conditions that may merit separate analysis: (1) already-expressed severe genetic conditions, such as symptomatic Huntington's disease, (2) already-expressed minor genetic conditions, such as polydactyly expressed in an extra finger or toe, (3) unexpressed late-onset genetic conditions, such as presymptomatic Huntington's disease detected through genetic testing, (4) genetic mutations associated with increased risk of disease (predispositions), such as BRCA1 or BRCA2 mutations detected through genetic testing, (5) unaffected carriers of recessive and X-linked disorders, such as carrier-status for cystic fibrosis, (6) genetic conditions that are cured or kept under control through treatment, such as phenylketonuria controlled through diet, and (7) conditions with a genetic basis that do not limit major life activities but are stigmatized or misunderstood, such as Down syndrome or Tourette syndrome (18). Category 1 conditions should satisfy the first prong of the definition of disability. Category 2 conditions would appear to fall outside the definition of disability (unless they are stigmatized or misunderstood). For the other categories, the outcome is uncertain, although the legislative history and the opinions of administrative agencies and the courts can be mined for insight.

Legislative History and Agency Interpretations

The legislative history of the ADA concerning genetic conditions is scanty at best. On the day the House of Representatives voted on the final conference report, three congressmen entered in the record statements that the law would protect "carriers of a disease-associated gene" from employment discrimination based on speculation about future illness or increased health care costs for carriers or their dependents (18). There is no record of debate on this point. To the extent the issue was thought of at all, then, the cases that came readily to mind for the few genetically minded legislators were autosomal recessive conditions such as sickle cell anemia. This is not surprising, since presymptomatic predispositional testing for diseases such

as breast cancer and colon cancer has only recently become widely available.

Reliance on agency interpretations is complicated by the fact that the introductory sections of ADA, unlike Titles I through IV, contain no delegation of authority to a particular agency. EEOC, which has charge of Title I, has produced several documents that elaborate on the definition of disability. EEOC's Title I regulations attempt to clarify certain aspects of the definition of disability, although they do not specifically mention genetic conditions. According to the Title I regulations, a "physical or mental impairment" includes "any physiological disorder, or condition, cosmetic disfigurement, or anatomical loss" affecting at least one of the major body systems. Some examples of major life activities are given, such as caring for oneself, walking, and working. To "substantially limit" is to significantly restrict as to condition, manner, or duration of performance relative to the performance of the average person. Factors to be considered include the nature and severity of the impairment, its duration or expected duration, and "the permanent or long term impact, or the expected permanent or long term impact of or resulting from the impairment" (19). The Title I regulations also present several possible variants of "regarded as" disability; the unifying element is the focus on what the third party's treatment of the individual suggests, rather than on the physical or mental state of the individual. The Title III regulations issued by the Department of Justice are similar to the Title I regulations on these points (20).

Interpretive guidelines receive considerably less deference from the courts than regulations but may still carry some weight. EEOC's interpretive guidance for Title I, published as an appendix to the regulations, affirms the importance of case-by-case determinations. However, it contains some rather unnuanced assertions. It states that the definition of impairment does not include "characteristic predisposition to illness or disease" (21). This suggests that regardless of contextual factors, genetic predispositions to cancer or heart disease are not disabilities protected under ADA—unless the "regarded as" prong of the definition fits the case. On the other hand, EEOC makes HIV infection an example of an impairment that is inherently substantially limiting. Hence, to the extent that a genetic mutation can be analogized to HIV infection, this language suggests a strong case can be made for recognizing the mutation not only as an impairment but also as a disability under the first prong of the ADA definition.

EEOC has also issued a compliance manual for Title I. On March 15, 1995, an amended manual was released that included the following language concerning "regarded as" disability: "This part of the definition of 'disability' applies to individuals who are subjected to discrimination on the basis of genetic information relating to illness, disease, or other disorders. Covered entities that discriminate against individuals on the basis of such genetic information are regarding the individuals as having impairments that substantially limit a major life activity" (22). In the compliance manual, EEOC gives the example of an asymptomatic individual with a genetic mutation conferring an increased risk of colon cancer; an employer discovers this information after making a conditional offer of employment and withdraws the offer due to concerns about attendance, productivity, and insurance costs. The example suggests that where action is taken on the basis of present fears, the definition of disability is satisfied, even if the fears are about future performance or future costs and even if the fears are not, strictly speaking, unfounded. Although the emphasis in EEOC's interpretation of "regarded as" disability is on myths, stereotypes, and misperceptions, EEOC states that the individual does not have to demonstrate that the employer's perception is wrong, for example, that health care costs will not increase if persons are hired who are at elevated risk of serious illness. (On the other hand, a finding of disability under ADA will not necessarily lead to a finding of liability. As discussed below, a prima facie case must include a showing that discrimination occurred because of or on the basis of the disability. Establishing a violation may be difficult, given that employers are unlikely to document that decisions are being made on the basis of genetic information.) The Department of Justice has not addressed genetic conditions in its interpretive guidance for Title III.

Judicial Opinions

No published judicial opinion addresses whether an individual with a genetic mutation associated with disease, but not yet expressed in symptoms, has a disability under ADA. A number of cases have raised somewhat similar issues. In *Bragdon v. Abbott*, the U.S. Supreme Court held that HIV infection, even in its early stages, is a disability under the first prong of the ADA definition (23). It is important to understand the court's reasoning, since an unexpressed genetic condition may or may not share the characteristics the court found significant in *Bragdon*. First, the court addressed whether HIV infection is an impairment. The court found that given the immediacy of the damage to the hemic and lymphatic systems, the predictable course of the disease, and its severity, an impairment exists from the moment of infection. Next the court concluded that as to the plaintiff reproduction was a major life activity. The court further concluded that the risk of infecting a partner and the risk of infecting a child could substantially limit this activity. Even assuming the risk of perinatal transmission could be lowered from 25 to 8 percent using antiretroviral therapies, the opinion stated that it is not possible to say as a matter of law "that an 8 percent risk of transmitting a dread and fatal disease to one's child does not represent a substantial limitation on reproduction."

Huntington's disease and other similar late-onset genetic conditions would appear to meet the *Bragdon* criteria of predictability and severity. Assuming some damage to a body system could be established prior to full expression, they would likely qualify as impairments from the moment of transmission. The case for genetic mutations associated with increased risk of disease is considerably weaker under the *Bragdon* criteria, since the element of predictability is missing; the same would be true for the category of unaffected carriers. (There is

some irony here. Many people are struck by the unfairness where a third-party treats a possibility of disease as if it were a certainty as a risk avoidance measure, but the case for protection under ADA appears stronger where the question is not whether but when a disease will develop.) Genetic conditions would pose risks of transmission to a child analogous to the risk of infection associated with HIV, although the risk to a partner would be absent. For monogenic genetic disorders, the risk of transmission will generally be 25 or 50 percent. Factors affecting the reproductive options of particular plaintiffs, such as the availability of preimplantation genetic diagnosis, would also appear relevant under the Supreme Court's framework for analysis.

The concurring and dissenting opinions in *Bragdon* are of interest as well. Justice Ginsburg, concurring, stated that "[n]o rational legislator ... would require nondiscrimination once symptoms become visible but permit discrimination when the disease, though present, is not yet visible." This reasoning lends further support to a distinction, in the assessment of impairment, between those genetic mutations that inevitably give rise to disease and those that simply increase the risk of disease. Writing for the dissenters, Chief Justice Rehnquist stated that taken to its extreme the logic of *Bragdon* would "render every individual with a genetic marker for some debilitating disease 'disabled' here and now because of some possible future effects." Justice Rehnquist meant this as a warning of a peril to be avoided, but the language could be used by those who favor just such an extension.

In *Sutton v. United Air Lines*, the Supreme Court returned to the definition of disability, focusing on the substantial limitation requirement (24). The case involved two women who suffered from severe myopia but had 20/20 vision with corrective lenses. The court held that what matters for purposes of determining disability is an individual's present state, which includes mitigating measures such as corrective lenses; being "potentially or hypothetically" substantially limited does not suffice. (The court was influenced by a congressional finding, in the introductory provisions of ADA, that 43 million Americans had one or more physical or mental disabilities. The court found this number hard to reconcile with a broad interpretation of disability.) While the holding in *Sutton* would not change the result in *Bragdon*, or an analogous case in which the potential for transmission of a genetic mutation would substantially limit current reproductive options, it does suggest that any argument that a mutation qualifies as a disability under the first prong of the definition due to its anticipated effects will fail. Certainly those with genetic conditions that are cured or kept under control through treatment will have a hard time establishing disability, unless they can show that the side effects of treatment are themselves disabling, or satisfy one of the other prongs of the definition.

Unfortunately for plaintiffs, *Sutton* also puts up a barrier to establishing "regarded as" disability, by suggesting that concerns about an impairment that are sufficient to prompt negative employment action may not be sufficient to establish that the employer is regarding the individual as disabled. Cases decided before *Sutton* had interpreted the definition of disability to encompass a third party's perception that disability was likely in the future, if that perception influenced present action. In *Doukas v. Metropolitan Life Insurance Company*, a federal district court noted that limiting ADA to perception of present disability would violate congressional intent and "allow an employer to refuse to hire an epileptic as long as the job applicant was not having a seizure at the time" (25). The court in *Winslow v. IDS Life Insurance Co.* found this reasoning persuasive (26). These cases, which concerned mental illness, and analogous cases involving genetic predisposition to disease, might be distinguished from *Sutton* by the unavailability of mitigating measures and the severity of the potential impairment. *Sutton* was, after all, a case about mitigating measures, despite language dismissive of probabilistic calculations and fears about the future as elements in the construction of disability. In *Cook v. State of Rhode Island Dept. of Mental Health*, a case decided under the Rehabilitation Act, the First Circuit Court of Appeals found that an employer's fears about risks associated with the plaintiff's morbid obesity were sufficient to establish perceived disability (27).

Is it good policy to adopt a generous interpretation of the ADA definition of disability and so extend ADA protections to unaffected carriers of recessive and X-linked disorders and individuals with unexpressed late-onset genetic conditions or genetic predispositions to disease? The answer to this question would appear to rest on a careful analysis of the fit between the purposes of ADA and the experience of persons falling within the particular category under consideration. Unaffected carriers of sickle cell trait can point to a history of isolation and segregation. Individuals with a genetic predisposition to cancer can point out that when employers turn them away, their contributions as productive members of society are diminished, perhaps unnecessarily and certainly prematurely. It is worth noting that state antidiscrimination laws may contain definitions of disability which are broader than the ADA's. For example, New York's highest state court has interpreted the New York State Human Rights Law to include "diagnosable medical anomalies which impair bodily integrity and thus may lead to more serious conditions in the future" (28).

COLLECTION OF INFORMATION BY EMPLOYERS

Title I of the ADA regulates medical examinations and inquiries conducted by employers (29). What is permissible varies according to the stage in the hiring process. To make sense of ADA, one needs to view hiring in terms of three stages: a pre-employment or interviewing stage prior to an offer of employment, a pre-placement or entrance examination stage after an offer of employment has been made but before commencement of employment duties, and a postemployment stage initiated with employment duties. The only acceptable *pre-employment* inquiries concern the ability of an applicant to perform job-related functions. Examinations and inquiries are generally prohibited *postemployment*, unless they can be shown to be "job-related and consistent with business necessity." However, two kinds of information-gathering activities relating to existing employees are expressly permitted: (1) voluntary medical examinations, including voluntary medical histories, as part of an employee health program, and (2) inquiries into the ability of an employee to perform job-related functions.

Employers have the most freedom at the *pre-placement* stage of employment. Title I states that after an offer of employment has been made, but prior to the commencement of employment duties, an employer may require a medical examination (which may include a review of medical records) and may condition the offer of employment on the results. The two limitations on employer discretion in this area are (1) all entering employees must be subjected to examination regardless of disability, and (2) the medical information obtained in this way must be collected and maintained on separate forms and in separate files, must be treated as confidential, and can be used only as permitted under Title I. Title I permits release of information to supervisors and managers where it concerns necessary work restrictions and accommodations, to first-aid and safety personnel when appropriate if emergency treatment may be required, and to government officials conducting compliance investigations. The same rules concerning separate forms and files, confidentiality, and use, apply to information obtained through examinations and inquiries made of existing employees.

The regulations issued by EEOC explicitly state that employment entrance examinations need not be job-related and consistent with business necessity (30). However, if an employer withdraws an offer of employment based on the results of an examination, the criteria used must not be of a kind to screen out or tend to screen out individuals with disabilities, or must be job-related and consistent with business necessity. This restriction on employer discretion may be hard to enforce, since job applicants who have received conditional offers of employment will often have a difficult time detecting illegal uses of information. An employer is generally not required to share the employer's reasons for withdrawing an offer of employment with the affected individual, and ADA does not alter this state of affairs (18). Some state genetic privacy laws require specific consent for genetic testing and disclosure of results, but absent such legislation, an individual may be completely in the dark concerning the nature or results of any tests conducted or information reviewed as part of an entrance examination.

Once in court, job applicants or employees face several hurdles. They may be met with the argument that only individuals who meet the statutory definition of disability are protected from inquiries and examinations. (As discussed at considerable length above, this test may be difficult to satisfy.) Federal appeals courts in the Eighth, Ninth, and Tenth Circuits (covering the western states and much of the midwest) have ruled that a plaintiff need not be disabled in order to state a claim for the unauthorized gathering or disclosure of confidential information by an employer (31). Plaintiffs alleging a violation of ADA's confidentiality protections may have difficulty showing that the violation resulted in some kind of tangible injury. Further ADA's confidentiality protections only apply to information collected through pre-placement medical examinations and the kinds of medical examinations and inquiries authorized under the ADA for existing employees. This leaves out medical information in benefit records, for example, medical information contained in benefit request forms. In *Yoder v. Ingersoll-Rand Company*, a federal district court ruled that the confidentiality provisions of ADA did not apply to a physician's statement confirming a diagnosis of HIV/AIDS in a disability benefit request form (32). The court rejected an argument for broader protection based on the general purposes of ADA. Medical information obtained before the effective date of ADA will also fall outside its protections (33). Finally, ADA does not prohibit employers from using general release forms, or soliciting consent to a broad battery of tests, at least at the conditional offer (pre-placement) and postemployment stages. Employers may compile extensive information in connection with voluntary wellness and employee assistance programs (34).

The Lawrence Berkeley Laboratory case, mentioned in the introduction, illustrates some of the difficulties associated with pursuing a remedy for unauthorized genetic testing. The employees in that case sued their employer under Title I of ADA and under other state and federal laws. The employees contended that testing for sickle cell trait and other sensitive medical conditions, allegedly without their knowledge or authorization, violated ADA because the testing was neither job related nor consistent with business necessity. They also advanced claims based on violations of privacy rights under federal and state constitutions. Finally, they argued that in singling out black employees for sickle cell testing (and female employees for pregnancy testing), the defendants violated Title VII of the Civil Rights Act. Title VII prohibits discrimination in employment based on race, color, religion, sex, or national origin. The employees did not allege that any employment-related action was taken on the basis of their test results or that their tests results were disclosed to third parties.

In *Norman-Bloodsaw v. Lawrence Berkeley Laboratory*, the Ninth Circuit Court of Appeals concluded that there is no remedy under ADA for unauthorized testing or testing lacking a job- or business-related justification at the pre-placement stage of employment (35). The court suggested that the only viable claim given the facts would concern a failure to properly maintain medical records according to ADA requirements but that something more than a general allegation of inadequate safeguards would be necessary for that purpose. The court found that the plaintiffs were entitled to a trial on their other claims.

It is important to note that ADA does not preempt state laws that provide greater or equal protections, and plaintiffs may have greater success in pursuing claims for breaches of confidentiality or unauthorized testing or inappropriate inquiries under state law. Title I of ADA must also be considered together with other federal laws affecting employment. For example, the Occupational Safety and Health Act (OSHA) requires medical monitoring of employees who may be exposed to hazardous chemicals. The implementing regulations require physical examinations before workers are assigned to certain sites, including a family history addressing genetic factors, but they do not require genetic testing (36). EEOC has stated that Title I does not halt the performance of such examinations, which might in any event be justified as job related and consistent with business necessity (37).

DISCRIMINATION IN EMPLOYMENT AND EMPLOYER-PROVIDED BENEFITS

Elements of Prima Facie Case and General Defenses

The general prohibition of discrimination in Title I is broadly stated to encompass all aspects of employment (38). So long as an individual with a disability is "qualified," that is, can perform all essential job functions, he or she is protected from discrimination because of or on the basis of the disability with respect to job application procedures, hiring, advancement, discharge, compensation, training, and other terms, conditions, and privileges of employment. Limiting, segregating, or classifying a job applicant or employee in a way that adversely affects his or her employment opportunities or status, and failing to make reasonable accommodations (i.e., accommodations that could be accomplished without undue hardship to the employer), are instances of discrimination, as are failures to abide by the rules concerning medical examinations and inquiries. Individuals with dependents with disabilities are also protected, because discrimination is defined to include the denial of equal jobs or benefits to a qualified individual because of the known disability of an individual with whom the qualified individual is known to have a relationship or association.

Even if a plaintiff makes a prima facie case, that is, establishes the elements of disability and disability-based discrimination, the employer can avoid liability by showing that a particular use of qualification standards, tests, or selection criteria was job-related and consistent with business necessity, and that no reasonable accommodation was possible under the circumstances. Qualification standards may include a requirement that an individual not pose a "direct threat" to the health or safety of other individuals. EEOC's Title I regulations specify that a direct threat is "a significant risk of substantial harm" (to self or others) that cannot be eliminated or reduced by reasonable accommodation (39). The assessment of risk must be based on "the most current medical knowledge and/or on the best available objective evidence." The likelihood and imminence of the potential harm are among the factors to be considered. This fairly stringent standard should preclude employers from making employment or job assignment decisions based on genetic susceptibilities or predispositions to disease, unless the science is good and tests of significance and substantiality are met. ADA does not prevent employers from seeking to understand and reduce hazards in the work environment. Nor does it prohibit the offer of accommodation to an employee with a condition that greatly increases the likelihood that the employee will suffer harm from a particular activity, or become incapacitated in a way that puts others at risk, assuming information about the condition is acquired by legal means. The fall-back defense of "undue hardship" requires a showing of "significant difficulty or expense," considering the nature and net cost of the accommodation, the financial resources of the facility and the covered entity, the impact of the accommodation on operations, and so on.

Employer-Provided Insurance and the Insurance Safe Harbor

While an employer cannot refuse to hire, or fire, a qualified individual with a disability due to fears about increased health care costs, or exclude the individual from benefit programs available to other employees, the employer is given considerable latitude in the area of insurance. The key provision in this area is what has become known as ADA's "insurance safe harbor" (40). The relevant subsection states that Titles I through IV should not be construed to prohibit or restrict (1) an insurer or other entity that administers benefit plans from underwriting risks, classifying risks, or administering such risks in a manner based on or not inconsistent with state law; (2) a person or organization from establishing, sponsoring, observing or administering the terms of a bona fide benefit plan based on underwriting risks, classifying risks, or administering such risks in a manner based on or not inconsistent with state law; or (3) a person or organization from establishing, sponsoring, observing or administering the terms of a bona fide benefit plan that is not subject to state laws that regulate insurance. (The Employee Retirement Income Security Act of 1974, known as ERISA, prevents the application of state insurance laws to employers' self-funded health plans.) However, ADA states that this provision cannot be used as a "subterfuge" to evade the purposes of Titles I and III.

EEOC's interpretive guidance on Title I addresses medical underwriting and preexisting condition clauses and benefit design issues. The guidance document states that medical underwriting and preexisting condition clauses included in health insurance policies offered by employers are not affected by ADA, except to the extent that practices are found to be inconsistent with applicable state law (21). In the area of health insurance, the Health Insurance Portability and Accountability Act of 1996 (HIPAA) may offer more extensive protections. (HIPAA focuses on group policies, but those with individual polices may benefit to a limited extent from provisions governing the transition from group to individual coverage and guaranteed renewability.) HIPAA *permits* issuers of group policies to impose limited preexisting condition exclusions *but only* if these relate to conditions for which medical advice, diagnosis, care, or treatment was recommended or received within the six-month period ending on the enrollment date. Genetic information cannot be treated as a condition in the absence of a diagnosis of the condition related to such information (41). HIPAA *prohibits* issuers of group policies from excluding an individual within the group from coverage on the basis of a health status-related factor relating to the individual or a dependent. The list of health status-related factors includes genetic information. HIPAA also prohibits variation in benefits, premiums, and contributions for similarly situated group members on the basis of these factors, but neither HIPAA nor the ADA requires that employers offer any insurance at all.

Employers can affect health care costs through benefit design as well as through medical underwriting and preexisting condition clauses. EEOC has concluded that ADA does not prohibit employers from placing limits on coverage for certain procedures or treatments (e.g., visit limits), even if these adversely affect individuals with disabilities, so long as the limits are applied equally to individuals with and without disabilities (21). EEOC offers more extensive comment on health insurance in its Interim Enforcement Guidance on Disability-Based Distinctions in Employer Provided Health Insurance (42), and in the Title I Technical Assistance Manual (37). The guidelines on benefit design would permit an employer to exclude all experimental drugs or procedures from coverage, so long as this restriction is applied evenhandedly to all insured individuals. It follows that employers would have no obligation to arrange for coverage of gene therapy and other interventions still in the research phase. Indeed, EEOC states that broad distinctions that apply to a range of dissimilar conditions and constrain individuals with and without disabilities are not distinctions based on disability. A term or provision is disability-based only if it singles out a particular disability or discrete group of disabilities or disability in general for inferior treatment. Cancers, muscular dystrophies, and kidney diseases are given as examples of discrete groups of disabilities. Genetic disorders, or inherited genetic disorders, would arguably constitute a discrete group of disabilities, meaning that it would not be permissible for an employer or plan administrator to single out interventions targeting genetic conditions for more limited coverage than other conditions. Picking and choosing among genetic conditions might also run afoul of the ADA. In *Henderson v. Bodine Aluminum, Inc.*, a woman with breast cancer argued that an insurer's policy of paying for bone marrow transplants for some cancers, but not breast cancer, violated the ADA. The Eighth Circuit Court of Appeals ruled in her favor, stating that "if the evidence shows that a given treatment is non-experimental ... and the plan provides the treatment for other conditions directly comparable to the one at issue, the denial of that treatment arguably violates the ADA" (43).

Courts are still struggling with the "subterfuge" language in ADA. EEOC's Interim Enforcement Guidance states that this language refers to "disability-based disparate treatment that is not justified by the risks or costs associated with the disability" (42). If an employee can make a prima facie case of discrimination, then EEOC puts the burden on the employer to produce evidence that "the disparate treatment is justified by legitimate actuarial data, or by actual or reasonably anticipated experience, and that conditions with comparable actuarial data and/or experience are treated in the same fashion," or to offer some other acceptable justification for the practice (e.g., that there is no other way to ensure the solvency of the plan or prevent a drastic increase in premiums). This suggests that ADA can be used to challenge underwriting decisions that are based on outdated or inaccurate information about genetics in general or specific genetic disorders, or on myths, fears, or stereotypes that have no basis in science. However, the phrase "reasonably anticipated experience," common in state insurance laws, does appear to create some room for inference from available data. Also a few courts have adopted a restrictive interpretation of subterfuge, holding that a benefit plan cannot be a subterfuge unless the employer intended by virtue of the plan to discriminate in a non-fringe-benefit-related aspect of the employment relation (e.g., the employer set out to design a benefit plan that would discourage persons with disabilities from applying for jobs) (44).

DISCRIMINATION BY INSURERS

Threshold Question: Scope of Title III

As noted above, Title III of the ADA regulates public accommodations. Rather than a true definition of the term, the statute offers a laundry list of private entities that are covered if their operations affect interstate or foreign commerce. These include an "insurance office, professional office of a health care provider, hospital, or other service establishment" (45). Because insurance is mentioned, individuals who have experienced discrimination in insurance have turned to Title III for a remedy. Resort to Title III is most common where insurance is not provided through an employer (or relief under Title I is unavailable for some other reason). Under Title III, as under Title I, the individual seeking a remedy for discrimination must first establish that he or she is an individual with a disability within the meaning of ADA.

A major point of controversy at present is whether Title III extends beyond access to physical structures to address access to services such as insurance policies. The courts are divided. In its Title III Technical Assistance Manual, the Department of Justice assumes rather than argues for the broad view (46). The evidence offered in favor of the restrictive view includes the many references to "offices" in the list that defines public accommodation and the insurance safe harbor. The leading case for the restrictive view is *Parker v. Metropolitan Life Ins. Co.* (47). In *Parker*, an individual sued her employer and her insurer claiming that a shorter benefit period for mental disability than for physical disability under an employer-provided disability policy violated ADA. The Sixth Circuit Court of Appeals ruled that a public accommodation is a physical place and a disability policy not obtained in an office transaction is *not* a service or good offered by a place of public accommodation. Although it was unnecessary to the decision in the case, the court also concluded that Title III does not extend to the contents (terms and conditions) of insurance policies. *Parker* has been followed by the Third and Ninth Circuit Courts of Appeals, and the Seventh Circuit Court of Appeals has also taken the position that Title III does not apply to the contents of insurance policies (48).

The First Circuit Court of Appeals has presented the case for the broad view. In *Carparts Distribution Center v. Automotive Wholesaler's Ass'n*, the First Circuit reasoned that by including "travel service" in the list of examples of public accommodations, Congress signaled that commercial enterprises not requiring physical entry could be public accommodations (49). The court noted that neither Title III nor the implementing regulations makes any mention of physical boundaries or physical access. Further the court believed it would be irrational, and inconsistent with the purposes of ADA, to conclude that persons who enter an office are protected by ADA, but persons who purchase services over the telephone or by mail are not. As to whether ADA requires scrutiny of the contents of insurance policies, the court found that in some cases, meaningful access to a service requires a change in substance. The reasoning of *Carparts* has been adopted by the Second Circuit Court of Appeals (50).

Other Title III Issues

Even if Title III is found to apply to insurance policies, plaintiffs may have a difficult time prevailing on a claim. The antidiscrimination language in Title III is broad: "No individual shall be discriminated against on the basis of disability in the full and equal enjoyment of the goods, services, facilities, privileges, advantages, or accommodations of any place of public accommodation by any person who owns, leases (or leases to), or operates a place of public accommodation" (51). Administrative methods that have the effect of discriminating on the basis of disability are expressly included. And Title III, like Title I, addresses discrimination based on association with an individual with a disability. However, if an insurer can establish that the actions taken were in accordance with sound actuarial principles, reasonably anticipated experience, or bona fide risk classification, they will likely be sheltered by the insurance safe harbor (discussed above). Construing the Title III antidiscrimination provisions in light of the insurance safe harbor, the Department of Justice has concluded that "a public accommodation may offer a plan that limits certain kinds of coverage based on classification of risk, but may not refuse to insure, or refuse to continue to insure, or limit the amount, extent, or kind of coverage available to an individual, or charge a different rate for the same coverage solely because of a physical or mental impairment, except where the refusal, limitation, or rate differential is based on sound actuarial principles or is related to actual or reasonably anticipated experience" (46). In essence, the Department of Justice and the EEOC have chosen the same middle course.

As in the employment context, it may be difficult for plaintiffs to make a prima facie case because they lack access to key information. The Title III Technical Assistance Manual states that ADA does not require an insurer to provide a copy of the actuarial data on which its actions were based at the request of the applicant (46). Further several courts, including the Third Circuit Court of Appeals, have suggested that allegations of subterfuge do not compel insurers to come forward with evidence to justify their coverage or underwriting decisions (48).

Still a number of plaintiffs have prevailed in lawsuits against insurers brought under Title III. For example, in *Chabner v. United of Omaha Life Insurance Co.*, an individual with fascioscapulohumeral muscular dystropy sued a life insurer for issuing him a life insurance policy at a premium that was considerably higher than the standard premium (52). As a threshold matter, the court found that Title III applies to insurance underwriting practices. Next the court held that where it is undisputed that an individual was treated differently based solely on his disability, the insurer has the burden of coming forward with evidence that the differential treatment was based on sound actuarial principles or actual and reasonably anticipated experience. The court added that even though the legal standard refers to "anticipated experience," insurers may not engage in speculation; that is to say, underwriting must always have a basis in actuarial data. The court found that in this case the defendant had failed to satisfy its evidentiary burden. As a result the court entered summary judgment in favor of the plaintiff. A recent decision from the Ninth Circuit Court of Appeals may affect the continued validity of the legal analysis in *Chabner* (48), but the facts are representative of the type of case that may become increasingly common as genetic disorders are subject to medical underwriting.

DISCRIMINATION IN PUBLIC PROGRAMS

Although the focus of discussion has been on employment and private insurance, the potential exists for discrimination in many public sector programs. In the ADA framework, public programs fall under Title II, governing agencies of state and local government. Although the bulk of Title II is devoted to public transportation, this title contains a general prohibition of discrimination by reason of disability affecting participation in or receipt of benefits of services, programs, or activities (53). Professional licensing appears to be one area where new developments in genetic technology could give rise to discrimination. Some have noted that there is also considerable potential for genetic discrimination in the public schools. (Private schools would be public accommodations covered under Title III.) It is conceivable that genetic information will someday be used to make modifications to programs to better meet the needs of children with genetic conditions. Genetic information may also be used to segregate children, where an administrator or teacher is persuaded that a certain mutation is associated with behavioral and disciplinary problems, or to attach labels to children that may become self-fulfilling prophecies (54). In these circumstances Title II of ADA could be used to challenge segregation and to combat other damaging practices based on stereotypes or hypothetical risks rather than individualized assessment. The Rehabilitation Act, which preceded ADA, remains available as an additional source of protection against discrimination by entities receiving federal financial assistance.

CONCLUSION

In sum, ADA exhibits the usual limitations of legislation: failure to adequately address situations remote from the experience of its framers, and resort to ambiguous language to achieve consensus. Few of the lawmakers who debated ADA reflected on the significance of developments in the field of genetics for civil rights and disability

law, and the text of ADA contains no mention of genetic conditions or genetic testing. Accordingly it is uncertain whether genetic conditions that are known, but presently asymptomatic, are covered under the statute. The extent to which the new antidiscrimination law should change the rules for insurance companies was certainly debated, but the resolution of that debate allowed for a range of interpretations. Indeed, the language of the statute is sufficiently ambiguous to send courts in different directions on the question of whether ADA imposes any constraints on the substance of insurance policies, especially those purchased by individuals. At present, then, there is considerable uncertainty concerning the relevance of ADA to genetic discrimination. ADA is certainly not a comprehensive response to the problem of genetic discrimination. Nonetheless, unless and until comprehensive legislation is enacted at the federal level, ADA will have to serve as proxy for a more complete and considered response, supplemented by the Rehabilitation Act and HIPAA and other federal and state laws. As genetic knowledge increases, and with it the potential for genetic discrimination, the courts will inevitably have to address some of the areas of uncertainty described above. In the not-to-distant future we should have, if not more justice, then at least more clarity.

BIBLIOGRAPHY

1. Pub. L. No. 101-336 (codified at 42 U.S.C. §§12101-12213).
2. 42 U.S.C. §12101(b).
3. M.S. Marquis and S.H. Long, *Health Affairs* **18**, 161–166 (1999).
4. H.T. Greely, in D.J. Kevles and L. Hood, eds., *The Code of Codes*, Harvard University Press, Cambridge, MA, 1993, pp. 264–280.
5. Office of Technology Assessment, U.S. Congress, *Genetic Monitoring and Screening in the Workplace*, vol. 1, U.S. Government Printing Office, Washington, DC, 1990.
6. M.A. Rothstein, *Am. J. Law, Med. Ethics* **26**, 198–204 (1998).
7. L.C. Volpe, *Genet. Test.* **2**(1), 9–12 (1998).
8. M.A. Hall and S.S. Rich, *Am. J. Hum. Genet.* **66**, 293–307 (2000).
9. O. Carrasquillo et al., *N. Engl. J. Med.* **340**, 109–114 (1999).
10. H. Ostrer et al., *Am. J. Hum. Genet.* **52**, 565–568 (1993).
11. L. Goch, *Best's Review-Life-Health Insurance Ed.*, November 1, 1998, WL 11240436, Westlaw, ALLNEWS database.
12. P.S. Miller, *J. Law, Med. Ethics* **26**(3), 189–197 (1998).
13. J. Seltzer, *Hofstra Law Rev.* **27**, 411–471 (1998).
14. Pub. L. No. 92-112, as amended by Pub. L. No. 93-516 (codified at 29 U.S.C. §§701-796).
15. 29 U.S.C. §794.
16. 42 U.S.C. §§12112, 12132, 12182.
17. 42 U.S.C. §12102.
18. M.A. Rothstein, *Houston Law Rev.* **29**(1), 23–84 (1992).
19. 29 C.F.R. §1630.2(h).
20. 28 C.F.R. §36.104.
21. 29 C.F.R. Pt. 1630, App., Interpretive Guidance on Title I of the Americans with Disabilities Act, §1630.2(h).
22. Equal Employment Opportunity Commission, *Compliance Manual*, §902.8 (1995).

23. 524 U.S. 624 (1998).

24. 527 U.S. 471 (1999).

25. 7 A.D. Cases 848, 1997 WL 833134 (D.N.H. 1997).

26. 29 F. Supp. 2d 557 (D. Minn. 1998).

27. 10 F.3d 17 (1st Cir. 1993).

28. *State Div. of Human Rights v. Xerox Corp.*, 65 N.Y.2d 213, 491 N.Y.S.2d 106, 109, 480 N.E.2d 695 (N.Y. 1985); *Application of State Div. of Human Rights on Complaint of Granelle*, 504 N.Y.S.2d 92, 118 A.D.2d 3 (N.Y.A.D. 1 1986).

29. 42 U.S.C. §12112(d).

30. 29 C.F.R. §1630.14.

31. *Cossette v. Minnesota Power & Light*, 188 F.3d 964 (8th Cir. 1999); *Armstrong v. Turner Industries*, 141 F.3d 554 (5th Cir. 1998).

32. 31 F.Supp.2d 565 (N.D. Ohio 1997).

33. See *Buchanan v. City of San Antonio*, 85 F.3d 196 (5th Cir. 1996).

34. E.E. Schultz, *Wall Street J.* May 18 (1999).

35. 135 F.3d 1260 (9th Cir. 1998).

36. OSHA Instruction STD 1-23.4 (1980), 1 OSHR Ref. File 21:8212.

37. Equal Employment Opportunity Commission, *A Technical Assistance Manual on the Employment Provisions (Title I) of the Americans with Disabilities Act*, U.S. Government Printing Office, Washington, DC, 1992.

38. 42 U.S.C. §12112.

39. 29 CFR §1630.2(r).

40. 42 U.S.C. §12201(c).

41. 42 U.S.C. §300gg.

42. Equal Employment Opportunity Commission, *Interim Enforcement Guidance on the Application of the Americans with Disabilities Act of 1990 to Disability-Based Distinctions in Employer Provided Health Insurance*, U.S. Government Printing Office, Washington, DC, 1993.

43. 70 F.3d 958 (8th Cir. 1995).

44. *Krauel v. Iowa Methodist Medical Center*, 95 F.3d 674 (8th Cir. 1996).

45. 42 U.S.C. §12181.

46. U.S. Department of Justice, *Title III of the Americans with Disabilities Act Technical Assistance Manual*, U.S. Government Printing Office, Washington, DC, 1993.

47. 121 F.3d 1006 (6th Cir. 1997), *cert. denied*, 522 U.S. 1084 (1998).

48. *Ford v. Schering-Plough Corporation*, 145 F.3d 601 (3d Cir. 1998), *cert. denied*, 119 S.Ct. 850 (1999); *Weyer v. Twentieth Century Fox Film Corp.*, 198 F.3d 1104 (9th Cir. 2000); *Doe v. Mutual of Omaha Ins. Co.*, 179 F.3d 557 (7th Cir. 1999), *cert. denied*, 120 S.Ct. 845 (2000).

49. 37 F.3d 12 (1st Cir. 1994).

50. *Pallozzi v. Allstate Life Ins. Co.*, 198 F.3d 28 (2d Cir. 1999), *amended on denial of reh'g*, 204 F.3d 392, (2d Cir. 2000).

51. 42 U.S.C. §12182.

52. 994 F.Supp. 1185 (N.D. Cal. 1998).

53. 42 U.S.C. §12132.

54. L.F. Rothstein, in M. Rothstein, ed., *Genetic Secrets: Protecting Privacy and Confidentiality in the Genetic Era*, Yale University Press, New Haven, CT, 1997, pp. 317–331.

See also GENETIC INFORMATION, LEGAL, REGULATING GENETIC SERVICES.

GENETIC INFORMATION, LEGAL, REGULATING GENETIC SERVICES

GEORGE C. CUNNINGHAM
State of California-Department of Health Services
Berkeley, California

OUTLINE

INTRODUCTION

Genetic testing is expanding at an accelerating rate (1–5). Genetic tests were first used to screen for or diagnose a hereditary disorder in a given individual. This early testing was based on the determination of abnormal protein products or metabolites produced by the mutant genes, such as sickle hemoglobin in sickle cell anemia or phenylalanine in phenylketonuria. Subsequently this was expanded to include testing for carriers of mutations with potential expression in subsequent offspring by identification of sequences of deoxyribonucleic acid (DNA) representing the actual mutation, such as sickle cell and cystic fibrosis carrier screening. Most recently it now includes testing for genes which, based on other genetic and environmental interactions, predispose to a wide variety of disorders, such as breast/ovarian cancer (BRCA1 and 2), cardiovascular disease, schizophrenia, and obesity. Genetic tests now include tests for gene products, actual genes, that is, DNA sequences, and abnormalities of chromosomal number and morphology. The field of gene therapy and preventative genetic engineering is just beginning to develop effective interventions.

The broadening of the scope of genetic medicine has not yet been recognized by public policy makers in legislative bodies, public health agencies, and the courts. As a consequence there is only a relatively limited body of law and regulation that addresses the many unresolved problems this new technology presents.

The legitimate interests of government include protecting the liberty, privacy, health and safety of its citizens, promoting the welfare of the community, and arbitrating disputes between competing values and interests in the society so as to obtain the greatest good for the greatest number with due consideration of minority rights and values. A great deal of interest has been generated and legislation enacted that addresses the issues of the use of genetic information in insurance, employment, and research (6–8). Much of the preoccupation with this issue is due to the unique market-based philosophy applied to health care delivery in the United States. Discriminatory use of genetic information to deny access to or increase the cost of insurance is a symptom of this broader problem, and its importance diminishes greatly in other developed nations where health care is not dependent on employment and is available as a universal social benefit. Discrimination has been discussed in the press and in professional journals for some time and will not be discussed in this article. Rather, we will try to address the neglected issue of ensuring equitable access to genetic services that meet a minimal standard of quality at reasonable cost. The issues of how best to use this new knowledge to alleviate the burden of genetic disorders on individuals and the society requires a response irrespective of the problems associated with discrimination.

PUBLIC HEALTH ROLE

The core functions of public health have been listed by the National Academy of Science Institute of Medicine as assessment, policy development, and assurance (9). Federal and state public health agencies have made only a small beginning in fulfilling these functions in genetic services.

ASSURANCE OF QUALITY GENETIC SERVICES

Licensure

Federal and state healthcare agencies are charged with the general mission of protection of public health, safety, and welfare. One of the major administrative tools used to ensure this general objective is achieved is regulation. This includes regulation of persons and facilities providing services, that is, governmental licensure. A license is an official governmental document that allows the holder to perform certain actions that are prohibited to the general public. It is therefore a restriction of personal liberty. Any legislator who proposes a law that restricts personal liberty needs to convince the legislature and the public that the restriction is necessary to protect the public from a greater harm than the restrictions would impose and that there is no other effective or less restrictive way to prevent the harm. Based on these basic considerations, is there a case for governmental licensure of genetic personnel such as doctoral-level geneticists in cytogenetics, genetic

counselors, genetic laboratory technology, genetic nurse specialists, and genetic facilities such as laboratories?

There are different concepts of what evidence would be sufficient to demonstrate public harm. Some would espouse a proactive stance and would be satisfied with evidence that supports a reasonable probability of harm.

Others favor a reactive philosophy and want to "count the bodies," such as document-specific instances of real harm.

PUBLIC EDUCATION

If citizens can protect themselves, there is no need for governmental involvement. There is little reason to believe the public can protect it self. The National Academy of Sciences in 1975 concluded: "It is essential to begin the study of human biology, including genetics and probability, in primary school, continuing with a more health-related program in secondary school.... Sufficient knowledge of genetics, probability, and medicine leading to appropriate perceptions of susceptibility to the seriousness of genetic disease and of carrier status cannot be acquired as a consequence of incidental, accidental, or haphazard learning..." (10). This has been reinforced by the Presidential Commission, the Institute of Medicine, the National Science Foundation, and other public and private groups in the intervening years. However, studies continue to document the low level of scientific literacy in the United States. The public at large is not well informed about genetics and genetic disorders and needs to rely on the services of experts. Five out of six never heard of genetic engineering. Only one-third of college graduates can correctly describe deoxyribonucleic acid (DNA). Surveys of scientific literacy have validated this deficiency (11–16). Erroneous genetic information or misinterpretations can lead to decisions not to get pregnant, terminate a pregnancy, stigmatization, loss of self-esteem, marital or familial disruptions, inappropriate and risky interventions, and the like.

PRIMARY CARE

Can we rely on the usual sources of care to protect the public from unnecessary services while assuring access to appropriate, high-quality services? With respect to genetics, the high probability of harm resulting from services provided by persons without specific training and experience in genetics is documented in several studies of genetic knowledge and practice of nongeneticists, such as, primary care practitioners (17–22). The inclusion of modern human genetics in the training programs for physicians and nurses is a relatively recent event. The number of human chromosomes and their relationship to disorders such as Down syndrome was not known until 1959. Prenatal diagnosis was introduced in the early 1960s. Use of DNA based tests in clinical medicine only began in the late 1970s. There are a large number of practicing physicians who have had little or no genetics in their training. Even after medical schools began to introduce genetics, the hours were minimal and frequently elective. Few questions on genetics are included on medical licensing or specialty board examinations. The result is a generally unsatisfactory understanding and application of clinical genetics in primary care. This is not a reflection on the dedication and concern of primary care physicians, who could not reasonably be expected to keep up with the explosion of genetic knowledge, but simply a factual description of the current situation.

The occurrence or high probability of occurrence of harm in the absence of access to qualified genetic personnel has been demonstrated and accepted by most knowledgeable professionals who have studied this area, but the questions remains, Is licensing the only or best solution to the problem? What are the alternatives? One alternative solution would be improved education of primary care providers. This is certainly necessary in any event and could contribute to improvement of services but does not appear to be practical or effective as the only response. Primary care physicians have limited time for continuing education and many areas of clinical practice other than genetics are competing for this limited time. In addition they have limited time with their patients. Studies indicate an office visit includes approximately 11 minutes of face-to-face contact with the physician. The average genetic counseling visit is approximately 50 minutes (23–25).

PROFESSIONAL CERTIFICATION

Another alternative is self-certification by professional societies (26). The first effort in certification of genetic counselors was the American Society of Human Genetics (ASHG) program which began in 1979. This was followed in 1980 by the recognition by the American Board of Medical Specialties of the American Board of Medical Genetics (ABMG) with the first certificates issued in 1981 under the auspices of ASHG. Certificates were issued to genetic counselors, doctoral-level geneticists, and physicians. The American College of Medical Genetics (ACMG) was established in 1992 and recognized as a component society by the American Medical Association (AMA). In 1993 the American Board of Genetic Counselors (ABGC) was established to continue the certification of genetic counselors. There are now 1006 medical genetic specialists, 150 Ph.D. medical geneticists, and 779 genetic counselors with board certification in the United States. While this process was a major contribution to the resolution of the problem, it is again not completely satisfactory. The medical specialty is still not listed in telephone directories nor recognized by the majority of third-party payers as one to be included in panels of specialists or listed in directories of providers. Utilization guidelines and reimbursement for services for these specialists are still being developed. Failure to authorize or reimburse for genetic counseling done by nonphysicians has limited the use of genetic counselors and Ph.D. geneticists. Furthermore, in most states, any physician can legally provide genetic services without special training or qualifications or specialty board recognition. This situation allows any physician, nurse, or counselor to be self-designated as a genetic specialist or subspecialist, which is misleading to the public.

FUNDING

Another alternative to prevent harm is the use of federal and state payment for services as a means to require the use of qualified staff in delivery of high quality genetic services. However, this is only effective for those services eligible for state payment and leaves the citizens who are not eligible unprotected. The reimbursement requirements of Medicaid or children with special health care needs are examples of this approach.

LICENSURE

As a result of this analysis, the only effective solution is to legally recognize the training and expertise of genetic professionals by licensure. There are additional arguments for licensure in terms of development of quality genetic services. Frequently quality genetic services require consideration of complex questions of risks, benefits, conditional interpretation, various interventions of varying effectiveness, and the like, in brief, information that would be beyond the reasonable expectation of the scope of primary care. If genetic personnel were licensed, the state could require the use of licensed or otherwise qualified personnel for complex genetic problems in order to maintain the quality of services. The existence of a license would allow genetic counselors as recognized professionals to have some say in governmental policy affecting their field of interest. Licensure would also promote the creation and funding of positions needed for expansion of new genetic services. In order to expand training programs, it is necessary to create a defined pool of positions for the graduates. Licensure would allow the public to specify and request referral to these professionals as recognized by the state.

SCOPE OF PERSONNEL LICENSURE

What kind of genetic personnel need licensure? Physician geneticists are licensed as physicians and only require better recognition and utilization of their specialty board certification. There are a group of doctoral geneticists trained in human clinical genetics who are certified by ABMG as Ph.D. medical geneticists. The laws of most states do not allow such persons to obtain a physician's and surgeon's license, and therefore prohibit their clinical utilization. For example, a cytogeneticist is prohibited from diagnosing chromosome anomalies as the practice of medicine. Provided their clinically related services are limited to their area of training and expertise, provision should be made to legally recognize them as practitioners by certification or licensure. Masters and doctoral level genetic counselors are an essential part of quality genetic services and need to have their practice made legal, including registration or licensure. In the laboratory area there are four subspecialist certificates issued by ABMG: clinical cytogeneticist, clinical biochemical geneticist, clinical molecular geneticist, and clinical biochemical molecular geneticist. The 1997 directory lists 425 clinical cytogeneticists, 137 clinical biochemical geneticists, 143 clinical molecular geneticists, and 49 clinical biochemical molecular geneticists (27). These are doctoral level categories, and they require candidates to pass an examination in general medical genetics as a precondition to taking the specialty examination. Individuals in these classifications are intended to function as laboratory directors of genetic specialty laboratories. The actual bench-level performance of tests is the job of the laboratory technologist. The complex techniques used by these technologists are not a part of traditional laboratory technology training programs. Recognizing the need for special training and certification in the field of cytogenetics, five California cytogeneticists organized an Association of Cytogenetic Technologists (ACT) in 1975. In cooperation with the National Certifying Agency for Medical Laboratory Personnel (NCAMLP), a national technologist certification program was developed in cytogenetics, which issued its first certificates in 1981.

The field of molecular biology, namely DNA analysis, was largely research oriented until the early 1990s. The California Department of Health Services contacted ACT and NCAMLP and requested a certification process be established in molecular biology. As a result ACT expanded its area of interest and in 1996 changed its name to Association of Genetic Technologists (AGT). NCAMLP responded by establishing a certification program for Certified Laboratory Specialist in Molecular Biology with the first examination given in July 1997. There is no specialty certification for technologists in biochemical genetics at this time.

FACILITIES LICENSURE

Federal and State Licensing Responsibilities

The federal government has not assumed responsibility for licensure of personnel. The basic law regulating laboratory practices, the Clinical Laboratory Improvement Act of 1988 (CLIA), does provide for certain minimum educational and experience qualifications for laboratory directors, technical supervisors, and testing personnel and describes their responsibilities (28). The federal regulations first require a current license issued by the state in which the laboratory is located. The only specific reference to genetics is the area of cytogenetics where the technical supervisor must have four years of genetic training or experience, two of which must have been in cytogenetics. There is currently an advisory committee working on improving the coverage of genetic laboratory personnel. All laboratories must be in compliance with CLIA, including the genetic specialty laboratories. The federal regulations cover areas of staffing, patient test management, quality control, proficiency testing, inspections, and sanctions.

States generally have laws that require licenses for clinical laboratories and laboratory personnel. New York (1972) (29) and California (1995) (30) recognize genetics as a laboratory specialty area. New York defines the qualifications for laboratory director and makes the director responsible for using qualified technologists and maintaining quality control. Specific standards are detailed for cytogenetics (1972), and genetic testing (1990). Proficiency

testing and site visits are required. California is currently implementing a similar program. However, in addition California requires licensure of genetic technologists.

LICENSURE CONSIDERATIONS

Public agencies and legislatures, in considering a proposed licensure program, need to collect information in a variety of areas before the full societal impact can be assessed. This includes the numbers of personnel and their professional representation; what segment of the public is served, what is the position of public advocacy groups, is there duplication or competition with existing licenses, what is the nature and severity of harm to be prevented? Are there alternatives? What will licensure cost? What will be the effect on supply? What is the limit on the scope of practice? Are knowledge and skills testable? Are there approved schools to provide training? What is the economic impact?

LEGISLATION AND REGULATIONS

With or without licensure, laws can be passed regulating the provision of genetic services. Legislation has been used in California to regulate prenatal serum screening (31). The state law permits the Department of Health Services to specify standards for vendors participating in the statewide birth defect screening program, which is called the Expanded AFP program. This program is based on the well-documented association of specific patterns of analytes (alpha feto-protein, human chorionic gonadotropin, and unconjugated estriol) with increased risk of birth defects (neural tube defects, abdominal wall defects, Down syndrome, and other chromosomal defects). The law also requires that all women seen before the twentieth week of gestation be provided information about the screening and be offered an opportunity to be screened by the program. If the woman elects to be tested, she signs an informed consent. Specimens are collected and transmitted to the laboratory. Specimens are analyzed in one of eight regional private laboratories under contract to the department. These laboratories use uniform methodology and are subject to daily quality control by the state. All data and laboratory results are communicated to a central computer in Berkeley. All persons who are judged to be high risk by the central computer algorithm are authorized, at no additional charge, to receive follow-up diagnostic services at one of 29 state-approved Prenatal Diagnostic Centers. The follow-up services include genetic counseling, ultrasound examination and, if necessary, amniocentesis, amniotic fluid analysis, and karyotyping. Any facility that meets state standards can be designated an approved vendor. The standards require that the prenatal diagnostic center be directed by a board-certified medical geneticist, that genetic counseling be provided by board-certified genetic counselors, that ultrasound examinations be performed by specially skilled and experienced ultrasonologists, and that amniocentesis, if indicated, be performed by experienced perinatologists/obstetricians. The state collects a participation fee from third-party payers or from the participant, which covers all operating costs and is used to reimburse vendors. This public–private partnership design has succeeded in providing universal access to high-quality services.

TECHNOLOGY ASSESSMENT

In addition to prevention of adverse consequences of genetic disorders by promotion and regulation of personnel, the state is obligated to provide another kind of protection, namely protection from premature promotion of tests and substandard services that could adversely affect individual citizens or the community at large. This establishes the public health department as the primary technology assessment agency. Technology assessment really is a process of reviewing the scientific evidence and the information and opinions of experts to determine if a given technology should be applied in clinical practice and under what circumstances and conditions.

FEDERAL ROLE

Decisions on the appropriate implementation of any new genetic testing program are currently not centralized (32). There was an Office of Technology Assessment (OTA) established in 1972 to conduct assessments for the Congress. The mandate was broad, encompassing any technological problem, and OTA did publish some studies of heath technologies. OTA was abolished in 1995. In 1989 the Institute of Medicine published a monograph recommending a national technology assessment agency (33). The Congress established the Agency for Health Care Policy and Research (AHCPR) in 1989, but again, it has a broad area of responsibility and has not included many genetic technologies in its reviews. The National Institutes of Health (NIH) has established a mechanism called a consensus conference where a panel hears presentations from experts, reviews the literature, and publishes consensus statements on technologies. The federal Food and Drug Administration (FDA) has been proposed to play this role through their regulation of diagnostic kits and devices (34,35). While they could regulate clinical accuracy and utility, they do not have the authority to regulate the ancillary clinical setting in which the test is used.

STATE ROLE

Given the numbers of technologies being proposed or currently in use without rigorous analysis, all these efforts contribute useful information and should be encouraged. However, it is important to establish technology assessment capacity at the state level since the states play such a critical role in regulation and funding of health care. The legislature of the state of Maryland established a Maryland Commission of Hereditary Disorders in 1973, which reviews genetic tests, but this model has not been adopted by other states. There is a legitimate role for the state regulatory process using input from both experts and the public. The regulations can prohibit unvalidated testing,

when the preponderance of evidence indicates that the public either individually or collectively could be harmed, except as part of a research project. The state can impose conditions on genetic testing by regulation when the evidence indicates such conditions are necessary. These conditions might include specialized informed consent, confidentiality, pre- and/or post-test counseling, protective measures, quality assurance requirements, record keeping requirements, availability of diagnostic and intervention resources, and so on. This approach could include use of state accredited or registered personnel who had training essential for the proposed testing program. Finally, once a technology is accepted by the experts and the public using the evidentiary process, the state has an obligation to promote equitable access. This could involve mandating that all public and private payment sources pay for any cost-effective technology, funding screening centers, and conducting public and professional education or outreach with the at-risk public. One example of the states' carrying out this function is the newborn screening for genetic disorders.

PRIVATE SECTOR

In addition to state regulatory practices, most insurance companies and managed care organizations have technology assessment groups. The criteria used by these private groups can be unduly influenced by cost considerations and it is not unusual to find a test accepted by one payer and regarded as experimental and not accepted by another.

PROBLEMS WITH LEGISLATION AND REGULATION

While laws and regulations have undoubtedly saved lives, prevented disease and disability, and increased the value of the goods and services, the potential for regulatory abuse and damage is all too apparent to the public. The arguments made in favor of an expanded activist role for public health in regulation of genetic services could be seriously undermined by failure to avoid the situations that have contributed to the current low esteem accorded the use of this governmental tool. The first error is what is referred to as "agency capture" where the special interest group affected by the regulations controls directly or indirectly the governmental regulators. Regulation should not be used to increase incomes of specialists through unnecessary restriction of services. Licensure laws can be used to exclude qualified providers in order to maintain incomes. Facility standards can be used to monopolize services and improve their economic outcomes. The public and the regulators should be aware of this tendency and should maintain an open public process that remains focused on the goal of assuring universal access to comprehensive high-quality, cost-effective services. While recognizing the contributions and qualifications of such professional groups as ACMG and the National Society of Genetic Counselors (NSGC), the state should remain open to including others, such as pathologists with subspecialty training in cytogenetics or molecular biology and nurses with subspecialty training in genetics as qualified providers. Special certification of training in hereditary cancer counseling, sickle cell counseling, and cystic fibrosis genetics can be used to create a pool of qualified personnel to implement specific screening programs. Finally, involvement in policy development of the increasingly active groups of individuals and families affected by genetic disorders and the general public can exert a corrective action.

Part of the process of effective regulation is the follow-up of enforcement and monitoring. Simply putting a requirement in a statute or regulation is no guarantee that the system will implement the requirement as intended. Failure to enforce regulations is another factor undermining public confidence in government's interest and ability to represent the public interest. On-site monitoring, chart reviews, records maintenance, reporting outcomes or events, and the like, are required on a continuous basis to ensure uniform application of the law and community-wide implementation. While these add to the burden of regulation they are essential if the benefits of regulation are to be real instead of imagined.

Another problem is the incompetence of some of the regulators. It is difficult to provide in the public sector the kinds of salaries that will be guaranteed to attract the kind of genetic expertise needed. It is important that regulators have access to expert consultants and adopt policies and processes that permit input from a broad variety of genetic and nongenetic professionals, as well as affected members of the public. Qualifications of regulators should include familiarity with the field of genetics, public health, law and administration, and the way the health care system operates. The regulatory agency should have sufficient resources and visibility to be able to develop and implement effective programs and formulate and enforce regulatory standards. The regulations should reflect a consensus of affected parties as to the minimum requirements of currently accepted standards of care, and not utopian efforts to provide cutting-edge technologies to anyone who might possibly benefit.

CONCLUSION

The rapid development of genetic knowledge and technology poses problems, both familiar and novel, for the society. As the representative of the public, the federal and state public health agencies need to be prepared to ensure equitable access to quality genetic testing for high-risk populations. Regulation of personnel and facilities providing testing can play a constructive role in assuring that the inevitable adverse consequences of testing are minimized and benefits are maximized with fair treatment of all the involved parties and interests.

BIBLIOGRAPHY

1. F. Collins, *N. Engl. J. Med.* **341**, 28–37 (1999).
2. M. Khoury, *Am. J. Pub. Health.* **86**, 1717–1722 (1996).
3. *Science*, The Genome Issue **286**, 443–491 (1999).
4. J. Bell, *Br. Med. J.* **316**, 618–620 (1998).
5. *Time*, January 11 (1999).
6. K. Rothenberg, *J. Law Med. Ethics* **312**, 312 (1995).
7. K. Rothenberg et al., *Science* **275**, 1755–1757 (1997).
8. A.M. Capron, in *Proc. Conf. on The Implications of Genetics for Health Professional Education*, 1999, pp. 123–162.
9. Institute of Medicine, *The Future of Public Health*, National Academy Press, Washington, DC, 1988.
10. L.B. Andrews et al., eds., *Assessing Genetic Risks: Implications for Health and Social Policy*, National Academy Press, Washington, DC, 1994, p. 186.
11. J. Miller, *Report to the National Science Foundation*, NSF, Washington, DC, 1992.
12. March of Dimes Birth Defects Foundation, *Genetic Testing and Gene Therapy*, Louis Harris and Association, White Plains, NY.
13. March of Dimes Birth Defect Foundation, *News Release*, September 29, White Plains, NY, 1992.
14. K.J. Hodgkinson, *J. Med. Genet.* **27**, 552–558 (1990).
15. L.B. Andrews et al., eds., *Assessing Genetic Risks: Implications for Health and Social Policy*, National Academy Press, Washington, DC, 1994, p. 186.
16. K.J. Hofman et al., *Acad. Med.* **68**, 625–631 (1993).
17. N. Holtzman, *J. Clin. Ethics* **2**, 1–5 (1991).
18. N. Holtzman, *J. Clin. Ethics* **2**, 1–2 (1992).
19. E.S. Tambor et al., *Am. J. Pub. Health* **83**, 1599–1603 (1993).
20. D.C. Wertz, *Am. J. Hum. Genet.* **61**, A193 (1997).
21. A.D. Kline, G. Halpin, and I.S. Mittman, *Am. J. Hum. Genet.* **61**, A189 (1997).
22. L.B. Andrews et al., eds., *Assessing Genetic Risks: Implications for Health and Social Policy*, National Academy Press, Washington, DC, 1994, pp. 116, 119.
23. B.A. Bernhardt et al., *Obs. Gyn.* **91**, 648–655 (1998).
24. N.C. Blumenthal et al., *J. Family Pract.* **48**, 264–271 (1999).
25. T.G. Ferns et al., *Arch. Ped. Adoles. Med.* **152**, 227–233.
26. V.A. McKusick, *J. Am. Med. Assoc.* **270**, 2351–2356 (1993).
27. American Board of Medical Genetics, Available at: *http//www/faseb.org/genetics/abmg/stats.allyears.htm*
28. Title 42 code of Federal Regulations Part 405 et seq., Section 493.1449.
29. New York Public Health Law, Title V, Article 5.
30. California Business and Professions Code Section 1202.5 et seq.
31. California Health and Safety Code Part 5, Article 1, Hereditary Disorders Act, Section 124975 et seq.
32. N. Holtzman, *Science* **286**, 409 (1999).
33. L.B. Andrews et al., eds., *Assessing Genetic Risks: Implications for Health and Social Policy*, National Academy Press, Washington, DC, 1994, p. 13.
34. B.S. Wilford and K. Nolan, *J. Am. Med. Assoc.* **270**, 2948–2954 (1993).
35. National Academy of Science Institute of Medicine, *Assessment of Diagnostic Technology in Health Care*, National Academy Press, Washington, DC, 1989.

See other entries GENETIC INFORMATION, ETHICS, AND INFORMATION RELATING TO BIOLOGICAL PARENTHOOD; GENETIC INFORMATION, ETHICS, INFORMED CONSENT TO TESTING AND SCREENING; GENETIC INFORMATION, LEGAL, FDA REGULATION OF GENETIC TESTING; GENETIC INFORMATION, LEGAL, GENETIC PRIVACY LAWS; GENETIC INFORMATION, LEGAL, GENETICS AND THE AMERICANS WITH DISABILITIES ACT.

H

HUMAN ENHANCEMENT USES OF BIOTECHNOLOGY, ETHICS, COGNITIVE ENHANCEMENT

Peter J. Whitehouse
Case Western Reserve University
Cleveland, Ohio

Cynthia R. Marling
Ohio University
Athens, Ohio

OUTLINE

INTRODUCTION

Have you ever smoked a cigarette or drunk a cup of coffee and felt more alert? Do you know someone who wears a hearing aid and attends better to oral language, thus remembering things better? Do you want your surgeon to be using the latest technological equipment in treating your condition? Should Indian chess players participating in international competition be permitted to consume bramin, which is used widely in their country for cognitive enhancement? Do you believe that electronic technology will play an increasing role in education? Do you think that biological and information sciences will lead to new knowledge that will enhance our ability to pay attention and remember? Do you believe that, particularly at this time in the history of the human race, such enhancements in our ability to think may be especially critical? Do you think wisdom is a desirable individual and social goal?

This article presumes that many readers will answer some of these questions in the affirmative. People already consume products and participate in activities that they believe will enhance their thinking abilities, and they are inclined to seek out the latest advancements in these areas (1). Perhaps this interest reflects the fast pace of life in modern times and the desire for individuals and groups to obtain competitive advantage in the worlds of business and education.

This article examines some of the underlying ethical issues that relate to cognitive enhancement. We will explore these issues after a discussion of what we mean by cognition and by enhancement (2). We believe that interest and activity in this area of life will only increase in intensity in the future.

Cognition

The term "cognition" could perhaps be replaced with the word "thinking." However, clinicians and researchers use the term to refer to a variety of intellectual skills, including attention, learning, memory, language, skilled motor behaviors, and perceptual abilities. In addition it often encompasses so-called executive functions, such as goal setting, planning, judgment, problem-solving, and decision-making.

Most of the literature on cognitive enhancement addresses methods of enhancing either attention or memory through medications. Thus drugs may help one stay awake and attend to stimuli in the environment or remember past or upcoming events better. Undoubtedly, we would be even more excited by cognitive enhancers that improve so-called higher level thinking, such as the executive functions and even wisdom (3). All intellectual abilities depend on adequate arousal. (Note that it is difficult to learn new material when in a coma.) Arousal is intimately related to attentional mechanisms. Another fundamental intellectual task is to selectively attend to important stimuli in the environment and to avoid distraction by those less critical. Interventions that affect the speed of processing, allocation of attention, and accomplishment of complex tasks might be expected to improve higher level decision-making and problem-solving as well.

Cognitive Enhancement

The term cognitive enhancement is usually used to differentiate this concept from enhancement of noncognitive or emotional abilities. The ability to appreciate and detect the wide range of human emotion both in oneself and in others is also a critical ability that deserves consideration for enhancement. Clearly, alterations in mood, such as depression and anxiety, can impair decision-making and problem-solving. Drugs such as Fluoxetine (Prozac) can help clinically depressed patients, but whether interventions can enhance emotional abilities in normal individuals remains uncertain (4,5). Nevertheless, presumably normal grocery shoppers snap up herbal teas and substances like St. John's Wort in hopes that they can.

The distinction between cognitive and emotional capabilities is somewhat arbitrary. Fundamental psychological functions, such as arousal and motivation, suggest that basic distinctions between thought and feeling deserve scrutiny. However, for this article we will not focus on enhancement of noncognitive abilities, although we will make brief mention of the possibilities of biological intervention in this domain. We should also point out that one of the ultimate goals of enhancement technologies, that is, to improve human wisdom, would undoubtedly involve improvement in both cognitive and emotional skills, as wisdom represents that integration of high-level processing of both thoughts and feelings (6).

Biological Substrates

We are beginning to understand more about the biological substrates of human cognition, opening this as an avenue for the development of enhancement technologies through molecular biology and neurochemistry. Much of this understanding has come from the study of diseases such as Alzheimer disease and other dementias (7,8). Dementia is the medical term for loss of cognitive abilities in more than one domain, and Alzheimer disease is the most common dementia. Alzheimer disease is biologically characterized by the loss of specific populations of nerve cells in association with specific pathological features observable under the microscope. Relating the loss of particular populations of cells to clinical symptoms has been the Holy Grail of clinical pathological correlation in Alzheimer disease and related disorders.

One of the biological systems underlying learning and memory is the cholinergic basal forebrain (9). Loss of nerve cells in this structure, located deep in the brain underneath the basal ganglia, is a substrate for the cognitive impairment found in Alzheimer disease and some other disorders. The basic scientific evidence for this is fairly convincing in terms of the effects of damage to the structure in animals on learning and memory. Drugs that block cholinergic systems can cause memory problems in normal human beings. Most important from a therapeutic point of view, drugs such as Donepezil, which enhance cholinergic function (they are cholinesterase inhibitors that work by blocking the enzyme that breaks down acetylcholine), improve attention and memory in these conditions. Although the effects are moderate in size, they have been definitely and conclusively demonstrated in double-blind placebo controlled studies (8).

There is no specific biological marker to differentiate normal aging from Alzheimer disease. Some degree of dysfunction to the basal forebrain occurs as we all age.

A variety of labels (10,11) have been applied to this condition, ranging from benign senile forgetfulness to aging associated memory impairment to the term most commonly used today, mild cognitive impairment. These labels are applied to individuals who are usually older but do not have significant enough cognitive impairment to be considered demented. For example, memory problems rarely affect function in daily life. Thus a variety of trials are underway to enhance memory in individuals who are labeled with these conditions. They are by definition not demented and hence, normal. Therefore these trials really represent attempts to enhance cognition in normal individuals.

In addition to interventions that enhance cognition symptomatically, science is attempting to develop therapies that may slow the progression of conditions like Alzheimer disease that are due to gradually increasing loss of nerve cells in structures such as the cholinergic basal forebrain. A variety of approaches are being used, ranging from antioxidants such as vitamin E, antiinflammatory agents, and compounds that act to enhance the viability of nerve cells, such as nerve growth factors. Trials are also underway to treat patients who have mild cognitive impairment with these agents to delay the onset of Alzheimer disease, which would occur in a significant number of these individuals. Hence two forms of treatment are currently being used in normal people to try to enhance cognition, one symptomatical and the other preventative.

We should hasten to add that the therapeutic targets biologically are more than just the cholinergic system. Neuronal loss occurs in the locus ceruleus, which uses the neurotransmitter noradrenalin and the raphae nuclei, which use serotonin. The clinical consequences of loss of cells in these populations are less clear. However, drugs that act on these transmitters can affect cognitive abilities such as attention, as well as mood (4,12). Antidepressant medications work to enhance neurogenetic and serotonergic functions. For example, drugs such as amphetamine and Ritalin can affect mood and attention in normal individuals, and are used to treat Attention Deficit Disorder in children and adults.

Cholinergic medications may improve behavioral symptoms, namely noncognitive symptoms in dementia. Thus, again, we are reminded that the distinction between things cognitive and things affective is not always easy to determine either at a clinical or biological level. Although biological approaches are promising and growing increasingly so, most of human history has focused on nonbiological approaches to enhancing thinking abilities. Schools have been widely used throughout history to enhance thinking ability in children and adults, although their effectiveness has been under increasing scrutiny. Various assistive devices have been employed, such as the Chinese abacus, which was one of the earliest. Varieties of memory assistive devices have been used probably since the advent of commerce. However, the power of information sciences is increasing as rapidly or perhaps more rapidly than that of the biological sciences, offering other forms of enhancement possibilities for cognitive abilities. As computers and personal digital assistants proliferate and become more intelligent, the symbiosis between individual people and their computers becomes greater.

We have chosen here to focus on both biological and information systems enhancement and their ethical implications, although at first glance they would appear to be quite different. However, devices are currently used clinically in which microcomputers embedded in people control the infusion of biological substances designed to improve cognition (e.g., insulin pumps in diabetics). In principal, as biochips become increasingly sophisticated, enhancement technologies would likely include both silicon- and carbon-based approaches.

THE NATURE OF ENHANCEMENT

The article by Eric Juengst (13) in this encyclopedia provides the framework for our considerations of enhancement. Moreover we have benefited from work focusing on enhancement in sports (14). The area of genetic enhancement and physical enhancement has received more attention (15), for example, in the area of cosmetic surgery (16). Enhancement has been useful to limit the domain of medical practice, as it is usually considered to lie at the fringe of the scope of medicine. However, we will focus on the second meaning that Juengst gives to enhancement, which is the notion of self-improvement. The limiting of the domain of medicine is an interesting area in cognitive enhancement. The success of the medical establishment in identifying Alzheimer disease as a treatable condition requiring medical research and intervention has been noted. Interestingly the nature of Alzheimer disease as a disease is being challenged by the approach to try to enhance cognition in people with so-called mild cognitive impairment. However, we will focus principally on the notion of cognitive enhancement as improvement rather than treatment.

IMPORTANCE OF COGNITIVE ENHANCEMENT

Individual Use Pattern

It seems evident that individual human beings have decided that cognitive enhancement is a goal worth pursuing. Not only do we put considerable energy into going to school, but we also pursue increasingly faster individual computers and frequently take biological interventions ranging from coffee to complementary alternative measures in seeking out enhanced thinking ability. Global expenditures of billions of dollars, far exceeding expenditures on prescription drugs for all conditions, let alone those designed to enhance cognition, demonstrate the importance that individuals assign to this area, as this money is spent out of pocket (17). The range of biological products that have claimed to enhance cognition is enormous (1). Perhaps the most commonly used drug for this purpose worldwide is ginkgo biloba. The scientific evidence that ginkgo biloba helps in any disease state is inconclusive. The evidence that it enhances normal cognition is even weaker.

An entire class of drugs has been referred to as nootropics, meaning mind growth. The original compound, piracetam, has been demonstrated to improve learning and memory in a variety of animal models. Its effects in human beings are limited, however. Yet compounds in the same class such as nefiracetam are under active investigation for treating stroke and dementia. The power of genomics, combinatory chemistry, high throughput screening, and a variety of other approaches available in the industry, make it reasonable to think that more effective medications will be developed in the future, not only for diseased conditions, but for enhancing normal thinking.

Understanding Complex Systems

The stakes for cognitive enhancement go far beyond the individual economic performance of human beings. The human race is facing complex challenges to its very survival (18,19). For example, it seems evident that human beings have had significant impact on their environment and that of other species. A topic such as global warming and the controversy surrounding it illustrates the difficulties that human beings have in understanding the behavior of complex systems, such as our biosphere, and projecting the effects of our behavior in the present onto the state of our biosphere in the future. Computer models can be used to make projections about the viability of life on this planet, and thus we are already using information technologies to help us analyze complex system behavior. It is obvious that if we had biological and information science interventions that could enhance our ability to understand the consequences of our own behavior in the present, this would be a tremendous advancement for future generations. This would be particularly important if the cognitive abilities enhanced were, in fact, executive functions and even wisdom, improving the human brain's ability to model the consequences of present behavior on future states.

Thus we believe that the current interest expressed by individual human beings in enhancement could be reason enough to consider the ethical issues seriously. However, the need to enhance our cognitive abilities to help ensure the sustainability of life on this planet raises the stakes even further. It seems clear to us that thinking through the ethical issues surrounding cognitive enhancement warrants serious consideration.

ETHICAL ISSUES COMMON TO BIOLOGICAL AND INFORMATION INTERVENTIONS

Goals

If the goal of cognitive enhancement using either drugs or computer systems were self-evidently a desirable outcome, then the enhancement of higher level thinking, such as wisdom, would seem to be especially desirable. However, it is quite possible that enhancement of selective areas of cognition would not necessarily result in overall improvement. We already exist in a world in which people find the pace of life rapid. Would a drug that merely improved the ability of an individual to think more things be a desirable product? Would this merely focus attention on quantitative rather than qualitative outcomes? Would we create increased unhappiness by driving people forward to greater and greater productivity? Enhancing cognition might have detrimental effects on the broader personality. One is reminded of the warning of Spock on *Star Trek*, that there might be sacrifices to be made in the emotional life by enhancement in the cognitive sphere. Surely this is not necessarily a consequence of cognitive enhancement, but it is one worth being aware of. After all, we do live at a time that still celebrates the rational values of the enlightenment. Focusing on enhancing thought without feeling, that is, knowledge without wisdom, might in fact contribute to further self-destructive and society-damaging behaviors.

Costs

Another ethical issue common to enhancement technologies, be they drugs or computers, represents the issue of costs. How much energy should human society place into trying to develop interventions to improve cognition? Of course, this is a difficult question to answer because one can never predict the results of scientific research. If very effective interventions were developed at low costs, this might be desirable. However, it is a daunting challenge to enhance human cognition in any way, and thus societal resources could be invested out of proportion to the likelihood of success.

Justice

The fact that enhancement technologies already do cost considerable amounts of money raises the ethical issues surrounding justice and access (20,21). This topic is already of considerable interest to health care professionals and individuals who have thought about universal access to computer technology. Would the availability of even more effective enhancement technologies increase the already growing distance between the have and have not countries, as well as the have and have not populations within countries?

Risks and Benefits

The introduction of new technologies raises the issue of risks and benefits. Although we have talked mostly about the positive outcomes of enhancement technology, all technologies have potential downsides. Medications to enhance cognition that are given to healthy people would need to have a low chance of significant side effects in order to justify their use. Who would decide what level of enhancement is worth what level of risk? In the United States and in most countries, the regulatory authorities in governments attend to the issues of risk benefit and disease but have limited jurisdiction over enhancement technology designed for individuals who suffer from no illness. Clearly, at a more macro level, the Y2K problem illustrates the risk of dependence on information technology. Will we someday regret that we have become so dependent on computers that human lives can be lost as a result of power failures or other system crashes?

ETHICAL ISSUES RELATING PRINCIPALLY TO BIOLOGICAL INTERVENTIONS

Interventions designed to change the biology of an individual human being present some ethical issues, which

if not unique, are at least more obvious in relationship to this form of intervention. We might have included the issue of risk-benefit in this discussion, as various adverse events are more likely to be associated with medications than with the use of computers. However, pills seem to create the ethical issue of an artificial road to enhancement. Steroids taken by athletes are viewed as creating an unfair competitive advantage. While improving athletic prowess through diet, exercise, videotape feedback, and even computer analysis of physical motion is viewed as laudatory, taking pills to improve athletic prowess is not. Thus drug tests are now de rigueur at sporting events, and athletes who are found to have taken performance enhancing drugs are disqualified. Are drugs equivalent to hours of training? Will a cognitive enhancing pill that replaced hours of toil and sweat in a classroom similarly be viewed as some kind of inauthentic perversion (22)? However, distinctions between artificial and natural are actually difficult to make and, if the pills were relatively safe, most individuals might believe that this form of enhancement would be appropriate.

ETHICAL ISSUES RELATIVELY UNIQUE TO COGNITIVE ENHANCEMENT WITH INFORMATION TECHNOLOGY

Clearly computers are a more evident technology that enhances human beings' ability to remember and problem solve. What is clear also is the rapid advance in the intelligence of computers. What are the ethical issues that relate to enhancing human cognition through the use of computers?

Confidentiality

The first issue relates to that of confidentiality (23). Computers enhance human cognition in part by allowing information to be shared more quickly between different individuals, for example, through the use of e-mail. Yet this easy distribution also allows for the possibility that more information becomes available essentially to the entire world. Who has not received an e-mail entitled something like "learn everything about everybody"? Therefore violation of privacy becomes a major concern in using information technology to enhance human thinking abilities.

Computer-Assisted Information Processing

As mentioned in the beginning, in terms of environmental issues, solving complex health and resource problems is becoming increasingly difficult. Within the area of medicine, for example, it is becoming increasingly difficult to know which medical interventions to offer which patients. It is ironic that almost a century after the beginning of so-called scientific medicine, we are now promoting the notion of evidence-based medicine and health care. We need to take seriously the moral obligation to use optimally the information that has been collected to make individual and population health decisions. One approach that has been used as part of that evidence-based medicine is meta-analysis (24). Individual clinical trials often provide useful information that affects the behavior of clinicians, for example, which drugs to use, in which quantities, and for what conditions. However, many studies are equivocal in their interpretation. Hence the notion arises of examining conclusions that might be drawn by reviewing the entire body of information available about a particular intervention. In other words, meta-analysis synthesizes the results of studies, viewing them not just as individual protocols, but as a sum total acknowledgment in a particular domain.

Meta-analysis has, however, engendered considerable controversy, being described variously as obvious, necessary and wise, or statistical fakery. The controversy can be seen, for example, in the meta-analysis of the effects of secondary smoking, namely smoke inhaled by individuals who do not themselves smoke but who are in the environment of smokers. It seems evident that it is worthwhile examining all the evidence available on a particular topic, such as the effect of passive cigarette smoking, but this must be done in a rigorous way.

Missing Data

A further ethical issue has to do with the availability of all the information about a particular intervention. Academics and drug companies, for different but related reasons, do not like to make so-called negative studies available through publication. Reporting on something when a study does not work cannot advance one's career or one's bottom line. Yet a systematic and important bias is introduced into the domain of knowledge about an intervention if only positive trials are reported. Thus it seems apparent that there should be ethical obligations to publish negative results as well as positive results, or else clinical decision makers will be misled when they review the available literature.

Cognitive Prostheses

The branch of computer science most intimately concerned with cognitive enhancement is artificial intelligence (AI). A recent trend in AI research is toward building "cognitive prostheses," systems that amplify a human problem solver's own thought processes. As used in this context, the term "prosthesis" includes enhancement as well as treatment so that both hearing aids and stethoscopes would qualify as prostheses. The goal of building cognitive prostheses is to enhance the power of even the finest human mind (25). The vision is not to build an all purpose problem solver that makes its user generally smarter but to build a series of special purpose computer tools, each of which symbiotically interacts with its user to solve one particular type of problem. Thus one tool might help a lawyer to plan a more brilliant defense, and another might help a professor to write a more interesting lecture. There is no theoretical limit to the type of cognitive activity that might be enhanced. Clearly, if that activity were itself unethical, such as plotting a perfect crime, then using cognitive prostheses to support it would also be unethical. However, not all of the ethical issues involved are so obvious.

Computer Program Error

One new issue is that of determining who is responsible if the computer makes a mistake. Programming errors in devices designed to administer doses of radiation to treat cancer have occurred and led to human death (26). AI systems are not infallible, and they might conceivably offer bad advice or focus a user's attention away from critical information. While legal responsibility is still solely assigned to human beings, other possibilities for moral responsibility have been proposed for the case in which computers autonomously make bad decisions. One possibility is to think of a computer as an agent, which can be liable for harm, just as human medical assistants can be. Another possibility is to allow that no one may be responsible for faulty judgment rendered by machines. The rationale for the latter position is that the use of truly life-enhancing technology should be encouraged, and blame is clearly discouraging (27).

It has been suggested that the reason responsibility is so hard to assign when computers make mistakes is that the norms for building good computer decision systems are not well understood (28). There are no accepted standard practice guidelines, as there are in medicine, to help determine if system designers and programmers have done everything reasonable to ensure the goodness of a computer system. We must depend on the integrity and skill of system developers to ensure dependable, accurate systems. Fortunately, in symbiotic systems, in which people and computers work together, the human partners may serve as valuable safeguards, recognizing when computer outputs seem dubious.

Symbiosis

Still other issues arise from the symbiotic nature of the relationship between the human user and the computer system. A central goal of AI research is to create machines that think like people. Whether or not that goal is ever fully achieved is open to technical and ethical debate. However, there can be no doubt that great strides have already been made. So it is entirely possible that a cognitive prosthesis could function like a virtual colleague. How will we relate to our virtual colleagues? Will we come to depend on them, feel emotionally attached to them, and even debate ethical issues with them, as we might with real colleagues? If so, will this enhance our professional lives, or merely reduce the social interaction we might otherwise have with real colleagues? What roles will we allow our virtual colleagues to play, and what types of activities will we reserve for human beings?

CONCLUSIONS AND FUTURE DIRECTIONS

In this article, we have reviewed a broad collection of issues relating to human values and hence ethics surrounding the use of biological and information technologies to enhance cognition in normal people. We have tried to illustrate our case by examples of behavior and practice in evidence today. It seems that there are important reasons to try to enhance individual and social intelligence using both drugs and information systems. What issues might we see emerging in the future if we momentarily take the viewpoint of the science fiction writer?

Cyberspace

First, recall that it was William Gibson who introduced the notion of cyberspace some years ago in his book *Neuromancer* (29). The overlap between biological and computer enhancement was evident in this work, as the protagonists would choose from a wide assortment of stimulants and related biological compounds before "jacking in," that is, creating a direct biological link to the computer network before they immerse themselves in the shared mental space known as "cyberspace." To some extent we are already involved in cyberspace. Multiuser domains are spaces created electronically in which people can interact with so-called avatars, where a visual image can be created and observed to interact with other individuals in that space. It is quite possible to adopt a false identity, even to change age and gender, and to interact in social circumstances that can cause benefit or harm to other participants in this space. Admittedly, the power of interaction in this kind of space is more limited than that envisioned in *Neuromancer*, but it is certainly a start. We are all aware of stories of people meeting on the Internet who either marry or kill each other.

The notion that drug enhancement can be combined with computer enhancement is clearly evident in those of us who brew a cup of coffee before answering our e-mail. The likelihood of direct biological connection to the Internet is also not so farfetched. Already we can wear headphones and goggles that permit us to enter the world of virtual reality generated by computers. From its earliest applications in space exploration and flight simulation, virtual reality has grown to encompass surgical simulation, virtual anatomy for medical education, and artificial threatening environments for use in the psychiatric treatment of phobias (30). Yet the Internet is also full of virtual reality "games" featuring countless fictitious creatures, as well as representations of real people, who have been killed, maimed, or destroyed. Ensuring that this powerful technology is used only for societal benefit becomes a new moral imperative.

It is a small step to recognize in an individual with a cochlear implant or some visual assistive device that involves the interaction between their own individual assistive device and their nervous system to directly connect this device to the Internet. It is not so farfetched to imagine a plug on a cochlear implant that would allow a direct connection to music obtained and downloaded from the Web.

Computers as Moral Players

We will conclude on one, perhaps most distant and yet provocative issue that exists in *Neuromancer* but not in reality yet. We do already meet intelligent entities on the World Wide Web that are not human beings or even the manifestations of live human beings. Many have played a game of chess where the opponent is a computer. At what level of intellectual capacity would a computer have to exist before we would give it moral status? Perhaps the answer to that question is an infinite amount, since moral status is not granted on the basis of cognitive abilities alone.

Yet other types of human abilities are already being envisioned for computers and built into working computer systems. Kurzweil has written a book called *The Age of Spiritual Machines* (31,32), which asks whether emotional and other more affable human abilities can be programmed into a computer. AI researchers are currently building systems that can understand and model human emotions (33). Initial results in the field of affective computing may seem modest. Computers sense human emotions, like frustration, in computer users so that they may better respond to user needs. Virtual animals and cartoon characters are set in virtual worlds, where they act in accordance with their levels of hunger, fear, playfulness, aggression, and desire for affection, rather than according to programmed scripts. However, the research goals are far from modest. Following neurological findings that rational thinking may be affected by too much or too little emotion, AI researchers seek to enhance computers with the abilities to recognize, possess, and express human emotion. Will these new capabilities enhance computers to the point where we might afford them moral status?

Certainly science fiction writers have already addressed this topic. We could be forming the moral relationships between silicon-based information entities and carbon-based entities. Isaac Asimov developed an entire world based on the three principles of robotics that define the moral obligation of intelligent robots towards human beings (34). The first principal of robotics is that robots must not harm human beings nor allow them to come to harm through inaction. Who would determine the three laws of human beings that would govern their behavior toward complex computer systems? Destroying a computer system would certainly cause significant moral harm to human beings dependent on it, but at what point would we raise concern about destroying the computer itself?

Yes, this does seem farfetched, but it is not unrelated to growing concerns in bioethics about the moral relationships between human beings and other biological information processing entities. As concern about the environment continues and the relationships between human beings and other life forms become more fully understood, it seems evident that we should have moral responsibilities to other creatures in our biosphere and to the biosphere itself. We can ask whether we have a greater moral responsibility to a chimpanzee than an amoeba. We can ask whether we have a greater moral responsibility to a chimpanzee than an anencephalic child. We can ask at what point do we have a moral responsibility to a highly complex information system compared to a simple biological entity. Would you be willing to consider that the Internet has distributed intelligence and good moral purpose in its own right, beyond the effects it has on human beings? Would you sacrifice a single amoeba and give it less moral status than a distributed information system being considered for termination? If you are willing to take this moral step, then when in the process of the evolution of a biological and a computer information entity does this moral shift occur, if ever?

Importance Revisited

Life will be very different in the future because biological life itself will be changed, and there will be change in large part because of the availability of human beings to manipulate and create life forms that are biological. The future will be also dramatically different because of our ability to create different information processing entities. In fact the very survival of the human race depends on the responsible use of powerful biological and information technology. Clearly, we should give thought to the ethics of cognitive enhancement now and hope that some of these improvements in biology and information technology will someday assist us in being wiser about the use of technology to enhance human thinking.

BIBLIOGRAPHY

1. D.J. Ward, *Smart Drugs II: The Next Generation*, Health Freedom Publications, Menlo Park, CA, 1993.
2. P.J. Whitehouse, E. Juengst, M. Mehlman, and T.H. Murray, *Hastings Center Rep.* **27**(3), 14–21 (1997).
3. P. Whitehouse, J. Ballenger, and J. Gute, The Neuropsychology of Wisdom. I Virtual Congress of Neuropsychology in Internet (http://www.uninet.edu/union99). *Journal Revista de Neuropsichogia Espanola.* (in press).
4. P.D. Kramer, *Listening to Prozac*, Penguin, New York, 1993.
5. G. Klerman, *Hastings Center Rep.* **2**(4), 1–3 (1972).
6. J.E. Birren and L.M. Fisher, in R.J. Sternberg, ed., *WISDOM — Its Nature, Origins, and Development*, Cambridge University Press, New York, 1990, pp. 317–332.
7. P.J. Whitehouse, *Dementia*, FA Davis, Philadelphia, PA, 1993.
8. M.J. Knapp, *J. Am. Med. Assoc.* **271**, 985–991 (1994).
9. P.J. Whitehouse et al., *Science* **215**, 1237–1239 (1982).
10. T. Crook et al., *Developmental Neuropsychol.* **2**, 261–276 (1986).
11. R. Levy et al., *Int. Psychogeriatrics* **6**, 63–68 (1994).
12. L. Diller, *Hastings Center Rep.* **26**(2), 12–18 (1996).
13. E. Juengst, in T.H. Murray and M.J. Mehlman, (eds.), *The Encyclopedia of Ethical, Legal and Policy Issues in Biotechnology*, Wiley, New York, 2000.
14. T.H. Murray, *Drugs, Sports and Ethics in Feeling Good and Doing Better: Ethics and Nontherapeutic Drug Use*, Humana Press, Clifton, NJ, 1984.
15. Ad Hoc Committee on Growth Hormone Usage, *Pediatrics* **72**, 891–894 (1984).
16. E. Parens, ed., *Enhancing Human Traits: Ethical and Social Implications*, Georgetown University Press, Washington, DC, 1998.
17. Y. Ku and P.J. Whitehouse, *Alzheimer Disease Assoc. Disorders* **10**(2), 61–62 (1996).
18. V. Potter, *Global Bioethics Building on the Leopold Legacy*, Michigan State University Press, East Lansing, 1998.
19. V. Potter and P. Whitehouse, *Scientist* **12**(1), 9 (1998).
20. D. Norman, *Just Health Care*, Cambridge University Press, New York, 1986.
21. J. Sabin and N. Daniels, *Hastings Center Rep.* **24**(6), 5–13 (1994).
22. C. Taylor, *The Ethics of Authenticity*, Harvard University Press, Cambridge, MA, 1991.
23. S.A. Alpert, in K.W. Goodman, ed., *Ethics, Computing, and Medicine informatics and the Transformation of Health Care*, Cambridge University Press, Cambridge, UK, 1998.
24. K.W. Goodman, in K.W. Goodman, ed., *Ethics, Computing, and Medicine informatics and the Transformation of Health Care*, Cambridge University Press, Cambridge, UK, 1998.
25. W.E.L Grimson, in Anonymous, 1999.
26. N.G. Leveson and C.S. Turner, *IEEE Comp.* **26**(7), 18–41 (1993).
27. J.W. Snapper, in K.W. Goodman, ed., *Ethics, Computing and Medicine informatics and the Transformation of Health Care*, Cambridge University Press, Cambridge, UK, 1998.
28. D.G. Johnson and J.M. Mulvey, *Comm. ACM* **38**(12), 58–64 (1995).
29. W. Gibson, *Neuromancer*, Ace Books, New York, 1984.
30. R. Zajtchuk and R.M. Satava, *Comm. ACM* **40**(9), 63–64 (1997).
31. R. Kurzweil, *When Computers Exceed Human Intelligence: The Age of Spiritual Machines*, Viking, New York, 1999.
32. D. Gelertner, *The Muse in the Machine: Computerizing the Poetry of Human Thought*, Free Press, New York, 1994.
33. R.W. Picard, *Affective Computing*, MIT Press, Cambridge, MA, 1997.
34. I. Asimov, *The Foundation Trilogy: Three Classics of Science Fiction*, Doubleday, Garden City, 1951.

See other entries Behavioral genetics, human; see also Human enhancement uses of biotechnology entries.

HUMAN ENHANCEMENT USES OF BIOTECHNOLOGY, ETHICS, HUMAN GROWTH HORMONE

Carol A. Tauer
College of St. Catherine
St. Paul, Minnesota

OUTLINE

INTRODUCTION

The appropriate use of drugs that supplement endogenous production of growth hormone (GH) in children and adolescents is the subject of both ethical and medical controversy. From a scientific standpoint, questions about the long-term efficacy of GH treatment for children who are not GH-deficient (GHD) remain unresolved (1–8). Yet GH supplementation is proposed for a variety of indications, ranging from the child who lacks naturally produced GH to the average-height child whose parents want him to star in basketball. A regimen that is clearly therapeutic for some indications is proposed for questionably therapeutic or enhancement purposes in other situations. Can we draw a line between applications that are directed toward appropriate goals for the practice of medicine and those that fall outside its scope? This question raises conceptual issues because of the difficulty of distinguishing "therapy" from "enhancement," professional ethics questions related to defining the goals of medicine, and social ethics questions about what is a just or fair allocation of medical resources. Growth hormone therapy has become a paradigm case for testing theoretical analyses of issues like these.

THE THERAPY/ENHANCEMENT DISTINCTION

Debates on the ethical scope for genetic manipulation often cite the insertion or correction of the growth hormone gene as a test case. In discussions of therapeutic versus enhancement applications of genetic engineering, the manipulation of the GH gene in short to average healthy children is used as a typical example of enhancement engineering. W. French Anderson, pioneer in gene therapy research who advocates drawing a line to preclude enhancement applications, states:

> The most obvious [enhancement] example at the moment would be the insertion of a growth hormone gene into a normal child in the hope that this would make the child grow larger (9, p. 22).

While insertion of the GH gene itself is still a hypothetical possibility, the modification of height that is sought through biochemical GH raises similar questions. Insertion of the gene is a more drastic measure and most likely carries more risks, yet the administration of the drug also leads to questions about appropriate uses of medical technology for purposes that go beyond therapy, or for enhancement purposes. In fact John Robertson cites the presumed right of parents to increase a child's height through injections of growth hormone to support the claim that parents have a similar right to enhance a child's height through gene insertion at the time of conception (10).

Yet while conceptual and ethical debates about the goals of medicine and therapeutic versus enhancement applications of biotechnology continue to cite growth hormone as a test case (10–13), the literature of pediatric endocrinology has become increasingly skeptical of the potential success of expanded uses of GH. As studies of long-term height gains from GH treatment appear to show less promising results than were anticipated from short-term studies, researchers are showing decreased optimism and professional societies are advising great caution in the prescription of GH (2–5,15–19). If these unpromising results are confirmed through long-term controlled studies, then the ethical questions may become moot, at least for the specific case of enhancing height through GH. But the questions will remain, even if transferred to other technologies.

HISTORY OF GROWTH HORMONE

Treatment of children who are deficient in GH began in 1958, when hormone taken from the pituitaries of human cadavers was shown to significantly increase the growth of treated children. A child who does not secrete GH is described as a pituitary dwarf, and if untreated, is not likely to reach an adult height greater than 4 ft 6 in. However, a two-year course of treatment with natural GH required hormone from 50 to 100 pituitary glands, so dosage was limited and selection criteria for treatment were stringent.

In 1985, after some natural GH was found to be contaminated with an infectious agent or the prion that causes Creutzfeldt-Jakob disease, the FDA halted its sale. Two companies, Genentech and Eli Lilly, had developed biosynthetic versions of GH through recombinant DNA techniques. Genentech's drug Protropin was approved in 1985, Lilly's Humatrope soon after. FDA approval, however, was limited to the population of children who had been treated with cadaveric hormone, that is, children with classical GH deficiency or pituitary dwarfism.

The cost of treatment with biosynthetic GH is very high; it is currently estimated to average $18,000 a year, varying with weight of child, dosage, and frequency of injections (range estimated from $10,000 to $50,000 a year) (19,20). The usual course of treatment is four to five years, yielding an average cost per child of $80,000 to $100,000 (21,22).

From a situation of scarcity we have moved to a situation of highly plentiful (but very costly) growth hormone. Thus it has become possible to consider higher dosages, more frequent injections, and a more extended course of treatment, as well as extension to populations other than children with classical GH deficiency, such as girls with Turner's syndrome and children with chronic renal failure, or previous cranial irradiation or craniopharyngioma of the pituitary. In addition GH has been prescribed for children who are very short but who exhibit no definable medical condition or deficiency. While this usage is both ethically and medically controversial, if it were successful it could result in an even more extensive use of GH for healthy children of average height whose parents for some reason wanted them to be taller.

CONTROVERSY ABOUT APPROPRIATE USE OF GROWTH HORMONE

In the United States prescription of GH is primarily in the hands of pediatric endocrinologists, to whom referral is made when there is a concern about growth (20).

But endocrinologists disagree on appropriate criteria for prescribing GH beyond the population with demonstrated GH deficiency (GHD). Some reject the limitation to children with GHD, since scientifically the line between GHD and non-GHD children is by no means bright and clear, and may not even be meaningful (23).

Indications for Prescription of Growth Hormone

Clinical Criteria. It is very difficult to measure endogenous production of GH because secretion is variable or pulsatile rather than continuous. The usual method is provocative stimulatory testing through administration of substances like arginine, insulin, or Levodopa. An alternate approach is labeled physiological: sampling at frequent intervals over a 24-hour period or in a particular situation. There is no agreement on the best method; no measurement is viewed as completely reliable (24). Some authors describe three different diagnoses: complete or classic GHD, partial GHD or GH insufficiency, and non-GHD (25). Others hold that there is no discrete condition that can be labeled GHD but rather a spectrum of disorders involving GH secretion and utilization (23).

Given the pitfalls of trying to differentiate GHD from non-GHD children, some endocrinologists say the problem to be treated is short stature or low growth rate, not simply GH deficiency (26). The latter should not be the only, or even the primary, indication for GH treatment. In this view, the criterion for treating a very short child should be actual responsiveness to administration of GH, demonstrated through significantly increased growth velocity over a trial period.

For short but otherwise normal children, such use is presently experimental. An "off-label" use of a drug approved by the FDA for another population is not illegal. However, short-term acceleration of growth in normal children may not actually be a predictor of long-term gains. The drug might accelerate bone aging and result in a child's reaching the same adult height more quickly (or even a lesser height). Thus an enormous cost and the burdens of three to seven shots a week for four to five years might yield little or no long-term increase in height (2–5,15–17,27).

Current Practice Patterns. A recent national study of pediatric endocrinologists regarding their prescribing patterns showed a wide range of variation in the indications for which these doctors would recommend GH treatment (28). With a response rate of 81.3 percent (434 out of a possible 534), this study provides broad information about current practice. Physicians were given eight scenarios, all representing variations in the clinical status of a 10-year-old boy or girl (sex alternated randomly) whose current height was either 2 or 3 standard deviations below the mean. Other factors that were varied were growth rate, whether bone age was normal or delayed, and predicted adult height. Three price variations were added to each scenario (cost of $13,000/year, drop to $2000/year, drop to $100/year), as well as intensity of family desire or pressure, resulting in 32 different decision situations.

Respondents generally agreed that GH use for non-GHD children has been increasing over the last five years, and almost unanimously agreed that "short stature matters and has dysfunctional emotional impact." But there was lack of consensus regarding the perceived efficacy (both expected adult height and long-term adverse effects) of GH treatment for non-GHD children. Thus there was more agreement on the psychosocial ramifications of shortness than on the scientific and medical evidence, which is presumably the area of these professionals' expertise.

The investigators concluded that GH recommendations are currently based at least partly "on a desire to address perceived impairment rather than on a clear knowledge of patient response [i.e., medical evidence]" (28). External nonphysiological variables such as strength of family wishes and cost are apparently also significant in relation to prescribing decisions. A commentator on this study noted that until we have validation of GH treatment from controlled prospective trials, "the use of GH treatment in non-GHD children will continue to be based on anecdote and emotion rather than fact" (29).

Forces Driving Interest in Expanded Uses of GH

In the United States, thousands of short healthy children are currently receiving GH treatment which is unvalidated (30). A large number of them are part of a study that follows them at least through the period while they are receiving the drug. Genentech, whose drugs Protropin and Nutropin dominated the U.S. market in the early years of biosynthetic GH, provides for all recipients to be part of its postmarketing surveillance program, the National Cooperative Growth Study (31). While approximately two-thirds of the 12,000 children enrolled in the program by 1992 had GH deficiency or another definable medical condition, the remaining third is believed to represent primarily healthy short children (30,31).

The National Institutes of Health (NIH) are conducting several studies, mainly directed toward subject populations that have medical diagnoses, but one to test the long-term effectiveness of GH therapy in very short healthy children (27). (A study similar to the latter is ongoing in Great Britain, the Wessex Growth Study.) The NIH study calls for 80 "short stature" subjects to participate in a double-blind controlled clinical trial, with half the children receiving GH injections three times a week, the other half placebo injections. All children undergo extensive tests and examinations and will be followed to adult height, when the mean heights of the two groups will be compared.

Because of complaints that the NIH study violated ethical and regulatory requirements for research with children, an independent Review Committee was convened in late 1992. While this committee determined that continuation of the study was ethically acceptable, commentators pointed out that this conclusion was reached only by extending ethical norms on research involving sick children to research involving healthy children who were short (30,32,33). Thus the Review Committee's approval of the NIH study suggests that it is appropriate for medical science to seek remedies for the condition of short stature, even in the absence of an identifiable medical condition.

Four types of forces are driving research programs on expanded uses of growth hormone: desire to resolve uncertainties in clinical medicine, economic interests of the drug companies, consumer demand and family autonomy, and ethical claims related to equality and justice.

Desire to Resolve Uncertainties in Clinical Medicine. The question of whether growth hormone prescribed to short normal children actually results in an increase in final adult height is considered one of the most pressing problems in pediatric endocrinology today. Thousands of short children with no diagnosable medical conditions are receiving GH, but their physicians do not know what the treatment is actually achieving, even at a purely physiological level. This widespread prescription of a nonvalidated (and costly) therapy mandates a research program of the highest rigor, according to many proponents including the NIH Review Committee (1,30).

In addition there is some concern about the safety of long-term use of GH in children who are not measurably deficient. While a history of administration to children with GHD has revealed no significant problems, we are now dealing with a different population (34,35). The dosages being administered are higher than when GH was scarce; and in order to sustain increases in growth rate in non-GHD children, it appears necessary to administer even larger dosages. Two organizations that filed court complaints opposing the NIH "short stature" study focused on what they regard as real risks of harm to healthy children (36,37). While the NIH Committee assessed these risks as hypothetical or insignificant (30), Arthur Levine of NIH defended the necessity for the research by stating that GH "could be dangerous," and its safety in short healthy children needed to be proved (38).

Many clinicians would argue that until the clinical uncertainties about benefits and risks have been resolved, it is premature to worry about what they view as theoretical ethical questions about the goals of medicine and enhancement therapies.

Economic Interests of the Drug Companies. The rationale for the NIH "short stature" protocol states that the prospect of a plentiful supply of GH leads to "the need to evaluate other potential uses for this hormone" (27). Genentech had an "orphan drug" permit for GH which expired in 1992, and Eli Lilly's expired in February 1994. After these two companies lost their protected market in the United States, three other companies, Novo Nordisk, Pharmacia and Upjohn, and Serono developed biosynthetic GH drugs that have been approved for use in the United States. All five companies manufacture drugs approved for treatment of pediatric GHD, and some are also approved for other indications (39). In addition several European manufacturers have shown interest in obtaining FDA approval to market GH in the United States.

In the early 1990s market analysts suggested that extensive competition could drive the price of GH well below the cost of the protected Genentech and Eli Lilly

drugs (40). It was even suggested that the cost of human GH might eventually come near that of bovine GH, which is produced by a similar recombinant-DNA process. When the congressional ban on use of bovine GH expired in February 1994, the cost of a two-week injection was $5, or $130 a year (41). However, the price of human GH has remained relatively stable despite new manufacturers and loss of orphan drug status (20).

The threat of loss of market share and a potentially lower cash return on investment would necessarily influence a company to explore additional uses for a product it has developed at great expense. In the case of recombinant-GH, potential uses in adults have begun to be explored and are the focus of numerous research efforts. Studies indicate that adults who are GHD frequently exhibit increased fat mass, reduced muscle mass and strength, smaller hearts and lower cardiac output, lower bone density, and psychological problems, and they appear to have an increased risk of death from cardiovascular disease (42). While there are some risks of side effects, GH replacement therapy for GHD adults has been demonstrated to be beneficial through double-blind, placebo-controlled trials (42,43). This therapy has been approved by the FDA for use with GHD adults as well as to treat AIDS wasting. Estimates indicate that about 70,000 adults in the United States may be affected by GHD, while the current pediatric market includes approximately 40,000 patients (39). Thus expanding the prescription of GH to GHD adults offers the prospect for a highly lucrative market for drug companies.

Nevertheless, there will likely continue to be interest in exploring additional uses for GH with children and adolescents and in extending FDA approval beyond the current indications of GHD, Turner's syndrome, and chronic renal insufficiency (39). Since short stature and abnormally slow growth of a child are observable characteristics and are generally of concern to parents, consumer interest in remedies for short stature will continue to drive research on the use of GH for non-GHD short children.

Consumer Demand and Autonomy. As the public becomes aware of the availability of growth hormone treatment, parents of a short child increasingly take initiative in requesting that it be tried. These parents may be concerned about psychological and social difficulties that may confront a short child in later childhood and adolescence, or about the long-term economic and social disadvantages of being short. Such problems are perceived to be more serious for males than for females; approximately 75 percent of the children in the Genentech collaborative study who do not have specific medical diagnoses are male (31), as are 90 percent of the children enrolled in the NIH "short stature" study (30). But even when classical GHD is included, the percentage of black children treated is only a third of what would be expected given the percentage of black children in the U.S. population, and females with problems are identified only when their deficiency is more extreme than that of males. Moreover, as awareness of the availability of GH treatment increases, data show "an even greater tendency to refer or test males [than] females" (31).

Parental request for testing and for a trial of GH treatment is largely based on a subjective perception of inadequacy, to which the medical system responds. The result is a documented referral bias in favor of testing and treating white male children, even though females, and males of other races, may be able to demonstrate greater need. Moreover there is some evidence that treatment of non-GHD females may result in more significant long-term height gains than treatment of non-GHD males (6,8,16). Given that the risks of therapy appear to be slight and that a short-term growth benefit is likely, parental pressure for GH treatment of a short healthy child, particularly a short boy, may be difficult to resist. Note that the study of prescribing patterns described earlier found that strong family wishes for GH increased the likelihood that an endocrinologist would prescribe it (28).

In this era when patient or parental autonomy regarding medical treatment is a central or even overriding ethical value, the endocrinologist may consider the option of GH therapy a matter of parental choice. Provided that parents are well-informed that for short healthy children the treatment is still experimental with a remote possibility of risk, then their autonomy permits them to choose it. However, thoughtful physicians will also raise issues of possible psychological harm: What if expectations are not reached? Will the child be severely disappointed if the gain is only one or two inches? Does the use of drug treatment in itself suggest to the child that there is something wrong with him or her? (44) After the full range of risks and benefits has been explored, the principle of autonomy may appear to support the prerogative of parents to make the choice of treatment in the interests of this particular child, whom they presumably know better than anyone else does.

Justice: Two Aspects. As the previous section notes, the current allocation or distribution of GH treatment raises questions about fairness or justice in relation to race and gender. Prescribing patterns suggest that this very costly drug is currently prescribed in a way that is discriminatory. Even if the discrimination is purely de facto, resulting from which particular parents and children pursue GH treatment most energetically, still the discrimination and its effects remain.

But a different interpretation of justice has been invoked to support a principled extension of GH treatment and to justify providing it to short healthy children who are not GHD. David Allen and Norman Fost have devised the cases of Johnny and Billy to persuade us that a short non-GHD child has as much right to GH treatment as a GHD child:

> Johnny is a short 11-year-old-boy with documented GH deficiency.... His predicted adult height without GH treatment is approximately 160 cm (5 ft 3 in.). Billy is a short 11-year-old boy with normal secretion according to current testing methods.... He has a predicted adult height of 160 cm (5 ft 3 in.) (26, p. 18).

Allen and Fost argue that it is unjust and discriminatory to provide treatment to Johnny but not to Billy, solely on grounds that Johnny has an identifiable medical deficiency

and Billy does not. If the two boys can be expected to experience equivalent psychosocial problems and to be similarly disadvantaged, both as children and adults, then it seems arbitrary to treat one with GH but not the other.

The NIH Review Committee gave a great deal of weight to the principle of equal treatment, arguing that short non-GHD children suffer the same "functional impairment and psychosocial stigmatization" as GHD children. Therefore it could be unjust to deny them access to treatment simply on the basis of an "imprecise definition of 'deficiency'" (30). On this interpretation of justice it could be discriminatory to deny GH treatment to a child, no matter what his or her medical condition, if the child's short stature is perceived to be disadvantaging, disabling, or otherwise problematic. Authors who have written in defense of the treatment/enhancement distinction acknowledge that this example presents a hard case for them (13).

GOALS OF GROWTH HORMONE THERAPY

Importance of Identifying the Goal of Therapy

In the biomedical ethics literature, attempts to discern a conceptual and ethical distinction between therapeutic and enhancement uses of biotechnology often cite the case of Johnny and Billy, sometimes with discomfort that Billy appears to be as entitled to treatment as Johnny. In the literature of pediatric medicine, the current focus is on formulating a standard of care that has a rational connection with research results (18,19). Here the aim is to develop criteria for the classes of children for whom GH treatment is appropriate, based on studies that demonstrate significant benefits to that class in proportion to risks and costs, both personal and financial.

For either enterprise, the goals for GH treatment must be explicitly identified. Possible goals cover a variety of statistical, therapeutic, pychosocial, and enhancement ends or purposes. Each of these goals presumes a different understanding of the benefit to be provided, and thus each one points to a different way of evaluating whether treatment has been successful. Some of the goals appear to be only instrumental ends, where success requires that their attainment be a means to the attainment of some other more ultimate goal. Thus a rather lengthy list of goals collapses into a shorter set of ultimate goals.

Goal 1. To Increase Growth Velocity or Growth Rate. Some clinicians emphasize potential benefits of an increased growth rate, whether or not GH treatment produces a significant augmentation of eventual adult height (45). A child with measurable GH deficiency (GHD) has a pathology of growth and a physiologically abnormal growth rate. Similarly, it may be argued that a child without measurable GHD but whose growth velocity is measurably abnormal also has a pathology of growth or a growth disorder (46,47). If such a pathology is regarded as a medical condition, then its remediation is therapeutic and falls under traditional goals of medicine.

In their clinical discussion of "disorders of stature," Hindmarsh and Brook advocate using abnormality of growth velocity as the main criterion for considering

GH treatment. It is clinically diagnosable by objective standards. It circumvents the relationship of height to societal biases and is independent of racial or ethnic height differences (48). It is also independent of male–female height differences. This view maintains that amelioration of abnormal growth rate is the correction of a physiological deficiency.

But providing GH treatment to increase growth velocity in a short child may also be regarded as a means to some other goal, for example, to prevent or remedy psychosocial or behavioral problems that short children are thought to experience. However, if increased growth velocity is perceived as instrumental to the achievement of other goals, then its success and legitimacy must be evaluated in relation to those other goals (49). Only if studies demonstrate that psychosocial gains are actually achieved could an increase in growth rate directed toward such goals be defended.

Goal 2. To Increase Eventual Adult Height. The NIH "short stature" study, the Wessex Growth Study, and a number of already-completed studies aim to determine whether short-term gains from GH treatment translate into long-term height gains (2–8,15). For many physicians, parents, and researchers, the increase in growth rate sought in goal 1 is only a means to an increased adult height.

However, while medical science can assess normal growth (or growth velocity), there is really no medical criterion for normal height per se. Height is partly genetic, and is sex, race and ethnicity dependent (as growth rate is not). While it is within the competence of medicine to investigate why a child is not growing and to ameliorate that deficiency, medical science lacks objective criteria for what is a "good" or "healthy" height (13,50,51).

Height is not the sort of thing where one necessarily desires more rather than less, and some people (given goals such as gymnastics or riding race horses) may prefer to be statistically quite short. An increase in adult height through GH treatment, if achievable, is a goal that is necessarily only a means to the attainment of other desired goals. Increased height may be sought in order to avoid perceived psychosocial problems, or because extreme shortness could be functionally disabling, or because shortness is economically and socially disadvantaging within certain contexts. Thus the success of treatment must again be evaluated in terms of whether it achieves its ultimate goals.

Goal 3. To Ameliorate a Perceived Deformity. The NIH study protocol states that extreme short stature is perceived as "a major developmental abnormality" and the Review Committee cites evidence that it results in "psychosocial stigmatization." In relating this problem to traditional medical practice, Lantos, Siegler, and Cuttler note that surgery for deforming congenital anomalies is considered proper and is generally reimbursable. They believe that the short stature resulting from classical GHD is so severe that it is generally viewed as a major deformity, and hence is treatable (21). In most cases, however, the shortness of children who are not GHD is not as extreme.

Goal 4. To Prevent or Treat Psychosocial, Learning, and Behavioral Problems. Studies have suggested that children with very short stature may experience failure in school and have psychosocial and behavioral problems (52,53). For some endocrinologists, preventing or ameliorating these problems is the major goal of GH therapy. Underwood and Rieser recommend that therapeutic goals be defined "in terms of how short-term acceleration of growth would benefit the patient socially and psychologically" (45). Increasing the rate of growth could be instrumental for this short-term purpose even if significant gains in adult height could not be demonstrated.

If the actual or potential psychosocial problems of the patient are definable under the umbrella of mental health treatment (see Ref. 54 for difficulties in such definition), then GH treatment might perhaps be viewed as a therapeutic or preventive means for dealing with them (49). There are precedents in the practice of medicine for the use of drug therapy for behavioral or psychological disorders in children. However, it would be highly unusual to prescribe drug injections to children merely because they were believed to be at risk for such problems, namely for prevention (21). But what about very short children who are actually displaying these problems?

Recent studies show that psychosocial problems are not correlated to shortness per se, but that they are more likely to be related to hormonal deficiencies (2,55,57). Thus in assessing the appropriateness of GH therapy as a means for treating mental health problems, the two populations of GHD and non-GHD children must be considered separately. However, even with GHD children, no one yet is able to claim that we have scientific evidence for the effectiveness of GH therapy in improving their psychosocial functioning (52,58). "There have been no placebo-controlled evaluations of the behavioral effects of GH treatment," according to Richard Clopper (58). Some clinicians decide to prescribe GH largely on the basis of a perception that short children experience psychosocial problems, although there is no solid evidence from research to support GH as a remedy (28,29). If the reason for GH therapy is its supposedly beneficial effect on psychosocial adjustment and school performance, then studies must measure whether these results are actually achieved.

Goal 5. To Correct a Functional Disability. Advocates of GH treatment for short stature stress its functionally handicapping effects, particularly if the shortness is extreme. The NIH Review Committee cited "functional impairment" and "difficulty with physical aspects of the culture" as problems of short people that justified a research program (30). In providing examples to show that very short people are functionally handicapped, authors frequently cite problems in driving a car, and inability to reach shelves, light switches, and elevator buttons. Since children who are still growing are not regarded as handicapped simply because they are not tall enough to do these things, attaining this goal requires that GH treatment achieve an increase in final adult height.

Medical treatment is typically provided when it is effective in remedying a functional disability that interferes with activities of daily living or that prevents one from earning a living or living independently. The question is: what level of short adult stature can realistically be defined as that sort of disability? Often it would be simpler (and less costly) to modify the environment or the vehicle than to attempt to modify the body of a completely healthy person by means of GH injections. Moreover the criteria currently used to define "extreme shortness" are statistical and have no logical relationship to standard definitions of functional disability. Thus, while the goal of remedying a disability is consistent with therapeutic goals for the practice of medicine, its application to GH treatment of non-GHD children is questionable.

Goal 6. To Remove the Economic and Social Disadvantage of Short Stature. Data show that our society has a bias against short people, particularly short males. Allen and Fost note that "discrimination based on height—heightism—pervades American life," and that shortness "imposes a disadvantage in the competition for schools, jobs, income and mates" (26).

Society's attitude toward short people is a prejudice that often leads to discriminatory consequences. Lantos et al. note that "we do not usually call prejudice-induced conditions ... diseases" (21). In his analysis of prevention as a legitimate goal of medicine, Eric Juengst argues that this goal should be limited to efforts to defend people from "robust pathological entities [i.e., genuine disease entities], rather than changing their bodies to evade social injustices" (51). In other words, it is not appropriate for medicine to respond to discrimination by modifying the physical characteristics against which society is prejudiced. Other authors describe the use of medicine to respond to society's preference for taller people as social engineering. In attempting to make some people taller, medicine may simply be reinforcing "heightism" in our society.

Goal 7. To Enable Parents to Seek a Preferred Height for Their Children. This goal suggests that any height could be achievable, which is almost certainly incorrect. Ann Johanson, former Director of Clinical Affairs for Genentech, acknowledges that children with normal stature and growth rate are unlikely to gain significant additional height or growth velocity from GH treatment (59). Recently completed studies confirm this hypothesis (2–7,15,60). However, if a safe method of supplementing GH, either biochemically or through genetic manipulation, were shown to cause significant height increases in normal-height children, this goal would represent a pure enhancement use of GH. In the literature specifically focused on growth hormone, no one appears to advocate such enhancement use, even though parents may request it. But discussions of enhancement therapies in general sometimes defend providing them, based on the constitutional right of parents to autonomy and discretion in rearing their children (10).

GH TREATMENT FOR CHILDREN WITHOUT MEDICAL CONDITIONS

The Argument Invoking Equal Treatment

The Case of Johnny and Billy. The only clearly therapeutic goal for GH treatment is to increase or normalize growth rate in children who have a definable medical condition that is related to a pathology of growth. Secondarily this increased growth rate ought to produce an increment in final adult height over what was predicted. But what of Allen and Fost's case of Johnny and Billy? Recall that Johnny and Billy are both short 11-year-old boys, and that each has a predicted adult height of 160 cm (5 ft 3 in.). The difference between them is that Johnny has a documented GH deficiency, while Billy's tests show normal GH secretion. Allen and Fost argue that it is discriminatory to offer GH treatment to Johnny but not to Billy (26).

In applying the concept of justice to this situation, we are invoking the principle that equal cases should be treated equally. Since no two cases are ever exactly alike or equal, we must identify the relevant factors in which similarity is required. With Johnny and Billy, the similarities stated are short stature at present and equal predicted adult heights. Similarities assumed are that shortness will cause them similar psychosocial problems and will disadvantage them equally, and that GH treatment will benefit them equally and carry equivalent risks. If these assumptions are correct, then fairness seems to mandate that the two cases be treated equally in terms of provision of GH. In brief, a non-GHD child predicted to reach the same adult height as a GHD child has an equal right to GH treatment.

Justice entails this conclusion, however, only if the two cases are truly similar in relevant respects. But actual data show that they are different in important and often overlooked ways, so that the principle of equal treatment of similar cases does not apply to these two situations.

Dissimilarities between Johnny and Billy. First, a lack of endogenous GH appears to be linked to physiological and functional consequences in addition in growth. While earlier studies of short-statured children had shown them more susceptible to learning difficulties and academic failure than children of normal height, these studies did not distinguish according to the etiology of the short stature. Newer studies which concentrate on short children with GHD suggest that these children "show significant deficits in several specialized cognitive domains including those requiring complex visuoconstructional skills, orientation in space, long-term memory, and attention span" (61).

Some of this research fails to distinguish children with isolated GHD from children who lack not only growth hormone but are multiply hormone deficient or even panhypopituitary (lacking all pituitary hormones). When these distinctions are made, children with more severe endocrine problems do demonstrate more severe cognitive deficits and psychological disturbances (61,62). But authors of the Michigan longitudinal study, which followed a group of GHD children for seven years, reached a preliminary conclusion that even some children with isolated GHD "may have cognitive profiles similar to those described as learning disabled" (61,63).

Two studies of the psychological adjustment of GHD adults show that their profiles differ from those of matched controls of equally short stature, and that they are at risk for anxiety and depressive disorders, frequently displaying symptoms seen in patients with clinically diagnosed social phobia (64,65). Again, subjects with multiple endocrine deficiencies appear more vulnerable than those with isolated GHD. But Pine, Cohen, and Brook found that "A blunted growth hormone response to physiologic challenges remains perhaps the best-replicated biological correlate of emotional disorder," while not claiming which is cause and which is effect (66).

In summarizing the results of these studies, Brian Stabler observes that short stature alone is not responsible for the low quality of life of many GHD patients, but that "neuropsychologic functioning is fundamentally impaired in many GHD children." The more deficient in GH a child is, the greater the cognitive and behavioral difficulties, suggesting a "relationship between psychosocial functioning and degree of endocrine deficiency" (62).

On the other hand, studies of the adjustment of short children in general show that their level of social and academic functioning is "reasonably indistinguishable from that of average-statured peers" (55–57). After reporting their study that showed minimal long-term height gains from GH treatment of short normal children, Hindmarsh and Brook concluded:

> It has been alleged that short stature adversely affects children and the short-term effects of r-hGH on height ... might therefore find advocates for this therapy. Neither we nor others have been able to document markers of adverse psychological effects in normal short children (2, p. 16).

Thus short stature is not the primary factor underlying the behavioral, academic, or emotional difficulties of some short individuals. Rather, these problems appear linked to an underlying medical condition of which short stature is only one feature (55). While it may be reasonable to provide GH to Johnny to help him overcome these types of deficits in addition to growth failure, it is unlikely that merely increasing Billy's stature will offer comparable psychosocial benefits to him.

Second, there is skepticism as to whether a non-GHD child could make comparable height gains on comparable doses of GH. While some authors believe that cumulative growth response of non-GHD subjects is comparable to that of GHD patients (67), others cite strongly conflicting data (68). The latter argue that in order to achieve comparable height gains you would have to treat over a longer time period, at higher dosages, with correspondingly higher costs. At some higher dosage toxicity could be expected, possibly leading to side effects such as diabetes.

Recent studies that compare two groups of non-GHD children, one group treated with GH and the other untreated, also come to differing conclusions (2,3,6,7,17). Hindmarsh and Brook found that both treated and observation groups had some increases in final height over that predicted, but that the change in the treated group was not significantly greater than that in the observation group (2).

While studies show that GHD children like Johnny can be expected to make significant height gains through GH treatment, there is no comparable consensus among researchers as to whether Billy could anticipate similar gains, especially in the long term. It is unlikely that an expenditure on Billy equivalent to that on Johnny would bring Billy equal benefits, even in relation to height itself.

Third, the case description makes no mention of the two boys' growth velocities. It is almost certain that Johnny has subnormal growth velocity, while Billy's may well be within the normal range. If so, then the two boys differ significantly with respect to pathological versus normal growth, and Billy would be regarded as developmentally normal.

Fourth, because Billy does not have an identifiable medical or developmental problem, he is apt to be at greater psychological risk as a result of prescription of drug therapy. In his article "Is Taller Really Better?" Douglas Diekema argues that treatment itself may have a stigmatizing effect:

> When we seek to change the height (or physical appearance) of a child, he may perceive that he is incomplete and unacceptable. His peers may have suggested this on the playground. Now his parents seem to have confirmed it through their efforts to make him taller (69, p. 114).

Cosmetic interventions after an illness or accident are "less likely to make the child perceive himself as undesirable," since they restore something that was taken away. Interventions which repair abnormalities that interfere with functional capacity offer the child a benefit for him or herself, not merely in relation to how the child is perceived by others or in comparison with others. But GH treatment aimed solely at changing a child's body or appearance could have adverse psychological effects and do more harm than good (69).

C.G.D. Brook of the London Centre for Paediatric Endocrinology warns doctors who are pressured into providing GH for non-GHD children to be mindful of "the problems of stigmatising otherwise normal children" (60). Billy is at more risk of this stigmatization through the medicalization of his stature than is Johnny, who has a definable medical problem.

Fifth, while Johnny and Billy as individuals may both experience psychosocial problems, Billy's are more apt to be treatable without use of GH. Stabler notes that short individuals who do not have neuroendocrine impairment are very adaptable, and he cannot say whether GH or psychotherapy is more effective in treating them (70). Since Stabler advises that GH should never be given without supplementary counseling or psychotherapy, it would be significantly more cost effective to choose psychotherapy rather than GH plus psychotherapy for Billy. Brook, in noting the lack of evidence that short children suffer psychosocial disadvantage solely because of their stature, as well as the impossibility of testing whether an increase in growth rate would provide psychosocial benefits, concludes:

It is much more important for a short child to acquire coping skills than to buy [socially unimportant] inches through pharmacological means (60, p. 692).

These five arguments show that in the real world of GH treatment, the given data about Johnny and Billy are entirely consistent with the conclusion that they differ in a number of ways that are relevant with respect to treatment with GH. The fact that GH is provided for GHD children does not entail that it must be provided for equally short non-GHD children.

Short Stature as Disability or Disadvantage

Even if Johnny and Billy are not completely equivalent with regard to the provision of GH therapy, still it may be argued that because of the handicapping effects of very short stature, or the social and economic disadvantages of shortness, it follows that increased height (and hence GH treatment) is a legitimate medical goal for Billy. Even advocates of a restrictive definition of legitimate goals for the practice of medicine generally support treatment to remedy or ameliorate a disability. For example, Sabin and Daniels hold that "the central purpose of health care is to maintain, restore, or compensate for the restricted opportunity and loss of function caused by disease and disability" (54). The application of medical science in order to level social and economic disparities is more controverted. But even the goal of treating "disability" has problems when invoked to justify the extension of GH treatment.

Is Short Stature a Handicapping Condition? Allen and Fost classify extreme short stature as a handicap, that is, "a physical ... disability that prevents or restricts normal achievement." They suggest that height below the 1st percentile is "likely to be handicapping" (26). (For North American adult males this would be a height about 5 ft 3 in.; for adult females about 4 ft 10.5 in.)

As noted earlier, difficulty in driving a car or in reaching shelves or switches usually can be ameliorated through modification of a vehicle or environment. While it is possible to imagine statures so short that a person is unable to function outside a radically adapted environment, the heights suggested do not appear to be that restricting.

Moreover the appeal to short stature as a functional disability is belied by height criteria that are significantly different for males and females. Since the mean height for adult males is about 4.5 in. greater than that for females, treatment criteria based on these mean heights cannot really be aimed at overcoming a functional handicap. Males and females do not drive different cars according to their sex, nor use different light switches. If a 5 ft 1 in. female is not functionally impaired in activities of daily living, why is a 5 ft 1 in. male?

Much publicity has been given to fatalities allegedly caused by air bags at the time of an automobile crash. More than half of such deaths have involved children, and almost all adult deaths have been women. Evidence indicates that "women of short stature are particularly susceptible to head and neck injuries from air bags" (71,72). Women under 5 ft 2 in. are statistically at particular risk; however, that height, while below the first percentile for males, is well above it for females. Moreover federal crash-worthiness standards call for testing that uses 5 ft 8 in. dummies, the height and weight of an average male. In September 1998 the National Highway Transportation Safety Administration proposed new rules that would require testing to minimize the risks of air bags to infants, children, and short adults. However, comment on these rules was extended to December 30, 1999, and as of early 2000, the rules had not yet been finalized (73). The reason that the driving environment is more hazardous to short people is that safety tests and modifications are premised on drivers' being average-sized males. This example illustrates how the "handicap" of shortness results from the way equipment is tested and designed, not from shortness per se.

Short people must live in the world as it exists at present, however. A careful study of daily life activities and a range of occupational choices might identify a height level below which a person really is handicapped, that is, unable to function without major environmental accommodations. (This height level might turn out to be around 4 ft 6 in.) Such a result would base the criteria for "handicapped by reason of stature" on actual data, rather than on statistical population norms. It might then be possible to consider offering GH treatment to remedy the disability of short stature.

In determining the allocation of treatment, however, the resources that are generally available for the treatment and rehabilitation of people with disabilities must also be taken into account. Fairness requires that equally debilitating handicaps be provided equal resources. Economist Mary Ann Baily suggests that money spent on expanding access to GH could better be used instead to help severely handicapped children through rehabilitative care, training, and high-tech devices, since at present these benefits are "not well-covered by private or public payers" (22).

If we apply Allen and Fost's suggestion that children in the first percentile be considered handicapped, then with about 39,000 U.S. children in the first percentile at any given age, to provide all of them with five years of GH treatment at $18,000 a year would cost over $3.5 billion a year. Compared with the defense budget this amount may seem small. However, total 1996 Medicaid payments for dependent children under 21 were only $17.5 billion, and the 1999 allocation to NIH for medical research for infants and children through the National Institute of Child Health and Development was only $752 million (74,75). Many children in the first percentile may have pathologies of growth or other medical conditions. But for those who do not, the use of scarce resources to provide GH treatment appears to be a low priority in relation to other possible expenditures related to the needs of disabled and sick children.

Should Medicine Remedy the Disadvantage of Short Stature?. Short stature has been shown to carry both economic and social disadvantages. One study suggests that there may be an increased income (height bonus) of roughly $1150 a year (1999 dollars) per inch of greater height (76). Over a work life of 40 years, the total increment would be $46,000 per inch. If GH treatment were evaluated purely on an economic basis, a gain of two inches or $92,000 through a five-year course of treatment costing $90,000 would just about balance the investment in GH treatment. However, the anticipated increment in income would be gained through an increase in height relative to others who would become comparably shorter and presumably earn less. Thus an individual's interest in greater height is not correlated to a societal interest when medical resources are expended for GH treatment.

Most discussions of the disadvantaging effects of short stature focus on males. Studies show that a 12-inch height reduction is a significant predictor of lower economic status among men but not among women (77,78). (Obesity is a significant predictor of lower economic status for women but not for men.) These data could be taken to mean that women suffer less economically from shortness, and hence the disproportionate treatment of males with GH is not a concern. On the other hand, there is evidence that the typical size difference between males and females plays a major role in the status and income differential between the sexes. One study found that women who are 5 ft 7 in. earn on average the same salary as men who are 5 ft 7 in. (79). A study of the effects of both sex and size on status ranking found, as expected, that both males and people of greater size are perceived as having higher status. More surprisingly, this study found that the correlation between sex and status rank is largely (65 percent) accounted for by the correlation of sex with size, rather than by socioculturally influenced factors (80).

When the practice of medicine engages in trying to make some people (mostly white males) relatively taller, it is not only contributing to the "heightism" of our society. Given the correlations among sex, size, and status, it is also reinforcing discriminatory attitudes in relation to women (and toward people of typically shorter racial and ethnic groups) (79).

Moreover we do not know whether persons whose height is increased through GH treatment acquire the same advantages as people who are taller to begin with (69). Demographic data on adults who were GHD and were treated with GH are limited and often disappointing. Subjects in some studies, while reaching educational levels comparable to the population as a whole, were not comparably employed, married, or living independently of parents (62,81,82). Other studies, however, have identified more positive outcomes (83,84). A well-controlled study at Children's Hospital of Buffalo found a high educational level in previously treated adults, and also a low unemployment rate of 8 percent, at a time when overall unemployment in the area was 9 to 10 percent. None of these adults were experiencing significant emotional or adjustment problems (84).

Thus the data on the effectiveness of GH treatment for overcoming economic and social disadvantages are mixed, and placebo-controlled studies are completely unavailable (58,85). Even for GHD individuals, studies do not provide convincing evidence that GH treatment results in alleviation of the economic and social disadvantages of short stature.

These arguments provide three reasons for questioning the use of GH treatment to overcome economic and social disadvantages: (1) We cannot show that it is effective for this purpose. (2) Even if it were, providing some persons a height advantage relative to others offers no net benefit to society. (3) Not only is there no societal benefit, there is a negative impact through the reinforcement of societal biases and prejudices.

Liberty Rights and Personal Preferences

It could be argued that if an individual has a particular personal goal for which greater height would be an asset, then that individual should have the freedom to seek professional assistance to achieve that increased height. Since physicians, and specifically pediatric endocrinologists, control the prescription of biochemical growth hormone, physicians would necessarily be involved, and their prescription of GH to satisfy personal preferences would be no different from other uses of medical expertise for cosmetic purposes. Arguments about a just or fair allocation of societal resources would carry less weight if the patient either paid for, or contracted with an insurer who would pay for, such treatment.

This argument overlooks the fact that GH treatment to increase height must be provided during the child's growing years, and is most effective when begun early and continued until growth is completed. Hence (almost) all treatment occurs while the patient is a minor and unable to give an autonomous consent. Given the uncertainties about the clinical outcomes and the psychosocial benefits of GH treatment for non-GHD children, as well as the unpromising results of long-term studies, even adults would find it difficult to weigh these supposed benefits against the risks, inconveniences, and costs of years of medical treatment.

The endocrinologist has been designated as the customary gatekeeper for prescription of GH because of the need for a well-informed person to weigh risks against benefits, and to safeguard against the possibility of nonbeneficial (or even harmful) usage (19). Pediatricians, of whom pediatric endocrinologists are a subset, are given authority regarding child health issues because they are regarded as "guardians of and spokespersons for the well-being of children" (86). This charge requires that at minimum they base prescription of GH on evidence from studies and on practice guidelines developed by professional societies (18,19,29). It also requires them to recognize that while adults may choose to utilize medicine to change their bodies according to their preferences, adults (even parents) do not have a right to impose arduous treatment on healthy children simply to change their appearance.

The use of GH to treat short stature per se is a "medicalization" of problems whose sources lie within society rather than the short individual. The search for drug treatment to ameliorate psychosocial problems experienced by the individual patient often overlooks a wide range of other therapies, resources, and supports that are apt to be more effective in dealing with these problems. In fact some practitioners argue that focus on short stature "is in some ways an obstacle to gaining truly comprehensive care for [short] children" (62).

Moreover treatment that is provided by physicians, even if paid for by individual parents, increases the pressure to expand reimbursement through public funds and through health plans. What is at first regarded as a consumer issue, namely providing what clients want and will pay for, eventually comes to be expressed in terms of fairness: Why should the wealthy have access to a medical treatment that most people cannot afford? History shows that entitlements are often expanded in this way, without other justification.

The use of medical technologies to modify people according to their preferences is apt to increase public concern about emerging genetic therapies. While most genetics researchers and clinicians stress their interest in applying gene therapy only to severe and lethal diseases (9), yet the public can easily be aroused to fear and oppose all interventions that involve genetics. Avoiding the use of GH for what can only be regarded as enhancement purposes upholds a consistent policy of focusing genetic science on therapeutic applications.

Finally, no patient or parent has a constitutional or liberty right to demand any particular medical treatment. Even the right to procreative liberty, a highly protected right in the United States, does not permit one to demand RU-486 as an abortion option, nor to insist on Depo-Provera as a contraceptive before it was approved for that purpose. While parents have a wide range of choices as to how they wish to rear their children, they cannot insist that the medical profession cooperate with whatever plan they may have.

CONCLUSION

The development of recombinant human growth hormone has resulted in a plentiful supply of the GH drug. It has been shown to be safe and effective in improving growth in children who are growth hormone deficient, and is also approved for treatment of girls with Turner's syndrome and children with chronic renal insufficiency. In addition administration of GH has been recognized as therapeutic for GHD adults and is effective to treat AIDS wasting. The question of whether the drug ought to be prescribed beyond these categories, particularly for children who are healthy but short, remains an unresolved question. For many endocrinologists, abnormally slow growth, but not simply shortness, is an indication for treatment, even in the absence of other medical conditions.

The prescription of GH to short but otherwise normal children represents the use of medicine for enhancement purposes. Beyond the as-yet-unanswered question of whether such treatment actually achieves a significant long-term gain in height, there are unresolved ethical questions regarding the appropriateness of directing medical and societal resources to the amelioration of short stature. If amelioration of shortness could be shown to achieve other appropriate goals, such as improvement in psychosocial functioning, there might be justification for the prescription of GH to short normal children. However, studies do not demonstrate that shortness in

itself is problematic, and other solutions for psychosocial and similar problems appear to be preferable, especially given the high cost and the burdensome administration of GH treatment.

Because of a variety of factors associated with GH treatment for idiopathic or unexplained short stature, many arguments can be marshaled against allocating medical resources to this form of enhancement. However, these arguments are not necessarily transferable to other biotechnologies in which the enhancement versus therapy debate arises. Exploring these biotechnologies on a case-by-case basis provides one means for clarifying the enhancement or therapy distinction and its ethical implications, or for supporting the position that the distinction is not a helpful one.

BIBLIOGRAPHY

1. M.D.C. Donaldson, *Lancet* **348**, 3–4 (1996).
2. P.C. Hindmarsh and C.G.D. Brook, *Lancet* **348**, 13–16 (1996).
3. S. Loche et al., *J. Pediatr.* **125**, 196–200 (1994).
4. J.M. Wit et al., *Clin. Endocrinol.* **42**, 365–372 (1995).
5. L.T.M. Rekers-Mombarg et al., *J. Pediatr.* **132**, 455–460 (1998).
6. J.G. Buchlis et al., *J. Clin. Endocrinol. Metab.* **83**, 1075–1079 (1998).
7. R.L. Hintz et al., *N. Engl. J. Med.* **340**, 502–507 (1999).
8. E.S. McCaughey et al., *Lancet* **351**, 940–944 (1998).
9. W.F. Anderson, *Hastings Center Rep.* **20**(1), 21–24 (1990).
10. J. Robertson, in J. Robertson, ed., *Children of Choice*, Princeton University Press, Princeton, NJ, 1994, pp. 149–172.
11. L. Walters and J.G. Palmer, *The Ethics of Human Gene Therapy*, Oxford University Press, New York, 1997, pp. 110–113.
12. N. Daniels, *GROWTH. Genet. Hormones* **8**(suppl. 1), 46–48 (1992).
13. E. Parens, in E. Parens, ed., *Enhancing Human Traits: Ethical and Social Implications*, Georgetown University Press, Washington, DC, 1998, pp. 1–28.
14. D.W. Brock, in E. Parens, ed., *Enhancing Human Traits: Ethical and Social Implications*, Georgetown University Press, Washington, DC, 1998, pp. 48–69.
15. J. Coste et al., *Br. Med. J.* **315**, 708–713 (1997).
16. M. Kawai et al., *J. Pediatr.* **130**, 205–209 (1997).
17. S.E. Oberfield, *N. Engl. J. Med.* **340**, 557–559 (1999).
18. Drug and Therapeutics Committee, Lawson Wilkins Pediatric Endocrine Society, *J. Pediatr.* **127**, 857–867 (1995).
19. Committee on Drugs and Committee on Bioethics, American Academy of Pediatrics, *Pediatrics* **99**, 122–129 (1997).
20. B.S. Finkelstein et al., *J. Am. Med. Assoc.* **279**, 663–668 (1998).
21. J. Lantos, M. Siegler, and L. Cuttler, *J. Am. Med. Assoc.* **261**, 1020–1024 (1989).
22. M.A. Baily, *GROWTH. Genet. Hormones* **8**(suppl. 1), 54–57 (1992).
23. R. Stanhope, *GROWTH. Genet. Hormones* **8**(suppl. 1), 6–8 (1992).
24. B. Lippe and S.D. Frasier, *J. Pediatr.* **115**, 585–587 (1989).
25. A.W. Root, *GROWTH. Genet. Hormones* **8**(suppl. 1), 1–5 (1992).
26. D.B. Allen and N.C. Fost, *J. Pediatr.* **117**, 16–21 (1990).

27. National Institutes of Health, Clinical Research Project No. 84-CH-148 (later renumbered 91-CH-0046), September 18, 1990.

28. L. Cuttler et al., *J. Am. Med. Assoc.* **276**, 531–537 (1996).

29. B.B. Bercu, *J. Am. Med. Assoc.* **276**, 567–568 (1996).

30. National Institutes of Health, Report of the NIH human growth hormone protocol review committee, October 2, 1992.

31. B.M. Lippe, *GROWTH. Genet. Hormones* **8**(suppl. 1), 31–34 (1992).

32. C.A. Tauer, *IRB* **16**(3), 1–9 (1994).

33. G.B. White, *Kennedy Inst. Ethics J.* **3**, 401–409 (1993).

34. J.M. Gertner, *GROWTH. Genet. Hormones* **8**(suppl. 1), 18–22 (1992).

35. S. Bertelloni et al., *J. Pediatr.* **135**, 367–370 (1999).

36. Physicians Committee for Responsible Medicine, *Concerns About Growth Hormone Experiments in Short Children*, Fact sheet, Washington, DC, 1993.

37. J. Mendelson, *Int. J. Risk Safety Med.* **4**, 61–70 (1993).

38. R. Herman, *Washington Post*, Health Section, 9, June 29 (1993).

39. M.J. Pramik, *Genet. Eng. News* **19**(1), 15, 27, 32 (1999).

40. M. Ho, *Genentech Inc.*, Series U.S. Research, Goldman Sachs, New York, October 20, 1993.

41. J. Walsh, Debate over BST expands, *Star Tribune* (Minneapolis), A1 and A9, February 5 (1994).

42. M.L. Vance and N. Mauras, *N. Engl. J. Med.* **341**, 1206–1216 (1999).

43. P.H. Sonksen and A.J. Weissberger, *GROWTH. Genet. Hormones* **14**, 41–48 (1998).

44. L.E. Underwood, *Hosp. Prac.* 192–198, April 15 (1992).

45. L.E. Underwood and P.A. Rieser, *Acta Paediatr. Scand.* **362**(suppl.), 18–23 (1989).

46. Discussion I: C & D, *GROWTH. Genet. Hormones* **8**(suppl. 1), 17 (1992).

47. Discussion III: A & B, *GROWTH. Genet. Hormones* **8**(suppl. 1), 41–45 (1992).

48. P.C. Hindmarsh and C.G.D. Brook, in A.B. Grossman, ed., *Clinical Endocrinology*, Blackwell, Oxford, UK, 1992, pp. 810–836.

49. M. Verweij and F. Kortmann, *J. Med. Ethics* **23**, 305–309 (1997).

50. E.T. Juengst, in E. Parens, ed., *Enhancing Human Traits: Ethical and Social Implications*, Georgetown University Press, Washington, DC, 1998, pp. 29–47.

51. E.T. Juengst, *J. Med. Phil.* **22**, 125–142 (1997).

52. D.L. Young-Hyman, in C.S. Holmes, ed., *Psychoneuroendocrinology: Brain, Behavior, and Hormonal Interactions*, Springer-Verlag, New York, 1990, pp. 40–55.

53. B. Stabler and L.E. Underwood, eds., *Slow Grows the Child*, Lawrence Erlbaum, Hillsdale, NJ, 1986.

54. J.E. Sabin and N. Daniels, *Hastings Center Rep.* **24**(6), 5–13 (1994).

55. D.E. Sandberg, *GROWTH. Genet. Hormones* **8**(4), 5–6 (1992).

56. D.E. Sandberg, A.E. Brook, and S.P. Campos, in B. Stabler and L.E. Underwood, eds., *Growth, Stature and Adaptation: Behavioral, Social, and Cognitive Aspects of Growth Delay*, University of North Carolina Press, Chapel Hill, NC, 1994, pp. 19–33.

57. L.D. Voss and J. Mulligan, in B. Stabler and L.E. Underwood, eds., *Growth, Stature and Adaptation: Behavioral, Social, and Cognitive Aspects of Growth Delay*, University of North Carolina Press, Chapel Hill, NC, 1994, pp. 47–64.

58. R.S. Clopper, *GROWTH. Genet. Hormones* **8**(suppl. 1), 27–30 (1992).

59. Discussion IV, C & D, *GROWTH. Genet. Hormones* **8**(suppl. 1), 64–67 (1992).

60. C.G.D. Brook, *Br. Med. J.* **315**, 692–693 (1997).

61. P.T. Siegel, in C.S. Holmes, ed., *Psychoneuroendocrinology: Brain, Behavior, and Hormonal Interactions*, Springer-Verlag, New York, 1990, pp. 17–39.

62. B. Stabler, *GROWTH. Genet. Hormones* **8**(suppl. 1), 24–26 (1992).

63. P.T. Siegel and N.J. Hopwood, in B. Stabler and L.E. Underwood, eds., *Slow Grows the Child*, Lawrence Erlbaum, Hillsdale, NJ, 1986, pp. 57–71.

64. B. Stabler et al., *Clin. Endocrinol.* **36**, 467–473 (1992).

65. B. Stabler et al., *Psychosom. Med.* **56**, 175 (1994).

66. D.S. Pine, P. Cohen, and J. Brook, *Pediatrics* **97**, 856–863 (1996).

67. R.L. Hintz, *GROWTH. Genet. Hormones* **8**(suppl. 1), 10–14 (1992).

68. S.D. Frasier, *GROWTH. Genet. Hormones* **8**(suppl. 1), 15–16 (1992).

69. D.S. Diekema, *Persp. Biol. Med.* **34**, 109–123 (1990).

70. Discussion II, A & B, *GROWTH. Genet. Hormones* **8**(suppl. 1), 30–31 (1992).

71. *Washington Post* Staff, reprinted *Star Tribune* (Minneapolis), A12, November 3 (1996).

72. Editorial Staff, *Star Tribune* (Minneapolis), November 27 (1996).

73. National Highway Transportation Safety Administration, Proposed rule for 49 CFR 552, Available at: *http://www.nhtsa.gov/cars/rules/rulings/AAirbaqSNPRM/Index.html*

74. Health Care Financing Administration, *1998 Data Compendium*, U.S. Department of Health and Human Services, August 1998, Available at: *http://www.hcfa.gov/stats/stats/htm*

75. National Institutes of Health, Press briefing, FY 2001 President's Budget, Available at: *http://www4.od.nih.gov/ofm/budget/fy2001/Pressbriefing.htm*

76. M. Benjamin, J. Muyskens, and P. Saenger, *Hastings Center Rep.* **14**(2), 5–9 (1984).

77. S.L. Gortmaker et al., *N. Engl. J. Med.* **329**, 1008–1012 (1993).

78. J.D. Sargent and D.G. Blanchflower, *Arch. Pediatr. Adolesc. Med.* **148**, 681–687 (1994).

79. Editorial Staff, *Psych. Today* **5**(3), 102, August (1971).

80. P.W. Crosbie, *Soc. Psychol. Q.* **42**, 340–354 (1979).

81. H.J. Dean et al., *Am. J. Dis. Child* **139**, 1105–1110 (1985).

82. H.J. Dean, in C.S. Holmes, ed., *Psychoneuroendocrinology: Brain, Behavior, and Hormonal Interactions*, Springer-Verlag, New York, 1990, pp. 79–91.

83. C.M. Mitchell et al., in B. Stabler and L.E. Underwood, eds., *Slow Grows the Child*, Lawrence Erlbaum, Hillsdale, NJ, 1986, pp. 97–109.

84. R.R. Clopper et al., in B. Stabler and L.E. Underwood, eds., *Slow Grows the Child*, Lawrence Erlbaum, Hillsdale, NJ, 1986, pp. 83–96.

85. D.B. Allen, *GROWTH. Genet. Hormones* **8**(suppl. 1), 24 (1992).

86. J.D. Lantos, *GROWTH. Genet. Hormones* **8**(suppl. 1), 68–70 (1992).

See other HUMAN ENHANCEMENT USES OF BIOTECHNOLOGY entries.

HUMAN ENHANCEMENT USES OF BIOTECHNOLOGY, ETHICS, THE ETHICS OF ENHANCEMENT

Eric T. Juengst
Case Western Reserve University
Cleveland, Ohio

OUTLINE

INTRODUCTION

In discussions of ethics in biotechnology one frequently encounters the claim that there is an important moral distinction between using biotechnological tools and products to combat human disease, and attempting to use them to "enhance" human traits. Thus people argue that using biosynthetic human growth hormone to treat an inborn growth hormone deficiency is praiseworthy, but not the use of the same product to increase the height of a hormonally normal short child (1). Similarly, while the use of human gene transfer techniques to treat disease enjoys widespread support from secular and religious moral authorities, a line is usually drawn at using the same protocols to attempt to improve upon otherwise healthy traits (2,3). Even those unwilling to condemn the enhancement uses of biotechnology outright almost all concur that ethics demands that therapeutic applications of these tools be given priority for research and development (4). As a result the distinction has been enshrined in biotechnology policies at both professional and governmental levels, and continues to inform much of the public discussion of new biotechnological advances.

Despite its widespread support as a moral divide, however, treatment–enhancement distinction is not easy to characterize conceptually. It often even seems in danger of evaporating entirely under conceptual critiques even before the question of its moral merits is entertained. If "enhancement" is to keep serving as a significant policy boundary, we should at least be clear about just what it demarcates. Examining the multiple ways the distinction is interpreted within the bioethical and science policy literature can help with this clarification, and that is the goal of this entry. Ultimately it will suggest a normative point: that the interpretations that most accurately identify the moral concerns at stake in the uses of biotechnology are those that focus on the uses that would serve to exacerbate, rather than reform, the social injustices that flow from our intolerance for human biological variation.

TREATMENT–ENHANCEMENT DISTINCTION

The treatment–enhancement distinction is usually used in bioethics to argue that curative or therapeutic uses of biotechnology fall within (and are protected by) the boundaries of medicine's traditional domain, while enhancement uses do not, and to that extent are more problematic as a professional medical practice or a legitimate health care need (5). Unfortunately, making the distinction between treatment and enhancement is not without its own complexities. The distinction is explicated in at least three distinctly different ways, which have different merits as boundary markers for medical research and practice. There are accounts that rely on medicine's own understanding of its professional goals, accounts which rely on theoretical measures of "species-typical functioning" that go well beyond medicine, and accounts that turn on particular concepts of disease.

Professional Domain Accounts

One approach to the enhancement/treatment distinction is to define it in terms of the accepted limits of professional medical practice. On this view, "treatments" are any interventions which physicians and their patients agree are useful and proper, while "enhancements" are simply interventions which are considered to fall beyond a physician's professional purview. Thus physician-prescribed physical therapy to improve muscle strength would be considered legitimate medical treatment, while weight-lifting under a coach's supervision to achieve a particular physique would be considered an enhancement. This view resonates well with a number of contemporary social scientific critiques of biomedicine, which suggest that medicine has no natural domain of practice beyond that which it negotiates with society (6). It also provides a simple normative lesson for professionals concerned about their obligations in specific cases: One takes one's cues from the patient's value system, and negotiates towards interventions that can help achieve the patient's vision of human flourishing (7).

Unfortunately, however, these same features also deny this approach the ability to be of help to those attempting to use the treatment/enhancement distinction in order to regulate gene transfer research. Relying on the conventions of professional practice provides no principled way to classify technological innovations as either within or outside medicine's proper domain until after the fact of their acceptance or rejection by the medical community. To the extent that useful "upper-boundary" concepts are required at the policy level — for societies making health care research allocation decisions, for example — this impotence is an important weakness.

Normalcy Accounts

Fortunately another approach to interpreting the treatment/enhancement distinction is framed explicitly as a policy tool for separating legitimate health care needs from luxury services. The most developed exposition of this view is Sabin and Daniel's endorsement of what they call the "normal function" standard for determining the limits of "medically necessary" (and therefore socially underwritten) health services (8). Sabin and Daniels argue that an appropriate boundary between medically necessary treatments and optional enhancements can be drawn by thinking about how to provide medical services fairly within a population. Following Daniels' earlier work (1,10), they construe health care as one of society's means for preserving equality of opportunity for its citizens, and define health care needs as those services that allow individuals to enjoy the portion of the society's "normal opportunity range" to which their full array of skills and talents would give them access, by restoring or improving their abilities to the range of functional capacities typical for members of their reference class (e.g., age and gender) within the human species. Daniels has specified this definition of health care needs further by saying that the notion of "species-typical functioning" it relies upon is not "merely a statistical notion," but implies "a theoretical account of the design of the organism," that describes the "natural functional organization of a typical member of the species." Any interventions that would take expand an individual's range of functional capacities beyond the range typical for his or her reference class would count as an (medically unnecessary) enhancement.

The "normal function" approach is a sophisticated attempt to define the limits of social obligations to provide health services for policy purposes, and comes close to accurately reconstructing the rationale behind many actual "line drawing" judgments by health care coverage plans and professional societies. Unfortunately, this approach is also semipermeable in an important way for our purposes.

The first serious problem is the problem of prevention. While efforts at generic "health promotion" straddle the border of biomedicine, efforts to prevent the manifestation of specific maladies in individuals are always accepted as legitimate parts of biomedicine, and would be automatically located on the "treatment" side of the enhancement boundary. On the other hand, one of the ways one can prevent a disease is to strengthen the body's ability to resist it long before any diagnosable problem appears. These forms of prevention attempt to elevate bodily functions above the normal range for the individual (and in some cases the species), and to that extent seem to slide into enhancement. Consider the case that LeRoy Walters and Julie Palmer make for including some genetic enhancements within the domain of legitimate medical needs. They start with the paradigm of a nongenetic preventive intervention — immunization against infectious disease — and then drive their genetic truck through the border-crossing it creates:

> In current medical practice, the best example of a widely-accepted health-related physical enhancement is immunization against infectious disease. With immunizations against diseases like polio or hepatitis B, what we are saying is in effect, "The immune system that we inherited from our parents may not be adequate to ward off certain viruses if we are exposed to them. Therefore, we will enhance the capabilities of our immune system by priming it to fight against these viruses.

> From the current practice of immunizations against particular diseases, it would seem to be only a small step to try to enhance the general function of the immune system by genetic means. ...In our view, the genetic enhancement of immune system function would be morally justifiable if this kind of enhancement assisted in preventing disease and did not cause offsetting harms to the people treated by the technique (4, pp. 13–14).

This argument is bolstered by the fact that the technical prospects for such preventive–enhancement interventions already look good, given gene transfer research now underway to treat ill patients in just those ways. Thus the gene therapist summarizes the prospects for using gene therapy in oncology this way:

> Over the next few years, it appears that the greatest application will be in the treatment of cancer, where a number of genes that have been isolated have the potential to *empower* the immune system to eliminate cancer cells. ...Human gene therapy cancer trials have also been initiated for insertion of the tumor necrosis factor (TNF) gene into T-lymphocytes in an effort *to enhance the ability* of T-lymphocytes to kill tumors. Another approach has been to insert the TNF gene into tumor cells in an effort *to induce a more vigorous immune response* against the tumor (emphasis added) (10).

Another gene transfer protocol already underway 'treats' people with an inherited high risk of heart disease by increasing the number of low density lipoprotein receptors their blood cells carry, enhancing their ability to clear their high levels of cholesterol from their blood before it causes heart disease (11). If it works to reduce their risk of heart disease, why not use it prophylacticly to reduce my more modest risk? Moreover, if human gene transfer protocols like these are acceptable as forms of preventive medicine, the critics ask, how can we claim that we should be "drawing the line" at enhancement?

Disease-Based Accounts

Probably the most common rejoinder to the problem of prevention is to distinquish the problems to which they respond. Treatments are interventions which address the health problems created by diseases and disabilities — "maladies" in the helpful language of Clouser, Culver, and Gert (12). Enhancements, on the other hand, are interventions aimed at healthy systems and normal traits. Thus, prescribing biosynthetic growth hormone to rectify a diagnosable growth hormone deficiency is legitimate treatment, while prescribing it for patients with normal growth hormone levels would be an attempt at "positive genetic engineering" or enhancement (13). On this account, to justify an intervention as appropriate medicine means to be able to identify a pathological problem in the patient; if no medically recognizable malady can be diagnosed, the intervention cannot be "medically necessary," and is thus suspect as an enhancement.

This interpretation has the advantages of being simple, intuitively appealing, and consistent with a good bit of biomedical behavior. Maladies are objectively observable phenomena and the traditional target of medical intervention. We can know maladies through diagnosis, and we can tell that we have gone beyond medicine when no pathology can be identified (14). Thus the pediatric endocrinologists discourage the enhancement uses of biosynthetic growth hormone by citing the old adage "If it ain't broke, don't fix it" (15). This interpretation is also the one at work in the efforts of professionals working at the boundary, like cosmetic surgeons, to justify their services in terms of relieving "diagnosable" psychological suffering rather than satisfying the aesthetic tastes of their clients (16), and in our insurance companies' insistence on being provided with that diagnosis before providing coverage for such surgeries.

Unfortunately, this interpretation does also face at least two major difficulties. The first problem that any disease-based interpretation of the enhancement boundary faces is, of course, biomedicine's infamous nosological elasticity. It is not that hard to coin new maladies for the purposes of justifying the use of enhancement interventions. By interpreting the boundary of medicine in terms of maladies, this approach puts the power for drawing that boundary squarely in the profession's hands, with the corresponding potential for abuse. Moreover, the preventive powers of a give enhancement intervention can be difficult to disprove, if the targeted disease never manifests itself. Enhancing interventions would have the advantage of the man who claimed his dance was keeping dragons out of Central Park: Until a dragon lands, it is hard to argue that he's not providing a preventive service.

The more important problem, however, is that for practical purposes no matter how the line is drawn, most biotechnological interventions that could become problematic as enhancement interventions would not have to cross that line in order to be developed and approved for clinical use, because they will also have legitimate therapeutic applications. In fact, most biosynthetic biologicals and gene transfer protocols with potential for enhancement uses will first emerge as therapeutic agents. General cognitive enhancement interventions, for example, are likely to be approved for use only in patients with neurological diseases (17). However, to the extent that they are in high demand by individuals who are merely suffering the effects of normal aging, the risk of unapproved or "off-label" uses of these products will be high (18). This risk poses unique regulatory challenges. Even if, for example, the U.S. Food and Drug Administration (FDA) vigorously attempted to regulate genetic enhancement technology, under its current legislative mandate the agency is unlikely to regulate off-label uses of approved products (19). Moreover, given the current regulatory vacuum surrounding the private practice of reproductive medicine, there is little to prohibit the application of these techniques to early human embryos as well, in hopes of effecting germ-line transformations.

This last point is critical for policy purposes, because it suggests that, in countries like the United States, the real challenge to regulation in this area may not be the development of enhancement interventions or "enhancement research" but about the downstream "off-label" uses of gene therapies for nonmedical enhancement

purposes. The policy problems then becomes one of controlling access and use of the technologies, not their research and development. Unless laws are changed, regulation of biotechnological enhancements ill fall under the common law—which, given the absence of FDA regulation of off-label uses remains the most potent potential source of legal regulation—will focus on physician malpractice and actions for lack of informed consent. This presents another set of challenges for the law, since the novelty of enhancement technologies will make it difficult for judges and juries to ascertain the reasonableness of physician behavior (19).

These realities have pressed those who would use the treatment/enhancement distinction for policy purposes to articulate the moral dangers of genetic enhancement more clearly. After all, personal improvement is praised in many spheres of human endeavor, and, as purely elective matter, biomedical interventions like cosmetic surgery are well accepted in our society as means to achieving personal improvement goals.

ENHANCEMENT AS A FORM OF CHEATING

There are two lines of thought that have emerged from this recent work. The first focuses on the idea that biomedical enhancements are a form of social cheating. This is the view that taking the biomedical shortcut erodes the specific social practices that would make the analogous human achievement valuable in the first place. Thus some people argue that it defeats the purpose of the contest for the marathon runner to gain endurance chemically rather than through training, and it misses the point of meditation to gain Nirvana through psychosurgery. In both cases the value of the improvements lie in the achievements they reward as well as the benefits they bring. The achievements—successful training or disciplined meditation—add value to the improvements because they are understood to be admirable social practices in themselves. Wherever a biomedical intervention is used to bypass an admirable social practice, then the improvement's social value—the value of a runner's physical endurance or a mystic's visions—is weakened accordingly. If we are to preserve the value of the social practices we count as "enhancing," it may be in society's interest to impose a means-based limit on biomedical enhancement efforts.

Interpreting enhancement interventions as those which short-circuit admirable human practices has special utility for policy analysis. To the extent that biomedical shortcuts increasingly allow specific accomplishments to be divorced from the admirable practices they were designed to signal, the social value of those accomplishments will be undermined. Not only will the intrinsic value be diminished for everyone that takes the shortcut, but the resulting disparity between the enhanced and unenhanced will call the fairness of the whole game (be it educational, recreational or professional) into question. If the extrinsic value of being causally responsible for certain accomplishments is high enough (like professional sports salaries), the intrinsic value of the admirable practices that a particular institution was designed to foster may even start to be called into question (20). For institutions interested in continuing to foster the social values for which they have traditionally been the guardians, this has two alternative policy implications. Either they must redesign the game (of education, sports, etc.) to find new ways to evaluate excellence in the admirable practices that are not affected by available enhancements, or they must prohibit the use of the enhancing shortcuts. Which route an institution should take depends on the possibility and practicality of taking either, because ethically they are equivalent.

ENHANCEMENT AS AN ABUSE OF MEDICINE

Unfortunately, some of the social games we can play (and cheat in) do not turn on participants' achievements at all but on traits over which individuals have little control, like stature, shape, and skin color. The social games of stigmatization, discrimination, and exclusion use these traits in the same manner that other practices use achievements: as intrinsically valuable keys to extrinsic goods. Now it is becoming increasingly possible to seek biomedical help in changing these traits in order to short-circuit these games as well. Here, the biomedical interventions involved, like skin lighteners or stature increasers, are "enhancements" because they serve to improve the recipient's social standing, but only by perpetuating the social bias under which they originally labored. When "enhancement" is understood in this way, it warns of still another set of moral concerns.

On this interpretation, what makes the provision of human growth hormone to a short child a morally suspicious enhancement is not the absence of a diagnosable disease or the "species atypical" hormone level that would result: Rather it is the intent to improve the child's social status by changing the child rather than by changing her social environment (21). Enhancement interventions are almost always wrongheaded under this account because the source of the social status they seek to improve is, by definition, the social group and not the individual. Attempting to improve that status in the individual amounts to a moral mistake akin to "blaming the victim": It misattributes causality, is ultimately futile, and can have harmful consequences. This is the interpretation of enhancement that seems to be at work when people argue that it inappropriately "medicalizes" a social problem to use Ritalin to induce cooperative behavior in the classroom. In such cases the critics dispute the assumption that the human need in question is one that is created by, and quenchable through, our bodies, and assert that both its source and solution really lie in quite a different sphere of human experience.

This interpretation of the enhancement concept is useful to those interested in the ethics of personal improvement because it warns of a number of moral pitfalls beyond the baseline considerations that the enhancement–treatment distinction provides. Attempting to improve social status by changing the individual risks being self-defeating (by inflating expectations), futile (if the individual's comparative gains are neutralized by the enhancement's availability to the whole social group), unfair (if the whole group does not have access to

the enhancement), or complicitous with unjust social prejudices (by forcing people into a range of variation dictated by biases that favor one group over others). For those faced with decisions about whether to attempt to enhance themselves or their children through gene transfer, this way of understanding enhancement is much more illuminating than attempts to distinquishing it from medical treatment, because it points to the real values at stake. Ideally, one should do no gene transfer that will make an existing social problem worse, even if exacerbating injustice would further one's own interests.

On the other hand, protecting these values is difficult in a pluralistic society like ours, since it means developing ways to policing individuals' complicity with suspect social norms (22). Under the historical shadow of state sponsored eugenics programs, our government is unlikely to promulgate lists of acceptable and unacceptable enhancements, even if the intent of the lists are to protect the interests of those who are unenhanced.

Moreover regulatory limits on access to genetic enhancements in the United States could be ineffectual if individuals could obtain enhancements abroad. In the past, we have seen a number of examples of persons circumventing U.S. laws to obtain medical care abroad, including seeking illegal abortions abroad prior to *Roe v. Wade*, purchasing unapproved AIDS drugs (RU486) and other pharmaceuticals abroad and returning with them to the United States (23), and traveling to foreign countries to obtain infertility treatments that were illegal or unavailable in the United States (24). Most nations' ability under domestic legal authority to control offshore access is extremely limited. Another approach would be to try to prevent people from travelling abroad in the first place in order to obtain contraband enhancements. Again, the United States arguably possesses the authority to restrict travel on grounds of national security (25). Would obtaining enhancements abroad amount to such a threat? Even if it did, travel restrictions have been imposed on travel to specific countries; it would be virtually impossible to restrict travel to a country for a specific purpose, if travel were permitted for other purposes.

CONCLUSION

Clearly, all of the ways of understanding "enhancement" as a moral concept that I've reviewed have limitations. However, all these interpretations do seem to be alive and well and mixed together in the literature on the topic. It is not possible to cleanly assign the different interpretations of "enhancement" to different spheres of ethical analysis. But there do seem to be some rough correlations that might be made. Thus the interpretations that contrast enhancement interventions with "treatments" seem most useful where it is the limits of medicine's expertise that is at issue. Whether medicine's boundary is defined in terms of concepts of disease, or in sociological terms as the scope of medical practice, or in terms of some theory of the human norm, this interpretation at least provides tools to draw that boundary. Moreover, all other considerations being equal, the line that it draws is the boundary of medical obligation, not the boundary of medical tolerance. Using this tool, enhancement interventions like cosmetic surgery can still be permissable to perform as phsysicians, but also permissable to deny. This has important implications for social policy making about health care coverage, to the extent that society relies on medicine's sense of the medically necessary to define the limits of its obligations to underwrite care. Again, all other considerations being equal, this interpretation of the concept suggests that few enhancement interventions should be actively prohibited by society or foregone by individuals, even when they are not underwritten as a part of health care, since there is nothing intrinsically wrong with seeking self-improvements beyond good health.

By contrast, the interpretations of enhancement that focus on the misuse of biomedical tools in efforts at self-improvement seem the most relevant to issues in the personal, rather than professional, ethics of enhancement. Concerns about the authenticity of particular accomplishments are moral challenges to the individual but find little purchase in the professional ethics of biomedicine, with its focus on the physical safety and efficacy of its tools. The primary policy implications of this interpretation are for the social institutions charged with fostering particular admirable practices: Enhancement interventions that offer biomedical shortcuts to achievement force reassessments within those institutions of the values they stand for and the practices they have designed to foster them.

Finally, at the other end of the spectrum, enhancement interventions that seem to commit the moral mistake of trying to address social problems through the bodies of the potentially oppressed do seem to mark a stronger set of moral boundaries for all concerned. For biomedicine, this concept marks an epistemic limit beyond which medical approaches to problem-solving are not only unnecessary but conceptually wrongheaded. For individuals, parents, and society, these kinds of enhancement interventions risk either backfiring by exacerbating the social problems they are intended to address, or being futile, if they merely result in a shift of the normal range for a given social trait. Where the medicalization account of enhancements fits a given intervention, there does seem to be more justification for stronger warnings, protections, or prohibitions across the board, whether the interventions falls within medicine's boundaries or not (26).

ACKNOWLEDGMENT

The research conducted for this article was supported in part by NIH grant R01 HG-1446-02. I am grateful to Tom Murray, Maxwell Mehlman, Erik Parens, and David Smith for useful discussions of the text and many of the ideas presented here.

BIBLIOGRAPHY

1. N. Daniels, *Growth, Genet. Hormones* **8**(suppl. 1), 46–48 (1992).
2. W.F. Anderson, *J. Med. Philos.* **14**, 681–693 (1989).
3. P. Baird, *Perspect. Biol. Med.* **37**, 566–575 (1994).
4. L. Walters and J. Palmer, *The Ethics of Human Gene Therapy*, Oxford University Press, Oxford, UK, 1996.
5. E. Parens, *Enhancing Human Traits: Ethical and Social Implications*, Georgetown University Press, Washington, DC, 1998.
6. B. Good, *Medicine, Rationality and Experience: An Anthropological Perspective*, Cambridge University Press, New York, 1994.
7. H.T. Engelhardt, *Soc. Philos. Policy* **8**, 180–191 (1990).
8. J. Sabin and N. Daniels, *Hastings Center Rep.* **24**, 5–13 (1994).
9. N. Daniels, *Just Health Care*, Cambridge University Press, New York, 1986.
10. K. Culver, *Genet. Resource* **7**, 5–10 (1993).
11. J. Wilson et al., *Hum. Gene Ther.* **3**, 179–222 (1992).
12. K. Clouser, C. Culver, and B. Gert, *Hastings Center Rep.* **11**, 29–37 (1981).
13. E. Berger and B. Gert, *J. Med. Philos.* **16**, 667–685 (1991).
14. E. Juengst, *J. Med. Philos.* **22**, 125–142 (1997).
15. Ad Hoc Committee on Growth Hormone Usage. *Pediatrics* **72**, 891–894 (1984).
16. K. Morgan, *Hypatia* **6**, 25–53 (1991).
17. P. Whitehouse et al., *Hastings Center Rep.* **27**(3), 14–22 (1997).
18. D. Kessler, *Harvard J. Legis.* **15**, 693 (1978).
19. M. Mehlman, *Wake Forest Law Rev.* **34**(Fall), 671–617 (1999).
20. T. Murray, in T. Murray, W. Gaylin, and R. Macklin, eds., *Feeling Good and Doing Better: Ethics and Nontherapeutic Drug Use*, Humana Press, Clifton, NJ, pp. 107–129.
21. G. White, *Kennedy Inst. Ethics J.* **3**, 401–409 (1993).
22. M. Little, in I. de Beaufort, M. Hilhorst, and S. Holm, eds., *In the Eye of the Beholder: Ethics and Medical Change of Appearance*, Scandanavian University Press, Stockholm, Sweden, 1997, pp. 151–167.
23. J.S. Batterman, *Hofstra Univ. Law Rev.* **19**, 191 (1990).
24. J. Sauer, *Hum. Reprod.* **12**, 1844–1845 (1997).
25. Kriemer, *N.Y. Univ. Law Rev.* **67**, 451 (1992).
26. E. Parens, *Kennedy Inst. Ethics J.* **5**, 141–153 (1995).

See other entries BEHAVIORAL GENETICS, HUMAN; see also HUMAN ENHANCEMENT USES OF BIOTECHNOLOGY entries.

HUMAN ENHANCEMENT USES OF BIOTECHNOLOGY, ETHICS, THERAPY VS. ENHANCEMENT

GERALD McKENNY
Rice University
Houston, Texas

OUTLINE

INTRODUCTION

In recent years ongoing practices as well as new or anticipated developments in reconstructive surgery, sports medicine, psychopharmacology, human gene therapy, and emerging areas of biomedical engineering aim at altering the appearance of the body; increasing the efficiency, capacity or productivity of various human functions or performances; and changing features of the personality or mood. Conveniently (if imprecisely) referred to as enhancements, these phenomena, insofar as they are implemented through the institutions and practices of medicine, raise the question of whether their pursuit can be distinguished from the treatment of diseases, and if so, what normative significance, if any, such a distinction has. At stake in this distinction are important ethical and policy issues. As various authors have argued, the distinction between therapy and enhancement is commonly invoked to determine the composition of a basic health care insurance package and to limit the medicalization of human life (1).

THEORIES OF HEALTH, DISEASE, AND ENHANCEMENT

It may seem obvious that, for example, a chemotherapy regimen to treat cancer and a pharmacological agent or (should it become possible) a gene transfer to increase one's powers of concentration fall under radically different categories of intervention. It is natural for many to think of the former as the treatment of a disease and the latter as the enhancement of a capacity. However, there has been significant disagreement over how to account for such a difference and even whether it can be maintained at all in the end. The controversy derives from rival theories of disease in the philosophy of medicine. These include naturalist theories for which disease is a value-free concept grounded in the biological and medical sciences, and normativist theories for which disease is a value-laden concept referring to certain states that individuals (or groups or societies) seek to avoid or overcome. There are strong and weak versions of each type of theory.

Strong Naturalism

The most influential strong naturalist theory is that of Christopher Boorse (2,3). Analyzing what he takes to be the traditional medical understanding of disease, Boorse argues that (1) health (or normality) is the absence of disease (or pathology); (2) because the medical sciences are based on physiology, pathology is defined with reference to the survival and reproductive competence of the organism as the goals physiology studies (recognizing that other goals, e.g., survival of genes or ecological equilibrium,

may be important for other branches of biology); (3) a pathology is a reduction of a part-function at any level of the interlocking hierarchy of functional processes (ranging from organelle to cell to tissue to organ to gross behavior) below its statistically species-typical range as determined with reference to the relevant age and sex class. While other theories of disease make use of a statistical range to define normality and pathology, Boorse's emphasis on physiology with its notion of a species design consisting of interlocking functional processes ultimately explained in terms of their contribution to survival and reproductive competence makes him a naturalist.

Boorse's theory has the advantage of enabling one clearly to distinguish pathologies from other characteristics. For example, to cite David Allen and Norman Fost's now-familiar scenario (4), one can distinguish short stature due to human growth hormone deficiency (a part-function operating below the normal range) from short stature as an inherited trait reflecting normal human genetic diversity (assuming that a currently unknown genetic dysfunction is not the cause). Similarly Boorse's theory enables one to distinguish, with James Sabin and Norman Daniels (5), a case of shyness characterized by interpersonal sensitivity and defensive withdrawal due to a bipolar disorder from shyness that simply falls on one end of a normal distribution of social adaptation. Moreover, because many deformities (including harelip and cleft palate) typically involve dysfunction as well as deformity, they may be distinguished from other features (ranging from male baldness to much more severe structural defects) that do not involve dysfunctions. And because age is used as a reference group to determine pathology, dementia among older adults (an abnormal condition) can be counted as a disease while osteoporosis among postmenopausal women (a normal condition) is not. Finally, part-functions operating below the normal range can be distinguished, using this theory, from those operating at the low end of the normal range. For each of these pairs, improvements of the first element in the pair would count as therapies, while improvements of the second element would be considered enhancements.

How would Boorse's theory fare in view of more complex cases? Actual and hypothetical developments in human gene transfer raise the question of how the theory would evaluate the enhancement of certain functions in the course of treating or preventing disease. For example, Juan Manuel Torres describes a multidrug resistance protocol that enhances the capacity of the bone-marrow cells of cancer patients to resist certain side effects of chemotherapy (6). Would this be considered therapy or enhancement? While the protocol involves raising a part-function beyond the normal range, this enhancement (as Torres argues) is simply a necessary part of a procedure aimed at the removal or mitigation of a severe pathology. So long as the various elements or phases of a course of treatment are explained and evaluated in terms of the overall process or its end, there is no reason not to describe the intervention as part of the treatment of a disease. The same reasoning would apply to measures such as immunizations, which enhance an immune function as a necessary condition of preventing specific pathologies.

But what about an intervention such as the one proposed by Leroy Walters and Julie Palmer, which would aim not at a specific pathology but at increasing the capacity of the immune system as a whole to resist pathology more generally (7)? Would this constitute treatment or enhancement?

Nothing in Boorse's theory suggests that the category of pathology in medical science is restricted by the level and scale of the intervention, the complexity of the part-function involved, or the nonspecificity of the pathology. Would enhancing the immune system in this way nevertheless fall under the somewhat dubious category of positive health, that is, a state beyond the mere absence of pathology? Probably not, since such an intervention would not seem to violate Boorse's three criticisms of notions of positive health, namely that such notions lack a clear limit (in this case, autoimmune disorders would establish the limit of improvement), involve trade-offs among incompatible positive goals (the goal in this case still refers to freedom from disease rather than to a positive ideal of somatic good), and are value laden in a way health as the absence of disease is not (8). Boorse's theory, then, seems capable of handling all of these difficult cases. However, in order to cover these cases, it would be more precise to follow Walters and Palmer in distinguishing between health-related and non-health-related enhancements.

A final question for Boorse concerns the distinction between genetic disease and genetic variation. Returning to an earlier example, how would his theory greet the discovery of one or more genetic patterns linked with short stature? The mere existence of such a pattern could simply be an instance of human diversity and would therefore be insufficient for designating that pattern as a pathology. But what if the pattern resulted in a low output or inhibited performance of some part-function that depressed the latter below a normal range? Since disrupted part-functions at any level of the organism count as pathologies, it would appear necessary to describe the condition as such and to consider an appropriate therapy as the treatment of a disease. This example illustrates how genetic knowledge could require one to reclassify as therapies many interventions that would now, on Boorse's theory, be considered enhancements.

Boorse distinguishes sharply between the theoretical level of medical science (specifically, pathology) and the clinical level of medical practice. This raises the question of what normative significance his theory has for policy and medical practice. Boorse recognizes that his is a theoretical and not a clinical concept of disease. On the one hand, many minor, merely local or compensated pathologies (e.g., small benign internal tumors, minor warts or scars, mild cirrhosis of the liver) are undiagnosable and/or have no gross effects on disability, deformity or distress; on the other hand, physicians legitimately perform functions (e.g., childbirth, male circumcision, treatment of osteoporosis) other than the treatment of disease. Moreover he recognizes that diagnostic and therapeutic acts are always subject to the moral criteria elaborated in the bioethics literature; clinical pathology, unlike pathology as a theoretical medical science, is not value

free. However, Boorse describes the removal, mitigation, and prevention of pathology as the *core* of medicine while other activities (including enhancement) are *peripheral*: not necessarily illegitimate, but more controversial and lacking the objectivity and urgency of the treatment of pathology. He argues that the treatment of disease constitutes a presumption for clinical medicine because health (biological normality, i.e., absence of pathology) is almost always in the interests of patients and is neutral to most choices of activity and lifestyle. When these conditions fail to hold — when health is not in the interest of a patient or is not neutral with regard to activity or lifestyle — the presumption is defeated and other values may take priority over health.

However, it is doubtful that the treatment of disease constitutes a presumption even on Boorse's own account. The interests or values of the patient, and not the concept of disease, appear to be the normative core of clinical medicine; they determine when and to what extent health is the legitimate goal of clinical practice. Of course, it would be wrong to conclude from this alone that enhancements are justified by the interests or values of patients — that health should not be pursued in some cases does not entail that enhancement should or may be pursued — but there are some cases, such as the treatment of osteoporosis in postmenopausal women, where enhancements clearly should be pursued. It would be odd and even irresponsible to describe these cases as the overriding of a presumption and lacking in urgency and objectivity.

For all of these reasons it is questionable even on Boorse's account whether the objectivity of pathology in the theoretical sphere can simply be carried over into the clinical sphere (even if, as seems likely, biological normality will be the highest priority of clinical practice in the majority of cases). Hence, while Boorse gives a convincing argument for a theoretical distinction between treating a disease and enhancing a trait, the usefulness for policy makers and clinicians of a theory that places treatment of osteoporosis in the same category as facelifts is minimal at best. It says nothing about which health care services should be covered or what limits should be placed on medicalization.

Weak Naturalism

Not all forms of naturalism assume that science requires value-free concepts of health and disease that leave a gulf between medical theory and medical practice. Leon Kass's naturalism differs from Boorse's in two crucial respects (9). First, health is not simply the absence of disease but the wholeness and well functioning of the organism. There are degrees of health, and health can be promoted and maintained even in the absence of disease. The notion of wholeness cannot be captured in a physiological description and, especially in Kass's later essays, involves an experiential oneness of the lived body.

Nevertheless, this concept is still biological. It is manifested in phenomena such as the self-healing and pain-response capacities of organisms, and while Kass describes health as the fitness of the organism for its characteristic kinds of activity, he does not, as normativists do, define health in terms of what individuals or societies value. The difference between health and mere absence of pathology, the emphasis on wholeness with its abstraction from specifiable part-functions, and the lack of a biostatistical normal range would all seem to encourage Kass to welcome health-related enhancements. His theory would seem especially hospitable to those health-related enhancements (e.g., improving the performance of the immune system) that target the organism as a whole. However, Kass never argues for categorizing any such enhancements under the pursuit of health, and his view of medicine as most properly cooperative with, rather than transformative of, the body's own processes seems to cast a general suspicion on the entire enterprise of technological enhancement (assuming that cooperation and transformation can be consistently distinguished).

Second, Kass formulates a biology grounded on reflection on formal and final causes, thus closing the gap Boorse opened between value-laden medical practice and the value-free science on which medicine is based. No longer limited to physiology with its narrowly focused concern with individual survival and reproductive competence, medicine for Kass is concerned with "powers and desires for work, friendship, love, learning, awareness, mobility, thought and memory, self-command, and the sheer enjoyment of life." Health is not the realization of these goods, but the fitness of the organism for pursuing them. The biology medicine presupposes thus constitutes a broad view of human fulfillment — one that would appear to authorize medicine to extend far beyond the domain of health and disease and would appear to endorse a wide range of non-health-related enhancements. Nevertheless, Kass is highly critical of the use of medicine to fulfill individual preferences or desires, to advance societal priorities or programs, or to alter or replace basic physical processes or social institutions.

The problems with Kass's naturalism follow from the indeterminacy, vagueness, and ambiguity of his major claims. First, it is unclear whether in the end Kass's emphases on degrees of health and on health as wholeness differ significantly from Boorse. Within a biostatistical normal range, there are degrees of difference between those at the higher and lower ends; in this sense health even for Boorse is not *merely* the absence of disease. Health maintenance and promotion, highly commended by Kass, either merely prevent or mitigate disease, or they advance one from a lower to a higher level on a normal range (i.e., they enhance one, in the health-related sense of enhancement). The difference with Boorse seems to be an optical illusion produced by fundamentally different theoretical orientations. Boorse exhibits the modern tendency to treat pathology as basic, with death as its ultimate epistemological and ontological condition. Health is defined negatively, as the absence of pathology, namely normality, a biostatistical notion. Kass aims to recover what he believes is a classical tradition in which health is basic and is not mere biostatistical normality but greater or lesser fitness for the characteristic activities of the organism. Hence what Boorse would simply describe as statistically diverse levels of a single state of being normal, Kass would describe as different degrees of health. Of course, Kass may still wish to distinguish

pursuits of higher degrees of health from enhancements by distinguishing those enhancements that cooperate with nature from those that transform nature. But it is not clear what would determine such a distinction. Does it ride on the degree of alteration, on the extent to which the altered capacity exceeds the human average for that capacity, on the type of intervention involved or the technology it employs?

Second, it is unclear whether Kass can distinguish his broad list of functions that constitute health from the desires, preferences, ideals, projects, and effects that for him fall outside the pursuit of health. Kass is aware of the difficulties involved in giving an account of what would constitute a healthy power or desire for, say, work or learning without specifying the particular kinds of work or learning that are the object of individual desires, aptitudes or commitments. But without such a distinction it seems impossible to distinguish the pursuit of health from what Kass must regard as the pursuit of idiosyncrasy, or health from enhancement. Health would be relative to the particular activities and goals of individuals, making Kass a normativist. One way out of this problem would be to try to distinguish general purpose interventions — those that improve capacities needed for all or nearly all ways of life — from idiosyncratic interventions that equip one for a limited range of ways of life at the expense of others. However, in addition to problems involved in making and defending this distinction beyond the obvious cases of treating serious disease, its definition of health in terms of ways of life would move Kass into the normativist camp once again.

Third, as Kass's list of human capacities and activities indicates, health can affect almost any human activity or practice. It is likely that enhancement technologies will be increasingly capable of improving "powers and desires" for these capacities and activities indefinitely. Unless, then, he can determine what levels of realization are appropriate for each capacity and what role, if any, medical intervention should play in their realization, the medicalization of these capacities and activities will go virtually unchecked. Finally, even when Kass's reflective biology leads him to approve certain procedures, such as in vitro fertilization (IVF), that are not strictly health-related (IVF does not treat infertility, and infertility, for complex reasons, is not actually a matter of disease and health for Kass) but that fulfill desires that are, according to his biology, natural, he is unsure whether physicians should perform them or not. He seems torn between his conviction that medicine should ideally restrict itself to the pursuit of health and his belief that bringing such procedures under the umbrella of medicine is the best strategy for curtailing their abuse.

In sum, despite his effort to close the gap between the natural and the normative, it is not possible to derive from Kass any clear guidelines regarding basic health care coverage or the limits of medicalization.

Strong Normativism

Strong normativist theories permit a much simpler treatment of the therapy–enhancement distinction because most of these theories deny that any such distinction can be made in a publicly binding way. H. Tristam Engelhardt, Jr. repeatedly claims that problems become medical problems when (1) they appear as a failure to achieve a valued state such as a certain level of freedom from pain or anxiety, a certain level of function, a particular realization of human form or grace, or an expected span of life, and when (2) they are problems of a sort that cannot be willed away and are embedded in a web of anatomical, physiological, or psychological causal forces that are open to medical explanation and manipulation (10). This does not prevent Engelhardt from making distinctions among such problems. First, some conditions are likely to be disvalued in nearly all cultures and environments and in light of nearly all human purposes, while others will be disvalued only under some such circumstances. Second, the judgment that a condition is a matter for medicine rather than, say, law or religion will for some conditions (e.g., appendicitis) be a virtually unavoidable judgment, while for other conditions (e.g., alcoholism or criminal behavior) the plausibility of such a judgment will be contested. Third, there are both negative senses of health (the absence of particular diseases, deformities, and dysfunctions) and positive senses (enhancement of capacities, augmentation of pleasures, etc.).

It is natural to refer to the former phenomena in each of these three pairs as treating, describing, or diagnosing a disease and the latter phenomena as involving the enhancement of a trait. But for Engelhardt this could be misleading. To group phenomena under one or the other of these categories merely reflects, respectively, the range of contingent agreements and disagreements concerning which states are and are not disvalued; the purposes and circumstances that make one or another judgment more useful and plausible; and a particular view of which among the diversity of human traits are species typical, normative, and natural. Only within a particular substantive view of the good is it possible to determine which states genuinely constitute values and disvalues; what properly comes under the domains of medicine, law, and religion; and where to draw the line between negative and positive health. It follows that the decisions regarding the composition of insurance coverage and the proper limits of medicalization can be made only within particular communities with their substantive views of the role of medicine in the overall human good.

The plausibility of Engelhardt's account follows from the insuperable difficulties both strong and weak naturalists face in arguing, from within naturalism, for any normatively binding force for their distinctions between therapy and enhancement. Its implausibility follows from its failure to understand the relation of medicine to the sciences of physiology and pathology. In his criticism of Boorse, Engelhardt argues that to define disease in terms of what affects the individual is arbitrary in light of what evolutionary biology tells us about nature's preference for species or genes over individuals. However, Boorse argues that disease is a concept in the medical science of pathology, itself based on physiology, a science which focuses on goals of individual organisms (specifically, on their survival and reproductive competence). It is therefore a mistake to argue that concepts of disease arbitrarily privilege the individual organism over the species. Of course, it

is true, as Boorse admits, that the choice medical practice (usually) makes to combat disease rather than to serve the evolutionary fitness of the human species is a normative choice. But this proves only that medicine (as a clinical practice) is normative; it says nothing about disease (11).

Similarly one may, as Engelhardt (on the basis of his views of the philosophy of science and the history of medicine) does, reject Boorse's claim that physiology and pathology are altogether value free, question whether the concept of disease they support can in difficult cases be distinguished as rigorously from other states, as Boorse assumes, and deprive the concept of disease of any normative significance for the practice of medicine — one may do all of this without denying that medicine, on the basis of physiology and pathology, can in principle distinguish disease from other states of the organism. Of course, Engelhardt could concede this and still argue that basic health care coverage and the limits of medicalization must both be established by particular communities. However, with regard to insurance coverage, Engelhardt himself admits that many conditions are regarded with near (though still contingent) universality as disvalues. If so, it is reasonable to assume for purposes of public policy that most serious pathologies would fall under this description and thus be eligible for coverage, while also allowing trade-offs for more highly contested conditions and leaving room for particular individuals and communities to determine the rest (perhaps even in the form of vouchers if one's theory of justice demands, as Engelhardt's does not, a level of public support of health coverage that extends beyond what most people regard as serious conditions).

Medicalization is more difficult. While particular communities could be left to determine for themselves what role medicine should play in their pursuit of what they understand is the human good, it will be necessary for many policy purposes and for purposes of criminal and tort law to make public decisions on the proper domain (whether that of law, medicine, or religion) of conditions such as alcoholism and socially disruptive behavior.

Weak Normativism

Not all normativist theories deny the possibility of public agreement in principle on the distinction between therapy and enhancement. K. Danner Clouser, Charles Culver, and Bernard Gert define a malady (an inclusive term covering what injury, illness, sickness, disease, trauma, wound, disorder, lesion, syndrome, etc. have in common) as a condition of an individual, other than his or her rational beliefs and desires, such that he or she is incurring or at significant risk of incurring a harm or evil (death, pain, disability, loss of freedom, or loss of pleasure) in the absence of a direct sustaining cause (12). Maladies are harms (or significant risks of harm) that are caused in a certain way and that rational persons want to avoid unless they have good reasons not to avoid them. The normative core of this concept is clear in the notion of a harm that one wants to avoid. But unlike strong normativism, the reference to what rational persons want in the absence of good reasons to the contrary stakes a claim to universality. Maladies are disvalued states, but the disvalue is objective, assuming that the underlying theory of rationality, which Gert develops elsewhere (13), is true.

The notions of harm, significant risk, and absence of a direct sustaining cause all have significance for the distinction between therapy and enhancement. Of their various categories of harm, Clouser, Culver, and Gert devote the most attention to disability. Here they differ from Boorse in two important respects. First, rather than specify age as a reference class against which to evaluate a level of function, they highlight for each ability the stage in normal development when that ability is at its peak. Prior to that stage, to lack the ability is to have an inability; after that stage, to lack it or lose it constitutes a malady. This means that when humans of very advanced age lack or lose an ability (e.g., to walk a certain distance), they suffer a malady even if, say, 98 percent of those in their reference class lack this ability. To attempt to return such individuals to, or to maintain them at, a level of ability that is normal for the species as a whole, regardless of age, is therefore to treat a malady, not to enhance a characteristic. This denial of the relevance of age leads Clouser, Culver, and Gert to classify menopause as a malady, since it constitutes the loss of an ability women normally have.

Second, what counts as lacking or losing an ability is determined by a statistical normal range, but without a concept of species design. This means that a person who develops an extraordinary ability, such as running a marathon in fewer than three hours, and then loses that ability does not suffer a malady (unless her ability is so compromised that it falls below the normal range). It also means that statistical normality, not function, determines which reactions to environmental factors and what levels of risk are maladies. Since nearly everyone tends to become short of breath in a smoke-filled room and to attract mosquitoes, these conditions are not maladies; for the same reason, improving the performance of the immune system would constitute an enhancement.

The concept of a distinct sustaining cause is also crucial for distinguishing maladies from other conditions. For example, individuals whose stature falls below the normal range suffer a variety of discriminatory attitudes and practices. These constitute a distinct sustaining cause of the suffering: when these attitudes and practices are not in effect, the harm disappears. Thus this condition is analogous to the pain that accompanies a wrestler's lock and disappears when the hold is released, and is distinct from a genuine malady such as an allergy, which also involves factors external to the individual but which persists for a period of time even after these external factors are removed.

The harms individuals suffer due to skin color also have a distinct sustaining cause in social attitudes and thus do not constitute a malady. What, then, of the harms individuals suffer due to extreme deformities or disfigurations? Clouser, Culver, and Gert argue that unlike responses to skin color, which are learned responses and vary among societies, responses to these conditions are "universal," spontaneous, and "natural," indeed "like the natural environment." This argument is confusing, but the point seems to be that one may distinguish allergies and

severe deformities from short stature and skin color by arguing that responses of the environment to the former are "normal" and intractable, while responses to the latter are variable and alterable. If this is so, the causes of the latter conditions are distinct from the individual in a way that the causes of the former conditions are not. Assuming that responses to deformity can be assimilated to nature in this way (a major assumption), gross deformities and allergies would be maladies while short stature and skin color would not.

While critics of the Clouser-Culver-Gert theory have attacked as counterintuitive or sexist its designation of pregnancy (which involves significant risks of various maladies) and menstruation (which involves pain and discomfort) as maladies (14), the features outlined above also have problematic implications. The problems follow from the effort to distinguish between maladies and normal human variation using the notions of a normal range and absence of a distinct sustaining cause, but without the notions of species design and references classes found in Boorse's strong naturalism. For this theory, procedures aimed at the postponement or reversal of menopause (perhaps no longer an outrageously unlikely prospect) would constitute treatment of a malady, but curing or relieving a mild asthmatic condition that prevents a marathon runner from competing at her previous level would not. A near-universal risk of succumbing to a virus, being statistically normal, apparently would not count as a malady. Gross deformities would count as maladies only if it could be proved that negative responses to them are natural to human beings.

In fairness to this theory, Clouser, Culver, and Gert do not recommend discarding all of the terms for which malady constitutes the commonality. It may be possible, then, to regard the marathon runner's asthma as a disease though not a malady, and its relief or cure as a therapy and not an enhancement. However, this simply raises the questions of how the theory *would* distinguish treating a disease and enhancing a trait and, more generally, what advantage the concept of malady has if it cannot cover admitted instances of diseases with significant effects.

Clouser, Culver, and Gert claim that the concept of malady does have several advantages over rival theories, but this is questionable. Two of the advantages claimed for it, namely the recognition that abnormality is neither a necessary nor a sufficient definition of disease (or related terms) and that malady applies equally to mental and physical conditions, are shared by all of the theories examined here. A third alleged advantage, that the concept of malady could be useful for setting precedents and negotiating borderline cases in determining insurance coverage, is doubtful: It would be an unusual insurance plan that would cover reversal of menopause but not asthmatic conditions of athletes. The fourth alleged advantage would occur if genetic interventions make it possible to enhance properties such as height, intelligence, memory, and strength: They argue that the concept of malady may help determine whether unenhanced (but still statistically normal) levels of these properties constitute deficiencies or not. But any theory that makes use of a normal range will offer this benefit. The more difficult question is whether these properties are candidates for disease or malady or simply represent human diversity. This question turns not on the differences between malady and disease but on whether it is more plausible to link the normal range to a concept of species design (as Boorse would) or to the absence of distinct sustaining causes.

Finally, Clouser, Culver, and Gert quite plausibly deny that any concept of malady or disease can establish the proper limits of medicalization by determining what medicine should and should not choose to provide. In sum, this theory does not seem to offer, in comparison with other theories, any advantages that would offset its disadvantages.

ARGUMENTS FOR THE NORMATIVE SIGNIFICANCE OF THE DISTINCTION

The foregoing discussion indicates the unlikelihood that a theory of health, disease, and enhancement will succeed in determining which health care services should be covered or what limits to medicalization should be observed. Other arguments begin with a normative principle determined independently of such theories and then attempt to show how the principle requires a distinction between therapy and enhancement, at least in some contexts.

Risk–Benefit, Just Allocation, Discrimination

Two common arguments for distinguishing between therapy and enhancement and for excluding the latter from the practice of medicine refer to the risks of enhancement technologies and the problems they pose for just allocation and discrimination. W. French Anderson (15) and Juan Manuel Torres (16) invoke both arguments to assert the normative force of the line between therapy and enhancement in gene transfer technology, though the arguments and their shortcomings are applicable in other areas of medicine as well. Anderson focuses on the different levels of medical risk involved, respectively, in adding a normal gene to overcome the effects of a nonfunctioning gene and adding a normal gene to increase the productivity of a gene functioning at a normal level, and on the relation of these risks to the likely benefits of each kind of intervention, concluding that only in cases of very serious disease do the benefits outweigh the risks. However, this is only a temporary rationale for a line between therapy and enhancement.

It is reasonable to expect that the enhancement of normally functioning genes will gradually become safe enough that the risk–benefit ratio will increasingly favor at least some uses of gene transfer for non-health-related enhancements, while some uses of gene transfer for the treatment of disease will have unfavorable risk–benefit ratios. In this regard gene therapy will likely resemble every other branch of medicine where lines drawn by risk–benefit ratios cut across the line between therapy and enhancement. Anderson also points to the important and difficult questions of who should benefit from enhancements and how to avoid potentially discriminatory uses of them, arguing that until we resolve these questions we should limit our genetic interventions to the treatment of serious diseases. However, while these are urgent

questions in nearly all areas of medicine, our inability to resolve them has not prevented us from intervening into less serious nongenetic conditions without a consensus on who should benefit from such interventions, or from carrying out other interventions and practices that could be accompanied by the discriminatory effects Anderson and Torres cite (i.e., pressure to undergo treatment or to adhere to eugenic goals, exacerbation of the gap between haves and have nots).

Genetic enhancement is neither a necessary nor a sufficient condition for moral problems of these kinds; such problems have occured in the past and present without genetic enhancements, and they need not occur, or even be exacerbated, because of genetic enhancements. It is true that gene transfer technologies could provide new and potentially dangerous occasions for such problems, resulting in discrimination of a much greater magnitude than at present. However, coerced treatment and eugenics would require extensive use of gene transfer in the population, which Torres himself considers highly unlikely, while the odds that gene transfer will expand the gap between haves and have nots more than a nonmedical enhancement such as education already does (with the enthusiastic complicity of many of those who argue against genetic enhancements on these grounds) is just as unlikely.

The Threat to Athletic Competition

Another context in which the distinction between therapy and enhancement is considered relevant is sports medicine. The issue, noted by Dan Brock (17), Eric Juengst (18), Thomas Murray (19), and others, is whether the use of sports medicine for purposes beyond the treatment of injuries destroys the very definition of the activity itself or the significance of the athlete's achievement. However, sports medicine appears to have developed to the point that the line is no longer drawn between therapy and enhancement but between different types and different levels of enhancement. For example, many nutritional supplements aimed at improving the performance of one or another somatic capacity are currently permitted in a number of sports. If effective, these supplements clearly enhance performance; they do not treat an injury or related condition. Of course, other nutritional supplements are banned. But usually this is either because they are deemed to be unsafe, or to have an unfavorable risk–benefit ratio, or because the level of advantage they offer is deemed to be unfair to other competitors or inconsistent with the meaning of the activity, not because they go beyond the treatment of injuries and related conditions. Clearly, then, the distinction between permitted and banned nutritional supplements occurs within the category of enhancements; few among those responsible for the oversight of these sports are advocating a ban on nutritional supplements altogether (if indeed the latter could even be consistently distinguished from ordinary dietary measures).

Fair Equality of Opportunity

Finally, the therapy-enhancement distinction is invoked to determine what medical services a just system of health care allocation is obligated to provide. Norman Daniels argues that because disease and disability significantly affect the opportunities open to individuals, health care has a central function for justice (20,21). He defines justice in Rawlsian terms of fair equality of opportunity, which requires that opportunity be equal for persons of similar skills and talents (although the resulting inequalities are mitigated by the difference principle in Rawls's theory, which holds that these inequalities must work to the advantage of the least well off) (22). Justice therefore must protect individuals against factors, such as race or sex discrimination and disease or disability, that restrict the range of opportunities that would otherwise be open to persons of similar skills and talents. Justice requires access to health care services that prevent, cure, palliate, or compensate for diseases and disabilities. However, fair equality of opportunity does not require provision of health care services for all conditions that create inequalities of opportunity but only for conditions of disability or disease (i.e., for pathologies).

Daniels's position turns on two points: the principle of fair equality of opportunity itself, and a rationale for including pathologies but not disadvantageous normal human traits in the set of conditions for which society is obligated to provide health care services. Daniels recognizes that his principle of fair equality of opportunity is vulnerable to those who point out that because both pathologies and skills and talents are due significantly, though not entirely, to the natural lottery, they should be treated identically so that either society is obligated to remedy restrictions of opportunity in both cases (social welfarists) or in neither case (libertarians). However, Daniels's arguments against these opponents are largely circular: he repeatedly appeals to certain actual beliefs and practices that reflect his intuitions to criticize other actual beliefs and practices that do not. Moreover, his theory lacks a convincing rationale for excluding disadvantageous traits from fair equality of opportunity. If justifiable, such an exclusion would serve two important purposes: it would keep health care expenses in check (as Daniels notes), and it would help preserve human diversity against the leveling effects of the quest for competitive advantage.

In presenting his case for this exclusion, Daniels quite plausibly draws on Boorse to distinguish pathologies from traits. However, it is not clear why this distinction should matter to fair equality of opportunity. Considering Allen and Fost's example, Daniels observes that short stature has roughly the same effect on equality of opportunity whether it is caused by human growth hormone deficiency (a pathology) or simply reflects human genetic variation. He notes further that in both cases it is equally the result of the natural lottery and equally the object of social prejudice.

Why then should this distinction count for purposes of justice? In defense of the distinction, Daniels simply reasserts his principle of fair equality of opportunity, which (he claims) recognizes from the outset that skills and talents and (he now adds) "other capabilities" are unequally distributed. Styling this view "the standard model" which (he alleges) reflects "our actual concerns" and "our consensus," Daniels in effect tries to salvage the

normative significance of the distinction between therapy and enhancement in these cases by pleading that the matter has already been settled in favor of the normativity of this distinction. However, even if this were in fact settled (which it is not), whether the settlement is defensible is precisely what the example calls into question.

Having appealed to an allegedly prevailing adherence to his rule of exclusion from basic coverage, Daniels goes on to undermine the credibility of his own adherence to it by arguing that if an inexpensive treatment for improving the cognitive capacities of children becomes available, there would be compelling reasons—he mentions the enhancement of education, the narrowing of the gap between children at the low end of the normal range and others, and the increase of social productivity—to seek enhancement in this way. Even if one ignores the unlikelihood that such a treatment would in fact realize the second benefit, one must wonder about a theory of fair equality of opportunity that invokes the therapy–enhancement distinction to treat as unequals two individuals who are equally the victims of the natural lottery and social discrimination while it violates that distinction to serve socioeconomic ends and advance the status of persons already in the normal range.

ARGUMENTS AGAINST THE NORMATIVE SIGNIFICANCE OF THE DISTINCTION

Critics such as Kathy Davis (23) and David Frankford (24) seek to show how the therapy–enhancement distinction could, if adopted in the formulation of health care policy, be used by insurers and bureaucrats to deny genuine health care needs on the grounds that they are enhancements rather than therapies. Neither Davis nor Frankford wishes to abandon the distinction, only to ensure that it is applied with context-sensitive judgment. Frankford questions whether such context sensitivity is possible in the policy arena, while the examples Davis draws from the Dutch health care system point to the same conclusion despite her hopes to the contrary.

CONCLUSION

The result of this critical survey is that distinguishing between therapy and enhancement is easier than it is often assumed to be while articulating and defending any normative force for this distinction is more difficult than it is often assumed to be. The implication is that the therapy–enhancement distinction has little if any relevance for determining basic health insurance coverage or the limits of medicalization. With regard to insurance coverage, some procedures that are technically enhancements according to current medical science (e.g., treatment of osteoporosis for postmenopausal women) are almost certainly of higher priority in many cases than some procedures that treat diseases. Fortunately, while the therapy–enhancement distinction does not help in this or in many other cases, it is possible to arrive at a rough agreement on which conditions seriously inhibit almost any way of life, and thus should, in principle, be covered in a basic plan while permitting trade-offs when there are reasonable disagreements even at this level. If one's theory of justice does not permit all of these conditions to be covered, one may be forced to prioritize these conditions, after the example of the Oregon Medicaid program. If one's theory of justice requires funding basic coverage beyond these conditions, vouchers would enable individuals or groups to determine the composition of this additional coverage in accordance with their views about the relation of various somatic conditions to valued activities or ways of life. Medicalization and the normalization that accompanies it present a more difficult problem. To determine the appropriate limits of existing and emerging technologies will require a view of the ethical significance of the body and its capacities and limitations, and of the place (if any) of the discourses and practices of biomedicine in realizing these capacities and responding to these limitations (25).

BIBLIOGRAPHY

1. E. Parens, ed., *Enhancing Human Traits: Ethical and Social Implications*, Georgetown University Press, Washington, DC, 1998.
2. C. Boorse, in D. VanDeVeer and T. Regan, eds., *Health Care Ethics: An Introduction*, Temple University Press, Philadelphia, 1987, pp. 359–393.
3. C. Boorse, in J.M. Humber and R.F. Almeder, eds., *What Is Disease?* Humana Press, Totowa, NJ, 1997, pp. 1–134.
4. D. Allen and N. Fost, *J. Pediatrics* 117, 16–21 (1990).
5. J.E. Sabin and N. Daniels, *Hastings Center Rep.* 24(6), 5–13 (1994).
6. J.M. Torres, *J. Med. Philos.* 22, 43–53 (1997).
7. L. Walters and J.G. Palmer, *The Ethics of Human Gene Therapy*, Oxford University Press, New York, 1997.
8. C. Boorse, *Philos. Sci.* 44, 542–573 (1977).
9. L. Kass, *Toward a More Natural Science*, Free Press, New York, 1985.
10. H.T. Engelhardt, Jr., *Foundations of Bioethics*, 2nd ed., Oxford University Press, New York, 1996.
11. C. Boorse, in J.M. Humber and R.F. Almeder, eds., *What Is Disease?* Humana Press, Totowa, NJ, 1997, pp. 1–134.
12. K.D. Clouser, C.M. Culver, and B. Gert, in J.M. Humber and R.F. Almeder, eds., *What Is Disease?* Humana Press, Totowa, NJ, 1997, pp. 173–217.
13. B. Gert, *Morality: A New Justification of the Moral Rules*, Oxford University Press, New York, 1988.
14. M. Martin, *J. Med. Philos.* 10, 329–337 (1985).
15. W.F. Anderson, *J. Med. Philos.* 14, 681–693 (1989).
16. J.M. Torres, *J. Med. Philos.* 22, 43–53 (1997).
17. D. Brock, in E. Parens, ed., *Enhancing Human Traits: Ethical and Social Implications*, Georgetown University Press, Washington, DC, 1998, pp. 48–69.
18. E. Juengst, in E. Parens, ed., *Enhancing Human Traits: Ethical and Social Implications*, Georgetown University Press, Washington, DC, 1998, pp. 48–69.
19. T.H. Murray, in T.H. Murray, W. Gaylin, and R. Macklin, eds., *Feeling Good and Doing Better: Ethics and Nontherapeutic Drug Use*, Humana Press, Clifton, NJ, 1984.
20. N. Daniels, *Just Health Care*, Cambridge University Press, New York, 1985.
21. N. Daniels, in T.F. Murphy and M.A. Lappe, eds., *Justice and the Human Genome Project*, University of California Press, Berkeley, CA, 1994, pp. 110–132.
22. J. Rawls, *A Theory of Justice*, Harvard University Press, Cambridge, MA, 1971.
23. K. Davis, in E. Parens, ed., *Enhancing Human Traits: Ethical and Social Implications*, Georgetown University Press, Washington, DC, 1998, pp. 124–134.
24. D. Frankford, in E. Parens, ed., *Enhancing Human Traits: Ethical and Social Implications*, Georgetown University Press, Washington, DC, 1998, pp. 70–94.
25. G.P. McKenny, *To Relieve the Human Condition: Bioethics, Technology, and the Body*, State University of New York Press, Albany, NY, 1997.

See other entries BEHAVIORAL GENETICS, HUMAN; see also HUMAN ENHANCEMENT USES OF BIOTECHNOLOGY entries.

HUMAN ENHANCEMENT USES OF BIOTECHNOLOGY, LAW, GENETIC ENHANCEMENT, AND THE REGULATION OF ACQUIRED GENETIC ADVANTAGES

MAXWELL J. MEHLMAN
Case Western Reserve University
Cleveland, Ohio

OUTLINE

INTRODUCTION

Genetic enhancement, whether in the form of somatic enhancement for adults or for children, genetic selection for enhancement, or germ cell enhancement, may give the recipients significant social advantages. It is impossible at this time to be certain which traits will prove susceptible to genetic enhancement, but they may include physical traits, such as beauty, stature, strength, and stamina, personality characteristics such as charm, cheerfulness, charisma, confidence, and energy, and mental capabilities, including memory, intelligence, and creativity.

These improvements obviously will be in great demand. But how widely available will the technologies be that make them possible? Some genetically engineered drugs that produce somatic enhancements may be relatively affordable. Others may not. Genetic selection for enhancement in which fetuses were tested in utero might not add much to the cost of performing genetic tests to detect abnormalities or disease (1). This might make it, and the accompanying abortions, the "poor person's" genetic enhancement technique. But any enhancements performed on embryos would be expensive, since they would include the costs of in vitro fertilization (IVF). Currently IVF costs average $25,000, and there would be added costs of the genetic manipulations, which are likely to be substantially greater, particularly when the technology is first introduced.

Genetic enhancement is not likely to be paid for by any public or private health insurance program or policy. This is evident from the lack of third-party payment for cosmetic medicine, which is perhaps the most analogous biomedical technology currently available. The legislation governing the Medicare program contains a general prohibition against paying for "items or services ... which are not reasonable and necessary for the treatment of illness, or to improve the functioning of a malformed body part" (2), and includes a specific exclusion for "cosmetic surgery" (3). States have adopted the same coverage exceptions under their Medicaid programs (4). Private health insurance plans also do not cover cosmetic medicine; the language in the author's high-option Blue Cross policy is typical: "Coverage is not provided for services and supplies ... for surgery and other services primarily to improve appearance or to treat a mental or emotional condition through a change in body form. ..."

Even if the government wanted to provide general access to genetic enhancement, the cost would prove prohibitive. Widespread access to enhancements such as preimplantation selection or manipulation that depends on access to IVF currently would amount to $120 billion per year for the IVF services alone (5). Somatic enhancement would not be cheaper. A single somatic enhancement in the form of a substance like human growth hormone, which currently costs about $30,000 per child per year (5), would amount to $22 billion a year just to provide to the 1.7 million children who were in the lowest 3 percent of the population in terms of height (6). The figure for multiple somatic enhancements over a number of years for the entire population would be astronomical.

The high cost and the lack of coverage by third-party payment plans, of course, does not mean that no one will access genetic enhancement but only that it will be limited to persons who can purchase enhancement with their own assets. This gives rise to two related problems. The first is inequality, which is also discussed in another entry; the second is unfairness. The issue of fairness would arise at the micro level if genetically enhanced individuals compete for scarce resources against, or find themselves in a conflict of interest with, those who are unenhanced.

Genetic enhancement could confer a decisive advantage in competitive circumstances. How should society respond to the potential unfairness?

SPECIAL NATURE OF GENETIC ENHANCEMENTS

In a society whose members believe in the possibility of upward social mobility, people seek to better themselves and their children. They educate themselves and try to obtain the best education they can for their children. They may try to marry "upward," hoping for a mate who will increase their opportunities, social standing, and wealth. They push themselves and their children to cultivate and make the utmost use of their talents.

Many of these efforts take the form of medical or pharmaceutical interventions. People take drugs to improve their athletic and cognitive performance. They subject themselves to surgery to improve their appearance. Some of these activities, such as selecting one's mate, have at least an indirect influence on the genetic makeup of succeeding generations.

Against this background of current enhancement practices, what is so exceptional about genetic enhancement? Society has had plenty of experience coping with the social implications of efforts at self-improvement. While society's response has not always been adequate or successful—witness the difficulties in trying to control the use of performance-enhancing drugs in sports—will the problems created by genetic enhancement be so different that they require special attention?

Of course, even if we felt that wealth-based access to genetic enhancement did not constitute a new kind of threat to social equality, society might still need to respond to it in a vigorous fashion. The additional inequality arising from genetic enhancement, when added to existing sources of inequality, might tip the scales in favor of social unrest. At least, we might well want to monitor the situation closely, and stand prepared to respond if necessary.

Yet in a number of important ways, genetic enhancement does differ from previous sources of inequality and unfairness. Taken together, these differences justify a significantly heightened level of concern, if not outright alarm:

1. The probable high cost of genetic enhancement means that fewer individuals will gain access to it compared to those who can avail themselves of other forms of self-improvement. Twenty million Americans are members of commercial health and sports clubs (7). In 1997, 480,588 persons obtained cosmetic surgery. In contrast, only 39,390 per year obtain IVF (8), and even fewer would be able to afford the additional cost of preimplantation genetic enhancement. Somatic enhancement might be cheaper, but it might still be beyond the reach of many who wished to enhance multiple traits or to produce long-term results.

2. The effects achievable with genetic enhancement are likely to affect some traits that are not highly susceptible to current forms of self-improvement, including

some that are fundamental to personal success. Current self-improvements are limited in scope. One can change one's weight (although usually not permanently); employ cosmetics and cosmetic medicine to improve appearance within certain limits; somewhat increase the ability to cope with loss, failure, and stress; build muscles; develop greater physical, mental, and social skills; and increase reading speed. Genetic enhancement, however, may improve intelligence, cognition, charisma, creativity, energy, cheerfulness, sense of humor, and other characteristics that are arguably central to success and well-being.

3. Persons who are fortunate enough to be able to gain access to genetic enhancements are likely to obtain a much greater and long-lasting advantage than those who employ more traditional forms of self-improvement. Performance-enhancing drugs in sports produce their effects on the basic human phenotype, and the effects, while perhaps enough to win competitions, are relatively modest (9). Although cosmetic interventions change the appearance, they rarely stray from "normal" ranges for physical traits, and with the exception of cosmetic surgery, are often transitory, as any dieter knows. To date, techniques for improving memory and other cognitive functioning do not appear able to significantly increase intelligence or to have a particular profound or permanent effect (10,11). But there is no telling how powerful genetic enhancement can be. It could stretch the limits of desirable human traits considerably, perhaps even indefinitely. For example, there may be no such thing as being "too intelligent." Moreover, enhanced persons can still employ traditional forms of self-improvement on top of their genetically enhanced starting point.

4. Current self-improvement techniques tend to affect at most only a few aspects of performance or appearance at one time. In most cases, people work on one trait — for example, their facial appearance, their weight, their ability to solve puzzles, or memorize facts. Cosmetic polymedicine, while not unknown, is rare (12). Even a professional athlete in full training mode can do no more than exercise and take performance-enhancing drugs to increase strength and stamina, hire a famous coach and perhaps a sports psychologist, and repeatedly practice a skill or routine. Genetic enhancement, on the other hand, may permit wholesale changes in characteristics. Parents with sufficient resources may engineer numerous improvements in their children, and they may purchase multiple somatic enhancements for themselves or their dependents (13).

5. As the result of the ability of genetic enhancement to alter multiple traits in significant ways, genetic enhancements may give people decisive advantages or major success not just in one or two spheres of social activity but in a broad range of social endeavors. This may enable them to cross what Michael Walzer calls "spheres of distributive justice." Imagine the following individual, he says:

Here is a person whom we have freely chosen (without reference to his family ties or his wealth) as our political representative. He is also a bold and inventive entrepreneur. When he was younger, he studied science, scored amazingly high grades in every exam, and made important discoveries. In war, he is surpassingly brave and wins the highest honors. Himself compassionate and compelling, he is loved by all who know him (14).

If genetic enhancement made such a person possible, he and his kind would be likely to dominate the rest of society.

6. Finally, unlike most advantages derived from self-improvement, some genetic enhancements — those achieved through genetic selection for enhancement or germ-line engineering — will be incorporated into the genetic makeup of future generations (15). Both the genetic enhancements and the societal advantages that they confer will be inherited, and those who obtain them will comprise a special class within society. Although initially defined by its wealth, this class eventually will come to be characterized by its superior genetic endowment.

In short, genetic enhancement possesses a number of characteristics that raise special concerns for society (16). Some of the objections, like playing God, are metaphysical. Others concern the serious practical consequences for the individual and for society. The rest of this article will concentrate on these two consequences, beginning with the implications for social equality.

THREAT TO EQUALITY FROM GENETIC ENHANCEMENT

Is it fair for some people to have greater genetic advantages than others? This is a question that forms the crux of the age-old problem of "natural inequality" which has plagued philosophers and social theorists at least since the ancient Greeks. If genetic enhancement is unfair, then presumably society should do what it can to rectify the situation, and this engenders the secondary, but equally vexing, problem of what form societal intervention should take and how feasible it would be.

Some philosophers tolerate natural inequality more than others. Meritocrats, for example, welcome substantial inequalities resulting from the distribution of natural talents, pointing to the benefits that accrue to society from the accomplishments of the gifted. John Gardner, for example, objects to what he calls "extreme equalitarianism," which, he states, "ignores differences in native capacity and achievement and eliminates incentives to individuals." In Gardner's opinion, this signifies "the end of that striving for excellence that has produced history's greatest achievements" (17). Others relish excellence as much for its own sake as for what it can achieve. According to Thomas Nagel, "[a] society should try to foster the creation and preservation of what is best, or as good as it possibly can be. . . . Such an aim can be pursued only by recognizing and exploiting the natural inequalities between persons, encouraging specialization and distinction of levels in education, and accepting the variation in accomplishment which results" (18). Robert Nozick even disputes the idea that the fact that natural assets are arbitrarily distributed in society means that they are not deserved (19).

Philosophers who are morally troubled by inequality, on the other hand, tend to regard it as unjust for some individuals to benefit by virtue of their genetic endowment compared to others who do not fare as well in the genetic lottery (20). These philosophers generally agree that unchosen and unearned advantages and disadvantages must be minimized in order to achieve a more just society. As Rawls states: "It seems to be one of the fixed points of our considered judgments that no one deserves his place in the distribution of native endowments, any more than one deserves one's initial starting place in society" (21; also see Ref. 22).

Although liberal philosophers agree on the goal of rectifying the injustices of the natural lottery, they disagree substantially on how this should be achieved. A basic dispute, for example, concerns just what is to be equalized: "welfare" — that is, some subjective measure of well-being — or "resources" (23,24). Another contentious issue is how much inequality society should tolerate, whether of welfare or resources, in order to assure the production of desired goods. For example, meritocratic, libertarian, and free-market theorists all justify their tolerance for inequality at least in part on the ground that permitting people to profit from the exercise of their natural talents is necessary to induce them to increase the total sum of societal goods. In reaction at least in part to these and other difficulties, many philosophers abandon the quest for absolute equality, whether of resources or welfare, in favor of providing everyone with a minimum level of assets or of well-being, or with "equality of opportunity" (18,25,26).

Despite the gaps and imprecisions in the theory of equality and its application, it does provide at least one clear imperative in regard to wealth-based genetic enhancement: If genetic enhancements are viewed as natural assets, distributed largely by chance, then the principle of equality requires that society attempt to rectify the advantages that they confer, except to the extent that it may be necessary to allow individuals to profit to some degree from their natural talents in order to secure benefits for society as a whole. On the other hand, if genetic enhancements are viewed as "earned" advantages, obtained through diligence and effort, then at least some theories of equality would permit the enhanced individual to retain the additional value created by enhancement.

The easiest case for saying that genetic enhancements were unearned is when they were obtained by children from their parents or other family members. The children have done nothing to entitle them to such advantages; from a moral standpoint, their enhancement is no more than the luck of the draw (27). The interest of equality therefore would seem to justify depriving enhanced individuals of the benefit of enhancements that were installed by their parents or that were purchased with inherited or unearned wealth. Another easy case would be when persons acquired the money to purchase enhancements by immoral means; they too have no moral claim to the benefits.

But what about the person who obtains the necessary funds by dint of the sweat of her brow, without exploiting others or behaving otherwise immorally? This person has a strong claim to be allowed to retain the benefit from her genetic enhancements as morally deserved. Similarly a parent who earned the wherewithal to purchase genetic enhancements for her children in morally acceptable ways may contend that her children ought to be entitled to enjoy the benefits (28).

If we feel obliged to level the genetic playing field even when the advantages of genetic enhancements have been acquired in morally deserving ways, we might base our action on the view that equality is a moral imperative that overrides desert. This rationale no doubt to some extent lies behind schemes that redistribute earned wealth, like progressive income taxation. But the techniques that would be required to level the genetic playing field, as we will see, are far more intrusive than progressive taxation — even when accompanied by aggressive government enforcement. If these methods are to be politically acceptable, they must be premised on more than an abstract belief in the value of equality. They must be based as well on the conviction that genetic enhancement, if left unchecked, would be a grave threat to society. Can this view be sustained? I believe it can.

One of the most important of our societal goals is maintaining a liberal democratic form of government. This goal is directly threatened by wealth-based genetic enhancement: The inequality of social opportunity that results may be so great that a liberal democratic form of government becomes unsustainable, and our political system instead becomes autocratic or oligarchic. This follows from the assumption that a minimum degree of equality is necessary for modern liberal democracy to exist (29). If social inequality becomes too pronounced, liberal democratic political systems become unstable. As one sociologist states:

> Inequality in the distribution of rewards is always a potential source of political and social instability. Because upper, relatively advantaged strata are generally fewer in number than disadvantaged lower strata, the former are faced with crucial problems of social control over the latter. One way of approaching this issue is to ask not why the disprivileged rebel against the privileged but why they do not rebel more often than they do (30).

The characteristics of genetic enhancement that threaten to destabilize liberal democratic government are the features mentioned earlier that distinguish genetic enhancement from other forms of self-improvement: its high cost which may place it beyond the reach of all but the very wealthy, the broad and fundamental nature of the traits that it could enhance, the magnitude of its effects; their multiplicity, the resulting ability to gain advantages in multiple spheres of social activity, and the possibility created by germ-line enhancement that these advantages would be passed on to successive generations.

These characteristics not only give rise to social inequality; more insidiously, they undermine the belief in equality of opportunity. A widespread belief in equality of opportunity is the principal manner in which liberal democracies accommodate the reality of inequality—that everyone is not equally endowed with natural assets, nor with the same luck or disposition to work hard. In the United States, although most people will tolerate large in equalities in the distribution of resources, the main principle is that everyone have an equal opportunity to these resources (31). As John Shaar notes, the belief in equal opportunity is instrumental in maintaining the prevailing social order:

> No policy formula is better designed to fortify the dominant institutions, values, and ends of the American social order than the formula of equality of opportunity, for it offers *everyone* a fair and equal chance to find a place within that order (32).

Genetic enhancement would create such profound, true differences in ability that they would endow the wealthy with opportunities that are irrevocably beyond the reach of the less fortunate. We have from history other societies with population characteristics similar to those that would be created by wealth-based genetic enhancement. In medieval Europe, individuals were born into their respective classes. Only in rare exceptions were peasants able to obtain education in religious institutions or become apprenticed to a trade, so, they remained found to their station in life (33). In slave-owning societies, people were born into bondage and could be freed only by escape (self-exile) or at the pleasure of their masters. In contempory India the caste system is an example of such a society, and it is a constant threat to the nation's democratic institutions (34).

In short, wealth-based access to genetic enhancement creates not only a moral challenge but a political threat. From a moral standpoint, those who gain enhancement may not have done anything to deserve it. Adults may have come by the means necessary to purchase enhancement in objectionable or morally irrelevant ways—through exploitation or the brute luck of inheritance. Children who are enhanced by their parents are unlikely to have done anything to earn it; this is patent in the case of more remote generations enhanced through prior germ cell manipulations. Yet genetic enhancement poses more than an ethical quandary. Even if the price of genetic enhancement had been earned in a moral sense, the social impact of wealth-based enhancement is likely to be severe. Somatic enhancement alone could so dramatically widen the gulf between have's and have-not's that class warfare would ensue, and the conflict could topple democratic government. Germ-line enhancement could create, quite literally, a master race. The question is whether there is any practical way to prevent this.

PROMOTING GENETIC EQUALITY

In the face of the serious threats to equality represented by wealth-based genetic enhancement, what options do we have to promote equality? One approach would be to "level up." An obvious example is to give everyone access to genetic enhancement regardless of wealth. As pointed out earlier, however, this would be prohibitively expensive.

Even if we decided to divert some enormous portion of the gross national product (GNP) to finance a massive enhancement entitlement program, what enhancement services would such an entitlement program provide? This raises the old argument over resource versus welfare equality. If the objective were to give everyone an equal amount of enhancement resources, those who started out with a more favorable distribution of natural assets would end up better off. If instead we attempted to give everyone an equal or minimum share of enhancement, or an equal or minimum share of enhancement-created opportunity, how would we measure equivalence? Would an extra inch of height be equal to an extra ten points of IQ? The problem would not be solved if we gave everyone an equal share of money and allowed them to purchase whatever enhancements they desired. Logically, unless all enhancements cost the same, those who desired expensive enhancements would be less advantaged than those who were content with cheaper ones—again, the problem of expensive tastes (32). Moreover, since we could not afford to provide everyone with access to the same enhancements that the wealthy could purchase, the wealthy always could stay ahead of the rest of the population. Now we could solve the problem if we gave everyone the maximum amount of enhancements available, but then again we would run into the problem of prohibitive cost. Such problems would plague any attempt to give the unenhanced some countervailing benefit other than genetic enhancements, like money, information, or political power which would level the playing field.

The fact that some people start out with a more favorable distribution of natural assets suggests another approach: subsidize access to enhancements, not for everyone, but for those who were genetically disadvantaged. In other words, bring everyone up to the same level of genetic well-being. This would comport with Rawls's difference principle by improving the fortunes of the least well-off (21). But it runs into the same problems that were just described in attempting to equalize access to enhancements. In addition, allocating enhancements to the genetically disadvantaged would necessitate identifying genetically disadvantaged individuals or groups within the population, and measuring their degree of disadvantage. This would raise serious practical, moral, and political objections. Determining what counts as a genetic disadvantage is similar to trying to identify whether or not someone has a disability—a determination that is controversial, and often appears arbitrary (35–38). Measuring the extent of disability is an even thornier enterprise: Witness the morass that the state of Oregon found itself in when it tried to ration Medicaid services on the basis in part of how much they alleviated disability. Even if we could identify and quantify genetic disadvantage, we would need to establish a "normal" degree of genetic well-being state that the disadvantaged could attain, so we could give them the correct amount of enhancements or money with which to purchase enhancements. But "normalcy," as noted earlier, is highly arbitrary, value laden, and subjective (11). Furthermore it can become a constantly moving target as the distribution of advantages and disadvantages within the population shifts and as the average level of advantage increases with the number of people becoming enhanced. Finally, any attempt by the government to identify and rectify genetic disadvantage smacks of eugenics, which is politically suspect if not unthinkable.

If leveling up is not a feasible response to genetic enhancement, the alternative is to level down (14). The most straightforward approach would be to prevent anyone from obtaining genetic enhancement. A ban on genetic enhancements could be aimed at a variety of targets. For example, purchasing or possessing enhancements could be made illegal, similar to laws punishing illegal drug use or rules prohibiting the use of performance-enhancing drugs in sports. Another target would be health care professionals and institutional providers such as hospitals and IVF clinics. Congress or state legislators could make it a crime for health care professionals to provide genetic enhancements. Violators would face disciplinary actions by state medical boards, including loss or suspension of their licenses (39). Hospitals and other facilities like IVF clinics that continue to offer enhancement services would lose their licenses, their accreditation, or their ability to receive reimbursements under Medicare and Medicaid. Finally, if genetic enhancements were proprietary products such as drugs or medical devices, the Food and Drug Administration (FDA) could deny marketing approval.

But why wait until enhancements are available before banning their use? Why not prohibit research aimed at developing enhancement technologies in the first place? An analogy is the federal government's ban on federal funding of research on embryos and fetuses (40). Privately funded research could be restricted by penalizing institutions such as hospitals that participate in clinical trials, and by FDA denying permission to ship experimental enhancement products across state lines for purposes of human testing (41).

All of these restrictive approaches have limitations, however. Penalizing people who genetically enhanced their children would trigger intense constitutional debate. Particularly in the case of passive enhancements involving traditional "coital" methods of reproduction, the Supreme Court is likely to apply a strict scrutiny standard under which the right to decide what type of child to conceive or bring to term can be overridden only by a compelling state interest, and then only if the state uses the least intrusive means of regulation. Genetic enhancement accompanying less traditional methods of reproduction, such as IVF, may be entitled to less constitutional protection. However, even then, the courts are likely to take a hard look at overly intrusive government regulation. Somatic self-enhancement, while not raising issues of reproductive freedom, would set the state's interest in promoting equality against the individual's constitutionally protected interest in personal liberty and autonomy, including the right to make life-style decisions that do not harm others.

Those FDA restrictions on the sale of enhancement drug products, biologics, or devices able to survive constitutional challenge as an appropriate regulation of interstate commerce would be hampered by the way in which they are likely to become commercially available: as unapproved or "off-label" uses of products approved for therapeutic rather than enhancement purposes. A genetically engineered drug that enhances cognition, for example, could be approved to treat cognitive impairment, such as the effects of Alzheimer's disease. After it is approved for a therapeutic purpose, people might begin to seek it for unapproved enhancement purposes. The experience with human growth hormone mentioned earlier is a prime example. This genetically engineered drug is approved for use in children with "a lack of adequate endogenous growth hormone secretion," causing short stature (42). Yet parents are reported to be asking doctors to prescribe it for children who are merely short, and there are anecdotal accounts of parents requesting the drug for children who are already tall, in order to enhance their chances of playing competitive basketball (43).

FDA does not effectively regulate off-label uses of unapproved drugs. It merely limits the ways in which the manufacture may promote the drug for unapproved uses. Even if the FDA attempted to prohibit manufacturers altogether from promoting drugs for an unapproved use, enhancement uses would become public knowledge through media reports, the Internet, and word of mouth. Targeting health professionals who provided enhancement products to their patients would present similar obstacles. The FDA presently has no authority to control the prescribing behavior of physicians, who are free to prescribe products for uses which are not approved (44). There is nothing unlawful about a physician prescribing human growth hormone for children for an enhancement purpose which is not indicated on the product labeling. The only effective action the agency can take now is to ban or limit sales of the product altogether — for both therapeutic and enhancement uses. Yet in the case of products approved to treat serious and especially popularized diseases, this would carry an intolerably high political price.

The same problem would beset efforts to prevent research on genetic enhancements from taking place. Consider a ban on research on genetically engineered drugs to enhance cognitive function. Such a ban would be justified, it might be argued, on the ground that developing such a product would give those who used them an unfair advantage in competitions for scarce resources like college acceptances or aptitude-based job slots. But these same products most likely would be useful in treating diseases of cognitive deficiency, such as Alzheimer's and dementia. It is extremely difficult to curtail research on a specific use of a product. In any event there is little point, since, as stated above, once the product is developed for therapeutic use, it can easily migrate to enhancement uses.

Moreover an effective ban on access to genetic enhancements, whether aimed at individuals obtaining them for themselves or their children, or at providers and manufacturers, would require an elaborate enforcement regime. The analogies that best describe what would be necessary are programs to control the use of performance enhancing drugs in sports and the use of illicit recreational drugs. Indeed, the most appropriate government agency for regulating genetic enhancements may not be FDA but rather the Drug Enforcement Agency (DEA). After all, DEA, pursuant to the Controlled Substances Act, is responsible for enforcing restrictions on access to physiologically active products stemming from societal objections to their use.

Like the war on drugs and the effort to ban drugs in sports, restricting access to genetic enhancements to promote equality is bound to be extremely intrusive and expensive. These precedents were not completely effective. Somatic enhancements in the form of drugs, although perhaps complicated to manufacture, may be easy to conceal. Even enhancements that depended on sophisticated medical procedures such as IVF might be procured if one "knew the right person," the way "back-alley" abortions could be obtained prior to *Roe v. Wade* (45). The overwhelming consumer demand for genetic enhancements is certain to spawn a robust black market. As the experience with abortions indicates, people who are prevented from obtaining genetic enhancements domestically simply will procure them abroad (46).

The most troublesome aspect of enforcing a ban on genetic enhancements is likely to be the difficulty of determining that someone has been illegally enhanced so that they, and/or the person who enhanced them, can be punished. In part, this is a technical problem of detecting the presence of enhancement products or enhanced DNA in the human body. A similar problem plagues attempts to ban performance-enhancing drugs in sports. Athletes and their coaches are becoming increasingly adept at deceiving drug-screening tests. The athletes may use substances such as erythropoeitin that are naturally occurring in the body so that the exogenous enhancement cannot be chemically distinguished (47). Furthermore the athlete may be able to use an enhancement substance to produce a benefit, such as increased muscle mass, and then stop taking the substance sufficiently in advance of a screening test so that its use cannot be detected.

In the case of genetic enhancements, the enforcement problem would be compounded by the difficulty of distinguishing between therapeutic and enhancement uses. As noted earlier, the difference between the two often is not clear. Someone could claim, for example, that an improvement in appearance was necessary to treat feelings of inadequacy, or that an increase in strength or dexterity was preventive therapy in that it enabled them to avoid injury. Furthermore, as has been noted, many genetic enhancements have lawful medical uses. Someone could take human growth hormone in an attempt to become tall enough to play professional basketball, but someone else could take the exact same substance to combat pituitary dwarfism. A ban on enhancements would require a complicated system for distinguishing between legitimate and prohibited activity involving the same products. In addition, banning genetic enhancement in conjunction with assisted reproductive technologies such as those delivered in IVF clinics would require a far more effective scheme of regulatory regime than is currently in place (48).

Yet the strongest objection to banning genetic enhancements has not been mentioned so far: Enhanced individuals not only may personally benefit from their advantages, but they may confer advantages on society. For example, a person whose science ability was enhanced (assuming that this collection of traits was amenable to genetic manipulation) might make important discoveries that would

be impossible, or take much longer, for an unenhanced scientist. As noted earlier, even proponents of equality recognize the need to permit a certain degree of inequality in order to increase social benefit. In short, we might want to permit an individual to be enhanced if we expected the ratio of societal to personal benefit to be favorable enough.

Together with the practical limits on the effectiveness of a complete ban on genetic enhancements discussed earlier, the social value of certain kinds of enhanced performance make the goal of a total ban both unrealistic and undesirable. Some people will manage to enhance themselves no matter what it takes, and in some cases we will want people to do so. This leads to several policy suggestions: (*1*) enhancement licensing, (*2*) establising an enhancement lottery, and (*3*) regulating germ-line enhancement.

Enhancement Licensing

In order to permit genetic enhancements to produce desirable social gains, as well as to take some of the pressure off of a regulatory embargo that attempted to prevent the wealthy from purchasing enhancements, we should institute a system for licensing individuals to obtain genetic enhancements on the condition that they employ their enhanced abilities in some predefined manner to benefit society. By reducing the number of people who were enhanced, a licensing program would reduce the degree of social inequality, and the threat that genetic inequality poses to democratic institutions.

The system would be similar to legally enforced professional licensing schemes that give their holders powers and privileges denied ordinary citizens in return for agreements to abide by rules designed to promote social goals and to refrain from behaving in socially undesirable ways. The system also would bear some resemblance to licensing ownership or use of dangerous products such as handguns or automobiles. The administrative costs could be financed by licensing fees.

Such a licensing requirement could be enforced in the first instance against providers of genetic enhancement products or services. They themselves would be required to be licensed as a supplier, which would carry with it restrictions and reporting requirements. (A similar program operates under the Controlled Substances Act to keep track of the prescribing of narcotics and other dangerous drugs.) Individuals who seek to purchase enhancements would apply to a licensing board and would be required to propose the socially desirable purposes for which they seek to be enhanced. Those whose applications are approved would report to the board periodically to provide assurance of satisfactory performance, and their reports would be carefully audited. Licensed enhancements that involved manipulation of DNA would be genetically "tagged" so that lawfully enhanced individuals could be distinguished from those who obtained enhancements on the black market (49,50). Failure to fulfill the terms of the license would be penalized by loss of access to the enhancement or to its benefits. Depending on the nature of the enhancement, the penalty could take the form of being deprived of supplies of the enhancement product, actual biological reversal of the enhancement, various forms of social handicapping, surtaxes or monetary penalties, and perhaps in cases of egregious violations, such as the use of enhancements to cause serious harm to others, imprisonment. Similar penalties would be imposed on persons who were discovered to have supplied or obtained enhancements without being licensed.

Establishing an Enhancement Lottery

The licensing scheme so far described would be open only to the wealthy, since they would be the only persons who could afford to purchase genetic enhancements. This would perpetuate the inequalities described earlier that would result from wealth-based access to enhancements described earlier. The solution would seem to be to provide some people with access to enhancements even if they could not afford it. One approach would be a government program that subsidizes enhancements for certain persons, perhaps those who, in return for their license, promise to provide the most desirable set of social benefits. But this would embroil the government in an enhancement-rationing program in which it was required to judge the relative merit of different proposals, a task that would raise objections similar to those that have been lodged in the past against health care rationing programs in general (51). On the other hand, such a licensing plan does not raise similar objections because the licensing authority would not compare individuals seeking enhancement but would allow anyone to purchase enhancements so long as they agreed to meet certain minimum social objectives.

A better solution than a rationing program would be to establish a national lottery for genetic enhancements (46). Everyone would be given one chance in each drawing. The winner or winners would be entitled to resources sufficient to enable them to purchase the maximum package of enhancements lawfully available in the private market, although in order to "cash in" their winnings, they would have to become licensed like everyone else who was enhanced. Like the licensing program itself, the lottery could be financed by license fees paid by those who purchased enhancements. Among the advantages of a lottery approach is that its randomness would give continued vitality to the concept of equality of opportunity.

Regulating Germ Line Enhancement

The greatest threat to social equality posed by genetic enhancements is the formation of a genobility — a class of related individuals who achieve and maintain an unassailable grip on wealth, power, and social privilege and who pass their advantages on to successive generations. As discussed earlier, a genetic aristocracy of this sort is antithetical to liberal democratic political systems. If genetic enhancements are obtainable at all, then to some extent the formation of an enhanced class cannot be prevented; persons who were wealthy enough to purchase enhancements presumably would be able to provide their children with greater material advantages than persons who were not enhanced, thereby making it more likely that these

children would be able to purchase genetic enhancements in their turn. Yet the formation of such a genobility is far more likely to occur if individuals were permitted to make enhancement changes in their germ lines. Their offspring would inherit these genetic advantages, which they would be able to supplement with additional germ-line enhancements that they purchased, which in turn would be passed on to their children, and so on (52).

The social threat created by the inequality that would result from germ-line enhancement may not readily be mitigated by the licensing requirement that would accompany the lawful acquisition of somatic enhancements. It is difficult to imagine how to ensure that a person's children would abide by the licensing conditions agreed to by their parents. The children could be required to become licensed in their turn (e.g., when they reached the age of majority) on penalty of forfeiting their enhancement advantages, but despite the stipulation that the enhanced individual devote some degree of his or her enhanced capabilities to the public good, being licensed at the age of majority may not be sufficient to counteract the inequality that germ-line enhancements would produce.

The solution then would seem to be to prohibit germ-line genetic enhancement altogether. Conceivably the threat in social equality could be met by banning only those forms of germ-line enhancement involving gene transfer, and not the passive sorts of germ-line enhancement that would occur with genetic selection for enhancement, selective abortion for enhancement, or preconception enhancement. Moreover laws that make it illegal for individuals for enhancement reasons to discover their genetic endowment, to select embryos for implantation, or to abort a fetus, might be more realistic politically than laws that prohibit the alteration of germ cells for enhancement purposes.

A ban on germ-line engineering would raise a host of problems. It might be challenged as an unconstitutional interference with procreative liberty, although the justification that it was necessary in order to preserve democratic liberties from being engulfed by a genetic aristocracy might be deemed a compelling state interest. Detecting when someone had altered germ cells would be difficult and intrusive (53). Nevertheless, a ban may be necessary to promote a minimum level of genetic equality.

UNFAIRNESS

Regardless of the manner in which we attempt to reduce the inequalities that may be created by wealth-based genetic enhancement, some people invariably will become enhanced. A licensing scheme that is vigorously and effectively enforced will go some distance toward offsetting the advantages enjoyed by enhanced persons but not far enough. Enhanced individuals still will be in a superior position compared to unenhanced persons. This raises the question of whether and in what ways society should respond in order to reduce the resulting unfairness.

This unfairness will be felt most acutely when the unenhanced compete with the enhanced for scarce societal resources or when an enhanced individual exerts power over an unenhanced person in a relationship in which their interests conflict. These circumstances can occur in a large number of settings: between rivals for someone's affection or in interpersonal relationships such as those between boyfriend and girlfriend (and similar relationships between members of the same sex); in contests, including sports, games, beauty pageants, and talent shows; in competition for access to limited privileges, such as admission to academic institutions; in fiduciary relationships, such as those between patients and health care professionals, trustees and beneficiaries, directors and shareholders, attorneys and clients, and insurers and insureds; and in ordinary business relationships, such as those between seller and buyer, landlord and tenant, realtor and purchaser, lender and debtor, manufacturer and consumer, and employer and employee. The object of the competition may be any desirable good: money, jobs, status, affection, sexual favors, political influence, or market power. The relative advantage conferred by genetic enhancement would depend on both the context and the nature of the enhancement: In a test of strength, for example, enhanced intelligence may be of little value. Unfairness could arise either in a zero-sum situation in which the enhanced person obtains benefit at the unenhanced person's expense, or in non-zero-sum situations in which, although both the enhanced and the unenhanced person gain, the share gained by the enhanced person is greater and the share gained by unenhanced person smaller than would be the case if the parties were equivalently advantaged.

All these situations are subject in some fashion to external rules of behavior. They may be formal public laws; legally enforceable private law, such as the bylaws and other governing principles adopted by corporations, partnerships and unincorporated associations; or social norms or customs. How should these systems of rules respond to the potential unfairness created by genetic enhancement? Should the rules treat these differences as if they do not exist or do not bear on the activity? Or should the rules attempt, in some fashion, to level the playing field? If one person possesses an advantage over another, should the rules permit her to profit from it at the other's expense? Although it would be fascinating to consider nonlegal responses based on social norms and customs, the focus of the rest of this article will be on legally enforceable rules, that is, on public and private law.

If we attempt to level the genetic playing field in response to genetic enhancement, what would our options be? Basically there are the same two approaches that we examined in the previous section in discussing how to reduce genetic inequality: Either we decrease the advantages of those persons who were genetically enhanced, or we improve the lot of those who were not. In short, once again, we can level "up" or "down."

Leveling up would entail giving those who were not genetically enhanced some countervailing benefit. This could be money, professional advice, information that was hard to come by, or any other desirable resource that would help level the playing field. It could be a preference in access to a scarce resource, such as an affirmative action program. Yet, it is difficult to conceive of how this approach would work in the context of personal interactions. Would an unadvantaged person be permitted to draw on some public store of resources to place her on the same level as the enhanced person? Obviously this could not be a store of genetic enhancements, since that would contradict the basic assumption that we cannot afford to provide genetic enhancement to everybody. Yet, the same problem of scarce resources would plague any other subsidy, monetary or otherwise: It would cost too much to put the unadvantaged on the same level as the enhanced.

A less expensive alternative might be to level up only those who were the most disadvantaged relative to the enhanced. This would resemble laws prohibiting employment discrimination against persons with disabilities (54). The effect of these laws is to require employers to subsidize persons with disabilities so that in competitions for employment they can match persons who are not disabled. Only disadvantaged employees or applicants for employment receive this benefit, thus leveling the employment playing field.

This approach is intuitively appealing. By focusing on improving the lot of the worst off, it moves in the same basic direction as Rawls's difference principle. Yet it would produce odd results if it were applied to a more realistically complex society in which some people are enhanced, some (the unenhanced) are merely "normal," and some are disadvantaged: If through access to countervailing benefits, the disadvantaged are truly brought up to the level of the enhanced, they would pass those who previously had been neither advantaged nor disadvantaged. The formerly unadvantaged now would constitute the disadvantaged. In short, unless everyone is made equal, or the distribution of countervailing benefits is a once-only event, a policy of benefiting the worst-off would create an infinite regression. There will always be a group that is disadvantaged or unadvantaged and that riskes being treated unfairly by the enhanced — and also, under a genetic "affirmative action program," by the formerly disadvantaged who have been leveled up.

This leaves the other option of "leveling down." Since we cannot prevent some people from obtaining genetic enhancements for themselves or their children, unfairness might be avoided by preventing them from taking advantage of their enhancements when competing with the unenhanced or exerting power over them.

Some idea of the ways in which we might level down the genetic playing field can be obtained by reviewing how rules currently respond to the potential unfairness inherent in interactions between advantaged and unadvantaged individuals. Here, instead of advantages conferred by genetic enhancement, society is concerned with natural or acquired advantages such as youth, beauty, size, strength, endurance, intelligence, memory, creativity, information and knowledge, experience, social status, money, and personal power. If we examine current public and private law rules, we see a number of ways in which they attempt to level the playing field by leveling down these sorts of advantages:

1. Competition that is arguably unfair is sometimes prohibited. A private law example is the ban on the use of performance-enhancing drugs in sports

competitions (55). Another sports example is weight classes in certain competitions such as rowing and wrestling. In these competitions, athletes who have an advantage in weight are precluded from competing with those who weigh less.

Banning competitions between advantaged and unadvantaged individuals is not confined to sports. A public law example is the prohibition against insider trading in securities. Here the advantage is information that is not available to the public about a corporation whose stock is publicly traded. The law attempts to deny those who possess this information any financial gain from it. The advantaged individual is given the choice of either disclosing the information or not trading stock in the company.

2. The rules permit a transaction to take place only if the person with the advantage forfeits it by sharing it with the unadvantaged. An obvious example is information possessed by one party in certain business transactions, such as when the advantaged person knows that "disclosure of the fact would correct a mistake of the other party as to a basic assumption on which that party is making the contract" and nondisclosure would be a failure to act in good faith and "with reasonable standards of fair dealing" (56). Presumably such transactions are not prohibited altogether because it is sufficiently inexpensive to enforce the forfeiture rule and there is a sufficiently high possibility that, given adequate enforcement of the rule, the result will be fair.

Similarly in some cases the person with the advantage is handicapped so that the advantage is removed. This occurs, for example, in horse racing where jockeys who weigh relatively little are deprived of their advantage literally by having to carry weights. Better golfers are also deprived of their advantage by removing strokes from the score of other golfers.

3. The rules do not prohibit the competition but allow the unadvantaged to avoid the outcome if it seems too unfair. The doctrine of unconscionability in contracts is such a rule, which applies to advantages in the form of information or market power (57). Another example is the fiduciary rules that permit a court to a void a transaction by a trustee of a trust if the result would be unfair to the beneficiaries (58).

4. The rules sometimes level the playing field by eliminating the arm's-length nature of the transaction. The advantaged person is permitted to engage in the transaction but not allowed to employ the advantage in such a way as to take advantage of the other party. This is the result, for example, of fiduciary rules that mandate the fiduciary's undivided loyalty towards the entrustor and prevent the fiduciary from serving an interest other than the beneficiary's (59).

On the other hand, the rules could make no effort to level the playing field, and legislators could ignore or even to celebrate the advantages that some people have over others. With the exception of affirmative action programs,

for example, admissions criteria at selective educational institutions do not adjust applicants' accomplishments in light of their background or abilities. A person applying to Harvard with an IQ of 120 competes with applicants with IQ's of 160; the fact that an A in AP Calculus or a high score on the Scholastic Aptitude Test (SAT) achieved by the person with the 120 IQ is a far greater accomplishment than the same grade achieved by the person with the 160 IQ is irrelevant. Many athletic competitions force athletes to compete with those who are advantaged by being younger: older baseball and basketball players must compete with those considerably younger, some straight out of high school. Shorter basketball players are not allowed to shoot from stepladders, and there are no professional leagues for players of "normal" height. In football the slight take the field at their own peril.

This raises the question of whether the unfairness problem raised by wealth-based genetic enhancement simply should be ignored, as it seems to be in the case of college entrance criteria and in certain sports settings. What would justify ignoring the problem?

In many cases the fact that the rules ignore advantages, or certain advantages, is probably arbitrary, coincidental, or an historical artifact of no theoretical significance. In horse racing, for example, jockeys' weights are equalized on the premise that it is the quality of the horse and the jockey's horsemanship that should matter. There are no weight categories in football because that is just not how the game was conceived. Organized chess competition does not prohibit the use of cognitive enhancers such as nicotine or stimulants because the organizers simply never thought of it (E.C. Johnson, Assistant Director, U.S. Chess Federation, personal communication, October 4, 1995).

Nevertheless, we can posit several principled reasons why it may be inappropriate to deprive genetically enhanced individuals of their advantage in specific transactions or relationships: (1) loss of societal benefit from the enhancement, (2) difficulty in detecting enhancement, (3) difficulty in distinguishing between enhancement and effort, (4) nonenhancement advantages, (5) public intrusion into private affaires, and (6) transaction costs.

Preventing the Loss of Societal Benefit from the Transaction

An enhanced scientist, presumably enjoys personal advantages by virtue of being enhanced; she otherwise might not have been admitted to MIT, for example, or be able to earn a fortune from her patents. But despite the unfairness to unenhanced persons who applied to MIT or tried to develop patentable inventions, we might forgo trying to strip her of her personal benefits. By allowing her to benefit personally, we encourage people like her to purchase scientific enhancements so that society could reap the benefits. (This might well be the justification for not leveling the playing field in terms of intelligence in the case of admissions to institutions of higher learning.)

An example of a societal benefit that might be a sufficient reason to permit enhanced individuals to retain personal benefit are reductions in the costs of accidents. A naturally talented automobile mechanic might be expected to make safer repairs than someone with less talent, and therefore might be entitled to a hiring preference over someone who lacked her natural talents. The same might be said for an enhanced automobile mechanic. The argument becomes even more compelling in the case of persons responsible for the safety of large numbers of people: airline pilots, railroad engineers, operators of nuclear power plants, and the like.

Difficulty of Detecting Enhancement

Systems of rules might have no choice but to ignore the unfairness created by competitions involving enhanced and unenhanced individuals if enhancements cannot be detected. This is a severe problem in attempting to prohibit the use of performance-enhancing drugs in sports. The earlier discussion of licensing catalogued the difficulties of detecting enhancement and the potential solutions.

One further approach to the detection problem might be to permit unenhanced individuals to assert a rebuttable presumption against persons they interacted or competed with whom they believed were enhanced. Unless persons against whom the presumption was asserted could establish that they were not in fact enhanced, the rules would proceed to level the playing field (e.g., by having courts undo the deal, or penalizing the person presumed to be enhanced for participating in a prohibited competition). Inability to produce proof of lack of enhancement would satisfy the burden of proof that the person was enhanced. This would encourage the enhancement industry itself to develop a workable tracking and record-keeping system.

Distinguishing Between Enhancement and Effort

Arguably society should focus its leveling efforts on advantages derived from genetic enhancement, rather than on advantages obtained through personal effort. Otherwise, the effect will be to discourage effort, leading to sloth and loss of social benefit. As noted earlier, however, it may be difficult if not impossible to distinguish between achievements that are earned and achievements that result from enhancement. Accordingly, it might be argued, the rules ought to ignore genetic enhancement.

Yet genetic enhancement may be a sufficient social threat that it is appropriate to level playing fields regardless of this risk. Indeed, society often deprives people of plainly earned advantages in order to promote equality or fairness. Though not without its critics, progressive income taxation transfers earned wealth to achieve a more just distribution of resources. Weight categories in sports are enforced regardless of whether an athlete's size is the product of diet and exercise or steroids. Fiduciary law requires individuals with superior information to disclose it to beneficiaries, clients, and patients, even though the information may have been obtained through great effort. Similarly an enhanced person automatically might be required to disgorge her advantage (of information, market power, etc.), even though she obtained some or all of it through her own efforts. Alternatively, as mentioned before, the fact that someone was enhanced could establish a rebuttable presumption that any advantage related to the enhancement was due to the enhancement rather than to effort.

Nonenhancement Advantages

Just because someone is not genetically enhanced does not mean that they lack sufficient resources or talents to compete fairly with someone who is enhanced. The unenhanced person may possess great wealth, or have some special store of knowledge, or have access to the best advisors. Leveling the genetic playing field may exacerbate unfairness if it focuses on genetic enhancements to the exclusion of these other types of advantages. If society attempted to correct for all differences between people, however, there would be no end to societal interference.

On the other hand, as discussed earlier, the advantages conferred by genetic enhancement could be so great that it would be appropriate to single them out for remediation. Moreover, where enhancement merely creates a rebuttable presumption of unfairness, the enhanced individual would be free to prove that her enhancement advantages are equaled or outweighed by nonenhancement advantages possessed by the complaining party.

Intrusiveness

Given the difficulties of detection and differentiation described above, any attempt to level the playing field would invite public intrusion into highly personal affairs. To rebut an inference of enhancement, for example, people would have to reveal their personal and medical history, including particularly sensitive information relating to their genetic makeup and their reproductive activities.

If the stakes are high enough, however, we seem to be willing to require people to compromise their privacy rights. For example, athletes must submit to physical examination and to yield samples of bodily fluids for testing, often under nearly public conditions. Given sufficient concern for maintaining privacy and the confidentiality of sensitive personal information, and so long as the least intrusive means were employed to decide if someone were enhanced, the cost may be justifiable.

Transaction Costs

Leveling the genetic playing field is liable to be costly. Forums, advocates, and referees would be required to resolve fairness disputes. Black markets, both domestic and foreign, would need to be policed. The specter of a "war on genes" is not an attractive one. Yet again, the threats posed by genetic enhancement might well be worth the cost of leveling.

In short, there seems to be no obvious reason why we would ignore the unfairness created by wealth-based genetic enhancement, except in situations like preventing accidents or achieving scientific breakthroughs in which the ratio of social to personal benefit clearly demonstrated a substantial net benefit to society, or in situations in which the costs of leveling were deemed to be greater than the costs of unfairness. In all other cases, one or more of the leveling techniques listed earlier would be appropriate, depending on the circumstances.

LEVELING THE GENETIC PLAYING FIELD

Although our overall objective is to minimize individual unfairness caused by interactions between enhanced and unenhanced individuals, at same time, we want to maximize the societal benefit from individual enhancements. This raises the question of how to respond when the two objectives are incompatible, namely when individual unfairness can be prevented only by sacrificing societal benefit.

The answer depends on the nature and magnitude of the unfairness and of the forgone societal benefit. Ultimately public policy should favor preventing unfairness if the cost of unfairness is deemed to exceed the expected societal benefit. Conversely, a substantial amount of societal benefit should be sought at the expense of a relatively small amount of individual unfairness. The more substantial the unfairness, and the more equal the costs of unfairness and societal benefits tend to be, the more emphasis should be given to correcting the unfairness of the transaction.

For example, suppose that we are reviewing applicants for scarce medical research funding. Successful research is expected to yield significant societal benefits. Applications are submitted by both unenhanced individuals and individuals enhanced in ways that significantly increase their chances of research success. All other things being equal, we ought to award funding to the enhanced individuals. If the impact on the unenhanced individuals' careers is deemed significant enough, however, consideration might be given to mitigating the unfairness, such as by allocating a certain amount of funding for them alone (a sort of "unenhanced persons' affirmative action program"), or by favoring applications involving both enhanced and unenhanced investigators.

As suggested by the approaches described above to leveling the playing fields, there are a number of techniques for mitigating genetic unfairness. Some of these techniques are more costly than others, both in terms of implementation costs and in terms of forgone social benefit. The enhanced individual might be required to share with the unadvantaged person the advantage created by enhancement. At a minimum, enhanced individuals would have to disclose that they were enhanced. In a business transaction an enhanced party who by virtue of their enhancement has obtained superior information could be required to disclose that information to the unenhanced party. Sharing might be a preferred mitigation technique where a transaction is expected to yield societal benefit because it encourages the enhanced party to engage in the transaction by allowing that party some degree of personal benefit.

If sharing is impractical, such as in zero-sum situations, or if the implementation costs of sharing are too great compared with the expected societal gain, unfairness might be mitigated instead by handicapping the enhanced party. For instance, in contests, including athletic competitions, the enhanced individual could be put at a disadvantage, such as being given a longer distance or a harder question.

Another technique worth considering is allowing the interaction to take place but permitting the unadvantaged

party to apply to a court or an administrative agency to challenge and overturn or adjust a result if it is too unfair. This flexible, posthoc approach might be appropriate where the unfairness costs and societal benefits of a transaction were difficult to predict in advance. Business deals might be candidates for this approach, for example, particularly those in which the advantages enjoyed by the enhanced party, such as market power, could not, like information, be shared, and in which the particularities of transactions made the application of a priori handicapping rules too inexact. Making outcomes voidable also saves the costs of intervening in every transaction; only those results that seem too unfair will be reviewed.

An interesting option is to eliminate the arm's length nature of transactions between enhanced and unenhanced individuals. Like fiduciaries, the enhanced would be made responsible for the welfare of the unenhanced, a sort of genetic noblesse oblige. As in true fiduciary relationships, this could decrease the costs of monitoring the behavior of the enhanced by substituting a system of sanctioned trust for a regime of direct external controls. It also would encourage the unenhanced to interact for their benefit with the enhanced, facilitating resulting societal benefits (60).

Finally, if no significant social benefit were expected from an interaction, it could be prohibited. An alternative to handicapping enhanced athletes, for example, would be to forbid them from competing against athletes who were unenhanced. Such a competition might be allowed only if the costs of enforcing such a prohibition were great compared to the unfairness.

BIBLIOGRAPHY

1. C. Tucker, *The Weekly Standard*, December 2, p. 20 (1996); J. Ketelsen, *Forbes*, May 9, p. 184 (1994).

2. 42 USC §1395y(a)(1)(A).

3. 42 USC §1395y(9)(10).

4. 42 C.F.R. §440.230(d).

5. U.S. Department of Health and Human Services, *Health United States 1995*, Table 3, USDHHS, Washington, DC, 1995, p. 82.

6. L. Cuttler et al., *JAMA, J. Am. Med. Assoc.* **276**, 531, 532 (1996).

7. *U.S. News & World Rep.*, May 13, p. 20 (1996).

8. R.J. Paulson, *West. J. Med.*, December 1, p. 377 (1996).

9. D.R. Lamb, *Am. J. Sports Med.* **12**, 31–34 (1984); N.A. Ghaphery, *Sports Med.* **26**, 433 (1995); D.A. Smith and P.J. Perry, *Ann. Pharmocother.* **26**, 653–658 (1992).

10. G. Cowley and A. Underwood, *Newsweek*, June 15, p. 54 (1998).

11. P.J. Whitehouse, E. Juengst, M.J. Melhman, and T.H. Murray, *Hastings Cent. Rep.* **27**(3), 14–22 (1997).

12. *Orlando Sentinel*, October 12, p. E2 (1991).

13. E. Parens, *Hastings Cent. Rep.*, January, February, p. S8 (1998).

14. M. Walzer, in M. Robertson, ed., *Spheres of Justice*, Oxford University press, Oxford, UK, 1983, p. 11.

15. D.S. Diekema, *Perspect. Biol. Med.* **34**(1), 109, 112, 113 (1990); T.H. Murray, *The Worth of a Child*, University of California Press, Berkeley, 1996, pp. 90–91.

16. E. Parens, *Hastings Cent. Rep.* **28**(1), S1, S11 (1998); G. McGee, *ibid.* **27**, 16 (1997).

17. J.W. Gardner, *Excellence: Can We Be Equal and Excellent Too?* Norton, New York, 1984, p. 30.

18. T. Nagel, *Equality and Partiality*, Oxford University Press, New York, 1991, p. 135.

19. R. Nozick, *Anarchy State and Utopia*, Basic Books, New York, 1974, pp. 223–224.

20. Id. at 347, n.41.

21. J. Rawls, *A Theory of Justice*, Belknap Press of Harvard University Press, Cambridge, MA, 1971, p. 104.

22. R. Dworkin, *A Matter of Principle*, Harvard University Press, Cambridge, MA, 1985, p. 207.

23. R. Dworkin, *Philos. Public Affairs* **10**, 185, 283 (1981).

24. J.E. Roemer, *Theories of Distributive Justice*, Harvard University Press, Cambridge, MA, 1996, pp. 237–252.

25. H. Frankfurt, *Ethics* **98**, 21 (1987).

26. R. Arneson, *Philos. Stud.* **56**, 83, 84 (1989); M. Rosenfeld, *Cal. K. Rev.* **74**, 1687–1699 (1986).

27. E. Rakowski, *Equal Justice*, Oxford University Press, New York, 1991, p. 159.

28. E. Rakowski, *Tax Law Rev.* **51**, 419 (1996).

29. J. Riedinger, *Cap. Law. Rev.* **22**, 893, 895–897 (1993).

30. F. Parkin, *Class Inequality and Political Order: Social Stratification in Capitalist and Communist Societies*, Praeger, New York, 1971.

31. D.B. Grusky and A.A. Takata, in E.F. Borgatta and M.L. Borgatta, eds., *Encyclopedia of Sociology*, Macmillan, New York, 1992.

32. J.H. Schaar, *Legitimacy in the Modern State*, Transaction Books, New Brunswick, NJ, 1981, p. 195.

33. G.A. Hedger, ed., *An Introduction to Western Civilization*, Odyssey Press, New York, 1949, p. 217.

34. A. Bonner et al., *Democracy in India: A Hollow Shell*, American University Press, Washington, DC, 1994.

35. M. Mehlman, M. Durchslag, and D. Neuhauser, *J. Health Politics, Policy and Law* **22**, 1385–1391 (1997).

36. A.M. Capron, *Hastings Cent. Rep.*, November 22, pp. 18–20 (1992).

37. D.C. Hadorn, *JAMA, J. Am. Med. Assoc.* **268**, 1454–1459.

38. D. Orentlicher, *JAMA, J. Am. Med. Assoc.* **271**, 308–314.

39. AMA Council on Judicial and Ethical Affairs, *Arch. Fam. Med.* **3**, 633 (1994).

40. R.M. Cook-Deegan, *Cloning Human Beings*, vol. II, National Bioethics Advisory Commission, Washington, DC, 1997, p. H8.

41. R.A. Merrill, *Va. Law. Rev.* **82**, 1753, 1777–1782, 1821 (1996).

42. *Physician's Desk Reference*, 1998, p. 993.

43. R. Rubin, *Dallas Morning News*, July 7, p. A1 (1986).

44. 21 U.S.C.A. §396 (Supp. 1998).

45. Z. Leavy and J. Kummer, *S. Calif. Law. Rev.* **35**, 123 (1962).

46. M.J. Mehlman and J.R. Botkin, *Access to the Genome: The Challenge to Equality*, Georgetown University Press, Washington, DC, 1998, p. 119.

47. J. Maher, *Austin Am.-Statesman*, July 26, p. E3 (1992).

48. Fertility Clinic Success Rate and Certification Act of 1992, 42 U.S.C. §263a-1 to 263a-7 (1994).

49. M. Pollan, *N.Y. Times Mag.* October 25 (1998).

50. R.F. Service, *Science* **474** (1997).

51. M.J. Mehlman, *Wis. Law. Rev.* **239**, 256–260 (1985).

52. D.S. Karjala, *Jurimetrics* **32**, 121, 136 (1992).

53. R. Cooke, *Newsday*, August 30, p. A06 (1998).

54. Rehabilitation Act of 1973, 29 U.S.C. §706. Americans with Disabilities Act, 42 U.S.C. §12102.

55. U.S. Olympic Committee, *Drug Control Education*. Available at: <*http://www.olympic-usa.org/inside/in_1_3_7_1.html*>

56. Restatement (Second) of Contracts §161(d).

57. Restatement (Second) of Contracts §153 (1981), §208.

58. A. Scott and W. Fratcher, *The Law of Trusts*, 4th ed., Little, Brown, Boston, MA, 1987.

59. S.P. Shapiro, *Am. J. Soc.* **93**, 623 (1987); M.J. Mehlman, *Conn. Law. Rev.* **25**, 349, 368–371 (1993).

See other entries BEHAVIORAL GENETICS, HUMAN; see also HUMAN ENHANCEMENT USES OF BIOTECHNOLOGY entries.

HUMAN ENHANCEMENT USES OF BIOTECHNOLOGY, POLICY, TECHNOLOGICAL ENHANCEMENT, AND HUMAN EQUALITY

MICHAEL H. SHAPIRO
University of Southern California Law School
Los Angeles, California

OUTLINE

Introduction

Meanings of Enhancement

Mapping Meanings of "Equality" onto Meanings of "Enhancement," and Vice Versa

Technological Expectations; Germ-Line and Somatic Enhancements

The Idea of Enhancement

Competing Versions of (In)Equality

A Thought Experiment Not Far Removed from Reality

Equality Wars: Conflicting and Concurring Versions of Equality and Inequality; Remedies for Inequality; Equality, Enhancement, and Respect for Persons

Equality and Other Values: Conflicts and Connections

Distributional Equality Generally; Distribution That Transforms the Distributees; Distributional and Nondistributional Equalities

Enhancement and Its Effects on (In)Equality; (Non)Distribution of Enhancement Resources; Regulatory Choices

Nondistribution Options: Nonallocation at the Macro Level; Restrictions on Manufacture, Distribution, and Use; Black Markets; Paternalism and Community Self-Protection

Equality Impacts of Technological Enhancement: More on Distributional Options

Constitutional Considerations in Brief

Constitutional Frameworks

Paths of Constitutional Interpretation

INTRODUCTION

This article outlines some of the moral, legal, and general policy difficulties societies and individuals will face if technological enhancements via germ-line and somatic mechanisms become possible (1). It identifies and analyzes some of the conceptual structures necessary to explain the nature of these difficulties, suggests some alternative basic scenarios—such as greater or lesser scarcity of technological enhancement resources, impacts on how we perceive each other, different remediation patterns—and then maps and reverse-maps the projected technological developments against the value and legal structures. It also describes and comments on what many see as the most critical threats and promises, from our present value standpoints, of the anticipated changes, as well as on what might be the fate of these very standpoints themselves. The idea of enhancement is compared to other processes of human change, principally to the familiar forms of self-progress and the practices of treating disorders, injuries, and the like. Questions are raised about the very significance of these distinctions as rational *authorizing* and *limiting* tools that might guide us in distinguishing among permissible and impermissible interventions, and among obligatory and nonobligatory ones.

The moral and legal issues are explored primarily by way of the concept of equality. Some of the classic and (possibly) novel difficulties in the equality analyses include matters of access to and distribution of technological resources; the possibility of increased socioeconomic and political stratification that may be irreversible; the effects of a technological enhancement regime on the ways in which we view each other (as planned and assembled objects or as persons? some blend of these or other attitudes?) and on the viability of present views about equality and its relationship to justice, fairness, autonomy, utility, and ideas of merit, virtue, and desert. Considered in the discussion of our notions of merit, virtue, and desert is whether they are to be reconstructed or abandoned. Distributional criteria and, more generally, different egalitarian arguments based on different visions of equality are sorted, and there is a brief exploration of how the structure of our democratic institutions might be altered by responses to particular distribution patterns of merit attribute enhancements—in particular, by adopting a plural voting system of the sort envisioned by John Stuart Mill. Different forms of remediation or prevention of inequality and of affirmative promotion of equality are briefly touched on. At the end there is a brief review of some issues arising under the United States Constitution: If technology changes as many anticipate, the acute moral and policy issues will eventually be vetted and disputed within the legal system.

MEANINGS OF ENHANCEMENT

Mapping Meanings of "Equality" onto Meanings of "Enhancement," and Vice Versa

To explain how equality may be affected by technological enhancement requires some account of the meaning of "enhancement"; a review of the intimidating complexities of the equality analysis, including equality's relationship to other basic values; and an examination of how varying understandings of the concepts of enhancement and equality affect each other.

This article outlines a bi-directional mapping of differing versions of equality and projected forms of enhancement against each other. Different ideas of equality may lead to different valuations of enhancement, and the reverse. An initial task is to distinguish different equality arguments, and this rests, in part, on asking the now-familiar question, *What* is supposed to be equal to *what?* An obvious example of variant meanings of equality is suggested by the tension between equality of opportunity in its several forms (2) (certain ex ante positions are to be set equal) against equal-outcome standards (certain ex post positions are to be set equal). These opposing pulls are especially vivid when considering the possibility of, say, major enhancement of intellectual abilities. Equal opportunity understood as rights against interference by others with access to enhancement resources may yield unequal outcomes that track and intensify existing inequalities in wealth, income, status, and power. Diminishing returns in the value of increments in ability may set in slowly, thus prolonging the incentive to continue adding increments to intellectual talents, further deepening inequalities in power and social status. As a given form of enhancement becomes widespread, its value to a particular individual may shift from enabling her to tower over others to enabling her to avoid being towered over. For some traits, then, the more widespread the enhancement, the more urgently the less able need it in order to avoid losing more and more ground to more and more persons. At some later stage, relative interpersonal positions may be unchanged, although "absolute" performance capacities are amplified. An equality of outcome standard, on the other hand, would require major centralized intervention either to narrow the ability gaps among persons, or at least to preserve their relative standing. In the latter case, equality of outcome would encompass—not flat-out equal abilities—but preserving the status quo ante concerning the relative "distance" between persons. Of course, egalitarian maneuvers might involve redistribution of traditional goods and services, either in addition to or instead of enhancement opportunities (3).

Technological Expectations; Germ-Line and Somatic Enhancements

Current directions in technology clearly justify assuming for argument's sake that we will be able to influence significantly the development of our targeted traits as compared with what life's lottery (genetic or environmental) might otherwise have presented. The apparently successful germ-line alteration that resulted in superior learning ability in mice illustrates the point nicely (4). In earlier experiments, mouse embryos assimilated rat genes coding for growth hormone, producing some large mice that themselves bred several hefty offspring. The important point to take from these results is that even complex polygenic and multifactorial traits such as intellectual ability or size may be heavily influenced by a given gene (5): Not all genes and environmental factors are equal—some may have outsize or disproportionate effects (6). The accompanying point, of course, is that similar outcomes in human beings may be quite far off, if they are possible at all.

One distinction requires immediate attention—that between *germ-line* and *non-germ-line* techniques for altering traits in a specific possible or existing person. (Selective breeding would alter the distribution of traits in a population but not in a particular individual.) The latter include genetic alteration of somatic (body) cells ("gene therapy"); such alteration does not affect gametes (mishaps aside) and so does not affect one's descendants. Nevertheless, because "gene therapy" and the development of substances directly affecting gene operations require extensive knowledge of genetic mechanisms, all these modes of enhancement will be mentioned here.

The Idea of Enhancement

In General. For convenience, "enhancement" and "augmentation" are used interchangeably, although the latter might also suggest "extension" or "supplementation" (e.g., a springier vaulting pole). Standing alone, the term "enhancement" will refer to technological enhancement, not to socially and legally accepted processes of self-improvement—say, gradually increasing one's strength by lifting weights, or improving analytical skills through study. Many observers view the results of such accepted measures as "internally" rather than "externally" or "unnaturally" generated changes that compromise claims of personal, meritorious achievement (7). In fact these varying paths toward superior traits will generally be intertwined, further complicating our analysis.

Here are three basic families of overlapping questions concerning the meaning of "enhancement." Who is to be enhanced and why? What is to be enhanced and why? What counts as enhancement and why?

What Is the "Unit of Enhancement"? We need to ask first *who* or *what* is to be enhanced. A possible person presently in the form of an early embryo, or as-yet-unjoined gametes? A particular living person or group? The present or future human race?

What Is Enhanced? Traits, attributes, and characteristics are the targets for enhancement, but to what do these terms refer? Behavior patterns? Physical appearance? Tissue structure? Molecular arrangements—such as genomic structure—and biochemical processes? Competitive performance? Predispositions to develop particular physical or mentational conditions? And which of these targets should be selected for improvement? What role does culture play in characterizing and valuing traits, and how might this track genetics? A culturally valued trait may rarely have a clear genomic correlate.

What Counts as Enhancement? The two most discussed issues are, first, whether and how to distinguish permissible forms of enhancement (practice, pumping iron) with impermissible forms (ingesting memory-aiding substances, altering the germ line of one's children—technological fixes or shortcuts generally); and, second, whether and how to distinguish between enhancement, on the one hand, and repair of "defects" or injuries or control of disorder, on the other. As to the latter, a major reason for insisting on the treatment/augmentation distinction is the belief that it provides limits on an enhancement imperative that might seem to be utterly unbounded. This is not without cost, however: We risk devaluing and stigmatizing those with "imperfections" of various sorts. In any case, resource constraints may impose severe limits on the extent to which the "treatment model" and the "enhancement model" will be "lumped."

Enhancement of Merit and Wealth-Attracting Attributes (Resource Attractors). Traits plainly vary in importance. Those strongly favored for whatever reason—special abilities, health, appearance, personality, culturally preferred predispositions—are critical variables affecting the distribution of life's rewards, including social and political status, income and wealth, praise, mating opportunities, and prizes. The moral and conceptual foundations of merit and desert judgments are complex (8) and cannot be plumbed here, but it is vital to see that "merit attributes" are often (but not necessarily) "resource attractors"—they are distributional criteria of sorts, whether in market, centrally directed, or other economic systems. The close coincidence of these ideas permits some interchangeable use. Suppose, now, that we can enhance these distributional criteria through technological alteration. Those persons already in a position to draw substantial resources may sharply augment their resource-attractiveness—possibly in a self-accelerating cycle that draws increasing wealth and power to the enhanced persons. In at least a metaphoric sense, then, one's very "merit" is increasing—one's "merit basis" is "stepped up"—thus amplifying one's claim for still more of everything, including still more merit. The resulting risks of increased and more inflexible social stratification are obvious (9). (The risks seem lessened when enhancement is temporary and must be repeatedly renewed—a point to retain throughout this discussion.) One existing parallel is the distribution of educational resources, particularly advanced and specialized higher education. Another is wealth itself: One needs it to get more of it, and even to keep what one has. Here the intersection of equality with justice and fairness considerations is obvious. (A close comparison and ranking of these values is not possible here.)

"Repair of Disorder" as Distinguished from "Enhancement"; Disorder versus Enhancement Models for Justifying Trait Alteration. If a pathological condition is successfully treated, we are unlikely to describe the restorative process or its result as "enhancement" unless the intervention appears to go beyond "canceling out" the disorder and induces a "nonnatural" condition that masks or displaces the impairment rather than restoring the patient's ex ante personal baseline. If the improvement is justified by a supposed medical need, however, complaints of unequal distribution of enhancement resources may be blunted, although complaints about unequal access to medical resources may continue. In many cases the two models are partially "merged"—for example, measures to increase immunity or other resistance to certain disorders or conditions.

More generally, equality issues concerning enhancement will not just vanish when a disorder model is invoked. For one thing, reliance on a disorder model will not go unopposed. One can question the moral relevance of the distinction between disorder- and enhancement-based justifications for distribution: If an egalitarian imperative requires remedial measures, then—resource scarcity aside—what difference does the therapy-enhancement contrast make if one's relative position will be improved either way? One can also note the difficulties in distinguishing between the two in various other cases. For example, repair of a fracture might make one less vulnerable to future fractures, leaving one in effect stronger than before the injury. And, as we saw, germ-line or somatic manipulations that make one less vulnerable to, say, infectious diseases straddle the treatment/repair versus enhancement distinction. An oft-mentioned example of these difficulties is the use of human growth hormone. Its administration as treatment for short stature caused by pituitary or other disease is more readily accepted than its use on short persons not suffering from a height-impairing disorder—although from the short person's viewpoint, it may make little difference what accounts for his fate (10). The latter use is often seen as technological enhancement—although of a socially handicapping trait (11)—rather than treatment. Of course, whether something is a treatment at all rests on whether it is directed at a disorder, disease, injury and the like, and this may again depend in part on cultural habits and existing environmental conditions: What are socially (un)acceptable moods, or prevailing attitudes toward persons of very short/tall stature? Does the society's current "physical plant" (e.g., lots of stairs, few elevators) contribute to the limitations of persons with particular conditions? Still more, the treatment/enhancement distinction might be viewed as immaterial to the more general goal of "normalization" (12)—a concept overlapping, but distinct from, that of enhancement. But the standard of "normalization" may rise with enhancement or treatment, and the notion of relative handicap—or even disorder—may thus expand, with unfortunate consequences for the "handicapped."

Despite problems with the systematically murky treatment/enhancement distinction, however, it is far from meaningless (12).

Enhancement, Illicit Transformations, and Compromise of Identity, Merit, and Desert; the Paradox of Perfectionism; Effort and Merit. The traits a given culture most values at a particular time—say, intelligence (13), strength, and the capacity for diligent effort—are (1) the main targets of traditional improvement efforts such as training and practice and (2) arguably the most sacrosanct against technological tampering (14).

So the very traits selected for improvement are precisely those whose "artificial" ("nonnatural," "identity-compromising," "externally induced") augmentation is the most suspect. Indeed, if the wrong paths are taken, we may not even *count* the result as "improvement" or the accomplishment as (fully) merited. The very status of merit attributes may be impaired when they are technologically refashioned to extend beyond one's preexisting natural baseline, as augmented by traditional effort. But traditional baseline methods of self-alteration are not only *not* banned, they are *required* by perfectionist/progress ideals. (Whether such personal obligations accompany social obligations to assist individual perfectionism, of course, depends on the content of the ideal, and in turn on underlying basic value conceptions.) The upshot is that bettering ourselves in *inappropriate* ways does not "perfect" us—it lessens us. Thus, more is less. A possibly connected idea is that dispensing with effort as a critical component of achievements and improvements might "cheapen their value and cheat the social practices in which they play a role" (15).

As a rough intuitive matter, enhancement also raises troubling images of compromised personal identity, and thus of assignments of credit or rewards. If technology threatens identity, it also threatens the moral and political relevance of merit and desert. In turn, where merit/desert ascriptions are undercut, equality constraints become more muddled than usual. To assign greater rewards to those of greater merit than to others does not—on *some* views of equality—breach equality standards, and indeed may be required by them. If we cannot say who won the race, we cannot fully justify our assignment of prizes.

The capacity for effort at self-improvement or anything else seems to be a merit attribute, so this deserves some additional comments. We often prize *trying*, which we commonly view as under our control, as much or more than native endowments, many of which seem arbitrarily fixed. The results of traditional forms of striving are thought to be consistent with a stable identity. Now, the capacity to try is thought to be influenced by genetics and noncontrollable aspects of environment, as are other merit attributes. The talent for struggle is itself subject both to technological and nontechnological improvement (an infinite regress of trying?). How should we morally rate an increased capacity to exert diligent effort when the capacity is itself altered technologically? Isn't this as questionable as alterations of supposedly "fixed" traits such as intelligence, and of incrementally improvable traits such as strength? (One thinks of steroids in athletics here.)

It is also possible, in context, to view technological enhancement of endowments—including the capacity for effort—as itself reflecting a kind of praiseworthy effort. An increment in powers of memory, for example, may be unearned but nevertheless possess intrinsic value and instrumental value, as where it aids air traffic controllers in keeping up with the ever-increasing flood of data.

Demand for Enhancement Resources; Economics. The scale of demand for such resources depends on many variables that cannot now be clearly identified and measured. These variables include the nature of the enhancement, its monetary costs and perceived medical risks, the deterrence or incentive effects of gatekeepers and their standards (physicians as well as bureaucrats may keep the gates because of medical risks), cultural variables (whether technological enhancement is (dis)favored in general or for specific traits will obviously affect its level of use), interpersonal pressures (also influenced by culture), links and interactions among different traits, and personal preferences. Different forms of technological enhancement may of course fare quite differently in the market (16).

Any stable economic unit requires supply and demand equilibrium for any commodity, and this presupposes diminishing returns for incremental distributions. Diminishing returns will no doubt set in at some point for distributions of "increments in merit," but as noted, this onset may be quite late in the game (as with many medical resources generally). You may not value yet another hotdog, but you can always stand to be smarter. There may even be expanding returns at various distributional stages. To make modeling and prediction still more difficult, different assemblies of merit traits may interact in unpredictable ways in the market; some traits will reinforce or "potentiate" each other, some will impair each other, some increments can substitute for others, and so on. Finally, recall the impact of extent of social use: Being highly intelligent is less valuable when all are highly intelligent, yet there may be as much or more pressure to consume enhancement resources even if solely to maintain one's position.

COMPETING VERSIONS OF (IN)EQUALITY

A Thought Experiment Not Far Removed from Reality

This exercise is meant to illustrate differences in specifying what is to be "equalized" through distribution (2,17). If the egalitarian goal is to attain $X = Y$, what might X and Y be and what does "=" mean?

Suppose we have a mechanism (e.g., drugs, somatic gene "therapy," or germ-line alteration) that can significantly enhance one's mental abilities. There are many possible distributional schemes for the enhancing techniques. The distributees might be individuals or possible individuals in early embryonic or even dissociated gametic form. The distributive mechanism might be central direction by government, markets, or kinship and other interpersonal relations. The effects of the technology, for good or ill, are likely to vary among persons. Sophisticated models will take account of the variability, but simplifying assumptions concerning uniform efficacy are appropriate for now. For example, one might assume that we will see the same per-dose linear increments in a given ability for all persons, or that effectiveness is a direct or inverse function of preexisting ability—the abler one is, the greater or lesser the increment. It is also helpful to leave aside the fact that mental abilities come in many varieties as well as strengths and that their recognition and status may vary among cultures. The particular regulatory or licensing schemes that would implement the distributive plan are not discussed here. Whatever the schemes are, for oneself, one's children, or some group, the licensing procedure must embrace either a substantive criterion or some objective mechanism such as a lottery or queuing (neither of which is entirely "objective").

Market Distribution. Free market exchange implements a sort of equality of opportunity based on ability to pay (including insurance and borrowing power). There is an extensive literature on the moral foundations of markets, including commentaries on the role of characterizing existing distributions of goods and bads as just or unjust, and on the impact of "preinstitutional desert" and fairness in promoting realization of legitimate expectations, but this is left aside here (18). "Enhancement" might be financed through health insurance mechanisms, particularly where the procedure can be viewed as treatment for disorder, defect, or injury, or at least as ambiguous (recall the example of immunological augmentation). (19) Here, a *ratio* is equalized across persons: *dose/financial resources-economic power*. As for nonmarket distribution via central direction, we might consider—

Centrally Directed Distribution of Equal "Doses" to Everyone. This is a simple, ham-fisted sort of equality. It suppresses individual variations and thus bypasses questions of need, merit, and utility. The ratio of dose to threshold status as a person is the same for everyone.

Distribution in Proportion to Need. What is equalized is the ratio of *dose to need*. "Need" itself is a disputed concept for several reasons, including the fact that many asserted needs are based on one's *relative* status within a population, that need exists in degrees, that it is afflicted with the difficulties within the treatment/enhancement distinction, that it may be linked to nothing more than one's preferences or goals, and that it may be unclear what follows from an ascription of need: Are there duties not to interfere with anyone's trying to meet their needs, or government duties to provide assistance? If need rests on having a recognized disorder or injury, then only the afflicted receive doses—e.g., the demented or persons with Down syndrome. It is unclear how to apply a "need" standard to statistical "outliers" who are not disordered but nevertheless are handicapped by their distance from the median. And need may, as suggested, be task related: Did Einstein "need" enhancement to make progress on a unified field theory?

The fact that one's needs may be based on being relatively worse off, whether in natural endowments or in environmental circumstances, requires special attention (20). Well-known political and moral theories call for measures of "redress" because many of the worse off are seriously disadvantaged. Rawls's difference principle, for example, suggests distribution of resource-attractors to equalize the dose/need ratio—where need is linked to relatively low status (21). The difference principle, however, can also be viewed as threatening other visions of equality, as well as values of autonomy, justice and fairness. Redistribution entails interference with autonomy, and is arguably unfair in allowing some persons (the worse off) to reap the full benefits of their native abilities while preventing others (the better off) from doing so. Of course, a central question is "redistribution of *what?* The existing stock of wealth? Opportunities for enhancement?

Distribution in Proportion to "Social Utility." The ratio equalized here is dose to social utility. The social utility of some distribution pattern might be inversely *or* directly related to the distributee's relative ability, without regard to whether disorder underlies his or her low-end status. Distribution to those handicapped by low intelligence might reduce the need for social services. As for the very talented, think of encryption specialists trying to break an enemy's code (suppose that the British "Enigma" program hadn't cracked the German code in World War II?). Only those on the "edges" of human ability would become licensees. One might even expect pressures on government to require certain workforce groups, if not everyone, to use enhancement resources, though enforcement might be quite unpleasant—not to mention immoral and unconstitutional (e.g., violations of the Thirteenth Amendment).

Distribution in Proportion to Preexisting Merit. There are "native" ("endowed") merit attributes such as mental abilities, physical agility, and the capacity for diligent effort. There is also "acquired" merit based on accomplishment, good works, and developed skills and aptitudes. [The division between the two is hazy (8,53–68,109–131).] Here the ratio equalized is *dose to merit.* On this standard, the answer to "Who merits (more) merit?" is simply: those who *already* are highly meritorious. This sharply contrasts with a view of equality that sees natural variation in aptitudes as something to be overcome rather than presupposed as a suitable basis for distributing life's rewards (21).

Distribution in Proportion to Intensity of Personal Preference. The criterion here is how badly one wants something—including enhancement itself as a major facilitator for success generally. Extremely (pathologically?) intense preferences might be viewed as needs. Preferences and their intensities may also be regarded as a form of merit. We admire persons whose "desire to win" is strong enough to overcome serious odds.

Distribution to Achieve Equality of Outcome. Here, doses are distributed so that all have equal intelligence, however this level is chosen. There is no unitary concept of intelligence, and the task of equalizing all recognized forms of intelligence (22) seems far-fetched, but this is a thought experiment, after all. This rather open-ended outcome standard may entail that everyone be as intelligent as the previously most intelligent, or that the more intelligent are affirmatively impaired (they receive "negative doses") to reduce their status (23) while the less intelligent are upgraded. If the desired uniform ability level is Φ, the ratio equalized is the absolute value of *dose effectiveness to distance from* Φ (including positive and negative doses).

The driving force here might be the alarming idea that equality requires or is aided by making persons as identical as possible. On the other hand, "equality of outcome" may refer to identical proportional increases, leaving everyone's position in the "pecking order" the same. Of course, these different forms of equal outcome are in general quite different, and they bear only an uncertain relationship to more comprehensive forms of equality of outcome—income or wealth, social standing, political power, and so on.

Distribution by Lottery. In an effort to bypass the immense difficulties in applying the ideas of equality, fairness, and justice, some have recommended distribution of scarce resources via lottery (24). Perhaps in this sense lotteries represent a form of being "unprincipled on principle" (25). In any case, the suppression of interpersonal differences entailed by lotteries (once the lottery's constituency is defined) is both the *point* of resorting to them and the chief *objection* to them.

Randomization schemes cannot be properly denounced on egalitarian grounds without a theory of equality that explains why equalizing over one field rather than others (e.g., doses, dose/merit, or other ratios) reflects or produces "true equality," or at least a preferable form of equality. If a satisfactory equality theory is unavailable, values other than equality must be invoked. As it stands, lotteries serve some visions of equality and rationality and contravene others. On one view, lotteries promote equality because all who qualify for the lottery (qualifying itself raises serious equality issues) have an equal chance of winning it, *despite* their varying personal characteristics. Indeed, it is precisely the attention to these varying individual traits that constitutes for lottery supporters a violation of equality: These interpersonal variations—rather than basic personhood itself—are to be *suppressed.* On the other hand, some will receive the resource and others will not, *without a "substantive" reason.* This situation is arguably irrational and thus a violation of equality standards, possibly under prevailing views of personhood and its entailments.

Equality Wars: Conflicting and Concurring Versions of Equality and Inequality; Remedies for Inequality; Equality, Enhancement, and Respect for Persons

In General: Equality of Whom or What and with Respect to What? What *do* we assert in saying that $X = Y$? 'X' and 'Y' might designate persons, groups, opportunities or prospects held by persons or groups; means for taking advantage of opportunities to achieve one's goals; specific outcomes (wealth, victories, etc.); social or moral status; political power; rights as persons (26), without regard to differing traits; traits characterizing different persons; ideas, conceptual systems and philosophies; and overall ("net") personal or group merit or social worth despite differing traits.

Each possibility rests on concepts that are themselves difficult to penetrate, and are likely to reflect serious political and philosophical differences. To assert equality of noninterference rights—such as free speech, free exercise of religion—is far from asserting equality of means, opportunity, or prospect in securing audiences or places of worship (2). And these differing equalities may hold drastically different positions of respect and commitment among different persons and groups within a society, and from society to society. "Fundamental" noninterference rights are protected under the U.S. Constitution. Affirmative ("welfare") rights generally are not, even when directed toward increasing or preserving equality. The rejection seems to be founded partly on autonomy grounds, and partly on rival views of equality: Redistribution entails that some receive unearned rewards and others do not, and different persons will be allowed to keep different proportions of their wealth or income. The flags of equality, fairness, and justice are carried by all sides here—a point that is retained throughout this discussion.

One could raise parallel questions by asking about the meaning of the equality operator "=." Is it an assertion of fact (Arnold's strength is equal to Sylvester's), and if so of what sort? Is it a moral or political claim about equal rights or entitlements, and if so to what? Does it reflect an ideal both of the *threshold* equality of persons without regard to their differences and of how they should be treated? If we say persons are equal because they are all equally persons, why is undifferentiated personhood the right level of abstraction rather than personhood qualified by particular (dis)favored traits? If we say that the political power of (person)(group) X "equals" that of (person)(group) Y, we may mean they have equal numbers of votes, or equal power to elect candidates of their choice, or equal power to influence government policies (a particularly obscure claim), or any of several other options. It is not clear that we can justify our choice of "meaning" here via reference to equality alone, without reference to justice, fairness, autonomy, and utility—even if equality is not fully "reducible" to any of these other values or to some subset of them.

Equality and the Special Status of Merit Attributes. Judgments about an individual's merit are often relied on as a fundamental ground for sorting people—specifying certain (in)equalities among them—and for acting on these characterizations. The governing moral intuition (perhaps not in all cultures) is that outside the domain where only threshold personhood counts, persons are to be judged on their relative merits, and not on "arbitrary" personal characteristics or relationships. It is difficult to formulate a coherent theory for sound application of the epithet "arbitrary." For example, is it arbitrary—and thus perhaps morally improper—for individuals to search for a mate solely within their principal social group(s), which may be defined in part on the basis of ethnicity, race, religion, or national origin?

But comparative merit judgments are also criticized because, among other things, they produce unequal outcomes and may rest on unjust features of the status quo. More distribution of resources that strengthens one's measure of merit may expand and reify existing inequalities—even if it remains unclear whether artificial enhancements would be recognized as merit claims.

Valuation of Equality. As Temkin asks: "Is equality really desirable? And what kind of equality should we seek—that is, insofar as we are egalitarians, should we want equality of opportunity, primary goods, need satisfaction, welfare, or what? ... When is one situation *worse* than another regarding inequality?" (17, p.3).

The question of what "equality" means is distinct from the question whether it is desirable or valuable—although the two inquiries are linked in complex ways: Assignment or recognition of meaning often involves value analysis. There is, to be sure, an oddity in asking about the value of equality or of any "basic" value. How can one "value" basic values when these basic values represent the very terms in which value is defined and assessed? However paradoxical this may seem, we characteristically rank-order our values and assess them with respect to each other. But this is a matter for a comprehensive enterprise in moral theory.

Rectifying Inequalities. Plainly a major issue in genetic enhancement is whether it should be used to rectify inequalities by affirmatively creating equalities—and if so, how. Suppose that we reach a rough consensus on the preferred meaning of "equality" in various situations. There nevertheless may remain significant differences over appropriate measures to rectify or prevent inequalities. For example, if A's cache of goods is v but B has more, holding w, is A intrinsically worse off when B acquires still more but A continues to hold v? (17) From an equality standpoint, is it better to achieve equality by raising A's holdings from some outside source or transferring some of B's holdings to A? Or to enhance A's aptitudes and let him, on his own, try to overtake B? Should we worry more about inequality between certain groups than inequality within those groups (17)? Rectifying existing or past inequalities may implicate procedures that *themselves* may violate specific conceptions of equality—such as transfer payments. Such redistributions arguably impair the right to reap the benefits of one's natural gifts, as amplified by skills acquired through effort. As mentioned, they entail that some persons—those less well off—can acquire additional resources earned by others, and perhaps can keep a larger proportion of what they earn than can others.

"Rectifying" Differences. A population consisting of a single human clone (in the collective sense) might have equality problems, but the problems would surely be rather different from (but not necessarily lesser than) ours. A rather drastic (and perhaps technologically impossible maneuver) would be to make as many persons as identical in major respects as possible. The costs (from our present framework) in reduction of cultural and physical diversity and the loss of multiple perspectives in human endeavors seem very difficult to bear, although the radical transfiguration of human life makes them hard to assess. And, of course, the resource costs might be prohibitive. The point here is simply to observe that "difference" does not entail "inequality" in any sense relevant here, and few are on the stump for technological erasure of human variation (27).

Equality and the Morality of Inclusion and Exclusion. One might select among competing ideas of equality by appealing to a preferred morality of inclusion — of lumping by appealing to commonalities. The point of this brief reference is to emphasize two observations. Some may see the "technology of perfection" as allowing displacement of chance variations by planned similarities: We will all be perfectly equal because we will all be equally perfect. However, the very emphasis on perfection may impose serious burdens on those viewed as disabled or handicapped (3).

Is Equality "Empty"? Perhaps the indeterminacies (a term left undefined here) or conflicts *within* the idea of equality cannot be eased by further analysis of equality. There seems to be no overarching notion of equality to appeal to in all contested cases. The tensions may be irresolvable (17), though occasional consensus on certain matters may be attainable.

This is the central idea behind the claim that equality, at least in many important circumstances, is "empty" — a vacuous concept (28). The emptiness claim is roughly that the egalitarian maxim, "treat persons (dis)similarly situated in (dis)similar ways," cannot be understood and followed without a substantive moral/political theory of (dis)similarity that cannot itself depend on equality. We need, on this view, a theory with normative content to tell us what difference a difference ought to make. Equality alone does not tell what characteristics or actions (or anything) to "lump" as relevantly similar, nor what to "split" or suppress as relevantly different. For example, if government action permits some speech and restricts other speech on the basis of content, we cannot tell whether a constitutional or moral equality principle has been breached without a substantive free speech theory.

Equality and Other Values: Conflicts and Connections

It is often said that in many circumstances, equality, autonomy, fairness, justice, and utility (or any subgrouping) conflict. The nature of the conflict of course depends on the versions of equality and other values under review (29). A standard example is affirmative action. Distributing benefits on the basis of racial, ethnic, or gender criteria entails reduction of opportunities — a form of reduction of liberty and autonomy — for those without the relevant characteristics, and imposes forms of personal association on unwilling persons. These processes and outcomes conflict not only with particular views of equality, but with those of fairness, justice, autonomy, and utility (29,31). On any given set of views, justice may dictate what egalitarian maneuvers to prefer — say, that equality of opportunity, in justice, requires some degree of access to enhancement resources either via noninterference rights or via positive entitlements. Or some states of affairs or actions may be viewed as unjust because of a violation of some equality standard. (There are level-of-category problems in listing "basic" values. Not everyone would place these values on the same plane of moral reality or discourse. "Justice as fairness," for example, presupposes that at least certain versions of fairness are criteria for a higher-order concept, justice. Such difficulties cannot be further addressed here.)

Distributional Equality Generally; Distribution That Transforms the Distributees; Distributional and Nondistributional Equalities

Distribution and Personal Transformation. Distributional equality concerns who gets what, why, when, and how under any given system for distributing scarce resources. It addresses matters both ex ante (e.g., who gets the "merit-enhancing" commodity) and ex post (e.g., who gets what rewards — including still "more merit" — after the distribution and, at least in part, as a result of it). This ex ante, ex post distinction is particularly important given the possible "transformative" effects (left undefined here) of the distribution of enhancement resources. Augmentation may change the structure of the distributional game by disproportionately enlarging the distributee's resource-drawing power — ratcheting it up so that it is hard to undo. One might argue that all distributions "transform" the recipients and that there is no sharp distinction between the transformative effects of education or training, on the one hand, and of technologically augmented intellectual or physical functions, on the other. This is obviously true, but the absence of clear borders marking a distinction does not of itself trash the distinction.

Equality and Reduction, Mere Use of Persons, and Objectification: Some (Largely) Nondistributional Problems. Suppose that we believe a practice of enhancement reflects and generates excessive concern with the measures of specific traits and thus "reduces" persons to the (often commercial) value of these traits. This reduction is intimately connected with the processes of "mere use" and "objectification" of persons — their devaluation or descent from persons to "objects." A person who is (at least partially) objectified, reduced, and subject to mere use has thus suffered an egalitarian loss. (If everyone were reduced or objectified, however, there might be equality among the objects.)

ENHANCEMENT AND ITS EFFECTS ON (IN)EQUALITY; (NON)DISTRIBUTION OF ENHANCEMENT RESOURCES; REGULATORY CHOICES

Nondistribution Options: Nonallocation at the Macro Level; Restrictions on Manufacture, Distribution, and Use; Black Markets; Paternalism and Community Self-Protection

In General. There are many commodities that we think, for whatever reason, should not be distributed widely, if at all. To limit distribution, we can avoid allocating resources to the creation of such evils. If this fails, we can enact prohibitions or lesser regulations concerning distribution and ultimate use, although this may risk greater loss of control because of the rise of black markets. For example, prohibitions or severe restraints on use of enhancement resources may compound their risks by inhibiting safety controls such as physician guidance.

Whatever we decide about use of the commodity, the selection of the best regulatory mechanisms to implement our preferences remains open (31). Resolving this may require empirical inquiries to inform the moral, legal and policy options. Certain arguments frequently offered to justify nondistribution require attention because they bear on equality.

Nondistribution to Protect Those Who Prefer Non-use: The Perceived Risk of Greater Inequality as a "Coercive" Factor; Technology-Driven Demand for Greater Skills and Thus for Enhancement. People often do things they would rather avoid because, if they don't, they fear others will gain advantages over them. One might say that doing so reflects a "straitened preference"—it is what they want only under adverse, dispreferred circumstances. This is not necessarily bad: A child averse to learning to read may find a "second-best" reason to do so when advised that "all the other kids are doing it." Similarly the risk of falling behind in athletic activities may drive some otherwise unwilling competitors to steroids or other supposed enhancers. Although one can dispute the claim that this is rightly called "coercion" via excessively strong incentives, a strong sense of pressure in some form on the unwilling is likely. In some contexts this matter is easily put aside: Many athletes and students do not wish to practice, study, train hard, or diet, but few speak of coercion in these contexts. One must compare such standard efforts with, say, ingesting memory-or muscle-enhancers to shorten the task. Some pressure to do the former is widely considered desirable—possibly obligatory under some circumstances; not so with the other. Few complain of coercion in these contexts, partly because the general endorsement of such self-improvement counts against applying pejoratives such as "coercion" or even "undue influence." Still the wide use of these characterizations suggests that something is believed to be amiss in the choice situations in question.

More generally, pressures favoring enhancement of living persons, fetuses, and possible persons are likely to grow. There is some evidence of technology-driven increases in the demand for "human capital" as reflected in investment in education and training (32).

Paternalistic Nondistribution to Protect Persons against Physical or Psychological Harm; Autonomy versus Autonomy. Enhancement measures, whatever their efficacy, carry risks of adverse effects. Of course, the nature, incidence, and seriousness of such effects—perhaps even whether they are in context adversities at all—is largely unknown. One justification for nondistribution may thus be pure paternalism (a term left undefined here) (33). Another justification, suggested above, reveals autonomy's internal tensions: promoting autonomy by reducing "coercive incentives" (34) to use disfavored commodities. But a broad interpretation of "coercive" may impair the autonomy of those wanting to use the suspect resources and who knowingly assume their risks.

An Equality Argument against the Preceding Nondistribution Arguments. Those seeking access to enhancement resources may of course also offer an important equality argument: They are denied equality of opportunity to better themselves and are relegated to an inferior status as against their superiors, whose natural gifts are as arbitrarily deemed meritorious as are the enhanced attributes.

Nondistribution to Reinforce Equality Values and to Avoid Devaluation of Life and of Effort; Ambiguities of Identity.
The Lombardi Effect: Winning Isn't Everything—it's the Only Thing (35); Paradoxes of Reduction and Valuation; Reduction as Compromising Equality. We learn in part from observing and interpreting social practices, and the societal risk here lies in the "lesson" or "message" that athletic victory is worth serious bodily or mentational harm. [A 1984 poll of uncertain rigor reported that Olympic athletes would accept death at an early age in a Faustian exchange for guaranteed gold (36).] If the practice is banned, getting caught and sanctioned is also perceived as a risk. However, it is the ban itself that is in question here. Note also that to characterize the risk as "mere" (as in "risking life and limb for mere athletic victory or for show") presupposes certain value premises. One might urge, for example, that in a nation besieged by enemies on all sides, the supposedly adverse lesson about winning at all costs is useful for national self-defense by reinforcing a warrior state ideal.

To the extent the victory-is-all lesson is learned (by observers and the competitors themselves), it may reflect and constitute a reduction of human value to a single function or goal—athletic or other competitive success. Such devaluation bears on equality. Those persons who are reduced (a concept strongly linked to "mere use" and "objectification") are seen as less worthy than others, and thus unequal to full-valued persons. The enhanced—if that appellation survives—may be viewed as (partial) artifacts of *lesser* merit who leapfrogged over their associates and competitors. If so, it may be the *enhanced* who require protection from the unenhanced (37). (In that event, resulting inequalities of distribution would not be perceived as reflecting true differences in personal worth, and artificially enhanced talents would not be seen as lessening or devaluing the natural talents or acquired skills of others.) The value of enhanced persons will have collapsed into a narrow range of traits based on their prospects of victory. This supposed reductive risk carries us to the next difficulty.

Troublesome Links Among Enhancement, Value Reduction, and Positive Valuations; Role of Risk-Taking; "Person Perception". Value reduction by focus on specific traits or accomplishments is intricately linked to valuing persons positively—viewing them as meritorious and deserving of our high regard. We value persons not only because of the traits that define their threshold value as persons, but because the strength of those traits distinguishes them from others. How does this differ from "reducing" them to their traits? If we cannot say, the contrast between reduction and positive valuation seems empty. Suppose that an athlete says "winning is the only thing and is worth my life." Does this reflect "reduction" and lesser status—in her own eyes or the eyes of others? Or, on the contrary, supervaluation and greater status? Or some combination? The more general question concerns how we indeed perceive each other as particular persons—an issue not only for moral and political theory and practice, but for continuing work in cognitive psychology.

The answers, if any, may depend partly on cultural baselines. The United States for example, is not about to ban football or boxing because they risk severe permanent injuries or death. But *incremental* risks beyond a traditional baseline may be rejected because they reflect inappropriate trade-offs: All risks are justified if winning is everything or the only thing. (Think of escalating boxing to fights to the death.) Such a view arguably reflects a debasement of the value of life in the eyes of the audience and in the competitors' own eyes. (This particular argument would of course not directly apply to "magic bullets" — zero risk but effective enhancers. But there may nevertheless be possible adverse effects of extreme focus on the traits needed to win.)

As for risks undertaken for intellectual enhancement — such as a dangerous, possibly fatal drug that greatly augments memory — the situation is unclear and may depend on cultural circumstances affecting the comparative valuation of traits (e.g., intellectual as against athletic abilities). In Frank Herbert's *Dune*, rival feudal houses far in the future relied on resident "Mentats" — "wizards" of a sort — who amplified their preexisting exceptional intelligence with an addicting "spice." They evidently were judged *both* on their initial baseline abilities and the level of their enhanced abilities — the latter probably being a partial function of the former (38). Perhaps they were also judged on the skill with which they enhanced themselves and on their courage in facing risks. In this sense, at least a sliver of their "endowed" merit endured. Given their culturally imposed duties, what they did can be characterized in context as praiseworthy effort rather than merely as deriving an unearned benefit. Moreover their work would continue to reflect effort and struggle if, partly as a result of their enhancement, the complexity of their tasks increased (14); those who can do more complex work are likely to have it presented to them. Still, as suggested, artificial enhancement may not be viewed as meritorious enhancement at all. Indeed, from this standpoint, the more revered an attribute is, the more it is corrupted through technology's manipulations, and the more degraded is the "enhanced" person.

Avoiding the Social Devaluation of Effort. This category embraces both paternalistic and nonpaternalistic reasons for nondistribution of "elite-creating resources." As suggested, a possible (perhaps inaccurate) impression conveyed by a visible practice of enhancement is that the enhanced are getting a free ride (or at least a reduced-cost one). So, it is argued, enhancement entails getting too much bang for the buck — cheating of sorts — even if done in the open. If effort is devalued, merit judgments are distorted, and so are our judgments concerning the (in)equalities holding among competitors. (The claim that merit evaluations are "distorted" presupposes some preferred baseline from which to measure distortion.) What are we to make of the fact, for example, that some students, relying on their own gifts as aided by work, are disadvantaged in college-entrance examinations as against "artificially" or "unnaturally" able persons? To avoid this inegalitarian disadvantage and the disruption of prevailing norms, would a student need a license reflecting that she had not been enhanced before the exam? (Or

might it be the other way around — "nonenhanced need not apply"?) Of course, one may question the sanctity of prevailing norms, but within reason, communities have some defeasible moral and legal right to maintain the major features of their normative systems (39).

Devaluation of effort might also (paradoxically?) arise from enhancing the very capacity or inclination to make efforts: Some may view the result as external to one's character — an outside supplement that carries no merit with it. Some authorities, for example, say that steroids may expand one's capacity to exert efforts before reaching exhaustion (40).

More on Threats to Identity; Confused Attributions; Effects on Equality Judgments.

1. *Altering living persons.* Recall the distinction between germ-line alteration and gene therapy or other somatic treatment on living persons, fetuses, and embryos. In dealing with living persons, some aspects of equality concern assessing the fairness of returns on effort and of rewards for one's native endowments — particularly when effort and endowment are combined. Such judgments obviously require identifying and comparing persons. Enhancement may confuse notions of personal identity in at least two ways. It may create ambiguity as to the "source" of one's performance — whether it is "internally" generated and thus causally attributable to that person, or the result of "external" artificial augmentation and thus "attributable" to an outside source. If the latter, then the person in question "didn't do it." The judgments of merit and desert underlying the distribution of rewards will also be confused.

   Moreover, enhancement may, by altering personal traits, appear to interrupt the continuity of human identity, which in other circumstances endures despite gradual, historically acceptable change. In both situations equality appraisals — at least in extreme cases (not yet at hand) — may be distorted or even meaningless. Who or what is equal to whom or what? Who won the fight? Should an enhanced Mentat transplanted from a Dune world be eligible for a chaired professorship or the Nobel Prize? (Data, an android in *Star Trek: The Next Generation*, was ultimately awarded the Lucasian Chair — the same Cambridge chair held by Sir Isaac Newton.) Should he be rewarded only if all other candidates had similar enhancement opportunities?

   Do major trait changes truly compromise personal identity — say, a sudden escalation of intelligence from average to extraordinary (41). After all, one's "baseline" identity doesn't really disappear: The new one is built on it, and all new identities will continue to differ sharply from each other. One can well imagine, as suggested, educational institutions debating whether their admissions criteria should exclude augmented persons as not "truly meritorious" unless their natural abilities ex ante — their "enduring merit" — would have secured their admission. Should we compare and rate persons only on the basis of endowments pre-enhancement? Or is this irrelevant history? Even if pre-enhancement merit remains relevant, the "locked-in" resource-accumulations made possible by accelerating returns might be viewed as going far beyond fair rewards for ability — assuming that *that* distributional criterion is believed sound.

2. *Altering germ lines.* Identity problems may take on a different form when one considers the genomically enhanced — through alteration of early embryos or of gametes — as well as those who were enhanced somatically. The latter can be further divided into persons enhanced in utero, as embryos in vitro, or as young children, or adults. Our sense of self-identity, autonomy, and personal worth may differ sharply depending on our knowledge of the nature and timing of our enhancement, and on the reasons for it. Knowing that one's genome was altered will not necessarily have the same impact as knowing that one's physiology was altered. There will also be differences depending on the timing of the somatic changes — principally whether pre-memory (where the person would know herself only in her enhanced — and, to her, "native" form) or within memory. In any of these variations, think of children asking their parents just why their "natural identities" were tampered with or even changed, whether genomically or somatically. Was a potential person (the *un*enhanced entity) adversely affected because its existence was blocked and replaced by "another" person deriving from the altered entity (42)?

Equality Impacts of Technological Enhancement: More on Distributional Options

In General: Enhancement That Alters the Bases for Distributing Benefits and Burdens. As we saw earlier, the "equality impacts" of distributing anything depend not only on matters of fact but on what notions of equality are used. In turn this may determine what "units" or entities are being compared and targeted for what forms of "equalizing" with respect to what distributable entities. We may be addressing existing or future persons, families, groups, and so on, with respect to equalizing income, wealth, social status, legal and political rights, and opportunities of many sorts, and so on (2).

We now need also to distinguish between distribution of the resources needed for enhancement, and distribution of all other commodities. Distribution of enhancement resources changes the game by altering the criteria for resolving distributive claims; this "feedback" may be far more striking than the distribution of education and wealth.

Do Disorder/Treatment Models Blunt Equality-Based Objections to Enhancement Distributions? — Treatment as Restoring Equality Rather Than Distorting It through Enhancement. The connection between disorder models and equality was suggested earlier. If one's relative incapacity is disorder based, health care may restore normality. (It may also create it for those congenitally

disordered by raising them to a "normality baseline.") This is less likely to be viewed as a suspect form of enhancement, at least where the disorder and matched treatment are well-recognized as such. Indeed, such medical intervention may be thought to *promote equality* by reinstating equality of opportunity, undistorted by adverse medical conditions (3,15). In other respects, it may worsen equality conditions through pressures to move affected persons to normality. This may downgrade "alternative lifestyles," with adverse effects on various groups — those within the deaf culture, for example. There will, nevertheless, be strong pressures to expand the boundaries of the disorder/treatment model in order to secure insurance or other forms of payment. Of course, the greater the expansion of coverage of various "medical conditions" and of persons, the higher the price of insurance, and the greater the exclusion of lower income groups.

But enhancement not justified within a disorder model is likely to be seen as impairing equality in several ways. One is by distorting "nature-based" equality of opportunity resting on native endowments as elevated by customary forms of self-improvement. Another is by interfering with the unequal but arguably justified outcomes of competitive pursuits (43). (This is of course heavily dependent on the reigning political philosophy.) On the other hand, suppose that enhancement becomes legal, its use disclosed, its price relatively low, and its efficacy roughly the same for all. Then, whatever other objections would remain, inequality concerns, though remaining important because of existing and enduring positional differences, would be partially muted. Of course, other moral issues about enhancement would endure.

However, if variations in natural endowments were thought irrelevant to merit and desert, the point of distinguishing treatment from augmentation would largely be lost, except for clear medical need. As Daniels puts it, if one rejects the "standard model" that takes the distribution of abilities as given, then "the distinction between treatment and enhancement has no point, at least where enhancement is aimed at equalizing capabilities (43, pp. 124–125)."

Enhancement and the Demise of Merit; Person Perception Again; Interpersonal Comparisons of Merit, Desert, and Equality; Entrenchment of Elite Blocs (New or Old); Racial, Ethnic and Gender Dimensions of Enhancement.

In General. Perhaps the very idea of merit would largely "drop out" if the use of enhancement resources were widespread and comprehensive, surviving, if at all, only when applied to judging skill in arranging for and using such techniques. (If we also do not deserve the traits we each received From Above, what meaning does "merit" have other than a thin estimate of economic worth?) Assuming such "no-merit" assessments are made, they are likeliest when living persons are augmented by medical/surgical means, including somatic cell gene therapy or "genetic pharmacology." But our more immediate target is to trace possible effects of such enhancements on different forms of equality: social equality; political equality; equality of opportunity (broken down into matters of means, prospect, and so on (2,44)); group equality; and the roles of merit and need in making equality judgments. Although these broad and rather clumsy categorizations of equality can carry us only so far, they are useful starting points.

"Social equality" rests partly on the differing frameworks for "person-perception" (44) we use to appraise each other—and ourselves. Perhaps the genetically enhanced would perceive themselves—and be perceived by others—as "superior" in any of several senses: possessing greater intrinsic merit (even if artificially elevated by humans) and hence desert for various rewards; being more useful to society—and so more worthy, in both moral and nonmoral senses; and belonging to an elite group holding substantial political power. Perhaps this elite group would be the successor to an established powerful group; or it might constitute a new kind of elite based on genetic or other augmentation.

Another concern relates to the formation of blocs defined by the particular nature of the enhancement. People regularly sort themselves into groups defined roughly by the strength of particular traits: the more intelligent, the more physically fit, the more nerdy and so on. Enhancement of these traits might solidify these groups and strengthen their political and economic power. Their continued existence as discrete and enduring entities is suggested by how they differ from, say, political parties in the United States.

More generally, if distribution of expensive enhancement resources followed a market or preexisting merit path, existing socioeconomic distances would be enlarged and less bridgeable. This might reinforce adverse views about various ethnic and racial characteristics. Because of the self-reinforcing nature of distributions of "merit"—of the very grounds for distribution generally—the creation of entrenched elites may be hard to reverse. (If traits involving regard for others were so distributed, it might seem odd to speak of unbridgeable distances.)

Threats to Equal Respect for Everyone's Common Personhood: Enhancement as Intensifying Concern with the Strength of Specific Traits, Leading to Reductionism Generally and to Devaluation Resulting from One's Reduced Standing. Some equality judgments may involve a sort of "suspended belief" concerning the extent of interpersonal differences. One reason technological enhancement seems more unsettling than familiar forms of self-improvement is that it calls attention more specifically to the enhancing agent's target traits. Why such enhancement is more salient than long-term, gradual advances is an issue for cognitive psychology, particularly as it bears on person-perception. In any event, our common personhood may be overshadowed by a more intense focus on interpersonal differences. Our moral value *as persons* may be partially displaced by our increased value *as bearers of certain traits in certain measures* (14). To *plan* a person's traits suggests that those traits, as augmented, reflect his primary or even his only value—"value" here meaning social utility. This reduction, as suggested, is affiliated with the ideas of mere use of persons (in violation of the second formulation of Kant's categorical imperative) (45), objectification (46), and related processes. (The nature of this affiliation will

not be investigated; it is not necessary to specify which of these notions are criteria for the others or inferences from the others, and they are taken as more or less substitutable here.)

Threats to the Valuation of Persons with Conditions Generally Viewed as Disabling, Particularly Those Who Decline Measures—"Therapeutic" or "Enhancing"—to Eliminate (or Improve) the Condition or Prevent It in Others. There are conditions that, in some cases, are not viewed as disabilities but as enablers. The best-known example is that of deaf persons, at least within what has come to be known as the deaf culture. If they decline measures to enable them to hear—assuming effective measures became available—they might be severely called to task, particularly if they continue to press for special social services. If parents, deaf or hearing, decline these measures for their children, they may be criticized even more severely. These measures—the limiting case of "normalization" (as compared to "assimilating" or "mainstreaming")—may, as Silvers puts it, "devalu[e] alternative or adaptive modes of functioning" (12, p. 112). Still more, there is controversy concerning the moral propriety of terminating pregnancies when the developing fetus is believed *not* to be affected by a form of hereditary deafness. Moves to make this more difficult for deaf couples may also be viewed as devaluing deaf culture.

Forms of Regulation; Markets and Other Procedures; More on Equality's Internal Conflicts.

Natural Differences. There are obvious natural and acquired differences among persons, although their significance and even their recognition may rest both on competing moral frameworks and on cultural variables (18). Not all differences—assuming they are perceived at all—count as inequalities. In some cases, cultural variation may be of modest significance. Persons born without limbs, for example, will have difficulty in moving independently from place to place, and this is obviously a crucial ability for most persons in most cultures. But just how well such persons fare depends heavily on variations in familial and general social practices affecting those impaired in this and other ways.

Where differences are recognized as significant, however, one must inquire into their moral status how they should be dealt with. The differences might be taken as given and their effects left to the workings of decentralized market, kinship, or other private arrangements. Or, communities might try to improve matters from an egalitarian perspective, viewing the fact of major interpersonal differences as "natural wrongs" or injustices (17,18). Any effort to displace the market would of course take us into a different phase of moral and policy analysis: establishing criteria for distributing resources, including enhancement resources, and specifying procedures to verify that the criteria have been satisfied by prospective recipients (native endowments, accomplishments, prospects, interpersonal connections, etc.). Any choice of distributional regime—market, nonmarket or mixed—necessarily involves contested moral issues. The "genetic supermarket" (Nozick) may be efficient (21,47) in some sense, but it does not bypass foundational problems.

Recall that the developing technologies involved here are "reflexive" in the sense that they are meant to alter and enhance its consumers, and access to such resources is likely to roughly track prevailing distributions of economic or political power. This changes the normative terrain considerably because each distribution changes any static models we have been dealing with by (possibly) sharply changing the criteria for each successive set of distributions. We can no longer rest on assumptions that major human traits change only gradually, if at all (48).

Janus-Faced Equality. Technological enhancement provides some opportunities to even out nature's hierarchical roughness. It also creates the possibility of worsening it, as emphasized above (1,37). If such leveling is indeed a community moral obligation on egalitarian (and other) grounds, a ban on enhancement—when used to promote rather than impair equality—might violate a principle (corollary to some forms of equality) mandating rectification of specified inequalities through certain mechanisms.

But such apparently egalitarian rectification efforts would require centralized intervention into distribution of enhancement resources. Enabling the have-lesses to move closer to the have-mores might thus *itself* violate some aspects of equality—and of autonomy, fairness, justice, and utility—through coercive redistribution. As we saw earlier, some would lose more of what they earn than others, and some would receive unearned benefits while others would not.

In any case, it is unlikely that any distributive scheme would "level out" human traits. And few—from current perspectives—would think it desirable, morally or otherwise.

A Review: Inequalities Compounded; the "Matthew Effect" and Terminal Social Stratification: The Problem of "Who Merits Merit?" Again.

The rich get richer, the poor get poorer, ... and the smart get smarter? Why not? The well-educated already more easily qualify for still more education, often to the exclusion of the less-educated (49). "For unto every one that hath shall be given, and he shall have abundance: But from him that hath not shall be taken away even that which he hath" (Matthew, 25:14–30). (Merton coined the phrase "Matthew Effect" in referring to allocation of resources in scientific research.)

Distribution of scarce resources is of course a classic problem for economics, ethics, political theory, public policy, and just plain politics. But the distribution of enhancement resources, as we saw, raises special issues. Enhancement almost inevitably targets merit attributes—which are generally wealth-attracting resources.

The relevance of the Matthew Effect is obvious. The distribution of resources for enhancing "merit" claims for distribution, as we saw, involves a sort of feedback loop: It alters the very ground on which the initial distribution is made, generating a multiplier effect. Under such conditions Thomas Jefferson's "natural aristocracy" of "virtue and talents" is replaced by an artificial aristocracy of technologically enhanced abilities (50).

Of course, if the idea of merit does not survive the new age of technological alteration, "Who merits more 'merit'?" becomes doubly a nonsense question. Not only are we unable to increase our merit artificially, but merit itself is gone as a relevant moral category. Even if all or some characteristics lose all or some of their status as merit attributes, it seems likely that enhanced intellectual and physical powers, unevenly distributed, will continue to attract wealth and resources. Business, after all, is business: Intelligence counts for scientific research; heft counts for football; attractive faces and bodies draw attention and money—whether or not we talk about merit. The demise of merit, moreover, may not dispatch the view that we are nevertheless "entitled" to the fruits of our varying natural *or* enhanced abilities—whether we "deserve" any of them or not. In any case, the outcome of decentralized distribution of resource-attractors, as suggested, might ratchet up social, economic, and political stratification, and the hierarchical structure of community life generally. At least this is a potential outcome if distribution is based largely on decentralized mechanisms, such as markets, kinship, or old-boy/girl networks.

Enhancement and Interpersonal Desert: Time Scales, Life Plans, and Social Stability.

Enhancement of Living Persons within Their Respective Memories. We are accustomed to the gradual acquisition of merit earned by effort, resulting in gently escalating desert Sudden, major alterations in attributes, particularly merit attributes that help define one's identity, aren't associated with ordinary persons; Western culture links "shape-shifting" to mythological para-human creatures. But part of the very point of technological enhancement is to shorten the time span and reduce the effort needed to strengthen one's attributes beyond their endowed "maximums," and to gain the resulting incremental rewards. Such sudden changes in individual capacities may present major difficulties to a transformed person, to those around her, and to society generally (41). Our choices about life style and life plan have always depended strongly on presuppositions about our attributes—both assets and deficiencies—and their general stability (a stability consistent with their gradual elevation or deterioration). Suppose, however, that someone of modest talents and accomplishments rapidly becomes abler. Would she think that she *deserves* more of life's rewards because her abilities have sharply increased? How would she acquire these rewards? The newly intelligent or memorious (51) can't just saunter onto the grounds of Acme University and demand entry and possibly displacement of their (new) inferiors. Or can they?

Perhaps the spreading self-awareness of new powers—and the spreading fears of those stuck where they are—will provoke political and social instability. A somewhat distant analogy would be the sudden emancipation of large numbers of slaves or indentured servants who had been denied education and other resources needed to flourish as free persons. Think also of the comparatively rapid (if incomplete) change in the status of women in the United States and elsewhere. Virtually every aspect of equality would be challenged by technological enhancement. In particular, the nature of the contests between different forms of equality may also change. The perennial war between forms of equality of opportunity and forms of equality of outcome may be intensified by the limited availability of enhancement resources that greatly enlarge one's prospects, and by the growing "distances" between the enhanced and the unenhanced.

There is thus a two-stage egalitarian problem: determining who gets "merit-enhancing" resources, and determining what collective or individual responses to make when faced with the escalating demands of the newly enhanced. These *nouveau intelligent* do not suddenly enter the fabled set of fully qualified rocket scientists. (Memory and skills transfer is not discussed here.) But they will argue that they have joined the set of persons immediately entitled to further education and training, and, within a short time, to appropriate forms of employment and their attendant rewards. It is too early to say whether it will make any difference whether they frame their claims by relying on merit and desert or on economic and social utility.

Other Enhancements. Questions parallel to those just raised arise with those whose genomes were altered, or possibly whose traits were revised during embryonic development (but without genome changes) or fetal development or in early childhood. Our responses, however, may be different. All of these persons are likely, in different ways, to look upon themselves as identified with traits they have had "wired in" for as long as they can remember. Moreover, for whatever it is worth, they cannot themselves be accused of having tried to evade or soften the struggle for self-improvement and to reap unearned benefits. Those enhanced as adults or older children, however, will be able to compare their attributes "before and after" enhancement, and those who affirmatively opted for enhancement might be blamed for such (partial) evasions.

Again, it is unclear how limited-access institutions such as educational facilities and desirable employment opportunities could quickly adjust, even over one or two generations, to a sharp escalation of merit claims for entry. A further complication is that some forms of labor may become even more disfavored than they are now among more educated groups — cleaning/sanitation, simple but hard labor, some forms of blue-collar work, and various low-skilled personal-service functions. Other things remaining fixed, however, the shortage of supply for such labor would raise its wage rate, which presumably would draw applicants willing to trade (temporary?) embarrassment for an enlarged income. (But then, their services might then be too expensive for many consumers, especially among the unenhanced.)

Both Groups in the Long Run. The questions just raised also apply to long-run considerations, and here matters become still more speculative. How would we forecast shifts in attitudes and beliefs about interpersonal comparative valuations? We do not know how we will value (or reduce) each other when merit traits are significantly malleable.

Still we are not entirely at sea and can make at least minimalist projections. One would think that with escalating demand for the (possibly) superior services produced through stronger merit attributes applied to increasingly more demanding tasks, investment would gradually yield institutional responses: more educational facilities, more complex mental and physical competitions, and new technologies enabling disfavored lines of work to be done more by machines and less by persons. This would generate greater incentives and pressures for still further personal enhancement, new stages of institutional response, and so on. It is hard to say where diminishing returns and equilibrium would set in. Recall also that one effect of the greater salience and strength of merit attributes might be to amplify the social, economic, and political importance of the enhanced traits: enhancement efforts would be likely to require major investments of all kinds, both financial and emotional — and people want returns on their investments. The result would be still greater emphasis on interpersonal differences.

One theoretical possibility should be kept in mind for analytical purposes, however unlikely it may be: With broad access to similarly effective agents, there might be little *relative* interpersonal change, even though everyone's individual performance level was raised. But this would also represent a major source of pressure for social and economic revision: Nearly everyone will be abler, more insistent on appropriate rewards for ability, and more concerned about the responsive formation of new institutions to satisfy their new levels of talent. The sluggishness of social and political responses to the claims of those with newly enhanced attributes may contribute heavily to various forms of social instability. And some instabilities might well arise because of negative shifts of attitude both toward those with clearly defined disabilities or handicaps and those who are unenhanced (or not successfully enhanced) and find themselves ever-lower in relative standing.

Equality of Groups and Blocs: More on Social Stability. One uncontroversial point is that groups and communities play major roles in social and political life and, partly as a result of this, in the formation of one's sense of identity and self-regard. Matters of interpersonal equality are thus conceptually linked both to intergroup and intragroup equality — and both realms of equality are affected by prevalent views on merit and desert.

Humanity has generally sorted itself into groups, and some existing groupings are defined by observed or supposed differences in merit traits and accomplishments. Indeed, in a distributional system based entirely on the purest notions of merit, with the arbitrariness of prejudice, stereotyping, corruption, fraud, and coercion largely absent, one could infer that resulting differences in attainments, rewards, and status are based entirely on differences in abilities or other merit or wealth-attracting resources. Perhaps this just replaces one set of "arbitrary" criteria (old-boy/girl networks, ability to pay, kinship preferences) with another (genetic and environmental lotteries), but this is another issue. In any case, this somewhat intimidating prospect has been addressed in several well-known (and controversial) works (52). If realized, we would in theory lose our excuses for failure (e.g., "politics did you in"). Our relative status would rest on "the merits" and unambiguously reflect our attributes, perhaps as in Mensa, whose membership is chosen (in theory) on the basis of pure ability rather than interests or accomplishments. Once again, we face a raising of the borders between existing groups, and possibly the creation of new entrenched factions.

Still, we do not know whether any given pattern of enhancement would inspire social/political instability. Much may depend on whether, despite the greater gaps between individual and groups, the lot of the worst off is nevertheless improved (53). If the size of the gap between the better and worse off is great enough, the overall risks of instability may go up even if the resources of the less well off increase. Our notions of poverty seem to involve ordinal rankings as well as the cardinal value of one's holdings.

Political Equality Imperiled: In General.

Shifts in Political and Moral Ideals; Widespread Use and Low-Cost Access. It seems that many prospective parents now prefer the genetic lottery and are eager, or at least willing, to accept whatever they receive from it. We often eschew planning even where it is possible to plan, preferring vagueness and uncertainty to precision and predictability. Perhaps a partial explanation for this is fear of responsibility when things go awry or simply not as planned, or confusion over what to select. Beyond this, such preferences seem linked to what is perceived as a defining element of personhood: We envision persons as creative and autonomous, not bounded by fixed life plans imposed on them. Nonpersons have none of these attributes. Neither natural nor assembled objects possess them, and other living things seem too far off the mark to justify such characterizations (54). But competitive pressures may inspire many parents-to-be to seek greater precision of outcome in their reproductive plans, and (possibly) to rigorously enforce these plans on their offspring. Fear of an unenhanced child's eventual reaction to discovering that she is disadvantaged might well play a role here. (The parents in most cases cannot respond by telling the child that she had no alternative existence. The selfsame embryo from which she developed could have been isolated and altered, or her traits could have been changed after birth.)

Assume that the longer-term results of these pressures and of economies of scale are nearly universal low-cost successful efforts to enhance. Equality complications attributable to enhancement would be then be greatly attenuated, though not entirely removed (and standard equality problems would likely endure). From this particular egalitarian perspective, if not others, the more technology and the wider its use, the better.

However, where there are large-scale distributional inequalities, there is a risk of (irreversible?) erosion of equality's status. Equality could be adhered to (if at all) only in the sense of preserving the abstract idea of equality of opportunity: no affirmative blockade interfering with one's right to use her preexisting intelligence and wealth to secure more intelligence and wealth, for herself and for her existing or future offspring, and to reap the benefits of her enriched capacities—and so on down the dynastic generations.

How might this shift our ideals? Institutions and practices, by their very existence and visibility, "communicate" ideas and impressions, and these may have learning effects (55). Of course, what is "learned" depends on what is perceived or understood, and might be reshaped by responsive public debate.

Segmented Society. One feature of a world with both genetically and nongenetically enhanced persons might be a more rigorous division of labor, perhaps of the sort envisioned by Plato in his *Republic* (56). After all, if we take the trouble to (re)assemble our offspring with certain "engineered" traits, they had better do what we planned, right? Equality analysis here is of course beset with factual and normative/conceptual uncertainty. On the one hand, the escalation of technological complexity combined with enhancement might lead to greater division of labor and social stratification. On the other hand, enhanced persons might form a world with less rigorous division of labor because they become polymaths and jacks-of-more-than-one-profession.

Still, it is conceivable that regardless of how rigorous the division of labor is, political and social equality of a sort may hold. Different professions, trades, and occupations, and the varying aptitudes underlying them, might be viewed as equally worthy—an "equality of the enhanced." The "alphas" may be viewed as equal to the "betas," though their augmentations (via the germ line or the living body) and life work may be entirely different.

But this is nothing to count on. It is also plausible to expect that equality is largely "read out" where (from our present perspective) it is most applicable and most needed. The more entrenched the social stratification becomes, the greater will be the need for corrective notions of equality and of "remediation," but the less likely it is that there will be influential partisans for equality in any sense.

A More Equalized Society Instead? As suggested, enhancement resources might be distributed in ways that promote equality in several forms, consistently with whatever divisions of labor are implemented. Distribution might rest on need, where "need" is linked to enhancing equality of opportunity, perhaps vindicated by some degree of social assistance. Moreover every person might be considered to have a stronger claim to augmentation than his immediate "superiors" in preexisting attributes, giving him the right of first refusal for the next set of resources. And, as we saw earlier, where different traits are enhanced, the "net equality of the differently enhanced" may hold—as equality of "overall" merit. Finally, greater equality might be pursued by familiar redistributive or other social measures that are not directed toward trait alteration. So there is a slight possibility of a "more equalized" society. But this possibility should now be vetted through the lens of democratic theory.

More on Political Equality Imperiled: Democracy and Governance.

Enhancement and Democratic Theory: Millian Plural Voting and the Attenuation of Democracy

1. *Kinds of democracy: Is one-person, one-vote a defining characteristic of democracy?* What are and should be the effects of sharp differences in human

characteristics on matters of political governance? If we are not in fact equal to each other in deliberative ability, judgment, and drive, why do we all have equal voting power in the sense that when casting ballots in general elections, no one's vote counts for more than another's? We are not equal in our knowledge of the issues, our abilities to assess competing arguments, the nature and intensities of our preferences, our capacities to contribute to our social and economic system, our stakes in the outcomes of particular government policies, or even in our interest in participating in public affairs. And enhancement technologies may amplify these differences.

Yet for most of us, "democracy" seems to be all but definitionally connected with the maxim "one person, one vote." Unless this maxim holds, there is on this view no true democracy. Is this definitional link indeed appropriate given our vast interpersonal differences? Not all political thinkers have thought so. As Thompson summarizes John Stuart Mill's discussion of plural voting:

> The principle of competence expresses Mill's belief that a democracy should give as much weight as possible to superior intelligence and virtue in the political process (57).

Mill thus did not think that equal votes among electors was essential to democracy or for promoting the public good—quite the contrary. He endorsed plural voting (though perhaps with later reservations and possibly as a temporary measure) in which individual citizens had votes proportional to their "individual mental superiority" (58,60). The number of votes per elector would thus be a function of his or her revealed competence. Mill discussed occupational success, test results, and educational status as criteria for assigning more than one vote (59, pp. 475–476)

For Mill, plural voting is one method, among others, for furthering the principle of competence (57). His idea of competence is complex, however. It is not addressed simply to intelligence, but to skills (57), since highly intelligent persons might lack skills needed for sound governance. His vision of ideal competence also includes "moral competence." Education seems to be not just a proxy for competence, but partly constitutive of it. (One supposes that it might be a proxy for native ability.) Finally, Mill qualified his recommendations by recognizing that participation values were in tension with competence values (59,60).

The link between Mill's competence principle and human enhancement is clear. Genetic engineering, for example, has long stimulated fears that enhancement would threaten democracy—at least in forms demanding equal votes for all electors. Sinsheimer has asked, "Could … deeper knowledge of the realities of human genetics affect our commitment to democracy (61)?" If mere knowledge of the physical bases (both genetic and nongenetic) of the differing endowments underlying Millian competence can threaten democracy, one might well expect that the vivid reality of the huge gulfs between the enhanced and the unenhanced would represent an even greater threat.

But the nature of the threat to democracy must be specified. This is no simple task, given the fact that "democracy" may take quite different forms, and that the status of the one-person, one-vote standard as the premier form of democracy is a question at issue. All forms of democracy are linked by the idea that the governed—or some portion of them—are to have a significant say in what affects them, and that this voice is to be broadcast by some form of majoritarian aggregating of votes on important matters. This voice is an obvious component of autonomy, which is in turn an essential ground of democracy, and it is not simply advisory or merely a request for redress coming From Above. It is to be decisive within significant domains, although it may be subject to principled constraints derived from constitutions or other sources of law. What constitutes an "important matter" is of course hugely uncertain, but clarifying it is unnecessary here.

Return now to the question that opened this section: Why is the political equality that is implemented by one-person, one-vote accepted in the face of individual differences? Dahl raises a parallel question:

> [I]f income, wealth, and economic position are also political resources, and if they are distributed unequally, then how can citizens be political equals? And if citizens cannot be political equals, how is democracy to exist (30, p. 326)?

(Dahl is describing conflicting theoretical perspectives, not necessarily endorsing any.)

2. *Applications to an age of enhancement; one-person, one-vote.* A system of equal votes at the ballot box is far from ensuring "equal" political influence or equality in anything. Think, for example, of the suppression of group preferences in at-large voting districts (62). Nevertheless, in its own way, one-person, one-vote implements equality both practically and symbolically. If effective enhancement is feasible, might equal-vote democracy (somewhat paradoxically) be the preferred form of political governance *because of*—rather than despite—greater interpersonal differences? After all, even though we are not equally *able*, we may be more or less equally *affected* by particular government policies, and, whatever the unequal impacts, they are not uniquely correlated to ability. To respond that impacts on the less able count for less than impacts on the more able is to presuppose a far different theory of the equality of persons *as* persons than is now held, at least in many quarters. Still, the idea that all persons are equally affected by a given kind and degree of adversity might itself be under siege in an enhancement age.

It seems unlikely that unequal allocation of votes would be seen as a realistic, efficient, and benign recognition of differences in ability, native or augmented, or of the varying impacts of political policies. It will probably be taken, correctly, as reflecting deep disrespect for those allotted fewer votes (63). And no doubt, many of those with more votes—and some with fewer—will believe that disrespect is justified given the substantial gulfs in resource-attractive or merit traits.

There are, of course, conceptual issues and "exceptions" to the one-person, one-vote standard. That standard is arguably attenuated by many institutions, such as the U.S. Senate, where states have equal votes whatever their respective populations, or special voting units such as water districts, where votes are allocated on the basis of varying rates of use or on other variables.

But in general elections, at the ballot box level, plural voting is excluded from most modern ideas of democracy. Mill himself did not necessarily endorse it over other techniques for enhancing the influence of competent elites (57,59). He seemed well aware of the substance of the Matthew Effect: Those with excess voting power may draw increasingly disproportionate shares of rewards (48,65) and possibly still more voting power, in an extended cycle. He did not endorse the ' "blind submission of dunces to men of knowledge." ' (57, p. 85) As mentioned, he also strongly emphasized participation values in democracy, which help to control government and to educate the participants, making them more competent (57). As Thompson describes Mill's resolution of the tension:

> Just as the educative benefits of participation partly justify the extension of participation, so the educative value of superior competence partly justifies the influence of a competent minority (57, p. 79).

But participation to promote competence will not necessarily save the day for one-person, one vote—particularly in an age of enhancement, with its increased and entrenched gulfs in ability, as discussed next.

Enhancement and Democratic Governance.
Plural voting is a long way from dictatorship or other autocracy, but it is nevertheless likely to be taken as inconsistent with the idea of equality of persons *as* persons (60). One might thus question the seriousness of enhancement's challenge to democracy by recalling that we now maintain democratic ideals *notwithstanding* the *present* perception of *very* wide interpersonal differences. A major rationale for maintaining the one-person, one-vote regime is to prevent further consolidations of power that leave persons with inadequate access to basic commodities and opportunities.

But our commitment to democracy—and/or to various sociopolitical conditions that enable democrats to implement their commitment—might be fragile nonetheless

Equality of control is an unstable equilibrium. Differences in knowledge, skill, opportunity and activity create inequalities of control; these in turn tend to generate further differences, which create further inequalities. [Note how this may be compounded in still further cycles by enhancement.] Hence the struggle to maintain a polyarchal organization ["[t]he main sociopolitical process for approximating (although not achieving) democracy ..."] is never won; indeed, it is always on the verge of being lost (65, pp. 41, 282).

Enhancement and Participation.
Representative democracy is not just a matter of voting rights and voting power, and elections do not confer unreviewable, irreversible delegations of authority to representatives or officials. Ideally it entails genuine opportunities for participation, in order to promote its underlying ideals of autonomy and equality, in one form or another. The sort of narrowly defined "efficiency" promoted by restricting voting and governing to superior elites is not part of the democratic canon (47). Participation is a troublesome concept to interpret: There is speaking one's piece before the appropriate representatives and government officials, there is *influencing* their exercises of power, there is having access to relevant information and ability to comprehend it, there is being a plausible candidate for office as a representative or for appointment to public office, and so on. All these aspects of democratic participation may be affected by enhancement, whatever mode of distribution of enhancement resources is selected. One's greater or lesser abilities may expand or contract one's audience, or ability to communicate with politically powerful persons and groups, or relative deliberative skills, or capacity to quickly grasp the issues of the day, or ultimate influence. Moreover, in republics we delegate responsibility for governing to others, partly from the sheer need for division of labor, partly because we want government to be run by persons capable of doing so soundly. To be nonenhanced—that is, to be relatively less capable—may be to risk exclusion from government office.

As suggested, however, the prospect of enhancement may not be fatal to egalitarian democracy, either in political theory or in fact.

First, the arguments about allocating votes as a function of competence may be somewhat misdirected. Democracy, again, is in part about having a say in what affects one. But as we saw, how much something affects you may have little or no connection to your varying competences. Moreover, to justify plural voting on our understanding of democracy, we need a moral premise concerning the proper relationship between one's political power and one's particular circumstances—including not only one's competence but one's vulnerabilities to harm under government policies. Within our present political framework, the premise is not confirmed. Representative democracy may contemplate an ideal of superbly qualified electors and even more superbly qualified representatives, but the ground for democracy is not the superior decision-making competence of the people and their delegates, as opposed to despotic rulers or elites. It rests *generally* on the unfairness and injustice of impairing autonomy by subjecting people to policies, conditions, and interactions that seriously affect them when they do not have a voice in the matter, at *some* important level of choice. We cannot order the President to cease bombing the principality of Lower Paregoric, but we can select the President—via electors we vote for. And the ground for *equal-vote* democracy, as we saw earlier, rests partly on the unfairness of giving unequal power to persons whose vulnerabilities are likely to be quite similar, whatever their mental and physical aptitudes. Thus, the "equally affected" argument may overpower the "superior contribution" argument. Nevertheless, the possibility remains that our current notions of equal vulnerability and impact will change as enhancement technologies develop.

To turn matters around, one might urge that under given circumstances it is the *enhanced* whose participation is at risk—particularly if they are a numerical minority. But even if they are endangered by their "suspect" status, plural voting may not be the best mechanism for protecting them as compared with a strong regime of individual rights. Of course, that regime may also be impaired by a hostile majority of the nonenhanced.

Second, egalitarian democracy might survive even within the Millian framework because the available enhancements might not be seen as affecting competences relevant to democratic governance. It is not clear, for example, that moral competence can be affected in any but the most slapdash way by genetic engineering—although the possibility of doing so should not be entirely dismissed (66).

Third, as a matter of theory, it is unclear how a competence criterion for ballot power can be assessed entirely independently of certain background moral issues concerning, say, the fair/just/egalitarian distribution of goods and services. With enhancement, the "ability gulfs" between persons are themselves a partial function of preexisting wealth differences. Depending on what forms of distribution of enhancement resources were in place, these wealth differences would be unjustifiably ratified and reified by plural voting. Because Millian competence is empirically tied to wealth, which may be morally irrelevant, to defend plural voting on competence grounds thus begs some questions of moral evaluation concerning distribution and its underlying issues of equality and fairness.

Superior competence, in this context, thus remains a murky concept. Indeed, as Singer observes, "Mill himself said, later in life, that [plural voting] was a proposal which found favour with no one. The reason, I think, is not that it would obviously be unfair to give more votes to better qualified people, but rather that it would be impossible to get everyone to agree on who was to have the extra votes" (67).

Fourth, even if technological enhancement did affect relevant forms of competence, those who remain unenhanced are not "incompetent" in any sense, including Mill's. A loss of relative standing in ability or depth of learning does not entail deliberative incompetence. "Competence," at least for present purposes, arguably concerns attaining a certain threshold at least as much as it concerns the distance between oneself and others, though the two are connected. In this respect it is similar in structure to "personhood." Here a Millian might respond

that enhancement could simply elevate the accepted competence threshold for qualifying as a voter, establishing a new minimal baseline for competence—but this still would not make the case for supernumerary votes.

Fifth, far from being inconsistent with equal-vote democracy, the increasing gaps between persons make it all the more desirable to retain that voting system, as suggested earlier. The less endowed and less enhanced are not likely to suspend pursuit of their own interests, despite their new relative dimness. Although the better endowed might be better able to protect themselves, given their superiority, a possible result of plural voting might be dangerous instabilities, partly because of the perceived risk of—and actual—aggrandizement of resources by the elites. The greater the fear of such risks, the more that departure from equal voting will be seen as sending us down a steep, greasy slope emptying into an abusive oligarchy—run either by the numerically inferior enhanced or by the unenhanced, each fearing domination by the other. In such a world, not only is equality compromised, but so also are all other basic values.

Turn now from equality to autonomy. (This in turn will shortly return us to equality.) What will become of *it* if enhancement is institutionalized to some degree? In parallel to the dismissal of the respect owed to the less gifted, one might urge that not only do they deserve fewer rewards, their autonomy is of lesser worth. From contemporary liberal perspectives, however, basic autonomy is not tied to one's measure of abilities, unless it falls below the general competence threshold, however defined. Yet, just as we make interpersonal comparisons of "worth" in various senses, we may in fact think that autonomy as exercised by different persons may decline in value with the declining *relative* competence of these actors. This view may have still greater pull where enhancement is practiced. Perhaps if autonomy for all is to be protected, some sort of equal-vote democracy is necessary to preserve it. As we saw, democracy might remain preferred partly because of the posited inequalities, not despite them. Still, defenders of plural voting or rule by an elite are likely to suggest that, precisely because of the elite group's superior competence, autonomy and even equality itself are better promoted by what seems like an inegalitarian system (29). It is thus hard to deny that participatory/autonomy values are at elevated risk in an enhancement context.

Sixth, perhaps the most obvious defense of equal-vote democracy is that it may be instrumental in promoting opportunities to obtain the very enhancement resources that inspired this debate about democracy's requirements—a continuation of enhancement's potential role as "remediation" of natural inequalities. There is certainly no assurance that the elites will look out for anyone's interests but their own, except on the doubtful assumption that they will also be moral elites with a strong egalitarian or altruistic bent. It bears mention at least once in this entry, despite the point's familiarity, that the result of superior competence may be greater and more successful evil.

Finally, plural voting defenders will, sooner or later, make the simple-sounding argument that there is *no* threat to equality in an enhancement age. Equality, after all, concerns the similar treatment of similarly situated persons and the dissimilar treatment of dissimilarly situated persons. If the more able are relevantly different from the less able, treating them differently is not only not inconsistent with equality, it is required by it. The obvious response, which can only be summarily stated here, is that this claim presupposes a large set of unconfirmed and strongly contested moral propositions.

Social Changes in Attitudes Concerning Equality, Self-Regard, and Community: Symbols, Communication, and Learning. The operation and observation of our social institutions and practices generate learning effects. Present conceptions of equality and other values may eventually confront a world where long-standing assumptions about the relative stability of traits and character will be loosened. This emergence of a world in which human traits are far more controllable than now may, as suggested, drive changes in our attitudes about the demands of equality and fairness generally, and merit and desert in particular. These value shifts may occur for several reasons. For example, the consolidation of political power into hierarchies (whether or not reflected in plural voting) may result from the distribution of enhancement opportunities to those already holding wealth and power. Hierarchical institutions and practices may generate self-perpetuating learning effects through citizen participation or observation. People may come to perceive themselves and their social stations differently, perhaps as fully locked in. Enhancement may spur an increasingly intense focus on traits and their comparative measures, and magnify their apparent social and commercial value. True, we might still think that traditional enhancement enhances but that technological enhancement reduces. But whether the latter will indeed *reduce* persons to the social value of their enhanced traits or *elevate* them in a morally relevant sense is not now predictable.

CONSTITUTIONAL CONSIDERATIONS IN BRIEF

Constitutional Frameworks

In the United States, government regulation of use and distribution of enhancement technologies must be tested against claims of violating implied "fundamental liberty interests" under the due processes clauses of the Fifth and Fourteenth Amendments. (As for equal protection considerations, see the discussion below.) Federal action must also be tested against express and implied limitations on the powers of the federal government. Mention of some constitutional considerations is thus called for, and constitutional argument structures are in any event useful in discerning and addressing some of the most important issues generated by enhancement.

The right to procreate, as articulated in *Skinner v. Oklahoma* (68), might be taken to encompass at least certain forms of germ-line engineering or fetal manipulation, although the strength of such rights is open to serious question; the U.S. Supreme Court may be forced to consider a hierarchy of procreational liberty interests,

each imposing a greater or lesser burden of justification for government regulatory maneuvers. The recently emerged practice of prenatal and preconception screening for disorders, which is likely to be protected as a major adjunct to procreational autonomy, might suggest parallel protection of affirmative intervention to forestall the disorders via germ-line or fetal alteration. Nevertheless, the issue is uncertain, partly because of the differences between "standard" (if technologized) reproduction and anticipated future forms: *having children at all* is not the same as *having children in certain ways, or of certain (arranged) sorts.* It is one thing to leave matters to unrevised sexual recombination, and another to affirmatively determine the traits of a specific individual.

Somatic trait augmentation—at least for competent adults—arguably ought to have greater constitutional protection than parental choice to manipulate the germ line or alter fetal development because the affected party is the decision maker. Nevertheless, it is more challenging to describe the constitutional terrain because there is no clear, recognized conceptual bin in which to place it. (Compare "procreational autonomy.") There is no general constitutional liberty or "privacy" interest embracing a right to do what you will with your body, although some commentaries, scholarly and nonscholarly, might suggest otherwise. There are recognized liberty interests of sorts in refusing various forms of medical treatment and in "personal security," which are likely to extend to forced administration of enhancement techniques, medical or nonmedical. But these doctrines do not settle matters of noninterference with voluntary use or positive assistance in securing access.

This is not to say, however, that a persuasive case cannot be made for protecting the decision whether to enhance one's basic merit attributes as an important feature of the liberty protected by the Fifth and Fourteenth Amendments; it is much too simple to assert that such textually unmentioned rights are impossible because no long-standing "tradition" protects it. There is in fact a tradition of substantially free choice in making use of changing methods of instruction, training and general pedagogy for the purpose, among others, of self-improvement. It might well be thought to extend presumptively to control of mental functions and of bodily physiology generally. [Because of the logical link between mental functioning and communication, a First Amendment argument for fair access to intellectual enhancement resources—as well as the right to refuse such resources—might also be crafted (69).] One might also urge that the liberty interest in shaping the nurture and education of one's children encompasses enhancement. The interpretive maneuvers underlying these constitutional arguments are complex and entertaining; they are described briefly below. If serious enhancement arrives on the scene, however, arguments of these sorts are certain to be offered in opposition to restricting access to augmentation services.

Under the logic of constitutional protection of liberty interests, if any of these characterizations of a right to noninterference with enhancement decisions are successful, governments will have to justify their prohibitions and their regulatory systems generally. The weight assigned to the liberty interest will, in theory, determine how heavy these burdens of justification will be. Government will at a minimum have to identify serious interests that may be compromised by attempted or successful augmentation—for example, avoiding injuries to existing or possible persons. It will also, in theory, have to defend the precision of its means for protecting these interests. Imposing a major burden of justification on government action, for whatever reason, is the core component of a judicial decision path known as "heightened scrutiny." In recent years the U.S. Supreme Court has recognized liberty interests that apparently draw an "intermediate" level of scrutiny, rather than the maximum "strict scrutiny" standard (70), and with technological change it may well have to construct still more levels of calibrated protection.

Paths of Constitutional Interpretation

It is especially difficult to project future constitutional analysis when the transformative processes in question seem so far removed from traditional paradigms and historical understandings—assuming these matters remain constitutionally relevant. Tradition, history, original intent, and lexical understandings at the time of framing continue to be viewed as important and perhaps decisive interpretive criteria (separately or in some combination) for both explicit and implicit liberty interests, although some cases have been offered as counterexamples (71). Those arguing that enhancement falls within a strongly protected liberty interest—whether as an aspect of procreational liberty, a right of personal development, or a right to control our mental and physical functions—will have varying difficulties making their case. If their characterization is rejected by the courts, then the government's burden of justification is very weak—a minimal rationality test that constrains far less than use of the term "rational" would suggest in everyday language. The difficulty in constitutional characterization of an interest is greatly compounded when the asserted interest reflects an innovation that does not seem to fit existing categories. Thus, determining whether reproductive ventures involving germ-line engineering are included within a strongly protected liberty interest will, as with any form of legal characterization, involve (inter alia) comparisons to exemplars of what is or is not protected. Partisans then characteristically state whether the interest proposed for special protection is "too far removed" or "distant" from the archetype. (This vastly oversimplifies huge interpretive issues.) The problem, however, is that the supposedly defining features of the models offered may be contested. Is the process of creating a person who didn't exist before a sufficient condition for calling the process procreation—either in common discourse or in constitutionalese? Or must the person have been created by human sexual recombination rather than asexually? Or must the person not only be the result of sexual recombination but of sexual recombination *simpliciter* (where we simply rely on the genetic lottery and avoid affirmative trait-changing,

though possibly using prenatal and preconception screening, possibly followed by abortion or nonconception)? If we do not know what defines the standard example, we cannot tell how "far" we are from it. As things stand, prohibiting prenatal or preconception screening would seem to impermissibly burden procreational rights, but this does not show that forbidding germ line alteration—even for enhancement rather than disorder prevention—is also be impermissible. How "far" is affirmative genetic change in persons-to-be from prenatal or preconception testing in aid of deciding upon abortion or nonconception? All are forms of "genetic control," but germ-line alternation is vastly different, at least when viewed through prevailing constitutional frameworks. When biological technologies separate and rearrange life processes in ways not contemplated by our existing concepts, the interpretive difficulties we already face may be greatly amplified.

For now, it is enough to say that human procreation has come to vary along several overlapping axes. They concern technological facilitation of gamete union (e.g., in vitro fertilization); social arrangements (within or outside marriage; collaborative—e.g., surrogacy; use of gamete banks, whether or not for eugenic purposes); whether the efforts involve asexual methods (cloning); technological mechanisms for trait prediction; and technological mechanisms for positive control or influence over traits. It seems plausible to think that procreational autonomy extends as a *presumptive* protection—in full or near full strength to technological facilitation of gamete union (subject to limited health and kinship regulations), to procreation regardless of marital status (subject to certain protection-of-marriage limitations), and to prenatal or preconception screening for disorders, defects, or injuries. This presumptive protection might be overcome by compelling or important countervailing interests. Broader protection of collaborative procreation, and for which participants, is less certain, although strong coverage is likely for gamete donation or sale (at least for a modest price to avoid charges of "economic coercion") by persons within or outside the intended nuclear family. The constitutional fate of human cloning is seriously in doubt because of the perception by many that asexual reproduction is a truly radical departure from standard procreation, and does not belong within protected constitutional categories. Although germ line enhancement within sexual reproduction is a striking departure from standard reproduction because of its partial nullification of the genetic lottery, it seems likelier than cloning to be assigned some serious presumptive protection, at least within a disorder model. (Cloning may not go entirely unprotected, however.) As noted, however, the strength of "compelling" or "important" governmental interests can in theory override the individual rights claim, if the regulations are carefully tailored to further those interests so as to reduce intrusions on constitutionally protected interests.

Constitutional Equality Standards

There may also be questions concerning the status of the enhanced or the nonenhanced as members of discrete, identifiable groups at risk for discrimination and exploitation. If so identified, classifications concerning the group may be treated as "suspect" to some degree under the Fourteenth Amendment's equal protection clause and the Fifth Amendment's implied parallel protection. This will again trigger heightened scrutiny and, in theory, impose a nontrivial burden of justification on government action. If this "suspectness" characterization fails, the government is likely (but not certain) to prevail (72). The point here is that many egalitarian claims find little or no purchase within the constitutional framework of equality, which offers strong protection against certain forms of discrimination (racial, ethnic, gender, etc.), modest protection against certain forms of classificational irrationality involving vulnerable groups (it is hard to predict which groups will be considered vulnerable), and for all practical purposes no protection for any other form of classification. In egregious cases of abuse or manipulative control over enhanced or unenhanced persons, however, one might claim violation of the Thirteenth Amendment (banning slavery) or the Nobility Clause (Article I, §9) (73). Both provisions are heavily inspired by considerations of equality.

Congressional Powers

Congress has an uncertain range of powers to promote constitutional rights under Section 5 of the Fourteenth Amendment (and parallel provisions in other amendments), subject to Supreme Court control. If any group — nonenhanced or enhanced — seems especially put upon, Congress may consider remedial legislation (perhaps as a form of "affirmative action") (37). Congress also retains considerable powers to protect or promote constitutional rights under the commerce clause, and the taxing and spending powers.

CONCLUSION

Dealing with technological enhancement and its impact on basic values is beset with the usual problems associated with value analysis — vagueness, ambiguity, "open texture," indeterminacy, and collision with other values. But these problems are aggravated because the new powers seem to undermine assumptions concerning our understanding of these values. Determining just how crucial these assumptions are to the tasks of moral and legal evaluation of enhancement technologies forms a major portion of the analytical work required.

The most obvious assumption being tested, of course, is that we are severely limited in altering native traits — including our most valued merit attributes and resource attractors — by the constraints of our individual genetic endowments and by the very nature of familiar and slow-working tools of self-improvement: study, training, practice, effort, self-discipline.

It now appears, however, that technological intervention via the germ-line and somatic mechanisms will eventually allow us to alter, at least in certain ways, the limits of what we now view as relatively fixed potentials for improvement. Today, with extended study and practice as he grows up, Forrest Gump can learn to make change,

to balance checking accounts, and to do some algebra, but quantum gravity will forever elude him. Later, perhaps such limits will no longer hold: evidence of the possibility of serious and accelerated trait changes seems to be growing. Does one's merit, virtue, and ultimate desert rest only on traditional paths toward personal progress?

BIBLIOGRAPHY

1. M.H. Shapiro, *Wake Forest Law Rev.* **34**, 561 (1999).
   M.H. Shapiro, *S. Cal. Law Rev.* **65**, 11 (1991).
   M.H. Shapiro, *S. Cal. Law Rev.* **48**, 318 (1974).
2. D. Rae, *Equalities*, Harvard University Press, Cambridge, 1981, Chap. 4.
3. D.W. Brock, A. Buchanan, N. Daniels, and D. Wikler, *From Chance to Choice: Genes and the Just Society*, Cambridge University Press, Cambridge, 2000.
4. J. Travis, *Science News* **156**, 149 (1999).
   K. Howard, *Sci. Am.* **281**, 30 (1999).
5. Associated Press, *New York Times*, June 3, 1998, A19; T.H. Maugh II, Los Angels Times, February 13, 1997, B2; 1997 WL 2181814.
6. C. Holden, *Science* **280**, 681 (1998).
7. E. Parens, ed., *Enhancing Human Traits: Ethical and Social Implications*, Georgetown University Press, Washington, 1998, p. 66.
8. G. Sher, *Desert*, Princeton University Press, Princeton, 1987.
9. L.M. Fleck, in T.F. Murphy and M.A. Lappé, eds., *Justice and the Human Genome Project*, University of California Press, Berkeley, 1994, pp. 133, 143.
10. D.W. Brock, in E. Parens, ed., *Enhancing Human Traits: Ethical and Social Implications*, Georgetown University Press, Washington, 1998.
11. D.B. Allen and N.C. Fost, *J. Pediatrics* **117**, 16, 19 (1990).
12. A. Silvers, in E. Parens, ed., *Enhancing Human Traits: Ethical and Social Implications*, Georgetown University Press, Washington, 1998.
13. Times Wire Reports, *Los Angeles Times*, September 9, 1999, A14; 1999 WL 26173732.
14. R. Cole-Turner, in E. Parens, ed., *Enhancing Human Traits: Ethical and Social Implications*, Georgetown University Press, Washington, 1998.
15. E.T. Juengst, in E. Parens, ed., *Enhancing Human Traits. Ethical and Social Implications*, Georgetown University Press, Washington, 1998, pp. 29, 38.
    D.M. Frankford, in E. Parens, ed., *Enhancing Human Traits*, Georgetown University Press, Washington, 1998.
16. P.H. Huang, *Wake Forest Law Rev.* **34**, 639 (1999).
17. L.S. Temkin, *Inequality*, Oxford University Press, New York, 1993.
18. N.H. Hsieh, *J. Polit. Phil.* **7**, 7–4–1.1 (1999).
19. M.J. Mehlman and J.R. Botkin, *Access to the Genome: The Challenge to Equality*, Georgetown University Press, Washington, 1998, pp. 62–85.
20. Americans With Disabilities Act of 1990, 42 U.S.C. §§12101–12213 (1994 & Supp. II 1996).
21. J. Rawls, *A Theory of Justice*, Harvard University Press, Cambridge, 1971.
    R. Nozick, *Anarchy, State, and Utopia*, Basic Books, New York, 1974.

22. R.J. Sternberg, *Beyond IQ: A Triarchic Theory of Human Intelligence*, Cambridge University Press, New York, 1985.

23. K. Vonnegut, *Welcome to the Monkey House*, Dell Publishing, New York, 1950, p. 7.

24. B. Goodwin, *Justice by Lottery*, University of Chicago Press, Chicago, 1992.

    M. Waldholz, *Wall St. J.*, June 21, 1995 (1995 WL-WSJ 8730611).

    G.J. Annas, *Am. J. Law Med.* **3**, 59–76 (1977).

25. A.M. Bickel, *Harv. Law Rev.* **75**, 40, 76 (1961).

26. S. Darwall, *Philosophical Ethics*, 1998.

27. E. Parens, *Kennedy Inst. Ethics. J.* **5**, 141, 146–147 (1995).

28. P. Westen, *Harv. Law Rev.* **95**, 537 (1982).

29. I. Berlin, in A. Quinton, ed., *Political Philosophy*, Oxford University Press, London, 1967.

    G. Dworkin, *The Theory and Practice of Autonomy*, Cambridge University Press, Cambridge, 1988.

30. R.A. Dahl, *Democracy and Its Critics*, Yale University Press, New Haven, 1989.

31. M.J. Mehlman, *Wake Forest Law Rev.* **34**, 671 (1999).

32. J. Mincer, in C. Kerr and P.D. Staudohar, eds., *Labor Economics and Industrial Relations: Markets and Institutions*, Harvard University Press, Cambridge, 1994.

33. R. Sartorius, ed., *Paternalism*, University of Minnesota Press, Minneapolis, 1983.

34. A. Wertheimer, *Coercion*, Princeton University Press, Princeton, 1987.

35. J. Bartlett, *Familiar Quotations*, Little Brown, Boston, MA, 1980.

36. B. Edlund, *Reuters Library Report*, September 27, 1988, p. 1.

37. J.B. Attanasio, *U. Chi. Law Rev.* **53**, 1274 (1986).

38. F. Herbert, *Dune*, Chilton Books, Philadelphia, 1965.

39. S. Gardbaum, *Harv. Law Rev.* **104**, 1350 (1991).

40. E.J. Keenan, in J.A. Thomas, ed., *Drugs, Athletes, and Physical Performance*, Plenum Medical Book, New York, 1988.

41. D. Keyes, *Flowers for Algernon*, Harcourt Brace and World, New York, 1966.

    P. Anderson, *Brain Wave*, Ballantine Books, New York, 1954.

42. R.M. Berry, *Wake Forest Law Rev.* **34**, 715, 729–731 (1999).

43. N. Daniels, in T.F. Murphy and M.A. Lappe, eds., *Justice and the Human Genome Project*, University of California Press, Berkeley, 1994.

44. D.J. Schneider, A.H. Hastorf, and P.C. Ellsworth, *Person Perception*, 2nd ed., McGraw-Hill, New York, 1979, pp. 166–169, 267–269.

45. T.E. Hill, Jr., *Dignity and Practical Reason in Kant's Moral Theory*, Cornell University Press, Ithaca, 1992, pp. 38–39.

46. M.J. Radin, *Harv. Law. Rev.* **100**, 1849, 1933–1936 (1987). Contested Commodities 102–114 (1996).

    M.H. Shapiro, *Hastings Law. J.* **47**, 1081, 1180–1199 (1996).

47. M.H. Shapiro, *U. Pitt. Law. Rev.* **55**, 681, 686–687, 765–769 (1994).

48. H.L.A. Hart, *Harv. Law. Rev.* **71**, 593, 622 (1958).

    C.R. Beitz, *Political Equality: An Essay in Democratic Theory*, 1989, p. 35.

49. R.K. Merton, *Science* **159**, 56 (1968).

50. L.J. Cappon, ed., *The Adams-Jefferson Letters*, University of North Carolina, Chapel Hill, 1959, p. 388.

51. J.L. Borges, *Ficciones*, Grove/Atlantic, New York, 1962, pp. 107, 115.

52. R.J. Herrnstein, *I.Q. in the Meritocracy*, Atlantic Monthly Press, Boston, 1973.

    R. Herrnstein and C. Murray, *The Bell Curve: Intelligence and Class Life in American Life*, Free Press, New York, 1974.

    Richard Stone Science, ed., **267**, 779 (1995).

    G. Leach, *The Biocrats*, rev. ed., Pelican, Harmondsworth, 1972, pp. 221–223.

    M. Young, *The Rise of the Meritocracy*, Transaction, New Brunswick, 1994.

53. A. Posner, *J. Law Econ. Org.* **13**, 344 (1997).

54. T.L. Beauchamp and J.F. Childress, *Principles of Biomedical Ethics*, 4th ed., Oxford University Press, New York, 1994, pp. 120–188.

55. M.H. Shapiro, *U. Pitt. Law. Rev.* **55**, 681, 772–774 (1994).

56. Plato, *The Republic*, Benjamin Jowett, trans., Heritage Press, New York, 1944.

57. D.F. Thompson, *John Stuart Mill and Representative Government*, Princeton University Press, Princeton, 1976.

58. M.H. Morales, *Perfect Equality: John Stuart Mill on Well-Constituted Communities*, Rowman and Littlefield, Lanham, 1996, p. 86.

59. J. Stuart Mill, in J.M. Robson, ed., *Considerations on Representative Government*, University of Toronto Press, Toronto, Buffalo, London, 1977, pp. 371, 475.

60. J. Peters, *Colum. Law. Rev.* **97**, 312, 334–336 (1997).

    J. Waldron, *Geo. Law. J.* **84**, 2185, 2211–2212 (1996).

    R.J. Arneson, *J. Hist. Phil.* **20**, 43, 59–62 (1982).

61. R. Sinsheimer, in *Limits of Scientific Inquiry*, Daedalus, Spring, 1978, p. 34.

62. *Rogers v. Lodge*, 458 U.S. 613 (1982).

63. J. Cohen and C. Sabel, *Eur. Law. J.* **3**, 313 (1997).

64. N. Lemann, *New York Times*, April 26, 1998, §4, p. 15.

    N. Lemann, *The Big Test: The Secret History of the American Meritocracy*, Farrar, Straus and Giroux, New York, 1999.

65. R.A. Dahl and C.E. Lindblom, *Politics, Economics, and Welfare*, University of Chicago Press, Chicago, 1976, p. 282.

66. H.T. Engelhardt, Jr., *Soc. Phil. Pol.* **8**, 180, 186–189 (1990).

67. P. Singer, *Democracy and Disobedience*, Oxford University Press, New York, 1973, pp. 34–35.

68. *Skinner v. Oklahoma*, 316 U.S. 535, 541 (1942).

69. M.H. Shapiro, *S. Cal. Law. Rev.* **47**, 237, 256–257 (1974).

70. E. Chemerinsky, *Constitutional Law: Principles and Policies*, Panel Publishers, New York, §6.5, 1997.

71. *Roe v. Wade*, 410 U.S. 113 (1973).

    Brown v. Board of Education, 347 U.S. 483 (1954).

72. *Plyler v. Doe*, 457 U.S. 202 (1982).

73. F.C. Pizzulli, *S. Cal. Law. Rev.* **47**, 476 (1974).

See other entries BEHAVIORAL GENETICS, HUMAN; see also HUMAN ENHANCEMENT USES OF BIOTECHNOLOGY entries.